Integrated SCIENCE

Fifth Edition

Bill W. Tillery
Arizona State University

Eldon D. Enger
Delta College

Frederick C. Ross
Delta College

Mc Graw Hill

Connect
Learn
Succeed™

INTEGRATED SCIENCE, FIFTH EDITION

Published by McGraw-Hill, a business unit of The McGraw-Hill Companies, Inc., 1221 Avenue of the Americas, New York, NY 10020. Copyright © 2011 by The McGraw-Hill Companies, Inc. All rights reserved. Previous editions © 2008, 2007, and 2004.

Some ancillaries, including electronic and print components, may not be available to customers outside the United States.

This book is printed on acid-free paper containing 10% postconsumer waste.

1 2 3 4 5 6 7 8 9 0 DOW/DOW 1 0 9 8 7 6 5 4 3 2 1 0

ISBN 978-0-07-351219-8
MHID 0-07-351219-2

Vice President & Editor-in-Chief: *Martin Lange*
Vice President, EDP: *Kimberly Meriwether David*
Publisher: *Ryan Blankenship*
Sponsoring Editor: *Debra B. Hash*
Director of Developmental: *Kristine Tibbets*
Senior Developmental Editor: *Mary E. Hurley*
Senior Marketing Manager: *Lisa Nicks*
Senior Project Manager: *Joyce Watters*
Senior Production Supervisor: *Laura Fuller*
Lead Media Project Manager: *Judi David*
Design Coordinator: *Brenda A. Rolwes*
Cover Designer: *Studio Montage, St. Louis, Missouri*
(USE) Cover Image: © *Royalty-Free/CORBIS*
Senior Photo Research Coordinator: *John C. Leland*
Photo Research: *David Tietz/Editorial Image, LLC*
Compositor: *Lachina Publishing Services*
Typeface: *10/12 Minion*
Printer: *R. R. Donnelley*

Library of Congress Cataloging-in-Publication Data

Tillery, Bill W.
   Integrated science / Bill W. Tillery, Eldon D. Enger, Frederick C. Ross.—5th ed.
      p. cm.
   Includes index.
   ISBN-13 978-0-07-3512192-8 (alk. paper)
      1. Science—Textbooks. I. Enger, Eldon D. II. Ross, Frederick C. III. Title.
Q161.2.T54 2010
500—dc22
                                                                        2009038109

www.mhhe.com

Contents

CHAPTER 6

Electricity   125

CHAPTER 7

Light   152

CHAPTER 8

Atoms and Periodic Properties   178

CHAPTER 9

Chemical Reactions   196

CHAPTER 10

Water and Solutions   222

CHAPTER 11

Nuclear Reactions   240

Chapter Overviews

Each chapter begins with an introductory overview. The overview previews the chapter's contents and what students can expect to learn from reading the chapter. It adds to the general outline of the chapter by introducing students to the concepts to be covered. It also expands upon the core concept map, facilitating in the integration of topics. Finally, the overview will help students to stay focused and organized while reading the chapter for the first time. After reading this introduction, students should browse through the chapter, paying particular attention to the topic headings and illustrations so that they get a feel for the kinds of ideas included within the chapter.

APPLYING SCIENCE TO THE REAL WORLD

Concepts Applied

As students look through each chapter, they will find one or more Concepts Applied boxes. These activities are simple exercises that students can perform at home or in the classroom to demonstrate important concepts and reinforce their understanding of them. This feature also describes the application of those concepts to their everyday lives.

Science and Society

These readings relate the chapter's content to current societal issues. Many of these boxes also include Questions to Discuss that provide students an opportunity to discuss issues with their peers.

Myths, Mistakes, and Misunderstandings

These brief boxes provide short, scientific explanations to dispel a societal myth or a home experiment or project that enables students to dispel the myth on their own.

People Behind the Science

Many chapters also have one or two fascinating biographies that spotlight well-known scientists, past and present. From these People Behind the Science biographies, students learn about the human side of science: science is indeed relevant, and real people do the research and make the discoveries. These readings present the sciences in real-life terms that students can identify with and understand.

Closer Look and Connections

Each chapter of *Integrated Science* also includes one or more **Closer Look** readings that discuss topics of special human or environmental concern, topics concerning interesting technological applications, or topics on the cutting edge of scientific research. These readings enhance the learning experience by taking a more detailed look at related topics and adding con-

Preface

WHAT SETS THIS BOOK APART?

Creating Informed Citizens

Integrated Science is a straightforward, easy-to-read, but substantial introduction to the fundamental behavior of matter and energy in living and nonliving systems. It is intended to serve the needs of nonscience majors who must complete one or more science courses as part of a general or basic studies requirement.

Integrated Science provides an introduction to a scientific way of thinking as it introduces fundamental scientific concepts, often in historical context. Several features of the text provide opportunities for students to experience the methods of science by evaluating situations from a scientific point of view. While technical language and mathematics are important in developing an understanding of science, only the language and mathematics needed to develop central concepts are used. No prior work in science is assumed.

Many features, such as Science and Society readings, as well as basic discussions of the different branches of science help students understand how the branches relate. This allows students to develop an appreciation of the major developments in science and an ability to act as informed citizens on matters that involve science and public policy.

> *"I especially like the application of the concepts and the connections in this text. We try very hard to show that science has connections to the everyday world and why it's important to see those connections. I don't think science can be taught to nonscience majors unless this type of approach is taken."*
>
> —Richard L. Kopec, St. Edward's University

Flexible Organization

The *Integrated Science* sequence of chapters is flexible, and the instructor can determine topic sequence and depth of coverage as needed. The materials are also designed to support a conceptual approach or a combined conceptual and problem-solving approach. The *Integrated Science* ARIS Instructor's Resources offer suggestions for integrating the text's topics around theme options. With laboratory studies, the text contains enough material for the instructor to select a sequence for a one- or two-semester course.

THE GOALS OF *INTEGRATED SCIENCE*

1. **Create an introductory science course aimed at the nonscience major.** The origin of this book is rooted in our concern for the education of introductory-level students in the field of science. Historically, nonscience majors had to

enroll in courses intended for science or science-related majors such as premeds, architects, or engineers. Such courses are important for these majors but are mostly inappropriate for introductory-level nonscience students. To put a nonscience student in such a course is a mistake. Few students will have the time or background to move through the facts, equations, and specialized language to gain any significant insights into the logic or fundamental understandings; instead, they will leave the course with a distaste for science. Today, society has a great need for a few technically trained people but a much larger need for individuals who understand the process of science and its core concepts.

2. **Introduce a course that presents a coherent and clear picture of all science disciplines through an interdisciplinary approach.** Recent studies and position papers have called for an interdisciplinary approach to teaching science to nonmajors. For example, the need is discussed in *Science for All Americans—Project 2061* (American Association for the Advancement of Science), *National Science Education Standards* (National Research Council, 1994), and *Science in the National Interest* (White House, 1994). Interdisciplinary science is an attempt to broaden and humanize science education by reducing and breaking down the barriers that enclose traditional science disciplines as distinct subjects.

> *"The authors obviously feel that emphasizing the interconnectedness of nature should be studied by integrating all of the sciences into a coherent, understandable network of facts, concepts, and interpretations that lead the student to view the universe in a new and enlightened perspective. This philosophy is particularly important in the education of nonscience majors who may never again formally study science."*
>
> —Jay R. Yett, Orange Coast College

3. **Help instructors build their own mix of descriptive and analytical aspects of science, arousing student interest and feelings as they help students reach the educational goals of their particular course.** The spirit of interdisciplinary science is sometimes found in courses called "General Science," "Combined Science," or "Integrated Science." These courses draw concepts from a wide range of the traditional fields of science but are not concentrated around certain problems or questions. For example, rather than just dealing with the physics of energy, an interdisciplinary approach might consider broad aspects of energy—dealing with potential problems of an energy crisis—including social and ethical issues. A number of approaches can be used in interdisciplinary science, including the teaching of science in a *social, historical, philosophical,* or *problem-solving* context, but there is

no single best approach. One of the characteristics of interdisciplinary science is that it is not constrained by the necessity of teaching certain facts or by traditions. It likewise cannot be imposed as a formal discipline, with certain facts to be learned. It is justified by its success in attracting and holding the attention and interest of students, making them a little wiser as they make their way toward various careers and callings.

4. **Humanize science for nonscience majors.** Each chapter presents historical background where appropriate, uses everyday examples in developing concepts, and follows a logical flow of presentation. A discussion of the people and events involved in the development of scientific concepts puts a human face on the process of science. The use of everyday examples appeals to the nonscience major, typically accustomed to reading narration, not scientific technical writing, and also tends to bring relevancy to the material being presented. The logical flow of presentation is helpful to students not accustomed to thinking about relationships between what is being read and previous knowledge learned, a useful skill in understanding the sciences.

VALUED INPUT WENT INTO STRIVING TO MEET YOUR NEEDS

Text development today involves a team that includes authors and publishers and valuable input from instructors who share their knowledge and experience with publishers and authors through reviews and focus groups. Such feedback has shaped this edition, resulting in reorganization of existing content and expanded coverage in key areas. This text has continued to evolve as a result of feedback from instructors actually teaching integrated science courses in the classroom. Reviewers point out that current and accurate content, a clear writing style with concise explanations, quality illustrations, and dynamic presentation materials are important factors considered when evaluating textbooks. Those criteria have guided the revision of the *Integrated Science* text and the development of its ancillary resources.

New to This Edition

Chapter 1 What Is Science?: All English units have been converted to metric. A new example of the scientific method has also been added.

Chapter 2 Motion: All English units have been converted to metric, and new information on GPS appears.

Chapter 4 Heat and Temperature: Revised Example 4.5 provides additional explanation about the change of units.

Chapter 5 Wave Motions and Sound: A new A Closer Look on "Hearing Problems" appears.

Chapter 7 Light: A new discussion on the application of relativity has been added. The A Closer Look: The Compact Disc (CD) has been revised and updated.

Chapter 8 Atoms and Periodic Properties: The section on electron configuration has been moved to the Integrated

Science companion website to simplify the text discussion for nonscience majors. However, the material is still available online for those who would like to include this detailed discussion in their courses.

Chapter 9 Chemical Reactions: The description of chemical change has been revised.

Chapter 10 Water and Solutions: A new A Closer Look: Decompression Sickness appears. The section on concentration of solutions has been moved to the Integrated Science companion website to simplify the text discussion for nonscience majors. However, the material is still available online for those who would like to include this detailed discussion in their courses.

Chapter 12 The Universe: A new Myth, Mistakes, and Misunderstandings: Stars Visible from Bottom of a Well? has been added.

Chapter 13 The Solar System: The information on and names of dwarf planets has been updated, along with the information on space probes.

Chapter 14 Earth in Space: A new table of future eclipses now appears. There is also a new Myths, Mistakes, and Misunderstandings: Moon Mistakes.

Chapter 15 Earth: A new Myths, Mistakes, and Misunderstandings: Bye-Bye California? has been included.

Chapter 16 Earth's Surface: Two new Myths, Mistakes, and Misunderstandings: Do Earthquakes Swallow People? and Unchanging as the Hills? are now included. The revised chapter also includes a rewritten description of the Richter scale and updated information on Mount St. Helens.

Chapter 17 Earth's Weather: The revised chapter includes a section on Climate Change and Global Warming, along with a table showing greenhouse gas concentrations. A new Science and Society box on climate change by Congressman John Shadegg has been added. There is also a new section on Milankovitch cycles and revised sections on the nitrogen cycle, low pressure centers and stormy weather, and the description of how tornadoes move.

Chapter 18 Earth's Waters: A new section on the Ocean Conveyer Belt appears.

Chapter 19 Organic and Biochemistry: The chapter has been reorganized and divided into two parts: Part I: The Nature of Organic Compounds, and Part II: Organic Compounds of Life to make the chapter more digestible for nonscience students. There are also twelve new figures to add to the visual explanation of the content. New A Closer Look boxes on Nonpersistent and Persistent Organic Pollutants, and Fat and Your Diet appear along with a new People Behind the Science: Watson, Crick, and Wilson, and Connections: Biomass to Biofuels. New questions were added to the Applying the Concepts material at the end of the chapter.

Chapter 20 The Nature of Living Things: This chapter has also been reorganized and divided into three parts: Part I: Life's Characteristics and Cells, Part II: Energy Transformations in Cells, and Part III: Cellular Reproduction to make the chapter more digestible for nonscience students. There are also sixteen new or updated figures to add to the visual ex-

planation of the content. A new A Closer Look Antibiotics and Cell Structural Differences, and Science and Society: Abnormal Cell Division: Cancer have also been included.

Chapter 21 The Origin and Evolution of Life: The chapter was divided into three distinct parts and reorganized to help instructors and students better distinguish the three major topics discussed in the chapter: Part I: How Did Life Originate?, Part II: The Process of Evolution, and Part III: Speciation. There is a new A Closer Look: Evolution Without Natural Selection. The term *panspermia* and the idea that planets have been discovered around other stars are newly introduced. The material on the nature of Archaea was updated throughout the chapter based on recent discoveries. The nature of levorotatory and dextrorotatory molecules was rewritten to better explain the concept. Several new questions were added to the Applying the Concepts material at the end of the chapter.

Chapter 22 The History of Life on Earth: The chapter was divided into two parts: Part I: The Geologic History on Earth, and Part II: Kinds of Organisms to help instructors and students better distinguish the major topics discussed in the chapter. The terms *catastrophism* and *gradualism* were introduced. Material on the copper age man found in the Alps was added to the discussion on preserved fossils. The possibility of a current human-caused mass extinction was added to the section on mass extinctions. The term *Eubacteria* has been replaced with the now more commonly used name *Bacteria*, for the domain in which bacteria are classified. The section on Archaea was extensively revised to include more recent information about the nature of Archaea. The section on prions was also revised. A section on the recent discovery of tiny hominins from Indonesia was added to discussion of human evolution.

Chapter 23 Ecology and Environment: A new section on Nutrient Cycles and Geologic Time was added. A new section on Bioaccumulation and Biomagnification along with a new illustration of the biomagnification of DDT was added. The concept of a deceleration phase in the population growth curve is introduced. A Closer Look: Carbon Dioxide and Global Warming was updated with information from the latest publication of the Intergovernmental Panel on Climate Change. Subheadings were added to the section on Population Characteristics to clarify the discussion. Numerous new photos and figures have been added to improve the visual explanation of the chapter. Table 23.2 on Comparisons of the Population Characteristics of Developed and Less Developed Nations was updated with 2008 data. Science and Society: Government Policy and Population Control was updated with 2008 data and information on the Chinese government's efforts to encourage births of girls.

Chapter 24 Human Biology: Materials Exchange and Control Mechanism: The section Exchanging Materials: Basic Principles was rewritten to focus on organisms rather than cells. The section on obesity was substantially rewritten to include a clearer description on body mass index and to enumerate the various factors correlated with the high incidence of obesity. Numerous new photos and figures have been added to improve the visual explanation of the chapter.

Chapter 25 Human Biology: Reproduction: There a figures to add to the visual explanation of the c new Science and Society boxes: A Link Betwee tion and Endometriosis? and Multiple Births, mones have been added. Updated informatio also appears in this revised chapter.

Chapter 26 Mendelian and Molecular Geneti has been reorganized and divided into t Mendelian Genetics Updated, and Part Basis of Genetics to make the chapte for nonscience students. There are als updated figures to add to the visual content. The concepts of epigenetics two new boxes appear: A Closer L Connections: Astronauts, Cells, and

Appendix D Solutions for Second Ex is a new appendix providing solut follow-up Example exercises.

THE LEARNING SYSTEM

To achieve the goals stated, this te tures that should make student's more effective and enjoyable. Th you clearly understand the conc as the foundation of the integra

OVERVIEW TO INTEGF

Chapter 1 provides an ove science in general and this the fundamental methods study and understand the

MULTIDISCIPLIN/

Chapter Openir

Core Concept a

NEW! Core and Su cepts and the chapt outline and emph porting concepts by identifying th

Connectio

NEW! The r the text are map, integr list, the co big pictur how that crete examples to help students better appreciate the real-world applications of science.

In addition to the **Closer Look** readings, each chapter contains concrete interdisciplinary **Connections** that are highlighted. **Connections** will help students better appreciate the interdisciplinary nature of the sciences. The **Closer Look** and **Connections** readings are informative materials that are supplementary in nature. These boxed features highlight valuable information beyond the scope of the text and relate intrinsic concepts discussed to real-world issues, underscoring the relevance of integrated science in confronting the many issues we face in our day-to-day lives. They are identified with the following icons:

"A Closer Look: The Compact Disc was, again, an excellent application of optics to everyday life and to something modern students thrive on—CDs and DVDs."

—Treasure Brasher, West Texas A&M University

"Connections—wonderful!!!. . . . . . A Closer Look . . . excellent. Clear, interesting, good figures. You have presented crucial information in a straightforward and uncompromising way."

—Megan M. Hoffman, Berea College

General: This icon identifies interdisciplinary topics that cross over several categories; for example, life sciences and technology.

Life: This icon identifies interdisciplinary life science topics, meaning connections concerning all living organisms collectively: plant life, animal life, marine life, and any other classification of life.

Technology: This icon identifies interdisciplinary technology topics, that is, connections concerned with the application of science for the comfort and well being of people, especially through industrial and commercial means.

Measurement, Thinking, Scientific Methods: This icon identifies interdisciplinary concepts and understandings concerned with people trying to make sense out of their surroundings by making observations, measuring, thinking, developing explanations for what is observed, and experimenting to test those explanations.

Environmental Science: This icon identifies interdisciplinary concepts and understandings about the problems caused by human use of the natural world and remedies for those problems.

END-OF-CHAPTER FEATURES

At the end of each chapter are the following materials:

- *Summary:* highlights the key elements of the chapter
- *Summary of Equations* (chapters 1–9, 11): highlights the key equations to reinforce retention of them
- *Key Terms:* page-referenced where students will find the terms defined in context

- *Applying the Concepts:* a multiple choice quiz to test students' comprehension of the material covered
- *Questions for Thought:* designed to challenge students to demonstrate their understandings of the topic
- *For Further Analysis:* exercises include analysis or discussion questions, independent investigations, and activities intended to emphasize critical thinking skills and societal issues, and develop a deeper understanding of the chapter content
- *Invitation to Inquiry:* exercises that consist of short, open-ended activities that allow students to apply investigative skills to the material in the chapter
- *Parallel Exercises* (chapters 1–9, 11): There are two groups of parallel exercises, Group A and Group B. The Group A parallel exercises have complete solutions worked out, along with useful comments in appendix D. The Group B parallel exercises are similar to those in Group A but do not contain answers in the text. By working through the Group A parallel exercises and checking the solution in appendix D, students will gain confidence in tackling the parallel exercises in Group B and thus reinforce their problem-solving skills.

"I like the key terms with the page numbers with each one. I always like to see more conceptual- and synthesis-type questions, which is why I like the 'Questions for Thought' and 'For Further Analysis' parts. . . . Exercises such as 'Questions for Thought' number 7, having students think about why oxygen is in Earth's atmosphere but not in Venus or Mars' atmosphere, is a valuable sort of question, because it requires students to know something and apply it."

—Jim Hamm, Big Bend Community College

END-OF-TEXT MATERIAL

At the back of the text are appendices that give additional background details, charts, and answers to chapter exercises. A new Appendix D provides solutions for each chapter's follow-up Example exercises. There are also a glossary of all key terms, an index organized alphabetically by subject matter, and special tables printed on the inside covers for reference use.

> "... many books addressing similar disciplines have a tendency to talk over a student's head, making a student frustrated further in a class they do not want to be attending.... Personally, I would admit that Integrated Science has a slight edge. The glossary seems up-to-date and centers in on words many nonscience majors may not understand."
>
> —David J. DiMattio, St. Bonaventure University

MULTIMEDIA SUPPLEMENTS

McGraw-Hill Connect Physics

McGraw-Hill Connect Physics is a web-based assignment and assessment platform that gives students the means to better connect with their coursework, with their instructors, and with the important concepts that they will need to know for success now and in the future. With Connect Physics, instruc-tors can deliver assignments, quizzes, and tests easily online. Students can practice important skills at their own pace and on their own schedule. Connect Physics includes all Parallel Exercises from the text (in open-ended, numerical entry, and algorithmic format whenever possible) as well as all Questions for Thought in multiple-choice format (written by the authors of the text).

With the *Integrated Science* companion site, instructors also have access to PowerPoint lecture outlines, an Instructor's Manual, an Instructor's Edition Lab Manual, electronic images from the text, clicker questions, quizzes, animations, and many other resources directly tied to text-specific materials in *Integrated Science*. Students have access to self-quizzing, animations, and more.

Presentation Tools

Presentation Tools offer the ultimate multimedia resource center for the instructor. Graphics from the textbook are available in electronic format to create customized classroom presentations, visually based tests and quizzes, dynamic course website content, or attractive printed support materials.

The following assets are available in digital formats, for full-text download and also grouped by chapter:

- **Art and Photo Library:** Full-color digital files of all of the illustrations and many of the photos in the text can be readily incorporated into lecture presentations, exams, or custom-made classroom materials.

- **Worked Example Library and Table Library:** Access the worked examples and tables from the text in electronic format for inclusion in your classroom resources.
- **Animations Library:** Files of animations and videos covering various topics are included so that you can easily make use of these animations in a lecture or classroom setting.
- **Lecture Outlines:** Lecture notes, incorporating illustrations and animated images, have been written to the fifth edition text. They are provided in PowerPoint format so that you may use these lectures as written or customize them to fit your lecture.

Personal Response Systems

Personal response systems bring interactivity into the classroom or lecture hall. A wireless response system gives the instructor and students immediate feedback from the entire class. The wireless response pads are essentially easy-to-use remotes that engage students, allowing instructors to motivate student preparation, interactivity, and active learning. Instructors receive immediate feedback to gauge which concepts students understand. Questions covering the content of the *Integrated Science* text and formatted in PowerPoint are available on the *Integrated Science* companion site.

Computerized Test Bank Online

A comprehensive bank of over nine hundred test questions written by the authors of the text, in multiple-choice format at a variety of difficulty levels is provided within a computerized test bank powered by McGraw-Hill's flexible electronic testing program—EZ Test Online (www.eztestonline.com). EZ Test Online allows you to create the paper and online tests or quizzes in this easy-to-use program!

Imagine being able to create and access your test or quiz anywhere, at any time without installing the testing software. Now, with EZ Test Online, instructors can select questions from multiple McGraw-Hill test banks or author their own, and then either print the test for paper distribution or give it online.

Electronic Books

If you or your students are ready for an alternative version of the traditional textbook, McGraw-Hill brings you innovative and inexpensive electronic textbooks. By purchasing E-books from McGraw-Hill, students can save as much as 50 percent on selected titles delivered on the most advanced E-book platforms available.

E-books from McGraw-Hill are smart, interactive, searchable, and portable, with such powerful built-in tools as detailed searching, highlighting, note taking, and student-to-student or instructor-to-student note sharing. E-books from McGraw-Hill will help students to study smarter and quickly find the information they need. Students will also save money. Contact your McGraw-Hill sales representative to discuss E-book packaging options.

PRINTED SUPPLEMENTARY MATERIALS

Laboratory Manual

The laboratory manual, written and classroom-tested by the authors, presents a selection of laboratory exercises specifically written for the interest and abilities of nonscience majors. Each lab begins with an open-ended *Invitations to Inquiry,* designed to pique student interest in the lab concept. This is followed by laboratory exercises that require measurement and data analysis for work in a more structured learning environment. When the laboratory manual is used with *Integrated Science,* students will have an opportunity to master basic scientific principles and concepts, learn new problem-solving and thinking skills, and understand the nature of scientific inquiry from the perspective of hands-on experiences. There is also an **instructor's edition lab manual** available on the *Integrated Science* companion site.

ACKNOWLEDGMENTS

This revision of *Integrated Science* has been made possible by the many users and reviewers of its fourth edition. The authors are indebted to the fourth edition reviewers for their critical reviews, comments, and suggestions. The reviewers were:

Moses Adeniyi, *Savannah State University*
Karen Chassereau, *Georgia Southern University*
Sarah Cooper, *Arcadia University*
Andrew Evans, Jr., *Metropolitan State College of Denver*
Laura Frost, *Point Park University*
John R. Hoffman, *Arcadia University, College of Graduate Studies*
David King, *Auburn University*
Matthew Opdyke, *Point Park University*
Barbara Rendleman, *Ivy Tech Community College, Evansville Campus*
Amy Rollins, *Mercer University, Henry County Campus*
Paramasivam Sivapatham, *Savannah State University*
Emily Willingham, *St. Edwards University*

The authors would also like to thank the theme integration authors for their contributions to the Instructor's Resources on the companion site. Those contributors include:

Mary Brown, *Lansing Community College*
David J. DiMattio, *St. Bonaventure University*
Tasneem F. Khaleel, *Montana State University–Billings*
G. A. Nixon and Mary Ellen Teasdale, *Texas A&M–Commerce*
Thad Zaleskiewicz and Jennifer Siegert, *University of Pittsburgh–Greensburg*

The authors would also like to thank the following media ancillary authors for their contributions of the PowerPoint Lecture Outlines and the clicker questions, respectively:

Christine McCreary, *University of Pittsburgh–Greensburg*
Jeffrey J. Miller, *Metropolitan State College of Denver*

MEET THE AUTHORS

Bill W. Tillery

Bill W. Tillery is professor emeritus of Physics at Arizona State University. He earned a bachelor's degree at Northeastern State University (1960) and master's and doctorate degrees from the University of Northern Colorado (1967). Before moving to Arizona State University, he served as director of the Science and Mathematics Teaching Center at the University of Wyoming (1969–73) and as an assistant professor at Florida State University (1967–69). Bill has served on numerous councils, boards, and committees and was honored as the "Outstanding University Educator" at the University of Wyoming in 1972. He was elected the "Outstanding Teacher" in the Department of Physics and Astronomy at Arizona State University in 1995.

During his time at Arizona State, Bill has taught a variety of courses, including general education courses in science and society, physical science, and introduction to physics. He has received more than forty grants from the National Science Foundation, the U. S. Office of Education, private industry (Arizona Public Service), and private foundations (Flinn Foundation) for science curriculum development and science teacher inservice training. In addition to teaching and grant work, Bill has authored or co-authored more than sixty textbooks and many monographs, and has served as editor of three newsletters and journals between 1977 and 1996.

Eldon D. Enger

Eldon D. Enger is professor emeritus of biology at Delta College, a community college near Saginaw, Michigan. He received his B.A. and M.S. degrees from the University of Michigan. Professor Enger has over thirty years of teaching experience, during which he has taught biology, zoology, environmental science, and several other courses. He has been very active in curriculum and course development.

Professor Enger is an advocate for variety in teaching methodology. He feels that if students are provided with varied experiences, they are more likely to learn. In addition to the standard textbook assignments, lectures, and laboratory activities, his classes are likely to include writing assignments, student presentation of lecture material, debates by students on controversial issues, field experiences, individual student projects, and discussions of local examples and relevant current events. Textbooks are very valuable for presenting content, especially if they contain accurate, informative drawings and visual examples. Lectures are best used to help students see themes and make connections, and laboratory activities provide important hands-on activities.

Professor Enger has been a Fulbright Exchange Teacher to Australia and Scotland, received the Bergstein Award for Teaching Excellence and the Scholarly Achievement Award from Delta College, and participated as a volunteer in Earthwatch Research Programs in Costa Rica, the Virgin Islands and Australia. During 2001, he was a member of a People to People delegation to South Africa.

Professor Enger is married, has two adult sons, and enjoys a variety of outdoor pursuits such as cross-country skiing, hiking, hunting, kayaking, fishing, camping, and gardening. Other interests include reading a wide variety of periodicals, beekeeping, singing in a church choir, and preserving garden produce.

Frederick C. Ross

Fred Ross is professor emeritus of biology at Delta College, a community college near Saginaw, Michigan. He received his B.S. and M.S. from Wayne State University, Detroit, Michigan, and has attended several other universities and institutions. Professor Ross has over thirty years' teaching experience, including junior and senior high school, during which he has taught biology, cell biology and biological chemistry, microbiology, environmental science, and zoology. He has been very active in curriculum and course development. These activities included the development of courses in infection control and microbiology, and AIDS and infectious diseases, and a PBS ScienceLine course for elementary and secondary education majors in cooperation with Central Michigan University. In addition, he was involved in the development of the wastewater microbiology technician curriculum offered by Delta College.

He was also actively involved in the National Task Force of Two Year College Biologists (American Institute of Biological Sciences) and in the National Science Foundation College Science Improvement Program, and has been an evaluator for science and engineering fairs, Michigan Community College Biologists, a judge for the Michigan Science Olympiad and the Science Bowl, a member of a committee to develop and update blood-borne pathogen standards protocol, and a member of Topic Outlines in Introductory Microbiology Study Group of the American Society for Microbiology.

Professor Ross involves his students in a variety of learning techniques and has been a prime advocate of the writing-to-learn approach. Besides writing, his students are typically engaged in active learning techniques including use of inquiry-based learning, the Internet, e-mail communications, field experiences, classroom presentation, as well as lab work. The goal of his classroom presentations and teaching is to actively engage the minds of his students in understanding the material, not just memorization of "scientific facts." Professor Ross is married and recently a grandfather. He enjoys sailing, horseback riding, and cross-country skiing.

1

What Is Science?

Science is concerned with your surroundings and your concepts and understanding of these surroundings.

CORE **CONCEPT**

Science is a way of thinking about and understanding your surroundings.

OUTLINE

Measurement is used to accurately describe properties and events (p. 4).

An equation is a statement of a relationship between variables (p. 10).

Objects and Properties
Quantifying Properties
Measurement Systems
Standard Units for the Metric System
    Length
    Mass
    Time
Metric Prefixes
Understandings from Measurements
    Data
    Ratios and Generalizations
    The Density Ratio
    Symbols and Equations

The Nature of Science
    The Scientific Method
    Explanations and Investigations
    Scientific Laws
    Models and Theories
Science, Nonscience, and Pseudoscience
    From Experimentation to Application
    Science and Nonscience
    Pseudoscience
    Limitations of Science
People Behind the Science:
    Florence Bascom

Science investigations include collecting observations, developing explanations, and testing explanations (p. 12).

Scientific laws describe relationships between events that happen time after time (p. 14).

in and out of your surroundings (Ch. 2–7).

...try

...atter is composed of atoms that interact on several different levels (Ch. 8–11).

Earth Science

▷ Earth is matter and energy that interact through cycles of change (Ch. 14–18).

Astronomy

▷ The stars and solar system are matter and energy that interact through cycles of change (Ch. 12–13).

OVERVIEW

Have you ever thought about your thinking and what you know? On a very simplified level, you could say that everything you know came to you through your senses. You see, hear, and touch things of your choosing, and you can smell and taste things in your surroundings. Information is gathered and sent to your brain by your sense organs. Somehow, your brain processes all this information in an attempt to find order and make sense of it all. Finding order helps you understand the world and what may be happening at a particular place and time. Finding order also helps you predict what may happen next, which can be very important in a lot of situations.

This is a book on thinking about and understanding your surroundings. These surroundings range from the obvious, such as the landscape and the day-to-day weather, to the not so obvious, such as how atoms are put together. Your surroundings include natural things as well as things that people have made and used (figure 1.1). You will learn how to think about your surroundings, whatever your previous experience with thought-demanding situations. This first chapter is about "tools and rules" that you will use in the thinking process.

OBJECTS AND PROPERTIES

Science is concerned with making sense out of the environment. The early stages of this "search for sense" usually involve *objects* in the environment, things that can be seen or touched. These could be objects you see every day, such as a glass of water, a moving automobile, or a running dog. They could be quite large, such as the Sun, the Moon, or even the solar system, or invisible to the unaided human eye. Objects can be any size, but people are usually concerned with objects that are larger than a pinhead and smaller than a house. Outside these limits, the actual size of an object is difficult for most people to comprehend.

As you were growing up, you learned to form a generalized mental image of objects called a *concept*. Your concept of an object is an idea of what it is, in general, or what it should be according to your idea (figure 1.2). You usually have a word stored away in your mind that represents a concept. The word *chair*, for example, probably evokes an idea of "something to sit on." Your generalized mental image for the concept that goes with the word *chair* probably includes a four-legged object with a backrest. Upon close inspection, most of your (and everyone else's) concepts are found to be somewhat vague. For example, if the word *chair* brings forth a mental image of something with four legs and a backrest (the concept), what is the difference between a "high chair" and a "bar stool"? When is a chair a chair and not a stool? These kinds of questions can be troublesome for many people.

Not all of your concepts are about material objects. You also have concepts about intangibles such as time, motion, and relationships between events. As was the case with concepts of material objects, words represent the existence of intangible concepts. For example, the words *second, hour, day,* and *month* represent concepts of time. A concept of the pushes and pulls that come with changes of motion during an airplane flight might be represented with such words as *accelerate* and *falling*. Intangible concepts might seem to be more abstract since they do not represent material objects.

By the time you reach adulthood, you have literally thousands of words to represent thousands of concepts. But most,

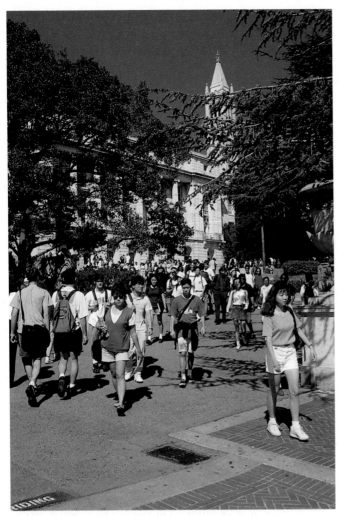

FIGURE 1.1 Your surroundings include naturally occurring objects and manufactured objects such as sidewalks and buildings.

you would find on inspection, are somewhat ambiguous and not at all clear-cut. That is why you find it necessary to talk about certain concepts for a minute or two to see if the other person has the same "concept" for words as you do. That is why when one person says, "Boy, was it hot!" the other person may respond, "How hot was it?" The meaning of *hot* can be quite different for two people, especially if one is from Arizona and the other from Alaska!

The problem with words, concepts, and mental images can be illustrated by imagining a situation involving you and another person. Suppose that you have found a rock that you believe would make a great bookend. Suppose further that you are talking to the other person on the telephone, and you want to discuss the suitability of the rock as a bookend, but you do not know the name of the rock. If you knew the name, you would simply state that you found a "_____." Then you would probably discuss the rock for a minute or so to see if the other person really understood what you were talking about. But not knowing the name of the rock and wanting to communicate about the suitability of the object as a bookend, what would you do? You would probably describe the characteristics, or **properties,** of the rock. Properties are the qualities or attributes that, taken together, are usually peculiar to an object. Since you commonly determine properties with your senses (smell, sight, hearing, touch, and taste), you could say that the properties of an object are the effect the object has on your senses. For example, you might say that the rock is a "big, yellow, smooth rock with shiny gold cubes on one side." But consider the mental image that the other person on the telephone forms when you describe these properties. It is entirely possible that the other person is thinking of something very different from what you are describing (figure 1.3)!

FIGURE 1.2 What is your concept of a chair? Are all of these pieces of furniture chairs? Most people have concepts, or ideas of what things in general should be, that are loosely defined. The concept of a chair is one example of a loosely defined concept.

FIGURE 1.3 Could you describe this rock to another person over the telephone so that the other person would know *exactly* what you see? This is not likely with everyday language, which is full of implied comparisons, assumptions, and inaccurate descriptions.

As you can see, the example of describing a proposed bookend by listing its properties in everyday language leaves much to be desired. The description does not really help the other person form an accurate mental image of the rock. One problem with the attempted communication is that the description of any property implies some kind of *referent*. The word **referent** means that you *refer to*, or think of, a given property in terms of another, more familiar object. Colors, for example, are sometimes stated with a referent. Examples are "sky blue," "grass green," or "lemon yellow." The referents for the colors blue, green, and yellow are, respectively, the sky, living grass, and a ripe lemon.

Referents for properties are not always as explicit as they are with colors, but a comparison is always implied. Since the comparison is implied, it often goes unspoken and leads to assumptions in communications. For example, when you stated that the rock was "big," you assumed that the other person knew that you did not mean as big as a house or even as big as a bicycle. You assumed that the other person knew that you meant that the rock was about as large as a book, perhaps a bit larger.

Another problem with the listed properties of the rock is the use of the word *smooth*. The other person would not know if you meant that the rock *looked* smooth or *felt* smooth. After all, some objects can look smooth and feel rough. Other objects can look rough and feel smooth. Thus, here is another assumption, and probably all of the properties lead to implied comparisons, assumptions, and a not very accurate communication. This is the nature of your everyday language and the nature of most attempts at communication.

QUANTIFYING PROPERTIES

Typical day-to-day communications are often vague and leave much to be assumed. A communication between two people, for example, could involve one person describing some person, object, or event to a second person. The description is made by using referents and comparisons that the second person may or may not have in mind. Thus, such attributes as "long"

fingernails or "short" hair may have entirely different meanings to different people involved in a conversation. Assumptions and vagueness can be avoided by using **measurement** in a description. Measurement is a process of comparing a property to a well-defined and agreed-upon referent. The well-defined and agreed-upon referent is used as a standard called a **unit.** The measurement process involves three steps: (1) *comparing* the referent unit to the property being described, (2) following a *procedure*, or operation, which specifies how the comparison is made, and (3) *counting* how many standard units describe the property being considered.

The measurement process thus uses a defined referent unit, which is compared to a property being measured. The *value* of the property is determined by counting the number of referent units. The name of the unit implies the procedure that results in the number. A measurement statement always contains a *number* and *name* for the referent unit. The number answers the question "How much?" and the name answers the question "Of what?" Thus a measurement always tells you "how much of what." You will find that using measurements will sharpen your communications. You will also find that using measurements is one of the first steps in understanding your physical environment.

MEASUREMENT SYSTEMS

Measurement is a process that brings precision to a description by specifying the "how much" and "of what" of a property in a particular situation. A number expresses the value of the property, and the name of a unit tells you what the referent is as well as implying the procedure for obtaining the number. Referent units must be defined and established, however, if others are to understand and reproduce a measurement. It would be meaningless, for example, for you to talk about a length in "clips" if other people did not know what you meant by a "clip" unit. When standards are established, the referent unit is called a **standard unit** (figure 1.4). The use of standard units makes it possible to communicate and duplicate measurements. Standard units are usually defined and established by governments and their agencies that are created for that purpose. In the United States, the agency concerned with measurement standards is the National Institute of Standards and Technology. In Canada, the Standards Council of Canada oversees the National Standard System.

50 leagues
130 nautical miles
150 miles
158 Roman miles
1,200 furlongs
12,000 chains
48,000 rods
452,571 cubits
792,000 feet

FIGURE 1.4 Any of these units and values could have been used at some time or another to describe the same distance between these hypothetical towns. Any unit could be used for this purpose, but when one particular unit is officially adopted, it becomes known as the *standard unit.*

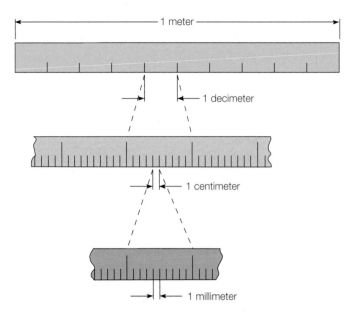

FIGURE 1.7 Compare the units shown above. How many millimeters fit into the space occupied by 1 centimeter? How many millimeters fit into the space of 1 decimeter? Can you express this as multiples of ten?

FIGURE 1.8 A cubic decimeter of water (1,000 cm³) has a liquid volume of 1 L (1,000 mL) and a mass of 1 kg (1,000 g). Therefore, 1 cm³ of water has a liquid volume of 1 mL and a mass of 1 g.

The volume of 1,000 cm³ also defines a metric unit that is commonly used to measure liquid volume, the *liter* (L). For smaller amounts of liquid volume, the *milliliter* (mL) is used. The relationship between liquid volume, volume, and mass of water is therefore

$$1.0 \text{ L} \rightarrow 1.0 \text{ dm}^3 \text{ and has a mass of } 1.0 \text{ kg}$$

or, for smaller amounts,

$$1.0 \text{ mL} \rightarrow 1.0 \text{ cm}^3 \text{ and has a mass of } 1.0 \text{ g}$$

UNDERSTANDINGS FROM MEASUREMENTS

One of the more basic uses of measurement is to *describe* something in an exact way that everyone can understand. For example, if a friend in another city tells you that the weather

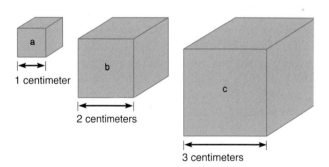

FIGURE 1.9 A weather report gives exact information, data that describe the weather by reporting numerically specified units for each condition.

has been "warm," you might not understand what temperature is being described. A statement that the air temperature is 70°F carries more exact information than a statement about "warm weather." The statement that the air temperature is 70°F contains two important concepts: (1) the numerical value of 70 and (2) the referent unit of degrees Fahrenheit. Note that both a numerical value and a unit are necessary to communicate a measurement correctly. Thus, weather reports describe weather conditions with numerically specified units; for example, 70° Fahrenheit for air temperature, 5 miles per hour for wind speed, and 0.5 inch for rainfall (figure 1.9). When such numerically specified units are used in a description, or a weather report, everyone understands *exactly* the condition being described.

Data

Measurement information used to describe something is called **data.** Data can be used to describe objects, conditions, events, or changes that might be occurring. You really do not know if the weather is changing much from year to year until you compare the yearly weather data. The data will tell you, for example, if the weather is becoming hotter or dryer or is staying about the same from year to year.

Let's see how data can be used to describe something and how the data can be analyzed for further understanding. The cubes illustrated in figure 1.10 will serve as an example. Each cube can be described by measuring the properties of size and surface area.

FIGURE 1.10 Cube *a* is 1 centimeter on each side, cube *b* is 2 centimeters on each side, and cube *c* is 3 centimeters on each side. These three cubes can be described and compared with data, or measurement information, but some form of analysis is needed to find patterns or meaning in the data.

First, consider the size of each cube. Size can be described by *volume*, which means *how much space something occupies.* The volume of a cube can be obtained by measuring and multiplying the length, width, and height. The data is

volume of cube *a*   1 cm^3
volume of cube *b*   8 cm^3
volume of cube *c*   27 cm^3

Now consider the surface area of each cube. *Area* means *the extent of a surface,* and each cube has six surfaces, or faces (top, bottom, and four sides). The area of any face can be obtained by measuring and multiplying length and width. The data for the three cubes thus describes them as follows:

	Volume	Surface Area
cube *a*	1 cm^3	6 cm^2
cube *b*	8 cm^3	24 cm^2
cube *c*	27 cm^3	54 cm^2

Ratios and Generalizations

Data on the volume and surface area of the three cubes in figure 1.10 describes the cubes, but whether it says anything about a relationship between the volume and surface area of a cube is difficult to tell. Nature seems to have a tendency to camouflage relationships, making it difficult to extract meaning from raw data. Seeing through the camouflage requires the use of mathematical techniques to expose patterns. Let's see how such techniques can be applied to the data on the three cubes and what the pattern means.

One mathematical technique for reducing data to a more manageable form is to expose patterns through a *ratio.* A ratio is a relationship between two numbers obtained when one number is divided by another number. Suppose, for example, that an instructor has 50 sheets of graph paper for a laboratory group of 25 students. The relationship, or ratio, between the number of sheets and the number of students is 50 papers to 25 students, and this can be written as 50 papers/25 students. This ratio is *simplified* by dividing 25 into 50, and the ratio becomes 2 papers/1 student. The 1 is usually understood (not stated), and the ratio is written as simply 2 papers/student. It is read as 2 papers "for each" student, or 2 papers "per" student. The concept of simplifying with a ratio is an important one, and you will see it time and time again throughout science. It is important that you understand the meaning of *per* and *for each* when used with numbers and units.

Applying the ratio concept to the three cubes in figure 1.10, the ratio of surface area to volume for the smallest cube, cube *a,* is 6 cm^2 to 1 cm^3, or

$$\frac{6 \text{ cm}^2}{1 \text{ cm}^3} = 6 \frac{\text{cm}^2}{\text{cm}^3}$$

meaning there are 6 square centimeters of area *for each* cubic centimeter of volume.

The middle-sized cube, cube *b,* had a surface area of 24 cm^2 and a volume of 8 cm^3. The ratio of surface area to volume for this cube is therefore

$$\frac{24 \text{ cm}^2}{8 \text{ cm}^3} = 3 \frac{\text{cm}^2}{\text{cm}^3}$$

meaning there are 3 square centimeters of area *for each* cubic centimeter of volume.

The largest cube, cube *c,* had a surface area of 54 cm^2 and a volume of 27 cm^3. The ratio is

$$\frac{54 \text{ cm}^2}{27 \text{ cm}^3} = 2 \frac{\text{cm}^2}{\text{cm}^3}$$

or 2 square centimeters of area *for each* cubic centimeter of volume. Summarizing the ratio of surface area to volume for all three cubes, you have

small cube   *a*–6:1
middle cube   *b*–3:1
large cube   *c*–2:1

Now that you have simplified the data through ratios, you are ready to generalize about what the information means. You can generalize that the surface-area-to-volume ratio of a cube *decreases* as the volume of a cube becomes larger. Reasoning from this generalization will provide an explanation for a number of related observations. For example, why does crushed ice melt faster than a single large block of ice with the same volume? The explanation is that the crushed ice has a larger surface-area-to-volume ratio than the large block, so more surface is exposed to warm air. If the generalization is found to be true for shapes other than cubes, you could explain why a log chopped into small chunks burns faster than the whole log. Further generalizing might enable you to predict if large potatoes would require more or less peeling than the same weight of small potatoes. When generalized explanations result in predictions that can be verified by experience, you gain confidence in the explanation. Finding patterns of relationships is a satisfying intellectual adventure that leads to understanding and generalizations that are frequently practical.

The Density Ratio

The power of using a ratio to simplify things, making explanations more accessible, is evident when you compare the simplified ratio 6 to 3 to 2 with the hodgepodge of numbers that you would have to consider without using ratios. The power of using the ratio technique is also evident when considering other properties of matter. Volume is a property that is sometimes confused with mass. Larger objects do not necessarily contain more matter than smaller objects. A large balloon, for example, is much larger than this book, but the book is much more massive than the balloon. The simplified way of comparing the mass of a particular volume is to find the ratio of mass to volume. This ratio is called **density,** which is defined as *mass per unit volume.* The *per* means "for each," as previously

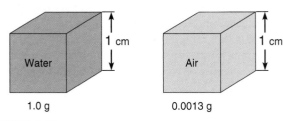

FIGURE 1.11 Equal volumes of different substances do not have the same mass, as these cube units show. Calculate the densities in g/m³. Do equal volumes of different substances have the same density? Explain.

discussed, and *unit* means "one," or "each." Thus "mass per unit volume" literally means the "mass of one volume" (figure 1.11). The relationship can be written as

$$\text{density} = \frac{\text{mass}}{\text{volume}}$$

or

$$\rho = \frac{m}{V} \qquad \text{equation 1.1}$$

(ρ is the symbol for the Greek letter *rho*.)

As with other ratios, density is obtained by dividing one number and unit by another number and unit. Thus, the density of an object with a volume of 5 cm³ and a mass of 10 g is

$$\text{density} = \frac{10 \text{ g}}{5 \text{ cm}^3} = 2 \frac{\text{g}}{\text{cm}^3}$$

The density in this example is the ratio of 10 g to 5 cm³, or 10 g/5 cm³, or 2 g to 1 cm³. Thus, the density of the example object is the mass of *one* volume (a unit volume), or 2 g *for each cm³*.

TABLE 1.3   Densities of Some Common Substances

Substance	Density (ρ) (g/cm³)
Aluminum	2.70
Copper	8.96
Iron	7.87
Lead	11.4
Water	1.00
Seawater	1.03
Mercury	13.6
Gasoline	0.680

Connections . . .

Fish Density

Sharks and rays are marine animals that have an internal skeleton made entirely of cartilage. These animals have no swim bladder to adjust their body density in order to maintain their position in the water; therefore, they must constantly swim or they will sink. The bony fish, on the other hand, have a skeleton composed of bone and most also have a swim bladder. These fish can regulate the amount of gas in the bladder to control their density. Thus, the fish can remain at a given level in the water without expending large amounts of energy.

Any unit of mass and any unit of volume may be used to express density. The densities of solids, liquids, and gases are usually expressed in grams per cubic centimeter (g/cm³), but the densities of liquids are sometimes expressed in grams per milliliter (g/mL). Using SI standard units, densities are expressed as kg/m³. Densities of some common substances are shown in table 1.3.

CONCEPTS APPLIED

Density Examples

1. What is the density of this book? Measure the length, width, and height of this book in cm, then multiply to find the volume in cm³. Use a balance to find the mass of this book in grams. Compute the density of the book by dividing the mass by the volume. Compare the density in g/cm³ with other substances listed in table 1.3.

2. Compare the densities of some common liquids. Pour a cup of vinegar in a large bottle. Carefully add a cup of corn syrup, then a cup of cooking oil. Drop a coin, tightly folded pieces of aluminum foil, and toothpicks into the bottle. Explain what you observe in terms of density. Take care not to confuse the property of *density*, which describes the compactness of matter, with *viscosity*, which describes how much fluid resists flowing under normal conditions. (Corn syrup has a greater viscosity than water—is this true of density, too?)

Myths, Mistakes, and Misunderstandings

Tap a Can?

Some people believe that tapping on the side of a can of carbonated beverage will prevent it from foaming over when the can is opened. Is this true or a myth? Set up a controlled experiment to compare opening cold cans of a carbonated beverage that have been tapped with cans that have not been tapped. Are you sure you have controlled all the other variables?

If matter is distributed the same throughout a volume, the *ratio* of mass to volume will remain the same no matter what mass and volume are being measured. Thus, a teaspoonful, a cup, or a lake full of freshwater at the same temperature will all have a density of about 1 g/cm³ or 1 kg/L.

EXAMPLE 1.1   *(Optional)*

Two blocks are on a table. Block A has a volume of 30.0 cm³ and a mass of 81.0 g. Block B has a volume of 50.0 cm³ and a mass of 135 g. Which block has the greater density? If the two blocks have the same density, what material are they? (See table 1.3.)

SOLUTION

Density is defined as the ratio of the mass of a substance per unit volume. Assuming the mass is distributed equally throughout the volume, you could assume that the ratio of mass to volume is the same no matter what quantity of mass and volume are measured. If you can accept this assumption, you can use equation 1.1 to determine the density:

Block A

$$\text{mass } (m) = 81.0 \text{ g}$$
$$\text{volume } (V) = 30.0 \text{ cm}^3 \qquad \rho = \frac{m}{V}$$
$$\text{density } (\rho) = ?$$
$$= \frac{81.0 \text{ g}}{30.0 \text{ cm}^3}$$
$$= \frac{81.0}{30.0} \frac{\text{g}}{\text{cm}^3}$$
$$= 2.70 \frac{\text{g}}{\text{cm}^3}$$

Block B

$$\text{mass } (m) = 135 \text{ g}$$
$$\text{volume } (V) = 50.0 \text{ cm}^3 \qquad \rho = \frac{m}{V}$$
$$\text{density } (\rho) = ?$$
$$= \frac{135 \text{ g}}{50.0 \text{ cm}^3}$$
$$= \frac{135}{50.0} \frac{\text{g}}{\text{cm}^3}$$
$$= 2.70 \frac{\text{g}}{\text{cm}^3}$$

As you can see, both blocks have the same density. Inspecting table 1.3, you can see that aluminum has a density of 2.70 g/cm³, so both blocks must be aluminum.

EXAMPLE 1.2   *(Optional)*

A rock with a volume of 4.50 cm³ has a mass of 15.0 g. What is the density of the rock? (Answer: 3.33 g/cm³)

Symbols and Equations

In the previous section, the relationship of density, mass, and volume was written with symbols. Density was represented by ρ, the lowercase letter rho in the Greek alphabet, mass was represented by m, and volume by V. The use of such symbols is established and accepted by convention, and these symbols are like the vocabulary of a foreign language. You learn what the symbols mean by use and practice, with the understanding that *each symbol stands for a very specific property or concept.* The symbols actually represent **quantities,** or *measured properties.* The symbol m thus represents a quantity of mass that is specified by a number and a unit, for example, 16 g. The symbol V represents a quantity of volume that is specified by a number and a unit, such as 17 cm³.

Symbols

Symbols usually provide a clue about which quantity they represent, such as m for mass and V for volume. However, in some cases, two quantities start with the same letter, such as volume and velocity, so the uppercase letter is used for one (V for volume) and the lowercase letter is used for the other (v for velocity). There are more quantities than upper- and lowercase letters, however, so letters from the Greek alphabet are also used, for example, ρ for mass density. Sometimes a subscript is used to identify a quantity in a particular situation, such as v_i for initial, or beginning, velocity and v_f for final velocity. Some symbols are also used to carry messages; for example, the Greek letter delta (Δ) is a message that means "the change in" a value. Other message symbols are the symbol $\therefore$, which means "therefore," and the symbol $\propto$, which means "is proportional to."

Equations

Symbols are used in an **equation,** a statement that describes a relationship where *the quantities on one side of the equal sign are identical to the quantities on the other side. Identical* refers to both the numbers and the units. Thus, in the equation describing the property of density, $\rho = m/V$, the numbers on both sides of the equal sign are identical (e.g., 5 = 10/2). The units on both sides of the equal sign are also identical (e.g., g/cm³ = g/cm³).

Equations are used to (1) *describe a property,* (2) *define a concept,* or (3) *describe how quantities change relative to each other.* Understanding how equations are used in these three classes is basic to comprehension of physical science. Each class of uses is considered separately in the following discussion.

Describing a property. You have already learned that the compactness of matter is described by the property called density. Density is a ratio of mass to a unit volume, or $\rho = m/V$. The key to understanding this property is to understand the meaning of a ratio and what *per* or *for each* means. Other examples of properties that will be defined by ratios are how fast something is moving (speed) and how rapidly a speed is changing (acceleration).

Defining a concept. A physical science concept is sometimes defined by specifying a measurement procedure. This is called an *operational definition* because a procedure is established that defines a concept as well as telling you how to measure it. Concepts of what is meant by force, mechanical work, and mechanical power and concepts involved in electrical and magnetic interactions can be defined by measurement procedures.

Describing how quantities change relative to each other. Nature is full of situations where one or more quantities change in value, or vary in size, in response to changes in other quantities. Changing quantities are called **variables.** Your weight, for example, is a variable that changes in size in response to changes in another variable; for example, the amount of food you eat. You already know about the pattern, or relationship, between these two variables. With all other factors being equal, an increase in the amount of food you eat results in an increase in your weight. When two variables increase (or decrease) together in the same ratio, they are said to be in *direct proportion.* When two variables are in direct proportion, *an increase or decrease in one variable results in the same relative increase or decrease in a second variable.*

Variables do not always increase or decrease in direct proportion. Sometimes one variable *increases* while a second variable *decreases* in the same ratio. This is an *inverse proportion* relationship. Other common relationships include one variable increasing in proportion to the *square* or to the *inverse square* of a second variable. Here are the forms of these four different types of proportional relationships:

Direct	$a \propto b$
Inverse	$a \propto 1/b$
Square	$a \propto b^2$
Inverse square	$a \propto 1/b^2$

FIGURE 1.12 The volume of fuel you have added to the fuel tank is directly proportional to the amount of time that the fuel pump has been running. This relationship can be described with an equation by using a proportionality constant.

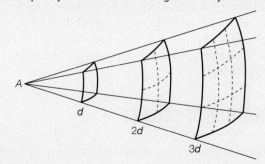
Proportionality Statements

Proportionality statements describe in general how two variables change relative to each other, but a proportionality statement is *not* an equation. For example, consider the last time you filled your fuel tank at a service station (figure 1.12). You could say that the volume of gasoline in an empty tank you are filling is directly proportional to the amount of time that the fuel pump was running, or

$$\text{volume} \propto \text{time}$$

This is not an equation because the numbers and units are not identical on both sides. Considering the units, for example, it should be clear that minutes do not equal gallons; they are two different quantities. To make a statement of proportionality into an equation, you need to apply a *proportionality constant,* which is sometimes given the symbol k. For the fuel pump example, the equation is

$$\text{volume} = (\text{time})(\text{constant})$$

or

$$V = tk$$

In this example, the constant is the flow of gasoline from the pump in L/min (a ratio). Assume the rate of flow is 10 L/min. In units, you can see why the statement is now an equality.

$$L = (\text{min})\left(\frac{L}{\text{min}}\right)$$

$$L = \frac{\text{min} \times L}{\text{min}}$$

$$L = L$$

Here are some tips to consider if your instructor uses a problem-solving approach. You could think of an equation as a *set of instructions.* The density equation, for example, is $\rho = m/V$. The equation tells you that mass density is a ratio of mass to volume, and you can find the density by dividing the mass by the volume. If you have difficulty, you either do not know the instructions or do not know how to follow the instructions (algebraic rules). Appendix D has worked examples that can help you with both the instructions and how to follow them. See also appendix E, which deals with problem solving.

THE NATURE OF SCIENCE

Most humans are curious, at least when they are young, and are motivated to understand their surroundings. These traits have existed since antiquity and have proven to be a powerful motivation. In recent times, the need to find out has motivated the launching of space probes to learn what is "out there," and humans have visited the Moon to satisfy their curiosity. Curiosity and the motivation to understand nature were no less powerful in the past than today. Over two thousand years ago, the ancient Greeks lacked the tools and technology of today and could only make conjectures about the workings of nature. These early seekers of understanding are known as *natural philosophers,* and they observed, thought, and wrote about the workings of all of nature. They are called philosophers because their understandings came from reasoning only, without experimental evidence. Nonetheless, some of their ideas were essentially correct and are still in use today. For example, the idea of matter being composed of *atoms* was first reasoned by certain ancient Greeks in the fifth century B.C. The idea of *elements,* basic components that make up matter, was developed much earlier but refined by the ancient Greeks in the fourth century B.C. The concept of what the elements are and the concept of the nature of atoms have changed over time, but the ideas first came from ancient natural philosophers.

The Scientific Method

Some historians identify the time of Galileo and Newton, approximately three hundred years ago, as the beginning of modern science. Like the ancient Greeks, Galileo and Newton were interested in studying all of nature. Since the time of Galileo and Newton, the content of physical science has increased in scope and specialization, but the basic means of acquiring understanding, the scientific investigation, has changed little. A *scientific investigation* provides understanding through *experimental evidence,* as opposed to the conjectures based on thinking only of the ancient natural philosophers. In chapter 2, for example, you will learn how certain ancient Greeks described how objects fall toward Earth with a thought-out, or reasoned, explanation. Galileo, on the other hand, changed how people thought of falling objects by developing explanations from both creative thinking and precise measurement of physical quantities, providing experimental evidence for his explanations. Experimental evidence provides explanations today, much as it did for Galileo, as relationships are found from precise measurements of physical quantities. Thus, scientific knowledge about nature has grown as measurements and investigations have led to understandings that lead to further measurements and investigations.

What is a scientific investigation and what methods are used to conduct one? Attempts have been made to describe scientific methods in a series of steps (define problem, gather data, make hypothesis, test, make conclusion), but no single description has ever been satisfactory to all concerned. Scientists do similar things in investigations, but there are different approaches and different ways to evaluate what they find. Overall, the similar things might look like this:

1. Observe some aspect of nature.
2. Propose an explanation for something observed.
3. Use the explanation to make predictions.
4. Test predictions by doing an experiment or by making more observations.
5. Modify explanation as needed.
6. Return to step 3.

The exact approach a scientist uses depends on the individual doing the investigation as well as the particular field of science being studied.

Another way to describe what goes on during a scientific investigation is to consider what can be generalized. At least three separate activities seem to be common to scientists in different fields as they conduct scientific investigations, and these generalized activities are:

- Collecting observations.
- Developing explanations.
- Testing explanations.

No particular order or routine can be generalized about these common elements. In fact, individual scientists might not even be involved in all three activities. Some, for example, might spend all of their time out in nature, "in the field" collecting data and generalizing about their findings. This is an acceptable means of scientific investigation in some fields of science. Yet, other scientists might spend all of their time indoors, at computer terminals, developing theoretical equations that offer explanations for generalizations made by others. Again, the work at a computer terminal is an acceptable means of scientific investigation. Thus, there is not an order of five steps that are followed, particularly by today's specialized scientists. This is one reason why many philosophers of science argue there is no such thing as *the* scientific method. There are common activities of observing, explaining, and testing in scientific investigations in different fields, and these activities will be discussed next.

Explanations and Investigations

Explanations in the natural sciences are concerned with things or events observed, and there can be several different means of developing or creating explanations. In general, explanations can come from the results of experiments, from an educated guess, or just from imaginative thinking. In fact, there are several examples in the history of science of valid explanations being developed even from dreams.

Explanations go by various names, each depending on intended use or stage of development. For example, an explanation in an early stage of development is sometimes called a *hypothesis.* A **hypothesis** is a tentative thought- or experiment-derived explanation. It must be compatible with observations and provide understanding of some aspect of nature, but the key word here is *tentative.* A hypothesis is tested by experiment and is rejected, or modified, if a single observation or test does not fit.

The successful testing of a hypothesis may lead to the design of experiments, or it could lead to the development of another hypothesis, which could, in turn, lead to the design of yet more experiments, which could lead to. . . . As you can see, this is a branching, ongoing process that is very difficult to describe in specific terms. In addition, it can be difficult to identify a conclusion, an endpoint in the process. The search for new concepts to explain experimental evidence may lead from a hypothesis to a new theory, which results in more new hypotheses. This is why one of the best ways to understand scientific methods is to study the history of science. Or you can conduct a scientific investigation yourself.

Testing a Hypothesis

In some cases, a hypothesis may be tested by simply making additional observations. For example, if you hypothesize that a certain species of bird uses cavities in trees as places to build nests, you could observe several birds of the species and record the kinds of nests they build and where they are built.

Another common method for testing a hypothesis involves devising an experiment. An experiment is a re-creation of an event or occurrence in a way that enables a scientist to support or disprove a hypothesis. This can be difficult since a particular event may be influenced by a great many separate things. For example, the growth of a crop, such as corn, is influenced by a variety of factors including sunlight, soil moisture, air and soil temperature, soil nutrients, competition with other plants, and damage caused by insects and diseases. It might seem that developing an understanding of the factors that influence the growth of corn would be an impossible task. To help unclutter such situations, scientists have devised what is known as a *controlled experiment.* A **controlled experiment** compares two situations that have all the influencing factors identical except one. The situation used as the basis of comparison is called the *control group* and the other is called the *experimental group.* The single influencing factor that is allowed to be different in the experimental group is called the *experimental variable.*

The importance of various factors in corn growth would have to be broken down into a large number of simple questions, as previously mentioned. Each question would provide the basis on which experimentation would occur. Each experiment would provide information about one factor involved in corn growth. For example, to test the hypothesis that nitrogen is an important soil nutrient for corn growth, an experiment could be performed in which two identical fields would be prepared and planted with corn. The two fields would need to have the same type of soil, exposure to sunlight, moisture, pest-control treatment, and history of previous crops. In practice in agricultural experiments, the two fields are typically side by side. One corn field would be planted with a known quantity nitrogen-containing fertilizer added to the soil (the experimental group), while the other field (control group) would be grown without any nitrogen-containing fertilizer. Data on growth would need to be gathered in a standard way. One way to do this would be to harvest all the corn plants and weigh them. Another way could be to determine the average height of the plants. After the experiment, the new data (facts) gathered would be analyzed. If there were no differences between the two groups, scientists could conclude that the variable (amount of nitrogen in the soil) evidently did not influence corn growth. However, if there was a difference, it would be likely that the variable was responsible for the difference between the control and experimental groups. In the case of corn growth in this kind of experiment, the presence of nitrogen fertilizer enhances growth.

Accept Results?

Scientists are not likely to accept the results of a single experiment, since it is possible that a variable was overlooked in the process of designing and carrying out the experiment. For example, it is possible that a fungus pest could have affected one field more than the other, or the machinery used to spray for pest control may have malfunctioned on one field, or a local rain shower could have provided more rain to one of the fields. Therefore, scientists repeat experiments with exactly the same conditions and compare the results of several replicates (copies) of the experiment before arriving at a conclusion.

Basic and Applied Research

Science is the process of understanding your environment. It begins with making observations, creating explanations, and conducting research experiments. New information and conclusions are based on the results of the research.

There are two types of scientific research: basic and applied. *Basic research* is driven by a search for understanding and may or may not have practical applications. Examples of basic research include seeking understandings about how the solar system was created, finding new information about matter by creating a new element in a research lab, or mapping temperature variations on the bottom of the Chesapeake Bay. Such basic research expands our knowledge but will not lead to practical results.

Applied research has a goal of solving some practical problem rather than just looking for answers. Examples of applied research include the creation and testing of a highly efficient fuel cell to run cars on hydrogen fuel, improving the energy efficiency of the refrigerator, or creating a faster computer chip from new materials.

Whether research is basic or applied depends somewhat on the time frame. If a practical use cannot be envisioned in the future, then it is definitely basic research. If a practical use is immediate, then the work is definitely applied research. If a practical use is developed some time in the future, then the research is partly basic and partly practical. For example, when the laser was invented, there was no practical use for it. It was called "an answer waiting for a question." Today, the laser has many, many practical applications.

Knowledge gained by basic research has sometimes resulted in the development of technological breakthroughs. On the other hand, other basic research—such as learning how the solar system formed—has no practical value other than satisfying our curiosity.

Questions to Discuss

1. Should funding priorities go to basic research, applied research, or both?
2. Should universities concentrate on basic research and industries concentrate on applied research, or should both do both types of research?
3. Should research-funding organizations specify which types of research should be funded?

The results of an experiment are considered convincing only when there is just one variable, many replicates of the same experiment have been conducted, and the results are consistent.

Furthermore, scientists often apply statistical tests to the results to help decide in an impartial manner if the results obtained are *valid* (meaningful; fit with other knowledge), *reliable* (give the same results repeatedly), and show cause-and-effect, or if they are just the result of random events.

During experimentation, scientists learn new information and formulate new questions that can lead to yet more experiments. One good experiment can result in a hundred new questions and experiments. The discovery of the structure of the DNA molecule by Watson and Crick resulted in thousands of experiments and stimulated the development of the entire field of molecular biology. Similarly, the discovery of molecules that regulate the growth of plants resulted in much research about how the molecules work and which molecules might be used for agricultural purposes.

If the processes of questioning and experimentation continue, and evidence continually and consistently supports the original hypothesis and other closely related hypotheses, the scientific community will begin to see how these hypotheses and facts fit together into a broad pattern.

Scientific Laws

Sometimes you can observe a series of relationships that seem to happen over and over again. There is a popular saying, for example, that "if anything can go wrong, it will." This is called Murphy's law. It is called a *law* because it describes a relationship between events that seems to happen time after time. If you drop a slice of buttered bread, for example, it can land two ways, butter side up or butter side down. According to Murphy's law, it will land butter side down. With this example, you know at least one way of testing the validity of Murphy's law.

Another "popular saying" type of relationship seems to exist between the cost of a houseplant and how long it lives. You could call it the "law of houseplant longevity." The relationship is that the life of a houseplant is inversely proportional to its purchase price. This "law" predicts that a $10 houseplant will wilt and die within a month, but a 50 cent houseplant will live for years. The inverse relationship is between the variables of (1) cost and (2) life span, meaning the more you pay for a plant, the shorter the time it will live. This would also mean that inexpensive plants will live for a long time. Since the relationship seems to occur time after time, it is called a law.

A **scientific law** describes an important relationship that is observed in nature to occur consistently time after time. Basically, scientific laws describe *what* happens in nature. The law is often identified with the name of a person associated with the formulation of the law. For example, with all other factors being equal, an increase in the temperature of the air in a balloon results in an increase in its volume. Likewise, a decrease in the temperature results in a decrease in the total volume of the balloon.

The volume of the balloon varies directly with the temperature of the air in the balloon, and this can be observed to occur consistently time after time. This relationship was first discovered in the latter part of the eighteenth century by two French scientists, A. C. Charles and Joseph Gay-Lussac. Today, the relationship is sometimes called *Charles' law.* When you read about a scientific *law,* you should remember that a law is a statement that means something about a relationship that you can observe time after time in nature.

Have you ever heard someone state that something behaved a certain way *because* of a scientific law? For example, a big truck accelerated slowly *because* of Newton's laws of motion. Perhaps this person misunderstands the nature of scientific laws. Scientific laws do not dictate the behavior of objects; they simply describe it. They do not say how things ought to act but rather how things *do* act. A scientific law is *descriptive;* it describes how things act.

Models and Theories

Often the part of nature being considered is too small or too large to be visible to the human eye and the use of a *model* is needed. A **model** (figure 1.13) is a description of a theory or idea that accounts for all known properties. The description can come in many different forms, such as an actual physical model, a computer model, a sketch, an analogy, or an equation. No one has ever seen the whole solar system, for example, and all you can see in the real world is the movement of the Sun, Moon, and planets against a background of stars. A physical model or sketch of the solar system, however, will give you a pretty good idea of what the solar system might look like. The physical model and the sketch are both models since they give you a mental picture of the solar system.

At the other end of the size scale, models of atoms and molecules are often used to help us understand what is happening in this otherwise invisible world. Also, a container of small, bouncing rubber balls can be used as a model to explain the relationships of Charles' law. This model helps you see what happens to invisible particles of air as the temperature, volume, and pressure of the gas change. Some models are better than others, and models constantly change along with our understanding about nature. Early twentieth-century models of atoms, for example, were based on a "planetary model," which had electrons in the role of planets moving around the nucleus, which played the role of the Sun. Today, the model has changed as our understandings about the nature of the atom have changed. Electrons are now pictured as vibrating with certain wavelengths, which can make standing waves only at certain distances from the nucleus. Thus, the model of the atom changed from one with electrons viewed as solid particles to one that views them as vibrations on a string.

The most recently developed scientific theory was refined and expanded during the 1970s. This theory concerns the surface of Earth, and it has changed our model of what the Earth is like. At first, however, the basic idea of today's accepted theory was pure and simple conjecture. The term *conjecture* usually means an explanation or idea based on speculation, or one based on trivial grounds without any real evidence. Scientists would look at a map of Africa and South America, for example, and mull over how the two continents seem

to be as pieces of a picture puzzle that had moved apart (figure 1.14). Any talk of moving continents was considered conjecture because it was not based on anything acceptable as real evidence.

Many years after the early musings about moving continents, evidence was collected from deep-sea drilling rigs that the ocean floor becomes progressively older toward the African and South American continents. This was good enough evidence to establish the "seafloor spreading hypothesis" that described the two continents moving apart.

If a hypothesis survives much experimental testing and leads, in turn, to the design of new experiments with the generation of new hypotheses that can be tested, you now have a working *theory.* A **theory** is defined as a broad, working hypothesis that is based on extensive experimental evidence. A scientific theory is a well-tested hypothesis that tells you *why* something happens. For example, the "seafloor spreading hypothesis" did survive requisite experimental testing and, together with other working hypotheses, is today found as part of the *plate tectonic theory.* The plate tectonic theory describes how the continents have moved apart, just like pieces of a picture puzzle. Is this the same idea that was once considered conjecture? Sort of, but this time it is supported by experimental evidence.

The term *scientific theory* is reserved for historic schemes of thought that have survived the test of detailed examination for long periods of time. The *atomic theory,* for example, was developed in the late 1800s and has been the subject of extensive investigation and experimentation over the last century. The atomic theory and other scientific theories form the framework of scientific thought and experimentation today. Scientific theories point to new ideas about the behavior of nature, and these ideas result in more experiments, more data to collect, and more explanations to develop. All of this may lead to a slight modification of an existing theory, a major modification, or perhaps the creation of an entirely new one. These activities continue in an ongoing attempt to satisfy the curiosity of people by understanding nature.

SCIENCE, NONSCIENCE, AND PSEUDOSCIENCE

As you can see from the discussion of the nature of science, a scientific approach to the world requires a certain way of thinking. There is an insistence on ample supporting evidence by numerous studies rather than easy acceptance of strongly stated opinions. Scientists must separate opinions from statements of fact. A scientist is a healthy skeptic.

Careful attention to detail is also important. Since scientists publish their findings and their colleagues examine their work, there is a strong desire to produce careful work that can be easily defended. This does not mean that scientists do not speculate and state opinions. When they do, however, they should take great care to clearly distinguish fact from opinion.

There is also a strong ethic of honesty. Scientists are not saints, but the fact that science is conducted out in the open in

A

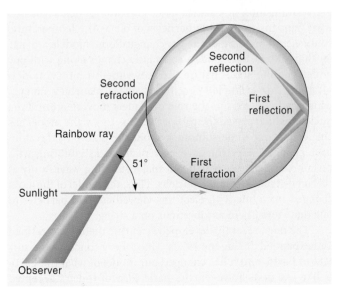

B

FIGURE 1.13 A model helps you visualize something that cannot be observed. You cannot observe what is making a double rainbow, for example, but models of light entering the upper and lower surface of a raindrop help you visualize what is happening. The drawings in (B) serve as a model that explains how a double rainbow is produced. (Also, see "The Rainbow" in chapter 7.)

A

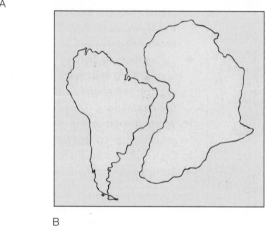

B

FIGURE 1.14 (A) Normal position of the continents on a world map. (B) A sketch of South America and Africa, suggesting that they once might have been joined together and subsequently separated by a continental drift.

front of one's peers tends to reduce the incidence of dishonesty. In addition, the scientific community strongly condemns and severely penalizes those who steal the ideas of others, perform shoddy science, or falsify data. Any of these infractions could lead to the loss of one's job and reputation.

From Experimentation to Application

The scientific method has helped us to understand and control many aspects of our natural world. Some information is extremely important in understanding the structure and functioning of things in our world but at first glance appears to have little practical value. For example, understanding the life cycle of a star may be important for people who are trying to answer questions about how the universe is changing, but it seems of little practical value to the average citizen. However, as our knowledge has increased, the time between first discovery to practical application has decreased significantly.

For example, scientists known as *genetic engineers* have altered the chemical code system of small organisms (microorganisms) so that they may produce many new drugs such as antibiotics, hormones, and enzymes. The ease with which these complex chemicals are produced would not have been possible had it not been for the information gained from the basic, theoretical sciences of microbiology, molecular biology, and genetics. Our understanding of how organisms genetically control the manufacture of proteins has led to the large-scale production of enzymes. Some of these chemicals can remove stains from clothing, deodorize, clean contact lenses, remove damaged skin from burn patients, and "stone wash" denim for clothing.

Science and Nonscience

The differences between science and nonscience are often based on the assumptions and methods used to gather and organize information and, most important, the testing of these assumptions. The difference between a scientist and a nonscientist is that a scientist continually challenges and tests principles and assumptions to determine a cause-and-effect relationship, whereas a nonscientist may not feel that this is important.

Once you understand the nature of science, you will not have any trouble identifying astronomy, chemistry, physics, and biology as sciences. But what about economics, sociology, anthropology, history, philosophy, and literature? All of these fields may make use of certain central ideas that are derived in a logical way, but they are also nonscientific in some ways. Some things cannot be approached using the scientific method. Art, literature, theology, and philosophy are rarely thought of as sciences. They are concerned with beauty, human emotion, and speculative thought rather than with facts and verifiable laws. On the other hand, physics, chemistry, geology, and biology are always considered sciences.

Music is an area of study in a middle ground where scientific approaches may be used to some extent. "Good" music is certainly unrelated to science, but the study of how the human larynx generates the sound of a song is based on scientific principles. Any serious student of music will study the physics of sound and how the vocal cords vibrate to generate sound waves.

Similarly, economists use mathematical models and established economic laws to make predictions about future economic conditions. However, the regular occurrence of unpredicted economic changes indicates that economics is far from scientific, since the reliability of predictions is a central criterion of science. Anthropology and sociology are also scientific in nature in many respects, but they cannot be considered true sciences because many of the generalizations they have developed cannot

Science and Society

Herbal Medicine, Legislation, and Pseudoscience

The Dietary Supplement Health and Education Act of 1994 effectively removed herbal medicines from regulation by the U.S. Food and Drug Administration (FDA). The law, which amended the federal Food, Drug, and Cosmetic Act, defined "dietary supplements" as a separate regulatory category with very few regulations.

Following passage of the Dietary Supplement Health and Education Act,

▶ Manufacturers no longer needed to ask for permission from the FDA to produce and distribute products unless it is a "new product." There is no useful definition of a "new product." Since herbal medicines have been used for centuries, manufacturers could show that most of the things they would market would have been used previously somewhere and would not be new products.

▶ Manufacturers are responsible for determining the safety of the product. The FDA can only act if it finds a product is unsafe.

▶ Manufacturers determine serving size or quantity to be taken.

▶ Manufacturers do not need to prove that the herbal product actually works. The public is not protected from the sale of useless but harmless products.

Box Figure 1.2 This is Saint-John's-wort, an herbal medicine claimed to relieve depression.

▶ Manufacturers may not make claims about curing particular disease conditions but may make vague claims such as "improves mood," "boosts immune system," or "supports liver function."

▶ If a manufacturer makes a claim about how a product affects the structure or function of the body, it must include the following disclaimer:

"These statements have not been evaluated by the Food and Drug Administration. This product is not intended to diagnose, treat, cure, or prevent any disease."

One outcome of this change in law was a continued growth in the amount of misinformation provided by marketers of herbal medicines under the guise of scientific statements—pseudoscience.

Let's examine an example of such misinformation.

The following is a typical information statement supporting the use of Saint-John's-wort to treat depression.

▶ Provides support for a positive mood balance.*

▶ May help promote general well-being.*

▶ Grows in Europe and the United States.

▶ The botanical species is positively identified by the sophisticated thin layer chromatography (TLC) technology.

▶ TLC verification method is as accurate and reliable for identifying true herbal species as human fingerprinting.

▶ Whole ground herb is minimally processed, dried, and pulverized.

▶ Each capsule contains 500 mg of Saint-John's-wort herb.

* These statements have not been evaluated by the Food and Drug Administration. This product is not intended to diagnose, treat, cure, or prevent any disease.

---

be tested by repeated experimentation. They also do not show a significantly high degree of cause-and-effect, or they have poor predictive value.

Pseudoscience

Pseudoscience (*pseudo* = false) is a deceptive practice that uses the appearance or language of science to convince, confuse, or mislead people into thinking that something has scientific validity when it does not. When pseudoscientific claims are closely examined, they are not found to be supported by unbiased tests. For example, although nutrition is a respected scientific field, many individuals and organizations make claims about their nutritional products and diets that cannot be supported. Because of nutritional research, we all know that we must obtain certain nutrients such as amino acids, vitamins, and minerals

from the food that we eat or we may become ill. Many scientific experiments reliably demonstrate the validity of this information. However, in most cases, it has not been proven that the nutritional supplements so vigorously promoted are as useful or desirable as advertised. Rather, selected bits of scientific information (amino acids, vitamins, and minerals are essential to good health) have been used to create the feeling that additional amounts of these nutritional supplements are necessary or that they can improve your health. In reality, the average person eating a varied diet will obtain all of these nutrients in adequate amounts, and nutritional supplements are not required.

Another related example involves the labeling of products as organic or natural. Marketers imply that organic or natural products have greater nutritive value because they are organically grown (grown without pesticides or synthetic fertilizers) or because they come from nature. Although there

Typical Herbal Supplements and Claims

Common Name of Plant	Scientific Name of Plant	Claimed Benefit
St.-John's-wort	*Hypericum perforatum*	Relieves mild or moderate depression
Gingko	*Ginkgo biloba*	Improves memory
Garlic	*Allium sativum*	Improves immune function; prevents atherosclerosis
Ginseng	*Panax quinquefolius* or *P. ginseng*	Elevates energy level
Green tea	*Camellia sinensis*	Prevents certain cancers, atherosclerosis, and tooth decay
Echinacea (purple coneflower)	*Echinacea pupurea*	Improves immune system
Black cohosh	*Cimicifuga racemosa* or *Actaea racemosa*	Relieves menopause symptoms or painful menstruation

How does this information exemplify pseudoscience?

1. The first two statements provide vague statements about mental health. They do not say that it is used to treat depression because that link has not been scientifically established to most scientists' satisfaction. The manufacturer relies on scientifically unverified statements in popular literature and on the Internet, and propagated by word of mouth to provide the demand for the product.

2. The two statements about thin layer chromatography are meant to suggest accuracy and purity. In fact, thin layer chromatography is a poor way to identify plants. It simply looks at the pigments present in plants. The use of this technique is unnecessary, since Saint-John's-wort is a plant that is easily identified by amateurs.

3. The statement about the herb being "minimally processed" is meant to suggest that nothing was lost from the plant during the preparation of the capsule and implies a high level of quality control and purity. Although the statement may be true, it is quite possible that the plant material collected contained dust, insects, and other contaminants. It is also highly likely that individual plants differ greatly in the amount of specific chemicals they contain.

Questions to Discuss

1. Select an herbal medicine and examine the claims made about this herb. Design a means of experimentally testing these claims. Decide with your group what would be acceptable evidence and what would not be acceptable.

2. Discuss with your group why people tend to ignore experimentally testing a claim.

3. Discuss with your group the possibility of a "placebo effect," that is, if someone strongly believes a claim, it might come true.

are questions about the health effects of trace amounts of pesticides in foods, no scientific study has shown that a diet of natural or organic products has any benefit over other diets. The poisons curare, strychnine, and nicotine are all organic molecules that are produced in nature by plants that could be grown organically, but we would not want to include them in our diet.

Absurd claims that are clearly pseudoscience sometimes appear to gain public acceptance because of promotion in the media. Thus, some people continue to believe stories that psychics can really help solve puzzling crimes, that perpetual energy machines exist, or that sources of water can be found by a person with a forked stick. Such claims could be subjected to scientific testing and disposed of if they fail the test, but this process is generally ignored. In addition to experimentally testing such a claim that appears to be pseudoscience, here are some questions that you should consider when you suspect something is pseudoscience:

1. What is the background and scientific experience of the person promoting the claim?
2. How many articles have been published by the person in peer-reviewed scientific journals?
3. Has the person given invited scientific talks at universities and national professional organization meetings?
4. Has the claim been researched and published by the person in a peer-reviewed scientific journal *and* have other scientists independently validated the claim?
5. Does the person have something to gain by making the claim?

Limitations of Science

By definition, science is a way of thinking and seeking information to solve problems. Therefore, scientific methods can be

People Behind the Science

Florence Bascom (1862–1945)

Florence Bascom, a U.S. geologist, was an expert in the study of rocks and minerals and founded the geology department at Bryn Mawr College, Pennsylvania. This department was responsible for training the foremost women geologists of the early twentieth century.

Born in Williamstown, Massachusetts, in 1862, Bascom was the youngest of the six children of suffragist and schoolteacher Emma Curtiss Bascom and William Bascom, professor of philosophy at Williams College. Her father, a supporter of suffrage and the education of women, later became president of the University of Wisconsin, to which women were admitted in 1875. Florence Bascom enrolled there in 1877 and with other women was allowed limited access to the facilities but was denied access to classrooms filled with men. In spite of this, she earned a B.A. in 1882, a B.Sc. in 1884, and an M.S. in 1887. When Johns Hopkins University graduate school opened to women in 1889, Bascom was allowed to enroll to study geology on the condition that she sit behind a screen to avoid distracting the male students. With the support of her advisor, George Huntington Williams, and her father, she managed in 1893 to become the second woman to gain a Ph.D. in geology (the first being Mary Holmes at the University of Michigan in 1888).

Bascom's interest in geology had been sparked by a driving tour she took with her father and his friend Edward Orton, a geology professor at Ohio State University. It was an exciting time for geologists with new areas opening up all the time. Bascom was also inspired by her teachers at Wisconsin and Johns Hopkins, who were experts in the new fields of metamorphism and crystallography. Bascom's Ph.D. thesis was a study of rocks that had previously been thought to be sediments but which she proved to be metamorphosed lava flows.

While studying for her doctorate, Bascom became a popular teacher, passing on her enthusiasm and rigor to her students. She taught at the Hampton Institute for Negroes and American Indians and at Rockford College before becoming an instructor and associate professor of geology at Ohio State University from 1892 to 1895. Moving to Bryn Mawr College, where geology was considered subordinate to the other sciences, she spent two years teaching in a storeroom while building a considerable collection of fossils, rocks, and minerals. While at Bryn Mawr, she took great pride in passing on her knowledge and training to a generation of women who would become successful. At Bryn Mawr, she rose rapidly, becoming reader (1898), associate professor (1903), professor (1906), and finally professor emerita from 1928 till her death in 1945 in Northampton, Massachusetts.

Bascom became, in 1896, the first woman to work as a geologist on the U.S. Geological Survey, spending her summers mapping formations in Pennsylvania, Maryland, and New Jersey and her winters analyzing slides. Her results were published in Geographical Society of America bulletins. In 1924, she became the first woman to be elected a fellow of the Geographical Society and went on, in 1930, to become the first woman vice president. She was associate editor of the *American Geologist* (1896–1905) and achieved a four-star place in the first edition of *American Men and Women of Science* (1906), a sign of how highly regarded she was in her field.

Bascom was the author of over forty research papers. She was an expert on the crystalline rocks of the Appalachian Piedmont, and she published her research on Piedmont geomorphology. Geologists in the Piedmont area still value her contributions, and she is still a powerful model for women seeking status in the field of geology today.

Source: Modified from the *Hutchinson Dictionary of Scientific Biography*. © RM, 2009. All rights reserved. Helicon Publishing is a division of RM.

---

applied only to questions that have a factual basis. Questions concerning morals, value judgments, social issues, and attitudes cannot be answered using the scientific methods. What makes a painting great? What is the best type of music? Which wine is best? What color should I paint my car? These questions are related to values, beliefs, and tastes; therefore, scientific methods cannot be used to answer them.

Science is also limited by the ability of people to pry understanding from the natural world. People are fallible and do not always come to the right conclusions, because information is lacking or misinterpreted, but science is self-correcting. As new information is gathered, old incorrect ways of thinking must be changed or discarded. For example, at one time, people were sure that the Sun went around Earth. They observed that the Sun rose in the east and traveled across the sky to set in the west. Since they could not feel Earth moving, it seemed perfectly logical that the sun traveled around Earth. Once they understood that Earth rotated on its axis, people began to understand that the rising and setting of the Sun could be explained in other ways. A completely new concept of the relationship between the Sun and Earth developed.

Although this kind of study seems rather primitive to us today, this change in thinking about the Sun and Earth was a very important step in understanding the universe and how the various parts are related to one another. This background information was built upon by many generations of astronomers and space scientists, and finally led to space exploration.

People also need to understand that science cannot answer all the problems of our time. Although science is a powerful tool, there are many questions it cannot answer and many problems it cannot solve. The behavior and desires of people generate most of the problems societies face. Famine, drug abuse, and pollution are human-caused and must be resolved by humans. Science may provide some tools for social planners, politicians, and ethical thinkers, but science does not have, nor does it attempt to provide, all the answers to the problems of the human race. Science is merely one of the tools at our disposal.

SUMMARY

Science is a search for order in our surroundings. People have *concepts,* or mental images, about material *objects* and intangible *events* in their surroundings. Concepts are used for thinking and communicating. Concepts are based on *properties,* or attributes that describe a thing or event. Every property implies a *referent* that describes the property. Referents are not always explicit, and most communications require assumptions.

Measurement is a process that uses a well-defined and agreed upon *referent* to describe a *standard unit.* The unit is compared to the property being defined by an *operation* that determines the *value* of the unit by *counting.* Measurements are always reported with a *number,* or value, and a *name* for the unit.

The two major *systems* of standard units are the *English system* and the *metric system.* The English system uses standard units that were originally based on human body parts, and the metric system uses standard units based on referents found in nature. The metric system also uses a system of prefixes to express larger or smaller amounts of units. The metric standard units for length, mass, and time are the *meter, kilogram,* and *second.*

Measurement information used to describe something is called *data.* One way to extract meanings and generalizations from data is to use a *ratio,* a simplified relationship between two numbers. Density is a ratio of mass to volume, or $\rho = m/V$.

Symbols are used to represent *quantities,* or measured properties. Symbols are used in *equations,* which are shorthand statements that describe a relationship where the quantities (both number values and units) are identical on both sides of the equal sign. Equations are used to (1) *describe* a property, (2) *define* a concept, or (3) *describe* how *quantities change* together.

Quantities that can have different values at different times are called *variables.* Variables that increase or decrease together in the same ratio are said to be in *direct proportion.* If one variable increases while the other decreases in the same ratio, the variables are in *inverse*

proportion. Proportionality statements are not necessarily equations. A *proportionality constant* can be used to make such a statement into an equation. Proportionality constants might have numerical value only, without units, or they might have both value and units.

Modern science began about three hundred years ago during the time of Galileo and Newton. Since that time, *scientific investigation* has been used to provide *experimental evidence* about nature. The investigations provide *accurate, specific,* and *reliable* data that are used to develop and test *explanations.* A *hypothesis* is a tentative explanation that is accepted or rejected from experimental data. An accepted hypothesis may result in a *scientific law,* an explanation concerned with important phenomena. Laws are sometimes identified with the name of a scientist and can be expressed verbally, with an equation, or with a graph.

A *model* is used to help understand something that cannot be observed directly, explaining the unknown in terms of things already understood. Physical models, mental models, and equations are all examples of models that explain how nature behaves. A *theory* is a broad, detailed explanation that guides development and interpretations of experiments in a field of study.

Science and *nonscience* can be distinguished by the kinds of laws and rules that are constructed to unify the body of knowledge. Science involves the continuous *testing* of rules and principles by the collection of new facts. If the rules are not testable, or if no rules are used, it is not science. *Pseudoscience* uses scientific appearances to mislead.

Summary of Equations

1.1
$$\text{density} = \frac{\text{mass}}{\text{volume}}$$
$$\rho = \frac{m}{V}$$

KEY TERMS

controlled experiment (p. **13**)

data (p. **7**)

density (p. **8**)

equation (p. **10**)

hypothesis (p. **13**)

measurement (p. **4**)

model (p. **15**)

properties (p. **3**)

pseudoscience (p. **18**)

quantities (p. **10**)

referent (p. **4**)

scientific law (p. **14**)

standard unit (p. **4**)

theory (p. **15**)

unit (p. **4**)

variables (p. **11**)

APPLYING THE CONCEPTS

Additional multiple choice and conversion exercise questions are on the website: **www.mhhe.com/tillery.**

1. The process of comparing a property of an object to a well-defined and agreed-upon referent is called the process of
   a. generalizing.
   b. measurement.
   c. graphing.
   d. scientific investigation.

2. The height of an average person is closest to
   a. 1.0 m.
   b. 1.5 m.
   c. 2.5 m.
   d. 3.5 m.

3. Which of the following standard units is defined in terms of an object as opposed to an event?
   a. kilogram
   b. meter
   c. second
   d. none of the above

4. One-half liter of water has a mass of
   a. 0.5 g.
   b. 5 g.
   c. 50 g.
   d. 500 g.

5. A cubic centimeter (cm^3) of water has a mass of about 1
   a. mL.
   b. kg.
   c. g.
   d. dm.

6. Measurement information that is used to describe something is called
   a. referents.
   b. properties.
   c. data.
   d. a scientific investigation.

7. The property of volume is a measure of
   a. how much matter an object contains.
   b. how much space an object occupies.
   c. the compactness of matter in a certain size.
   d. the area on the outside surface.

8. As the volume of a cube becomes larger and larger, the surface area-to-volume ratio
   a. increases.
   b. decreases.
   c. remains the same.
   d. sometimes increases and sometimes decreases.

9. If you consider a very small portion of a material that is the same throughout, the density of the small sample will be
   a. much less.
   b. slightly less.
   c. the same.
   d. greater.

10. A scientific law can be expressed as
    a. a written concept.
    b. an equation.
    c. a graph.
    d. Any of the above is correct.

Answers

1. b 2. b 3. a 4. d 5. c 6. c 7. b 8. b 9. c 10. d

QUESTIONS FOR THOUGHT

1. What is a concept?
2. What are two components of a measurement statement? What does each component tell you?
3. Other than familiarity, what are the advantages of the English system of measurement?
4. Define the metric standard units for length, mass, and time.
5. Does the density of a liquid change with the shape of a container? Explain.
6. Does a flattened pancake of clay have the same density as the same clay rolled into a ball? Explain.
7. Compare and contrast a scientific hypothesis and a scientific law.
8. What is a model? How are models used?
9. Are all theories always completely accepted or completely rejected? Explain.
10. What is pseudoscience and how can you always recognize it?

FOR FURTHER ANALYSIS

1. Select a statement that you feel might represent pseudoscience. Write an essay supporting *and* refuting your selection, noting facts that support one position or the other.
2. Evaluate the statement that science cannot solve human-produced problems such as pollution. What does it mean to say pollution is caused by humans and can only be solved by humans? Provide evidence that supports your position.
3. Make an experimental evaluation of what happens to the density of a substance at larger and larger volumes.
4. If your wage were dependent on your work-time squared, how would it affect your pay if you double your hours?
5. *Merriam-Webster's 11th Collegiate Dictionary* defines science, in part, as "knowledge or a system of knowledge covering general truths or the operation of general laws especially as obtained and tested through scientific method." How would you define science?
6. Are there any ways in which scientific methods differ from common-sense methods of reasoning?
7. The United States is the only country in the world that does not use the metric system of measurement. With this understanding, make a list of advantages and disadvantages for adopting the metric system in the United States.

INVITATION TO INQUIRY

Paper Helicopters

Construct paper helicopters and study the effects that various variables have on their flight. After considering the size you wish to test, copy the pattern shown in figure 1.15A on a sheet of notebook paper. Note that solid lines are to be cut and dashed lines are to be folded. Make three scissor cuts on the solid lines. Fold A toward you and B away from you to form the wings. Then fold C and D inward to overlap, forming the body. Finally, fold up the bottom on the dashed line and hold it together with a paper clip. Your finished product should look like the helicopter in figure 1.15B. Try a preliminary flight test by standing on a chair or stairs and dropping your helicopter.

Decide what variables you would like to study to find out how they influence the total flight time. Consider how you will hold everything else constant while changing one variable at a time. You can change the wing area by making new helicopters with more or less area in the A and B flaps. You can change the weight by adding more paper clips. Study these and other variables to find out who can design a helicopter that will remain in the air the longest. Who can design a helicopter that is most accurate in hitting a target?

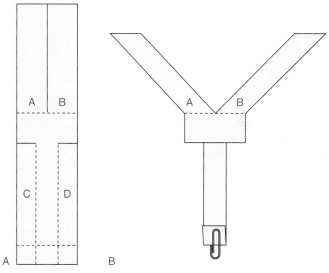

FIGURE 1.15 Pattern for a paper helicopter.

PARALLEL EXERCISES

The exercises in groups A and B cover the same concepts. Solutions to group A exercises are located in appendix D.

Group A

Note: You will need to refer to table 1.3 to complete some of the following exercises.

1. What is your height in meters? In centimeters?
2. What is the mass density of mercury if 20.0 cm³ has a mass of 272 g?
3. What is the mass of a 10.0 cm³ cube of lead?
4. What is the volume of a rock with a mass density of 3.00 g/cm³ and a mass of 600 g?
5. If you have 34.0 g of a 50.0 cm³ volume of one of the substances listed in table 1.3, which one is it?
6. What is the mass of water in a 40 L aquarium?

7. A 2.1 kg pile of aluminum cans is melted, then cooled into a solid cube. What is the volume of the cube?
8. A cubic box contains 1,000 g of water. What is the length of one side of the box in meters? Explain your reasoning.
9. A loaf of bread (volume 3,000 cm³) with a density of 0.2 g/cm³ is crushed in the bottom of a grocery bag into a volume of 1,500 cm³. What is the density of the mashed bread?
10. According to table 1.3, what volume of copper would be needed to balance a 1.00 cm³ sample of lead on a two-pan laboratory balance?

Group B

Note: You will need to refer to table 1.3 to complete some of the following exercises.

1. What is your mass in kilograms? In grams?
2. What is the mass density of iron if 5.0 cm³ has a mass of 39.5 g?
3. What is the mass of a 10.0 cm³ cube of copper?
4. If ice has a mass density of 0.92 g/cm³, what is the volume of 5,000 g of ice?
5. If you have 51.5 g of a 50.0 cm³ volume of one of the substances listed in table 1.3, which one is it?
6. What is the mass of gasoline ($\rho = 0.680$ g/cm³) in a 94.6 L gasoline tank?
7. What is the volume of a 2.00 kg pile of iron cans that are melted, then cooled into a solid cube?
8. A cubic tank holds 1,000.0 kg of water. What are the dimensions of the tank in meters? Explain your reasoning.
9. A hot dog bun (volume 240 cm³) with a density of 0.15 g/cm³ is crushed in a picnic cooler into a volume of 195 cm³. What is the new density of the bun?
10. According to table 1.3, what volume of iron would be needed to balance a 1.00 cm³ sample of lead on a two-pan laboratory balance?

2

Motion

Information about the mass of a hot air balloon and forces on the balloon will enable you to predict if it is going to move up, down, or drift across the river. This chapter is about such relationships between force, mass, and changes in motion.

CORE **CONCEPT**

A net force is required for any change in a state of motion.

OUTLINE

Physics

▷ Mechanical work is the product of a force and the distance an object moves as a result of the force (Ch. 3).

▷ Resonance occurs when the frequency of an applied force matches a natural frequency (Ch. 5).

Life Science

▷ Biomechanics is the application of principles of motion to living things (Ch. 20).

Astronomy

▷ Gravity pulls clouds of gas together in space to form stars (Ch. 12).

▷ The solar system may have formed when gas, dust, and elements from a previously existing star were pulled together by gravity into a large disk (Ch. 12).

Earth Science

▷ Earth's surface is made up of large, rigid plates that move from applied forces (Ch. 15).

OVERVIEW

In chapter 1, you learned some "tools and rules" and some techniques for finding order in your surroundings. Order is often found in the form of patterns, or relationships between quantities that are expressed as equations. Equations can be used to (1) describe properties, (2) define concepts, and (3) describe how quantities change relative to one another. In all three uses, patterns are quantified, conceptualized, and used to gain a general understanding about what is happening in nature.

In the study of science, certain parts of nature are often considered and studied together for convenience. One of the more obvious groupings involves *movement*. Most objects around you appear to spend a great deal of time sitting quietly without

motion. Buildings, rocks, utility poles, and trees rarely, if ever, move from one place to another. Even things that do move from time to time sit still for a great deal of time. This includes you, automobiles, and bicycles (figure 2.1). On the other hand, the Sun, the Moon, and starry heavens always seem to move, never standing still. Why do things stand still? Why do things move?

Questions about motion have captured the attention of people for thousands of years. But the ancient people answered questions about motion with stories of mysticism and spirits that lived in objects. It was during the classic Greek culture, between 600 B.C. and 300 B.C., that people began to look beyond magic and spirits. One particular Greek philosopher, Aristotle, wrote a theory

about the universe that offered not only explanations about things such as motion but also offered a sense of beauty, order, and perfection. The theory seemed to fit with other ideas that people had and was held to be correct for nearly two thousand years after it was written. It was not until the work of Galileo and Newton during the 1600s that a new, correct understanding about motion was developed. The development of ideas about motion is an amazing and absorbing story. You will learn in this chapter how to describe and use some properties of motion. This will provide some basic understandings about motion and will be very helpful in understanding some important aspects of astronomy and the earth sciences, as well as the movement of living things.

DESCRIBING MOTION

Motion is one of the more common events in your surroundings. You can see motion in natural events such as clouds moving, rain and snow falling, and streams of water, all moving in a never-ending cycle. Motion can also be seen in the activities of people who walk, jog, or drive various machines from place to place. Motion is so common that you would think everyone would intuitively understand the concepts of motion, but history indicates that it was only during the past three hundred years or so that people began to understand motion correctly. Perhaps the correct concepts are subtle and contrary to common sense,

requiring a search for simple, clear concepts in an otherwise complex situation. The process of finding such order in a multitude of sensory impressions by taking measurable data and then inventing a concept to describe what is happening is the activity called *science*. We will now apply this process to motion.

What is motion? Consider a ball that you notice one morning in the middle of a lawn. Later in the afternoon, you notice that the ball is at the edge of the lawn, against a fence, and you wonder if the wind or some person moved the ball. You do not know if the wind blew it at a steady rate, if many gusts of wind moved it, or even if some children kicked it all over the yard. All you know for sure is that the ball has been moved because are moving across the land outside the car since your location on the highway changes from one moment to another. Observing your fellow passenger, however, reveals no change of position. You are in motion relative to the highway outside the car. You are not in motion relative to your fellow passenger. Your motion, and the motion of any other object or body, is the process of a change in position *relative* to some reference object or location. Thus, *motion* can be defined as the act or process of changing position relative to some reference during a period of time.

MEASURING MOTION

You have learned that objects can be described by measuring certain fundamental properties such as mass and length. Since motion involves (1) a change of position and (2) the passage of *time*, the motion of objects can be described by using combinations of the fundamental properties of length and time. Combinations of these measurements describe three properties of motion: *speed, velocity,* and *acceleration.*

Speed

Suppose you are in a car that is moving over a straight road. How could you describe your motion? You need at least two measurements: (1) the distance you have traveled and (2) the time that has elapsed while you covered this distance. Such a distance and time can be expressed as a ratio that describes your motion. This ratio is a property of motion called **speed,** which is a measure of how fast you are moving. Speed is defined as distance per unit of time, or

$$\text{speed} = \frac{\text{distance}}{\text{time}}$$

The units used to describe speed are usually miles/hour (mi/h), kilometers/hour (km/h), or meters/second (m/s).

Let's go back to your car that is moving over a straight highway and imagine you are driving to cover equal distances in equal periods of time (figure 2.2). If you use a stopwatch to measure the time required to cover the distance between highway mile markers (those little signs with numbers along major highways), the time intervals will all be equal. You might find, for example, that one minute elapses between each mile marker. Such a uniform straight-line motion that covers equal distances in equal periods of time is the simplest kind of motion.

If your car were moving over equal distances in equal periods of time, it would have a *constant speed.* This means that the car is neither speeding up nor slowing down. It is usually difficult to maintain a constant speed. Other cars and distractions such as interesting scenery cause you to reduce your speed. At other times you increase your speed. If you calculate your speed over an entire trip, you are considering a large distance between two places and the total time that elapsed. The increases and

FIGURE 2.1 The motion of this snowboarder, and of other moving objects, can be described in terms of the distance covered during a certain time period.

it is in a different position after some time passed. These are the two important aspects of motion: (1) a change of position and (2) the passage of time.

If you did happen to see the ball rolling across the lawn in the wind, you would see more than the ball at just two locations. You would see the ball moving continuously. You could consider, however, the ball in continuous motion to be a series of individual locations with very small time intervals. Moving involves a change of position during some time period. Motion is the act or process of something changing position.

The motion of an object is usually described with respect to something else that is considered to be not moving. (Such a stationary object is said to be "at rest.") Imagine that you are traveling in an automobile with another person. You know that you

FIGURE 2.2 If you know the value of any two of the three variables of distance, time, and speed, you can find the third. What is the average speed of this car?

decreases in speed would be averaged. Therefore, most speed calculations are for an *average speed.* The speed at any specific instant is called the *instantaneous speed.* To calculate the instantaneous speed, you would need to consider a very short time interval—one that approaches zero. An easier way would be to use the speedometer, which shows the speed at any instant.

It is easier to study the relationships between quantities if you use symbols instead of writing out the whole word. The letter *v* can be used to stand for speed when dealing with straight-line motion, which is the only kind of motion that will be considered in the problems in this text. A bar over the *v* ($\bar{v}$) is a symbol that means average (it is read "*v*-bar" or "*v*-average"). The letter *d* can be used to stand for distance and the letter *t* to stand for time. The relationship between average speed, distance, and time is therefore

$$\bar{v} = \frac{d}{t} \qquad \text{equation 2.1}$$

This is one of the three types of equations that were discussed earlier, and in this case, the equation defines a motion property.

Constant, instantaneous, or average speeds can be measured with any distance and time units. Common units in the English system are miles/hour and feet/second. Metric units for speed are commonly kilometers/hour and meters/second. The ratio of any distance/time is usually read as distance per time, such as miles per hour.

Velocity

The word *velocity* is sometimes used interchangeably with the word *speed,* but there is a difference. **Velocity** describes the *speed and direction* of a moving object. For example, a speed might be described as 60 km/h. A velocity might be described as 60 km/h to the west. To produce a change in velocity, either the speed or the direction is changed (or both are changed). A satellite moving with a constant speed in a circular orbit around Earth does not have a constant velocity since its direction of movement is constantly changing. Velocities can be represented graphically with arrows. The lengths of the arrows are proportional to the speed, and the arrowheads indicate the direction (figure 2.3).

EXAMPLE 2.1 *(Optional)*

The driver of a car moving at 72.0 km/h drops a road map on the floor. It takes her 3.00 seconds to locate and pick up the map. How far did she travel during this time?

SOLUTION

The car has a speed of 72.0 km/h and the time factor is 3.00 s, so km/h must be converted to m/s. From inside the front cover of this book, the conversion factor is 1 km/h = 0.2778 m/s, so

$$\bar{v} = \frac{0.2778 \frac{m}{s}}{\frac{km}{h}} \times 72.0 \frac{km}{h}$$

$$= (0.2778)(72.0) \frac{m}{s} \times \frac{h}{km} \times \frac{km}{h}$$

$$= 20.0 \frac{m}{s}$$

The relationship between the three variables, $\bar{v}$, *t,* and *d,* is found in equation 2.1: $\bar{v} = d/t$

$$\bar{v} = 20.0 \frac{m}{s} \qquad \bar{v} = \frac{d}{t}$$

$$t = 3.00 \text{ s} \qquad \bar{v}t = \frac{d\cancel{t}}{\cancel{t}}$$

$$d = ? \qquad d = \bar{v}t$$

$$= \left(20.0 \frac{m}{s}\right)(3.00 \text{ s})$$

$$= (20.0)(3.00) \frac{m}{\cancel{s}} \times \frac{\cancel{s}}{1}$$

$$= \boxed{60.0 \text{ m}}$$

EXAMPLE 2.2 *(Optional)*

A bicycle has an average speed of 8.00 km/h. How far will it travel in 10.0 seconds? (Answer: 22.2 m)

FIGURE 2.3 Velocity can be presented graphically with arrows. Here are three different velocities represented by three different arrows. The length of each arrow is proportional to the speed, and the arrowhead shows the direction of travel.

Acceleration

Motion can be changed in three different ways: (1) by changing the speed, (2) by changing the direction of travel, or (3) by changing both the speed and the direction of travel. Since velocity describes both the speed and the direction of travel, any of these three changes will result in a change of velocity. You need at least one additional measurement to describe a change of motion, which is how much time elapsed while the change was taking place. The change of velocity and time can be combined to define the *rate* at which the motion was changed. This rate is called **acceleration.** Acceleration is defined as a change of velocity per unit time, or

$$\text{acceleration} = \frac{\text{change of velocity}}{\text{time elapsed}}$$

Another way of saying "change in velocity" is the final velocity minus the initial velocity, so the relationship can also be written as

$$\text{acceleration} = \frac{\text{final velocity} - \text{initial velocity}}{\text{time elapsed}}$$

Acceleration due to a change in speed only can be calculated as follows. Consider a car that is moving with a constant, straight-line velocity of 60 km/h when the driver accelerates

to 80 km/h. Suppose it takes 4 s to increase the velocity of 60 km/h to 80 km/h. The change in velocity is therefore 80 km/h minus 60 km/h, or 20 km/h. The acceleration was

$$\text{acceleration} = \frac{80 \frac{\text{km}}{\text{h}} - 60 \frac{\text{km}}{\text{h}}}{4\text{s}}$$

$$= \frac{20 \frac{\text{km}}{\text{h}}}{4\text{s}}$$

$$= 5 \frac{\text{km/h}}{\text{s}} \text{ or,}$$

$$= 5 \text{ km/h/s}$$

The average acceleration of the car was 5 km/h for each ("per") second. This is another way of saying that the velocity increases an average of 5 km/h in each second. The velocity of the car was 60 km/h when the acceleration began (initial velocity). At the end of 1 s, the velocity was 65 km/h. At the end of 2 s, it was 70 km/h; at the end of 3 s, 75 km/h; and at the end of 4 s (total time elapsed), the velocity was 80 km/h (final velocity). Note how fast the velocity is changing with time. In summary,

start (initial velocity)	60 km/h
end of first second	65 km/h
end of second second	70 km/h
end of third second	75 km/h
end of fourth second (final velocity)	80 km/h

As you can see, acceleration is really a description of how fast the speed is changing; in this case, it is increasing 5 km/h each second.

Usually, you would want all the units to be the same, so you would convert km/h to m/s. A change in velocity of 5.0 km/h converts to 1.4 m/s and the acceleration would be 1.4 m/s/s. The units "m/s per s" mean what change of velocity (1.4 m/s) is occurring every second. The combination "m/s/s" is rather cumbersome, so it is typically treated mathematically to simplify the expression (to simplify a fraction, invert the divisor and multiply, or m/s × 1/s = m/s²). Remember that the expression 1.4 m/s² means the same as 1.4 m/s per s, a change of velocity in a given time period.

The relationship among the quantities involved in acceleration can be represented with the symbols a for average acceleration, v_f for final velocity, v_i for initial velocity, and t for time. The relationship is

$$a = \frac{v_f - v_i}{t} \qquad \qquad \textbf{equation 2.2}$$

There are also other changes in the motion of an object that are associated with acceleration. One of the more obvious is a change that results in a decreased velocity. Your car's brakes, for example, can slow your car or bring it to a complete stop. This is *negative acceleration,* which is sometimes called *deceleration.* Another change in the motion of an object is a change of direction. Velocity encompasses both the rate of motion as well as direction, so a change of direction is an acceleration. The satellite

---

CONCEPTS APPLIED

Style Speeds

Observe how many different styles of walking you can identify in students walking across the campus. Identify each style with a descriptive word or phrase.

Is there any relationship between any particular style of walking and the speed of walking? You could find the speed of walking by measuring a distance, such as the distance between two trees, then measuring the time required for a student to walk the distance. Find the average speed for each identified style of walking by averaging the walking speeds of ten people.

Report any relationships you find between styles of walking and the average speed of people with each style. Include any problems you found in measuring, collecting data, and reaching conclusions.

CONCEPTS APPLIED

How Fast Is a Stream?

A stream is a moving body of water. How could you measure the speed of a stream? Would timing how long it takes a floating leaf to move a measured distance help?

What kind of relationship, if any, would you predict for the speed of a stream and a recent rainfall? Would you predict a direct relationship? Make some measurements of stream speeds and compare your findings to recent rainfall amounts.

Connections ...

Travel

The super-speed magnetic levitation (maglev) train is a completely new technology based on magnetically suspending a train 3 to 10 cm (about 1 to 4 in) above a monorail, then moving it along with a magnetic field that travels along the monorail guides. The maglev train does not have friction between wheels and the rails since it does not have wheels. This lack of resistance and the easily manipulated magnetic fields makes very short acceleration distances possible. For example, a German maglev train can accelerate from 0 to 300 km/h (about 185 mi/h) over a distance of just 5 km (about 3 mi). A conventional train with wheels requires about 30 km (about 19 mi) in order to reach the same speed from a standing start. The maglev is attractive for short runs because of its superior acceleration. It is also attractive for longer runs because of its high top speed—up to about 500 km/h (about 310 mi/h). Today, only an aircraft can match such a speed.

moving with a constant speed in a circular orbit around Earth is constantly changing its direction of movement. It is therefore constantly accelerating because of this constant change in its motion. Your automobile has three devices that could change the state of its motion. Your automobile therefore has three accelerators—the gas pedal (which can increase magnitude of

---

CONCEPTS APPLIED

Acceleration Patterns

Suppose the radiator in your car has a leak and drops of coolant fall constantly, one every second. What pattern would the drops make on the pavement when you accelerate the car from a stoplight? What pattern would they make when you drive a constant speed? What pattern would you observe as the car comes to a stop? Use a marker to make dots on a sheet of paper that illustrate (1) acceleration, (2) constant speed, and (3) negative acceleration. Use words to describe the acceleration in each situation.

A Constant direction increase speed

B Constant direction decrease speed

C Change direction constant speed

D Change direction change speed

FIGURE 2.4 Four different ways (A–D) to accelerate a car.

velocity), the brakes (which can decrease magnitude of velocity), and the steering wheel (which can change direction of velocity). (See figure 2.4.) The important thing to remember is that acceleration results from any *change* in the motion of an object.

EXAMPLE 2.3 *(Optional)*

A bicycle moves from rest to 5 m/s in 5 s. What was the acceleration?

SOLUTION

$$v_i = 0 \text{ m/s}$$
$$v_f = 5 \text{ m/s}$$
$$t = 5 \text{ s}$$
$$a = ?$$

$$a = \frac{v_f - v_i}{t}$$
$$= \frac{5 \text{ m/s} - 0 \text{ m/s}}{5 \text{ s}}$$
$$= \frac{5}{5} \frac{\text{m/s}}{\text{s}}$$
$$= 1 \frac{\text{m}}{\text{s}} \times \frac{1}{\text{s}}$$
$$= \boxed{1 \frac{\text{m}}{\text{s}^2}}$$

EXAMPLE 2.4 *(Optional)*

An automobile uniformly accelerates from rest at 5 m/s² for 6 s. What is the final velocity in m/s? (Answer: 30 m/s)

FORCES

The Greek philosopher Aristotle considered some of the first ideas about the causes of motion back in the fourth century B.C. However, he had it all wrong when he reportedly stated that a dropped object falls at a constant speed that is determined by its weight. He also incorrectly thought that an object moving across Earth's surface requires a continuously applied force to continue moving. These ideas were based on observing and thinking, not measurement, and no one checked to see if they were correct. It took about two thousand years before people began to correctly understand motion.

Aristotle did recognize an association between force and motion, and this much was acceptable. It is partly correct because a force is closely associated with *any* change of motion, as you will see. This section introduces the concept of a force, which will be developed more fully when the relationship between forces and motion is considered later.

A **force** is a push or a pull that is capable of changing the state of motion of an object. Consider, for example, the movement of a ship from the pushing of two tugboats (figure 2.5). Tugboats can vary the strength of the force exerted on a ship, but they can also push in different directions. What effect does direction have on two forces acting on an object? If the tugboats were side by side, pushing in the same direction, the overall force is the sum of the two forces. If they act in exactly opposite directions, one pushing on each side of the ship, the overall force is the difference between the strength of the two forces. If they have the same strength, the overall effect is to cancel each other without producing any motion. The **net force** is the sum of all the forces acting on an object. Net force means "final," after the forces are added (figure 2.6).

FIGURE 2.5 The rate of movement and the direction of movement of this ship are determined by a combination of direction and size of force from each of the tugboats. Which direction are the two tugboats pushing? What evidence would indicate that one tugboat is pushing with a greater force? If the tugboat by the numbers is pushing with a greater force and the back tugboat is keeping the back of the ship from moving, what will happen?

FIGURE 2.6 (*A*) When two parallel forces are acting on the ship in the same direction, the net force is the two forces added together. (*B*) When two forces are opposite and of equal size, the net force is zero. (*C*) When two parallel forces are not of equal size, the net force is the difference in the direction of the larger force.

Connections . . .

Weather

Classification schemes are imaginative mental constructions used to show similarities and differences in objects or events. For example, the following describes two schemes used to help us classify storms that are not associated with weather fronts.

What is the difference between a tropical depression, tropical storm, and a hurricane? They are classified according to the *speed* of the maximum sustained surface winds. In the United States, the maximum sustained surface wind is measured by averaging the wind speed over a 1-minute period. Here is the classification scheme:

Tropical Depression. This is a center of low pressure around which the winds are generally moving 55 km/h (about 35 mi/h) or less. The tropical depression might dissolve into nothing, or it might develop into a more intense disturbance.

Tropical Storm. This is a more intense, highly organized center of low pressure with winds between 56 and 120 km/h (about 35 to 75 mi/h).

Hurricane. This is an intense low-pressure center with winds greater than 120 km/h (about 75 mi/h). A strong storm of this type is called a "hurricane" if it occurs over the Atlantic Ocean or the Pacific Ocean east of the international date line. It is called a "typhoon" if it occurs over the North Pacific Ocean west of the international date line. Hurricanes are further classified according to category and damage to be expected. Here is the classification scheme:

Category	Damage	Winds
1	minimal	120–153 km/h (75–95 mi/h)
2	moderate	154–177 km/h (96–110 mi/h)
3	extensive	178–210 km/h (111–130 mi/h)
4	extreme	211–250 km/h (131–155 mi/h)
5	catastrophic	>250 km/h (>155 mi/h)

When two parallel forces act in the same direction, they can be simply added. In this case, there is a net force that is equivalent to the sum of the two forces. When two parallel forces act in opposite directions, the net force is the difference between the two forces and is in the direction of the larger force. When two forces act neither in a way that is exactly together nor exactly opposite each other, the result will be like a new, different force having a new direction and strength.

Forces have a strength and direction that can be represented by force arrows. The tail of the arrow is placed on the object that feels the force, and the arrowhead points in the direction in which the force is exerted. The length of the arrow is proportional to the strength of the force. The use of force arrows helps you visualize and understand all the forces and how they contribute to the net force.

There are four **fundamental forces** that *cannot* be explained in terms of any other force. They are gravitational, electromagnetic, weak, and the strong nuclear force. Gravitational forces act between all objects in the universe—between you and Earth, between Earth and the Sun, between the planets in the solar systems, and, in fact, hold stars in large groups called galaxies. Switching scales from the very large galaxy to inside an atom, we find electromagnetic forces acting between electrically charged parts of atoms, such as electrons and protons. Electromagnetic forces are responsible for the structure of atoms, chemical change, and electricity and magnetism. Weak and strong forces act inside the nucleus of an atom, so they are not as easily observed at work as are gravitational and electromagnetic forces. The weak force is involved in certain nuclear reactions. The strong nuclear force is involved in close-range holding of the nucleus together. In general, the strong nuclear force between particles inside a nucleus is about 10^2 times stronger than the electromagnetic force and about 10^{39} times stronger than the gravitational force. The fundamental forces are responsible for everything that happens in the universe, and we will learn more about them in chapters on electricity, light, nuclear energy, chemistry, geology, and astronomy.

HORIZONTAL MOTION ON LAND

Everyday experience seems to indicate that Aristotle's idea about horizontal motion on Earth's surface is correct. After all, moving objects that are not pushed or pulled do come to rest in a short period of time. It would seem that an object keeps moving only if a force continues to push it. A moving automobile will slow and come to rest if you turn off the ignition. Likewise, a ball that you roll along the floor will slow until it comes to rest. Is the natural state of an object to be at rest, and is a force necessary to keep an object in motion? This is exactly what people thought until Galileo (figure 2.7) published his book *Two New Sciences* in 1638, which described his findings about motion.

The book had three parts that dealt with uniform motion, accelerated motion, and projectile motion. Galileo described details of simple experiments, measurements, calculations, and thought experiments as he developed definitions and concepts of motion. In one of his thought experiments, Galileo presented an argument against Aristotle's view that a force is needed to keep an object in motion. Galileo imagined an object (such as a ball) moving over a horizontal surface without the force of friction. He concluded that the object would move forever with a constant velocity as long as there was no unbalanced force acting to change the motion.

Why does a rolling ball slow to a stop? You know that a ball will roll farther across a smooth, waxed floor such as a bowling lane than it will across floor covered with carpet. The rough carpet offers more resistance to the rolling ball. The resistance of the floor friction is shown by a force arrow, F_{floor}, in figure 2.8. This force, along with the force arrow for air resistance, F_{air}, oppose the forward movement of the ball. Notice the dashed line arrow in part A of figure 2.8. There is no other force applied to the ball, so the rolling speed decreases until the ball finally comes to a complete stop. Now imagine what force you would need to exert by pushing with your hand, moving along with the ball to keep it rolling at a uniform rate. An examination of the forces in part B of figure 2.8 can help you determine the amount of force. The force you apply, $F_{applied}$, must counteract the resistance forces. It opposes

FIGURE 2.7 Galileo (left) challenged the Aristotelian view of motion and focused attention on the concepts of distance, time, velocity, and acceleration.

the forces that are slowing down the ball as illustrated by the direction of the arrows. To determine how much force you should apply, look at the arrow equation. $F_{applied}$ has the same length as the sum of the two resistance forces, but it is in the opposite direction of the resistance forces. Therefore, the overall force, F_{net}, is zero. The ball continues to roll at a uniform rate when you *balance* the force opposing its motion. It is reasonable, then, that if there were no opposing forces, you would not need to apply a force to keep it rolling. This was the kind of reasoning that Galileo did when he discredited the Aristotelian view that a force was necessary to keep an object moving. Galileo concluded that a moving object would continue moving with a constant velocity if no unbalanced forces were applied; that is, if the net force were zero.

It could be argued that the difference in Aristotle's and Galileo's views of forced motion is really a degree of analysis. After all, moving objects on Earth do come to rest unless continuously pushed or pulled. But Galileo's conclusion describes *why* they must be pushed or pulled and reveals the true nature of the motion of objects. Aristotle argued that the natural state of objects is to be at rest and attempted to explain why objects move. Galileo, on the other hand, argued that it is just as natural for an object to be moving and attempted to explain why they come to rest. The behavior of matter to persist in its state of motion is called **inertia.** Inertia is the *tendency of an object to remain in unchanging motion whether actually moving or at rest, when the net force is zero.* The development of this concept changed the way people viewed the natural state of an object and opened the way for further understandings about motion. Today, it is understood that a satellite moving through free space will continue to do so with no unbalanced forces acting on it (figure 2.9A). An unbalanced force is needed to slow the satellite (figure 2.9B), increase its speed (figure 2.9C), or change its direction of travel (figure 2.9D).

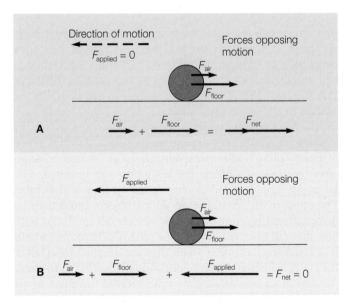

FIGURE 2.8 (A) This ball is rolling to your left with no forces in the direction of motion. The sum of the force of floor friction (F_{floor}) and the force of air friction (F_{air}) results in a net force opposing the motion, so the ball slows to a stop. (B) A force is applied to the moving ball, perhaps by a hand that moves along with the ball. The force applied ($F_{applied}$) equals the sum of the forces opposing the motion, so the ball continues to move with a constant velocity.

A No force—constant speed in straight line

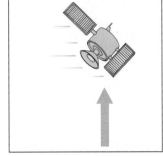

B Force applied against direction of motion

C Force applied in same direction as motion

D Force applied sideways to direction of motion

FIGURE 2.9 Explain how the combination of drawings (A–D) illustrates inertia.

A Bicycle Racer's Edge

Galileo was one of the first to recognize the role of friction in opposing motion. As shown in figure 2.8, friction with the surface and air friction combine to produce a net force that works against anything that is moving on the surface. This article is about air friction and some techniques that bike riders use to reduce that opposing force—perhaps giving them an edge in a close race.

The bike riders in box figure 2.1 are forming a single-file line, called a *paceline*, because the slip-stream reduces the air resistance for a closely trailing rider. Cyclists say that riding in the slipstream of another cyclist will save much of their energy. In fact, the cyclists will be able to move 5 mi/h faster than they would expending the same energy while riding alone.

In a sense, riding in a slipstream means that you do not have to push as much air out of your way. It has been estimated that at 20 mi/h a cyclist must move a little less than half a ton of air out of the way every minute.

Along with the problem of moving about a half-ton of air out of the way every

Box Figure 2.1 The object of the race is to be in the front, to finish first. If this is true, why are these racers forming a single-file line?

minute, there are two basic factors related to air resistance. These are (1) a turbulent versus a smooth flow of air and (2) the problem of frictional drag. A turbulent flow of air contributes to air resistance because it causes the air to separate slightly on the back side, which increases the pressure on the front of the moving object. This is why racing cars, airplanes, boats, and other racing vehicles are streamlined to a tear-droplike shape. This shape is not as likely to have the lower-pressure-producing air turbulence behind (and resulting greater

pressure in front) because it smoothes or streamlines the air flow.

The frictional drag of air is similar to the frictional drag that occurs when you push a book across a rough tabletop. You know that smoothing the rough tabletop will reduce the frictional drag on the book. Likewise, the smoothing of a surface exposed to moving air will reduce air friction. Cyclists accomplish this "smoothing" by wearing smooth spandex clothing and by shaving hair from arm and leg surfaces that are exposed to moving air. Each hair contributes to the overall frictional drag, and removal of the arm and leg hair can thus result in seconds saved. This might provide enough of an edge to win a close race. Shaving legs and arms, together with the wearing of spandex or some other tight, smooth-fitting garments, are just a few of the things a cyclist can do to gain an edge. Perhaps you will be able to think of more ways to reduce the forces that oppose motion.

Walk or Run in Rain?

Is it a mistake to run in rain if you want to stay drier? One idea is that you should run because you spend less time in the rain so you will stay drier. On the other hand, this is true only if the rain lands only on the top of your head and shoulders. If you run, you will end up running into more raindrops on the larger surface area of your face, chest, and front of your legs.

Two North Carolina researchers looked into this question with one walking and the other running over a measured distance while wearing cotton sweatsuits. They then weighed their clothing and found that the walking person's sweatsuit weighed more. This means you should run to stay drier.

FALLING OBJECTS

Did you ever wonder what happens to a falling rock during its fall? Aristotle reportedly thought that a rock falls at a uniform speed that is proportional to its weight. Thus, a heavy

rock would fall at a faster uniform speed than a lighter rock. As stated in a popular story, Galileo discredited Aristotle's conclusion by dropping a solid iron ball and a solid wooden ball simultaneously from the top of the Leaning Tower of Pisa (figure 2.10). Both balls, according to the story, hit the ground nearly at the same time. To do this, they would have to fall with the same velocity. In other words, the velocity of a falling object does not depend on its weight. Any difference in freely falling bodies is explainable by air resistance. Soon after the time of Galileo, the air pump was invented. The air pump could be used to remove the air from a glass tube. The effect of air resistance on falling objects could then be demonstrated by comparing how objects fall in the air with how they fall in an evacuated glass tube. You know that a coin falls faster than a feather when they are dropped together in the air. A feather and heavy coin will fall together in the near vacuum of an evacuated glass tube because the effect of air resistance on the feather has been removed. When objects fall toward Earth without considering air resistance, they are said to be in **free fall**. Free fall considers only gravity and neglects air resistance.

Galileo concluded that light and heavy objects fall together in free fall, but he also wanted to know the details of what was

FIGURE 2.10 According to a widespread story, Galileo dropped two objects with different weights from the Leaning Tower of Pisa. They reportedly hit the ground at about the same time, discrediting Aristotle's view that the speed during the fall is proportional to weight.

going on while they fell. He now knew that the velocity of an object in free fall was *not* proportional to the weight of the object. He observed that the velocity of an object in free fall *increased* as the object fell and reasoned from this that the velocity of the falling object would have to be (1) somehow proportional to the *time* of fall and (2) somehow proportional to the *distance* the object fell. If the time and distance were both related to the velocity of a falling object at a given time and distance, how were they related to one another?

Galileo reasoned that a freely falling object should cover a distance *proportional to the square of the time of the fall ($d \propto t^2$)*. In other words, the object should fall 4 times as far in 2 s as in 1 s ($2^2 = 4$), 9 times as far in 3 s ($3^2 = 9$), and so on. Galileo checked this calculation by rolling balls on an inclined board with a smooth groove in it. He used the inclined board to slow

There are two different meanings for the term *free fall*. In physics, *free fall* means the unconstrained motion of a body in a gravitational field, without considering air resistance. Without air resistance, all objects are assumed to accelerate toward the surface at 9.8 m/s^2.

In the sport of skydiving, *free fall* means falling within the atmosphere without a drag-producing device such as a parachute. Air provides a resisting force that opposes the motion of a falling object, and the net force is the difference between the downward force (weight) and the upward force of air resistance. The weight of the falling object depends on the mass and acceleration from gravity, and this is the force downward. The resisting force is determined by at least two variables: (1) the area of the object exposed to the airstream and (2) the speed of the falling object. Other variables such as streamlining, air temperature, and turbulence play a role, but the greatest effect seems to be from exposed area and the increased resistance as speed increases.

A skydiver's weight is constant, so the downward force is constant. Modern skydivers typically free-fall from about 3,650 m (about 12,000 ft) above the ground until about 750 m (about 2,500 ft), where they open their parachutes. After jumping from the plane, the diver at first accelerates toward the surface, reaching speeds up to about 185–210 km/h (about 115–130 mi/h). The air resistance increases with increased speed and the net force becomes less and less. Eventually, the downward weight force will be balanced by the upward air resistance force, and the net force becomes zero. The person now falls at a constant speed, and we say the terminal velocity has been reached. It is possible to change your body position to vary your rate of fall up or down by 32 km/h (about 20 mi/h). However, by diving or "standing up" in free fall, experienced skydivers can reach speeds of up to 290 km/h (about 180 mi/h). The record free fall speed, done without any special equipment, is 517 km/h (about 321 mi/h). Once the parachute opens, a descent rate of about 16 km/h (about 10 mi/h) is typical.

Falling Bodies

Galileo concluded that all objects fall together, with the same acceleration, when the upward force of air resistance is removed. It would be most difficult to remove air from the room, but it is possible to do some experiments that provide some evidence of how air influences falling objects.

1. Take a sheet of paper and your textbook and drop them side by side from the same height. Note the result.
2. Place the sheet of paper on top of the book and drop them at the same time. Do they fall together?
3. Crumple the sheet of paper into a loose ball and drop the ball and book side by side from the same height.
4. Crumple a sheet of paper into a very tight ball and again drop the ball and book side by side from the same height.

Explain any evidence you found concerning how objects fall.

the motion of descent in order to measure the distance and time relationships, a necessary requirement since he lacked the accurate timing devices that exist today. He found, as predicted, that the falling balls moved through a distance proportional to the square of the time of falling. This also means that the *velocity of the falling object increased at a constant rate*. Recall that a change of velocity during some time period is called *acceleration*. In other words, a falling object *accelerates* toward the surface of Earth.

Since the velocity of a falling object increases at a constant rate, this must mean that falling objects are *uniformly accelerated* by the force of gravity. *All objects in free fall experience a constant acceleration.* During each second of fall, the object on Earth gains 9.8 m/s (32 ft/s) in velocity. This gain is the acceleration of the falling object, 9.8 m/s^2 (32 ft/s^2).

The acceleration of objects falling toward Earth varies slightly from place to place on Earth's surface because of Earth's shape and spin. The acceleration of falling objects decreases from the poles to the equator and also varies from place to place because Earth's mass is not distributed equally. The value of 9.8 m/s^2 (32 ft/s^2) is an approximation that is fairly close to, but not exactly, the acceleration due to gravity in any particular location. The acceleration due to gravity is important in a number of situations, so the acceleration from this force is given a special symbol, *g.*

COMPOUND MOTION

So far we have considered two types of motion: (1) the horizontal, straight-line motion of objects moving on the surface of Earth and (2) the vertical motion of dropped objects that accelerate toward the surface of Earth. A third type of motion occurs when an object is thrown, or projected, into the air. Essentially, such a projectile (rock, football, bullet, golf ball, or whatever) could be directed straight upward as a vertical projection, directed straight out as a horizontal projection, or directed at some angle between the vertical and the horizontal. Basic to understanding such compound motion is the observation that (1) gravity acts on objects *at all times,* no matter where they are, and (2) the acceleration due to gravity (*g*) is *independent of any motion* that an object may have.

Vertical Projectiles

Consider first a ball that you throw straight upward, a vertical projection. The ball has an initial velocity but then reaches a maximum height, stops for an instant, then accelerates back toward Earth. Gravity is acting on the ball throughout its climb, stop, and fall. As it is climbing, the force of gravity is continually reducing its velocity. The overall effect during the climb is deceleration, which continues to slow the ball until the instantaneous stop. The ball then accelerates back to the surface just like a ball that has been dropped. If it were not for air resistance, the ball would return with the same velocity that it had initially. The velocity arrows for a ball thrown straight up are shown in figure 2.11.

Horizontal Projectiles

Horizontal projectiles are easier to understand if you split the complete motion into vertical and horizontal parts. Consider, for example, an arrow shot horizontally from a bow. The force of gravity accelerates the arrow downward, giving it an increasing downward velocity as it moves through the air. This increasing downward velocity is shown in figure 2.12 as increasingly longer velocity arrows (v_v). There are no forces in the horizontal direction if you can ignore air resistance, so the horizontal velocity of the arrow remains the same as shown by the v_h velocity arrows. The combination of the increasing

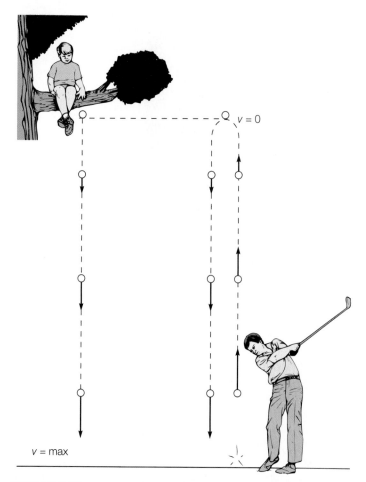

FIGURE 2.11 On its way up, a vertical projectile such as this misdirected golf ball is slowed by the force of gravity until an instantaneous stop; then it accelerates back to the surface, just as another golf ball does when dropped from the same height. The straight up- and down-moving golf ball has been moved to the side in the sketch so we can see more clearly what is happening.

vertical (v_v) motion and the unchanging horizontal (v_h) motion causes the arrow to follow a curved path until it hits the ground.

An interesting prediction that can be made from the shot arrow analysis is that an arrow shot horizontally from a bow will hit the ground at the same time as a second arrow that is simply dropped from the same height (figure 2.12). Would this be true of a bullet dropped at the same time as one fired horizontally from a rifle? The answer is yes; both bullets would hit the ground at the same time. Indeed, without air resistance, all the bullets and arrows should hit the ground at the same time if dropped or shot from the same height.

Golf balls, footballs, and baseballs are usually projected upward at some angle to the horizon. The horizontal motion of these projectiles is constant as before because there are no horizontal forces involved. The vertical motion is the same as that of a ball projected directly upward. The combination of these two motions causes the projectile to follow a curved path called a *parabola,* as shown in figure 2.13. The next time you have the opportunity, observe the path of a ball that has been

FIGURE 2.12 A horizontal projectile has the same horizontal velocity throughout the fall as it accelerates toward the surface, with the combined effect resulting in a curved path. Neglecting air resistance, an arrow shot horizontally will strike the ground at the same time as one dropped from the same height above the ground, as shown here by the increasing vertical velocity arrows.

FIGURE 2.13 A football is thrown at some angle to the horizon when it is passed downfield. Neglecting air resistance, the horizontal velocity is a constant, and the vertical velocity decreases, then increases, just as in the case of a vertical projectile. The combined motion produces a parabolic path. Contrary to statements by sportscasters about the abilities of certain professional quarterbacks, it is impossible to throw a football with a "flat trajectory" because it begins to accelerate toward the surface as soon as it leaves the quarterback's hand.

projected at some angle (figure 2.14). Note that the second half of the path is almost a reverse copy of the first half. If it were not for air resistance, the two values of the path would be exactly the same. Also note the distance that the ball travels as compared to the angle of projection. An angle of projection of 45° results in the maximum distance of travel.

THREE LAWS OF MOTION

In the previous sections, you learned how to describe motion in terms of distance, time, velocity, and acceleration. In addition, you learned about different kinds of motion, such as straight-line motion, the motion of falling objects, and the compound motion of objects projected up from the surface of Earth. You were also introduced, in general, to two concepts closely associated with motion: (1) that objects have inertia, a tendency to resist a change in motion, and (2) that forces are involved in a change of motion.

Transportation and the Environment

Environmental science is an interdisciplinary study of Earth's environment. The concern of this study is the overall problem of human degradation of the environment and remedies for that damage. As an example of an environmental topic of study, consider the damage that results from current human activities involving the use of transportation. Researchers estimate that overall transportation activities are responsible for about one-third of the total U.S. carbon emissions that are added to the air every day. Carbon emissions are a problem because they are directly harmful in the form of carbon monoxide. They are also indirectly harmful because of the contribution of carbon dioxide to global warming and the consequences of climate change.

Here is a list of things that people might do to reduce the amount of environmental damage from transportation:

A. Use a bike, carpool, walk, or take public transportation wherever possible.

B. Plan to combine trips to the store, mall, and work, leaving the car parked whenever possible.

C. Purchase hybrid electric or fuel cell-powered cars and vehicles whenever possible.

D. Move to a planned community that makes the use of cars less necessary and less desirable.

Questions to Discuss

Discuss with your group the following questions concerning connections between thought and feeling:

1. What are your positive or negative feelings associated with each item in the list?
2. Would you feel differently if you had a better understanding of the global problem?
3. Do your feelings mean that you have reached a conclusion?
4. What new items could be added to the list?

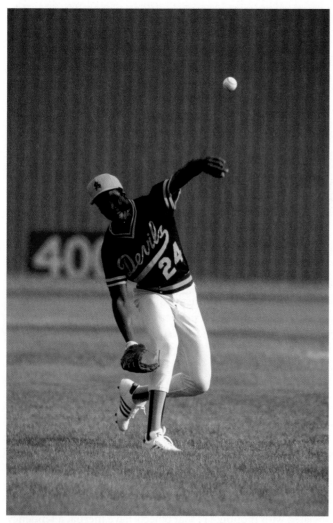

FIGURE 2.14 Without a doubt, this baseball player is aware of the relationship between the projection angle and the maximum distance acquired for a given projection velocity.

The relationship between forces and a change of motion is obvious in many everyday situations. When a car, bus, or plane in which you are riding starts moving, you feel a force on your back. Likewise, you feel a force on the bottoms of your feet when an elevator you are in starts moving upward. On the other hand, you seem to be forced toward the dashboard if a car stops quickly, and it feels as if the floor pulls away from your feet when an elevator drops rapidly. These examples all involve patterns between forces and motion, patterns that can be quantified, conceptualized, and used to answer questions about why things move or stand still. These patterns are the subject of Newton's three laws of motion.

Newton's First Law of Motion

Newton's first law of motion is also known as the *law of inertia* and is very similar to one of Galileo's findings about motion. Recall that Galileo used the term *inertia* to describe the tendency of an object to resist changes in motion. Newton's first law describes this tendency more directly. In modern terms (not Newton's words), the **first law of motion** is as follows:

> **Every object retains its state of rest or its state of uniform straight-line motion unless acted upon by an unbalanced force.**

This means that an object at rest will remain at rest unless it is put into motion by an unbalanced force; that is, the net force must be greater than zero. Likewise, an object moving with uniform straight-line motion will retain that motion unless a net force causes it to speed up, slow down, or change its direction of travel. Thus, Newton's first law describes the tendency of an object to resist *any* change in its state of motion.

Think of Newton's first law of motion when you ride standing in the aisle of a bus. The bus begins to move, and you, being an independent mass, tend to remain at rest. You take a

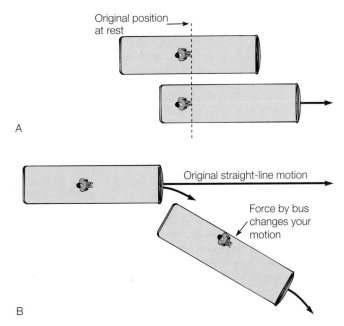

Original position at rest

Original straight-line motion

Force by bus changes your motion

A

B

FIGURE 2.15 Top view of a person standing in the aisle of a bus. (A) The bus is at rest and then starts to move forward. Inertia causes the person to remain in the original position, appearing to fall backward. (B) The bus turns to the right, but inertia causes the person to retain the original straight-line motion until forced in a new direction by the side of the bus.

few steps back as you tend to maintain your position relative to the ground outside. You reach for a seat back or some part of the bus. Once you have a hold on some part of the bus, it supplies the forces needed to give you the same motion as the bus and you no longer find it necessary to step backward. You now have the same motion as the bus, and no forces are involved, at least until the bus goes around a curve. You now feel a tendency to move to the side of the bus. The bus has changed its straight-line motion, but you, again being an independent mass, tend to move straight ahead. The side of the seat forces you into following the curved motion of the bus. The forces you feel when the bus starts moving or turning are a result of your tendency to remain at rest or follow a straight path until forces correct your motion so that it is the same as that of the bus (figure 2.15).

Newton's Second Law of Motion

Newton successfully used Galileo's ideas to describe the nature of motion. Newton's first law of motion explains that any object, once started in motion, will continue with a constant velocity in a straight line unless a force acts on the moving object. This law not only describes motion but establishes the role of a force as well. A change of motion is therefore *evidence* of the action of a net force. The association of forces and a change of motion is common in your everyday experience. You have felt forces on your back in an accelerating automobile, and you have felt other forces as the automobile turns or stops. You have also learned about gravitational forces that accelerate objects toward the surface of Earth. Unbalanced forces and acceleration are involved in any change of motion. Newton's second law of motion is a relationship between *net force, acceleration,* and *mass* that describes the cause of a change of motion.

Consider the motion of you and a bicycle you are riding. Suppose you are riding your bicycle over level ground in a straight line at 10 mi/h. Newton's first law tells you that you will continue with a constant velocity in a straight line as long as no external, unbalanced force acts on you and the bicycle. The force that you *are* exerting on the pedals seems to equal some external force that moves you and the bicycle along (more on this later). The force exerted as you move along is needed to *balance* the resisting forces of tire friction and air resistance. If these resisting forces were removed, you would not need to exert any force at all to continue moving at a constant velocity. The net force is thus the force you are applying minus the forces from tire friction and air resistance. The *net force* is therefore zero when you move at a constant speed in a straight line (figure 2.16).

If you now apply a greater force on the pedals, the *extra* force you apply is unbalanced by friction and air resistance. Hence, there will be a net force greater than zero, and you will accelerate. You will accelerate during, and *only* during, the time that the net force is greater than zero. Likewise, you will slow down if you apply a force to the brakes, another kind of resisting friction. A third way to change your velocity is to apply a force on the handlebars, changing the direction of

F_2

F_1 F_a

$$F_2 \quad F_1 \atop F_a \; = F_{net} = 0$$

FIGURE 2.16 At a constant velocity, the force of tire friction (F_1) and the force of air resistance (F_2) have a sum that equals the force applied (F_a). The net force is therefore zero.

your velocity. Thus, *unbalanced forces* on you and your bicycle produce an *acceleration*.

Starting a bicycle from rest suggests a relationship between force and acceleration. You observe that the harder you push on the pedals, the greater your acceleration. If you double the net force, then you will also double the acceleration, reaching the same velocity in half the time. Likewise, if you triple the net force, you will increase the acceleration threefold. Recall that when quantities increase or decrease together in the same ratio, they are said to be *directly proportional*. The acceleration is therefore directly proportional to the force.

Suppose that your bicycle has two seats, and you have a friend who will ride (but not pedal) with you. Suppose also that the addition of your friend on the bicycle will double the mass of the bike and riders. If you use the same net force as before, the bicycle will undergo a much smaller acceleration. In fact, with all other factors equal, doubling the mass and applying the same extra force will produce an acceleration of only half as much (figure 2.17). An even more massive friend would reduce the acceleration even more. If you triple the mass and apply the same extra force, the acceleration will be one-third as much. Recall that when a relationship between two quantities shows that one quantity increases as another decreases, in the same ratio, the quantities are said to be *inversely proportional*.

The acceleration of an object depends on *both* the *net force applied* and the *mass* of the object. The **second law of motion** is as follows:

The acceleration of an object is directly proportional to the net force acting on it and inversely proportional to the mass of the object.

If we express force in appropriate units, we can write this statement as an equation,

$$a = \frac{F}{m}$$

By solving for *F*, we rearrange the equation into the form it is most often expressed,

$$F = ma \qquad \qquad \textbf{equation 2.3}$$

In the metric system, you can see that the units for force will be the units for mass (*m*) times acceleration (*a*). The unit for mass is kg and the unit for acceleration is m/s^2. The combination of these units, (kg) (m/s^2) is a unit of force called the **newton** (N), in honor of Isaac Newton. So,

$$1 \text{ newton} = 1 \text{ N} = 1 \frac{\text{kg} \cdot \text{m}}{\text{s}^2}$$

Newton's second law of motion is the essential idea of his work on motion. According to this law, there is always a relationship between the acceleration, a net force, and the mass of an object. Implicit in this statement are three understandings: (1) that we are talking about the net force, meaning total external force acting on an object, (2) that the motion statement is concerned with acceleration, not velocity, and (3) that the mass does not change unless specified.

EXAMPLE 2.5   *(Optional)*

A 60 kg bicycle and rider accelerate at 0.5 m/s^2. How much extra force was applied?

SOLUTION

The mass (*m*) of 60 kg and the acceleration (*a*) of 0.5 m/s^2 are given. The problem asked for the extra force (*F*) needed to give the mass the acquired acceleration. The relationship is found in equation 2.3, $F = ma$.

$$m = 60 \text{ kg} \qquad F = ma$$
$$a = 0.5 \frac{\text{m}}{\text{s}^2} \qquad = (60 \text{ kg})\left(0.5 \frac{\text{m}}{\text{s}^2}\right)$$
$$F = ? \qquad = (60)(0.5)\text{kg} \times \frac{\text{m}}{\text{s}^2}$$
$$= 30 \frac{\text{kg} \cdot \text{m}}{\text{s}^2}$$
$$= \boxed{30 \text{ N}}$$

An extra force of 30 N beyond that required to maintain constant speed must be applied to the pedals for the bike and rider to maintain an acceleration of 0.5 m/s^2. (Note that the units kg·m/s^2 form the definition of a newton of force, so the symbol N is used.)

FIGURE 2.17   More mass results in less acceleration when the same force is applied. With the same force applied, the riders and bike with twice the mass will have half the acceleration, with all other factors constant. Note that the second rider is not pedaling.

EXAMPLE 2.6 (Optional)

What is the acceleration of a 20 kg cart if the net force on it is 40 N? (Answer: 2 m/s^2)

Weight and Mass

What is the meaning of weight—is it the same concept as mass? Weight is a familiar concept to most people, and in everyday language, the word is often used as having the same meaning as mass. In physics, however, there is a basic difference between weight and mass, and this difference is very important in Newton's explanation of motion and the causes of motion.

Mass is defined as the property that determines how much an object resists a change in its motion. The greater the mass the greater the *inertia,* or resistance to change in motion. Consider, for example, that it is easier to push a small car into motion than to push a large truck into motion. The truck has more mass and therefore more inertia. Newton originally defined mass as the "quantity of matter" in an object, and this definition is intuitively appealing. However, Newton needed to measure inertia because of its obvious role in motion and redefined mass as a measure of inertia.

You could use Newton's second law to measure a mass by exerting a force on the mass and measuring the resulting acceleration. This is not very convenient, so masses are usually measured on a balance by comparing the force of gravity acting on a standard mass compared to the force of gravity acting on the unknown mass.

The force of gravity acting on a mass is the *weight* of an object. Weight is a force and has different units (N) from those of mass (kg). Since weight is a measure of the force of gravity acting on an object, the force can be calculated from Newton's second law of motion,

$$F = ma$$

or

$$\text{downward force} = (\text{mass})(\text{acceleration due to gravity})$$

or

$$\text{weight} = (\text{mass})(g)$$

$$w = mg \qquad \text{equation 2.4}$$

You learned previously that g is the symbol used to represent acceleration due to gravity. Near Earth's surface, g has an approximate value of 9.8 m/s^2. To understand how g is applied to an object not moving, consider a ball you are holding in your hand. By supporting the weight of the ball, you hold it stationary, so the upward force of your hand and the downward force of the ball (its weight) must add to a net force of zero. When you let go of the ball, the gravitational force is the only force acting on the ball. The ball's weight is then the net force that accelerates it at g, the acceleration due to gravity. Thus, $F_{net} = w = ma = mg$. The weight of the ball never changes in a given location, so its weight is always equal to $w = mg$, even if the ball is not accelerating.

An important thing to remember is that *pounds* and *newtons* are units of *force* (table 2.1). A *kilogram,* on the other hand, is a measure of *mass.* Thus, the English unit of 1.0 lb is comparable to the metric unit of 4.5 N (or 0.22 lb is equivalent to 1.0 N). Conversion tables sometimes show how to convert from pounds (a unit of weight) to kilograms (a unit of mass). This is possible because weight and mass are proportional in a given location on the surface of Earth. Using conversion factors from inside the front cover of this book, see if you can express your weight in pounds and newtons and your mass in kilograms.

EXAMPLE 2.7 (Optional)

What is the weight of a 60.0 kg person on the surface of Earth?

SOLUTION

A mass (m) of 60.0 kg is given, and the acceleration due to gravity (g) 9.8 m/s^2 is implied. The problem asked for the weight (w). The relationship is found in equation 2.4, $w = mg$, which is a form of $F = ma$.

$$m = 60.0 \text{ kg}$$
$$g = 9.8 \text{ m/s}^2 \quad w = mg$$
$$w = ? \qquad = (60.0 \text{ kg})\left(9.8 \frac{\text{m}}{\text{s}^2}\right)$$
$$= (60.0)(9.8) \text{ kg} \times \frac{\text{m}}{\text{s}^2}$$
$$= 588 \frac{\text{kg} \cdot \text{m}}{\text{s}^2}$$
$$= 588 \text{ N}$$
$$= \boxed{590 \text{ N}}$$

EXAMPLE 2.8 (Optional)

A 60.0 kg person weighs 100.0 N on the Moon. What is the value of g on the Moon? (Answer: 1.67 m/s^2)

Connections . . .

Weight on Different Planets*

Planet	Acceleration of Gravity	Approximate Weight(N)	Approximate Weight (lb)
Mercury	3.72 m/s²	223 N	50 lb
Venus	8.92 m/s²	535 N	120 lb
Earth	9.80 m/s²	588 N	132 lb
Mars	3.72 m/s²	223 N	50 lb
Jupiter	24.89 m/s²	1,493 N	336 lb
Saturn	10.58 m/s²	635 N	143 lb
Uranus	8.92 m/s²	535 N	120 lb
Neptune	11.67 m/s²	700 N	157 lb

*For a 60.0 kg person

TABLE 2.1 **Units of Mass and Weight in the Metric and English Systems of Measurement**

	Mass	×	Acceleration	=	Force
Metric system	kg	×	$\dfrac{m}{s^2}$	=	$\dfrac{kg \cdot m}{s^2}$ (newton)
English system	$\left(\dfrac{lb}{ft/s^2}\right)$	×	$\dfrac{ft}{s^2}$	=	lb (pound)

Newton's Third Law of Motion

Newton's first law of motion states that an object retains its state of motion when the net force is zero. The second law states what happens when the net force is *not* zero, describing how an object with a known mass moves when a given force is applied. The two laws give one aspect of the concept of a force; that is, if you observe that an object starts moving, speeds up, slows down, or changes its direction of travel, you can conclude that an unbalanced force is acting on the object. Thus, any change in the state of motion of an object is *evidence* that an unbalanced force has been applied.

Newton's third law of motion is also concerned with forces. First, consider where a force comes from. A force is always produced by the interaction of two objects. Sometimes we do not know what is producing forces, but we do know that they always come in pairs. Anytime a force is exerted, there is always a matched and opposite force that occurs at the same time. For example, if you push on the wall, the wall pushes back with an equal and opposite force. The two forces are opposite and balanced, and you know this because $F = ma$ and neither you nor the wall accelerated. If the acceleration is zero, then you know from $F = ma$ that the net force is zero (zero equals zero). Note also that the two forces were between two different objects, you and the wall. Newton's third law always describes what happens between two different objects. To simplify the many interactions that occur on Earth, consider a satellite freely floating in space.

FIGURE 2.18   Forces occur in matched pairs that are equal in strength and opposite in direction.

According to Newton's second law ($F = ma$), a force must be applied to change the state of motion of the satellite. What is a possible source of such a force? Perhaps an astronaut pushes on the satellite for 1 second. The satellite would accelerate *during* the application of the force, then move away from the original position at some constant velocity. The astronaut would also move away from the original position, but in the opposite direction (figure 2.18). A *single force does not exist* by itself. There is always a matched and opposite force that occurs at the same time. Thus, the astronaut exerted a momentary force on the satellite, but the satellite evidently exerted a momentary force back on the astronaut as well, for the astronaut moved away from the original position in the opposite direction. Newton did not have astronauts and satellites to think about, but this is the kind of reasoning he did when he concluded that forces always occur in matched pairs that are equal and opposite. Thus, the **third law of motion** is as follows:

> **Whenever two objects interact, the force exerted on one object is equal in strength and opposite in direction to the force exerted on the other object.**

The third law states that forces always occur in matched pairs that act in opposite directions and on two *different* bodies. Sometimes the third law of motion is expressed as follows: "For every action there is an equal and opposite reaction," but this can be misleading. Neither force is the cause of the other. The forces are at every instant the cause of each other, and they appear and disappear at the same time. If you are going to describe the force exerted on a satellite by an astronaut, then you must realize that there is a simultaneous force exerted on the astronaut by the satellite. The forces (astronaut on satellite and satellite on astronaut) are equal in magnitude but opposite in direction.

Perhaps it would be more common to move a satellite with a small rocket. A satellite is maneuvered in space by firing a rocket in the direction opposite to the direction someone wants to move the satellite. Exhaust gases (or compressed gases) are accelerated in one direction and exert an equal but opposite force on the satellite that accelerates it in the opposite direction. This is another example of the third law.

Consider how the pairs of forces work on Earth's surface. You walk by pushing your feet against the ground (figure 2.19). Of course you could not do this if it were not for friction. You would slide as on slippery ice without friction. But since friction does exist, you exert a backward horizontal force on the ground, and, as the third law explains, the ground exerts an equal and opposite force on you. You accelerate forward from the net force as explained by the second law. If Earth had the same mass as you, however, it would accelerate backward at the same rate that you were accelerated forward. Earth is much more massive than you, however, so any acceleration of Earth is a vanishingly small amount. The overall effect is that you are accelerated forward by the force the ground exerts on you.

Return now to the example of riding a bicycle that was discussed previously. What is the source of the *external* force that accelerates you and the bike? Pushing against the pedals is not external to you and the bike, so that force will *not* accelerate you and the bicycle forward. This force is transmitted through the bike mechanism to the rear tire, which pushes against the ground. It is the ground exerting an equal and opposite force against the system of you and the bike that accelerates you forward. You must consider the forces that act on the system of you and the bike before you can apply $F = ma$. The only forces that will affect the forward motion of the bike system are the force of the ground pushing it forward and the frictional forces that oppose the forward motion. This is another example of the third law.

MOMENTUM

Sportscasters often refer to the *momentum* of a team, and newscasters sometimes refer to an election where one of the candidates has *momentum*. Both situations describe a competition where one side is moving toward victory and it is difficult to stop. It seems appropriate to borrow this term from the physical sciences because momentum is a property of movement. It takes a longer time to stop something from moving when it has a lot of momentum. The physical science concept of momentum is closely related to Newton's laws of motion. **Momentum** (p) is defined as the product of the mass (m) of an object and its velocity (v),

$$\text{momentum} = \text{mass} \times \text{velocity}$$

or

$$p = mv \qquad \textbf{equation 2.5}$$

The astronaut in figure 2.20 has a mass of 60.0 kg and a velocity of 0.750 m/s as a result of the interaction with the satellite. The resulting momentum is therefore (60.0 kg) (0.750 m/s), or 45.0 kg·m/s. As you can see, the momentum would be greater if the astronaut had acquired a greater velocity or if the astronaut had a greater mass and acquired the same velocity. Momentum involves both the inertia and the velocity of a moving object.

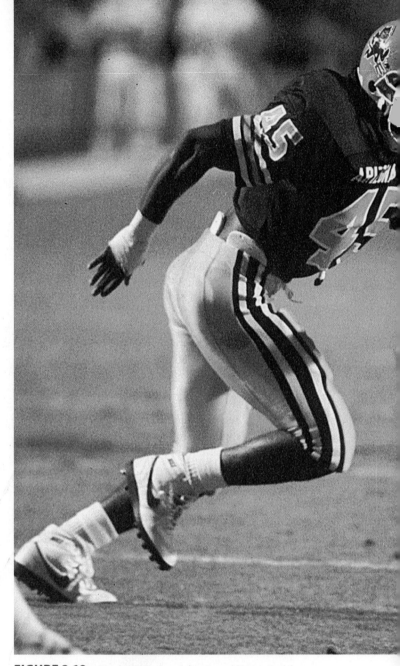

FIGURE 2.19 The football player's foot is pushing against the ground, but it is the ground pushing against the foot that accelerates the player forward to catch a pass.

Conservation of Momentum

Notice that the momentum acquired by the satellite in figure 2.20 is *also* 45.0 kg·m/s. The astronaut gained a certain momentum in one direction, and the satellite gained the *very same momentum in the opposite direction*. Newton originally defined the second law in terms of a change of momentum being proportional to the net force acting on an object. Since the third law explains that the forces exerted on both the astronaut and satellite were equal and opposite, you would expect both objects to acquire equal momentum in the opposite direction. This result is observed any time objects in a system interact and the only forces involved are those between the interacting objects (figure 2.20). This statement leads to a

Connections . . .

Swimming Scallop

Newton's laws of motion apply to animal motion as well as that of satellites and automobiles. Consider, for example, the dilemma of a growing scallop. A scallop is the shell often seen as a logo for a certain petroleum company, a fan-shaped shell with a radiating fluted pattern (box figure 2.2). The scallop is a marine mollusk that is most unusual since it is the only clamlike mollusk that is capable of swimming. By opening and closing its shell, it is able to propel itself by forcing water from the interior of the shell in a jetlike action. The popular seafood called "scallops" is the edible muscle that the scallop uses to close its shell.

A scallop is able to swim by orienting its shell at a proper angle and maintaining a minimum acceleration to prevent sinking. For example, investigations have found that one particular species of scallop must force enough water backward to move about six body lengths per second with a 10 degree angle of attack to maintain level swimming. Such a swimming effort can be maintained for up to about 20 seconds, enabling the scallop to escape predation or some other disturbing condition.

A more massive body limits the swimming ability of the scallop, as a greater force is needed to give a greater mass the same acceleration (as you would expect from Newton's second law of motion). This problem becomes worse as the scallop grows larger and larger without developing a greater and greater jet force.

Box Figure 2.2 A scallop shell.

particular kind of relationship called a *law of conservation*. In this case, the law applies to momentum and is called the **law of conservation of momentum**:

> **The total momentum of a group of interacting objects remains the same in the absence of external forces.**

Conservation of momentum, energy, and charge are among examples of conservation laws that apply to everyday situations. These situations always illustrate two understandings: (1) each conservation law is an expression that describes a physical principle that can be observed, and (2) each law holds regardless of the details of an interaction or how it took place. Since the conservation laws express something that always occurs, they tell us what might be expected to happen and what might be expected not to happen in a given situation. The conservation laws also allow unknown quantities to be found by analysis. The law of conservation of momentum, for example, is useful in analyzing motion in simple systems of collisions such as those of billiard balls, automobiles, or railroad cars. It is also useful in measuring action and reaction interactions, as in rocket propulsion, where the backward momentum of the exhaust gases equals the momentum given to the rocket in the opposite direction (figure 2.21). When this is done, momentum is always found to be conserved.

$F = 30.0$ N
$t = 1.50$ s

$F = 30.0$ N
$t = 1.50$ s

$m = 60.0$ kg
$v = 0.750$ m/s
$p = mv$

$= (60.0 \text{ kg}) (0.750 \frac{\text{m}}{\text{s}})$

$= 45.0 \frac{\text{kg} \cdot \text{m}}{\text{s}}$

$m = 120.0$ kg
$v = 0.375$ m/s
$p = mv$

$= (120.0 \text{ kg}) (0.375 \frac{\text{m}}{\text{s}})$

$= 45.0 \frac{\text{kg} \cdot \text{m}}{\text{s}}$

FIGURE 2.20 Both the astronaut and the satellite receive a force of 30.0 N for 1.50 s when they push on each other. Both then have a momentum of 45.0 kg·m/s in the opposite direction. This is an example of the law of conservation of momentum.

Impulse

Have you ever heard that you should "follow through" when hitting a ball? When you follow through, the bat is in contact with the ball for a longer period of time. The force of the hit is important, of course, but both the force and how long the force is applied determine the result. The product of the force and the time of application is called **impulse.** This quantity can be expressed as

$$\text{impulse} = Ft$$

where F is the force applied during the time of contact (t). The impulse you give the ball determines how fast the ball will move and thus how far it will travel.

Impulse is related to the change of motion of a ball of a given mass, so the change of momentum (mv) is brought about by the impulse. This can be expressed as

Change of momentum = (applied force) (time of contact)
$$\Delta p = Ft$$

equation 2.6

FIGURE 2.21 According to the law of conservation of momentum, the momentum of the expelled gases in one direction equals the momentum of the rocket in the other direction in the absence of external forces.

Contact time is also important in safety. Automobile airbags, the padding in elbow and knee pads, and the plastic barrels off the highway in front of overpass supports are examples of designs intended to increase the contact time. Again, increasing the contact time reduces the force, since $\Delta p = Ft$. The impact force is reduced and so are the injuries. Think about this the next time you see a car that was crumpled and bent by a collision. The driver and passengers were probably saved from injuries that are more serious since more time was involved in stopping the car that crumpled. A car that crumples is a safer car in a collision.

FORCES AND CIRCULAR MOTION

Consider a communications satellite that is moving at a uniform speed around Earth in a circular orbit. According to the first law of motion, there *must be* forces acting on the satellite, since it does *not* move off in a straight line. The second law of motion also indicates forces, since an unbalanced force is required to change the motion of an object.

Recall that acceleration is defined as a rate of change in velocity and that velocity has both strength and direction. Velocity is changed by a change in speed, direction, or both speed and direction. The satellite in a circular orbit is continuously being accelerated. This means that there is a continuously acting unbalanced force on the satellite that pulls it out of a straight-line path.

The force that pulls an object out of its straight-line path and into a circular path is called a **centripetal** (center-seeking) **force.** Perhaps you have swung a ball on the end of a string in a horizontal circle over your head. Once you have the ball moving, the only unbalanced force (other than gravity) acting on the ball is the centripetal force your hand exerts on the ball

where Δp is a change of momentum. You "follow through" while hitting a ball in order to increase the contact time. If the same force is used, a longer contact time will result in a greater impulse. A greater impulse means a greater change of momentum, and since the mass of the ball does not change, the overall result is a moving ball with a greater velocity. This means following through will result in more distance from hitting the ball with the same force. That's why it is important to follow through when you hit the ball.

Now consider bringing a moving object to a stop by catching it. In this case, the mass and the velocity of the object are fixed at the time you catch it, and there is nothing you can do about these quantities. The change of momentum is equal to the impulse, and the force and time of force application *can* be manipulated. For example, consider how you would catch a raw egg that is tossed to you. You would probably move your hands with the egg as you catch it, increasing the contact time. Increasing the contact time has the effect of reducing the force, since $\Delta p = Ft$. You changed the force applied by increasing the contact time, and hopefully, you reduced the force sufficiently so the egg does not break.

Connections . . .

Circular Fun

Amusement park rides are designed to accelerate your body, sometimes producing changes in the acceleration (jerk) as well. This is done by changes in speed, changes in the direction of travel, or changes in both direction and speed. Many rides move in a circular path, since such movement is a constant acceleration.

Why do people enjoy amusement park rides? It is not the high speed, since your body is not very sensitive to moving at a constant speed. Moving at a steady 600 mi/h in an airplane, for example, provides little sensation when you are seated in an aisle seat in the central cabin.

Your body is not sensitive to high-speed traveling, but it is sensitive to acceleration and changes of acceleration. Acceleration affects the fluid in your inner ear, which controls your sense of balance. In most people, acceleration also produces a reaction that results in the release of the hormones epinephrine and norepinephrine from the adrenal medulla, located near the kidney. The heart rate increases, blood pressure rises, blood is shunted to muscles, and the breathing rate increases. You have probably experienced this reaction many times in your life, as when you nearly have an automobile accident or slip and nearly fall. In the case of an amusement park ride, your body adapts and you believe you enjoy the experience.

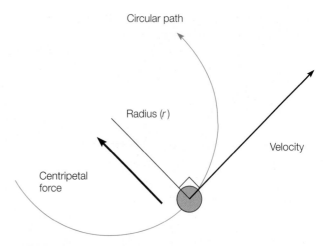

FIGURE 2.22 Centripetal force on the ball causes it to change direction continuously, or accelerate into a circular path. Without the force acting on it, the ball would continue in a straight line.

orbit of a radius (r) can be found by substituting equation 2.7 in $F = ma$, or

$$F = \frac{mv^2}{r}$$ **equation 2.8**

EXAMPLE 2.9 *(Optional)*

A 0.25 kg ball is attached to the end of a 0.5 m string and moved in a horizontal circle at 2.0 m/s. What net force is needed to keep the ball in its circular path?

SOLUTION

$m = 0.25$ kg

$r = 0.5$ m

$v = 2.0$ m/s

$F = ?$

$$F = \frac{mv^2}{r}$$

$$= \frac{(0.25 \text{ kg})(2.0 \text{ m/s})^2}{0.5 \text{ m}}$$

$$= \frac{(0.25 \text{ kg})(4.0 \text{ m}^2/\text{s}^2)}{0.5 \text{ m}}$$

$$= \frac{(0.25)(4.0)}{0.5} \frac{\text{kg} \cdot \text{m}^2}{\text{s}^2} \times \frac{1}{\text{m}}$$

$$= 2 \frac{\text{kg} \cdot \text{m}^2}{\text{m} \cdot \text{s}^2}$$

$$= 2 \frac{\text{kg} \cdot \text{m}}{\text{s}^2}$$

$$= \boxed{2 \text{ N}}$$

EXAMPLE 2.10 *(Optional)*

Suppose you make the string in example 2.9 half as long, 0.25 m. What force is now needed? (Answer: 4.0 N)

through the string. This centripetal force pulls the ball from its natural straight-line path into a circular path. There are no outward forces acting on the ball. The force that you feel on the string is a consequence of the third law; the ball exerts an equal and opposite force on your hand. If you were to release the string, the ball would move away from the circular path in a *straight line* that has a right angle to the radius at the point of release (figure 2.22). When you release the string, the centripetal force ceases, and the ball then follows its natural straight-line motion. If other forces were involved, it would follow some other path. Nonetheless, the apparent outward force has been given a name just as if it were a real force. The outward tug is called a *centrifugal force.*

The magnitude of the centripetal force required to keep an object in a circular path depends on the inertia, or mass, of the object and the acceleration of the object, just as you learned in the second law of motion. The acceleration of an object moving in a circle can be shown to be directly proportional to the square of the speed around the circle (v^2) and inversely proportional to the radius of the circle (r). (A smaller radius requires a greater acceleration.) Therefore, the acceleration of an object moving in uniform circular motion (a_c) is

$$a_c = \frac{v^2}{r}$$ **equation 2.7**

The magnitude of the centripetal force of an object with a constant mass (m) that is moving with a velocity (v) in a circular

NEWTON'S LAW OF GRAVITATION

You know that if you drop an object, it always falls to the floor. You define *down* as the direction of the object's movement and *up* as the opposite direction. Objects fall because of the force of gravity, which accelerates objects on Earth at $g = 9.8$ m/s^2 (32 ft/s^2) and gives them weight, $w = mg$.

Gravity is an attractive force, a pull that exists between all objects in the universe. It is a mutual force that, just like all other forces, comes in matched pairs. Since Earth attracts you with a certain force, you must attract Earth with an exact opposite force. The magnitude of this force of mutual attraction depends on several variables. These variables were first described by Newton in *Principia,* his famous book on motion that was printed in 1687. Newton had, however, worked out his ideas much earlier, by the age of twenty-four, along with ideas about his laws of motion and the formula for centripetal acceleration. In a biography written by a friend in 1752, Newton stated that the notion of gravitation came to mind during a time of thinking that "was occasioned by the fall of an apple." He was thinking about why the Moon stays in orbit around Earth rather than moving off in a straight line as would be predicted by the first law of motion. Perhaps the same force that attracts the Moon toward Earth, he thought, attracts the apple to Earth. Newton developed a theoretical equation for gravitational force that explained not only the motion of the Moon but the motion of the whole solar system. Today, this relationship is known as the **universal law of gravitation:**

> **Every object in the universe is attracted to every other object with a force that is directly proportional to the product of their masses and inversely proportional to the square of the distances between them.**

In symbols, m_1 and m_2 can be used to represent the masses of two objects, d the distance between their centers, and G a constant of proportionality. The equation for the law of universal gravitation is therefore

$$F = G\frac{m_1m_2}{d^2} \qquad \textbf{equation 2.9}$$

This equation gives the magnitude of the attractive force that each object exerts on the other. The two forces are oppositely directed. The constant G is a universal constant, since the law applies to all objects in the universe. It was first measured experimentally by Henry Cavendish in 1798. The accepted value today is $G = 6.67 \times 10^{-11}$ N·m^2/kg^2. Do not confuse G, the universal constant, with g, the acceleration due to gravity on the surface of Earth.

Thus, the magnitude of the force of gravitational attraction is determined by the mass of the two objects and the distance between them (figure 2.23). The law also states that *every* object is attracted to every other object. You are attracted to all the objects around you—chairs, tables, other people, and so forth. Why don't you notice the forces between you and other objects? One or both of the interacting objects must be quite massive before a noticeable force results from the interaction. That is why you do not notice the force of gravitational attraction between you and objects that are not very massive compared to Earth. The attraction between you and Earth overwhelmingly predominates, and that is all you notice.

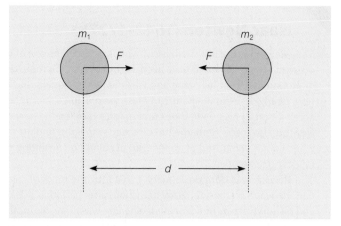

FIGURE 2.23 The variables involved in gravitational attraction. The force of attraction (F) is proportional to the product of the masses (m_1, m_2) and inversely proportional to the square of the distance (d) between the centers of the two masses.

EXAMPLE 2.11 *(Optional)*

What is the force of gravitational attraction between two 60.0 kg (132 lb) students who are standing 1.00 m apart?

SOLUTION

$G = 6.67 \times 10^{-11}$ Nm^2/kg^2

$m_1 = 60.0$ kg

$m_2 = 60.0$ kg $\qquad F = G\dfrac{m_1m_2}{d^2}$

$d = 1.00$ m

$F = ?$

$$= \frac{(6.67 \times 10^{-11}\,\text{Nm}^2/\text{kg}^2)(60.0\,\text{kg})(60.0\,\text{kg})}{(1.00\,\text{m})^2}$$

$$= (6.67 \times 10^{-11})(3.60 \times 10^3)\frac{\text{Nm}^2\text{kg}^2}{\dfrac{\text{kg}^2}{\text{m}^2}}$$

$$= 2.40 \times 10^{-7}(\text{Nm}^2)\left(\frac{1}{\text{m}^2}\right)$$

$$= 2.40 \times 10^{-7}\frac{\text{Nm}^2}{\text{m}^2}$$

$$= \boxed{2.40 \times 10^{-7}\ \text{N}}$$

(Note: A force of 2.40×10^{-7} [0.00000024] N is equivalent to a force of 5.40×10^{-8} [0.00000005] lb, a force that you would not notice. In fact, it would be difficult to measure such a small force.)

EXAMPLE 2.12 *(Optional)*

What would be the value of g if Earth were less dense, with the same mass and double the radius? (Answer: $g = 2.44$ m/s^2)

The acceleration due to gravity, g, is about 9.8 m/s^2 on Earth and is practically a constant for relatively short distances above the surface. Notice, however, that Newton's law of gravitation is an inverse square law. This means if you double the distance, the force is $1/(2)^2$ or 1/4 as great. If you triple the distance, the force is $1/(3)^2$ or 1/9 as great. In other words, the force of

People Behind the Science

Isaac Newton (1642–1727)

Isaac Newton was a British physicist who is regarded as one of the greatest scientists ever to have lived. He discovered the three laws of motion that bear his name and was the first to explain gravitation, clearly defining the nature of mass, weight, force, inertia, and acceleration. In his honor, the SI unit of force is called the newton. Newton also made fundamental discoveries in light, finding that white light is composed of a spectrum of colors and inventing the reflecting telescope.

Newton was born on January 4, 1643 (by the modern calendar). He was a premature, sickly baby born after his father's death, and his survival was not expected. When he was three, his mother remarried, and the young Newton was left in his grandmother's care. He soon began to take refuge in things mechanical, making water clocks, kites bearing fiery lanterns aloft, and a model mill powered by a mouse, as well as innumerable drawings and diagrams. When Newton was twelve, his mother withdrew him from school with the intention of making him into a farmer. Fortunately, his uncle recognized Newton's ability and managed to get him back into school to prepare for college.

Newton was admitted to Trinity College, Cambridge, and graduated in 1665, the same year that the university was closed because of the plague. Newton returned to his boyhood farm to wait out the plague, making only an occasional visit back to Cambridge. During this period, he performed his first prism experiments and thought about motion and gravitation.

Newton returned to study at Cambridge after the plague had run its course, receiving a master's degree in 1668 and becoming a professor at the age of only twenty-six. Newton remained at Cambridge almost thirty years, studying alone for the most part, though in frequent contact with other leading scientists by letter and through the Royal Society in London. These were Newton's most fertile years. He labored day and night, thinking and testing ideas with calculations.

In Cambridge, he completed what may be described as his greatest single work, *Philosophae Naturalis Principia Mathematica* (*Mathematical Principles of Natural Philosophy*). This was presented to the Royal Society in 1686, which subsequently withdrew from publishing it because of a shortage of funds. The astronomer Edmund Halley (1656–1742), a wealthy man and friend of Newton, paid for the publication of the *Principia* in 1687. In it, Newton revealed his laws of motion and the law of universal gravitation.

Newton's greatest achievement was to demonstrate that scientific principles are of universal application. In *Principia Mathematica,* he built the evidence of experiment and observation to develop a model of the universe that is still of general validity. "If I have seen further than other men," he once said, "it is because I have stood on the shoulders of giants"; and Newton was certainly able to bring together the knowledge of his forebears in a brilliant synthesis.

No knowledge can ever be total, but Newton's example brought about an explosion of investigation and discovery that has never really abated. He perhaps foresaw this when he remarked, "To myself, I seem to have been only like a boy playing on the seashore, and diverting myself in now and then finding a smoother pebble or a prettier shell than ordinary, whilst the great ocean of truth lay all undiscovered before me."

With his extraordinary insight into the workings of nature and rare tenacity in wresting its secrets and revealing them in as fundamental and concise a way as possible, Newton stands as a colossus of science. In physics, only Archimedes (287–212 B.C.) and Albert Einstein (1879–1955), who also possessed these qualities, may be compared to him.

Source: Modified from the *Hutchinson Dictionary of Scientific Biography.* © RM, 2009. All rights reserved. Helicon Publishing is a division of RM.

---

gravitational attraction and *g* decrease inversely with the square of the distance from Earth's center.

Newton was able to calculate the acceleration of the Moon toward Earth, about 0.0027 m/s². The Moon "falls" toward Earth because it is accelerated by the force of gravitational attraction. This attraction acts as a *centripetal force* that keeps the Moon from following a straight-line path as would be predicted from the first law. Thus, the acceleration of the Moon keeps it in a somewhat circular orbit around Earth. Figure 2.24 shows that the Moon would be in position A if it followed a straight-line path instead of "falling" to position B as it does. The Moon thus "falls" around Earth. Newton was able to analyze the motion of the Moon quantitatively as evidence that it is gravitational force that keeps the Moon in its orbit. The law of gravitation was extended to the Sun, other planets, and eventually the universe. The quantitative predictions of observed relationships

CONCEPTS APPLIED

Apparent Weightlessness

Use a sharp pencil to make a small hole in the bottom of a Styrofoam cup. The hole should be large enough for a thin stream of water to flow from the cup, but small enough for the flow to continue for 3 or 4 seconds. Test the water flow over a sink.

Hold a finger over the hole on the cup as you fill it with water. Stand on a ladder or outside stairwell as you hold the cup out at arm's length. Move your finger, allowing a stream of water to flow from the cup as you drop it. Observe what happens to the stream of water as the cup is falling. Explain your observations. Also predict what you would see if you were falling with the cup.

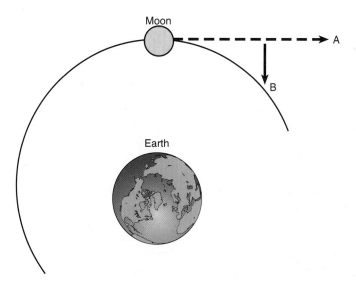

FIGURE 2.24 Gravitational attraction acts as a centripetal force that keeps the Moon from following the straight-line path shown by the dashed line to position A. It was pulled to position B by gravity (0.0027 m/s²) and thus "fell" toward Earth the distance from the dashed line to B, resulting in a somewhat circular path.

FIGURE 2.25 This photograph of Earth from space shows that it is nearly spherical.

among the planets were strong evidence that all objects obey the same law of gravitation. In addition, the law provided a means to calculate the mass of Earth, the Moon, the planets, and the Sun. Newton's law of gravitation, laws of motion, and work with mathematics formed the basis of most physics and technology for the next two centuries, as well as accurately describing the world of everyday experience.

Earth Satellites

As you can see in figure 2.25, Earth is round-shaped and nearly spherical. The curvature is obvious in photographs taken from space but not so obvious back on the surface because Earth is so large. However, you can see evidence of the curvature in places on the surface where you have unobstructed vision for long distances. For example, a tall ship appears to "sink" on the horizon as it sails away, following Earth's curvature below your line of sight. The surface of Earth curves away from your line of sight or any other line tangent to the surface, dropping at a rate of about 4.9 m for every 8 km (16 ft in 5 mi). This means that a ship 8 km away will appear to drop about 5 m below the horizon and anything less than about 5 m tall at this distance will be out of sight, below the horizon.

Recall that a falling object accelerates toward Earth's surface at *g*, which has an average value of 9.8 m/s². Ignoring air resistance, a falling object will have a speed of 9.8 m/s at the end of 1 second and will fall a distance of 4.9 m. If you wonder why the object did not fall 9.8 m in 1 s, recall that the object starts with an initial speed of zero and has a speed of 9.8 m/s only during the last instant. The average speed was an average of the initial and final speeds, which is 4.9 m/s. An average

speed of 4.9 m/s over a time interval of 1 second will result in a distance covered of 4.9 m.

Did you know that Newton was the first to describe how to put an artificial satellite into orbit around Earth? He did not discuss rockets, however, but described in *Principia* how to put a cannonball into orbit. He described how a cannonball shot with sufficient speed straight out from a mountaintop would go into orbit around Earth. If it had less than the sufficient speed, it would fall back to Earth following the path of projectile motion, as discussed in the section on "Compound Motion." What speed does it need to go into orbit? Earth curves away from a line tangent to the surface at 4.9 m per 8 km. Any object falling from a resting position will fall a distance of 4.9 m during the first second. Thus, a cannonball shot straight out from a mountaintop with a speed of 8 km/s (nearly 18,000 mi/hr, or 5 mi/s) will fall toward the surface, dropping 4.9 m during the first second. But the surface of Earth drops, too, curving away below the falling cannonball. So the cannonball is still moving horizontally, no closer to the surface than it was a second ago. As it falls 4.9 m during the next second, the surface again curves away 4.9 m over the 8 km distance. This repeats again and again, and the cannonball stays the same distance from the surface, and we say it is now an *artificial satellite* in orbit. The satellite requires no engine or propulsion as it continues to fall toward the surface, with Earth curving away from it continuously. This assumes, of course, no loss of speed from air resistance.

Today, an artificial satellite is lofted by rocket or rockets to an altitude of more than 320 km (about 200 mi), above the

Gravity Problems

Gravity does act on astronauts in spacecraft that are in orbit around Earth. Since gravity is acting on the astronaut and spacecraft, the term *zero gravity* is not an accurate description of what is happening. The astronaut, spacecraft, and everything in it are experiencing *apparent weightlessness* because they are continuously falling toward the surface. Everything seems to float because everything is falling together. But, strictly speaking, everything still has weight, because weight is defined as a gravitational force acting on an object ($w = mg$).

Whether weightlessness is apparent or real, however, the effects on people are the same. Long-term orbital flights have provided evidence that the human body changes from the effect of weightlessness. Bones lose calcium and other minerals, the heart shrinks to a much smaller size, and leg muscles shrink so much on prolonged flights that astronauts cannot walk when they return to the surface. These changes occur because on Earth, humans are constantly subjected to the force of gravity. The nature of the skeleton and the strength of the muscles are determined by how the body reacts to this force. Metabolic pathways and physiological processes that maintain strong bones and muscles evolved having to cope with a specific gravitational force. When we are suddenly subjected to a place where gravity is significantly different, these processes result in weakened systems. If we lived on a planet with a different gravitational force, we would have muscles and bones that were adapted to the gravity on that planet. Many kinds of organisms have been used in experiments in space to try to develop a better understanding of how their systems work without gravity.

The problems related to prolonged weightlessness must be worked out before long-term weightless flights can take place. One solution to these problems might be a large, uniformly spinning spacecraft. The astronauts would tend to move in a straight line, and the side of the turning spacecraft (now the "floor") would exert a force on them to make them go in a curved path. This force would act as an artificial gravity.

---

air friction of the atmosphere, before being aimed horizontally. The satellite is then "injected" into orbit by giving it the correct tangential speed. This means it has attained an orbital speed of at least 8 km/s (5 mi/s) but less than 11 km/s (7 mi/s). At a speed less than 8 km/s, the satellite would fall back to the surface in a parabolic path. At a speed more than 11 km/s, it will move faster than the surface curves away and will escape from Earth into space. But with the correct tangential speed, and above the atmosphere and air friction, the satellite will follow a circular orbit for long periods of time without the need for any more propulsion. An orbit injection speed of more than 8 km/s (5 mi/s) would result in an elliptical rather than a circular orbit.

A satellite could be injected into orbit near the outside of the atmosphere, closer to Earth but outside the air friction that might reduce its speed. The satellite could also be injected far away from Earth, where it takes a longer time to complete one orbit. Near the outer limits of the atmosphere—that is, closer to the surface—a satellite might orbit Earth every 90 minutes or so. A satellite as far away as the Moon, on the other hand, orbits Earth in a little less than 28 days. A satellite at an altitude of 36,000 km (a little more than 22,000 mi) has a period of 1 day. In the right spot over the equator, such a satellite is called a **geosynchronous satellite,** since it turns with Earth and does not appear to move across the sky (figure 2.26). The photographs of the cloud cover you see in weather reports were taken from one or more geosynchronous weather satellites. Communications networks are also built around geosynchronous satellites. One way to locate one of these geosynchronous satellites is to note the aiming direction of backyard satellite dishes that pick up television signals.

Weightlessness

News photos sometimes show astronauts "floating" in the Space Shuttle or next to a satellite (figure 2.27). These astronauts appear to be weightless but technically are no more weightless than a skydiver in free fall or a person in a falling elevator. Recall that weight is a gravitational force, a measure of the gravitational attraction between Earth and an object (mg). The weight of a cup of coffee, for example, can be measured by placing the cup on a scale. The force the cup of coffee exerts against the scale

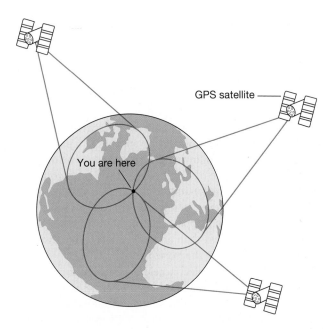

FIGURE 2.26 In the Global Positioning System (GPS), each of a fleet of orbiting satellites sends out coded radio signals that enable a receiver on the earth to determine both the exact position of the satellite in space and its exact distance from the receiver. Given this information, a computer in the receiver then calculates the circle on the Earth's surface on which the receiver must lie. Data from three satellites give three circles and the receiver must be located at the one point where all three intersect (also see pages 173–74).

FIGURE 2.27 Astronauts in an orbiting space station may appear to be weightless. Technically, however, they are no more weightless than a skydiver in free fall or a person in a falling elevator near or on the surface of Earth.

is its weight. You also know that the scale pushes back on the cup of coffee since it is not accelerating, which means the net force is zero.

Now consider what happens if a skydiver tries to pour a cup of coffee while in free fall. Even if you ignore air resistance, you can see that the skydiver is going to have a difficult time, at best. The coffee, the cup, and the skydiver will all be falling together. Gravity is acting to pull the coffee downward, but gravity is also acting to pull the cup from under it at the same rate. The coffee, the cup, and the skydiver all fall together, and the skydiver will see the coffee appear to "float" in blobs. If the diver lets go of the cup, it too will appear to float as everything continues to fall together. However, this is only *apparent weightlessness,* since gravity is still acting on everything; the coffee, the cup, and the skydiver only *appear* to be weightless because they are all accelerating at *g*.

Astronauts in orbit are in free fall, falling toward Earth just as the skydiver, so they too are undergoing apparent weightlessness. To experience true weightlessness, the astronauts would have to travel far from Earth and its gravitational field, and far from the gravitational fields of other planets.

SUMMARY

Motion can be measured by speed, velocity, and acceleration. *Speed* is a measure of how fast something is moving. It is a ratio of the distance covered between two locations to the time that elapsed while moving between the two locations. The *average speed* considers the distance covered during some period of time, while the *instantaneous speed* is the speed at some specific instant. *Velocity* is a measure of the speed and direction of a moving object. *Acceleration* is a change of velocity per unit of time.

A *force* is a push or a pull that can change the motion of an object. The *net force* is the sum of all the forces acting on an object.

Galileo determined that a continuously applied force is not necessary for motion and defined the concept of *inertia:* an object remains in unchanging motion in the absence of a net force. Galileo also determined that falling objects accelerate toward Earth's surface independent of the weight of the object. He found the acceleration due to gravity, *g*, to be 9.8 m/s^2(32 ft/s^2), and the distance an object falls is proportional to the square of the time of free fall ($d \propto t^2$).

Compound motion occurs when an object is projected into the air. Compound motion can be described by splitting the motion into vertical and horizontal parts. The acceleration due to gravity, *g*, is a constant that is acting at all times and acts independently of any motion that an object has. The path of an object that is projected at some angle to the horizon is therefore a parabola.

Newton's *first law of motion* is concerned with the motion of an object and the lack of a net force. Also known as the *law of inertia,* the first law states that an object will retain its state of straight-line motion (or state of rest) unless a net force acts on it.

The *second law of motion* describes a relationship between net force, mass, and acceleration. A *newton* of force is the force needed to give a 1.0 kg mass an acceleration of 1.0 m/s^2.

Weight is the downward force that results from Earth's gravity acting on the mass of an object. Weight is measured in *newtons* in the metric system and *pounds* in the English system.

Newton's *third law of motion* states that forces are produced by the interaction of *two different* objects. These forces always occur in matched pairs that are equal in size and opposite in direction.

Momentum is the product of the mass of an object and its velocity. In the absence of external forces, the momentum of a group of interacting objects always remains the same. This relationship is the *law of conservation of momentum. Impulse* is a change of momentum equal to a force times the time of application.

An object moving in a circular path must have a force acting on it, since it does not move in a straight line. The force that pulls an object out of its straight-line path is called a *centripetal force.* The centripetal force needed to keep an object in a circular path depends on the mass of the object, its velocity, and the radius of the circle.

The *universal law of gravitation* is a relationship between the masses of two objects, the distance between the objects, and a proportionality constant. Newton was able to use this relationship to show that gravitational attraction provides the centripetal force that keeps the Moon in its orbit.

Summary of Equations

2.1

$$\text{average speed} = \frac{\text{distance}}{\text{time}}$$

$$\overline{v} = \frac{d}{t}$$

2.2

$$\text{acceleration} = \frac{\text{change of velocity}}{\text{time}}$$

$$= \frac{\text{final velocity} - \text{initial velocity}}{\text{time}}$$

$$a = \frac{v_f - v_i}{t}$$

2.3

$$\text{force} = \text{mass} \times \text{acceleration}$$

$$F = ma$$

2.4

$$\text{weight} = \text{mass} \times \text{acceleration due to gravity}$$

$$w = mg$$

2.5

$$\text{momentum} = \text{mass} \times \text{velocity}$$

$$p = mv$$

2.6

$$\text{change of momentum} = \text{force} \times \text{time}$$

$$\Delta p = Ft$$

2.7

$$\text{centripetal acceleration} = \frac{\text{velocity squared}}{\text{radius of circle}}$$

$$a_c = \frac{v^2}{r}$$

2.8

$$\text{centripetal force} = \frac{\text{mass} \times \text{velocity squared}}{\text{radius of circle}}$$

$$F = \frac{mv^2}{r}$$

2.9

$$\text{gravitational force} = \text{constant} \times \frac{\text{one mass} \times \text{another mass}}{\text{distance squared}}$$

$$F = G\frac{m_1 m_2}{d^2}$$

KEY TERMS

acceleration (p. **29**)
centripetal force (p. **45**)
first law of motion (p. **38**)
force (p. **30**)
free fall (p. **34**)
fundamental forces (p. **32**)

g (p. **36**)
geosynchronous satellite (p. **50**)
impulse (p. **44**)
inertia (p. **33**)
law of conservation of
  momentum (p. **44**)

mass (p. **41**)
momentum (p. **43**)
net force (p. **30**)
newton (p. **40**)
second law of motion (p. **40**)
speed (p. **27**)

third law of motion (p. **42**)
universal law of gravitation
  (p. **47**)
velocity (p. **28**)

APPLYING THE CONCEPTS

1. A quantity of 5 m/s^2 is a measure of
   a. metric area.
   b. acceleration.
   c. speed.
   d. velocity.

2. An automobile has how many different devices that can cause it to undergo acceleration?
   a. none
   b. one
   c. two
   d. three or more

3. Ignoring air resistance, an object falling toward the surface of Earth has a *velocity* that is
   a. constant.
   b. increasing.
   c. decreasing.
   d. acquired instantaneously but dependent on the weight of the object.

4. Ignoring air resistance, an object falling near the surface of Earth has an *acceleration* that is
   a. constant.
   b. increasing.
   c. decreasing.
   d. dependent on the weight of the object.

5. Two objects are released from the same height at the same time, and one has twice the weight of the other. Ignoring air resistance,
   a. the heavier object hits the ground first.
   b. the lighter object hits the ground first.
   c. they both hit at the same time.
   d. whichever hits first depends on the distance dropped.

6. A ball rolling across the floor slows to a stop because
   a. there is a net force acting on it.
   b. the force that started it moving wears out.
   c. the forces are balanced.
   d. the net force equals zero.

FIGURE 3.1 Glen Canyon Dam on the Colorado River between Utah and Arizona is 216 m (710 ft) tall, dropping 940 m³ of water per second (33,000 ft³/s) through eight generating units.

Mechanical work is the product of a force and the distance an object moves as a result of the force. There are two important considerations to remember about this definition: (1) something *must move* whenever work is done, and (2) the movement must be in the *same direction* as the direction of the force. When you move a book to a higher shelf in a bookcase, you are doing work on the book. You apply a vertically upward force equal to the weight of the book as you move it in the same direction as the direction of the applied force. The work done on the book can therefore be calculated by multiplying the weight of the book by the distance it was moved (figure 3.2).

If you simply stand there holding the book, however, you are doing no work on the book. Your arm may become tired from holding the book, since you must apply a vertically upward force equal to the weight of the book. But this force is not acting through a distance, since the book is not moving. According to equation 3.1, a distance of zero results in zero work. Only a force that results in motion in the same direction results in work.

Units of Work

The units of work can be obtained from the definition of work, $W = Fd$. In the metric system, a force is measured in newtons (N), and distance is measured in meters (m), so the unit of work is

$$W = Fd$$
$$W = (\text{newton})(\text{meter})$$
$$W = (\text{N})(\text{m})$$

The newton-meter is therefore the unit of work. This derived unit has a name. The newton-meter is called a **joule** (J) (pronounced "jool").

$$1 \text{ joule} = 1 \text{ newton-meter}$$

The units for a newton are kg·m/s², and the unit for a meter is m. It therefore follows that the units for a joule are kg·m²/s².

In the English system, the force is measured in pounds (lb), and the distance is measured in feet (ft). The unit of work in the English system is therefore the ft·lb. The ft·lb does not have a name of its own as the N·m does (figure 3.3). (Note that although the equation is $W = Fd$, and this means = (pounds)(feet), the unit is called the ft·lb.)

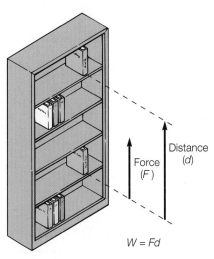

FIGURE 3.2 The force on the book moves it through a vertical distance from the second shelf to the fifth shelf, and work is done, $W = Fd$.

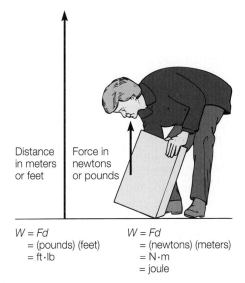

FIGURE 3.3 Work is done against gravity when lifting an object. Work is measured in joules or foot-pounds.

EXAMPLE 3.1   (Optional)

How much work is needed to lift a 5.0 kg backpack to a shelf 1.0 m above the floor?

SOLUTION

The backpack has a mass (m) of 5.0 kg, and the distance (d) is 1.0 m. To lift the backpack requires a vertically upward force equal to the weight of the backpack. Weight can be calculated from $w = mg$:

$$m = 5.0 \text{ kg} \quad w = mg$$
$$g = 9.8 \text{ m/s}^2$$
$$w = ? \quad\quad = (5.0 \text{ kg})\left(9.8 \frac{\text{m}}{\text{s}^2}\right)$$
$$= (5.0)(9.8) \text{ kg} \times \frac{\text{m}}{\text{s}^2}$$
$$= 49 \frac{\text{kg}\cdot\text{m}}{\text{s}^2}$$
$$= 49 \text{ N}$$

The definition of work is found in equation 3.1,

$$F = 49 \text{ N} \quad W = Fd$$
$$d = 1.0 \text{ m} \quad\quad = (49 \text{ N})(1.0 \text{ m})$$
$$W = ? \quad\quad = (49)(1.0) \text{ N} \times \text{m}$$
$$= 49 \text{ N}\cdot\text{m}$$
$$= \boxed{49 \text{ J}}$$

EXAMPLE 3.2   (Optional)

How much work is required to lift a 50 lb box vertically a distance of 2 ft? (Answer: 100 ft·lb)

Power

You are doing work when you walk up a stairway, since you are lifting yourself through a distance. You are lifting your weight (force exerted) the *vertical* height of the stairs (distance through which the force is exerted). Consider a person who weighs 120 lb and climbs a stairway with a vertical distance of 10 ft. This person will do (120 lb)(10 ft) or 1,200 ft·lb of work. Will the amount of work change if the person were to run up the stairs? The answer is no, the same amount of work is accomplished. Running up the stairs, however, is more tiring than walking up the stairs. You use the same amount of energy but at a greater *rate* when running. The rate at which energy is transformed or the rate at which work is done is called **power** (figure 3.4). Power is defined as work per unit of time,

$$\text{power} = \frac{\text{work}}{\text{time}}$$
$$P = \frac{W}{t} \quad\quad \textbf{equation 3.2}$$

Book Work

Place a tied loop of string between the center pages of a small book. Pick up the loop so the string lifts the book, supporting it open pages down. Use a spring scale to find the weight of the book in newtons. Measure the work done in lifting the book 1 meter. Use the spring scale to measure the work done in pulling the book along a tabletop for 1 meter. Is the amount of work done to lift the book the same as the amount of work done pulling the book along the tabletop? Why or why not?

Considering just the work and time factors, the 120 lb person who ran up the 10 ft height of stairs in 4 seconds would have a power rating of

$$P = \frac{W}{t} = \frac{(120 \text{ lb})(10 \text{ ft})}{4 \text{ s}} = 300 \frac{\text{ft}\cdot\text{lb}}{\text{s}}$$

If the person had a time of 3 seconds on the same stairs, the power rating would be greater, 400 ft·lb/s. This is a greater *rate* of energy use, or greater power.

When the steam engine was first invented, there was a need to describe the rate at which the engine could do work. Since people at this time were familiar with using horses to do their

FIGURE 3.4   (*A*) The work accomplished in climbing a stairway is the person's weight times the vertical distance. (*B*) The power level is the work accomplished per unit of time.

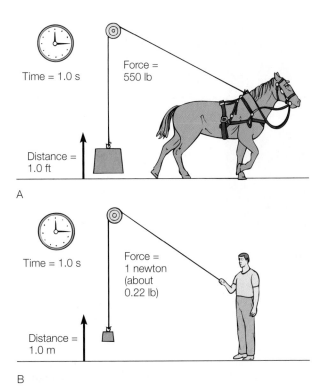

FIGURE 3.5 (A) A horsepower is defined as a power rating of 550 ft·lb/s. (B) A watt is defined as a newton-meter per second, or joule per second.

and the power would be $P = mgh/t$. Note that h is for height, a vertical distance (d).

$$
\begin{aligned}
m &= 500.0 \text{ kg} \\
g &= 9.8 \text{ m/s}^2 \\
h &= 10.0 \text{ m} \\
t &= 5.0 \text{ s}
\end{aligned}
\qquad
\begin{aligned}
P &= \frac{mgh}{t} \\[2mm]
&= \frac{(500.0 \text{ kg})\left(9.8\,\dfrac{\text{m}}{\text{s}^2}\right)(10.0 \text{ m})}{5.0 \text{ s}} \\[2mm]
&= \frac{(500.0)(9.8)(10.0)}{5.0}\,\frac{\text{kg} \times \dfrac{\text{m}}{\text{s}^2} \times \text{m}}{\text{s}} \\[2mm]
&= 9{,}800\,\frac{\text{N·m}}{\text{s}} \\[2mm]
&= 9{,}800\,\frac{\text{J}}{\text{s}} \\[2mm]
&= 9{,}800 \text{ W} \\[2mm]
&= \boxed{9.8 \text{ kW}}
\end{aligned}
$$

The power in horsepower (hp) units would be

$$9{,}800 \text{ W} \times \frac{\text{hp}}{746 \text{ W}} = 13 \text{ hp}$$

EXAMPLE 3.4 *(Optional)*

A 150 lb person runs up a 15 ft stairway in 10.0 s. What is the horsepower rating of the person? (Answer: 0.41 horsepower)

MOTION, POSITION, AND ENERGY

Closely related to the concept of work is the concept of **energy.** Energy can be defined as the *ability to do work.* This definition of energy seems consistent with everyday ideas about energy and physical work. After all, it takes a lot of energy to do a lot of work. In fact, one way of measuring the energy of something is to see how much work it can do. Likewise, when work is done *on* something, a change occurs in its energy level. The following examples will help clarify this close relationship between work and energy.

Potential Energy

Consider a book on the floor next to a bookcase. You can do work on the book by vertically raising it to a shelf. You can measure this work by multiplying the vertical upward force applied times the distance that the book is moved. You might find, for example, that you did an amount of work equal to 10 J on the book.

Suppose that the book has a string attached to it, as shown in figure 3.6. The string is threaded over a frictionless pulley and attached to an object on the floor. If the book is caused to fall from the shelf, the object on the floor will be vertically lifted through some distance by the string. The falling book exerts a force on the object through the string, and the object is moved

work, the steam engines were compared to horses. James Watt, who designed a workable steam engine, defined **horsepower** as a power rating of 550 ft·lb/s (figure 3.5A). To convert a power rating in the English units of ft·lb/s to horsepower, divide the power rating by 550 ft·lb/s/hp. For example, the 120 lb person who had a power rating of 400 ft·lb/s had a horsepower of 400 ft·lb/s ÷ 550 ft·lb/s/hp, or 0.7 hp.

In the metric system, power is measured in joules per second. The unit J/s, however, has a name. A J/s is called a **watt** (W). The watt (figure 3.5B) is used with metric prefixes for large numbers: 1,000 W = 1 kilowatt (kW) and 1,000,000 W = 1 megawatt (MW). It takes 746 W to equal 1 horsepower. One kilowatt is equal to about 1 1/3 horsepower. The electric utility company charges you for how much electrical energy you have used. Electrical energy is measured by power (kW) times the time of use (h). Thus, electrical energy is measured in kWh. We will return to kilowatts and kilowatt-hours later when we discuss electricity.

EXAMPLE 3.3 *(Optional)*

An electric lift can raise a 500.0 kg mass a distance of 10.0 m in 5.0 s. What is the power of the lift?

SOLUTION

Power is the work per unit time ($P = W/t$), and work is force times distance ($W = Fd$). The vertical force required is the weight lifted, and $w = mg$. Therefore, the work accomplished would be $W = mgh$,

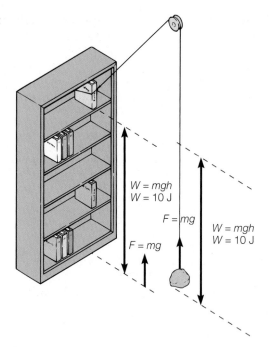

FIGURE 3.6 If moving a book from the floor to a high shelf requires 10 J of work, then the book will do 10 J of work on an object of the same mass when the book falls from the shelf.

through a distance. In other words, the *book* did work on the object through the string, $W = Fd$.

The book can do more work on the object if it falls from a higher shelf, since it will move the object a greater distance. The higher the shelf, the greater the *potential* for the book to do work. The ability to do work is defined as energy. The energy that an object has because of its position is called **potential energy** (*PE*). Potential energy is defined as *energy due to position.* This type of potential energy is called *gravitational potential energy,* since it is a result of gravitational attraction. There are other types of potential energy, such as that in a compressed or stretched spring.

The gravitational potential energy of an object can be calculated from the work done *on* the object to change its position. You exert a force equal to its weight as you lift it some height above the floor, and the work you do is the product of the weight and height. Likewise, the amount of work the object *could* do because of its position is the product of its weight and height. For the metric unit of mass, weight is the product of the mass of an object times *g*, the acceleration due to gravity, so

gravitational potential energy = weight × height

$$PE = mgh \qquad \text{equation 3.3}$$

For English units, the pound *is* the gravitational unit of force, or weight, so equation 3.3 becomes $PE = (w)(h)$.

Under what conditions does an object have zero potential energy? Considering the book in the bookcase, you could say that the book has zero potential energy when it is flat on the floor. It can do no work when it is on the floor. But what if that floor happens to be the third floor of a building? You could, after all, drop the book out of a window. The answer is that it makes no difference. The same results would be obtained in either case since it is the *change of position* that is important

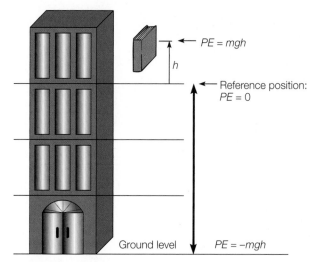

FIGURE 3.7 The zero reference level for potential energy is chosen for convenience. Here the reference position chosen is the third floor, so the book will have a negative potential energy at ground level.

in potential energy. The zero reference position for potential energy is therefore arbitrary. A zero reference point is chosen as a matter of convenience. Note that if the third floor of a building is chosen as the zero reference position, a book on ground level would have negative potential energy. This means that you would have to do work on the book to bring it back to the zero potential energy position (figure 3.7). You will learn more about negative energy levels later in the chapters on chemistry.

EXAMPLE 3.5 *(Optional)*

What is the potential energy of a 2.14 kg book that is on a bookshelf 1.0 m above the floor?

SOLUTION

Equation 3.3, $PE = mgh$, shows the relationship between potential energy (*PE*), weight (*mg*), and height (*h*).

$$
\begin{aligned}
m &= 2.14 \text{ kg} \qquad PE = mgh \\
h &= 1.0 \text{ m} \qquad\quad = (2.14 \text{ kg})(9.8 \tfrac{\text{m}}{\text{s}^2})(1.0 \text{ m}) \\
PE &= ? \qquad\qquad = (2.14)(9.8)(1.0) \, \frac{\text{kg} \cdot \text{m} \cdot \text{m}}{\text{s}^2} \\
&\qquad\qquad\quad = 21 \frac{\text{kg} \cdot \text{m}}{\text{s}^2} \times \text{m} \\
&\qquad\qquad\quad = 21 \text{ Nm} \\
&\qquad\qquad\quad = \boxed{21 \text{ J}}
\end{aligned}
$$

EXAMPLE 3.6 *(Optional)*

How much work can a 5.00 kg mass do if it is 5.00 m above the ground? (Answer: 250 J)

Kinetic Energy

Moving objects have the ability to do work on other objects because of their motion. A rolling bowling ball exerts a force on the bowling pins and moves them through a distance, but the ball loses speed as a result of the interaction (figure 3.8). A moving car has the ability to exert a force on a tree and knock

FIGURE 3.8 (A) Work is done on the bowling ball as a force (F_B) moves it through a distance. (B) This gives the ball a kinetic energy equal in amount to the work done on it. (C) The ball does work on the pins and has enough remaining energy to crash into the wall behind the pins.

it down, again with a corresponding loss of speed. Objects in motion have the ability to do work, so they have energy. The energy of motion is known as **kinetic energy.** Kinetic energy can be measured (1) in terms of the work done to put the object in motion or (2) in terms of the work the moving object will do in coming to rest. Consider objects that you put into motion by throwing. You exert a force on a football as you accelerate it through a distance before it leaves your hand. The kinetic energy that the ball now has is equal to the work (force times distance) that you did on the ball. You exert a force on a baseball through a distance as the ball increases its speed before it leaves your hand. The kinetic energy that the ball now has is equal to the work that you did on the ball.

CONCEPTS APPLIED

Work and Power

Power is the rate of expending energy or of doing work. You can find your power output by taking a few measurements.

First, let's find how much work you do in walking up a flight of stairs. Your work output will be approximately equal to the change in your potential energy (mgh), so you will need (1) to measure the vertical height of a flight of stairs in metric units and (2) to calculate or measure your mass (conversion factors are located inside the front cover). Record your findings in your report.

Second, find your power output by climbing the stairs as fast as you can while someone measures the time with a stopwatch. Find your power output in watts. Convert this to horsepower by consulting the conversion factors inside the front cover. Did you develop at least 1 horsepower? Does a faster person always have more horsepower?

$$\text{walking power} = \frac{(\quad \text{kg})(9.8 \text{ m/s}^2)(\quad \text{m})}{(\quad \text{s})}$$

$$\text{running power} = \frac{(\quad \text{kg})(9.8 \text{ m/s}^2)(\quad \text{m})}{(\quad \text{s})}$$

The ball exerts a force on the hand of the person catching the ball and moves it through a distance. The net work done on the hand is equal to the kinetic energy that the ball had.

A baseball and a bowling ball moving with the same velocity do not have the same kinetic energy. You cannot knock down many bowling pins with a slowly rolling baseball. Obviously, the more-massive bowling ball can do much more work than a less-massive baseball with the same velocity. Is it possible for the bowling ball and the baseball to have the same kinetic energy? The answer is yes, if you can give the baseball sufficient velocity. This might require shooting the baseball from a cannon, however. Kinetic energy is proportional to the mass of a moving object, but velocity has a greater influence. Consider two balls of the same mass, but one is moving twice as fast as the other. The ball with twice the velocity will do *four* times as much work as the slower ball. A ball with three times the velocity will do *nine* times as much work as the slower ball. Kinetic energy is proportional to the square of the velocity ($2^2 = 4$; $3^2 = 9$). The kinetic energy (*KE*) of an object is

$$\text{kinetic energy} = \frac{1}{2}(\text{mass})(\text{velocity})^2$$

$$KE = \frac{1}{2}mv^2 \qquad \textbf{equation 3.4}$$

Kinetic energy is measured in joules, as is work ($F \times d$ or N·m) and potential energy (mgh or N·m).

EXAMPLE 3.7 *(OPTIONAL)*

A 7.00 kg bowling ball is moving in a bowling lane with a velocity of 5.00 m/s. What is the kinetic energy of the ball?

SOLUTION

The relationship between kinetic energy (*KE*), mass (*m*), and velocity (*v*) is found in equation 3.4, $KE = 1/2mv^2$:

$$
\begin{aligned}
m &= 7.00 \text{ kg} \\
v &= 5.00 \text{ m/s} \\
KE &= ?
\end{aligned}
$$

$$
\begin{aligned}
KE &= \frac{1}{2}mv^2 \\
&= \frac{1}{2}(7.00 \text{ kg})\left(5.00\frac{\text{m}}{\text{s}}\right)^2 \\
&= \frac{1}{2}(7.00 \text{ kg})\left(25.0\frac{\text{m}^2}{\text{s}^2}\right) \\
&= \frac{1}{2}(7.00)(25.0) \text{ kg} \times \frac{\text{m}^2}{\text{s}^2} \\
&= \frac{1}{2} \times 175\frac{\text{kg}\cdot\text{m}^2}{\text{s}^2} \\
&= 87.5\frac{\text{kg}\cdot\text{m}}{\text{s}^2} \times \text{m} \\
&= 87.5 \text{ N}\cdot\text{m} \\
&= \boxed{87.5 \text{ J}}
\end{aligned}
$$

EXAMPLE 3.8 *(Optional)*

A 100.0 kg football player moving with a velocity of 6.0 m/s tackles a stationary quarterback. How much work was done on the quarterback? (Answer: 1,800 J)

ENERGY FLOW

The key to understanding the individual concepts of work and energy is to understand the close relationship between the two. When you do work on something, you give it energy of position (potential energy) or you give it energy of motion (kinetic energy). In turn, objects that have kinetic or potential energy can now do work on something else as the transfer of energy continues. Where does all this energy come from and where does it go? The answer to this question is the subject of this section on energy flow.

Energy Forms

Energy comes in various forms, and different terms are used to distinguish one form from another. Although energy comes in various *forms,* this does not mean that there are different *kinds* of energy. The forms are the result of the more common fundamental forces—gravitational, electromagnetic, and nuclear—and objects that are interacting. Energy can be categorized into five forms: (1) *mechanical,* (2) *chemical,* (3) *radiant,* (4) *electrical,* and (5) *nuclear.* The following is a brief discussion of each of the five forms of energy.

Mechanical energy is the form of energy of familiar objects and machines (figure 3.9). A car moving on a highway has kinetic mechanical energy. Water behind a dam has potential mechanical energy. The spinning blades of a steam turbine have kinetic mechanical energy. The form of mechanical energy is usually associated with the kinetic energy of everyday-sized objects and the potential energy that results from gravity. There are other possibilities (e.g., sound), but this description will serve the need for now.

Chemical energy is the form of energy involved in chemical reactions (figure 3.10). Chemical energy is released in the chemical reaction known as *oxidation.* The fire of burning wood is an example of rapid oxidation. A slower oxidation

A

B

FIGURE 3.10   Chemical energy is a form of potential energy that is released during a chemical reaction. Both (*A*) wood and (*B*) coal have chemical energy that has been stored through the process of photosynthesis. The pile of wood might provide fuel for a small fireplace for several days. The pile of coal might provide fuel for a power plant for a hundred days.

releases energy from food units in your body. As you will learn in the chemistry unit, chemical energy involves electromagnetic forces between the parts of atoms. Until then, consider the following comparison. Photosynthesis is carried on in green plants. The plants use the energy of sunlight to rearrange carbon dioxide and water into plant materials and oxygen. Leaving out many steps and generalizing, this could be represented by the following word equation:

$$energy + carbon\ dioxide + water = wood + oxygen$$

The plant took energy and two substances and made two different substances. This is similar to raising a book to a higher shelf in a bookcase. That is, the new substances have more energy than the original ones did. Consider a word equation for the burning of wood:

$$wood + oxygen = carbon\ dioxide + water + energy$$

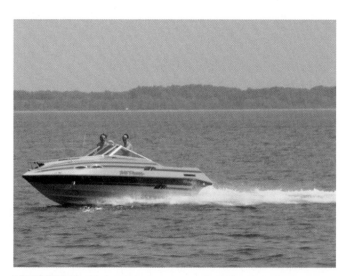

FIGURE 3.9   Mechanical energy is the energy of motion, or the energy of position, of many familiar objects. This boat has energy of motion.

FIGURE 3.11 This demonstration solar cell array converts radiant energy from the Sun to electrical energy, producing an average of 200,000 watts of electric power (after conversion).

Notice that this equation is exactly the reverse of photosynthesis. In other words, the energy used in photosynthesis was released during oxidation. Chemical energy is a kind of potential energy that is stored and later released during a chemical reaction.

Radiant energy is energy that travels through space (figure 3.11). Most people think of light or sunlight when considering this form of energy. Visible light, however, occupies only a small part of the complete electromagnetic spectrum, as shown in figure 3.12. Radiant energy includes light and all other parts of the spectrum (see chapter 7). Infrared radiation is sometimes called "heat radiation" because of the association with heating when this type of radiation is absorbed. For example, you feel the interaction of infrared radiation when you hold your hand near a warm range element. However, infrared radiation is another type of radiant energy. In fact, some snakes, such as the sidewinder, have pits between their eyes and nostrils that can detect infrared radiation emitted from warm animals. Microwaves are another type of radiant energy that is used in cooking. As with other forms of energy, light, infrared, and microwaves will be considered in more detail later. For now, consider all types of radiant energy to be forms of energy that travel through space.

Electrical energy is another form of energy from electromagnetic interactions that will be considered in detail later. You are familiar with electrical energy that travels through wires to your home from a power plant (figure 3.13), electrical energy that is generated by chemical cells in a flashlight, and electrical energy that can be "stored" in a car battery.

Nuclear energy is a form of energy often discussed because of its use as an energy source in power plants. Nuclear energy is another form of energy from the atom, but this time the energy involves the nucleus, the innermost part of an atom, and nuclear interactions.

Energy Conversion

Potential energy can be converted to kinetic energy and vice versa. The simple pendulum offers a good example of this

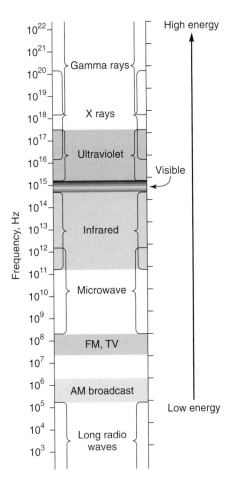

FIGURE 3.12 The electromagnetic spectrum includes many categories of radiant energy. Note that visible light occupies only a tiny part of the entire spectrum.

FIGURE 3.13 The blades of a steam turbine. In a power plant, chemical or nuclear energy is used to heat water to steam, which is directed against the turbine blades. The mechanical energy of the turbine turns an electric generator. Thus, a power plant converts chemical or nuclear energy to mechanical energy, which is then converted to electrical energy.

Connections . . .

Energy Converter Coaster

A roller coaster is an energy converter that swaps kinetic and potential energy back and forth. An outside energy source is used only to move the cars to the top of the first hill. The first hill is always the highest hill above the ground, and here the cars will have the most potential energy they will have for the entire ride.

When the coaster reaches the top of the first hill, it begins to move down a sloping track with increasing speed as potential energy is converted to kinetic energy. At the bottom of this hill, the track then starts up a second hill, and the cars this time convert kinetic energy back to potential energy. Ideally, all the potential energy is converted to kinetic energy as the cars move down a hill, and all the kinetic energy is converted to potential energy as the cars move up a hill. There is no perfect energy conversion, and some

energy is lost to air resistance and friction. The roller coaster design allows for these losses and leaves room for a slight surplus of kinetic energy. You know this is true because the operator must apply brakes at the end of the ride.

Is the speed of the moving roller coaster at the bottom of a hill directly proportional to the height of the hill? Ignoring friction, the speed of a coaster at the bottom of a hill is proportional to the *square root of the height* of the hill. This means that to double the speed of the coaster at the bottom of the hill, you would need to build the hill more than four times higher. It would need to be more than four times higher because the coaster does not drop straight down. Doubling the height increases the theoretical speed only 40 percent.

What is the relationship between the weight of the people on a roller coaster

and the speed achieved at the bottom of the hill? The answer is that the weight of the people does not matter. Ignoring air resistance and friction, a heavy roller coaster full of people and a lighter one with just a few people will both have the same speed at the bottom of the hill.

Draw a profile of a roller coaster ride to find out about potential and kinetic energy exchanges and the height of each hill. The profile should show the relative differences in height between the crest of each hill and the bottom of each upcoming dip. From such a profile, you could find where in a ride the maximum speed would occur, as well as what speed to expect. As you study a profile keep in mind that the roller coaster is designed to produce many changes of speed—accelerations—rather than high speed alone.

---

conversion. A simple pendulum is an object, called a bob, suspended by a string or wire from a support. If the bob is moved to one side and then released, it will swing back and forth in an arc. At the moment that the bob reaches the top of its swing, it stops for an instant, then begins another swing. At the instant of stopping, the bob has 100 percent potential energy and no kinetic energy. As the bob starts back down through the swing, it is gaining kinetic energy and losing potential energy. At the instant the bob is at the bottom of the swing, it has 100 percent kinetic energy and no potential energy. As the bob now climbs through the other half of the arc, it is gaining potential energy and losing kinetic energy until it again reaches an instantaneous stop at the top, and the process starts over. The kinetic energy of the bob at the bottom of the arc is equal to the potential energy it had at the top of the arc (figure 3.14). Disregarding friction, the sum of the potential energy and the kinetic energy remains constant throughout the swing.

Any *form* of energy can be converted to another form. In fact, most technological devices that you use are nothing more than *energy-form converters* (figure 3.15). A lightbulb, for example, converts electrical energy to radiant energy. A car converts chemical energy to mechanical energy. A solar cell converts radiant energy to electrical energy, and an electric motor converts electrical energy to mechanical energy. Each technological device converts some form of energy (usually chemical or electrical) to another form that you desire (usually mechanical or radiant).

It is interesting to trace the *flow of energy* that takes place in your surroundings. Suppose, for example, that you are riding a bicycle. The bicycle has kinetic mechanical energy as it moves along. Where did the bicycle get this energy? It came from you, as you use the chemical energy of food units to contract your muscles and move the bicycle along. But where did your chemical energy come from? It came from your food, which consists of plants, animals who eat plants, or both plants and animals. In any case, plants are at the bottom of your food chain. Plants convert radiant energy from the Sun into chemical

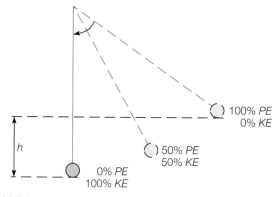

FIGURE 3.14 This pendulum bob loses potential energy (*PE*) and gains an equal amount of kinetic energy (*KE*) as it falls through a distance *h*. The process reverses as the bob moves up the other side of its swing.

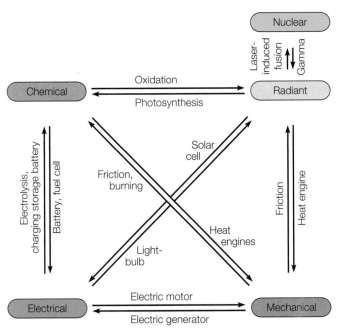

FIGURE 3.15 The energy forms and some conversion pathways.

energy. Radiant energy comes to the plants from the Sun because of the nuclear reactions that took place in the core of the Sun. Your bicycle is therefore powered by nuclear energy that has undergone a number of form conversions!

Energy Conservation

Energy can be transferred from one object to another, and it can be converted from one form to another form. If you make a detailed accounting of all forms of energy before and after a transfer or conversion, the total energy will be *constant*. Consider your bicycle coasting along over level ground when you apply the brakes. What happened to the kinetic mechanical energy of the bicycle? It went into heating the rim and brakes of your bicycle, then eventually radiated to space as infrared radiation. All radiant energy that reaches Earth is eventually radiated back to space (figure 3.16). Thus, throughout all the

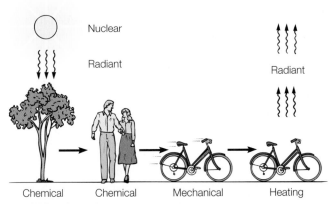

FIGURE 3.16 Energy arrives from the Sun, goes through a number of conversions, then radiates back into space. The total sum leaving eventually equals the original amount that arrived.

form conversions and energy transfers that take place, the total sum of energy remains constant.

The total energy is constant in every situation that has been measured. This consistency leads to another one of the conservation laws of science, the **law of conservation of energy:**

> **Energy is never created or destroyed. Energy can be converted from one form to another but the total energy remains constant.**

See also "Thermodynamics" in chapter 4. You may be wondering about the source of nuclear energy. Does a nuclear reaction create energy? Albert Einstein answered this question back in the early 1900s when he formulated his now-famous relationship between mass and energy, $E = mc^2$. This relationship will be discussed in detail in chapter 11. Basically, the relationship states that mass *is* a form of energy, and this has been experimentally verified many times.

Energy Transfer

Earlier it was stated that when you do work on something, you give it energy. The result of work could be increased kinetic mechanical energy, increased gravitational potential energy, or an increase in the temperature of an object. You could summarize this by stating that either *working* or *heating* is always involved any time energy is transformed. This is not unlike your financial situation. To increase or decrease your financial status, you need some mode of transfer, such as cash or checks, as a means of conveying assets. Just as with cash flow from one individual to another, energy flow from one object to another requires a mode of transfer. In energy matters, the mode of transfer is working or heating. Any time you see working or heating occurring, you know that an energy transfer is taking place. The next time you see heating, think about what energy form is being converted to what new energy form. (The final form is usually radiant energy.) Heating is the topic of chapter 4, where you will consider the role of heat in energy matters.

ENERGY SOURCES TODAY

Prometheus, according to ancient Greek mythology, stole fire from heaven and gave it to humankind. Fire has propelled human advancement ever since. All that was needed was something to burn—fuel for Prometheus's fire.

Any substance that burns can be used to fuel a fire, and various fuels have been used over the centuries as humans advanced. First, wood was used as a primary source for heating. Then coal fueled the Industrial Revolution. Eventually, humankind roared into the twentieth century burning petroleum. Today, petroleum is the most widely used source of energy (figure 3.17). It provides about 40 percent of the total energy used by the United States. Natural gas contributes about 23 percent of the total energy used today. The use of coal also provides about 23 percent of the total. Biomass—any material formed by photosynthesis—contributes about 3 percent of the total energy used today. Note that petroleum, coal, biomass, and natural gas are all chemical sources of energy, sources that are mostly burned for their energy. These chemical sources supply about 87 percent of the total energy consumed. About a third of this is burned for heating, and the rest is burned to drive engines or generators.

Nuclear energy and hydropower are the nonchemical sources of energy. These sources are used to generate electrical energy. The alternative sources of energy, such as solar, wind, and geothermal, provide about 1 percent of the total energy consumed in the United States today.

The energy-source mix has changed from past years, and it will change in the future. Wood supplied 90 percent of the energy until the 1850s, when the use of coal increased. Then, by 1910, coal was supplying about 75 percent of the total energy needs. Then petroleum began making increased contributions to the energy supply. Now increased economic and environmental constraints and a decreasing supply of petroleum are

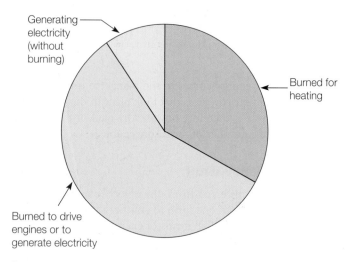

FIGURE 3.17 (A) Primary energy consumed in the United States, 2005,* and (B) uses of this energy.
*Source: Energy Information Administration
(www.eia.doe.gov/emeu/aer/pdf/pages/sec1.pdf).

producing another supply shift. The present petroleum-based energy era is about to shift to a new energy era.

About 92 percent of the total energy consumed today is provided by four sources: (1) petroleum (including natural gas), (2) coal, (3) moving water, and (4) nuclear. The following is a brief introduction to these four sources.

Petroleum

The word *petroleum* is derived from the word *petra*, meaning rock, and the word *oleum*, meaning oil. Petroleum is oil that comes from oil-bearing rock. Natural gas is universally associated with petroleum and has similar origins. Both petroleum and natural gas form from organic sediments, materials that have settled out of bodies of water. Sometimes a local condition permits the accumulation of sediments that are exceptionally rich in organic material. This could occur under special conditions in a freshwater lake, or it could occur on shallow ocean basins. In either case, most of the organic material is from plankton—tiny free-floating animals and plants such as algae. It is from such accumulations of buried organic material that petroleum and natural gas are formed. The exact process by which these materials become petroleum and gas is not understood. It is believed that bacteria, pressure, appropriate temperatures, and time are all important. Natural gas is formed at higher temperatures than is petroleum. Varying temperatures over time may produce a mixture of petroleum and gas or natural gas alone.

Petroleum forms a thin film around the grains of the rock where it formed. Pressure from the overlying rock and water moves the petroleum and gas through the rock until they reach a rock type or structure that stops them. If natural gas is present, it occupies space above the accumulating petroleum. Such accumulations of petroleum and natural gas are the sources of supply for these energy sources.

Discussions about the petroleum supply and the cost of petroleum usually refer to a "barrel of oil." The *barrel* is an accounting device of 42 U.S. gallons. Such a 42-gallon barrel does not exist. When or if oil is shipped in barrels, each drum holds 55 U.S. gallons.

The supply of petroleum and natural gas is limited. Most of the continental drilling prospects appear to be exhausted, and the search for new petroleum supplies is now offshore. In general, over 25 percent of the United States' petroleum is estimated to come from offshore wells. The amount of imported petroleum has ranged from 30 to 50 percent over the years, with most imported oil coming from Canada, Saudi Arabia, Venezuela, Mexico, and Nigeria.

Petroleum is used for gasoline (about 45 percent), diesel (about 40 percent), and heating oil (about 15 percent). Petroleum is also used in making medicine, clothing fabrics, plastics, and ink.

Coal

Petroleum and natural gas formed from the remains of tiny organisms that lived millions of years ago. Coal, on the other hand, formed from an accumulation of plant materials that collected under special conditions millions of years ago. Thus, petroleum, natural gas, and coal are called **fossil fuels.** Fossil fuels contain the stored radiant energy of organisms that lived millions of years ago.

The first thing to happen in the formation of coal was for plants in swamps to die and sink. Stagnant swamp water protected the plants and plant materials from consumption by animals and decomposition by microorganisms. Over time, chemically altered plant materials collected at the bottom of pools of water in the swamp. This carbon-rich material is *peat* (not to be confused with peat moss). Peat is used as a fuel in many places in the world. The flavor of Scotch (whisky) is the result of the peat fires used to brew the liquor. Peat is still being produced naturally in swampy areas today. Under pressure and at high temperatures, peat will eventually be converted to coal.

There are several stages, or *ranks,* in the formation of coal. The lowest rank is lignite (brown coal), then subbituminous, then bituminous (soft coal), and the highest rank is anthracite (hard coal).

Each rank of coal has different burning properties and a different energy content. Coal also contains impurities of clay, silt, iron oxide, and sulfur. The mineral impurities leave an ash when the coal is burned, and the sulfur produces sulfur dioxide, a pollutant.

Most of the coal mined today is burned by utilities to generate electricity (about 80 percent). The coal is ground to a face-powder consistency and blown into furnaces. This greatly increases efficiency but produces *fly ash,* ash that "flies" up the chimney. Industries and utilities are required by the federal Clean Air Act to remove sulfur dioxide and fly ash from plant emissions. About 20 percent of the cost of a new coal-fired power plant goes into air pollution control equipment. Coal is an abundant but dirty energy source.

Moving Water

Moving water has been used as a source of energy for thousands of years. It is considered a renewable energy source, inexhaustible as long as the rain falls. Today, hydroelectric plants generate about 3 percent of the United States' *total* energy consumption at about 2,400 power-generating dams across the nation. Hydropower furnished about 40 percent of the United States' electric power in 1940. Today, dams furnish 9 percent of the electric power. Energy consumption has increased, but hydropower production has not kept pace because geography limits the number of sites that can be built.

Water from a reservoir is conducted through large pipes called penstocks to a powerhouse, where it is directed against turbine blades that turn a shaft on an electric generator.

Nuclear

Nuclear power plants use nuclear energy to produce electricity. Energy is released as the nuclei of uranium and plutonium atoms split, or undergo fission. The fissioning takes place in a large steel vessel called a *reactor.* Water is pumped through the reactor to produce steam, which is used to produce electrical energy, just as in the fossil fuel power plants. The nuclear processes are described in detail in chapter 11, and the process of producing electrical energy is described in detail in chapter 6. Nuclear power plants use nuclear energy to produce electricity, but there are opponents of this process. The electric utility companies view nuclear energy as *one* energy source used to produce electricity. They state that they have no allegiance to any one energy source but are seeking to utilize the most reliable and dependable of several energy sources. Petroleum, coal, and hydropower are also presently utilized as energy sources for electric power production. The electric utility companies are concerned that petroleum and natural gas are becoming increasingly expensive, and there are questions about long-term supplies. Hydropower has limited potential for growth, and solar energy is prohibitively expensive today. Utility companies

People Behind the Science

James Prescott Joule (1818–1889)

James Joule was a British physicist who helped develop the principle of conservation of energy by experimentally measuring the mechanical equivalent of heat. In recognition of Joule's pioneering work on energy, the SI unit of energy is named the joule.

Joule was born on December 24, 1818, into a wealthy brewing family. He and his brother were educated at home between 1833 and 1837 in elementary math, natural philosophy, and chemistry, partly by the English chemist John Dalton (1766–1844). Joule was a delicate child and very shy, and apart from his early education, he was entirely self-taught in science. He does not seem to have played any part in the family brewing business, although some of his first experiments were done in the laboratory at the brewery.

Joule had great dexterity as an experimenter, and he could measure temperatures very precisely. At first, other scientists could not believe such accuracy and were skeptical about the theories that Joule developed to explain his results. The encouragement of Lord Kelvin from 1847 on changed these attitudes, however, and Kelvin subsequently used Joule's practical ability to great advantage. By 1850, Joule was highly regarded by other scientists and was elected a fellow of the Royal Society. Joule's own wealth funded his scientific career, and he never took an academic post. His funds eventually ran out, however. He was awarded a pension in 1878 by Queen Victoria, but by that time, his mental powers were going. He suffered a long illness and died on October 11, 1889.

Joule realized the importance of accurate measurement very early on, and exact data became his hallmark. His most active research period was between 1837 and 1847. In a long series of experiments, he studied the relationship between electrical, mechanical, and chemical effects and heat, and in 1843, he was able to announce his determination of the amount of work required to produce a unit of heat. This is called the mechanical equivalent of heat (4.184 joules per calorie).

One great value of Joule's work was the variety and completeness of his experimental evidence. He showed that the same relationship could be examined experimentally and that the ratio of equivalence of the different forms of energy did not depend on how one form was converted into another or on the materials involved. The principle that Joule had established is that energy cannot be created or destroyed but only transformed.

Joule lives on in the use of his name to measure energy, supplanting earlier units such as the erg and calorie. It is an appropriate reflection of his great experimental ability and his tenacity in establishing a basic law of science.

Source: Modified from the *Hutchinson Dictionary of Scientific Biography*. © RM, 2009. All rights reserved. Helicon Publishing is a division of RM.

---

see two major energy sources that are available for growth: coal and nuclear. There are problems and advantages to each, but the utility companies feel they must use coal and nuclear power until the new technologies, such as solar power, are economically feasible.

Conserving Energy

Conservation is not a way of generating energy, but it is a way of reducing the need for additional energy consumption and it saves money for the consumer. Some conservation technologies are sophisticated, while others are quite simple. For example, if a small, inexpensive wood-burning stove were developed and used to replace open fires in the less-developed world, energy consumption in these regions could be reduced by 50 percent.

Many observers have pointed out that demanding more energy while failing to conserve is like demanding more water to fill a bathtub while leaving the drain open. To be sure, conservation and efficiency strategies by themselves will not eliminate demands for energy, but they can make the demands much easier to meet, regardless of what options are chosen to provide the primary energy. Energy efficiency improvements have significantly reduced the need for additional energy sources. Consider these facts, which are based primarily on data published by the U.S. Energy Information Administration:

- Total primary energy use per capita in the United States in 2003 was almost identical to that in 1973. Over the same thirty-year period, economic output (gross domestic product, or GDP) per capita increased 74 percent.
- National energy intensity (energy use per unit of GDP) fell 43 percent between 1973 and 2002. About 60 percent of this decline is attributable to energy efficiency improvements.
- If the United States had not dramatically reduced its energy intensity over the past thirty years, consumers and businesses would have spent at least $430 billion more on energy purchases in 2003.

Even though the United States is much more energy-efficient today than it was thirty years ago, the potential is still enormous for additional cost-effective energy savings. Some newer energy efficiency measures have barely begun to be adopted. Other efficiency measures could be developed and commercialized in coming years.

Much of the energy we consume is wasted. This statement is not meant as a reminder to simply turn off lights and lower furnace thermostats; it is a technological challenge. Our use of energy is so inefficient that most potential energy in fuel is lost as waste heat, becoming a form of environmental pollution.

The amount of energy wasted through poorly insulated windows and doors alone is about as much energy as the United States receives from the Alaskan pipeline each year. It is estimated that by using inexpensive, energy-efficient measures, the average energy bills of a single home could be reduced by 10 percent to 50 percent and could help reduce the emissions of carbon dioxide into the atmosphere.

Many conservation techniques are relatively simple and highly cost-effective. More efficient and less energy-intensive industry and domestic practices could save large amounts of energy. Improved automobile efficiency, better mass transit, and increased railroad use for passenger and freight traffic are simple and readily available means of conserving transportation energy. In response to the 1970s' oil price shocks, automobile mileage averages in the United States more than doubled, from 5.55 km/L (13 mi/gal) in 1975 to 12.3 km/L (28.8 mi/gal) in 1988. Unfortunately, the oil glut and falling fuel prices of the late 1980s discouraged further conservation. Between 1990 and 1997, the average slipped to only 11.8 km/L (27.6 mi/gal). It remains to be seen if the sharp increase of gasoline prices in the early years of the twenty-first century will translate into increased miles per gallon in new car design.

Several technologies that reduce energy consumption are now available. Highly efficient fluorescent lightbulbs (CFL) that can be used in regular incandescent fixtures give the same amount of light for 25 percent of the energy and produce less heat. Since lighting and air conditioning (which removes the heat from inefficient incandescent lighting) account for 25 percent of U.S. electricity consumption, widespread use of these lights could significantly reduce energy consumption. Low-emissive glass for windows can reduce the amount of heat entering a building while allowing light to enter. The use of this type of glass in new construction and replacement windows could have a major impact on the energy picture. Many other technologies, such as automatic light-dimming or shutoff devices, are being used in new construction.

The shift to more efficient use of energy needs encouragement. Often, poorly designed, energy-inefficient buildings and machines can be produced inexpensively. The short-term cost is low, but the long-term cost is high. The public needs to be educated to look at the long-term economic and energy costs of purchasing poorly designed buildings and appliances.

Electric utilities have recently become part of the energy conservation picture. In some states, they have been allowed to make money on conservation efforts; previously, they could make money only by building more power plants. This encourages them to become involved in energy conservation education, because teaching their customers how to use energy more efficiently allows them to serve more people without building new power plants.

ENERGY TOMORROW

An *alternative source of energy* is one that is different from the typical sources used today. The sources used today are fossil fuels (coal, petroleum, and natural gas), nuclear, and falling water. Alternative sources could be solar, geothermal, hydrogen gas, fusion, or any other energy source that a new technology could utilize.

Solar Technologies

The term *solar energy* is used to describe a number of technologies that directly or indirectly utilize sunlight as an alternative energy source (figure 3.18). There are eight main categories of these solar technologies:

1. **Solar cells.** A solar cell is a thin crystal of silicon, gallium, or some polycrystalline compound that generates electricity when exposed to light. Also called photovoltaic devices, solar cells have no moving parts and produce electricity directly, without the need for hot fluids or intermediate conversion states. Solar cells have been used extensively in space vehicles and satellites. Here on Earth, however, use has been limited to demonstration projects, remote-site applications, and consumer specialty items such as solar-powered watches and calculators. The problem with solar cells today is that the manufacturing cost is too high (they are essentially handmade). Research is continuing on the development of highly efficient, affordable solar cells that could someday produce electricity for the home. See page 147 to find out how a solar cell is able to create a current.

2. **Power tower.** This is another solar technology designed to generate electricity. One type of planned power tower will have a 171 m (560 ft) tower surrounded by some 9,000 special mirrors called heliostats. The heliostats will focus sunlight

FIGURE 3.18 Wind is another form of solar energy. This wind turbine generates electrical energy for this sailboat, charging batteries for backup power when the wind is not blowing. In case you are wondering, the turbine cannot be used to make a wind to push the boat. In accord with Newton's laws of motion, this would not produce a net force on the boat.

on a boiler at the top of the tower where salt (a mixture of sodium nitrate and potassium nitrate) will be heated to about 566°C (about 1,050°F). This molten salt will be pumped to a steam generator, and the steam will be used to drive a generator, just like other power plants. Water could be heated directly in the power tower boiler. Molten salt is used because it can be stored in an insulated storage tank for use when the Sun is not shining, perhaps for up to 20 hours.

3. **Passive application.** In passive applications, energy flows by natural means, without mechanical devices such as motors, pumps, and so forth. A passive solar house would include considerations such as its orientation to the Sun, the size and positioning of windows, and a roof overhang that lets sunlight in during the winter but keeps it out during the summer. There are different design plans to capture, store, and distribute solar energy throughout a house, and some of these designs are described on pages 88–89.

4. **Active application.** An active solar application requires a solar collector in which sunlight heats air, water, or some liquid. The liquid or air is pumped through pipes in a house to generate electricity, or it is used directly for hot water. Solar water heating makes more economic sense today than the other solar applications.

5. **Wind energy.** The wind has been used for centuries to move ships, grind grain into flour, and pump water. The wind blows, however, because radiant energy from the Sun heats some parts of Earth's surface more than others. This differential heating results in pressure differences and the horizontal movement of air, which is called wind. Thus, wind is another form of solar energy. Wind turbines are used to generate electrical energy or mechanical energy. The biggest problem with wind energy is the inconsistency of the wind. Sometimes the wind speed is too great, and other times it is not great enough. Several methods of solving this problem are being researched. (See "Wind Power" in chapter 17 for more on wind energy.)

6. **Biomass.** Biomass is any material formed by photosynthesis, including small plants, trees, and crops, and any garbage, crop residue, or animal waste. Biomass can be burned directly as a fuel, converted into a gas fuel (methane), or converted into liquid fuels such as alcohol. The problems with using biomass include the energy expended in gathering the biomass, as well as the energy used to convert it to a gaseous or liquid fuel.

7. **Agriculture and industrial heating.** This is a technology that simply uses sunlight to dry grains, cure paint, or do anything that can be done with sunlight rather than using traditional energy sources.

8. **Ocean thermal energy conversion (OTEC).** This is an electric generating plant that would take advantage of the approximately 22°C (about 40°F) temperature difference between the surface and the depths of tropical, subtropical, and equatorial ocean waters. Basically, warm water is drawn into the system to vaporize a fluid, which expands through a turbine generator. Cold water from the depths condenses the vapor back to a liquid form, which is then cycled back to the warm-water side. The concept has been tested and found to be technically successful. The greatest interest seems to be from islands that have warm surface waters and cold depths, such as Hawaii, Puerto Rico, Guam, and the Virgin Islands.

Geothermal Energy

Geothermal energy is energy from beneath Earth's surface. The familiar geysers, hot springs, and venting steam of Yellowstone National Park are clues that this form of energy exists. There is substantially more geothermal energy than is revealed in Yellowstone, however, and geothermal resources are more widespread than once thought. Earth has a high internal temperature, and recoverable geothermal resources may underlie most states. These resources occur in four broad categories of geothermal energy: (1) dry steam, (2) hot water, (3) hot, dry rock, and (4) geopressurized resources. Together, the energy contained in these geothermal resources represents about 15,000 times more than is consumed in the United States in a given year. The only problem is getting to the geothermal energy, then using it in a way that is economically attractive.

Most geothermal energy occurs as *hot, dry rock,* which accounts for about 85 percent of the total geothermal resource. Hot, dry rock is usually in or near an area of former volcanic activity. The problem of utilizing this widespread resource is how to get the energy to the surface. Research has been conducted by drilling wells, then injecting water into one well and extracting energy from the heated water pumped from the second well. There is more interest in the less widespread but better understood geothermal systems of hot water and steam.

Geopressurized resources are trapped underground reservoirs of hot water that contain dissolved natural gas. The water temperature is higher than the boiling point, so heat could be used as a source of energy as well as the dissolved natural gas. Such geopressurized reservoirs make up about 14 percent of the total accessible geothermal energy found on Earth. They are being studied in some areas to determine whether they are large enough to be economically feasible as an energy source. More is known about recovering energy from other types of hot water and steam resources, so these seem more economically attractive.

Hot water and steam make up the smallest geothermal resource category, together comprising only about 1 percent of the total known resource. However, more is known about the utilization and recovery of these energy sources, which are estimated to contain an amount of energy equivalent to about half of the known reserves of petroleum in the United States. *Steam* is very rare, occurring in only three places in the United States. Two of these places are national parks (Lassen and Yellowstone), so this geothermal steam cannot be used as an energy source. The third place is at the Geysers, an area of fumaroles near San Francisco, California. Steam from the Geysers is used to generate a significant amount of electricity.

Hot water systems make up most of the *recoverable* geothermal resources. Heat from deep volcanic or former volcanic sources create vast, slow-moving convective patterns in groundwater. If the water circulating back near the surface is hot enough, it can be used for generating electricity, heating buildings, or many other possible applications. Worldwide, geothermal energy is used to operate pulp and paper mills, cool hotels, raise fish, heat greenhouses, dry crops, desalt water, and dozens of other things. Thousands of apartments, homes, and businesses are today heated geothermally in Oregon and Idaho in the United States, as well as in Hungary, France, Iceland, and New Zealand. Today, each British thermal unit (Btu) of heat supplied by geothermal energy does not have to be supplied by fossil fuels. Tomorrow, geothermal resources will become more and more attractive as the price and the consequences of using fossil fuels continue to increase.

Hydrogen

Hydrogen is the lightest and simplest of all the elements, occurring as a diatomic gas that can be used for energy directly in a fuel cell or burned to release heat. Hydrogen could be used to replace natural gas with a few modifications of present natural gas burners. A big plus in favor of hydrogen as a fuel is that it produces no pollutants. In addition to the heat produced, the only emission from burning hydrogen is water, as shown in the following equation:

$$\text{Hydrogen} + \text{oxygen} \rightarrow \text{water} + 68{,}300 \text{ calories}$$

The primary problem with using hydrogen as an energy source is that *it does not exist* on or under Earth's surface in any but trace amounts! Hydrogen must therefore be obtained by a chemical reaction from such compounds as water. Water is a plentiful substance on Earth, and an electric current will cause decomposition of water into hydrogen and oxygen gas. Measurement of the electric current and voltage will show that:

$$\text{Water} + 68{,}300 \text{ calories} \rightarrow \text{hydrogen} + \text{oxygen}$$

Thus, assuming 100 percent efficiency, the energy needed to obtain hydrogen gas from water is exactly equal to the energy released by hydrogen combustion. So hydrogen cannot be used to produce energy, since hydrogen gas is not available, but it can be used as a means of storing energy for later use. Indeed, hydrogen may be destined to become an effective solution to the problems of storing and transporting energy derived from solar energy sources. In addition, hydrogen might serve as the transportable source of energy, such as that needed for cars and trucks, replacing the fossil fuels. In summary, hydrogen has the potential to provide clean, alternative energy for a number of uses, including lighting, heating, cooling, and transportation.

SUMMARY

Work is defined as the product of an applied force and the distance through which the force acts. Work is measured in newton-meters, a metric unit called a *joule*. *Power* is work per unit of time. Power is measured in *watts*. One watt is 1 joule per second. Power is also measured in *horsepower*. One horsepower is 550 ft·lb/s.

Energy is defined as the ability to do work. An object that is elevated against gravity has a potential to do work. The object is said to have *potential energy*, or *energy of position*. Moving objects have the ability to do work on other objects because of their motion. The *energy of motion* is called *kinetic energy*.

Work is usually done *against inertia, fundamental forces, friction, shape,* or *combinations of these*. As a result, there is a gain of *kinetic energy, potential energy, an increased temperature,* or *any combination of these*. Energy comes in the *forms* of *mechanical, chemical, radiant, electrical,* or *nuclear*. Potential energy can be *converted* to kinetic and kinetic can be *converted* to potential. Any form of energy can be *converted* to any other form. Most technological devices are *energy-form converters* that do work for you. Energy flows into and out of the surroundings, but the amount of energy is always constant. The *law of conservation of energy* states that *energy is never created or destroyed*. Energy conversion always takes place through *heating* or *working*.

The basic energy sources today are the chemical *fossil fuels* (petroleum, natural gas, and coal), *nuclear energy,* and *hydropower*. *Petroleum* and *natural gas* were formed from organic material of plankton, tiny free-floating plants and animals. A barrel of petroleum is 42 U.S. gallons, but such a container does not actually exist. *Coal* formed from plants that were protected from consumption by falling into a swamp.

The decayed plant material, *peat,* was changed into the various *ranks* of coal by pressure and heating over some period of time. Coal is a dirty fuel that contains impurities and sulfur. Controlling air pollution from burning coal is costly. Moving water and nuclear energy are used for the generation of electricity. An *alternative* source of energy is one that is different from the typical sources used today. Alternative sources could be *solar, geothermal, or hydrogen*.

Summary of Equations

3.1

$$\text{work} = \text{force} \times \text{distance}$$
$$W = Fd$$

3.2

$$\text{power} = \frac{\text{work}}{\text{time}}$$
$$P = \frac{W}{t}$$

3.3

$$\text{potential energy} = \text{weight} \times \text{height}$$
$$PE = mgh$$

3.4

$$\text{kinetic energy} = \frac{1}{2}(\text{mass})(\text{velocity})^2$$
$$KE = \frac{1}{2}mv^2$$

KEY TERMS

energy (p. **59**)
fossil fuels (p. **67**)
geothermal energy (p. **70**)

horsepower (p. **59**)
joule (p. **57**)
kinetic energy (p. **61**)

law of conservation of
   energy (p. **65**)
potential energy (p. **60**)

power (p. **58**)
watt (p. **59**)
work (p. **56**)

APPLYING THE CONCEPTS

1. The metric unit of a joule (J) is a unit of
   a. potential energy.
   b. work.
   c. kinetic energy.
   d. Any of the above is correct.

2. Power is
   a. the rate at which work is done.
   b. the rate at which energy is expended.
   c. work per unit time.
   d. Any of the above is correct.

3. Which of the following is a combination of units called a watt?
   a. N·m/s
   b. kg·m²/s²/s
   c. J/s
   d. All of the above are correct.

4. About how many watts are equivalent to 1 horsepower?
   a. 7.5
   b. 75
   c. 750
   d. 7,500

5. A kilowatt-hour is a unit of
   a. power.
   b. work.
   c. time.
   d. electrical charge.

6. The potential energy of a box on a shelf, relative to the floor, is a measure of
   a. the work that was required to put the box on the shelf from the floor.
   b. the weight of the box times the distance above the floor.
   c. the energy the box has because of its position above the floor.
   d. All of the above are correct.

7. A rock on the ground is considered to have zero potential energy. In the bottom of a well, then, the rock would be considered to have
   a. zero potential energy, as before.
   b. negative potential energy.
   c. positive potential energy.
   d. zero potential energy but will require work to bring it back to ground level.

8. Which quantity has the greatest influence on the amount of kinetic energy that a large truck has while moving down the highway?
   a. mass
   b. weight
   c. velocity
   d. size

9. Most energy comes to and leaves Earth in the form of
   a. nuclear energy.
   b. chemical energy.
   c. radiant energy.
   d. kinetic energy.

10. The law of conservation of energy is basically that
   a. energy must not be used up faster than it is created, or the supply will run out.
   b. energy should be saved because it is easily destroyed.
   c. energy is never created or destroyed.
   d. you are breaking a law if you needlessly destroy energy.

11. The most widely used source of energy today is
   a. coal.
   b. petroleum.
   c. nuclear.
   d. moving water.

12. The accounting device of a "barrel of oil" is defined to hold how many U.S. gallons of petroleum?
   a. 24
   b. 42
   c. 55
   d. 100

Answers

1. d 2. d 3. d 4. c 5. b 6. d 7. b 8. c 9. c 10. c 11. b 12. b

QUESTIONS FOR THOUGHT

1. How is work related to energy?

2. What is the relationship between the work done while moving a book to a higher bookshelf and the potential energy that the book has on the higher shelf?

3. Does a person standing motionless in the aisle of a moving bus have kinetic energy? Explain.

4. A lamp bulb is rated at 100 W. Why is a time factor not included in the rating?

5. Is a kWh a unit of work, energy, power, or more than one of these? Explain.

6. If energy cannot be destroyed, why do some people worry about the energy supplies?

7. A spring clamp exerts a force on a stack of papers it is holding together. Is the spring clamp doing work on the papers? Explain.

8. Why are petroleum, natural gas, and coal called *fossil fuels?*

9. From time to time, people claim to have invented a machine that will run forever without energy input and develops more energy than it uses (perpetual motion). Why would you have reason to question such a machine?

10. Define a joule. What is the difference between a joule of work and a joule of energy?

11. Compare the energy needed to raise a mass 10 m on Earth to the energy needed to raise the same mass 10 m on the Moon. Explain the difference, if any.

12. What happens to the kinetic energy of a falling book when the book hits the floor?

FOR FURTHER ANALYSIS

1. Evaluate the requirement that something must move whenever work is done. Why is this a requirement?

2. What are the significant similarities and differences between work and power?

3. Whenever you do work on something, you give it energy. Analyze how you would know for sure that this is a true statement.

4. Simple machines are useful because they are able to trade force for distance moved. Describe a conversation between yourself and another person who believes that you do less work when you use a simple machine.

5. Use the equation for kinetic energy to prove that speed is more important than mass in bringing a speeding car to a stop.

6. Describe at least several examples of negative potential energy and how each shows a clear understanding of the concept.

7. The five forms of energy are the result of the fundamental forces—gravitational, electromagnetic, and nuclear—and objects that are interacting. Analyze which force is probably involved with each form of energy.

8. Most technological devices convert one of the five forms of energy into another. Try to think of a technological device that does not convert an energy form to another. Discuss the significance of your finding.

9. Are there any contradictions to the law of conservation of energy in any area of science?

INVITATION TO INQUIRY

New Energy Source?

Is waste paper a good energy source? There are 103 waste-to-energy plants in the United States that burn solid garbage, so waste paper would be a good source, too. The plants burn solid garbage to make steam that is used to heat buildings and generate electricity. Schools might be able to produce a pure waste paper source because waste paper accumulates near computer print stations and in offices. Collecting waste paper from such sources would yield 6,800 Btu/lb, which is about half the heat value of coal.

If you accept this invitation, start by determining how much waste paper is created per month in your school. Would this amount produce enough energy to heat buildings or generate electricity?

PARALLEL EXERCISES

The exercises in groups A and B cover the same concepts. Solutions to group A exercises are located in appendix D.
Note: Neglect all frictional forces in all exercises.

Group A

1. A force of 200 N is needed to push a table across a level classroom floor for a distance of 3 m. How much work was done on the table?

2. An 880 N box is pushed across a level floor for a distance of 5.0 m with a force of 440 N. How much work was done on the box?

3. How much work is done in raising a 10.0 kg backpack from the floor to a shelf 1.5 m above the floor?

Group B

1. Work of 1,200 J is done while pushing a crate across a level floor for a distance of 1.5 m. What force was used to move the crate?

2. How much work is done by a hammer that exerts a 980.0 N force on a nail, driving it 1.50 cm into a board?

3. A 5.0 kg textbook is raised a distance of 30.0 cm as a student prepares to leave for school. How much work did the student do on the book?

—continued

Group A

4. If 5,000 J of work is used to raise a 102 kg crate to a shelf in a warehouse, how high was the crate raised?

5. A 60.0 kg student runs up a 5.00 m high stairway in a time of 3.92 s. (a) How many watts of power did she develop? (b) How many horsepower is this?

6. What is the kinetic energy of a 2,000 kg car moving at 72 km/h?

7. How much work is needed to stop a 1,000.0 kg car that is moving straight down the highway at 54.0 km/h?

8. A 1,000 kg car stops on top of a 51.02 m hill. (a) How much energy was used in climbing the hill? (b) How much potential energy does the car have?

9. A horizontal force of 10.0 lb is needed to push a bookcase 5 ft across the floor. (a) How much work was done on the bookcase? (b) How much did the gravitational potential energy change as a result?

10. (a) How much work is done in moving a 2.0 kg book to a shelf 2.00 m high? (b) What is the change in potential energy of the book as a result? (c) How much kinetic energy will the book have as it hits the ground as it falls?

11. A 60.0 kg jogger moving at 2.0 m/s decides to double the jogging speed. How did this change in speed change the kinetic energy?

12. A 170.0 lb student runs up a stairway to a classroom 25.0 ft above ground level in 10.0 s. (a) How much work did the student do? (b) What was the average power output in horsepower?

Group B

4. An electric hoist does 196,000 J of work in raising a 250.0 kg load. How high was the load lifted?

5. What is the horsepower of a 1,500.0 kg car that can go to the top of a 360.0 m high hill in exactly one minute?

6. What is the kinetic energy of a 30.0 g bullet that is traveling at 200.0 m/s?

7. How much work will be done by a 30.0 g bullet traveling at 200 m/s?

8. A 10.0 kg box is lifted 15 m above the ground by a construction crane. (a) How much work did the crane do on the box? (b) How much potential energy does the box have relative to the ground?

9. A force of 50.0 lb is used to push a box 10.0 ft across a level floor. (a) How much work was done on the box? (b) What is the change of potential energy as a result of this move?

10. (a) How much work is done in raising a 50.0 kg crate a distance of 1. 5 m above a storeroom floor? (b) What is the change of potential energy as a result of this move? (c) How much kinetic energy will the crate have as it falls and hits the floor?

11. The driver of an 800.0 kg car decides to double the speed from 20.0 m/s to 40.0 m/s. What effect would this have on the amount of work required to stop the car, that is, on the kinetic energy of the car?

12. A 70.0 kg student runs up the stairs of a football stadium to a height of 10.0 m above the ground in 10.0 s. (a) What is the power of the student in kilowatts? (b) in horsepower?

4

Heat and Temperature

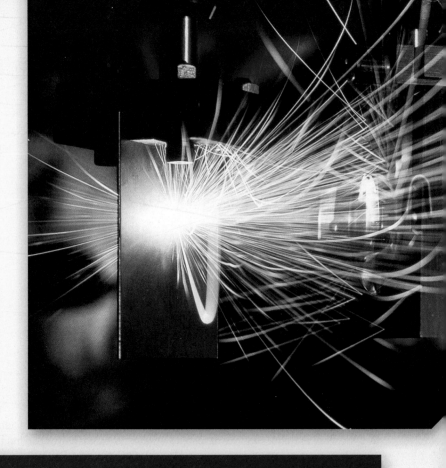

Sparks fly from a plate of steel as it is cut by an infrared laser. Today, lasers are commonly used to cut as well as weld metals, so the cutting and welding is done by light, not by a flame or electric current.

CORE CONCEPT

A relationship exists between heat, temperature, and the motion and position of the molecules.

OUTLINE

All matter is made of molecules that move and interact (p. 76).

Temperature is a measure of the average kinetic energy of molecules (p. 79).

Heat is a measure of internal energy that has been transferred or absorbed (p. 82).

The Kinetic Molecular Theory
    Molecules
    Molecules Interact
    Phases of Matter
    Molecules Move
Temperature
    Thermometers
    Thermometer Scales
Heat
    Heat as Energy Transfer
    Measures of Heat
    Specific Heat
    Heat Flow
A Closer Look: Passive Solar Design

Energy, Heat, and Molecular Theory
    Phase Change
    Evaporation and Condensation
    Relative Humidity
Thermodynamics
    The First Law of
        Thermodynamics
    The Second Law of
        Thermodynamics
    The Second Law and Natural
        Processes
People Behind the Science:
    Count Rumford (Benjamin
    Thompson)

The laws of thermodynamics describe a relationship between changes of internal energy, work, and heat (p. 94).

Chemistry

▷ Chemical reactions involve changes in the internal energy of molecules (Ch. 9).

Astronomy

▷ Stars form from compressional heating (Ch. 12).

Earth Science

▷ Earth's interior is hot, making possible the movement of plates on the surface (Ch. 15–16).

▷ Movements of air, air masses, and ocean currents are driven in cycles involving temperature differences (Ch. 17–18).

Life Science

▷ The process of aerobic respiration results in a release of heat (Ch. 20).

OVERVIEW

Heat has been closely associated with the comfort and support of people throughout history. You can imagine the appreciation when your earliest ancestors first discovered fire and learned to keep themselves warm and cook their food. You can also imagine the wonder and excitement about 3000 B.C., when people put certain earthlike substances on the hot, glowing coals of a fire and later found metallic copper, lead, or iron. The use of these metals for simple tools followed soon afterwards. Today, metals are used to produce complicated engines that use heat for transportation and do the work of moving soil and rock, construction, and agriculture. Devices made of heat-extracted metals are also used to control the temperature of structures, heating or cooling the air as necessary. Thus, the production and control of heat gradually built the basis of civilization today (figure 4.1).

The sources of heat are the energy forms that you learned about in chapter 3. The fossil fuels are *chemical* sources of heat. Heat is released when oxygen is combined with these fuels. Heat also results when *mechanical* energy does work against friction, such as in the brakes of a car coming to a stop. Heat also appears when *radiant* energy is absorbed. This is apparent when solar energy heats water in a solar collector or when sunlight melts snow. The transformation of *electrical* energy to heat is apparent in toasters, heaters, and ranges. *Nuclear* energy provides the heat to make steam in a nuclear power plant. Thus, all the energy forms can be converted to heat.

The relationship between energy forms and heat appears to give an order to nature, revealing patterns that you will want to understand. All that you need is some kind of explanation for the relationships—a model or theory that helps make sense of it all. This chapter is concerned with heat and temperature and their relationship to energy. It begins with a simple theory about the structure of matter and then uses the theory to explain the concepts of heat, energy, and temperature changes.

THE KINETIC MOLECULAR THEORY

The idea that substances are composed of very small particles can be traced back to certain early Greek philosophers. The earliest record of this idea was written by Democritus during the fifth century B.C. He wrote that matter was empty space filled with tremendous numbers of tiny, indivisible particles called *atoms*. This idea, however, was not acceptable to most of the ancient Greeks, because matter seemed continuous, and empty space was simply not believable. The idea of atoms was rejected by Aristotle as he formalized his belief in continuous matter composed of the earth, air, fire, and water elements.

Aristotle's belief about matter, like his beliefs about motion, predominated through the 1600s. Some people, such as Galileo and Newton, believed the ideas about matter being composed of tiny particles, or atoms, since this theory seemed to explain the behavior of matter. Widespread acceptance of the particle model did not occur, however, until strong evidence was developed through chemistry in the late 1700s and early 1800s. The experiments finally led to a collection of assumptions about the small particles of matter and the space around them. Collectively, the assumptions could be called the **kinetic molecular theory.** The following is a general description of some of these assumptions.

FIGURE 4.1 Heat and modern technology are inseparable. These glowing steel slabs, at over 1,100°C (about 2,000°F), are cut by an automatic flame torch. The slab caster converts 300 tons of molten steel into slabs in about 45 minutes. The slabs are converted to sheet steel for use in the automotive, appliance, and building industries.

Molecules

The basic assumption of the kinetic molecular theory is that all matter is made up of tiny, basic units of structure called *atoms*. Atoms are neither divided, created, nor destroyed during any type of chemical or physical change. There are similar groups of atoms that make up the pure substances known as chemical *elements*. Each element has its own kind of atom, which is different from the atoms of other elements. For example, hydrogen, oxygen, carbon, iron, and gold are chemical elements, and each has its own kind of atom.

In addition to the chemical elements, there are pure substances called *compounds* that have more complex units of structure. Pure substances, such as water, sugar, and alcohol, are composed of atoms of two or more elements that join together in definite proportions. Water, for example, has structural units that are made up of two atoms of hydrogen tightly bound to one atom of oxygen (H_2O). These units are not easily broken apart and stay together as small physical particles of which water is composed. Each is the smallest particle of water that can exist, a molecule of water. A *molecule* is generally defined as a tightly bound group of atoms in which the atoms maintain their identity. How atoms become bound together to form molecules is discussed in chapter 9.

Some elements exist as gases at ordinary temperatures, and all elements are gases at sufficiently high temperatures. At ordinary temperatures, the atoms of oxygen, nitrogen, and other gases are paired in groups of two to form *diatomic molecules*. Other gases, such as helium, exist as single, unpaired atoms at ordinary temperatures. At sufficiently high temperatures, iron, gold, and other metals vaporize to form gaseous, single,

unpaired atoms. In the kinetic molecular theory, the term *molecule* has the additional meaning of the smallest, ultimate particle of matter that can exist. Thus, the ultimate particle of a gas, whether it is made up of two or more atoms bound together or of a single atom, is conceived of as a molecule. A single atom of helium, for example, is known as a *monatomic molecule*. For now, a **molecule** is defined as the smallest particle of a compound, or a gaseous element, that can exist and still retain the characteristic properties of that substance.

Molecules Interact

Some molecules of solids and liquids interact, strongly attracting and clinging to each other. When this attractive force is between the same kind of molecules, it is called *cohesion*. It is a stronger cohesion that makes solids and liquids different from gases, and without cohesion, all matter would be in the form of gases. Sometimes one kind of molecule attracts and clings to a different kind of molecule. The attractive force between unlike molecules is called *adhesion*. Water wets your skin because the adhesion of water molecules and skin is stronger than the cohesion of water molecules. Some substances, such as glue, have a strong force of adhesion when they harden from a liquid state, and they are called adhesives.

Phases of Matter

Three phases of matter are common on Earth under conditions of ordinary temperature and pressure. These phases—or forms of existence—are solid, liquid, and gas. Each of these has a different molecular arrangement (figure 4.2). The different characteristics of each phase can be attributed to the molecular arrangements and the strength of attraction between the molecules (table 4.1).

Solids have definite shapes and volumes because they have molecules that are fixed distances apart and bound by relatively strong cohesive forces. Each molecule is a nearly fixed distance from the next, but it does vibrate and move around an equilibrium position. The masses of these molecules and the spacing between them determine the density of the solid. The hardness of a solid is the resistance of a solid to forces that tend to push its molecules further apart.

Liquids have molecules that are not confined to an equilibrium position as in a solid. The molecules of a liquid are close together and bound by cohesive forces that are not as strong as in a solid. This permits the molecules to move from place to place within the liquid. The molecular forces are strong enough to give the liquid a definite volume but not strong enough to give it a definite shape. Thus, a pint of milk is always a pint of milk (unless it is under tremendous pressure) and takes the shape of the container holding it. Because the forces between the molecules of a liquid are weaker than the forces between the molecules of a solid, a liquid cannot support the stress of a rock placed on it as a solid does. The liquid molecules *flow*, rolling over each other as the rock pushes its way between the

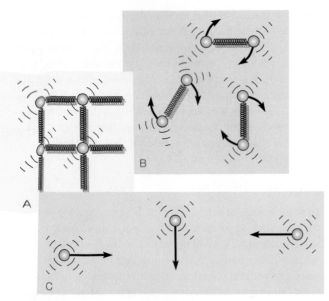

FIGURE 4.2 (A) In a solid, molecules vibrate around a fixed equilibrium position and are held in place by strong molecular forces. (B) In a liquid, molecules can rotate and roll over each other because the molecular forces are not as strong. (C) In a gas, molecules move rapidly in random, free paths.

TABLE 4.1 The Shape and Volume Characteristics of Solids, Liquids, and Gases Are Reflections of Their Molecular Arrangements*

	Solids	Liquids	Gases
Shape	Fixed	Variable	Variable
Volume	Fixed	Fixed	Variable

*These characteristics are what would be expected under ordinary temperature and pressure conditions on the surface of Earth.

molecules. Yet, the molecular forces are strong enough to hold the liquid together, so it keeps the same volume.

Gases are composed of molecules with weak cohesive forces acting between them. The gas molecules are relatively far apart and move freely in a constant, random motion that is changed often by collisions with other molecules. Gases therefore have neither fixed shapes nor fixed volumes.

Gases that are made up of positive ions and negative electrons are called *plasmas*. Plasmas have the same properties as gases but also conduct electricity and interact strongly with magnetic fields. Plasmas are found in fluorescent and neon lights on Earth, and in the Sun and other stars. Nuclear fusion occurs in plasmas of stars (see chapter 11), producing starlight as well as sunlight. Plasma physics is studied by scientists in their attempt to produce controlled nuclear fusion.

There are other distinctions between the phases of matter. The term *vapor* is sometimes used to describe a gas that is usually in the liquid phase. Water vapor, for example, is the gaseous form of liquid water. Liquids and gases are collectively called *fluids* because of their ability to flow, a property that is lacking in most solids.

Molecules Move

Suppose you are in an evenly heated room with no air currents. If you open a bottle of ammonia, the odor of ammonia is soon noticeable everywhere in the room. According to the kinetic molecular theory, molecules of ammonia leave the bottle and bounce around among the other molecules making up the air until they are everywhere in the room, slowly becoming more evenly distributed. The ammonia molecules *diffuse,* or spread, throughout the room. The ammonia odor diffuses throughout the room faster if the air temperature is higher and slower if the air temperature is lower. This would imply a relationship between the temperature and the speed at which molecules move about.

The relationship between the temperature of a gas and the motion of molecules was formulated in 1857 by Rudolf Clausius. He showed that the temperature of a gas is proportional to the average kinetic energy of the gas molecules. This means that ammonia molecules have a greater average velocity at a higher temperature and a slower average velocity at a lower temperature. This explains why gases diffuse at a greater rate at higher temperatures. Recall, however, that kinetic energy involves the mass of the molecules as well as their velocity ($KE = 1/2\,mv^2$). It is the *average kinetic energy* that is proportional to the temperature, which involves the molecular mass as well as the molecular velocity. Whether the kinetic energy is jiggling, vibrating, rotating, or moving from place to place, the **temperature** of a substance is a *measure of the average kinetic energy of the molecules making up the substance* (figure 4.3).

FIGURE 4.3 The number of oxygen molecules with certain velocities that you might find in a sample of air at room temperature. Notice that a few are barely moving and some have velocities over 1,000 m/s at a given time, but the *average* velocity is somewhere around 500 m/s.

TEMPERATURE

If you ask people about the temperature, they usually respond with a referent ("hotter than the summer of '89") or a number ("68°F or 20°C"). Your response, or feeling, about the referent or number depends on a number of factors, including a *relative* comparison. A temperature of 20°C (68°F), for example, might seem cold during the month of July but warm during the month of January. The 20°C temperature is compared to what is expected at the time, even though 20°C is 20°C, no matter what month it is.

When people ask about the temperature, they are really asking *how hot or how cold something is.* Without a thermometer, however, most people can do no better than *hot* or *cold,* or perhaps *warm* or *cool,* in describing a relative temperature. Even then, there are other factors that confuse people about temperature. Your body judges temperature on the basis of the net *direction* of energy flow. You sense situations in which heat is flowing into your body as *warm* and situations in which heat is flowing from your body as *cool.* Perhaps you have experienced having your hands in snow for some time, then washing your hands in cold water. The cold water feels warm. Your hands are colder than the water, energy flows into your hands, and they communicate "warm."

Thermometers

The human body is a poor sensor of temperature, so a device called a *thermometer* is used to measure the hotness or coldness of something. Most thermometers are based on the relationship between some property of matter and changes in temperature. Almost all materials expand with increasing temperatures. A strip of metal is slightly longer when hotter and slightly shorter when cooler, but the change of length is too small to be useful in a thermometer. A more useful, larger change is obtained when two metals that have different expansion rates are bonded together in a strip. The bimetallic (*bi* = two; *metallic* = metal) strip will bend toward the metal with less expansion when the strip is heated (figure 4.4). Such a bimetallic strip is formed into a coil and used in thermostats and dial thermometers (figure 4.5).

The common glass thermometer is a glass tube with a bulb containing a liquid, usually mercury or colored alcohol, that expands up the tube with increases in temperature and contracts back toward the bulb with decreases in temperature. The height of this liquid column is used with a referent scale

FIGURE 4.4 (*A*) A bimetallic strip is two different metals, such as iron and brass, bonded together as a single unit, shown here at room temperature. (*B*) Since one metal expands more than the other, the strip will bend when it is heated. In this example, the brass expands more than the iron, so the bimetallic strip bends away from the brass.

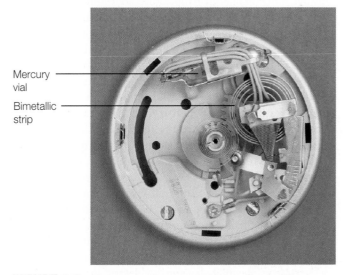

FIGURE 4.5 This thermostat has a coiled bimetallic strip that expands and contracts with changes in the room temperature. The attached vial of mercury is tilted one way or the other, and the mercury completes or breaks an electric circuit that turns the heating or cooling system on or off.

to measure temperature. Some thermometers, such as a fever thermometer, have a small constriction in the tube so the liquid cannot normally return to the bulb. Thus, the thermometer shows the highest reading, even if the temperature it measures has fluctuated up and down during the reading. The liquid must be forced back into the bulb by a small swinging motion, bulb-end down, then sharply stopping the swing with a snap of the wrist. The inertia of the mercury in the bore forces it past the constriction and into the bulb. The fever thermometer is then ready to use again.

Today, scientists have developed a different type of thermometer and a way around the problems of using a glass-mercury fever thermometer. This new approach measures the internal core temperature by quickly reading infrared radiation from the eardrum. All bodies with a temperature above absolute zero emit radiation, including your body. The intensity of the radiation is a sensitive function of body temperature, so reading the radiation emitted will tell you about the temperature of that body.

The human eardrum is close to the hypothalamus, the body's thermostat, so a temperature reading taken here must be close to the temperature of the internal core. You cannot use a mercury thermometer in the ear because of the very real danger of puncturing the eardrum, along with obtaining doubtful readings from a mercury bulb. You can use a pyroelectric material to measure the infrared radiation coming from the entrance to the ear canal, however, to quickly obtain a temperature reading. A pyroelectric material is a polarized crystal that generates an electric charge in proportion to a temperature change. The infrared fever thermometer has a short barrel that is inserted in the ear canal opening. A button opens a shutter inside the battery-powered device, admitting infrared radiation for about 300 milliseconds. Infrared radiation from the ear canal increases the temperature of a thin pyroelectric crystal, which develops an electric charge. A current spike from the pyroelectric sensor now moves through some filters and converters and into a microprocessor chip. The chip is programmed with the relationship between the body temperature and the infrared radiation emitted. Using this information, it calculates the temperature by measuring the current spike produced by the infrared radiation falling on the pyroelectric crystal. The microprocessor now sends the temperature reading to an LCD display on the outside of the device, where it can be read almost immediately.

Thermometer Scales

Several referent scales are used to define numerical values for measuring temperatures (figure 4.6). The **Fahrenheit scale** was developed by the German physicist Gabriel D. Fahrenheit

in about 1715. Fahrenheit invented a mercury-in-glass thermometer with a scale based on two arbitrarily chosen reference points. The original Fahrenheit scale was based on the temperature of an ice and salt mixture for the lower reference point (0°) and the temperature of the human body as the upper reference point (about 100°). Thus, the original Fahrenheit scale was a centigrade scale with 100 divisions between the high and the low reference points. The distance between the two reference points was then divided into equal intervals called *degrees*. There were problems with identifying a "normal" human body temperature as a reference point, since body temperature naturally changes during a given day and from day to day. However, some people "normally" have a higher body temperature than others. Some may have a normal body temperature of 99.1°F, while others have a temperature of 97°F. The average for a large population is 98.6°F. The only consistent thing about the human body temperature is constant change. The standards for the Fahrenheit scale were eventually changed to something more consistent, the freezing point and the boiling point of water at normal atmospheric pressure. The original scale was retained with the new reference points, however, so the "odd" numbers of 32°F (freezing point of water) and 212°F (boiling point of water under normal pressure) came to be the reference points. There are 180 equal intervals, or degrees, between the freezing and boiling points on the Fahrenheit scale.

The **Celsius scale** was invented by Anders C. Celsius, a Swedish astronomer, in about 1735. The Celsius scale uses the freezing point and the boiling point of water at normal atmospheric pressure, but it has different arbitrarily assigned values. The Celsius scale identifies the freezing point of water as 0°C and the boiling point as 100°C. There are 100 equal intervals, or degrees, between these two reference points, so the Celsius scale is sometimes called the *centigrade* scale.

There is nothing special about either the Celsius scale or the Fahrenheit scale. Both have arbitrarily assigned numbers, and one is no more accurate than the other. The Celsius scale is more convenient because it is a decimal scale and because it has a direct relationship with a third scale to be described shortly,

FIGURE 4.6 The Fahrenheit, Celsius, and Kelvin temperature scales.

Goose Bumps and Shivering

For an average age and minimal level of activity, many people feel comfortable when the environmental temperature is about 25°C (77°F). Comfort at this temperature probably comes from the fact that the body does not have to make an effort to conserve or get rid of heat.

Changes in the body that conserve heat occur when the temperature of the air and clothing directly next to a person becomes less than 20°C, or if the body senses rapid heat loss. First, blood vessels in the skin are constricted. This slows the flow of blood near the surface, which reduces heat loss by conduction. Constriction of skin blood vessels reduces body heat loss but may also cause the skin and limbs to become significantly cooler than the body core temperature (producing cold feet, for example).

Sudden heat loss, or a chill, often initiates another heat-saving action by the body. Skin hair is pulled upright, erected to slow heat loss to cold air moving across the skin. Contraction of a tiny muscle attached to the base of the hair shaft makes a tiny knot, or bump, on the skin. These are sometimes called "goose bumps" or "chill bumps." Although "goose bumps" do not significantly increase insulation in humans, the equivalent response in birds and many mammals elevates hairs or feathers and greatly enhances insulation.

Further cooling after the blood vessels in the skin have been constricted results in the body taking yet another action. The body now begins to produce *more* heat, making up for heat loss through involuntary muscle contractions called "shivering." The greater the need for more body heat, the greater the activity of shivering.

If the environmental temperatures rise above about 25°C (77°F), the body triggers responses that cause it to *lose* heat. One response is to make blood vessels in the skin larger, which increases blood flow in the skin. This brings more heat from the core to be conducted through the skin, then radiated away. It also causes some people to have a red blush from the increased blood flow in the skin. This action increases conduction through the skin, but radiation alone provides insufficient cooling at environmental temperatures above about 29°C (84°F). At about this temperature, sweating begins and perspiration pours onto the skin to provide cooling through evaporation. The warmer the environmental temperature, the greater the rate of sweating and cooling through evaporation.

The actual responses to a cool, cold, warm, or hot environment will be influenced by a person's level of activity, age, and gender, and environmental factors such as the relative humidity, air movement, and combinations of these factors. Temperature is the single most important comfort factor. However, when the temperature is high enough to require perspiration for cooling, humidity also becomes an important factor in human comfort.

---

the Kelvin scale. Both scales have arbitrarily assigned reference points and an arbitrary number line that indicates *relative* temperature changes. Zero is simply one of the points on each number line and does *not* mean that there is no temperature. Likewise, since the numbers are relative measures of temperature change, 2° is not twice as hot as a temperature of 1° and 10° is not twice as hot as a temperature of 5°. The numbers simply mean some measure of temperature *relative to* the freezing and boiling points of water under normal conditions.

You can convert from one temperature to the other by considering two differences in the scales: (1) the difference in the degree size between the freezing and boiling points on the two scales, and (2) the difference in the values of the lower reference points.

The Fahrenheit scale has 180° between the boiling and freezing points (212°F − 32°F) and the Celsius scale has 100° between the same two points. Therefore, each Celsius degree is 180/100 or 9/5 as large as a Fahrenheit degree. Each Fahrenheit degree is 100/180 or 5/9 of a Celsius degree. In addition, considering the difference in the values of the lower reference points (0°C and 32°F) gives the equations for temperature conversion.

$$T_F = \frac{9}{5}T_C + 32°$$

equation 4.1

$$T_C = \frac{5}{9}(T_F - 32°)$$

equation 4.2

EXAMPLE 4.1 *(Optional)*

The average human body temperature is 98.6°F. What is the equivalent temperature on the Celsius scale?

SOLUTION

$$T_C = \frac{5}{9}(T_F - 32°)$$
$$= \frac{5}{9}(98.6° - 32°)$$
$$= \frac{5}{9}(66.6°)$$
$$= \frac{333°}{9}$$
$$= \boxed{37°C}$$

EXAMPLE 4.2 *(Optional)*

A temperature display outside of a bank indicates 20°C (room temperature). What is the equivalent temperature on the Fahrenheit scale? (Answer: 68°F)

There is a temperature scale that does not have arbitrarily assigned reference points and zero *does* mean nothing. This is not a relative scale but an absolute temperature scale called the **absolute scale,** or **Kelvin scale.** The zero point on the absolute scale is thought to be the lowest limit of temperature.

Absolute zero is the *lowest temperature possible,* occurring when all random motion of molecules has ceased. Absolute zero is written as 0 K. A degree symbol is not used, and the K stands for the SI standard scale unit kelvin. The absolute scale uses the same degree size as the Celsius scale and $-273°C = 0$ K. Note in figure 4.6 that 273 K is the freezing point of water and 373 K is the boiling point. You could think of the absolute scale as a Celsius scale with the zero point shifted by 273°. Thus, the relationship between the absolute and Celsius scales is

$$T_K = T_C + 273 \qquad \textbf{equation 4.3}$$

A temperature of absolute zero has never been reached, but scientists have cooled a sample of sodium to 700 nanokelvins, or 700 billionths of a kelvin above absolute zero.

EXAMPLE 4.3 *(Optional)*

A science article refers to a temperature of 300.0 K. (a) What is the equivalent Celsius temperature? (b) What is the equivalent Fahrenheit temperature?

SOLUTION

(a) The relationship between the absolute scale and Celsius scale is found in equation 4.3, $T_K = T_C + 273$. Solving this equation for Celsius yields $T_C = T_K - 273$.

$$T_C = T_K - 273$$
$$= 300.0 - 273$$
$$= \boxed{27°C}$$

(b)
$$T_F = \frac{9}{5}T_C + 32°$$
$$= \frac{9}{5}27° + 32°$$
$$= \frac{243°}{5} + 32°$$
$$= 48.6° + 32°$$
$$= \boxed{81°F}$$

HEAT

Suppose you have a bowl of hot soup or a cup of hot coffee that is too hot. What can you do to cool it? You can blow across the surface, which speeds evaporation and therefore results in cooling, but this is a slow process. If you were in a hurry, you would probably add something cooler, such as ice. Adding a cooler substance will cool the hot liquid.

You know what happens when you mix fluids or objects with a higher temperature with fluids or objects with a lower temperature. The warmer temperature object becomes cooler and the cooler temperature object becomes warmer. Eventu-

ally, both will have a temperature somewhere between the warmer and the cooler. This might suggest that something is moving between the warmer and cooler objects, changing the temperature.

The relationship that exists between energy and temperature will help explain the concept of heat, so we will consider it first. If you rub your hands together a few times, they will feel a little warmer. If you rub them together vigorously for a while, they will feel a lot warmer, maybe hot. A temperature increase takes place anytime mechanical energy causes one surface to rub against another (figure 4.7). The two surfaces could be solids, such as the two blocks, but they can also be the surface of a solid and a fluid, such as air. A solid object moving through the air encounters compression, which results in a higher temperature of the surface. A high-velocity meteor enters Earth's atmosphere and is heated so much from the compression that it begins to glow, resulting in the fireball and smoke trail of a "falling star."

To distinguish between the energy of the object and the energy of its molecules, we use the terms *external* and *internal* energy. **External energy** is the total potential and kinetic energy of an everyday-sized object. All the kinetic and potential energy considerations discussed in previous chapters were about the external energy of an object.

Internal energy is the total kinetic and potential energy of the *molecules* of an object. The kinetic energy of a molecule can be much more complicated than straight-line velocity might suggest, however, as a molecule can have many different types of motion at the same time (pulsing, twisting, turning, etc.). Overall, internal energy is characterized by properties such as temperature, density, heat, volume, pressure of a gas, and so forth.

When you push a table across the floor, the observable *external* kinetic energy of the table is transferred to the *internal*

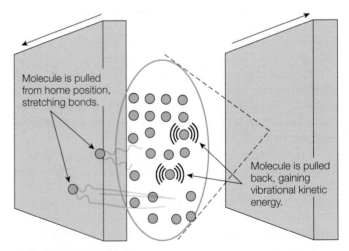

FIGURE 4.7   Here is how friction results in increased temperatures: Molecules on one moving surface will catch on another surface, stretching the molecular forces that are holding it. The molecules are pulled back to their home position with a snap, resulting in a gain of vibrational kinetic energy.

FIGURE 4.8 *External energy* is the kinetic and potential energy that you can see. *Internal energy* is the total kinetic and potential energy of molecules. When you push a table across the floor, you do work against friction. Some of the external mechanical energy goes into internal kinetic and potential energy, and the bottom surface of the legs becomes warmer.

1 liter of water at 90°C

250 milliliter of water at 90°C

FIGURE 4.9 Heat and temperature are different concepts, as shown by a liter of water (1,000 mL) and a 250 mL cup of water, both at the same temperature. You know the liter of water contains more internal energy since it will require more ice cubes to cool it to, say, 25°C than will be required for the cup of water. In fact, you will have to remove 48,750 *additional* calories to cool the liter of water.

kinetic energy of the molecules between the table legs and the floor, resulting in a temperature increase (figure 4.8). The relationship between external and internal kinetic energy explains why the heating is proportional to the amount of mechanical energy used.

Heat as Energy Transfer

Temperature is a measure of the degree of hotness or coldness of a body, a measure that is based on the average molecular kinetic energy. Heat, on the other hand, is based on the *total internal energy* of the molecules of a body. You can see one difference in heat and temperature by considering a cup of water and a pitcher of water. If both the small and the large amount of water have the same temperature, both must have the same average molecular kinetic energy. Now, suppose you wish to cool both by, say, 20°. The pitcher of water would take much longer to cool, so it must be that the large amount of water has more internal energy (figure 4.9). Heat is a measure based on the *total* internal energy of the molecules of a body, and there is more total energy in a pitcher of water than in a cup of water at the same temperature.

How can we measure heat? Since it is difficult to see molecules, internal energy is difficult to measure directly. Thus, heat is nearly always measured during the process of a body gaining or losing energy. This measurement procedure will also give us a working definition of **heat:**

> **Heat is a measure of the internal energy that has been absorbed or transferred from one body to another.**

The *process* of increasing the internal energy is called "heating," and the *process* of decreasing internal energy is called "cooling." The word *process* is italicized to emphasize that heat is energy in transit, not a material thing you can add or take away. Heat is understood to be a measure of internal energy that can be measured as energy flows in or out of an object.

There are two general ways that heating can occur. These are (1) from a temperature difference, with energy moving from the region of higher temperature, and (2) from an object gaining energy by way of an energy-form conversion.

When a *temperature difference* occurs, energy is transferred from a region of higher temperature to a region of lower temperature. Energy flows from a hot range element, for example, to a pot of cold water on a range. It is a natural process for energy to flow from a region of a higher temperature to a region of a lower temperature just as it is natural for a ball to roll downhill. The temperature of an object and the temperature of the surroundings determine if heat will be transferred to or from an object. The terms *heating* and *cooling* describe the direction of energy flow, naturally moving from a region of higher energy to one of lower energy.

Measures of Heat

Since heating is a method of energy transfer, a quantity of heat can be measured just like any quantity of energy. The metric unit for measuring work, energy, or heat is the *joule*. However, the separate historical development of the concepts of heat and the concepts of motion resulted in separate units, some based on temperature differences.

The metric unit of heat is called the **calorie** (cal). A calorie is defined as the *amount of energy (or heat) needed to increase the temperature of 1 gram of water 1 degree Celsius*. A more precise definition specifies the degree interval from 14.5°C to 15.5°C because the energy required varies slightly at different temperatures. This precise definition is not needed for a general discussion. A **kilocalorie** (kcal) is the *amount of energy (or heat) needed to increase the temperature of 1 kilogram of water 1 degree Celsius*. The measure of the energy released by the oxidation of food is the kilocalorie, but it is called the Calorie (with a capital C) by nutritionists (figure 4.10). Confusion can be avoided by making sure that the scientific calorie is never capitalized (cal) and the dieter's Calorie is always capitalized.

Connections . . .

Energy of Some Foods

Note: Typical basal metabolic rate equals 1,200–2,200 Calories per day.

Food	Nutritional Calories	Kilojoules
Candy bar, various	120 to 230	502 to 963
Beer, regular	150	628
Beer, lite	100	419
Carbonated drink	150	628
Arby's roast beef	380	1,590
Whopper, w/cheese	705	2,951
DQ large cone	340	1,423
Domino's deluxe pizza	225 per slice (16")	942
Hardee's big deluxe	500	2,093
Jumbo Jack w/cheese	630	2,638
KFC breast	300	1,256
Big Mac	560	2,344
Pizza Hut supreme	140 per slice (10")	586
Taco Bell taco	160	670
Wendy's big classic	570	2,386

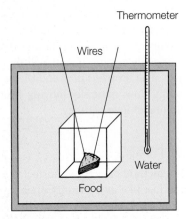

FIGURE 4.10 The Calorie value of food is determined by measuring the heat released from burning the food. If there is 10.0 kg of water and the temperature increases from 10°C to 20°C, the food contains 100 Calories (100,000 calories). The food illustrated here would release much more energy than this.

The best solution would be to call the Calorie what it is, a kilocalorie (kcal).

The English system's measure of heating is called the **British thermal unit** (Btu). A Btu is *the amount of energy (or heat) needed to increase the temperature of 1 pound of water 1 degree Fahrenheit*. The Btu is commonly used to measure the heating or cooling rates of furnaces, air conditioners, water heaters, and so forth. The rate is usually expressed or understood to be in Btu per hour. A much larger unit is sometimes mentioned in news reports and articles about the national energy consumption. This unit is the *quad,* which is 1 quadrillion Btu (a million billion or 10^{15} Btu).

In metric units, the mechanical equivalent of heat is

$$4.184 \text{ J} = 1 \text{ cal}$$

or

$$4,184 \text{ J} = 1 \text{ kcal}$$

The establishment of this precise proportionality means that, fundamentally, mechanical energy and heat are different forms of the same thing.

EXAMPLE 4.4 *(Optional)*

A 1,000.0 kg car is moving at 90.0 km/h (25.0 m/s). How many kilocalories are generated when the car brakes to a stop?

SOLUTION

The kinetic energy of the car is

$$
\begin{aligned}
KE &= \frac{1}{2}mv^2 \\
&= \frac{1}{2}(1,000.0 \text{ kg})\left(25.0 \frac{\text{m}}{\text{s}}\right)^2 \\
&= \frac{1}{2}(1,000.0 \text{ kg})\left(625 \frac{\text{m}^2}{\text{s}^2}\right) \\
&= (500.0)(625)\frac{\text{kg} \cdot \text{m}^2}{\text{s}^2} \\
&= 312,500 \text{ J}
\end{aligned}
$$

You can convert this to kcal by using the relationship between mechanical energy and heat:

$$
\begin{aligned}
&(312,500 \text{ J})\frac{1 \text{ kcal}}{4,184 \text{ J}} \\
&\frac{312,500}{4,184}\frac{\cancel{\text{J}} \cdot \text{kcal}}{\cancel{\text{J}}} \\
&\boxed{74.7 \text{ kcal}}
\end{aligned}
$$

(Note: The temperature increase from this amount of heating could be calculated from equation 4.4.)

Specific Heat

You can observe a relationship between heat and different substances by doing an experiment in "kitchen physics." Imagine that you have a large pot of liquid to boil in preparing a meal. Three variables influence how much heat you need:

1. The initial temperature of the liquid;
2. How much liquid is in the pot; and,
3. The nature of the liquid (water or soup, for example).

What this means specifically, is

1. **Temperature change.** The amount of heat needed is proportional to the temperature change. It takes more heat to raise the temperature of cool water, so this relationship could be written as $Q \propto \Delta T$.
2. **Mass.** The amount of heat needed is also proportional to the amount of the substance being heated. A larger mass requires more heat to go through the same temperature change than a smaller mass. In symbols, $Q \propto m$.
3. **Substance.** Different materials require different amounts of heat to go through the same temperature range when their masses are equal (figure 4.11). This property is called the **specific heat** of a material, which is defined as the amount of heat needed to increase the temperature of 1 gram of a substance 1 degree Celsius.

Considering all the variables involved in our kitchen physics cooking experience, we find the heat (Q) need is described by the relationship of

$$Q = mc\Delta T \qquad \textbf{equation 4.4}$$

where c is the symbol for specific heat.

Specific heat is responsible for the larger variation of temperature observed over land than near a large body of water. Table 4.2 gives the specific heat of soil as 0.200 cal/gC° and the specific heat of water as 1.00 cal/gC°. Since specific

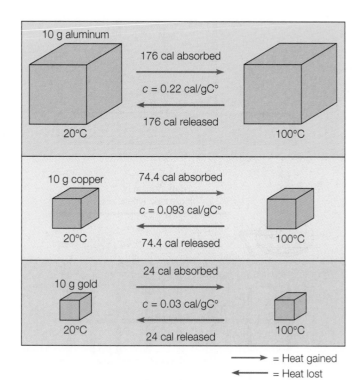

FIGURE 4.11 Of these three metals, aluminum needs the most heat per gram per degree when warmed and releases the most heat when cooled. Why are the cubes different sizes?

TABLE 4.2 The Specific Heat of Selected Substances

Substance	Specific Heat (cal/gC° or kcal/kgC°)
Air	0.17
Aluminum	0.22
Concrete	0.16
Copper	0.093
Glass (average)	0.160
Gold	0.03
Ice	0.500
Iron	0.11
Lead	0.0305
Mercury	0.033
Seawater	0.93
Silver	0.056
Soil (average)	0.200
Steam	0.480
Water	1.00

Note: To convert to specific heat in J/kgC°, multiply each value by 4,184. Also note that 1 cal/gC° = 1 kcal/kgC°.

CONCEPTS APPLIED

More Kitchen Physics

Consider the following information as it relates to the metals of cooking pots and pans:

1. It is easier to change the temperature of metals with low specific heats.
2. It is harder to change the temperature of metals with high specific heats.

Look at the list of metals and specific heats listed in table 4.2 and answer the following questions:

1. Considering specific heat alone, which metal could be used for making practical pots and pans that are the most energy efficient to use?
2. Again considering specific heat alone, would certain combinations of metals provide any advantages for rapid temperature changes?

heat is defined as the amount of heat needed to increase the temperature of 1 gram of a substance 1 degree, this means 1 gram of water exposed to 1 calorie of sunlight will warm 1°C. One gram of soil exposed to 1 calorie of sunlight, on the other hand, will be warmed to 5°C since it only takes 0.2 calories to warm the soil 1°C. Thus, the temperature is more even near large bodies of water since it is harder to change the temperature of the water.

EXAMPLE 4.5 *(Optional)*

How much heat must be supplied to a 500.0 g pan to raise its temperature from 20.0°C to 100.0°C if the pan is made of (a) iron and (b) aluminum?

SOLUTION

The relationship between the heat supplied (Q), the mass (m), and the temperature change (ΔT) is found in equation 4.4. The specific heats (c) of iron and aluminum can be found in table 4.2.

(a) Iron:

$m = 500.0$ g

$c = 0.11$ cal/gC°

$T_f = 100.0°C$

$T_i = 20.0°C$

$Q = ?$

$$Q = mc\Delta T$$
$$= (500.0 \text{ g})\left(0.11\frac{\text{cal}}{\text{gC}°}\right)(80.0°C)$$
$$= (500.0)(0.11)(80.0) \text{ g} \times \frac{\text{cal}}{\text{g} \cdot \text{C}°} \times °C$$
$$= 4,400 \frac{\text{g} \cdot \text{cal} \cdot °C}{\text{g} \cdot \text{C}°}$$
$$= \boxed{4.4 \text{ kcal}}$$

(b) Aluminum:

$m = 500.0$ g

$c = 0.22$ cal/gC°

$T_f = 100.0°C$

$T_i = 20.0°C$

$Q = ?$

$$Q = mc\Delta T$$
$$= (500.0 \text{ g})\left(0.22\frac{\text{cal}}{\text{gC}°}\right)(80.0°C)$$
$$= (500.0)(0.22)(80.0) \text{ g} \times \frac{\text{cal}}{\text{g} \cdot \text{C}°} \times °C$$
$$= 8,800 \frac{\text{g} \cdot \text{cal} \cdot °C}{\text{g} \cdot \text{C}°}$$
$$= \boxed{8.8 \text{ kcal}}$$

It takes twice as much heat energy to warm the aluminum pan through the same temperature range as the iron pan. Thus, with equal rates of energy input, the iron pan will warm twice as fast as the aluminum pan. Note that using kcal rather than the cal unit results in a more compact number. Either unit is correct, but the more compact number is easier to work with, and this makes it more desirable.

EXAMPLE 4.6 *(Optional)*

What is the specific heat of a 2 kg metal sample if 1.2 kcal are needed to increase the temperature from 20.0°C to 40.0°C? (Answer: 0.03 kcal/kgC°)

Heat Flow

In a previous section, you learned that heating is a transfer of energy that involves (1) a temperature difference or (2) energy-form conversions. Heat transfer that takes place because of a temperature difference takes place in three different ways: by conduction, convection, or radiation.

Conduction

Anytime there is a temperature difference, there is a natural transfer of heat from the region of higher temperature to the region of lower temperature. In solids, this transfer takes place as heat is *conducted* from a warmer place to a cooler one. Recall that the molecules in a solid vibrate in a fixed equilibrium position and that molecules in a higher temperature region have more kinetic energy, on the average, than those in a lower temperature region. When a solid, such as a metal rod, is held in a flame, the molecules in the warmed end vibrate violently. Through molecular interaction, this increased energy of vibration is passed on to the adjacent, slower-moving molecules, which also begin to vibrate more violently. They, in turn, pass on more vibrational energy to the molecules next to them. The increase in activity thus moves from molecule to molecule, causing the region of increased activity to extend along the rod. This is called **conduction,** the transfer of energy from molecule to molecule (figure 4.12).

Most insulating materials are good insulators because they contain many small air spaces (figure 4.13). The small air spaces are poor conductors because the molecules of air are far apart, compared to a solid, making it more difficult to pass the increased vibrating motion from molecule to molecule. Styrofoam, glass wool, and wool cloth are good insulators because they have many small air spaces, not because of the material they are made of. The best insulator is a vacuum, since there are no molecules to pass on the vibrating motion.

Transmission of increased kinetic energy

FIGURE 4.12 Thermometers placed in holes drilled in a metal rod will show that heat is conducted from a region of higher temperature to a region of lower temperature. The increased molecular activity is passed from molecule to molecule in the process of conduction.

FIGURE 4.13 Fiberglass insulation is rated in terms of R-value, a ratio of the conductivity of the material to its thickness.

Can you spend too much for home insulation? Should the local government require maximum insulation in new homes to save energy? Research economic limits to insulation: when does the cost of installing insulation exceed any returns in energy saved?

Questions to Discuss

Write a script designed to inform a city planning board about the benefits, advantages, and disadvantages of requiring maximum insulation in all new homes.

Wooden and metal parts of your desk have the same temperature, but the metal parts will feel cooler if you touch them. Metal is a better conductor of heat than wood and feels cooler because it conducts heat from your finger faster. This is the same reason that a wood or tile floor feels cold to your bare feet. You use an insulating rug to slow the conduction of heat from your feet.

Convection

Convection is the transfer of heat by a large-scale displacement of groups of molecules with relatively higher kinetic energy. In conduction, increased kinetic energy is passed from molecule to molecule. In convection, molecules with higher kinetic energy are moved from one place to another place. Conduction happens primarily in solids, but convection happens only in liquids and gases, where fluid motion can carry molecules with higher kinetic energy over a distance. When molecules gain energy, they move more rapidly and push more vigorously against their surroundings. The result is an expansion as the region of heated molecules pushes outward and increases the volume. Since the same amount of matter now occupies a larger volume, the overall density has been decreased.

In fluids, expansion sets the stage for convection. Warm, less-dense fluid is pushed upward by the cooler, more-dense fluid around it. In general, cooler air is more dense; it sinks and flows downhill. Cold air, being more dense, flows out near the bottom of an open refrigerator. You can feel the cold, dense air pouring from the bottom of an open refrigerator to your toes on the floor. On the other hand, you hold your hands *over* a heater because the warm, less-dense air is pushed upward. In a room, warm air is pushed upward from a heater. The warm air spreads outward along the ceiling and is slowly displaced as newly warmed air is pushed upward to the ceiling. As the air cools it sinks over another part of the room, setting up a circulation pattern known as a *convection current* (figure 4.14). Convection

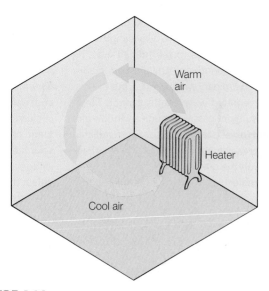

FIGURE 4.14 Convection currents move warm air throughout a room as the air over the heater becomes warmed, expands, and is moved upward by cooler air.

currents can also be observed in a large pot of liquid that is heating on a range. You can see the warmer liquid being forced upward over the warmer parts of the range element, then sink over the cooler parts. Overall, convection currents give the pot of liquid an appearance of turning over as it warms.

Radiation

The third way that heat transfer takes place because of a temperature difference is called **radiation.** Radiation involves the form of energy called *radiant energy,* energy that moves through space. As you will learn in chapter 7, radiant energy includes visible light and many other forms as well. All objects with a temperature above absolute zero give off radiant energy. The absolute temperature of the object determines the rate, intensity, and kinds of radiant energy emitted. You know that visible light is emitted if an object is heated to a certain temperature. A heating element on an electric range, for example, will glow with a reddish-orange light when at the highest setting, but it produces no visible light at lower temperatures, although you

A Closer Look

Passive Solar Design

Passive solar application is an economically justifiable use of solar energy today. Passive solar design uses a structure's construction to heat a living space with solar energy. Few electric fans, motors, or other energy sources are used. The passive solar design takes advantage of free solar energy; it stores and then distributes this energy through natural conduction, convection, and radiation.

Sunlight that reaches Earth's surface is mostly absorbed. Buildings, the ground, and objects become warmer as the radiant energy is absorbed. Nearly all materials, however, reradiate the absorbed energy at longer wavelengths, wavelengths too long to be visible to the human eye. The short wavelengths of sunlight pass readily through ordinary window glass, but the longer, reemitted wavelengths cannot. Therefore, sunlight passes through a window and warms objects inside a house. The reradiated longer wavelengths cannot pass readily back through the glass but are absorbed by certain molecules in the air. The temperature of the air is thus increased. This is called the "greenhouse effect." Perhaps you have experienced the effect when you left your car windows closed on a sunny, summer day.

In general, a passive solar home makes use of the materials from which it is constructed to capture, store, and distribute solar energy to its occupants. Sunlight enters the house through large windows facing south and warms a thick layer of concrete, brick, or stone. This energy "storage mass" then releases energy during the day and, more important, during the night. This release of energy can be by direct radiation to occupants, by conduction to adjacent air, or by convection of air across the surface of the storage mass. The living space is thus heated without special plumbing or forced air circulation. As you can imagine, the key to a successful passive solar home is to consider every detail of natural energy flow, including the materials of which floors and walls are constructed, convective air circulation patterns, and the size and placement of windows. In addition, a passive solar home requires a different lifestyle and living patterns. Carpets, for example, would defeat the purpose of a storage-mass floor, since they would insulate the storage mass from sunlight. Glass is not a good insulator, so windows must have curtains or movable insulation panels to slow energy loss

Box Figure 4.1 The direct solar gain design collects and stores solar energy in the living space.

at night. This requires the daily activity of closing curtains or moving insulation panels at night and then opening curtains and moving panels in the morning. Passive solar homes, therefore, require a high level of personal involvement by the occupants.

There are three basic categories of passive solar design: (1) direct solar gain, (2) indirect solar gain, and (3) isolated solar gain.

A *direct solar gain* home is one in which solar energy is collected in the actual living space of the home (box figure 4.1). The advantage of this design is the large, open window space with a calculated overhang

---

CONCEPTS APPLIED

Candle Wax, Bubbles, and a Lightbulb

Here are three experiments you can do on heat flow:

Conduction. Use melted candle wax to stick thumbtack heads to a long metal rod. Heat one end of the rod with a flame. Record any evidence you observe that heat moves across the rod by conduction.

Convection. Choose a calm, warm day with plenty of strong sunlight. Make soap bubbles to study convection currents between a grass field and an adjacent asphalt parking lot. Find other adjacent areas where you think you might find convection currents and study them with soap bubbles, too. Record your experiments, findings, and explanations for what you observed.

Radiation. Hold your hand under an unlighted electric lightbulb and then turn on the bulb. Describe evidence that what you feel traveled to you by radiation, not conduction or convection. Describe any experiments you can think of to prove you felt radiant energy.

feel warmth in your hand when you hold it near the element. Your hand absorbs the nonvisible radiant energy being emitted from the element. The radiant energy does work on the molecules of your hand, giving them more kinetic energy. You sense this as an increase in temperature, that is, warmth.

All objects above absolute zero emit radiant energy, but all objects also absorb radiant energy. A hot object, however, emits more radiant energy than a cold object. The hot object will emit more energy than it absorbs from the colder object, and the colder object will absorb more energy from the hot object than it emits. There is, therefore, a net energy transfer that will take place by radiation as long as there is a temperature difference between the two objects.

ENERGY, HEAT, AND MOLECULAR THEORY

The kinetic molecular theory of matter is based on evidence from different fields of science, not just one subject area. Chemists and physicists developed some convincing conclusions about the structure of matter over the past 150 years, using carefully designed experiments and mathematical calculations

Box Figure 4.2 The indirect solar gain design uses a Trombe wall to collect, store, and distribute solar energy.

Box Figure 4.3 The isolated solar gain design uses a separate structure to collect and store solar energy.

that admits maximum solar energy in the winter but prevents solar gain in the summer. The disadvantage is that the occupants are living in the collection and storage components of the design and can place nothing (such as carpets and furniture) that would interfere with warming the storage mass in the floors and walls.

An *indirect solar gain* home uses a massive wall inside a window that serves as a storage mass. Such a wall, called a *Trombe wall,* is shown in box figure 4.2. The Trombe wall collects and stores solar energy, then warms the living space with radiant energy and convection currents. The disadvantage

to the indirect solar gain design is that large windows are blocked by the Trombe wall. The advantage is that the occupants are not in direct contact with the solar collection and storage area, so they can place carpets and furniture as they wish. Controls to prevent energy loss at night are still necessary with this design.

An *isolated solar gain* home uses a structure that is separated from the living space to collect and store solar energy. Examples of an isolated gain design are an attached greenhouse or sun porch (box figure 4.3). Energy flow between the attached structure and the living space can

be by conduction, convection, and radiation, which can be controlled by opening or closing off the attached structure. This design provides the best controls, since it can be completely isolated, opened to the living space as needed, or directly used as living space when the conditions are right. Additional insulation is needed for the glass at night, however, and for sunless winter days.

It has been estimated that building a passive solar home would cost about 10 percent more than building a traditional home of the same size. Considering the possible energy savings, you might believe that most homes would now have a passive solar design. They do not, however, as most new buildings require technology and large amounts of energy to maximize comfort. Yet, it would not require too much effort to consider where to place windows in relation to the directional and seasonal intensity of the Sun and where to plant trees. Perhaps in the future you will have an opportunity to consider using the environment to your benefit through the natural processes of conduction, convection, and radiation.

that explained observable facts about matter. Step by step, the detailed structure of this submicroscopic, invisible world of particles became firmly established. Today, an understanding of this particle structure is basic to physics, chemistry, biology, geology, and practically every other science subject. This understanding has also resulted in present-day technology.

Phase Change

Solids, liquids, and gases are the three common phases of matter, and each phase is characterized by different molecular arrangements. The motion of the molecules in any of the three common phases can be increased through (1) the addition of heat through a temperature difference or (2) the absorption of one of the five forms of energy, which results in heating. In either case, the temperature of the solid, liquid, or gas increases according to the specific heat of the substance, and more heating generally means higher temperatures.

More heating, however, does not always result in increased temperatures. When a solid, liquid, or gas changes from one phase to another, the transition is called a **phase change.** A phase change always absorbs or releases energy, *a quantity of*

heat that is not associated with a temperature change. Since the quantity of heat associated with a phase change is not associated with a temperature change, it is called *latent heat.* Latent heat refers to the "hidden" energy of phase changes, which is energy (heat) that goes into or comes out of *internal potential energy* (figure 4.15).

Three kinds of major phase changes can occur: (1) *solid-liquid,* (2) *liquid-gas,* and (3) *solid-gas.* In each case, the phase change can go in either direction. For example, the solid-liquid phase change occurs when a solid melts to a liquid or when a liquid freezes to a solid. Ice melting to water and water freezing to ice are common examples of this phase change and its two directions. Both occur at a temperature called the *freezing point* or the *melting point,* depending on the direction of the phase change. In either case, however, the freezing and melting points are the same temperature.

The liquid-gas phase change also occurs in two different directions. The temperature at which a liquid boils and changes to a gas (or vapor) is called the *boiling point.* The temperature at which a gas or vapor changes back to a liquid is called the *condensation point.* The boiling and condensation points are

FIGURE 4.15 Each phase change absorbs or releases a quantity of latent heat, which goes into or is released from molecular potential energy.

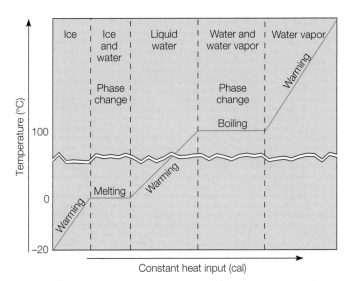

FIGURE 4.16 This graph shows three warming sequences and two phase changes with a constant input of heat. The ice warms to the melting point, then absorbs heat during the phase change as the temperature remains constant. When all the ice has melted, the now-liquid water warms to the boiling point, where the temperature again remains constant as heat is absorbed during this second phase change from liquid to gas. After all the liquid has changed to gas, continued warming increases the temperature of the water vapor.

the same temperature. There are conditions other than boiling under which liquids may undergo liquid-gas phase changes, and these conditions are discussed in the next section.

You probably are not as familiar with solid-gas phase changes, but they are common. A phase change that takes a solid directly to a gas or vapor is called *sublimation*. Mothballs and dry ice (solid CO_2) are common examples of materials that undergo sublimation, but frozen water, meaning common ice, also sublimates under certain conditions. Perhaps you have noticed ice cubes in a freezer become smaller with time as a result of sublimation. The frost that forms in a freezer, on the other hand, is an example of a solid-gas phase change that takes place in the other direction. In this case, water vapor forms the frost without going through the liquid state, a solid-gas phase change that takes place in an opposite direction to sublimation.

For a specific example, consider the changes that occur when ice is subjected to a constant source of heat (figure 4.16). Starting at the left side of the graph, you can see that the temperature of the ice increases from the constant input of heat. The ice warms according to $Q = mc\Delta T$, where c is the specific heat of ice. When the temperature reaches the melting point (0°C), it stops increasing as the ice begins to melt. More and more liquid water appears as the ice melts, but the temperature *remains* at 0°C even though heat is still being added at a constant rate. It takes a certain amount of heat to melt all of the ice. Finally, when all the ice is completely melted, the temperature again increases at a constant rate between the melting and boiling points. Then, at constant temperature, the addition of heat produces another phase change, from liquid to gas. The quantity of heat involved in

this phase change is used in doing the work of breaking the molecule-to-molecule bonds in the solid, making a liquid with molecules that are now free to move about and roll over one another. Since the quantity of heat (Q) is absorbed without a temperature change, it is called the **latent heat of fusion** (L_f). The latent heat of fusion is *the heat involved in a solid-liquid phase change in melting or freezing.* For water, the latent heat of fusion is 80.0 cal/g (144.0 Btu/lb). This means that every gram of ice that melts *absorbs* 80.0 cal of heat. Every gram of water that freezes *releases* 80.0 cal. The total heat involved in a solid-liquid phase change depends on the mass of the substance involved, so

$$Q = mL_f \qquad \textbf{equation 4.5}$$

where L_f is the latent heat of fusion for the substance involved.

Refer again to figure 4.16. After the solid-liquid phase change is complete, the constant supply of heat increases the temperature of the water according to $Q = mc\Delta T$, where c is now the specific heat of liquid water. When the water reaches the boiling point, the temperature again remains constant even though heat is still being supplied at a constant rate. The quantity of heat involved in the liquid-gas phase change again goes into doing the work of overcoming the attractive molecular forces. This time the molecules escape from the liquid state to become single, independent molecules of gas. The quantity of heat (Q) absorbed or released during this phase change is called the **latent heat of vaporization** (L_v). The latent heat of vaporization is *the heat involved in a liquid-gas phase change where there is evaporation or condensation.* For water, the latent heat of vaporization is 540.0 cal/g (970.0 Btu/lb).

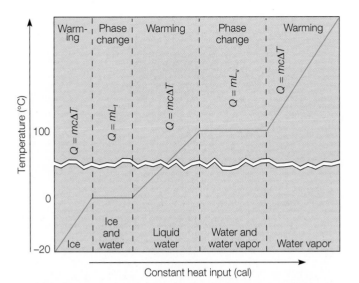

FIGURE 4.17 Compare this graph to the one in figure 4.16. This graph shows the relationships between the quantity of heat absorbed during warming and phase changes as water is warmed from ice at $-20°C$ to water vapor at some temperature above 100°C. Note that the specific heat for ice, liquid water, and water vapor (steam) has different values.

TABLE 4.3   Some Physical Constants for Water and Heat

Specific Heat (c)

Water	$c = 1.00$ cal/gC°
Ice	$c = 0.500$ cal/gC°
Steam	$c = 0.480$ cal/gC°

Latent Heat of Fusion

L_f (water)	$L_f = 80.0$ cal/g

Latent Heat of Vaporization

L_v (water)	$L_v = 540.0$ cal/g

Mechanical Equivalent of Heat

1 kcal	4,184 J

This means that every gram of water vapor that condenses on your bathroom mirror releases 540.0 cal, which warms the bathroom. The total heating depends on how much water vapor condensed, so

$$Q = mL_v \qquad \textbf{equation 4.6}$$

where L_v is the latent heat of vaporization for the substance involved. The relationships between the quantity of heat absorbed during warming and phase changes are shown in figure 4.17. Some physical constants for water and heat are summarized in table 4.3.

EXAMPLE 4.7   *(Optional)*

How much energy does a refrigerator remove from 100.0 g of water at 20.0°C to make ice at −10.0°C?

SOLUTION

This type of problem is best solved by subdividing it into smaller steps that consider (1) the heat added or removed and the resulting temperature changes for each phase of the substance, and (2) the heat flow resulting from any phase change that occurs within the ranges of changes as identified by the problem (see figure 4.17). The heat involved in each phase change and the heat involved in the heating or cooling of each phase are identified as Q_1, Q_2, and so forth. Temperature readings are calculated with absolute values, so you ignore any positive or negative signs.

1. Water in the liquid state cools from 20.0°C to 0°C (the freezing point) according to the relationship $Q = mc\Delta T$, where c is the specific heat of water, and

$$\begin{aligned} Q_1 &= mc\Delta T \\ &= (100.0 \text{ g})\left(1.00\frac{\text{cal}}{\text{gC°}}\right)(0°C - 20.0°C) \\ &= (100.0)(1.00)(20.0) \text{ g} \times \frac{\text{cal}}{\text{g} \cdot \text{C°}} \times °C \\ &= 2,000 \frac{\text{g} \cdot \text{cal} \cdot °\cancel{C}}{\cancel{g} \cdot \cancel{C°}} \\ &= 2.00 \text{ kcal} \end{aligned}$$

2. The latent heat of fusion must now be removed as water at 0°C becomes ice at 0°C through a phase change, and

$$\begin{aligned} Q_2 &= mL_f \\ &= (100.0 \text{ g})\left(80.0\frac{\text{cal}}{\text{g}}\right) \\ &= (100.0)(80.0) \text{ g} \times \frac{\text{cal}}{\text{g}} \\ &= 8,000 \text{ cal} \\ &= 8.00 \text{ kcal} \end{aligned}$$

3. The ice is now at 0°C and is cooled to −10°C, as specified in the problem. The ice cools according to $Q = mc\Delta T$, where c is the specific heat of ice. The specific heat of ice is 0.500 cal/gC°, and

$$\begin{aligned} Q_3 &= mc\Delta T \\ &= (100.0 \text{ g})\left(0.500\frac{\text{cal}}{\text{gC°}}\right)(10°C - 0°C) \\ &= (100.0)(0.500)(10.0) \text{ g} \times \frac{\text{cal}}{\text{g} \cdot \text{C°}} \times °C \\ &= 500 \text{ cal} \\ &= 0.500 \text{ kcal} \end{aligned}$$

4. The total energy removed is then

$$\begin{aligned} Q_T &= Q_1 + Q_2 + Q_3 \\ &= (2.00 \text{ kcal}) + (8.00 \text{ kcal}) + (0.500 \text{ kcal}) \\ &= \boxed{10.50 \text{ kcal}} \end{aligned}$$

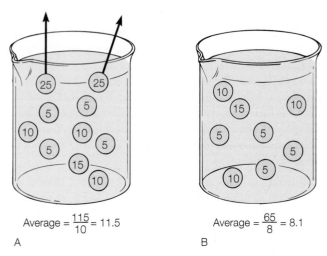

$$\text{Average} = \frac{115}{10} = 11.5$$

A

$$\text{Average} = \frac{65}{8} = 8.1$$

B

FIGURE 4.18 Temperature is associated with the average energy of the molecules of a substance. These numbered circles represent arbitrary levels of molecular kinetic energy that, in turn, represent temperature. The two molecules with the higher kinetic energy values [25 in (A)] escape, which lowers the average values from 11. 5 to 8. 1 (B). Thus, evaporation of water molecules with more kinetic energy contributes to the cooling effect of evaporation, in addition to the absorption of latent heat.

Evaporation and Condensation

Liquids do not have to be at the boiling point to change to a gas and, in fact, tend to undergo a phase change at any temperature when left in the open. The phase change occurs at any temperature but does occur more rapidly at higher temperatures. The temperature of the water is associated with the *average* kinetic energy of the water molecules. The word *average* implies that some of the molecules have a greater energy and some have less. If a molecule of water that has an exceptionally high energy is near the surface and is headed in the right direction, it may overcome the attractive forces of the other water molecules and escape the liquid to become a gas. This is the process of *evaporation.* Evaporation reduces a volume of liquid water as water molecules leave the liquid state to become water vapor in the atmosphere (figure 4.18).

Water molecules that evaporate move about in all directions, and some will return, striking the liquid surface. The same forces that they escaped from earlier capture the molecules, returning them to the liquid state. This is called the process of condensation. Condensation is the opposite of evaporation. In *evaporation,* more molecules are leaving the liquid state than are returning. In *condensation,* more molecules are returning to the liquid state than are leaving. This is a dynamic, ongoing process with molecules leaving and returning continuously. The net number leaving or returning determines if evaporation or condensation is taking place (figure 4.19).

When the condensation rate *equals* the evaporation rate, the air above the liquid is said to be *saturated.* The air immediately next to a surface may be saturated, but the condensation of water molecules is easily moved away with air movement. There is no net energy flow when the air is saturated, since the heat carried away by evaporation is returned by condensation.

FIGURE 4.19 The inside of this closed bottle is isolated from the environment, so the space above the liquid becomes saturated. While it is saturated, the evaporation rate equals the condensation rate. When the bottle is cooled, condensation exceeds evaporation and droplets of liquid form on the inside surfaces.

This is why you fan your face when you are hot. The moving air from the fanning action pushes away water molecules from the air near your skin, preventing the adjacent air from becoming saturated, thus increasing the rate of evaporation. Think about this process the next time you see someone fanning his or her face.

There are four ways to increase the rate of evaporation. (1) An increase in the temperature of the liquid will increase the average kinetic energy of the molecules and thus increase the number of high-energy molecules able to escape from the liquid state. (2) Increasing the surface area of the liquid will also increase the likelihood of molecular escape to the air. This is why you spread out wet clothing to dry or spread out a puddle you want to evaporate. (3) Removal of water vapor from near the surface of the liquid will prevent the return of the vapor molecules to the liquid state and thus increase the net rate of evaporation. This is why things dry more rapidly on a windy day. (4) *Pressure* is defined as *force per unit area,* which can be measured in lb/in^2 or N/m^2. Gases exert a pressure, which is interpreted in terms of the kinetic molecular theory.

Evaporative Coolers

Evaporative cooling is one of the earliest forms of mechanical air conditioning. An evaporative cooler (sometimes called a "swamp cooler") works by cooling outside air by evaporation, then blowing the cooler but now more humid air through a house. Usually, an evaporative cooler moves a sufficient amount of air through a house to completely change the air every two or three minutes.

An evaporative cooler is a metal or fiberglass box with louvers. Inside the box is a fan and motor, and there might be a small pump to recycle water. Behind the louvers are loose pads made of wood shavings (excelsior) or some other porous material. Water is pumped from the bottom of the cooler and trickles down through the pads, thoroughly wetting them. The fan forces air into the house and dry, warm outside air moves through the wet pads into the house in a steady flow. Windows or a door in the house must be partly open, or the air will not be able to flow through the house.

The water in the pads evaporates, robbing heat from the air moving through them. Each liter of water that evaporates could take over 540 kcal of heat from the air. The actual amount removed depends on the temperature of the water, the efficiency of the cooler, and the relative humidity of the outside air.

An electric fan cools you by evaporation, but an evaporative cooler cools the air that blows around you. Relative humidity is the main variable that determines how much an evaporative cooler will cool the air. The following data show the cooling power of a typical evaporative cooler at various relative humidities when the outside air is 38°C (about 100°F).

The advantage of an evaporative cooler is the low operating cost compared to refrigeration. The disadvantages are that it doesn't cool the air that much when the humidity is high, it adds even more humidity to the air, and outside air with its dust, pollen, and pollution is continually forced through the house. Another disadvantage is that mineral deposits left by evaporating hard water could require a frequent maintenance schedule.

At This Humidity:	Air at 38°C (100°F) Will Be Cooled to:
10%	22°C (72°F)
20%	24°C (76°F)
30%	27°C (80°F)
40%	29°C (84°F)
50%	31°C (88°F)
60%	33°C (91°F)
70%	34°C (93°F)
80%	35°C (95°F)
90%	37°C (98°F)

Atmospheric pressure is discussed in detail in chapter 17. For now, consider that the atmosphere exerts a pressure of about 10.0 N/cm² (14.7 lb/in²) at sea level. The atmospheric pressure, as well as the intermolecular forces, tend to hold water molecules in the liquid state. Thus, reducing the atmospheric pressure will reduce one of the forces holding molecules in a liquid state. Perhaps you have noticed that wet items dry more quickly at higher elevations, where the atmospheric pressure is less.

Relative Humidity

There is a relationship between evaporation-condensation and the air temperature. If the air temperature is decreased, the average kinetic energy of the molecules making up the air is decreased. Water vapor molecules condense from the air when they slow enough that molecular forces can pull them into the liquid state. Fast-moving water vapor molecules are less likely to be captured than slow-moving ones. Thus, as the air temperature increases, there is less tendency for water vapor molecules to return to the liquid state. Warm air can therefore hold more water vapor than cool air. In fact, air at 38°C (100°F) can hold five times as much water vapor as air at 10°C (50°F) (figure 4.20).

The ratio of how much water vapor *is* in the air to how much water vapor *could be* in the air at a certain temperature is called **relative humidity.** This ratio is usually expressed as a percentage, and

$$\text{relative humidity} = \frac{\text{water vapor in air}}{\text{capacity at present temperature}} \times 100\%$$

$$R.H. = \frac{\text{g/m}^3 \text{(present)}}{\text{g/m}^3 \text{(max)}} \times 100\% \qquad \textbf{equation 4.7}$$

Figure 4.20 shows the maximum amount of water vapor that can be in the air at various temperatures. Suppose that the air contains 10 g/m³ of water vapor at 10°C (50°F). According to figure 4.20, the maximum amount of water vapor that *can be* in the air when the air temperature is 10°C (50°F) is 10 g/m³. Therefore, the relative humidity is (10 g/m³) ÷ (10 g/m³) × 100%, or 100 percent. This air is therefore saturated. If the air held only 5 g/m³ of water vapor at 10°C, the relative humidity would be 50 percent, and 2 g/m³ of water vapor in the air at 10°C is 20 percent relative humidity.

As the air temperature increases, the capacity of the air to hold water vapor also increases. This means that if the *same amount* of water vapor is in the air during a temperature increase, the relative humidity will decrease. Thus, the relative humidity increases every night because the air temperature decreases, not because water vapor has been added to the air. The relative humidity is important because it is one of the things that controls the rate of evaporation, and the evaporation rate is one of the variables involved in how well you can cool yourself in hot weather.

FIGURE 4.20 The curve shows the *maximum* amount of water vapor in g/m³ that can be in the air at various temperatures.

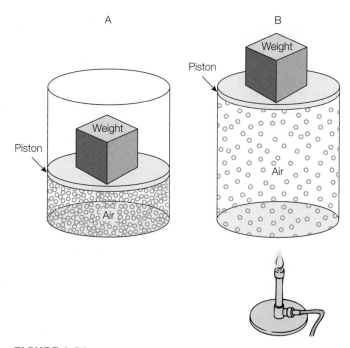

FIGURE 4.21 A very simple heat engine. The air in (*B*) has been heated, increasing the molecular motion and thus the pressure. Some of the heat is transferred to the increased gravitational potential energy of the weight as it is converted to mechanical energy.

CONCEPTS APPLIED

Why It Is Called a Pop Can!

Obtain two empty, clean pop (soft drink) cans, a container of ice water with ice cubes, and a container of boiling water. You might want to "dry run" this experiment to make sure of the procedure before actually doing it.

Place about 2 cm of water in a pop can and heat it on a stove until the water boils and you see evidence of steam coming from the opening. Using tongs, quickly invert the can half way into a container of ice water. Note how much water runs from the can as you remove it from the ice water.

Repeat this procedure, this time inverting the can half way into a container of boiling water. Note how much water runs from the can as you remove it from the boiling water.

Explain your observations in terms of the kinetic molecular theory, evaporation, and condensation. It is also important to explain any differences observed between what happened to the two pop cans.

THERMODYNAMICS

The branch of physical science called *thermodynamics* is concerned with the study of heat and its relationship to mechanical energy, including the science of heat pumps, heat engines, and the transformation of energy in all its forms. The *laws of*

thermodynamics describe the relationships concerning what happens as energy is transformed to work and the reverse, also serving as useful intellectual tools in meteorology, chemistry, and biology.

Mechanical energy is easily converted to heat through friction, but a special device is needed to convert heat to mechanical energy. A *heat engine* is a device that converts heat into mechanical energy. The operation of a heat engine can be explained by the kinetic molecular theory, as shown in figure 4.21. This illustration shows a cylinder, much like a big can, with a closely fitting piston that traps a sample of air. The piston is like a slightly smaller cylinder and has a weight resting on it, supported by the trapped air. If the air in the large cylinder is now heated, the gas molecules will acquire more kinetic energy. This results in more gas molecule impacts with the enclosing surfaces, which results in an increased pressure. Increased pressure results in a net force, and the piston and weight move upward, as shown in figure 4.21B. Thus, some of the heat has now been transformed to the increased gravitational potential energy of the weight.

Thermodynamics is concerned with the *internal energy* (U), the total internal potential and kinetic energies of molecules making up a substance, such as the gases in the simple heat engine. The variables of temperature, gas pressure, volume, heat, and so forth characterize the total internal energy, which is called the *state* of the system. Once the system is identified, everything else is called the *surroundings*. A system can exist in a number of states since the variables that characterize a state can have any number of values and combinations of values. Any two systems that have the same values of variables that characterize internal energy are said to be in the same state.

The First Law of Thermodynamics

Any thermodynamic system has a unique set of properties that will identify the internal energy of the system. This state can be changed two ways: (1) by heat flowing into (Q_{in}) or out (Q_{out}) of the system, or (2) by the system doing work (W_{out}) or work being done on the system (W_{in}). Thus, work (W) and heat (Q) can change the internal energy of a thermodynamic system according to

$$JQ - W = U_2 - U_1 \qquad \textbf{equation 4.8}$$

where J is the mechanical equivalence of heat ($J = 4.184$ joule/calorie) and ($U_2 - U_1$) is the internal energy difference between two states. This equation represents the **first law of thermodynamics,** which states that the energy supplied to a thermodynamic system in the form of heat minus the work done by the system is equal to the change in internal energy. The first law of thermodynamics is an application of the *law of conservation of energy,* which applies to all energy matters. The first law of thermodynamics is concerned specifically with a thermodynamic system. As an example, consider energy conservation that is observed in the thermodynamic system of a heat engine (see figure 4.22). As the engine cycles to the original state of internal energy ($U_2 - U_1 = 0$), all the external work accomplished must be equal to all the heat absorbed in the cycle. The heat supplied to the engine from a high temperature source (Q_H) is partly converted to work (W), and the rest is rejected in the lower-temperature exhaust (Q_L). The work accomplished is therefore the difference in the heat input and the heat output ($Q_H - Q_L$), so the work accomplished represents the heat used,

$$W = J(Q_H - Q_L) \qquad \textbf{equation 4.9}$$

where J is the mechanical equivalence of heat ($J = 4.184$ joules/calorie). A schematic diagram of this relationship is shown in figure 4.22. You can increase the internal energy (produce heat) as long as you supply mechanical energy (or do work). The first law of thermodynamics states that the conversion of work to heat is reversible, meaning that heat can be changed to work. There are several ways of converting heat to work, for example, the use of a steam turbine or gasoline automobile engine.

The Second Law of Thermodynamics

A heat pump is the opposite of a heat engine, as shown schematically in figure 4.23. The heat pump does work (W) in compressing vapors and moving heat from a region of lower temperature (Q_L) to a region of higher temperature (Q_H). That

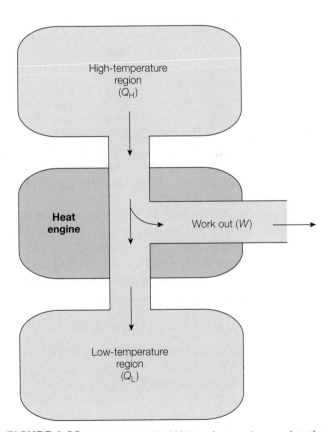

FIGURE 4.22 The heat supplied (Q_H) to a heat engine goes into the mechanical work (W), and the remainder is expelled in the exhaust (Q_L). The work accomplished is therefore the difference in the heat input and output ($Q_H - Q_L$), so the work accomplished represents the heat used, $W = J(Q_H - Q_L)$.

FIGURE 4.23 A heat pump uses work (W) to move heat from a low temperature region (Q_L) to a high temperature region (Q_H). The heat moved (Q_L) requires work (W), so $JQ_L = W$. A heat pump can be used to chill things at the Q_L end or warm things at the Q_H end.

work is required to move heat this way is in accord with the observation that heat naturally flows from a region of higher temperature to a region of lower temperature. Energy is required for the opposite, moving heat from a cooler region to a warmer region. The natural direction of this process is called the **second law of thermodynamics,** which is that heat flows from objects with a higher temperature to objects with a cooler temperature. In other words, if you want heat to flow from a colder region to a warmer one, you must *cause* it to do so by using energy. And if you do, such as with the use of a heat pump, you necessarily cause changes elsewhere, particularly in the energy sources used in the generation of electricity. Another statement of the second law is that it is impossible to convert heat completely into mechanical energy. This does not say that you cannot convert mechanical energy completely into heat, for example, in the brakes of a car when the brakes bring it to a stop. The law says that the reverse process is not possible, that you cannot convert 100 percent of a heat source into mechanical energy. Both of the preceding statements of the second law are concerned with a *direction* of thermodynamic processes, and the implications of this direction will be discussed next.

The Second Law and Natural Processes

Energy can be viewed from two considerations of scale: (1) the observable *external energy* of an object, and (2) the *internal energy* of the molecules or particles that make up an object. A ball, for example, has kinetic energy after it is thrown through the air, and the entire system of particles making up the ball acts like a single massive particle as the ball moves. The motion and energy of the single system can be calculated from the laws of motion and from the equations representing the concepts of work and energy. All of the particles are moving together in *coherent motion* when the external kinetic energy is considered.

But the particles making up the ball have another kind of kinetic energy, with the movements and vibrations of internal kinetic energy. In this case, the particles are not moving uniformly together but are vibrating with motions in many different directions. Since there is a lack of net motion and a lack of correlation, the particles have a jumbled *incoherent motion*, which is often described as chaotic. This random, chaotic motion is sometimes called *thermal motion*.

Thus, there are two kinds of motion that the particles of an object can have: (1) a coherent motion where they move together, in step, and (2) an incoherent, chaotic motion of individual particles. These two types of motion are related to the two modes of energy transfer, working and heating. The relationship is that *work* on an object is associated with its *coherent motion,* while *heating* an object is associated with its internal *incoherent motion.*

The second law of thermodynamics implies a direction to the relationship between work (coherent motion) and heat (incoherent motion), and this direction becomes apparent as you analyze what happens to the motions during energy conversions. Some forms of energy, such as electrical and mechanical, have a greater amount of order since they involve particles moving together in a coherent motion. The term *quality of energy* is used to identify the amount of coherent motion. Energy with high order and coherence is called *high-quality energy.* Energy with less order and less coherence, on the other hand, is called *low-quality energy.* In general, high-quality energy can be easily converted to work, but low-quality energy is less able to do work.

High-quality electrical and mechanical energy can be used to do work but then become dispersed as heat through energy-form conversions and friction. The resulting heat can be converted to do more work only if there is a sufficient temperature difference. The temperature differences do not last long, however, as conduction, convection, and radiation quickly disperse the energy even more. Thus, the transformation of high-quality energy into lower-quality energy is a natural process. Energy tends to disperse, both from the conversion of an energy form to heat and from the heat flow processes of conduction, convection, and radiation. Both processes flow in one direction only and cannot be reversed. This is called the *degradation of energy,* which is the transformation of high-quality energy to lower-quality energy. In every known example, it is a natural process of energy to degrade, becoming less and less available to do work. The process is *irreversible,* even though it is possible to temporarily transform heat to mechanical energy through a heat engine or to upgrade the temperature through the use of a heat pump. Eventually, the upgraded mechanical energy will degrade to heat and the increased heat will disperse through the processes of heat flow.

The apparent upgrading of energy by a heat pump or heat engine is always accompanied by a greater degrading of energy someplace else. The electrical energy used to run the heat pump, for example, was produced by the downgrading of chemical or nuclear energy at an electrical power plant. The overall result is that the *total* energy was degraded toward a more disorderly state.

A *thermodynamic measure of disorder* is called **entropy.** Order means patterns and coherent arrangements. Disorder means dispersion, no patterns, and a randomized or spread-out arrangement. Entropy is therefore a measure of chaos, and this leads to another statement about the second law of thermodynamics and the direction of natural change, that

the total entropy of the universe continually increases.

Note the use of the words *total* and *universe* in this statement of the second law. The entropy of a system can decrease (more order); for example, when a heat pump cools and condenses the random, chaotically moving water vapor molecules into the more ordered state of liquid water. When the energy source for the production, transmission, and use of electrical energy is considered, however, the *total* entropy will be seen

Thermodynamics in Action

The laws of thermodynamics are concerned with changes in energy and heat. This application explores some of these relationships.

Obtain an electric blender and a thermometer. Fill the blender half way with water, then let it remain undisturbed until the temperature is constant as shown by two consecutive temperature readings.

Run the blender at the highest setting for a short time, then stop and record the temperature of the water. Repeat this procedure several times.

Explain your observations in terms of thermodynamics. See if you can think of other experiments that show relationships between changes in energy and heat.

It Makes Its Own Fuel?

Have you ever heard of a perpetual motion machine? A perpetual motion machine is a hypothetical device that would produce useful energy out of nothing. This is generally accepted as being impossible according to laws of physics. In particular, perpetual motion machines would violate either the first or the second law of thermodynamics.

A perpetual motion machine that violates the first law of thermodynamics is called a "machine of the first kind." In general, the first law says that you can never get something for nothing. This means that without energy input, there can be no change in internal energy, and without a change in internal energy, there can be no work output. Machines of the first kind typically use no fuel or make their own fuel faster than they use it. If this type of machine appears to work, look for some hidden source of energy.

A "machine of the second kind" does not attempt to make energy out of nothing. Instead, it tries to extract random molecular motion into useful work or extract useful energy from some degraded source such as outgoing radiant energy. The second law of thermodynamics says this just cannot happen any more than rocks can roll uphill on their own. This just does not happen.

The American Physical Society states, "The American Physical Society deplores attempts to mislead and defraud the public based on claims of perpetual motion machines or sources of unlimited useful free energy, unsubstantiated by experimentally tested established physical principles."

Visit www.phact.org/e/dennis4.html to see a historical list of perpetual motion and free energy machines.

as increasing. Likewise, the total entropy increases during the growth of a plant or animal. When all the food, waste products, and products of metabolism are considered, there is again an increase in *total* entropy.

Thus, the *natural process* is for a state of order to degrade into a state of disorder with a corresponding increase in entropy. This means that all the available energy of the universe is gradually diminishing, and over time, the universe should therefore approach a limit of maximum disorder called the *heat death* of the universe. The heat death of the universe is the theoretical limit of disorder, with all molecules spread far, far apart, vibrating slowly with a uniform low temperature.

The heat death of the universe seems to be a logical consequence of the second law of thermodynamics, but scientists are not certain if the second law should apply to the whole universe. What do you think? Will the universe end as matter becomes spread out with slowly vibrating molecules? As has been said, nature is full of symmetry. So why should the universe begin with a bang and end with a whisper?

EXAMPLE 4.8   *(Optional)*

A heat engine operates with 65.0 kcal of heat supplied and exhausts 40.0 kcal of heat. How much work did the engine do?

SOLUTION

Listing the known and unknown quantities:

> heat input $Q_H = 65.0 \text{ kcal}$
> heat rejected $Q_L = 40.0 \text{ kcal}$
> mechanical equivalent of heat 1 kcal = 4,184 J

The relationship between these quantities is found in equation 4.9 $W = J(Q_H - Q_L)$. This equation states a relationship between the heat supplied to the engine from a high-temperature source (Q_H), which is partly converted to work (W), with the rest rejected in a lower-temperature exhaust (Q_L). The work accomplished is therefore the difference in the heat input and the heat output ($Q_H - Q_L$), so the work accomplished represents the heat used, where J is the mechanical equivalence of heat (1 kcal = 4,184 J). Therefore,

$$W = J(Q_H - Q_L)$$

$$= 4{,}184 \ \frac{J}{kcal} \ (65.0 \text{ kcal} - 40.0 \text{ kcal})$$

$$= 4{,}184 \ \frac{J}{kcal} \ (25.0 \text{ kcal})$$

$$= (4{,}184)(25.0) \ \frac{J \, kcal}{kcal}$$

$$= 104{,}600 \text{ J}$$

$$= \boxed{105 \text{ kJ}}$$

People Behind the Science

Count Rumford (Benjamin Thompson) (1753–1814)

Count Rumford was a U.S.-born physicist who first demonstrated conclusively that heat is not a fluid but a form of motion. He was born Benjamin Thompson in Woburn, Massachusetts, on March 26, 1753. At the age of nineteen, he became a schoolmaster as a result of much self-instruction and some help from local clergy. He moved to Rumford (now named Concord), New Hampshire, and almost immediately married a wealthy widow many years his senior.

Thompson's first activities seem to have been political. When the War of Independence broke out, he remained loyal to the British crown and acted as some sort of secret agent. Because of these activities, he had to flee to London in 1776 (having separated from his wife the year before). He was rewarded with government work and an appointment as a lieutenant colonel in a British regiment in New York.

In 1791, Thompson was made a count of the Holy Roman Empire in recognition of his work in Bavaria. He took his title from Rumford in his homeland, and it is by this name that we know him today—Count Rumford.

Rumford's early work in Bavaria combined social experiments with his lifelong interests concerning heat in all its aspects. When he employed beggars from the streets to manufacture military uniforms, he faced the problem of feeding them. A study of nutrition led him to recognize the importance of water and vegetables, and Rumford decided that soups would fit his requirements. He devised many recipes and developed cheap food emphasizing the potato. Soldiers were employed in gardening to produce the vegetables. Rumford's enterprise of manufacturing military uniforms led to a study of insulation and to the conclusion that heat was lost mainly through convection. Therefore, he designed clothing to inhibit convection—sort of the first thermal clothing.

No application of heat technology was too humble for Rumford's experiments. He devised the domestic range—the "fire in a box"—and special utensils to go with it. In the interest of fuel efficiency, he devised a calorimeter to compare the heats of combustion of various fuels. Smoky fireplaces also drew his attention, and after a study of the various air movements, he produced designs incorporating all the features now considered essential in open fires and chimneys, such as the smoke shelf and damper. His search for an alternative to alcoholic drinks led to the promotion of coffee and the design of the first percolator.

The work for which Rumford is best remembered took place in 1798. As military commander for the elector of Bavaria, he was concerned with the manufacture of cannons. These were bored from blocks of iron with drills, and it was believed that the cannons became hot because as the drills cut into the iron, heat was escaping in the form of a fluid called caloric. However, Rumford noticed that heat production increased as the drills became blunter and cut less into the metal. If a very blunt drill was used, no metal was removed, yet the heat output appeared to be limitless. Clearly, heat could not be a fluid in the metal but must be related to the work done in turning the drill. Rumford also studied the expansion of liquids of different densities and different specific heats, and showed by careful weighing that the expansion was not due to caloric taking up the extra space.

Rumford's contribution to science in demolishing the caloric theory of heat was very important, because it paved the way to the realization that heat is related to energy and work and that all forms of energy can be converted to heat. However, it took several decades to establish the understanding that caloric does not exist and there was no basis for the caloric theory of heat.

Source: Modified from the *Hutchinson Dictionary of Scientific Biography*. © RM, 2009. All rights reserved. Helicon Publishing is a division of RM.

SUMMARY

The kinetic theory of matter assumes that all matter is made up of tiny, ultimate particles of matter called *molecules*. A molecule is defined as the smallest particle of a compound or a gaseous element that can exist and still retain the characteristic properties of that substance. Molecules interact, attracting each other through a force of *cohesion*. Liquids, solids, and gases are the *phases of matter* that are explained by the molecular arrangements and forces of attraction between their molecules. A *solid* has a definite shape and volume because it has molecules that vibrate in a fixed equilibrium position with strong cohesive forces. A *liquid* has molecules that have cohesive forces strong enough to give it a definite volume but not strong enough to give it a definite shape. The molecules of a liquid can flow, rolling over each other. A *gas* is composed of molecules that are far apart, with weak cohesive forces. Gas molecules move freely in a constant, random motion.

The *temperature* of an object is related to the *average kinetic energy* of the molecules making up the object. A measure of temperature tells how hot or cold an object is on two arbitrary scales, the *Fahrenheit scale* and the *Celsius scale*. The *absolute scale*, or *Kelvin scale*, has the coldest temperature possible ($-273°C$) as zero (0 K).

The observable potential and kinetic energy of an object is the *external energy* of that object, while the potential and kinetic energy of the molecules making up the object is the *internal energy* of the object. Heat refers to the total internal energy and is a transfer of energy that takes place (1) because of a *temperature difference* between two objects or

(2) because of an *energy-form conversion*. An energy-form conversion is actually an energy conversion involving work at the molecular level, so all energy transfers involve *heating* and *working*.

A quantity of heat can be measured in *joules* (a unit of work or energy) or *calories* (a unit of heat). A *kilocalorie* is 1,000 calories, another unit of heat. A *Btu,* or *British thermal unit,* is the English system unit of heat. The *mechanical equivalent of heat* is 4,184 J = 1 kcal.

The *specific heat* of a substance is the amount of energy (or heat) needed to increase the temperature of 1 gram of a substance 1 degree Celsius. The specific heat of various substances is not the same because the molecular structure of each substance is different.

Energy transfer that takes place because of a temperature difference does so through conduction, convection, or radiation. *Conduction* is the transfer of increased kinetic energy from molecule to molecule. Substances vary in their ability to conduct heat, and those that are poor conductors are called *insulators*. Gases, such as air, are good insulators. The best insulator is a vacuum. *Convection* is the transfer of heat by the displacement of large groups of molecules with higher kinetic energy. Convection takes place in fluids, and the fluid movement that takes place because of density differences is called a *convection current. Radiation* is radiant energy that moves through space. All objects with an absolute temperature above zero give off radiant energy, but all objects absorb it as well. Energy is transferred from a hot object to a cold one through radiation.

The transition from one phase of matter to another is called a *phase change*. A phase change always absorbs or releases a quantity of *latent heat* not associated with a temperature change. Latent heat is energy that goes into or comes out of *internal potential energy*. The *latent heat of fusion* is absorbed or released at a solid-liquid phase change. The latent heat of fusion for water is 80.0 cal/g (144.0 Btu/lb). The *latent heat of vaporization* is absorbed or released at a liquid-gas phase change. The latent heat of vaporization for water is 540.0 cal/g (970.0 Btu/lb).

Molecules of liquids sometimes have a high enough velocity to escape the surface through the process called *evaporation*. Evaporation is a cooling process, since the escaping molecules remove the latent heat of vaporization in addition to their high molecular energy. Vapor molecules return to the liquid state through the process called *condensation*. Condensation is the opposite of evaporation and is a warming process. When the condensation rate equals the evaporation rate, the air is said to be *saturated*. The rate of evaporation can be *increased* by (1) increased temperature, (2) increased surface area, (3) removal of evaporated molecules, and (4) reduced atmospheric pressure.

Warm air can hold more water vapor than cold air, and the ratio of how much water vapor is in the air to how much could be in the air at that temperature (saturation) is called *relative humidity*.

Thermodynamics is the study of heat and its relationship to mechanical energy, and the *laws of thermodynamics* describe these relationships: The *first law of thermodynamics* is a thermodynamic statement of the law of conservation of energy. The *second law of thermodynamics* states that heat flows from objects with a higher temperature to objects with a cooler temperature. The second law implies a *degradation of energy* as *high-quality* (more ordered) energy sources undergo *degradation* to *low-quality* (less ordered) sources. *Entropy* is a thermodynamic measure of disorder, and entropy is seen as continually increasing in the universe and may result in the maximum disorder called the *heat death* of the universe.

Summary of Equations

4.1
$$T_F = \frac{9}{5} T_C + 32°$$

4.2
$$T_C = \frac{5}{9}(T_F - 32°)$$

4.3
$$T_K = T_C + 273$$

4.4 Quantity of heat = (mass)(specific heat)(temperature change)
$$Q = mc\Delta T$$

4.5 Heat absorbed or released = (mass)(latent heat off usion)
$$Q = mL_f$$

4.6 Heat absorbed or released = (mass)(latent heat of vaporization)
$$Q = mL_v$$

4.7 relative humidity $= \dfrac{\text{water vapor in air}}{\text{capacity at present temperature}} \times 100\%$

$$R.H. = \frac{g/m^3(\text{present})}{g/m^3(\text{max})} \times 100\%$$

4.8 (mechanical equivalence of heat) − (work) = internal energy difference between two states
$$JQ - W = U_2 - U_1$$

4.9 work = (mechanical equivalence ofh eat)(difference between heat input and output)
$$W = J(Q_H - Q_L)$$

KEY TERMS

absolute scale (p. **81**)
absolute zero (p. **82**)
British thermal unit (p. **84**)
calorie (p. **83**)
Celsius scale (p. **80**)
conduction (p. **86**)
convection (p. **87**)

entropy (p. **96**)
external energy (p. **82**)
Fahrenheit scale (p. **80**)
first law of thermodynamics (p. **95**)
heat (p. **83**)
internal energy (p. **82**)

Kelvin scale (p. **81**)
kilocalorie (p. **83**)
kinetic molecular theory (p. **76**)
latent heat of fusion (p. **90**)
latent heat of vaporization (p. **90**)
molecule (p. 77)

phase change (p. **89**)
radiation (p. **87**)
relative humidity (p. **93**)
second law of thermodynamics (p. **96**)
specific heat (p. **85**)
temperature (p. **78**)

APPLYING THE CONCEPTS

1. The kinetic molecular theory explains the expansion of a solid material with increases of temperature as basically the result of
   a. individual molecules expanding.
   b. increased translational kinetic energy.
   c. molecules moving a little farther apart.
   d. heat taking up the spaces between molecules.

2. Using the absolute temperature scale, the freezing point of water is correctly written as
   a. 0 K.
   b. 0°K.
   c. 273 K.
   d. 273°K.

3. The metric unit of heat called a *calorie* is
   a. the specific heat of water.
   b. the energy needed to increase the temperature of 1 gram of water 1 degree Celsius.
   c. equivalent to a little over 4 joules of mechanical work.
   d. All of the above are correct.

4. Table 4.2 lists the specific heat of soil as 0.20 kcal/kgC° and the specific heat of water as 1.00 kcal/kgC°. This means that if 1 kg of soil and 1 kg of water each receive 1 kcal of energy, ideally
   a. the water will be warmer than the soil by 0.8°C.
   b. the soil will be 4°C warmer than the water.
   c. the water will be 4°C warmer than the soil.
   d. the water will warm by 1°C, and the soil will warm by 0.2°C.

5. The heat transfer that takes place by energy moving directly from molecule to molecule is called
   a. conduction.
   b. convection.
   c. radiation.
   d. None of the above is correct.

6. The heat transfer that does not require matter is
   a. conduction.
   b. convection.
   c. radiation.
   d. impossible, for matter is always required.

7. Styrofoam is a good insulating material because
   a. it is a plastic material that conducts heat poorly.
   b. it contains many tiny pockets of air.
   c. of the structure of the molecules making up the Styrofoam.
   d. it is not very dense.

8. The transfer of heat that takes place because of density difference in fluids is
   a. conduction.
   b. convection.
   c. radiation.
   d. None of the above is correct.

9. As a solid undergoes a phase change to a liquid state, it
   a. releases heat while remaining at a constant temperature.
   b. absorbs heat while remaining at a constant temperature.
   c. releases heat as the temperature decreases.
   d. absorbs heat as the temperature increases.

10. The condensation of water vapor actually
    a. warms the surroundings.
    b. cools the surroundings.
    c. sometimes warms and sometimes cools the surroundings, depending on the relative humidity at the time.
    d. neither warms nor cools the surroundings.

11. No water vapor is added to or removed from a sample of air that is cooling, so the relative humidity of this sample of air will
    a. remain the same.
    b. be lower.
    c. be higher.
    d. be higher or lower, depending on the extent of change.

12. Compared to cooler air, warm air can hold
    a. more water vapor.
    b. less water vapor.
    c. the same amount of water vapor.
    d. less water vapor, the amount depending on the humidity.

Answers

1. c 2. c 3. d 4. b 5. a 6. c 7. b 8. b 9. b 10. a 11. c 12. a

QUESTIONS FOR THOUGHT

1. What is temperature? What is heat?

2. Explain why most materials become less dense as their temperature is increased.

3. Would the tight packing of more insulation, such as glass wool, in an enclosed space increase or decrease the insulation value? Explain.

4. A true vacuum bottle has a double-walled, silvered bottle with the air removed from the space between the walls. Describe how this design keeps food hot or cold by dealing with conduction, convection, and radiation.

5. Why is cooler air found in low valleys on calm nights?

6. Why is air a good insulator?

7. A piece of metal feels cooler than a piece of wood at the same temperature. Explain why.

8. What is condensation? Explain, on a molecular level, how the condensation of water vapor on a bathroom mirror warms the bathroom.

9. Which Styrofoam cooler provides more cooling, one with 10 lb of ice at 0°C or one with 10 lb of ice water at 0°C? Explain your reasoning.

10. Explain why a glass filled with a cold beverage seems to sweat. Would you expect more sweating inside a house during the summer or during the winter? Explain.

11. Explain why a burn from 100°C steam is more severe than a burn from water at 100°C.

12. The relative humidity increases almost every evening after sunset. Explain how this is possible if no additional water vapor is added to or removed from the air.

FOR FURTHER ANALYSIS

1. Considering the criteria for determining if something is a solid, liquid, or a gas, what is table salt, which can be poured?

2. What are the significant similarities and differences between heat and temperature?

3. Gas and plasma are phases of matter, yet gas runs a car and plasma is part of your blood. Compare and contrast these terms and offer an explanation for the use of similar names.

4. Analyze the table of specific heats (table 4.2) and determine which metal would make an energy-efficient and practical pan, providing more cooking for less energy.

5. This chapter contains information about three types of passive solar home design. Develop criteria or standards of evaluation that would help someone decide which design is right for their local climate.

6. Could a heat pump move heat without the latent heat of vaporization? Explain.

7. Explore the assumptions on which the "heat death of the universe" idea is based. Propose and evaluate an alternative idea for the future of the universe.

INVITATION TO INQUIRY

Who Can Last Longest?

How can we be more energy efficient? Much of our household energy consumption goes into heating and cooling, and much is lost through walls and ceilings. This invitation is about the insulating properties of various materials and their arrangement.

The challenge of this invitation is to create an insulated container that can keep an ice cube from melting. Decide on a maximum size for the container, and then decide what materials to use. Consider how

you will use the materials. For example, if you are using aluminum foil, should it be shiny side in or shiny side out? If you are using newspapers, should they be folded flat or crumpled loosely?

One ice cube should be left outside the container to use as a control. Find out how much longer your insulated ice cube will outlast the control.

PARALLEL EXERCISES

The exercises in groups A and B cover the same concepts. Solutions to group A exercises are located in appendix D.
Note: Neglect all frictional forces in all exercises.

Group A

1. The average human body temperature is 98.6°F. What is the equivalent temperature on the Celsius scale?

2. An electric current heats a 221 g copper wire from 20.0°C to 38.0°C. How much heat was generated by the current? (c_{copper} = 0.093 kcal/kgC°)

3. A bicycle and rider have a combined mass of 100.0 kg. How many calories of heat are generated in the brakes when the bicycle comes to a stop from a speed of 36.0 km/h?

4. A 15.53 kg bag of soil falls 5.50 m at a construction site. If all the energy is retained by the soil in the bag, how much will its temperature increase? (c_{soil} = 0.200 kcal/kgC°)

5. A 75.0 kg person consumes a small order of french fries (250.0 Cal) and wishes to "work off" the energy by climbing a 10.0 m stairway. How many vertical climbs are needed to use all the energy?

6. A 0.5 kg glass bowl (c_{glass} = 0.2 kcal/kgC°) and a 0.5 kg iron pan (c_{iron} = 0.11 kcal/kgC°) have a temperature of 68°F when placed in a freezer. How much heat will the freezer have to remove from each to cool them to 32°F?

7. A sample of silver at 20.0°C is warmed to 100.0°C when 896 cal is added. What is the mass of the silver? (c_{silver} = 0.056 kcal/kgC°)

Group B

1. The Fahrenheit temperature reading is 98° on a hot summer day. What is this reading on the Kelvin scale?

2. A 0.25 kg length of aluminum wire is warmed 10.0°C by an electric current. How much heat was generated by the current? ($c_{aluminum}$ = 0.22 kcal/kgC°)

3. A 1,000.0 kg car with a speed of 90.0 km/h brakes to a stop. How many calories of heat are generated by the brakes as a result?

4. A 1.0 kg metal head of a geology hammer strikes a solid rock with a velocity of 5.0 m/s. Assuming all the energy is retained by the hammer head, how much will its temperature increase? (c_{head} = 0.11 kcal/kgC°)

5. A 60.0 kg person will need to climb a 10.0 m stairway how many times to "work off" each excess Cal (kcal) consumed?

6. A 50.0 g silver spoon at 20.0°C is placed in a cup of coffee at 90.0°C. How much heat does the spoon absorb from the coffee to reach a temperature of 89.0°C?

7. If the silver spoon placed in the coffee in problem 6 causes it to cool 0.75°C, what is the mass of the coffee? (Assume c_{coffee} = 1.0 cal/gC°)

—continued

PARALLEL EXERCISES—*Continued*

Group A

8. A 300.0 W immersion heater is used to heat 250.0 g of water from 10.0°C to 70.0°C. About how many minutes did this take?

9. A 100.0 g sample of metal is warmed 20.0°C when 60.0 cal is added. What is the specific heat of this metal?

10. How much heat is needed to change 250.0 g of water at 80.0°C to steam at 100.0°C?

11. A 100.0 g sample of water at 20.0°C is heated to steam at 125.0°C. How much heat was absorbed?

12. In an electric freezer, 400.0 g of water at 18.0°C is cooled, frozen, and the ice is chilled to −5.00°C. How much total heat was removed from the water?

Group B

8. How many minutes would be required for a 300.0 W immersion heater to heat 250.0 g of water from 20.0°C to 100.0°C?

9. A 200.0 g china serving bowl is warmed 65.0°C when it absorbs 2.6 kcal of heat from a serving of hot food. What is the specific heat of the china dish?

10. A 500.0 g pot of water at room temperature (20.0°C) is placed on a stove. How much heat is required to change this water to steam at 100.0°C?

11. Spent steam from an electric generating plant leaves the turbines at 120.0°C and is cooled to 90.0°C liquid water by water from a cooling tower in a heat exchanger. How much heat is removed by the cooling tower water for each kilogram of spent steam?

12. Lead is a soft, dense metal with a specific heat of 0.028 kcal/kgC°, a melting point of 328.0°C, and a heat of fusion of 5.5 kcal/kg. How much heat must be provided to melt a 250.0 kg sample of lead with a temperature of 20.0°C?

5

Wave Motions and Sound

Compared to the sounds you hear on a calm day in the woods, the sounds from a waterfall can carry up to a million times more energy.

CORE **CONCEPT**

Sound is transmitted as increased and decreased pressure waves that carry energy.

OUTLINE

Mechanical waves are longitudinal or transverse; sound waves are longitudinal (p. 106).

Forces and Elastic Materials
    Forces and Vibrations
    Describing Vibrations
Waves
    Kinds of Waves
    Waves in Air
    Hearing Waves in Air
Describing Waves
A Closer Look: Hearing Problems

All sounds originate from vibrating matter (p. 111).

Sound Waves
    Velocity of Sound in Air
    Refraction and Reflection
    Interference

Energy and Sound
    Loudness
    Resonance
Sources of Sounds
    Vibrating Strings
People Behind the Science: *Johann Christian Doppler*
    Sounds from Moving Sources

An external force that matches the natural frequency results in resonance (p. 116).

The Doppler effect is an apparent change of pitch brought about from motion (p. 119).

Life Science

▷ The brain perceives sounds based on signals from the ear (Ch. 24).

OVERVIEW

Sometimes you can feel the floor of a building shake for a moment when something heavy is dropped. You can also feel prolonged vibrations in the ground when a nearby train moves by. The floor of a building and the ground are solids that transmit vibrations from a disturbance. Vibrations are common in most solids because the solids are elastic, having a tendency to rebound, or snap back, after a force or an impact deforms them. Usually you cannot see the vibrations in a floor or the ground, but you sense they are there because you can feel them.

There are many examples of vibrations that you can see. You can see the rapid blur of a vibrating guitar string (figure 5.1). You can see the vibrating up-and-down movement of a bounced-upon diving board. Both the vibrating guitar string and the diving board set up a vibrating motion of air that you identify as a sound. You cannot see the vibrating motion of the air, but you sense it is there because you hear sounds.

There are many kinds of vibrations that you cannot see but can sense. Heat, as you have learned, is associated with molecular vibrations that are too rapid and too tiny for your senses to detect other than as an

increase in temperature. Other invisible vibrations include electrons that vibrate, generating spreading electromagnetic radio waves or visible light. Thus, vibrations take place as an observable motion of objects but are also involved in sound, heat, electricity, and light. The vibrations involved in all these phenomena are alike in many ways and all involve energy. Therefore, many topics of science are concerned with vibrational motion. In this chapter, you will learn about the nature of vibrations and how they produce waves in general. These concepts will be applied to sound in this chapter and to electricity and electromagnetic radiation in later chapters.

FORCES AND ELASTIC MATERIALS

If you drop a rubber ball, it bounces because it is capable of recovering its shape when it hits the floor. A ball of clay, on the other hand, does not recover its shape and remains a flattened blob on the floor. An *elastic* material is one that is capable of recovering its shape after a force deforms it. A rubber ball is elastic and a ball of clay is not elastic. You know a metal spring is elastic because you can stretch it or compress it and it always recovers its shape.

A direct relationship exists between the extent of stretching or compression of a spring and the amount of force applied to it. A large force stretches a spring a lot; a small force stretches it a little. As long as the applied force does not exceed the elastic limit of the spring, it will always return to its original shape when you remove the applied force. There are three important considerations about the applied force and the response of the spring:

1. The greater the applied force, the greater the compression or stretch of the spring from its original shape.
2. The spring appears to have an *internal restoring force*, which returns it to its original shape.
3. The farther the spring is pushed or pulled, the *stronger* the restoring force that returns the spring to its original shape.

Forces and Vibrations

A **vibration** is a back-and-forth motion that repeats itself. Such a motion is not restricted to any particular direction and can be in many different directions at the same time. Almost any solid can be made to vibrate if it is elastic. To see how forces are involved in vibrations, consider the spring and mass in figure 5.2. The spring and mass are arranged so that the mass can freely move back and forth on a frictionless surface. When the mass has not been disturbed, it is at rest at an *equilibrium position* (figure 5.2A). At the equilibrium position, the spring is not compressed or stretched, so it applies no force on the mass. If, however, the mass is pulled to the right (figure 5.2B), the spring is stretched and applies a restoring force on the mass toward the left. The farther the mass is displaced, the greater the stretch of the spring and thus the greater the restoring force. The restoring force is proportional to the displacement and is in the opposite direction of the applied force.

If the mass is now released, the restoring force is the only force acting (horizontally) on the mass, so it accelerates back toward the equilibrium position. This force will continuously decrease until the moving mass arrives back at the equilibrium

FIGURE 5.1 Vibrations are common in many elastic materials, and you can see and hear the results of many in your surroundings. Other vibrations in your surroundings, such as those involved in heat, electricity, and light, are invisible to the senses.

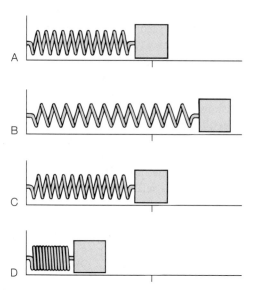

FIGURE 5.2 A mass on a frictionless surface is at rest at an equilibrium position (*A*) when undisturbed. When the spring is stretched (*B*) or compressed (*D*), then released (*C*), the mass vibrates back and forth because restoring forces pull opposite to and proportional to the displacement.

position, where the force is zero. The mass will have a maximum velocity when it arrives, however, so it overshoots the equilibrium position and continues moving to the left (figure 5.2C). As it moves to the left of the equilibrium position, it compresses the spring, which exerts an increasing force on the mass. The moving mass comes to a temporary halt (figure 5.2D), but now the restoring force again starts it moving back toward the equilibrium position. The whole process repeats itself again and again as the mass moves back and forth over the same path.

The periodic vibration, or oscillation, of the mass is similar to many vibrational motions found in nature called *simple harmonic motion*. Simple harmonic motion is defined as the vibratory motion that occurs when there is a restoring force opposite to and proportional to a displacement.

The vibrating mass and spring system will continue to vibrate for a while, slowly decreasing with time until the vibrations stop completely. The slowing and stopping is due to air resistance and internal friction. If these could be eliminated or compensated for with additional energy, the mass would continue to vibrate with a repeating, or *periodic*, motion.

Describing Vibrations

A vibrating mass is described by measuring several variables (figure 5.3). The extent of displacement from the equilibrium position is called the **amplitude.** A vibration that has a mass displaced a greater distance from equilibrium thus has a greater amplitude than a vibration with less displacement.

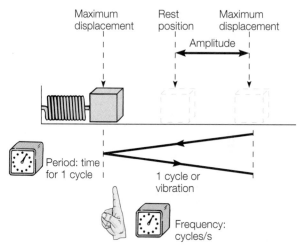

FIGURE 5.3 A vibrating mass attached to a spring is displaced from rest, or equilibrium position, and then released. The maximum displacement is called the *amplitude* of the vibration. A cycle is one complete vibration. The period is the time required for one complete cycle. The frequency is a count of how many cycles it completes in 1 s.

A complete vibration is called a **cycle.** A cycle is the movement from some point, say the far left, all the way to the far right, and back to the same point again, the far left in this example. The **period** (*T*) is simply the time required to complete one cycle. For example, suppose 0.1 s is required for an object to move through one complete cycle, to complete the back-and-forth motion from one point, then back to that point. The period of this vibration is 0.1 s.

Sometimes it is useful to know how frequently a vibration completes a cycle every second. The number of cycles per second is called the **frequency** (f). For example, a vibrating object moves through 10 cycles in 1 s. The frequency of this vibration is 10 cycles per second. Frequency is measured in a unit called a **hertz** (Hz). The unit for a hertz is 1/s because a cycle does not have dimensions. Thus, a frequency of 10 cycles per second is referred to as 10 hertz or 10 1/s.

The period and frequency are two ways of describing the time involved in a vibration. Since the period (T) is the total time involved in one cycle and the frequency (f) is the number of cycles per second, the relationship is

$$T = \frac{1}{f} \qquad \text{equation 5.1}$$

or

$$f = \frac{1}{T} \qquad \text{equation 5.2}$$

EXAMPLE 5.1 *(Optional)*

A vibrating system has a period of 0.5 s. What is the frequency in Hz?

SOLUTION

$$T = 0.5 \text{ s} \qquad f = \frac{1}{T}$$
$$f = ?$$
$$= \frac{1}{0.5 \text{ s}}$$
$$= \frac{1}{0.5} \frac{1}{\text{s}}$$
$$= 2 \frac{1}{\text{s}}$$
$$= \boxed{2 \text{ Hz}}$$

You can obtain a graph of a vibrating object, which makes it easier to measure the amplitude, period, and frequency. If a pen is fixed to a vibrating mass and a paper is moved beneath it at a steady rate, it will draw a curve, as shown in figure 5.4. The greater the amplitude of the vibrating mass, the greater the height of this curve. The greater the frequency, the closer together the peaks and valleys. Note the shape of this curve. This shape is characteristic of simple harmonic motion and is called a *sinusoidal,* or sine, graph. It is so named because it is the same shape as a graph of the sine function in trigonometry.

WAVES

A vibration can be a repeating, or *periodic,* type of motion that disturbs the surroundings. A *pulse* is a disturbance of a single event of short duration. Both pulses and periodic vibrations can create a physical *wave* in the surroundings. A wave is a disturbance that moves through a medium such as a solid or the air. A heavy object dropped on the floor, for example, makes a pulse that sends a mechanical wave that you feel. It might also make a sound wave in the air that you hear. In either case, the medium that transported a wave (solid floor or air) returns to its normal state after the wave has passed. The medium does not travel from place to place, the wave does. Two major considerations about a wave are that (1) a wave is a traveling disturbance and (2) a wave transports energy.

You can observe waves when you drop a rock into a still pool of water. The rock pushes the water into a circular mound as it enters the water. Since it is forcing the water through a distance, it is doing work to make the mound. The mound starts to move out in all directions, in a circle, leaving a depression behind. Water moves into the depression, and a circular wave—mound and depression—moves from the place of disturbance outward (figure 5.5). Any floating object in the path of the wave, such as a leaf, exhibits an up-and-down motion as the mound and

FIGURE 5.5 A water wave moves across the surface. How do you know for sure that it is energy and not water that is moving across the surface?

FIGURE 5.4 A graph of simple harmonic motion is described by a sinusoidal curve.

(Figure 5.4 labels: 1 cycle, Maximum displacement, Rest position, Amplitude, Maximum displacement, Paper pulled this way →)

depression of the wave pass. But the leaf merely bobs up and down and after the wave has passed, it is much in the same place as before the wave. Thus, it was the disturbance that traveled across the water, not the water itself. If the wave reaches a leaf floating near the edge of the water, it may push the leaf up and out of the water, doing work on the leaf. Thus, the wave is a moving disturbance that transfers energy from one place to another.

Kinds of Waves

If you could see the motion of an individual water molecule near the surface as a water wave passed, you would see it trace out a circular path as it moves up and over, down and back. This circular motion is characteristic of the motion of a particle reacting to a water wave disturbance. There are other kinds of waves, and each involves particles in a characteristic motion.

A **longitudinal wave** is a disturbance that causes particles to move closer together or farther apart in the same direction that the wave is moving. If you attach one end of a coiled spring to a wall and pull it tight, you will make longitudinal waves in the spring if you grasp the spring and then move your hand back and forth parallel to the spring. Each time you move your hand toward the length of the spring, a pulse of closer-together coils will move across the spring (figure 5.6A). Each time you pull your hand back, a pulse of farther-apart coils will move across the spring. The coils move back and forth in the same direction that the wave is moving, which is the characteristic movement in reaction to a longitudinal wave.

You will make a different kind of wave in the stretched spring if you now move your hand up and down perpendicular to the length of the spring. This creates a **transverse wave.** A transverse wave is a disturbance that causes motion perpendicular to the direction that the wave is moving. Particles

responding to a transverse wave do not move closer together or farther apart in response to the disturbance; rather, they vibrate back and forth or up and down in a direction perpendicular to the direction of the wave motion (see figure 5.6B).

Whether you make mechanical longitudinal or transverse waves depends not only on the nature of the disturbance creating the waves but also on the nature of the medium. Mechanical transverse waves can move through a material only if there is some interaction, or attachment, between the molecules making up the medium. In a gas, for example, the molecules move about freely without attachments to one another. A pulse can cause these molecules to move closer together or farther apart, so a gas can carry a longitudinal wave. But if a gas molecule is caused to move up and then down, there is no reason for other molecules to do the same, since they are not attached. Thus, a gas will carry mechanical longitudinal waves but not mechanical transverse waves. Likewise a liquid will carry mechanical longitudinal waves but not mechanical transverse waves since the liquid molecules simply slide past one another. The surface of a liquid, however, is another story because of surface tension. A surface water wave is, in fact, a combination of longitudinal and transverse wave patterns that produce the circular motion of a disturbed particle. Solids can and do carry both mechanical longitudinal and transverse waves because of the strong attachments between the molecules.

Waves in Air

Waves that move through the air are longitudinal, so sound waves must be longitudinal waves. A familiar situation will be used to describe the nature of a longitudinal wave moving through air before considering sound specifically. The situation involves a small room with no open windows and two doors that open into the room. When you open one door into the room, the other door closes. Why does this happen? According to the kinetic molecular theory, the room contains many tiny, randomly moving gas molecules that make up the air. As you opened the door, it pushed on these gas molecules, creating a jammed-together zone of molecules immediately adjacent to the door. This jammed-together zone of air now has a greater density and pressure, which immediately spreads outward from

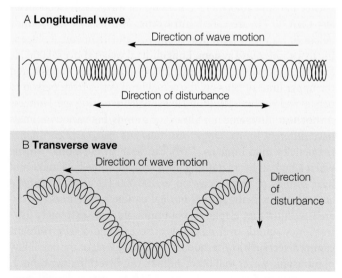

FIGURE 5.6 (*A*) Longitudinal waves are created in a spring when the free end is moved back and forth parallel to the spring. (*B*) Transverse waves are created in a spring when the free end is moved up and down.

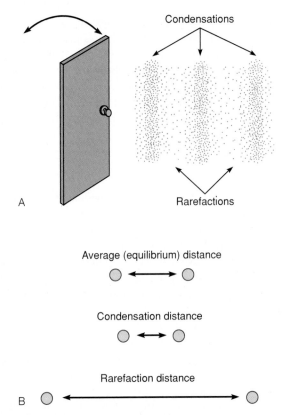

FIGURE 5.7   (A) Swinging the door inward produces pulses of increased density and pressure called *condensations*. Pulling the door outward produces pulses of decreased density and pressure called *rarefactions*. (B) In a condensation, the average distance between gas molecules is momentarily decreased as the pulse passes. In a rarefaction, the average distance is momentarily increased.

the door as a pulse. The disturbance is rapidly passed from molecule to molecule, and the pulse of compression spreads through the room. The pulse of greater density and increased pressure of air reached the door at the other side of the room, and the composite effect of the molecules impacting the door, that is, the increased pressure, caused it to close.

If the door at the other side of the room does not latch, you can probably cause it to open again by pulling on the first door quickly. By so doing, you send a pulse of thinned-out molecules of lowered density and pressure. The door you pulled quickly pushed some of the molecules out of the room. Other molecules quickly move into the region of less pressure, then back to their normal positions. The overall effect is the movement of a thinned-out pulse that travels through the room. When the pulse of slightly reduced pressure reaches the other door, molecules exerting their normal pressure on the other side of the door cause it to move. After a pulse has passed a particular place, the molecules are very soon homogeneously distributed again due to their rapid, random movement.

If you were to swing a door back and forth, it would be a vibrating object. As it vibrates back and forth, it would have a certain frequency in terms of the number of vibrations per second. As the vibrating door moves toward the room, it creates a pulse of jammed-together molecules called a **condensation** (or compression) that quickly moves throughout the room. As the vibrating door moves away from the room, a pulse of thinned-out molecules called a **rarefaction** quickly moves throughout the room. The vibrating door sends repeating pulses of condensation (increased density and pressure) and rarefaction (decreased density and pressure) through the room as it moves back and forth (figure 5.7). You know that the pulses transmit energy because they produce movement, or do work on, the other door. Individual molecules execute a harmonic motion about their equilibrium position and can do work on a movable object. Energy is thus transferred by this example of longitudinal waves.

Hearing Waves in Air

You cannot hear a vibrating door because the human ear normally hears sounds originating from vibrating objects with a frequency between 20 and 20,000 Hz. Longitudinal waves with frequencies less than 20 Hz are called **infrasonic.** You usually *feel* sounds below 20 Hz rather than hear them, particularly if you are listening to a good sound system. Longitudinal waves above 20,000 Hz are called **ultrasonic.** Although 20,000 Hz is usually considered the upper limit of hearing, the actual limit varies from person to person and becomes lower and lower with increasing age. Humans do not hear infrasonic nor ultrasonic sounds, but various animals have different limits. Dogs, cats, rats, and bats can hear higher frequencies than humans. Dogs can hear an ultrasonic whistle when a human hears nothing, for example. Some bats make and hear sounds of frequencies up to 100,000 Hz as they navigate and search for flying insects in total darkness. Scientists discovered recently that elephants communicate with extremely low-frequency sounds over distances of several kilometers. Humans cannot detect such low-frequency sounds. This raises the possibility of infrasonic waves that other animals can detect that we cannot.

A tuning fork that vibrates at 260 Hz makes longitudinal waves much like the swinging door, but these longitudinal waves are called *sound waves* because they are within the frequency

Box figure 5.1 shows the anatomy of the ear. The sound that arrives at the ear is first funneled by the external ear to the tympanum, also known as the eardrum. The cone-shaped nature of the external ear focuses sound on the tympanum and causes it to vibrate at the same frequency as the sound waves reaching it. Attached to the tympanum are three tiny bones known as the malleus (hammer), incus (anvil), and stapes (stirrup). The malleus is attached to the tympanum, the incus is attached to the malleus and stapes, and the stapes is attached to a small, membrane-covered opening called the oval window in a snail-shaped structure known as the cochlea. The vibration of the tympanum causes the tiny bones (malleus, incus, and stapes) to vibrate, and they, in turn, cause a corresponding vibration in the membrane of the oval window.

The cochlea of the ear is the structure that detects sound, and it consists of a snail-shaped set of fluid-filled tubes. When the oval window vibrates, the fluid in the cochlea begins to move, causing a membrane in the cochlea, called the basilar membrane, to vibrate. High-pitched, short-wavelength sounds cause the basilar membrane to vibrate at the base of the cochlea near the oval window. Low-pitched, long-wavelength sounds vibrate the basilar membrane far from the oval window. Loud sounds cause the basilar membrane to vibrate more vigorously than do faint sounds. Cells on this membrane depolarize when they are stimulated by its vibrations. Since they synapse with neurons, messages can be sent to the brain.

Because sounds of different wavelengths stimulate different portions of the cochlea, the brain is able to determine the pitch of a sound. Most sounds consist of a mixture of pitches that are heard. Louder sounds stimulate the membrane more forcefully, causing the sensory cells in the cochlea to send more nerve impulses per second. Thus, the brain is able to perceive the loudness of various sounds, as well as the pitch.

Associated with the cochlea are two fluid-filled chambers and a set of fluid-filled tubes called the semicircular canals. These structures are not involved in hearing but are involved in maintaining balance and posture. In the walls of these canals and chambers are cells similar to those found on the basilar membrane. These cells are stimulated by movements of the head and by the position of the head with respect to the force of gravity. The constantly changing position of the head results in sensory input that is important in maintaining balance.

Box Figure 5.1 A schematic sketch of the human ear.

range of human hearing. The prongs of a struck tuning fork vibrate, moving back and forth. This is more readily observed if the prongs of the fork are struck, then held against a sheet of paper or plunged into a beaker of water. In air, the vibrating prongs first move toward you, pushing the air molecules into a condensation of increased density and pressure. As the prongs then move back, a rarefaction of decreased density and pressure is produced. The alternation of increased and decreased pressure pulses moves from the vibrating tuning fork and spreads outward equally in all directions, much like the surface of a rapidly expanding balloon (figure 5.8). Your eardrum is forced in and out by the pulses it receives. It now vibrates with the same frequency as the tuning fork. The vibrations of the eardrum are transferred by three tiny bones to a fluid in a coiled chamber. Here, tiny hairs respond to the frequency and size of the disturbance, activating nerves that transmit the information to the brain. The brain interprets a frequency as a sound with a certain **pitch.** High-frequency sounds are interpreted as high-pitched musical notes, for example, and low-frequency sounds are interpreted as low-pitched musical notes. The brain then selects certain sounds from all you hear,

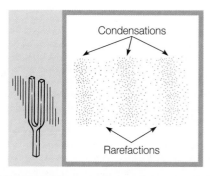

FIGURE 5.8 A vibrating tuning fork produces a series of condensations and rarefactions that move away from the tuning fork. The pulses of increased and decreased pressure reach your ear, vibrating the eardrum. The ear sends nerve signals to the brain about the vibrations, and the brain interprets the signals as sounds.

and you "tune" to certain ones, enabling you to listen to whatever sounds you want while ignoring the background noise, which is made up of all the other sounds.

DESCRIBING WAVES

A tuning fork vibrates with a certain frequency and amplitude, producing a longitudinal wave of alternating pulses of increased-pressure condensations and reduced-pressure rarefactions. A graph of the frequency and amplitude of the vibrations is shown in figure 5.9A, and a representation of the condensations and rarefactions is shown in figure 5.9B. The wave pattern can also be represented by a graph of the changing air pressure of the traveling sound wave, as shown in figure 5.9C. This graph can be used to define some interesting concepts associated with sound waves. Note the correspondence between the (1) amplitude, or displacement, of the vibrating prong, (2) the pulses of condensations and rarefactions, and (3) the changing air pressure. Note also the correspondence between the frequency of the vibrating prong and the frequency of the wave cycles.

Figure 5.10 shows the terms commonly associated with waves from a continuously vibrating source. The wave *crest* is the maximum disturbance from the undisturbed (rest) position. For a sound wave, this would represent the maximum increase of air pressure. The wave *trough* is the maximum disturbance in the opposite direction from the rest position. For a sound wave, this would represent the maximum decrease of air pressure. The *amplitude* of a wave is the displacement from rest to the crest *or* from rest to the trough. The time required for a wave to repeat itself is the *period* (*T*). To repeat itself means the time required to move through one full wave, such as from the crest of one wave to the crest of the next wave. This length in which the wave repeats itself is called the **wavelength** (the symbol is λ, which is the Greek letter lambda). Wavelength is measured in centimeters or meters just like any other length.

There is a relationship between the wavelength, period, and speed of a wave. The relationship is

$$v = \lambda f \qquad \textbf{equation 5.3}$$

which we will call the *wave equation*. This equation tells you that the velocity of a wave can be obtained from the product of the wavelength and the frequency. Note that it also tells you that the wavelength and frequency are inversely proportional at a given velocity.

EXAMPLE 5.2 *(Optional)*

A sound wave with a frequency of 260 Hz has a wavelength of 1.27 m. With what speed would you expect this sound wave to move?

SOLUTION

$$f = 260 \text{ Hz}$$
$$\lambda = 1.27 \text{ m}$$
$$v = ?$$

$$v = \lambda f$$
$$= (1.27 \text{ m})\left(260 \, \frac{1}{\text{s}}\right)$$
$$= (1.27)(260) \text{ m} \times \frac{1}{\text{s}}$$
$$= \boxed{330 \, \frac{\text{m}}{\text{s}}}$$

EXAMPLE 5.3 *(Optional)*

In general, the human ear is most sensitive to sounds at 2,500 Hz. Assuming that sound moves at 330 m/s, what is the wavelength of sounds to which people are most sensitive? (Answer: 13 cm)

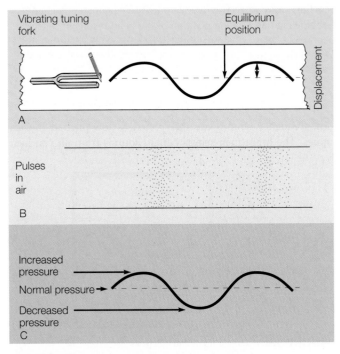

FIGURE 5.9 Compare the (*A*) back-and-forth vibrations of a tuning fork with (*B*) the resulting condensations and rarefactions that move through the air and (*C*) the resulting increases and decreases of air pressure on a surface that intercepts the condensations and rarefactions.

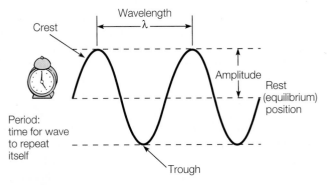

FIGURE 5.10 Here are some terms associated with periodic waves. The *wavelength* is the distance from a part of one wave to the same part in the next wave, such as from one crest to the next. The *amplitude* is the displacement from the rest position. The *period* is the time required for a wave to repeat itself, that is, the time for one complete wavelength to move past a given location.

There are three general areas of hearing problems that are related to your age and the intensity and duration of sounds falling on your ears. These are (1) middle ear infections of young children, (2) loss of ability to hear higher frequencies because of aging, and (3) ringing and other types of noise heard in the head or ear.

Middle ear infections are one of the most common illnesses of young children. The middle ear is a small chamber behind the eardrum that has three tiny bones that transfer sound vibrations from the eardrum to the inner ear (see box figure 5.1). The middle ear is connected to the throat by a small tube named the eustachian tube. This tube allows air pressure to be balanced behind the eardrum. A middle ear infection usually begins with a cold. Small children

have short eustachian tubes, which become swollen from a cold, and this traps fluid in the middle ear. Fluid buildup causes pain and discomfort, as well as reduced hearing ability. This condition often clears on its own in several weeks or more. For severe and recurring cases, small drainage tubes are sometimes inserted through the eardrum to allow fluid drainage. These tubes eventually fall out, and the eardrum heals. Middle ear infections are less of a problem by the time a child reaches school age.

There is a normal loss of hearing because of aging. This loss is more pronounced for higher frequencies and is also greater in males than females. The normal loss of hearing begins in the early twenties, and then increases through the sixties and seventies. However, this process is accelerated by

spending a lot of time in places with very loud sounds. Loud concerts or riding in a "boom car" are two examples that will speed the loss of ability to hear higher frequencies. A "boom car" is one with loud music you can hear "booming" a half a block or more away.

Tinnitus is a sensation of sound, as a ringing or swishing that seems to originate in the ears or head and can only be heard by the person affected. There are several different causes, but the most common is nerve damage in the inner ear. Exposure to loud noises, explosions, firearms, and loud bands are common causes of tinnitus. Advancing age and certain medications can also cause tinnitus. Medications, such as aspirin, can be stopped to end the tinnitus, but there is no treatment for nerve damage to the inner ear.

SOUND WAVES

The transmission of a sound wave requires a medium, that is, a solid, liquid, or gas to carry the disturbance. Therefore, sound does not travel through the vacuum of outer space, since there is nothing to carry the vibrations from a source. The nature of the molecules making up a solid, liquid, or gas determines how well or how rapidly the substance will carry sound waves. The two variables are (1) the inertia of the molecules and (2) the strength of the interaction, if the molecules are attached to one another. Thus, hydrogen gas, with the least massive molecules with no interaction or attachments, will carry a sound wave at 1,284 m/s (4,213 ft/s) when the temperature is 0°C. More massive helium gas molecules have more inertia and carry a sound wave at only 965 m/s (3,166 ft/s) at the same temperature. A solid, however, has molecules that are strongly attached so vibrations are passed rapidly from molecule to molecule. Steel, for example, is highly elastic, and sound will move through a steel rail at 5,940 m/s (19,488 ft/s). Thus, there is a reason for the old saying, "Keep your ear to the ground," because sounds move through solids more rapidly than through a gas (table 5.1).

Velocity of Sound in Air

Most people have observed that sound takes some period of time to move through the air. If you watch a person hammering on a roof a block away, the sounds of the hammering are not in sync with what you see. Light travels so rapidly that you can consider what you see to be simultaneous with what

is actually happening for all practical purposes. Sound, however travels much more slowly and the sounds arrive late in comparison to what you are seeing. This is dramatically illustrated by seeing a flash of lightning, then hearing thunder seconds later. Perhaps you know of a way to estimate the distance to a lightning flash by timing the interval between the flash and boom.

The air temperature influences how rapidly sound moves through the air. The gas molecules in warmer air have a greater kinetic energy than those of cooler air. The molecules of warmer air therefore transmit an impulse from molecule to molecule more rapidly. More precisely, the speed of a sound wave increases 0.60 m/s (2.0 ft/s) for *each* Celsius degree increase in temperature. In *dry* air at sea-level density (normal pressure) and 0°C (32°F), the velocity of sound is about 331 m/s (1,087 ft/s).

TABLE 5.1   Speed of Sound in Various Materials

Medium	m/s	ft/s
Carbon dioxide (0°C)	259	850
Dry air (0°C)	331	1,087
Helium (0°C)	965	3,166
Hydrogen (0°C)	1,284	4,213
Water (25°C)	1,497	4,911
Seawater (25°C)	1,530	5,023
Lead	1,960	6,430
Glass	5,100	16,732
Steel	5,940	19,488

Refraction and Reflection

When you drop a rock into a still pool of water, circular patterns of waves move out from the disturbance. These water waves are on a flat, two-dimensional surface. Sound waves, however, move in three-dimensional space like a rapidly expanding balloon. Sound waves are *spherical waves* that move outward from the source. Spherical waves of sound move as condensations and rarefactions from a continuously vibrating source at the center. If you identify the same part of each wave in the spherical waves, you have identified a *wave front*. For example, the crests of each condensation could be considered as a wave front. From one wave front to the next, therefore, identifies one complete wave or wavelength. At some distance from the source, a small part of a spherical wave front can be considered a *linear wave front* (figure 5.11).

Waves move within a homogeneous medium such as a gas or a solid at a fairly constant rate but gradually lose energy to friction. When a wave encounters a different condition (temperature, humidity, or nature of material), however, drastic changes may occur rapidly. The division between two physical conditions is called a *boundary*. Boundaries are usually encountered (1) between different materials or (2) between the same materials with different conditions. An example of a wave moving between different materials is a sound made in the next room that moves through the air to the wall and through the wall to the air in the room where you are. The boundaries are air-wall and wall-air. If you have ever been in a room with "thin walls," it is obvious that sound moved through the wall and air boundaries.

An example of sound waves moving through the same material with different conditions is found when a wave front moves through air of different temperatures. Since sound travels faster in warm air than in cold air, the wave front becomes bent. The bending of a wave front between boundaries is called *refraction*. Refraction changes the direction of travel of a wave front. Consider, for example, that on calm, clear nights, the air near Earth's surface is cooler than air farther above the surface. Air at rooftop height above the surface might be four or five degrees warmer under such ideal conditions. Sound will travel faster in the higher, warmer air than it will in the lower, cooler air close to the surface. A wave front will therefore become bent, or refracted, toward the ground on a cool night and you will be able to hear sounds from farther away than on warm nights (figure 5.12A). The opposite process occurs during the day as Earth's surface becomes warmer from sunlight (figure 5.12B). Wave fronts are refracted upward because part of the wave front travels faster in the warmer air near the surface. Thus, sound does not seem to carry as far in the summer as it does in the winter. What is actually happening is that during the summer, the wave fronts are refracted away from the ground before they travel very far.

When a wave front strikes a boundary that is parallel to the front the wave may be absorbed, transmitted, or undergo *reflection*, depending on the nature of the boundary medium, or the wave may be partly absorbed, partly transmitted, partly reflected, or any combination thereof. Some materials, such as hard, smooth surfaces, reflect sound waves more than they absorb them. Other materials, such as soft, ruffly curtains, absorb sound waves more than they reflect them. If you have

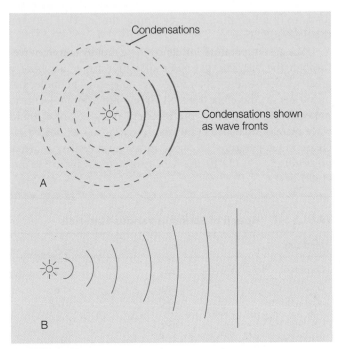

FIGURE 5.11 (*A*) Spherical waves move outward from a sounding source much as a rapidly expanding balloon. This two-dimensional sketch shows the repeating condensations as spherical wave fronts. (*B*) Some distance from the source, a spherical wave front is considered a linear, or plane, wave front.

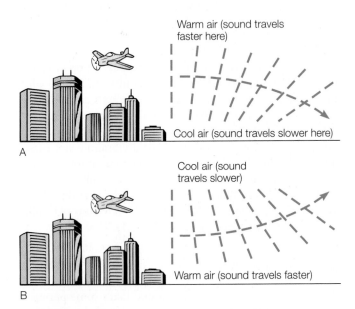

FIGURE 5.12 (*A*) Since sound travels faster in warmer air, a wave front becomes bent, or refracted, toward Earth's surface when the air is cooler near the surface. (*B*) When the air is warmer near the surface, a wave front is refracted upward, away from the surface.

FIGURE 5.13 This closed-circuit TV control room is acoustically treated by covering the walls with sound-absorbing baffles.

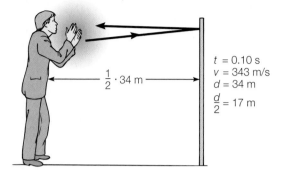

A Echo

$t = 0.10$ s
$v = 343$ m/s
$d = 34$ m
$\frac{d}{2} = 17$ m

$\frac{1}{2} \cdot 34$ m

B Sonar

$t = 1.0$ s
$v = 5{,}023$ ft/s
$d = 5{,}023$ ft
depth $= \frac{1}{2} \cdot d$

$\frac{1}{2} \cdot 5{,}023$ ft

FIGURE 5.14 (*A*) At room temperature, sound travels at 343 m/s. In 0.10 s, sound would travel 34 m. Since the sound must travel to a surface and back in order for you to hear an echo, the distance to the surface is one-half the total distance. (*B*) Sonar measures a depth by measuring the elapsed time between an ultrasonic sound pulse and the echo. The depth is one-half the round trip.

ever been in a room with smooth, hard walls and with no curtains, carpets, or furniture, you know that sound waves may be reflected several times before they are finally absorbed.

Do you sing in the shower? Many people do because the tone is more pleasing than singing elsewhere. The walls of a shower are usually hard and smooth, reflecting sounds back and forth several times before they are absorbed. The continuation of many reflections causes a tone to gain in volume. Such mixing of reflected sounds with the original is called **reverberation.** Reverberation adds to the volume of a tone, and it is one of the factors that determines the acoustical qualities of a room, lecture hall, or auditorium. An open-air concert sounds flat without the reverberation of an auditorium and is usually enhanced electronically to make up for the lack of reflected sounds. Too much reverberation in a room or classroom is not good since the spoken word is not as sharp. Sound-absorbing materials are therefore used on the walls and floors where clear, distinct speech is important (figure 5.13). The carpet and drapes you see in a movie theater are not decorator items but are there to absorb sounds.

If a reflected sound arrives after 0.10 s, the human ear can distinguish the reflected sound from the original sound. A reflected sound that can be distinguished from the original is called an **echo.** Thus, a reflected sound that arrives before 0.10 s is perceived as an increase in volume and is called a reverberation, but a sound that arrives after 0.10 s is perceived as an echo.

Sound wave echoes are measured to determine the depth of water or to locate underwater objects by a *sonar* device. The word *sonar* is taken from *so*und *na*vigation *r*anging. The device generates an underwater ultrasonic sound pulse, then measures the elapsed time for the returning echo. Sound waves travel at about 1,531 m/s (5,023 ft/s) in seawater at 25°C (77°F). A 1 s lapse between the ping of the generated sound and the echo return would mean that the sound traveled 5,023 ft for the round trip. The bottom would be half this distance below the surface (figure 5.14).

Myths, Mistakes, and Misunderstandings

A Duck's Quack Doesn't Echo?

You may have heard the popular myth that "a duck's quack doesn't echo, and no one knows why." An acoustic research experiment was carried out at the University of Salford in Greater Manchester, England, to test this myth. Acoustic experts first recorded a quacking duck in a special chamber that was constructed to produce no sound reflections. Simulations were then done in a reverberation chamber to match the effect of the duck quacking when flying past a cliff face. The tests found that a duck's quack indeed does echo, just like any other sound.

The quack researchers speculated that the myth might have resulted because:

1. The quack does echo, but it is usually too quiet to hear. A duck quacks too quietly, so the echo is too quiet to hear.
2. Ducks quack near water, not near reflecting surfaces such as a mountain or building that would make an echo.
3. It is hard to hear the echo of a sound that fades in and out, as a duck quack does.

EXAMPLE 5.4   (Optional)

The human ear can distinguish a reflected sound pulse from the original sound pulse if 0.10 s or more elapses between the two sounds. What is the minimum distance to a reflecting surface from which we can hear an echo (see figure 5.14A) if the speed of sound is 343 m/s?

SOLUTION

$$t = 0.10 \text{ s (minimum)} \qquad v = \frac{d}{t} \therefore d = vt$$
$$v = 343 \text{ m/s}$$
$$d = ?$$
$$= \left(343 \frac{\text{m}}{\text{s}}\right)(0.10 \text{ s})$$
$$= (343)(0.10)\frac{\text{m}}{\text{s}} \times s$$
$$= 34.3 \frac{\text{m} \cdot \text{s}}{\text{s}}$$
$$= 34 \text{ m}$$

Since the sound pulse must travel from the source to the reflecting surface, then back to the source,

$$34 \text{ m} \times 1/2 = \boxed{17 \text{ m}}$$

The minimum distance to a reflecting surface from which we hear an echo when the air is at room temperature is therefore 17 m (about 56 ft).

EXAMPLE 5.5   (Optional)

An echo is heard exactly 1.00 s after a sound when the speed of sound is 1,147 ft/s. How many feet away is the reflecting surface? (Answer: 574 ft)

Interference

Waves interact with a boundary much as a particle would, reflecting or refracting because of the boundary. A moving ball, for example, will bounce from a surface at the same angle it strikes the surface, just as a wave does. A particle or a ball, however, can be in only one place at a time, but waves can be spread over a distance at the same time. You know this since many different people in different places can hear the same sound at the same time.

Another difference between waves and particles is that two or more waves can exist in the same place at the same time. When two patterns of waves meet, they pass through each other without refracting or reflecting. However, at the place where they meet, the waves interfere with each other, producing a *new* disturbance. This new disturbance has a different amplitude, which is the algebraic sum of the amplitudes of the two separate wave patterns. If the wave crests or wave troughs arrive at the same place at the same time, the two waves are said to be *in phase*. The result of two waves arriving in phase is a new disturbance with a crest and trough that has greater displacement than either of the two separate waves. This is called *constructive interference* (figure 5.15A). If the trough of one wave arrives at the same place and time as the crest of another

wave, the waves are completely *out of phase*. When two waves are completely out of phase, the crest of one wave (positive displacement) will cancel the trough of the other wave (negative displacement), and the result is zero total disturbance, or no wave. This is called *destructive interference* (figure 5.15B). If the two sets of wave patterns do not have the exact same amplitudes or wavelengths, they will be neither completely in phase nor completely out of phase. The result will be partly constructive or destructive interference, depending on the exact nature of the two wave patterns.

Suppose that two vibrating sources produce sounds that are in phase, equal in amplitude, and equal in frequency. The resulting sound will be increased in volume because of constructive interference. But suppose the two sources are slightly different in frequency, for example, 350 Hz and 352 Hz. You will hear a regularly spaced increase and decrease of sound known as **beats**. Beats occur because the two sound waves experience alternating constructive and destructive interferences (figure 5.16). The phase relationship changes because of the difference in frequency, as you can see in the illustration. These alternating constructive and destructive interference zones are moving from the source to the receiver, and the receiver hears the results as a rapidly rising and falling sound level. The beat frequency is the difference between the frequencies of the two sources. A 352 Hz source and 350 Hz source sounded together would result in a beat frequency of 2 Hz. Thus, as two frequencies are closer and closer together, fewer beats will be heard per second. You may be familiar with the phenomenon of beats if

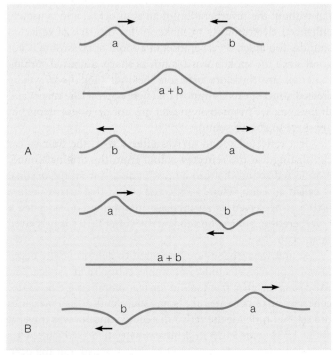

FIGURE 5.15   (A) Constructive interference occurs when two equal, in-phase waves meet. (B) Destructive interference occurs when two equal, out-of-phase waves meet. In both cases, the wave displacements are superimposed when they meet, but they then pass through one another and return to their original amplitudes.

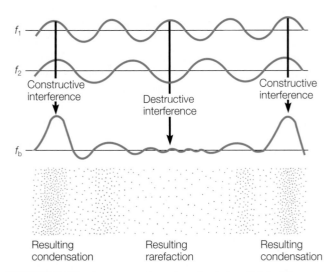

FIGURE 5.16 Two waves of equal amplitude but slightly different frequencies interfere destructively and constructively. The result is an alternation of loudness called a *beat*.

you have ever flown in an airplane with two engines. If one engine is running slightly faster than the other, you hear a slow beat. The beat frequency (f_b) is equal to the absolute difference in frequency of two interfering waves with slightly different frequencies, or

$$f_b = f_2 - f_1 \qquad \textbf{equation 5.4}$$

ENERGY AND SOUND

All waves involve the transportation of energy, including sound waves. The vibrating mass and spring in figure 5.2 vibrate with an amplitude that depends on how much work you did on the mass in moving it from its equilibrium position. More work on the mass results in a greater displacement and a greater amplitude of vibration. A vibrating object that is producing sound waves will produce more intense condensations and rarefactions if it has a greater amplitude. The intensity of a sound wave is a measure of the energy the sound wave is carrying (figure 5.17). **Intensity** is defined as the power (in watts) transmitted by a wave to a unit area (in square meters) that is perpendicular to the waves.

Loudness

The *loudness* of a sound is a subjective interpretation that varies from person to person. Loudness is also related to (1) the energy of a vibrating object, (2) the condition of the air the sound wave travels through, and (3) the distance between you and the vibrating source. Furthermore, doubling the amplitude of the vibrating source will quadruple the *intensity* of the resulting sound wave, but the sound will not be perceived as four times as loud. The relationship between perceived loudness and the intensity of a sound wave is not a linear relationship.

In fact, a sound that is perceived as twice as loud requires ten times the intensity, and quadrupling the loudness requires a one-hundredfold increase in intensity.

The human ear is very sensitive, capable of hearing sounds with intensities as low as 10^{-12} W/m², and is not made uncomfortable by sound until the intensity reaches about 1 W/m². The second intensity is a million million (10^{12}) times greater than the first. Within this range, the subjective interpretation of intensity seems to vary by powers of ten. This observation led to the development of the **decibel scale** to measure the intensity level. The scale is a ratio of the intensity level of a given sound to the threshold of hearing, which is defined as 10^{-12} W/m² at 1,000 Hz. In keeping with the power-of-ten subjective interpretations of intensity, a logarithmic scale is used rather than a linear scale. Originally, the scale was the logarithm of the ratio of the intensity level of a sound to the threshold of hearing. This definition set the zero point at the threshold of human hearing. The unit was named the *bel* in honor of Alexander Graham Bell. This unit was too large to be practical, so it was reduced by one-tenth and called a *decibel*. The intensity level of a sound is therefore measured in decibels (table 5.2).

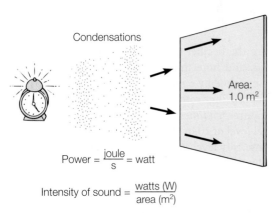

FIGURE 5.17 The intensity of a sound wave is the rate of energy transferred to an area perpendicular to the waves. Intensity is measured in watts per square meter, W/m².

TABLE 5.2   Comparison of Noise Levels in Decibels with Intensity

Example	Response	Decibels	Intensity W/m²
Least needed for hearing	Barely perceived	0	1×10^{-12}
Calm day in woods	Very, very quiet	10	1×10^{-11}
Whisper (15 ft)	Very quiet	20	1×10^{-10}
Library	Quiet	40	1×10^{-8}
Talking	Easy to hear	65	3×10^{-6}
Heavy street traffic	Conversation difficult	70	1×10^{-5}
Pneumatic drill (50 ft)	Very loud	95	3×10^{-3}
Jet plane (200 ft)	Discomfort	120	1

Noise Pollution Solution?

Bells, jet planes, sirens, motorcycles, jackhammers, and construction noises all contribute to the constant racket that has become commonplace in many areas. Such noise pollution is everywhere, and it is a rare place where you can escape the ongoing din. Earplugs help reduce the noise level, but they also block sounds that you want to hear, such as music and human voices. In addition, earplugs are more effective against high-frequency sounds than they are against lower ones such as aircraft engines or wind noise.

A better solution for noise pollution might be the relatively new "antinoise" technology that cancels sound waves before they reach your ears. A microphone detects background noise and transmits information to microprocessors about the noise waveform. The microprocessors then generate an "antinoise" signal that is 180 degrees out of phase with the background noise. When the noise and antinoise meet, they undergo destructive interference and significantly reduce the final sound loudness.

The noise-canceling technology is available today as a consumer portable electronic device with a set of headphones, looking much like a portable tape player. The headphones can be used to combat steady, ongoing background noise such as you might hear from a spinning computer drive, a vacuum cleaner, or while inside the cabin of a jet plane. Some vendors claim their device cancels up to 40 percent of whirring air conditioner noise, 80 percent of ongoing car noise, or up to 95 percent of constant airplane cabin noise. You hear all the other sounds—people talking, warning sounds, and music—as the microprocessors are not yet fast enough to match anything but a constant sound source.

Noise-canceling microphones can also be used to limit background noise that muddles the human voice in teleconferencing or during cellular phone use from a number of high-noise places. With improved microprocessor and new digital applications, noise-canceling technology will help bring higher sound quality to voice-driven applications. It may also find extended practical use in helping to turn down the volume on our noisy world.

Compare the decibel noise level of familiar sounds listed in table 5.2, and note that each increase of 10 on the decibel scale is matched by a *multiple* of 10 on the intensity level. For example, moving from a decibel level of 10 to a decibel level of 20 requires *ten times* more intensity. Likewise, moving from a decibel level of 20 to 40 requires a one-hundredfold increase in the intensity level. As you can see, the decibel scale is not a simple linear scale.

Resonance

You know that sound waves transmit energy when you hear a thunderclap rattle the windows. In fact, the sharp sounds from an explosion have been known not only to rattle but also break windows. The source of the energy is obvious when thunderclaps or explosions are involved. But sometimes energy transfer occurs through sound waves when it is not clear what is happening. A truck drives down the street, for example, and one window rattles but the others do not. A singer shatters a crystal water glass by singing a single note, but other objects remain undisturbed. A closer look at the nature of vibrating objects and the transfer of energy will explain these phenomena.

Almost any elastic object can be made to vibrate and will vibrate freely at a constant frequency after being sufficiently disturbed. Entertainers sometimes discover this fact and appear on late-night talk shows playing saws, wrenches, and other odd objects as musical instruments. All material objects have a **natural frequency** of vibration determined by the materials and shape of the objects. The natural frequencies of different wrenches enable an entertainer to use the suspended tools as if they were the bars of a xylophone.

If you have ever pumped a swing, you know that small forces can be applied at any frequency. If the frequency of the applied forces matches the natural frequency of the moving swing, there is a dramatic increase in amplitude. When the two frequencies match, energy is transferred very efficiently. This condition, when the frequency of an external force matches the natural frequency, is called **resonance**. The natural frequency of an object is thus referred to as the *resonant frequency*, that is, the frequency at which resonance occurs.

A silent tuning fork will resonate if a second tuning fork with the same frequency is struck and vibrates nearby (figure 5.18). You will hear the previously silent tuning fork sounding if you stop the vibrations of the struck fork by touching it. The waves of condensations and rarefactions produced by the struck tuning fork produce a regular series of impulses

Struck tuning fork

Not struck, but vibrating, tuning fork

FIGURE 5.18 When the frequency of an applied force, including the force of a sound wave, matches the natural frequency of an object, energy is transferred very efficiently. The condition is called *resonance.*

CONCEPTS APPLIED

A Singing Glass

Did you ever hear a glass "sing" when the rim is rubbed? The trick to making the glass sing is to remove as much oil from your finger as possible. Then you lightly rub around and on the top of the glass rim at the correct speed. Without oil, your finger will imperceptibly catch on the glass as you rub the rim. With the appropriate pressure and speed, your catching finger might match the natural frequency of the glass. The resonate vibration will cause the glass to "sing" with a high-pitched note.

Laser Bug

Hold a fully inflated balloon lightly between your fingertips and talk. You will be able to feel the slight vibrations from your voice. Likewise, the sound waves from your voice will cause a nearby window to vibrate slightly. If a laser beam is bounced off the window, the reflection will be changed by the vibrations. The incoming laser beam is coherent; all the light has the same frequency and amplitude (see p. 169). The reflected beam, however, will have different frequencies and amplitudes from the window pane vibrating in and out. The changes can be detected by a receiver and converted into sound in a headphone.

You cannot see an infrared laser beam because infrared is outside the frequencies that humans can see. Any sound-sensitive target can be used by the laser bug, including a windowpane, inflated balloon, hanging picture, or the glass front of a china cabinet.

Questions to Discuss

1. Is it legal for someone to listen in on your private conversations?
2. Should the sale of technology such as the laser bug be permitted? What are the issues?

that match the natural frequency of the silent tuning fork. This illustrates that at resonance, relatively little energy is required to start vibrations.

A truck causing vibrations as it is driven past a building may cause one window to rattle while others do not. Vibrations caused by the truck have matched the natural frequency of this window but not the others. The window is undergoing resonance from the sound wave impulses that matched its natural frequency. It is also resonance that enables a singer to break a water glass. If the tone is at the resonant frequency of the glass, the resulting vibrations may be large enough to shatter it.

Resonance considerations are important in engineering. A large water pump, for example, was designed for a nuclear power plant. Vibrations from the electric motor matched the resonant frequency of the impeller blades, and they shattered after a short period of time. The blades were redesigned to have a different natural frequency when the problem was discovered. Resonance vibrations are particularly important in the design of buildings.

SOURCES OF SOUNDS

All sounds have a vibrating object as their source. The vibrations of the object send pulses or waves of condensations and rarefactions through the air. These sound waves have physical properties that can be measured, such as frequency and intensity. Subjectively, your response to frequency is to identify a certain pitch. A high-frequency sound is interpreted as a high-pitched sound, and a low-frequency sound is interpreted as a low-pitched sound. Likewise, a greater intensity is interpreted as increased loudness, but there is not a direct relationship between intensity and loudness as there is between frequency and pitch.

There are other subjective interpretations about sounds. Some sounds are bothersome and irritating to some people but go unnoticed by others. In general, sounds made by brief, irregular vibrations such as those made by a slamming door, dropped book, or sliding chair are called *noise*. Noise is characterized by sound waves with mixed frequencies and jumbled intensities (figure 5.19). On the other hand, there are sounds made by very regular, repeating vibrations such as those made by a tuning fork. A tuning fork produces a *pure tone* with a sinusoidal curved

FIGURE 5.19 Different sounds that you hear include (*A*) noise, (*B*) pure tones, and (*C*) musical notes.

pressure variation and regular frequency. Yet a tuning fork produces a tone that most people interpret as bland. You would not call a tuning fork sound a musical note! Musical sounds from instruments have a certain frequency and loudness, as do noise and pure tones, but you can readily identify the source of the very same musical note made by two different instruments. You recognize it as a musical note, not noise and not a pure tone. You also recognize if the note was produced by a violin or a guitar. The difference is in the wave form of the sounds made by the two instruments, and the difference is called the *sound quality*. How does a musical instrument produce a sound of a characteristic quality? The answer may be found by looking at instruments that make use of vibrating strings.

Vibrating Strings

A stringed musical instrument, such as a guitar, has strings that are stretched between two fixed ends. When a string is plucked, waves of many different frequencies travel back and forth on the string, reflecting from the fixed ends. Many of these waves quickly fade away, but certain frequencies resonate, setting up patterns of waves. Before considering these resonant patterns in detail, keep in mind that (1) two or more waves can be in the same place at the same time, traveling through one another from opposite directions; (2) a confined wave will be reflected at a boundary, and the reflected wave will be inverted (a crest becomes a trough); and (3) reflected waves interfere with incoming waves of the same frequency to produce **standing waves.** Figure 5.20 is a graphic "snapshot" of what happens when reflected wave patterns meet incoming wave patterns. The incoming wave is shown as a solid line, and the reflected wave is shown as a dotted line. The result is (1) places of destructive interference, called *nodes,* which show no disturbance, and (2) loops of constructive interference, called *antinodes,* which take place where the crests and troughs of the two wave patterns produce a disturbance that rapidly alternates upward and downward. This pattern of alternating nodes and antinodes does not move along the string and is thus called a *standing wave.* Note that the standing wave for *one wavelength* will have a node at both ends and in the center, as well as two antinodes. Standing waves occur at the natural, or resonant, frequencies of the string, which are a consequence of the nature of the string, the string length, and the tension in the string. Since the standing waves are resonant vibrations, they continue as all other waves quickly fade away.

Since the two ends of the string are not free to move, the ends of the string will have nodes. The *longest* wave that can

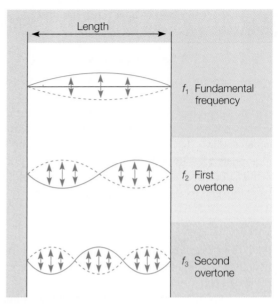

FIGURE 5.21 A stretched string of a given length has a number of possible resonant frequencies. The lowest frequency is the fundamental, f_1; the next higher frequencies, or overtones, shown are f_2 and f_3.

make a standing wave on such a string has a wavelength (λ) that is twice the length (L) of the string. Since frequency (f) is inversely proportional to wavelength ($f = v/\lambda$ from equation 5.3), this longest wavelength has the lowest frequency possible, called the **fundamental frequency.** The fundamental frequency has one antinode, which means that the length of the string has one-half a wavelength. The fundamental frequency (f_1) determines the pitch of the *basic* musical note being sounded and is called the first harmonic. Other resonant frequencies occur at the same time, however, since other standing waves can also fit onto the string. A higher frequency of vibration (f_2) could fit two half-wavelengths between the two fixed nodes. An even higher frequency (f_3) could fit three half-wavelengths between the two fixed nodes (figure 5.21). Any whole number of halves of the wavelength will permit a standing wave to form. The frequencies (f_2, f_3, etc.) of these wavelengths are called the *overtones,* or harmonics. It is the presence and strength of various overtones that give a musical note from a certain instrument its characteristic quality. The fundamental and the overtones add together to produce the characteristic *sound quality,* which is different for the same-pitched note produced by a violin and by a guitar (figure 5.22).

The vibrating string produces a waveform with overtones, so instruments that have vibrating strings are called *harmonic instruments.* Instruments that use an air column as a sound maker are also harmonic instruments. These include all the wind instruments such as the clarinet, flute, trombone, trumpet, pipe organ, and many others. The various wind instruments have different ways of making a column of air vibrate. In the flute, air vibrates as it moves over a sharp edge, while in the clarinet, saxophone, and other reed instruments, it vibrates through fluttering thin reeds. The air column in brass instruments, on the other hand, is vibrated by the tightly fluttering lips of the player.

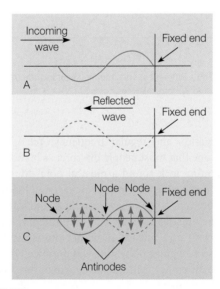

FIGURE 5.20 An incoming wave on a cord with a fixed end (*A*) meets a reflected wave (*B*) with the same amplitude and frequency, producing a standing wave (*C*). Note that a standing wave of one wavelength has three nodes and two antinodes.

People Behind the Science

Johann Christian Doppler (1803–1853)

Johann Doppler was an Austrian physicist who discovered the Doppler effect, which relates the observed frequency of a wave to the relative motion of the source and the observer. The Doppler effect is readily observed in moving sound sources, producing a fall in pitch as the source passes the observer, but it is of most use in astronomy, where it is used to estimate the velocities and distances of distant bodies.

Doppler was born in Salzburg, Austria, on November 29, 1803, the son of a stonemason. He showed early promise in mathematics and attended the Polytechnic Institute in Vienna from 1822 to 1825. Despairing of ever obtaining an academic post, he decided to emigrate to the United States. Then, on the point of departure, he was offered a professorship of mathematics at the State Secondary School in Prague and changed his mind. He subsequently obtained professorships in mathematics at the State Technical Academy in Prague in 1841 and at the Mining Academy in Schemnitz in 1847. Doppler returned to Vienna the following year and, in 1850, became director of the new Physical Institute and professor of experimental physics at the Royal Imperial University of Vienna. He died from a lung disease in Venice on March 17, 1853.

Doppler explained the effect that bears his name by pointing out that sound waves from a source moving toward an observer will reach the observer at a greater frequency than if the source is stationary, thus increasing the observed frequency and raising the pitch of the sound. Similarly, sound waves from a source moving away from the observer reach the observer more slowly, resulting in a decreased frequency and a lowering of pitch. In 1842, Doppler

put forward this explanation and derived the observed frequency mathematically in Doppler's principle.

The first experimental test of Doppler's principle was made in 1845 at Utrecht in Holland. A locomotive was used to carry a group of trumpeters in an open carriage to and fro past some musicians able to sense the pitch of the notes being played. The variation of pitch produced by the motion of the trumpeters verified Doppler's equations.

Doppler correctly suggested that his principle would apply to any wave motion and cited light as an example as well as sound. He believed that all stars emit white light and that differences in color are observed on Earth because the motion of stars affects the observed frequency of the light and hence its color. This idea was not universally true, as stars vary in their basic color. However, Armand Fizeau (1819–1896) pointed out in 1848 that shifts in the spectral lines of stars could be observed and ascribed to the Doppler effect and hence enable their motion to be determined. This idea was first applied in 1868 by William Huggins (1824–1910), who found that Sirius is moving away from the solar system by detecting a small redshift in its spectrum. With the linking of the velocity of a galaxy to its distance by Edwin Hubble (1889–1953) in 1929, it became possible to use the redshift to determine the distances of galaxies. Thus, the principle that Doppler discovered to explain an everyday and inconsequential effect in sound turned out to be of truly cosmological importance.

Source: Modified from the *Hutchinson Dictionary of Scientific Biography*. © RM, 2009. All rights reserved. Helicon Publishing is a division of RM.

The length of the air column determines the frequency, and woodwind instruments have holes in the side of a tube that are opened or closed to change the length of the air column. The resulting tone depends on the length of the air column and the resonate overtones.

Sounds from Moving Sources

When the source of a sound is stationary, equally spaced sound waves expand from a source in all directions. But if the sounding source starts moving, then successive sound waves become displaced in the direction of movement and this changes the pitch. For example, the siren of an approaching ambulance seems to change pitch when the ambulance passes you. The sound wave is "squashed" as the ambulance approaches you and you hear a higher-frequency siren than the people inside the ambulance. When the ambulance passes you, the sound

FIGURE 5.22   A combination of the fundamental and overtone frequencies produces a composite waveform with a characteristic sound quality.

Connections...

Doppler Radar

The Doppler effect was named after the Austrian scientist Christian Doppler, who first demonstrated the effect using sound waves back in 1842. The same principle applies to electromagnetic radiation as well as sound, but now the shifts are in frequency of the radiation. A lower frequency is observed when a source of light is moving away, and this is called a "redshift." Also, a "blueshift" toward a higher frequency occurs when a source of light is moving toward an observer. Radio waves will also experience such shifts of frequency, and weather radar that measures frequency changes as a result of motion is called *Doppler radar*.

Weather radar broadcasts short radio waves from an antenna. When directed at a storm, the waves are reflected back to the antenna by rain, snow, and hail. Reflected radar waves are electronically converted and displayed on a monitor, showing the location and intensity of precipitation. A Doppler radar also measures frequency shifts in the reflected radio waves. Waves from objects moving toward the antenna show a higher frequency, and waves from objects moving away from the antenna show a lower frequency. These shifts of frequency are measured, then displayed as the speed and direction of winds, moving raindrops, and other objects in the storm.

Weather forecasters can direct a Doppler radar machine to measure different elevations of a storm system. This shows highly accurate information that can be used to identify, for example, where and when a tornado might form, as well as the intensity of storm winds in a given area, and even provide an estimate of how much precipitation fell from the storm.

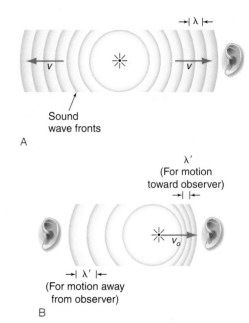

FIGURE 5.23 (*A*) Sound waves emitted by a stationary source and observed by a stationary observer. (*B*) Sound waves emitted by a source in motion toward the right. An observer on the right receives wavelengths that are shortened; an observer on the left receives wavelengths that are lengthened.

waves are "stretched" and you hear a lower-frequency siren. The overall effect of a higher pitch as a source approaches and then a lower pitch as it moves away is called the **Doppler effect.** The Doppler effect is evident if you stand by a street and an approaching car sounds its horn as it drives by you. You will hear a higher-pitched horn as the car approaches, which shifts to a lower-pitched horn as the waves go by you. The driver of the car, however, will hear the continual, true pitch of the horn since the driver is moving with the source (figure 5.23).

A Doppler shift is also noted if the observer is moving and the source of sound is stationary. When the observer moves toward the source, the wave fronts are encountered more frequently than if the observer were standing still. As the observer moves away from the source, the wave fronts are encountered less frequently than if the observer were not moving. An observer on a moving train approaching a crossing with a sounding bell thus hears a high-pitched bell that shifts to a lower-pitched bell as the train whizzes by the crossing.

When an object moves through the air at the speed of sound, it keeps up with its own sound waves. All the successive wave fronts pile up on one another, creating a large wave disturbance called a *shock wave*. The shock wave from a supersonic airplane is a cone-shaped wave of intense condensations trailing backward at an angle dependent on the speed of the aircraft. Wherever this cone of superimposed crests passes, a **sonic boom** occurs (figure 5.24). The many crests have been added together, each contributing to the pressure increase. The human ear cannot differentiate between such a pressure wave

Shock waves are produced when a projectile travels faster than the speed of sound in air.

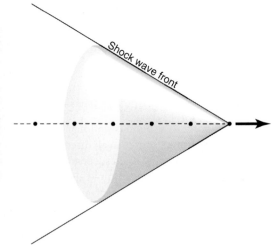

FIGURE 5.24 A sound source moves with velocity *greater* than the speed of sound in the medium. The envelope of spherical wave front forms the conical shock wave.

created by a supersonic aircraft and a pressure wave created by an explosion.

Does a sonic boom occur only when an airplane breaks the sound barrier? The answer is no, an airplane traveling at or faster than the speed of sound produces the shock wave continuously, and a sonic boom will be heard everywhere the plane drags its cone-shaped shock wave. Can you find evidence of shock waves associated with projections on the airplane pictured in figure 5.25?

The Austrian physicist Ernst Mach published a paper in 1877 laying out the principles of supersonics. He also came up with the idea of using a ratio of the velocity of an object to the velocity of sound. Today, this ratio is called the Mach number. A plane traveling at the speed of sound has a Mach number of 1, a plane traveling at twice the speed of sound has a Mach number of 2, and so on. Ernst Mach was also the first to describe what

FIGURE 5.25 A cloud sometimes forms just as a plane accelerates to break the sound barrier. Moist air is believed to form the cloud water droplets as air pressure drops behind the shock wave.

is happening to produce a sonic boom, and he observed the existence of a conical shock wave formed by a projectile as it approached the speed of sound.

SUMMARY

Elastic objects *vibrate,* or move back and forth, in a repeating motion when disturbed by some external force. They are able to do this because they have an *internal restoring force* that returns them to their original positions after being deformed by some external force. If the internal restoring force is opposite to and proportional to the deforming displacement, the vibration is called a *simple harmonic motion.* The extent of displacement is called the *amplitude,* and one complete back-and-forth motion is one *cycle.* The time required for one cycle is a *period.* The *frequency* is the number of cycles per second, and the unit of frequency is the *hertz.* A *graph* of the displacement as a function of time for a simple harmonic motion produces a *sinusoidal* graph.

Periodic, or repeating, vibrations or the *pulse* of a single disturbance can create *waves,* disturbances that carry energy through a medium. A wave that disturbs particles in a back-and-forth motion in the direction of the wave travel is called a *longitudinal wave.* A wave that disturbs particles in a motion perpendicular to the direction of wave travel is called a *transverse wave.* The nature of the medium and the nature of the disturbance determine the type of wave created.

Waves that move through the air are longitudinal and cause a back-and-forth motion of the molecules making up the air. A zone of molecules forced closer together produces a *condensation,* a pulse of increased density and pressure. A zone of reduced density and pressure is a *rarefaction.* A vibrating object produces condensations and rarefactions that expand outward from the source. If the frequency is between 20 Hz and 20,000 Hz, the human ear perceives the waves as *sound* of a certain *pitch.* High frequency is interpreted as high-pitched sound and low frequency as low-pitched sound.

A graph of pressure changes produced by condensations and rarefactions can be used to describe sound waves. The condensations produce *crests,* and the rarefactions produce *troughs.* The *amplitude* is the maximum change of pressure from the normal. The *wavelength* is the distance between any two successive places on a wave train, such as the distance from one crest to the next crest. The *period* is the time required for a wave to repeat itself. The *velocity* of a wave is how quickly a wavelength passes. The *frequency* can be calculated from the *wave equation,* $v = \lambda f$.

Sound waves can move through any medium but not a vacuum. The velocity of sound in a medium depends on the molecular inertia and strength of interactions. Sound, therefore, travels most rapidly through a solid, then a liquid, then a gas. In air, sound has a greater velocity in warmer air than in cooler air because the molecules of air are moving about more rapidly, therefore transmitting a pulse more rapidly.

Sound waves are *reflected* or *refracted* from a *boundary,* which means a change in the transmitting medium. Reflected waves that are *in phase* with incoming waves undergo *constructive interference,* and waves that are *out of phase* undergo *destructive interference.* Two waves that are otherwise alike but with slightly different frequencies produce an alternating increasing and decreasing of loudness called *beats.*

The *energy* of a sound wave is called the wave *intensity,* which is measured in watts per square meter. The intensity of sound is expressed on the *decibel scale,* which relates it to changes in loudness as perceived by the human ear.

All elastic objects have *natural frequencies* of vibration that are determined by the materials they are made of and their shapes. When energy is transferred at the natural frequencies, there is a dramatic increase of amplitude called *resonance.* The natural frequencies are also called *resonant frequencies.*

Sounds are compared by pitch, loudness, and *quality.* The quality is determined by the instrument sounding the note. Each instrument has its own characteristic quality because of the resonant frequencies that it produces. The basic, or *fundamental, frequency* is the longest standing wave that it can make. The fundamental frequency determines the basic note being sounded, and other resonant frequencies, or standing waves called *overtones* or *harmonics,* combine with the fundamental to give the instrument its characteristic quality.

A moving source of sound or a moving observer experiences an apparent shift of frequency called the *Doppler effect*. If the source is moving as fast or faster than the speed of sound, the sound waves pile up into a *shock wave* called a *sonic boom*. A sonic boom sounds very much like the pressure wave from an explosion.

Summary of Equations

5.1

$$\text{period} = \frac{1}{\text{frequency}}$$

$$T = \frac{1}{f}$$

5.2

$$\text{frequency} = \frac{1}{\text{period}}$$

$$f = \frac{1}{T}$$

5.3

$$\text{velocity} = (\text{wavelength})(\text{frequency})$$

$$v = \lambda f$$

5.4

$$\text{beat frequency} = \text{one frequency} - \text{other frequency}$$

$$f_b = f_2 - f_1$$

KEY TERMS

amplitude (p. **105**)
beats (p. **114**)
condensation (p. **108**)
cycle (p. **105**)
decibel scale (p. **115**)
Doppler effect (p. **120**)
echo (p. **113**)

frequency (p. **106**)
fundamental frequency
  (p. **118**)
hertz (p. **106**)
infrasonic (p. **108**)
intensity (p. **115**)
longitudinal wave (p. **107**)

natural frequency (p. **116**)
period (p. **105**)
pitch (p. **109**)
rarefaction (p. **108**)
resonance (p. **116**)
reverberation (p. **113**)
sonic boom (p. **120**)

standing waves (p. **118**)
transverse wave (p. **107**)
ultrasonic (p. **108**)
vibration (p. **104**)
wavelength (p. **110**)

APPLYING THE CONCEPTS

1. The time required for a vibrating object to complete one full cycle is the
   a. frequency.
   b. amplitude.
   c. period.
   d. hertz.

2. The unit of cycles per second is called a
   a. hertz.
   b. lambda.
   c. wave.
   d. watt.

3. The period of a vibrating object is related to the frequency, since they are
   a. directly proportional.
   b. inversely proportional.
   c. frequently proportional.
   d. not proportional.

4. A longitudinal mechanical wave causes particles of a material to move
   a. back and forth in the same direction the wave is moving.
   b. perpendicular to the direction the wave is moving.
   c. in a circular motion in the direction the wave is moving.
   d. in a circular motion opposite the direction the wave is moving.

5. Transverse mechanical waves will move only through
   a. solids.
   b. liquids.
   c. gases.
   d. All of the above are correct.

6. Longitudinal mechanical waves will move only through
   a. solids.
   b. liquids.
   c. gases.
   d. All of the above are correct.

7. The characteristic of a wave that is responsible for what you interpret as pitch is the wave
   a. amplitude.
   b. shape.
   c. frequency.
   d. height.

8. The number of cycles that a vibrating tuning fork experiences each second is related to the resulting sound wave characteristic of
   a. frequency.
   b. amplitude.
   c. wave height.
   d. quality.

Lemon Battery

1. You can make a simple compass galvanometer that will detect a small electric current (box figure 6.5). All you need is a magnetic compass and some thin insulated wire (the thinner the better).

2. Wrap the thin insulated wire in parallel windings around the compass. Make as many parallel windings as you can, but leave enough room to see both ends of the compass needle. Leave the wire ends free for connections.

3. To use the galvanometer, first turn the compass so the needle is parallel to the wire windings. When a current passes through the coil of wire, the magnetic field produced will cause the needle to move from its north-south position, showing the presence of a current. The needle will deflect one way or the other, depending on the direction of the current.

4. Test your galvanometer with a "lemon battery." Roll a soft lemon on a table while pressing on it with the palm of your hand. Cut two slits in the lemon about 1 cm apart. Insert a 8 cm (approximate) copper wire in one slit and a same-sized length of a straightened paper clip in the other slit, making sure the metals do not touch inside the lemon. Connect the galvanometer to the two metals. Try the two metals in other fruits, vegetables, and liquids. Can you find a pattern?

Box Figure 6.5 You can use the materials shown here to create and detect an electric current.

moved away from the coil. The same effect occurs if you move the coil back and forth over a stationary magnet. Furthermore, no current is detected when the magnetic field and the coil of wire are not moving. Thus, electromagnetic induction depends on the relative motion of the magnetic field and the coil of wire. It does not matter which one moves or changes, but one

Swinging Coils

The interactions between moving magnets and moving charges can be easily demonstrated with two large magnets, two coils of wire, and a galvanometer.

1. Make a coil of insulated bell wire (#18 copper wire) by wraping fifty windings around a narrow jar. Tape the coil at several places so it does not come apart.

2. Connect the coil to a galvanometer (see "Concepts Applied: Lemon Battery" to make your own).

3. Move a strong bar magnet into and out of the stationary coil of wire and observe the galvanometer. Note the magnetic pole, direction of movement, and direction of current for both in and out movements.

4. Move the coil of wire back and forth over the stationary bar magnet.

5. Now make a second coil of wire from insulated bell wire and tape the coil as before.

6. Suspend the coil of wire on a ring stand or some other support on a table top. The coil should hang so it will swing with the broad circle of the coil moving back and forth. Place a large magnet on supports so it is near the center of the coil.

7. Set up an identical coil of wire, ring stand or support, and magnet on another table. Connect the two coils of wire.

8. Move one of the coils of wire and observe what happens to the second coil. The second coil should move, mirroring the movements of the first coil. (If it does not move, find some stronger magnets.)

9. Explain what happens at the first coil and the second coil.

FIGURE 6.22 A current is induced in a coil of wire moved through a magnetic field. The direction of the current depends on the direction of motion.

Connections . . .

Current War

Thomas Edison built the first electric generator and electrical distribution system to promote his new long-lasting lightbulbs. The DC generator and distribution system was built in lower Manhattan, New York City, and was switched on September 4, 1882. It supplied 110 volts DC to fifty-nine customers. Edison studied both AC and DC systems and chose DC because of the advantages it offered at the time. Direct current was used because batteries are DC, and batteries were used as a system backup. Also, DC worked fine with electric motors, and AC motors were not yet available.

George Westinghouse was in the business of supplying gas for gas lighting, and he could see that electric lighting would soon be replacing all the gaslights. After studying the matter, he decided that Edison's low-voltage system was not efficient enough. In 1885, he began experimenting with AC generators and transformers in Pittsburgh. In 1886, he installed a 500-volt AC system in Great Barrington, Massachusetts. The system stepped up the voltage to 3,000 volts for transmission, then back down to 110 volts, 60 Hz AC to power lightbulbs in homes and businesses.

Westinghouse's promotion of AC led to direct competition with Edison and his DC electrical systems. A "war of currents" resulted, with Edison claiming that transmission of such high voltage was dangerous. He emphasized this point by recommending the use of high-voltage AC in an electric chair as the best way to execute prisoners.

The advantages of AC were greater since you could increase the voltage, transmit for long distances at a lower cost, and then decrease the voltage to a safe level. Eventually, even Edison's own General Electric company switched to producing AC equipment. Westinghouse turned his attention to the production of large steam turbines for producing AC power and was soon setting up AC distribution systems across the nation.

down the AC voltage. It has two basic parts: (1) a *primary* or "input" coil, and (2) a *secondary* or "output" coil, which is close by. Both coils are often wound on a single iron core but are always fully insulated from each other. When an alternating current flows through the primary coil, a magnetic field grows around the coil to a maximum size, collapses to zero, then grows to a maximum size with an opposite polarity. This happens 120 times a second as the alternating current oscillates at 60 hertz. The growing and collapsing magnetic field moves across the wires in the secondary coil, inducing a voltage in the secondary coil.

The size of the induced voltage in the secondary coil is proportional to the number of wire loops in the two coils. Each turn or loop of the output coil has the same voltage induced in it, so more loops in an output coil means higher voltage output. If the output coil has the same number of loops as the input coil, the induced voltage in the secondary coil will be the same as the voltage in the primary coil. If the secondary coil has one-tenth as many loops as the primary coil, then the induced voltage in the secondary coil will be one-tenth the voltage in the primary coil. This is called a **step-down transformer** since the voltage was stepped down in the secondary coil. On the other hand, if the secondary coil has ten times more loops than the primary coil, then the voltage will be *increased* by a factor of 10. This is a **step-up transformer.** How much the voltage is stepped up or stepped down depends on the ratio of wire loops in the primary and secondary coils (figure 6.23).

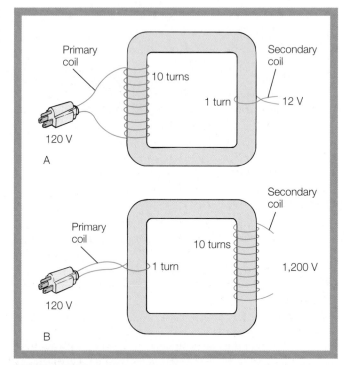

FIGURE 6.23 (*A*) This step-down transformer has ten turns on the primary for each turn on the secondary and reduces the voltage from 120 V to 12 V. (*B*) This step-up transformer increases the voltage from 120 V to 1,200 V, since there are ten turns on the secondary to each turn on the primary.

must move or change relative to the other for electromagnetic induction to occur.

Electromagnetic induction occurs when the loop of wire moves across magnetic field lines or when magnetic field lines move across the loop. The size of the induced voltage is proportional to (1) the number of wire loops passing through the magnetic field lines, (2) the strength of the magnetic field, and (3) the rate at which magnetic field lines pass through the wire.

An **electric generator** is a device that converts mechanical energy into electrical energy. A simple generator is built much like an electric motor, with an axle with many loops in a wire coil that rotates. The coil is turned by some form of mechanical energy, such as a water turbine or a steam turbine, which uses steam generated from fossil fuels or nuclear energy. As the coil rotates in a magnetic field, a current is induced in the coil.

Current from a power plant goes to a transformer to step up the voltage. A **transformer** is a device that steps up or steps

PARALLEL EXERCISES

The exercises in groups A and B cover the same concepts. Solutions to group A exercises are located in appendix D.

Group A

1. A rubber balloon has become negatively charged from being rubbed with a wool cloth, and the charge is measured as 1.00×10^{-14} C. According to this charge, the balloon contains an excess of how many electrons?

2. An electric current through a wire is 6.00 C every 2.00 s. What is the magnitude of this current?

3. A current of 4.00 A flows through a toaster connected to a 120.0 V circuit. What is the resistance of the toaster?

4. What is the current in a 60.0 Ω resistor when the potential difference across it is 120.0 V?

5. A lightbulb with a resistance of 10.0 Ω allows a 1.20 A current to flow when connected to a battery. (a) What is the voltage of the battery? (b) What is the power of the lightbulb?

6. A small radio operates on 3.00 V and has a resistance of 15.0 Ω. At what rate does the radio use electric energy?

7. A 1,200 W hair dryer is operated on a 120 V circuit for 15 min. If electricity costs $0.10/kWh, what was the cost of using the blow dryer?

8. An automobile starter rated at 2.00 hp draws how many amps from a 12.0 V battery?

9. An average-sized home refrigeration unit has a 1/3 hp fan motor for blowing air over the inside cooling coils, a 1/3 hp fan motor for blowing air over the outside condenser coils, and a 3.70 hp compressor motor. (a) All three motors use electric energy at what rate? (b) If electricity costs $0.10/kWh, what is the cost of running the unit per hour? (c) What is the cost for running the unit 12 hours a day for a 30-day month?

10. A 15 ohm toaster is turned on in a circuit that already has a 0.20 hp motor, three 100 W lightbulbs, and a 600 W electric iron that are on. Will this trip a 15 A circuit breaker? Explain.

11. A power plant generator produces a 1,200 V, 40 A alternating current that is fed to a step-up transformer before transmission over the high lines. The transformer has a ratio of 200 to 1 wire loops. (a) What is the voltage of the transmitted power? (b) What is the current?

12. A step-down transformer has an output of 12 V and 0.5 A when connected to a 120 V line. Assuming no losses: (a) What is the ratio of primary to secondary loops? (b) What current does the transformer draw from the line? (c) What is the power output of the transformer?

Group B

1. An inflated rubber balloon is rubbed with a wool cloth until an excess of a billion electrons is on the balloon. What is the magnitude of the charge on the balloon?

2. A wire carries a current of 2.0 A. At what rate is the charge flowing?

3. A current of 0.83 A flows through a lightbulb in a 120 V circuit. What is the resistance of this lightbulb?

4. What is the voltage across a 60.0 Ω resistor with a current of 3⅓ amp?

5. A 10.0 Ω lightbulb is connected to a 12.0 V battery. (a) What current flows through the bulb? (b) What is the power of the bulb?

6. A lightbulb designed to operate in a 120.0 V circuit has a resistance of 192 Ω. At what rate does the bulb use electric energy?

7. What is the monthly energy cost of leaving a 60 W bulb on continuously if electricity costs $0.10 per kWh?

8. An electric motor draws a current of 11.5 A in a 240 V circuit. (a) What is the power of this motor in W? (b) How many horsepower is this?

9. A swimming pool requiring a 2.0 hp motor to filter and circulate the water runs for 18 hours a day. What is the monthly electrical cost for running this pool pump if electricity costs $0.10 per kWh?

10. Is it possible for two people to simultaneously operate 1,300 W hair dryers on the same 120 V circuit without tripping a 15 A circuit breaker? Explain.

11. A step-up transformer has a primary coil with 100 loops and a secondary coil with 1,500 loops. If the primary coil is supplied with a household current of 120 V and 15 A, (a) what voltage is produced in the secondary circuit? (b) What current flows in the secondary circuit?

12. The step-down transformer in a local neighborhood reduces the voltage from a 7,200 V line to 120 V. (a) If there are 125 loops on the secondary, how many are on the primary coil? (b) What current does the transformer draw from the line if the current in the secondary is 36 A? (c) What are the power input and output?

7

Light

This fiber optics bundle carries pulses of light from an infrared laser to carry much more information than could be carried by electrons moving through wires. This is part of a dramatic change underway, a change that will first find a hybrid "optoelectronics" replacing the more familiar "electronics" of electrons and wires.

CORE **CONCEPT**

Light is electromagnetic radiation—energy—that interacts with matter.

OUTLINE

Light that interacts with matter can be reflected, absorbed, or transmitted (p. 156).

Light moving from one transparent material to another undergoes a change of direction called refraction (p. 158).

Sources of Light
Properties of Light
  Light Interacts with Matter
  Reflection
  Refraction
A Closer Look: Optics
  Dispersion and Color

Evidence for Waves
  Interference
  Polarization
A Closer Look: The Rainbow
Evidence for Particles
  Photoelectric Effect
  Quantization of Energy
The Present Theory
A Closer Look: The Compact Disc (CD)
Relativity
  Special Relativity
People Behind the Science: James Clerk Maxwell
  General Relativity
  Relativity Theory Applied

Interference and polarization provide evidence that light is a wave (pp. 164–68).

The photoelectric effect and the quantization of energy provide evidence that light is a particle (pp. 168–70).

Physics

▷ Light is a form of energy (Ch. 3–4).
▷ Light energy can be converted to electricity (Ch. 3).

Chemistry

▷ Light is emitted from an atom when an electron jumps from a higher to a lower orbital (Ch. 8).
▷ Gamma radiation is electromagnetic radiation (Ch. 11).

Astronomy

▷ Light from a star will tell you about the star's temperature and motion (Ch. 12).

Life Science

▷ Sunlight provides energy flow in ecosystems (Ch. 22).

OVERVIEW

You use light and your eyes more than any other sense to learn about your surroundings. All of your other senses—touch, taste, sound, and smell—involve matter, but the most information is provided by light. Yet, light seems more mysterious than matter. You can study matter directly, measuring its dimensions, taking it apart, and putting it together to learn about it. Light, on the other hand, can only be studied indirectly in terms of how it behaves (figure 7.1). Once you understand its behavior, you know everything there is to know about light. Anything else is thinking about what the behavior means.

The behavior of light has stimulated thinking, scientific investigations, and debate for hundreds of years. The investigations and debate have occurred because light cannot be directly observed, which makes the exact nature of light very difficult to pin down. For example, you know that light moves energy from one place to another place. You can feel energy from the Sun as sunlight warms you, and you know that light has carried this energy across millions of miles of empty space. The ability of light to move energy like this could be explained (1) as energy transported by waves, just as sound waves carry energy from a source, or (2) as the kinetic energy of a stream of moving particles, which give up their energy when they strike a surface. The movement of energy from place to place could be explained equally well by a wave model of light or by a particle model of light. When two possibilities exist like this in science, experiments are designed and measurements are made to support one model and reject the other. Light, however, presents a baffling dilemma. Some experiments provide evidence that light consists of waves and not a stream of moving particles. Yet other experiments provide evidence of just the opposite, that light is a stream of particles and not a wave. Evidence for accepting a wave or particle model seems to depend on which experiments are considered.

The purpose of using a model is to make new things understandable in terms of what is already known. When these new things concern light, three models are useful in visualizing separate behaviors. Thus, the electromagnetic wave model will be used to describe how light is created at a source. Another model, a model of light as a ray, a small beam of light, will be used to discuss some common properties of light such as reflection and the refraction, or bending, of light. Finally, properties of light that provide evidence for a particle model will be discussed before ending with a discussion of the present understanding of light.

SOURCES OF LIGHT

The Sun, and other stars, lightbulbs, and burning materials all give off light. When something produces light, it is said to be **luminous.** The Sun is a luminous object that provides almost all of the *natural* light on Earth. A small amount of light does reach Earth from the stars but not really enough to see by on a moonless night. The Moon and planets shine by reflected light and do not produce their own light, so they are not luminous.

Burning has been used as a source of *artificial* light for thousands of years. A wood fire and a candle flame are luminous because of their high temperatures. When visible light is given off as a result of high temperatures, the light source is said to be **incandescent.** A flame from any burning source, an ordinary lightbulb, and the Sun are all incandescent sources because of high temperatures.

How do incandescent objects produce light? One explanation is given by the electromagnetic wave model. This model describes a relationship between electricity, magnetism, and light. The model pictures an electromagnetic wave as forming whenever an electric charge is *accelerated* by some external force. The acceleration produces a wave consisting of electrical

FIGURE 7.2 The electric and magnetic fields in an electromagnetic wave vary together. Here the fields are represented by arrows that indicate the strength and direction of the fields. Note the fields are perpendicular to each other and to the direction of the wave.

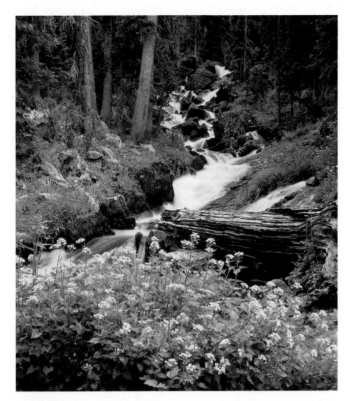

FIGURE 7.1 Light, sounds, and odors can identify the pleasing environment of this garden, but light provides the most information. Sounds and odors can be identified and studied directly, but light can only be studied indirectly, that is, in terms of how it behaves. As a result, the behavior of light has stimulated thinking, scientific investigations, and debate for hundreds of years. Perhaps you have wondered about light and its behaviors. What is light?

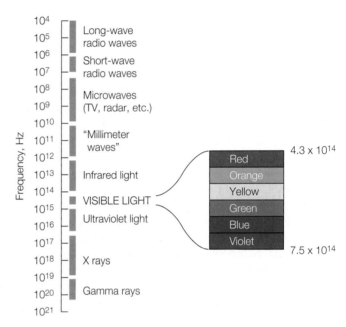

FIGURE 7.3 The electromagnetic spectrum. All electromagnetic waves have the same fundamental character and the same speed in a vacuum, but many aspects of their behavior depend on their frequency.

and magnetic fields that become separated from the accelerated charge, moving off into space (figure 7.2). As the wave moves through space, the two fields exchange energy back and forth, continuing on until they are absorbed by matter and give up their energy.

The frequency of an electromagnetic wave depends on the acceleration of the charge; the greater the acceleration, the higher the frequency of the wave that is produced. The complete range of frequencies is called the *electromagnetic spectrum* (figure 7.3). The spectrum ranges from radio waves at the low frequency end of the spectrum to gamma rays at the high frequency end. Visible light occupies only a small part of the middle portion of the complete spectrum.

Visible light is emitted from incandescent sources at high temperatures, but actually electromagnetic radiation is given off from matter at *any* temperature. This radiation is called **blackbody radiation,** which refers to an idealized material (the *blackbody*) that perfectly absorbs and perfectly emits electromagnetic radiation. From the electromagnetic wave model, the radiation originates from the acceleration of charged particles near the surface of an object. The frequency of the black-body radiation is determined by the energy available for accelerating

charged particles, that is, the temperature of the object. Near absolute zero, there is little energy available and no radiation is given off. As the temperature of an object is increased, more energy is available, and this energy is distributed over a range of values, so more than one frequency of radiation is emitted. A graph of the frequencies emitted from the range of available energy is thus somewhat bell-shaped. The steepness of the curve and the position of the peak depend on the temperature (figure 7.4). As the temperature of an object increases, there is an increase in the *amount* of radiation given off, and the peak radiation emitted progressively *shifts* toward higher and higher frequencies.

At room temperature, the radiation given off from an object is in the infrared region, invisible to the human eye. When the temperature of the object reaches about 700°C (about 1,300°F), the peak radiation is still in the infrared region, but the peak has shifted enough toward the higher frequencies that a little visible light is emitted as a dull red glow. As the

Ultraviolet Light and Life

Ultraviolet (UV) light is a specific portion of the electromagnetic spectrum that has a significant impact on living things. UV light has been subdivided into different categories based on wavelength—the shorter the wavelength, the higher the energy level. Ultraviolet light causes mutations when absorbed by DNA.

▶ Ultraviolet light is important in forming vitamin D in the skin. When cholesterol in the skin is struck by ultraviolet light, some of the cholesterol is converted to vitamin D. Therefore, some exposure to ultraviolet light is good. However, since large amounts of ultraviolet light damage the skin, the normal tanning reaction seen in the skin of light-skinned people is a protective reaction to the damaging effects of ultraviolet light. The darker skin protects the deeper layers of the skin from damage.

▶ Tanning booths have mostly UV-A with a small amount of UV-B. Dosages in tanning booths are usually higher than in sunlight.

▶ Since UV-C disrupts DNA, it can be used to sterilize medical instruments, drinking water, and wastewater. A standard mercury vapor light emits much of its light in the UV-C range, but the lamp is coated to prevent light in that range from leaving the lamp. Specially manufactured lamps without the coating and ultraviolet transparent glass are used to provide UV-C for germicidal purposes.

▶ Some animals are able to see light in the ultraviolet spectrum. In particular, bees do not see the color red but are able to see ultraviolet. When flowers are photographed with filters that allow ultraviolet light to be recorded on the film, flowers often have a different appearance from what we see. The pattern of ultraviolet light assists the bees in finding the source of nectar within the flower.

Class	Wavelength (nanometers)	Characteristics	Biological Effects
UV-A	320–400	Most passes through the ozone layer; can pass through window glass	Causes skin damage and promotes skin cancer
UV-B	280–320	90% blocked by ozone layer; amount is highest at midday	Causes sunburn, skin damage, and skin cancer
UV-C	100–280	Nearly all blocked by ozone layer	Kills bacteria and viruses

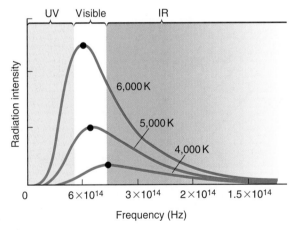

FIGURE 7.4 Blackbody spectra for several different temperatures. The frequency of the peak of the curve (shown by a dot) shifts to higher frequency at higher temperatures.

temperature of the object continues to increase, the amount of radiation increases, and the peak continues to shift toward shorter wavelengths. Thus, the object begins to glow brighter, and the color changes from red, to orange, to yellow, and eventually to white. The association of this color change with temperature is noted in the referent description of an object being "red hot," "white hot," and so forth.

The incandescent flame of a candle or fire results from the blackbody radiation of carbon particles in the flame. At a black-body temperature of 1,500°C (about 2,700°F), the carbon particles emit visible light in the red to yellow frequency range. The tungsten filament of an incandescent lightbulb is heated to about 2,200°C (about 4,000°F) by an electric current. At this temperature, the visible light emitted is in the reddish, yellow-white range.

The radiation from the Sun, or sunlight, comes from the Sun's surface, which has a temperature of about 5,700°C (about 10,000°F). As shown in figure 7.5, the Sun's radiation has a broad spectrum centered near the yellow-green frequency. Your eye is most sensitive to this frequency of sunlight. The spectrum of sunlight before it travels through Earth's atmosphere is infrared (about 51 percent), visible light (about 40 percent), and ultraviolet (about 9 percent). Sunlight originated as energy released from nuclear reactions in the Sun's core (see chapter 12). This energy requires about a million years to work its way up to the surface. At the surface, the energy from the core accelerates charged particles, which then emit light like tiny antennas. The sunlight requires about eight minutes to travel the distance from the Sun's surface to Earth.

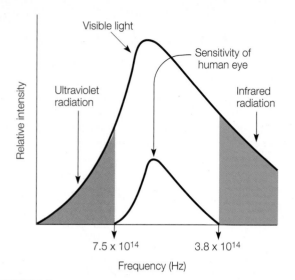

FIGURE 7.5 Sunlight is about 9 percent ultraviolet radiation, 40 percent visible light, and 51 percent infrared radiation before it travels through Earth's atmosphere.

PROPERTIES OF LIGHT

You can see luminous objects from the light they emit, and you can see nonluminous objects from the light they reflect, but you cannot see the path of the light itself. For example, you cannot see a flashlight beam unless you fill the air with chalk dust or smoke. The dust or smoke particles reflect light, revealing the path of the beam. This simple observation must be unknown to the makers of science fiction movies, since they always show visible laser beams zapping through the vacuum of space.

Some way to represent the invisible travels of light is needed in order to discuss some of its properties. Throughout history, a **light ray model** has been used to describe the travels of light. The meaning of this model has changed over time, but it has always been used to suggest that "something" travels in *straight-line paths.* The light ray is a line that is drawn to represent the straight-line travel of light. A line is drawn to represent this imaginary beam to illustrate the law of reflection (as from a mirror) and the law of refraction (as through a lens). There are limits to using a light ray for explaining some properties of light, but it works very well in explaining mirrors, prisms, and lenses.

Light Interacts with Matter

A ray of light travels in a straight line from a source until it encounters some object or particles of matter (figure 7.6). What happens next depends on several factors, including (1) the smoothness of the surface, (2) the nature of the material, and (3) the angle at which the light ray strikes the surface.

The *smoothness* of the surface of an object can range from perfectly smooth to extremely rough. If the surface is perfectly smooth, rays of light undergo *reflection,* leaving the surface parallel to each other. A mirror is a good example of a very smooth surface that reflects light this way (figure 7.7A).

If a surface is not smooth, the light rays are reflected in many random directions as *diffuse reflection* takes place (figure 7.7B). Rough and irregular surfaces and dust in the air make diffuse reflections. It is diffuse reflection that provides light in places not in direct lighting, such as under a table or a tree. Such shaded areas would be very dark without diffuse reflection of light.

Some materials allow much of the light that falls on them to move through the material without being reflected. Materials that allow transmission of light through them are called *transparent.* Glass and clear water are examples of transparent materials. Many materials do not allow transmission of any light and are called *opaque.* Opaque materials reflect light,

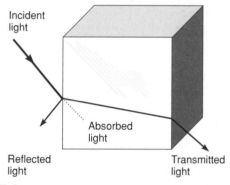

FIGURE 7.6 Light that interacts with matter can be reflected, absorbed, or transmitted through transparent materials. Any combination of these interactions can take place, but a particular substance is usually characterized by what it mostly does to light.

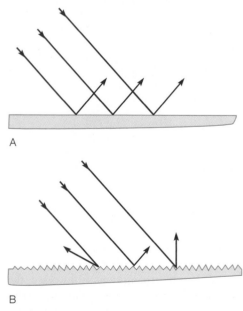

FIGURE 7.7 (A) Rays reflected from a perfectly smooth surface are parallel to each other. (B) Diffuse reflection from a rough surface causes rays to travel in many random directions.

FIGURE 7.8 Light travels in a straight line, and the color of an object depends on which wavelengths of light the object reflects. Each of these flowers absorbs the colors of white light and reflects the color that you see.

absorb light, or some combination of partly absorbing and partly reflecting light (figure 7.8). The light that is reflected varies with wavelength and gives rise to the perception of color, which will be discussed shortly. Absorbed light gives up its energy to the material and may be reemitted at a different wavelength, or it may simply show up as a temperature increase.

The *angle* of the light ray to the surface and the nature of the material determine if the light is absorbed, transmitted through a transparent material, or reflected. Vertical rays of light, for example, are mostly transmitted through a transparent material with some reflection and some absorption. If the rays strike the surface at some angle, however, much more of the light is reflected, bouncing off the surface. Thus, the glare of reflected sunlight is much greater around a body of water in the late afternoon than when the Sun is directly overhead.

Light that interacts with matter is reflected, transmitted, or absorbed, and all combinations of these interactions are possible. Materials are usually characterized by which of these interactions they *mostly* do, but this does not mean that other interactions are not occurring too. For example, a window glass is usually characterized as a transmitter of light. Yet, the glass always *reflects* about 4 percent of the light that strikes it. The reflected light usually goes unnoticed during the day because of the bright light that is transmitted from the outside. When it is dark outside, you notice the reflected light as the window glass now appears to act much like a mirror. A one-way mirror is another example of both reflection and transmission occurring. A mirror is usually characterized as a reflector of light. A one-way mirror, however, has a very thin silvering that reflects most of the light but still transmits a little. In a lighted room, a one-way mirror appears to reflect light just as any other mirror does. But a person behind the mirror in a dark room can see into the lighted room by means of the transmitted light. Thus, you know that this mirror transmits as well as reflects light.

One-way mirrors are used to unobtrusively watch for shoplifters in many businesses.

Reflection

Most of the objects that you see are visible from diffuse reflection. For example, consider some object such as a tree that you see during a bright day. Each *point* on the tree must reflect light in all directions, since you can see any part of the tree from any angle (figure 7.9). As a model, think of bundles of light rays entering your eye, which enable you to see the tree. This means that you can see any part of the tree from any angle because different bundles of reflected rays will enter your eye from different parts of the tree.

Light rays that are diffusely reflected move in all possible directions, but rays that are reflected from a smooth surface, such as a mirror, leave the mirror in a definite direction. Suppose you look at a tree in a mirror. There is only one place on the mirror where you look to see any one part of the tree. Light is reflecting off the mirror from all parts of the tree, but the only rays that reach your eye are the rays that are reflected at a certain angle from the place where you look. The relationship between the light rays moving from the tree and the direction in which they are reflected from the mirror to reach your eyes can be understood by drawing three lines: (1) a line representing an original ray from the tree, called the *incident ray,* (2) a line representing a reflected ray, called the *reflected ray,* and (3) a reference line that is perpendicular to the reflecting surface and is located at the point where the incident ray struck the surface. This line is called the *normal.* The angle between the incident

FIGURE 7.9 Bundles of light rays are reflected diffusely in all directions from every point on an object. Only a few light rays are shown from only one point on the tree in this illustration. The light rays that move to your eyes enable you to see the particular point from which they were reflected.

ray and the normal is called the *angle of incidence,* θ_i, and the angle between the reflected ray and the normal is called the *angle of reflection,* θ_r (figure 7.10). The *law of reflection,* which was known to the ancient Greeks, is that the *angle of incidence equals the angle of reflection,* or

$$\theta_i = \theta_r \qquad \textbf{equation 7.1}$$

Figure 7.11 shows how the law of reflection works when you look at a flat mirror. Light is reflected from all points on the box, and of course only the rays that reach your eyes are detected. These rays are reflected according to the law of reflection, with the angle of reflection equaling the angle of incidence. If you move your head slightly, then a different bundle of rays reaches your eyes. Of all the bundles of rays that reach your eyes, only two rays from a point are shown in the illustration. After these two rays are reflected, they continue to spread apart at the same rate that they were spreading before reflection. Your eyes and brain do not know that the rays have been reflected, and the diverging rays appear to come from behind the mirror, as the dashed lines show. The image, therefore, appears to be the same distance *behind* the mirror as the box is from the front of the mirror. Thus, a mirror image is formed where the rays of

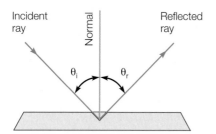

FIGURE 7.10 The law of reflection states that the angle of incidence (θ_i) is equal to the angle of reflection (θ_r). Both angles are measured from the *normal,* a reference line drawn perpendicular to the surface at the point of reflection.

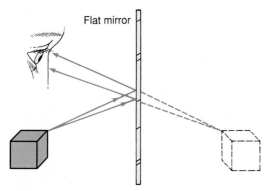

FIGURE 7.11 Light rays leaving a point on the block are reflected according to the law of reflection, and those reaching your eye are seen. After reflecting, the rays continue to spread apart at the same rate. You interpret this to be a block the same distance behind the mirror. You see a virtual image of the block, because light rays do not actually move from the image.

light *appear* to originate. This is called a **virtual image.** A virtual image is the result of your eyes' and brain's interpretations of light rays, not actual light rays originating from an image. Light rays that do originate from the other kind of image are called a **real image.** A real image is like the one displayed on a movie screen, with light originating from the image. A virtual image cannot be displayed on a screen, since it results from an interpretation.

Curved mirrors are either *concave,* with the center part curved inward, or *convex,* with the center part bulging outward. A concave mirror can be used to form an enlarged virtual image, such as a shaving or makeup mirror, or it can be used to form a real image, as in a reflecting telescope. Convex mirrors, for example, the mirrors on the sides of trucks and vans, are often used to increase the field of vision. Convex mirrors are also placed above an aisle in a store to show a wide area.

Refraction

You may have observed that an object that is partly in the air and partly in water appears to be broken, or bent, where the air and water meet. When a light ray moves from one transparent material to another, such as from water through air, the ray undergoes a change in the direction of travel at the boundary between the two materials. This change of direction of a

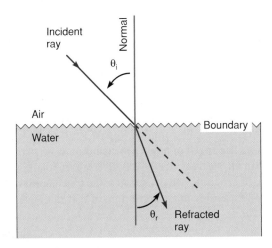

FIGURE 7.12 A ray diagram shows refraction at the boundary as a ray moves from air through water. Note that θ_i does not equal θ_r in refraction.

A

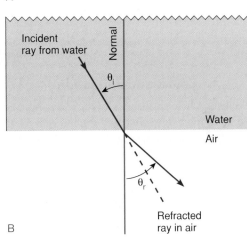

B

FIGURE 7.13 (A) A light ray moving to a new material with a slower speed of light is refracted toward the normal ($\theta_i > \theta_r$). (B) A light ray moving to a new material with a faster speed is refracted away from the normal ($\theta_i < \theta_r$).

light ray at the boundary is called **refraction.** The amount of change can be measured as an angle from the normal, just as it was for the angle of reflection. The incoming ray is called the *incident ray,* as before, and the new direction of travel is called the *refracted ray.* The angles of both rays are measured from the normal (figure 7.12).

Refraction results from a *change in speed* when light passes from one transparent material into another. The speed of light in a vacuum is 3.00×10^8 m/s, but it is slower when moving through a transparent material. In water, for example, the speed of light is reduced to about 2.30×10^8 m/s. The speed of light has a magnitude that is specific for various transparent materials.

When light moves from one transparent material to another transparent material with a *slower* speed of light, the ray is refracted *toward* the normal (figure 7.13A). For example, light travels through air faster than through water. Light traveling from air into water is therefore refracted toward the normal as it enters the water. On the other hand, if light has a *faster* speed in the new material, it is refracted *away* from the normal. Thus, light traveling from water into the air is refracted away from the normal as it enters the air (figure 7.13B).

The magnitude of refraction depends on (1) the angle at which light strikes the surface and (2) the ratio of the speed of light in the two transparent materials. An incident ray that is perpendicular (90°) to the surface is not refracted at all. As the angle of incidence is increased, the angle of refraction is also increased. There is a limit, however, that occurs when the angle of refraction reaches 90°, or along the water surface. Figure 7.14 shows rays of light traveling from water to air at various angles. When the incident ray is about 49°, the angle of refraction that results is 90°, along the water surface. This limit to the angle of incidence that results in an angle of refraction of 90° is called the *critical angle* for a water-to-air surface (figure 7.14). At any incident angle greater than the critical angle, the light ray does not move from the water to the air but

is *reflected* back from the surface as if it were a mirror. This is called **total internal reflection** and implies that the light is trapped inside if it arrived at the critical angle or beyond. Faceted transparent gemstones such as the diamond are brilliant because they have a small critical angle and thus reflect much light internally. Total internal reflection is also important in fiber optics.

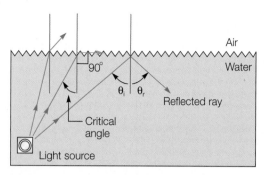

FIGURE 7.14 When the angle of incidence results in an angle of refraction of 90°, the refracted light ray is refracted along the water surface. The angle of incidence for a material that results in an angle of refraction of 90° is called the *critical angle*. When the incident ray is at this critical angle or greater, the ray is reflected internally. The critical angle for water is about 49°, and for a diamond it is about 25°.

EXAMPLE 7.1 *(Optional)*

What is the speed of light in a diamond?

SOLUTION

The relationship between the speed of light in a material (v), the speed of light in a vacuum ($c = 3.00 \times 10^8$ m/s), and the index of refraction is given in equation 7.2. The index of refraction of a diamond is found in table 7.1 ($n = 2.42$).

$$n_{\text{diamond}} = 2.42 \qquad n = \frac{c}{v} \therefore v = \frac{c}{n}$$
$$c = 3.00 \times 10^8 \text{ m/s}$$
$$v = ? \qquad\qquad = \frac{3.00 \times 10^8 \text{ m/s}}{2.42}$$
$$= \boxed{1.24 \times 10^8 \text{ m/s}}$$

As was stated earlier, refraction results from a change in speed when light passes from one transparent material into another. The ratio of the speeds of light in the two materials determines the magnitude of refraction at any given angle of incidence. The greatest speed of light possible, according to current theory, occurs when light is moving through a vacuum. The speed of light in a vacuum is accurately known to nine decimals but is usually rounded to 3.00×10^8 m/s for general discussion. The speed of light in a vacuum is a very important constant in physical science, so it is given a symbol of its own, c. The ratio of c to the speed of light in some transparent material, v, is called the **index of refraction**, n, of that material or

$$n = \frac{c}{v}$$

$\qquad\qquad\qquad\qquad\qquad\qquad\qquad$ **equation 7.2**

The indexes of refraction for some substances are listed in table 7.1. The values listed are constant physical properties and can be used to identify a specific substance. Note that a larger value means a greater refraction at a given angle. Of the materials listed, diamond refracts light the most and air the least. The index for air is nearly 1, which means that light is slowed only slightly in air.

Note that table 7.1 shows that colder air at 0°C (32°F) has a higher index of refraction than warmer air at 30°C (86°F), which means that light travels faster in warmer air. This difference explains the "wet" highway that you sometimes see at a distance in the summer. The air near the road is hotter on a clear, calm day. Light rays traveling toward you in this hotter air are refracted upward as they enter the cooler air. Your brain interprets this refracted light as *reflected* light, but no reflection is taking place. Light traveling downward from other cars is also refracted upward toward you, and you think you are seeing cars "reflected" from the wet highway (figure 7.15).

TABLE 7.1 **Index of Refraction**

Substance	$n = c/v$
Glass	1.50
Diamond	2.42
Ice	1.31
Water	1.33
Benzene	1.50
Carbon tetrachloride	1.46
Ethyl alcohol	1.36
Air (0°C)	1.00029
Air (30°C)	1.00026

FIGURE 7.15 Mirages are caused by hot air near the ground refracting, or bending, light rays upward into the eyes of a distant observer. The observer believes he is seeing an upside-down image reflected from water on the highway.

Historians tell us there are many early stories and legends about the development of ancient optical devices. The first glass vessels were made about 1500 B.C., so it is possible that samples of clear, transparent glass were available soon after. One legend claimed that the ancient Chinese invented eyeglasses as early as 500 B.C. A burning glass (lens) was mentioned in a Greek play written about 424 B.C. Several writers described how Archimedes saved his hometown of Syracuse with a burning glass in about 214 B.C. Syracuse was besieged by Roman ships when Archimedes supposedly used the burning glass to focus sunlight on the ships, setting them on fire. It is not known if this story is true or not, but it is known that the Romans indeed did have burning glasses. Glass spheres, which were probably used to start fires, have been found in Roman ruins, including a convex lens recovered from the ashes of Pompeii.

Today, lenses are no longer needed to start fires, but they are common in cameras, scanners, optical microscopes, eyeglasses, lasers, binoculars, and many other optical devices. Lenses are no longer just made from glass, and today many are made from a transparent, hard plastic that is shaped into a lens.

The primary function of a lens is to form an image of a real object by refracting incoming parallel light rays. Lenses have two basic shapes, with the center of a surface either bulging in or bulging out. The outward bulging shape is thicker at the center than around the outside edge and is called a *convex lens* (box figure 7.1A). The other basic lens shape is just the opposite—thicker around the outside edge than at the center—and is called a *concave lens* (box figure 7.1B).

Convex lenses are used to form images in magnifiers, cameras, eyeglasses, projectors, telescopes, and microscopes (box figure 7.2). Concave lenses are used in some eyeglasses and in combinations with the convex lens to correct for defects. The convex lens is the most commonly used lens shape.

Your eyes are optical devices with convex lenses. Box figure 7.3 shows the basic

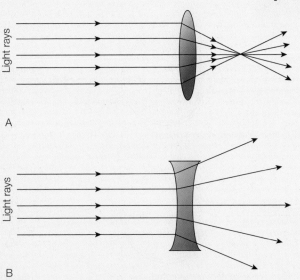

Box Figure 7.1 (*A*) Convex lenses are called converging lenses since they bring together, or converge, parallel rays of light. (*B*) Concave lenses are called diverging lenses since they spread apart, or diverge, parallel rays of light.

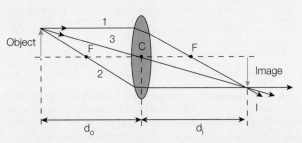

Box Figure 7.2 A convex lens forms an inverted image from refracted light rays of an object outside the focal point. Convex lenses are mostly used to form images in cameras, film or overhead projectors, magnifying glasses, and eyeglasses.

structure. First, a transparent hole called the *pupil* allows light to enter the eye. The size of the pupil is controlled by the *iris,* the colored part that is a muscular diaphragm. The *lens* focuses a sharp image on the back surface of the eye, the *retina.* The retina is made up of millions of light-sensitive structures, and nerves carry signals (impulses) from the retina through the optic nerve to the brain.

The lens is a convex, pliable material held in place and changed in shape by the attached *ciliary muscle.* When the eye is focused on a distant object, the ciliary muscle is completely relaxed. Looking at a closer object requires the contraction of the ciliary muscles to change the curvature of the lens. This adjustment of focus by action of the

Box Figure 7.3 Light rays from a distant object are focused by the lens onto the retina, a small area on the back of the eye.

ciliary muscle is called *accommodation.* The closest distance an object can be seen without a blurred image is called the *near point,* and this is the limit to accommodation.

The near point moves outward with age as the lens becomes less pliable. By

—Continued top of next page

Continued—

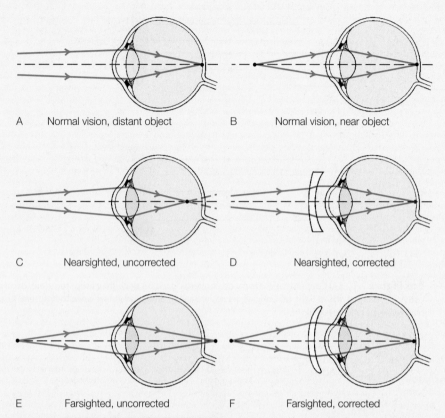

A   Normal vision, distant object

B   Normal vision, near object

C   Nearsighted, uncorrected

D   Nearsighted, corrected

E   Farsighted, uncorrected

F   Farsighted, corrected

Box Figure 7.4 (*A*) The relaxed, normal eye forms an image of distant objects on the retina. (*B*) For close objects, the lens of the normal eye changes shape to focus the image on the retina. (*C*) In a nearsighted eye, the image of a distant object forms in front of the retina. (*D*) A diverging lens corrects for nearsightedness. (*E*) In a farsighted eye, the image of a nearby object forms beyond the retina. (*F*) A converging lens corrects for farsightedness.

Box Figure 7.5 A simple microscope uses a system of two lenses, which are an objective lens that makes an enlarged image of the specimen and an eyepiece lens that makes an enlarged image of that image.

Box Figure 7.6 This illustrates how the path of light moves through a simple reflecting astronomical telescope. Several different designs and placement of mirrors are possible.

middle age, the near point may be twice this distance or greater, creating the condition known as farsightedness. The condition of farsightedness, or *hyperopia,* is a problem associated with aging (called presbyopia). Hyperopia can be caused at an early age by an eye that is too short or by problems with the cornea or lens that focuses the image behind the retina. Farsightedness can be corrected with a convex lens as shown in box figure 7.4A.

Nearsightedness, or *myopia,* is a problem caused by an eye that is too long or problems with the cornea or lens that focus the image in front of the retina. Nearsightedness can be corrected with a concave lens, as shown in box figure 7.4B.

The microscope is an optical device used to make things look larger. It is essentially a system of two lenses, one to produce an image of the object being studied and the other to act as a magnifying glass and enlarge that image. The power of the microscope is basically determined by the *objective lens,* which is placed close to the specimen on the stage of the microscope. Light is projected up through the specimen, and the objective lens makes an enlarged image of the specimen inside the tube between the two lenses. The *eyepiece lens* is positioned so that it makes a sharp enlarged image of the image produced by the objective lens (box figure 7.5).

Telescopes are optical instruments used to provide enlarged images of near and distant objects. There are two major types of telescopes, *refracting* telescopes that use two lenses and *reflecting* telescopes that use combinations of mirrors or a mirror and a lens. The refracting telescope has two lenses, with the objective lens forming a reduced image, which is viewed with an eyepiece lens to enlarge that image. In reflecting telescopes, mirrors are used instead of lenses to collect the light (box figure 7.6).

Finally, the *digital camera* is one of the more recently developed light-gathering and photograph-taking optical instruments. This camera has a group of small photocells, with perhaps thousands lined up on the focal plane behind a converging lens. An image falls on the array, and each photocell stores a charge that is proportional to the amount of light falling on the cell. A microprocessor measures the amount of charge registered by each photocell and considers it as a pixel, a small bit of the overall image. A shade of gray or a color is assigned to each pixel, and the image is ready to be enhanced, transmitted to a screen, printed, or magnetically stored for later use.

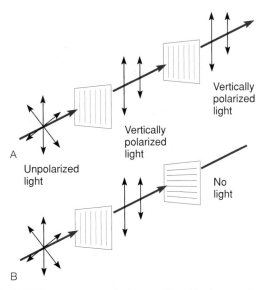

FIGURE 7.18 (A) Unpolarized light has transverse waves vibrating in all possible directions perpendicular to the direction of travel. (B) Polarized light vibrates only in one plane. In this illustration, the wave is vibrating in a vertical direction only.

FIGURE 7.19 (A) Two crystals that are aligned both transmit vertically polarized light that looks like any other light. (B) When the crystals are crossed, no light is transmitted.

little absorption. When the crystals are *crossed* at 90°, the light transmitted by the first is vibrating in a plane that is absorbed by the second crystal, and practically all the light is absorbed. At some other angle, only a fraction of the polarized light from the first crystal is transmitted by the second.

You can verify whether or not a pair of sunglasses is made of polarizing material by rotating a lens of one pair over a lens of a second pair. Light is transmitted when the lenses are aligned but mostly absorbed at 90° when the lenses are crossed.

Light is completely polarized when all the waves are removed except those vibrating in a single direction. Light is partially polarized when some of the waves are in a particular orientation, and any amount of polarization is possible. There are several means of producing partially or completely polarized light, including (1) selective absorption, (2) reflection, and (3) scattering.

Selective absorption is the process that takes place in certain crystals, such as tourmaline, where light in one plane is transmitted and all the other planes are absorbed. A method of manufacturing a polarizing film was developed in the 1930s by Edwin H. Land. The film is called *Polaroid*. Today, Polaroid is made of long chains of hydrocarbon molecules that are aligned in a film. The long-chain molecules ideally absorb all light waves that are parallel to their lengths and transmit light that is perpendicular to their lengths. The direction that is *perpendicular* to the oriented molecular chains is thus called the polarization direction or the *transmission axis*.

Reflected light with an angle of incidence between 1° and 89° is partially polarized as the waves parallel to the reflecting surface are reflected more than other waves. Complete polarization, with all waves parallel to the surface, occurs at a particular angle of incidence. This angle depends on a number of variables, including the nature of the reflecting material. Figure 7.20 illustrates polarization by reflection. Polarizing sunglasses reduce the glare of reflected light because they have vertically oriented transmission axes. This absorbs the

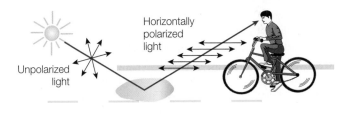

FIGURE 7.20 Light that is reflected becomes partially or fully polarized in a horizontal direction, depending on the incident angle and other variables.

Connections . . .

Lasers

A laser is a device that produces a coherent beam of single-frequency, in-phase light. The beam comes from atoms that have been stimulated by electricity. Most ordinary light sources produce incoherent light; light that is emitted randomly and at different frequencies. The coherent light from a laser has the same frequency, phase, and direction, so it does not tend to spread out and can be very intense (box figure 7.8). This has made possible a number of specialized applications, and the list of uses continues to grow. The word *laser* is from *l*ight *a*mplification by *s*timulated *e*mission of *r*adiation.

There are different kinds of lasers in use and new ones are under development. One common type of laser is a gas-filled tube with mirrors at both ends. The mirror at one end is only partly silvered, which allows light to escape as the laser beam. The distance between the mirrors matches the resonate frequency of the light produced, so the trapped light will set up an optical standing wave. An electric discharge produces fast electrons that raise the energy level of the electrons of the specific gas atoms in the tube. The electrons of the energized gas atoms emit a particular

Ordinary light contains many frequencies

Single-frequency incoherent light

Single-frequency coherent light

Box Figure 7.8   A laser produces a beam of light whose waves all have the same frequency and are in step with one another (coherent). The beam is also very narrow and spreads out very little even over long distances.

frequency of light as they drop back to the original level, and this emitted light sets up the standing wave. The standing wave stimulates other atoms of the gas, resulting in the emission of more light at the same frequency and phase.

Lasers are everywhere today and have connections with a wide variety of technologies. At the supermarket, a laser and detector unit reads the bar code on each grocery item. The laser sends the pattern to a computer, which sends a price to the

register while also tracking the store inventory. A low-powered laser and detector also reads your CD music disk or MP3 disk and can be used to make a three-dimensional image. Most laser printers use a laser, and a laser makes the operational part of a fiber optics communication system. Stronger lasers are used for cutting, drilling, and welding. Lasers are used extensively in many different medical procedures, from welding a detached retina to bloodless surgery.

horizontally oriented reflected light. If you turn your head from side to side so as to rotate your sunglasses while looking at a reflected glare, you will see the intensity of the reflected light change. This means that the reflected light is partially polarized.

The phenomenon called *scattering* occurs when light is absorbed and reradiated by particles about the size of gas molecules that make up the air. Sunlight is initially unpolarized. When it strikes a molecule, electrons are accelerated and vibrate horizontally and vertically. The vibrating charges reradiate polarized light. Thus, if you look at the blue sky with a pair of polarizing sunglasses and rotate them, you will observe that light from the sky is polarized. Bees are believed to be able to detect polarized skylight and use it to orient the direction of their flights. Violet and blue light have the shortest wavelengths of visible light, and red and orange light have the longest. The violet and blue rays of sunlight are scattered the most. At sunset, the path of sunlight through the atmosphere is much longer than when the Sun is more directly overhead. Much of the blue and violet have been scattered away as a result of the longer path through the atmosphere at sunset. The remaining

light that comes through is mostly red and orange, so these are the colors you see at sunset.

EVIDENCE FOR PARTICLES

In 1850, J. L. Foucault was able to prove that light travels much slower in transparent materials than it does in air. This was in complete agreement with the wave theory and completely opposed to the particle theory. By the end of the 1800s, James Maxwell's theoretical concept of electric and magnetic fields changed the concept of light from mechanical waves to waves of changing electric and magnetic fields. Further evidence removed the necessity for ether, the material supposedly needed for waves to move through. Light was now seen as electromagnetic waves that could move through empty space. By this time, it was possible to explain all behaviors of light moving through empty space or through matter with a wave theory. Yet, there were nagging problems that the wave theory could not explain. In general, these problems concerned light that is absorbed by or emitted from matter.

Photoelectric Effect

Light is a form of energy, and it gives its energy to matter when it is absorbed. Usually, the energy of absorbed light results in a temperature increase, such as the warmth you feel from absorbed sunlight. Sometimes, however, the energy from absorbed light results in other effects. In some materials, the energy is acquired by electrons, and some of the electrons acquire sufficient energy to jump out of the material. The movement of electrons as a result of energy acquired from light is known as the **photoelectric effect.** The photoelectric effect is put to a practical use in a solar cell, which transforms the energy of light into an electric current (figure 7.21).

The energy of light can be measured with great accuracy. The kinetic energy of electrons after they absorb light can also be measured with great accuracy. When these measurements were made of the light and electrons involved in the photoelectric effect, some unexpected results were observed. Monochromatic light, that is, light of a single, fixed frequency, was used to produce the photoelectric effect. First, a low-intensity, or dim, light was used, and the numbers and energy of the ejected electrons were measured. Then a high-intensity light was used, and the numbers and energy of the ejected electrons were again measured. Measurement showed that (1) low-intensity light caused fewer electrons to be ejected, and high-intensity light caused many to be ejected, and (2) all electrons ejected from low- or high-intensity light ideally had the *same* kinetic energy. Surprisingly, the kinetic energy of the ejected electrons was found to be *independent* of the light intensity. This was contrary to what the wave theory of light would predict, since a stronger light should mean that waves with more energy have more energy to give to the electrons. Here is a behavior involving light that the wave theory could not explain.

Quantization of Energy

In addition to the problem of the photoelectric effect, there were problems with blackbody radiation, light emitted from hot objects. The experimental measurements of light emitted through blackbody radiation did not match predictions made from theory. In 1900, Max Planck (pronounced *plonk*), a German physicist, found that he could fit the experimental measurements and theory together by assuming that the vibrating molecules that emitted the light could only have a *discrete amount* of energy. Instead of energy existing through a continuous range of amounts, Planck found that the vibrating molecules could only have energy in multiples of energy in certain amounts, or **quanta** (meaning "discrete amounts"; *quantum* is singular and *quanta*, plural).

Planck's discovery of quantized energy states was a radical, revolutionary development, and most scientists, including Planck, did not believe it at the time. Planck, in fact, spent considerable time and effort trying to disprove his own discovery. It was, however, the beginning of the quantum theory, which was eventually to revolutionize physics.

Five years later, in 1905, Albert Einstein applied Planck's quantum concept to the problem of the photoelectric effect. Einstein described the energy in a light wave as quanta of energy called **photons.** Each photon has an energy E that is related to the frequency f of the light through Planck's constant h, or

$$E = hf \qquad \text{equation 7.4}$$

The value of Planck's constant is 6.63×10^{-34} J·s. This relationship says that higher-frequency light (e.g., blue light at 6.50×10^{14} Hz) has more energy than lower-frequency light (e.g., red light at 4.00×10^{14} Hz). The energy of such high- and low-frequency light can be verified by experiment.

The photon theory also explained the photoelectric effect. According to this theory, light is a stream of moving photons. It is the number of photons in this stream that determines if the light is dim or intense. A high-intensity light has many, many photons, and a low-intensity light has only a few photons. At any particular fixed frequency, all the photons would have the same energy, the product of the frequency and Planck's constant (hf). When a photon interacts with matter, it is absorbed and gives up all of its energy. In the photoelectric effect, this interaction takes place between photons and electrons. When an intense light is used, there are more photons to interact with the electrons, so more electrons are ejected. The energy given up by each photon is a function of the frequency of the light, so at a fixed frequency, the energy of each photon, *hf,* is the same, and the acquired kinetic energy of each ejected electron is the same. Thus, the photon theory explains the measured experimental results of the photoelectric effect.

Light

FIGURE 7.21 A setup for observing the photoelectric effect. Light strikes the negatively charged plate, and electrons are ejected. The ejected electrons move to the positively charged plate and can be measured as a current in the circuit.

EXAMPLE 7.3   (Optional)

What is the energy of a photon of red light with a frequency of 4.00×10^{14} Hz?

SOLUTION

The relationship between the energy of a photon (E) and its frequency (f) is found in equation 7.4. Planck's constant is given as 6.63×10^{-34} J·s.

$$f = 4.00 \times 10^{14} \text{ Hz} \qquad E = hf$$
$$h = 6.63 \times 10^{-34} \text{ J·s}$$
$$E = ? \qquad\qquad = (6.63 \times 10^{-34} \text{ J·s})\left(4.00 \times 10^{14} \frac{1}{\text{s}}\right)$$

$$= (6.63 \times 10^{-34})(4.00 \times 10^{14}) \text{J·s} \times \frac{1}{\text{s}}$$

$$= 2.65 \times 10^{-19} \frac{\text{J·s}}{\text{s}}$$

$$= \boxed{2.65 \times 10^{-19} \text{ J}}$$

EXAMPLE 7.4   (Optional)

What is the energy of a photon of violet light with a frequency of 7.00×10^{14} Hz? (Answer: 4.64×10^{-19} J)

The photoelectric effect is explained by considering light to be photons with quanta of energy, not a wave of continuous energy. This is not the only evidence about the quantum nature of light, and more will be presented in chapter 8. But, as you can see, there is a dilemma. The electromagnetic wave theory and the photon theory seem incompatible. Some experiments cannot be explained by the wave theory and seem to support the photon theory. Other experiments are contradictions, providing seemingly equal evidence to reject the photon theory in support of the wave theory.

THE PRESENT THEORY

Today, light is considered to have a dual nature, sometimes acting like a wave and sometimes acting like a particle. A wave model is useful in explaining how light travels through space and how it exhibits such behaviors as refraction, interference, and diffraction. A particle model is useful in explaining how light is emitted from and absorbed by matter, exhibiting such behaviors as blackbody radiation and the photoelectric effect. Together, both of these models are part of a single theory of light, a theory that pictures light as having both particle and wave properties. Some properties are more useful when explaining some observed behaviors, and other properties are more useful when explaining other behaviors.

Frequency is a property of a wave, and the energy of a photon is a property of a particle. Both frequency and the energy of a photon are related in equation 7.4, $E = hf$. It is thus possible to describe light in terms of a frequency (or wavelength) or in terms of a quantity of energy. Any part of the electromagnetic spectrum can thus be described by units of frequency, wavelength, or energy, which are alternative means of describing light. The radio radiation parts of the spectrum are low-frequency, low-energy, and long-wavelength radiations. Radio radiations have more wave properties and practically no particle properties, since the energy levels are low. Gamma radiation, on the other hand, is high-frequency, high-energy, and short wavelength radiation. Gamma radiation has more particle properties, since the extremely short wavelengths have very high energy levels. The more familiar part of the spectrum, visible light, is between these two extremes and exhibits both wave and particle properties, but it never exhibits both properties at the same time in the same experiment.

Part of the problem in forming a concept or mental image of the exact nature of light is understanding this nature in terms of what is already known. The things you already know about are observed to be particles, or objects, or they are observed to be waves. You can see objects that move through the air, such as baseballs or footballs, and you can see waves on water or in a field of grass. There is nothing that acts like a moving object in some situations but acts like a wave in other situations. Objects are objects, and waves are waves, but objects do not become waves, and waves do not become objects. If this dual nature did exist, it would seem very strange. Imagine, for example, holding a book at a certain height above a lake (figure 7.22). You can make measurements and calculate the kinetic energy the book will have when dropped into the lake. When it hits the water, the book disappears, and water waves move away from the point of impact in a circular pattern that moves across the water. When the waves reach another person across the lake, a book identical to the one you dropped pops up out of the water as the waves disappear. As it leaves the water across the lake, the book has the same kinetic energy that your book had when it hit the water in front of you. You and the other person could measure things about either book, and you could measure things about the waves, but you could not measure both at the same time. You might say that this

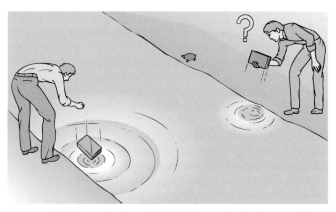

FIGURE 7.22   It would seem very strange if there were not a sharp distinction between objects and waves in our everyday world. Yet this appears to be the nature of light.

The Compact Disc (CD)

A compact disc (CD) is a laser-read (also called *optically read*) data storage device. There are a number of different formats in use today, including music CDs, DVD movies, Blu-Ray DVD, and CDs for storing computer data. All of these utilize the general working principles described below, but some have different refinements. Some, for example, can fit much more data on a disc by utilizing smaller recording tracks.

The CD disc rotates between 200 and 500 revolutions per minute, but the drive changes speed to move the head at a constant linear velocity over the recording track, faster near the inner hub and slower near the outer edge of the disc. Furthermore, the drive reads from the inside out, so the disc will slow as it is played.

The CD disc is a 12 cm diameter, 1.3 mm thick sandwich of a hard plastic core, a mirrorlike layer of metallic aluminum, and a tough, clear plastic overcoating that protects the thin layer of aluminum. The CD records digitized data: music, video, or computer data that have been converted into a string of binary numbers. First, a master disc is made. The binary numbers are translated into a series of pulses that are fed to a laser. The laser is focused onto a photosensitive material on a spinning master disc. Whenever there is a pulse in the signal, the laser burns a small oval pit into the surface, making a pattern of pits and bumps on the track of the master disc. The laser beam is incredibly small, making marks about a micron or so in diameter. A micron is one-millionth of a meter, so you can fit a tremendous number of data tracks onto the disc, which has each track spaced 1.6 microns apart. Next, commercial CD discs are made by using the master disc as a mold. Soft plastic is pressed against the master disc in a vacuum-forming machine so the small physical marks—the pits and bumps made by the laser—are pressed into the plastic. This makes a record of the strings of binary numbers that were etched into the master disc by the strong but tiny laser beam. During playback, a low-powered laser beam is reflected off the track to read the binary marks on it. The optical sensor head contains a tiny diode laser, a lens, mirrors, and tracking devices that can move the head in three directions. The head moves side to side to keep the head over a single track (within 1.6 micron), it moves up and down to keep the laser beam in focus, and it moves forward and backward as a fine adjustment to maintain a constant linear velocity.

The disadvantage of the commercial CD is the lack of ability to do writing or rewriting. Writing and rewritable optical media are available, and these are called CD-R and CD-RW.

A CD-R records data to a disc by using a laser to burn spots into an organic dye. Such a "burned" spot reflects less light than an area that was not heated by the laser. This is designed to mimic the way light reflects from pits and bumps of a commercial CD, except this time the string of binary numbers are burned (nonreflective) and not burned areas (reflective). Since this is similar to how data on a commercial CD is represented, a CD-R disc can generally be used in a CD player as if it were a commercial CD. The dyes in a CD-R disc are photosensitive organic compounds that are similar to those used in making photographs. The color of a CD-R disc is a result of the kind of dye that was used in the recording layer combined with the type of reflective coating used. Some of these dye and reflective coating combinations appear green, some appear blue, and others appear to be gold. Once a CD-R disc is burned, it cannot be rewritten or changed.

The CD-RW is designed to have the ability to do writing or rewriting. It uses a different technology but again mimics the way light reflects from the pits and bumps of a pressed commercial CD. Instead of a dye-based recording layer, the CD-RW uses a compound made from silver, indium, antimony, and tellurium. This layer has a property that permits rewriting the information on a disc. The nature of this property is that when it is heated to a certain temperature and cooled, it becomes crystalline. However, when it is heated to a higher temperature and cooled, it becomes noncrystalline. A crystalline surface reflects a laser beam while a noncrystalline surface absorbs the laser beam. The CD-RW is again designed to mimic the way light reflects from the pits and bumps of a commercial CD, except this time the string of binary numbers are noncrystalline (nonreflective) and crystalline areas (reflective). In order to write, erase, and read, the CD-RW recorder must have three different laser powers. It must have (1) a high power to heat spots to about 600°C, which cool rapidly and make noncrystalline spots that are less reflective. It must have (2) a medium power to erase data by heating the media to about 200°C, which allows the media to crystallize and have a uniform reflectivity. Finally, it must have (3) a low setting that is used for finding and reading nonreflective and the more reflective areas of a disc. The writing and rewriting of a CD-RW can be repeated hundreds of times.

---

behavior is not only strange but impossible. Yet, it is an analogy to the observed behavior of light.

As stated, light has a dual nature, sometimes exhibiting the properties of a wave and sometimes exhibiting the properties of moving particles but never exhibiting both properties at the same time. Both the wave and the particle nature are accepted as being part of one model today, with the understanding that the exact nature of light is not describable in terms of anything that is known to exist in the everyday-sized world. Light is an extremely small-scale phenomenon that must be different, without a sharp distinction between a particle and a wave. Evidence about this strange nature of an extremely small-scale phenomenon will be considered again in chapter 8 as a basis for introducing the quantum theory of matter.

RELATIVITY

The electromagnetic wave model brought together and explained electric and magnetic phenomena, and explained that light can be thought of as an electromagnetic wave (see figure 7.2). There remained questions, however, that would not be answered until Albert Einstein developed a revolutionary new theory. Even at the age of seventeen, Einstein was already thinking about ideas that would eventually lead to his new theory. For example, he wondered about chasing a beam of light if you were also moving at the speed of light. Would you see the light as an oscillating electric and magnetic field at rest? He realized there was no such thing, either on the basis of experience or according to Maxwell's theory of electromagnetic waves.

In 1905, at the age of twenty-six, Albert Einstein published an analysis of how space and time are affected by motion between an observer and what is being measured. This analysis is called the *special theory of relativity*. Eleven years later, Einstein published an interpretation of gravity as distortion of the structure of space and time. This analysis is called the *general theory of relativity*. A number of remarkable predictions have been made based on this theory, and all have been verified by many experiments.

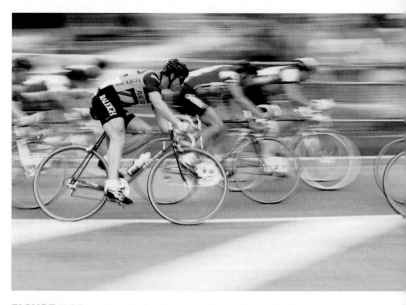

FIGURE 7.23   All motion is relative to a chosen frame of reference. Here the photographer has turned the camera to keep pace with one of the cyclists. Relative to him, both the road and the other cyclists are moving. There is no fixed frame of reference in nature and, therefore, no such thing as "absolute motion"; all motion is relative.

Special Relativity

The **special theory of relativity** is concerned with events as observed from different points of view, or different "reference frames." Here is an example: You are on a bus traveling straight down a highway at a constant 100 km/h. An insect is observed to fly from the back of the bus at 5 km/h. With respect to the bus, the insect is flying at 5 km/h. To someone observing from the ground, however, the speed of the insect is 100 km/h plus 5 km/h, or 105 km/h. If the insect is flying toward the back of the bus, its speed is 100 km/h minus 5 km/h, or 95 km/h with respect to Earth. Generally, the reference frame is understood to be Earth, but this is not always stated. Nonetheless, we must specify a reference frame whenever a speed or velocity is measured.

Einstein's special theory is based on two principles. The first concerns frames of reference and the fact that all motion is relative to a chosen frame of reference. This principle could be called the **consistent law principle:**

> **The laws of physics are the same in all reference frames that are moving at a constant velocity with respect to each other.**

Ignoring vibrations, if you are in a windowless bus, you will not be able to tell if the bus is moving uniformly or if it is not moving at all. If you were to drop something—say, your keys—in a moving bus, they would fall straight down, just as they would in a stationary bus. The keys fall straight down with respect to the bus in either case. To an observer outside the bus, in a different frame of reference, the keys would appear to take a curved path because they have an initial velocity. Moving objects follow the same laws in a uniformly moving bus or any other uniformly moving frame of reference (figure 7.23).

The second principle concerns the speed of light and could be called the **constancy of speed principle:**

> **The speed of light in empty space has the same value for all observers regardless of their velocity.**

The speed of light in empty space is 3.00×10^8 m/s (186,000 mi/s). An observer traveling toward a source would measure the speed of light in empty space as 3.00×10^8 m/s. An observer not moving with respect to the source would measure this very same speed. This is not like the insect moving in a bus—you do not add or subtract the velocity of the source from the velocity of light. The velocity is always 3.00×10^8 m/s for all observers, regardless of the velocity of the observers and regardless of the velocity of the source of light. Light behaves differently than anything in our everyday experience.

The special theory of relativity is based solely on the consistent law principle and the constancy of speed principle. Together, these principles result in some very interesting results if you compare measurements from the ground of the length, time, and mass of a very fast airplane with measurements made by someone moving with the airplane. You, on the ground, would find that

- The length of an object is shorter when it is moving.
- Moving clocks run more slowly.
- Moving objects have increased mass.

James Clerk Maxwell (1831–1879)

James Maxwell was a British physicist who discovered that light consists of electromagnetic waves and established the kinetic theory of gases. He also proved the nature of Saturn's rings and demonstrated the principles governing color vision.

Maxwell was born at Edinburgh, Scotland, on November 13, 1831. He was educated at Edinburgh Academy from 1841 to 1847, then entered the University of Edinburgh. He then went on to study at Cambridge University in 1850, graduating in 1854.

He became professor of natural philosophy at Marischal College, Aberdeen, in 1856 and moved to London in 1860 to take up the post of professor of natural philosophy and astronomy at King's College. On the death of his father in 1865, Maxwell returned to his family home in Scotland and devoted himself to research. However, in 1871, he was persuaded to move to Cambridge, where he became the first professor of experimental physics and set up the Cavendish Laboratory, which opened in 1874.

Maxwell continued in this position until 1879, when he contracted cancer. He died at Cambridge on November 5, 1879, at the age of 48.

Maxwell's development of the electromagnetic theory of light took many years. It began with the paper *On Faraday's Lines of Force*, in which Maxwell built on the views of Michael Faraday (1791–1867) that electric and magnetic effects result from fields of lines of force that surround conductors and magnets. Maxwell drew an analogy between the behavior of the lines of force and the flow of an incompressible liquid, thereby deriving equations that represented known electric and magnetic effects. The next step toward the electromagnetic theory took place with the publication of the paper *On Physical Lines of Force* (1861–1862). In it, Maxwell developed a model for the medium in which electric and magnetic effects could occur. In *A Dynamical Theory of the Electromagnetic Field* (1864), Maxwell developed the fundamental equations that describe the electromagnetic field. These showed that light is propagated in two waves, one magnetic and the other electric, which vibrate perpendicular to each other and to the direction of propagation. This was confirmed in Maxwell's *Note on the Electromagnetic Theory of Light* (1868), which used an electrical derivation of the theory instead of the dynamical formulation, and Maxwell's whole work on the subject was summed up in *Treatise on Electricity and Magnetism* in 1873.

The treatise also established that light has a radiation pressure and suggested that a whole family of electromagnetic radiations must exist, of which light was only one. This was confirmed in 1888 with the sensational discovery of radio waves by Heinrich Hertz (1857–1894). Sadly, Maxwell did not live long enough to see this triumphant vindication of his work.

Maxwell is generally considered to be the greatest theoretical physicist of the 1800s, as his forebear Michael Faraday was the greatest experimental physicist. His rigorous mathematical ability was combined with great insight to enable him to achieve brilliant syntheses of knowledge in the two most important areas of physics at that time. In building on Faraday's work to discover the electromagnetic nature of light, Maxwell not only explained electromagnetism but also paved the way for the discovery and application of the whole spectrum of electromagnetic radiation that has characterized modern physics.

The special theory of relativity shows that measurements of length, time, and mass are different in different moving reference frames. Einstein developed equations that describe each of the changes described above. These changes have been verified countless times with elementary particle experiments, and the data always fit Einstein's equations with predicted results.

General Relativity

Einstein's **general theory of relativity** could also be called Einstein's geometric theory of gravity. According to Einstein, a gravitational interaction does not come from some mysterious force called gravity. Instead, the interaction is between a mass and the geometry of space and time where the mass is located. Space and time can be combined into a fourth-dimensional "spacetime" structure. A mass is understood to interact with the spacetime, telling it how to curve. Spacetime also interacts with a mass, telling it how to move. A gravitational interaction is considered to be a local event of movement along a geodesic (shortest distance between two points on a curved surface) in curved spacetime (figure 7.24). This different viewpoint has led to much more accurate measurements and has been tested by many events in astronomy (see page 297 for one example).

Relativity Theory Applied

You use the Global Positioning System (GPS) when you use the locator in your cell phone or trace your route on a car navigation system. The GPS consists of a worldwide network of 24 satellites, each with an atomic clock that keeps accurate time to within three nanoseconds (0.000000003 of a second). The satellites broadcast a radio signal with the position and time of transmission. A receiver on the surface of Earth, for example your navigation system, uses signals from four satellites to determine location, speed, and time. A computer chip in the receiver uses such data from the satellites to calculate latitude and longitude, which can be used in a mapping application.

GPS satellites move with high velocity at a high altitude above the surface. This results in a combination of errors from the satellite velocity (special relativity error) and from the high location in Earth's gravitational field (general relativity error). Clocks moving with high velocity run slower compared to an identical clock on Earth's surface. For a satellite moving at 14,000 km/h this amounts to a slowing of 7,200 nanoseconds/day.

Clocks located at a higher altitude run faster than an identical clock on Earth's surface. For a satellite at 26,000 km this amounts to running fast by 45,900 nanoseconds/day. Combining special and general relativity errors results in a GPS clock

FIGURE 7.24 General relativity pictures gravity as a warping of the structure of space and time due to the presence of a body of matter. An object nearby experiences an attractive force as a result of this distortion in spacetime, much as a marble rolls toward the bottom of a saucer-shaped hole in the ground.

running fast by 38,700 nanoseconds/day. This would result in a position error of more than 10 km/day. GPS satellite clocks correct relativity errors by adjusting the rate so the fast moving, high altitude clocks tick at the same rate as an identical clock on Earth's surface.

SUMMARY

Electromagnetic radiation is emitted from all matter with a temperature above absolute zero, and as the temperature increases, more radiation and shorter wavelengths are emitted. Visible light is emitted from matter hotter than about 700°C, and this matter is said to be *incandescent.* The Sun, a fire, and the ordinary lightbulb are incandescent sources of light.

The behavior of light is shown by a light ray model that uses straight lines to show the straight-line path of light. Light that interacts with matter is *reflected* with parallel rays, moves in random directions by *diffuse reflection* from points, or is *absorbed,* resulting in a temperature increase. Matter is *opaque,* reflecting light, or *transparent,* transmitting light.

In reflection, the incoming light, or *incident ray,* has the same angle as the *reflected ray* when measured from a perpendicular from the point of reflection, called the *normal.* That the two angles are equal is called the *law of reflection.* The law of reflection explains how a flat mirror forms a *virtual image,* one from which light rays do not originate. Light rays do originate from the other kind of image, a *real image.*

Light rays are bent, or *refracted,* at the boundary when passing from one transparent media to another. The amount of refraction depends on the *incident angle* and the *index of refraction,* a ratio of the speed of light in a vacuum to the speed of light in the media. When the refracted angle is 90°, *total internal reflection* takes place. This limit to the angle of incidence is called the *critical angle,* and all light rays with an incident angle at or beyond this angle are reflected internally.

Each color of light has a range of wavelengths that forms the *spectrum* from red to violet. A glass prism has the property of *dispersion,* separating a beam of white light into a spectrum. Dispersion occurs because the index of refraction is different for each range of colors, with short wavelengths refracted more than larger ones.

A wave model of light can be used to explain diffraction, interference, and polarization, all of which provide strong evidence for the wavelike nature of light. *Interference* occurs when light passes through two small slits or holes and produces an *interference pattern* of bright lines and dark zones. *Polarized light* vibrates in one direction only, in a plane. Light can be polarized by certain materials, by reflection, or by scattering. Polarization can only be explained by a transverse wave model.

A wave model fails to explain observations of light behaviors in the *photoelectric effect* and *blackbody radiation.* Max Planck found that he could modify the wave theory to explain blackbody radiation by assuming that vibrating molecules could only have discrete amounts, or *quanta,* of energy and found that the quantized energy is related to the frequency and a constant known today as *Planck's constant.* Albert Einstein applied Planck's quantum concept to the photoelectric effect and described a light wave in terms of quanta of energy called *photons.* Each photon has an energy that is related to the frequency and Planck's constant.

Today, the properties of light are explained by a model that incorporates both the wave and the particle nature of light. Light is considered to have both wave and particle properties and is not describable in terms of anything known in the everyday-sized world.

The *special theory of relativity* is an analysis of how space and time are affected by motion between an observer and what is being measured. The *general theory of relativity* relates gravity to the structure of space and time.

Summary of Equations

7.1 angle of incidence = angle of reflection

$$\theta_i = \theta_r$$

7.2 index of refraction = $\dfrac{\text{speed of light in vacuum}}{\text{speed of light in material}}$

$$n = \frac{c}{v}$$

7.3 speed of light in vacuum = (wavelength)(frequency)

$$c = \lambda f$$

7.4 $\begin{matrix}\text{energy of} \\ \text{photon}\end{matrix} = \left(\begin{matrix}\text{Planck's} \\ \text{constant}\end{matrix}\right)(\text{frequency})$

$$E = hf$$

KEY TERMS

blackbody radiation (p. **154**)
consistent law principle (p. **172**)
constancy of speed principle
  (p. **172**)
general theory of relativity
  (p. **173**)

incandescent (p. **153**)
index of refraction (p. **160**)
interference (p. **165**)
light ray model (p. **156**)
luminous (p. **153**)
photoelectric effect (p. **169**)

photons (p. **169**)
polarized (p. **166**)
quanta (p. **169**)
real image (p. **158**)
refraction (p. **159**)
special theory of relativity (p. **172**)

total internal reflection
  (p. **159**)
unpolarized light (p. **166**)
virtual image (p. **158**)

APPLYING THE CONCEPTS

1. An object is hot enough to emit a dull red glow. When this object is heated even more, it will emit
   a. shorter-wavelength, higher-frequency radiation.
   b. longer-wavelength, lower-frequency radiation.
   c. the same wavelengths as before but with more energy.
   d. more of the same wavelengths with more energy.

2. The difference in the light emitted from a candle, an incandescent lightbulb, and the Sun is basically from differences in
   a. energy sources.
   b. materials.
   c. temperatures.
   d. phases of matter.

3. Before it travels through Earth's atmosphere, sunlight is mostly
   a. infrared radiation.
   b. visible light.
   c. ultraviolet radiation.
   d. blue light.

4. You are able to see in shaded areas, such as under a tree, because light has undergone
   a. refraction.
   b. incident bending.
   c. a change in speed.
   d. diffuse reflection.

5. The ratio of the speed of light in a vacuum to the speed of light in some transparent materials is called
   a. the critical angle.
   b. total internal reflection.
   c. the law of reflection.
   d. the index of refraction.

6. Any part of the electromagnetic spectrum, including the colors of visible light, can be measured in units of
   a. wavelength.
   b. frequency.
   c. energy.
   d. Any of the above is correct.

7. A prism separates the colors of sunlight into a spectrum because
   a. each wavelength of light has its own index of refraction.
   b. longer wavelengths are refracted more than shorter wavelengths.
   c. red light is refracted the most, violet the least.
   d. All of the above are correct.

8. Which of the following can only be explained by a wave model of light?
   a. reflection
   b. refraction
   c. interference
   d. photoelectric effect

9. The polarization behavior of light is best explained by considering light to be
   a. longitudinal waves.
   b. transverse waves.
   c. particles.
   d. particles with ends, or poles.

10. Max Planck made the revolutionary discovery that the energy of vibrating molecules involved in blackbody radiation existed only in
   a. multiples of certain fixed amounts.
   b. amounts that smoothly graded one into the next.
   c. the same, constant amount of energy in all situations.
   d. amounts that were never consistent from one experiment to the next.

11. Einstein applied Planck's quantum discovery to light and found
    a. a direct relationship between the energy and frequency of light.
    b. that the energy of a photon divided by the frequency of the photon always equaled a constant known as Planck's constant.
    c. that the energy of a photon divided by Planck's constant always equaled the frequency.
    d. All of the above are correct.

12. Today, light is considered to be
    a. tiny particles of matter that move through space, having no wave properties.
    b. electromagnetic waves only, with no properties of particles.
    c. a small-scale phenomenon without a sharp distinction between particle and wave properties.
    d. something that is completely unknown.

Answers

1. a **2.** c **3.** a **4.** d **5.** d **6.** d **7.** a **8.** c **9.** b **10.** a **11.** d **12.** c

QUESTIONS FOR THOUGHT

1. What determines if an electromagnetic wave emitted from an object is a visible light wave or a wave of infrared radiation?

2. What model of light does the polarization of light support? Explain.

3. Which carries more energy, red light or blue light? Should this mean anything about the preferred color of warning and stop lights? Explain.

4. What model of light is supported by the photoelectric effect? Explain.

5. What happens to light that is absorbed by matter?

6. One star is reddish, and another is bluish. Do you know anything about the relative temperatures of the two stars? Explain.

7. When does total internal reflection occur? Why does this occur in the diamond more than other gemstones?

8. Why does a highway sometimes appear wet on a hot summer day when it is not wet?

9. How can you tell if a pair of sunglasses is polarizing or not?

10. Explain why the intensity of reflected light appears to change if you tilt your head from side to side while wearing polarizing sunglasses.

11. What was so unusual about Planck's findings about blackbody radiation? Why was this considered revolutionary?

12. Why are both the photon model and the electromagnetic wave model accepted today as a single theory? Why was this so difficult for people to accept at first?

FOR FURTHER ANALYSIS

1. Clarify the distinction between light reflection and light refraction by describing clear, obvious examples of each.

2. Describe how you would use questions alone to help someone understand that the shimmering they see above a hot pavement is not heat.

3. Use a dialogue as you "think aloud" in considering the evidence that visible light is a wave, a particle, or both.

4. Compare and contrast the path of light through a convex and a concave lens. Give several uses for each lens, and describe how the shape of the lens results in that particular use.

5. Analyze how the equation $E = hf$ could mean that visible light is a particle and a wave at the same time.

6. How are visible light and a radio wave different? How are they the same?

INVITATION TO INQUIRY

Best Sunglasses?

Obtain several different types of sunglasses. Design experiments to determine which combination of features will be found in the best pair of sunglasses. First, design an experiment to determine which type reduces reflected glare the most. Find out how sunglasses are able to block ultraviolet radiation. According to your experiments and research, describe the "best" pair of sunglasses.

PARALLEL EXERCISES

The exercises in groups A and B cover the same concepts. Solutions to group A exercises are located in appendix D.

Group A

1. What is the speed of light while traveling through (a) water and (b) ice?

2. How many minutes are required for sunlight to reach Earth if the Sun is 1.50×10^8 km from Earth?

3. How many hours are required before a radio signal from a space probe near the dwarf planet Pluto reaches Earth, 6.00×10^9 km away?

4. A light ray is reflected from a mirror with an angle 10° to the normal. What was the angle of incidence?

5. Light travels through a transparent substance at 2.20×10^8 m/s. What is the substance?

6. The wavelength of a monochromatic light source is measured to be 6.00×10^{-7} m in an experiment. (a) What is the frequency? (b) What is the energy of a photon of this light?

7. At a particular location and time, sunlight is measured on a 1 m^2 solar collector with a power of 1,000.0 W. If the peak intensity of this sunlight has a wavelength of 5.60×10^{-7} m, how many photons are arriving each second?

8. A light wave has a frequency of 4.90×10^{14} cycles per second. (a) What is the wavelength? (b) What color would you observe (see table 7.2)?

9. What is the energy of a gamma photon of frequency 5.00×10^{20} Hz?

10. What is the energy of a microwave photon of wavelength 1.00 mm?

Group B

1. (a) What is the speed of light while traveling through a vacuum? (b) While traveling through air at 30°C? (c) While traveling through air at 0°C?

2. How much time is required for reflected sunlight to travel from the Moon to Earth if the distance between Earth and the Moon is 3.85×10^5 km?

3. How many minutes are required for a radio signal to travel from Earth to a space station on Mars if Mars is 7.83×10^7 km from Earth?

4. An incident light ray strikes a mirror with an angle of 30° to the surface of the mirror. What is the angle of the reflected ray?

5. The speed of light through a transparent substance is 2.00×10^8 m/s. What is the substance?

6. A monochromatic light source used in an experiment has a wavelength of 4.60×10^{-7} m. What is the energy of a photon of this light?

7. In black-and-white photography, a photon energy of about 4.00×10^{-19} J is needed to bring about the changes in the silver compounds used in the film. Explain why a red light used in a darkroom does not affect the film during developing.

8. The wavelength of light from a monochromatic source is measured to be 6.80×10^{-7} m. (a) What is the frequency of this light? (b) What color would you observe?

9. How much greater is the energy of a photon of ultraviolet radiation ($\lambda = 3.00 \times 10^{-7}$ m) than the energy of an average photon of sunlight ($\lambda = 5.60 \times 10^{-7}$ m)?

10. At what rate must electrons in a wire vibrate to emit microwaves with a wavelength of 1.00 mm?

8

Atoms and Periodic Properties

This is a picture of pure zinc, one of the eighty-nine naturally occurring elements found on Earth.

CORE CONCEPT

Different fields of study contributed to the model of the atom.

OUTLINE

The electron was discovered from experiments with electricity (p. 180).

The nucleus and proton were discovered from experiments with radioactivity (p. 181).

Experiments with light and line spectra and application of the quantum concept led to the Bohr model of the atom (pp. 184–85).

Atomic Structure Discovered
- Discovery of the Electron
- The Nucleus

The Bohr Model
  The Quantum Concept
  Atomic Spectra
- Bohr's Theory

Quantum Mechanics
The Periodic Table
A Closer Look: The Rare Earths
Metals, Nonmetals, and Semiconductors
People Behind the Science: Dmitri Ivanovich Mendeleyev

Application of the wave properties of electrons led to the quantum mechanics model of the atom (pp. 185–87).

Chemistry

▷ Electron configuration can be used to explain how atoms join together (Ch. 9).

▷ Water and its properties can be explained by considering the electron structure of hydrogen and oxygen atoms (Ch. 10).

▷ The nature of the atomic nucleus will explain radioactivity and nuclear energy (Ch. 11).

Astronomy

▷ The energy from stars originates in nuclear reactions in their core (Ch. 12).

▷ Stars are nuclear reactors (Ch. 12).

Earth Science

▷ Materials that make up Earth can be understood by considering their atomic structures (Ch. 15–16).

▷ Earth cycles materials through change (Ch. 15–18).

Life Science

▷ Living things use energy and materials in complex interactions (Ch. 19–26).

OVERVIEW

The development of the modern atomic model illustrates how modern scientific understanding comes from many different fields of study. For example, you will learn how studies of electricity led to the discovery that atoms have subatomic parts called *electrons*. The discovery of radioactivity led to the discovery of more parts, a central nucleus that contains protons and neutrons. Information from the absorption and emission of light was used to construct a model of how these parts are put together, a model resembling a miniature solar system with electrons circling the nucleus. The solar system model had initial, but limited, success and was inconsistent with other understandings about matter and energy. Modifications of this model were attempted, but none solved the problems. Then the discovery of wave properties of matter led to an entirely new model of the atom (figure 8.1).

The atomic model will be put to use in later chapters to explain the countless varieties of matter and the changes that matter undergoes. In addition, you will learn how these changes can be manipulated to make new materials, from drugs to ceramics. In short, you will learn how understanding the atom and all the changes it undergoes not only touches your life directly but shapes and affects all parts of civilization.

ATOMIC STRUCTURE DISCOVERED

Did you ever wonder how scientists could know about something so tiny that you cannot see it, even with the most powerful optical microscope? The atom is a tiny unit of matter, so small that 1 gram of hydrogen contains about 600,000,000,000,000,000,000,000 (six-hundred-thousand-billion billion or 6×10^{23}) atoms. Even more unbelievable is that atoms are not individual units but are made up of even smaller particles. How is it possible that scientists are able to tell you about the parts of something so small that it cannot be seen? The answer is that these things cannot be observed directly, but their existence can be inferred from experimental evidence. The following story describes the evidence and how scientists learned about the parts—electrons, the nucleus, protons, neutrons, and others—and how they are all arranged in the atom.

The atomic concept is very old, dating back to ancient Greek philosophers some 2,500 years ago. The ancient Greeks also reasoned about the way that pure substances are put together. A glass of water, for example, appears to be the same throughout. Is it the same? Two plausible, but conflicting, ideas were possible as an intellectual exercise. The water could have a *continuous* structure; that is, it could be completely homogeneous throughout. The other idea was that the water only appears to be continuous but is actually *discontinuous*. This means that if you continue to divide the water into smaller and smaller volumes, you would eventually reach a limit to this dividing, a particle that could not be further subdivided. The Greek philosopher Democritus (460–362 B.C.) developed this model in the fourth century B.C., and he called the indivisible particle an *atom*, from a Greek word meaning "uncuttable." However, neither Plato nor Aristotle accepted the atomic theory of matter, and it was not until about two thousand years later that the atomic concept of matter was reintroduced. In the early 1800s, the English chemist John Dalton brought back the ancient Greek idea of hard, indivisible atoms to explain

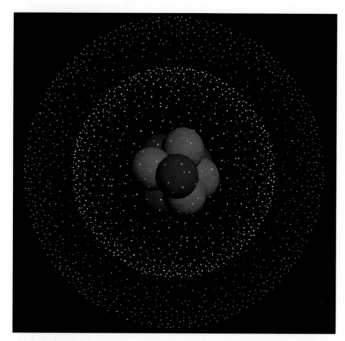

FIGURE 8.1 This is a computer-generated model of a beryllium atom, showing the nucleus and electron spacing relationships. This configuration can also be predicted from information on a periodic table. (Not to scale.)

FIGURE 8.2 A vacuum tube with metal plates attached to a high-voltage source produces a greenish beam called *cathode rays*. These rays move from the cathode (negative charge) to the anode (positive charge).

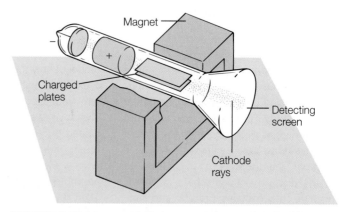

FIGURE 8.3 A cathode ray passed between two charged plates is deflected toward the positively charged plate. The ray is also deflected by a magnetic field. By measuring the deflection by both, J. J. Thomson was able to calculate the ratio of charge to mass. He was able to measure the deflection because the detecting screen was coated with zinc sulfide, a substance that produces a visible light when struck by a charged particle.

chemical reactions. Five statements will summarize his theory. As you will soon see, today we know that statement 2 is not strictly correct:

1. Indivisible minute particles called atoms make up all matter.
2. All the atoms of an element are exactly alike in shape and mass.
3. The atoms of different elements differ from one another in their masses.
4. Atoms chemically combine in definite whole-number ratios to form chemical compounds.
5. Atoms are neither created nor destroyed in chemical reactions.

During the 1800s, Dalton's concept of hard, indivisible atoms was familiar to most scientists. Yet, the existence of atoms was not generally accepted by all scientists. There was skepticism about something that could not be observed directly. Strangely, full acceptance of the atom came in the early 1900s with the discovery that the atom was not indivisible after all. The atom has parts that give it an internal structure. The first part to be discovered was the *electron,* a part that was discovered through studies of electricity.

Discovery of the Electron

Scientists of the late 1800s were interested in understanding the nature of the recently discovered electric current. To observe a current directly, they tried to produce one by itself, away from wires, by removing the air from a tube and then running electricity through the vacuum. When metal plates inside a tube were connected to the negative and positive terminals of a high-voltage source (figure 8.2), a greenish beam was observed that seemed to move from the cathode (negative terminal) through the empty tube and collect at the anode (positive terminal). Since this mysterious beam seemed to come out of the cathode, it was said to be a *cathode ray.*

The English physicist J. J. Thomson figured out what the cathode ray was in 1897. He placed charged metal plates on each side of the beam (figure 8.3) and found that the beam was deflected away from the negative plate. Since it was known that like charges repel, this meant that the beam was composed of negatively charged particles.

The cathode ray was also deflected when caused to pass between the poles of a magnet. By balancing the deflections made by the magnet with the deflections made by the electric field, Thomson could determine the ratio of the charge to mass for an individual particle. Today, the charge-to-mass ratio is considered to be 1.7584×10^{11} coulomb/kilogram. A significant

FIGURE 8.4 Millikan measured the charge of an electron by balancing the pull of gravity on oil droplets with an upward electrical force. Knowing the charge-to-mass ratio that Thomson had calculated, Millikan was able to calculate the charge on each droplet. He found that all the droplets had a charge of 1.60×10^{-19} coulomb or multiples of that charge. The conclusion was that this had to be the charge of an electron.

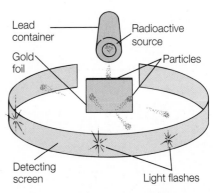

FIGURE 8.5 Rutherford and his co-workers studied particle scattering from a thin metal foil. The particles struck the detecting screen, producing a flash of visible light. Measurements of the angles between the flashes, the metal foil, and the source of the particles showed that they were scattered in all directions, including straight back toward the source.

part of Thomson's experiments was that he found the charge-to-mass ratio was the same no matter what gas was in the tube and of what materials the electrodes were made. Thomson had discovered the **electron,** a fundamental particle of matter.

A method for measuring the charge and mass of the electron was worked out by an American physicist, Robert A. Millikan, around 1906. Millikan used an apparatus like the one illustrated in figure 8.4 to measure the charge on tiny droplets of oil. Millikan found that none of the droplets had a charge less than one particular value (1.60×10^{-19} coulomb) and that larger charges on various droplets were always multiples of this unit of charge. Since all of the droplets carried the single unit of charge or multiples of the single unit, the unit of charge was understood to be the charge of a single electron.

Knowing the charge of a single electron and knowing the charge-to-mass ratio that Thomson had measured now made it possible to calculate the mass of a single electron. The mass of an electron was thus determined to be about 9.11×10^{-31} kg, or about 1/1,840 of the mass of the lightest atom, hydrogen.

Thomson had discovered the negatively charged electron, and Millikan had measured the charge and mass of the electron. But atoms themselves are electrically neutral. If an electron is part of an atom, there must be something else that is positively charged, canceling the negative charge of the electron. The next step in the sequence of understanding atomic structure would be to find what is neutralizing the negative charge and to figure out how all the parts are put together.

Thomson had proposed a model for what was known about the atom at the time. He suggested that an atom could be a blob of massless, positively charged stuff in which electrons were stuck like "raisins in plum pudding." If the mass of a hydrogen atom is due to the electrons embedded in a massless, positively charged matrix, then 1,840 electrons would be needed together with sufficient positive matter to make the atom electrically neutral.

The Nucleus

The nature of radioactivity and matter were the research interests of a British physicist, Ernest Rutherford (see page 242). In 1907, Rutherford was studying the scattering of radioactive particles directed toward a thin sheet of metal. As shown in figure 8.5, the particles from a radioactive source were allowed to move through a small opening in a lead container, so only a narrow beam of the massive, fast-moving particles would penetrate a very thin sheet of gold. The particles were detected by plates, which produced a small flash of light when struck by the particles.

Rutherford found that most of the particles went straight through the foil. However, he was astounded to find that some were deflected at very large angles and some were even reflected backward. He could account for this only by assuming that the massive, positively charged particles were repelled by a massive positive charge concentrated in a small region of the atoms in the gold foil (figure 8.6). He concluded that an atom must have a tiny, massive, and positively charged **nucleus** surrounded by electrons.

From measurements of the scattering, Rutherford estimated electrons must be moving around the nucleus at a distance 100,000 times the radius of the nucleus. This means the volume of an atom is mostly empty space. A few years later, Rutherford was able to identify the discrete unit of positive charge, which we now call a **proton.** Rutherford also speculated about the existence of a neutral particle in the nucleus, a neutron. The **neutron** was eventually identified in 1932 by James Chadwick.

Today, the number of protons in the nucleus of an atom is called the **atomic number.** All the atoms of a particular element have the same number of protons in their nucleus, so all atoms of an element have the same atomic number. Hydrogen has an atomic number of 1, so any atom that has one proton in its nucleus is an atom of the element hydrogen. Today, scientists have identified 117 different kinds of elements, each with a different number of protons.

The neutrons of the nucleus, along with the protons, contribute to the mass of an atom. Although all the atoms of an

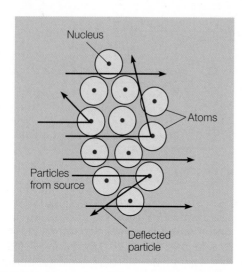

FIGURE 8.6 Rutherford's nuclear model of the atom explained the particle-scattering results as positive particles experiencing a repulsive force from the positive nucleus. Measurements of the percent of particles passing straight through and of the various angles of scattering of those coming close to the nuclei gave Rutherford a means of estimating the size of the nucleus.

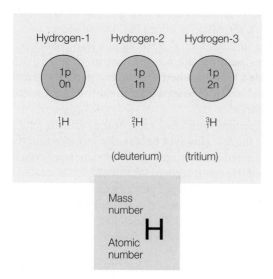

FIGURE 8.7 The three isotopes of hydrogen have the same number of protons but different numbers of neutrons. Hydrogen-1 is the most common isotope. Hydrogen-2, with an additional neutron, is named *deuterium,* and hydrogen-3 is called *tritium.*

element must have the same number of protons in their nuclei, the number of neutrons may vary. Atoms of an element that have different numbers of neutrons are called **isotopes.** There are three isotopes of hydrogen illustrated in figure 8.7. All three isotopes have the same number of protons and electrons, but one isotope has no neutrons, one isotope has 1 neutron (deuterium), and one isotope has 2 neutrons (tritium).

An atom is very tiny, and it is impossible to find the mass of a given atom. It is possible, however, to compare the mass of one atom to another. The mass of any atom is compared to the mass of an atom of a particular isotope of carbon. This particular carbon isotope is assigned a mass of exactly

12.00 . . . units called **atomic mass units** (u). Since this isotope is *defined* to be exactly 12 u, it can have an infinite number of significant figures. This isotope, called *carbon-12,* provides the standard to which the masses of all other isotopes are compared. The relative mass of any isotope is based on the mass of a carbon-12 isotope.

The relative mass of the hydrogen isotope without a neutron is 1.007 when compared to carbon-12. The relative mass of the hydrogen isotope with 1 neutron is 2.0141 when compared to carbon-12. Elements occur in nature as a mixture of isotopes, and the contribution of each is calculated in the atomic weight. **Atomic weight** is a weighted average of the isotopes based on their mass compared to carbon-12 and their relative abundance found on Earth. Of all the hydrogen isotopes, for example, 99.985 percent occurs as the isotope without a neutron and 0.015 percent is the isotope with 1 neutron (the other isotope is not considered because it is radioactive). The fractional part of occurrence is multiplied by the relative atomic mass for each isotope, and the results are summed to obtain the atomic weight. Table 8.1 gives the atomic weight of hydrogen as 1.0079 as a result of this calculation.

The sum of the number of protons and neutrons in a nucleus of an atom is called the **mass number** of that atom. Mass numbers are used to identify isotopes. A hydrogen atom with 1 proton and 1 neutron has a mass number of 1 + 1, or 2, and is referred to as hydrogen-2. A hydrogen atom with 1 proton and 2 neutrons has a mass number of 1 + 2, or 3, and is referred to as hydrogen-3. Using symbols, hydrogen-3 is written as

$$^{3}_{1}\text{H}$$

where H is the chemical symbol for hydrogen, the subscript to the bottom left is the atomic number, and the superscript to the top left is the mass number.

How are the electrons moving around the nucleus? It might occur to you, as it did to Rutherford and others, that an atom might be similar to a miniature solar system. In this analogy, the nucleus is in the role of the Sun, electrons in the role of moving planets in their orbits, and electrical attractions between the nucleus and electrons in the role of gravitational attraction. There are, however, big problems with this idea. If electrons were moving in circular orbits, they would continually change

TABLE 8.1 Selected Atomic Weights Calculated from Mass and Abundance of Isotopes

Stable Isotopes	Mass of Isotope Compared to C-12		Abundance		Atomic Weight
$^{1}_{1}\text{H}$	1.007	×	99.985%		
$^{2}_{1}\text{H}$	2.0141	×	0.015%	=	1.0079
$^{9}_{4}\text{Be}$	9.01218	×	100%	=	9.01218
$^{14}_{7}\text{N}$	14.00307	×	99.63%		
$^{15}_{7}\text{N}$	15.00011	×	0.37%	=	14.0067

Atomic Parts

Identify the number of protons, neutrons, and electrons in an atom of $^{16}_{8}O$. Write your answer before reading the solution in the next paragraph.

The subscript to the bottom left is the atomic number. Atomic number is defined as the number of protons in the nucleus, so this number identifies the number of protons as 8. Any atom with 8 protons is an atom of oxygen, which is identified with the symbol O. The superscript to the top left identifies the mass number of this isotope of oxygen, which is 16. The mass number is defined as the sum of the number of protons and the number of neutrons in the nucleus. Since you already know the number of protons is 8 (from the atomic number), then the number of neutrons is 16 minus 8, or 8 neutrons. Since a neutral atom has the same number of electrons as protons, an atom of this oxygen isotope has 8 protons, 8 neutrons, and 8 electrons.

Now, can you describe how many protons, neutrons, and electrons are found in an atom of $^{17}_{8}O$? Compare your answer with a classmate's to check.

their direction of travel and would therefore be accelerating. According to the Maxwell model of electromagnetic radiation, an accelerating electric charge emits electromagnetic radiation such as light. If an electron gave off light, it would lose energy. The energy loss would mean that the electron could not maintain its orbit, and it would be pulled into the oppositely charged nucleus. The atom would collapse as electrons spiraled into the nucleus. Since atoms do not collapse like this, there is a significant problem with the solar system model of the atom.

THE BOHR MODEL

Niels Bohr was a young Danish physicist who visited Rutherford's laboratory in 1912 and became very interested in questions about the solar system model of the atom. He wondered what determined the size of the electron orbits and the energies of the electrons. He wanted to know why orbiting electrons did not give off electromagnetic radiation. Seeking answers to questions such as these led Bohr to incorporate the *quantum concept* of Planck and Einstein with Rutherford's model to describe the electrons in the outer part of the atom. This quantum concept will be briefly reviewed before proceeding with the development of Bohr's model of the hydrogen atom.

The Quantum Concept

In 1900, Max Planck introduced the idea that matter emits and absorbs energy in discrete units that he called **quanta.** Planck had been trying to match data from spectroscopy experiments with data that could be predicted from the theory of electromagnetic radiation. In order to match the experimental findings

with the theory, he had to assume that specific, discrete amounts of energy were associated with different frequencies of radiation. In 1905, Albert Einstein extended the quantum concept to light, stating that light consists of discrete units of energy that are now called **photons.** The energy of a photon is directly proportional to the frequency of vibration, and the higher the frequency of light, the greater the energy of the individual photons. In addition, the interaction of a photon with matter is an "all-or-none" affair; that is, matter absorbs an entire photon or none of it. The relationship between frequency (f) and energy (E) is

$$E = hf \qquad \text{equation 8.1}$$

where h is the proportionality constant known as *Planck's constant* (6.63×10^{-34} J·s). This relationship means that higher-frequency light, such as ultraviolet, has more energy than lower-frequency light, such as red light.

EXAMPLE 8.1 *(Optional)*

What is the energy of a photon of red light with a frequency of 4.60×10^{14} Hz?

SOLUTION

$f = 4.60 \times 10^{14}$ Hz

$h = 6.63 \times 10^{-34}$ J·s

$E = ?$

$$E = hf$$
$$= (6.63 \times 10^{-34} \text{ J·s})\left(4.60 \times 10^{14}\,\frac{1}{\text{s}}\right)$$
$$= (6.63 \times 10^{-34})(4.60 \times 10^{14})\text{ J·s} \times \frac{1}{\text{s}}$$
$$= \boxed{3.05 \times 10^{-19} \text{ J}}$$

EXAMPLE 8.2 *(Optional)*

What is the energy of a photon of violet light with a frequency of 7.30×10^{14} Hz? (Answer: 4.84×10^{-19} J)

Atomic Spectra

Planck was concerned with hot solids that emit electromagnetic radiation. The nature of this radiation, called *blackbody radiation,* depends on the temperature of the source. When this light is passed through a prism, it is dispersed into a *continuous spectrum,* with one color gradually blending into the next as in a rainbow. Today, it is understood that a continuous spectrum comes from solids, liquids, and dense gases because the atoms interact, and all frequencies within a temperature-determined range are emitted. Light from an incandescent gas, on the other hand, is dispersed into a **line spectrum,** narrow lines of colors with no light between the lines (figure 8.8). The atoms in the incandescent gas are able to emit certain characteristic frequencies, and each frequency is a line of color that represents a definite value of energy. The line spectra are specific for a substance, and increased or decreased temperature changes only the intensity of the lines of colors. Thus, hydrogen always produces the same colors of lines in

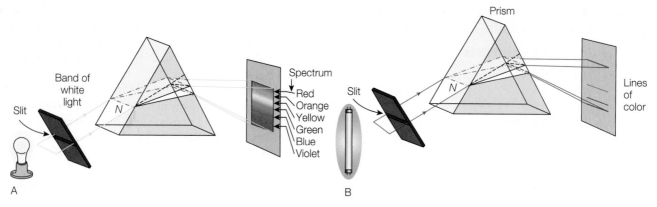

FIGURE 8.8 (A) Light from incandescent solids, liquids, or dense gases produces a continuous spectrum as atoms interact to emit all frequencies of visible light. (B) Light from an incandescent gas produces a line spectrum as atoms emit certain frequencies that are characteristic of each element.

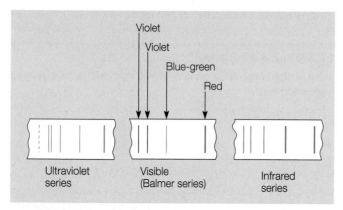

FIGURE 8.9 Atomic hydrogen produces a series of characteristic line spectra in the ultraviolet, visible, and infrared parts of the total spectrum. The visible light spectra always consist of two violet lines, a blue-green line, and a bright red line.

the same position. Helium has its own specific set of lines, as do other substances. Line spectra are a kind of fingerprint that can be used to identify a gas. A line spectrum might also extend beyond visible light into ultraviolet, infrared, and other electromagnetic regions.

In 1885, a Swiss mathematics teacher named J. J. Balmer was studying the regularity of spacing of the hydrogen line spectra. Balmer was able to develop an equation that fit all the visible lines. These four lines became known as the *Balmer series*. Other series were found later, outside the visible part of the spectrum (figure 8.9).

Such regularity of observable spectral lines must reflect some unseen regularity in the atom. At this time it was known that hydrogen had only one electron. How could one electron produce series of spectral lines with such regularity?

Bohr's Theory

An acceptable model of the hydrogen atom would have to explain the characteristic line spectra and their regularity as described by Balmer. In fact, a successful model should be able to predict the occurrence of each color line as well as account for its origin. By 1913, Bohr was able to do this by applying

the quantum concept to a solar system model of the atom. He began by considering the single hydrogen electron to be a single "planet" revolving in a circular orbit around the nucleus. Three sets of rules described this electron:

1. **Allowed orbits.** An electron can revolve around an atom only in specific allowed orbits. Bohr considered the electron to be a particle with a known mass in motion around the nucleus and used Newtonian mechanics to calculate the distances of the allowed orbits. According to the Bohr model, electrons can exist only in one of these allowed orbits and nowhere else.

2. **Radiationless orbits.** An electron in an allowed orbit does not emit radiant energy as long as it remains in the orbit. According to Maxwell's theory of electromagnetic radiation, an accelerating electron should emit an electromagnetic wave, such as light, which would move off into space from the electron. Bohr recognized that electrons moving in a circular orbit are accelerating, since they are changing direction continuously. Yet, hydrogen atoms did not emit light in their normal state. Bohr decided that the situation must be different for orbiting electrons and that electrons could stay in their allowed orbits and not give off light. He postulated this rule as a way to make his theory consistent with other scientific theories.

3. **Quantum leaps.** An electron gains or loses energy only by moving from one allowed orbit to another (figure 8.10). In the Bohr model, the energy an electron has depends on which allowable orbit it occupies. The only way that an electron can change its energy is to jump from one allowed orbit to another in quantum "leaps." An electron must acquire energy to jump from a lower orbit to a higher one. Likewise, an electron gives up energy when jumping from a higher orbit to a lower one. Such jumps must be all at once, not part way and not gradual. An electron acquires energy from high temperatures or from electrical discharges to jump to a higher orbit. An electron jumping from a higher to a lower orbit gives up energy in the form of light. A single photon is emitted when a downward jump occurs, and the energy of the photon is *exactly* equal to the difference in the energy level of the two orbits.

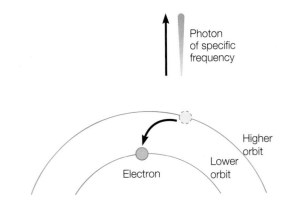

Photon
of specific
frequency

Higher
orbit

Lower
orbit

Electron

FIGURE 8.10 Each time an electron makes a "quantum leap," moving from a higher-energy orbit to a lower-energy orbit, it emits a photon of a specific frequency and energy value.

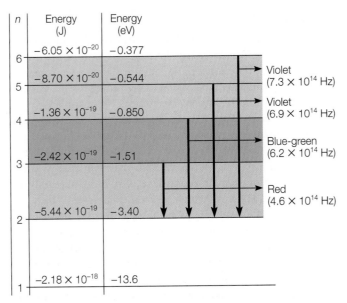

n	Energy (J)	Energy (eV)	
6	-6.05×10^{-20}	-0.377	
5	-8.70×10^{-20}	-0.544	Violet (7.3×10^{14} Hz)
4	-1.36×10^{-19}	-0.850	Violet (6.9×10^{14} Hz)
3	-2.42×10^{-19}	-1.51	Blue-green (6.2×10^{14} Hz)
			Red (4.6×10^{14} Hz)
2	-5.44×10^{-19}	-3.40	
1	-2.18×10^{-18}	-13.6	

FIGURE 8.11 An energy level diagram for a hydrogen atom, not drawn to scale. The energy levels (n) are listed on the left side, followed by the energies of each level in J and eV. The color and frequency of the visible light photons emitted are listed on the right side, with the arrow showing the orbit moved from and to.

The energy level diagram in figure 8.11 shows the energy states for the orbits of a hydrogen atom. The lowest energy state is the **ground state** (or normal state). The higher states are the **excited states.** The electron in a hydrogen atom would normally occupy the ground state, but high temperatures or electric discharge can give the electron sufficient energy to jump to one of the excited states. Once in an excited state, the electron immediately jumps back to a lower state, as shown by the arrows in the figure. The length of the arrow represents the frequency of the photon that the electron emits in the process. A hydrogen atom can give off only one photon at a time, and the many lines of a hydrogen line spectrum come from many atoms giving off many photons at the same time.

As you can see, the energy level diagram in figure 8.11 shows how the change of known energy levels from known orbits results in the exact frequencies of the color lines in the Balmer series. Bohr's theory did explain the lines in the hydrogen spectrum

with a remarkable degree of accuracy. However, the model did not have much success with larger atoms. The Bohr model could not explain the spectra line of larger atoms with its single quantum number. A German physicist, A. Sommerfeld, tried to modify Bohr's model by adding elliptical orbits in addition to Bohr's circular orbits. It soon became apparent that the "patched up" model, too, was not adequate. Bohr had made the rule that there were radiationless orbits without an explanation, and he did not have an explanation for the quantized orbits. There was something fundamentally incomplete about the model.

QUANTUM MECHANICS

The Bohr model of the atom successfully accounted for the line spectrum of hydrogen and provided an understandable mechanism for the emission of photons by atoms. However, the model did not predict the spectra of any atom larger than hydrogen, and there were other limitations. A new, better theory was needed. The roots of a new theory would again come from experiments with light. Experiments with light had established that sometimes light behaves like a stream of particles and at other times like a wave. Eventually, scientists began to accept that light has both wave properties and particle properties, which is now referred to as the *wave-particle duality of light*. In 1923, Louis de Broglie, a French physicist, reasoned that symmetry is usually found in nature, so if a particle of light has a dual nature, then particles such as electrons should too.

Recall that waves confined on a fixed string establish resonant modes of vibration called *standing waves* (see chapter 5). Only certain fundamental frequencies and harmonics can exist on a string, and the combination of the fundamental and overtones gives the stringed instrument its particular quality. The same result of resonant modes of vibrations is observed in *any* situation where waves are confined to a fixed space. Characteristic standing wave patterns depend on the wavelength and wave velocity for waves formed on strings, in enclosed columns of air, or for any kind of wave in a confined space. Electrons are confined to the space near a nucleus, and electrons have wave properties, so an electron in an atom must be a confined wave. Does an electron form a characteristic wave pattern? This was the question being asked in about 1925 when Heisenberg, Schrödinger, Dirac, and others applied the wave nature of the electron to develop a new model of the atom based on the mechanics of electron waves. The new theory is now called *wave mechanics*, or quantum mechanics.

Erwin Schrödinger, an Austrian physicist, treated the atom as a three-dimensional system of waves to derive the *Schrödinger equation*. Instead of the simple circular planetary orbits of the Bohr model, solving the Schrödinger equation results in a description of three-dimensional shapes of the patterns that develop when electron waves are confined by a nucleus. Schrödinger first considered the hydrogen atom, calculating the states of vibration that would be possible for an electron wave confined by a nucleus. He found that the frequency of these vibrations, when multiplied by Planck's constant, matched exactly, to the last decimal point, the observed energies of the

Science and Society

Atomic Research

There are two types of scientific research: basic and applied. Basic research is driven by a search for understanding and may or may not have practical applications. Applied research has a goal of solving some practical problems rather than just looking for understanding.

Some people feel that all research should result in something practical, so all research should be applied. Hold that thought while considering if the following research discussed in this chapter is basic or applied:

1. J. J. Thomson investigates cathode rays.
2. Robert Millikan measures the charge of an electron.
3. Ernest Rutherford studies radioactive particles striking gold foil.
4. Niels Bohr proposes a solar system model of the atom by applying the quantum concept.
5. Erwin Schrödinger proposes a model of the atom based on the wave nature of the electron.

Questions to Discuss:

1. Were the five research topics basic or applied research (see page 14)?

2. Would we ever have developed a model of the atom if all research had to be practical?

quantum states of the hydrogen atom. The conclusion is that the wave nature of the electron is the important property to consider for a successful model of the atom.

The quantum mechanics theory of the atom proved to be very successful; it confirmed all the known experimental facts and predicted new discoveries.

Quantum mechanics does share at least one idea with the Bohr model. This is that an electron emits a photon when jumping from a higher state to a lower one. The Bohr model, however, considered the *particle* nature of an electron moving in a circular orbit with a definitely assigned position at a given time. Quantum mechanics considers the *wave* nature, with the electron as a confined wave with well-defined shapes and frequencies. A wave is not localized like a particle and is spread out in space. The quantum mechanics model is, therefore, a series of orbitlike smears, or fuzzy statistical representations, of where the electron might be found.

The quantum mechanical theory is not an extension or refinement of the Bohr model. The Bohr model considered electrons as particles in circular orbits that could be only certain distances from the nucleus. The quantum mechanical model, on the other hand, considers the electron as a wave and considers the energy of its harmonics, or modes, of standing waves. In the Bohr model, the location of an electron was certain—in an orbit. In the quantum mechanical model, the electron is a spread-out wave.

Quantum mechanics describes the energy state of an electron wave with four *quantum numbers:*

1. **Distance from the nucleus.** The *principal quantum number* describes the *main energy level* of an electron in terms of its most probable distance from the nucleus. The lowest energy state possible is closest to the nucleus and is assigned the principal quantum number of 1 ($n = 1$). Higher states are assigned progressively higher positive whole numbers of $n = 2$, $n = 3$, $n = 4$, and so on. Electrons with higher principal quantum numbers have higher energies and are located farther from the nucleus.

2. **Energy sublevel.** The *angular momentum quantum number* defines energy sublevels within the main energy levels. Each sublevel is identified with a letter. The first four of these letters, in order of increasing energy, are s, p, d, and f. The letter s represents the lowest sublevel, and the letter f represents the highest sublevel. A principal quantum number and a letter indicating the angular momentum quantum number are combined to identify the main energy state and energy sublevel of an electron. For an electron in the lowest main energy level, $n = 1$, and in the lowest sublevel, s, the number and letter are 1s (read as "one-s"). Thus, 1s indicates an electron that is as close to the nucleus as possible in the lowest energy sublevel possible.

   There are limits to how many sublevels can occupy each of the main energy levels. Basically, the lowest main energy level can have only the lowest sublevel, and another sublevel is added as you move up through the main energy levels. Thus, the lowest main energy level, $n = 1$, can have only the s sublevel. The $n = 2$ can have s and p sublevels. The $n = 3$ main energy level can have the s, p, and the d sublevels. Finally, the $n = 4$ main energy level can have all four sublevels, with s, p, d, and f. Therefore, the number of possible sublevels is the same as the principal quantum number.

   The Bohr model considered the location of an electron as certain, like a tiny shrunken marble in an orbit. The quantum mechanics model considers the electron as a wave, and knowledge of its location is very uncertain. The **Heisenberg uncertainty principle** states that you cannot measure the exact position of a wave because a wave is spread out. One cannot specify the position and the momentum of a spread-out electron. The location of the electron can be described only in terms of *probabilities* of where it might be at a given instant. The probability of location is described by a fuzzy region of space called an **orbital.** An orbital defines the space where an electron is likely to be found. Orbitals have characteristic three-dimensional shapes and sizes and are identified with electrons of characteristic energy levels (figure 8.12). An orbital shape represents where an electron could probably be located at any particular instant. This "probability cloud" could likewise have any particular orientation in space, and the direction of this orientation is uncertain.

3. **Orientation in space.** An external magnetic field applied to an atom produces different energy levels that are related to the orientation of the orbital to the magnetic field. The orientation of an orbital in space is described by the *magnetic quantum number.* This number is related to the energies of orbitals as they are oriented in space relative to an external

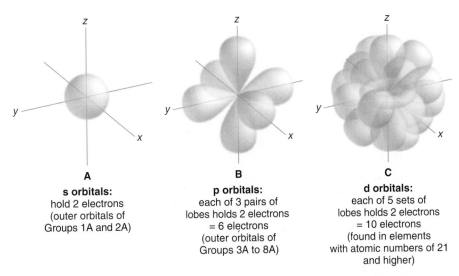

A	B	C
s orbitals:	**p orbitals:**	**d orbitals:**
hold 2 electrons (outer orbitals of Groups 1A and 2A)	each of 3 pairs of lobes holds 2 electrons = 6 electrons (outer orbitals of Groups 3A to 8A)	each of 5 sets of lobes holds 2 electrons = 10 electrons (found in elements with atomic numbers of 21 and higher)

FIGURE 8.12 The general shapes of s, p, and d orbitals, the regions of space around the nuclei of atoms in which electrons are likely to be found. (The f orbital is too difficult to depict.)

TABLE 8.2   Quantum Numbers and Electron Distribution to $n = 4$

Main Energy Level	Energy Sublevels	Maximum Number of Electrons	Maximum Number of Electrons per Main Energy Level
$n = 1$	s	2	2
$n = 2$	s	2	
	p	6	8
$n = 3$	s	2	
	p	6	
	d	10	18
$n = 4$	s	2	
	p	6	
	d	10	
	f	14	32

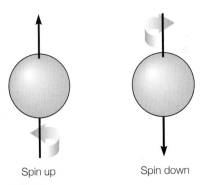

Spin up   Spin down

FIGURE 8.13 Experimental evidence supports the concept that electrons can be considered to spin one way or the other as they move about an orbital under an external magnetic field.

magnetic field, a kind of energy sub-sublevel. In general, the lowest-energy sublevel (s) has only one orbital orientation (figure 8.12A). The next higher-energy sublevel (p) can have three orbital orientations (figure 8.12B). The d sublevel can have five orbital orientations, and the highest sublevel, f, can have a total of seven different orientations (figure 8.12C) (also see table 8.2).

4. **Direction of spin.**   Detailed studies have shown that an electron spinning one way, say clockwise, in an external magnetic field would have a different energy than one spinning the other way, say counterclockwise. The *spin quantum number* describes these two spin orientations (figure 8.13).

Electron spin is an important property that helps determine the electronic structure of an atom. As it turns out, two electrons spinning in opposite directions produce unlike magnetic fields that are attractive, balancing some of the normal repulsion from two like charges. Two electrons of op-

posite spin, called an electron pair, can thus occupy the same orbital. Wolfgang Pauli, a German physicist, summarized this idea in 1924. His summary, now known as the **Pauli exclusion principle,** states that *no two electrons in an atom can have the same four quantum numbers.* This principle provides the key for understanding the electron structure of atoms. (For a detailed discussion of electron configuration, see the chapter 8 resources on www.mhhe.com/tillery.)

THE PERIODIC TABLE

The periodic table is made up of rows and columns of cells, with each element having its own cell in a specific location. The cells are not arranged symmetrically. The arrangement has a meaning, both about atomic structure and about chemical behaviors. It will facilitate your understanding of the code if you refer frequently to a periodic table during the following discussion (figure 8.14).

An element is identified in each cell with its chemical symbol. The number above the symbol is the atomic number of the element, and the number below the symbol is the rounded

Periodic Table of the Elements

Period	IA (1)	IIA (2)	IIIB (3)	IVB (4)	VB (5)	VIB (6)	VIIB (7)	VIIIB (8)	VIIIB (9)	VIIIB (10)	IB (11)	IIB (12)	IIIA (13)	IVA (14)	VA (15)	VIA (16)	VIIA (17)	VIIIA (18)
1	Hydrogen 1 **H** 1.008																	Helium 2 **He** 4.003
2	Lithium 3 **Li** 6.941	Beryllium 4 **Be** 9.012											Boron 5 **B** 10.81	Carbon 6 **C** 12.01	Nitrogen 7 **N** 14.01	Oxygen 8 **O** 16.00	Fluorine 9 **F** 19.00	Neon 10 **Ne** 20.18
3	Sodium 11 **Na** 22.99	Magnesium 12 **Mg** 24.31											Aluminum 13 **Al** 26.98	Silicon 14 **Si** 28.09	Phosphorus 15 **P** 30.97	Sulfur 16 **S** 32.07	Chlorine 17 **Cl** 35.45	Argon 18 **Ar** 39.95
4	Potassium 19 **K** 39.10	Calcium 20 **Ca** 40.08	Scandium 21 **Sc** 44.96	Titanium 22 **Ti** 47.88	Vanadium 23 **V** 50.94	Chromium 24 **Cr** 52.00	Manganese 25 **Mn** 54.94	Iron 26 **Fe** 55.85	Cobalt 27 **Co** 58.93	Nickel 28 **Ni** 58.69	Copper 29 **Cu** 63.55	Zinc 30 **Zn** 65.39	Gallium 31 **Ga** 69.72	Germanium 32 **Ge** 72.61	Arsenic 33 **As** 74.92	Selenium 34 **Se** 78.96	Bromine 35 **Br** 79.90	Krypton 36 **Kr** 83.80
5	Rubidium 37 **Rb** 85.47	Strontium 38 **Sr** 87.62	Yttrium 39 **Y** 88.91	Zirconium 40 **Zr** 91.22	Niobium 41 **Nb** 92.91	Molybdenum 42 **Mo** 95.94	Technetium 43 **Tc** (98)	Ruthenium 44 **Ru** 101.1	Rhodium 45 **Rh** 102.9	Palladium 46 **Pd** 106.4	Silver 47 **Ag** 107.9	Cadmium 48 **Cd** 112.4	Indium 49 **In** 114.8	Tin 50 **Sn** 118.7	Antimony 51 **Sb** 121.8	Tellurium 52 **Te** 127.6	Iodine 53 **I** 126.9	Xenon 54 **Xe** 131.3
6	Cesium 55 **Cs** 132.9	Barium 56 **Ba** 137.3	Lanthanum 57 **La** 138.9 †	Hafnium 72 **Hf** 178.5	Tantalum 73 **Ta** 180.9	Tungsten 74 **W** 183.8	Rhenium 75 **Re** 186.2	Osmium 76 **Os** 190.2	Iridium 77 **Ir** 192.2	Platinum 78 **Pt** 195.1	Gold 79 **Au** 197.0	Mercury 80 **Hg** 200.6	Thallium 81 **Tl** 204.4	Lead 82 **Pb** 207.2	Bismuth 83 **Bi** 209.0	Polonium 84 **Po** (209)	Astatine 85 **At** (210)	Radon 86 **Rn** (222)
7	Francium 87 **Fr** (223)	Radium 88 **Ra** (226)	Actinium 89 **Ac** (227) ‡	Rutherfordium 104 **Rf** (261)	Dubnium 105 **Db** (262)	Seaborgium 106 **Sg** (266)	Bohrium 107 **Bh** (264)	Hassium 108 **Hs** (277)	Meitnerium 109 **Mt** (268)	Ununnilium 110 **Uun** (281)	Unununium 111 **Uuu** (272)	Ununbium 112 **Uub** (285)	Ununtrium 113 **Uut** (284)	Ununquadium 114 **Uuq** (289)	Ununpentium 115 **Uup** (288)	Ununhexium 116 **Uuh** (292)		Ununoxtium 118 **Uuo** (294)

Alkali Metals — IA (1)
Alkaline Earth Metals — IIA (2)
Transition Elements
Inner Transition Elements

Metals
Semiconductors
Nonmetals

Noble gases — VIIIA (18)
Halogens — VIIA (17)

†Lanthanides 6

	Cerium 58 **Ce** 140.1	Praseodymium 59 **Pr** 140.9	Neodymium 60 **Nd** 144.2	Promethium 61 **Pm** (145)	Samarium 62 **Sm** 150.4	Europium 63 **Eu** 152.0	Gadolinium 64 **Gd** 157.3	Terbium 65 **Tb** 158.9	Dysprosium 66 **Dy** 162.5	Holmium 67 **Ho** 164.9	Erbium 68 **Er** 167.3	Thulium 69 **Tm** 168.9	Ytterbium 70 **Yb** 173.0	Lutetium 71 **Lu** 175.0

‡Actinides 7

	Thorium 90 **Th** 232.0	Protactinium 91 **Pa** 231.0	Uranium 92 **U** 238.0	Neptunium 93 **Np** (237)	Plutonium 94 **Pu** (244)	Americium 95 **Am** (243)	Curium 96 **Cm** (247)	Berkelium 97 **Bk** (247)	Californium 98 **Cf** (251)	Einsteinium 99 **Es** (252)	Fermium 100 **Fm** (257)	Mendelevium 101 **Md** (258)	Nobelium 102 **No** (259)	Lawrencium 103 **Lr** (262)

Values in parentheses are the mass numbers of the most stable or best-known isotopes.

Names and symbols for elements 110–118 are under review.

Key

element name — Hydrogen
1
atomic number

symbol of element — **H**

1.008 — atomic weight

FIGURE 8.14 The periodic table of the elements.

atomic weight of the element. Horizontal rows of elements run from left to right with increasing atomic numbers. Each row is called a *period.* The periods are numbered from 1 to 7 on the left side. A vertical column of elements is called a *family* (or group) of elements. Elements in families have similar properties, but this is more true of some families than others. The table is subdivided into A and B groups. The members of the A-group families are called the *main group,* or **representative elements.** The members of the B-group are called the **transition elements** (or metals). Some science organizations use a "1 to 18" designation for the A and B groups, as shown in figure 8.14. The A and B designations will be used throughout this text.

As shown in table 8.3, all of the elements in the first column have an outside electron configuration of one electron. With the exception of hydrogen, the elements of the first column are shiny, low-density metals that are so soft you can cut them easily with a knife. These metals are called the *alkali metals* because they react violently with water to form an alkaline solution. The alkali metals do not occur in nature as free elements because they are so reactive. Hydrogen is a unique element in the periodic table. It is not an alkali metal, but it is placed in the group because it seems to fit there because it has one electron in its outer orbital.

The elements in the second column all have an outside configuration of two electrons and are called the *alkaline earth metals.* The alkaline earth metals are soft, reactive metals but not as reactive or soft as the alkali metals. Calcium and magnesium, in the form of numerous compounds, are familiar examples of this group.

The elements in group VIIA all have an outside configuration of seven electrons, needing only one more electron to completely fill the outer orbitals. These elements are called the *halogens.* The halogens are very reactive nonmetals. The halogens fluorine and chlorine are pale yellow and yellowish green gases. Bromine is a reddish brown liquid and iodine is a dark purple solid. Halogens are used as disinfectants, bleaches, and combined with a metal as a source of light in halogen lights. Halogens react with metals to form a group of chemicals called *salts,* such as sodium chloride. In fact, the word *halogen* is Greek, meaning "salt former."

As shown in table 8.4, the elements in group VIIIA have orbitals that are filled to capacity. These elements are colorless, odorless gases that almost never react with other elements to form compounds. Sometimes they are called the *noble gases*

because they are chemically inert, perhaps indicating they are above the other elements. They have also been called the *rare gases* because of their scarcity and *inert gases* because they are mostly chemically inert, not forming compounds. The noble gases are inert because they have filled outer electron configurations, a particularly stable condition.

Each period *begins* with a single electron in a new orbital. Second, each period *ends* with the filling of an orbital, completing the maximum number of electrons that can occupy that main energy level. Since the first A family is identified as IA, this means that all the atoms of elements in this family have one electron in their outer orbitals. All the atoms of elements in family IIA have two electrons in their outer orbitals. This pattern continues on to family VIIIA, in which all the atoms of elements have eight electrons in their outer orbitals except helium. Thus, the number identifying the A families *also identifies the number of electrons in the outer orbitals* (also called *valence electrons*), with the exception of helium. Helium is nonetheless similar to the other elements in this family, since all have filled outer orbitals. The electron theory of chemical bonding, which is discussed in chapter 9, states that only the electrons in the outermost orbitals of an atom are involved in chemical reactions. Thus, *the outer orbital electrons are mostly responsible for the chemical properties of an element.* Since the members of a family all have similar outer configurations, you would expect them to have similar chemical behaviors, and they do.

TABLE 8.3 Electron Structures of the Alkali Metal Family

Element	Number of Electrons in Main Energy Level						
	1ST	2ND	3RD	4TH	5TH	6TH	7TH
Lithium (Li)	2	1	—	—	—	—	—
Sodium (Na)	2	8	1	—	—	—	—
Potassium (K)	2	8	8	1	—	—	—
Rubidium (Rb)	2	8	18	8	1	—	—
Cesium (Cs)	2	8	18	18	8	1	—
Francium (Fr)	2	8	18	32	18	8	1

TABLE 8.4 Electron Structures of the Noble Gas Family

Element	Number of Electrons in Main Energy Level						
	1ST	2ND	3RD	4TH	5TH	6TH	7TH
Helium (He)	2	—	—	—	—	—	—
Neon (Ne)	2	8	—	—	—	—	—
Argon (Ar)	2	8	8	—	—	—	—
Krypton (Kr)	2	8	18	8	—	—	—
Xenon (Xe)	2	8	18	18	8	—	—
Radon (Rn)	2	8	18	32	18	8	

A Closer Look

The Rare Earths

Compounds of the rare earths were first identified when they were isolated from uncommon minerals in the late 1700s. The elements are very reactive and have similar chemical properties, so they were not recognized as elements until some fifty years later. Thus, they were first recognized as earths, that is, non-metal substances, when in fact they are metallic elements. They were also considered to be rare since, at that time, they were known to occur only in uncommon minerals. Today, these metallic elements are known to be more abundant in the Earth than gold, silver, mercury, or tungsten. The rarest of the rare earths, thulium, is twice as abundant as silver. The rare earth elements are neither rare nor earths, and they are important materials in glass, electronic, and metallurgical industries.

You can identify the rare earths in the two lowermost rows of the periodic table. These rows contain two series of elements that actually belong in periods 6 and 7, but they are moved below so that the entire table is not so wide. Together, the two series are called the inner transition elements. The top series is fourteen elements wide from elements 58 through 71. Since this series belongs next to element 57, lanthanum, it is sometimes called the *lanthanide series*. This series is also known as the rare earths. The second series of fourteen elements is called the *actinide series*. These are mostly the artificially prepared elements that do not occur naturally.

You may never have heard of the rare earth elements, but they are key materials in many advanced or high-technology products. Lanthanum, for example, gives glass special refractive properties and is used in optic fibers and expensive camera lenses. Samarium, neodymium, and dysprosium are used to manufacture crystals used in lasers. Samarium, ytterbium, and terbium have special magnetic properties that have made possible new electric motor designs, magnetic-optical devices in computers, and the creation of a ceramic superconductor. Other rare earth metals are also being researched for use in possible high-temperature superconductivity materials. Many rare earths are also used in metal alloys; for example, an alloy of cerium is used to make heat-resistant jet-engine parts. Erbium is also used in high-performance metal alloys. Dysprosium and holmium have neutron-absorbing properties and are used in control rods to control nuclear fission. Europium should be mentioned because of its role in making the red color of color television screens. The rare earths are relatively abundant metallic elements that play a key role in many common and high-technology applications. They may also play a key role in superconductivity research.

CONCEPTS APPLIED

Outer Orbitals

How many outer orbital electrons are found in an atom of (a) oxygen, (b) calcium, and (c) aluminum? Write your answers before reading the answers in the next paragraph.

(a) According to the list of elements on the inside back cover of this text, oxygen has the symbol O and an atomic number of 8. The square with the symbol O and the atomic number 8 is located in the column identified as VIA. Since the A family number is the same as the number of electrons in the outer orbital, oxygen has six outer orbital electrons. (b) Calcium has the symbol Ca (atomic number 20) and is located in column IIA, so a calcium atom has two outer orbital electrons. (c) Aluminum has the symbol Al (atomic number 13) and is located in column IIIA, so an aluminum atom has three outer orbital electrons.

METALS, NONMETALS, AND SEMICONDUCTORS

As indicated earlier, chemical behavior is mostly concerned with the outer shell electrons. The outer shell electrons, that is, the highest energy level electrons, are conveniently represented with an **electron dot notation,** made by writing the chemical symbol with dots around it indicating the number of outer shell electrons. Electron dot notations are shown for the representative elements in figure 8.15. Again, note the pattern in figure 8.15—all the noble gases are in group VIIIA, and all (except helium) have eight outer electrons. All the group IA elements (alkali metals) have one dot, all the IIA elements have two dots, and so on. This pattern will explain the difference in metals, nonmetals, and a third group of in-between elements called semiconductors.

There are several ways to classify matter according to properties. One example is to group substances according to the physical properties of metals and nonmetals—luster, conductivity, malleability, and ductility. Metals and nonmetals

FIGURE 8.15   Electron dot notation for the representative elements.

Nutrient Elements

All molecules required to support living things are called nutrients. Some nutrients are inorganic molecules such as calcium, iron, or potassium, and others are organic molecules such as carbohydrates, proteins, fats, and vitamins. All minerals are elements and cannot be synthesized by the body. Because they are elements, they cannot be broken down or destroyed by metabolism or cooking. They commonly occur in many foods and in water. Minerals retain their characteristics whether they are in foods or in the body, and each plays a different role in metabolism.

Minerals* can function as regulators, activators, transmitters, and controllers of various enzymatic reactions. For example, sodium ions (Na^+) and potassium ions (K^+) are important in the transmission of nerve impulses, while magnesium ions (Mg^{2+}) facilitate energy release during reactions involving ATP. Without iron, not enough hemoglobin would be formed to transport oxygen, a condition called anemia, and a lack of calcium may result in osteoporosis. Osteoporosis is a condition that results from calcium loss leading to painful, weakened bones. There are many minerals that are important in your diet. In addition to those just mentioned, you need chlorine, cobalt, copper, iodine, phosphorus, potassium, sulfur, and zinc to remain healthy. With few exceptions, adequate amounts of minerals are obtained in a normal diet. Calcium and iron supplements may be necessary, particularly in women.

Osteoporosis is a nutritional deficiency disease that results in a change in the density of the bones as a result of the loss of bone mass. Bones that have undergone this change look lacy or like Swiss cheese, with larger than normal holes. A few risk factors found to be associated with this disease are being female and fair skinned; having a sedentary lifestyle; using alcohol, caffeine, and tobacco; being anorexic; and having reached menopause.

*Note that the term *mineral* has a completely different meaning in geology. See page 328 in chapter 15.

---

also have certain chemical properties that are related to their positions in the periodic table. Figure 8.16 shows where the *metals, nonmetals,* and *semiconductors* are located. Note that about 80 percent of all the elements are metals.

The noble gases have completely filled outer orbitals in their highest energy levels, and this is a particularly stable arrangement. Other elements react chemically, either *gaining or losing electrons to attain a filled outermost energy level like the noble gases.* When an atom loses or gains electrons, it acquires an unbalanced electron charge and is called an **ion.** An atom of lithium, for example, has three protons (plus charges) and three electrons (negative charges). If it loses the outermost electron, it now has an outer filled orbital structure like helium, a noble gas. It is also now an ion, since it has three protons (3+) and two electrons (2−), for a net charge of 1+. A lithium ion thus has a 1+ charge.

Elements with one, two, or three outer electrons tend to lose electrons to form positive ions. The metals lose electrons like this, and the *metals are elements that lose electrons to form positive ions* (figure 8.17). Nonmetals, on the other hand, are elements with five to seven outer electrons that tend to acquire electrons to fill their outer orbitals. *Nonmetals are elements that gain electrons to form negative ions.* In general, elements located in the left two-thirds or so of the periodic table are metals. The nonmetals are on the right side of the table.

The dividing line between the metals and nonmetals is a steplike line from the left top of group IIIA down to the bottom left of group VIIA. This is not a line of sharp separation between the metals and nonmetals, and elements *along* this line sometimes act like metals, sometimes like nonmetals, and sometimes like both. These hard-to-classify elements are called **semiconductors** (or *metalloids*). Silicon, germanium, and arsenic have physical properties of nonmetals; for example, they are brittle materials that cannot be hammered into a new shape. Yet these elements conduct electric currents under certain conditions. The ability to conduct

FIGURE 8.16 The location of metals, nonmetals, and semiconductors in the periodic table.

People Behind the Science

Dmitri Ivanovich Mendeleyev (1834–1907)

Dmitri Mendeleyev was a Russian chemist whose name will always be linked with his outstanding achievement, the development of the periodic table. He was the first chemist to understand that all elements are related members of a single, ordered system. He converted what had been a highly fragmented and speculative branch of chemistry into a true, logical science. The spelling of his name has been a source of confusion for students and frustration for editors for more than a century, and the forms Mendeléeff, Mendeléev, and even Mendelejeff can all be found in print.

Mendeleyev was born in Tobol'sk, Siberia, on February 7, 1834, the youngest of the seventeen children of the head of the local high school. His father went blind when Mendeleyev was a child, and the family had to rely increasingly on their mother for support. He was educated locally but could not gain admission to any Russian university because of prejudice toward the supposedly backward attainments of those educated in the provinces. In 1855, he finally qualified as a teacher at the Pedagogical Institute in St. Petersburg. He took an advanced-degree course in chemistry. In 1859, he was sent by the government for further study at the University of Heidelberg, and in 1861, he returned to St. Petersburg and became professor of general chemistry at the Technical Institute in 1864. He could find no textbook adequate for his students' needs, so he decided to produce his own. The resulting *Principles of Chemistry* (1868–1870) won him international renown; it was translated into English in 1891 and 1897.

Before Mendeleyev produced his periodic law, understanding of the chemical elements had long been an elusive and frustrating task. According to Mendeleyev, the properties of the elements are periodic functions of their atomic weights. In 1869, he stated that "the elements arranged according to the magnitude of atomic weights show a periodic change of properties." Other chemists, notably Lothar Meyer in Germany, had meanwhile come to similar conclusions, with Meyer publishing his findings independently.

Mendeleyev compiled the first true periodic table, listing all the sixty-three elements then known. Not all elements would "fit" properly using the atomic weights of the time, so he altered indium from 76 to 114 (modern value 114.8) and beryllium from 13.8 to 9.2 (modern value 9.013). In 1871, he produced a revisionary paper showing the correct repositioning of seventeen elements.

To make the table work, Mendeleyev also had to leave gaps, and he predicted that further elements would eventually be discovered to fill them. These predictions provided the strongest endorsement of the periodic law. Three were discovered in Mendeleyev's lifetime: gallium (1871), scandium (1879), and germanium (1886), all with properties that tallied closely with those he had assigned to them.

Farsighted though Mendeleyev was, he had no notion that the periodic recurrences of similar properties in the list of elements reflected anything in the structures of their atoms. It was not until the 1920s that it was realized that the key parameter in the periodic system is not the atomic weight but the atomic number of the elements—a measure of the number of protons in the atom. Since then, great progress has been made in explaining the periodic law in terms of the electronic structures of atoms and molecules.

Source: Modified from the *Hutchinson Dictionary of Scientific Biography.* © RM, 2009. All rights reserved. Helicon publishing is a division of RM.

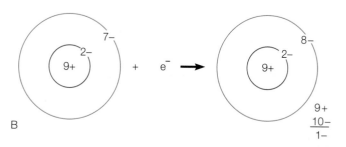

FIGURE 8.17 (*A*) Metals lose their outer electrons to acquire a noble gas structure and become positive ions. Lithium becomes a 1+ ion as it loses its one outer electron. (*B*) Nonmetals gain electrons to acquire an outer noble gas structure and become negative ions. Fluorine gains a single electron to become a 1− ion.

CONCEPTS APPLIED

Metals and Charge

Is strontium a metal, nonmetal, or semiconductor? What is the charge on a strontium ion?

The list of elements inside the back cover identifies the symbol for strontium as Sr (atomic number 38). In the periodic table, Sr is located in family IIA, which means that an atom of strontium has two electrons in its outer shell. For several reasons, you know that strontium is a metal: (1) An atom of strontium has two electrons in its outer shell, and atoms with one, two, or three outer electrons are identified as metals; (2) strontium is located in the IIA family, the alkaline earth metals; and (3) strontium is located on the left side of the periodic table, and in general, elements located in the left two-thirds of the table are metals.

Elements with one, two, or three outer electrons tend to lose electrons to form positive ions. Since strontium has an atomic number of 38, you know that it has thirty-eight protons (38+) and thirty-eight electrons (38−). When it loses its two outer shell electrons, it has 38+ and 36− for a charge of 2+ .

an electric current is a property of a metal, and nonmalleability is a property of nonmetals, so as you can see, these semiconductors have the properties of both metals and nonmetals.

The transition elements, which are all metals, are located in the B-group families. Unlike the representative elements, which form vertical families of similar properties, the transition elements tend to form horizontal groups of elements with similar properties. Iron (Fe), cobalt (Co), and nickel (Ni) in group VIIIB, for example, are three horizontally arranged metallic elements that show magnetic properties.

A family of representative elements all form ions with the same charge. Alkali metals, for example, all lose an electron to form a 1+ ion. The transition elements have *variable charges.* Some transition elements, for example, lose their one outer electron to form 1+ ions (copper, silver). Copper, because of its special configuration, can also lose an additional electron to form a 2+ ion. Thus, copper can form either a 1+ ion or a 2+ ion. Most transition elements have two outer s orbital electrons and lose them both to form 2+ ions (iron, cobalt, nickel), but some of these elements also have special configurations that permit them to lose more of their electrons. Thus, iron and cobalt, for example, can form either a 2+ ion or a 3+ ion. Much more can be interpreted from the periodic table, and more generalizations will be made as the table is used in the following chapters.

SUMMARY

Attempts at understanding matter date back to ancient Greek philosophers, who viewed matter as being composed of *elements,* or simpler substances. Two models were developed that considered matter to be (1) *continuous,* or infinitely divisible, or (2) *discontinuous,* made up of particles called *atoms.*

In the early 1800s, Dalton published an *atomic theory,* reasoning that matter was composed of hard, indivisible atoms that were joined together or dissociated during chemical change.

When a good air pump to provide a vacuum was invented in 1885, *cathode rays* were observed to move from the negative terminal in an evacuated glass tube. The nature of cathode rays was a mystery. The mystery was solved in 1887 when Thomson discovered they were negatively charged particles now known as *electrons.* Thomson had discovered the first elementary particle of which atoms are made and measured their charge-to-mass ratio.

Rutherford developed a solar system model based on experiments with alpha particles scattered from a thin sheet of metal. This model had a small, massive, and positively charged *nucleus* surrounded by moving electrons. These electrons were calculated to be at a distance from the nucleus of 100,000 times the radius of the nucleus, so the volume of an atom is mostly empty space. Later, Rutherford proposed that the nucleus contained two elementary particles: *protons* with a positive charge and *neutrons* with no charge. The *atomic number* is the number of protons in an atom. Atoms of elements with different numbers of neutrons are called *isotopes.* The mass of each isotope is compared to the mass of carbon-12, which is assigned a mass of exactly 12.00 *atomic mass units.* The mass contribution of the isotopes of an element according to their abundance is called the *atomic weight* of an element. Isotopes are identified by their *mass number,* which is the sum of the number of protons and neutrons in the nucleus. Isotopes are identified by their chemical symbol, with the atomic number as a subscript and the mass number as a superscript.

Bohr developed a model of the hydrogen atom to explain the characteristic *line spectra* emitted by hydrogen. His model specified that (1) electrons can move only in allowed orbits, (2) electrons do not emit radiant energy when they remain in an orbit, and (3) electrons move from one allowed orbit to another when they gain or lose energy. When an electron jumps from a higher orbit to a lower one, it gives up energy in the form of a single photon. The energy of the photon corresponds to the difference in energy between the two levels. The Bohr model worked well for hydrogen but not for other atoms.

Schrödinger and others used the wave nature of the electron to develop a new model of the atom called *wave mechanics,* or *quantum mechanics.* This model was found to confirm exactly all the experimental data as well as predict new data. The quantum mechanical model describes the energy state of the electron in terms of quantum numbers based on the wave nature of the electron. The quantum numbers defined the *probability* of the location of an electron in terms of fuzzy regions of space called *orbitals.*

The *periodic table* has horizontal rows of elements called *periods* and vertical columns of elements called *families.* Members of a given family have the same outer orbital electron configurations, and it is the electron configuration that is mostly responsible for the chemical properties of an element.

Summary of Equations

8.1

energy = (Planck's constant)(frequency)

$$E = hf$$

where $h = 6.63 \times 10^{-34}$ J · s

KEY TERMS

atomic mass units (p. **182**)
atomic number (p. **181**)
atomic weight (p. **182**)

electron (p. **181**)
electron dot notation (p. **190**)
excited states (p. **185**)

ground state (p. **185**)
Heisenberg uncertainty principle (p. **186**)

ion (p. **191**)
isotope (p. **182**)
line spectrum (p. **183**)

mass number (p. 182) orbital (p. 187) proton (p. 181) semiconductors (p. 191)
neutron (p. 181) Pauli exclusion principle (p. 187) quanta (p. 183) transition elements (p. 189)
nucleus (p. 181) photons (p. 183) representative elements (p. 189)

APPLYING THE CONCEPTS

1. The electron was discovered through experiments with
   a. radioactivity.
   b. light.
   c. matter waves.
   d. electricity.

2. Thomson was convinced that he had discovered a subatomic particle, the electron, from the evidence that
   a. the charge-to-mass ratio was the same for all materials.
   b. cathode rays could move through a vacuum.
   c. electrons were attracted toward a negatively charged plate.
   d. the charge was always 1.60×10^{-19} coulomb.

3. The existence of a tiny, massive, and positively charged nucleus was deduced from the observation that
   a. fast, massive, and positively charged alpha particles all move straight through metal foil.
   b. alpha particles were deflected by a magnetic field.
   c. some alpha particles were deflected by metal foil.
   d. None of the above is correct.

4. According to Rutherford's calculations, the volume of an atom is mostly
   a. occupied by protons and neutrons.
   b. filled with electrons.
   c. occupied by tightly bound protons, electrons, and neutrons.
   d. empty space.

5. Hydrogen, with its one electron, produces a line spectrum in the visible light range with
   a. one color line.
   b. two color lines.
   c. three color lines.
   d. four color lines.

6. According to the Bohr model, an electron gains or loses energy only by
   a. moving faster or slower in an allowed orbit.
   b. jumping from one allowed orbit to another.
   c. being completely removed from an atom.
   d. jumping from one atom to another atom.

7. When an electron in a hydrogen atom jumps from an orbit farther from the nucleus to an orbit closer to the nucleus, it
   a. emits a single photon with an energy equal to the energy difference of the two orbits.
   b. emits four photons, one for each of the color lines observed in the line spectrum of hydrogen.
   c. emits a number of photons dependent on the number of orbit levels jumped over.
   d. None of the above is correct.

8. The quantum mechanics model of the atom is based on
   a. the quanta, or measured amounts of energy of a moving particle.
   b. the energy of a standing electron wave that can fit into an orbit.
   c. calculations of the energy of the three-dimensional shape of a circular orbit of an electron particle.
   d. Newton's laws of motion but scaled down to the size of electron particles.

9. The Bohr model of the atom described the energy state of electrons with one quantum number. The quantum mechanics model uses how many quantum numbers to describe the energy state of an electron?
   a. one
   b. two
   c. four
   d. ten

10. The space in which it is probable that an electron will be found is described by a(an)
    a. circular orbit.
    b. elliptical orbit.
    c. orbital.
    d. geocentric orbit.

Answers

1. d 2. a 3. c 4. d 5. d 6. b 7. a 8. b 9. c 10. c

QUESTIONS FOR THOUGHT

1. What was the experimental evidence that Thomson had discovered the existence of a subatomic particle when working with cathode rays?

2. Describe the experimental evidence that led Rutherford to the concept of a nucleus in an atom.

3. What is the main problem with a solar system model of the atom?

4. Compare the size of an atom to the size of its nucleus.

5. What does *atomic number* mean? How does the atomic number identify the atoms of a particular element? How is the atomic number related to the number of electrons in an atom?

6. An atom has 11 protons in the nucleus. What is the atomic number? What is the name of this element? What is the electron configuration of this atom?

7. How is the atomic weight of an element determined?

8. Describe the three main points in the Bohr model of the atom.

9. Why do the energies of electrons in an atom have negative values? (*Hint:* It is *not* because of the charge of the electron.)

10. Which has the lowest energy, an electron in the first energy level ($n = 1$) or an electron in the third energy level ($n = 3$)? Explain.

11. What is similar about the Bohr model of the atom and the quantum mechanical model? What are the fundamental differences?

12. What is the difference between a hydrogen atom in the ground state and one in the excited state?

FOR FURTHER ANALYSIS

1. Evaluate Millikan's method for finding the charge of an electron. Are there any doubts about the results of using this technique?

2. What are the significant similarities and differences between the isotopes of a particular element?

3. Thomson's experiments led to the discovery of the electron. Analyze how you know for sure that he discovered the electron.

4. Describe a conversation between yourself and another person as you correct his or her belief that atomic weight has something to do with gravity.

5. Analyze the significance of the observation that matter only emits and absorbs energy in discrete units.

6. Describe at least several basic differences between the Bohr and quantum mechanics models of the atom.

INVITATION TO INQUIRY

Too Small to See?

As Rutherford knew when he conducted his famous experiment with radioactive particles and gold foil, the structure of an atom is too small to see, so it must be inferred from other observations. To illustrate this process, pour 50 mL of 95 percent ethyl alcohol (or some other almost pure alcohol) into a graduated cylinder. In a second graduated cylinder, measure 50 mL of water. Mix the alcohol and water together thoroughly and record the volume of the combined liquids. Assuming no evaporation took place, is the result contrary to your expectation? What question could be asked about the result?

Answers to questions about things that you cannot see often require a model that can be observed. For example, a model might represent water and alcohol molecules by using beans for alcohol molecules and sand for water molecules. Does mixing 50 mL of beans and 50 mL of sand result in 100 mL of mixed sand and beans? Explain the result of mixing alcohol and water based on your observation of mixing beans and sand.

To continue this inquiry, fill a water glass to the brim with water. Add a small amount of salt to the water. You know that two materials cannot take up the same space at the same time, so what happened to the salt? How can you test your ideas about things that are too tiny to see?

PARALLEL EXERCISES

The exercises in groups A and B cover the same concepts. Solutions to group A exercises are located in appendix D.

Group A

1. How much energy is needed to move an electron in a hydrogen atom from $n = 2$ to $n = 6$? (See figure 8.11 for needed values.)

2. What frequency of light is emitted when an electron in a hydrogen atom jumps from $n = 6$ to $n = 2$? What color would you see?

3. How much energy is needed to completely remove the electron from a hydrogen atom in the ground state?

4. Thomson determined the charge-to-mass ratio of the electron to be -1.76×10^{11} coulomb/kilogram. Millikan determined the charge on the electron to be -1.60×10^{-19} coulomb. According to these findings, what is the mass of an electron?

Group B

1. How much energy is needed to move an electron in a hydrogen atom from the ground state ($n = 1$) to $n = 3$?

2. What frequency of light is emitted when an electron in a hydrogen atom jumps from $n = 2$ to the ground state ($n = 1$)?

3. How much energy is needed to completely remove an electron from $n = 2$ in a hydrogen atom?

4. If the charge-to-mass ratio of a proton is 9.58×10^7 coulomb/ kilogram and the charge is 1.60×10^{-19} coulomb, what is the mass of the proton?

9

Chemical Reactions

A chemical change occurs when iron rusts, and rust is a different substance with physical and chemical properties different from iron. This rusted anchor makes a colorful display on the bow of a grain ship.

CORE **CONCEPT**

Electron structure will explain how and why atoms join together in certain numbers.

OUTLINE

Chemical reactions are changes in matter in which different substances are created by forming or breaking chemical bonds (p. 199).

A chemical bond is an attractive force that holds atoms together in a compound (p. 202).

An ionic bond is a chemical bond of electrostatic attraction between ions (p. 203).

Compounds
Elements
Chemical Change
Valence Electrons and Ions
Chemical Bonds
  Ionic Bonds
  Covalent Bonds
A Closer Look: Name That Compound
Composition of Compounds
Chemical Equations

A Closer Look: How to Write a Chemical Formula
A Closer Look: On Balancing Equations
Types of Chemical Reactions
  Combination Reactions
  Decomposition Reactions
  Replacement Reactions
  Ion Exchange Reactions
People Behind the Science: Linus Carl Pauling

A covalent bond is a chemical bond formed by the sharing of electrons (p. 204).

Chemistry

▷ The electron structure of water will explain its properties and the properties of solutions (Ch. 10 and 17).

Earth Science

▷ Chemical change will explain one way solid rocks on Earth's surface are weathered (Ch. 16).

Life Science

▷ Organic chemistry and biochemistry consider chemical changes in the important organic compounds of life (Ch. 19).

OVERVIEW

In chapter 8, you learned how the modern atomic theory is used to describe the structures of atoms of different elements. The electron structures of different atoms successfully account for the position of elements in the periodic table as well as for groups of elements with similar properties. On a large scale, all metals were found to have a similarity in electron structure, as were nonmetals. On a smaller scale, chemical families such as the alkali metals were found to have the same outer electron configurations. Thus, the modern atomic theory accounts for observed similarities between elements in terms of atomic structure.

So far, only individual, isolated atoms have been discussed; we have not considered how atoms of elements join together to produce compounds. There is a relationship between the electron structure of atoms and the reactions they undergo to produce specific compounds. Understanding this relationship will explain the changes that matter itself undergoes. For example, hydrogen is a highly flammable, gaseous element that burns with an explosive reaction. Oxygen, on the other hand, is a gaseous element that supports burning. As you know, hydrogen and oxygen combine to form water. Water is a liquid that neither burns nor supports burning. What happens when atoms of elements such as hydrogen and oxygen join to form molecules such as water? Why do such atoms join and why do they stay together? Why does water have properties different from the elements that combine to produce it? And finally, why is water H_2O and not H_3O or H_4O?

Answers to questions about why and how atoms join together in certain numbers are provided by considering the electronic structures of the atoms. Chemical substances are formed from the interactions of electrons as their structures merge, forming new patterns that result in molecules with new properties. It is the new electron pattern of the water molecule that gives water properties that are different from the oxygen or hydrogen from which it formed (figure 9.1). Understanding how electron structures of atoms merge to form new patterns is understanding the changes that matter itself undergoes, the topic of this chapter.

COMPOUNDS

Matter that you see around you seems to come in a wide variety of sizes, shapes, forms, and kinds. Are there any patterns in the apparent randomness that will help us comprehend matter? Yes, there are many patterns, and one of the more obvious is that all matter occurs as either a mixture or as a pure substance (figure 9.2). A *mixture* has unlike parts and a composition that varies from sample to sample. For example, sand from a beach is a variable mixture of things such as bits of rocks, minerals, and sea shells.

There are two distinct ways in which mixtures can have unlike parts with a variable composition. A *heterogeneous mixture* has physically distinct parts with different properties. Beach sand, for example, is usually a heterogeneous mixture of tiny pieces of rocks and tiny pieces of shells that you can see. It is said to be heterogeneous because any two given samples will have a different composition with different kinds of particles.

A solution of salt dissolved in water also meets the definition of a mixture since it has unlike parts and can have a variable composition. A solution, however, is different from a sand mixture since it is a *homogeneous mixture*, meaning it is the same throughout a given sample. A homogeneous mixture, or solution, is the same throughout. The key to understanding that a solution is a mixture is found in its variable composition; that is, a given solution might be homogeneous, but one solution can vary from the next. Thus, you can have a salt solution with a 1 percent concentration, another with a 7 percent

FIGURE 9.1 Water is the most abundant liquid on Earth and is necessary for all life. Because of water's great dissolving properties, any sample is a solution containing solids, other liquids, and gases from the environment. This stream also carries suspended, ground-up rocks, called *rock flour*, from a nearby glacier.

A   B

FIGURE 9.3 Sugar (*A*) is a compound that can be easily decomposed to simpler substances by heating. (*B*) One of the simpler substances is the black element carbon, which cannot be further decomposed by chemical or physical means.

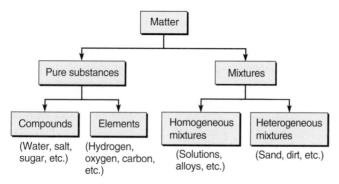

FIGURE 9.2 A classification scheme for matter.

concentration, yet another with a 10 percent concentration, and so on. Solutions, mixed gases, and metal alloys are all homogeneous mixtures since they are made of unlike parts and do not have a fixed, definite composition.

Mixtures can be separated into their component parts by physical means. For example, you can physically separate the parts making up a sand mixture by using a magnifying glass and tweezers to move and isolate each part. A solution of salt in water is a mixture since the amount of salt dissolved in water can have a variable composition. But how do you separate the parts of a solution? One way is to evaporate the water, leaving the salt behind. There are many methods for separating mixtures, but all involve a **physical change.** A physical change does not alter the *identity* of matter. When water is evaporated from the salt solution, for example, it changes to a different state of matter (water vapor) but is still recognized as water. Physical changes involve physical properties only; no new substances

are formed. Examples of physical changes include evaporation, condensation, melting, freezing, and dissolving, as well as reshaping processes such as crushing or bending.

Mixtures can be physically separated into *pure substances,* materials that are the same throughout and have a fixed, definite composition. If you closely examine a sample of table salt, you will see that it is made up of hundreds of tiny cubes. Any one of these cubes will have the same properties as any other cube, including a salty taste. Sugar, like table salt, has all of its parts alike. Unlike salt, sugar grains have no special shape or form, but each grain has the same sweet taste and other properties as any other sugar grain.

If you heat salt and sugar in separate containers, you will find very different results. Salt, like some other pure substances, undergoes a physical change and melts, changing back to a solid upon cooling with the same properties that it had originally. Sugar, however, changes to a black material upon heating, while it gives off water vapor. The black material does not change back to sugar upon cooling. The sugar has *decomposed* to a new substance, while the salt did not. The new substance is carbon, and it has properties completely different from sugar. The sugar has gone through a **chemical change.** A chemical change alters the identity of matter, producing new substances with different properties. In this case, the chemical change was one of decomposition. Heat produced a chemical change by decomposing the sugar into carbon and water vapor.

The decomposition of sugar always produces the same mass ratio of carbon to water, so sugar has a fixed, definite composition. A **compound** is a pure substance that can be decomposed by a chemical change into simpler substances with a fixed mass ratio. This means that sugar is a compound (figure 9.3).

A pure substance that cannot be broken down into anything simpler by chemical or physical means is an **element.**

Sugar is decomposed by heating into carbon and water vapor. Carbon cannot be broken down further, so carbon is an element. It has been known since about 1800 that water is a compound that can be broken down by electrolysis into hydrogen and oxygen, two gases that cannot be broken down to anything simpler. So sugar is a compound made from the elements of carbon, hydrogen, and oxygen.

But what about the table salt? Is table salt a compound? Table salt is a stable compound that is not decomposed by heating. It melts at a temperature of about 800°C and then returns to the solid form with the same salty properties upon cooling. Electrolysis—the splitting of a compound by means of electricity—can be used to decompose table salt into the elements sodium and chlorine, positively proving that it is a compound.

Pure substances are either compounds or elements. Decomposition through heating and decomposition through electrolysis are two means of distinguishing between compounds and elements. If a substance can be decomposed into something simpler, you know for sure that it is a compound. If the substance cannot be decomposed, it might be an element, or it might be a stable compound that resists decomposition. More testing would be necessary before you can be confident that you have identified an element. Most pure substances are compounds. There are millions of different compounds but only 113 known elements at the present time. These elements are the fundamental materials of which all matter is made.

ELEMENTS

Elements are not equally abundant and only a few are common. In table 9.1, for example, you can see that only eight elements make up about 99 percent of the solid surface of Earth. Oxygen is most abundant, making up about 50 percent of the weight of Earth's crust. Silicon makes up more than 25 percent, so these two nonmetals alone make up about 75 percent of Earth's solid surface. Almost all the rest is made up of just six metals, as shown in the table.

The number of common elements is limited elsewhere, too. Only two elements make up about 99 percent of the atmospheric

TABLE 9.1 **Elements Making up 99 Percent of Earth's Crust**

Element (Symbol)	Percent by Weight
Oxygen (O)	46.6
Silicon (Si)	27.7
Aluminum (Al)	8.1
Iron (Fe)	5.0
Calcium (Ca)	3.6
Sodium (Na)	2.8
Potassium (K)	2.6
Magnesium (Mg)	2.1

air around Earth. Air is mostly nitrogen (about 78 percent) and oxygen (about 21 percent), with traces of five other elements and compounds. Water is hydrogen and oxygen, of course, but seawater also contains elements in solution. These elements are chlorine (55 percent), sodium (31 percent), sulfur (8 percent), and magnesium (4 percent). Only three elements make up about 97 percent of your body. These elements are hydrogen (60 percent), oxygen (26 percent), and carbon (11 percent). Generally, all of this means that the elements are not equally distributed or equally abundant in nature (figure 9.4).

CHEMICAL CHANGE

The air you breathe, the liquids you drink, and all the things around you are elements, compounds, or mixtures. Most are compounds, however, and very few are pure elements. Water, sugar, gasoline, and chalk are examples of compounds. Each can be broken down into the elements that make it up. Recall that elements are basic substances that cannot be broken down into simpler substances. Examples of elements are hydrogen, carbon, and calcium. Why and how these elements join together in different ways to form different compounds is the subject of this chapter.

You have already learned that elements are made up of atoms that can be described by the modern atomic theory. You can also consider an **atom** to be *the smallest unit of an element that can exist alone or in combination with other elements.* Compounds are formed when atoms are held together by an

FIGURE 9.4 The elements of aluminum, iron, oxygen, and silicon make up about 88 percent of Earth's solid surface. Water on the surface and in the air as clouds and fog is made up of hydrogen and oxygen. The air is 99 percent nitrogen and oxygen. Hydrogen, oxygen, and carbon make up 97 percent of a person. Thus, almost everything you see in this picture is made up of just six elements.

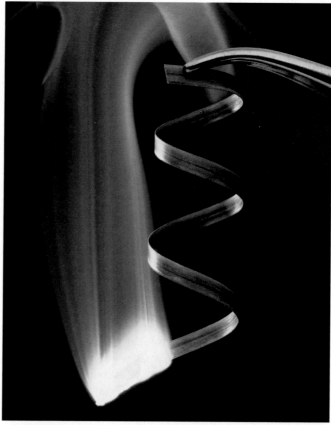

FIGURE 9.5 Magnesium is an alkaline earth metal that burns brightly in air, releasing heat and light. As chemical energy is released, a new chemical substance is formed. The new chemical material is magnesium oxide, a soft powdery material that forms an alkaline solution in water (called *milk of magnesia*).

attractive force called a *chemical bond.* The chemical bond binds individual atoms together in a compound. A molecule is generally thought of as a tightly bound group of atoms that maintains its identity. More specifically, a **molecule** is defined as *the smallest particle of a compound, or a gaseous element, that can exist and still retain the characteristic chemical properties of a substance.* Compounds with one type of chemical bond, as you will see, have molecules that are electrically neutral groups of atoms held together strongly enough to be considered independent units. For example, water is a compound. The smallest unit of water that can exist alone is an electrically neutral unit made up of two hydrogen atoms and one oxygen atom held together by chemical bonds. The concept of a molecule will be expanded as chemical bonds are discussed.

Compounds occur naturally as gases, liquids, and solids. Many common gases occur naturally as molecules made up of two or more atoms. For example, at ordinary temperatures, hydrogen gas occurs as molecules of two hydrogen atoms bound together. Oxygen gas also usually occurs as molecules of two oxygen atoms bound together. Both hydrogen and oxygen occur naturally as *diatomic molecules* (*di-* means "two"). Oxygen sometimes occurs as molecules of three oxygen atoms bound together. These *triatomic molecules* (*tri-* means "three") are called *ozone.*

When molecules of any size are formed or broken down into simpler substances, new materials with new properties are produced. This kind of a change in matter is called a chemical change, and the process is called a chemical reaction. A **chemical reaction** is defined as

> **a change in matter in which different chemical substances are created by forming or breaking chemical bonds.**

In general, chemical bonds are formed when atoms of elements are bound together to form compounds. Chemical bonds are broken when a compound is decomposed into simpler substances. Chemical bonds are electrical in nature, formed by electrical attractions, as discussed in chapter 6.

Chemical reactions happen all the time, all around you. A growing plant, burning fuels, and your body's utilization of food all involve chemical reactions. These reactions produce different chemical substances with greater or smaller amounts of internal potential energy (see chapter 4 for a discussion of internal potential energy). Energy is *absorbed* to produce new chemical substances with more internal potential energy. Energy is *released* when new chemical substances are produced with less internal potential energy (figure 9.5). In general, changes in internal potential energy are called **chemical energy.** For example, new chemical substances are produced in green plants through the process called *photosynthesis.* A green plant uses radiant energy (sunlight), carbon dioxide, and water to produce new chemical materials and oxygen. These new chemical materials, the stuff that leaves, roots, and wood are made of, contain more chemical energy than the carbon dioxide and water they were made from.

A **chemical equation** is a way of describing what happens in a chemical reaction. The chemical reaction of photosynthesis can be described by using words in an equation:

$$\underset{\text{(sunlight)}}{\text{energy}} + \underset{\text{molecules}}{\underset{\text{dioxide}}{\text{carbon}}} + \underset{\text{molecules}}{\text{water}} \rightarrow \underset{\text{molecules}}{\underset{\text{material}}{\text{plant}}} + \underset{\text{molecules}}{\text{oxygen}}$$

The substances that are changed are on the left side of the word equation and are called *reactants*. The reactants are carbon dioxide molecules and water molecules. The equation also indicates that energy is absorbed, since the term *energy* appears on the left side. The arrow means *yields*. The new chemical substances are on the right side of the word equation and are called *products*. Reading the photosynthesis reaction as a sentence, you would say, "Carbon dioxide and water absorb energy to react, yielding plant materials and oxygen."

The plant materials produced by the reaction have more internal potential energy, also known as *chemical energy,* than the reactants. You know this from the equation because the term *energy* appears on the left side but not the right. This means that the energy on the left went into internal potential energy on the right. You also know this because the reaction can be reversed to release the stored energy (figure 9.6). When plant materials (such as wood) are burned, the materials react with oxygen, and chemical energy is released in the form of radiant energy (light) and high kinetic energy of the newly formed gases and vapors. In words,

$$\underset{\text{molecules}}{\underset{\text{material}}{\text{plant}}} + \underset{\text{molecules}}{\text{oxygen}} \rightarrow \underset{\text{molecules}}{\underset{\text{dioxide}}{\text{carbon}}} + \underset{\text{molecules}}{\text{water}} + \text{energy}$$

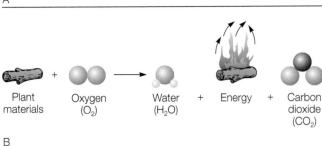

A

B

FIGURE 9.6 (*A*) New chemical bonds are formed as a green plant makes new materials and stores solar energy through the photosynthesis process. (*B*) The chemical bonds are later broken, and the same amount of energy and the same original materials are released. The same energy and the same materials are released rapidly when the plant materials burn, and they are released slowly when the plant decomposes.

If you compare the two equations, you will see that burning is the opposite of the process of photosynthesis! The energy released in burning is exactly the same amount of solar energy that was stored as internal potential energy by the plant. Such chemical changes, in which chemical energy is stored in one reaction and released by another reaction, are the result of the making, then the breaking, of chemical bonds. Chemical bonds were formed by utilizing energy to produce new chemical substances. Energy was released when these bonds were broken then re-formed to produce the original substances. In this example, chemical reactions and energy flow can be explained by the making and breaking of chemical bonds. Chemical bonds can be explained in terms of changes in the electron structures of atoms. Thus, the place to start in seeking understanding about chemical reactions is the electron structure of the atoms themselves.

VALENCE ELECTRONS AND IONS

As discussed in chapter 8, it is the number of electrons in the outermost orbital that usually determines the chemical properties of an atom. These outer electrons are called **valence electrons,** and it is the valence electrons that participate in chemical bonding. The inner electrons are in stable, fully occupied orbitals and do not participate in chemical bonds. The representative elements (the A-group families) have valence electrons in the outermost orbitals, which contain from one to eight valence electrons. Recall that you can easily find the number of valence electrons by referring to a periodic table. The number at the top of each representative family is the same as the number of outer orbital electrons (with the exception of helium).

The noble gases have filled outer orbitals and do not normally form compounds. Apparently, half-filled and filled orbitals are particularly stable arrangements. Atoms have a tendency to seek such a stable, filled outer orbital arrangement such as the one found in the noble gases. For the representative elements, this tendency is called the **octet rule.** The octet rule states that *atoms attempt to acquire an outer orbital with eight electrons* through chemical reactions. This rule is a generalization, and a few elements do not meet the requirement of eight electrons but do seek the same general trend of stability. There are a few other exceptions, and the octet rule should be considered a generalization that helps keep track of the valence electrons in most representative elements.

The family number of the representative element in the periodic table tells you the number of valence electrons and what the atom must do to reach the stability suggested by the octet rule. For example, consider sodium (Na). Sodium is in family IA, so it has one valence electron. If the sodium atom can get rid of this outer valence electron through a chemical reaction, it will have the same outer electron configuration as an atom of the noble gas neon (Ne) (compare figure 9.7B and 9.7C).

When a sodium atom (Na) loses an electron to form a sodium ion (Na^+), it has the same, stable outer electron configuration as a neon atom (Ne). The sodium ion (Na^+) is still a form of sodium since it still has eleven protons. But it is now

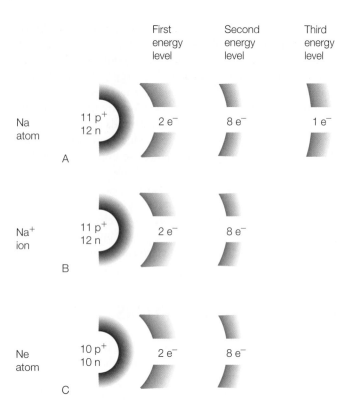

First energy level	Second energy level	Third energy level

Na atom
$11\,p^+$
$12\,n$
$2\,e^-$ $8\,e^-$ $1\,e^-$

A

Na$^+$ ion
$11\,p^+$
$12\,n$
$2\,e^-$ $8\,e^-$

B

Ne atom
$10\,p^+$
$10\,n$
$2\,e^-$ $8\,e^-$

C

FIGURE 9.7 (*A*) A sodium atom has two electrons in the first energy level, eight in the second energy level, and one in the third level. (*B*) When it loses its one outer, or valence, electron, it becomes a positively charged sodium ion with the same electron structure as an atom of neon (*C*).

a sodium *ion,* not a sodium *atom,* since it has eleven protons (eleven positive charges) and now has ten electrons (ten negative charges) for a total of

$$11 + (\text{protons})$$
$$\underline{10 - (\text{electrons})}$$
$$1 + (\text{net charge on sodium ion})$$

CONCEPTS APPLIED

Calcium

What is the symbol and charge for a calcium ion?

From the list of elements on the inside back cover, the symbol for calcium is Ca, and the atomic number is 20. The periodic table tells you that Ca is in family IIA, which means that calcium has two valence electrons. According to the octet rule, the calcium ion must lose two electrons to acquire the stable outer arrangement of the noble gases. Since the atomic number is 20, a calcium atom has twenty protons (20 +) and twenty electrons (20 −). When it is ionized, the calcium ion will lose two electrons for a total charge of (20 +) + (18 −), or 2 +. The calcium ion is represented by the chemical symbol for calcium and the charge shown as a superscript: Ca^{2+}. What is the symbol and charge for an aluminum ion? (Answer: Al^{3+})

This charge is shown on the chemical symbol of Na$^+$ for the *sodium ion.* Note that the sodium nucleus and the inner orbitals do not change when the sodium atom is ionized. The sodium ion is formed when a sodium atom loses its valence electron, and the process can be described by

where Na is the electron dot symbol for sodium, and the e$^-$ is the electron that has been pulled off the sodium atom.

CHEMICAL BONDS

Atoms gain or lose electrons through a chemical reaction to achieve a state of lower energy, the stable electron arrangement of the noble gas atoms. Such a reaction results in a **chemical bond,** an *attractive force that holds atoms together in a compound.* There are three general classes of chemical bonds: (1) ionic bonds, (2) covalent bonds, and (3) metallic bonds.

Ionic bonds are formed when atoms *transfer* electrons to achieve the noble gas electron arrangement. Electrons are given up or acquired in the transfer, forming positive and negative ions. The electrostatic attraction between oppositely charged ions forms ionic bonds, and ionic compounds are the result. In general, ionic compounds are formed when a metal from the left side of the periodic table reacts with a nonmetal from the right side.

Covalent bonds result when atoms achieve the noble gas electron structure by *sharing* electrons. Covalent bonds are generally formed between the nonmetallic elements on the right side of the periodic table.

Metallic bonds are formed in solid metals such as iron, copper, and the other metallic elements that make up about 80 percent of all the elements. The atoms of metals are closely packed and share many electrons in a "sea" that is free to move throughout the metal, from one metal atom to the next. Metallic bonding accounts for metallic properties such as high electrical conductivity.

Ionic, covalent, and metallic bonds are attractive forces that hold atoms or ions together in molecules and crystals. There are two ways to describe what happens to the electrons when one of these bonds is formed: by considering (1) the new patterns formed when atomic orbitals overlap to form a combined orbital called a *molecular orbital* or (2) the atoms in a molecule as *isolated atoms* with changes in their outer shell arrangements. The molecular orbital description considers that the electrons belong to the whole molecule and form a molecular orbital with its own shape, orientation, and energy levels. The isolated atom description considers the electron energy levels as if the atoms in the molecule were alone, isolated from the molecule. The isolated atom description is less accurate than the molecular orbital description, but it is less complex and more easily understood. Thus, the following details about chemical bonding will mostly consider individual atoms and ions in compounds.

Ionic Bonds

An **ionic bond** is defined as the *chemical bond of electrostatic attraction* between negative and positive ions. Ionic bonding occurs when an atom of a metal reacts with an atom of a nonmetal. The reaction results in a transfer of one or more valence electrons from the metal atom to the valence shell of the nonmetal atom. The atom that loses electrons becomes a positive ion, and the atom that gains electrons becomes a negative ion. Oppositely charged ions attract one another, and when pulled together, they form an ionic solid with the ions arranged in an orderly geometric structure (figure 9.8). This results in a crystalline solid that is typical of salts such as sodium chloride (figure 9.9).

As an example of ionic bonding, consider the reaction of sodium (a soft reactive metal) with chlorine (a pale yellow-green gas). When an atom of sodium and an atom of chlorine collide, they react violently as the valence electron is transferred from the sodium to the chlorine atom. This produces a sodium ion and a chlorine ion. The reaction can be illustrated with electron dot symbols as follows:

$$Na \cdot \ + \ \cdot \overset{\cdot \cdot}{\underset{\cdot \cdot}{Cl}} : \ \longrightarrow \ Na^+ \ \left(: \overset{\cdot \cdot}{\underset{\cdot \cdot}{Cl}} : \right)^-$$

As you can see, the sodium ion transferred its valence electron, and the resulting ion now has a stable electron configuration. The chlorine atom accepted the electron in its outer orbital to acquire a stable electron configuration. Thus, a stable positive ion and a stable negative ion are formed. Because of opposite electrical charges, the ions attract each other to produce an ionic bond. When many ions are involved, each Na^+ ion is surrounded by six Cl^- ions, and each Cl^- ion is surrounded by six Na^+ ions. This gives the resulting solid NaCl its crystalline cubic structure, as shown in figure 9.9. In the solid state, all the sodium ions and all the chlorine ions are bound together in one giant unit. Thus, the term *molecule* is not really appropriate for ionic solids such as sodium chloride. But the term is sometimes used anyway, since any given sample will have the same number of Na^+ ions as Cl^- ions.

The sodium-chlorine reaction can be represented with electron dot notation as occurring in three steps:

1. $\quad$ energy $\quad + \quad Na \cdot \longrightarrow \ Na^+ \ + \ e^-$

2. $\quad \cdot \overset{\cdot \cdot}{\underset{\cdot \cdot}{Cl}} : \ + \ e^- \ \longrightarrow \ \left(: \overset{\cdot \cdot}{\underset{\cdot \cdot}{Cl}} : \right)^- + $ energy

3. $\quad Na^+ \ + \ \left(: \overset{\cdot \cdot}{\underset{\cdot \cdot}{Cl}} : \right)^- \longrightarrow \ Na^+ \ \left(: \overset{\cdot \cdot}{\underset{\cdot \cdot}{Cl}} : \right)^- + $ energy

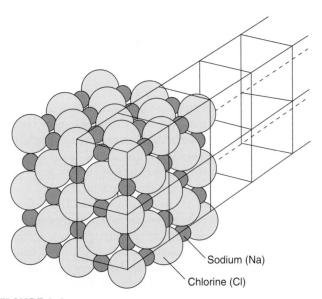

FIGURE 9.8 Sodium chloride crystals are composed of sodium and chlorine ions held together by electrostatic attraction. A crystal builds up, giving the sodium chloride crystal a cubic structure.

The energy released in steps 2 and 3 is greater than the energy absorbed in step 1, and an ionic bond is formed. The energy released is called the **heat of formation.** It is also the amount of energy required to decompose the compound (sodium chloride) into its elements. The reaction does not take place in steps as described, however, but occurs all at once. Note again, as in the photosynthesis-burning reactions described earlier, that the total amount of chemical energy is conserved. The energy released by the formation of the sodium chloride compound is the *same* amount of energy needed to decompose the compound.

Ionic bonds are formed by electron transfer, and electrons are conserved in the process. This means that electrons are not created or destroyed in a chemical reaction. The same total number of electrons exists after a reaction that existed before the reaction. There are two rules you can use for keeping track of electrons in ionic bonding reactions:

1. Ions are formed as atoms gain or lose valence electrons to achieve the stable noble gas structure.

FIGURE 9.9 You can clearly see the cubic structure of these ordinary table salt crystals because they have been magnified about ten times.

2. There must be a balance between the number of electrons lost and the number of electrons gained by atoms in the reaction.

The sodium-chlorine reaction follows these two rules. The loss of one valence electron from a sodium atom formed a stable sodium ion. The gain of one valence electron by the chlorine atom formed a stable chlorine ion. Thus, both ions have noble gas configurations (rule 1), and one electron was lost and one was gained, so there is a balance in the number of electrons lost and the number gained (rule 2).

The **formula** of a compound *describes what elements are in the compound and in what proportions.* Sodium chloride contains one positive sodium ion for each negative chlorine ion. The formula of the compound sodium chloride is NaCl. If there are no subscripts at the lower right part of each symbol, it is understood that the symbol has a number "1." Thus, NaCl indicates a compound made up of the elements sodium and chlorine, and there is one sodium atom for each chlorine atom.

Calcium (Ca) is an alkaline metal in family IIA, and fluorine (F) is a halogen in family VIIA. Since calcium is a metal and fluorine is a nonmetal, you would expect calcium and fluorine atoms to react, forming a compound with ionic bonds. Calcium must lose two valence electrons to acquire a noble gas configuration. Fluorine needs one valence electron to acquire a noble gas configuration. So calcium needs to lose two electrons and fluorine needs to gain one electron to achieve a stable configuration (rule 1). Two fluorine atoms, each acquiring one electron, are needed to balance the number of electrons lost and the number of electrons gained. The compound formed from the reaction, calcium fluoride, will therefore have a calcium ion with a charge of plus two for every two fluorine ions with a charge of minus one. Recalling that electron dot symbols show only the outer valence electrons, you can see that the reaction is

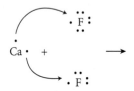

which shows that a calcium atom transfers two electrons, one each to two fluorine atoms. Now showing the results of the reaction, a calcium ion is formed from the loss of two electrons (charge 2+) and two fluorine ions are formed by gaining one electron each (charge 1−):

$$Ca \cdot \; + \; \cdot \ddot{\underset{..}{F}} \colon \; \longrightarrow \; Ca^{2+} \quad (\colon \ddot{\underset{..}{F}} \colon)^{-}$$
$$\cdot \ddot{\underset{..}{F}} \colon \qquad\qquad (\colon \ddot{\underset{..}{F}} \colon)^{-}$$

The formula of the compound is therefore CaF$_2$, with the subscript 2 for fluorine and the understood subscript 1 for calcium. This means that there are two fluorine atoms for each calcium atom in the compound.

Sodium chloride (NaCl) and calcium fluoride (CaF$_2$) are examples of compounds held together by ionic bonds. Such compounds are called **ionic compounds.** Ionic compounds of the representative elements are generally white, crystalline solids that form colorless solutions. Sodium chloride, the most common example, is common table salt. Many of the transition elements form colored compounds that make colored solutions. Ionic compounds dissolve in water, producing a solution of ions that can conduct an electric current.

In general, the elements in families IA and IIA of the periodic table tend to form positive ions by losing electrons. The ion charge for these elements equals the family number of these elements. The elements in families VIA and VIIA tend to form negative ions by gaining electrons. The ion charge for these elements equals their family number minus 8. The elements in families IIIA and VA have less of a tendency to form ionic compounds, except for those in higher periods. Common ions of some representative elements are given in table 9.2. The transition elements form positive ions of several different charges. Common ions of some transition elements are listed in table 9.3.

Covalent Bonds

Most substances do not have the properties of ionic compounds since they are not composed of ions. Most substances are molecular, composed of electrically neutral groups of atoms that are tightly bound together. As noted earlier, many gases are diatomic, occurring naturally as two atoms bound together as an electrically neutral molecule. Hydrogen, for example, occurs as molecules of H$_2$ and no ions are involved. The hydrogen atoms are held together by a covalent bond. A **covalent bond** is a *chemical bond formed by the sharing of at least a pair of electrons.* In the diatomic hydrogen molecule, each hydrogen atom contributes a single electron to the shared pair. Both hydrogen atoms count the shared pair of electrons in achieving

TABLE 9.2 Common Ions of Some Representative Elements

Element	Symbol	Ion
Lithium	Li	1+
Sodium	Na	1+
Potassium	K	1+
Magnesium	Mg	2+
Calcium	Ca	2+
Barium	Ba	2+
Aluminum	Al	3+
Oxygen	O	2−
Sulfur	S	2−
Hydrogen	H	1+, 1−
Fluorine	F	1−
Chlorine	Cl	1−
Bromine	Br	1−
Iodine	I	1−

TABLE 9.3 Common Ions of Some Transition Elements

Single-Charge Ions

Element	Symbol	Charge
Zinc	Zn	2+
Tungsten	W	6+
Silver	Ag	1+
Cadmium	Cd	2+

Variable-Charge Ions

Element	Symbol	Charge
Chromium	Cr	2+, 3+, 6+
Manganese	Mn	2+, 4+, 7+
Iron	Fe	2+, 3+
Cobalt	Co	2+, 3+
Nickel	Ni	2+, 3+
Copper	Cu	1+, 2+
Tin	Sn	2+, 4+
Gold	Au	1+, 3+
Mercury	Hg	1+, 2+
Lead	Pb	2+, 4+

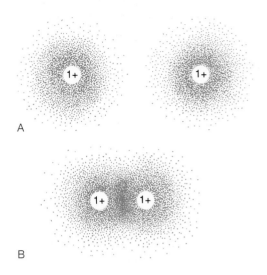

FIGURE 9.10 (*A*) Two hydrogen atoms, each with its own probability distribution of electrons about the nucleus. (*B*) When the hydrogen atoms bond, a new electron distribution pattern forms around the entire molecule, and both atoms share the two electrons.

their noble gas configuration. Hydrogen atoms both share one pair of electrons, but other elements might share more than one pair to achieve a noble gas structure.

Consider how the covalent bond forms between two hydrogen atoms by imagining two hydrogen atoms moving toward one another. Each atom has a single electron. As the atoms move closer and closer together, their orbitals begin to overlap. Each electron is attracted to the oppositely charged nucleus of the other atom and the overlap tightens. Then the repulsive forces from the like-charged nuclei will halt the merger. A state of stability is reached between the two nuclei and two electrons, and an H_2 molecule has been formed. The two electrons are now shared by both atoms, and the attraction of one nucleus for the other electron and vice versa holds the atoms together (figure 9.10).

Electron dot notation can be used to represent the formation of covalent bonds. For example, the joining of two hydrogen atoms to form an H_2 molecule can be represented as

$$H\cdot + H\cdot \longrightarrow H\!:\!H$$

Since an electron pair is *shared* in a covalent bond, the two electrons move throughout the entire molecular orbital. Since each hydrogen atom now has both electrons on an equal basis, each can be considered now to have the noble gas configuration of helium. A dashed circle around each symbol shows that both atoms have two electrons:

$$H\cdot + H\cdot \longrightarrow (H\!:\!H)$$

Hydrogen and fluorine react to form a covalent molecule (how this is known will be discussed shortly), and this bond can be represented with electron dots. Fluorine is in the VIIA family, so you know an atom of fluorine has seven valence electrons in the outermost energy level. The reaction is

$$H\cdot + \cdot \ddot{\underset{..}{F}}\!: \longrightarrow (H\!:\!\ddot{\underset{..}{F}}\!:)$$

Each atom shares a pair of electrons to achieve a noble gas configuration. Hydrogen achieves the helium configuration, and fluorine achieves the neon configuration. All the halogens have seven valence electrons, and all need to gain one electron (ionic bond) or share an electron pair (covalent bond) to achieve a noble gas configuration. This also explains why the halogen gases occur as diatomic molecules. Two fluorine atoms can achieve a noble gas configuration by sharing a pair of electrons:

$$\cdot \ddot{\underset{..}{F}}\!: + \cdot \ddot{\underset{..}{F}}\!: \longrightarrow (\ddot{\underset{..}{F}}\!:\!\ddot{\underset{..}{F}}\!:)$$

The diatomic hydrogen (H_2) and fluorine (F_2), hydrogen fluoride (HF), and water (H_2O) are examples of compounds held together by covalent bonds. A compound held together by covalent bonds is called a **covalent compound.** In general, covalent compounds form from nonmetallic elements on the right side of the periodic table. For elements in families IVA through VIIA, the number of unpaired electrons (and thus the number of covalent bonds formed) is eight minus the family number. You can get a lot of information from the periodic table from generalizations like this one. For another generalization, compare table 9.4 with

Aluminum Fluoride

Use electron dot notation to predict the formula of a compound formed when aluminum (Al) combines with fluorine (F).

Aluminum, atomic number 13, is in family IIIA, so it has three valence electrons and an electron dot notation of

Al ·

According to the octet rule, the aluminum atom would need to lose three electrons to acquire the stable noble gas configuration. Fluorine, atomic number 9, is in family VIIA, so it has seven valence electrons and an electron dot notation of

· F :

Fluorine would acquire a noble gas configuration by accepting one electron. Three fluorine atoms, each acquiring one electron, are needed to balance the three electrons lost by aluminum. The reaction can be represented as

The ratio of aluminum atoms to fluorine atoms in the compound is 1:3. The formula for aluminum fluoride is therefore AlF_3.

Predict the formula of the compound formed between aluminum and oxygen using electron dot notation. (Answer: Al_2O_3)

TABLE 9.4 Structures and Compounds of Nonmetallic Elements Combined with Hydrogen

Nonmetallic Element	Elements (E represents any element of family)	Compound
Family IVA: C, Si, Ge	· E ·	H : E : H (with H above and below)
Family VA: N, P, As, Sb	· E ·	H : E : H (with H below)
Family VIA: O, S, Se, Te	· E ·	H : E : H
Family VIIA: F, Cl, Br, I	· E ·	H : E

---

the periodic table. The table gives the structures of nonmetals combined with hydrogen and the resulting compounds.

Two dots can represent a lone pair of valence electrons, or they can represent a bonding pair, a single pair of electrons being shared by two atoms. Bonding pairs of electrons are often represented by a simple line between two atoms. For example,

H : H   is shown as   H — H

and

: O :
H   H   is shown as   O with H and H branching below

Note that the line between the two hydrogen atoms represents an electron pair, so each hydrogen atom has two electrons in the outer shell, as does helium. In the water molecule, each hydrogen atom has two electrons as before. The oxygen atom has two lone pairs (a total of four electrons) and two bonding pairs (a total of four electrons) for a total of eight electrons. Thus, oxygen has acquired a stable octet of electrons.

A covalent bond in which a single pair of electrons is shared by two atoms is called a *single covalent bond,* or simply a **single bond.** Some atoms have two unpaired electrons and can share more than one electron pair. A **double bond** is a covalent bond formed when *two pairs* of electrons are shared by two atoms. This happens mostly in compounds involving atoms of the elements C, N, O, and S. Ethylene, for example, is a gas given off from ripening fruit. The electron dot formula for ethylene is

H   H   H   H
: C :: C :   or   \   /
H   H   C = C
                   /   \
                  H   H

The ethylene molecule has a double bond between two carbon atoms. Since each line represents two electrons, you can simply count the lines around each symbol to see if the octet rule has been satisfied. Each H has one line, so each H atom is sharing two electrons. Each C has four lines so each C atom has eight electrons, satisfying the octet rule.

A **triple bond** is a covalent bond formed when *three pairs* of electrons are shared by two atoms. Triple bonds occur mostly in compounds with atoms of the elements C and N. Acetylene, for example, is a gas often used in welding torches (figure 9.11). The electron dot formula for acetylene is

H : C ::: C : H   or   H — C ≡ C — H

Name That Compound

Ionic Compound Names

Ionic compounds formed by metal ions are named by stating the name of the metal (positive ion), then the name of the nonmetal (negative ion). Ionic compounds formed by variable-charge ions have an additional rule to identify which variable-charge ion is involved. There was an old way of identifying the charge on the ion by adding either *-ic* or *-ous* to the name of the metal. The suffix *-ic* meant the higher of two possible charges, and the suffix *-ous* meant the lower of two possible charges. For example, iron has two possible charges, 2+ or 3+. The old system used the Latin name for the root. The Latin name for iron is *ferrum,* so a higher charged iron ion (3+) was named a ferric ion. The lower charged iron ion (2+) was called a ferrous ion. You still hear the old names sometimes, but chemists now have a better way to identify the variable-charge ion. The newer system uses the English name of the metal with Roman numerals in parentheses to indicate the charge number. Thus, an iron ion with a charge of 2+ is called an iron(II) ion and an iron ion with a charge of 3+ is an iron(III) ion. Box table 9.1 gives some of the modern names for variable-charge ions. These names are used with the name of a nonmetal ending in *-ide,* just like the single-charge ions in ionic compounds made up of two different elements.

Some ionic compounds contain three or more elements, and so are more complex than a combination of a metal ion and a nonmetal ion. This is possible because they have **polyatomic ions,** groups of two or more atoms that are bound together tightly and behave very much like a single monatomic ion (box table 9.2). For example, the OH^- ion is an oxygen atom bound to a hydrogen atom with a net charge of $1-$. This polyatomic ion is called a *hydroxide ion.* The hydroxide compounds make up one of the main groups of ionic compounds, the *metal hydroxides.* A metal hydroxide is an ionic compound consisting of a metal with the hydroxide ion. Another main group consists of the salts with polyatomic ions.

Box Table 9.1

Modern Names of Some Variable-Charge Ions	
Ion	Name of Ion
Fe^{2+}	Iron(II) ion
Fe^{3+}	Iron(III) ion
Cu^+	Copper(I) ion
Cu^{2+}	Copper(II) ion
Pb^{2+}	Lead(II) ion
Pb^{4+}	Lead(IV) ion
Sn^{2+}	Tin(II) ion
Sn^{4+}	Tin(IV) ion
Cr^{2+}	Chromium(II) ion
Cr^{3+}	Chromium(III) ion
Cr^{6+}	Chromium(VI) ion

The metal hydroxides are named by identifying the metal first and the term *hydroxide* second. Thus, NaOH is named sodium hydroxide and KOH is potassium hydroxide. The salts are similarly named, with the metal (or ammonium ion) identified first, then the name of the polyatomic ion. Thus, $NaNO_3$ is named sodium nitrate and $NaNO_2$ is sodium nitrite. Note that the suffix *-ate* means the polyatomic ion with one more oxygen atom than the *-ite* ion. For example, the chlor*ate* ion is $(ClO_3)^-$ and the chlor*ite* ion is $(ClO_2)^-$. Sometimes more than two possibilities exist, and more oxygen atoms are identified with the prefix *per-* and less with the prefix *hypo-*. Thus, the *per*chlor*ate* ion is $(ClO_4)^-$ and the *hypo*chlor*ite* ion is $(ClO)^-$.

Covalent Compound Names

Covalent compounds are molecular, and the molecules are composed of two *nonmetals,* as opposed to the metal and nonmetal elements that make up ionic compounds. The combinations of nonmetals do not present simple names as the ionic compounds did, so a different set of rules for naming and formula writing is needed.

Ionic compounds were named by stating the name of the positive metal ion, then the name of the negative nonmetal ion with an *-ide* ending. This system is not adequate for naming the covalent com-

Box Table 9.2

Some Common Polyatomic Ions	
Ion Name	Formula
Acetate	$(C_2H_3O_2)^-$
Ammonium	$(NH_4)^+$
Borate	$(BO_3)^{3-}$
Carbonate	$(CO_3)^{2-}$
Chlorate	$(ClO_3)^-$
Chromate	$(CrO_4)^{2-}$
Cyanide	$(CN)^-$
Dichromate	$(Cr_2O_7)^{2-}$
Hydrogen carbonate (or bicarbonate)	$(HCO_3)^-$
Hydrogen sulfate (or bisulfate)	$(HSO_4)^-$
Hydroxide	$(OH)^-$
Hypochlorite	$(ClO)^-$
Nitrate	$(NO_3)^-$
Nitrite	$(NO_2)^-$
Perchlorate	$(ClO_4)^-$
Permanganate	$(MnO_4)^-$
Phosphate	$(PO_4)^{3-}$
Phosphite	$(PO_3)^{3-}$
Sulfate	$(SO_4)^{2-}$
Sulfite	$(SO_3)^{2-}$

pounds. To begin, covalent compounds are composed of two or more nonmetal atoms that form a molecule. It is possible for some atoms to form single, double, or even triple bonds with other atoms, including atoms of the same element. The net result is that the same two elements can form more than one kind of covalent compound. Carbon and oxygen, for example, can combine to form the gas released from burning and respiration, carbon dioxide (CO_2). Under certain conditions, the very same elements combine to produce a different gas, the poisonous carbon monoxide (CO). Similarly, sulfur and oxygen can combine differently to produce two different covalent compounds. A successful system for naming covalent compounds must therefore provide a means of identifying different compounds made of the same elements. This is accomplished by using a system of Greek prefixes (see box table 9.3). The rules are as follows:

—Continued top of next page

Continued—

1. The first element in the formula is named first with a prefix indicating the number of atoms if the number is greater than one.

2. The stem name of the second element in the formula is next. A prefix is used with the stem if two elements form more than one compound. The suffix *-ide* is again used to indicate a compound of only two elements. For example, CO is carbon monoxide and CO_2 is carbon dioxide. The compound BF_3 is boron trifluoride and N_2O_4 is dinitrogen tetroxide. Knowing the for-

mula and the prefix and stem information in box table 9.3, you can write the name of any covalent compound made up of two elements by ending it with *-ide*. Conversely, the name will tell you the formula. However, there are a few polyatomic ions with *-ide* endings that are compounds made up of more than just two elements (hydroxide and cyanide). Compounds formed with the ammonium ion will also have an *-ide* ending, and these are also made up of more than two elements.

Box Table 9.3

Prefixes and Element Stem Names

Prefixes		Stem Names	
PREFIX	MEANING	ELEMENT	STEM
Mono-	1	Hydrogen	Hydr-
Di-	2	Carbon	Carb-
Tri-	3	Nitrogen	Nitr-
Tetra-	4	Oxygen	Ox-
Penta-	5	Fluorine	Fluor-
Hexa-	6	Phosphorus	Phosph-
Hepta-	7	Sulfur	Sulf-
Octa-	8	Chlorine	Chlor-
Nona-	9	Bromine	Brom-
Deca-	10	Iodine	Iod-

Note: The *a* or *o* ending on the prefix is often dropped if the stem name begins with a vowel, for example, "tetroxide," not "tetraoxide."

FIGURE 9.11 Acetylene is a hydrocarbon consisting of two carbon atoms and two hydrogen atoms held together by a triple covalent bond between the two carbon atoms. When mixed with oxygen gas (the tank to the right), the resulting flame is hot enough to cut through most metals.

The acetylene molecule has a triple bond between two carbon atoms. Again, note that each line represents two electrons. Each C atom has four lines, so the octet rule is satisfied.

COMPOSITION OF COMPOUNDS

As you can imagine, there are literally millions of different chemical compounds from all the possible combinations of over ninety natural elements held together by ionic or covalent bonds. Each of these compounds has its own name, so there are millions of names and formulas for all the compounds. In the early days, compounds were given *common names* according to how they were used, where they came from, or some other means of identifying them. Thus, sodium carbonate was called soda, and closely associated compounds were called baking soda (sodium bicarbonate), washing soda (sodium carbonate), and caustic soda

CONCEPTS APPLIED

Household Chemicals

Pick a household product that has a list of ingredients with names of covalent compounds or of ions you have met in this chapter. Write the brand name of the product and the type of product (example: Sani-Flush; toilet-bowl cleaner), then list the ingredients as given on the label, writing them one under the other (column 1). Beside each name put the formula, if you can figure out what it should be (column 2). Also, in a third column, put whatever you know or can guess about the function of that substance in the product. (Example: This is an acid; helps dissolve mineral deposits.)

Myths, Mistakes, and Misunderstandings

Ban DHMO?

"Dihydrogen monoxide (DHMO) is colorless, odorless, tasteless, and kills uncounted thousands of people every year. Most of these deaths are caused by accidental inhalation of DHMO, but the dangers do not end there. Prolonged exposure to its solid form causes severe tissue damage. Symptoms of DHMO ingestion can include excessive sweating and urination, and possibly a bloated feeling, nausea, vomiting, and body electrolyte imbalance. For those who have become dependent, DHMO withdrawal means certain death."

The above is part of a hoax that was recently circulated on the Internet. The truth is that dihydrogen monoxide is the chemical name of H_2O—water.

FIGURE 9.12 These substances are made up of sodium and some form of a carbonate ion. All have common names with the term *soda* for this reason. Soda water (or "soda pop") was first made by reacting soda (sodium carbonate) with an acid, so it was called "soda water."

(sodium hydroxide), and the bubbly drink made by reacting soda with acid was called soda water, later called soda pop (figure 9.12). Potassium carbonate was extracted from charcoal by soaking in water and came to be called potash. Such common names are colorful, and some are descriptive, but it was impossible to keep up with the names as the number of known compounds grew. So a systematic set of rules was developed to determine the name and formula of each compound. Once you know the rules, you can write the formula when you hear the name. Conversely, seeing the formula will tell you the systematic name of the compound. This can be an interesting intellectual activity and can also be important when reading the list of ingredients to understand the composition of a product.

A different set of systematic rules is used with ionic compounds and covalent compounds, but there are a few rules in common. For example, a compound made of only two different elements always ends with the suffix *-ide*. So when you hear the name of a compound ending with *-ide,* you automatically know that the compound is made up of only two elements. Sodium chlor*ide* is an ionic compound made up of sodium and chlorine ions. Carbon diox*ide* is a covalent compound with carbon and oxygen atoms. Thus, the systematic name tells you what elements are present in a compound with an *-ide* ending.

CHEMICAL EQUATIONS

Chemical reactions occur when bonds between the outermost parts of atoms are formed or broken. Bonds are formed, for example, when a green plant uses sunlight—a form of energy—to create molecules of sugar, starch, and plant fibers. Bonds are broken and energy is released when you digest the sugars and starches or when plant fibers are burned. Chemical reactions thus involve changes in matter, the creation of new materials with new properties, and energy exchanges. So far, you have considered chemical symbols as a concise way to represent elements and formulas as a concise way to describe what a compound is made of. There is also a concise way to describe a chemical reaction, the *chemical equation.*

Word equations are useful in identifying what has happened before and after a chemical reaction. The substances that existed before a reaction are called *reactants,* and the substances that exist after the reaction are called the *products.* The equation has a general form of

$$\text{reactants} \rightarrow \text{products}$$

where the arrow signifies a separation in time; that is, it identifies what existed before the reaction and what exists after the reaction. For example, the charcoal used in a barbecue grill is carbon (figure 9.13). The carbon reacts with oxygen while burn-

FIGURE 9.13 The charcoal used in a grill is basically carbon. The carbon reacts with oxygen to yield carbon dioxide. The chemical equation for this reaction, $C + O_2 \rightarrow CO_2$, contains the same information as the English sentence but has quantitative meaning as well.

CONCEPTS APPLIED

Name and Write

Name the following compounds: (a) LiF and (b) PbF_2. Write the formulas for the following compounds: (c) potassium bromide and (d) copper(I) sulfide.

(a) The formula LiF means that the positive metal ions are lithium, the negative nonmetal ions are fluorine, and there are only two elements in the compound. Lithium ions are Li^{1+} (family IA), and fluorine ions are F^{1-} (family VIIA). The name is lithium fluoride.

(b) Lead is a variable-charge transition element (table 9.3), and fluorine ions are F^{1-}. The lead ion must be Pb^{2+} because the compound PbF_2 is electrically neutral. Therefore, the name is lead(II) fluoride.

(c) The ions are K^{1+} and Br^{1-}. Crossing over the charge numbers and dropping the signs gives the formula KBr.

(d) The Roman numeral tells you the charge on the copper ion, so the ions are Cu^{1+} and S^{2-}. The formula is Cu_2S.

A Closer Look

How to Write a Chemical Formula

Ionic Compound Formulas

The formulas for ionic compounds are easy to write. There are two rules:

1. **The symbols.** Write the symbol for the positive element first, followed by the symbol for the negative element (same order as in the name).
2. **The subscripts.** Add subscripts to indicate the numbers of ions needed to produce an electrically neutral compound.

As an example, let us write the formula for the compound calcium chloride. The name tells you that this compound consists of positive calcium ions and negative chlorine ions. The suffix -ide tells you there are only two elements present. Following rule 1, the symbols would be CaCl.

For rule 2, note the calcium ion is Ca^{2+} and the chlorine ion is Cl^-. You know the calcium is plus two and chlorine is negative one by applying the atomic theory, knowing their positions in the periodic table, or by using a table of ions and their charges. To be electrically neutral, the compound must have an equal number of pluses and minuses. Thus, you will need two negative chlorine ions for every calcium ion with its 2+ charge. Therefore, the formula is $CaCl_2$.

The total charge of two chlorines is thus 2−, which balances the 2+ charge on the calcium ion.

One easy way to write a formula showing that a compound is electrically neutral is to cross over the absolute charge numbers (without plus or minus signs) and use them as subscripts. For example, the symbols for the calcium ion and the chlorine ion are

$$Ca^{2+}Cl^{1-}$$

Crossing the absolute numbers as subscripts, as follows

and then dropping the charge numbers gives

$$Ca_1 Cl_2$$

No subscript is written for 1; it is understood. The formula for calcium chloride is thus

$$CaCl_2$$

The crossover technique works because ionic bonding results from a transfer of electrons, and the net charge is conserved. A calcium ion has a 2+ charge because the atom lost two electrons and two chlorine atoms gain one electron each, for a total of two electrons gained. Two electrons lost equals two electrons gained, and the net charge on calcium chloride is zero, as it has to be.

When using the crossover technique, it is sometimes necessary to reduce the ratio to the lowest common multiple. Thus, Mg_2O_2 means an equal ratio of magnesium and oxygen ions, so the correct formula is MgO.

The formulas for variable-charge ions are easy to write, since the Roman numeral tells you the charge number. The formula for tin(II) fluoride is written by crossing over the charge numbers (Sn^{2+}, F^{1-}), and the formula is SnF_2.

The formulas for ionic compounds with polyatomic ions are written from combinations of positive metal ions or the ammonium ion with the polyatomic ions, as listed in box table 9.2. Since the polyatomic ion is a group of atoms that has a charge and stays together in a unit, it is sometimes necessary to indicate this with parentheses. For example, magnesium hydroxide is composed of Mg^{2+} ions and $(OH)^{1-}$ ions.

---

ing, and the reaction (1) releases energy and (2) forms carbon dioxide. The reactants and products for this reaction can be described as

$$carbon + oxygen \rightarrow carbon\ dioxide$$

The arrow means *yields,* and the word equation is read as, "Carbon reacts with oxygen to yield carbon dioxide." This word equation describes what happens in the reaction but says nothing about the quantities of reactants or products.

Chemical symbols and formulas can be used in the place of words in an equation and the equation will have a whole new meaning. For example, the equation describing carbon reacting with oxygen to yield carbon dioxide becomes

$$C + O_2 \rightarrow CO_2 \qquad \textbf{(balanced)}$$

The new, added meaning is that one atom of carbon (C) reacts with one molecule of oxygen (O_2) to yield one molecule of carbon dioxide (CO_2). Note that the equation also shows one atom of carbon and two atoms of oxygen (recall that oxygen occurs as a diatomic molecule) as reactants on the left side and one atom of carbon and two atoms of oxygen as products on the right side.

Since the same number of each kind of atom appears on both sides of the equation, the equation is said to be *balanced.*

You would not want to use a charcoal grill in a closed room because there might not be enough oxygen. An insufficient supply of oxygen produces a completely different product, the poisonous gas carbon monoxide (CO). An equation for this reaction is

$$C + O_2 \rightarrow CO \qquad \textbf{(not balanced)}$$

As it stands, this equation describes a reaction that violates the **law of conservation of mass,** that matter is neither created nor destroyed in a chemical reaction. From the point of view of an equation, this law states that

$$mass\ of\ reactants = mass\ of\ products$$

Mass of reactants here means all that you start with, including some that might not react. Thus, elements are neither created nor destroyed, and this means the elements present and their mass. In any chemical reaction, the kind and mass of the reactive elements are identical to the kind and mass of the product elements.

From the point of view of atoms, the law of conservation of mass means that *atoms are neither created nor destroyed*

Using the crossover technique to write the formula, you get

The parentheses are used and the subscript is written *outside* the parenthesis to show that the entire hydroxide unit is taken twice. The formula $Mg(OH)_2$ means

which shows that the pluses equal the minuses. Parentheses are not used, however, when only one polyatomic ion is present. Sodium hydroxide is NaOH, not $Na(OH)_1$.

Covalent Compound Formulas

The systematic name tells you the formula for a covalent compound. The gas that dentists use as an anesthetic, for example, is dinitrogen monoxide. This tells you there are two nitrogen atoms and one oxygen atom in the molecule, so the formula is N_2O. A different molecule composed of the very same elements is nitrogen dioxide. Nitrogen dioxide is the pollutant responsible for the brownish haze of smog. The formula for nitrogen dioxide is NO_2. Other examples of formulas from systematic names are carbon dioxide (CO_2) and carbon tetrachloride (CCl_4).

Formulas of covalent compounds indicate a pattern of how many atoms of one element combine with atoms of another. Carbon, for example, combines with no more than two oxygen atoms to form carbon dioxide. Carbon combines with no more than four chlorine atoms to form carbon tetrachloride. Electron dot formulas show these two molecules as

$$\ddot{O} :: C :: \ddot{O} \qquad :\ddot{Cl} : \overset{:\ddot{Cl}:}{\underset{:\ddot{Cl}:}{C}} : \ddot{Cl}:$$

Using a dash to represent bonding pairs, we have

$$O = C = O \qquad \overset{\displaystyle Cl}{\underset{\displaystyle Cl}{Cl - C - Cl}}$$

In both of these compounds, the carbon atom forms four covalent bonds with another atom. The number of covalent bonds that an atom can form is called its *valence*. Carbon has a valence of four and can form single, double, or triple bonds. Here are the possibilities for a single carbon atom (combining elements not shown):

$$-\overset{|}{\underset{|}{C}}- \qquad -\overset{|}{C}=$$

$$=C= \qquad -C\equiv$$

Hydrogen has only one unshared electron, so the hydrogen atom has a valence of one. Oxygen has a valence of two, and nitrogen has a valence of three. Here are the possibilities for hydrogen, oxygen, and nitrogen:

$$H- \qquad -\ddot{O}- \qquad :\ddot{O}=$$

$$-\overset{..}{\underset{|}{N}}- \qquad -\overset{..}{N}= \qquad :N\equiv$$

in the chemical reaction. A chemical reaction is the making or breaking of chemical bonds between atoms or groups of atoms. Atoms are not lost or destroyed in the process, nor are they changed to a different kind. The equation for the formation of carbon monoxide has two oxygen atoms in the reactants (O_2) but only one in the product (CO). An atom of oxygen has disappeared somewhere, and that violates the law of conservation of mass. You cannot fix the equation by changing the CO to a CO_2, because this would change the identity of the compounds. Carbon monoxide is a poisonous gas that is different from carbon dioxide, a relatively harmless product of burning and respiration. *You cannot change the subscript in a formula* because that would change the formula. A different formula means a different composition and thus a different compound.

You cannot change the subscripts of a formula, but you can place a number called a *coefficient* in *front* of the formula. Changing a coefficient changes the *amount* of a substance, not the identity. Thus 2 CO means two molecules of carbon monoxide and 3 CO means three molecules of carbon monoxide. If there is no coefficient, 1 is understood, as with subscripts. The meaning of coefficients and subscripts is illustrated in figure 9.14.

C	means	⚪	One atom of carbon
O	means	⚪	One atom of oxygen
O_2	means	⚪⚪	One molecule of oxygen consisting of two atoms of oxygen
CO	means	⚪⚪	One molecule of carbon monoxide consisting of one atom of carbon attached to one atom of oxygen
CO_2	means	⚪⚪⚪	One molecule of carbon dioxide consisting of one atom of carbon attached to two atoms of oxygen
3 CO_2	means	⚪⚪⚪ ⚪⚪⚪ ⚪⚪⚪	Three molecules of carbon dioxide, each consisting of one atom of carbon attached to two atoms of oxygen

FIGURE 9.14 The meaning of subscripts and coefficients used with a chemical formula. The subscripts tell you how many atoms of a particular element are in a compound. The coefficient tells you about the quantity, or number, of molecules of the compound.

On Balancing Equations

Natural gas is mostly methane, CH_4, which burns by reacting with oxygen (O_2) to produce carbon dioxide (CO_2) and water vapor (H_2O). Write a balanced chemical equation for this reaction by following a procedure of four steps.

Step 1. Write the correct formulas for the reactants and products in an unbalanced equation. For the burning of methane, the unbalanced, but otherwise correct, formula equation would be

$$CH_4 + O_2 \rightarrow CO_2 + H_2O \quad \textbf{(not balanced)}$$

Step 2. Inventory the number of each kind of atom on both sides of the unbalanced equation. In the example, there are

Reactants	Products
1 C	1 C
4 H	2 H
2 O	3 O

This step shows that the H and O are unbalanced.

Step 3. Determine where to place coefficients in front of formulas to balance the equation. It is often best to focus on the simplest thing you can do with whole number ratios. The H and the O are unbalanced, for example, and there are 4 H atoms on the left and 2 H atoms on the right. Placing a coefficient 2 in front of H_2O will balance the H atoms:

$$CH_4 + O_2 \rightarrow CO_2 + 2\,H_2O \quad \textbf{(not balanced)}$$

Now take a second inventory:

Reactants	Products
1 C	1 C
4 H	4 H
2 O	4 O (O_2+2 O)

This shows the O atoms are still unbalanced with 2 on the left and 4 on the right. Placing a coefficient of 2 in front of O_2 will balance the O atoms.

$$CH_4 + 2\,O_2 \rightarrow CO_2 + 2\,H_2O \quad \textbf{(balanced)}$$

Step 4. Take another inventory to determine if the numbers of atoms on both sides are now equal. If they are, determine if the coefficients are in the lowest possible whole-number ratio. The inventory is now

Reactants	Products
1 C	1 C
4 H	4 H
4 O	4 O

The number of each kind of atom on each side of the equation is the same, and the ratio of 1:2 → 1:2 is the lowest possible whole-number ratio. The equation is balanced, which is illustrated with sketches of molecules in box figure 9.1.

Balancing chemical equations is mostly a trial-and-error procedure. But with practice, you will find there are a few generalized "role models" that can be useful in balancing equations for many simple reactions. The key to success at balancing equations is to think it out step-by-step while remembering the following:

1. Atoms are neither lost nor gained nor do they change their identity in a chemical reaction. The same kind and number of atoms in the reactants must appear in the products, meaning atoms are conserved.

2. A correct formula of a compound cannot be changed by altering the number or placement of subscripts. Changing subscripts changes the identity of a compound and the meaning of the entire equation.

3. A coefficient in front of a formula multiplies everything in the formula by that number.

Here are a few generalizations that can be helpful in balancing equations:

1. Look first to formulas of compounds with the most atoms and try to balance the atoms or compounds from which they were formed or to which they decomposed.

2. You should treat polyatomic ions that appear on both sides of the equation as independent units with a charge. That is, consider the polyatomic ion as a unit while taking an inventory rather than the individual atoms making up the polyatomic ion. This will save time and simplify the procedure.

3. Both the "crossover technique" and the use of "fractional coefficients" can be useful in finding the least common multiple to balance an equation.

4. The physical state of reactants and products in a reaction is often identified by the symbols (g) for gas, (l) for liquid, (s) for solid, and (aq) for an aqueous solution, which means water. If a gas escapes, this is identified with an arrow pointing up (↑). A solid formed from a solution is identified with an arrow pointing down (↓). The Greek symbol delta (Δ) is often used under or over the yield sign to indicate a change of temperature or other physical values.

Reaction:
    Methane reacts with oxygen to yield carbon dioxide and water

Balanced equation:
$$CH_4 + 2\,O_2 \longrightarrow CO_2 + 2\,H_2O$$

Sketches representing molecules:

Meaning:

| 1 molecule of methane | + | 2 molecules of oxygen | → | 1 molecule of carbon dioxide | + | 2 molecules of water |

Box Figure 9.1 Compare the numbers of each kind of atom in the balanced equation with the numbers of each kind of atom in the sketched representation. Both the equation and the sketch have the same number of atoms in the reactants and in the products.

Placing a coefficient of 2 in front of the C and a coefficient of 2 in front of the CO in the equation will result in the same numbers of each kind of atom on both sides:

$$2 \, C + O_2 \rightarrow 2 \, CO$$

Reactants: 2 C   Products: 2 C

2 O   2 O

The equation is now balanced.

Generalizing from groups of chemical reactions also makes it possible to predict what will happen in similar reactions. For example, the combustion of methane (CH_4), propane (C_3H_8), and octane (C_8H_{18}) all involve a *hydrocarbon*, a compound of the elements hydrogen and carbon. Each hydrocarbon reacts with O_2, yielding CO_2 and releasing the energy of combustion. Generalizing from these reactions, you could predict that the combustion of *any* hydrocarbon would involve the combination of atoms of the hydrocarbon molecule with O_2 to produce CO_2 and H_2O with the release of energy. Such reactions could be analyzed by chemical experiments, and the products could be identified by their physical and chemical properties. You would find your predictions based on similar reactions would be correct, thus justifying predictions from such generalizations. Butane, for example, is a hydrocarbon with the formula C_4H_{10} The balanced equation for the combustion of butane is

$$2 \, C_4H_{10(g)} + 13 \, O_{2(g)} \rightarrow 8 \, CO_{2(g)} + 10 \, H_2O_{(g)}$$

You could extend the generalization further, noting that the combustion of compounds containing oxygen as well as carbon and hydrogen also produces CO_2 and H_2O (figure 9.15). These compounds are *carbohydrates,* composed of carbon and water. Glucose, for example, is a compound with the formula $C_6H_{12}O_6$.

FIGURE 9.15 *Hydrocarbons* are composed of the elements hydrogen and carbon. Propane (C_3H_8) and gasoline, which contains a mixture of molecules such as octane (C_8H_{18}), are examples of hydrocarbons. *Carbohydrates* are composed of the elements of hydrogen, carbon, and oxygen. Table sugar, for example, is the carbohydrate $C_{12}H_{22}O_{11}$. Generalizing, all hydrocarbons and carbohydrates react completely with oxygen to yield CO_2 and H_2O.

Glucose combines with oxygen to produce CO_2 and H_2O, and the balanced equation is

$$C_6H_{12}O_{6(s)} + 6 \, O_{2(g)} \rightarrow 6 \, CO_{2(g)} + 6 \, H_2O_{(g)}$$

Note that three molecules of oxygen were not needed from the O_2 reactant since the other reactant, glucose, contains six oxygen atoms per molecule. An inventory of atoms will show that the equation is thus balanced.

Combustion is a rapid reaction with O_2 that releases energy, usually with a flame. A very similar, although much slower, reaction takes place in plant and animal respiration. In respiration, carbohydrates combine with O_2 and release energy used for biological activities. This reaction is slow compared to combustion and requires enzymes to proceed at body temperature. Nonetheless, CO_2 and H_2O are the products.

TYPES OF CHEMICAL REACTIONS

The reactions involving hydrocarbons and carbohydrates with oxygen are examples of an important group of chemical reactions called *oxidation-reduction* reactions. When the term *oxidation* was first used, it specifically meant reactions involving the combination of oxygen with other atoms. But fluorine, chlorine, and other nonmetals were soon understood to have similar reactions as oxygen, so the definition was changed to one concerning the shifts of electrons in the reaction.

An **oxidation-reduction reaction** (or **redox reaction**) is broadly defined as a reaction in which electrons are transferred from one atom to another. As is implied by the name, such a reaction has two parts and each part tells you what happens to the electrons. *Oxidation* is the part of a redox reaction in which there is a loss of electrons by an atom. *Reduction* is the part of a redox reaction in which there is a gain of electrons by an atom. The name also implies that in any reaction in which oxidation

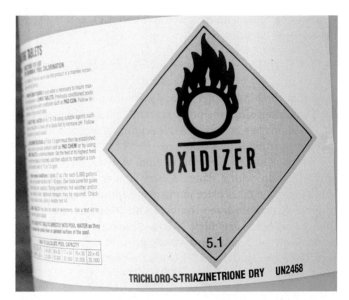

FIGURE 9.16 Oxidizing agents take electrons from other substances that are being oxidized. Oxygen and chlorine are commonly used strong oxidizing agents.

occurs, reduction must take place, too. One cannot take place without the other.

Substances that take electrons from other substances are called **oxidizing agents.** Oxidizing agents take electrons from the substances being oxidized. Oxygen is the most common oxidizing agent, and several examples have already been given about how it oxidizes foods and fuels. Chlorine is another commonly used oxidizing agent, often for the purposes of bleaching or killing bacteria (figure 9.16).

A **reducing agent** supplies electrons to the substance being reduced. Hydrogen and carbon are commonly used reducing agents. Carbon is commonly used as a reducing agent to extract metals from their ores. For example, carbon (from coke, which is coal that has been baked) reduces Fe_2O_3, an iron ore, in the reaction

$$2\,Fe_2O_{3(s)} + 3\,C_{(s)} \rightarrow 4\,Fe_{(s)} + 3\,CO_2 \uparrow$$

The Fe in the ore gained electrons from the carbon, the reducing agent in this reaction.

Many chemical reactions can be classified as redox or nonredox reactions. Another way to classify chemical reactions is to consider what is happening to the reactants and products. This type of classification scheme leads to four basic categories of chemical reactions, which are (1) *combination,* (2) *decomposition,* (3) *replacement,* and (4) *ion exchange* reactions. The first three categories are subclasses of redox reactions. It is in the ion exchange reactions that you will find the first example of a reaction that is not a redox reaction.

Combination Reactions

A **combination reaction** is a synthesis reaction in which two or more substances combine to form a single compound. The combining substances can be (1) elements, (2) compounds, or

(3) combinations of elements and compounds. In generalized form, a combination reaction is

$$X + Y \rightarrow XY$$

Many redox reactions are combination reactions. For example, metals are oxidized when they burn in air, forming a metal oxide. Consider magnesium, which gives off a bright white light as it burns:

$$2\,Mg_{(s)} + O_{2(g)} \rightarrow 2\,MgO_{(s)}$$

The rusting of metals is oxidation that takes place at a slower pace than burning, but metals are nonetheless oxidized in the process. Again noting the generalized form of a combination reaction, consider the rusting of iron:

$$4\,Fe_{(s)} + 3\,O_{2(g)} \rightarrow 2\,Fe_2O_{3(s)}$$

Nonmetals are also oxidized by burning in air, for example, when carbon burns with a sufficient supply of O_2:

$$C_{(s)} + O_{2(g)} \rightarrow CO_{2(g)}$$

Note that all the combination reactions follow the generalized form of $X + Y \rightarrow XY$.

Decomposition Reactions

A **decomposition reaction,** as the term implies, is the opposite of a combination reaction. In decomposition reactions, a compound is broken down (1) into the elements that make up the compound, (2) into simpler compounds, or (3) into elements and simpler compounds. Decomposition reactions have a generalized form of

$$XY \rightarrow X + Y$$

Decomposition reactions generally require some sort of energy, which is usually supplied in the form of heat or electrical energy. An electric current, for example, decomposes water into hydrogen and oxygen:

$$2\,H_2O_{(l)} \xrightarrow{\text{electricity}} 2\,H_{2(g)} + O_{2(g)}$$

Mercury(II) oxide is decomposed by heat, an observation that led to the discovery of oxygen:

$$2\,HgO_{(s)} \xrightarrow{\Delta} 2\,Hg_{(s)} + O_2 \uparrow$$

Note that all the decomposition reactions follow the generalized form of $XY \rightarrow X + Y$.

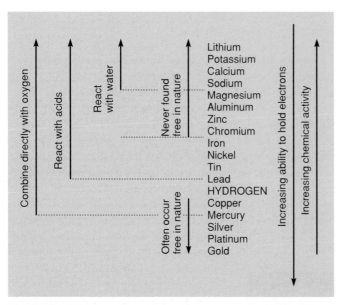

FIGURE 9.17 The activity series for common metals, together with some generalizations about the chemical activities of the metals. The series is used to predict which replacement reactions will take place and which reactions will not occur. (Note that hydrogen is not a metal and is placed in the series for reference to acid reactions.)

Replacement Reactions

In a **replacement reaction,** an atom or polyatomic ion is replaced in a compound by a different atom or polyatomic ion. The replaced part can be either the negative or positive part of the compound. In generalized form, a replacement reaction is

$$XY + Z \rightarrow XZ + Y$$

(negative part replaced)

or

$$XY + A \rightarrow AY + X$$

(positive part replaced)

Replacement reactions occur because some elements have a stronger electron-holding ability than other elements. Elements that have the least ability to hold on to their electrons are the most chemically active. Figure 9.17 shows a list of chemical activity of some metals, with the most chemically active at the top. Hydrogen is included because of its role in acids. Take a few minutes to look over the generalizations listed in figure 9.17. The generalizations apply to combination, decomposition, and replacement reactions.

Replacement reactions take place as more active metals give up electrons to elements lower on the list with a greater electron-holding ability. For example, aluminum is higher on the activity series than copper. When aluminum foil is placed in a solution of copper(II) chloride, aluminum is oxidized, losing electrons to the copper. The loss of electrons from metallic aluminum forms aluminum ions in solution, and the copper comes out of solution as a solid metal (figure 9.18).

$$2\,Al_{(s)} + 3\,CuCl_{2(aq)} \rightarrow 2\,AlCl_{3(aq)} + 3\,Cu_{(s)}$$

FIGURE 9.18 This shows a reaction between metallic aluminum and the blue solution of copper(II) chloride. Aluminum is above copper in the activity series, and aluminum replaces the copper ions from the solution as copper is deposited as a metal. The aluminum loses electrons to the copper and forms aluminum ions in solution.

A metal will replace any metal ion in solution that it is above in the activity series. If the metal is listed below the metal ion in solution, no reaction occurs. For example, $Ag_{(s)} + CuCl_{2(aq)} \rightarrow$ no reaction.

The very active metals (lithium, potassium, calcium, and sodium) react with water to yield metal hydroxides and hydrogen. For example,

$$2\,Na_{(s)} + 2\,H_2O_{(1)} \rightarrow 2\,NaOH_{(aq)} + H_2\uparrow$$

Acids yield hydrogen ions in solution, and metals above hydrogen in the activity series will replace hydrogen to form a metal salt. For example,

$$Zn_{(s)} + H_2SO_{4(aq)} \rightarrow ZnSO_{4(aq)} + H_2\uparrow$$

In general, the energy involved in replacement reactions is less than the energy involved in combination or decomposition reactions.

Ion Exchange Reactions

An **ion exchange reaction** is a reaction that takes place when the ions of one compound interact with the ions of another compound, forming (1) a solid that comes out of solution (a precipitate), (2) a gas, or (3) water.

A water solution of dissolved ionic compounds is a solution of ions. For example, solid sodium chloride dissolves in water to become ions in solution,

$$NaCl_{(s)} \rightarrow Na^+_{(aq)} + Cl^-_{(aq)}$$

If a second ionic compound is dissolved with a solution of another, a mixture of ions results. The formation of a precipitate, a gas, or water, however, removes ions from the solution,

Science and Society

The Catalytic Converter

The modern automobile produces two troublesome products in the form of (1) nitrogen monoxide and (2) hydrocarbons from the incomplete combustion of gasoline. These products from the exhaust enter the air to react in sunlight, eventually producing an irritating haze known as photochemical smog. To reduce photochemical smog, modern automobiles are fitted with a catalytic converter as part of their exhaust system (box figure 9.2).

Molecules require a certain amount of energy to change chemical bonds. This certain amount of energy is called the *activation energy,* and it represents an energy barrier that must be overcome before a chemical reaction can take place. This explains why chemical reactions proceed at a faster rate at higher temperatures. At higher temperatures, molecules have greater average kinetic energies; thus, they already have part of the minimum energy needed for a reaction to take place.

The rate at which a chemical reaction proceeds is affected by a *catalyst,* a material that speeds up a chemical reaction without being permanently changed by the reaction. A catalyst appears to speed a chemical reaction by lowering the activation energy. Molecules become temporarily attached to the surface of the catalyst, which weakens the chemical bonds holding the molecule together. The weakened molecule is easier to break apart and the activation energy is lowered. Some catalysts do this better with some specific compounds than others, and extensive chemical research programs are devoted to finding new and more effective catalysts.

Automobile catalytic converters use metals such as platinum and transition metal oxides such as copper(II) oxide and chromium(III) oxide. Catalytic reactions that occur in the converter can reduce or oxidize about 90 percent of the hydrocarbons, 85 percent of the carbon monoxide, and 40 percent of the nitrogen monoxide from exhaust gases. Other controls, such as exhaust gas recirculation, are used to reduce further nitrogen monoxide formation.

Box Figure 9.2   This silver-colored canister is the catalytic converter. The catalytic converter is located between the engine and the muffler, which is farther back toward the rear of the car.

and this must occur before you can say that an ionic exchange reaction has taken place. For example, water being treated for domestic use sometimes carries suspended matter that is removed by adding aluminum sulfate and calcium hydroxide to the water. The reaction is

$$3\,Ca(OH)_{2(aq)} + Al_2(SO_4)_{3(aq)} \rightarrow 3\,CaSO_{4(aq)} + 2\,Al(OH)_3 \downarrow$$

The aluminum hydroxide is a jellylike solid that traps the suspended matter for sand filtration. The formation of the insoluble aluminum hydroxide removed the aluminum and hydroxide ions from the solution, so an ion exchange reaction took place.

In general, an ion exchange reaction has the form

$$AX + BY \rightarrow AY + BX$$

where one of the products removes ions from the solution. The calcium hydroxide and aluminum sulfate reaction took place as the aluminum and calcium ions traded places. A solubility table such as the one in appendix B will tell you if an ionic exchange reaction has taken place. Aluminum hydroxide is

insoluble, according to the table, so the reaction did take place. No ionic exchange reaction occurred if the new products are both soluble.

Another way for an ion exchange reaction to occur is if a gas or water molecule forms to remove ions from the solution. When an acid reacts with a base (an alkaline compound), a salt and water are formed

$$HCl_{(aq)} + NaOH_{(aq)} \rightarrow NaCl_{(aq)} + H_2O_{(l)}$$

The reactions of acids and bases are discussed in chapter 10.

Linus Carl Pauling (1901–1994)

Linus Pauling was a U.S. theoretical chemist and biologist whose achievements ranked among the most important of any in twentieth-century science. His main contribution was to the understanding of molecular structure and chemical bonding. He was one of the very few people to have been awarded two Nobel prizes: he received the 1954 Nobel Prize for chemistry (for his work on intermolecular forces) and the 1962 Peace Prize. Throughout his career, his work was noted for the application of intuition and inspiration, assisted by his phenomenal memory; he often carried over principles from one field of science to another.

Pauling was born in Portland, Oregon, on February 28, 1901, the son of a pharmacist. He began his scientific studies at Oregon State Agricultural College, from which he graduated in chemical engineering in 1922. He then began his research at the California Institute of Technology, Pasadena, gaining his Ph.D. in 1925. He became a full professor at Pasadena in 1931 and left there in 1936 to take up the post of director of the Gates and Crellin Laboratories, which he held for the next twenty-two years. He also held appointments at the University of California, San Diego, and Stanford University. His last appointment was as director of the Linus Pauling Institute of Science and Medicine at Menlo Park, California.

In 1931, Pauling published a classic paper, "The Nature of the Chemical Bond," in which he used quantum mechanics to explain that an electron-pair bond is formed by the interaction of two unpaired electrons, one from each of two atoms, and that once paired, these electrons cannot take part in the formation of other bonds. It was followed by the book *Introduction to Quantum Mechanics* (1935), of which he was coauthor. He was a pioneer in the application of quantum mechanical principles to the structures of molecules.

It was Pauling who introduced the concept of hybrid orbitals in molecules to explain the symmetry exhibited by carbon atoms in most of its compounds. Pauling also investigated electronegativity of atoms and polarization in chemical bonds. He assigned electronegativities on a scale up to 4.0. A pair of electrons in a bond is pulled preferentially toward an atom with a higher electronegativity. In hydrogen chloride (HCl), for example, hydrogen has an electronegativity of 2.1 and chlorine of 3.5. The bonding electrons are pulled toward the chlorine atom, giving it a small excess negative charge (and leaving the hydrogen atom with a small excess positive charge), polarizing the hydrogen-chlorine bond.

Pauling's ideas on chemical bonding are fundamental to modern theories of molecular structure. Much of this work was consolidated in his book *The Nature of the Chemical Bond, The Structure of Molecules and Crystals* (1939). In the 1940s, Pauling turned his attention to the chemistry of living tissues and systems. He applied his knowledge of molecular structure to the complexity of life, principally to proteins in blood. With Robert Corey, he worked on the structures of amino acids and polypeptides. They proposed that many proteins have structures held together with hydrogen bonds, giving them helical shapes. This concept assisted Francis Crick and James Watson in their search for the structure of DNA, which they eventually resolved as a double helix.

In his researches on blood, Pauling investigated immunology and sickle-cell anemia. Later work confirmed his hunch that the disease is genetic and that normal hemoglobin and the hemoglobin in abnormal "sickle" cells differ in electrical charge. Throughout the 1940s, he studied living materials; he also carried out research on anesthesia. At the end of this period, he published two textbooks, *General Chemistry* (1948) and *College Chemistry* (1950), which became best-sellers.

Source: Modified from the *Hutchinson Dictionary of Scientific Biography*. © RM, 2009. All rights reserved. Helicon Publishing is a division of RM.

SUMMARY

Mixtures are made up of *unlike parts* with a *variable composition. Pure substances* are the *same throughout* and have a *definite composition.* Mixtures can be separated into their components by *physical changes,* changes that do not alter the identity of matter. Some pure substances can be broken down into simpler substances by a *chemical change,* a change that *alters the identity of matter as it produces new substances with different properties.* A pure substance that can be decomposed by chemical change into simpler substances with a definite composition is a *compound.* A pure substance that cannot be broken down into anything simpler is an *element.*

A *chemical change* produces new substances by making or breaking chemical bonds. The process of chemical change is called a *chemical reaction.* During a chemical reaction, different chemical substances with greater or lesser amounts of internal potential energy are produced. *Chemical energy* is the change of internal potential energy during a chemical reaction. A *chemical equation* is a short-hand way of describing a chemical reaction. An equation shows the substances that are changed, the *reactants,* on the left side and the new substances produced, the *products,* on the right side.

Chemical reactions involve *valence electrons,* the electrons in the outermost energy level of an atom. Atoms tend to lose or acquire electrons to achieve the configuration of the noble gases with stable, filled outer orbitals. This tendency is generalized as the *octet rule,* that atoms lose or gain electrons to acquire the noble gas structure of eight electrons in the outer orbital. Atoms form negative or positive *ions* in the process.

A chemical bond is an attractive force that holds atoms together in a compound. Chemical bonds formed when atoms transfer electrons to become ions are *ionic bonds*. An ionic bond is an electrostatic attraction between oppositely charged ions. Chemical bonds formed when ions share electrons are *covalent bonds*.

Ionic bonds result in *ionic compounds* with a crystalline structure. The energy released when an ionic compound is formed is called the *heat of formation*. It is the same amount of energy that is required to decompose the compound into its elements.

A *formula* of a compound uses symbols to tell what elements are in a compound and in what proportions. Ions of representative elements have a single, fixed charge, but many transition elements have variable charges. Electrons are conserved when ionic compounds are formed, and the ionic compound is electrically neutral. A formula shows the overall balance of charges.

Covalent compounds are molecular, composed of electrically neutral groups of atoms bound together by *covalent bonds*. The sharing of a pair of electrons, with each atom contributing a single electron to the shared pair, forms a *single covalent bond*. Covalent bonds formed when two pairs of electrons are shared are called *double bonds*, and a *triple bond* is the sharing of three pairs of electrons.

Compounds are named with different rules for ionic and covalent compounds. Both ionic and covalent compounds that are made up of only two different elements always end with an *-ide* suffix, but there are a few exceptions.

The rule for naming variable-charge ions states the English name and gives the charge with Roman numerals in parentheses.

Ionic compounds are electrically neutral, and formulas must show a balance of charge. The *crossover technique* is an easy way to write formulas that show a balance of charge.

Covalent compounds are molecules of two or more nonmetal atoms held together by a covalent bond. The system for naming covalent compounds uses Greek prefixes to identify the numbers of atoms, since more than one compound can form from the same two elements (CO and CO_2, for example).

A concise way to describe a chemical reaction is to use formulas in a *chemical equation*. A chemical equation with the same number of each kind of atom on both sides is called a *balanced equation*. A balanced equation is in accord with the *law of conservation of mass,* which states that atoms are neither created nor destroyed in a chemical reaction. To balance a chemical equation, *coefficients* are placed in front of chemical formulas. Subscripts of formulas may not be changed since this would change the formula, meaning a different compound.

One important group of chemical reactions is called *oxidation reduction reactions,* or *redox* reactions for short. Redox reactions are reactions where shifts of electrons occur. The process of losing electrons is called *oxidation,* and the substance doing the losing is said to be *oxidized.* The process of gaining electrons is called *reduction,* and the substance doing the gaining is said to be *reduced.* Substances that take electrons from other substances are called *oxidizing agents.* Substances that supply electrons are called *reducing agents.*

Chemical reactions can also be classified as (1) *combination,* (2) *decomposition,* (3) *replacement,* or (4) *ion exchange.* The first three of these are redox reactions, but ion exchange is not.

KEY TERMS

atom (p. **199**)	covalent compound (p. **205**)	ionic compounds (p. **204**)	polyatomic ion (p. **207**)
chemical bond (p. **202**)	decomposition reaction	law of conservation of mass	redox reaction (p. **213**)
chemical change (p. **198**)	(p. **214**)	(p. **211**)	reducing agent (p. **214**)
chemical energy (p. **200**)	double bond (p. **206**)	molecule (p. **200**)	replacement reaction (p. **215**)
chemical equation (p. **201**)	element (p. **198**)	octet rule (p. **201**)	single bond (p. **206**)
chemical reaction (p. **200**)	formula (p. **204**)	oxidation-reduction reaction	triple bond (p. **206**)
combination reaction (p. **214**)	heat of formation (p. **203**)	(p. **213**)	valence electrons (p. **201**)
compound (p. **198**)	ion exchange reaction (p. **215**)	oxidizing agents (p. **214**)	
covalent bond (p. **204**)	ionic bond (p. **203**)	physical change (p. **198**)	

APPLYING THE CONCEPTS

1. Which of the following represents a chemical change?
   a. heating a sample of ice until it melts
   b. tearing a sheet of paper into tiny pieces
   c. burning a sheet of paper
   d. All of the above are correct.

2. A pure substance that cannot be decomposed into anything simpler by chemical or physical means is called a (an)
   a. element.
   b. compound.
   c. mixture.
   d. isotope.

3. The electrons that participate in chemical bonding are (the)
   a. valence electrons.
   b. electrons in fully occupied orbitals.
   c. stable inner electrons.
   d. All of the above are correct.

4. Which type of chemical bond is formed between two atoms by the sharing of electrons?
   a. ionic
   b. covalent
   c. metallic
   d. None of the above is correct.

5. Which combination of elements forms crystalline solids that will dissolve in water, producing a solution of ions that conduct an electric current?
   a. metal and metal
   b. metal and nonmetal
   c. nonmetal and nonmetal
   d. All of the above are correct.

6. An inorganic compound made of only two different elements has a name that always ends with the suffix
   a. *-ite.*
   b. *-ate.*
   c. *-ide.*
   d. *-ous.*

7. Dihydrogen monoxide is a compound with the common name of
   a. laughing gas.
   b. water.
   c. smog.
   d. rocket fuel.

8. A chemical equation is balanced by changing (the)
   a. subscripts.
   b. superscripts.
   c. coefficients.
   d. any of the above as necessary to achieve a balance.

9. Since wood is composed of carbohydrates, you should expect what gases to exhaust from a fireplace when complete combustion takes place?
   a. carbon dioxide, carbon monoxide, and pollutants
   b. carbon dioxide and water vapor
   c. carbon monoxide and smoke
   d. It depends on the type of wood being burned.

10. When carbon burns with an insufficient supply of oxygen, carbon monoxide is formed according to the following equation: $2\ C + O_2 \rightarrow 2\ CO$. What category of chemical reaction is this?
    a. combination
    b. ion exchange
    c. replacement
    d. None of the above is correct.

11. Of the elements listed below, the one with the greatest chemical activity is
    a. aluminum.
    b. zinc.
    c. iron.
    d. mercury.

12. A balanced chemical equation has
    a. the same number of molecules on both sides of the equation.
    b. the same kinds of molecules on both sides of the equation.
    c. the same number of each kind of atom on both sides of the equation.
    d. All of the above are correct.

Answers

1. c 2. a 3. a 4. b 5. b 6. c 7. b 8. c 9. b 10. a 11. a 12. c

QUESTIONS FOR THOUGHT

1. What is the difference between a chemical change and a physical change? Give three examples of each.

2. Describe how the following are alike and how they are different: (a) a sodium atom and a sodium ion, and (b) a sodium ion and a neon atom.

3. What is the difference between an ionic and covalent bond? What do atoms forming the two bond types have in common?

4. What is the octet rule?

5. What is a polyatomic ion? Give the names and formulas for several common polyatomic ions.

6. Write the formula for magnesium hydroxide. Explain what the parentheses mean.

7. What is the basic difference between a single bond and a double bond?

8. What is the law of conservation of mass? How do you know if a chemical equation is in accord with this law?

9. Describe in your own words how a chemical equation is balanced.

10. How is the activity series for metals used to predict if a replacement reaction will occur or not?

11. What must occur in order for an ion exchange reaction to take place? What is the result if this does not happen?

12. Predict the products for the following reactions: (a) the combustion of ethyl alcohol (C_2H_5OH), and (b) the rusting of aluminum (Al).

FOR FURTHER ANALYSIS

1. What are the significant similarities and differences between a physical change and a chemical change?

2. Analyze how you would know for sure that a pure substance you have is a compound and not an element.

3. Analyze how you would know for sure that a pure substance you have is an element and not a compound.

4. Make up an explanation for why ionic compounds are formed when a metal from the left side of the periodic table reacts with a nonmetal from the right side, but covalent bonds are formed between nonmetallic elements on the right side of the table.

5. What are the advantages and disadvantages to writing a chemical equation with chemical symbols and formulas rather than just words?

6. Provide several examples of each of the four basic categories of chemical reactions and describe how each illustrates a clear representation of the category.

7. Summarize for another person the steps needed for successfully writing a balanced chemical equation.

INVITATION TO INQUIRY

Rate of Chemical Reactions

Temperature is one of the more important factors that influence the rate of a chemical reaction. You can use a "light stick" or "light tube" to study how temperature can influence a chemical reaction. Light sticks and tubes are devices that glow in the dark and are very popular on July 4 and at other times when people might be outside after sunset. They work from a chemical reaction that is similar to the chemical reaction that produces light in a firefly. Design an experiment that uses light sticks to find out the effect of temperature on the brightness of light and how long the device will continue providing light. Perhaps you will be able to show by experimental evidence that use at a particular temperature produces the most light for the longest period of time.

PARALLEL EXERCISES

The exercises in groups A and B cover the same concepts. Solutions to group A exercises are located in appendix D.

Group A

1. How many outer orbital electrons are found in an atom of:
   a. Li
   b. N
   c. F
   d. Cl
   e. Ra
   f. Be

2. Write electron dot notations for the following elements:
   a. Boron
   b. Bromine
   c. Calcium
   d. Potassium
   e. Oxygen
   f. Sulfur

3. Identify the charge on the following ions:
   a. Boron
   b. Bromine
   c. Calcium
   d. Potassium
   e. Oxygen
   f. Nitrogen

4. Name the following polyatomic ions:
   a. $(OH)^-$
   b. $(SO_3)^{2-}$
   c. $(ClO)^-$
   d. $(NO_3)^-$
   e. $(CO_3)^{2-}$
   f. $(ClO_4)^-$

5. Use the crossover technique to write formulas for the following compounds:
   a. Iron(III) hydroxide
   b. Lead(II) phosphate
   c. Zinc carbonate
   d. Ammonium nitrate
   e. Potassium hydrogen carbonate
   f. Potassium sulfite

6. Write formulas for the following covalent compounds:
   a. Carbon tetrachloride
   b. Dihydrogen monoxide
   c. Manganese dioxide
   d. Sulfur trioxide
   e. Dinitrogen pentoxide
   f. Diarsenic pentasulfide

Group B

1. How many outer orbital electrons are found in an atom of:
   a. Na
   b. P
   c. Br
   d. I
   e. Te
   f. Sr

2. Write electron dot notations for the following elements:
   a. Aluminum
   b. Fluorine
   c. Magnesium
   d. Sodium
   e. Carbon
   f. Chlorine

3. Identify the charge on the following ions:
   a. Aluminum
   b. Chlorine
   c. Magnesium
   d. Sodium
   e. Sulfur
   f. Hydrogen

4. Name the following polyatomic ions:
   a. $(C_2H_3O_2)^-$
   b. $(HCO_3)^-$
   c. $(SO_4)^{2-}$
   d. $(NO_2)^-$
   e. $(MnO_4)^-$
   f. $(CO_3)^{2-}$

5. Use the crossover technique to write formulas for the following compounds:
   a. Aluminum hydroxide
   b. Sodium phosphate
   c. Copper(II) chloride
   d. Ammonium sulfate
   e. Sodium hydrogen carbonate
   f. Cobalt(II) chloride

6. Write formulas for the following covalent compounds:
   a. Silicon dioxide
   b. Dihydrogen sulfide
   c. Boron trifluoride
   d. Dihydrogen dioxide
   e. Carbon tetrafluoride
   f. Nitrogen trihydride

—continued

Group A

7. Name the following covalent compounds:
   a. CO
   b. CO_2
   c. CS_2
   d. N_2O
   e. P_4S_3
   f. N_2O_3

8. Write balanced chemical equations for each of the following unbalanced reactions:
   a. $SO_2 + O_2 \rightarrow SO_3$
   b. $P + O_2 \rightarrow P_2O_5$
   c. $Al + HCl \rightarrow AlCl_3 + H_2$
   d. $NaOH + H_2SO_4 \rightarrow Na_2SO_4 + H_2O$
   e. $Fe_2O_3 + CO \rightarrow Fe + CO_2$
   f. $Mg(OH)_2 + H_3PO_4 \rightarrow Mg_3(PO_4)_2 + H_2O$

9. Identify the following as combination, decomposition, replacement, or ion exchange reactions:
   a. $NaCl_{(aq)} + AgNO_{3(aq)} \rightarrow NaNO_{3(aq)} + AgCl\downarrow$
   b. $H_2O_{(l)} + CO_{2(g)} \rightarrow H_2CO_{3(l)}$
   c. $2\,NaHCO_{3(s)} \rightarrow Na_2CO_{3(s)} + H_2O_{(g)} + CO_{2(g)}$
   d. $2\,Na_{(s)} + Cl_{2(g)} \rightarrow 2\,NaCl_{(s)}$
   e. $Cu_{(s)} + 2\,AgNO_{3(aq)} \rightarrow Cu(NO_3)_{2(aq)} + 2\,Ag_{(s)}$
   f. $CaO_{(s)} + H_2O_{(l)} \rightarrow Ca(OH)_{2(aq)}$

10. Write complete, balanced equations for each of the following reactions:
   a. $C_5H_{12(g)} + O_{2(g)} \rightarrow$
   b. $HCl_{(aq)} + NaOH_{(aq)} \rightarrow$
   c. $Al_{(s)} + Fe_2O_{3(s)} \rightarrow$
   d. $Fe_{(s)} + CuSO_{4(aq)} \rightarrow$
   e. $MgCl_{(aq)} + Fe(NO_3)_{2(aq)} \rightarrow$
   f. $C_6H_{10}O_{5(s)} + O_{2(g)} \rightarrow$

Group B

7. Name the following covalent compounds:
   a. N_2O
   b. SO_2
   c. SiC
   d. PF_5
   e. $SeCl_6$
   f. N_2O_4

8. Write balanced chemical equations for each of the following unbalanced reactions:
   a. $NO + O_2 \rightarrow NO_2$
   b. $KClO_3 \rightarrow KCl + O_2$
   c. $NH_4Cl + Ca(OH)_2 \rightarrow CaCl_2 + NH_3 + H_2O$
   d. $NaNO_3 + H_2SO_4 \rightarrow Na_2SO_4 + HNO_3$
   e. $PbS + H_2O_2 \rightarrow PbSO_4 + H_2O$
   f. $Al_2(SO_4)_3 + BaCl_2 \rightarrow AlCl_3 + BaSO_4$

9. Identify the following as combination, decomposition, replacement, or ion exchange reactions:
   a. $ZnCO_{3(s)} \rightarrow ZnO_{(s)} + CO_2\uparrow$
   b. $2\,NaBr_{(aq)} + Cl_{2(g)} \rightarrow 2\,NaCl_{(aq)} + Br_{2(g)}$
   c. $2\,Al_{(s)} + 3\,Cl_{2(g)} \rightarrow 2\,AlCl_{3(s)}$
   d. $Ca(OH)_{2(aq)} + H_2SO_{4(aq)} \rightarrow CaSO_{4(aq)} + 2\,H_2O_{(l)}$
   e. $Pb(NO_3)_{2(aq)} + H_2S_{(g)} \rightarrow 2\,HNO_{3(aq)} + PbS\downarrow$
   f. $C_{(s)} + ZnO_{(s)} \rightarrow Zn_{(s)} + CO\uparrow$

10. Write complete, balanced equations for each of the following reactions:
   a. $C_3H_{6(g)} + O_{2(g)} \rightarrow$
   b. $H_2SO_{4(aq)} + KOH_{(aq)} \rightarrow$
   c. $C_6H_{12}O_{6(s)} + O_{2(g)} \rightarrow$
   d. $Na_3PO_{4(aq)} + AgNO_{3(aq)} \rightarrow$
   e. $NaOH_{(aq)} + Al(NO_3)_{3(aq)} \rightarrow$
   f. $Mg(OH)_{2(aq)} + H_3PO_{4(aq)} \rightarrow$

10

Water and Solutions

Water is often referred to as the *universal solvent* because it makes so many different kinds of solutions. Eventually, moving water can dissolve solid rock, carrying it away in solution.

CORE **CONCEPT**
Water and solutions of water have unique properties.

OUTLINE

Earth Science

▷ Water cycles in and out of the atmosphere (Ch. 7).

▷ Less than 1 percent of all water on Earth is fit for human consumption or agriculture (Ch. 18).

Life Science

▷ Metabolic processes require water for all aspects (Ch. 20).

▷ Water and temperature ranges determine biomes (Ch. 23).

OVERVIEW

What do you think about when you see a stream (figure 10.1)? Do you wonder about the water quality and what might be dissolved in the water? Do you wonder where the stream comes from and if it will ever run out of water?

Many people can look at a stream, but they might think about different things. A farmer might think about how the water could be diverted and used for his crops.

A city planner might wonder if the water is safe for domestic use, and if not, what it would cost to treat the water. Others might wonder if the stream has large fish they could catch. Many large streams can provide water for crops, domestic use, and recreation, and still meet the requirements for a number of other uses.

It is the specific properties of water that make it important for agriculture, domestic use, and recreation. Living things evolved in a watery environment, so water and its properties are essential to life on Earth. Some properties of water, such as the ability to dissolve almost anything, also make water very easy to pollute. This chapter is concerned with some of the unique properties of water, water solutions, and household use of water.

HOUSEHOLD WATER

Water is an essential resource, not only because it is required for life processes but also because of its role in a modern society. (See chapter 18.) Water is used in the home for drinking and cooking (2 percent), washing dishes (6 percent), laundry (11 percent), bathing (23 percent), toilets (29 percent), and maintaining lawns and gardens (29 percent).

The water supply is obtained from streams, lakes, and reservoirs on the surface or from groundwater pumped from below the surface. Surface water contains more sediments, bacteria, and possible pollutants than water from a well because it is exposed to the atmosphere and water runs off the land into streams and rivers. Surface water requires filtering to remove suspended particles, treatment to kill bacteria, and sometimes processing to remove pollution. Well water is generally cleaner but still might require treatment to kill bacteria and remove pollution that has seeped through the ground from waste dumps, agricultural activities, or industrial sites.

Most pollutants are usually too dilute to be considered a significant health hazard, but there are exceptions. There are five types of contamination found in U.S. drinking water that are responsible for the most widespread danger, and these are listed in table 10.1. In spite of these general concerns and other occasional local problems, the U.S. water supply is considered to be among the cleanest in the world.

Demand for domestic water sometimes exceeds the immediate supply in some growing metropolitan areas. This is most common during the summer, when water demand is

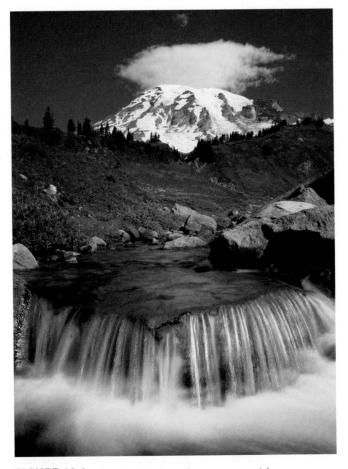

FIGURE 10.1 A freshwater stream has many potential uses.

Science and Society

Who Has the Right?

As the population grows and new industries develop, more and more demands are placed on the water supply. This raises some issues about how water should be divided among agriculture, industries, and city domestic use. Agricultural interests claim they should have the water because they produce the food and fibers that people must have. Industrial interests claim they should have the water because they create the jobs and the products that people must have. Cities, on the other hand, claim that domestic consumption is the most important because people cannot survive without water. Yet, others claim that no group has a right to use water when it is needed to maintain habitats.

Questions to Discuss:

1. Who should have the first priority for water use?
2. Who should have the last priority for water use?
3. What determined your answers to questions 1 and 2?

TABLE 10.1 Possible Pollution Problems in the U.S. Water Supply

Pollutant	Source	Risk
Lead	Lead pipes in older homes; solder in copper pipes, brass fixtures	Nerve damage, miscarriage, birth defects, high blood pressure, hearing problems
Chlorinated solvents	Industrial pollution	Cancer
Trihalomethanes	Chlorine disinfectant reacting with other pollutants	Liver damage, kidney damage, possibly cancer
PCBs	Industrial waste, older transformers	Liver damage, possibly cancer
Bacteria and viruses	Septic tanks, outhouses, overflowing sewer lines	Gastrointestinal problems, serious diseases

high and rainfall is often low. Communities in these areas often have public education campaigns designed to help reduce the demand for water. For example, did you know that taking a tub bath can use up to 135 liters (about 36 gal) of water compared to only 95 liters (about 25 gal) for a regular shower? Even more water is saved by a shower that does not run continuously—wetting down, soaping up, and rinsing off uses only 15 liters (about 4 gal) of water. You can also save about 35 liters (about 9 gal) of water by not letting the water run continuously while brushing your teeth.

It is often difficult to convince people to conserve water when it is viewed as an inexpensive, limitless supply. However, efforts to conserve water increase dramatically as the cost to the household consumer increases.

The issues involved in maintaining a safe water supply are better understood by considering some of the properties of water and water solutions. These are the topics of the following sections.

PROPERTIES OF WATER

Water is essential for life since living organisms are made up of cells filled with water and a great variety of dissolved substances. Foods are mostly water, with fruits and vegetables containing up to 95 percent water and meat consisting of about 50 percent water. Your body is over 70 percent water by weight. Since water is such a large component of living things, understanding the properties of water is important to understanding life. One important property is water's un-

usual ability to act as a solvent. Water is called a "universal solvent" because of its ability to dissolve most molecules. In living things, these dissolved molecules can be transported from one place to another by diffusion or by some kind of a circulatory system.

The usefulness of water does not end with its unique abilities as a solvent and transporter; it has many more properties that are useful, although unusual. For example, unlike other liquids, water in its liquid phase has a greater density than solid water (ice). This important property enables solid ice to float on the surface of liquid water, insulating the water below and permitting fish and other water organisms to survive the winter. If ice were denser than water, it would sink, freezing all lakes and rivers from the bottom up. Fish and most organisms that live in water would not be able to survive in a lake or river of solid ice.

As described in chapter 4, water is also unusual because it has a high specific heat. The same amount of sunlight falling on equal masses of soil and water will warm the soil 5°C for each 1°C increase in water temperature. Thus, it will take five times more sunlight to increase the temperature of the water to as much as the soil. This enables large bodies of water to moderate the temperature, making it more even.

A high latent heat of vaporization is yet another unusual property of water. This particular property enables people to dissipate large amounts of heat by evaporating a small amount of water. Since people carry this evaporative cooling system with them, they can survive some very warm desert temperatures, for example.

Finally, other properties of water are not crucial for life but are interesting nonetheless. For example, why do all snowflakes have six sides? Is it true that no two snowflakes are alike? The unique structure of the water molecule will explain water's unique solvent abilities, why solid water is less dense than liquid water, its high specific heat, its high latent heat of vaporization, and perhaps why no two snowflakes seem to be alike.

Structure of the Water Molecule

In chapter 9, you learned that atoms combine two ways. Atoms from opposite sides of the periodic table form ionic bonds after transferring one or more electrons. Atoms from the right side of the periodic table form covalent bonds by sharing one or more pairs of electrons. This distinction is clear-cut in many compounds but not in water. The way atoms share electrons in a water molecule is not exactly covalent, but it is not ionic either.

In a molecule of water, an oxygen atom shares a pair of electrons with two hydrogen atoms. Oxygen has six outer electrons and needs two more to satisfy the octet rule, achieving the noble gas structure of eight. Each hydrogen atom needs one more electron to fill its outer orbital with two. Therefore, one oxygen atom bonds with two hydrogen atoms, forming H_2O. Both oxygen and hydrogen are more stable with the outer orbital configuration of the noble gases (neon and helium in this case).

Electrons are shared in a water molecule but not equally. Oxygen, with its eight positive protons, has a greater attraction for the shared electrons than do either of the hydrogens with a single proton. Therefore, the shared electrons spend more time around the oxygen part of the molecule than they do around the hydrogen part. This results in the oxygen end of the molecule being more negative than the hydrogen end. When electrons in a covalent bond are not equally shared, the molecule is said to be polar. A **polar molecule** has a *dipole* (*di* = two; *pole* = side or end), meaning it has a positive end and a negative end.

A water molecule has a negative center at the oxygen end and a positive center at the hydrogen end. The positive charges on the hydrogen end are separated, giving the molecule a bent arrangement rather than a straight line. Figure 10.2A shows a model of a water molecule showing its polar nature.

It is the polar structure of the water molecule that is responsible for many of the unique properties of water. Polar molecules of any substances have attractions between the positive end of a molecule and the negative end of another molecule. When the polar molecule has hydrogen at one end and fluorine, oxygen, or nitrogen on the other part, the attractions are strong enough to make a type of bonding called **hydrogen bonding.** Hydrogen bonding is a bond that occurs between the hydrogen end of a molecule and the fluorine, oxygen, or nitrogen end of other similar molecules. A better name for this would be a hydrogen-fluorine bond, a hydrogen-oxygen bond, or a hydrogen-nitrogen bond. However, for brevity, the second part of the bond is not named and all the hydrogen-something bonds are simply known of as "hydrogen" bonds. The dotted line between the hydrogen and oxygen molecules in figure 10.2B represents a hydrogen bond. A dotted line is used to represent a bond that is not as strong as the bond represented by the solid line of a covalent compound.

FIGURE 10.2 (A) The water molecule is polar, with centers of positive and negative charges. (B) Attractions between these positive and negative centers establish hydrogen bonds between adjacent molecules.

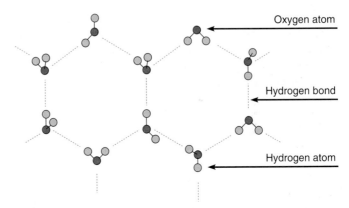

FIGURE 10.3 The hexagonal structure of ice. Hydrogen bonding between the oxygen atom and two hydrogen atoms of other water molecules results in an arrangement, which forms the open, hexagonal structure of ice. Note the angles of the water molecules do not change but have different orientations.

Hydrogen bonding accounts for the physical properties of water, including its unusual density changes with changes in temperature. Figure 10.3 shows the hydrogen-bonded structure of ice. Water molecules form a six-sided hexagonal structure that extends out for billions of molecules. The large channels, or holes, in the structure result in ice being less dense than water. The shape of the hexagonal arrangement also suggests why snowflakes always have six sides. Why does it seem like no two snowflakes are alike? Perhaps the answer can be found in the almost infinite variety of shapes that can be built from billions and billions of tiny hexagons of ice crystals.

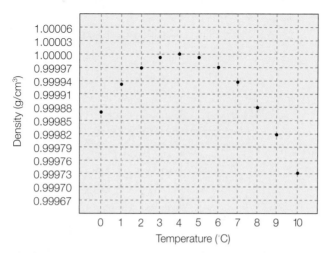

FIGURE 10.4 The density of water from 0°C to 10°C. The density of water is at a maximum at 4°C, becoming less dense as it is cooled or warmed from this temperature. Hydrogen bonding explains this unusual behavior.

Myths, Mistakes, and Misunderstandings

Teardrops Keep Falling?

It is a mistake to represent raindrops or drops of falling water with teardrop shapes. Small raindrops are pulled into a spherical shape by surface tension. Larger raindrops are also pulled into a spherical shape, but the pressure of air on the bottom of the falling drop somewhat flattens the bottom. If the raindrop is too large, the pressure of air on the falling drop forms a concave depression on the bottom, which grows deeper and deeper until the drop breaks up into smaller spherical drops.

When ice is warmed, the increased vibrations of the molecules begin to expand and stretch the hydrogen bond structure. When ice melts, about 15 percent of the hydrogen bonds break and the open structure collapses into the more compact arrangement of liquid water. As the liquid water is warmed from 0°C, still more hydrogen bonds break down, and the density of the water steadily increases. At 4°C, the expansion of water from the increased molecular vibrations begins to predominate, and the density decreases steadily with further warming (figure 10.4). Thus, water has its greatest density at a temperature of 4°C.

The heat of fusion, specific heat, and heat of vaporization of water are unusually high when compared to other chemically similar substances. These high values are accounted for by the additional energy needed to break hydrogen bonds.

The Dissolving Process

A **solution** is a homogeneous mixture of ions or molecules of two or more substances. *Dissolving* is the process of making a solution. During dissolving, the different components that make up the solution become mixed. For example, when sugar dissolves in water, molecules of sugar become uniformly dispersed throughout the molecules of water. The uniform taste of sweetness of any part of the sugar solution is a result of this uniform mixing.

The general terms of *solvent* and *solute* identify the components of a solution. The solvent is the component present in the larger amount. The solute is the component that dissolves in the solvent. Atmospheric air, for example, is about 78 percent nitrogen, so nitrogen is considered the solvent. Oxygen (about 21 percent), argon (about 0.9 percent), and other gases make up the solutes. If one of the components of a solution is a liquid, it is usually identified as the solvent. An *aqueous solution* is a solution of a solid, a liquid, or a gas in water.

A solution is formed when the molecules or ions of two or more substances become homogeneously mixed. But the process of dissolving must be more complicated than the simple mixing together of particles because (1) solutions become saturated, meaning there is a limit on solubility, and (2) some substances are *insoluble,* not dissolving at all or at least not noticeably. In general, the forces of attraction between molecules or ions of the solvent and solute determine if something will dissolve and if there are limits on the solubility. These forces of attraction and their role in the dissolving process will be considered in the following examples.

First, consider the dissolving process in gaseous and liquid solutions. In a gas, the intermolecular forces are small, so gases can mix in any proportion. Fluids that can mix in any proportion like this are called **miscible fluids.** Fluids that do not mix are called *immiscible fluids.* Air is a mixture of gases, so gases (including vapors) are miscible.

Liquid solutions can dissolve a gas, another liquid, or a solid. Gases are miscible in liquids, and a carbonated beverage (your favorite cola) is the common example, consisting of carbon dioxide dissolved in water. Whether or not two given liquids form solutions depends on some similarities in their molecular structures. The water molecule, for example, is a polar molecule with a negative end and a positive end. On the other hand, carbon tetrachloride (CCl_4) is a molecule with polar bonds that are symmetrically arranged. Because of the symmetry, CCl_4 has no negative or positive ends, so it is nonpolar. Thus, some liquids have polar molecules, and some have nonpolar molecules. The general rule for forming solutions is *like dissolves like.* A nonpolar compound, such as carbon tetrachloride, will dissolve oils and greases because they are nonpolar compounds. Water, a polar compound, will not dissolve the nonpolar oils and greases. Carbon tetrachloride was at one time used as a cleaning solvent because of its oil and grease dissolving abilities. Its use is no longer recommended because it causes liver damage.

Some molecules, such as soap, have a part of the molecule that is polar and a part that is nonpolar. Washing with water alone will not dissolve oils because water and oil are immiscible. When soap is added to the water, however, the polar end of the soap molecule is attracted to the polar water molecules, and the nonpolar end is absorbed into the oil. A particle (larger than a molecule) is formed, and the oil is washed away with the water.

The "like dissolves like" rule applies to solids and liquid solvents as well as liquids and liquid solvents. Polar solids, such as salt, will readily dissolve in water, which has polar molecules,

Decompression Sickness

Decompression sickness (DCS) is a condition caused by the formation of nitrogen bubbles in the blood and tissues of a scuba diver who surfaces too quickly, causing a rapid drop in pressure. This condition is usually marked by joint pain but can include chest pain, skin irritation, and muscle cramps. This may be barely noticed in mild cases, but severe cases can be fatal. The joint pain is often called *bends*.

DCS can occur in any situation where a person is subjected to a higher air pressure for some time and then experiences a rapid decompression. Here is what happens: Air is about 79 percent nitrogen and 21 percent oxygen. Nitrogen is inert to the human body, and what we inhale is exhaled, but some becomes dissolved in the blood. Dissolved

nitrogen is a normal occurrence and does not present a problem. However, when a person is breathing higher-pressure air, more nitrogen is breathed in and more becomes dissolved in the blood and other tissues. If the person returns to normal pressure slowly, the extra dissolved nitrogen is expelled by the lungs, and this is not a problem. If the return to normal pressure is too rapid, however, nitrogen bubbles form in the blood and other tissues, and this causes DCS. The nitrogen bubbles can cause pressure on nerves, block circulation, and cause joint pain. These symptoms usually appear when the diver returns to the surface or within eight hours of returning to normal air pressure.

One way to prevent DCS is for a diver to make "decompression stops" as he or

she returns to the surface. These stops allow the dissolved nitrogen to diffuse into the lungs rather than making bubbles in tissues and the bloodstream. Also, it is not a good idea for a diver to fly on a commercial airliner for a day or so. Commercial airliners are pressurized to an altitude of about 2,500 m (about 8,200 ft). This is safe for the normal passenger but can cause more decompression problems for the recent diver.

DCS can be treated by placing the affected person in a special sealed chamber. Pressure is slowly increased in the chamber to cause nitrogen bubbles to go back into solution, and then slowly decreased to allow the dissolved nitrogen to be expelled by the lungs.

---

but do not dissolve readily in oil, grease, or other nonpolar solvents. Polar water readily dissolves salt because the charged polar water molecules are able to exert an attraction on the ions, pulling them away from the crystal structure. Thus, ionic compounds dissolve in water.

Ionic compounds vary in their solubilities in water. This difference is explained by the existence of two different forces involved in an ongoing "tug of war." One force is the attraction between an ion on the surface of the crystal and a water molecule, an *ion-polar molecule force*. When solid sodium chloride and water are mixed together, the negative ends of the water molecules (the oxygen ends) become oriented toward the positive sodium ions on the crystal. Likewise, the positive ends of water molecules (the hydrogen ends) become oriented toward the negative chlorine ions. The attraction of water molecules for ions is called **hydration.** If the force of hydration is greater than the attraction between the ions in the solid, they are pulled away from the solid, and dissolving occurs (figure 10.5). Considering sodium chloride only, the equation is

$$Na^+Cl^-_{(s)} \rightarrow Na^+_{(aq)} + Cl^-_{(aq)}$$

which shows that the ions were separated from the solid to become a solution of ions. In other compounds, the attraction between the ions in the solid might be greater than the energy of hydration. In this case, the ions of the solid would win the "tug of war," and the ionic solid is insoluble.

The saturation of soluble compounds is explained in terms of hydration eventually occupying a large number of the polar water molecules. Fewer available water molecules means less attraction on the ionic solid, with more solute ions being pulled back to the surface of the solid. The tug of war continues back

and forth as an equilibrium condition is established. (For a detailed discussion of different measures of solution concentration, see the chapter 10 Resources on www.mhhe.com/tillery.)

Solubility

Gases and liquids appear to be soluble in all proportions, but there is a limit to how much solid can be dissolved in a liquid. You may have noticed that a cup of hot tea will dissolve several

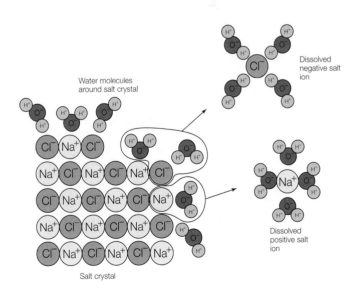

FIGURE 10.5 An ionic solid dissolves in water because the number of water molecules around the surface is greater than the number of other ions of the solid. The attraction between polar water molecules and a charged ion enables the water molecules to pull ions away from the crystal so the salt crystals dissolve in the water.

teaspoons of sugar, but the limit of solubility is reached quickly in a glass of iced tea. The limit of how much sugar will dissolve seems to depend on the temperature of the tea. More sugar added to the cold tea after the limit is reached will not dissolve, and solid sugar granules begin to accumulate at the bottom of the glass. At this limit, the sugar and tea solution is said to be *saturated*. Dissolving does not actually stop when a solution becomes saturated and undissolved sugar continues to enter the solution. However, dissolved sugar is now returning to the undissolved state at the same rate as it is dissolving. The overall equilibrium condition of sugar dissolving as sugar is coming out of solution is called a **saturated solution.** A saturated solution is a *state of equilibrium that exists between dissolving solute and solute coming out of solution.* You actually cannot see the dissolving and coming out of solution that occurs in a saturated solution because the exchanges are taking place with particles the size of molecules or ions.

Not all compounds dissolve as sugar does, and more or less of a given compound may be required to produce a saturated solution at a particular temperature. In general, the difficulty of dissolving a given compound is referred to as *solubility.* More specifically, the **solubility** of a solute is defined as the *concentration that is reached in a saturated solution at a particular temperature.* Solubility varies with the temperature, as the sodium and potassium salt examples show in figure 10.6. These solubility curves describe the amount of solute required to reach the saturation equilibrium at a particular temperature. In general, the solubilities of most ionic solids increase with temperature, but there are exceptions. In addition, some salts release heat when dissolved in water, and other salts absorb heat when dissolved. The "instant cold pack" used for first aid is a bag of water containing a second bag of ammonium nitrate (NH_4NO_3). When the bag of ammonium nitrate is broken, the compound dissolves and absorbs heat.

You can usually dissolve more of a solid, such as salt or sugar, as the temperature of the water is increased. Contrary to what you might expect, gases usually become *less* soluble in water as the temperature increases. As a glass of water warms, small bubbles collect on the sides of the glass as dissolved air comes out of solution. The first bubbles that appear when warming a pot of water to boiling are also bubbles of dissolved air coming out of solution. This is why water that has been boiled usually tastes "flat." The dissolved air has been removed by the heating. The "normal" taste of water can be restored by pouring the boiled water back and forth between two glasses. The water dissolves more air during this process, restoring the usual taste.

Changes in pressure have no effect on the solubility of solids in liquids but greatly affect the solubility of gases. The release of bubbles (fizzing) when a bottle or can of soda is opened occurs because pressure is reduced on the beverage and dissolved carbon dioxide comes out of solution. In general, *gas solubility decreases with temperature and increases with pressure.* As usual, there are exceptions to this generalization.

PROPERTIES OF WATER SOLUTIONS

Pure solvents have characteristic physical and chemical properties that are changed by the addition of a solute. Following are some of the more interesting changes.

Electrolytes

Water solutions of ionic substances will conduct an electric current, so they are called **electrolytes.** Ions must be present and free to move in a solution to carry the charge, so electrolytes are solutions containing ions. Pure water will not conduct an electric current as it is a covalent compound, which ionizes only very slightly. Water solutions of sugar, alcohol, and most other covalent compounds are nonconductors, so they are called *nonelectrolytes.* Nonelectrolytes are covalent compounds that form molecular solutions, so they cannot conduct an electric current.

Some covalent compounds are nonelectrolytes as pure liquids but become electrolytes when dissolved in water. Pure hydrogen chloride (HCl), for example, does not conduct an electric current, so you can assume that it is a molecular substance. When dissolved in water, hydrogen chloride does conduct a current, so it must now contain ions. Evidently, the hydrogen chloride has become *ionized* by the water. The process of forming ions from molecules is called *ionization.* Hydrogen chloride, just as water, has polar molecules. The positive hydrogen atom on the HCl molecule is attracted to the negative oxygen end of a water molecule, and the force of attraction is

FIGURE 10.6 Approximate solubility curves for sodium nitrate, potassium nitrate, potassium chloride, and sodium chloride.

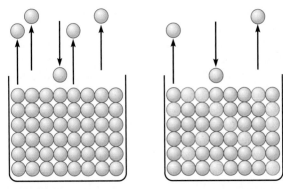

FIGURE 10.7 Three representations of water and hydrogen chloride in an ionizing reaction. (*A*) Sketches of molecules involved in the reaction. (*B*) Electron dot equation of the reaction. (*C*) The chemical equation for the reaction. Each of these representations shows the hydrogen being pulled away from the chlorine atom to form H_3O^+, the hydronium ion.

FIGURE 10.8 The rate of evaporation, and thus the vapor pressure, is less for a solution than for a solvent in the pure state. The greater the solute concentration, the less the vapor pressure.

strong enough to break the hydrogen-chlorine bond, forming charged particles (figure 10.7). The reaction is

$$H_2O_{(1)} + HCl_{(1)} \rightarrow H_3O^+_{(aq)} + Cl^-_{(aq)}$$

The H_3O^+ ion is called a **hydronium ion.** A hydronium ion is basically a molecule of water with an attached hydrogen ion. The presence of the hydronium ion gives the solution new chemical properties; the solution is no longer hydrogen chloride but is *hydrochloric acid.* Hydrochloric acid, and other acids, will be discussed shortly.

Boiling Point

Boiling occurs when the pressure of the vapor escaping from a liquid (vapor pressure) is equal to the atmospheric pressure on the liquid. The *normal* boiling point is defined as the temperature at which the vapor pressure is equal to the average atmospheric pressure at sea level. For pure water, this temperature is 100°C (212°F). It is important to remember that boiling is a purely physical process. No bonds within water molecules are broken during boiling.

The vapor pressure over a solution is *less* than the vapor pressure over the pure solvent at the same temperature. Molecules of a liquid can escape into the air only at the surface of the liquid, and the presence of molecules of a solute means that fewer solvent molecules can be at the surface to escape. Thus, the vapor pressure over a solution is less than the vapor pressure over a pure solvent (figure 10.8).

Because the vapor pressure over a solution is less than that over the pure solvent, the solution boils at a higher temperature. A higher temperature is required to increase the vapor pressure to that of the atmospheric pressure. Some cooks have been observed to add a "pinch" of salt to a pot of water before boiling. Is this to increase the boiling point and therefore cook the food more quickly? How much does a pinch of salt increase the boiling temperature? The answers are found in the relationship between the concentration of a solute and the boiling point of the solution.

It is the number of solute particles (ions or molecules) at the surface of a solution that increases the boiling point. A mole is a measure that can be defined as a number of particles called Avogadro's number. Since the number of particles at the surface is proportional to the ratio of particles in the solution, the concentration of the solute will directly influence the increase in the boiling point. In other words, the boiling point of any dilute solution is increased proportional to the concentration of the solute. For water, the boiling point is increased 0.521°C for every mole of solute dissolved in 1,000 g of water. Thus, any water solution will boil at a higher temperature than pure water. Since it boils at a higher temperature, it also takes a longer time to reach the boiling point.

It makes no difference what substance is dissolved in the water; 1 mole of solute in 1,000 g of water will elevate the boiling point by 0.521°C. A mole contains Avogadro's number of particles, so a mole of any solute will lower the vapor pressure by the same amount. Sucrose, or table sugar, for example, is $C_{12}H_{22}O_{11}$ and has a gram-formula weight of 342 g. Thus, 342 g of sugar in 1,000 g of water (about a liter) will increase the boiling point by 0.521°C. Therefore, if you measure the boiling point of a sugar solution, you can determine the concentration of sugar in the solution. For example, pancake syrup that boils at 100.261°C (sea-level pressure) must contain 171 g of sugar dissolved in 1,000 g of water. You know this because the increase of 0.261°C over 100°C is one-half of 0.521°C. If the boiling point were increased by 0.521°C over 100°C, the syrup would have the full gram-formula weight (342 g) dissolved in a kg of water.

Since it is the number of particles of solute in a specific sample of water that elevates the boiling point, different effects are observed in dissolved covalent and dissolved ionic compounds (figure 10.9). Sugar is a covalent compound, and the solute is molecules of sugar moving between the water molecules. Sodium chloride, on the other hand, is an ionic compound and dissolves by the separation of ions, or

$$Na^+Cl^-_{(s)} \rightarrow Na^+_{(aq)} + Cl^-_{(aq)}$$

This equation tells you that 1 mole of NaCl separates into 1 mole of sodium ions and 1 mole of chlorine ions for a total of *2* moles of solute. The boiling point elevation of a solution made from 1 mole of NaCl (58.5 g) is therefore multiplied by two, or $2 \times 0.521°C = 1.04°C$. The boiling point of a solution made by adding 58.5 g of NaCl to 1,000 g of water is therefore 101.04°C at normal sea-level pressure.

Now back to the question of how much a pinch of salt increases the boiling point of a pot of water. Assuming the pot contains about a liter of water (about a quart) and assuming that a "pinch" of salt has a mass of about 0.2 gram, the boiling point will be increased by 0.0037°C. Thus, there must be some reason other than increasing the boiling point that a cook adds a pinch of salt to a pot of boiling water. Perhaps the salt is for seasoning?

Freezing Point

Freezing occurs when the kinetic energy of molecules has been reduced sufficiently so the molecules can come together, forming the crystal structure of the solid. Reduced kinetic energy of the molecules, that is, reduced temperature, results in a specific freezing point for each pure liquid. The *normal* freezing point for pure water, for example, is 0°C (32°F) under normal pressure. The presence of solute particles in a solution interferes with the water molecules as they attempt to form the six-sided hexagonal structure. The water molecules cannot get by the solute particles until the kinetic energy of the solute particles is reduced, that is, until the temperature is below the normal freezing point. Thus, the presence of solute particles lowers the freezing point, and solutions freeze at a lower temperature than the pure solvent.

The freezing-point depression of a solution has a number of interesting implications for solutions such as seawater. When seawater freezes, the water molecules must work their way around the salt particles as was described earlier. Thus, the solute particles are *not* normally included in the hexagonal structure of ice. Ice formed in seawater is practically pure water. Since the solute was *excluded* when the ice formed, the freezing of seawater increases the salinity. Increased salinity means increased concentration, so the freezing point of seawater is further depressed and more ice forms only at a lower temperature. When this additional ice forms, more pure water is removed and the process goes on. Thus, seawater does not have a fixed freezing point but has a lower and lower freezing point as more and more ice freezes.

The depression of the freezing point by a solute has a number of interesting applications in colder climates. Salt, for example, is spread on icy roads to lower the freezing point (and thus the melting point) of the ice. Calcium chloride, $CaCl_2$, is a salt that is often used for this purpose. Water in a car radiator would also freeze in colder climates if a solute, called antifreeze, were not added to the radiator water. Methyl alcohol has been used as an antifreeze because it is soluble in water and does not damage the cooling system. Methyl alcohol, however, has a low boiling point and tends to boil away. Ethylene glycol has a higher boiling point, so it is called a "permanent" antifreeze.

A

B

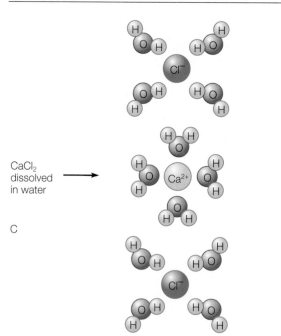

C

FIGURE 10.9   Since ionic compounds dissolve by the separation of ions, they provide more particles in solution than molecular compounds. (*A*) A mole of sugar provides Avogadro's number of particles. (*B*) A mole of NaCl provides two times Avogadro's number of particles. (*C*) A mole of $CaCl_2$ provides three times Avogadro's number of particles.

Like other solutes, ethylene glycol also raises the boiling point, which is an added benefit for summer driving.

ACIDS, BASES, AND SALTS

The electrolytes known as *acids, bases,* and *salts* are evident in environmental quality, foods, and everyday living. Environmental quality includes the hardness of water, which is determined by the presence of certain salts, the acidity of soils, which determines how well plants grow, and acid rain, which is a by-product of industry and automobiles. Many concerns about air and water pollution are often related to the chemistry concepts of acids, bases, and salts.

Properties of Acids and Bases

Acids and bases are classes of chemical compounds that have certain characteristic properties. These properties can be used to identify if a substance is an acid or a base (tables 10.2 and 10.3). The following are the properties of *acids* dissolved in water:

1. Acids have a sour taste, such as the taste of citrus fruits.
2. Acids change the color of certain substances; for example, litmus changes from blue to red when placed in an acid solution (figure 10.10A).
3. Acids react with active metals, such as magnesium or zinc, releasing hydrogen gas.
4. Acids *neutralize* bases, forming water and salts from the reaction.

Likewise, *bases* have their own characteristic properties. Bases are also called alkaline substances, and the following are the properties of bases dissolved in water:

1. Bases have a bitter taste, for example, the taste of caffeine.
2. Bases reverse the color changes that were caused by acids. Red litmus is changed back to blue when placed in a solution containing a base (figure 10.10B).
3. Basic solutions feel slippery on the skin. They have a *caustic* action on plant and animal tissue, converting tissue into soluble materials. A strong base, for example, reacts with

fat to make soap and glycerine. This accounts for the slippery feeling on the skin.
4. Bases *neutralize* acids, forming water and salts from the reaction.

Tasting an acid or base to see if it is sour or bitter can be hazardous, since some are highly corrosive or caustic. Many organic acids are not as corrosive and occur naturally in foods. Citrus fruit, for example, contains citric acid, vinegar is a solution of acetic acid, and sour milk contains lactic acid. The stings or bites of some insects (bees, wasps, and ants) and some plants (stinging nettles) are painful because an organic acid, formic acid, is injected by the insect or plant. Your stomach contains a solution of hydrochloric acid. In terms of relative strength, the hydrochloric acid in your stomach is about ten times stronger than the carbonic acid (H_2CO_3) of carbonated beverages.

Examples of bases include solutions of sodium hydroxide (NaOH), which has a common name of lye or caustic soda, and potassium hydroxide (KOH), which has a common name of caustic potash. These two bases are used in products known as drain cleaners. They open plugged drains because of their caustic action, turning grease, hair, and other organic "plugs" into soap and other soluble substances that are washed away. A weaker base is a solution of ammonia (NH_3), which is often used as a household cleaner. A solution of magnesium hydroxide, $Mg(OH)_2$, has a common name of milk of magnesia and is sold as an antacid and laxative.

Many natural substances change color when mixed with acids or bases. You may have noticed that tea changes color slightly, becoming lighter, when lemon juice (which contains citric acid) is added. Some plants have flowers of one color when grown in acidic soil and flowers of another color when grown in basic soil. A vegetable dye that changes color in the presence of acids or bases can be used as an *acid-base indicator*. An indicator is simply a vegetable dye that is used to distinguish between acid and base solutions by a color change. Litmus, for example, is an acid-base indicator made from a dye extracted from certain species of lichens. The dye is applied to paper strips, which turn red in acidic solutions and blue in basic solutions.

TABLE 10.2 Some Common Acids

Name	Formula	Comment
Acetic acid	CH_3COOH	A weak acid found in vinegar
Boric acid	H_3BO_3	A weak acid used in eyedrops
Carbonic acid	H_2CO_3	The weak acid of carbonated beverages
Formic acid	$HCOOH$	Makes the sting of insects and certain plants
Hydrochloric acid	HCl	Also called muriatic acid; used in swimming pools, soil acidifiers, and stain removers
Lactic acid	$CH_3CHOHCOOH$	Found in sour milk, sauerkraut, and pickles; gives tart taste to yogurt
Nitric acid	HNO_3	A strong acid
Phosphoric acid	H_3PO_4	Used in cleaning solutions; added to carbonated beverages for tartness
Sulfuric acid	H_2SO_4	Also called oil of vitriol; used as battery acid and in swimming pools

TABLE 10.3 Some Common Bases

Name	Formula	Comment
Sodium hydroxide	NaOH	Also called lye or caustic soda; a strong base used in oven cleaners and drain cleaners
Potassium hydroxide	KOH	Also called caustic potash; a strong base used in drain cleaners
Ammonia	NH_3	A weak base used in household cleaning solutions
Calcium hydroxide	$Ca(OH)_2$	Also called slaked lime; used to make brick mortar
Magnesium hydroxide	$Mg(OH)_2$	Solution is called milk of magnesia; used as antacid and laxative

A

B

FIGURE 10.10 (*A*) Acid solutions will change the color of blue litmus to red. (*B*) Solutions of bases will change the color of red litmus to blue.

Explaining Acid-Base Properties

Comparing the lists in tables 10.2 and 10.3, you can see that acids and bases appear to be chemical opposites. Notice in table 10.2 that the acids all have an H, or hydrogen atom, in their formulas. In table 10.3, most of the bases have a hydroxide ion, OH⁻, in their formulas. Could this be the key to acid-base properties?

The modern concept of an acid considers the properties of acids in terms of the hydronium ion, H_3O^+. As was mentioned earlier, the hydronium ion is a water molecule to which an H^+ ion is attached. Since a hydrogen ion is a hydrogen atom without its single electron, it could be considered as an ion consisting of a single proton. Thus, the H^+ ion can be called a *proton*. An **acid** is defined as any substance that is a *proton donor* when dissolved in water, increasing the hydronium ion concentration. For example, hydrogen chloride dissolved in water has the following reaction:

$$(H)Cl_{(aq)} \quad + \quad H_2O_{(l)} \quad \longrightarrow \quad H_3O^+_{(aq)} \quad + \quad Cl^-_{(aq)}$$

The dotted circle and arrow were added to show that the hydrogen chloride donated a proton to a water molecule. The resulting solution contains H_3O^+ ions and has acid properties, so the solution is called hydrochloric acid. It is the H_3O^+ ion that is responsible for the properties of an acid.

The bases listed in table 10.3 all appear to have a hydroxide ion, OH^-. Water solutions of these bases do contain OH^- ions, but the definition of a base is much broader. A **base** is defined as any substance that is a *proton acceptor* when dissolved in water, increasing the hydroxide ion concentration. For example, ammonia dissolved in water has the following reaction:

$$NH_{3(g)} \quad + \quad (H_2)O_{(l)} \quad \longrightarrow \quad (NH_4)^+ \quad + \quad OH^-$$

The dotted circle and arrow show that the ammonia molecule accepted a proton from a water molecule, providing a hydroxide ion. The resulting solution contains OH^- ions and has basic properties, so a solution of ammonium hydroxide is a base.

Carbonates, such as sodium carbonate (Na_2CO_3), form basic solutions because the carbonate ion reacts with water to produce hydroxide ions.

$$(CO_3)^{2-}_{(aq)} + H_2O_{(l)} \rightarrow (HCO_3)^-_{(aq)} + OH^-_{(aq)}$$

Thus, sodium carbonate produces a basic solution.

Acids could be thought of as simply solutions of hydronium ions in water, and bases could be considered solutions of hydroxide ions in water. The proton donor and proton acceptor definition is much broader, and it does include the definition of acids and bases as hydronium and hydroxide compounds. The broader, more general definition covers a wider variety of reactions and is therefore more useful.

The modern concept of acids and bases explains why the properties of acids and bases are **neutralized,** or lost, when acids and bases are mixed together. For example, consider the hydronium ion produced in the hydrochloric acid solution and the hydroxide ion produced in the ammonia solution. When these solutions are mixed together, the hydronium ion reacts with the hydroxide ion, and

$$H_3O^+{}_{(aq)} + OH^+{}_{(aq)} \rightarrow H_2O_{(l)} + H_2O_{(l)}$$

Thus, a proton is transferred from the hydronium ion (an acid), and the proton is accepted by the hydroxide ion (a base). Water is produced, and both the acid and base properties disappear or are neutralized.

Strong and Weak Acids and Bases

Acids and bases are classified according to their degree of ionization when placed in water. *Strong acids* ionize completely in water, with all molecules dissociating into ions. Nitric acid, for example, reacts completely in the following equation:

$$HNO_{3(aq)} + H_2O_{(l)} \rightarrow H_3O^+{}_{(aq)} + (NO_3)^-{}_{(aq)}$$

Nitric acid, hydrochloric acid (figure 10.11), and sulfuric acid are common strong acids.

Acids that ionize only partially produce fewer hydronium ions, so they are *weak acids.* Vinegar, for example, contains acetic acid that reacts with water in the following reaction:

$$HC_2H_3O_2 + H_2O \rightarrow H_3O^+ + (C_2H_3O_2)^-$$

Only about 1 percent or less of the acetic acid molecules ionize, depending on the concentration.

Bases are also classified as strong or weak. A *strong base* is completely ionized in solution. Sodium hydroxide, or lye, is the most common example of a strong base. It dissolves in water to form a solution of sodium and hydroxide ions:

$$^+OH^-{}_{(s)} \rightarrow Na^+{}_{(aq)} + OH^-{}_{(aq)}$$

A *weak base* is only partially ionized. Ammonia, magnesium hydroxide, and calcium hydroxide are examples of weak bases. Magnesium and calcium hydroxide are only slightly soluble in water, and this reduces the *concentration* of hydroxide ions in a solution.

The pH Scale

The strength of an acid or a base is usually expressed in terms of a range of values called a **pH scale.** The pH scale is based on the concentration of the hydronium ion (in moles/L) in an acidic or a basic solution. To understand how the scale is able to express both acid and base strength in terms of the hydronium ion, first note that pure water is very slightly ionized in the reaction:

$$H_2O_{(l)} + H_2O_{(l)} \rightarrow H_3O^+{}_{(aq)} OH^-{}_{(aq)}$$

The amount of self-ionization by water has been determined through measurements. In pure water at 25°C or any neutral water solution at that temperature, the H_3O^+ concentration is 1×10^{-7} moles/L, and the OH^- concentration is also 1×10^{-7} moles/L.

Since both ions are produced in equal numbers, then the H_3O^+ concentration equals the OH^- concentration, and pure water is neutral, neither acidic nor basic.

In general, adding an acid substance to pure water increases the H_3O^+ concentration. Adding a base substance to pure water increases the OH^- concentration. Adding a base also *reduces* the H_3O^+ concentration as the additional OH^- ions are able to combine with more of the hydronium ions to produce union-ized water. Thus, at a given temperature, an increase in OH^- concentration is matched by a *decrease* in H_3O^+ concentration. The concentration of the hydronium ion can be used as a measure of acidic, neutral, and basic solutions. In general, (1) acidic solutions have H_3O^+ concentrations above 1×10^{-7} moles/L, (2) neutral solutions have H_3O^+ concentrations equal to 1×10^{-7} moles/L, and (3) basic solutions have H_3O^+ concentrations less than 1×10^{-7} moles/L. These three statements lead directly to the pH scale, which is named from the French *pouvoir hydrogene,* meaning "hydrogen power." Power refers to the exponent of the hydronium ion concentration, and the pH is a *power of ten notation that expresses the H_3O^+ concentration.*

A neutral solution has a pH of 7.0. Acidic solutions have pH values below 7, and smaller numbers mean greater acidic properties. Increasing the OH^- concentration decreases the

---

CONCEPTS APPLIED

Acid or Base?

Pick some household product that probably has an acid or base character (Example: pH increaser for aquariums). On a separate paper, write the listed ingredients and identify any you believe would be distinctly acidic or basic in a water solution. Tell whether you expect the product to be an acid or a base. Describe your findings of a litmus paper test.

FIGURE 10.11 Hydrochloric acid (HCl) has the common name of *muriatic* acid. Hydrochloric acid is a strong acid used in swimming pools, soil acidifiers, and stain removers.

H_3O^+ Concentration (moles/liters)	pH	Meaning
1×10^{-0} (=1)	0	
1×10^{-1}	1	
1×10^{-2}	2	
1×10^{-3}	3	Increasing acidity
1×10^{-4}	4	
1×10^{-5}	5	
1×10^{-6}	6	
1×10^{-7}	7	Neutral
1×10^{-8}	8	
1×10^{-9}	9	
1×10^{-10}	10	Increasing basicity
1×10^{-11}	11	
1×10^{-12}	12	
1×10^{-13}	13	
1×10^{-14}	14	

FIGURE 10.12 The pH scale.

TABLE 10.4 The Approximate pH of Some Common Substances

Substance	pH (or pH Range)
Hydrochloric acid (4%)	0
Gastric (stomach) solution	1.6–1.8
Lemon juice	2.2–2.4
Vinegar	2.4–3.4
Carbonated soft drinks	2.0–4.0
Grapefruit	3.0–3.2
Oranges	3.2–3.6
Acid rain	4.0–5.5
Tomatoes	4.2–4.4
Potatoes	5.7–5.8
Natural rainwater	5.6–6.2
Milk	6.3–6.7
Pure water	7.0
Seawater	7.0–8.3
Blood	7.4
Sodium bicarbonate solution	8.4
Milk of magnesia	10.5
Ammonia cleaning solution	11.9
Sodium hydroxide solution	13.0

H_3O^+ concentration, so the strength of a base is indicated on the same scale with values greater than 7. Note that the pH scale is logarithmic, so a pH of 2 is ten times as acidic as a pH of 3. Likewise, a pH of 10 is one hundred times as basic as a pH of 8. Figure 10.12 is a diagram of the pH scale, and table 10.4 compares the pH of some common substances (figure 10.13).

Properties of Salts

Salt is produced by a neutralization reaction between an acid and a base. A **salt** is defined as any ionic compound except those with hydroxide or oxide ions. Table salt, NaCl, is but one example of this large group of ionic compounds. As an example of a salt produced by a neutralization reaction, consider the reaction of HCl (an acid in solution) with $Ca(OH)_2$ (a base in solution). The reaction is

$$2\,HCl_{(aq)} + Ca(OH)_{2(aq)} \rightarrow CaCl_{2(aq)} + 2\,H_2O_{(l)}$$

This is an ionic exchange reaction that forms molecular water, leaving Ca^{2+} and Cl^- in solution. As the water is evaporated, these ions begin forming ionic crystal structures as the solution concentration increases. When the water is all evaporated, the white crystalline salt of $CaCl_2$ remains.

If sodium hydroxide had been used as the base instead of calcium hydroxide, a different salt would have been produced:

$$HCl_{(aq)} + NaOH_{(aq)} \rightarrow NaCl_{(aq)} + H_2O_{(l)}$$

Salts are also produced when elements combine directly, when an acid reacts with a metal, and by other reactions.

Salts are essential in the diet both as electrolytes and as a source of certain elements, usually called *minerals* in this context. Plants must have certain elements that are derived from water-soluble salts. Potassium, nitrates, and phosphate salts are often used to supply the needed elements. There is no

FIGURE 10.13 The pH increases as the acidic strength of these substances decreases from left to right. Did you know that lemon juice is more acidic than vinegar? That a soft drink is more acidic than orange juice or grapefruit juice?

scientific evidence that plants prefer to obtain these elements from natural sources, as compost, or from chemical fertilizers. After all, a nitrate ion is a nitrate ion, no matter what its source. Table 10.5 lists some common salts and their uses.

Hard and Soft Water

Salts vary in their solubility in water, and a solubility chart appears in appendix B. Table 10.6 lists some generalizations

TABLE 10.5 **Some Common Salts and Their Uses**

Common Name	Formula	Use
Alum	$KAl(SO_4)_2$	Medicine, canning, baking powder
Baking soda	$NaHCO_3$	Fire extinguisher, antacid, deodorizer, baking powder
Bleaching powder (chlorine tablets)	$CaOCl_2$	Bleaching, deodorizer, disinfectant in swimming pools
Borax	$Na_2B_4O_7$	Water softener
Chalk	$CaCO_3$	Antacid tablets, scouring powder
Chile saltpeter	$NaNO_3$	Fertilizer
Cobalt chloride	$CoCl_2$	Hygrometer (pink in damp weather, blue in dry weather)
Epsom salt	$MgSO_4 \cdot 7\,H_2O$	Laxative
Fluorspar	CaF_2	Metallurgy flux
Gypsum	$CaSO_4 \cdot 2\,H_2O$	Plaster of Paris, soil conditioner
Lunar caustic	$AgNO_3$	Germicide and cauterizing agent
Niter (or saltpeter)	KNO_3	Meat preservative, makes black gunpowder (75 parts KNO_3, 15 of carbon, 10 of sulfur)
Potash	K_2CO_3	Makes soap, glass
Rochelle salt	$KNaC_4H_4O_6$	Baking powder ingredient
TSP	Na_3PO_4	Water softener, fertilizer

TABLE 10.6 **Generalizations About Salt Solubilities**

Salts	Solubility	Exceptions
Sodium Potassium Ammonium	Soluble	None
Nitrate Acetate Chlorate	Soluble	None
Chlorides	Soluble	Ag and Hg(I) are insoluble
Sulfates	Soluble	Ba, Sr, and Pb are insoluble
Carbonates Phosphates Silicates	Insoluble	Na, K, and NH_4 are soluble
Sulfides	Insoluble	Na, K, and NH_4 are soluble: Mg, Ca, Sr, and Ba decompose

FIGURE 10.14 A water softener exchanges sodium ions for the calcium and magnesium ions of hard water. Thus, the water is now soft, but it contains the same number of ions as before.

concerning the various common salts. Some of the salts are dissolved by water that will eventually be used for domestic supply. When the salts are soluble calcium or magnesium compounds, the water will contain calcium or magnesium ions in solution. A solution of Ca^{2+} or Mg^{2+} ions is said to be *hard water* because it is hard to make soap lather in the water. "Soft" water, on the other hand, makes a soap lather easily. The difficulty occurs because soap is a sodium or potassium compound that is soluble in water. The calcium or magnesium ions, when present, replace the sodium or potassium ions in the soap compound, forming an insoluble compound. It is this insoluble compound that forms a "bathtub ring" and also collects on clothes being washed, preventing cleansing.

The key to "softening" hard water is to remove the troublesome calcium and magnesium ions (figure 10.14). If the hardness is caused by magnesium or calcium *bicarbonates*, the removal is accomplished by simply heating the water. Upon heating, they decompose, forming an insoluble compound that effectively removes the ions from solution. The decomposition reaction for calcium bicarbonate is

$$Ca^{2+}(HCO_3)^-_{2(aq)} \rightarrow CaCO_{3(s)} + H_2O_{(l)} + CO_2\uparrow$$

The reaction is the same for magnesium bicarbonate. As the solubility chart in appendix B shows, magnesium and calcium

carbonates are insoluble, so the ions are removed from solution in the solid that is formed. Perhaps you have noticed such a white compound forming around faucets if you live where bicarbonates are a problem. Commercial products to remove such deposits usually contain an acid, which reacts with the carbonate to make a new, soluble salt that can be washed away.

Water hardness is also caused by magnesium or calcium *sulfate*, which requires a different removal method. Certain chemicals such as sodium carbonate (washing soda), trisodium phosphate (TSP), and borax will react with the troublesome ions,

A Closer Look

Acid Rain

Acid rain is a general term used to describe any acidic substances, wet or dry, that fall from the atmosphere. Wet acidic deposition could be in the form of rain, but snow, sleet, and fog could also be involved. Dry acidic deposition could include gases, dust, or any solid particles that settle out of the atmosphere to produce an acid condition.

Pure, unpolluted rain is naturally acidic. Carbon dioxide in the atmosphere is absorbed by rainfall, forming carbonic acid (H_2CO_3). Carbonic acid lowers the pH of pure rainfall to a range of 5.6 to 6.2. Decaying vegetation in local areas can provide more CO_2, making the pH even lower. A pH range of 4.5 to 5.0, for example, has been measured in remote areas of the Amazon jungle. Human-produced exhaust emissions of sulfur and nitrogen oxides can lower the pH of rainfall even more, to a 4.0 to 5.5 range. This is the pH range of acid rain.

The sulfur and nitrogen oxides that produce acid rain come from exhaust emissions of industries and electric utilities that burn coal and from the exhaust of cars, trucks, and buses (box figure 10.1). The emissions are sometimes called "SO_x" and "NO_x", which is read "socks" and "knox." The x subscript implies the variable presence of any or all of the oxides, for example, nitrogen monoxide (NO), nitrogen dioxide (NO_2), and dinitrogen tetroxide (N_2O_4) for NO_x.

SO_x and NO_x are the raw materials of acid rain and are not themselves acidic. They react with other atmospheric chemicals to form sulfates and nitrates, which combine with water vapor to form sulfuric acid (H_2SO_4) and nitric acid (HNO_3). These are the chemicals of concern in acid rain.

Many variables influence how much and how far SO_x and NO_x are carried in the atmosphere and if they are converted to acid rain or simply return to the surface as a dry gas or particles. During the 1960s and 1970s, concerns about local levels of pollution led to the replacement of short smokestacks of about 60 m (about 200 ft) with taller smokestacks of about 200 m (about 650 ft). This reduced the local levels of pollution by dumping the exhaust higher in the atmosphere where winds could carry it away. It also set the stage for longer-range transport of SO_x and NO_x and their eventual conversion into acids.

There are two main reaction pathways by which SO_x and NO_x are converted to acids: (1) reactions in the gas phase and (2) reactions in the liquid phase, such as in water droplets in clouds and fog. In the gas phase, SO_x and NO_x are oxidized to acids, mainly by hydroxyl ions and ozone, and the acid is absorbed by cloud droplets and precipitated as rain or snow. Most of the nitric acid in acid rain and about one-fourth of the sulfuric acid is formed in gas-phase reactions. Most of the liquid-phase reactions that produce sulfuric acid involve the absorbed SO_x and hydrogen peroxide (H_2O_2), ozone, oxygen, and particles of carbon, iron oxide, and manganese oxide particles. These particles also come from the exhaust of fossil fuel combustion.

Acid rain falls on land, bodies of water, forests, crops, buildings, and people. Concerns about acid rain center on its environmental impact on lakes, forests, crops, materials, and human health. Lakes in different parts of the world, for example, have been increasing in acidity over the past fifty years. Lakes in northern New England, the Adirondacks, and parts of Canada now have a pH of less than 5.0, and correlations have been established between lake acidity and decreased fish populations. Trees, mostly conifers, are dying at unusually rapid rates in the northeastern United States. Red spruce in Vermont's Green Mountains and the mountains of New York and New Hampshire have been affected by acid rain, as have pines in New Jersey's Pine Barrens. It is believed

Box Figure 10.1   Natural rainwater has a pH of 5.6 to 6.2. Exhaust emissions of sulfur and nitrogen oxides can lower the pH of rainfall to a range of 4.0 to 5.5. The exhaust emissions come from industries, electric utilities, and automobiles. Not all emissions are as visible as those pictured in this illustration.

that acid rain leaches essential nutrients, such as calcium, from the soil and mobilizes aluminum ions. The aluminum ions disrupt the water equilibrium of fine root hairs, and when the root hairs die, so do the trees.

Human-produced emissions of sulfur and nitrogen oxides from burning fossil fuels are the cause of acid rain. The heavily industrialized northeastern part of the United States, from the Midwest through New England, releases sulfur and nitrogen emissions that result in a precipitation pH of 4.0 to 4.5. This region is the geographic center of the nation's acid rain problem. The solution to the problem is found in (1) using fuels other than fossil fuels and (2) reducing the thousands of tons of SO_x and NO_x that are dumped into the atmosphere per day when fossil fuels are used.

---

forming an insoluble solid that removes them from solution. For example, washing soda and calcium sulfate react as follows:

$$Na_2CO_{3(aq)} + CaSO_{4(aq)} \rightarrow Na_2SO_{4(aq)} + CaCO_3\downarrow$$

Calcium carbonate is insoluble; thus, the calcium ions are removed from solution before they can react with the soap. Many laundry detergents have Na_2CO_3, TSP, or borax ($Na_2B_4O_7$) added to soften the water. TSP causes other problems, however, as the additional phosphates in the waste water can act as a fertilizer, stimulating the growth of algae to such an extent that other organisms in the water die.

A water softener unit is an ion exchanger. The unit contains a mineral that exchanges sodium ions for calcium and magnesium

People Behind the Science

Johannes Nicolaus Brönsted (1879–1947)

Johannes Brönsted was a Danish physical chemist whose work in solution chemistry, particularly electrolytes, resulted in a new theory of acids and bases.

Brönsted was born on February 22, 1879, in Varde, Jutland, the son of a civil engineer. He was educated at local schools before going to study chemical engineering at the Technical Institute of the University of Copenhagen in 1897. He graduated two years later and then turned to chemistry, in which he qualified in 1902. After a short time in industry, he was appointed an assistant in the university's chemical laboratory in 1905, becoming professor of physical and inorganic chemistry in 1908.

Brönsted's early work was wide-ranging, particularly in the fields of electrochemistry, the measurement of hydrogen ion concentrations, amphoteric electrolytes, and the behavior of indicators. He discovered a method of eliminating potentials in the measurement of hydrogen ion concentrations and devised a simple equation that connects the activity and osmotic coefficients of an electrolyte, as well as another that relates activity coefficients to reaction velocities. From the absorption spectra of chromium(III) salts, he concluded that strong electrolytes are completely dissociated and that the changes of molecular conductivity and freezing point that accompany changes in concentration are caused by the electrical forces between ions in solution.

In 1887, Svante Arrhenius had proposed a theory of acidity that explained its nature on an atomic level. He defined an acid as a compound that could generate hydrogen ions in aqueous solution and an alkali as a compound that could generate hydroxyl ions. A strong acid is completely ionized (dissociated) and produces many hydrogen ions, whereas a weak acid is only partly dissociated and produces few hydrogen ions. Conductivity measurements confirm the theory, as long as the solutions are not too concentrated.

In 1923, Brönsted published (simultaneously with Thomas Lowry in Britain) a new theory of acidity, which has certain important advantages over that of Arrhenius. Brönsted defined an acid as a proton donor and a base as a proton acceptor. The definition applies to all solvents, not just water. It also explains the different behavior of pure acids in solution. Pure liquid sulfuric acid or acetic acid does not change the color of indicators nor does it react with carbonates or metals. But as soon as water is added, all of these reactions occur.

Source: Modified from the *Hutchinson Dictionary of Scientific Biography.* © RM, 2009. All rights reserved. Helicon Publishing is a division of RM.

ions as water is run through it. The softener is regenerated periodically by flushing with a concentrated sodium chloride solution. The sodium ions replace the calcium and magnesium ions, which are carried away in the rinse water. The softener is then ready for use again. The frequency of renewal cycles depends on the water hardness, and each cycle can consume from 4 to 20 pounds of sodium chloride per renewal cycle. In general, water with less than 75 ppm calcium and magnesium ions is called soft water; with greater concentrations, it is called hard water. The greater the concentration above 75 ppm, the harder the water.

SUMMARY

A water molecule consists of two hydrogen atoms and an oxygen atom with bonding and electron pairs in a tetrahedral arrangement. Electrons spend more time around the oxygen, producing a *polar molecule,* with centers of negative and positive charge. Polar water molecules interact. The force of attraction is called a *hydrogen bond.* The hydrogen bond accounts for the decreased density of ice, the high heat of fusion, and the high heat of vaporization of water. The hydrogen bond is also involved in the *dissolving* process.

A *solution* is a homogeneous mixture of ions or molecules of two or more substances. The substance present in the large amount is the *solvent,* and the *solute* is dissolved in the solvent. If one of the components is a liquid, however, it is called the solvent.

Fluids that mix in any proportion are called *miscible fluids,* and *immiscible fluids* do not mix. Polar substances dissolve in polar solvents but not nonpolar solvents, and the general rule is *like dissolves like.* Thus, oil, a nonpolar substance, is immiscible in water, a polar substance.

A limit to dissolving solids in a liquid occurs when the solution is *saturated.* A *saturated solution* is one with equilibrium between solute dissolving and solute coming out of solution. The *solubility* of a solid is the concentration of a saturated solution at a particular temperature.

Water solutions that carry an electric current are called *electrolytes,* and nonconductors are called *nonelectrolytes.* In general, ionic substances make electrolyte solutions, and molecular substances make nonelectrolyte solutions. Polar molecular substances may be *ionized* by polar water molecules, however, making an electrolyte from a molecular solution.

The *boiling point of a solution* is greater than the boiling point of the pure solvent, and the increase depends only on the concentration of the solute (at a constant pressure). For water, the boiling point is increased 0.521°C for each mole of solute in each kg of water. The *freezing point of a solution* is lower than the freezing point of the pure solvent, and the depression also depends on the concentration of the solute.

Acids, bases, and salts are chemicals that form ionic solutions in water, and each can be identified by simple properties. These properties are accounted for by the modern concepts of each. *Acids* are *proton donors* that form *hydronium ions* (H_3O^+) in water solutions. *Bases* are *proton acceptors* that form *hydroxide ions* (OH⁻) in water solutions. The strength of an acid or base is measured on the *pH scale,* a power of ten notation of the hydronium ion concentration. On the scale, numbers from 0 up to 7 are acids, 7 is neutral, and numbers above 7 and up to 14 are bases. Each unit represents a tenfold increase or decrease in acid or base properties.

A *salt* is any ionic compound except those with hydroxide or oxide ions. Salts provide plants and animals with essential elements. The solubility of salts varies with the ions that make up the compound. Solutions of magnesium or calcium produce *hard water,* water in which it is hard to make soap lather. Hard water is softened by removing the magnesium and calcium ions.

KEY TERMS

acid (p. **232**)
base (p. **232**)
electrolytes (p. **228**)
hydration (p. **227**)

hydrogen bonding (p. **225**)
hydronium ion (p. **229**)
miscible fluids (p. **226**)
neutralized (p. **232**)

pH scale (p. **233**)
polar molecule (p. **225**)
salt (p. **234**)
saturated solution (p. **228**)

solubility (p. **228**)
solution (p. **226**)

APPLYING THE CONCEPTS

1. Which of the following is *not* a solution?
   a. seawater
   b. carbonated water
   c. sand
   d. brass

2. Atmospheric air is a homogeneous mixture of gases that is mostly nitrogen gas. The nitrogen is therefore (the)
   a. solvent.
   b. solution.
   c. solute.
   d. None of the above is correct.

3. Water has the greatest density at what temperature?
   a. 100°C
   b. 20°C
   c. 4°C
   d. 0°C

4. A solid salt is insoluble in water, so the strongest force must be the
   a. ion-water molecule force.
   b. ion-ion force.
   c. force of hydration.
   d. polar molecule force.

5. The ice that forms in freezing seawater is
   a. pure water.
   b. the same salinity as liquid seawater.
   c. more salty than liquid seawater.
   d. more dense than liquid seawater.

6. Which of the following would have a pH of *less* than 7?
   a. a solution of ammonia
   b. a solution of sodium chloride
   c. pure water
   d. carbonic acid

7. Which of the following would have a pH of *more* than 7?
   a. a solution of ammonia
   b. a solution of sodium chloride
   c. pure water
   d. carbonic acid

8. Substance A has a pH of 2 and substance B has a pH of 3. This means that
   a. substance A has more basic properties than substance B.
   b. substance B has more acidic properties than substance A.
   c. substance A is ten times more acidic than substance B.
   d. substance B is ten times more acidic than substance A.

Answers

1. c 2. a 3. c 4. b 5. a 6. d 7. a 8. c

QUESTIONS FOR THOUGHT

1. How is a solution different from other mixtures?

2. Explain why some ionic compounds are soluble while others are insoluble in water.

3. Explain why adding salt to water increases the boiling point.

4. A deep lake in Minnesota is covered with ice. What is the water temperature at the bottom of the lake? Explain your reasoning.

5. Explain why water has a greater density at 4°C than at 0°C.

6. What is hard water? How is it softened?

7. According to the definition of an acid and the definition of a base, would the pH increase, decrease, or remain the same when NaCl is added to pure water? Explain.

8. What is a hydrogen bond? Explain how a hydrogen bond forms.

9. What feature of a soap molecule gives it cleaning ability?

10. What ion is responsible for (a) acidic properties? (b) for basic properties?

FOR FURTHER ANALYSIS

1. Analyze the basic reason that water is a universal solvent, becomes less dense when it freezes, has a high heat of fusion, has a high specific heat, and has a high heat of vaporization.

2. What is the same and what is different between a salt that will dissolve in water and one that is insoluble?

3. There are at least three ways to change the boiling point of water, so describe how you know for sure that 100°C (212°F) is the boiling point.

4. What are the significant similarities and differences between an acid, a base, and a salt?

5. Describe how you would teach someone why the pH of an acid is a low number (less than 7), while the pH of a base is a larger number (greater than 7).

6. Describe at least four different examples of how you could make hard water soft.

INVITATION TO INQUIRY

Which Freezes Faster?

Is it true that hot water that has boiled freezes faster than fresh cold water from the tap? Investigate freezing of hot, boiled water and fresh cold water. Make sure you control all the variables in your experimental design. Compare your findings to those of classmates.

11
Nuclear Reactions

With the top half of the steel vessel and control rods removed, fuel rod bundles can be replaced in the water-flooded nuclear reactor.

CORE **CONCEPT**
Nuclear reactions involve changes in the nucleus of the atom.

OUTLINE

Natural radioactivity is the spontaneous emission of particles or energy from a disinte-grating nucleus (p. 241).

Natural Radioactivity
  Nuclear Equations
  The Nature of the Nucleus
  Types of Radioactive Decay
  Radioactive Decay Series
A Closer Look: How Is Half-Life Determined?

Measurement of Radiation
  Measurement Methods
  Radiation Units
  Radiation Exposure

Nuclear instability results from an imbalance between the attractive nuclear force and the repulsive electromagnetic force (pp. 243–45).

Nuclear Energy
  Nuclear Fission
  Nuclear Power Plants
  Nuclear Fusion
A Closer Look: Three Mile Island and Chernobyl
A Closer Look: Nuclear Waste
People Behind the Science: Marie Curie

The relationship between energy and mass changes is $E = mc^2$ (p. 250).

An unstable nucleus becomes more stable by emitting alpha, beta, or gamma radiation from the nucleus (p. 246).

Astronomy

▷ Nuclear fusion provides energy for the stars (Ch. 12).

Earth Science

▷ Radiometric dating is a powerful tool in Earth sciences (Ch. 22).

OVERVIEW

The ancient alchemist dreamed of changing one element into another, such as lead into gold. The alchemist was never successful, however, because such changes were attempted with chemical reactions. Chemical reactions are reactions that involve only the electrons of atoms. Electrons are shared or transferred in chemical reactions, and the internal nucleus of the atom is unchanged. Elements thus retain their identity during the sharing or transferring of electrons. This chapter is concerned with a different kind of reaction, one that involves the *nucleus* of the atom. In nuclear reactions, the nucleus of the atom is often altered, changing the identity of the elements involved. The ancient alchemist's dream of changing one element into another was actually a dream of achieving a nuclear change, that is, a nuclear reaction.

Understanding nuclear reactions is important because although fossil fuels are the major source of energy today, there are growing concerns about (1) air pollution from fossil fuel combustion, (2) increasing levels of CO_2 from fossil fuel combustion, which contributes to the warming of Earth (the greenhouse effect), and (3) the dwindling fossil fuel supply, which cannot last forever. Energy experts see nuclear energy as a means of meeting rising energy demands in an environmentally acceptable way. However, the topic of nuclear energy is controversial, and discussions of it often result in strong emotional responses. Decisions about the use of nuclear energy require some understandings about nuclear reactions and some facts about radioactivity and radioactive materials (figure 11.1). These understandings and facts are the topics of this chapter.

NATURAL RADIOACTIVITY

Natural **radioactivity** is the spontaneous emission of particles or energy from an atomic nucleus as it disintegrates. It was discovered in 1896 by Henri Becquerel, a French scientist who was very interested in the recent discovery of X rays. Becquerel was experimenting with fluorescent minerals, minerals that give off visible light after being exposed to sunlight. He wondered if fluorescent minerals emitted X rays in addition to visible light. From previous work with X rays, Becquerel knew that they would penetrate a wrapped, light-tight photographic plate, exposing it as visible light exposes an unprotected plate. Thus, Becquerel decided to place a fluorescent uranium mineral on a protected photographic plate while the mineral was exposed to sunlight. Sure enough, he found a silhouette of the mineral on the plate when it was developed. Believing the uranium mineral emitted X rays, he continued his studies until the weather turned cloudy. Storing a wrapped, protected photographic plate and the uranium mineral together during the cloudy weather, Becquerel returned to the materials later and developed the photographic plate to again find an image of the mineral (figure 11.2). He concluded that the mineral was emitting an "invisible radiation" that was not induced by sunlight. The emission of invisible radiation was later named *radioactivity*. Materials that have the property of radioactivity are called *radioactive* materials.

Becquerel's discovery led to the beginnings of the modern atomic theory and to the discovery of new elements. Ernest Rutherford studied the nature of radioactivity and found that there are three kinds, which are today known by the first three letters of the Greek alphabet—alpha (α), beta (β), and gamma (γ). These Greek letters were used at first before the nature of the radiation was known. Today, an **alpha particle** (sometimes called an alpha ray) is known to be the nucleus of a helium atom, that is, two protons and two neutrons. A **beta particle** (or beta ray) is a high-energy electron. A **gamma ray** is electromagnetic radiation, as is light, but of very short wavelength (figure 11.3).

It was Rutherford's work with alpha particles that resulted in the discovery of the nucleus and the proton (see chapter 8). At Becquerel's suggestion, Marie Curie searched for other radioactive materials and, in the process, discovered two new elements, polonium and radium. More radioactive elements have been discovered since that time, and, in fact, all the isotopes of all the elements with an atomic number greater than 83 (bismuth) are radioactive. As a result of radioactive disintegration, the nucleus of an atom often undergoes a change of identity, becoming a simpler nucleus. The spontaneous

FIGURE 11.1 Decisions about nuclear energy require some understanding of nuclear reactions and the nature of radioactivity. This is one of the three units of the Palo Verde Nuclear Generating Station in Arizona. With all three units running, enough power is generated to meet the electrical needs of nearly 4 million people.

disintegration of a given nucleus is a purely natural process and cannot be controlled or influenced. The natural spontaneous disintegration or decomposition of a nucleus is also called **radioactive decay.** Although it is impossible to know *when* a given nucleus will undergo radioactive decay, as you will see later, it is possible to deal with the *rate* of decay for a given radioactive material with precision.

Nuclear Equations

There are two main subatomic particles in the nucleus, the proton and the neutron. The proton and neutron are called

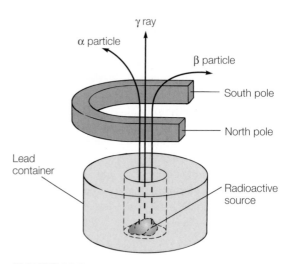

FIGURE 11.3 Radiation passing through a magnetic field shows that massive, positively charged alpha particles are deflected one way and less massive beta particles with their negative charge are greatly deflected in the opposite direction. Gamma rays, like light, are not deflected.

A

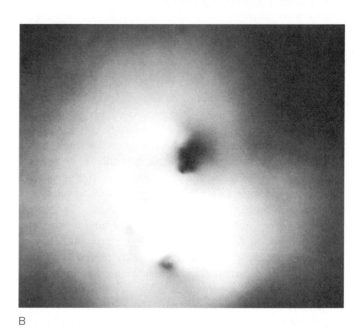

B

FIGURE 11.2 Radioactivity was discovered by Henri Becquerel when he exposed a light-tight photographic plate to a radioactive mineral, then developed the plate. (*A*) A photographic film is exposed to a uraninite ore sample. (*B*) The film, developed normally after a four-day exposure to uraninite. Becquerel found an image like this one and deduced that the mineral gave off invisible radiation.

nucleons. Recall that the number of protons, the *atomic number,* determines what element an atom is and that all atoms of a given element have the same number of protons. The number of neutrons varies in *isotopes,* which are atoms with the same atomic number but different numbers of neutrons. The number of protons and neutrons together determines the *mass number,* so different isotopes of the same element are identified with their mass numbers. Thus, the two most common, naturally occurring isotopes of uranium are referred to as uranium-238 and uranium-235, and the 238 and 235 are the mass numbers of these isotopes. Isotopes are also represented by the following symbol:

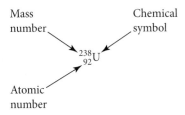

Subatomic particles involved in nuclear reactions are represented by symbols with the following form:

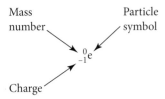

Symbols are used in an equation for a nuclear reaction that is written much like a chemical reaction with reactants and products. When a uranium-238 nucleus emits an alpha particle ($_2^4\text{He}$), for example, it loses two protons and two neutrons. The nuclear reaction is written in equation form as

$$_{92}^{238}\text{U} \rightarrow _{90}^{234}\text{Th} + _2^4\text{He}$$

The *products* of this nuclear reaction from the decay of a uranium-238 nucleus are (1) the alpha particle ($_2^4\text{He}$) given off and (2) the nucleus, which remains after the alpha particle leaves the original nucleus. What remains is easily determined since all nuclear equations must show conservation of charge and conservation of the total number of nucleons. Therefore, (1) the number of protons (positive charge) remains the same, and the sum of the subscripts (atomic number, or numbers of protons) in the reactants must equal the sum of the subscripts in the products; and (2) the total number of nucleons remains the same, and the sum of the superscripts (atomic mass, or number of protons plus neutrons) in the reactants must equal the sum of the superscripts in the products. The new nucleus remaining after the emission of an alpha particle, therefore, has an atomic number of $90 (92 - 2 = 90)$. According to the table of atomic numbers on the inside back cover

Name	Symbol	Mass Number	Charge
Proton	$_1^1\text{H}$ (or $_1^1\text{p}$)	1	1+
Electron	$_{-1}^0\text{e}$ (or $_{-1}^0\beta$)	0	1−
Neutron	$_0^1\text{n}$	1	0
Gamma photon	$_0^0\gamma$	0	0

of this text, this new nucleus is thorium (Th). The mass of the thorium isotope is 238 minus 4, or 234. The emission of an alpha particle thus decreases the number of protons by 2 and the mass number by 4. From the subscripts, you can see that the total charge is conserved ($92 = 90 + 2$). From the superscripts, you can see that the total number of nucleons is also conserved ($238 = 234 + 4$). The mass numbers (superscripts) and the atomic numbers (subscripts) are *balanced* in a correctly written nuclear equation. Such nuclear equations are considered to be independent of any chemical form or chemical reaction. Nuclear reactions are independent and separate from chemical reactions, whether or not the atom is in the pure element or in a compound. Each particle that is involved in nuclear reactions has its own symbol with a superscript indicating mass number and a subscript indicating the charge. These symbols, names, and numbers are given in table 11.1.

The Nature of the Nucleus

The modern atomic theory does not picture the nucleus as a group of stationary protons and neutrons clumped together by some "nuclear glue." The protons and neutrons are understood to be held together by a *nuclear force,* a strong fundamental force of attraction that is functional only at very short distances, on the order of 10^{-15} m or less. At distances greater than about 10^{-15} m, the nuclear force is negligible, and the weaker *electromagnetic force,* the force of repulsion between like charges, is the operational force. Thus, like-charged protons experience a repulsive force when they are farther apart than about 10^{-15} m. When closer together than 10^{-15} m, the short-range, stronger nuclear force predominates, and the protons experience a strong attractive force. This explains why the like-charged protons of the nucleus are not repelled by their like electric charges.

Observations of radioactive decay reactants and products and experiments with nuclear stability have led to a *shell model of the nucleus.* This model considers the protons and neutrons moving in energy levels in the nucleus analogous to the orbital structure of electrons in the outermost part of the atom. As in the electron orbitals, there are certain configurations of nuclear shells that have a greater stability than others. Considering electrons, filled and half-filled orbitals are more stable than other arrangements, and maximum stability occurs with the noble gases and their 2, 10, 18, 36, 54, and 86 electrons. Considering the nucleus, atoms with 2, 8, 20, 28, 50, 82, or 126 protons or

Plutonium Decay Equation

A plutonium-242 nucleus undergoes radioactive decay, emitting an alpha particle. Write the nuclear equation for this nuclear reaction.

Step 1: The table of atomic weights on the inside back cover gives the atomic number of plutonium as 94. Plutonium-242 therefore has a symbol of $^{242}_{94}\text{Pu}$. The symbol for an alpha particle is ($^{4}_{2}\text{He}$), so the nuclear equation so far is

$$^{242}_{94}\text{Pu} \rightarrow {}^{4}_{2}\text{He} + ?$$

Step 2: From the subscripts, you can see that $94 = 2 + 92$, so the new nucleus has an atomic number of 92. The table of atomic weights identifies element 92 as uranium with a symbol of U.

Step 3: From the superscripts, you can see that the mass number of the uranium isotope formed is $242 - 4 = 238$, so the product nucleus is $^{238}_{92}\text{U}$ and the complete nuclear equation is

$$^{242}_{94}\text{Pu} \rightarrow {}^{4}_{2}\text{He} + {}^{238}_{92}\text{U}$$

Step 4: Checking the subscripts ($94 = 2 + 92$) and the superscripts ($242 = 4 + 238$), you can see that the nuclear equation is balanced.

What is the product nucleus formed when radium emits an alpha particle? (Answer: Radon-222, a chemically inert, radioactive gas.)

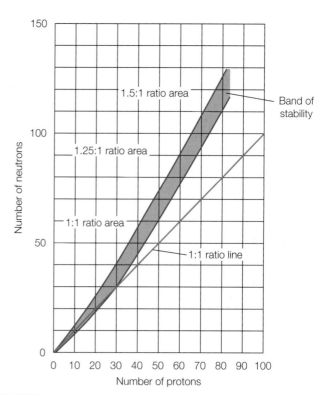

FIGURE 11.4 The shaded area indicates stable nuclei, which group in a band of stability according to their neutron-to-proton ratio. As the size of nuclei increases, so does the neutron-to-proton ratio that represents stability. Nuclei outside this band of stability are radioactive.

neutrons have a maximum nuclear stability. The stable numbers are not the same for electrons and nucleons because of differences in nuclear and electromagnetic forces.

Isotopes of uranium, radium, and plutonium, as well as other isotopes, emit an alpha particle during radioactive decay to a simpler nucleus. The alpha particle is a helium nucleus, ($^{4}_{2}\text{He}$). The alpha particle contains two protons as well as two neutrons, which is one of the nucleon numbers of stability, so you would expect the helium nucleus (or alpha particle) to have a stable nucleus, and it does. *Stable* means it does not undergo radioactive decay. Pairs of protons and pairs of neutrons have increased stability, just as pairs of electrons in a molecule do. As a result, nuclei with an *even number* of both protons and neutrons are, in general, more stable than nuclei with odd numbers of protons and neutrons. There are a little more than 150 stable isotopes with an even number of protons and an even number of neutrons, but there are only 4 stable isotopes with odd numbers of each. Just as in the case of electrons, other factors come into play as the nucleus becomes larger and larger with increased numbers of nucleons.

The results of some of these factors are shown in figure 11.4, which is a graph of the number of neutrons versus the number of protons in nuclei. As the number of protons increases, the neutron-to-proton ratio of the *stable nuclei* also

increases in a **band of stability.** Within the band, the neutron-to-proton ratio increases from about 1:1 at the bottom left to about 1½:1 at the top right. The increased ratio of neutrons is needed to produce a stable nucleus as the number of protons increases. Neutrons provide additional attractive *nuclear* (not electrical) forces, which counter the increased electrical repulsion from a larger number of positively charged protons. Thus, more neutrons are required in larger nuclei to produce a stable nucleus. However, there is a limit to the additional attractive forces that can be provided by more and more neutrons, and all isotopes of all elements with more than 83 protons are unstable and thus undergo radioactive decay.

The generalizations about nuclear stability provide a means of predicting if a particular nucleus is radioactive. The generalizations are as follows:

1. All isotopes with an atomic number greater than 83 have an unstable nucleus.
2. Isotopes that contain 2, 8, 20, 28, 50, 82, or 126 protons or neutrons in their nucleus occur in more stable isotopes than those with other numbers of protons or neutrons.
3. Pairs of protons and pairs of neutrons have increased stability, so isotopes that have nuclei with even numbers of both protons and neutrons are generally more stable than nuclei with odd numbers of both protons and neutrons.
4. Isotopes with an atomic number less than 83 are stable when the ratio of neutrons to protons in the nucleus

is about 1:1 in isotopes with up to 20 protons, but the ratio increases in larger nuclei in a band of stability (see figure 11.4). Isotopes with a ratio to the left or right of this band are unstable and thus will undergo radioactive decay.

Types of Radioactive Decay

Through the process of radioactive decay, an unstable nucleus becomes a more stable one with less energy. The three more familiar types of radiation emitted—alpha, beta, and gamma—were introduced earlier.

1. **Alpha emission.** Alpha (α) emission is the expulsion of an alpha particle ($_2^4\text{He}$) from an unstable, disintegrating nucleus. The alpha particle, a helium nucleus, travels from 2 to 12 cm through the air, depending on the energy of emission from the source. An alpha particle is easily stopped by a sheet of paper close to the nucleus. As an example of alpha emission, consider the decay of a radon-222 nucleus,

$$_{86}^{222}\text{Rn} \rightarrow {}_{84}^{218}\text{Po} + {}_2^4\text{He}$$

The spent alpha particle eventually acquires two electrons and becomes an ordinary helium atom.

2. **Beta emission.** Beta (β^-) emission is the expulsion of a different particle, a beta particle, from an unstable disintegrating nucleus. A beta particle is simply an electron ($_{-1}^0e$) ejected from the nucleus at a high speed. The emission of a beta particle *increases the number of protons* in a nucleus. It is as if a neutron changed to a proton by emitting an electron, or

$$_0^1\text{n} \rightarrow {}_1^1\text{p} + {}_{-1}^0\text{e}$$

Carbon-14 is a carbon isotope that decays by beta emission:

$$_6^{14}\text{C} \rightarrow {}_7^{14}\text{N} + {}_{-1}^0\text{e}$$

Note that the number of protons increased from six to seven, but the mass number remained the same. The mass number is unchanged because the mass of the expelled electron (beta particle) is negligible.

Beta particles are more penetrating than alpha particles and may travel several hundred centimeters through the air. They can be stopped by a thin layer of metal close to the emitting nucleus, such as a 1 cm thick piece of aluminum. A spent beta particle may eventually join an ion to become part of an atom, or it may remain a free electron.

3. **Gamma emission.** Gamma (γ) emission is a high-energy burst of electromagnetic radiation from an excited nucleus. It is a burst of light (photon) of a wavelength much too short to be detected by the eye. Other types of radioactive decay, such as alpha or beta emission, sometimes leave the nucleus with an excess of energy, a condition called an *excited state*. As in the case of excited electrons, the nucleus returns to a lower energy state by emitting electromagnetic radia-

tion. From a nucleus, this radiation is in the high-energy portion of the electromagnetic spectrum. Gamma is the most penetrating of the three common types of nuclear radiation. Like X rays, gamma rays can pass completely through a person, but all gamma radiation can be stopped by a 5 cm thick piece of lead close to the source. As with other types of electromagnetic radiation, gamma radiation is absorbed by and gives its energy to materials. Since the product nucleus changed from an excited state to a lower energy state, there is no change in the number of nucleons. For example, radon-222 is an isotope that emits gamma radiation:

$$_{86}^{222}\text{Rn}^* \rightarrow {}_{86}^{222}\text{Rn} + {}_0^0\gamma$$

($*$ denotes excited state)

Radioactive decay by alpha, beta, and gamma emission is summarized in table 11.2, which also lists the unstable nuclear conditions that lead to the particular type of emission. Just as electrons seek a state of greater stability, a nucleus undergoes radioactive decay to achieve a balance between nuclear attractions, electromagnetic repulsions, and a low quantum of nuclear shell energy. The key to understanding the types of reactions that occur is found in the band of stable nuclei illustrated in figure 11.4. The isotopes within this band have achieved the state of stability, and other isotopes above, below, or beyond the band are unstable and thus radioactive.

Nuclei that have a neutron-to-proton ratio beyond the upper right part of the band are unstable because of an imbalance between the proton-proton electromagnetic repulsions and all the combined proton and neutron nuclear attractions. Recall that the neutron-to-proton ratio increases from about 1:1 to about 1½:1 in the larger nuclei. The additional neutron provided additional nuclear attractions to hold the nucleus together, but atomic number 83 appears to be the upper limit to this additional stabilizing contribution. Thus, all nuclei with an atomic number greater than 83 are outside the upper right limit of the band of stability. Emission of an alpha particle reduces the number of protons by two and the number of neutrons by two, moving the nucleus more toward the band of stability. Thus, you can expect a nucleus that lies beyond the upper right part of the band of stability to be an alpha emitter (figure 11.5).

TABLE 11.2 Radioactive Decay

Unstable Condition	Type of Decay	Emitted	Product Nucleus
More than 83 protons	Alpha emission	$_2^4\text{He}$	Lost 2 protons and 2 neutrons
Neutron-to-proton ratio too large	Beta emission	$_{-1}^0\text{e}$	Gained 1 proton, no mass change
Excited nucleus	Gamma emission	$_0^0\gamma$	No change
Neutron-to-proton ratio too small	Other emission	$_1^0\text{e}$	Lost 1 proton, no mass change

Connections . . .

Radiation and Food Preservation

Radiation can be used to delay food spoilage and preserve foods by killing bacteria and other pathogens, just as heat is used to pasteurize milk. Foods such as wheat, flour, fruits, vegetables, pork, chicken, turkey, ground beef, and other uncooked meats are exposed to gamma radiation from cobalt-60 or cesium-137 isotopes, X rays, and electron beams. This kills insects, parasites such as *Trichinella spiralis* and tapeworms, and bacteria such as *E. coli, Listeria, Salmonellae,* and *Staphylococcus.* The overall effect is that many food-borne causes of human disease are eliminated and it is possible to store foods longer.

Food in the raw state, processed, or frozen is passed through a machine where it is irradiated. This process does not make the food radioactive since the food does not touch any radioactive substance. In addition, the radiation used in the process is not strong enough to disrupt the nucleus of atoms in food molecules, so it does not produce radioactivity, either.

In addition to killing the parasites and bacteria, the process might result in some nutritional loss but no more than that which normally occurs in canning. Some new chemical products may be formed by the exposure to radiation, but studies in several countries have not been able to identify any health problems or ill effects from these compounds.

Treatment with radiation works better for some foods than others. Dairy products undergo some flavor changes that are undesirable, and some fruits such as peaches become soft. Irradiated strawberries, on the other hand, remain firm and last for weeks instead of a few days in the refrigerator. Foods that are sterilized with stronger doses of irradiation can be stored for years without refrigeration just like canned foods that have undergone heat pasteurization.

In the United States, the Food and Drug Administration regulates which products can be treated by radiation and the dosages used in the treatment. The U.S. Department of Agriculture is responsible for the inspection of irradiated meat and poultry products. All foods that have undergone a radiation treatment must show the international logo for this process and a statement. The logo, called a radura, is a stylized flower inside a circle with five openings on the top part (box figure 11.1).

Box Figure 11.1

A nucleus with a neutron-to-proton ratio that is too large will be on the left side of the band of stability. Emission of a beta particle decreases the number of neutrons and increases the number of protons, so a beta emission will lower the

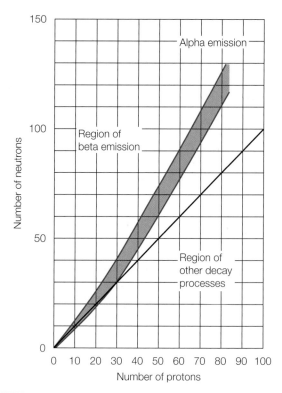

FIGURE 11.5 Unstable nuclei undergo different types of radioactive decay to obtain a more stable nucleus. The type of decay depends, in general, on the neutron-to-proton ratio, as shown.

neutron-to-proton ratio. Thus, you can expect a nucleus with a large neutron-to-proton ratio, that is, one to the left of the band of stability, to be a beta emitter.

A nucleus that has a neutron-to-proton ratio that is too small will be on the right side of the band of stability. These nuclei can increase the number of neutrons and reduce the number of protons in the nucleus by other types of radioactive decay. As usual when dealing with broad generalizations and trends, there are exceptions to the summarized relationships between neutron-to-proton ratios and radioactive decay.

Radioactive Decay Series

A radioactive decay reaction produces a simpler and eventually more stable nucleus than the reactant nucleus. As discussed in the previous section, large nuclei with an atomic number greater than 83 decay by alpha emission, giving up two protons and two neutrons with each alpha particle. A nucleus with an atomic number greater than 86, however, will emit an alpha particle and *still* have an atomic number greater than 83, which means the product nucleus will also be radioactive. This nucleus will also undergo radioactive decay, and the process will continue through a series of decay reactions until a stable nucleus is achieved. Such a series of decay reactions that (1) begins with one radioactive nucleus, which (2) decays to a second nucleus, which (3) then decays to a third nucleus, and so on until (4) a stable nucleus is reached is called a *radioactive decay series.* There are three naturally occurring radioactive decay series. One begins with thorium-232 and ends with lead-208, another begins with uranium-235 and ends with lead-207, and the third series begins with uranium-238 and ends with lead-206. Figure 11.6 shows the uranium-238 radioactive decay series.

As figure 11.6 illustrates, the uranium-238 begins with uranium-238 decaying to thorium-234 by alpha emission. Thorium has a new position on the graph because it now has a new atomic number and a new mass number. Thorium-234 is unstable and decays to protactinium-234 by beta emission, which is also unstable and decays by beta emission to uranium-234. The process continues with five sequential alpha emissions, then two beta-beta-alpha decay steps before the series terminates with the stable lead-206 nucleus.

The rate of radioactive decay is usually described in terms of its *half-life*. The **half-life** is the time required for one-half of the unstable nuclei to decay. Since each isotope has a characteristic decay constant, each isotope has its own characteristic half-life. Half-lives of some highly unstable isotopes are measured in fractions of seconds, and other isotopes have half-lives measured in seconds, minutes, hours, days, months, years, or billions of years. Table 11.3 lists half-lives of some of the isotopes, and the process is illustrated in figure 11.7.

As an example of the half-life measure, consider a hypothetical isotope that has a half-life of one day. The half-life is independent of the amount of the isotope being considered,

Radioactive Predictions

Refer to figure 11.5 and predict the type of radioactive decay for each of the following unstable nuclei:

(a) $^{131}_{53}$I

(b) $^{241}_{94}$Pu

(a) Iodine-131 has a nucleus with 53 protons and 131 minus 53, or 78 neutrons, so it has a neutron-to-proton ratio of 1.47:1. This places iodine-131 on the left side of the band of stability, with a high neutron-to-proton ratio that can be reduced by beta emission. The nuclear equation is

$$^{131}_{53}\text{I} \rightarrow {}^{131}_{54}\text{Xe} + {}^{0}_{-1}\text{e}$$

(b) Plutonium-241 has 94 protons and 241 minus 94, or 147 neutrons, in the nucleus. This nucleus is to the upper right, beyond the band of stability. It can move back toward stability by emitting an alpha particle, losing 2 protons and 2 neutrons from the nucleus. The nuclear equation is

$$^{242}_{94}\text{Pu} \rightarrow {}^{238}_{92}\text{U} + {}^{4}_{2}\text{He}$$

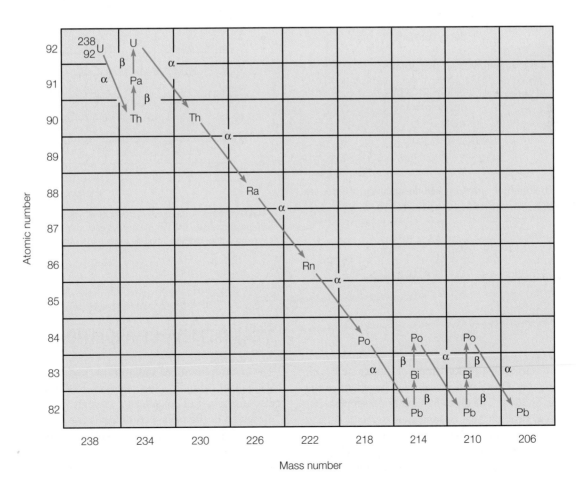

FIGURE 11.6 The radioactive decay series for uranium-238. This is one of three naturally occurring series.

A Closer Look

How Is Half-Life Determined?

It is not possible to predict when a radioactive nucleus will decay because it is a random process. It is possible, however, to deal with nuclear disintegration statistically since the rate of decay is not changed by any external conditions of temperature, pressure, or any chemical state. When dealing with a large number of nuclei, the ratio of the rate of nuclear disintegration per unit of time to the total number of radioactive nuclei is a constant, or

$$\text{radioactive decay constant} = \frac{\text{decay rate}}{\text{number of nuclei}}$$

The radioactive decay constant is a specific constant for a particular isotope, and each isotope has its own decay constant that can be measured. For example, a 238 g sample of uranium-238 (1 mole) that has 2.93×10^6 disintegrations per second would have a decay constant of

$$\begin{aligned}\text{radioactive decay constant} &= \frac{\text{decay rate}}{\text{number of nuclei}} \\ &= \frac{2.93 \times 10^6 \text{ nuclei/s}}{6.01 \times 10^{23} \text{ nuclei}} \\ &= 4.87 \times 10^{-18} \text{ 1/s}\end{aligned}$$

The half-life of a radioactive nucleus is related to its radioactive decay constant by

$$\text{half-life} = \frac{\text{a mathematical constant}}{\text{decay constant}}$$

The half-life of uranium-238 is therefore

$$\begin{aligned}\text{half-life} &= \frac{\text{a mathematical constant}}{\text{decay constant}} \\ &= \frac{0.693}{4.87 \times 10^{-18} \text{ 1/s}} \\ &= 1.42 \times 10^{17} \text{ s}\end{aligned}$$

This is the half-life of uranium-238 in seconds. There are $60 \times 60 \times 24 \times 365$, or 3.15×10^7 s in a year, so

$$\frac{1.42 \times 10^{17} \text{ s}}{3.15 \times 10^7 \text{ s/yr}} = 4.5 \times 10^9 \text{ yr}$$

The half-life of uranium-238 is thus 4.5 billion years.

TABLE 11.3   Half-Lives of Some Radioactive Isotopes

Isotope	Half-Life	Mode of Decay
$^{3}_{1}\text{H}$ (tritium)	12.26 years	Beta
$^{14}_{6}\text{C}$	5,730 years	Beta
$^{90}_{38}\text{Sr}$	28 years	Beta
$^{131}_{53}\text{I}$	8 days	Beta
$^{133}_{54}\text{Xe}$	5.27 days	Beta
$^{238}_{92}\text{U}$	4.51×10^9 years	Alpha
$^{242}_{94}\text{Pu}$	3.79×10^5 years	Alpha
$^{240}_{94}\text{Pu}$	6,760 years	Alpha
$^{239}_{94}\text{Pu}$	24,360 years	Alpha

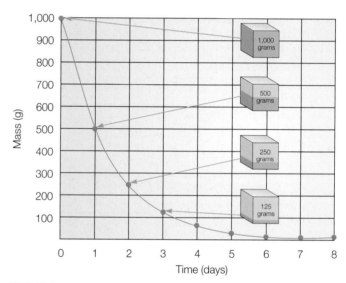

FIGURE 11.7   Radioactive decay of a hypothetical isotope with a half-life of one day. The sample decays each day by one-half to some other element. Actual half-lives may be in seconds, minutes, or any time unit up to billions of years.

but suppose you start with a 1.0 kg sample of this element with a half-life of one day. One day later, you will have half of the original sample, or 500 g. The other half did not disappear, but it is now the decay product, that is, some new element. During the next day, half of the remaining nuclei will disintegrate, and only 250 g of the initial sample is still the original element. One-half of the remaining sample will disintegrate each day until the original sample no longer exists.

MEASUREMENT OF RADIATION

The measurement of radiation is important in determining the half-life of radioactive isotopes. Radiation measurement is also important in considering biological effects, which will be discussed in the next section. As is the case with electricity, it is not possible to make direct measurements on things as small as electrons and other parts of atoms. Indirect measurement methods are possible, however, by considering the effects of the radiation.

Measurement Methods

As Becquerel discovered, radiation affects photographic film, exposing it as visible light does. Since the amount of film exposure is proportional to the amount of radiation, photographic film can be used as an indirect measure of radiation. Today, people who work around radioactive materials or X rays carry light-tight film badges. The film is replaced periodically and developed. The optical density of the developed film provides a record of the worker's exposure to radiation since the darkness of the developed film is proportional to the exposure.

There are also devices that indirectly measure radiation by measuring an effect of the radiation. An *ionization counter* is one type of device that measures ions produced by radiation. A second type of device is called a *scintillation counter*. *Scintillate* is a word meaning "sparks or flashes," and a scintillation counter measures the flashes of light produced when radiation strikes a phosphor.

The most common example of an ionization counter is known as a *Geiger counter*. The working components of a Geiger counter are illustrated in figure 11.8. Radiation is received in a metal tube filled with an inert gas, such as argon. An insulated wire inside the tube is connected to the positive terminal of a direct current source. The metal cylinder around the insulated wire is connected to the negative terminal. There is not a current between the center wire and the metal cylinder because the gas acts as an insulator. When radiation passes through the window, however, it ionizes some of the gas atoms, releasing free electrons. These electrons are accelerated by the field between the wire and cylinder, and the accelerated electrons ionize more gas molecules, which results in an *avalanche* of free electrons. The avalanche creates a pulse of current that is amplified and then measured. More radiation means more avalanches, so the pulses are an indirect means of measuring radiation. When connected to a speaker or earphone, each avalanche produces a "pop" or "click."

Some materials are *phosphors*, substances that emit a flash of light when excited by radiation. Zinc sulfide, for example, gives off a tiny flash of light when struck by radiation from a disintegrating radium nucleus. A scintillation counter measures the flashes of light through the photoelectric effect, producing free electrons that are accelerated to produce a pulse of current. Again, the pulses of current are used as an indirect means to measure radiation.

Radiation Units

You have learned that *radioactivity* is a property of isotopes with unstable, disintegrating nuclei and *radiation* is emitted particles (alpha or beta) or energy traveling in the form of photons (gamma). Radiation can be measured (1) at the source of radioactivity or (2) at a place of reception, where the radiation is absorbed.

The *activity* of a radioactive source is a measure of the number of nuclear disintegrations per unit of time. The unit of activity at the source is called a **curie** (Ci), which is defined as 3.70×10^{10} nuclear disintegrations per second. Activities are usually expressed in terms of fractions of curies, for example, a *picocurie* (pCi), which is a millionth of a millionth of a curie. Activities are sometimes expressed in terms of so many picocuries per liter (pCi/L).

The International System of Units (SI) unit for radioactivity is the *Becquerel* (Bq), which is defined as one nuclear disintegration per second. The unit for reporting radiation in the United States is the curie, but the Becquerel is the internationally accepted unit. Table 11.4 gives the names, symbols, and conversion factors for units of radioactivity.

As radiation from a source moves out and strikes a material, it gives the material energy. The amount of energy released by radiation striking living tissue is usually very small, but it can cause biological damage nonetheless because chemical bonds are broken and free polyatomic ions are produced by radiation.

The amount of radiation received by a human is expressed in terms of radiological dose. Radiation dose is usually written in units of a **rem,** which takes into account the possible biological damage produced by different types of radiation. Doses are usually expressed in terms of fractions of the rem, for example, a *millirem* (mrem). A millirem is 1/1,000 of a rem and is the unit of choice when low levels of radiation are discussed. The SI unit for radiation dose is the *millisievert* (mSv). Both the millirem and the millisievert relate ionizing radiation and biological effect to humans. The natural radiation that people receive from nature in one day is less than 1 millirem (0.01 millisievert). A single dose of 100,000 to 200,000 millirem (1,000 to 2,000 millisievert) can cause radiation sickness in humans (table 11.5). A single dose of 500,000 millirem (5,000 millisievert) results in death about 50 percent of the time.

Another measure of radiation received by a material is the **rad.** The term *rad* is from *rad*iation *a*bsorbed *d*ose. The SI unit

FIGURE 11.8 The working parts of a Geiger counter.

TABLE 11.4 Names, Symbols, and Conversion Factors for Radioactivity

Name	Symbol	To Obtain	Multiply By
becquerel	Bq	Ci	2.7×10^{11}
gray	Gy	rad	100
sievert	Sv	rem	100
curie	Ci	Bq	3.7×10^{10}
rem	rem	Sv	0.01
millirem	millirem	rem	0.001
rem	rem	millirem	1,000

TABLE 11.5 Approximate Single-Dose, Whole-Body Effects of Radiation Exposure

Level	Comment
0.240 rem	Average annual exposure to natural background radiation
0.500 rem	Upper limit of annual exposure to general public
25.0 rem	Threshold for observable effects such as blood count changes
100.0 rem	Fatigue and other symptoms of radiation sickness
200.0 rem	Definite radiation sickness, bone marrow damage, possibility of developing leukemia
500.0 rem	Lethal dose for 50 percent of individuals
1,000.0 rem	Lethal dose for all

for radiation received by a material is the *gray* (Gy). One gray is equivalent to an exposure of 100 rad.

Radiation Exposure

Natural radioactivity is a part of your environment, and you receive between 100 and 500 millirems each year from natural sources. This radiation from natural sources is called **background radiation.** Background radiation comes from outer space in the form of cosmic rays and from unstable isotopes in the ground, building materials, and foods. Many activities and situations will increase your yearly exposure to radiation. For example, the atmosphere absorbs some of the cosmic rays from space, so the less atmosphere above you, the more radiation you will receive. You are exposed to one additional millirem per year for each 100 feet you live above sea level. You receive approximately 0.3 millirem for each hour spent on a jet flight. Airline crews receive an additional 300 to 400 millirems per year because they spend so much time high in the atmosphere. Additional radiation exposure comes from medical X rays and television sets. In general, the worldwide average background radiation exposure for the average person is about 240 millirems per year.

What are the consequences of radiation exposure? Radiation can be a hazard to living organisms because it produces ionization along its path of travel. This ionization can (1) disrupt chemical bonds in essential macromolecules such as DNA and (2) produce molecular fragments, which are free polyatomic ions that can interfere with enzyme action and other essential cell functions. Tissues with highly active cells are more vulnerable to radiation damage than others, such as blood-forming tissue. Thus, one of the symptoms of an excessive radiation exposure is an altered blood count. Table 11.5 compares the estimated results of various levels of acute radiation exposure.

Radiation is not a mysterious, unique health hazard. It is a hazard that should be understood and neither ignored nor exaggerated. Excessive radiation exposure should be avoided, just as you avoid excessive exposure to other hazards such as certain chemicals, electricity, or even sunlight. Everyone agrees that *excessive* radiation exposure should be avoided, but there is some controversy about long-term, low-level exposure and

its possible role in cancer. Some claim that tolerable low-level exposure does not exist because that is not possible. Others point to many studies comparing high and low background radioactivity with cancer mortality data. For example, no cancer mortality differences could be found between people receiving 500 or more millirems a year and those receiving fewer than 100 millirems a year. The controversy continues, however, because of lack of knowledge about long-term exposure. Two models of long-term, low-level radiation exposure have been proposed: (1) a linear model and (2) a threshold model. The *linear model* proposes that any radiation exposure above zero is damaging and can produce cancer and genetic damage. The *threshold model* proposes that the human body can repair damage and get rid of damaging free polyatomic ions up to a certain exposure level called the threshold (figure 11.9). The controversy over long-term, low-level radiation exposure will probably continue until there is clear evidence about which model is correct. Whichever is correct will not lessen the need for rational risks versus cost-benefit analyses of all energy alternatives.

NUCLEAR ENERGY

Some nuclei are unstable because they are too large or because they have an unstable neutron-to-proton ratio. These unstable nuclei undergo radioactive decay, eventually forming products

Myths, Mistakes, and Misunderstandings

Antiradiation Pill?

It is a myth that an antiradiation pill exists that will protect you from ionizing radiation. There is a pill, an iodine supplement, that is meant to saturate your thyroid with a nonradioactive isotope of iodine. Once saturated, your thyroid will not absorb radioactive isotopes for storage in the gland, which could be dangerous.

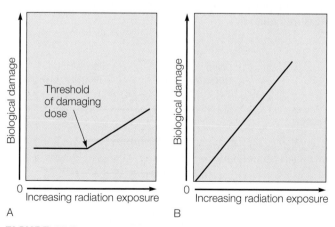

FIGURE 11.9 Graphic representation of the (*A*) threshold model and (*B*) linear model of low-level radiation exposure. The threshold model proposes that the human body can repair damage up to a threshold. The linear model proposes that any radiation exposure is damaging.

of greater stability. An example of this radioactive decay is the alpha emission reaction of uranium-238 to thorium-234,

$$^{238}_{92}U \rightarrow \ ^{234}_{90}Th + \ ^{4}_{2}He$$

$$238.0003 \ u \rightarrow 233.9942 \ u + 4.00150 \ u$$

The numbers below the nuclear equation are the *nuclear* masses (u) of the reactant and products. As you can see, there seems to be a loss of mass in the reaction,

$$233.9942 + 4.00150 - 238.0003 = -0.0046 \ u$$

This change in mass is related to the energy change according to the relationship that was formulated by Albert Einstein in 1905. The relationship is

$$E = mc^2 \qquad \textbf{equation 11.1}$$

where E is a quantity of energy, m is a quantity of mass, and c is a constant equal to the speed of light in a vacuum, $3.00 \times 10^8 \text{m/s}$. According to this relationship, matter and energy are the same thing, and energy can be changed to matter and vice versa.

The relationship between mass and energy explains why the mass of a nucleus is always *less* than the sum of the masses of the individual particles of which it is made.

The difference between the mass of the individual nucleons making up a nucleus and the actual mass of the nucleus is called the **mass defect** of the nucleus. The explanation for the mass defect is again found in $E = mc^2$. When nucleons join to make a nucleus, energy is released as the more stable nucleus is formed.

The energy equivalent released when a nucleus is formed is the same as the **binding energy,** the energy required to break the nucleus into individual protons and neutrons. The binding energy of the nucleus of any isotope can be calculated from the mass defect of the nucleus.

The ratio of binding energy to nucleon number is a reflection of the stability of a nucleus (figure 11.10). The greatest binding energy per nucleon occurs near mass number 56, then decreases for both more massive and less massive nuclei. This means that more massive nuclei can gain stability by splitting into smaller nuclei with the release of energy. It also means that less massive nuclei can gain stability by joining together with the release of energy. The slope also shows that more energy is released in the coming-together process than in the splitting process.

The nuclear reaction of splitting a massive nucleus into more stable, less massive nuclei with the release of energy is **nuclear fission** (figure 11.11). Nuclear fission occurs rapidly in an atomic bomb explosion and occurs relatively slowly in a nuclear reactor. The nuclear reaction of less massive nuclei, coming together to form more stable, and more massive, nuclei with the release of energy is **nuclear fusion.** Nuclear fusion occurs rapidly in a hydrogen bomb explosion and occurs continually in the Sun, releasing the energy essential for the continuation of life on Earth.

Nuclear Fission

Nuclear fission was first accomplished in the late 1930s when researchers were attempting to produce isotopes by bombarding massive nuclei with neutrons. In 1938, two German scien-

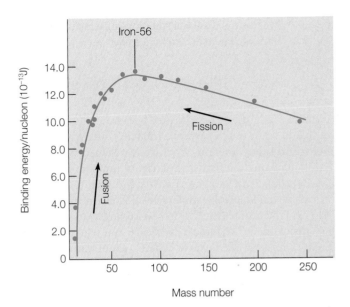

FIGURE 11.10 The maximum binding energy per nucleon occurs around mass number 56, then decreases in both directions. As a result, fission of massive nuclei and fusion of less massive nuclei both release energy.

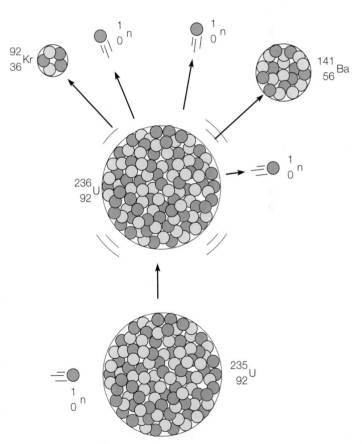

FIGURE 11.11 The fission reaction occurring when a neutron is absorbed by a uranium-235 nucleus. The deformed nucleus splits any number of ways into lighter nuclei, releasing neutrons in the process.

tists, Otto Hahn and Fritz Strassman, identified the element barium in a uranium sample that had been bombarded with neutrons. Where the barium came from was a puzzle at the time, but soon afterward, Lise Meitner, an associate who had moved to Sweden, deduced that the uranium nuclei had split, producing barium. The reaction might have been

$$\frac{1}{0}n + \frac{235}{92}U \rightarrow \frac{141}{56}Ba + \frac{92}{36}Kr + 3\frac{1}{0}n$$

The phrase "might have been" is used because a massive nucleus can split in many different ways, producing different products. About thirty-five different, less massive elements have been identified among the fission products of uranium-235. Some of these products are fission fragments, and some are produced by unstable fragments that undergo radioactive decay. Selected fission fragments are listed in table 11.6 below, together with their major modes of radioactive decay and half-lives. Some of the isotopes are the focus of concern about nuclear wastes, the topic of the "Science and Society" reading at the end of this chapter.

The fission of a uranium-235 nucleus produces two or three neutrons along with other products. These neutrons can each collide with other uranium-235 nuclei, where they are absorbed, causing fission with the release of more neutrons, which collide with other uranium-235 nuclei to continue the process. A reaction where the products are able to produce more reactions in a self-sustaining series is called a **chain reaction.** A chain reaction is self-sustaining until all the uranium-235 nuclei have fissioned or until the neutrons fail to strike a uranium-235 nucleus (figure 11.12).

You might wonder why all the uranium in the universe does not fission in a chain reaction. Natural uranium is mostly uranium-238, an isotope that does not fission easily. Only about 0.7 percent of natural uranium is the highly fissionable uranium-235. This low ratio of readily fissionable uranium-235 nuclei makes it unlikely that a stray neutron would be able to achieve a chain reaction.

To achieve a chain reaction, there must be (1) a sufficient mass with (2) a sufficient concentration of fissionable nuclei. When the mass and concentration are sufficient to sustain a chain reaction, the amount is called a **critical mass.** Likewise, a mass too small to sustain a chain reaction is called a *subcritical mass.* A mass of sufficiently pure uranium-235 (or plutonium-239) that is large enough to produce a rapidly accelerating chain reaction is called a *supercritical mass.* An atomic bomb is simply a device that uses a small, conventional explosive to push subcritical masses of fissionable material into a supercritical mass. Fission occurs almost instantaneously in the supercritical mass, and tremendous energy is released in a violent explosion.

Nuclear Power Plants

The nuclear part of a nuclear power plant is the *nuclear reactor,* a steel vessel in which a controlled chain reaction of fissionable material releases energy (figure 11.13). In the most popular design, called a pressurized light-water reactor, the fissionable material is enriched 3 percent uranium-235 and 97 percent

TABLE 11.6   Fragments and Products from Nuclear Reactors Using Fission of Uranium-235

Isotope	Major Mode of Decay	Half-Life	Isotope	Major Mode of Decay	Half-Life
Tritium	Beta	12.26 years	Cerium-144	Beta, gamma	285 days
Carbon-14	Beta	5,730 years	Promethium-147	Beta	2.6 years
Argon-41	Beta, gamma	1.83 hours	Samarium-151	Beta	90 years
Iron-55	Electron capture	2.7 years	Europium-154	Beta, gamma	16 years
Cobalt-58	Beta, gamma	71 days	Lead-210	Beta	22 years
Cobalt-60	Beta, gamma	5.26 years	Radon-222	Alpha	3.8 days
Nickel-63	Beta	92 years	Radium-226	Alpha, gamma	1,620 years
Krypton-85	Beta, gamma	10.76 years	Thorium-229	Alpha	7,300 years
Strontium-89	Beta	5.4 days	Thorium-230	Alpha	26,000 years
Strontium-90	Beta	28 years	Uranium-234	Alpha	2.48×10^5 years
Yttrium-91	Beta	59 days	Uranium-235	Alpha, gamma	7.13×10^8 years
Zirconium-93	Beta	9.5×10^5 years	Uranium-238	Alpha	4.51×10^9 years
Zirconium-95	Beta, gamma	65 days	Neptunium-237	Alpha	2.14×10^6 years
Niobium-95	Beta, gamma	35 days	Plutonium-238	Alpha	89 years
Technetium-99	Beta	2.1×10^5 years	Plutonium-239	Alpha	24,360 years
Ruthenium-106	Beta	1 year	Plutonium-240	Alpha	6,760 years
Iodine-129	Beta	1.6×10^7 years	Plutonium-241	Beta	13 years
Iodine-131	Beta, gamma	8 days	Plutonium-242	Alpha	3.79×10^5 years
Xenon-133	Beta, gamma	5.27 days	Americium-241	Alpha	458 years
Cesium-134	Beta, gamma	2.1 years	Americium-243	Alpha	7,650 years
Cesium-135	Beta	2×10^6 years	Curium-242	Alpha	163 days
Cesium-137	Beta	30 years	Curium-244	Alpha	18 years
Cerium-141	Beta	32.5 days			

FIGURE 11.12 A schematic representation of a chain reaction. Each fissioned nucleus releases neutrons, which move out to fission other nuclei. The number of neutrons can increase quickly with each series.

uranium-238 that has been fabricated in the form of small ceramic pellets (figure 11.14A). The pellets are encased in a long zirconium alloy tube called a *fuel rod*. The fuel rods are locked into a *fuel rod assembly* by locking collars, arranged to permit pressurized water to flow around each fuel rod (figure 11.14B) and to allow the insertion of *control rods* between the fuel rods. *Control rods* are constructed of materials, such as cadmium, that absorb neutrons. The lowering or raising of control rods within the fuel rod assemblies slows or increases the chain reaction by varying the amount of neutrons absorbed. When they are lowered completely into the assembly, enough neutrons are absorbed to stop the chain reaction.

It is physically impossible for the low-concentration fuel pellets to form a supercritical mass. A nuclear reactor in a power plant can only release energy at a comparatively slow rate, and it is impossible for a nuclear power plant to produce a nuclear explosion. In a pressurized water reactor, the energy released is carried away from the reactor by pressurized water in a closed pipe called the *primary loop* (figure 11.15). The water is pressurized at about 150 atmospheres (about 2,200 lb/in^2) to keep it from boiling, since its temperature may be 350°C (about 660°F).

In the pressurized light-water (ordinary water) reactor, the circulating pressurized water acts as a coolant, carrying heat away from the reactor. The water also acts as a *moderator*, a substance that slows neutrons so they are more readily absorbed by uranium-235 nuclei. Other reactor designs use heavy water (deuterium dioxide) or graphite as a moderator.

FIGURE 11.13 A schematic representation of the basic parts of a nuclear reactor. The largest commercial nuclear power plant reactors are 23 to 28 cm (9 to 11 in) thick steel vessels with a stainless steel liner, standing about 12 m (40 ft) high with a diameter of about 5 m (16 ft). Such a reactor has four pumps, which move 1,665,581 liters (440,000 gallons) of water per minute through the primary loop.

Water from the closed primary loop is circulated through a heat exchanger called a *steam generator* (figure 11.15). The pressurized high-temperature water from the reactor moves

A

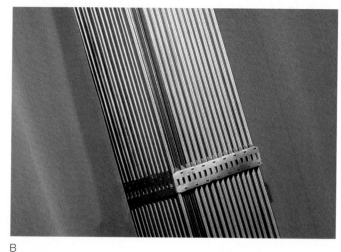

B

FIGURE 11.14 (*A*) These are uranium oxide fuel pellets that are stacked inside fuel rods, which are then locked together in a fuel rod assembly. (*B*) A fuel rod assembly. See also figure 11.17, which shows a fuel rod assembly being loaded into a reactor.

through hundreds of small tubes inside the generator as *feedwater* from the *secondary loop* flows over the tubes. The water in the primary loop heats feedwater in the steam generator and then returns to the nuclear reactor to become heated again. The feedwater is heated to steam at about 235°C (455°F) with a pressure of about 68 atmospheres (1,000 lb/in²). This steam is piped to the turbines, which turn an electric generator (figure 11.16).

After leaving the turbines, the spent steam is condensed back to liquid water in a second heat exchanger receiving water from the cooling towers. Again, the cooling water does not mix with the closed secondary loop water. The cooling-tower water

enters the condensing heat exchanger at about 32°C (90°F) and leaves at about 50°C (about 120°F) before returning to a cooling tower, where it is cooled by evaporation. The feedwater is preheated, then recirculated to the steam generator to start the cycle over again. The steam is condensed back to liquid water because of the difficulty of pumping and reheating steam.

After a period of time, the production of fission products in the fuel rods begins to interfere with effective neutron transmission, so the reactor is shut down annually for refueling. During refueling, about one-third of the fuel that had the longest exposure in the reactor is removed as "spent" fuel. New fuel rod assemblies are inserted to make up for the part

FIGURE 11.15 A schematic general system diagram of a pressurized water nuclear power plant, not to scale. The containment building is designed to withstand an internal temperature of 149°C (300°F) at a pressure of 414 kilopascals (60 lbs/in²) and still maintain its leak-tight integrity.

FIGURE 11.16 The turbine deck of a nuclear generating station. One large generator is in line with four steam turbines in this non-nuclear part of the plant. The large silver tanks are separators that remove water from the steam after it has left the high-pressure turbine and before it is recycled back into the low-pressure turbines.

removed (figure 11.17). However, only about 4 percent of the "spent" fuel is unusable waste, about 94 percent is uranium-238, 0.8 percent is uranium-235, and about 0.9 percent is plutonium (figure 11.18). Thus, "spent" fuel rods contain an appreciable amount of usable uranium and plutonium. For now, spent reactor fuel rods are mostly stored in cooling pools at the nuclear plant sites. In the future, a decision will be made either to reprocess the spent fuel, recovering the uranium and plutonium through chemical reprocessing, or put the fuel in terminal storage. Concerns about reprocessing are based on the fact that plutonium-239 and uranium-235 are fissionable and could possibly be used by terrorist groups to construct nuclear explosive devices. Six other countries do have reprocessing plants, however, and the spent fuel rods represent a significant energy source. Some energy experts say that it would be inappropriate to dispose of such an energy source.

The technology to dispose of fuel rods exists if the decision is made to do so. The longer half-life waste products are mostly alpha emitters. These metals could be converted to oxides, mixed with powdered glass (or a ceramic), melted, and then poured into stainless steel containers. The solidified canisters would then be buried in a stable geologic depository. The glass technology is used in France for disposal of high-level wastes. Buried at 610 to 914 m (2,000–3,000 ft) depths in solid granite, the only significant means of the radioactive wastes reaching the surface would be through groundwater dissolving the stainless steel, glass, and waste products and then transporting them back to the surface. Many experts believe that if such groundwater dissolving were to take place, it would require thousands of years. The radioactive isotopes would thus undergo natural radioactive decay by the time they could reach the surface. Nonetheless, research is continuing on nuclear waste and its disposal. In the meantime, the question of whether it is best to reprocess fuel rods or place them in permanent storage remains unanswered.

What is the volume of nuclear waste under question? If all the spent fuel rods from all the commercial nuclear plants accumulated since coming online were reprocessed, then mixed with glass, the total amount of glassified waste would make a pile on one football field an estimated 4 m (about 13 ft) high.

Nuclear Fusion

As the graph of nuclear binding energy versus mass numbers shows (see figure 11.10), nuclear energy is released when (1) massive nuclei such as uranium-235 undergo fission and (2) when less massive nuclei come together to form more massive nuclei through nuclear fusion. Nuclear fusion is responsible for the energy released by the Sun and other stars. At the present halfway point in the Sun's life—with about 5 billion years to go—the core is now 35 percent hydrogen and 65 percent

FIGURE 11.17 Spent fuel rod assemblies are removed and new ones are added to a reactor head during refueling. This shows an initial fuel load to a reactor, which has the upper part removed and set aside for the loading.

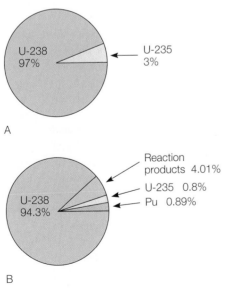

FIGURE 11.18 The composition of the nuclear fuel in a fuel rod (A) before and (B) after use over a three-year period in a nuclear reactor.

Three Mile Island and Chernobyl

Three Mile Island (downriver from Harrisburg, Pennsylvania) and Chernobyl (former USSR, now Ukraine) are two nuclear power plants that became famous because of accidents. Here is a brief accounting of what happened.

Three Mile Island

It was March 28, 1979, and the 880-megawatt Three Mile Island Nuclear Plant, operated by Metropolitan Edison Company, was going full blast. At 4:00 A.M. that morning the main feedwater pump that pumps water to the steam generator failed for some unexplained reason (follow this description in figure 11.15). Backup feedwater pumps kicked in, but the valves that should have been open were closed for maintenance, blocking the backup source of water for the steam generator. All of a sudden there was not a source of water for the steam generator that removes heat from the primary loop. Events began to happen quickly at that point.

The computer sensors registered that the steam generator was not receiving water, and the computer began to follow a shutdown procedure. First, the turbine was shut down as steam was vented from the steam line out through the turbine building, sounding much like a large jet plane. Within six seconds, the reactor was shut down (scrammed) with the control rods dropping down between the fuel rods in the reactor vessel. Fissioning began to slow, but the reactor was still hot.

Between three and six seconds, a significant event occurred when a pressure relief valve opened on the primary loop, relieving the excess pressure that was generated because feedwater was not entering the steam generator to remove heat from the primary loop. The valve should have closed when the excess pressure was released, but it did not. It was stuck in an open position. Pressurized water and steam was pouring from the primary loop into the containment building. As water was lost from the primary loop, temperatures inside the reactor began to climb. The loss of pressure resulted in high-pressure water flashing into steam. If an operator had

pressed the right button in the control room, it would have closed the open valve, but this did not happen until thirty-two minutes later.

At this point, the reactor could have recovered from the events. Two minutes after the initial shutdown, the computer sensors noted the loss of pressure from the open valve and kicked in the emergency core cooling system, which pumps more water into the reactor vessel. However, for some unknown reason, the control room operators shut down one pump four and a half minutes after the initial event and the second pump six minutes later.

Water continued to move through the open pressure relief valve into the containment building. At seven and a half minutes after the start of the event, the radioactive water on the floor was 2 feet deep and the sump pumps started pumping water into tanks in the auxiliary building. This water would become the source of the radioactivity that did escape the plant. It escaped because the pump seals leaked and spilled radioactive water. The filters had been removed from the auxiliary building air vents, allowing radioactive gases to escape.

Eleven minutes after the start of the event, an operator restarted the emergency core cooling system that had been turned off. With the cooling water flowing again, the pressure in the reactor stopped falling. The fuel rods, some thirty-six thousand in this reactor, had not yet suffered any appreciable damage. This would be taken care of by the next incredible event.

What happened next was that operators began turning off the emergency cooling pumps, perhaps because they were vibrating too much. In any case, with the pumps off, the water level in the reactor fell again, this time uncovering the fuel rods. Within minutes, the temperature was high enough to rupture fuel rods, dropping radioactive oxides into the bottom of the reactor vessel. The operators now had a general emergency.

Eleven hours later, the operators decided to start the main reactor coolant pump. This pump had shut down at the beginning of the series of events. Water again covered

the fuel rods, and the pressure and temperature stabilized.

The consequences of this series of events were as follows:

1. Local residences did receive some radiation from the release of gases. They received 10 millirem (0.1 millisievert) in a low-exposure area and up to 25 millirem (0.25 millisievert) in a high-exposure area.
2. Cleaning up the damaged reactor vessel and core required more than ten years to cut it up, pack it into canisters, and ship everything to the Federal Nuclear Reservation at Idaho Falls.
3. The cost of the cleanup was more than $1 billion.
4. Changes were implemented at other nuclear power plants as a consequence: Pressure relief valves have been removed, operators can no longer turn off the emergency cooling system, and operators must now spend about one-fourth of their time in training.

Chernobyl

The Soviet-designed Chernobyl reactor was a pressurized water reactor with individual fuel channels, which is very different from the pressurized water reactors used in the United States. The Chernobyl reactor was constructed with each fuel assembly in an individual pressure tube with circulating pressurized water. This heated water was circulated to a steam separator, and the steam was directed to turbines, which turned a generator to produce electricity. Graphite blocks surrounded the pressure tubes, serving as moderators to slow the neutrons involved in the chain reaction. The graphite was cooled by a mixture of helium and nitrogen. The reactor core was located in a concrete bunker that acted as a radiation shield. The top part was a steel cap and shield that supported the fuel assemblies. There were no containment buildings around the Soviet reactors as there are in the United States.

The Chernobyl accident was the result of a combination of a poorly engineered reactor design, poorly trained reactor operators, and serious mistakes made by the operators on the day of the accident.

The reactor design was flawed because at low power, steam tended to form pockets in the water-filled fuel channels, creating a condition of instability. Instability occurred because (1) steam is not as efficient at cooling as is liquid water, and (2) liquid water acts as a moderator and neutron absorber while steam does not. Excess steam therefore leads to overheating and increased power generation. Increased power can lead to increased steam generation, which leads to further increases in power. This coupled response is very difficult to control because it feeds itself.

On April 25, 1986, the operators of Chernobyl unit 4 undertook a test to find out how long the turbines would spin and supply power following the loss of electrical power. The operators disabled the automatic shutdown mechanisms and then started the test early on April 26. The plan was to stabilize the reactor at 1,000 MW, but an error was made and the power fell to about 30 MW and pockets of steam became a problem. They tried to increase the power by removing all the control rods. At 1:00 A.M., they were able to stabilize the reactor at 200 MW. Then instability returned and the operators were making continuous adjustments to maintain a constant power. They reduced the feedwater to maintain steam pressure, and this created even more steam voids in the fuel channels. Power surged to a very high level and fuel elements ruptured. A steam explosion moved the reactor cap, exposing individual fuel channels and releasing fission products to the environment. A second explosion knocked a hole in the roof, exposed more of the reactor core, and the reactor graphite burst into flames. The graphite burned for nine days, releasing about 324 million Ci $(12 \times 10^{18} \text{Bq})$ into the environment.

The fiery release of radioactivity was finally stopped by using a helicopter to drop sand, boron, lead, and other materials onto the burning graphite reactor. After the fire was out, the remains of the reactor were covered with a large concrete shelter.

In addition to destroying the reactor, the accident killed 30 people, 28 of whom died from radiation exposure. Another 134 people were treated for acute radiation poisoning and all recovered from the immediate effects.

Cleanup crews over the next year received about 10 rem (100 millisievert) to 25 rem (250 millisievert) and some received as much as 50 rem (500 millisievert). In addition to this direct exposure, large expanses of Belarus, Ukraine, and Russia were contaminated by radioactive fallout from the reactor fire. Hundreds of thousands of people have been resettled into less contaminated areas. The World Health Organization and other international agencies have studied the data to understand the impact of radiation-related disease. These studies do confirm a rising incidence of thyroid cancer but no increases in leukemia so far.

helium. Through fusion, the Sun converts about 650 million tons of hydrogen to 645 million tons of helium every second. The other roughly 5 million tons of matter are converted into energy. Even at this rate, the Sun has enough hydrogen to continue the process for an estimated 5 billion years. Several fusion reactions take place between hydrogen and helium isotopes, including the following:

$$ {}_1^1\text{H} + {}_1^1\text{H} \rightarrow {}_1^2\text{H} + {}_1^0\text{e} $$

$$ {}_1^2\text{H} + {}_1^2\text{H} \rightarrow {}_2^3\text{He} + {}_0^1\text{n} $$

$$ {}_2^3\text{He} + {}_2^3\text{He} \rightarrow {}_2^4\text{He} + 2\,{}_1^1\text{H} $$

The fusion process would seem to be a desirable energy source on Earth because (1) two isotopes of hydrogen, deuterium (${}_1^2\text{H}$) and tritium (${}_1^3\text{H}$), undergo fusion at a relatively low temperature; (2) the supply of deuterium is practically unlimited, with each gallon of seawater containing about a teaspoonful of deuterium dioxide; and (3) enormous amounts of energy are released with no radioactive by-products.

The oceans contain enough deuterium to generate electricity for the entire world for millions of years, and tritium can be constantly produced by a fusion device. Researchers know what needs to be done to tap this tremendous energy source. The problem is *how* to do it in an economical, continuous energy-producing fusion reactor. The problem, one of the most difficult engineering tasks ever attempted, is meeting three basic fusion reaction requirements of (1) temperature, (2) density, and (3) time (figure 11.19):

1. **Temperature.**   Nuclei contain protons and are positively charged, so they experience the electromagnetic repulsion of like charges. This force of repulsion can be overcome, moving the nuclei close enough to fuse together, by giving the nuclei sufficient kinetic energy. The fusion reaction of deuterium and tritium, which has the lowest temperature requirements of any fusion reaction known at the present time, requires temperatures on the order of 100 million degrees Celsius.
2. **Density.**   There must be a sufficiently dense concentration of heavy hydrogen nuclei, on the order of $10^{14}/\text{cm}^3$, so many reactions occur in a short time.
3. **Time.**   The nuclei must be confined at the appropriate density up to a second or longer at pressures of at least 10 atmospheres to permit a sufficient number of reactions to take place.

The temperature, density, and time requirements of a fusion reaction are interrelated. A short time of confinement, for example, requires an increased density, and a longer confinement time requires less density. The primary problems of fusion research are the high-temperature requirements and confinement. No material in the world can stand up to a temperature of

A Closer Look

Nuclear Waste

There are two general categories of nuclear wastes: (1) low-level wastes and (2) high-level wastes. *Low-level wastes* are produced by the normal operation of a nuclear reactor. Radioactive isotopes sometimes escape from fuel rods in the reactor and in the spent fuel storage pools. These isotopes are removed from the water by ion-exchange resins and from the air by filters. The used resins and filters will contain the radioactive isotopes and will become low-level wastes. In addition, any contaminated protective clothing, tools, and discarded equipment also become low-level wastes.

Low-level liquid wastes are evaporated, mixed with cement, then poured into 55 gallon steel drums. Solid wastes are compressed and placed in similar drums. The drums are currently disposed of by burial in government-licensed facilities. In general, low-level waste has an activity of less than 1.0 curie per cubic foot. Contact with the low-level waste could expose a person to up to 20 millirems per hour of contact.

High-level wastes from nuclear power plants are spent nuclear fuel rods. At the present time, most of the commercial nuclear power plants have these rods in temporary storage at the plant site. These

rods are "hot" in the radioactive sense, producing about 100,000 curies per cubic foot. They are also hot in the thermal sense, continuing to generate heat for months after removal from the reactor. The rods are cooled by heat exchangers connected to storage pools; they could otherwise achieve an internal temperature as high as 800°C for several decades. In the future, these spent fuel rods will be reprocessed or disposed of through terminal storage.

Agencies of the United States government have also accumulated millions of gallons of high-level wastes from the manufacture of nuclear weapons and nuclear research programs. These liquid wastes are stored in million-gallon stainless steel containers that are surrounded by concrete. The future of this large amount of high-level wastes may be evaporation to a solid form or mixture with a glass or ceramic matrix, which is melted and poured into stainless steel containers. These containers would be buried in solid granite rock in a stable geologic depository. Such high-level wastes must be contained for thousands of years as they undergo natural radioactive decay (box figure 11.2). Burial at a depth of 610 to 914 m (2,000–3,000 ft) in solid

Box Figure 11.2 This is a standard warning sign for a possible radioactive hazard. Such warning signs would have to be maintained around a nuclear waste depository for thousands of years.

granite would provide protection from exposure by explosives, meteorite impact, or erosion. One major concern about this plan is that a hundred generations later, people might lose track of what is buried in the nuclear garbage dump.

---

100 million degrees Celsius, and any material container would be instantly vaporized. Thus, research has centered on meeting the fusion reaction requirements without a material container. Two approaches are being tested, *magnetic confinement* and *inertial confinement.*

Magnetic confinement utilizes a very hot *plasma*, a gas consisting of atoms that have been stripped of their electrons because of the high kinetic energies. The resulting positively and negatively charged particles respond to electrical and magnetic forces, enabling researchers to develop a "magnetic bottle," that is, magnetic fields that confine the plasma and avoid the problems of material containers that would vaporize. A magnetically confined plasma is very unstable, however, and researchers have compared the problem to trying to carry a block of Jell-O on a pair of rubber bands. Different magnetic field geometries and magnetic "mirrors" are the topics of research in attempts to stabilize the hot, wobbly plasma. Electric currents, injection of fast ions, and radio frequency (microwave) heating methods are also being studied.

Inertial confinement is an attempt to heat and compress small frozen pellets of deuterium and tritium with energetic laser beams or particle beams, producing fusion. The focus

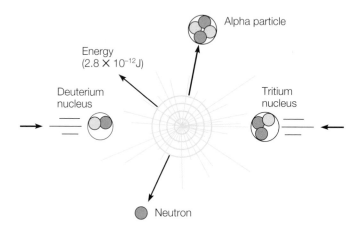

FIGURE 11.19 A fusion reaction between a tritium nucleus and a deuterium nucleus requires a certain temperature, density, and time of containment to take place.

of this research is new and powerful lasers, light ion and heavy ion beams. If successful, magnetic or inertial confinement will provide a long-term solution for future energy requirements.

High-Level Nuclear Waste

In 1982, the U.S. Congress established a national policy to solve the problem of nuclear waste disposal. This policy is a federal law called the Nuclear Waste Policy Act. Congress based this policy on what most scientists worldwide agree is the best way to dispose of nuclear waste.

The Nuclear Waste Policy Act made the U.S. Department of Energy (DOE) responsible for finding a site, building, and operating an underground disposal facility called a geologic repository.

In 1983, the DOE selected nine locations in six states for consideration as potential repository sites. This selection was based on data collected for nearly ten years. The nine sites were studied and results of these preliminary studies were reported in 1985. Based on these reports, President Ronald Reagan approved three sites for intensive scientific study called site characterization. The three sites were Hanford, Washington; Deaf Smith County, Texas; and Yucca Mountain, Nevada.

In 1987, Congress amended the Nuclear Waste Policy Act and directed the Department of Energy to study only Yucca Mountain.

On July 9, 2002, the U.S. Senate cast the final legislative vote approving the development of a repository at Yucca Mountain. On July 23, 2002, President George W. Bush signed House Joint Resolution 87, allowing the DOE to take the next step in establishing a safe repository in which to store our nation's nuclear waste.

In July 2004, the DOE issued a report that describes how commercial spent nuclear fuel and high-level waste will be acquired, transported to federal facilities, and disposed of in the Yucca Mountain Geologic Repository. The plan is to accept scheduled allocations from nuclear power plant owners beginning in 2010. In 2010, 400 metric tons of uranium will be accepted. This number will grow to 3,000 metric tons per year in 2014 and for each of the next five years.

The DOE is currently preparing an application to obtain a license from the U.S. Nuclear Regulatory Commission to construct a repository.

Source: www.ocrwm.doe.gov/repository/index.shtml

People Behind the Science

Marie Curie (1867–1934)

Marie Curie was a Polish-born French scientist who, with her husband, Pierre Curie (1859–1906), was an early investigator of radioactivity. From 1896, the Curies worked together, building on the results of Henri Becquerel, who had discovered radioactivity from uranium salts. Marie Curie discovered that thorium also emits radiation and found that the mineral pitchblende was even more radioactive than could be accounted for by any uranium and thorium content. The Curies then carried out an exhaustive search and in July 1898 announced the discovery of polonium, followed in December of that year with the discovery of radium. They shared the 1903 Nobel Prize for physics with Becquerel for the discovery of radioactivity. The Curies did not participate in Becquerel's discovery but investigated radioactivity and gave the phenomenon its name. Marie Curie went on to study the chemistry and medical applications of radium, and was awarded the 1911 Nobel Prize for chemistry in recognition of her work in isolating the pure metal.

At the outbreak of World War I in 1914, Marie Curie helped to equip ambulances with X-ray equipment and drove the ambulances to the front lines. The International Red Cross made her head of its Radiological Service. She taught medical orderlies and doctors how to use the new technique. By the late 1920s, her health began to deteriorate: continued exposure to high-energy radiation had given her leukemia. She entered a sanatorium and died on July 4, 1934.

Throughout much of her life, Marie Curie was poor, and the painstaking radium extractions were carried out in primitive conditions. The Curies refused to patent any of their discoveries, wanting them to benefit everyone freely. They used the Nobel Prize money and other financial rewards to finance further research. One of the outstanding applications of their work has been the use of radiation to treat cancer, one form of which cost Marie Curie her life.

Source: Modified from the Hutchinson Dictionary of Scientific Biography. © RM, 2009. All rights reserved. Helicon Publishing is a division of RM.

SUMMARY

Radioactivity is the spontaneous emission of particles or energy from an unstable atomic nucleus. The modern atomic theory pictures the nucleus as protons and neutrons held together by a short-range *nuclear force* that has moving *nucleons* (protons and neutrons) in *energy shells* analogous to the shell structure of electrons. A graph of the number of neutrons to the number of protons in a nucleus reveals that stable nuclei have a certain neutron-to-proton ratio in a *band of stability*. Nuclei that are above or below the band of stability and nuclei that are beyond atomic number 83 are radioactive and undergo *radioactive decay*.

Three common examples of radioactive decay involve the emission of an *alpha particle,* a *beta particle,* and a *gamma ray.* An alpha particle is a helium nucleus, consisting of two protons and two neutrons. A beta particle is a high-speed electron that is ejected from the nucleus. A gamma ray is a short-wavelength electromagnetic radiation from an excited nucleus. In general, nuclei with an atomic number of 83 or larger become more stable by alpha emission. Nuclei with a neutron-to-proton ratio that is too large become more stable by beta emission. Gamma ray emission occurs from a nucleus that was left in a high-energy state by the emission of an alpha or beta particle.

Each radioactive isotope has its own specific *radioactive decay rate.* This rate is usually described in terms of *half-life,* the time required for one-half the unstable nuclei to decay.

Radiation is measured by (1) its effects on photographic film, (2) the number of ions it produces, or (3) the flashes of light produced on a phosphor. It is measured at a source in units of a *curie,* defined as 3.70×10^{10} nuclear disintegrations per second. It is measured where received in units of a *rad.* A *rem* is a measure of radiation that takes into account the biological effectiveness of different types of radiation damage. In general, the natural environment exposes everyone to 100 to 500 millirems per year, an exposure called *background radiation.* Lifestyle and location influence the background radiation received, but the worldwide average is 240 millirems per year.

Energy and mass are related by Einstein's famous equation of $E = mc^2$, which means that *matter can be converted to energy and energy to matter.* The mass of a nucleus is always less than the sum of the masses of the individual particles of which it is made. This *mass defect* of a nucleus is equivalent to the energy released when the nucleus was formed according to $E = mc^2$. It is also the *binding energy,* the energy required to break the nucleus apart into nucleons.

When the binding energy is plotted against the mass number, the greatest binding energy per nucleon is seen to occur for an atomic number near that of iron. More massive nuclei therefore release energy by fission, or splitting to more stable nuclei. Less massive nuclei release energy by fusion, the joining of less massive nuclei to produce a more stable, more massive nucleus. Nuclear fission provides the energy for atomic explosions and nuclear power plants. Nuclear fusion is the energy source of the Sun and other stars and also holds promise as a future energy source for humans.

Summary of Equations

11.1   energy = mass × the speed of light squared

$$E = mc^2$$

KEY TERMS

alpha particle (p. **241**)

background radiation (p. **250**)

band of stability (p. **244**)

beta particle (p. **241**)

binding energy (p. **251**)

chain reaction (p. **252**)

critical mass (p. **252**)

curie (p. **249**)

gamma ray (p. **241**)

half-life (p. **247**)

mass defect (p. **251**)

nuclear fission (p. **251**)

nuclear fusion (p. **251**)

nucleons (p. **243**)

rad (p. **249**)

radioactive decay (p. **242**)

radioactivity (p. **241**)

rem (p. **249**)

APPLYING THE CONCEPTS

1. A high-speed electron ejected from a nucleus during radioactive decay is called a (an)
   a. alpha particle.
   b. beta particle.
   c. gamma ray.
   d. None of the above is correct.

2. The ejection of an alpha particle from a nucleus results in
   a. an increase in the atomic number by one.
   b. an increase in the atomic mass by four.
   c. a decrease in the atomic number by two.
   d. None of the above is correct.

3. An atom of radon-222 loses an alpha particle to become a more stable atom of
   a. radium.
   b. bismuth.
   c. polonium.
   d. radon.

4. A sheet of paper will stop a (an)
   a. alpha particle.
   b. beta particle.
   c. gamma ray.
   d. None of the above is correct.

5. The most penetrating of the three common types of nuclear radiation is the
   a. alpha particle.
   b. beta particle.
   c. gamma ray.
   d. All have equal penetrating ability.

6. An atom of an isotope with an atomic number greater than 83 will probably emit a (an)
   a. alpha particle.
   b. beta particle.
   c. gamma ray.
   d. None of the above is correct.

7. An atom of an isotope with a large neutron-to-proton ratio will probably emit a (an)
   a. alpha particle.
   b. beta particle.
   c. gamma ray.
   d. None of the above is correct.

8. The rate of radioactive decay can be increased by increasing the
   a. temperature.
   b. pressure.
   c. size of the sample.
   d. None of the above is correct.

9. Isotope A has a half-life of seconds, and isotope B has a half-life of millions of years. Which isotope is more radioactive?
   a. It depends on the sample size.
   b. isotope A
   c. isotope B
   d. Unknown from the information given.

10. A measure of radioactivity at the *source* is (the)
    a. curie..
    b. rad.
    c. rem.
    d. Any of the above is correct.

11. A measure of radiation received that considers the biological effect resulting from the radiation is (the)
    a. curie.
    b. rad.
    c. rem.
    d. Any of the above is correct.

12. Used fuel rods from a nuclear reactor contain about
    a. 96% usable uranium and plutonium.
    b. 33% usable uranium and plutonium.
    c. 4% usable uranium and plutonium.
    d. 0% usable uranium and plutonium.

Answers

1. b 2. c 3. c 4. a 5. c 6. a 7. b 8. c 9. b 10. a 11. c 12. a

QUESTIONS FOR THOUGHT

1. How is a radioactive material different from a material that is not radioactive?

2. What is radioactive decay? Describe how the radioactive decay rate can be changed if this is possible.

3. Describe three kinds of radiation emitted by radioactive materials. Describe what eventually happens to each kind of radiation after it is emitted.

4. How are positively charged protons able to stay together in a nucleus since like charges repel?

5. What is half-life? Give an example of the half-life of an isotope, describing the amount remaining and the time elapsed after five half-life periods.

6. Would you expect an isotope with a long half-life to be more, the same, or less radioactive than an isotope with a short half-life? Explain.

7. What is meant by background radiation? What is the normal radiation dose for the average person from background radiation?

8. Why is there controversy about the effects of long-term, low levels of radiation exposure?

9. What is a mass defect? How is it related to the binding energy of a nucleus? How can both be calculated?

10. Compare and contrast nuclear fission and nuclear fusion.

FOR FURTHER ANALYSIS

1. What are the significant differences between a radioactive isotope and an isotope that is not radioactive?

2. Analyze the different types of radioactive decay to explain how each is a hazard to living organisms.

3. Make up a feasible explanation for why some isotopes have half-lives of seconds, yet other kinds of isotopes have half-lives in the billions of years.

4. Suppose you believe the threshold model of radiation exposure is correct. Describe a conversation between yourself and another person who feels strongly that the linear model of radiation exposure is correct.

5. Explain how the fission of heavy elements and the fusion of light elements both release energy.

6. Write a letter to your congressional representative describing why used nuclear fuel rods should be reprocessed rather than buried as nuclear waste.

7. What are the similarities and differences between a nuclear fission power plant and a nuclear fusion power plant?

INVITATION TO INQUIRY

How Much Radiation?

Ionizing radiation is understood to be potentially harmful if certain doses are exceeded. How much ionizing radiation do you acquire from the surroundings where you live, from your lifestyle, and from medical procedures? Investigate radiation from cosmic sources, the Sun, television sets, time spent in jet airplanes, and dental or other X-ray machines. What are other sources of ionizing radiation in your community? How difficult is it to find relevant information and make recommendations? Does any agency monitor the amount of radiation that people receive? What are the problems and issues with such monitoring?

PARALLEL EXERCISES

The exercises in groups A and B cover the same concepts. Solutions to group A exercises are located in appendix D.

Group A

Note: You will need the table of atomic weights inside the back cover of this text.

1. Give the number of protons and the number of neutrons in the nucleus of each of the following isotopes:
   a. cobalt-60
   b. potassium-40
   c. neon-24
   d. lead-208

2. Write the nuclear symbols for each of the nuclei in exercise 1.

3. Predict if the nuclei in exercise 1 are radioactive or stable, giving your reasoning behind each prediction.

4. Write a nuclear equation for the decay of the following nuclei as they give off a beta particle:
   a. $^{56}_{26}\text{Fe}$
   b. $^{24}_{11}\text{Na}$
   c. $^{64}_{29}\text{Cu}$
   d. $^{24}_{11}\text{Na}$
   e. $^{214}_{82}\text{Pb}$
   f. $^{32}_{15}\text{P}$

5. Write a nuclear equation for the decay of the following nuclei as they undergo alpha emission:
   a. $^{235}_{92}\text{U}$
   b. $^{226}_{88}\text{Ra}$
   c. $^{239}_{94}\text{Pu}$
   d. $^{214}_{83}\text{Bi}$
   e. $^{230}_{90}\text{Th}$
   f. $^{210}_{84}\text{Po}$

6. The half-life of iodine-131 is 8 days. How much of a 1.0 oz sample of iodine-131 will remain after 32 days?

7. How much energy must be supplied to break a single iron-56 nucleus into separate protons and neutrons? (The mass of an iron-56 nucleus is 55. 9206 u, one proton is 1.00728 u, and one neutron is 1.00867 u.)

Group B

Note: You will need the table of atomic weights inside the back cover of this text.

1. Give the number of protons and the number of neutrons in the nucleus of each of the following isotopes:
   a. aluminum-25
   b. technetium-95
   c. tin-120
   d. mercury-200

2. Write the nuclear symbols for each of the nuclei in exercise 1.

3. Predict if the nuclei in exercise 1 are radioactive or stable, giving your reasoning behind each prediction.

4. Write a nuclear equation for the beta emission decay of each of the following:
   a. $^{14}_{6}\text{C}$
   b. $^{60}_{27}\text{Co}$
   c. $^{24}_{11}\text{Na}$
   d. $^{241}_{94}\text{Pu}$
   e. $^{131}_{53}\text{I}$
   f. $^{210}_{82}\text{Pb}$

5. Write a nuclear equation for each of the following alpha emission decay reactions:
   a. $^{241}_{95}\text{Am}$
   b. $^{232}_{90}\text{Th}$
   c. $^{223}_{88}\text{Ra}$
   d. $^{234}_{92}\text{U}$
   e. $^{242}_{96}\text{Cm}$
   f. $^{237}_{93}\text{Np}$

6. If the half-life of cesium-137 is 30 years, how much time will be required to reduce a 1.0 kg sample to 1.0 g?

7. How much energy is needed to separate the nucleons in a single lithium-7 nucleus? (The mass of a lithium-7 nucleus is 7.01435 u, one proton is 1.00728 u, and one neutron is 1.00867 u.)

12

The Universe

This is a planetary nebula in the constellation Aquarius. Planetary nebulae are clouds of ionized gases with no relationship to any planet. They were named long ago, when they appeared similar to the planets Neptune and Uranus when viewed through early telescopes.

CORE **CONCEPT**

The night sky is filled with billions of stars, and the Sun is an ordinary star with an average brightness.

OUTLINE

OVERVIEW

Astronomy is an exciting field of science that has fascinated people since the beginnings of recorded history. Ancient civilizations searched the heavens in wonder, some recording on clay tablets what they observed. Many religious and philosophical beliefs were originally based on interpretations of these ancient observations. Today, we are still awed by space, but now we are fascinated with ideas of space travel, black holes, and the search for extraterrestrial life. Throughout history, people have speculated about the universe and their place in it and watched the sky and wondered (figure 12.1). What is out there and what does it all mean? Are there other people on other planets, looking at the star in their sky that is our Sun, wondering if we exist?

Until about thirty years ago, progress in astronomy was limited to what could be observed and photographed. Developments in technology then began to provide the details of what is happening in the larger expanses of space away from Earth. This included new data made available from the development of infrared, radio, and X-ray telescopes and the Hubble Space Telescope. New data and the discovery of pulsars, neutron stars, and black holes began to fit together like the pieces of a puzzle. Theoretical models emerged about how stars, galaxies, and the universe have evolved. This chapter is concerned with these topics and how the stars are arranged in space. The chapter concludes with theoretical models of how the universe began and what may happen to it in the future.

THE NIGHT SKY

Early civilizations had a much better view of the night sky before city lights, dust, and pollution obscured much of it. Today, you must travel far from cities, perhaps to a remote mountain-top, to see a clear night sky as early people observed it. Back then, people could clearly see the motion of the moon and stars night after night, observing recurring cycles of motion. These cycles became important as people associated them with certain events. Thus, watching the Sun, Moon, and star movements became a way to identify when to plant crops, when to harvest, and when it was time to plan for other events. Observing the sky was an important activity, and many early civilizations built observatories with sighting devices to track and record astronomical events. Stonehenge, for example, was an ancient observatory built in England around 2600 B.C. by Neolithic people (figure 12.2).

Light from the stars and planets must pass through Earth's atmosphere to reach you and this affects the light. Stars appear as point *sources* of light, and each star generates its own light. The stars seem to twinkle because density differences in the atmosphere refract the point of starlight one way, then the other, as the air moves. The result is the slight dancing about and change in intensity called twinkling. The points of starlight are much steadier when viewed on a calm night or from high in the mountains, where there is less atmosphere for the starlight to pass through. Astronauts outside the atmosphere see no twinkling, and the stars appear as steady point sources of light.

Back at ground level, within the atmosphere, the *reflected* light from a planet does not seem to twinkle. A planet appears as a disk of light rather than a point source, so refraction from moving air of different densities does not affect the image as much. Sufficient air movement can cause planets to appear to shimmer, however, just as a road appears to shimmer on a hot summer day.

How far away is a star? When you look at the sky, it appears that all the stars are at the same distance. It seems impossible to know anything about the actual distance to any given star.

A

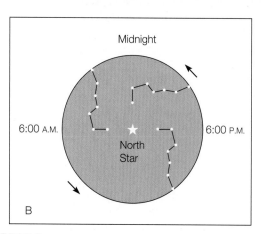

B

FIGURE 12.1 Ancient civilizations used celestial cycles of motion as clocks and calendars. (*A*) This photograph shows the path of stars around the North Star. (*B*) A "snapshot" of the position of the Big Dipper over a period of twenty-four hours as it turns around the North Star. This shows how the Big Dipper can be used to help you keep track of time.

FIGURE 12.2 The stone pillars of Stonehenge were positioned so that they could be used to follow the movement of the Sun and Moon with the seasons of the year.

Standard referent units of length such as kilometers or miles have little meaning in astronomy since there are no referent points of comparison. Distance without a referent point can be measured in terms of angles or time. The unit of astronomical distance that uses time is the **light-year** (ly). A light-year is the distance that light travels in one year, about 9.5×10^{12} km (about 6×10^{12} mi).

If you could travel by spaceship a few hundred light-years from Earth, you would observe the Sun shrink to a bright point of light among the billions and billions of other stars. The Sun is just an ordinary star with an average brightness. Like the other stars, the Sun is a massive, dense ball of gases with a surface heated to incandescence by energy released from fusion reactions deep within. Since the Sun is an average star, it can be used as a reference for understanding all the other stars.

ORIGIN OF STARS

Theoretically, stars are born from swirling clouds of hydrogen gas in the deep space between other stars. Such interstellar (between stars) clouds are called **nebulae.** These clouds consist of random, swirling atoms of gases that have little gravitational attraction for one another because they have little mass. Complex motion of stars, however, can produce a shock wave that causes particles to move closer together, making local compressions. Their mutual gravitational attraction then begins to pull them together into a cluster. The cluster grows as more atoms are pulled in, which increases the mass and thus the gravitational attraction, and still more atoms are pulled in from farther away. Theoretical calculations indicate that on the order of 1×10^{57} atoms are necessary, all within a distance of 3 trillion km (about 1.9 trillion mi). When these conditions occur, the cloud of gas atoms begins to condense by gravitational attraction to a **protostar,** an accumulation of gases that will become a star.

Gravitational attraction pulls the average protostar from a cloud with a diameter of trillions of kilometers (trillions of miles)

down to a dense sphere with a diameter of 2.5 million km (1.6 million mi) or so. As gravitational attraction accelerates the atoms toward the center, they gain kinetic energy, and the interior temperature increases. Over a period of some 10 million years of contracting and heating, the temperature and density conditions at the center of the protostar are sufficient to start nuclear fusion reactions. Pressure from hot gases and energy from increasing fusion reactions begin to balance the gravitational attraction over the next 17 million years, and the newborn, average star begins its stable life, which will continue for the next 10 billion years.

The interior of an average star, such as the Sun, is modeled after the theoretical pressure, temperature, and density conditions that would be necessary to produce the observed energy and light from the surface. This model describes the interior as a set of three shells: (1) the core, (2) a radiation zone, and (3) the convection zone (figure 12.3).

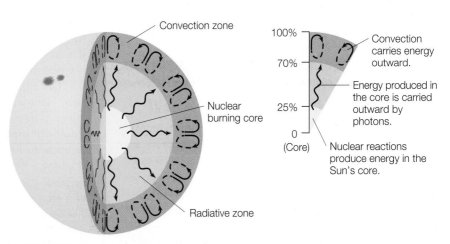

FIGURE 12.3 Energy-producing nuclear reactions occur only within the inner 25 percent of the Sun's radius. The energy produced by these reactions is carried outward by photons to 70 percent of the Sun's radius. From that distance outward, convection carries most of the Sun's energy.

Our model describes the **core** as a dense, very hot region where nuclear fusion reactions release gamma and X-ray radiation. The density of the core is about twelve times that of solid lead. Because of the exceedingly hot conditions, however, the core remains in a plasma state even at this density.

The model describes the **radiation zone** as less dense than the core, having a density about the same as the density of water. Energy in the form of gamma and X rays from the core is absorbed and reemitted by collisions with atoms in this zone. The radiation slowly diffuses outward because of the countless collisions over a distance comparable to the distance between Earth and the Moon. It could take millions of years before this radiation finally escapes the radiation zone.

In the model, the **convection zone** begins about seven-tenths of the way to the surface, where the density of the plasma is about 1 percent the density of water. Plasma at the bottom of this zone is heated by radiation from the radiation zone below, expands from the heating, and rises to the surface by convection. At the surface, the plasma emits energy in the form of visible light, ultraviolet radiation, and infrared radiation, which moves out into space. As it loses energy, the plasma contracts in volume and sinks back to the radiation zone to become heated again, continuously carrying energy from the radiation zone to the surface in convection cells. The surface is continuously heated by the convection cells as it gives off energy to space, maintaining a temperature of about 5,800 K (about 5,500°C).

Myths, Mistakes, and Misunderstandings

Stars Visible from the Bottom of a Well?

Perhaps you have heard that stars are visible from the bottom of a well, even in full daylight. This is a myth, as you cannot see any stars from the bottom of a well in full daylight. This idea is based on the belief that the brightness of the Sun, which obliterates the light from stars during the day, is somehow less in the bottom of a well. This idea might be based on the observation that looking at a star that you cannot see well is aided by looking at the star through a long, dark tube (a telescope). This somehow overlooks the role of lenses or mirrors in a telescope.

As an average star, the Sun converts about 1.4×10^{17} kg of matter to energy every year as hydrogen nuclei are fused to produce helium. The Sun was born about 5 billion years ago and has sufficient hydrogen in the core to continue shining for another 4 or 5 billion years. Other stars, however, have masses that are much greater or much less than the mass of the Sun so they have different life spans. More massive stars generate higher temperatures in the core because they have a greater gravitational contraction from their greater masses. Higher temperatures mean increased kinetic energy, which results in increased numbers of collisions between hydrogen nuclei with the end result an increased number of fusion reactions. Thus, a more massive star uses up its hydrogen more rapidly than a less massive star. On the other hand, stars that are less massive than the Sun use their hydrogen at a slower rate so they have longer life spans. The life spans of the stars range from a few million years for large, massive stars, to 10 billion years for average stars such as the Sun, to trillions of years for small, less massive stars.

BRIGHTNESS OF STARS

Stars generate their own light, but some stars appear brighter than others in the night sky. As you can imagine, this difference in brightness could be related to (1) the amount of light produced by the stars, (2) the size of each star, or (3) the distance to a particular star. A combination of these factors is responsible for the brightness of a star as it appears to you in the night sky. A classification scheme for different levels of brightness that you see is called the **apparent magnitude** scale (table 12.1).

The apparent magnitude scale is based on a system established by the Greek astronomer Hipparchus over two thousand years ago. Hipparchus made a catalog of stars he could see and assigned a numerical value to each to identify its relative brightness. The brightness values ranged from 1 to 6, with the number 1 assigned to the brightest star and the number 6 assigned to the faintest star that could be seen. Later, some stars were found to be brighter than the apparent magnitude of +1, which extended the scale into negative numbers. The brightest star in the night sky is Sirius, for example, with an apparent magnitude of −1.42 (see table 12.1).

The apparent magnitude of a star depends on how far away stars are in addition to differences in the stars themselves. Stars at a farther distance will appear fainter, and those closer will appear brighter, just as any other source of light. To compensate for distance differences, astronomers calculate the brightness that stars would appear to have if they were all at a defined, standard distance (32.6 light-years). The brightness of a star at this distance is called the **absolute magnitude.** The Sun, for example, is the closest star and has an apparent magnitude of −26.7 at an average distance from Earth. When viewed from the standard distance, the Sun would have an absolute magnitude of +4.8, which is about the brightness of a faint star.

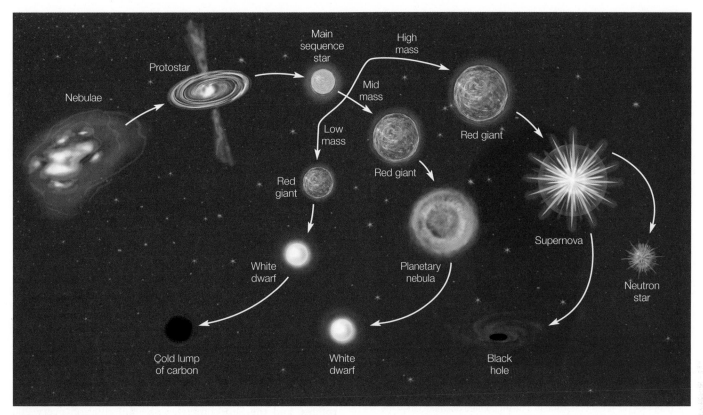

FIGURE 12.9 This shows some of the possible stages in the birth and aging of a star. The differences are determined by the mass of the star.

the explosion. All the elements heavier than iron were created as some less massive nuclei disintegrated in the explosion, joining with each other and with lighter nuclei to produce the nuclei of the elements from iron to uranium. As you will see in chapter 13, these newly produced, scattered elements will later become the building blocks for new stars and planets such as the Sun and Earth.

The remains of the compressed core after the supernova have yet another fate if the core has a remaining mass greater than 1.4 solar masses. The gravitational forces on the remaining matter, together with the compressional forces of the supernova explosion, are great enough to collapse nuclei, forcing protons and electrons together into neutrons, forming the core of a **neutron star.** A neutron star is the very small (10 to 20 km diameter), superdense (10^{11} kg/cm^3 or greater) remains of a supernova with a center core of pure neutrons.

Because it is a superdense form of matter, the neutron star also has an extremely powerful magnetic field, capable of becoming a pulsar. A **pulsar** is a very strongly magnetized neutron star that emits a uniform series of equally spaced electromagnetic pulses. Evidently, the magnetic field of a rotating neutron star makes it a powerful electric generator, capable of accelerating charged particles to very high energies. These accelerated charges are responsible for emitting a beam of electromagnetic radiation, which sweeps through space with amazing regularity (figure 12.10). The pulsating radio signals from a pulsar were a big mystery when first discovered. For a time, extraterrestrial life was considered as the source of the

signals, so they were jokingly identified as LGM (for "little green men"). Over three hundred pulsars have been identified, and most emit radiation in the form of radio waves. Two, however, emit visible light, two emit beams of gamma radiation, and one emits X-ray pulses.

Another theoretical limit occurs if the remaining core has a mass of about 3 solar masses or more. At this limit, the force of gravity overwhelms *all* nucleon forces, including the repulsive forces between like charged particles. If this theoretical limit is reached, nothing can stop the collapse, and the collapsed star will become so dense that even light cannot escape. The star is now a **black hole** in space. Since nothing can stop the collapsing star, theoretically a black hole would continue to collapse to a pinpoint and then to a zero radius called a *singularity*. This event seems contrary to anything that can be directly observed in the physical universe, but it does agree with the general theory of relativity and concepts about the curvature of space produced by such massively dense objects. Black holes are theoretical and none has been seen, of course, because a black hole theoretically pulls in radiation of all wavelengths and emits nothing. Evidence for the existence of a black hole is sought by studying X rays that would be given off by matter as it is accelerated into a black hole.

Another form of evidence for the existence of a black hole has now been provided by the Hubble Space Telescope. Hubble pictured a disk of gas only about 60 light-years out from the center of a galaxy (M87), moving at more than 1.6 million km/h (about 1 million mi/h). The only known possible

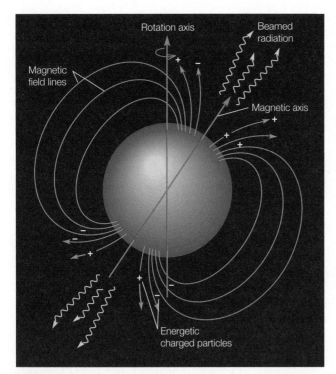

FIGURE 12.10 The magnetic axis of the pulsar is inclined with respect to the rotation axis. Rapidly moving electrons in the regions near the magnetic poles emit radiation in a beam pointed outward. When the beam sweeps past Earth, a pulse is detected.

explanation for such a massive disk of gas moving with this velocity at the distance observed would require the presence of a 1–2 billion solar-mass black hole. This gas disk could only be resolved by the Hubble Space Telescope, so this telescope has provided the first observational evidence of a black hole.

GALAXIES

Stars are associated with other stars on many different levels, from double stars that orbit a common center of mass, to groups of tens or hundreds of stars that have gravitational links and a common origin, to the billions and billions of stars that form the basic unit of the universe, a **galaxy.** The Sun is but one of an estimated four hundred billion stars that are held together by gravitational attraction in the Milky Way galaxy (figure 12.11). The numbers of stars and vastness of the Milky Way galaxy alone seem almost beyond comprehension, but there is more to come. The Milky Way is but one of *billions* of galaxies that are associated with other galaxies in clusters, and these clusters are associated with one another in superclusters. Through a large telescope, you can see more galaxies than individual stars in any direction, each galaxy with its own structure of billions of stars. Yet, there are similarities that point to a common origin. Some of the similarities and associations of stars will be introduced in this section along with the Milky Way galaxy, the vast, flat, spiraling arms of stars, gas, and dust where the Sun is located.

The Milky Way Galaxy

Away from city lights, you can clearly see the faint, luminous band of the Milky Way galaxy on a moonless night. Through a telescope or a good pair of binoculars, you can see that the luminous band is made up of countless numbers of stars. You may also be able to see the faint glow of nebulae, concentrations of gas and dust. There are dark regions in the Milky Way that also give an impression of something blocking starlight, such as dust. You can also see small groups of stars called **galactic clusters.** Galactic clusters are gravitationally bound subgroups of as many as one thousand stars that move together within the Milky Way. Other clusters are more symmetrical and tightly packed, containing as many as a million stars, and are known as **globular clusters.**

Viewed from a distance in space, the Milky Way would appear to be a huge, flattened cloud of spiral arms radiating out from the center. There are three distinct parts: (1) the spherical concentration of stars at the center of the disk called the *galactic nucleus;* (2) the rotating *galactic disk,* which contains most of the bright, blue stars along with much dust and gas; and (3) a spherical *galactic halo,* which contains some 150 globular clusters located outside the galactic disk (figure 12.12). The Sun is located in one of the arms of the galactic disk, some 25,000 to 30,000 light-years from the center. The galactic disk rotates, and the Sun completes one full rotation every 200 million years.

The diameter of the *galactic disk* is about 100,000 light-years. Yet, in spite of the recent estimate of four hundred billion stars in the Milky Way, it is mostly full of emptiness. By way of analogy, imagine reducing the size of the Milky Way disk until stars like the Sun were reduced to the size of tennis balls. The distance between two of these tennis-ball-sized stars would now compare to the distance across the state of Texas. The space between the stars is not actually empty since

FIGURE 12.11 A wide-angle view toward the center of the Milky Way galaxy. Parts of the white, milky band are obscured from sight by gas and dust clouds in the galaxy.

it contains a thin concentration of gas, dust, and molecules of chemical compounds. The gas particles outnumber the dust particles about 10^{12} to 1. The gas is mostly hydrogen, and the dust is mostly solid iron, carbon, and silicon compounds. Over forty different chemical molecules have been discovered in the space between the stars, including many organic molecules. Some nebulae consist of clouds of molecules with a maximum density of about 10^6 molecules/cm^3. The gas, dust, and chemical compounds make up part of the mass of the galactic disk, and the stars make up the remainder. The gas plays an important role in the formation of new stars, and the dust and chemical compounds play an important role in the formation of planets.

Other Galaxies

Outside the Milky Way is a vast expanse of emptiness, lacking even the few molecules of gas and dust spread thinly through the

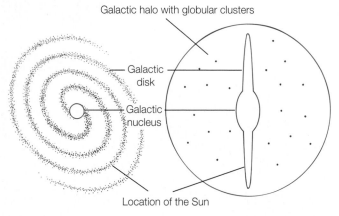

Galactic halo with globular clusters

Galactic disk

Galactic nucleus

Location of the Sun

FIGURE 12.12 The structure of the Milky Way galaxy.

galactic nucleus. There is only the light from faraway galaxies and the time that it takes for this light to travel across the vast vacuum of intergalactic space. How far away is the nearest galaxy? Recall that the Milky Way is so large that it takes light 100,000 years to travel the length of its diameter.

The nearest galactic neighbor to the Milky Way is a dwarf spherical galaxy only 80,000 light-years from the solar system. The nearby galaxy is called a dwarf because it has a diameter of only about 1,000 light-years. It is apparently in the process of being pulled apart by the gravitational pull of the Milky Way, which now is known to have eleven satellite galaxies.

The nearest galactic neighbor similar to the Milky Way is Andromeda, about 2 million light-years away. Andromeda is similar to the Milky Way in size and shape, with about four hundred billion stars, gas, and dust turning in a giant spiral pinwheel (figure 12.13). Other galaxies have other shapes and other characteristics. The American astronomer Edwin Hubble developed a classification scheme for the structure of galaxies based on his 1926 study of some six hundred different galaxies. The basic galactic structures were identified as elliptical, spiral, barred, and irregular.

The Life of a Galaxy

Hubble's classification of galaxies into distinctly different categories of shape was an exciting accomplishment because it suggested that some relationship or hidden underlying order might exist in the shapes. Finding underlying order is important because it leads to the discovery of the physical laws that govern the universe. Soon after Hubble published his classification results in 1926, two models of galactic evolution were proposed. One model, which was suggested by Hubble, had extremely slowly spinning spherical galaxies forming first, which gradually flattened out as their rate of spin increased while they condensed. This is a model of spherical galaxies flattening out to increasingly elliptical shapes, eventually spinning off spirals until they finally broke up into irregular shapes over a long period of time.

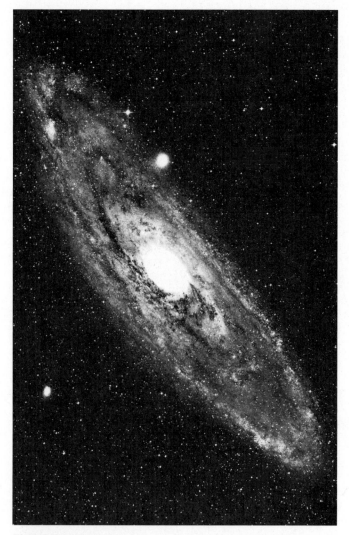

FIGURE 12.13 The Andromeda galaxy, which is believed to be similar in size, shape, and structure to the Milky Way galaxy.

Among many uses, the Hubble Space Telescope is used to study young galaxies and galaxies on collision courses. Based on these studies, astronomers today recognize that the different shapes of galaxies do not represent an evolutionary sequence. Their shapes are understood to be a result of the various conditions under which the galaxies were formed.

The current model of how galaxies form is based on the **big bang theory** of the creation of the universe. The big bang theory considers the universe to have had an explosive beginning. According to this theory, all matter in the universe was located together in an arbitrarily dense state from which it began to expand, an expansion that continues today. Evidence that supports the big bang theory comes from Albert Einstein's theory of general relativity as well as three areas of physical observations:

1. **Expansion of the universe.** The initial evidence for the big bang theory came from Edwin Hubble and his earlier work with galaxies. Hubble had determined the distances

Extraterrestrials?

Extraterrestrial is a descriptive term, meaning a thing or event outside Earth or its atmosphere. The term is also used to describe a being, a life-form that originated away from Earth. This reading is concerned with the search for extraterrestrials, intelligent life that might exist beyond Earth and outside the solar system.

Why do people believe that extraterrestrials might exist? The affirmative answer comes from a mixture of current theories about the origin and development of stars, statistical odds, and faith. Considering the statistical odds, note that our Sun is one of the some 400 billion stars that make up our Milky Way galaxy. The Milky Way galaxy is one of some 10 billion galaxies in the observable universe. Assuming an average of about 400 billion stars per galaxy, this means there are some 400 billion times 10 billion stars, or 4×10^{21} stars in the observable universe. There is nothing special or unusual about our Sun, and astronomers believe it to be quite ordinary among all the other stars (all 4,000,000,000,000,000,000,000 or so).

So the Sun is an ordinary star, but what about our planet? Not too long ago, most people, including astronomers, thought our solar system with its life-supporting planet (Earth) to be unique. Evidence collected over the past decade or so, however, has strongly suggested that this is not so. Planets are now believed to be formed as a natural part of the star-forming process. Evidence of planets around other stars has also been detected by astronomers. One of the stars with planets is "only" 53 light-years from Earth.

Even with a very low probability of planetary systems forming with the development of stars, a population of 4×10^{21} stars means there are plenty of planetary

Box Figure 12.1   The 300 m (984 ft) diameter radio telescope at Arecibo, Puerto Rico, is the largest fixed-dish radio telescope in the world.

systems in existence, some with the conditions necessary to support life. (Note: If 1 percent have planetary systems, this means 4×10^{19} stars have planets.) Thus, it is a statistical observation that suitable planets for life are very likely to exist. In addition, radio astronomers have found that many organic molecules exist, even in the space between the stars. Based on statistics alone, there should be life on other planets, life that may have achieved intelligence and developed into a technological civilization.

If extraterrestrial life exists, why have we not detected them or why have they not contacted us? The answer to this question is found in the unbelievable distances involved in interstellar space. For example, a logical way to search for extraterrestrial intelligence is to send radio signals to space and analyze those coming from space. Modern radio telescopes can send powerful radio beams and present-day computers and data processing techniques can now search through incoming radio

signals for the patterns of artificially generated radio signals (box figure 12.1). Radio signals, however, travel through space at the speed of light. The diameter of our Milky Way galaxy is about 100,000 light-years, which means 100,000 years would be required for a radio transmission to travel across the galaxy. For this discussion, assume Earth is located on one edge of the Milky Way. If we were to transmit a super, super strong radio beam from Earth, it would travel at the speed of light and cross the distance of our galaxy in 100,000 years. If some extraterrestrials on the other side of the Milky Way galaxy did detect the message and send a reply, it could not arrive at Earth until 200,000 years after the message was sent. Now consider the fact that of all the 10 billion other galaxies in the observable universe, our nearest galactic neighbor similar to the Milky Way is Andromeda. Andromeda is 2 million light-years from the Milky Way galaxy.

In addition to problems with distance and time, there are questions about in which part of the sky you should send and look for radio messages, questions about which radio frequency to use, and problems with the power of present-day radio transmitters and detectors. Realistically, the hope for any exchange of radio-transmitted messages would be restricted to within several hundred light-years of Earth.

Considering all the limitations, what sort of signals should we expect to receive from extraterrestrials? Probably a series of pulses that somehow indicate counting, such as 1, 2, 3, 4, and so on, repeated at regular intervals. This is the most abstract, while at the same time the simplest, concept that an intelligent being anywhere would have. It could provide the foundation for communications between the stars.

---

to some of the galaxies that had redshifted spectra. From this expansive redshift, it was known that these galaxies were moving away from the Milky Way. Hubble found a relationship between the distance to a galaxy and the velocity with which it was moving away. He found the velocity to be directly proportional to the distance; that is,

the greater the distance to a galaxy, the greater the velocity. This means that a galaxy twice as far from the Milky Way is moving away from the Milky Way at twice the speed. Since this relationship was seen in all directions, it meant that the universe is expanding uniformly. The same effect would be viewed from any particular galaxy; that is, all

A Closer Look

Redshift and Hubble's Law

As described in chapter 5, the Doppler effect tells us that the frequency of a wave depends on the relative motion of the source and observer. When the source and observer are moving toward each other, the frequency appears to be higher. If the source and observer are moving apart, the frequency appears to be lower.

Light from a star or galaxy is changed by the Doppler effect, and the frequency of the observed spectral lines depends on the relative motion. The Doppler effect changes the frequency from what it would be if the star or galaxy were motionless relative to the observer. If the star or galaxy is moving toward the observer, a shift occurs in the spectral lines toward a higher frequency (blueshift). If the star or galaxy is moving away from the observer, a shift occurs in the spectral lines toward a lower frequency (redshift). Thus, a redshift or blueshift in the spectral lines will tell you if a star or galaxy is moving toward or away from you.

One of the first measurements of the distance to other galaxies was made by Edwin Hubble at Mount Wilson Observatory in California. When Hubble compared the distance figures with the observed redshifts, he found that the recession speeds were proportional to the distance. Farther-away galaxies were moving away from the Milky Way, but galaxies that are more distant are moving away faster than closer galaxies. This proportional relationship between galactic speed and distances was discovered in 1929 by Hubble and today is known as *Hubble's law*. The conclusion was that all the galaxies are moving away from one another and an observer on any given galaxy would have the impression that all galaxies were moving away in all directions. In other words, the universe is expanding with component galaxies moving farther and farther apart.

the other galaxies are moving away with a velocity proportional to the distance to the other galaxies. This points to a common beginning, a time when all matter in the universe was together.

2. **Background radiation.** The big bang occurred as some unstable form of energy expanded and cooled, eventually creating matter and space. The initial temperature was 10 billion kelvins or so, and began to cool as the universe expanded. The afterglow of the big bang is called cosmic background radiation. A measurement of cosmic background radiation today agrees with the radiation that should be present according to an expanding model of the universe.

3. **Abundance of elements.** The proportion of helium in the universe should be about 24 percent, based on an expanding model of the universe. A measurement of the abundance of helium verifies the proportion as predicted by the big bang theory.

Cosmic Background Explorer (COBE) spacecraft studied diffuse cosmic background radiation to help answer such questions as how matter is distributed in the universe, whether the universe is uniformly expanding, and how and when galaxies first formed.

The 2003 results from NASA's orbiting *Wilkinson Microwave Anisotropy Probe (WMAP)* produced a precision map of the remaining cosmic microwave background from the big bang. *WMAP* surveyed the entire sky for a whole year with a resolution some 40 times greater than *COBE*. Analysis of *WMAP* data revealed that the universe is 13.7 billion years old, with a 1 percent margin of error. The *WMAP* data found strong support for the big bang and expanding universe theories. It also revealed that the content of the universe includes 4 percent ordinary matter, 23 percent of an unknown type of dark matter, and 73 percent of a mysterious dark energy.

How old is the universe? As mentioned earlier, astronomical and physical "clocks" indicate that the universe was created in a "big bang" some 13.7 billion years ago, expanding as an intense and brilliant fireball with a temperature of some 10 billion kelvins. This estimate of the age is based on precise measurements of the rate at which galaxies are moving apart, an expansion that started with the big bang. Astronomers used data on the expansion to back-calculate the age, much like running a movie backward, to arrive at the age estimate. At first, this technique found that the universe was about 18 to 20 billion years old. Then astronomers found that the universe is not expanding at a constant rate. Instead, the separation of galaxies is actually accelerating, pushed by a mysterious force known as "**dark energy.**" By adding in calculations for this poorly understood force, the estimate of 13 to 14 billion years was developed.

Other age-estimating techniques agree with this ballpark age. Studies of white dwarfs, for example, established their rate of cooling. By looking at the very faintest and oldest white dwarfs with the Hubble Space Telescope, astronomers were able to use the cooling rate to estimate the age of the universe. The result found the dimmest of the white dwarfs to be about 13 billion years old, plus or minus about half a billion years. More recently, data from the orbiting *WMAP* spacecraft has produced a precision map of the cosmic microwave background that shows the universe to be 13.7 billion years old. An age of

Dark Matter

Will the universe continue to expand forever, or will it be pulled back together into a really big crunch? Whether the universe will continue expanding or gravity will pull it back together again depends on the *critical density* of the universe. If there is enough matter, the actual density of the universe will be above the critical density, and gravity will stop the current expansion and pull everything back together again. On the other hand, if the actual density of the universe is less than or equal to the critical density, the universe will be able to escape its own gravity and continue expanding. Is the actual density small enough for the universe to escape itself?

All the detailed calculations of astrophysicists point to an actual density that is *less* than the critical density, meaning the universe will continue to expand forever. However, these calculations also show that there is more matter in the universe than can be accounted for in the stars and galaxies. This means there must be matter in the universe that is not visible, or not shining at least, so we cannot see it. Three examples follow that show why astrophysicists believe more matter exists than we can see.

1. There is a relationship between the light emitted from a star and the mass of that star. Thus, you can indirectly measure the amount of matter in a galaxy by measuring the light output of that galaxy. Clusters of galaxies have been observed moving, and the motion of galaxies within a cluster does not agree with matter information from the light. The motion suggests that galaxies are attracted by a gravitational force from about ten times more matter than can be accounted for by the light coming from the galaxies.

2. There are mysterious variations in the movement of stars within an individual galaxy. The rate of rotation of stars about the center of rotation of a galaxy is related to the distribution of matter in that galaxy. The outer stars are observed to rotate too fast for the amount of matter that can be seen in the galaxy. Again, measurements indicate there is about ten times more matter in each galaxy than can be accounted for by the light coming from the galaxies.

3. Finally, there are estimates of the total matter in the universe that can be made by measuring ratios such as deuterium to ordinary hydrogen, lithium−7 to helium−3, and helium to hydrogen. In general, theoretical calculations all seem to account for only about 10 percent of all the matter in the universe.

Calculations from a variety of sources all seem to agree that 90 percent or more of the matter that makes up the universe is missing. The missing matter and the mysterious variations in the orbits of galaxies and stars can be accounted for by the presence of **dark matter,** which is invisible and unseen. What is the nature of this dark matter? You could speculate that dark matter is simply normal matter that has been overlooked, such as dark galaxies, brown dwarfs, or planetary material such as rock, dust, and the like. You could also speculate that dark matter is dark because it is in the form of subatomic particles, too small to be seen. As you can see, scientists have reasons to believe that dark matter exists, but they can only speculate about its nature.

The nature of dark matter represents one of the major unsolved problems in astronomy today. There are really two dark matter problems: (1) the nature of the dark matter and (2) how much dark matter contributes to the actual density of the universe. Not everyone believes that dark matter is simply normal matter that has been overlooked. There are at least two schools of thought on the nature of dark matter and what should be the focus of research. One school of thought focuses on *particle physics* and contends that dark matter consists primarily of exotic undiscovered particles or known particles such as the neutrino. Neutrinos are electrically neutral, stable subatomic particles. This school of thought is concerned with forms of dark matter called WIMPs (after *weakly interacting massive particles*). In spite of the name, particles under study are not always massive. It is entirely possible that some low-mass species of neutrino—or some as-of-yet unidentified WIMP—will be experimentally discovered and found to have the mass needed to account for the dark matter and thus close the universe.

The other school of thought about dark matter focuses on *astrophysics* and contends that dark matter is ordinary matter in the form of brown dwarfs, unseen planets outside the solar system, and galactic halos. This school of thought is concerned with forms of dark matter called MACHOs (after *massive astrophysical compact halo objects*). Dark matter considerations in this line of reasoning consider massive objects, but ordinary matter (protons and neutrons) is also considered as the material making up galactic halos. Protons and neutrons belong to a group of subatomic particles called the *baryons,* so they are sometimes referred to as *baryonic dark matter*. Some astronomers feel that baryonic dark matter is the most likely candidate for halo dark matter.

In general, astronomers can calculate the probable cosmological abundance of WIMPs, but there is no proof of their existence. By contrast, MACHOs dark matter candidates are known to exist, but there is no way to calculate their abundance. MACHOs astronomers assert that halo dark matter and probably all dark matter may be baryonic. Do you believe WIMPs or MACHOs will provide an answer about the future of the universe? What is the nature of dark matter, how much dark matter exists, and what is the fate of the universe? Answers to these and more questions await further research.

Connections . . .

Dark Energy

The first evidence of dark energy came from astronomers trying to measure how fast the expanding universe is slowing. At the time, the generally accepted model of the universe was that it was still expanding but had been slowing from gravitational attraction since the big bang. The astronomers intended to measure the slowing rate to determine the average density of matter in the universe, providing a clue about the extent of dark matter. Their idea was to compare light traveling from a supernova in the distant universe with light traveling from a much closer supernova. Light from the distant universe is just now reaching us since it was emitted when the universe was very young. The brightness of the older and younger supernova would provide information about distance, and expansive redshift data would provide information about their speed of expansion. By comparing these two light sources, they would be able to calculate the expansion rate for now and in the distant past.

To their surprise, the astronomers did not find a rate of slowing expansion. Rather, they found that the expansion is speeding up. There was no explanation for this speeding up other than to assume that some unknown antigravity force was at work. This unknown force must be pushing the galaxies farther apart at the same time that gravity is trying to pull them together. The unknown repulsive force became known as "dark energy."

The idea of an unknown repulsive force is not new. It was a new idea when Einstein added a "cosmological constant" to the equations in the theory of general relativity. This constant represents a force that opposes gravity, and the force grows as a function of space. This means there was not much—shall we say, dark energy—when the universe was smaller, and gravity slowed the expansion. As the distance between galaxies increased, there was an increase of dark energy and the expansion accelerated. Einstein removed this constant when Edwin Hubble reported evidence that the universe is expanding. Removing the constant may have been a mistake. One of the problems in understanding what was happening to the early universe was a lack of information about the distant past as is represented by the distant universe of the present. Today, this is changing as more new technology is helping astronomers study the distant universe. Will they find the meaning of dark energy? What is dark energy? The answer is that no one knows. Stay tuned. . .this story is to be continued.

13.7 billion years may not be the final answer for the age of the universe, but when three independent measurement techniques agree closely, the answer becomes more believable.

An often-used analogy for the movement of galaxies after the big bang is a loaf of rising raisin bread. Consider galaxies as raisins in a loaf of raisin bread dough as it rises (figure 12.14). As the dough expands in size, it carries along the raisins, which are moved farther and farther apart by the expansion. If you were on a raisin, you would see all the other raisins moving away from you, and the speed of this movement would be directly proportional to their distance away. It would not matter which raisin you were on since all the raisins would appear to be moving away from all the other raisins. In other words, the galaxies are not expanding into space that is already there. It is space itself that is expanding, and the galaxies move with the expanding space (figure 12.15).

The ultimate fate of the universe will strongly depend on the mass of the universe and if the expansion is slowing or continuing. All the evidence tells us it began by expanding with a big bang. Will it continue to expand, becoming increasingly cold and diffuse? Will it slow to a halt, then start to contract to a fiery finale of a big crunch (figure 12.16)? Researchers continue searching for answers with experimental data and theoretical models, looking for clues in matches between the data and models.

A Raisin bread dough before rising

B Raisin bread dough after rising

FIGURE 12.14 As the universe (here represented by a loaf of raisin bread) expands, the expansion carries galaxies (represented by the raisins) away from each other at speeds that are proportional to their distances from each other. It doesn't matter within which galaxy an astronomer resides; the other galaxies all appear to be moving away.

People are fascinated with ideas of space travel as well as new discoveries about the universe and how it formed. Unmanned space probes, new kinds of instruments, and new kinds of telescopes have resulted in new information. We are learning what is happening in outer space away from Earth, as well as about the existence of other planets.

Few would deny that the space program has provided valuable information. However, some people wonder if it is worth the cost. They point to many problems here on Earth, such as growing energy and water needs, pollution problems, and ongoing health problems. They say that the money spent on exploring space could be better used for helping resolve problems on Earth.

In addition to new information and understanding coming out of the space program, supporters of space exploration point to new technology that helps people living on Earth. Satellites now provide valuable information for agriculture, including land use and weather monitoring. Untold numbers of lives have been saved thanks to storm warnings provided by weather satellites. There are many other spin-offs from the space program, including improvements in communication systems.

Questions to Discuss

1. What do you think: have the gain of knowledge and the spin-offs merited the expense? Or would the funds be better spent on solving other problems?

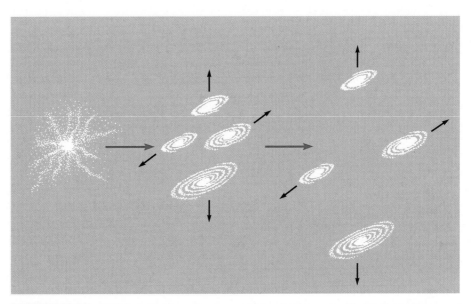

FIGURE 12.15   Will the universe continue expanding as the dust and gas in galaxies become locked up in white dwarf stars, neutron stars, and black holes?

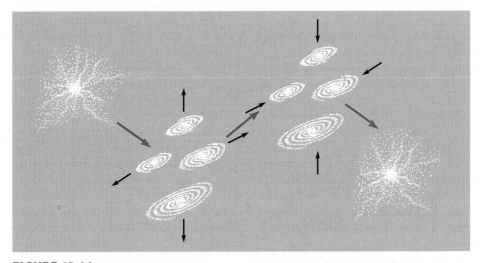

FIGURE 12.16   One theory of the universe assumes that the space between the galaxies is expanding, as does the big bang theory, but the galaxies gradually come back together to begin all over in another big bang.

People Behind the Science

Jocelyn (Susan) Bell Burnell (1943–)

Jocelyn Bell is a British astronomer who discovered pulsating radio stars—pulsars—an important astronomical discovery of the 1960s.

Bell was born in Belfast, Ireland, on July 15, 1943. The Armagh Observatory, of which her father was architect, was sited near her home, and the staff there was particularly helpful and offered encouragement when they learned of her early interest in astronomy. From 1956 to 1961, she attended the Mount School in York, England. She then went to the University of Glasgow, receiving her B.Sc. degree in 1965. In the summer of 1965, she began to work for her Ph.D. under the supervision of Anthony Hewish at the University of Cambridge. It was during the course of this work that she discovered pulsars.

She spent her first two years in Cambridge building a radio telescope that was specially designed to track quasars—her Ph.D. research topic. The telescope that she and her team built had the ability to record rapid variations in signals. It was also nearly 2 hectares (about 5 acres) in area, equivalent to a dish of 150 m (about 500 ft) in diameter, making it an extremely sensitive instrument.

The sky survey began when the telescope was finally completed in 1967, and Bell was given the task of analyzing the signals received. One day, while scanning the charts of recorded signals, she noticed a rather unusual radio source that had occurred during the night and had been picked up in a part of the sky that was opposite in direction to the Sun. This was curious because strong variations in the signals from quasars are caused by solar wind and are usually weak during the night. At first, she thought that the signal might be due to a local interference. After a month of further observations, it became clear that the position of the peculiar signals remained fixed with respect to the stars, indicating that it was neither terrestrial nor solar in origin. A more detailed examination of the signal showed that it was in fact composed of a rapid set of pulses that occurred precisely every 1.337 seconds. The pulsed signal was as regular as the most regular clock on Earth.

One attempted explanation of this curious phenomenon was that it represented an interstellar beacon sent out by extraterrestrial life on another star, so initially it was nicknamed LGM, for "little green men." Within a few months of noticing this signal, however, Bell located three other similar sources. They too pulsed at an extremely regular rate, but their periods varied over a few fractions of a second, and they all originated from widely spaced locations in our galaxy. Thus, it seemed that a more likely explanation of the signals was that they were being emitted by a special kind of star—a pulsar.

Since the astonishing discovery was announced, other observatories have searched the heavens for new pulsars. Some three hundred are now known to exist, their periods ranging from hundredths of a second to four seconds. It is thought that neutron stars are responsible for the signal. These are tiny stars, only about 7 km (about 4.3 mi) in diameter, but they are incredibly massive. The whole star and its associated magnetic field are spinning at a rapid rate, and the rotation produces the pulsed signal.

Source: Modified from the Hutchinson Dictionary of Scientific Biography. © RM, 2009. All rights reserved. Helicon Publishing is a division of RM.

SUMMARY

Stars are theoretically born in clouds of hydrogen gas and dust in the space between other stars. Gravity pulls huge masses of hydrogen gas together into a *protostar,* a mass of gases that will become a star. The protostar contracts, becoming increasingly hotter at the center, eventually reaching a temperature high enough to start *nuclear fusion* reactions between hydrogen atoms. Pressure from hot gases balances the gravitational contraction, and the average newborn star will shine quietly for billions of years. The average star has a dense, hot *core* where nuclear fusion releases radiation, a less dense *radiation zone* where radiation moves outward, and a thin *convection zone* that is heated by the radiation at the bottom, then moves to the surface to emit light to space.

The brightness of a star is related to the amount of energy and light it is producing, the size of the star, and the distance to the star. The *apparent magnitude* is the brightness of a star as it appears to you. To compensate for differences in brightness due to distance, astronomers calculate the brightness that stars would have at a standard distance. This standard-distance brightness is called the *absolute magnitude.*

Stars appear to have different colors because they have different surface temperatures. A graph of temperature by spectral types and brightness by absolute magnitude is called the *Hertzsprung-Russell diagram,* or H-R diagram for short. Such a graph shows that normal, mature stars fall on a narrow band called the *main sequence* of stars. Where a star falls on the main sequence is determined by its brightness and temperature, which in turn are determined by the mass of the star. Other groups of stars on the H-R diagram have different sets of properties that are determined by where they are in their evolution.

The life of a star consists of several stages, the longest of which is the *main sequence* stage after a relatively short time as a *protostar.* After using up the hydrogen in the core, a star with an average mass expands to a *red giant,* then blows off the outer shell to become a *white dwarf star,* which slowly cools to a black lump of carbon. The blown-off outer shell forms a *planetary nebula,* which disperses over time to become the gas and dust of interstellar space. More massive stars collapse into *neutron stars* or *black holes* after a violent *supernova* explosion.

Galaxies are the basic units of the universe. The Milky Way galaxy has three distinct parts: (1) the *galactic nucleus,* (2) a rotating *galactic disk,* and (3) a *galactic halo.* The galactic disk contains subgroups of stars that move together as *galactic clusters.* The halo contains symmetrical and tightly packed clusters of millions of stars called *globular clusters.*

Evidence from astronomical and physical "clocks" indicates that the galaxies formed some 13.7 billion years ago, expanding ever since from a common origin in a *big bang.* The *big bang theory* describes how the universe began by expanding.

KEY TERMS

absolute magnitude (p. **266**)
apparent magnitude (p. **266**)
big bang theory (p. **274**)
black hole (p. **271**)
Cepheid variable (p. **269**)
convection zone (p. **266**)

core (p. **266**)
dark energy (p. **276**)
dark matter (p. **277**)
galactic clusters (p. **272**)
galaxy (p. **272**)
globular clusters (p. **272**)

Hertzsprung-Russell diagram (p. **267**)
light-year (p. **265**)
main sequence stars (p. **268**)
nebulae (p. **265**)
neutron star (p. **271**)

protostar (p. **265**)
pulsar (p. **271**)
radiation zone (p. **266**)
red giant stars (p. **268**)
supernova (p. **270**)
white dwarf stars (p. **268**)

APPLYING THE CONCEPTS

1. Stars twinkle and planets do not twinkle because
   a. planets shine by reflected light, and stars produce their own light.
   b. all stars are pulsing light sources.
   c. stars appear as point sources of light, and planets are disk sources.
   d. All of the above are correct.

2. Which of the following stars would have the longer life spans?
   a. the less massive
   b. between the more massive and the less massive
   c. the more massive
   d. All have the same life span.

3. A bright blue star on the main sequence is probably
   a. very massive.
   b. less massive.
   c. between the more massive and the less massive.
   d. None of the above is correct.

4. The basic property of a main sequence star that determines most of its other properties, including its location on the H-R diagram, is
   a. brightness.   c. temperature.
   b. color.   d. mass.

5. All the elements that are more massive than the element iron were formed in a
   a. nova.   c. supernova.
   b. white dwarf.   d. black hole.

6. If the core remaining after a supernova has a mass between 1.5 and 3 solar masses, it collapses to form a
   a. white dwarf.   c. red giant.
   b. neutron star.   d. black hole.

7. The basic unit of the universe is a
   a. star.
   b. solar system.
   c. galactic cluster.
   d. galaxy.

8. The greater the distance to a galaxy, the greater a redshift in its spectral lines. This is known as
   a. Doppler's law.
   b. Cepheid's law.
   c. Hubble's law.
   d. Kepler's law.

9. Dark energy calculations and the age of cooling white dwarfs indicate that the universe is about how old?
   a. 6,000 years
   b. 4.5 billion years
   c. 13.7 billion years
   d. 100,000 billion years

10. Whether the universe will continue to expand or will collapse back into another big bang seems to depend on what property of the universe?
    a. the density of matter in the universe
    b. the age of galaxies compared to the age of their stars
    c. the availability of gases and dust between the galaxies
    d. the number of black holes

Answers

1. c 2. a 3. a 4. d 5. c 6. b 7. d 8. c 9. c 10. a

QUESTIONS FOR THOUGHT

1. What is a light-year and how is it defined?
2. Why are astronomical distances not measured with standard referent units of distance such as kilometers or miles?
3. Which size of star has the longest life span, a star sixty times more massive than the Sun, one just as massive as the Sun, or a star that has a mass of one-twenty-fifth that of the Sun? Explain.
4. What is the Hertzsprung-Russell diagram? What is the significance of the diagram?
5. Describe, in general, the life history of a star with an average mass like the Sun.
6. What, if anything, is the meaning of the Hubble classification scheme of the galaxies?
7. What is a nova? What is a supernova?
8. Describe the theoretical physical circumstances that lead to the creation of (a) a white dwarf star, (b) a red giant, (c) a neutron star, (d) a black hole, and (e) a supernova.
9. Describe the two forces that keep a star in a balanced, stable condition while it is on the main sequence. Explain how these forces are able to stay balanced for a period of billions of years or longer.

10. What is the source of all the elements in the universe that are more massive than helium but less massive than iron? What is the source of all the elements in the universe that are more massive than iron?

11. What is a red giant star? Explain the conditions that lead to the formation of a red giant. How can a red giant become brighter than it was as a main sequence star if it now has a lower surface temperature?

12. Describe the structure of the Milky Way galaxy. Where are new stars being formed in the Milky Way? Explain why they are formed in this part of the structure and not elsewhere.

FOR FURTHER ANALYSIS

1. A star is 518 light-years from Earth. During what event in history did the light now arriving at Earth leave the star?

2. What are the significant differences between the life and eventual fate of a massive star and an average-sized star such as the Sun?

3. Analyze when apparent magnitude is a better scale of star brightness and when absolute magnitude is a better scale of star brightness.

4. What is the significance of the Hertzsprung-Russell diagram?

5. The Milky Way galaxy is a huge, flattened cloud of spiral arms radiating out from the center. Describe several ideas that explain why it has this shape. Identify which idea you favor and explain why.

INVITATION TO INQUIRY

It Keeps Going, and Going, and . . .

Pioneer 10 was the first space probe to visit an outer planet of our solar system. It was launched March 2, 1972, and successfully visited Jupiter on June 13, 1983. After transmitting information and relatively close-up pictures of Jupiter, *Pioneer 10* continued on its trajectory, eventually becoming the first space probe to leave the solar system. It continued to move silently into deep space and sent the last signal on January 22, 2003, when it was 12.2 billion km (7.6 billion mi) from Earth. It will now continue to drift for the next 2 million years toward the star Aldebaran in the constellation Taurus.

As the first human-made object out of the solar system, *Pioneer 10* carries a gold-plated plaque with the image shown in box figure 12.2. Perhaps intelligent life will find the plaque and decipher the image to learn about us. What information is in the image? Try to do your own deciphering to reveal the information. When you have exhausted your efforts, see grin.hq.nasa.gov/ABSTRACTS/GPN-2000-001623.html.

For more on the *Pioneer 10* mission, see nssdc.gsfc.nasa.gov/nmc/mastercatalog.do?sc=1972-012A.

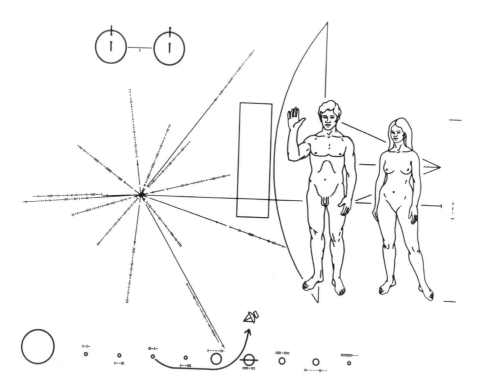

Box Figure 12.2   *Pioneer 10* plaque symbology.

13

The Solar System

Neptune, the most distant and smallest of the gas giant planets, is a cold and interesting place. It has a Great Dark Spot, as you can see in this photograph made by *Voyager*. This spot is about the size of Earth and is similar to the Great Red Spot on Jupiter. Neptune has the strongest winds of any planet of the solar system—up to 2,000 km/h (1,200 mi/h). Clouds were observed by *Voyager* to be "scooting" around Neptune every sixteen hours or so. *Voyager* scientists called these fast-moving clouds "scooters."

CORE **CONCEPT**

The solar system is composed of the Sun, a system of related planets, moons, comets, and asteroids.

OUTLINE

OVERVIEW

For generations, people have observed the sky in awe, wondering about the bright planets moving across the background of stars, but they could do no more than wonder. You are among the first generations on Earth to see close-up photographs of the planets, comets, and asteroids and to see Earth as it appears from space. Spacecrafts have now made thousands of photographs of the other planets and their moons, measured properties of the planets, and in some cases, studied their surfaces with landers. Astronauts have left Earth and visited the Moon, bringing back rock samples, data, and photographs of Earth as seen from the Moon (figure 13.1). All of these photographs and findings have given a new perspective of Earth, the planets, and the moons, comets, and asteroids that make up the solar system.

Viewed from the Moon, Earth is a spectacular blue globe with expanses of land and water covered by huge changing patterns of white clouds. Viewed from a spacecraft, other planets present a very different picture, each unique in its own way. Mercury has a surface that is covered with craters, looking very much like the surface of Earth's Moon. Venus is covered with clouds of sulfuric acid over an atmosphere of mostly carbon dioxide, which is under great pressures with surface temperatures hot enough to melt lead. The surface of Mars has great systems of canyons, inactive volcanoes, dry riverbeds and tributaries, and ice beneath the surface. The giant planets Jupiter and Saturn have orange, red, and white bands of organic and sulfur compounds and storms with gigantic lightning discharges much

larger than anything ever seen on Earth. One moon of Jupiter has active volcanoes spewing out liquid sulfur and gaseous sulfur dioxide. The outer giant planets Uranus and Neptune have moons and particles in rings that appear to be covered with powdery, black carbon.

These and many more findings, some fascinating surprises and some expected, have stimulated the imagination as well as added to our comprehension of the frontiers of space. The new information about the Sun's impressive system of planets, moons, comets, and asteroids has also added to speculations and theories about the planets and how they evolved over time in space. This information, along with the theories and speculations, will be presented in this chapter to give you a picture of the solar system.

PLANETS, MOONS, AND OTHER BODIES

The International Astronomical Union (IAU) is the governing authority over names of celestial bodies. At the August 24, 2006, meeting of the IAU, the definitions of planets, dwarf planets, and small solar system bodies were clarified and approved. To be a classical **planet,** an object must be orbiting the Sun, must be nearly spherical, and must be large enough to clear all matter from its orbital zone. A **dwarf planet** is defined as an object that is orbiting the Sun, is nearly spherical but has *not* cleared matter from its orbital zone and is *not* a satellite. All other objects orbiting the Sun are referred to collectively as **small solar system bodies.**

Some astronomers had dismissed Pluto as a true planet for years because it has properties that do not fit with those of

the other planets. The old definition of a planet was anything spherical that orbits the Sun, which resulted in nine planets. But in 2003, a new astronomical body was discovered. This body was named "Eris" after the Greek goddess of discord. Eris is larger than Pluto, round, and circles the Sun. Is it a planet? If so, the asteroid Ceres would also be a planet as would the fifty or so large, icy bodies believed to be orbiting the Sun far beyond Pluto. The idea of fifty or sixty planets in the solar system spurred astronomers to clarify the definition of a planet. However, Pluto does not clear its orbital zone, so it was downgraded to a dwarf planet. Today, there are eight planets, five dwarf planets (Pluto, Eris, Ceres, Makemake, and Haumea), and many, many small solar system bodies. These definitions may change again in the future as more is learned about our solar system.

FIGURE 13.1 This view of rising Earth was seen by the *Apollo 11* astronauts after they entered the orbit around the Moon. Earth is just above the lunar horizon in this photograph.

In this chapter, we will visit each of the planets, Earth's Moon, and other bodies of the solar system. (figure 13.2).

The Sun has seven hundred times the mass of all the planets, moons, and minor members of the solar system together. It is the force of gravitational attraction between the comparatively massive Sun and the rest of our solar system that holds it all together. The distance from Earth to the Sun is known as one **astronomical unit** (AU). One AU is about 1.5×10^8 km (about 9.3×10^7 mi). The astronomical unit is used to describe distances in the solar system, for example, Earth is 1 AU from the Sun.

Table 13.1 compares the basic properties of the eight planets. From this table, you can see that the planets can be classified into two major groups based on size, density, and nature of the atmosphere. The interior planets of Mercury, Venus, and Mars have densities and compositions similar to those of Earth, so these planets, along with Earth, are known as the **terrestrial planets.**

Outside the orbit of Mars are four **giant planets,** which are similar in density and chemical composition. The terrestrial planets are mostly composed of rocky materials and metallic nickel and iron. The giant planets of Jupiter, Saturn, Uranus, and Neptune, on the other hand, are massive giants mostly composed of hydrogen, helium, and methane. The density of the giant planets suggests the presence of rocky materials and iron as a core surrounded by a deep layer of compressed gases beneath a deep atmosphere of vapors and gases. Note that the terrestrial planets are separated from the giant planets by the asteroid belt.

We will start with the planet closest to the Sun and work our way outward, moving farther and farther from the Sun as we learn about our solar system.

Mercury

Mercury is the innermost planet, moving rapidly in a highly elliptical orbit that averages about 0.4 astronomical unit, or about 0.4 of the average distance of Earth from the Sun. Mercury is the smallest planet and is slightly larger than Earth's Moon. Mercury is very bright because it is so close to the Sun, but it is difficult to observe because it only appears briefly for a few hours immediately after sunset or before sunrise. This appearance,

FIGURE 13.2 The order of the planets out from the Sun. The planets are (1) Mercury, (2) Venus, (3) Earth, (4) Mars, (5) Jupiter, (6) Saturn, (7) Uranus, and (8) Neptune. The orbits and the planet sizes are not drawn to scale, and not all rings or moons are shown. Also, the planets are not in a line as shown.

TABLE 13.1 Properties of the Planets

	Mercury	Venus	Earth	Mars	Jupiter	Saturn	Uranus	Neptune
Average Distance from the Sun:								
in 10^6 km	58	108	150	228	778	1,400	3,000	4,497
in AU	0.38	0.72	1.0	1.5	5.2	9.5	19.2	30.1
Inclination to Ecliptic	7°	3.4°	0°	1.9°	1.3°	2.5°	0.8°	1.8°
Revolution Period								
(Earth years)	0.24	0.62	1.00	1.88	11.86	29.46	84.01	164.8
Rotation Period*	59 days	−243 days	23 h	24 h	9 h	10 h	−17 h	16 h
(Earth days,			56 min	37 min	50 min	39 min	14 min	6.7 min
min, and s)			4 s	23 s	30 s			
Mass (Earth = 1)	0.05	0.82	1.00	0.11	317.9	95.2	14.6	17.2
Equatorial Dimensions:								
diameter in km	4,880	12,104	12,756	6,787	142,984	120,536	57,118	49,528
in Earth radius = 1	0.38	0.95	1.00	0.53	11	9	4	4
Density (g/cm³)	5.43	5.25	5.52	3.95	1.33	0.69	1.29	1.64
Atmosphere								
(major compounds)	None	CO_2	N_2, O_2	CO_2	H_2, He	H_2, He	H_2, He, CH_4	H_2, He, CH_4
Solar Energy Received								
(cal/cm²/s)	13.4	3.8	2.0	0.86	0.08	0.02	0.006	0.002

*A negative value means spin is opposite to motion in orbit.

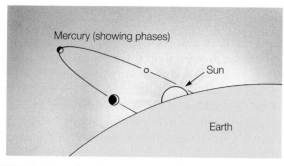

FIGURE 13.3 Mercury is close to the Sun and is visible from Earth only briefly before or after sunrise or sunset, showing phases. Mercury actually appears much smaller in an orbit that is not tilted as much as shown in this figure.

low on the horizon, means that Mercury must be viewed through more of Earth's atmosphere, making the study of such a small object difficult at best (figure 13.3).

Mercury moves around the Sun in about three Earth months, giving Mercury the shortest "year" of all the planets. With the highest orbital velocity of all the planets, Mercury was appropriately named after the mythical Roman messenger of speed. Oddly, however, this speedy planet has a rather long day in spite of its very short year. With respect to the stars, Mercury rotates once every fifty-nine days. This means that Mercury rotates on its axis three times every two orbits.

The long Mercury day with a nearby large, hot Sun means high temperatures on the surface facing the Sun. High temperatures mean higher gas kinetic energies, and with a low

gravity, gases easily escape from Mercury so it has only trace gases for an atmosphere. The lack of an atmosphere to even the heat gains from the long days and heat losses from the long nights results in some very large temperature differences. The temperature of the surface of Mercury ranges from above the melting point of lead on the sunny side to below the temperature of liquid oxygen on the dark side.

Mercury has been visited by *Mariner 10,* which flew by three times in 1973 and 1974. The photographs transmitted by *Mariner 10* revealed that the surface of Mercury is covered with craters and very much resembles the surface of Earth's Moon. There are large craters, small craters, superimposed craters, and craters with lighter colored rays coming from them just like the craters on the Moon. Also as on Earth's Moon, there are hills and smooth areas with light and dark colors that were covered by lava in the past, some time after most of the impact craters were formed (figure 13.4).

A new spacecraft, named *MESSENGER,* is now on its way to investigate the planet Mercury. Its name is an acronym for *MErcury Surface, Space ENvironment, GEochemistry and Ranging.* It was launched August 3, 2004, on a 7.9 billion km (about 4.9 billion mi) trip designed to slow the spacecraft as it falls toward the Sun. Overall, it looped by Earth in August 2005, then twice by Venus in October 2006 and June 2007, and then three times by Mercury in January 2008, October 2008, and September 2009. Eventually, it will be slow enough to be captured by the planet Mercury as it flies by in March 2011. It will then become the first spacecraft to orbit Mercury. Mercury has a high surface temperature, and the spacecraft instruments will be protected from high radiant energy from Mercury and the

FIGURE 13.4 A photomosaic of Mercury made from pictures taken by the *Mariner 10* spacecraft. The surface of Mercury is heavily cratered, looking much like the surface of Earth's Moon. All the interior planets and the Moon were bombarded early in the life of the solar system.

Sun by a sun-shade of heat-resistant ceramic fabric. (There is a NASA mission page at www.nasa.gov/mission_pages/messenger/main/index.html.)

Mercury has no natural satellites or moons, it has a weak magnetic field, and it has an average density more similar to that of Venus or Earth than to the Moon's. The presence of the magnetic field and the relatively high density for such a small body probably means that Mercury has a relatively large core of iron with at least part of the core molten. Because of its high density, Mercury is believed to have lost much of its less dense, outer layer of rock materials sometime during its formation.

Venus

Venus is the brilliant evening and morning "star" that appears near sunrise or sunset, sometimes shining so brightly that you can see it while it is still daylight. Venus orbits the Sun at an average distance of about 0.7 AU. Venus is sometimes to the left of the Sun, appearing as the evening star, and sometimes to the right of the Sun, appearing as the morning star. Venus also has phases just as the Moon does. When Venus is in the full phase, it is small and farthest away from Earth. A crescent Venus appears much larger and thus the brightest because it is closest to Earth when it is in its crescent phase. You can see the phases of Venus with a good pair of binoculars.

Venus shines brightly because it is covered with clouds that reflect about 80 percent of the sunlight, making it the brightest object in the sky after the Sun and the Moon. These same clouds prevented any observations of the surface of Venus until the early 1960s, when radar astronomers were able to penetrate the clouds and measure the planet's spin rate. Venus was found to spin slowly, so slowly that each day on Venus is longer than a Venus year! Also a surprise, Venus was found to spin in the *opposite* direction to its direction of movement in its orbit. On Venus, you would observe the Sun to rise in the west and set in the east, if you could see it, that is, through all the clouds.

In addition to early studies by radio astronomers, Venus has been the target of many American and Soviet spacecraft probes (Table 13.2).

Venus has long been called Earth's sister planet since its mass, size, and density are very similar. That is where the similarities with Earth end, however, as Venus has been found to have a very hostile environment. Spacecraft probes found a hot, dry surface under tremendous atmospheric pressure (figure 13.5). The atmosphere consists mostly of carbon dioxide, a few percent of nitrogen, and traces of water vapor and other gases. The atmospheric pressure at the surface of Venus is almost one hundred times the pressure at the surface of Earth, a pressure many times beyond what a human could tolerate. The average surface temperature is comparable to the surface

TABLE 13.2 Spacecraft Missions to Venus

Date	Name	Owner	Remark
Feb 12, 1961	*Venera 1*	USSR	Flyby
Aug 27, 1962	*Mariner 2*	U.S.	Flyby
Apr 2, 1964	*Zond 1*	USSR	Flyby
Nov 12, 1965	*Venera 2*	USSR	Flyby
Nov 16, 1965	*Venera 3*	USSR	Crashed on Venus
June 12, 1967	*Venera 4*	USSR	Impacted Venus
June 14, 1967	*Mariner 5*	U.S.	Flyby
Jan 5, 1969	*Venera 5*	USSR	Impacted Venus
Jan 10, 1969	*Venera 6*	USSR	Impacted Venus
Aug 17, 1970	*Venera 7*	USSR	Venus landing
Mar 27, 1972	*Venera 8*	USSR	Venus landing
Nov 3, 1973	*Mariner 10*	U.S.	Venus, Mercury flyby photos
June 8, 1975	*Venera 9*	USSR	Lander/orbiter
June 14, 1975	*Venera 10*	USSR	Lander/orbiter
May 20, 1978	*Pioneer 12* (also called *Pioneer Venus 1* or *Pioneer Venus*)	U.S.	Orbital studies of Venus
Aug 8, 1978	*Pioneer 13* (also called *Pioneer Venus 2* or *Pioneer Venus*)	U.S.	Orbital studies of Venus
Sept 9, 1978	*Venera 11*	USSR	Lander; sent photos
Sept 14, 1978	*Venera 12*	USSR	Lander; sent photos
Oct 30, 1981	*Venera 13*	USSR	Lander; sent photos
Nov 4, 1981	*Venera 14*	USSR	Lander; sent photos
June 2, 1983	*Venera 15*	USSR	Radar mapper
June 7, 1983	*Venera 16*	USSR	Radar mapper
Dec 15, 1984	*Vega 1*	USSR	Venus/Comet Halley probe
Dec 21, 1984	*Vega 2*	USSR	Venus/Comet Halley probe
May 4, 1989	*Magellan*	U.S.	Orbital radar mapper
Oct 18, 1989	*Galileo*	U.S.	Flyby measurements and photos
Apr 11, 2006	*Venus Express*	ESA	Orbital spacecraft
June 5, 2007	*Messenger*	U.S.	Flyby photos

FIGURE 13.5 This is an image of an 8 km (5 mile) high volcano on the surface of Venus. The image was created by a computer using *Magellan* radar data, simulating a viewpoint elevation of 1.7 km (1 mi) above the surface. The lava flows extend for hundreds of kilometers across the fractured plains shown in the foreground. The simulated colors are based on color images recorded by the Soviet *Venera 13* and *14* spacecraft.

atmosphere. Carbon dioxide molecules absorb the infrared radiation, increasing their kinetic energy and the temperature.

The surface of Venus is mostly a flat, rolling plain. There are several raised areas, or "continents," on about 5 percent of the surface. There is also a mountain larger than Mount Everest, a great valley deeper and wider than the Grand Canyon, and many large, old impact craters. In general, the surface of Venus appears to have evolved much as did the surface of Earth but without the erosion caused by ice, rain, and running water.

Venus, like Mercury, has no satellites. Venus also does not have a magnetic field, as might be expected. Two conditions seem to be necessary in order for a planet to generate a magnetic field: a molten center and a relatively rapid rate of rotation. Since Venus takes 243 days to complete one rotation, the slowest of all the planets, it does not have a magnetic field even if some of the interior of Venus is still liquid as in Earth.

The European Space Agency's (ESA) orbiting *Venus Express* is conducting an in-depth observation of the structure and chemistry of the atmosphere of Venus. One of the early findings was hot, extensive 21 km (13 mi) deep clouds of sulfuric acid moved by 334 km/h (220 mph) winds. The mission also plans to study the Venus greenhouse effect and volcanic activity of the past and present, if any.

Earth's Moon

Next to the Sun, the Moon is the largest, brightest object in the sky. The Moon is Earth's nearest neighbor and surface features can be observed with the naked eye. With the aid of a telescope or a good pair of binoculars, you can see light-colored mountainous regions called the *lunar highlands*, smooth dark areas called *maria*, and many sizes of craters, some with bright streaks extending from them (figure 13.6). The smooth dark

temperature on Mercury, which is hot enough to melt lead. The hot temperature on Venus, which is nearly twice the distance from the Sun as Mercury, is a result of the greenhouse effect. Sunlight filters through the atmosphere of Venus, warming the surface. The surface reemits the energy in the form of infrared radiation, which is absorbed by the almost pure carbon dioxide

FIGURE 13.6 You can easily see the light-colored lunar highlands, smooth and dark maria, and many craters on the surface of Earth's nearest neighbor in space.

areas are called maria after a Latin word meaning "sea." They acquired this name from early observers who thought the dark areas were oceans and the light areas were continents. Today, the maria are understood to have formed from ancient floods of molten lava that poured across the surface and solidified to form the "seas" of today. There is no water or atmosphere on the Moon.

Many facts known about the Moon were established during the *Apollo* missions, the first human exploration of a place away from Earth. A total of twelve *Apollo* astronauts walked on the Moon, taking thousands of photographs, conducting hundreds of experiments, and returning to Earth with over 380 kg (about 840 lb) of moon rocks. In addition, instruments were left on the Moon that continued to radio data back to Earth after the *Apollo* program ended in 1972. As a result of the *Apollo* missions, many questions were answered about the Moon, but unanswered questions still remain.

The *Apollo* astronauts found that the surface of the Moon is covered by a 3 m (about 10 ft) layer of fine gray dust that contains microscopic glass beads. The dust and beads were formed from millions of years of continuous bombardment of micrometeorites. These very small meteorites generally burn up in Earth's atmosphere. The Moon does not have an atmosphere, so meteorites have continually fragmented and pulverized the surface in a slow, steady rain. The glass beads are believed to have formed when larger meteorite impacts melted part of the surface, which was immediately forced into a fine spray that cooled rapidly while above the surface.

The rocks on the surface of the Moon were found to be mostly *basalts,* a type of rock formed on Earth from the cooling and solidification of molten lava. The dark-colored rocks from the maria are similar to Earth's basalts (see chapter 15) but contain greater amounts of titanium and iron oxides. The light-colored rocks from the highlands are mostly *breccias,* a kind of rock made up of rock fragments that have been compacted together. On the Moon, the compacting was done by meteorite impacts. The rocks from the highlands contain more aluminum and less iron than the maria basalts and thus have a lower density than the darker rocks.

All the moon rocks contained a substantial amount of radioactive elements, which made it possible to precisely measure their age. The light-colored rocks from the highlands were formed about 4 billion years ago. The dark-colored rocks from the maria were much younger, with ages ranging from 3.1 to 3.8 billion years. This indicates a period of repeated volcanic eruptions and lava flooding over a seven hundred-million-year period that ended about three billion years ago.

The moon rocks brought back to Earth, the results of the lunar seismographs, and all the other data gathered through the *Apollo* missions have increased our knowledge about the Moon, leading to new understandings of how it formed. This model pictures the present Moon developing through four distinct stages.

The *origin stage* describes how the moon originally formed. The Moon is believed to have originated from the impact of Earth with a very large object, perhaps as large as Mars or larger. The Moon formed from ejected material produced by this collision. The collision is believed to have vaporized the colliding body as well as part of Earth. Some of the debris condensed away from Earth to form the Moon.

The collision that resulted in the Moon took place after Earth's iron core formed, so there is not much iron in moon rocks. The difference in moon and earth rocks can be accounted for by the presence of materials from the impacting body.

The *molten surface stage* occurred during the first two hundred million years after the collision. Heating from a number of sources could have been involved in the melting of the entire lunar surface 100 km (about 60 mi) or so deep. The heating required to melt the surface is believed to have resulted from the impacts of rock fragments, which were leftover debris from the formation of the solar system that intensely bombarded the Moon. After a time, there were fewer fragments left to bombard the Moon, and the molten outer layer cooled and solidified to solid rock. The craters we see on the Moon today are the result of meteorites hitting the Moon between 3.9 and 4.2 billion years ago after the crust formed.

The *molten interior stage* involved the melting of the interior of the Moon. Radioactive decay had been slowly heating the interior, and 3.8 billion years ago, or about a billion years after the Moon formed, sufficient heat accumulated to melt the interior. The light and heavier rock materials separated during this period, perhaps producing a small iron core. Molten lava flowed into basins on the surface during this period, forming the smooth, darker maria seen today. The lava flooding continued for about 700 million years, ending about 3.1 billion years ago.

The *cold and quiet stage* began 3.1 billion years ago as the last lava flow cooled and solidified. Since that time, the surface of the Moon has been continually bombarded by micrometeorites and a few larger meteorites. With the exception of a few new craters, the surface of the Moon has changed little in the last 3 billion years.

Mars

Mars has always attracted attention because of its unique, bright reddish color. The properties and surface characteristics have also attracted attention, particularly since Mars seems to have similarities to Earth. It orbits the Sun at an average distance

A

B

FIGURE 13.12 (*A*) This image, made by the Hubble Space Telescope, clearly shows the large impact site made by fragment G of former comet Shoemaker-Levy 9 when it collided with Jupiter. (*B*) This is a picture of comet Shoemaker-Levy 9 after it broke into twenty-two pieces, lined up in this row, then proceeded to plummet into Jupiter during July 1994. The picture was made by the Hubble Space Telescope.

Unidentified flying objects (UFOs) are observed around the world. UFOs are generally sighted near small towns and out in the country, often near a military installation. Statistically, most sightings occur during the month of July at about 9:00 P.M., then at 3:00 A.M. UFOs can be grouped into three categories:

1. **Natural phenomena.** Most sightings of UFOs can be explained as natural phenomena. For example, a bright light that flashes red and green on the horizon could be a star viewed through atmospheric refraction. The vast majority of all UFO sightings are natural phenomena such as atmospheric refraction, ball lightning, swamp gas, or something simple such as a drifting weather balloon or military flares on parachutes, high in the atmosphere and drifting across the sky. This category of UFOs should also include exaggeration and fraud—such as balloons released with burning candles as a student prank.

2. **Aliens.** The idea that UFOs are alien spacecraft from other planets is very popular. However, there is no authentic, unambiguous evidence that would prove the existence of aliens. In fact, most unidentified flying objects are not objects at all. They are lights that can eventually be identified.

3. **Psychological factors.** This group of sightings includes misperceptions of natural phenomena resulting from unusual conditions that influence a person's perception and psychological health. This includes people who claim to receive information from aliens by "channeling" messages.

Questions to Discuss

Divide your group into three subgroups, with each subgroup selecting one of the three categories above. After preparing for a few minutes, have each group present reasons why we should understand UFOs to be natural phenomena, aliens from other planets, or psychological misperceptions. Then have the entire group discuss the three categories and try to come to a consensus.

(figure 13.12A) and then produced a once-in-a-lifetime spectacle as it proceeded to leave its imprint on Jupiter as well as people of Earth watching from the sidelines. The string of twenty-two comet fragments fell onto Jupiter during July 1994, creating a show eagerly photographed by viewers using telescopes around the world (figure 13.12B). The fragments impacted the upper atmosphere of Jupiter, producing visible, energetic fireballs. The aftereffects of these fireballs were visible for about a year. There are chains of craters on two of the Galilean moons that may have been formed by similar events.

Saturn

Saturn is slightly smaller and substantially less massive than Jupiter, and has similar features to Jupiter (see figure 13.9), but it is readily identified by its unique, beautiful system of rings. Saturn's rings consist of thousands of narrow bands of particles. Some rings are composed of particles large enough to be measured in meters and some have particles that are dust-sized (figure 13.13). Saturn is about 9.5 AU from the Sun, but its system of rings is easily spotted with a good pair of binoculars. Saturn also has the lowest average density of any of the planets, about 0.7 the density of water.

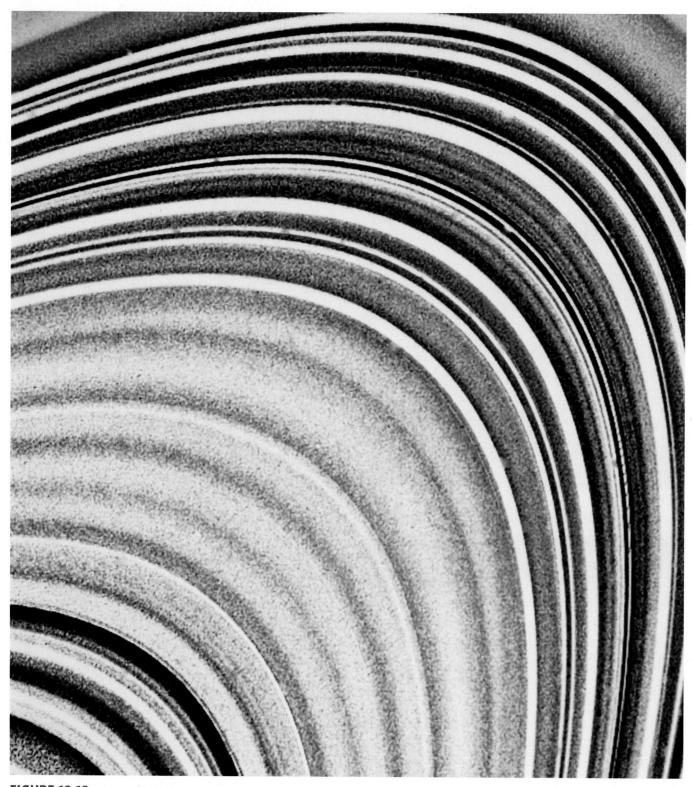

FIGURE 13.13 A part of Saturn's system of rings, pictured by *Voyager 2* from a distance of about 3 million km (about 2 million mi). More than sixty bright and dark ringlets are seen here; different colors indicate different surface compositions.

The surface of Saturn, like Jupiter's surface, has bright and dark bands that circle the planet parallel to the equator. Saturn also has a smaller version of Jupiter's Great Red Spot, but in general, the bands and spot are not as highly contrasted or brightly colored as they are on Jupiter. Saturn has thirty satel-lites, including Titan, the only moon in the solar system with a substantial atmosphere.

Titan is covered with clouds and impossible to observe. It is larger than the planet Mercury and is covered with a deep layer of reddish clouds. Titan's atmosphere is mostly nitrogen, with some hydrocarbons. This could be similar to Earth's atmosphere before life began adding oxygen to the atmosphere. The pressure on the surface is 1.5 atmospheres, but with a surface temperature of about −180°C (about −290°F), it is doubtful that life has developed on Titan.

The international *Cassini-Huygens* mission entered orbit around Saturn on July 1, 2004, after a 3.5 billion km (2.2 billion mi), seven-year voyage from Earth. Establishing orbit was the first step of a four-year study of Saturn and its rings and moons. So far, the mission has found that Saturn's largest moon, Titan, has a surface shaped by rock fracturing, winds, and erosion. Among the new discoveries is a 1,500 km (930 mi) long river of liquid methane. Titan's atmosphere is rich in organic (meaning carbon-containing) molecules. Most of the clouds on Titan occur over the south pole, and scientists believe this is where a cycle of methane rain, runoff with channel carving, and liquid methane evaporation is most active.

On January 14, 2005, *Cassini* delivered a detachable probe called *Huygens* to the moon Titan. The probe dropped by parachute, sampling the chemical composition of the atmosphere for the 2.5 hour descent and for 90 minutes after landing. It landed with more of a "splat" than a thud or splash, indicating that even though the surface temperature was –180°C, the surface was more like a thick organic stew than a solid or liquid.

Cassini has also discovered that Saturn's icy moon Enceladus has a significant atmosphere. Before this discovery, Titan was the only moon in the solar system known to have an atmosphere. Enceladus is too small to hold on to an atmosphere, so it must have a continuous source that may be volcanic, geysers, or gases escaping from the interior.

Cassini also provided a test of Einstein's general theory of relativity (p. 173). According to this theory, a massive object such as the Sun should cause spacetime to curve. This curve means that it should take longer for anything to travel by the Sun since a curved path is a greater distance. The general theory of relativity was tested by analyzing radio waves that traveled by the Sun on their way from *Cassini* to Earth. The researchers found that the radio waves traveling by the Sun were indeed delayed, and the amount of delay agreed precisely with predictions made from Einstein's theory.

The *Cassini-Huygens* mission is a cooperative project of NASA, the European Space Agency, and the Italian Space Agency.

Uranus and Neptune

Uranus and Neptune are two more giant planets that are far, far away from Earth. Uranus revolves around the Sun at an average distance of over 19 AU, taking about 84 years to circle the Sun once. Neptune is an average 30 AU from the Sun and takes about 165 years for one complete orbit. Uranus is about twice as far away from the Sun as Saturn, and Neptune is three times as far away. To give you an idea of these tremendous distances, consider that the time required for radio signals to travel from Uranus to Earth is more than 2.5 hours! It would be most difficult to carry on a conversation by radio with someone such a distance away. Even farther away, a radio signal from a

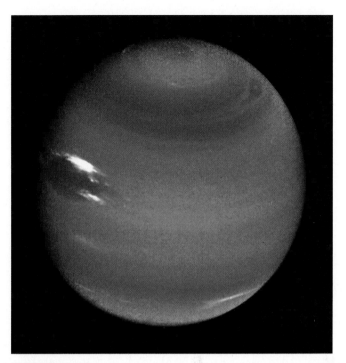

FIGURE 13.14 This is a photo image of Neptune taken by *Voyager 2*. Neptune has a turbulent atmosphere over a very cold surface of frozen hydrogen and helium.

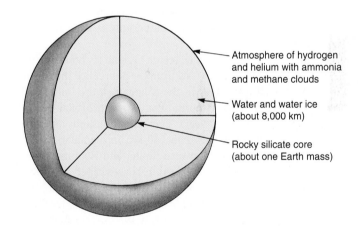

FIGURE 13.15 The interior structure of Uranus and Neptune.

transmitter near Neptune would require over 4 hours to reach Earth, which means 8 hours would be required for two people just to say "Hello" to each other!

Uranus and Neptune are more similar to each other than Saturn is to Jupiter (figure 13.14). Both are the smallest of the giant planets, with a diameter of about 50,000 km (about 30,000 mi) and about a third the mass of Jupiter. Both planets are thought to have similar interior structures (figure 13.15), which consist of water and water ice surrounding a rocky core with an atmosphere of hydrogen and helium. Because of their great distances from the Sun, both have very low average surface temperatures.

SMALL BODIES OF THE SOLAR SYSTEM

Comets, asteroids, and meteorites are the leftovers from the formation of the Sun and planets. Presently, the total mass of all these leftovers in and around the solar system may account for a significant fraction of the mass of the solar system, perhaps as much as two-thirds of the total mass. It must have been much greater in the past, however, as evidenced by the intense bombardment that took place on the Moon and other planets up to some four billion years ago.

Comets

A **comet** is known to be a relatively small, solid body of frozen water, carbon dioxide, ammonia, and methane, along with dusty and rocky bits of materials mixed in. Until the 1950s, most astronomers believed that comet bodies were mixtures of sand and gravel. Fred Whipple proposed what became known as the *dirty-snowball cometary model,* which was recently verified when spacecraft probes observed Halley's comet in 1986 (Table 13.4).

Based on calculation of their observed paths, comets are estimated to originate some 30 AU to a light-year or more from the Sun. Here, according to other calculations and estimates, is a region of space containing billions and billions of objects. There is a spherical "cloud" of the objects beyond the orbit of Pluto from about 30,000 AU out to a light-year or more from the Sun, called the **Oort cloud** (figure 13.16). The icy aggregates of the Oort cloud are understood to be the source of long-period comets, with orbital periods of more than two hundred years.

There is also a disk-shaped region of small icy bodies, which ranges from about 30 to 100 AU from the Sun, called the **Kuiper Belt.** The small icy bodies in the Kuiper Belt are understood to be the source of short-period comets, with orbital periods of less than two hundred years. There are thousands of Kuiper Belt objects that are larger than 100 km in diameter, and six are known to be orbiting between Jupiter and Neptune. Called *centaurs,* these objects are believed to have escaped the Kuiper Belt. Centaurs might be small, icy bodies similar to Pluto.

The current theory of the origin of comets was developed by the Dutch astronomer Jan Oort in 1950. According to the theory, the huge cloud and belt of icy, dusty aggregates are leftovers from the formation of the solar system and have been slowly circling the solar system ever since it formed. Something, perhaps a gravitational nudge from a passing star, moves one of the icy bodies enough that it is pulled toward the Sun in what will become an extremely elongated elliptical orbit. The icy, dusty body forms the only substantial part of a comet, and the body is called the comet *nucleus.*

Observations by the *Vega* and *Giotto* spacecrafts found the nucleus of Halley's comet to be an elongated mass of about 8 by 11 km (about 5 by 7 mi) with an overall density less than one-fourth that of solid water ice. As the comet nucleus moves toward the Sun, it warms from the increasing intense solar radiation. Somewhere between Jupiter and Mars, the ice and

TABLE 13.4   Completed Spacecraft Missions to Study Comets

Date	Name	Owner	Remark
Sep 11, 1985	ISSE 3 (or ICE)	U.S.	Studies of electric and magnetic fields around Giacobini-Zinner comet from 7,860 km (4,880 mi)
Mar 6, 1986	Vega 1	USSR	Photos and studies of nucleus of Halley's comet from 8,892 km (5,525 mi)
Mar 8, 1986	Suisei	Japan	Studied hydrogen halo of Halley's comet from 151,000 km (93,800 mi)
Mar 9, 1986	Vega 2	USSR	Photos and studies of nucleus of Halley's comet from 8,034 km (4,992 mi)
Mar 11, 1986	Sakigake	Japan	Studied solar wind in front of Halley's comet from 7.1 million km (4.4 million mi)
Mar 28, 1986	Giotto	ESA	Photos and studies of Halley's comet from 541 km (336 mi)
Mar 28, 1986	ISSE 3 (or ICE)	U.S.	Studies of electric and magnetic fields around Halley's comet from 32 million km (20 million mi)
Feb 17, 1996	NEAR Shoe-maker	U.S.	Studied the asteroid Eros
Oct 24, 1998	Deep Space 1	U.S.	Flyby of asteroid Braille and comet Borrelly
Feb 7, 1999	Stardust	U.S.	Studied comet Wild 2 and returned a sample of cosmic dust to Earth
Mar 2, 2004	Rosetta	ESA	Two probes will be launched into comet 67PChruyumov-Gerasimenko on Nov 2014
Jan 12, 2005	Deep Impact	U.S.	Studied comet Tempel 1 by sending impact probe on July 4, 2005
Jul 4, 2005	Deep Impact	U.S.	Projectile shot into comet Tempel 1
Sept 27, 2007	Dawn	U.S.	Study the two most massive asteroids Vesta (2011) and Ceres (2015)

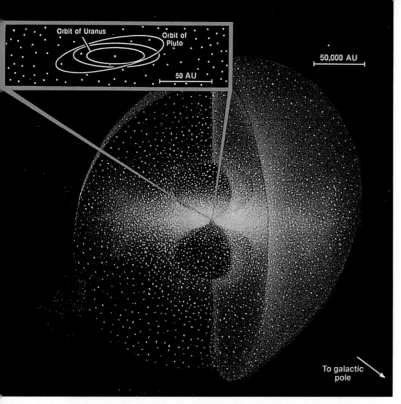

FIGURE 13.16 Although there may be as many as one trillion comets in the Oort cloud, the volume of space that the Oort cloud occupies is so immense that the comets are separated from one another by distances that are typically about 10 AU.

frozen gases begin to vaporize, releasing both grains of dust and evaporated ices. These materials form a large hazy head around the comet called a *coma*. The coma grows larger with increased vaporization, perhaps several hundred or thousands of kilometers across. The coma reflects sunlight as well as producing its own light, making it visible from Earth. The coma generally appears when a comet is within about 3 AU of the Sun. It reaches its maximum diameter about 1.5 AU from the Sun. The nucleus and coma together are called the *head* of the comet. In addition, a large cloud of invisible hydrogen gas surrounds the head, and this hydrogen *halo* may be hundreds of thousands of kilometers across.

As the comet nears the Sun, the solar wind and solar radiation ionize gases and push particles from the coma, pushing both into the familiar visible *tail* of the comet. Comets may have two types of tails: (1) ionized gases and (2) dust. The dust is pushed from the coma by the pressure from sunlight. It is visible because of reflected sunlight. The ionized gases are pushed into the tail by magnetic fields carried by the solar wind. The ionized gases of the tail are fluorescent, emitting visible light because they are excited by ultraviolet radiation from the Sun. The tail generally points away from the Sun, so it follows the comet as it approaches the Sun but leads the comet as it moves away from the Sun (figure 13.17).

Comets are not very massive or solid, and the porous, snow-like mass has a composition more similar to the giant planets than to the terrestrial planets in comparison. Each time a comet passes near the Sun, it loses some of its mass through evaporation of gases and loss of dust to the solar wind. After passing the Sun, the surface forms a thin, fragile crust covered with carbon and other dust particles. Each pass by the Sun means a loss of matter, and the coma and tail are dimmer with each succeeding pass. About 20 percent of the approximately six hundred comets that are known have orbits that return them to the Sun within a two-hundred-year period, some of which return as often as every five or ten years. The other 80 percent have long elliptical orbits that return them at intervals exceeding two hundred years. The famous Halley's comet has a smaller elliptical orbit and returns about every seventy-six years. Halley's comet, like all other comets, may eventually break up into a trail of gas and dust particles that orbit the Sun.

Asteroids

Between the orbits of Mars and Jupiter is a belt, or circular region of thousands of small rocky bodies called **asteroids** (figure 13.18). This belt contains thousands of *asteroids* that

FIGURE 13.17 As a comet nears the Sun it grows brighter, with the tail always pointing away from the Sun.

range in size from 1 km or less up to the largest asteroid, named Ceres, which has a diameter of about 1,000 km (over 600 mi). The asteroids are thinly distributed in the belt, 1 million km or so apart (about 600,000 mi), but there is evidence of collisions occurring in the past. Most asteroids larger than 50 km (about 30 mi) have been studied by analyzing the sunlight reflected from their surfaces. These spectra provide information about the composition of the asteroids. Asteroids on the inside of the belt are made of stony materials, and those on the outside of the belt are dark with carbon minerals. Still other asteroids are metallic, containing iron and nickel. These spectral composition studies, analyses of the orbits of asteroids, and studies of meteorites that have fallen to Earth all indicate that the asteroids are not the remains of a planet or planets that were broken up. The asteroids are now believed to have formed some 4.6 billion years ago from the original solar nebula. During their formation, or shortly thereafter, their interiors were partly melted, perhaps from the heat of short-lived radioactive decay reactions. Their location close to Jupiter, with its gigantic gravitational field, prevented the slow gravitational clumping-together process that would have formed a planet.

Jupiter's gigantic gravitational field also captured some of the asteroids, pulling them into its orbit. Today, there are two groups of asteroids, called the *Trojan asteroids,* which lead and follow Jupiter in its orbit. They lead and follow at a distance where the gravitational forces of Jupiter and the Sun balance to keep them in the orbit. A third group of asteroids, called the *Apollo asteroids,* has orbits that cross the orbit of Earth. It is possible that one of the Apollo asteroids could collide with Earth. One theory about what happened to the dinosaurs is based on evidence that such a collision indeed did occur some sixty-five million years ago. The chemical and physical properties of the two satellites of Mars, Phobos and Deimos, are more similar to the asteroids than to Mars. It is probable that the Martian satellites are captured asteroids.

Meteors and Meteorites

Comets leave trails of dust and rock particles after encountering the heat of the Sun, and collisions between asteroids in the past have ejected fragments of rock particles into space. In space, the remnants of comets and asteroids are called **meteoroids.** When a meteoroid encounters Earth moving through space, it accelerates toward the surface with a speed that depends on its direction of travel and the relative direction that Earth is moving. It soon begins to heat from air compression, melting into a visible trail of light and smoke. The streak of light and smoke in the sky is called a **meteor.** The "falling star" or "shooting star" is a meteor. Most meteors burn up or evaporate completely within seconds after reaching an altitude of about 100 km (about 60 mi) because they are nothing more than a speck of dust. A **meteor shower** occurs when Earth passes through a stream of particles left by a comet in its orbit. Earth might meet the stream of particles concentrated in such an orbit on a regular basis as it travels around the Sun, resulting in predictable meteor showers (Table 13.5). In the third

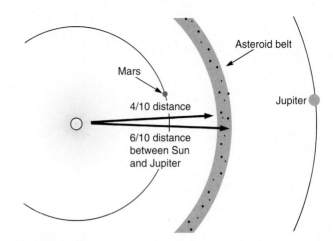

FIGURE 13.18   Most of the asteroids in the asteroid belt are about halfway between the Sun and Jupiter.

TABLE 13.5   Some Annual Meteor Showers

Name	Date of Maximum	Hour Rate
Quadrantid	January 3	30
Aquarid	May 4	5
Perseid	August 12	40
Orionid	October 22	15
Taurids	November 1, 16	5
Leonid	November 17	5
Geminid	December 12	55

week of October, for example, Earth crosses the orbital path of Halley's comet, resulting in a shower of some ten to fifteen meteors per hour. Meteor showers are named for the constellation in which they appear to originate. The October meteor shower resulting from an encounter with the orbit of Halley's comet, for example, is called the Orionid shower because it appears to come from the constellation Orion.

Did you know that atom-bomb-sized meteoroid explosions often occur high in Earth's atmosphere? Most smaller meteors melt into the familiar trail of light and smoke. Larger meteors may fragment upon entering the atmosphere, and the smaller fragments will melt into multiple light trails. Still larger meteors may actually explode at altitudes of about 32 km (about 20 mi) or so. Military satellites that watch Earth for signs of rockets blasting off or nuclear explosions record an average of eight meteor explosions a year. These are big explosions, with an energy equivalent estimated to be similar to a small nuclear bomb. Actual explosions, however, may be ten times larger than the estimation. Based on statistical data, scientists have estimated that every ten million years, Earth should be hit by a very, very large meteor. The catastrophic explosion and aftermath would devastate life over much of the planet, much like the theoretical dinosaur-killing impact of sixty-five million years ago.

If a meteoroid survives its fiery trip through the atmosphere to strike the surface of Earth, it is called a **meteorite.** Most meteors are from fragments of comets, but most meteorites generally come from particles that resulted from collisions between asteroids that occurred long ago. Meteorites are classified into three basic groups according to their composition: (1) *iron meteorites,* (2) *stony meteorites,* and (3) *stony-iron meteorites* (figure 13.19). The most common meteorites are stony, composed of the same minerals that make up rocks on Earth. The stony meteorites are further subdivided into two groups according to their structure, the *chondrites* and the *achondrites.* Chondrites have a structure of small spherical lumps of silicate minerals or glass, called *chondrules,* held together by a fine-grained cement. The achondrites do not have the chondrules, as their name implies, but have a homogeneous texture more like volcanic rocks such as basalt that cooled from molten rock.

The iron meteorites are about half as abundant as the stony meteorites. They consist of variable amounts of iron and nickel, with traces of other elements. In general, there is proportionally much more nickel than is found in the rocks of Earth. When cut, polished, and etched, beautiful crystal patterns are observed on the surface of the iron meteorite. The patterns mean that the iron was originally molten, then cooled very slowly over millions of years as the crystal patterns formed.

A meteorite is not, as is commonly believed, a ball of fire that burns up the landscape where it lands. The iron or rock has been in the deep freeze of space for some time, and it travels rapidly through Earth's atmosphere. The outer layers become hot enough to melt, but there is insufficient time for this heat to be conducted to the inside. Thus, a newly fallen iron meteorite will be hot since metals are good heat conductors, but it will not be hot enough to start a fire. A stone meteorite is a poor conductor of heat so it will be merely warm.

ORIGIN OF THE SOLAR SYSTEM

Any model of how the solar system originated presents a problem in testing or verification. This problem is that the solar system originated a long time ago, some five billion years ago, according to a number of different independent sources of evidence. Also, there are no other planetary systems that can be directly observed for comparison either in existence or in the process of being formed. At the distance they occur, even the Hubble Space Telescope would not be able to directly observe planets around their suns. Astronomers have identified hundreds of extrasolar planets. They identify the presence of planets by measuring the very slight wobble of a central star and then use the magnitude of this motion to determine the presence of orbiting planets, the size and shape of their orbits, and their mass. The technique works only for larger planets and cannot detect those much smaller than about half the mass of Saturn. The technique does not provide a

A

B

FIGURE 13.19 (*A*) A stony meteorite. The smooth, black surface was melted by friction with the atmosphere. (*B*) An iron meteorite that has been cut, polished, and etched with acid. The pattern indicates that the original material cooled from a molten material over millions of years.

visual image of the planets but only measures the gravitational effect of the planets on the star.

The basic idea behind how the solar system formed was put forth by both Immanuel Kant and Pierre Laplace back in the eighteenth century. Today this idea is called the **protoplanet nebular model.** A *protoplanet* is the earliest stage in the formation of a planet. The model can be considered in stages, which are not really a part of the model but are simply a convenient way to organize the total picture (figure 13.20).

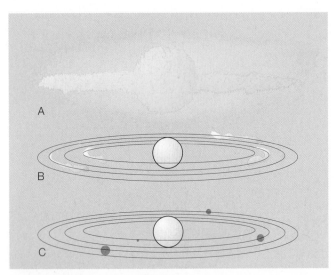

FIGURE 13.20 Formation of the solar system according to the protoplanet nebular model, not drawn to scale. (*A*) The process starts with a nebula of gas, dust, and chemical elements from previously existing stars. (*B*) The nebula is pulled together by gravity, collapsing into the protosun and protoplanets. (*C*) As the planets form, they revolve around the Sun in newly formed orbits.

Stage A

The first important event in the formation of our solar system involves stars that disappeared billions of years ago, long before the Sun was born. Earth, the other planets, and all the members of the solar system are composed of elements that were manufactured by these former stars. In a sequence of nuclear reactions, hydrogen fusion in the core of large stars results in the formation of the elements up to iron. Elements heavier than iron are formed in rare supernova explosions of dying massive stars. Thus, *stage A* of the formation of the solar system consisted of the formation of elements heavier than hydrogen in many, many previously existing stars, including the supernovas of more massive stars. Many stars had to live out their life cycles to provide the raw materials of the solar system. The death of each star, including supernovas, added newly formed elements to the accumulating gases and dust in interstellar space. Over a long period of time, these elements began to concentrate in one region of space as dust, gases, and chemical compounds, but hydrogen was still the most abundant element in the nebula that was destined to become the solar system.

Stage B

During *stage B,* the hydrogen gas, dust, elements, and chemical compounds from former stars began to form a large, slowly rotating nebula that was much, much larger than the present solar system. Under the influence of gravity, the large but diffuse, slowly rotating nebula began to contract, increasing its rate of spin. The largest mass pulled together in the center, contracting to the protostar, which eventually would become the Sun. The remaining gases, elements, and dusts formed an enormous, fat, bulging disk called an *accretion disk,* which would eventually form the planets and smaller bodies. The fragments of dust and other solid matter in the disk began to stick together in larger and larger accumulations from numerous collisions over the first million years or so. All of the present-day elements of the planets must have been present in the nebula along with the most abundant elements of hydrogen and helium. The elements and the familiar chemical compounds accumulated into basketball-sized or larger chunks of matter.

Did the planets have an icy slush beginning? Over a period of time, perhaps one-hundred million years or so, huge accumulations of frozen water, frozen ammonia, and frozen crystals of methane began to accumulate, together with silicon, aluminum, and iron oxide plus other metals in the form of rock and mineral grains. Such a slushy mixture would no doubt have been surrounded by an atmosphere of hydrogen, helium, and other vapors thinly interspersed with smaller rocky grains of dust. Local concentrations of certain minerals might have occurred throughout the whole accretion disk, with a greater concentration of iron, for example, in the disk where the protoplanet Mars was forming compared to where the protoplanet Earth was forming. Evidence for this part of the model is found in Mars today, with its greater abundance of iron. The abundant iron oxides are what color Mars the red planet.

All of the protoplanets might have started out somewhat similarly as huge accumulations of a slushy mixture with an atmosphere of hydrogen and helium gases. Gravitational attraction must have compressed the protoplanets as well as the protosun. During this period of contraction and heating, gravitational adjustments continue, and about a fifth of the disk nearest to the protosun must have been pulled into the central body of the protosun, leaving a larger accumulation of matter in the outer part of the accretion disk.

Stage C

During *stage C,* the warming protosun became established as a star, perhaps undergoing an initial flare-up that has been observed today in other newly forming stars. Such a flareup might have been of such a magnitude that it blasted away the hydrogen and helium atmospheres of the interior planets (Mercury, Venus, Earth, and Mars) out past Mars, but it did not reach far enough out to disturb the hydrogen and helium atmospheres of the outer planets. The innermost of the outer planets, Jupiter and Saturn, might have acquired some of the matter blasted away from the inner planets, becoming the giants of the solar system by comparison. This is just speculation, however, and the two giants may have simply formed from greater concentrations of matter in that part of the accretion disk.

The evidence, such as separation of heavy and light mineral matter, shows that the protoplanets underwent heating early in their formation. Much of the heating may have been provided by gravitational contraction, the same process that gave the protosun sufficient heat to begin its internal nuclear fusion reactions. Heat was also provided from radioactive decay processes inside the protoplanets, and the initial greater heating from the Sun may have played a role in the protoplanet heating process. Larger bodies were able to retain this heat better than smaller ones, which radiated it to space more readily. Thus, the

Percival Lowell (1855–1916)

Percival Lowell was an American astronomer, mathematician, and the founder of an important observatory in the United States whose main field of research was the planets of the solar system. Responsible for the popularization in his time of the theory of intelligent life on Mars, he also predicted the existence of a planet beyond Neptune that was later discovered and named Pluto. Today, Pluto is recognized as a dwarf planet.

Lowell was born in Boston, Massachusetts, on March 13, 1855. His interest in astronomy began to develop during his early school years. In 1876, he graduated from Harvard University, where he had concentrated on mathematics, and then traveled for a year before entering his father's cotton business. Six years later, Lowell left the business and went to Japan. He spent most of the next ten years traveling around the Far East, partly for pleasure, partly to serve business interests, but also holding a number of minor diplomatic posts.

Lowell returned to the United States in 1893 and soon afterward decided to concentrate on astronomy. He set up an observatory at Flagstaff, Arizona, at an altitude more than 2,000 m (6,564 ft) above sea level, on a site chosen for the clarity of its air and its favorable atmospheric conditions. He first used borrowed telescopes of 12- and 18-inch (30 and 45 cm) diameters to study Mars, which at that time was in a particularly suitable position. In 1896, he acquired a larger telescope and studied Mars by night and Mercury and Venus during the day. Overwork led to deterioration in Lowell's health, and from 1897 to 1901, he could do little research, although he was able to participate in an expedition to Tripoli in 1900 to study a solar eclipse.

He was made nonresident professor of astronomy at the Massachusetts Institute of Technology in 1902 and gave several lecture series in that capacity. He led an expedition to the Chilean Andes in 1907 that produced the first high-quality photographs of Mars. The author of many books and the holder of several honorary degrees, Lowell died in Flagstaff on November 12, 1916.

The planet Mars was a source of fascination for Lowell. Influenced strongly by the work of Giovanni Schiaparelli (1835–1910)—and possibly misled by the current English translation of "canals" for the Italian *canali* ("channels")—Lowell set up his observatory at Flagstaff originally with the sole intention of confirming the presence of advanced life forms on the planet. Thirteen years later, the expedition to South America was devoted to the study and photography of Mars. Lowell "observed" a complex and regular network of canals and believed that he detected regular seasonal variations that strongly indicated agricultural activity. He found darker waves that seemed to flow from the poles to the equator and suggested that the polar caps were made of frozen water. (The waves were later attributed to dust storms, and the polar caps are now known to consist not of ice but mainly of frozen carbon dioxide. Lowell's canal system also seems to have arisen mostly out of wishful thinking; part of the system does indeed exist, but it is not artificial and is apparent only because of the chance apposition of dark patches on the Martian surface.)

Lowell is remembered as a scientist of great patience and originality. He contributed to the advancement of astronomy through his observations and his establishment of a fine research center, and he did much to bring the excitement of the subject to the general public.

Source: Modified from the *Hutchinson Dictionary of Scientific Biography.* © RM, 2009. All rights reserved. Helicon Publishing is a division of RM.

larger bodies underwent a more thorough heating and melting, perhaps becoming completely molten early in their history. In the larger bodies, the heavier elements, such as iron, were pulled to the center of the now molten mass, leaving the lighter elements near the surface. The overall heating and cooling process took millions of years as the planets and smaller bodies were formed. Gases from the hot interiors formed secondary atmospheres of water vapor, carbon dioxide, and nitrogen on the larger interior planets.

Interestingly, the asteroid belt was discovered from a prediction made by the German astronomer Bode at the end of the eighteenth century. Bode had noticed a pattern of regularity in the spacing of the planets that were known at the time. He found that by expressing the distances of the planets from the Sun in astronomical units, these distances could be approximated by the relationship $(n + 4)/10$, where n is a number in the sequence 0, 3, 6, 12, and so on where each number (except the first) is doubled in succession. When these calculations were done, the distances turned out to be very close to the distances of all the planets known at that time, but the numbers also predicted a planet between Mars and Jupiter where there

Myths, Mistakes, and Misunderstandings

Solar Mistakes

The following are common misunderstandings about the Sun:

- The Sun is not a star; it is a unique object of our solar system.
- The Sun is solid.
- The Sun does not rotate.
- Sunlight is from burning gas on the Sun.
- The Sun will last forever.

was none. Later, a belt of asteroids was found where the Bode numbers predicted there should be a planet. This suggested to some people that a planet had existed between Mars and Jupiter in the past and somehow this planet was broken into pieces, perhaps by a collision with another large body (Table 13.6).

Such patterns of apparent regularity in the spacing of planetary orbits were of great interest because, if a true pattern existed, it could hold meaning about the mechanism that determined the location of planets at various distances from the Sun. Many attempts have been made to explain the mechanism of planetary spacing and why a belt of asteroids exists where the Bode numbers predict there should be a planet. The most successful explanations concern Jupiter and the influence of its gigantic gravitational field on the formation of clumps of matter at certain distances from the Sun. In other words, a planet does not exist today between Mars and Jupiter because there never was a planet there. The gravitational influence of Jupiter prevented the clumps of matter from joining together to form a planet, and a belt of asteroids formed instead.

TABLE 13.6   Distances from the Sun to Planets Known in the 1790s

Planet	n	Distance Predicted by $(n + 4)/10$ (AU)	Actual Distance (AU)
Mercury	0	0.4	0.39
Venus	3	0.7	0.72
Earth	6	1.0	1.0
Mars	12	1.6	1.5
(Asteroid belt)	24	2.8	—
Jupiter	48	5.2	5.2
Saturn	96	10.0	9.5
Uranus	192	19.6	19.2

SUMMARY

The planets can be classified into two major groups: (1) the *terrestrial planets* of Mercury, Venus, Mars, and Earth and (2) the *giant planets* of Jupiter, Saturn, Uranus, and Neptune.

Comets are porous aggregates of water ice, frozen methane, frozen ammonia, dry ice, and dust. The solar system is surrounded by the *Kuiper Belt* and the *Oort cloud* of these objects. Something nudges one of the icy bodies and it falls into a long elliptical orbit around the Sun. As it approaches the Sun, increased radiation evaporates ices and pushes ions and dust into a long visible tail. *Asteroids* are rocky or metallic bodies that are mostly located in a belt between Mars and Jupiter. The remnants of comets, fragments of asteroids, and dust are called *meteoroids*. A meteoroid that falls through Earth's atmosphere and melts to a visible trail of light and smoke is called a *meteor*. A meteoroid that survives the trip through the atmosphere to strike the surface of Earth is called a *meteorite*. Most meteors are fragments and pieces of dust from comets. Most meteorites are fragments that resulted from collisions between asteroids.

The *protoplanet nebular model* is the most widely accepted theory of the origin of the solar system, and this theory can be considered as a series of events, or stages. *Stage A* is the creation of all the elements heavier than hydrogen in previously existing stars. *Stage B* is the formation of a nebula from the raw materials created in stage A. The nebula contracts from gravitational attraction, forming the *protosun* in the center with a fat, bulging *accretion disk* around it. The Sun will form from the protosun, and the planets will form in the accretion disk. *Stage C* begins as the protosun becomes established as a star. The icy remains of the original nebula are the birthplace of *comets*. *Asteroids* are other remains that did undergo some melting.

KEY TERMS

asteroids (p. **299**)
astronomical unit (p. **285**)
comet (p. **298**)
dwarf planet (p. **284**)

giant planets (p. **285**)
Kuiper Belt (p. **298**)
meteor (p. **300**)
meteorite (p. **301**)

meteoroids (p. **300**)
meteor shower (p. **300**)
Oort cloud (p. **298**)
planet (p. **284**)

protoplanet nebular model (p. **301**)
small solar system bodies
 (p. **284**)
terrestrial planets (p. **285**)

APPLYING THE CONCEPTS

1. Earth, other planets, and all the members of the solar system
   a. have always existed.
   b. formed thousands of years ago from elements that have always existed.
   c. formed millions of years ago, when the elements and each body were created at the same time.
   d. formed billions of years ago from elements that were created in many previously existing stars.

2. The belt of asteroids between Mars and Jupiter is probably
   a. the remains of a planet that exploded.
   b. clumps of matter that condensed from the accretion disk but never got together as a planet.
   c. the remains of two planets that collided.
   d. the remains of a planet that collided with an asteroid or comet.

The Celestial Sphere

To locate the ecliptic, planets, or anything else in the sky, you need something to refer to, a referent system. A referent system is easily established by first imagining the sky to be a celestial sphere just as the ancient Greeks did. A coordinate system of lines can be visualized on this sphere. Imagine that you could inflate Earth until its surface touched the celestial sphere. If you now transfer lines to the celestial sphere, you will have a system of sky coordinates. From the surface of Earth, you can see that the *celestial equator* is a line on the celestial sphere directly above Earth's equator and the *north celestial pole* is a point directly above the North Pole of Earth. Likewise, the *south celestial pole* is a point directly above the South Pole of Earth.

You can only see half of the overall celestial sphere from any one place on the surface of Earth. Imagine a point on the celestial sphere directly above where you are located. An imaginary line that passes through this point, then passes north through the north celestial pole, continuing all the way around through the south celestial pole and back to the point directly above you makes a big circle called the *celestial meridian* (box figure 14.1). Note that the celestial meridian location is determined by where you are on Earth.

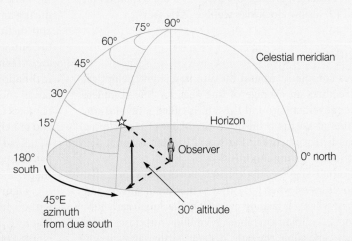

Box Figure 14.1 Once you have established the celestial equator, the celestial poles, and the celestial meridian, you can use a two-coordinate horizon system to locate positions in the sky. One popular method of using this system identifies the altitude angle (in degrees) from the horizon up to an object on the celestial sphere and the azimuth angle (again in degrees) the object on the celestial sphere is east or west of due south, where the celestial meridian meets the horizon. The illustration shows an altitude of 30° and an azimuth of 45° east of due south.

The celestial equator and the celestial poles, on the other hand, are always in the same place no matter where you are.

Overall, the celestial sphere appears to spin, turning on an axis through the celestial poles. A photograph made by pointing a camera at the north celestial pole and leaving the shutter open for several hours will show the apparent motion of the celestial sphere with star trails (see figure 12.1A).

The moderately bright star near the center is the North Star, Polaris. Polaris is almost, but not exactly, at the north celestial pole. If you observe the celestial sphere night after night, you will see that the stars maintain their positions relative to one another as they turn counterclockwise around Polaris. Those near Polaris pivot around it and are called "circumpolar." Those farther out rise in the east, move in an arc, then set in the west.

Thus, Earth is very close to, but not exactly, an oblate spheroid. The significance of this shape will become apparent when Earth's motions are discussed next (figure 14.4).

MOTIONS OF EARTH

Ancient civilizations had a fairly accurate understanding of the size and shape of Earth but had difficulty accepting the idea that Earth moves for at least two reasons: (1) they could not sense any motion of Earth, and (2) they had ideas about being at the center of a universe that was created for them. It was not until the 1700s that the concept of an Earth in motion became generally accepted. Today, Earth is understood to move a number of different ways, seven of which were identified in

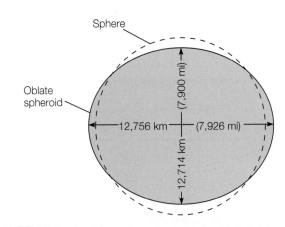

FIGURE 14.4 Earth has an irregular, slightly lopsided, slightly pear-shaped form. In general, it is considered to have the shape of an oblate spheroid, departing from a perfect sphere as shown here.

the introduction to this chapter. Three of these motions are independent of motions of the Sun and the galaxy. These are (1) a yearly revolution around the Sun, (2) a daily rotation on its axis, and (3) a slow clockwise wobble of its axis.

Revolution

Earth moves constantly around the Sun in a slightly elliptical orbit that requires an average of one year for one complete circuit. The movement around the Sun is called a **revolution,** and all points of Earth's orbit lie in the plane of the ecliptic. The average distance between Earth and the Sun is about 150 million km (about 93 million mi).

Earth's orbit is slightly elliptical, so it moves with a speed that varies. It moves fastest when it is closer to the Sun in January and moves slowest when it is farthest away from the Sun in early July. Earth is about 2.5 million km (about 1.5 million mi) closer to the Sun in January and about the same distance farther away in July than it would be if the orbit were a circle. This total difference of about 5 million km (about 3 million mi) results in a January Sun with an apparent diameter that is 3 percent larger than the July Sun, and Earth as a whole receives about 6 percent more solar energy in January. The effect of being closer to the Sun is much less than the effect of some other relationships, and winter occurs in the Northern Hemisphere when Earth is closest to the Sun. Likewise, summer occurs in the Northern Hemisphere when the Sun is at its greatest distance from Earth (figure 14.5).

The important directional relationships that override the effect of Earth's distance from the Sun involve the daily **rotation,** or spinning, of Earth around an imaginary line through the geographic poles called Earth's *axis.* The important directional relationships are a constant inclination of Earth's axis to the plane of the ecliptic and a constant orientation of the axis to the stars. The *inclination of Earth's axis* to the plane of the ecliptic is about 66.5° (or 23.5° from a line perpendicular to the plane). This relationship between the plane of Earth's orbit and the tilt of its axis is considered to be the same day after day throughout the year, even though small changes do occur in the inclination over time. Likewise, the *orientation of Earth's axis* to the stars is considered to be the same throughout the year as

Earth moves through its orbit. Again, small changes do occur in the orientation over time. Thus, in general, the axis points in the same direction, remaining essentially parallel to its position during any day of the year. The essentially constant orientation and inclination of the axis result in the axis pointing toward the Sun as Earth moves in one part of its orbit, then pointing away from the Sun six months later. The generally constant inclination and orientation of the axis, together with Earth's rotation and revolution, combine to produce three related effects: (1) days and nights that vary in length, (2) changing seasons, and (3) climates that vary with latitude.

Read about Milankovitch cycles on page 406 to learn how patterns of changes in the orientation of Earth's tilt, orbital shape, and axis wobble affect Earth's climate.

Figure 14.5 shows how the North Pole points toward the Sun on June 21 or 22, then away from the Sun on December 22 or 23 as it maintains its orientation to the stars. When the North Pole is pointed toward the Sun, it receives sunlight for a full twenty-four hours, and the South Pole is in Earth's shadow for a full twenty-four hours. This is summer in the Northern Hemisphere, with the longest daylight periods and the Sun at its maximum noon height in the sky. Six months later, on December 22 or 23, the orientation is reversed with winter in the Northern Hemisphere, the shortest daylight periods, and the Sun at its lowest noon height in the sky.

The beginning of a season can be recognized from any one of the three related observations: (1) the length of the daylight period, (2) the altitude of the Sun in the sky at noon, or (3) the length of a shadow from a vertical stick at noon. All of these observations vary with changes in the direction of Earth's axis of rotation relative to the Sun. On about June 22 and December 22, the Sun reaches its highest and lowest noon altitudes as Earth moves to point the North Pole directly toward the Sun (June 21 or 22) and directly away from the Sun (December 22 or 23). Thus, the Sun appears to stop increasing or decreasing its altitude in the sky, stop, then reverse its movement twice a year. These times are known as **solstices** after the Latin meaning "Sun stand still." The Northern Hemisphere's **summer solstice** occurs on about June 22 and identifies the beginning of the summer season. At the summer solstice, the Sun at noon has the highest altitude, and the shadow from a vertical stick is shorter than on any other day of the year. The Northern Hemisphere's **winter solstice** occurs on about December 22 and identifies the beginning of the winter season. At the winter solstice, the Sun at noon has the lowest altitude, and the shadow from a vertical stick is longer than on any other day of the year (figure 14.6).

As Earth moves in its orbit between pointing its North Pole toward the Sun on about June 22 and pointing it away on about December 22, there are two times when it is halfway between. At these times, Earth's axis is perpendicular to a line between the center of the Sun and Earth, and daylight and night are of equal length. These are called the **equinoxes** after the Latin meaning "equal nights." The **spring equinox** (also called the **vernal equinox**) occurs on about March 21 and identifies the beginning of the spring season. The **autumnal equinox** occurs on about September 23 and identifies the beginning of the fall season.

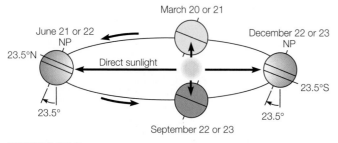

FIGURE 14.5 The consistent tilt and orientation of Earth's axis as it moves around its orbit is the cause of the seasons. The North Pole is pointing toward the Sun during the summer solstice and away from the Sun during the winter solstice.

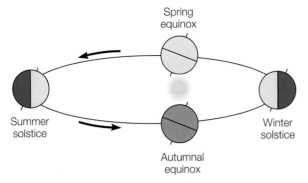

FIGURE 14.6 The length of daylight during each season is determined by the relationship of Earth's shadow to the tilt of the axis. At the equinoxes, the shadow is perpendicular to the latitudes, and day and night are of equal length everywhere. At the summer solstice, the North Pole points toward the Sun and is completely out of the shadow for a twenty-four-hour day. At the winter solstice, the North Pole is in the shadow for a twenty-four-hour night. The situation is reversed for the South Pole.

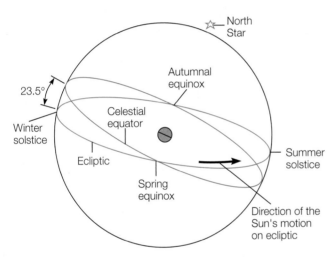

FIGURE 14.7 The position of the Sun on the celestial sphere at the solstices and the equinoxes.

The relationship between the apparent path of the Sun on the celestial sphere and the seasons is shown in figure 14.7. The celestial equator is a line on the celestial sphere directly above Earth's equator. The equinoxes are the points on the celestial sphere where the ecliptic, the path of the Sun, crosses the celestial equator. Note also that the summer solstice occurs when the ecliptic is 23.5° north of the celestial equator, and the winter solstice occurs when it is 23.5° south of the celestial equator.

Rotation

Observing the apparent turning of the celestial sphere once a day and seeing the east-to-west movement of the Sun, Moon, and stars, it certainly seems as if it is the heavenly bodies and not Earth doing the moving. You cannot sense any movement,

and there is little apparent evidence that Earth indeed moves. Evidence of a moving Earth comes from at least three different observations: (1) the observation that the other planets and the Sun rotate, (2) the observation of the changing plane of a long, heavy pendulum at different latitudes on Earth, and (3) the observation of the direction of travel of something moving across, but above, Earth's surface, such as a rocket.

Other planets, such as Jupiter, and the Sun can be observed to rotate by keeping track of features on the surface such as the Great Red Spot on Jupiter and sunspots on the Sun. While such observations are not direct evidence that Earth also rotates, they do show that other members of the solar system spin on their axes. As described earlier, Jupiter is also observed to be oblate, flattened at its poles with an equatorial bulge. Since Earth is also oblate, this is again indirect evidence that it rotates, too.

The most easily obtained and convincing evidence about Earth's rotation comes from a *Foucault pendulum,* a heavy mass swinging from a long wire. This pendulum is named after the French physicist Jean Foucault, who first used a long pendulum in 1851 to prove that Earth rotates. Foucault started a long, heavy pendulum moving just above the floor, marking the plane of its back-and-forth movement. Over some period of time, the pendulum appeared to slowly change its position, smoothly shifting its plane of rotation. Science museums often show this shifting plane of movement by setting up small objects for the pendulum to knock down. Foucault demonstrated that the pendulum actually maintains its plane of movement in space (inertia) while Earth rotates eastward (counterclockwise) under the pendulum. It is Earth that turns under the pendulum, causing the pendulum to appear to change its plane of rotation. It is difficult to imagine the pendulum continuing to move in a fixed direction in space while Earth, and everyone on it, turns under the swinging pendulum.

Figure 14.8 illustrates the concept of the Foucault pendulum. A pendulum is attached to a support on a stool that is free to rotate. If the stool is slowly turned while the pendulum is swinging, you will observe that the pendulum maintains its plane of rotation while the stool turns under it. If you were much smaller and looking from below the pendulum, it would appear to turn as you rotate with the turning stool. This is what happens on Earth. Such a pendulum at the North Pole would make a complete turn in about twenty-four hours. Moving south from the North Pole, the change decreases with latitude

FIGURE 14.8 The Foucault pendulum swings back and forth in the same plane while a stool is turned beneath it. Likewise, a Foucault pendulum on Earth's surface swings back and forth in the same plane while Earth turns beneath it. The amount of turning observed depends on the latitude of the pendulum.

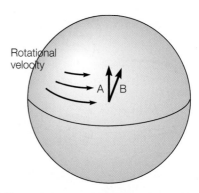

FIGURE 14.9 Earth has a greater rotational velocity at the equator and less toward the poles. As an object moves north or south (A), it passes over land with a different rotational velocity, which produces a deviation to the right in the Northern Hemisphere (B) and to the left in the Southern Hemisphere.

until, at the equator, the pendulum would not appear to turn at all. At higher latitudes, the plane of the pendulum appears to move clockwise in the Northern Hemisphere and counterclockwise in the Southern Hemisphere.

More evidence that Earth rotates is provided by objects that move above and across Earth's surface. As shown in figure 14.9, Earth has a greater rotational velocity at the equator than at the poles. As an object leaves the surface and moves north or south, the surface has a different rotational velocity, so it rotates beneath the object as it proceeds in a straight line. This gives the moving object an apparent deflection to the right of the direction of movement in the Northern Hemisphere and to the left in the Southern Hemisphere. The apparent deflection caused by Earth's rotation is called the **Coriolis effect.** The Coriolis effect will explain Earth's prevailing wind systems as well as the characteristic direction of wind in areas of high pressure and areas of low pressure (see chapter 17).

Precession

If Earth were a perfect spherically shaped ball, its axis would always point to the same reference point among the stars. The reaction of Earth to the gravitational pull of the Moon and the Sun on its equatorial bulge, however, results in a slow wobbling of Earth as it turns on its axis. This slow wobble of Earth's axis, called **precession,** causes it to swing in a slow circle like the wobble of a spinning top (figure 14.10). It takes Earth's axis about twenty-six thousand years to complete one turn, or wobble. Today, the axis points very close to the North Star, Polaris, but is slowly moving away to point to another star. In about twelve thousand years, the star Vega will appear to be in the position above the North Pole, and Vega will be the new North Star. The moving pole also causes changes over time in which particular signs of the zodiac appear with the spring equinox. Because of precession, the occurrence of the spring equinox has been moving backward (westward) through the zodiac constellations at about 1 degree every seventy-two years. Thus, after about twenty-six thousand years, the spring

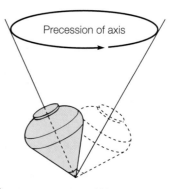

FIGURE 14.10 A spinning top wobbles as it spins, and the axis of the top traces out a small circle. The wobbling of the axis is called *precession*.

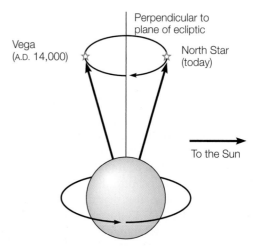

FIGURE 14.11 The slow, continuous precession of Earth's axis results in the North Pole pointing around a small circle over a period of about twenty-six thousand years.

equinox will have moved through all the constellations and will again approach the constellation of Aquarius for the next "age of Aquarius" (figure 14.11).

PLACE AND TIME

The continuous rotation and revolution of Earth establishes an objective way to determine direction, location, and time on Earth. If Earth were an unmoving sphere, there would be no side, end, or point to provide a referent for direction and location. Earth's rotation, however, defines an axis of rotation, which serves as a reference point for determination of direction and location on the entire surface. Earth's rotation and revolution together define cycles, which define standards of time. The following describes how Earth's movements are used to identify both place and time.

Identifying Place

A system of two straight lines can be used to identify a point, or position, on a flat, two-dimensional surface. The position of the letter X on this page, for example, can be identified by making a line a certain number of measurement units from the top of the page and a second line a certain number of measurement units from the left side of the page. Where the two lines intersect will identify the position of the letter X, which can be recorded or communicated to another person (figure 14.12).

A system of two straight lines can also be used to identify a point, or position, on a sphere, except this time the lines are circles. The reference point for a sphere is not as simple as in the flat, two-dimensional case, however, since a sphere does not have a top or side edge. Earth's axis provides the north-south reference point. The equator is a big circle around Earth that is exactly halfway between the two ends, or poles, of the rotational axis. An infinite number of circles are imagined to run around Earth parallel to the equator as shown in figure 14.13. The east- and west-running parallel circles are called **parallels.** Each parallel is the same distance between the equator and one of the poles all the way around Earth. The distance from the equator to a point on a parallel is called the **latitude** of that point. Latitude tells you how far north or south a point is from the equator by telling you the parallel the point is located on. The distance is measured northward from the equator (which is 0°) to the North Pole (90° north) or southward from the equator (0°) to the South Pole (90° south) (figure 14.14). If you are somewhere at a latitude of 35° north, you are somewhere on Earth on the 35° latitude line north of the equator.

Since a parallel is a circle, a location of 40°N latitude could be anyplace on that circle around Earth. To identify a location, you need another line, this time one that runs pole to pole and perpendicular to the parallels. These north-south running arcs that intersect at both poles are called **meridians** (figure 14.15). There is no naturally occurring, identifiable meridian that can be used as a point of reference such as the equator serves for parallels, so one is identified as the referent by international agreement. The referent meridian is the one that passes through the Greenwich Observatory near London, England, and this meridian is called the **prime meridian.** The distance from the prime meridian east or west is called the **longitude.** The degrees of longitude of a point on a parallel are measured to the east or to the west from

FIGURE 14.12 Any location on a flat, two-dimensional surface is easily identified with two references from two edges. This technique does not work on a motionless sphere because there are no reference points.

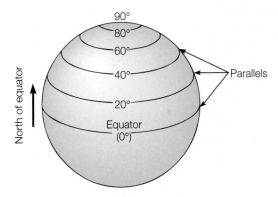

FIGURE 14.13 A circle that is parallel to the equator is used to specify a position north or south of the equator. A few of the possibilities are illustrated here.

the prime meridian up to 180° (figure 14.16). New Orleans, Louisiana, for example, has a latitude of about 30°N of the equator and a longitude of about 90°W of the prime meridian. The location of New Orleans is therefore described as 30°N, 90°W.

Locations identified with degrees of latitude north or south of the equator and degrees of longitude east or west of the prime meridian are more precisely identified by dividing each degree of latitude into subdivisions of 60 minutes (60') per degree, and each minute into 60 seconds (60"). On the other hand, latitudes near the equator are sometimes referred to in general as the *low latitudes,* and those near the poles are sometimes called the *high latitudes.*

In addition to the equator (0°) and the poles (90°), the parallels of 23.5°N and 23.5°S from the equator are important references for climatic consideration. The parallel of 23.5°N is called the **tropic of Cancer,** and 23.5°S is called the **tropic of Capricorn.** These two parallels identify the limits toward the poles within which the Sun appears directly overhead during the course of a year. The parallel of 66.5°N is called the **Arctic Circle,** and the parallel of 66.5°S is called the **Antarctic Circle.** These two parallels identify the limits toward the equator within which the Sun appears above the horizon all day during the summer (figure 14.17). This starts with six months of

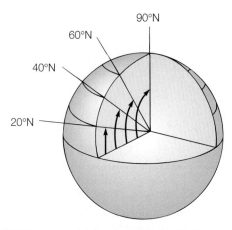

FIGURE 14.14 If you could see to Earth's center, you would see that latitudes run from 0° at the equator north to 90° at the North Pole (or to 90° south at the South Pole).

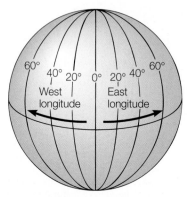

FIGURE 14.15 Meridians run pole to pole perpendicular to the parallels and provide a reference for specifying east and west directions.

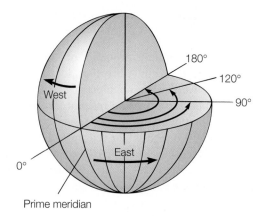

FIGURE 14.16 If you could see inside Earth, you would see 360° around the equator and 180° of longitude east and west of the prime meridian.

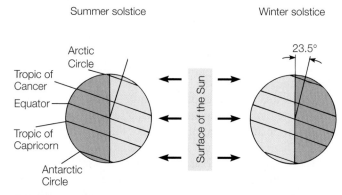

FIGURE 14.17 At the summer solstice, the noon Sun appears directly overhead at the tropic of Cancer (23.5°N) and twenty-four hours of daylight occurs north of the Arctic Circle (66.5°N). At the winter solstice, the noon Sun appears overhead at the tropic of Capricorn (23.5°S) and twenty-four hours of daylight occurs south of the Antarctic Circle (66.5°S).

daylight every day at the pole, then decreases as you get fewer days of full light until reaching the limit of one day of twenty-four-hour daylight at the 66.5° limit.

Measuring Time

Standards of time are determined by intervals between two successive events that repeat themselves in a regular way. Since ancient civilizations, many of the repeating events used to mark time have been recurring cycles associated with the rotation of Earth on its axis and its revolution around the Sun. Thus, the day, month, season, and year are all measures of time based on recurring natural motions of Earth. All other measures of time are based on other events or definitions of events. There are, however, several different ways to describe the day, month, and year, and each depends on a different set of events. These events are described in the following section.

Daily Time

The technique of using astronomical motions for keeping time originated some four thousand years ago with the Babylonian culture. The Babylonians marked the yearly journey of the Sun against the background of the stars, which was divided into twelve periods, or months, after the signs of the zodiac. Based on this system, the Babylonian year was divided into twelve months with a total of 360 days. In addition, the Babylonians invented the week and divided the day into hours, minutes, and seconds. The week was identified as a group of seven days, each based on one of the seven heavenly bodies that were known at the time. The hours, minutes, and seconds of a day were determined from the movement of the shadow around a straight, vertical rod.

As seen from a place in space above the North Pole, Earth rotates counterclockwise turning toward the east. On Earth, this motion causes the Sun to appear to rise in the east, travel across the sky, and set in the west. The changing angle between the tilt of Earth's axis and the Sun produces an apparent shift of the Sun's path across the sky, northward in the summer season and southward in the winter season. The apparent movement of the Sun across the sky was the basis for the ancient as well as the modern standard of time known as the day.

Today, everyone knows that Earth turns as it moves around the Sun, but it is often convenient to regard space and astronomical motions as the ancient Greeks did, as a celestial sphere that turns around a motionless Earth. Recall that the celestial meridian is a great circle on the celestial sphere that passes directly overhead where you are and continues around Earth through both celestial poles. The movement of the Sun across the celestial meridian identifies an event of time called **noon.** As the Sun appears to travel west, it crosses meridians that are farther and farther west, so the instant identified as noon moves west with the Sun. The instant of noon at any particular longitude is called the **apparent local noon** for that longitude because it identifies noon from the apparent position of the Sun in the sky. The morning hours before the Sun crosses the meridian are identified as *ante meridiem* (A.M.) hours, which is Latin for "before meridian." Afternoon hours are identified as *post meridiem* (P.M.) hours, which is Latin for "after the meridian."

There are several ways to measure the movement of the Sun across the sky. The ancient Babylonians, for example, used a vertical rod called a *gnomon* to make and measure a shadow that moved as a result of the apparent changes of the Sun's position. The gnomon eventually evolved into a *sundial,* a vertical or slanted gnomon with divisions of time marked on a horizontal plate beneath the gnomon. The shadow from the gnomon indicates the **apparent local solar time** at a given place and a given instant from the apparent position of the Sun in the sky. If you have ever read the time from a sundial, you know that it usually does not show the same time as a clock or a watch (figure 14.18). In addition, sundial time is nonuniform, fluctuating throughout the course of a year, sometimes running ahead of clock time and sometimes running behind clock time.

A sundial shows the apparent local solar time, but clocks are set to measure a uniform standard time based on **mean solar time.** Mean solar time is a uniform time averaged from the apparent solar time. The apparent solar time is nonuniform, fluctuating because (1) Earth moves sometimes faster and sometimes slower in its elliptical orbit around the Sun and (2) the equator of Earth is inclined to the ecliptic. The combined consequence of these two effects is a variable, nonuniform sundial time as compared to the uniform mean solar time, otherwise known as clock time.

A day is defined as the length of time required for Earth to rotate once on its axis. There are different ways to measure this rotation, however, which result in different definitions of the day. A **sidereal day** is the interval between two consecutive crossings of the celestial meridian by a particular star (*sidereal* means "star"). This interval of time depends only on the time Earth takes to rotate 360° on its axis. One sidereal day is practically the same length as any other sidereal day because Earth's rate of rotation is constant for all practical purposes.

An **apparent solar day** is the interval between two consecutive crossings of the celestial meridian by the Sun, for example, from one local solar noon to the next solar noon. Since Earth is moving in orbit around the Sun, it must turn a little bit farther to compensate for its orbital movement, bringing the Sun back to local solar noon (figure 14.19). As a consequence, the apparent solar day is about four minutes longer than the sidereal day. This additional time accounts for the observation that the stars and constellations of the zodiac rise about four minutes earlier every night, appearing higher in the sky at the

FIGURE 14.18 A sundial indicates the apparent local solar time at a given instant in a given location. The time read from a sundial, which is usually different from the time read from a clock, is based on an average solar time.

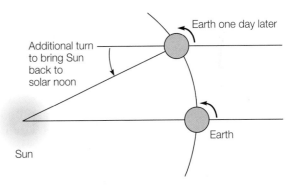

FIGURE 14.19 Because Earth is moving in orbit around the Sun, it must rotate an additional distance each day, requiring about four minutes to bring the Sun back across the celestial meridian (local solar noon). This explains why the stars and constellations rise about four minutes earlier every night.

same clock time until they complete a yearly cycle. A sidereal day is twenty-three hours, fifty-six minutes, and four seconds long. A **mean solar day** is twenty-four hours long, averaged from the mean solar time to keep clocks in closer step with the Sun than would be possible using the variable apparent solar day. Just how out of synchronization the apparent solar day can become with a clock can be illustrated with another ancient way of keeping track of the Sun's motions in the sky, the "hole in the wall" sun calendar and clock.

Variations of the "hole in the wall" sun calendar were used all over the world by many different ancient civilizations, including the early Native Americans of the American Southwest. More than one ancient Native American ruin has small holes in the western wall aligned in such a way as to permit sunlight to enter a chamber only on the longest and shortest days of the year. This established a basis for identifying the turning points in the yearly cycle of seasons.

A hole in the roof can be used as a sun clock, but it will require a whole year to establish the meaning of a beam of sunlight shining on the floor. Imagine a beam of sunlight passing through a small hole to make a small spot of light on the floor. For a year, you mark the position of the spot of light on the floor *each day* when your clock tells you the *mean solar time is noon*. You trace out an elongated, lopsided figure eight with the small end pointing south and the larger end pointing north (figure 14.20A). Note by following the monthly markings shown in figure 14.20B that the figure-eight shape is actually traced out by the spot of sunlight making two **S** shapes as the Sun changes its apparent position in the sky. Together, the two **S** shapes make the shape of the figure eight.

Why did the sunbeam trace out a figure eight over a year? The two extreme north-south positions of the figure are easy to understand because by December, Earth is in its orbit with the North Pole tilted away from the Sun. At this time, the direct rays of the Sun fall on the tropic of Capricorn (23.5° south of the equator), and the Sun appears low in the sky as seen from the Northern Hemisphere. Thus, on this date, the winter solstice, a beam of sunlight strikes the floor at its northernmost position beneath the hole. By June, Earth has moved halfway around its orbit, and the North Pole is now tilted toward the Sun. The direct rays of the Sun now fall on the tropic of Cancer (23.5° north of the equator), and the Sun appears high in the sky as seen from the Northern Hemisphere (figure 14.21). Thus, on this date, the summer solstice, a beam of sunlight strikes the floor at its southernmost position beneath the hole.

If everything else were constant, the path of the spot would trace out a straight line between the northernmost and southernmost positions beneath the hole. The east and west movements of the point of light as it makes an **S** shape on the floor must mean, however, that the Sun crosses the celestial meridian (noon) earlier one part of the year and later the other part. This

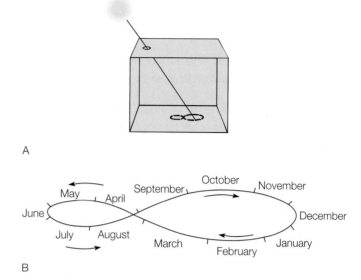

A

B

FIGURE 14.20 (*A*) During a year, a beam of sunlight traces out a lopsided figure eight on the floor if the position of the light is marked at noon every day. (*B*) The location of the point of light on the figure eight during each month.

early and late arrival is explained in part by Earth moving at different speeds in its orbit.

If changes in orbital speed were the only reason that the Sun does not cross the sky at the same rate during the year, the spot of sunlight on the floor would trace out an oval rather than a figure eight. The plane of the ecliptic, however, does not coincide with the plane of Earth's equator, so the Sun appears at different angles in the sky, and this makes it appear to change

FIGURE 14.21 The path of the Sun's direct rays during a year. The Sun is directly over the tropic of Cancer at the summer solstice and high in the Northern Hemisphere sky. At the winter solstice, the Sun is directly over the tropic of Capricorn and low in the Northern Hemisphere sky.

Saving Time?

The purpose of daylight saving time is to make better use of daylight during the summer by moving an hour of daylight from the morning to the evening. In the United States, daylight saving time is observed from the second Sunday in March to the first Sunday in November. Clocks are changed on these Sundays according to the saying, "Spring ahead, fall back." Arizona and Hawaii choose not to participate and stay on standard time all year.

Americans who say they like daylight saving time say they like it because it gives them more light in the evenings and it saves energy. Some people do not like daylight saving time because it requires them to reset all their clocks and adjust their sleep schedule twice a year. They also complain that the act of changing the clock is not saving daylight at all but it is sending them to bed an hour earlier. Farmers also complain that plants and animals are regulated by the Sun, not the clock, so they have to plan all their nonfarm interactions on a different schedule.

Questions to Discuss:

Divide your group into two subgroups, one representing those who like daylight saving time and the other representing those who do not. After a few minutes of preparation, have a short debate about the advantages and disadvantages of daylight saving time.

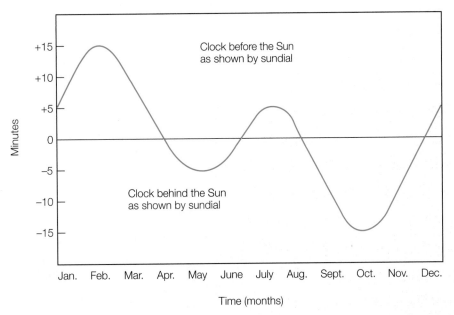

FIGURE 14.22 The equation of time, which shows how many minutes sundial time is faster or slower than clock time during different months of the year.

its speed during different times of the year. This effect changes the length of the apparent solar day by making the Sun up to ten minutes later or earlier than the mean solar time four times a year between the solstices and equinoxes.

The two effects add up to a cumulative variation between the apparent local solar time (sundial time) and the mean solar time (clock time) (figure 14.22). This cumulative variation is known as the **equation of time,** which shows how many minutes sundial time is faster or slower than clock time during different days of the year. The equation of time is often shown on globes in the figure-eight shape called an *analemma*, which also can be used to determine the latitude of direct solar radiation for any day of the year.

Since the local mean time varies with longitude, every place on an east-west line around Earth could possibly have clocks that were a few minutes ahead of those to the west and a few minutes behind those to the east. To avoid the confusion that would result from many clocks set to local mean solar time, Earth's surface is arbitrarily divided into one-hour **standard time zones** (figure 14.23). Since there are 360° around Earth and 24 hours in a day, this means that each time zone is 360° divided by 24, or 15° wide. These 15° zones are adjusted so that whole states are in the same time zone or for other political reasons. The time for each zone is defined as the mean solar time at the middle of each zone. When you cross a boundary between two zones, the clock is set ahead one hour if you are traveling east and back one hour if you are traveling west. Most states adopt **daylight saving time** during the summer, setting clocks ahead one hour in the spring and back one hour in the fall ("spring ahead and fall back"). Daylight saving time results in an extra hour of daylight during summer evenings.

The 180° meridian is arbitrarily called the **international date line,** an imaginary line established to compensate for cumulative time zone changes (figure 14.24). A traveler crossing

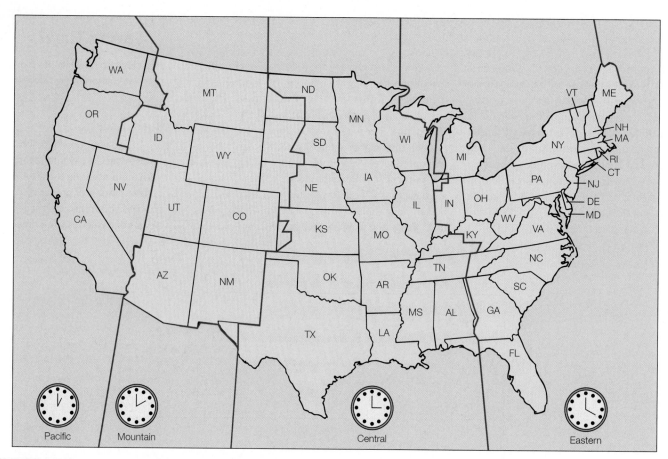

FIGURE 14.23   The standard time zones. Hawaii and most of Alaska are two hours earlier than Pacific Standard Time.

FIGURE 14.24   The international date line follows the 180° meridian but is arranged in a way that land areas and island chains have the same date.

the date line gains or loses a day just as crossing a time zone boundary results in the gain or loss of an hour. A person moving across the line while traveling westward gains a day; for example, the day after June 2 would be June 4. A person crossing the line while traveling eastward repeats a day; for example, the day after June 6 would be June 6. Note that the date line is curved around land masses to avoid local confusion.

Yearly Time

A *year* is generally defined as the interval of time required for Earth to make one complete revolution in its orbit. As was the case for definitions of a day, there are different definitions of what is meant by a year. The most common definition of a year is the interval between two consecutive spring equinoxes, which is known as the **tropical year** (*trope* is Greek for "turning"). The tropical year is 365 days, 5 hours, 48 minutes, and 46 seconds, or 365.24220 mean solar days.

A **sidereal year** is defined as the interval of time required for Earth to move around its orbit so the Sun is again in the same position relative to the stars. The sidereal year is slightly longer than the tropical year because Earth rotates more than 365.25 times during one revolution. Thus, the sidereal year is 365.25636 mean solar days, which is about 20 minutes longer than the tropical year.

The tropical and sidereal years would be the same interval of time if Earth's axis pointed in a consistent direction. The precession of the axis, however, results in the axis pointing in a slightly different direction with time. This shift of direction over the course of a year moves the position of the spring equinox westward, and the equinox is observed twenty minutes before the orbit has been completely circled. The position of the spring equinox against the background of the stars thus moves westward by some fifty seconds of arc per year.

It is the *tropical year* that is used as a standard time interval to determine the calendar year. Earth does not complete an exact number of turns on its axis while completing one trip around the Sun, so it becomes necessary to periodically adjust the calendar so it stays in step with the seasons. The calendar system that was first designed to stay in step with the seasons was devised by the ancient Romans. Julius Caesar reformed the calendar, beginning in 46 B.C., to have a 365-day year with a 366-day year (leap year) every fourth year. Since the tropical year of 365.24220 mean solar days is very close to 365¼ days, the system, called a *Julian calendar,* accounted for the ¼ day by adding a full day to the calendar every fourth year. The Julian calendar was very similar to the one now used, except the year began in March, the month of the spring equinox. The month of July was named in honor of Julius Caesar, and the following month was later named after his successor, Augustus.

There was a slight problem with the Julian calendar because it was longer than the tropical year by 365.25 minus 365.24220, or 0.0078 day per year. This small interval (which is 11 minutes, 14 seconds) does not seem significant when compared to the time in a whole year. But over the years, the error of minutes and seconds grew to an error of days. By 1582, when Pope Gregory XIII revised the calendar, the error had grown to 13 days but was corrected for 10 days of error. This revision resulted in the *Gregorian calendar,* which is the system used today. Since the accumulated error of 0.0078 day per year is almost 0.75 day per century, it follows that four centuries will have 0.75 times 4, or 3 days of error. The Gregorian system corrects for the accumulated error by dropping the additional leap year day three centuries out of every four. Thus, the century year of 2000 was a leap year with 366 days, but the century years of 2100, 2200, and 2300 will not be leap years. You will note that this approximation still leaves an error of 0.0003 day per century, so another calendar revision will be necessary in a few thousand years to keep the calendar in step with the seasons.

Monthly Time

In ancient times, people often used the Moon to measure time intervals that were longer than a day but shorter than a year. The word *month,* in fact, has its origins in the word *moon* and its period of revolution. The Moon revolves around Earth in an orbit that is inclined to the plane of Earth's orbit, the plane of the ecliptic, by about 5°. The Moon is thus never more than about ten apparent diameters from the ecliptic. It revolves in this orbit in about 27⅓ days as measured by two consecutive crossings of any star. This period is called a **sidereal month.** The Moon rotates in the same period as the time of revolution, so the sidereal month is also the time required for one rotation. Because the rotation and revolution rates are the same, you always see the same side of the Moon from Earth.

The ancient concept of a month was based on the **synodic month,** the interval of time from new moon to new moon (or any two consecutive identical phases). The synodic month is longer than a sidereal month at a little more than 29½ days. The Moon's phases (see "Phases of the Moon" in this chapter) are determined by the relative positions of Earth, Moon, and Sun. As shown in figure 14.25, the Moon moves with Earth in its orbit around the Sun. During one sidereal month, the Moon has to revolve a greater distance before the same phase is observed on Earth, and this greater distance requires 2.2 days. This makes the synodic month about 29½ days long, only a little less than ¹⁄₁₂ of a year, or the period of time the present calendar identifies as a "month."

THE EARTH-MOON SYSTEM

Earth and its Moon are unique in the solar system because of the size of the Moon. It is not the largest satellite, but the ratio of its mass to Earth's mass is greater than the mass ratio of any other moon to its planet. The Moon has a diameter of 3,476 km (about 2,159 mi), which is about one-fourth the diameter of Earth, and a mass of about 1/81 of Earth's mass. This is a small fraction of Earth's mass, but it is enough to affect Earth's motion as it revolves around the Sun.

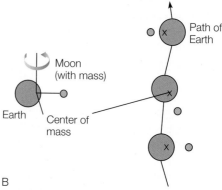

FIGURE 14.25 As the Moon moves in its orbit around Earth, it must revolve a greater distance to bring the same part to face Earth. The additional turning requires about 2.2 days, making the synodic month longer than the sidereal month.

If the Moon had a negligible mass, it would circle Earth with a center of rotation (center of mass) located at the center of Earth. In this situation, the center of Earth would follow a smooth path around the Sun (figure 14.26A). The mass of the Moon, however, is great enough to move the center of rotation away from Earth's center toward the Moon. As a result, both bodies act as a system, moving around a center of mass. The center of mass between Earth and the Moon follows a smooth orbit around the Sun. Earth follows a slightly wavy path around the Sun as it slowly revolves around the common center of mass (figure 14.26B).

FIGURE 14.26 (A) If the Moon had a negligible mass, the center of gravity between the Moon and Earth would be Earth's center, and Earth would follow a smooth orbit around the Sun. (B) The actual location of the center of mass between Earth and Moon results in a slightly in and out, or wavy, path around the Sun.

Phases of the Moon

The phases of the Moon are a result of the changing relative positions of Earth, the Moon, and the Sun as the Earth-Moon system moves around the Sun. Sunlight always illuminates half of the Moon, and half is always in shadow. As the Moon's path takes it between Earth and the Sun, then to the dark side of Earth, you see different parts of the illuminated half called *phases* (figure 14.27). When the Moon is on the dark side of Earth, you see the entire illuminated half of the Moon called the **full moon** (or the full phase). Halfway around the orbit, the lighted side of the Moon now faces away from Earth, and the unlighted side now faces Earth. This dark appearance is called the **new moon** (or the new phase). In the new phase, the Moon is not *directly* between Earth and the Sun, so it does not produce an eclipse (see "Eclipses of the Sun and Moon" in this chapter).

As the Moon moves from the new phase in its orbit around Earth, you will eventually see half the lighted surface, which is known as the **first quarter.** Often the unlighted part of the Moon shines with a dim light of reflected sunlight from Earth called *earthshine.* Note that the division between the lighted and un-lighted part of the Moon's surface is curved in an arc. A straight line connecting the ends of the arc is perpendicular to the direction of the Sun (figure 14.28). After the first quarter, the Moon moves to its full phase, then to the **last quarter** (see figure 14.27). The period of time between two consecutive phases, such as new moon to new moon, is the synodic month, or about 29.5 days.

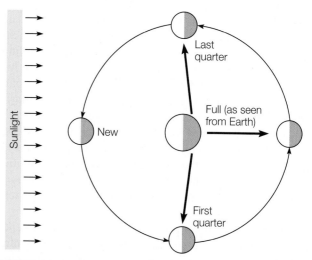

FIGURE 14.27 Half of the Moon is always lighted by the Sun, and half is always in the shadow. The Moon phases result from the view of the lighted and dark parts as the Moon revolves around Earth.

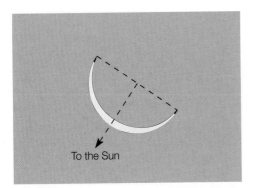

FIGURE 14.28 The cusps, or horns, of the Moon always point away from the Sun. A line drawn from the tip of one cusp to the other is perpendicular to a straight line between the Moon and the Sun.

Eclipses of the Sun and Moon

Sunlight is not visible in the emptiness of space because there is nothing to reflect the light, so the long conical shadow behind each spherical body is not visible, either. One side of Earth and one side of the Moon are always visible because they reflect sunlight. The shadow from Earth or from the Moon becomes noticeable only when it falls on the illuminated surface of the other body. This event of Earth's or the Moon's shadow falling on the other body is called an **eclipse** (table 14.1). Most of the time, eclipses do not occur because the plane of the Moon's orbit is inclined to Earth's orbit about 5° (figure 14.29). As a result, the shadow from the Moon or the shadow from Earth usually falls above or below the other body, too high or too low to produce an eclipse. An eclipse occurs only when the Sun, Moon, and Earth are in a line with each other.

The shadow from Earth and the shadow from the Moon are long cones that point away from the Sun. Both cones have two

TABLE 14.1 Total Eclipses in U.S.—2008–2025*

Total Solar Eclipses	
Aug 21, 2017	Path: Oregon to South Carolina
Apr 8, 2024	Path: Texas to Maine
Total Lunar Eclipses	
Feb 21, 2008	All visible eastern; moonrise western
Dec 21, 2010	All visible
Dec 10, 2011	Visible at moonset
Apr 15, 2014	Visible at moonset
Oct 8, 2014	All visible western; at moonset eastern
Apr 4, 2015	Visible at moonset
Sept 28, 2015	Western visible at moonrise; all eastern
Jan 31, 2018	Visible at moonset
Jan 21, 2019	All visible
May 26, 2021	Visible at moonset
May 16, 2022	Visible moonrise western; all eastern
Mar 14, 2025	All visible

*Eclipse predictions by Fred Espenak, NASA/GSFC See http://sunearth.gsfc.nasa.gov/eclipse/eclipse.html

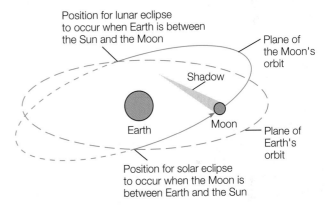

FIGURE 14.29 The plane of the Moon's orbit is inclined to the plane of Earth's orbit by about 5°. An eclipse occurs only where the two planes intersect and Earth, the Moon, and the Sun are in a line.

parts, an inner cone of a complete shadow called the **umbra** and an outer cone of partial shadow called the **penumbra.** When and where the umbra of the Moon's shadow falls on Earth, people see a **total solar eclipse.** During a total solar eclipse, the Moon completely covers the disk of the Sun. The total solar eclipse is preceded and followed by a partial eclipse, which is seen when the observer is in the penumbra. If the observer is in a location where only the penumbra passes, then only a partial eclipse will be observed (figure 14.30). More people see partial than full solar eclipses because the penumbra covers a larger area. The occurrence of a total solar eclipse is a rare event in a given location, occurring once every several hundred years and then lasting for less than seven minutes.

The Moon's cone-shaped shadow averages a length of 375,000 km (about 233,000 mi), which is less than the average distance between Earth and the Moon. The Moon's elliptical orbit brings it sometimes closer to and sometimes farther from

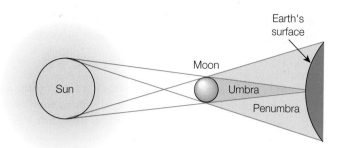

FIGURE 14.30 People in a location where the tip of the umbra falls on the surface of Earth see a total solar eclipse. People in locations where the penumbra falls on Earth's surface see a partial solar eclipse.

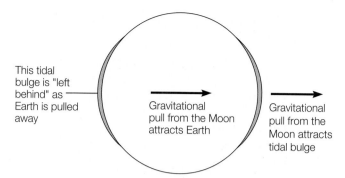

FIGURE 14.31 Gravitational attraction pulls on Earth's waters on the side of Earth facing the Moon, producing a tidal bulge. A second tidal bulge on the side of Earth opposite the Moon is produced when Earth, which is closer to the Moon, is pulled away from the waters.

Earth. A total solar eclipse occurs only when the Moon is close enough so at least the tip of its umbra reaches the surface of Earth. If the Moon's umbra fails to reach Earth, an **annular eclipse** occurs. Annular means "ring-shaped," and during this eclipse, the edge of the Sun is seen to form a bright ring around the Moon. As before, people located in the area where the penumbra falls will see a partial eclipse. The annular eclipse occurs more frequently than the total solar eclipse.

When the Moon is full and the Sun, Moon, and Earth are lined up so Earth's shadow falls on the Moon, a **lunar eclipse** occurs. Earth's shadow is much larger than the Moon's diameter, so a lunar eclipse is visible to everyone on the night side of Earth. This larger shadow also means a longer eclipse that may last for hours. As the umbra moves over the Moon, the darkened part takes on a reddish, somewhat copper-colored glow from light refracted and scattered into the umbra by Earth's atmosphere. This light passes through the thickness of Earth's atmosphere on its way to the eclipsed Moon, and it acquires the reddish color for the same reason that a sunset is red: much of the blue light has been removed by scattering in Earth's atmosphere.

Tides

If you live near or have ever visited a coastal area of the ocean, you are familiar with the periodic rise and fall of sea level known as **tides.** The relationship between the motions of the Moon and the magnitude and timing of tides has been known and studied since very early times. These relationships are that (1) the greatest range of tides occurs at full and new moon phases, (2) the least range of tides occurs at quarter moon phases, and (3) in most oceans, the time between two high tides or between two low tides is an average of twelve hours and twenty-five minutes. The period of twelve hours and twenty-five minutes is half the average time interval between consecutive passages of the Moon across the celestial meridian. A location on the surface of Earth is directly under the Moon when it crosses the meridian, and directly opposite it on the far side of Earth an average twelve hours and twenty-five minutes later. There are two *tidal bulges* that follow the Moon as it moves around Earth, one on the side facing the Moon and one on the opposite side. In general, tides are a result of these bulges moving westward around Earth.

A simplified explanation of the two tidal bulges involves two basic factors: the gravitational attraction of the Moon and the motion of the Earth-Moon system (figure 14.31). Water on Earth's surface is free to move, and the Moon's gravitational attraction pulls the water to the tidal bulge on the side of Earth facing the Moon. This tide-raising force directed toward the Moon bulges the water in mid-ocean some .75 m (about 2.5 ft), but it also bulges the land, producing a land tide. Since Earth is much more rigid than water, the land tide is much smaller at about 12 cm (about 4.5 in). Since all parts of the land bulge together, this movement is not evident without measurement by sensitive instruments.

The tidal bulge on the side of Earth opposite the Moon occurs as Earth is pulled away from the ocean by the Earth-Moon gravitational interaction. Between the tidal bulge facing the Moon and the tidal bulge on the opposite side, sea level is depressed across the broad surface. The depression is called a *tidal trough,* even though it does not actually have the shape of a trough. The two tidal bulges, with the trough between, move slowly eastward following the Moon. Earth turns more rapidly on its axis, however, which forces the tidal bulge to stay in front of the Moon moving through its orbit. Thus, the tidal axis is not aligned with the Earth-Moon gravitational axis.

The tides do not actually appear as alternating bulges that move around Earth. There are a number of factors that influence the making and moving of the bulges in complex interactions that determine the timing and size of a tide at a given time in a given location. Some of these factors include (1) the relative positions of Earth, Moon, and Sun, (2) the elliptical orbit of the Moon, which sometimes brings it closer to Earth, and (3) the size, shape, and depth of the basin holding the water.

The relative positions of Earth, Moon, and Sun determine the size of a given tide because the Sun, as well as the Moon, produces a tide-raising force. The Sun is much more massive than the Moon, but it is so far away that its tide-raising force is about half that of the closer Moon. Thus, the Sun basically modifies lunar tides rather than producing distinct tides of its own. For example, Earth, Moon, and Sun are nearly in line during the full and new moon phases. At these times, the lunar and

People Behind the Science

Carl Edward Sagan (1934–1996)

Carl Edward Sagan was an American astronomer and popularizer of astronomy whose main research was on planetary atmospheres, including that of the primordial Earth. His most remarkable achievement was to provide valuable insights into the origin of life on our planet.

Sagan was born on November 19, 1934, in New York City. Completing his education at the University of Chicago, he obtained his bachelor's degree in 1955 and his doctorate in 1960. Then, for two years, he was a research fellow at the University of California in Berkeley before he transferred to the Smithsonian Astrophysical Observatory in Cambridge, Massachusetts, lecturing also at Harvard University, where he became assistant professor. Finally, in 1968, Sagan moved to Cornell University in Ithaca (New York) to become director of the Laboratory for Planetary Studies; in 1970, he became professor of astronomy and space science there. He died on December 20, 1996.

The editor of the astronomical journal *Icarus*, Sagan wrote a number of popular books including *Broca's Brain: Reflections on the Romance of Science* (1979); *Cosmos* (1980), based on his television series of that name; and a science-fiction novel, *Contact*.

In the early 1960s, Sagan's first major research was into the planetary surface and atmosphere of Venus. At the time, although intense emission of radiation had shown that the dark-side temperature of Venus was nearly 600K, it was thought that the surface itself remained relatively cool—leaving open the possibility that there was some form of life on the planet. Various hypotheses were put forward to account for the strong emission actually observed: perhaps it was due to interactions between charged particles in Venus' dense upper atmosphere; perhaps it was glow discharge between positive and negative charges in the atmosphere; or perhaps emission was due to a particular radiation from charged particles trapped in the Venusian equivalent of a Van Allen Belt. Sagan showed that each of these hypotheses was incompatible with other observed characteristics or with implications of these characteristics. The positive part of Sagan's proposal was to show that all the observed characteristics were compatible with the straightforward hypothesis that the surface of Venus was very hot. On the basis of radar and optical observations, the distance between surface and clouds was calculated to be between 44 km (27 mi) and 65 km (40 mi); given the cloud-top temperature and Sagan's expectation

of a "greenhouse effect" in the atmosphere, surface temperature on Venus was computed to be between 500K (227°C/440°F) and 800K (527°C/980°F)—the range that would also be expected on the basis of emission rate.

Sagan then turned his attention to the early planetary atmosphere of Earth, with regard to the origins of life. One way of understanding how life began is to try to form the compounds essential to life in conditions analogous to those of the primeval atmosphere. Before Sagan, Stanley Miller and Harold Urey had used a mixture of methane, ammonia, water vapor, and hydrogen, sparked by a corona discharge that simulated the effect of lightning, to produce amino and hydroxy acids of the sort found in life-forms. Later experiments used ultraviolet light or heat as sources of energy, and even these had less energy than would have been available in Earth's primordial state. Sagan followed a similar method and, by irradiating a mixture of methane, ammonia, water, and hydrogen sulfide, was able to produce amino acids—and, in addition, glucose, fructose, and nucleic acids. Sugars can be made from formaldehyde under alkaline conditions and in the presence of inorganic catalysts. These sugars include five-carbon sugars, which are essential to the formation of nucleic acids, glucose, and fructose—all common metabolites found as constituents of present-day life-forms. Sagan's simulated primordial atmosphere not only showed the presence of those metabolites, it also contained traces of adenosine triphosphate (ATP)—the foremost agent used by living cells to store energy.

In 1966, in work done jointly with Pollack and Goldstein, Sagan was able to provide evidence supporting a hypothesis about Mars put forward by Wells, who observed that in regions on Mars where there were both dark and light areas, the clouds formed over the lighter areas aligned with boundaries of adjacent dark areas. Wells suggested that they were lee clouds formed by the Martian wind as it crossed dark areas. The implication, that dark areas mark the presence of ridges, was given support by Sagan's finding that dark areas had a high radar reflectivity that was slightly displaced in longitude. Sagan concluded that these dark areas were elevated areas with ridges of about 10 km (6 mi) and low slopes extending over long distances.

---

solar tide-producing forces act together, producing tides that are unusually high and corresponding low tides that are unusually low. The unusually high and low tides are called **spring tides** (figure 14.32A). Spring tides occur every two weeks and have nothing to do with the spring season. When the Moon is in its quarter phases, the Sun and Moon are at right angles to one another and the solar tides occur between the lunar tides, causing unusually less pronounced high and low tides called **neap tides** (figure 14.32B). Neap tides also occur every two weeks.

The size of the lunar-produced tidal bulge varies as the Moon's distance from Earth changes. The Moon's elliptical orbit brings it closest to Earth at a point called **perigee** and farthest from Earth at a point called **apogee.** At perigee, the Moon is about 44,800 km (about 28,000 mi) closer to Earth than at apogee, so its gravitational attraction is much greater. When perigee coincides with a new or full moon, especially high spring tides result.

The open basins of oceans, gulfs, and bays are all connected but have different shapes and sizes and have bordering

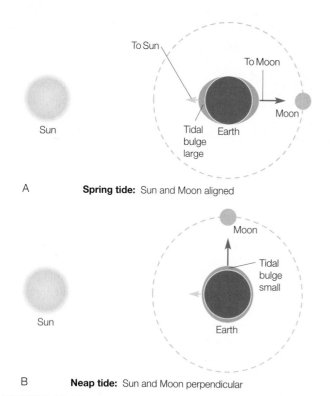

A   **Spring tide:** Sun and Moon aligned

B   **Neap tide:** Sun and Moon perpendicular

FIGURE 14.32 (A) When the Sun and Moon lie in the same or in opposite directions as seen from Earth, the tidal bulges they cause coincide, producing a large *spring* tide. (B) When the directions to the Sun and Moon are perpendicular, the tides of the Sun and Moon cancel to some extent, producing a weak *neap* tide.

landmasses in all possible orientations to the westward-moving tidal bulges. Water in each basin responds differently to the tidal forces, responding as periodic resonant oscillations that move back and forth much like the water in a bowl shifts when carried. Thus, coastal regions on open seas may experience tides that range between about 1 and 3 m (about 3 to 10 ft), but mostly enclosed basins such as the Gulf of Mexico have tides less than about 1/3 m (about 1 ft). The Gulf of Mexico, because of its size, depth, and limited connections with the open ocean, responds only to the stronger tidal attractions and has only one high and one low tide per day. Even lakes and ponds respond to tidal attractions, but the result is too small to be noticed. Other basins, such as the Bay of Fundy in Nova Scotia, are funnel-shaped and undergo an unusually high tidal range. The Bay of Fundy has experienced as much as a 15 m (about 50 ft) tidal range.

As the tidal bulges are pulled against a rotating Earth, friction between the moving water and the ocean basin tends to slow Earth's rotation over time. This is a very small slowing effect that is increasing the length of each day by about a second per year.* Evidence for this slowing comes from a number of sources, including records of ancient solar eclipses. The solar eclipses of two thousand years ago occurred three hours earlier than would be expected by using today's time but were on the mark if a lengthening day is considered. Fossils of a certain species of coral still living today provide further evidence of a lengthening day. This particular coral adds daily growth rings, and five-hundred-million-year-old fossils show that the day was about twenty-one hours long at that time. Finally, the Moon is moving away from Earth at a rate of about 4 cm (about 1.5 in) per year. This movement out to a larger orbit is a necessary condition to conserve angular momentum as Earth slows. As the Moon moves away from Earth, the length of the month increases. Some time in the distant future, both the day and the month will be equal, about fifty of the present days long.

*tycho.usno.navy.mil/leapsec.html

SUMMARY

Earth is an *oblate spheroid* that undergoes three basic motions: (1) a yearly *revolution* around the Sun, (2) a daily *rotation* on its axis, and (3) a slow wobble of its axis called *precession.*

As Earth makes its yearly *revolution* around the Sun, it maintains a generally *constant inclination of its axis* to the *plane of the ecliptic* of 66.5°, or 23.5° from a line perpendicular to the plane. In addition, Earth maintains a generally constant *orientation of its axis* to the stars, which always points in the same direction. The constant inclination and orientation of the axis, together with Earth's rotation and revolution, produce three effects: (1) days and nights that vary in length, (2) seasons that change during the course of a year, and (3) climates that vary with latitude. When Earth is at a place in its orbit so the axis points toward the Sun, the Northern Hemisphere experiences the longest days and the summer season. This begins on June 21 or 22, which is called the *summer solstice.* Six months later, the axis points away from the Sun and the Northern Hemisphere experiences the shortest days and the winter season. This begins on December 22 or 23 and is called the *winter solstice.* On March 20 or 21, Earth is halfway between the solstices and has days and nights of equal length, which is called the *spring* (or *vernal*) *equinox.* On September 22 or 23, the autumnal equinox, another period of equal nights and days, identifies the beginning of the fall season.

Precession is a slow wobbling of the axis as Earth spins. Precession is produced by the gravitational tugs of the Sun and Moon on Earth's equatorial bulge.

Lines around Earth that are parallel to the equator are circles called *parallels.* The distance from the equator to a point on a parallel is called the *latitude* of that point. North and south arcs that intersect at the poles are called *meridians.* The meridian that runs through the Greenwich Observatory is a reference line called the *prime meridian.* The distance of a point east or west of the prime meridian is called the *longitude* of that point.

The event of time called *noon* is the instant the Sun appears to move across the celestial meridian. The instant of noon at a particular location is called the *apparent local noon.* The time at a given place that is determined by a sundial is called the *apparent local solar time.* It is the basis for an averaged, uniform standard time called the *mean solar time.* Mean solar time is the time used to set clocks.

A *sidereal day* is the interval between two consecutive crossings of the celestial meridian by a star. An *apparent solar day* is the interval

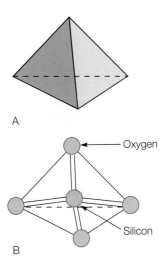

FIGURE 15.6 (*A*) The geometric shape of a tetrahedron with four equal sides. (*B*) A silicon and four oxygen atoms are arranged in the shape of a tetrahedron with the silicon in the center. This is the basic building block of all silicate minerals.

CONCEPTS APPLIED

Grow Your Own

Experiment with growing crystals from solutions of alum, copper sulfate, salt, or potassium permanganate. Write a procedure that will tell others what the important variables are for growing large, well-formed crystals.

The *nonferromagnesian silicates* have a light color and a low density compared to the ferromagnesians. This group includes the minerals *muscovite (white mica)*, the *feldspars*, and *quartz* (figure 15.9).

The remaining 8 percent of minerals making up Earth's crust that do not have silicon-oxygen tetrahedrons in their crystal structure are called *nonsilicates*. There are eight subgroups of nonsilicates: (1) carbonates, (2) sulfates, (3) oxides, (4) sulfides, (5) halides, (6) phosphates, (7) hydroxides, and (8) native elements. Some of these are identified in table 15.1. The carbonates are the most abundant of the nonsilicates, but others are important as fertilizers, sources of metals, and sources of industrial chemicals.

Rocks

Elements are *chemically* combined to make minerals. Minerals are *physically* combined to make rocks. A **rock** is defined as an aggregation of one or more minerals and perhaps other materials that have been brought together into a cohesive solid. These materials include volcanic glass, a silicate that is not considered a mineral because it lacks a crystalline structure. Thus, a rock can consist of one or more kinds of minerals that are somewhat

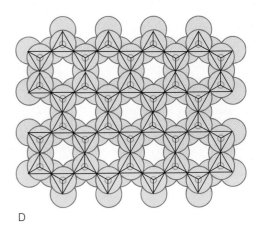

FIGURE 15.7 (*A*) Isolated silicon-oxygen tetrahedra do not share oxygens. This structure occurs in the mineral olivine. (*B*) Single chains of tetrahedra are formed by each silicon ion having two oxygens all to itself and sharing two with other silicons at the same time. This structure occurs in augite. (*C*) Double chains of tetrahedra are formed by silicon ions sharing either two or three oxygens. This structure occurs in hornblende. (*D*) The sheet structure in which each silicon shares three oxygens occurs in the micas, resulting in layers that pull off easily because of cleavage between the sheets.

A Closer Look

Asbestos

Asbestos is a common name for any of several minerals that can be separated into fireproof fibers, fibers that will not melt or ignite. The fibers can be woven into a fireproof cloth, used directly as fireproof insulation material, or mixed with plaster or other materials. People now have a fear of all asbestos because it is presumed to be a health hazard (box figure 15.1). However, there are about six commercial varieties of asbestos. Five of these varieties are made from an amphibole mineral and are commercially called "brown" and "blue" asbestos. The other variety is made from *chrysotile*, a *serpentine* family of minerals, and is commercially called "white" asbestos. White asbestos is the asbestos mined and most commonly used in North America. It is only the amphibole asbestos (brown and blue asbestos) that has been linked to cancer, even for a short exposure time. There is, however, no evidence that exposure to white asbestos results in an increased health hazard. It makes sense to ban the use of and remove all the existing amphibole asbestos from public buildings. It does not make sense to ban or remove the serpentine asbestos since it is not a proven health hazard.

Box Figure 15.1 Did you know there are different kinds of asbestos? Are all kinds of asbestos a health hazard?

FIGURE 15.8 Compare the dark colors of the ferromagnesian silicates augite (right), hornblende (left), and biotite to the light-colored nonferromagnesian silicates in figure 15.9.

FIGURE 15.9 Compare the light colors of the nonferromagnesian silicates mica (front center), white and pink orthoclase (top and center), and quartz (left center) to the dark-colored ferromagnesian silicates in figure 15.8.

Most people understand the fact that our mineral resources are limited, and when we use them, they are gone. Of course, some mineral resources can be recycled, reducing the need to mine increasingly more minerals. For example, aluminum can be recycled and used over and over again repeatedly. Glass, copper, iron, and other metals can similarly be recycled repeatedly. Other critical resources, however, cannot be recycled and, unfortunately, cannot be replaced. Crude oil, for example, is a dwindling resource that will eventually become depleted. Oil is not recyclable once it is burned, and no new supplies are being created, at least not at a rate that would make them available in the immediate future. Even if Earth were a hollow vessel completely filled with oil, it would eventually become depleted, perhaps sooner than you might think.

There is also another of our mineral resources that is critically needed for our survival but will probably soon be depleted. That resource is phosphorus derived from phosphate rock. Phosphorus is an essential nutrient required for plant growth, and if its concentration in soils is too low, plants grow poorly, if at all. Most agricultural soils are artificially fertilized with phosphate. Without this amendment, their productivity would decline and, in some cases, cease altogether.

Phosphate occurs naturally as the mineral apatite. Deposits of apatite were formed where ocean currents carried cold, deep water rich in dissolved phosphate ions to the upper continental slope and outer continental shelf. Here, phosphate ions replaced the carbonate ions in limestone, forming the mineral apatite. Apatite also occurs as a minor accessory mineral in most igneous, sedimentary, and metamorphic rock types. Some igneous rocks serve as a source of phosphate fertilizer, but most phosphate is mined from formerly submerged coastal areas of limestone, such as those found in Florida.

Trends in phosphate production and use over the past forty years suggest that the proved world reserves of phosphate rock will be exhausted within a few decades. New sources might be discovered, but some time soon, phosphate rock will be mined out and no longer available. When this happens, the food supply will have to be grown on lands that are naturally endowed with an adequate phosphate supply. Estimates are that this worldwide existing land area with adequate phosphate supply will supply food for only two billion people on the entire Earth. Phosphate is an essential element for all life on Earth, and no other element can function in its place.

Questions to Discuss:

Discuss with your group the following questions concerning the use of mineral resources:

1. Should the mining industry be permitted to exhaust an important mineral resource? Provide reasons with your answer.
2. What are the advantages and disadvantages of a controlled mining industry?
3. If phosphate minerals supplies become exhausted, who should be responsible for developing new supplies or substitutes: the mining industry or governments?

"glued" together by other materials such as glass. Most rocks are composed of silicate minerals, as you might expect since most minerals are silicates.

There is a classification scheme that is based on the way the rocks were formed. There are three main groups: (1) *igneous rocks* formed as a hot, molten mass of rock materials cooled and solidified; (2) *sedimentary rocks* formed from particles or dissolved materials from previously existing rocks; and (3) *metamorphic rocks* formed from rocks that were subjected to high temperatures and pressures that deformed or recrystallized the rock without complete melting.

Igneous Rocks

The word *igneous* comes from the Latin *ignis*, which means "fire." This is an appropriate name for **igneous rocks** that are defined as rocks that formed from a hot, molten mass of melted rock materials. The first step in forming igneous rocks is the creation of a very high temperature that is hot enough to melt rocks. A mass of melted rock materials is called *magma*. Magma may cool and crystallize to solid igneous rock either below or on the surface of Earth. Earth has a history of beginning as a molten material, all rocks of Earth were at one time igneous rocks. Today, about two-thirds of the outer layer, or crust, is made up of igneous rocks. This is not apparent in many loca-tions because the surface is covered by other kinds of rocks and rock materials (sand, soil, etc.). When they are found, igneous rocks are usually granite or basalt. *Granite* is a light-colored igneous rock that is primarily made up of three silicate minerals: quartz, mica, and feldspar. You can see the grains of these three minerals in a freshly broken surface of most samples of granite (figure 15.10). *Basalt* is a dark, more dense igneous rock with mineral grains too tiny to see. Basalt is the most common volcanic rock found on the surface of Earth.

Sedimentary Rocks

Sedimentary rocks are rocks that formed from particles or dissolved materials from previously existing rocks. Chemical reactions with air and water tend to break down and dissolve the less chemically stable parts of existing rocks, freeing more stable particles and grains in the process. The remaining particles are usually transported by moving water and deposited as sediments. *Sediments* are accumulations of silt, clay, sand, or other weathered materials. Weathered rock fragments and dissolved rock materials both contribute to sediment deposits (table 15.2).

Most sediments are deposited as many separate particles that accumulate as loose materials. Such accumulations of rock fragments, chemical deposits, or animal shells must become consolidated into a solid, coherent mass to become sedimentary rock.

FIGURE 15.10 Granite is a coarse-grained igneous rock composed mostly of light-colored, light-density, nonferromagnesian minerals. Earth's continental areas are dominated by granite and by rocks with the same mineral composition of granite.

TABLE 15.2 Classification Scheme for Sedimentary Rocks

Sediment Type	Particle or Composition	Rock
Fragment	Larger than sand	Conglomerate or breccia
Fragment	Sand	Sandstone
Fragment	Silt and clay	Siltstone, claystone, or shale
Chemical	Calcite	Limestone
Chemical	Dolomite	Dolomite
Chemical	Gypsum	Gypsum
Chemical	Halite (sodium chloride)	Salt

There are two main parts to this *lithification,* or rock-forming process: (1) compaction and (2) cementation (figure 15.11).

The weight of an increasing depth of overlying sediments causes an increasing pressure on the sediments below. This pressure squeezes the deeper sediments together, gradually reducing the pore space between the individual grains. This *compaction* of the grains reduces the thickness of a sediment deposit, squeezing out water as the grains are packed more tightly together. Compaction alone is usually not enough to make loose sediment into solid rock. Cementation is needed to hold the compacted grains together.

In *cementation,* the spaces between the sediment particles are filled with a chemical deposit. The chemical deposit binds the particles together into the rigid, cohesive mass of a sedimentary rock. Compaction and cementation may occur at the same time, but the cementing agent must have been introduced before compaction restricts the movement of the fluid through the open spaces. Many soluble minerals can serve as cementing agents, and calcite (calcium carbonate) and silica (silicon dioxide) are common.

Sediments accumulate from rocks that are in various stages of being broken down, so there is a wide range of sizes of particles in sediments. The size and shape of the particles are used as criteria to name the sedimentary rocks (table 15.3).

Metamorphic Rocks

The third group of rocks is called metamorphic. **Metamorphic rocks** are previously existing rocks that have been changed by heat, pressure, or hot solutions into a distinctly different rock. The heat, pressure, or hot solutions that produced the changes are associated with geologic events of (1) movement of the crust, which will be discussed in chapter 16, and with (2) heating and hot solutions from magma. Pressures from movement of the crust can change the rock texture by flattening, deforming, or realigning mineral grains. Temperatures from magma must be just right to

CONCEPTS APPLIED

Collect Your Own

Make a collection of rocks and minerals that can be found in your location, showing the name and location found for each. What will determine the total number of rocks and minerals it is possible to collect in your particular location?

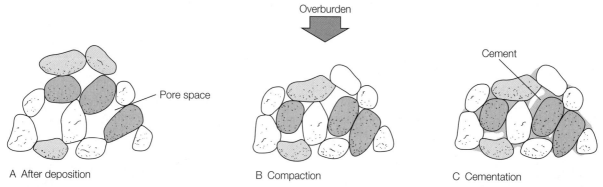

A After deposition   B Compaction   C Cementation

FIGURE 15.11 Cementation of sand grains to become sandstone. (*A*) Loose sand grains are deposited with open pore space between the grains. (*B*) The weight of overburden compacts the sand into a tighter arrangement, reducing pore space. (*C*) Precipitation of cement in the pores by groundwater binds the sand into the rock sandstone, which has a clastic texture.

TABLE 15.3 Simplified Classification Scheme for Sediment Fragments and Rocks

Sediment Name	Size Range	Rock
Boulder	Over 256 mm (10 in)	
Gravel	2 to 256 mm (0.08 – 10 in)	Conglomerate or breccia*
Sand	1/16 to 2 mm (0.025 – 0.08 in)	Sandstone
Silt (or dust)	1/256 to 1/16 mm (0.00015 – 0.025 in)	Siltstone**
Clay (or dust)	Less than 1/256 mm (less than 0.00015 in)	Claystone**

*Conglomerate has a rounded fragment; breccia has an angular fragment.
**Both are also known as mudstone; called shale if it splits along parallel planes.

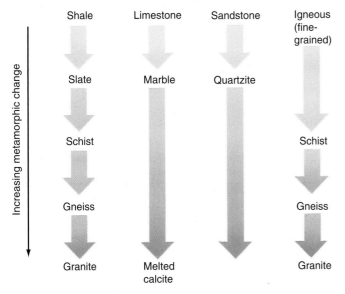

FIGURE 15.12 Increasing metamorphic change occurs with increasing temperatures and pressures. If the melting point is reached, the change is no longer metamorphic, and igneous rocks are formed.

CONCEPTS APPLIED

Sand Sort

Collect dry sand from several different locations. Use a magnifying glass to determine the minerals found in each sample.

---

produce a metamorphic rock. They must be high enough to disrupt the crystal structures to cause them to recrystallize but not high enough to melt the rocks and form igneous rocks (figure 15.12).

The Rock Cycle

Earth is a dynamic planet with a constantly changing surface and interior. As you will see in chapters 16, 17, and 18, internal changes alter Earth's surface by moving the continents and, for example, building mountains that are eventually worn away by weathering and erosion. Seas advance and retreat over the continents as materials are cycled from the atmosphere to the

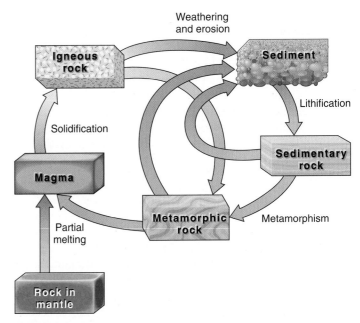

FIGURE 15.13   The rock cycle.

land and from the surface to the interior of Earth and then back again. Rocks are transformed from one type to another through this continual change. There is not a single rock on Earth's surface today that has remained unchanged through Earth's long history. The concept of continually changing rocks through time is called the *rock cycle* (figure 15.13). The rock cycle concept views an igneous, a sedimentary, or a metamorphic rock as the present but temporary stage in the ongoing transformation of rocks to new types. Any particular rock sample today has gone through countless transformations in the 4.6-billion-year history of Earth and will continue to do so in the future.

Chapter 22, The History of Life on Earth, is about how rocks are studied to find information about the history of Earth and the organisms on it. This is one of the main reasons that geologists are interested in rocks. For example, sedimentary rocks can be studied to learn about how the environment and the kinds of organisms found on Earth have changed over time.

EARTH'S INTERIOR

Earth is not a solid, stable sphere that is "solid as a rock." All rocks and rock materials can be made to flow, behaving as warm wax or putty that can be molded. Pressure and temperature just a few kilometers below the surface can be high enough so that rock flows very slowly. Rock flowage takes place as hot, buoyant material deep within Earth moves slowly upward toward the cooler surface. Elsewhere, cold, denser material moves downward.

Earth's internal heat and rock movement are related to some things that happen on the surface, such as moving continents, earthquakes, and volcanoes. Before considering how this happens, we will look at Earth's interior and some internal structures. Information about Earth's interior is deducted mostly from a study of earthquake waves as they move through different parts of Earth. Earthquake waves furnish a sort of "X-ray" machine to view Earth's interior.

The sudden movement of very large sections of rock can produce vibrations that move out as waves throughout the entire Earth. These vibrations are called *seismic waves*. Strong seismic waves are what people feel as a shaking, quaking, or vibrating during an earthquake. Seismic waves are generated when a huge mass of rock breaks and slides into a different position.

Seismic waves radiate outward from an earthquake, spreading in all directions through the solid Earth's interior like sound waves from an explosion. There are basically three kinds of waves:

1. A longitudinal (compressional) wave called a *P-wave*. P-waves are the fastest and move through surface rocks and solid and liquid materials below the surface.
2. A transverse (shear) wave called an *S-wave*. S-waves are second fastest after the P-wave. S-waves do not travel through liquids since liquids do not have the cohesion necessary to transmit a shear, or side-to-side, motion.
3. An up-and-down (crest and trough) wave that travels across the surface called a *surface wave* that is much like a wave on water that moves across the solid surface of Earth. Surface waves are the slowest and occur where S- or P-waves reach the surface. There are two important types of surface waves: *Love waves* and *Rayleigh waves*. Love waves are horizontal S-waves that move side to side. This motion knocks buildings off their foundations and can also destroy bridges and overpasses. Rayleigh waves are more like rolling water waves. Rayleigh waves are more destructive because they produce more up, down, and sideways ground movement for a longer time.

Using data from seismic waves, scientists were able to determine that the interior of Earth can be broken down into three parts (figure 15.14). The *crust* is the outer layer of rock that forms a thin shell around Earth. Below the crust is the *mantle*, a much thicker shell than the crust. The mantle separates the crust from the center part, which is called the *core*. The following section starts on Earth's surface, at the crust, and then digs deeper and deeper into Earth's interior.

The Crust

Earth's **crust** is a thin layer of rock that covers the entire Earth, existing below the oceans as well as making up the continents. According to seismic waves, there are differences in the crust making up the continents and the crust beneath the oceans (table 15.4). These differences are (1) the oceanic crust is much thinner than the continental crust and (2) seismic waves move through the oceanic crust faster than they do through the continental crust. The two types of crust vary because they are made up of different kinds of rock.

The boundary between the crust and the mantle is marked by a sharp increase in the velocity of seismic waves as they pass from the crust to the mantle. Today, this boundary is called the *Mohorovicic discontinuity,* or the "Moho" for short. The boundary is a zone where seismic P-waves increase in velocity because of changes in the composition of the materials. The increase occurs because the composition on both sides of the boundary is different. The mantle is richer in ferromagnesian minerals and poorer in silicon than the crust.

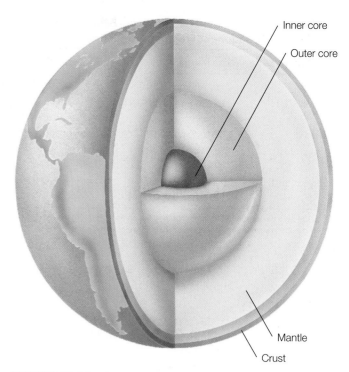

FIGURE 15.14 The structure of Earth's interior.

TABLE 15.4 **Comparison of Oceanic Crust and Continental Crust**

	Oceanic Crust	**Continental Crust**
Age	Less than 200 million years	Up to 3.8 billion years old
Thickness	5 to 8 km (3 to 5 mi)	10 to 70 km (6 to 45 mi)
Density	3.0 g/cm³	2.7 g/cm³
Composition	Basalt	Granite, schist, gneiss

Studies of the Moho show that the crust varies in thickness around Earth's surface. It is thicker under the continents and much thinner under the oceans.

The age of rock samples from Earth's continents has been compared with the age samples of rocks taken from the seafloor by oceanographic ships. This sampling has found the continental crust to be much older, with parts up to 3.8 billion years old. By comparison, the oldest oceanic crust is less than 200 million years old.

Comparative sampling also found that continental crust is a less dense, granite-type rock with a density of about 2.7 g/cm³. Oceanic crust, on the other hand, is made up of basaltic rock with a density of about 3.0 g/cm³. The less dense crust behaves as if it were floating on the mantle, much as less dense ice floats on water. There are exceptions, but in general, the thicker, less dense continental crust "floats" in the mantle above sea level and the thin, dense oceanic crust "floats" in the mantle far below sea level (figure 15.15).

The Mantle

The middle part of Earth's interior is called the **mantle.** The mantle is a thick shell between the core and the crust. This shell

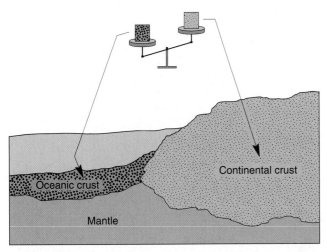

FIGURE 15.15 Continental crust is less dense, granite-type rock, while the oceanic crust is more dense, basaltic rock. Both types of crust behave as if they were floating on the mantle, which is more dense than either type of crust.

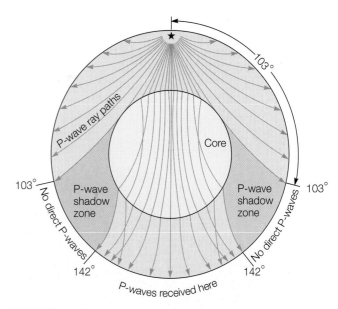

FIGURE 15.16 The P-wave shadow zone, caused by refraction of P-waves within Earth's core.

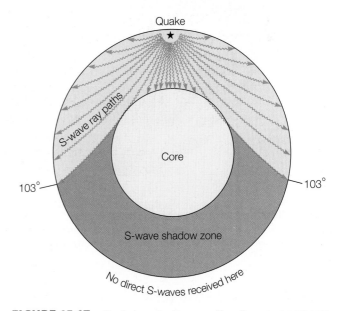

FIGURE 15.17 The S-wave shadow zone. Since S-waves cannot pass through a liquid, at least part of the core is either a liquid or has some of the same physical properties as a liquid.

takes up about 80 percent of the total volume of Earth and accounts for about two-thirds of Earth's total mass. Information about the composition and nature of the mantle comes from (1) studies of seismological data, (2) studies of the nature of meteorites, and (3) studies of materials from the mantle that have been ejected to Earth's surface by volcanoes. The evidence from these separate sources all indicates that the mantle is composed of silicates, predominantly the ferromagnesian silicate *olivine*. Meteorites, as mentioned in chapter 13, are basically either iron meteorites or stony meteorites. Most of the stony meteorites are silicates with a composition that would produce the chemical composition of olivine if they were melted and the heavier elements were separated by gravity. This chemical composition also agrees closely with the composition of basalt, the most common volcanic rock found on the surface of Earth.

The Core

Information about the nature of the **core,** the center part of Earth, comes from several sources of information. Seismological data provide the primary evidence for the structure of the core of Earth. Seismic P-waves spread through Earth from a large earthquake. Figure 15.16 shows how the P-waves spread out, soon arriving at seismic measuring stations all around the world. However, there are places between 103° and 142° of arc from the earthquake that do not receive P-waves. This region is called the *P-wave shadow zone*. The P-wave shadow zone is explained by P-waves being refracted by the core, leaving a shadow. The paths of P-waves can be accurately calculated, so the size and shape of Earth's core can also be accurately calculated.

Seismic S-waves leave a different pattern at seismic receiving stations around Earth. S- (sideways or transverse) waves can travel only through solid materials. An *S-wave shadow zone* also exists and is larger than the P-wave shadow zone (figure 15.17). S-waves are not recorded in the entire region more than 103° away from the epicenter. The S-wave shadow zone seems to indicate that S-waves do not travel through the core at all. If this is true, it implies that the core of Earth is a liquid, or at least acts like a liquid.

Analysis of P-wave data suggests that the core has two parts: a *liquid outer core* and a *solid inner core*. Both the P-wave and S-wave data support this conclusion. Overall, the core makes up about 15 percent of Earth's total volume and about one-third of its mass.

Evidence from the nature of meteorites indicates that Earth's core is mostly iron. Earth has a strong magnetic field that has its sources in the turbulent flow of the liquid part of its core. To produce such a field, the material of the core would have to be an electrical conductor, that is, a metal such as iron. There are two general kinds of meteorites that fall to Earth: (1) stony meteorites that are made of silicate minerals and (2) iron meteorites that are made of iron or of a nickel-iron alloy. Since Earth has a silicate-rich crust and mantle, by analogy, Earth's core must consist of iron or a nickel and iron alloy.

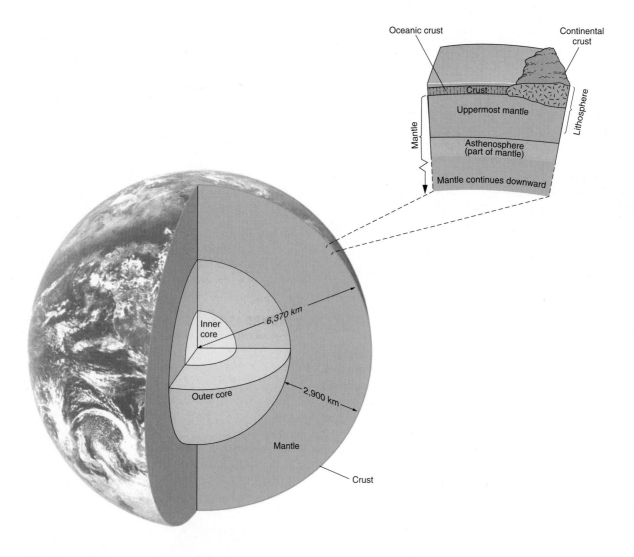

FIGURE 15.18 Earth's interior, showing the weak, plastic layer called the *asthenosphere*. The rigid, solid layer above the asthenosphere is called the *lithosphere*. The lithosphere is broken into plates that move on the upper mantle like giant ice sheets floating on water. This arrangement is the foundation for plate tectonics, which explains many changes that occur on Earth's surface such as earthquakes, volcanoes, and mountain building.

A More Detailed Structure

There is strong evidence that Earth has a layered structure with a core, mantle, and crust. This description of the structure is important for historical reasons and for understanding how Earth evolved over time. There is another, more detailed structure that can be described. This structure is far more important in understanding the history and present appearance of Earth's surface, including the phenomena of earthquakes and volcanoes.

The important part of this different detailed description of Earth's interior was first identified from seismic data. There is a thin zone in the mantle where seismic waves undergo a sharp decrease in velocity (figure 15.18). This low-velocity zone is evidently a hot, elastic semiliquid layer that extends around the entire Earth. It is called the **asthenosphere,** after the Greek word for "weak shell." The asthenosphere is weak because it is plastic, mobile, and yields to stresses. In some regions, the asthenosphere contains pockets of magma.

The rocks above and below the asthenosphere are rigid, solid, and brittle. The solid layer above the asthenosphere is called the **lithosphere,** after the Greek word for "stone shell." The lithosphere is also known as the "strong layer" in contrast to the "weak layer" of the asthenosphere. The lithosphere includes the entire crust, the Moho, and the upper part of the mantle. As you will see in the section on plate tectonics, the asthenosphere is one important source of magma that reaches Earth's surface. It is also a necessary part of the mechanism involved in the movement of the crust. The lithosphere is made up of comparatively rigid plates that are moving, floating in the upper mantle like giant ice sheets floating in the ocean.

PLATE TECTONICS

If you observe the shape of the continents on a world map or a globe, you will notice that some of the shapes look as if they would

■	0—5 million years
■	5—23 million years
■	23—38 million years
■	38—53 million years
■	53—65 million years
□	65—135 million years

FIGURE 15.22 The pattern of seafloor ages on both sides of the Mid-Atlantic Ridge reflects seafloor spreading activity. Younger rocks are found closer to the ridge.

or 2.5 m (about 8 ft). This slow rate is why most people do not recognize that the seafloor—and the continents—move.

Lithosphere Plates and Boundaries

The strong evidence for seafloor spreading soon led to the development of a new theory called **plate tectonics.** According to plate tectonics, the lithosphere is broken into a number of fairly rigid plates that move on the asthenosphere. Some plates, as shown in figure 15.24, contain continents and part of an ocean basin, while other plates contain only ocean basins. The plates move, and the movement is helping to explain why mountains form where they do, the occurrence of earthquakes and volcanoes, and in general, the entire changing surface of Earth.

Earthquakes, volcanoes, and most rapid changes in Earth's crust occur at the edge of a plate, which is called a *plate boundary.* Three general kinds of plate boundaries describe how one plate moves relative to another: divergent, convergent, and transform boundaries.

Divergent boundaries occur between two plates moving away from each other. Magma forms as the plates separate, decreasing pressure on the mantle below. This molten material from the asthenosphere rises, cools, and adds new crust to the edges of the separating plates. The new crust tends to move horizontally from both sides of the divergent boundary, usually known as an oceanic ridge. A divergent boundary is thus a **new crust zone.** Most new crust zones are presently on the seafloor, producing seafloor spreading (figure 15.25).

The Mid-Atlantic Ridge is a divergent boundary between the South American and African Plates, extending

FIGURE 15.23 Formation of magnetic strips on the seafloor. As each new section of seafloor forms at the ridge, iron minerals become magnetized in a direction that depends on the orientation of Earth's field at that time. This makes a permanent record of reversals of Earth's magnetic field.

north between the North American and Eurasian Plates (see figure 15.24). This ridge is one segment of the global mid-ocean ridge system that encircles Earth. The results of divergent plate movement can be seen in Iceland, where the Mid-Atlantic Ridge runs as it separates the North American and Eurasian Plates. In the northeastern part of Iceland, ground cracks are widening, often accompanied by volcanic activity. The movement was measured extensively between 1975 and 1984, when displacements caused a total separation of about 7 m (about 23 ft).

The measured rate of spreading along the Mid-Atlantic Ridge ranges from 1 to 6 centimeters per year. This may seem slow, but the process has been going on for millions of years and has caused a tiny inlet of water between the continents of Europe, Africa, and the Americas to grow into the vast Atlantic Ocean that exists today.

Another major ocean may be in the making in East Africa, where a divergent boundary has already moved Saudi Arabia away from the African continent, forming the Red Sea. If this spreading between the African Plate and the Arabian Plate

FIGURE 15.24 The major plates of the lithosphere that move on the asthenosphere.

Source: After W. Hamilton, U.S. Geological Survey.

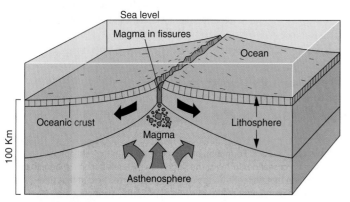

FIGURE 15.25 A diverging boundary at a mid-oceanic ridge. Hot asthenosphere wells upward beneath the ridge crest. Magma forms and squirts into fissures. Solid material that does not melt remains as mantle in the lower part of the lithosphere. As the lithosphere moves away from the spreading axis, it cools, becomes denser, and sinks to a lower level.

continues, the Indian Ocean will flood the area and the easternmost corner of Africa will become a large island.

Convergent boundaries occur between two plates moving toward each other. The creation of new crust at a divergent boundary means that old crust must be destroyed somewhere else at the same rate, or else Earth would have a continuously expanding diameter. Old crust is destroyed by returning to the asthenosphere at convergent boundaries. The collision produces an elongated belt of down-bending called a **subduction zone.** The lithosphere of one plate, which contains the crust, is subducted beneath the second plate and partially melts, then becoming part of the mantle. The more dense components of this may become igneous materials that remain in the mantle. Some of it may eventually migrate to a spreading ridge to make new crust

again. The less dense components may return to the surface as a silicon, potassium, and sodium-rich lava, forming volcanoes on the upper plate, or it may cool below the surface to form a body of granite. Thus, the oceanic lithosphere is being recycled through this process, which explains why ancient seafloor rocks do not exist. Convergent boundaries produce related characteristic geologic features depending on the nature of the materials in the plates, and there are three general possibilities: (1) converging continental and oceanic plates, (2) converging oceanic plates, and (3) converging continental plates.

As an example of *ocean-continent plate convergence,* consider the plate containing the South American continent (the South American Plate) and its convergent boundary with an oceanic plate (the Nazca Plate) along its western edge. Continent-oceanic plate convergence produces a characteristic set of geologic features as the oceanic plate of denser basaltic material is subducted beneath the less dense granite-type continental plate (figure 15.26). The subduction zone is marked by an oceanic trench (the Peru-Chile Trench), deep-seated earthquakes, and volcanic mountains on the continent (the Andes Mountains). The trench is formed from the down-bending associated with subduction and the volcanic mountains from subducted and melted crust that rise up through the overlying plate to the surface. The earthquakes are associated with the movement of the subducted crust under the overlying crust.

Ocean-ocean plate convergence produces another set of characteristics and related geologic features (figure 15.27). The northern boundary of the oceanic Pacific Plate, for example, converges with the oceanic part of the North American Plate near the Bering Sea. The Pacific Plate is subducted, forming the Aleutian oceanic trench with a zone of earthquakes that are shallow near the trench and progressively more deep-seated toward the continent. The deeper earthquakes are associated with the movement of more deeply subducted crust into the mantle. The Aleutian Islands are typical **island arcs,** curving chains of volcanic islands that occur over the belt of deep-seated earthquakes. These islands form where the melted subducted material rises up through the overriding plate above sea level. The Japanese, Marianas, and Indonesians are similar groups of arc islands associated with converging oceanic-oceanic plate boundaries.

During *continent-continent plate convergence,* subduction does not occur as the less dense, granite-type materials tend to resist subduction (figure 15.28). Instead, the colliding plates pile up into a deformed and thicker crust of the lighter material. Such a collision produced the thick, elevated crust known as the Tibetan Plateau and the Himalayan Mountains.

Transform boundaries occur between two plates sliding by each other. Crust is neither created nor destroyed at transform boundaries as one plate slides horizontally past another along a long, vertical fault. The movement is neither smooth nor equal along the length of the fault, however, as short seg-ments move independently with sudden jerks that are separated by periods without motion. The Pacific Plate, for example, is moving slowly to the northwest, sliding past the North American Plate. The San Andreas fault is one boundary along the California coastline. Vibrations from plate movements along this boundary are the famous California earthquakes.

Present-Day Understandings

The theory of plate tectonics, developed during the late 1960s and early 1970s, is new compared to most major scientific theories. Measurements are still being made, evidence is being gathered and evaluated, and the exact number of plates and their boundaries are yet to be determined with certainty. The major question that remains to be answered is what drives the plates, moving them apart, together, and past each other? One explanation is that slowly turning *convective cells* in the plastic asthenosphere drive the plates (figure 15.29). According to this hypothesis, hot fluid materials rise at the diverging boundaries. Some of the material escapes to form new crust, but most of it spreads out beneath the lithosphere. As it moves beneath the lithosphere, it drags the overlying plate with it. Eventually, it cools and sinks back inward under a subduction zone.

There is uncertainty about the existence of convective cells in the asthenosphere and their possible role because of a lack of clear evidence. Seismic data is not refined enough to show convective cell movement beneath the lithosphere. In addition, deep-seated earthquakes occur to depths of about 700 km (435 mi), which means that descending materials—parts of a subducted plate—must extend to that depth. This could mean

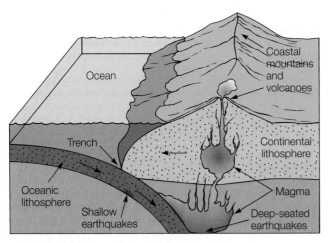

FIGURE 15.26 Ocean-continent plate convergence. This type of plate boundary accounts for shallow and deep-seated earthquakes, an oceanic trench, volcanic activity, and mountains along the coast.

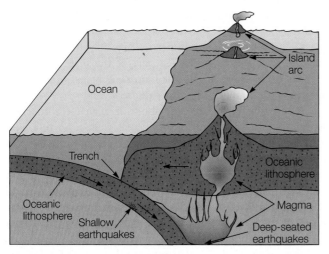

FIGURE 15.27 Ocean-ocean plate convergence. This type of plate convergence accounts for shallow and deep-focused earthquakes, an oceanic trench, and a volcanic arc above the subducted plate.

CONCEPTS APPLIED

New Earthquakes and Volcanoes

Locate and label the major plates of the lithosphere on an outline map of the world according to the most recent findings in plate tectonics. Show all types of boundaries and associated areas of volcanoes and earthquakes.

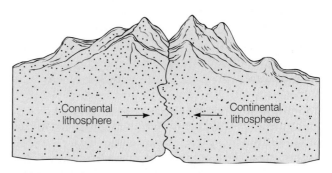

FIGURE 15.28 Continent-continent plate convergence. Rocks are deformed, and some lithosphere thickening occurs, but neither plate is subducted to any great extent.

A Closer Look

Seismic Tomography

The CAT scan is a common diagnostic imaging procedure that combines the use of X rays with computer technology. CAT stands for "computed axial tomography" and the word *tomography* means "drawing slices." The medical CAT scan shows organs of interest by making X-ray images from many different angles as the source of the X rays moves around the patient. CAT scan images are assembled by a computer into a three-dimensional picture that can show organs, bones, and tissues in great detail.

The CAT scan is applied to Earth's interior in a *seismic tomography* procedure. It works somewhat like the medical CAT scan, but seismic tomography uses seismic waves instead of X rays. The velocities of S- and P-seismic waves vary with depth, changing with density, temperature, pressure, and the composition of Earth's interior. Interior differences in temperature, pressure, and composition are what cause the motion of tectonic plates on Earth's surface. Thus,

a picture of Earth's interior can be made by mapping variations in seismic wave speeds.

Suppose an earthquake occurs and a number of seismic stations record when the S- and P- waves arrive. From these records, you could compare the seismic velocity data and identify if there were low seismic velocities between the source and some of the receivers. The late arrival of seismic waves could mean some difference in the structure of Earth between the source and the observing station. When an earthquake occurs in a new location, more data will be collected at different observing stations, which provides additional information about the shape and structure of whatever is slowing the seismic waves. Now, imagine repeating these measurements for many new earthquakes until you have enough data to paint a picture of what is beneath the surface.

Huge amounts of earthquake data—perhaps 10 million data points in 5 million

groups—are needed to construct a picture of Earth's interior. A really fast computer may take days to process this much data and construct cross sections through some interesting place, such as a subduction zone where an oceanic plate dives into the mantle.

Seismic tomography has also identified massive plumes of molten rock rising toward the surface from deep within Earth. These *superplumes* originate from the base of the mantle, rising to the lithosphere. The hot material was observed to spread out horizontally under the lithosphere toward midocean ridges. This may contribute to tectonic plate movement. Regions above the superplumes tend to bulge upward, and other indications of superplumes, such as variations in gravity, have been measured. Scientists continue to work with new seismic tomography data with more precise images and higher resolutions to better describe the interior structure of Earth.

Myths, Mistakes, and Misunderstandings

Bye-Bye California?

It is a myth that California will eventually fall off the continent into the ocean. The San Andreas fault is the boundary between the Pacific and North American Plates. The Pacific Plate is moving northwest along the North American Plate at 45 mm per year (about the rate your fingernails grow). The plates are moving horizontally by each other, so there is no reason to believe California will fall into the ocean. However, some 15 million years and millions of earthquakes from now, Los Angeles might be across the bay from San Francisco. See "Earthquakes, Megaquakes, and the Movies" at http://earthquake.usgs.gov/learning/topics/?topicID=36.

Heat and gravity are important in a proposed mechanism of plate motion called "*ridge-push*" (figure 15.30). This idea has a plate cooling and thickening as it moves away from a divergent boundary. As it subsides, it cools asthenospheric mantle to lithospheric mantle, forming a sloping boundary between the lithosphere and the asthenosphere. The plate slides down this boundary.

Another proposed mechanism of plate motion is called "*slab-pull*" (figure 15.31). In this mechanism, the subducting plate is colder and therefore denser than the surrounding hot mantle, so it pulls the surface part of the plate downward. Density of the slab may also be increased by loss of water and reforming of minerals into more dense forms.

What is generally accepted about the plate tectonic theory is the understanding that the solid materials of Earth are engaged in a continual cycle of change. Oceanic crust is subducted, melted, then partly returned to the crust as volcanic igneous rocks in island arcs and along continental plate boundaries. Other parts of the subducted crust become mixed with the upper mantle, returning as new crust at diverging boundaries. The materials of the crust and the mantle are thus cycled back and forth in a mixing that may include the deep mantle and the core as well. There is more to this story of a dynamic Earth that undergoes a constant change. The story continues in chapters 16, 17, and 18 with different cycles to consider.

that a convective cell might operate all the way down to the core-mantle boundary some 2,900 km (about 1,800 mi) below the surface. This presents another kind of problem because little is known about the lower mantle and how it interacts with the upper mantle. Theorizing without information is called speculation, and that is the best that can be done with existing data. The full answer may include the role of heat and the role of gravity.

"]

Measuring Plate Movement

According to the theory of plate tectonics, Earth's outer shell is made up of moving plates. The plates making up the continents are about 100 km (62 mi) thick and are gradually drifting at a rate of about 0.5 to 10 cm (0.2 to 4 in) per year. This reading is about one way that scientists know that Earth's plates are moving and how this movement is measured.

The very first human lunar landing mission took place in July 1969. Astronaut Neil Armstrong stepped onto the lunar surface, stating, "That's one small step for a man, one giant leap for mankind." In addition to fulfilling a dream, the *Apollo* project carried out a program of scientific experiments. The *Apollo 11* astronauts placed a number of experiments on the lunar surface in the Sea of Tranquility. Among the experiments was the first Laser Ranging Retro-Reflector

Experiment, which was designed to reflect pulses of laser light from Earth. Later, three more reflectors were placed on the Moon, including two by other *Apollo* astronauts and one by an unmanned Soviet *Lunakhod 2* lander.

The McDonald Observatory in Texas, the Lure Observatory on the island of Maui, Hawaii, and a third observatory in southern France have regularly sent laser beams through optical telescopes to the reflectors. The return signals, which are too weak to be seen with the unaided eye, are detected and measured by sensitive detection equipment at the observatories. The accuracy of these measurements, according to NASA reports, is equivalent to determining the distance between a point on the east and a point on the west coasts of the United States to an accuracy of 0.5 mm, about

the size of the period at the end of this sentence.

Reflected laser light experiments have found that the Moon is pulling away from Earth at about 4 cm/yr (about 1.6 in/yr), that the shape of Earth is slowly changing, undergoing isostatic adjustment from the compression by the glaciers during the last ice age, and that the observatory in Hawaii is slowly moving away from the one in Texas. This provides a direct measurement of the relative drift of two of Earth's tectonic plates. Thus, one way that changes on the surface of Earth are measured is through lunar ranging experiments. Results from lunar ranging, together with laser ranging to the artificial satellites in Earth's orbit, have revealed the small but constant drift rate of the plates making up Earth's dynamic surface.

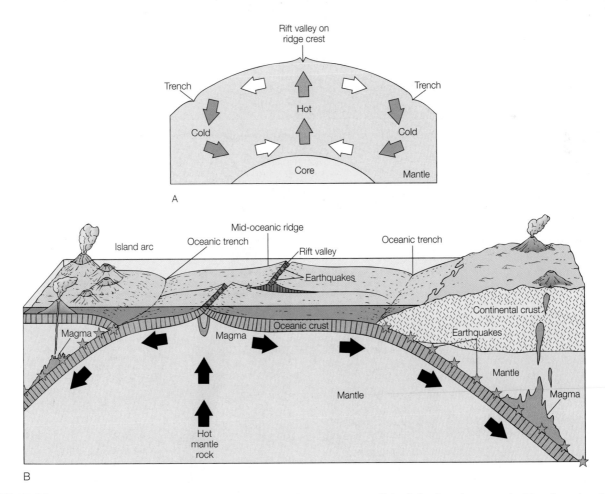

FIGURE 15.29   Not to scale. One idea about convection in the mantle has a convection cell circulating from the core to the lithosphere, dragging the overlying lithosphere laterally away from the oceanic ridge.

People Behind the Science

Frederick John Vine (1939–1988)

Frederick Vine was an English geophysicist whose work was an important contribution to the development of the theory of plate tectonics.

Vine was born in Brentford, Essex, England, and was educated at Latymer Upper School, London, and at St. John's College, Cambridge. From 1967 to 1970, he worked as assistant professor in the Department of Geological and Geophysical Sciences at Princeton University, before returning to England to become reader (1970) and then professor (1974) in the School of Environmental Sciences at the University of East Anglia.

In Cambridge in 1963, Vine collaborated with his supervisor, Drummond Hoyle Matthews (1931–), and wrote a paper, "Magnetic Anomalies over Ocean Ridges," that provided additional evidence for Harry H. Hess' (1962) seafloor spreading hypothesis. Alfred Wegener's original 1912 theory of continental drift (Wegener's hypothesis) had been met with hostility at the time because he could not explain why the continents had drifted apart, but Hess had continued his

work developing the theory of seafloor spreading to explain the fact that as the oceans grew wider, the continents drifted apart.

Following the work of Brunhes and Matonori Matuyama in the 1920s on magnetic reversals, Vine and Matthews predicted that new rock emerging from the oceanic ridges would intermittently produce material of opposing magnetic polarity to the old rock. They applied paleomagnetic studies to the ocean ridges in the North Atlantic and were able to argue that parallel belts of different magnetic polarities existed on either side of the ridge crests. This evidence was vital proof of Hess' hypothesis. Studies on ridges in other oceans also showed the existence of these magnetic anomalies. Vine and Matthews' hypothesis was widely accepted in 1966 and confirmed Dietz and Hess' earlier work. Their work was crucial to the development of the theory of plate tectonics, and it revolutionized the earth sciences.

Source: Modified from the *Hutchinson Dictionary of Scientific Biography.* © RM, 2009. All rights reserved. Helicon Publishing is a division of RM.

FIGURE 15.30 Ridge-push. A plate may slide downhill on the sloping boundary between the lithosphere and the asthenosphere at the base of the plate.

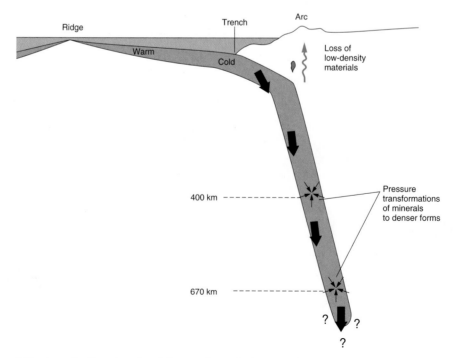

FIGURE 15.31 Slab-pull. The dense, leading edge of a subducting plate pulls the rest of the plate along. Plate density increases due to cooling, loss of low-density material, and pressure transformation of minerals to denser forms.

SUMMARY

The elements silicon and oxygen make up 75 percent of all the elements in the outer layer, or *crust,* of Earth. The elements combine to make crystalline chemical compounds called minerals. A *mineral* is defined as a naturally occurring, inorganic solid element or compound with a crystalline structure.

About 92 percent of the minerals of Earth's crust are composed of silicon and oxygen, the *silicate minerals.* The basic unit of the silicates is a *tetrahedral structure* that combines with positive metallic ions or with other tetrahedral units to form chains, sheets, or an interlocking framework. The *ferromagnesian silicates* are tetrahedral structures combined with ions of iron, magnesium, calcium, and other elements. The ferromagnesian silicates are darker in color and more dense than other silicates. The *nonferromagnesian silicates* do not have irons or magnesium ions, and they are lighter in color and less dense than the ferromagnesians. The *nonsilicate minerals* do not contain silicon and are carbonates, sulfates, oxides, halides, sulfides, and native elements.

A *rock* is defined as an aggregation of one or more minerals that have been brought together into a cohesive solid. *Igneous rocks* formed as hot, molten *magma* cooled and crystallized to firm, hard rocks. *Sedimentary rocks* are formed from *sediments,* accumulations of weathered rock materials that settle out of the atmosphere or out of water. Sediments become sedimentary rocks through a rock-forming process that involves both the *compaction* and *cementation* of the sediments. *Metamorphic rocks* are previously existing rocks that have been changed by heat, pressure, or hot solution into a different kind of rock without melting. The *rock cycle* is a concept that an igneous, a sedimentary, or a metamorphic rock is a temporary stage in the ongoing transformation of rocks to new types.

Earth has a layered interior that formed as Earth's materials underwent *differentiation,* the separation of materials while in the molten state. The center part, or *core,* is predominantly iron with a solid inner part and a liquid outer part. The core makes up about 15 percent of Earth's total volume and about a third of its total mass. The *mantle* is the middle part of Earth's interior that accounts for about two-thirds of Earth's total mass and about 80 percent of its total volume. The *Mohorovicic discontinuity* separates the outer layer, or *crust,* of Earth from the mantle. The crust of the *continents* is composed mostly of less dense granite-type rock. The crust of the *ocean basins* is composed mostly of the more dense basaltic rocks.

Another way to consider Earth's interior structure is to consider the weak layer in the upper mantle, the *asthenosphere* that extends around the entire Earth. The rigid, solid, and brittle layer above the asthenosphere is called the *lithosphere.* The lithosphere includes the entire crust, the Moho, and the upper part of the mantle.

Evidence from the ocean floor revived interest in the idea that continents could move. The evidence for *seafloor spreading* came from related observations concerning oceanic ridge systems, sediment and fossil dating of materials outward from the ridge, and magnetic patterns of seafloor rocks. Confirmation of seafloor spreading led to the *plate tectonic theory.* According to plate tectonics, new basaltic crust is added at *diverging boundaries* of plates, and old crust is *subducted* at *converging boundaries.* Mountain building, volcanoes, and earthquakes are seen as *related geologic features* that are caused by plate movements. The force behind the movement of plates is uncertain, but it may involve *convection* in the deep mantle.

KEY TERMS

asthenosphere (p. **338**)
convergent boundaries (p. **342**)
core (p. **337**)
crust (p. **336**)
divergent boundaries (p. **341**)
igneous rocks (p. **333**)

island arcs (p. **343**)
lithosphere (p. **338**)
mantle (p. **336**)
metamorphic rocks (p. **334**)
mineral (p. **328**)
new crust zone (p. **341**)

oceanic ridges (p. **340**)
oceanic trenches (p. **340**)
plate tectonics (p. **341**)
rift (p. **340**)
rock (p. **331**)
seafloor spreading (p. **340**)

sedimentary rocks (p. **333**)
silicates (p. **329**)
subduction zone (p. **342**)
transform boundaries (p. **343**)

APPLYING THE CONCEPTS

1. Sedimentary rocks are formed by the processes of compaction and
   a. pressurization.
   b. melting.
   c. cementation.
   d. heating but not melting.

2. Which type of rock probably existed first, starting the rock cycle?
   a. metamorphic
   b. igneous
   c. sedimentary
   d. All of the above are correct.

3. From seismological data, Earth's shadow zone indicates that part of Earth's interior must be
   a. liquid.
   b. solid throughout.
   c. plastic.
   d. hollow.

4. The Mohorovicic discontinuity is a change in seismic wave velocity that is believed to take place because of
   a. structural changes in minerals of the same composition.
   b. changes in the composition on both sides of the boundary.
   c. a shift in the density of minerals of the same composition.
   d. changes in the temperature with depth.

5. The oldest rocks are found in (the)
   a. continental crust.
   b. oceanic crust.
   c. neither, since both are the same age.

6. The least dense rocks are found in (the)
   a. continental crust.
   b. oceanic crust.
   c. neither, since both are the same density.

7. The idea of seafloor spreading along the Mid-Atlantic Ridge was supported by evidence from
   a. changes in magnetic patterns and ages of rocks moving away from the ridge.
   b. faulting and volcanoes on the continents.
   c. the observation that there was no relationship between one continent and another.
   d. All of the above are correct.

8. According to the plate tectonics theory, seafloor spreading takes place at a
   a. convergent boundary.   c. divergent boundary.
   b. subduction zone.   d. transform boundary.

9. The presence of an oceanic trench, a chain of volcanic mountains along the continental edge, and deep-seated earthquakes is characteristic of a (an)
   a. ocean-ocean plate convergence.
   b. ocean-continent plate convergence.
   c. continent-continent plate convergence.
   d. None of the above is correct.

10. The presence of an oceanic trench with shallow earthquakes and island arcs with deep-seated earthquakes is characteristic of a (an)
    a. ocean-ocean plate convergence.
    b. ocean-continent plate convergence.
    c. continent-continent plate convergence.
    d. None of the above is correct.

11. The ongoing occurrence of earthquakes without seafloor spreading, oceanic trenches, or volcanoes is most characteristic of a
    a. convergent boundary between plates.
    b. subduction zone.
    c. divergent boundary between plates.
    d. transform boundary between plates.

12. The evidence that Earth's core is part liquid or acts like a liquid comes from (the)
    a. P-wave shadow zone.
    b. S-wave shadow zone.
    c. meteorites.
    d. All of the above are correct.

Answers

1. c 2. b 3. a 4. b 5. a 6. a 7. a 8. c 9. b 10. a 11. d 12. b

QUESTIONS FOR THOUGHT

1. Briefly describe the rock-forming process that changes sediments into solid rock.

2. What are metamorphic rocks? What limits the maximum temperatures possible in metamorphism? Explain.

3. What evidence provides information about the nature of Earth's core?

4. What is the asthenosphere? Why is it important in modern understandings of Earth?

5. Describe the origin of the magnetic strip patterns found in the rocks along an oceanic ridge.

6. Explain why ancient rocks are not found on the seafloor.

7. Describe the three major types of plate boundaries and what happens at each.

8. Briefly describe the theory of plate tectonics and how it accounts for the existence of certain geologic features.

9. What is an oceanic trench? What is its relationship to major plate boundaries? Explain this relationship.

10. Describe the probable source of all the earthquakes that occur in southern California.

11. The northwestern coast of the United States has a string of volcanoes running along it. According to plate tectonics, what does this mean about this part of the North American Plate? What geologic feature would you expect to find on the seafloor off the northwestern coast? Explain.

12. Explain how the crust of Earth is involved in a dynamic, ongoing recycling process.

FOR FURTHER ANALYSIS

1. Is ice a mineral? Describe reasons to argue that ice is a mineral. Describe reasons to argue that ice is not a mineral.

2. What are the significant similarities and differences between igneous, sedimentary, and metamorphic rocks?

3. Why are there no active volcanoes in the eastern United States or Canada? Explain why you would or would not expect volcanoes there in the future.

4. Describe cycles that occur on Earth's surface and cycles that occur between the surface and the interior. Explain why these cycles do not exist on other planets of the solar system.

5. The rock cycle describes how igneous, metamorphic, and sedimentary rocks are changed into each other. If this is true, analyze why most of the rocks on Earth's surface are sedimentary.

6. If ice is a mineral, is a glacier a rock? Describe reasons to support or oppose calling a glacier a rock according to the definition of a rock.

7. Discuss evidence that would explain why plate tectonics occurs on Earth but not on other planets.

INVITATION TO INQUIRY

Measuring Plate Motion

Tectonic plate motion can be measured with several relatively new technologies, including satellite laser ranging and use of the Global Positioning System. Start your inquiry by visiting the Tectonic Plate Motion website at cddisa.gsfc.nasa.gov/926/slrtecto.html. Study the regional plate motion of the North American Plate, for example. Note the scale of 50 mm/yr, then measure to find the rate of movement at the different stations shown. Is this recent data on plate motion consistent with other available information on plate motion? How can you account for the different rates of motion at adjacent stations?

16

Earth's Surface

Folding, faulting, and lava flows, such as the one you see here, tend to build up, or elevate, Earth's surface.

CORE **CONCEPT**

The surface of the Earth is involved in an ongoing cycle of destruction and renewal.

OUTLINE

Earth Science

▷ Earth's surface is made up of a number of rigid plates that are moving (Ch. 15).

Life Science

▷ Erosion of rock releases phosphorus for use by plants (Ch. 23).

▷ Sedimentation traps organisms and forms fossils (Ch. 22)

▷ Layers of sedimentary rock reveal the history of life on Earth (Ch. 22).

OVERVIEW

The central idea of plate tectonics, which was discussed in chapter 15, is that Earth's surface is made up of rigid plates that are slowly moving. Since the plates and the continents riding on them are in constant motion, any given map of the world is only a snapshot that shows the relative positions of the continents at a given time. The continents occupied different positions in the distant past. They will occupy different positions in the distant future. The surface of Earth, which seems so solid and stationary, is in fact mobile.

Plate tectonics has changed the accepted way of thinking about the solid, stationary nature of Earth's surface and ideas about the permanence of the surface as well. The surface of Earth is no longer viewed as having a permanent nature but is understood to be involved in an ongoing cycle of destruction and renewal. Old crust is destroyed as it is plowed back into the mantle through subduction, becoming mixed with the mantle. New crust is created as molten materials move from the mantle through seafloor spreading and volcanoes. Over time, much of the crust must cycle into and out of the mantle.

The movement of plates, the crust-mantle cycle, and the rock cycle all combine to produce a constantly changing surface. There are basically two types of surface changes: (1) changes that originate within Earth, resulting in a building up of the surface (figure 16.1) and (2) changes that occur when rocks are exposed to the atmosphere and water, resulting in a sculpturing and tearing down of the surface. This chapter is about changes in the land. The concepts of this chapter will provide you with something far more interesting about Earth's surface than the scenic aspect. The existence of different features (such as mountains, folded hills, islands) and the occurrence of certain events (such as earthquakes, volcanoes, faulting) are all related. The related features and events also have a story to tell about Earth's past, a story about the here and now, and yet another story about the future.

INTERPRETING EARTH'S SURFACE

Because many geologic changes take place slowly, it is difficult for a person to see significant change occur to mountains, canyons, and shorelines in the brief span of a lifetime. Given a mental framework based on a lack of appreciation of change over geologic time, how do you suppose people interpreted the existence of features such as mountains and canyons? Some believed, as they had observed in their lifetimes, that the mountains and canyons had "always" been there. Statements such as "unchanging as the hills" or "old as the hills" illustrate this lack of appreciation of change over geologic time. Others did not believe the features had always been there but believed they were formed by a sudden, single catastrophic event (figure 16.2). A catastrophe created a feature of Earth's surface all at once, with little or no change occurring since that time. The Grand Canyon, for example, was not interpreted as the result of incomprehensibly slow river erosion but as the result of a giant crack or rip that appeared in the surface. The canyon that you see today was interpreted as forming when Earth split open and

the Colorado River fell into the split. This interpretation was used to explain the formation of major geologic features based on the little change observed during a person's lifetime.

About two hundred years ago, the idea of unchanging, catastrophically formed landscapes was challenged by James Hutton, a Scottish physician. Hutton, who is known today as the founder of modern geology, traveled widely throughout the British Isles. Hutton was a keen observer of rocks, rock structures, and other features of the landscape. He noted that sandstone, for example, was made up of rock fragments that appeared to be (1) similar to the sand being carried by rivers and (2) similar to the sand making up the beaches next to the sea. He also noted fossil shells of sea animals in sandstone on the land, while the living relatives of these animals were found in the shallow waters of the sea. This and other evidence led Hutton to realize that rocks were being ground into fragments, then carried by rivers to the sea. He surmised that these particles would be reformed into rocks later, then lifted and shaped into the hills and mountains of the land. He saw all this as quiet, orderly change that required only *time* and the

FIGURE 16.1 An aerial view from the south of the eruption of Mount St. Helens volcano on May 18, 1980.

FIGURE 16.2 Would you believe that this rock island has "always" existed where it is? Would you believe it was formed by a sudden, single event? What evidence would it take to convince you that the rock island formed ever so slowly, starting as a part of southern California and moving very slowly, at a rate of centimeters per year, to its present location near the coast of Alaska?

ongoing work of the water and some forces to make the sediments back into rocks. With Hutton's logical conclusion came the understanding that Earth's history could be interpreted by tracing it backwards, from the present to the past. This tracing required a frame of reference of slow, uniform change, not the catastrophic frame of reference of previous thinkers. The frame of reference of uniform changes is today called the **principle of uniformity** (also called *uniformitarianism*). The principle of uniformity is often represented by a statement that "the present is the key to the past." This statement means that the same geologic processes you see changing rocks today are the very same processes that changed them in the ancient past, although not necessarily at the same rate. The principle of uniformity does not *exclude* the happening of sudden or catastrophic events on the surface of Earth. A violent volcanic explosion, for example, is a catastrophic event that most certainly modifies the surface of Earth. What the principle of uniformity does state is that the physical and chemical laws we observe today operated exactly the same in the past. The rates of operation may or may not have been the same in the past, but the events you see occurring today are the same events that occurred in the past. Given enough time, you can explain the formation of the structures of Earth's surface with known events and concepts.

The principle of uniformity has been used by geologists since the time of Hutton. The concept of how the constant changes occur has evolved with the development of plate tectonics, but the basic frame of reference is the same. You will see how the principle of uniformity is applied by first considering what can happen to rocks and rock layers that are deeply buried.

PROCESSES THAT BUILD UP THE SURFACE

All the possible movements of Earth's plates, including drift toward or away from other plates, and any process that deforms Earth's surface are included in the term *diastrophism*. Diastrophism is the process of deformation that changes Earth's surface. It produces many of the basic structures you see on Earth's surface, such as plateaus, mountains, and folds in the crust. The movement of magma is called *vulcanism* or *volcanism*. Diastrophism, volcanism, and earthquakes are closely related, and their occurrence can usually be explained by events involving plate tectonics. The results of diastrophism are discussed in the section on stress and strain, which is followed by a discussion of earthquakes, volcanoes, and mountain chains.

Stress and Strain

Any solid material responds to a force in a way that depends on the extent of coverage (force per unit area, or pressure), the nature of the material, and other variables such as the temperature. Consider, for example, what happens if you place the point of a ballpoint pen on the side of an aluminum pop (soda) can and apply an increasing pressure. With increasing pressure, you can observe at least four different and separate responses:

1. At first, the metal successfully resists a slight pressure and *nothing happens.*
2. At a somewhat greater pressure, you will be able to deform, or bend, the metal into a concave surface. The metal will return to its original shape, however, when the pressure is removed. This is called an *elastic deformation* since the metal was able to spring back into its original shape.
3. At a still greater pressure, the metal is deformed to a concave surface, but this time, the metal does not return to its original shape. This means the *elastic limit* of the metal has been exceeded, and it has now undergone a *plastic deformation.* Plastic deformation permanently alters the shape of a material.
4. Finally, at some great pressure, the metal will rupture, resulting in a *break* in the material.

Many materials, including rocks, respond to increasing pressures in this way, showing (1) no change, (2) an elastic change with recovery, (3) a plastic change with no recovery, and (4) finally breaking from the pressure.

A **stress** is a force that tends to compress, pull apart, or deform a rock. Rocks in Earth's solid outer crust are subjected to forces as Earth's plates move into, away from, or alongside each other. However, not all stresses are generated directly by plate interaction. Three types of forces from plate interaction that cause rock stress are:

1. *Compressive stress* is caused by two plates moving together or by one plate pushing against another plate that is not moving.
2. *Tensional stress* is the opposite of compressional stress. It occurs when one part of a plate moves away, for example, and another part does not move.
3. *Shear stress* is produced when two plates slide past one another or by one plate sliding past another plate that is not moving.

Just like the metal in the soda can, a rock is able to withstand stress up to a limit. Then it might undergo elastic deformation, plastic deformation, or breaking with progressively greater pressures. The adjustment to stress is called **strain.** A rock unit might respond to stress by changes in volume, changes in shape, or by breaking. Thus, there are three types of strain: elastic, plastic, and fracture.

1. In *elastic strain,* rock units recover their original shape after the stress is released.
2. In *plastic strain,* rock units are molded or bent under stress and do not return to their original shape after the stress is released.
3. In *fracture strain,* rock units crack or break, as the name suggests.

The relationship between stress and strain, that is, exactly how the rock responds, depends on at least four variables. They are (1) the nature of the rock, (2) the temperature of the rock, (3) how slowly or quickly the stress is applied over time, and (4) the confining pressure on the rock. The temperature and con-

fining pressure are generally a function of how deeply the rock is buried. In general, rocks are better able to withstand compressional rather than pulling-apart stresses. Cold rocks are more likely to break than warm rocks, which tend to undergo plastic deformation. In addition, a stress that is applied quickly tends to break the rock, where stress applied more slowly over time, perhaps thousands of years, tends to result in plastic strain.

In general, rocks at great depths are under great pressure at higher temperatures. These rocks tend to undergo plastic deformation, then plastic flow, so rocks at great depths are bent and deformed extensively. Rocks closer to the surface can also bend, but they have a lower elastic limit and break more readily (figure 16.3). Rock deformation often results in recognizable surface features called folds and faults, the topics of the next sections.

Folding

Sediments that form most sedimentary rocks are deposited in nearly flat, horizontal layers at the bottom of a body of water. Conditions on the land change over time, and different mixtures of sediments are deposited in distinct layers of varying thickness. Thus, most sedimentary rocks occur naturally as structures of horizontal layers, or beds (figure 16.4).

A sedimentary rock layer that is not horizontal may have been subjected to some kind of compressive stress. The source of such a stress could be from colliding plates, from the intrusion of magma from below, or from a plate moving over a rising superplume of magma. Seismic tomography has identified massive plumes of molten rock rising toward the surface from deep within Earth. These superplumes originate from the base of the mantle, rising to the lithosphere. The hot material was observed to spread out horizontally, causing bulges as it exerts forces under the lithosphere.

Stress on buried layers of horizontal rocks can result in plastic strain, resulting in a wrinkling of the layers into *folds.* **Folds** are bends in layered bedrock (figure 16.5). They are analogous to layers of rugs or blankets that were stacked hori-

FIGURE 16.3 Stress and deformation relationships for deeply buried, warm rocks under high pressure (*A*) and for cooler rocks near the surface (*B*). Breaking occurs when stress exceeds rupture strength.

A

B

FIGURE 16.4 (*A*) Rock bedding on a grand scale in the Grand Canyon. (*B*) A closer example of rock bedding can be seen in this roadcut.

FIGURE 16.5 These folded rock layers are in the Calico Hills, California. Can you figure out what might have happened here to fold flat rock layers like this?

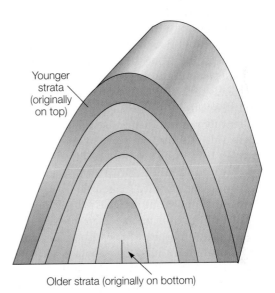

Younger strata (originally on top)

Older strata (originally on bottom)

FIGURE 16.6 An anticline, or arching fold, in layered sediments. Note that the oldest strata are at the center.

zontally, then pushed into a series of arches and troughs. Folds in layered bedrock of all shapes and sizes can occur from plastic strain, depending generally on the regional or local nature of the stress and other factors. Of course, when the folding occurred, the rock layers were in a plastic condition and were probably under considerable confining pressure from deep burial. However, you see the results of the folding when the rocks are under very different conditions at the surface.

The most common regional structures from deep plastic deformation are arch-shaped and trough-shaped folds. In general, an arch-shaped fold is called an **anticline** (figure 16.6). The corresponding trough-shaped fold is called a **syncline** (figure 16.7). Anticlines and synclines sometimes alternate across the land like waves on water. You can imagine that a great compressional stress must have been involved over a wide region to wrinkle the land like this.

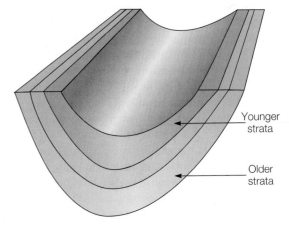

Younger strata

Older strata

FIGURE 16.7 A syncline, showing the reverse age pattern.

Anticlines, synclines, and other types of folds are not always visible as such on Earth's surface. The ridges of anticlines are constantly being weathered into sediments. The sediments, in turn, tend to collect in the troughs of synclines, filling them in. The Appalachian Mountains have ridges of rocks that are more resistant to weathering, forming hills and mountains. The San Joaquin Valley, on the other hand, is a very large syncline in California.

Note that any kind of rock can be folded. Sedimentary rocks are usually the best example of folding, however, since the fold structures of rock layers are easy to see and describe. Folding is much harder to see in igneous or metamorphic rocks that are blends of minerals without a layered structure.

Faulting

Rock layers do not always respond to stress by folding. Rocks near the surface are cooler and under less pressure, so they tend to be more brittle. A sudden stress on these rocks may reach the rupture point, resulting in a cracking and breaking of the rock structure. When there is relative movement between the rocks on either side of a fracture, the crack is called a **fault.** When faulting occurs, the rocks on one side move relative to the rocks on the other side along the surface of the fault, which is called the *fault plane.* Faults are generally described in terms of (1) the steepness of the fault plane, that is, the angle between the plane and imaginary horizontal plane, and (2) the direction of relative movement. There are basically three ways that rocks on one side of a fault can move relative to the rocks on the other side: (1) up and down (called "dip"), (2) horizontally, or sideways (called "strike"), and (3) with elements of both directions of movement (called "oblique").

One classification scheme for faults is based on an orientation referent borrowed from mining (many ore veins are associated with fault planes). Imagine a mine with a fault plane running across a horizontal shaft. Unless the plane is perfectly vertical, a miner would stand on the mass of rock below the fault plane and look up at the mass of rock above. Therefore, the mass of rock below is called the *footwall* and the mass of rock above is called the *hanging wall* (figure 16.8). How the footwall and hanging wall have moved relative to one another describes three basic classes of faults: (1) normal, (2) reverse, and (3) thrust faults. A **normal fault** is one in which the hanging wall has moved downward relative to the footwall. This seems "normal" in the sense that you would expect an upper block to slide *down* a lower block along a slope (figure 16.9A). Sometimes a huge block of rock bounded by normal faults will drop down, creating a *graben* (figure 16.9B). The opposite of a graben is a *horst,* which is a block bounded by normal faults that is uplifted (figure 16.9C). A very large block lifted sufficiently becomes a faultblock mountain. Many parts of the western United States are characterized by numerous fault-block mountains separated by adjoining valleys.

In a **reverse fault,** the hanging wall block has moved upward relative to the footwall block. As illustrated in figure 16.10A, a reverse fault is probably the result of horizontal compressive stress.

A reverse fault with a low-angle fault plane is also called a **thrust fault** (figure 16.10B). In some thrust faults, the hanging wall block has completely overridden the lower footwall

A

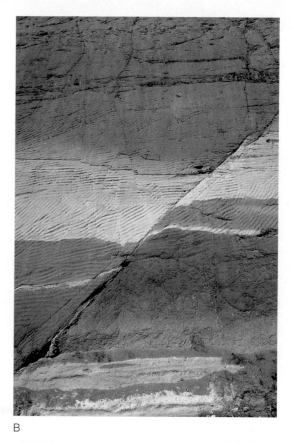

B

FIGURE 16.8 (*A*) Relationship between the hanging wall block and footwall block of a fault. (*B*) A photo of a fault near Kingman, Arizona, showing how the hanging wall has moved relative to the footwall.

for 10 to 20 km (6 to 12 mi). This is sometimes referred to as an "over-thrust."

As shown in figures 16.9 and 16.10, the relative movement of blocks of rocks along a fault plane provides information about the stresses that produced the movement. Reverse and thrust faulting result from compressional stress in the direction of the movement. Normal faulting, on the other hand, results from a pulling-apart stress that might be associated with diverging plates. It might also be associated with the stretching and bulging up of the crust over a hot spot.

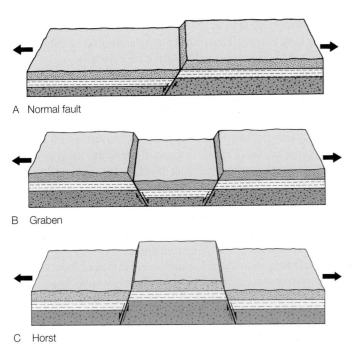

A Normal fault

B Graben

C Horst

FIGURE 16.9 How tensional stress could produce (A) a normal fault, (B) a graben, and (C) a horst.

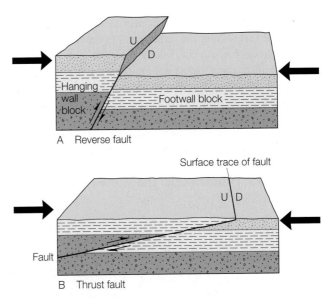

A Reverse fault

Surface trace of fault

B Thrust fault

FIGURE 16.10 How compressive stress could produce (A) a reverse fault and (B) a thrust fault.

EARTHQUAKES

What is an earthquake? An **earthquake** is a quaking, shaking, vibrating, or upheaval of the ground. Earthquakes are the result of the sudden release of energy that comes from *stress* on rock beneath Earth's surface. In the section on folding, you learned that rock units can bend and become deformed in response to stress, but there are limits as to how much stress rock can take before it fractures. When it does fracture, the sudden movement of blocks of rock produces vibrations that move out as waves throughout Earth. These vibrations are called **seismic waves.** It is strong seismic waves that people feel as a shaking, quaking, or vibrating during an earthquake.

Seismic waves are generated when a huge mass of rock breaks and slides into a different position. As you learned in the section on folding, the plane between two rock masses that have moved into new relative positions is called a *fault.* Major earthquakes occur along existing fault planes or when a new fault is formed by the fracturing of rock. In either case, most earthquakes occur along a fault plane when there is displacement of one side relative to the other.

Most earthquakes occur along a fault plane and near Earth's surface. You might expect this to happen since the rocks near the surface are brittle and those deeper are more plastic from increased temperature and pressure. Shallow earthquakes are typical of those that occur at the boundary of the North American Plate, which is moving against the Pacific Plate. In California, the boundary between these two plates is known as the *San Andreas fault* (figure 16.11). The San Andreas fault runs north-south through California, with the Pacific Plate moving on one side and the North American Plate moving on the other.

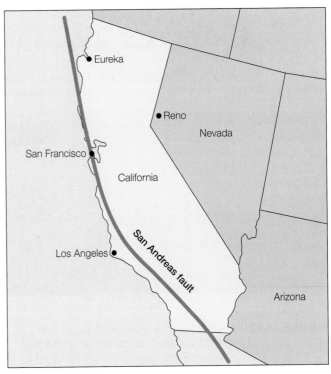

FIGURE 16.11 The San Andreas fault runs north-south for about 1,300 km (800 mi) through California.

The two plates are tightly pressed against each other, and friction between the rocks along the fault prevents them from moving easily. Stress continues to build along the entire fault as one plate attempts to move along the other. Some elastic deformation does occur from the stress, but eventually, the rupture strength of the rock (or the friction) is overcome. The stressed rock, now released of the strain, snaps suddenly into new positions in the phenomenon known as **elastic rebound** (figure 16.12). The rocks are displaced to new positions on either side of the fault, and the vibrations from the sudden movement are felt as an earthquake. The elastic rebound and movement tend to occur

along short segments of the fault at different times rather than along long lengths. Thus, the resulting earthquake tends to be a localized phenomenon rather than a regional one.

Most earthquakes occur near plate boundaries. They do happen elsewhere, but those are rare. The actual place where seismic waves originate beneath the surface is called the *focus* of the earthquake. The focus is considered to be the center of the earthquake and the place of initial rock movement on a fault. The point on Earth's surface directly above the focus is called the earthquake *epicenter* (figure 16.13).

Seismic waves radiate outward from an earthquake focus, spreading in all directions through the solid Earth's interior like the sound waves from an explosion. As introduced in chapter 15, there are three kinds of waves: *P-waves, S-waves,* and *surface waves.* A *seismometer* is an instrument used to measure seismic waves. Information about S- and P-waves as well as about the size of an earthquake can be read from seismometer recordings. S- and P-waves provide information about the location and magnitude of an earthquake as well as information about Earth's interior.

Seismic S- and P-waves leave the focus of an earthquake at essentially the same time. As they travel away from the focus, they gradually separate because the P-waves travel faster than the S-waves. To locate an epicenter, at least three recording stations measure the time lag between the arrival of the first P-waves and the first slower S-waves. The difference in the speed between the two waves is a constant. Therefore, the farther they travel, the greater the time lag between the arrival of the faster P-waves and the slower S-waves (figure 16.14A). By measuring the time lag and knowing the speed of the two waves, it is possible to calculate the distance to their source. However, the calculated distance provides no information about the direction or location of the source of the waves. The location is found by first using the calculated distance as the radius of a circle drawn on a map. The place where the circles from the three recording stations intersect is the location of the source of the waves (figure 16.14B).

Earthquakes range from the many that are barely detectable to the few that cause widespread damage. The *intensity* of an earthquake is a measure of the effect on people and buildings. Destruction is caused by the seismic waves, which cause

FIGURE 16.12 The elastic rebound theory of the cause of earthquakes. (*A*) Rock with stress acting on it. (*B*) Stress has caused strain in the rock. Strain builds up over a long period of time. (*C*) Rock breaks suddenly, releasing energy, with rock movement along a fault. Horizontal motion is shown; rocks can also move vertically. (*D*) Horizontal offset of rows in a lettuce field, 1979, El Centro, California.

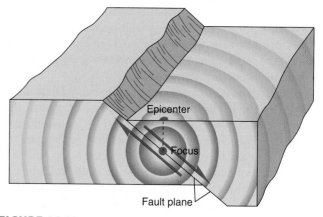

FIGURE 16.13 Simplified diagram of a fault, illustrating component parts and associated earthquake terminology.

the land and buildings to vibrate or shake. Vibrations during small quakes can crack windows and walls, but vibrations in strong quakes can topple bridges and buildings. Injuries and death are usually the result of falling debris, crumbling buildings, or falling bridges. Fire from broken gas pipes was a problem in the 1906 and 1989 earthquakes in San Francisco and the 1994 earthquake in Los Angeles. Broken water mains made it difficult to fight the 1906 San Francisco fires, but in 1989, fireboats and fire hoses using water pumped from the bay were able to extinguish the fires.

Other effects of earthquakes include landslides, displacement of the land surface, and tsunamis. Vertical, horizontal, or both vertical and horizontal displacement of the land can occur during a quake. People sometimes confuse cause and effect when they see a land displacement saying things like, "Look what the earthquake did!" The fact is that the movement of the land probably produced the seismic waves (the earthquake). The seismic waves did not produce the land displacement. Displacements from a single earthquake can be up to 10 to 15 m (about 30 to 50 ft), but such displacements rarely happen.

The size of an earthquake (if it was a "big one" or a "little one") can be measured in terms of vibrations, in terms of displacement, or in terms of the amount of energy released at the site of the earthquake. The larger the quake, the larger the waves recorded on a seismometer. From these recorded waves, scientists assign a number called the *magnitude*. Magnitude is a measure of the energy released during an earthquake. Earthquake magnitude is often reported using the *Richter scale* (table 16.1). The Richter scale was developed by Charles Richter, a seismologist at the California Institute of Technology in the early 1930s. The scale was based on the widest swing in the back-and-forth line traces of seismograph recording. The higher the magnitude of an earthquake, as measured by the Richter scale, the greater the (1) severity of the ground-shaking vibrations and (2) energy released by the earthquake. An increase of one on the Richter scale means that the amount of movement of the ground increased by a factor of ten, and the amount of energy released increased by a factor of thirty. An earthquake measuring below 3 on the scale is usually not felt by people near the epicenter. The largest earthquake measured so far had a magnitude over 9, but there is actually no upper limit to the scale. Today, professional seismologists rate the size of earthquakes in different ways, depending on what they are comparing and why, but each way results in logarithmic scales similar to the Richter scale.

A

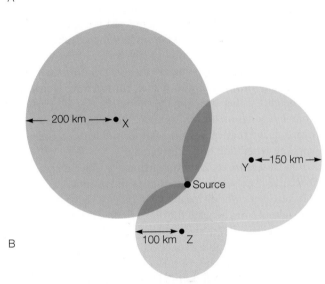

B

FIGURE 16.14 Use of seismic waves in locating the source of a disturbance. (*A*) Difference in times of first arrivals of P-waves and S-waves is a function of the distance from the source. (*B*) Triangulation using data from several seismograph stations allows location of the source.

Myths, Mistakes, and Misunderstandings

Do Earthquakes Swallow People?

It is a myth that the ground opens up, swallows people, then closes during an earthquake. There might be shallow crevasses that form during an earthquake, but faults do not open up and close during an earthquake. Movement occurs along a fault, not perpendicular to it. If faults opened up, no earthquake would occur because there would be no friction to lock the sides of the fault together.

TABLE 16.1   Effects of Earthquakes of Various Magnitudes

Richter Magnitudes	Description
0–2	Smallest detectable earthquake
2–3	Detected and measured but not generally felt
3–4	Felt as small earthquake but no damage occurs
4–5	Minor earthquake with local damage
5–6	Moderate earthquake with structural damage
6–7	Strong earthquake with destruction
7–8	Major earthquake with extensive damage and destruction
8–9	Great earthquake with total destruction

Tsunami is a Japanese term used to describe the very large ocean waves that can be generated by an earthquake, landslide, or volcanic explosion. Such large waves were formerly called "tidal waves." Since the large, fast waves were not associated with tides or tidal forces in any way, the term *tsunami* or *seismic sea wave* is preferred.

A tsunami, like other ocean waves, is produced by a disturbance. Most common ocean waves are produced by winds, travel at speeds of 90 km/h (55 mi/h), and produce a wave height of 0.6 to 3 m (2 to 10 ft) when they break on the shore. A tsunami, on the other hand, is produced by some strong disturbance in the seafloor, travels at speeds of 725 km/h (450 mi/h), and can produce a wave height of over 8 m (about 25 ft) when it breaks on the shore. A tsunami may have a very long wavelength of up to 200 km (about 120 mi) compared to the wavelength of ordinary wind-generated waves of 400 m (1,300 ft). Because of its great wavelength, a tsunami does not just break on the shore, then withdraw. Depending on the seafloor topography, the water from a tsunami may continue to rise for five to ten minutes, flooding the coastal region before the wave withdraws. A gently sloping seafloor and a funnel-shaped bay can force tsunamis to great heights as they break on the shore. The size of a particular tsunami depends on how the seafloor was disturbed. Generally, an earthquake that causes the seafloor suddenly to rise or fall favors the generation of a large tsunami.

On December 26, 2004, an earthquake with a magnitude of 9.0 occurred west of the Indonesian island of Sumatra. This was the largest quake worldwide in four decades. The focus was about 10 km (about 6 mi) beneath the ocean floor at the interface of the India and Burma Plates and was caused by the release of stresses that develop as the India Plate subducts beneath the overriding Burma Plate. A 100 km (62 mi) wide rupture occurred in the ocean floor, about 1,200 km (746 mi) long and parallel to the Sunda trench. The ocean floor was suddenly uplifted over 2 m (about 7 ft), with a movement about 10 m (about 33 ft) to the west-southwest. This displacement acted like a huge paddle at the bottom of the ocean, vertically displacing billions of tons of water and triggering a tsunami. The tsunami created a path of destruction across the 4,500 km (about 2,800 mi) wide Indian Ocean over the next seven hours. A series of very large waves struck the coast of the Indian Ocean, resulting in an estimated death toll of about 295,000 and over a million left homeless.

The world's largest recorded earthquakes have all occurred where one tectonic plate subducts beneath another. These include the magnitude-9.5 1960 Chile earthquake, the magnitude-9.2 1964 Prince William Sound, Alaska, earthquake, the magnitude-9.1 1957 Andreanof Islands, Alaska, earthquake, and the magnitude-9.0 1952 Kamchatka earthquake. Such earthquakes often create large tsunamis that result in death and destruction over a wide area.

ORIGIN OF MOUNTAINS

Folding and faulting have created most of the interesting features of Earth's surface, and the most prominent of these features are **mountains.** Mountains are elevated parts of Earth's crust that rise

FIGURE 16.15 The folded structure of the Appalachian Mountains, revealed by weathering and erosion, is obvious in this *Skylab* photograph of the Virginia-Tennessee-Kentucky boundary area. The clouds are over the Blue Ridge Mountains.

abruptly above the surrounding surface. Most mountains do not occur in isolation but rather in chains or belts. These long, thin belts are generally found along the edges of continents rather than in the continental interior. There are a number of complex processes involved in the origin of mountains and mountain chains, and no two mountains are exactly alike. For convenience, however, mountains can be classified according to three basic origins: (1) folding, (2) faulting, and (3) volcanic activity.

The major mountain ranges of Earth—the Appalachian, Rocky, and Himalayan Mountains, for example—have a great height that involves complex folding on a very large scale. The crust was thickened in these places as compressional forces produced tight, almost vertical folds. Thus, folding is a major feature of the major mountain ranges, but faulting and igneous intrusions are invariably also present. Differential weathering of different rock types produced the parallel features of the Appalachian Mountains that are so prominent in satellite photographs (figure 16.15). The folded sedimentary rocks of the Rockies are evident in the almost upright beds along the flanks of the front range.

A broad arching fold, which is called a dome, produced the Black Hills of South Dakota. The sedimentary rocks from the top of the dome have been weathered away, leaving a somewhat circular area of more resistant granite hills surrounded by

Oldest formation exposed on the surface

Youngest formation

FIGURE 16.16 A sketch of an eroded structural dome where all the rock layers dip away from the center.

Original sharply bounded fault blocks softened by erosion and sedimentation

Sediments

FIGURE 16.17 Faultblock mountains are weathered and eroded as they are elevated, resulting in a rounded shape and sedimentation rather than sharply edged fault blocks.

upward-tilting sedimentary beds (figure 16.16). The Adirondack Mountains of New York are another example of this type of mountain formed from folding, called domed mountains.

Compression and relaxation of compressional forces on a regional scale can produce large-scale faults, shifting large crustal blocks up or down relative to one another. Huge blocks of rocks can be thrust to mountainous heights, creating a series of faultblock mountains. Faultblock mountains rise sharply from the surrounding land along the steeply inclined fault plane. The mountains are not in the shape of blocks, however, as weathering has carved them into their familiar mountainlike shapes (figure 16.17). The Teton Mountains of Wyoming and the Sierra Nevadas of California are classic examples of faultblock mountains that rise abruptly from the surrounding land. The various mountain ranges of Nevada, Arizona, Utah,

Volcanoes Change the World

A volcanic eruption changes the local landscape; that much is obvious. What is not so obvious are the worldwide changes that can happen just because of the eruption of a single volcano. Perhaps the most discussed change brought about by a volcano occurred back in 1815–16 after the eruption of Tambora in Indonesia. The Tambora eruption was massive, blasting huge amounts of volcanic dust, ash, and gas high into the atmosphere. Most of the ash and dust fell back to Earth around the volcano, but some dust particles and sulfur dioxide gas were pushed high into the stratosphere.

It is known today that the sulfur dioxide from explosive volcanic eruptions reacts with water vapor in the stratosphere, forming tiny droplets of diluted sulfuric acid. In the stratosphere, there is no convection, so the droplets of acid and dust from the volcano eventually formed a layer around the entire globe. This formed a haze that remains in the stratosphere for years, reflecting and scattering sunlight.

What were the effects of haze from Tambora? There were fantastic, brightly colored sunsets from the added haze in the stratosphere. On the other hand, it was also cooler than usual, presumably because of the reflected sunlight that did not reach Earth's surface. It snowed in New England in June 1816, and the cold continued into July. Crops failed, and 1816 became known as the "year without summer."

More information is available about the worldwide effects of present-day volcanic eruptions because there are now instruments to make more observations in more places. However, it is still necessary to do a great deal of estimating because of the relative inaccessibility of the worldwide stratosphere. It was estimated, for example, that the 1982 eruption of El Chichon in Mexico created enough haze in the stratosphere to reflect 5 percent of the solar radiation away from Earth. Researchers also estimated that the effects of the El Chichon eruption cooled the global temperatures by a few tenths of a degree for two or three years. The cooling did take place, but the actual El Chichon contribution to the cooling is not clear because of other interactions. Earth may have been undergoing global warming from the greenhouse effect, for example, so the El Chichon cooling effect could have actually been much greater. Other complicating factors such as the effects of El Niño make changes difficult to predict.

In June 1991, the Philippine volcano Mount Pinatubo erupted, blasting twice as much gas and dust into the stratosphere as El Chichon had about a decade earlier. The haze from such eruptions has the potential to cool the climate about 0.5°C (1°F). The overall result, however, will always depend on a possible greenhouse effect, a possible El Niño effect, and other complications.

and southeastern California have large numbers of faultblock mountains that generally trend north and south.

Lava and other materials from volcanic vents can pile up to mountainous heights on the surface. These accumulations can form local volcano-formed mountains near mountains produced by folding or faulting. Such mixed-origin mountains are common in northern Arizona, New Mexico, and western Texas. The Cascade Mountains of Washington and Oregon are a series of towering volcanic peaks, most of which are not active today. As a source of mountains, volcanic activity has an overall limited impact on the continents. The major mountains built by volcanic activity are the mid-oceanic ridges formed at diverging plate boundaries.

Deep within Earth, previously solid rock melts at high temperatures to become *magma,* a pocket of molten rock. Magma is not just melted rock alone, however, as the melt contains variable mixtures of minerals (resulting in different types of lava flows). It also includes gases such as water vapor, sulfur dioxide, hydrogen sulfide, carbon dioxide, and hydrochloric acid. You can often smell some of these gases around volcanic vents and hot springs. Hydrogen sulfide smells like rotten eggs or sewer gas. The sulfur smells like a wooden match that has just been struck.

The gases dissolved in magma play a major role in forcing magma out of the ground. As magma nears the surface, it comes under less pressure, and this releases some of the dissolved gases from the magma. The gases help push the magma out of the ground. This process is similar to releasing the pressure on a can of warm soda, which releases dissolved carbon dioxide.

Magma works its way upward from its source below to Earth's surface, here to erupt into a lava flow or a volcano. A **volcano** is a hill or mountain formed by the extrusion of lava or rock fragments from magma below. Some lavas have a lower viscosity than others, are more fluid, and flow out over the land rather than forming a volcano. Such *lava flows* can accumulate into a plateau of basalt, the rock that the lava formed as it cooled and solidified. The Columbia Plateau of the states of Washington, Idaho, and Oregon is made up of layer after layer of basalt that accumulated from lava flows. Individual flows of lava formed basalt layers up to 100 m (about 330 ft) thick,

covering an area of hundreds of square kilometers. In places, the Columbia Plateau is up to 3 km (about 2 mi) thick from the accumulation of many individual lava flows.

The hill or mountain of a volcano is formed by ejected material that is deposited in a conical shape. The materials are deposited around a central *vent,* an opening through which an eruption takes place. The *crater* of a volcano is a basinlike depression over a vent at the summit of the cone. Figure 16.18 is an aerial view of Mount St. Helens, looking down into the crater at a volcanic dome that formed as magma periodically welled upward into the floor of the crater. This photo was taken several years after Mount St. Helens erupted in May 1980. The volcano was quiet between that time and late 2004, when thousands of small earthquakes preceded the renewed growth of the lava dome inside the crater. According to the U.S. Geological Survey, the new dome grew to more than 152 m (500 ft) above the old dome before the volcano entered a relatively quiet state. Go to http://vulcan.wr.usgs.gov/News?framework.html for a fact sheet on the latest activity of Mount St. Helens and links to current information.

A volcano forms when magma breaks out at Earth's surface. Only a small fraction of all the magma generated actually reaches Earth's surface. Most of it remains below the ground, cooling and solidifying to form igneous rocks that were described in chapter 15. A large amount of magma that has crystallized below the surface is known as a *batholith.* A small protrusion from a batholith is called a *stock.* By definition, a stock has less than 100 km^2 (40 mi^2) of exposed surface area, and a batholith is larger. Both batholiths and stocks become exposed at the surface through erosion of the overlying rocks and rock materials, but not much is known about their shape below. The sides seem to angle away with depth, suggesting that they become larger with depth. The intrusion of a batholith sometimes tilts rock layers upward, forming a *hogback,* a ridge with equal slopes on its sides (figure 16.19). Other forms of intruded rock were formed as moving magma took the paths of least resistance, flowing into joints, faults, and planes between

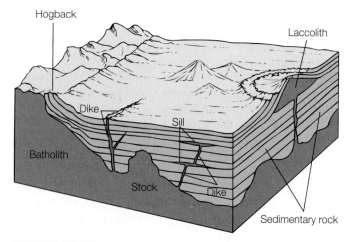

FIGURE 16.18   This is the top of Mount St. Helens several years after the 1980 explosive eruption.

FIGURE 16.19   Here are the basic intrusive igneous bodies that form from volcanic activity.

and other locations where repeated cycles of water freezing and thawing occur. The roots of trees and shrubs can also mechanically wedge rocks apart as they grow into cracks. You may have noticed the work of roots when trees or shrubs have grown next to a sidewalk for some period of time (figure 16.22).

The other example of mechanical weathering is believed to be caused by the reduction of pressure on rocks. As more and more weathered materials are removed from the surface, the downward pressure from the weight of the material on the rock below becomes less and less. The rock below begins to expand upward, fracturing into concentric sheets from the effect of reduced pressure. These curved, sheetlike plates fall away later in the mechanical weathering process called *exfoliation* (figure 16.21B). *Exfoliation* is the term given to the process of spalling off of layers of rock, somewhat analogous to peeling layers from an onion. Granite commonly weathers by exfoliation, producing characteristic dome-shaped hills and rounded boulders. Stone Mountain, Georgia, is a well-known example of an exfoliation-shaped dome. The onionlike structure of exfoliated granite is a common sight in the Sierras, Adirondacks, and any mountain range where older exposed granite is at the surface (figure 16.23).

A

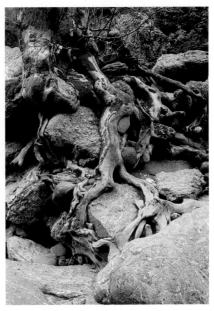
B

FIGURE 16.22 Growing trees can break, separate, and move solid rock. (*A*) Note how this tree has raised the sidewalk. (*B*) This tree is surviving by growing roots into tiny joints and cracks, which become larger as the tree grows.

Chemical Weathering

Examples of chemical weathering include (1) oxidation, (2) carbonation, and (3) hydration. *Oxidation* is a chemical reaction between oxygen and the minerals making up rocks. The ferromagnesian minerals contain iron, magnesium, and other metal ions in a silicate structure. Iron can react with oxygen to produce several different iron oxides, each with its own characteristic color. The most common iron oxide (hematite) has a deep red color. Other oxides of iron are brownish to yellow-brownish. It is the presence of such iron oxides that colors many sedimentary rocks and soils. The red soils of Oklahoma, Georgia, and many other places are colored by the presence of iron oxides produced by chemical weathering.

Carbonation is a chemical reaction between carbonic acid and the minerals making up rocks. Rainwater is naturally somewhat acidic because it dissolves carbon dioxide from the air. This forms a weak acid known as carbonic acid (H_2CO_3), the same acid that is found in carbonated soda. Carbonic acid rain falls on the land, seeping into cracks and crevices where it reacts with minerals. Limestone, for example, is easily weathered to a soluble form by carbonic acid. The limestone caves of Missouri, Kentucky, New Mexico, and elsewhere were produced by the chemical weathering of limestone by carbonation (figure 16.24). Minerals containing calcium, magnesium, sodium, potassium, and iron are chemically weathered by carbonation to produce salts that are soluble in water.

Hydration is a reaction between water and the minerals of rocks. The process of hydration includes (1) the dissolving of a

FIGURE 16.23 Spheroidal weathering of granite. The edges and corners of an angular rock are attacked by weathering from more than one side and retreat faster than flat rock faces. The result is rounded granite boulders, which often shed partially weathered minerals in onionlike layers.

mineral and (2) the combining of water directly with a mineral. Some minerals, for example, halite (which is sodium chloride), dissolve in water to form a solution. The carbonates formed from carbonation are mostly soluble, so they are easily leached from a rock by dissolving. Water also combines directly with some minerals to form new, different minerals. The feldspars, for example, undergo hydration and carbonation to produce (1) water-soluble potassium carbonate, (2) a chemical product that combines with water to produce a clay mineral, and (3) silica. The silica, which is silicon dioxide (SiO_2), may appear as a suspension of finely divided particles or in solution.

Mechanical and chemical weathering are interrelated, working together in breaking up and decomposing solid rocks of Earth's surface. In general, mechanical weathering results in cracks in solid rocks and broken-off coarse fragments. Chemical weathering results in finely pulverized materials and ions in solution, the ultimate decomposition of a solid rock. Consider, for example, a mountain of solid granite, the most common rock found on continents. In general, granite is made up of 65 percent feldspars, 25 percent quartz, and about 10 percent ferromagnesian minerals. Mechanical weathering begins the destruction process as exfoliation and frost wedging create cracks in the solid mass of granite. Rainwater, with dissolved oxygen and carbon dioxide, flows and seeps into the cracks and reacts with ferromagnesian minerals to form soluble carbonates and metal oxides. Feldspars undergo carbonation and hydration, forming clay minerals and soluble salts, which are washed away. Quartz is less susceptible to chemical weathering and remains mostly unchanged to form sand grains. The end products of the complete weathering of granite are quartz sand, clay minerals, metal oxides, and soluble salts.

Erosion

Weathering has prepared the way for erosion and for some agent of transportation to move or carry away the fragments, clays, and solutions that have been produced from solid rock. The weathered materials can be moved to a lower elevation by the direct result of gravity acting alone. They can also be moved to a lower elevation by gravity acting through some intermediate agent, such as running water, wind, or glaciers. The erosion of weathered materials as a result of gravity alone will be considered first.

Mass Movement

Gravity constantly acts on every mass of materials on the surface of Earth, pulling parts of elevated regions toward lower levels. Rocks in the elevated regions are able to temporarily resist this constant pull through their cohesiveness with a main rock mass or by the friction between the rock and the surface of the slope. Whenever anything happens to reduce the cohesiveness or to reduce the friction, gravity pulls the freed material to a lower elevation. Thus, gravity acts directly on individual rock fragments and on large amounts of surface materials as a mass, pulling all to a lower elevation. Erosion caused by gravity acting directly is called **mass movement** (also called mass wasting). Mass movement can be so slow that it is practically imperceptible. *Creep,* the slow downhill movement of soil down a steep slope, for example, is detectable only from the peculiar curved growth patterns of trees growing in the slowly moving soil (figure 16.25). At the other extreme, mass movement can be as sudden and swift as a rock bounding and clattering down a slope below a cliff. A *landslide* is a generic term used to describe any slow to rapid movement of any type or mass of materials, from the short slump of a hillside to the slide of a whole mountainside. Either slow or sudden, mass movement is a small victory for gravity in the ongoing process of leveling the landmass of Earth.

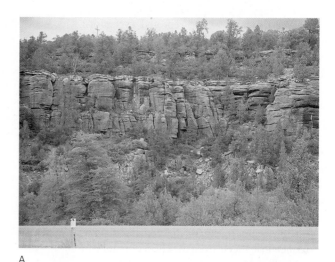

A

B

FIGURE 16.24   Limestone caves develop when slightly acidic groundwater dissolves limestone along joints and bedding planes, carrying away rock components in solution. (*A*) Joints and bedding planes in a limestone bluff. (*B*) This stream has carried away less-resistant rock components, forming a cave under the ledge.

Running Water

Running water is the most important of all the erosional agents of gravity that remove rock materials to lower levels. Erosion by

James Hutton (1726–1797)

James Hutton was a Scottish natural philosopher who pioneered uniformitarian geology. The son of an Edinburgh merchant, Hutton studied at Edinburgh University, Paris, and Leiden, training first for the law but taking his doctorate in medicine in 1749 (though he never practiced). He spent the next two decades traveling and farming in the southeast of Scotland. During this time, he cultivated a love of science and philosophy, developing a special taste for geology. About 1768, he returned to his native Edinburgh. A friend of Joseph Black, William Cullen, and James Watt, Hutton shone as a leading member of the scientific and literary establishment, playing a large role in the early history of the Royal Society of Edinburgh and in the Scottish Enlightenment.

Hutton wrote widely on many areas of natural science, including chemistry (where he opposed Lavoisier), but he is best known for his geology, set out in his *Theory of the Earth*, of which a short version appeared in 1788, followed by the definitive statement in 1795. In that work, Hutton attempted (on the basis of both theoretical considerations and personal fieldwork) to demonstrate that Earth formed a steady-state system in which terrestrial causes had always been of the same kind as at present, acting with comparable intensity (the principle later known as uniformitarianism). In Earth's economy, in the imperceptible creation and devastation of landforms, there was no vestige of a beginning nor prospect of an end. Continents were continually being gradually eroded by rivers and weather. Denuded debris accumulated on the seabed, to be consolidated into strata and subsequently thrust upward to form new continents, thanks to the action of Earth's central heat. Nonstratified rocks such as granite were of igneous origin. All Earth's processes were exceptionally leisurely, and hence, Earth must be incalculably old.

Though supported by the experimental findings of Sir James Hall, Hutton's theory was vehemently attacked in its day, partly because it appeared to point to an eternal Earth and hence to atheism. It found more favor when popularized by Hutton's friend, John Playfair, and later by Charles Lyell. The notion of uniformitarianism still forms the groundwork for much geological reasoning.

Source: Modified from the *Hutchinson Dictionary of Scientific Biography*. © RM, 2009. All rights reserved. Helicon Publishing is a division of RM.

A

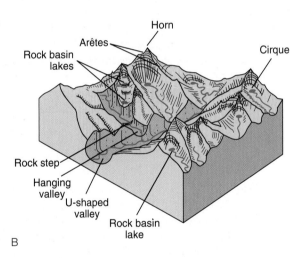

B

FIGURE 16.32   (*A*) A stream-carved mountainside before glaciation. (*B*) The same area after glaciation, with some of the main features of mountain glaciation labeled.

attack rocks chemically as water does through carbonation and other processes, and the wind cannot carry dissolved sediments in solution. Even in many deserts, more sediment is moved during the brief periods of intense surface runoff following the occasional rainstorms than is moved by wind during the prolonged dry periods.

Flowing air and moving water do have much in common as agents of erosion since both are fluids. Both can move larger particles by rolling them along the surface and can move finer particles by carrying them in suspension. Both can move larger and more massive particles with increased velocities. Water is denser and more viscous than air, so it is more efficient at transporting quantities of material than is the wind, but the processes are quite similar.

Two major processes of wind erosion are called (1) abrasion and (2) deflation. *Wind abrasion* is a natural sandblasting process that occurs when the particles carried along by the wind break off small particles and polish what they strike. Generally, the harder mineral grains such as quartz sand accomplish this best near the ground where the wind is bouncing them along. Wind abrasion can strip paint from a car exposed to the moving particles of a dust storm, eroding the paint along with rocks on the surface. Rocks and boulders exposed to repeated wind storms where the wind blows consistently from one or a few directions may be planed off from the repeated action of this natural sandblasting. Rocks sculptured by wind abrasion are called *ventifacts*, after the Latin meaning "wind-made" (figure 16.33).

Deflation, after the Latin meaning "to blow away," is the widespread picking up of loose materials from the surface. Deflation is naturally most active where winds are unobstructed and the materials are exposed and not protected by vegetation. These conditions are often found on deserts, beaches, and unplanted farmland between crops. During the 1930s, several years of drought killed the native vegetation in the Plains states during a period of increased farming activity. Unusually strong winds eroded the unprotected surface, removing and transporting hundreds of millions of tons of soil. This period of prolonged drought, dust storms, and general economic disaster for farmers in the area is known as the Dust Bowl episode.

The most common wind-blown deposits are (1) dunes and (2) loess. A **dune** is a low mound or ridge of sand or other sediments. Dunes form when sediment-bearing wind encounters an obstacle that reduces the wind velocity. With a slower velocity, the wind cannot carry as large a load, so sediments are deposited on the surface. This creates a windbreak, which results in a growing obstacle, a dune. Once formed, a dune tends to migrate, particularly if the winds blow predominantly from one direction. Dunes are commonly found in semiarid areas or near beaches.

Another common wind deposit is called *loess* (pronounced "luss"). Loess is a very fine dust, or silt, that has been deposited over a large area. One such area is located in the central part of the United States, particularly to the east sides of the major rivers of the Mississippi basin. Apparently, this deposit originated from the rock flour produced during the last great ice age. The rock flour was probably deposited along the major river valleys and later moved eastward by the prevailing westerly winds. Since rock flour is produced by the mechanical grinding action of

If wind is predominantly from one direction, rocks will be planed off or flattened on the upwind side.

With a persistent shift in wind direction, additional facets are cut in the rock.

FIGURE 16.33   Ventifact formation by abrasion from one or several directions.

glaciers, it has not been chemically broken down. Thus, the loess deposit contains many minerals that were not leached out of the deposit as typically occurs with chemical weathering. It also has an open, porous structure since it does not have as much of the chemically produced clay minerals. The good moisture-holding capacity from this open structure, together with the presence of minerals that serve as plant nutrients, make farming soils formed from loess deposits particularly productive.

SUMMARY

The *principle of uniformity* is the frame of reference that the same geologic processes you see changing rocks today are the same processes that changed them in the past.

Diastrophism is the process of deformation that changes Earth's surface, and the movement of magma is called *vulcanism.* Diastrophism, *vulcanism,* and *earthquakes* are closely related, and their occurrence can be explained most of the time by events involving *plate tectonics.*

Stress is a force that tends to compress, pull apart, or deform a rock, and the adjustment to stress is called *strain.* Rocks respond to stress by (1) withstanding the stress without change, (2) undergoing *elastic strain,* (3) undergoing *plastic strain,* or (4) by *breaking* in *fracture strain.* Exactly how a particular rock responds to stress depends on (1) the nature of the rock, (2) the temperature, and (3) how quickly the stress is applied over time.

Deeply buried rocks are at a higher temperature and tend to undergo *plastic deformation,* resulting in a wrinkling of the layers into *folds.* The most common are an arch-shaped fold called an *anticline* and a trough-shaped fold called a *syncline.* Anticlines and synclines are most easily observed in sedimentary rocks because they have bedding planes, or layers.

Rocks near the surface tend to break from a sudden stress. A break with movement on one side of the break relative to the other side is called a *fault.* The vibrations that move out as waves from the movement of rocks are called an *earthquake.* The actual place where an earthquake originates is called its *focus.* The place on the surface directly above a focus is called an *epicenter.* There are three kinds of waves that travel from the focus: *S-, P-,* and *surface waves.* The magnitude of earthquake waves is measured on the *Richter scale.*

Folding and faulting produce prominent features on the surface called mountains. *Mountains* can be classified as having an origin of *folding, faulting,* or *volcanic.* In general, mountains that occur in long narrow belts called *ranges* have an origin that can be explained by *plate tectonics.*

Weathering is the breaking up, crumbling, and destruction of any kind of solid rock. The process of physically picking up weathered rock materials is called *erosion.* After the eroded materials are picked up, they are removed by *transportation* agents. The combined action of weathering, erosion, and transportation wears away and lowers the surface of Earth.

The physical breaking up of rocks is called *mechanical weathering.* Mechanical weathering occurs by *wedging effects* and the *effects of reduced pressure. Frost wedging* is a wedging effect that occurs from repeated cycles of water freezing and thawing. The process of spalling off of curved layers of rock from reduced pressure is called *exfoliation*.

The breakdown of minerals by chemical reactions is called *chemical weathering*. Examples include *oxidation*, a reaction between oxygen and the minerals making up rocks; *carbonation*, a reaction between carbonic acid (carbon dioxide dissolved in water) and minerals making up rocks; and *hydration*, the dissolving or combining of a mineral with water. When the end products of complete weathering of rocks are removed directly by gravity, the erosion is called *mass movement*. Erosion and transportation also occur through the agents of *running water*, *glaciers*, or *wind*. Each creates their own characteristic features of erosion and deposition.

KEY TERMS

anticline (p. **353**)	fault (p. **354**)	mountains (p. **358**)	stress (p. **352**)
delta (p. **367**)	floodplain (p. **366**)	normal fault (p. **354**)	syncline (p. **353**)
dune (p. **370**)	folds (p. **352**)	principle of uniformity (p. **351**)	thrust fault (p. **354**)
earthquake (p. **355**)	frost wedging (p. **362**)	reverse fault (p. **354**)	tsunami (p. **358**)
elastic rebound (p. **356**)	glacier (p. **368**)	seismic waves (p. **355**)	volcano (p. **360**)
erosion (p. **362**)	mass movement (p. **364**)	strain (p. **352**)	weathering (p. **361**)

APPLYING THE CONCEPTS

1. The basic difference in the frame of reference called the principle of uniformity and the catastrophic frame of reference used by previous thinkers is
   a. the energy for catastrophic changes is much less.
   b. the principle of uniformity requires more time.
   c. catastrophic changes have a greater probability of occurring.
   d. None of the above is correct.

2. The difference between elastic deformation and plastic deformation of rocks is that plastic deformation
   a. permanently alters the shape of a rock layer.
   b. always occurs just before a rock layer breaks.
   c. returns to its original shape after the pressure is removed.
   d. All of the above are correct.

3. Whether a rock layer subjected to stress undergoes elastic deformation, plastic deformation, or rupture depends on
   a. the temperature of the rock.
   b. the confining pressure on the rock.
   c. how quickly or slowly the stress is applied over time.
   d. All of the above are correct.

4. When subjected to stress, rocks buried at great depths are under great pressure at high temperatures, so they tend to undergo
   a. no change because of the pressure.
   b. elastic deformation because of the high temperature.
   c. plastic deformation.
   d. breaking or rupture.

5. Earthquakes that occur at the boundary between two tectonic plates moving against each other occur along
   a. the entire length of the boundary at once.
   b. short segments of the boundary at different times.
   c. the entire length of the boundary at different times.
   d. None of the above is correct.

6. Each higher number of the Richter scale
   a. increases with the magnitude of an earthquake.
   b. means ten times more ground movement.
   c. indicates about thirty times more energy released.
   d. means all of the above.

7. Other than igneous activity, all mountain ranges have an origin resulting from
   a. folding.   c. stresses.
   b. faulting.   d. sedimentation.

8. The preferred name for the very large ocean waves that are generated by an earthquake, landslide, or volcanic explosion is
   a. tidal wave.   c. tidal bore.
   b. tsunami.   d. Richter wave.

9. Freezing water exerts pressure on the wall of a crack in a rock mass, making the crack larger. This is an example of
   a. mechanical weathering.
   b. chemical weathering.
   c. exfoliation.
   d. hydration.

10. Which of the following would have the greatest overall effect in lowering the elevation of a continent such as North America?
    a. continental glaciers
    b. alpine glaciers
    c. wind
    d. running water

11. Broad meanders on a very wide, gently sloping floodplain with oxbow lakes are characteristics you would expect to find in a river valley during what stage?
    a. newborn
    b. youth
    c. maturity
    d. old age

12. A glacier forms when
    a. the temperature does not rise above freezing.
    b. snow accumulates to form ice, which begins to flow.
    c. a summer climate does not occur.
    d. a solid mass of snow moves downhill under the influence of gravity.

Answers

1. b **2.** a **3.** d **4.** c **5.** b **6.** d **7.** c **8.** b **9.** a **10.** d **11.** d **12.** b

QUESTIONS FOR THOUGHT

1. What is the principle of uniformity? What are the underlying assumptions of this principle?

2. Describe the responses of rock layers to increasing compressional stress when it increases slowly on deeply buried, warm layers, when it increases slowly on cold rock layers, and when it is applied quickly to rock layers of any temperature.

3. Describe the conditions that would lead to faulting as opposed to folding of rock layers.

4. Where would the theory of plate tectonics predict that earthquakes would occur?

5. Identify some areas of probable active volcanic activity today in the United States. Explain your reasoning for selecting these areas.

6. Granite is the most common rock found on continents. What are the end products after granite has been completely weathered? What happens to these weathering products?

7. What other erosion processes are important as a stream of running water carves a valley in the mountains? Explain.

8. Describe the characteristic features associated with stream erosion as the stream valley passes through the stages of youth, maturity, and old age.

9. Could a stream erode the land lower than sea level? Explain.

10. Explain why glacial erosion produces a U-shaped valley but stream erosion produces a V-shaped valley.

11. Name and describe as many ways as you can think of that mechanical weathering occurs in nature. Do not restrict your thinking to those discussed in this chapter.

12. Compare the features caused by stream erosion, wind erosion, and glacial erosion.

FOR FURTHER ANALYSIS

1. Evaluate the statement "the present is the key to the past" as it represents the principle of uniformity. What evidence supports this principle?

2. Does the theory of plate tectonics support or not support the principle of uniformity? Provide evidence to support your answer.

3. What are the significant similarities and differences between elastic deformation and plastic deformation?

4. Explain the combination of variables that results in solid rock layers folding rather than faulting.

5. Analyze why you would expect most earthquakes to occur as localized, shallow occurrences near a plate boundary.

6. What are the significant similarities and differences between weathering and erosion?

7. Speculate if the continents will ever be weathered and eroded flat at sea level. Provide evidence to support your speculation.

8. Is it possible for any agent of erosion to erode the land to below sea level? Provide evidence or some observation to support your answer.

INVITATION TO INQUIRY

Building Rocks

Survey the use of rocks used in building construction in your community. Compare the type of rocks that are used for building interiors and those that are used for building exteriors. Where were the rocks quarried? Are any trends apparent for buildings constructed in the past and those built more recently? If so, are there reasons (cost, shipping, other limitations) underlying a trend, or is it simply a matter of style?

17

Earth's Weather

This cloud forms a thin covering over the mountaintop. Likewise, Earth's atmosphere forms a thin shell around Earth, with 99 percent of the mass within 32 km (about 20 mi) of the surface.

CORE **CONCEPT**

Solar radiation drives cycles in Earth's atmosphere, and some of these cycles determine weather and climate.

OUTLINE

Physics

▷ Sunlight is radiant energy (Ch. 3).

▷ Heat is energy in transit (Ch. 3).

Life Science

▷ Organisms show a high degree of adaptation to their local climate (Ch. 23).

▷ Through the process of photosynthesis, organisms remove carbon dioxide and add oxygen to the atmosphere (Ch. 23).

▷ Through the process of respiration, organisms remove oxygen and add carbon dioxide to the atmosphere (Ch. 23).

▷ Nitrogen is cycled between the atmosphere and the bodies of organisms (Ch. 23).

OVERVIEW

Earth's atmosphere has a unique composition because of the cyclic flow of materials. Some of these cycles involve the movement of materials in and out of Earth's atmosphere. Carbon dioxide, for example, is a very minor part of Earth's atmosphere. It has been maintained as a minor component in a mostly balanced state for about the past 570 million years, cycling into and out of the atmosphere.

Water is also involved in a global cyclic flow between the atmosphere and the surface. Water on the surface is mostly in the ocean, with lesser amounts in lakes, streams, and underground. Not much water is found in the atmosphere at any one time on a worldwide basis, but billions of tons are constantly evaporating into the atmosphere each year and returning as precipitation in an ongoing cycle.

The cycling of carbon dioxide and water to and from the atmosphere takes place in a dynamic system that is energized by the Sun. Radiant energy from the Sun heats some parts of Earth more than others. Winds redistribute this energy with temperature changes, rain, snow, and other changes that are generally referred to as the *weather*.

Understanding and predicting the weather is the subject of *meteorology*. Meteorology is the science of the atmosphere and weather phenomena, from understanding everyday rain and snow to predicting not-so-common storms and tornadoes (figure 17.1). Understanding weather phenomena depends on a knowledge of the atmosphere and the role of radiant energy on a spinning Earth that is revolving around the Sun. This chapter is concerned with understanding the atmosphere of Earth, its cycles, and the influence of radiant energy on the atmosphere.

THE ATMOSPHERE

The atmosphere is a relatively thin shell of gases that surrounds the solid Earth. If you could see the molecules making up the atmosphere, you would see countless numbers of rapidly moving particles, all undergoing a terrific jostling from the billions of collisions occurring every second. Since this jostling mass of tiny particles is pulled toward Earth by gravity, more are found near the surface than higher up. Thus, the atmosphere thins rapidly with increasing distance above the surface, gradually merging with the very diffuse medium of outer space.

To understand how rapidly the atmosphere thins with altitude, imagine a very tall stack of open boxes. At any given instant, each consecutively higher box would contain fewer of the jostling molecules than the box below it. Molecules in the lowest box on the surface, at sea level, might be able to move a distance of only 1×10^{-8} m (about 3×10^{-6} in) before colliding with another molecule. A box moved to an altitude of 80 km (about 50 mi) above sea level would have molecules that could move perhaps 10^{-2} m (about 1/2 in) before colliding with another molecule. At 160 km (about 100 mi), the distance traveled would be about 2 m (about 7 ft). As you can see, the distance between molecules increases rapidly with increasing altitude. Since air density is defined by the number of molecules in a unit volume, the density of the atmosphere decreases rapidly with increasing altitude (figure 17.2).

It is often difficult to imagine a distance above the surface because there is nothing visible in the atmosphere for comparison. Recall our stack of boxes from the previous example. Imagine that this stack of boxes is so tall that it reaches from the surface past the top of the atmosphere. Now imagine that this tremendously tall stack of boxes is tipped over and carefully laid out horizontally on the surface of Earth. How far would you have to move along these boxes to reach the box that was in

FIGURE 17.1 The probability of a storm can be predicted, but nothing can be done to stop or slow a storm. Understanding the atmosphere may help in predicting weather changes, but it is doubtful that weather will ever be controlled on a large scale.

FIGURE 17.2 At greater altitudes, the same volume contains fewer molecules of the gases that make up the air. This means that the density of air decreases with increasing altitude.

outer space, outside of the atmosphere? From the bottom box, you would cover a distance of only 5.6 km (about 3.5 mi) to reach the box that was above 50 percent of the mass of Earth's atmosphere. At 12 km (about 7 mi), you would reach the box that was above 75 percent of Earth's atmosphere. At 16 km (about 10 mi), you would reach the box that was above about 90 percent of the atmosphere. And, after only 32 km (about 20 mi), you would reach the box that was above 99 percent of Earth's atmosphere. The significance of these distances might be better appreciated if you can imagine the distances to some familiar locations; for example, from your campus to a store 16 km (about 10 mi) away would place you above 90 percent of the atmosphere if you were to travel this same distance straight up.

FIGURE 17.3 Earth's atmosphere thins rapidly with increasing altitude and is much closer to Earth's surface than most people realize.

Since the average radius of the solid Earth is about 6,373 km (3,960 mi), you can see that the atmosphere is a very thin shell with 99 percent of the mass within 32 km (about 20 mi) by comparison. The outer edge of the atmosphere is much closer to Earth than most people realize (figure 17.3).

Composition of the Atmosphere

A sample of pure, dry air is colorless, odorless, and composed mostly of the molecules of just three gases, nitrogen (N_2), oxygen (O_2), and argon (Ar). Nitrogen is the most abundant (about 78 percent of the total volume), followed by oxygen (about 21 percent), then argon (about 1 percent). The molecules of these three gases are well mixed, and this composition is nearly constant everywhere near Earth's surface (figure 17.4).

Nitrogen does not readily enter into chemical reactions with rocks, so it has accumulated in the atmosphere. Some nitrogen is removed from the atmosphere by certain bacteria in the soil and by lightning. The compounds formed are absorbed by plants and are consequently utilized by the food chain. Eventually, the nitrogen is returned to the atmosphere through the decay of plant and animal matter. Overall, these processes of nitrogen removal and release must be in balance

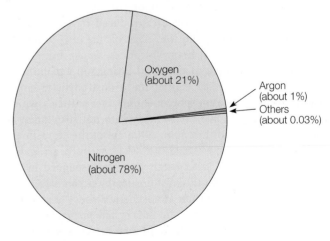

FIGURE 17.4 Earth's atmosphere has a unique composition of gases when compared to that of the other planets in the solar system.

since the amount of nitrogen in the atmosphere is essentially constant over time.

Oxygen gas also cycles into and out of the atmosphere in balanced processes of removal and release. Oxygen is removed (1) by living organisms as food is oxidized to carbon dioxide and water and (2) by chemical weathering of rocks as metals and other elements combine with oxygen to form oxides. Oxygen is released by green plants as a result of photosynthesis, and the amount released balances the amount removed by organisms and weathering. So oxygen, as well as nitrogen, is maintained in a state of constant composition through balanced chemical reactions.

The third major component of the atmosphere, argon, is inert and does not enter into any chemical reactions or cycles. It is produced as a product of radioactive decay and, once released, remains in the atmosphere as an inactive filler.

In addition to the relatively fixed amounts of nitrogen, oxygen, and argon, the atmosphere contains variable amounts of water vapor. Water vapor is the invisible, molecular form of water in the gaseous state, which should not be confused with fog or clouds. Fog and clouds are tiny droplets of liquid water, not water in the single molecular form of water vapor. The amount of water vapor in the atmosphere can vary from a small fraction of a percent composition by volume in cold, dry air to about 4 percent in warm, humid air. This small, variable percentage of water vapor is essential in maintaining life on Earth. It enters the atmosphere by evaporation, mostly from the ocean, and leaves the atmosphere as rain or snow.

Apart from the variable amounts of water vapor, the relatively fixed amounts of nitrogen, oxygen, and argon make up about 99.97 percent of the volume of a sample of dry air. The remaining gases are mostly carbon dioxide (CO_2) and traces of inert gases. Carbon dioxide makes up only 0.03 percent of the atmosphere, but it is important to life on Earth and it is constantly cycled through the atmosphere by the process of photo-synthesis that removes CO_2, and respiration that releases CO_2.

In addition to gases and water vapor, the atmosphere contains particles of dust, smoke, salt crystals, and tiny solid or liquid particles called *aerosols*. These particles become suspended and are dispersed among the molecules of the atmospheric gases. Aerosols are produced by combustion, often resulting in air pollution. Aerosols are also produced by volcanoes and forest fires. Volcanoes, smoke from combustion, and the force of the wind lifting soil and mineral particles into the air all contribute to dust particles larger than aerosols in the atmosphere. These larger particles, which range in size up to 500 micrometers, are not suspended as the aerosols are, and they soon settle out of the atmosphere as dust and soot.

Tiny particles of salt crystals that are suspended in the atmosphere come from the mist created by ocean waves and the surf. This mist forms an atmospheric aerosol of seawater that evaporates, leaving the solid salt crystals suspended in the air. The aerosol of salt crystals and dust becomes well mixed in the lower atmosphere around the globe, playing a large and important role in the formation of clouds.

Atmospheric Pressure

At Earth's surface (sea level), the atmosphere exerts a force of about 10.0 newtons on each square centimeter (14.7 lb/in²). As you go to higher altitudes above sea level the pressure rapidly decreases with increasing altitude. At an altitude of about 5.6 km (about 3.5 mi), the air pressure is about half of what it is at sea level, about 5.0 newtons/cm²(7.4 lb/in²). At 12 km (about 7 mi), the air pressure is about 2.5 newtons/cm²(3.7 lb/in²). Compare this decreasing air pressure at greater elevations to figure 17.3. Again, you can see that most of the atmosphere is very close to Earth, and it thins rapidly with increasing altitude. Even a short elevator ride takes you high enough that the atmospheric pressure on your eardrum is reduced. You equalize the pressure by opening your mouth, allowing the air under greater pressure inside the eardrum to move through the eustachian tube. This makes a "pop" sound that most people associate with changes in air pressure.

Atmospheric pressure is measured by an instrument called a **barometer.** The mercury barometer was invented in 1643 by an Italian named Torricelli. He closed one end of a glass tube and then filled it with mercury. The tube was then placed, open end down, in a bowl of mercury while holding the mercury in the tube with a finger. When Torricelli removed his finger with the open end below the surface in the bowl, a small amount of mercury moved into the bowl leaving a vacuum at the top end of the tube. The mercury remaining in the tube was supported by the atmospheric pressure on the surface of the mercury in the bowl. The pressure exerted by the weight of the mercury in the tube thus balanced the pressure exerted by the atmosphere. At sea level, Torricelli found that atmospheric pressure balanced a column of mercury about 76.00 cm (29.92 in) tall (figure 17.5).

As the atmospheric pressure increases and decreases, the height of the supported mercury column moves up and down. Atmospheric pressure can be expressed in terms of the height

FIGURE 17.5 The mercury barometer measures the atmospheric pressure from the balance between the pressure exerted by the weight of the mercury in a tube and the pressure exerted by the atmosphere. As the atmospheric pressure increases and decreases, the mercury rises and falls. This sketch shows the average height of the column at sea level.

of such a column of mercury. Public weather reports give the pressure by referring to such a mercury column; for example, "The pressure is 30 inches (about 76 cm) and rising." If the atmospheric pressure at sea level is measured many times over long periods of time, an average value of 76.00 cm (29.92 in) of mercury is obtained. This average measurement is called the **standard atmospheric pressure** and is sometimes referred to as the *normal pressure*. It is also called *one atmosphere of pressure*.

Warming the Atmosphere

Radiation from the Sun must pass through the atmosphere before reaching Earth's surface. The atmosphere filters, absorbs, and reflects incoming solar radiation, as shown in figure 17.6. On the average, Earth as a whole reflects about 30 percent of the total radiation back into space, with two-thirds of the reflection occurring from clouds. The amount reflected at any one time depends on the extent of cloud cover, the amount of dust in the atmosphere, and the extent of snow and vegetation on the surface. Substantial changes in any of these influencing variables could increase or decrease the reflectivity, leading to increased heating or cooling of the atmosphere.

As figure 17.6 shows, only about one-half of the incoming solar radiation reaches Earth's surface. The reflection and selective filtering by the atmosphere allow a global average of about 240 watts per square meter to reach the surface. Wide variations from the average occur with latitude as well as with the season.

The incoming solar radiation that does reach Earth's surface is absorbed. Rocks, soil, water, and the ground become warmer as a result. These materials emit the absorbed solar energy as infrared radiation, wavelengths longer than the visible part of the electromagnetic spectrum. This longer-wavelength infrared radiation has a frequency that matches some of the natural frequencies of vibration of carbon dioxide and water molecules. This match means that carbon dioxide and water

molecules readily absorb infrared radiation that is emitted from the surface of Earth. The absorbed infrared energy shows up as an increased kinetic energy of the molecules, which is indicated by an increase in temperature. Carbon dioxide and water vapor molecules in the atmosphere now emit infrared radiation of their own, this time in all directions. Some of this reemitted radiation is again absorbed by other molecules in the atmosphere, some is emitted to space, and significantly, some is absorbed by the surface to start the process all over again. The net result is that less of the energy from the Sun escapes immediately to space after being absorbed and emitted as infrared. It is retained through the process of being redirected to the surface, increasing the surface temperature more than it would have otherwise been. The more carbon dioxide that is present in the atmosphere, the more energy that will be bounced around and redirected back toward the surface, increasing the temperature near the surface. The process of heating the atmosphere in the lower parts by the absorption of solar radiation and reemission of infrared radiation is called the **greenhouse effect.** It is called the greenhouse effect because greenhouse glass allows the short wavelengths of solar radiation to pass into the greenhouse but does not allow all of the longer infrared radiation to leave. This analogy is misleading, however, because carbon dioxide and water vapor molecules do not "trap" infrared radiation, but they are involved in a dynamic absorption and downward reemission process that increases the surface temperature. The more carbon dioxide molecules that are involved in this dynamic process, the more infrared radiation will be directed back to Earth and the more the temperature will increase. More layers of glass on a greenhouse will not increase the temperature significantly. The significant heating factor in a real greenhouse is the blockage of convection by the glass, a process that does not occur from the presence of carbon dioxide and water vapor in the atmosphere.

Structure of the Atmosphere

Convection currents and the repeating absorption and reemission processes of the greenhouse effect tend to heat the atmosphere from the ground up. In addition, the higher-altitude gases lose radiation to space more readily than the lower-altitude gases. Thus, the lowest part of the atmosphere is warmer, and the temperature decreases with increasing altitude. On the average, the temperature decreases about 6.5°C for each kilometer of altitude (3.5°F/1,000 ft). This change of temperature with altitude is called the *observed lapse rate*. The observed lapse rate applies only to air that is not rising or sinking, and the actual change with altitude can be very different from this average value. For example, a stagnant mass of very cold air may settle over an area, producing colder temperatures near the surface than in the air layers above. Such a layer where the temperature increases with height is called an **inversion** (figure 17.7). Inversions often occur on calm winter days after the arrival of a cold front. They also occur on calm, clear, and cool nights ("C" nights), when the surface rapidly loses radiant energy to space. In either case, the situation results in a "cap" of

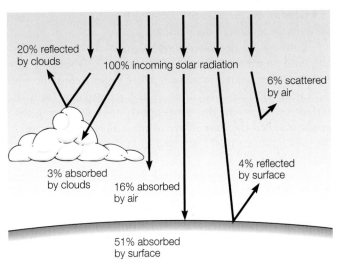

FIGURE 17.6 On the average, Earth's surface absorbs only 51 percent of the incoming solar radiation after it is filtered, absorbed, and reflected.

Hole in the Ozone Layer?

Ozone is triatomic oxygen (O_3) that is concentrated mainly in the upper portions of the stratosphere. Diatomic molecules of oxygen (O_2) are concentrated in the troposphere, and monatomic molecules of oxygen (O) are found in the outer edges of the atmosphere. Although the amount of ozone present in the stratosphere is not great, its presence is vital to life on Earth's surface. Ultraviolet (UV) radiation causes mutations to the DNA of organisms. Without the stratospheric ozone, much more ultraviolet radiation would reach the surface of Earth, causing mutations in all kinds of organisms. In humans, ultraviolet radiation is known to cause skin cancer and damage the cornea of eyes. It is believed that the incidence of skin cancer would rise dramatically without the protection offered by the ozone.

Here is how stratospheric ozone shields Earth from ultraviolet radiation. The ozone concentration is not static because there is an ongoing process of ozone formation and destruction. For ozone to form, diatomic oxygen (O_2) must first be broken down into the monatomic form (O). Shortwave ultraviolet radiation is absorbed by diatomic oxygen, breaking it down into the single-atom form. This reaction is significant because of (1) the high-energy ultraviolet radiation that is removed from the sunlight and (2) the monatomic oxygen that is formed, which will combine with diatomic oxygen to make triatomic oxygen (ozone) that will absorb even more ultraviolet radiation. This initial reaction is

$$O_2 + UV \rightarrow O + O$$

When the O molecule collides with an O_2 molecule and any third, neutral molecule (NM), the following reaction takes place:

$$O_2 + O + NM \rightarrow O_3 + NM$$

When O_3 is exposed to ultraviolet radiation, the ozone absorbs the UV radiation and breaks down to two forms of oxygen in the following reaction:

$$O_3 + UV \rightarrow O_2 + O$$

The monatomic molecule that is produced combines with an ozone molecule to produce two diatomic molecules,

$$O + O_3 \rightarrow 2\,O_2$$

and the process starts all over again.

Much concern exists about Freon (CF_2Cl_2) and other similar chemicals that make their way to the stratosphere. These chemicals are broken down by UV radiation, releasing chlorine (Cl), which reacts with ozone. The reaction might be

$$CF_2Cl_2 + UV \rightarrow CF_2Cl-^* + Cl-^*$$
$$Cl-^* + O_3 \rightarrow {}^*-ClO + O_2$$
$${}^*-ClO + O \rightarrow O_2 + Cl-^*$$

$-^*$ is an unattached bond

Note that ozone is decomposed in the second step, so it is removed from the UV radiation–absorbing process. Furthermore, the second and third steps repeat, so the destruction of one molecule of Freon can result in the decomposition of many molecules of ozone.

A regional zone of decreased ozone availability in the stratosphere is referred to as a "hole in the ozone layer," since much high-energy UV radiation can now reach the surface below. Concerns about the impact chlorine-containing chemicals such as Freon have on the ozone layer led to an international agreement known as the Montreal Protocol. The agreement has resulted in a phasing out of the use of Freon and similar compounds in most of the world. Subsequently, it appears that the sizes of the "holes in the ozone layer" are getting smaller.

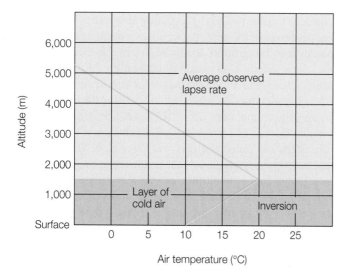

FIGURE 17.7 On the average, the temperature decreases about 6.5°C/km, which is known as the *observed lapse rate*. An inversion is a layer of air in which the temperature increases with height.

cooler, more dense air overlying the warmer air beneath. This often leads to an increase of air pollution because the inversion prevents dispersion of the pollutants.

Temperature decreases with height at the observed lapse rate until an average altitude of about 11 km (about 6.7 mi), where it then begins to remain more or less constant with increasing altitude. The layer of the atmosphere from the surface up to where the temperature stops decreasing with height is called the *troposphere*. Almost all weather occurs in the troposphere, which is derived from Greek words meaning "turning layer."

Above the troposphere is the second layer of the atmosphere called the *stratosphere*. This layer is derived from the Greek terms for "stratified layer." It is stratified, or layered, because the temperature increases with height. Cooler air below means that consecutive layers of air are denser on the bottom, which leads to a stable situation rather than the turning turbulence of the troposphere below. The stratosphere contains little moisture or dust and lacks convective turbulence, making it a

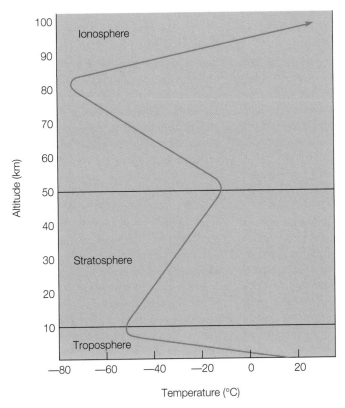

FIGURE 17.8 The structure of the atmosphere based on temperature differences.

desirable altitude for aircraft to fly. Temperature in the lower stratosphere increases gradually with increasing altitude to a height of about 48 km (about 30 mi) (figure 17.8).

The outermost layer is where the molecules merge with the diffuse vacuum of space. Molecules of this layer that have sufficient kinetic energy are able to escape and move off into space. This outer layer is sometimes called the *ionosphere* because of the free electrons and ions at this altitude. The electrons and ions here are responsible for reflecting radio waves around Earth and for the northern lights.

THE WINDS

The troposphere is heated from the bottom up as the surface of Earth absorbs sunlight. Uneven heating of Earth's surface sets the stage for *convection*. As a local region of air becomes heated, the air expands, reducing its density. This less dense air is pushed upward by nearby cooler, more dense air. This results in three general motions of air: (1) the upward movement of air over a region of greater heating, (2) the sinking of air over a cooler region, and (3) a horizontal air movement between the cooler and warmer regions. In general, a horizontal movement of air is called **wind,** and the direction of a wind is defined as the direction from which it blows.

Air in the troposphere rises, moves as the wind, and sinks. All three of these movements are related, and all occur at the same time over regions of the landscape. During a day with

gentle breezes on the surface, the individual, fluffy clouds you see are forming over areas where the air is moving upward. The clear air between the clouds is over areas where the air is moving downward. On a smaller scale, air can be observed moving from a field of cool grass toward an adjacent asphalt parking lot on a calm, sunlit day. Soap bubbles or smoke will often reveal the gentle air movement of this localized convection.

Local Wind Patterns

Considering average conditions, there are two factors that are important for a generalized model to help you understand local wind patterns. These factors are (1) the relationship between air temperature and air density, and (2) the relationship between air pressure and the movement of air.

The upward and downward movement of air leads to the second part of the generalized model, that (1) the upward movement produces a "lifting" effect on the surface that results in an area of lower atmospheric pressure and (2) the downward movement produces a "piling up" effect on the surface that results in an area of higher atmospheric pressure. On the surface, air is seen to move from the "piled up" area of higher pressure horizontally to the "lifted" area of lower pressure (figure 17.9). In other words, air generally moves from an area of higher pressure to an area of lower pressure. The movement of air and the pressure differences occur together, and neither is the cause of the other. This is an important relationship in a working model of air movement that can be observed and measured on a very small scale, such as between an asphalt parking lot and a grass field. It can also be observed and measured for local, regional wind patterns and for worldwide wind systems.

Adjacent areas of the surface can have different temperatures because of different heating or cooling rates. The difference is very pronounced between adjacent areas of land and water. Under identical conditions of incoming solar radiation,

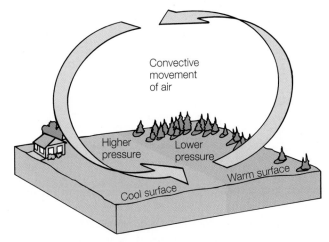

FIGURE 17.9 A model of the relationships between differential heating, the movement of air, and pressure difference in a convective cell. Cool air pushes the less-dense, warm air upward, reducing the surface pressure. As the uplifted air cools and becomes more dense, it sinks, increasing the surface pressure.

A Closer Look

The Wind Chill Factor

The term *wind chill* is attributed to the Antarctic explorer Paul A. Siple. During the 1940s, Siple and Charles F. Passel conducted experiments on how long it took a can of water to freeze at various temperatures and wind speeds at a height of 10 m (33 ft) above the ground (typical height of an anemometer). They found that the time depended on the air temperature and the wind speed. From these data, an equation was developed to calculate the **wind chill factor** for humans.

In 2001, the National Weather Service changed to a new method of computing wind chill temperatures. The new wind chill formula was developed by a year-long cooperative effort between the U.S. and Canadian governments and university scientists. The new standard is based on wind speeds at an average height of 1.5 m (5 ft) above the ground, which is closer to the height of a human face rather than the height of an anemometer and takes advantage of advances in science, technology, and computers. The new chart also highlights the danger of frostbite.

Here is the reason why the wind chill factor is an important consideration. The human body constantly produces heat to maintain a core temperature, and some of this heat is radiated to the surroundings. When the wind is not blowing (and you are not moving), your body heat is also able to warm some of the air next to your body. This warm blanket of air provides some insulation, protecting your skin from the colder air farther away. If the wind blows, however, it moves this air away from your body and you feel cooler. How much cooler depends on how fast the air is moving and on the outside temperature—which is what the wind chill factor tells you. Thus, wind chill is an attempt to measure the combined effect of low temperature and wind on humans (box figure 17.1). It is just one of the many factors that can affect winter comfort. Others include the type of clothes, level of physical exertion, amount of sunshine, humidity, age, and body type.

There is a wind chill calculator at the National Weather Service website, *www.nws.noaa.gov/om/windchill/*. All you need to do is enter the air temperature (in degrees Fahrenheit) and the wind speed (in miles per hour), and the calculator will give you the wind chill using both the old and the new formula.

Wind (mph)	Temperature (°F)																	
	40	35	30	25	20	15	10	5	0	−5	−10	−15	−20	−25	−30	−35	−40	−45
5	36	31	25	19	13	7	1	−5	−11	−16	−22	−28	−34	−40	−46	−52	−57	−63
10	34	27	21	15	9	3	−4	−10	−16	−22	−28	−35	−41	−47	−53	−59	−66	−72
15	32	25	19	13	6	0	−7	−13	−19	−26	−32	−39	−45	−51	−58	−64	−71	−77
20	30	24	17	11	4	−2	−9	−15	−22	−29	−35	−42	−48	−55	−61	−68	−74	−81
25	29	23	16	9	3	−4	−11	−17	−24	−31	−37	−44	−51	−58	−64	−71	−78	−84
30	28	22	15	8	1	−5	−12	−19	−26	−33	−39	−46	−53	−60	−67	−73	−80	−87
35	28	21	14	7	0	−7	−14	−21	−27	−34	−41	−48	−55	−62	−69	−76	−82	−89
40	27	20	13	6	−1	−8	−15	−22	−29	−36	−43	−50	−57	−64	−71	−78	−84	−91
45	26	19	12	5	−2	−9	−16	−23	−30	−37	−44	−51	−58	−65	−72	−79	−86	−93
50	26	19	12	4	−3	−10	−17	−24	−31	−38	−45	−52	−60	−67	−74	−81	−88	−95
55	25	18	11	4	−3	−11	−18	−25	−32	−39	−46	−54	−61	−68	−75	−82	−89	−97
60	25	17	10	3	−4	−11	−19	−26	−33	−40	−48	−55	−62	−69	−76	−84	−91	−98

Frostbite times: 30 minutes   10 minutes   5 minutes

Box Figure 17.1   Wind chill chart.

the temperature changes experienced by the water will be much less than the changes experienced by the adjacent land. There are three principal reasons for this difference: (1) The specific heat of water is about twice the specific heat of soil. This means that it takes more energy to increase the temperature of water than it does for soil. Equal masses of soil and water exposed to sunlight will result in the soil heating about 1°C while the water heats 1/2°C from absorbing the same amount of solar radiation. (2) Water is a transparent fluid that is easily mixed, so the incoming solar radiation warms a body of water throughout, spreading out the heating effect. Incoming solar radiation on land, on the other hand, warms a relatively thin layer on the top, concentrating the heating effect. (3) The water is cooled by evaporation, which helps keep a body of water at a lower temperature than an adjacent landmass under identical conditions of incoming solar radiation.

FIGURE 17.10 The land warms and cools more rapidly than an adjacent large body of water. During the day, the land is warmer, and air over the land expands and is buoyed up by cooler, more dense air from over the water. During the night, the land cools more rapidly than the water, and the direction of the breeze is reversed.

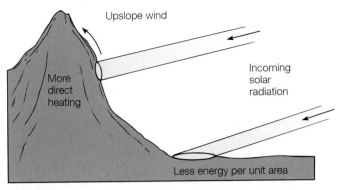

FIGURE 17.11 Incoming solar radiation falls more directly on the side of a mountain, which results in differential heating. The same amount of sunlight falls on the areas shown in this illustration, with the valley floor receiving a more spread-out distribution of energy per unit area. The overall result is an upslope mountain breeze during the day. During the night, dense cool air flows downslope for a reverse wind pattern.

A local wind pattern may result from the resulting temperature differences between a body of water and adjacent landmasses. If you have ever spent some time along a coast, you may have observed that a cool, refreshing gentle breeze blows from the water toward the land during the summer. During the day, the temperature of the land increases more rapidly than the water temperature. The air over the land is therefore heated more, expands, and becomes less dense. Cool, dense air from over the water moves inland under the air over the land, buoying it up. The air moving from the sea to the land is called a *sea breeze.* The sea breeze along a coast may extend inland several miles during the hottest part of the day in the summer. The same pattern is sometimes observed around the Great Lakes during the summer, but this breeze usually does not reach more than several city blocks inland. During the night, the land surface cools more rapidly than the water, and a breeze blows from the land to the sea (figure 17.10).

Another pattern of local winds develops in mountainous regions. If you have ever visited a mountain in the summer, you may have noticed that there is usually a breeze or wind blowing up the mountain slope during the afternoon. This wind pattern develops because the air over the mountain slope is heated more than the air in a valley. As shown in figure 17.11, the air over the slope becomes warmer because it receives more direct sunlight than the valley floor. Sometimes this air movement is so gentle that it would be unknown except for the evidence of clouds that form over the peaks during the day and evaporate at night. During the night, the air on the slope cools as the land loses radiant energy to space. As the air cools, it becomes denser and flows downslope, forming a reverse wind pattern to the one observed during the day.

During cooler seasons, cold, dense air may collect in valleys or over plateaus, forming a layer or "puddle" of cold air. Such an accumulation of cold air often results in some very cold nighttime temperatures for cities located in valleys, temperatures that are much colder than anywhere in the surrounding region. Some weather disturbance, such as an approaching front, can disturb such an accumulation of cold air and cause it to pour out of its resting place and through canyons or lower valleys. Air moving from a higher altitude like this becomes compressed as it moves to lower elevations under increasing atmospheric pressure. Compression of air increases the temperature by increasing the kinetic energy of the molecules. This creates a wind called a *Chinook,* which is common to mountainous and adjacent regions. A Chinook is a wind of compressed air with sharp temperature increases that can melt away any existing snow cover in a single day. The *Santa Ana* is a well-known compressional wind that occurs in southern California.

Global Wind Patterns

Local wind patterns tend to mask the existence of the overall global wind pattern that is also present. The global wind pattern is not apparent if the winds are observed and measured for a particular day, week, or month. It does become apparent when the records for a long period of time are analyzed. These records show that Earth has a large-scale pattern of atmospheric circulation that varies with latitude. There are belts in which the winds average an overall circulation in one direction, belts of higher atmospheric pressure averages, and belts of lower atmospheric pressure averages. This has led to a generalized pattern of atmospheric circulation and a global atmospheric model. This model, as you will see, today provides the basis for the daily weather forecast for local and regional areas.

As with local wind patterns, it is temperature imbalances that drive the global circulation of the atmosphere. Earth receives more direct solar radiation in the equatorial region than it does at higher latitudes (figure 17.12). As a result, the temperatures of the lower troposphere are generally higher in

Science and Society

Use Wind Energy?

Millions of windmills were installed in rural areas of the United States between the late 1800s and the late 1940s. These windmills used wind energy to pump water, grind grain, or generate electricity. Some are still in use today, but most were no longer needed after inexpensive electric power became generally available in those areas.

In the 1970s, wind energy began making a comeback as a clean, renewable energy alternative to fossil fuels. The windmills of the past were replaced by wind turbines of today. A wind turbine with blades that are rotated by the wind is usually mounted on a tower. The blades' rotary motion drives a generator that produces electricity. A location should have average yearly wind speeds of at least 19 km/h (12 mi/h) to provide enough wind energy for a turbine, and a greater yearly average means more energy is available. Farms, homes, and businesses in these locations can use smaller turbines, which are generally 50 kilowatts or less. Large turbines of 500 kilowatts or more are used

in "wind farms," which are large clusters of interconnected wind turbines connected to a utility power grid.

Many areas of the United States have a high potential for wind-power use. North Dakota, South Dakota, and Texas have enough wind resources to provide electricity for the entire nation. Today, only California has extensively developed wind farms, with more than thirteen thousand wind turbines on three wind farms in the Altamont Pass region (east of San Francisco), Tehachapi (southeast of Bakersfield), and San Gorgonio (near Palm Springs). With a total of 2,361 megawatts of installed capacity in California, wind energy generates enough electricity to more than meet the needs of a city the size of San Francisco. The wind farms in California have a rated capacity that is comparable to two large coal-fired power plants but without the pollution and limits of this nonrenewable energy source. Wind energy makes economic as well as environmental sense, and new wind farms are being developed in

Minnesota, Oregon, and Wyoming. Other states with a strong wind power potential include Kansas, Montana, Nebraska, Oklahoma, Iowa, Colorado, Michigan, and New York. All of these states, in fact, have a greater wind-energy potential than California.

Questions to Discuss

Discuss with your group the following questions concerning wind power:

1. Why have electric utilities not used much wind power as an energy source?
2. Should governments provide a tax break to encourage people to use wind power? Why or why not?
3. What are the advantages and disadvantages of using wind power in the place of fossil fuels?
4. What are the advantages and disadvantages of the government building huge wind farms in North Dakota, South Dakota, and Texas to supply electricity for the entire nation?

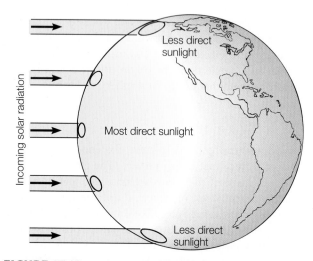

FIGURE 17.12 On a global, yearly basis, the equatorial region of Earth receives more direct incoming solar radiation than the higher latitudes. As a result, average temperatures are higher in the equatorial region and decrease with latitude toward both poles. This sets the stage for worldwide patterns of prevailing winds, high and low areas of atmospheric pressure, and climatic patterns.

the equatorial region, decreasing with latitude toward both poles. The lower troposphere from 10°N to 10°S of the equator is heated, expands, and becomes less dense. Hot air rises in this belt around the equator, known as the *intertropical convergence*

zone. The rising air cools because it expands as it rises, resulting in heavy average precipitation. The tropical rainforests of Earth occur in this zone of high temperatures and heavy rainfall. As the now dry, rising air reaches the upper parts of the troposphere, it begins to spread toward the north and the south, sinking back toward Earth's surface (figure 17.13). The descending air reaches the surface to form a high-pressure belt that is centered about 30°N and 30°S of the equator. Air moving on the surface away from this high-pressure belt produces the prevailing northeast trade winds and the prevailing westerly winds of the Northern Hemisphere. The great deserts of Earth are also located in this high-pressure belt of descending dry air.

Poleward of the belt of high pressure, the atmospheric circulation is controlled by a powerful belt of wind near the top of the troposphere called a **jet stream.** Jet streams are sinuous, meandering loops of winds that tend to extend all the way around Earth, moving generally from the west in both hemispheres at speeds of 160 km/h (about 100 mi/h) or more. A jet stream may occur as a single belt, or loop, of wind, but sometimes it divides into two or more parts. The jet stream develops north and south loops of waves much like the waves you might make on a very long rope. These waves vary in size, sometimes beginning as a small ripple but then growing slowly as the wave moves eastward. Waves that form on the jet stream bulge toward the poles (called a crest) or toward the equator (called a trough). Warm air masses move toward the poles

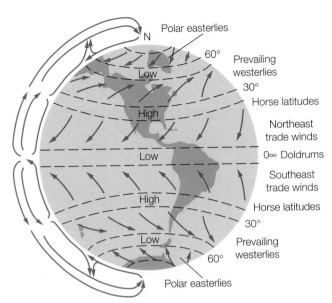

FIGURE 17.13 Simplified pattern of horizontal and vertical circulation in the actual atmosphere. Regions of high and low pressure are indicated.

ahead of a trough, and cool air masses move toward the equator behind a trough as it moves eastward. The development of a wave in the jet stream is understood to be one of the factors that influences the movement of warm and cool air masses, a movement that results in weather changes on the surface.

The intertropical convergence zone, the 30° belt of high pressure, and the northward and southward migration of a meandering jet stream all shift toward or away from the equator during the different seasons of the year. The troughs of the jet stream influence the movement of alternating cool and warm air masses over the belt of the prevailing westerlies, resulting in frequent shifts of fair weather to stormy weather, then back again. The average shift during the year is about 6° of latitude, which is sufficient to control the overall climate in some locations. The influence of this shift of the global circulation of Earth's atmosphere will be considered as a climatic influence after considering the roles of water and air masses in frequent weather changes.

WATER AND THE ATMOSPHERE

Water exists on Earth in all three states: (1) as a liquid when the temperature is generally above the freezing point of 0°C (32°F), (2) as a solid in the form of ice, snow, or hail when the temperature is generally below the freezing point, and (3) as the invisible, molecular form of water in the gaseous state, which is called *water vapor*.

Over 98 percent of all the water on Earth exists in the liquid state, mostly in the ocean, and only a small, variable amount of water vapor is in the atmosphere at any given time. Since so much water seems to fall as rain or snow at times, it may be a surprise that the overall atmosphere really does not contain very much water vapor. If the average amount of water vapor in Earth's atmosphere were condensed to liquid form, the vapor

and all the droplets present in clouds would form a uniform layer around Earth only 3 cm (about 1 in) thick. Nonetheless, it is this small amount of water vapor that is eventually responsible for (1) contributing to the greenhouse effect, which helps make Earth a warmer planet, (2) serving as one of the principal agents in the weathering and erosion of the land, which creates soils and sculptures the landscape, and (3) maintaining life, for all organisms (bacteria, protozoa, algae, fungi, plants, and animals) cannot survive without water. It is the ongoing cycling of water vapor into and out of the atmosphere that makes all this possible. Understanding this cycling process and the energy exchanges involved is also closely related to understanding Earth's weather patterns.

Evaporation and Condensation

Water tends to undergo a liquid-to-gas or a gas-to-liquid phase change at any temperature. The phase change can occur in either direction at any temperature.

In *evaporation*, more molecules are leaving the liquid state than are returning. In *condensation*, more molecules are returning to the liquid state than are leaving. This is a dynamic, ongoing process with molecules leaving and returning continuously (figure 17.14). If the air were perfectly dry and still, more molecules would leave (evaporate) the liquid state than would return (condense). Eventually, however, an equilibrium would be reached with as many molecules returning to the liquid state per unit of time as are leaving. An equilibrium condition between evaporation and condensation occurs in **saturated air.** Saturated air occurs when the processes of evaporation and condensation are in balance.

Air will remain saturated as long as (1) the temperature remains constant and (2) the processes of evaporation and condensation remain balanced. Temperature influences the equilibrium condition of saturated air because increases or decreases in the temperature mean increases or decreases

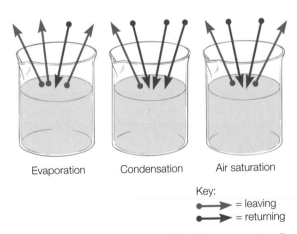

FIGURE 17.14 Evaporation and condensation are occurring all the time. If the number of molecules leaving the liquid state exceeds the number returning, the water is evaporating. If the number of molecules returning to the liquid state exceeds the number leaving, the water vapor is condensing. If both rates are equal, the air is saturated; that is, the relative humidity is 100 percent.

in the kinetic energy of water vapor molecules. Water vapor molecules usually undergo condensation when attractive forces between the molecules can pull them together into the liquid state. Lower temperature means lower kinetic energies, and slow-moving water vapor molecules spend more time close to one another and close to the surface of liquid water. Spending more time close together means an increased likelihood of attractive forces pulling the molecules together. On the other hand, higher temperature means higher kinetic energies, and molecules with higher kinetic energy are less likely to be pulled together. As the temperature increases, there is therefore less tendency for water molecules to return to the liquid state. If the temperature is increased in an equilibrium condition, more water vapor must be added to the air to maintain the saturated condition. Warm air can therefore hold more water vapor than cooler air. In fact, warm air on a typical summer day can hold five times as much water vapor as cold air on a cold winter day.

Humidity

The amount of water vapor in the air is referred to generally as **humidity.** Damp, moist air is more likely to have condensation than evaporation, and this air is said to have a *high humidity.* Dry air is more likely to have evaporation than condensation, on the other hand, and this air is said to have a *low humidity.* A measurement of the amount of water vapor in the air at a particular time is called the **absolute humidity** (figure 17.15). At room temperature, for example, humid air might contain 15 grams of water vapor in each cubic meter of air. At the same temperature, air of low humidity might have an absolute humidity of only 2 grams per cubic meter. The absolute humidity can range from near zero up to a maximum that is determined by the temperature at the time. Since the temperature of the water vapor present in the air is the same as the temperature of the air, the maximum absolute humidity is usually said to be determined by the air temperature. What this really means is that the maximum absolute humidity is determined by the temperature of the water vapor, that is, the average kinetic energy of the water vapor.

The relationship between the *actual* absolute humidity at a particular temperature and the *maximum* absolute humidity that can occur at that temperature is called the **relative humidity.** Relative humidity is a ratio between (1) the amount of water vapor in the air and (2) the amount of water vapor needed to saturate the air at that temperature. The relationship is

$$\text{relative humidity} = \frac{\substack{\text{absolute humidity} \\ \text{at present temperature}}}{\substack{\text{maximum absolute humidity} \\ \text{at present temperature}}} \times 100\%$$

For example, suppose a measurement of the water vapor in the air at 10°C (50°F) finds an absolute humidity of 5.0 g/m³. According to figure 17.15, the maximum amount of water vapor

FIGURE 17.15 The maximum amount of water vapor that can be in the air at different temperatures. The amount of water vapor in the air at a particular temperature is called the *absolute humidity.*

that can be in the air when the temperature is 10°C is about 10 g/m³. The relative humidity is then

$$\frac{5.0 \text{ g/m}^3}{10 \text{ g/m}^3} \times 100\% = 50\%$$

If the absolute humidity had been 10 g/m³, then the air would have all the water vapor it could hold, and the relative humidity would be 100 percent. A humidity of 100 percent means that the air is saturated at the present temperature.

The important thing to understand about relative humidity is that the capacity of air to hold water vapor changes with the temperature. Cold air cannot hold as much water vapor, and warming the air will increase its capacity. With the capacity *increased,* the relative humidity *decreases* because you can now add more water vapor to the air than you could before. Just warming the air, for example, can reduce the relative humidity from 50 percent to 3 percent. Lower relative humidity results because warming the air increases the capacity of air to hold water vapor. This explains the need to humidify a home in the winter. Evaporation occurs very rapidly when the humidity is low. Evaporation is a cooling process since the molecules with higher kinetic energy are the ones to escape, lowering the average kinetic energy as they evaporate. Dry air will therefore cause you to feel cool even though the air temperature is fairly high. Adding moisture to the air will enable you to feel warmer at lower air temperatures and thus lower your fuel bill.

The relationship between the capacity of air to hold water vapor and temperature also explains why the relative humidity increases in the evening after the Sun goes down. A cooler temperature means less capacity of air to hold water vapor. With the same amount of vapor in the air, a reduced capacity means a higher relative humidity.

The Condensation Process

Condensation depends on two factors: (1) the relative humidity and (2) the temperature of the air. During condensation, molecules of water vapor join together to produce liquid water on the surface as dew or in the air as the droplets of water making up fog or clouds. Water molecules may also join together to produce solid water in the form of frost or snow. Before condensation can occur, however, the air must be saturated, which means that the relative humidity must be 100 percent with the air containing all the vapor it can hold at the present temperature. A parcel of air can become saturated as a result of (1) water vapor being added to the air from evaporation, (2) cooling, which reduces the capacity of the air to hold water vapor and therefore increases the relative humidity, or (3) a combination of additional water vapor with cooling.

The process of condensation of water vapor explains a number of common observations. You are able to "see your breath" on a cold day, for example, because the high moisture content of your exhaled breath is condensed into tiny water droplets by cold air. The small fog of water droplets evaporates as it spreads into the surrounding air with a lower moisture content. The white trail behind a high-flying jet aircraft is also a result of condensation of water vapor. Water is one of the products of combustion, and the white trail is condensed water vapor, a trail of tiny droplets of water in the cold upper atmosphere. The trail of water droplets is called a *contrail* after "condensation trail." Back on the surface, a cold glass of beverage seems to "sweat" as water vapor molecules near the outside of the glass are cooled, moving more slowly. Slowly moving water vapor molecules spend more time closer together, and the molecular forces between the molecules pull them together, forming a thin layer of liquid water on the outside of the cold glass. This same condensation process sometimes results in a small stream of water from the cold air conditioning coils of an automobile or home mechanical air conditioner.

As air is cooled, its capacity to hold water vapor is reduced to lower and lower levels. Even without water vapor being added to the air, a temperature will eventually be reached at which saturation, 100 percent humidity, occurs. Further cooling below this temperature will result in condensation. The temperature at which condensation begins is called the **dew point temperature.** If the dew point is above 0°C (32°F), the water vapor will condense on surfaces as a liquid called **dew.** If the temperature is at or below 0°C, the vapor will condense on surfaces as a solid called **frost.** Note that dew and frost form on the tops, sides, and bottoms of objects. Dew and frost condense directly on objects and do not "fall out" of the air. Note also that the temperature that determines if dew or frost forms is the temperature of the object where they condense. This temperature near the open surface can be very different from the reported air temperature, which is measured more at eye level in a sheltered instrument enclosure.

Observations of where and when dew and frost form can lead to some interesting things to think about. Dew and frost, for example, seem to form on "C" nights, nights that can be described by the three "C" words of *clear, calm,* and *cool.* Dew and frost also seem to form more (1) in open areas rather than under trees or other shelters, (2) on objects such as grass rather than on the flat, bare ground, and (3) in low-lying areas before they form on slopes or the sides of hills. What is the meaning of these observations?

Dew and frost are related to clear nights and open areas because these are the conditions best suited for the loss of infrared radiation. Air near the surface becomes cooler as infrared radiation is emitted from the grass, buildings, streets, and everything else that absorbed the shorter-wavelength radiation of incoming solar radiation during the day. Clouds serve as a blanket, keeping the radiation from escaping to space so readily. So a clear night is more conducive to the loss of infrared radiation and therefore to cooling. On a smaller scale, a tree serves the same purpose, holding in radiation and therefore retarding the cooling effect. Thus, an open area on a clear, calm night would have cooler air near the surface than would be the case on a cloudy night or under the shelter of a tree.

The observation that dew and frost form on objects such as grass before forming on flat, bare ground is also related to loss of infrared radiation. Grass has a greater exposed surface area than the flat, bare ground. A greater surface area means a greater area from which infrared radiation can escape, so grass blades cool more rapidly than the flat ground. Other variables, such as specific heat, may be involved, but overall, frost and dew are more likely to form on grass and low-lying shrubs before they form on the flat, bare ground.

Dew and frost form in low-lying areas before forming on slopes and the sides of hills because of the density differences of cool and warm air. Cool air is more dense than warm air and is moved downhill by gravity, pooling in low-lying areas. You may have noticed the different temperatures of low-lying areas if you have ever driven across hills and valleys on a clear, calm, and cool evening. Citrus and other orchards are often located on slopes of hills rather than on valley floors because of the gravity drainage of cold air.

FIGURE 17.16 Fans like this one are used to mix the warmer, upper layers of air with the cooling air in the orchard on nights when frost is likely to form.

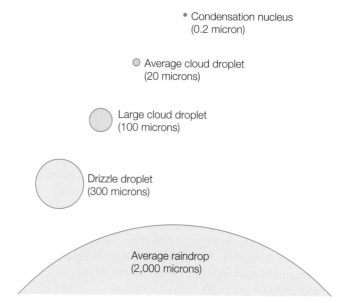

- Condensation nucleus (0.2 micron)
- Average cloud droplet (20 microns)
- Large cloud droplet (100 microns)
- Drizzle droplet (300 microns)
- Average raindrop (2,000 microns)

FIGURE 17.17 This figure compares the size of the condensation nuclei to the size of typical condensation droplets. Note that 1 micron is 1/1,000 mm.

It is air near the surface that is cooled first by the loss of radiation from the surface. Calm nights favor dew or frost formation because the wind mixes the air near the surface that is being cooled with warmer air above the surface. If you have ever driven near a citrus orchard, you may have noticed the huge, airplanelike fans situated throughout the orchard on poles. These fans are used on "C" nights when frost is likely to form to mix the warmer, upper layers of air with the cooling air in the orchard (figure 17.16).

Condensation occurs on the surface as frost or dew when the dew point is reached. When does condensation occur in the air? Water vapor molecules in the air are constantly colliding and banging into each other, but they do not just join together to form water droplets, even if the air is saturated. The water molecules need something to condense upon. Condensation of water vapor into fog or cloud droplets takes place on tiny particles present in the air. The particles are called **condensation nuclei.** There are hundreds of tiny dust, smoke, soot, and salt crystals suspended in each cubic centimeter of the air that serve as condensation nuclei. Tiny salt crystals, however, are particularly effective condensation nuclei because salt crystals attract water molecules. You may have noticed that salt in a salt shaker becomes moist on a humid day because of the way it attracts water molecules. Tiny salt crystals suspended in the air act the same way, serving as nuclei that attract water vapor into tiny droplets of liquid water.

After water vapor molecules begin to condense on a condensation nucleus, other water molecules will join the liquid water already formed, and the tiny droplet begins to increase in volume. The water droplets that make up a cloud are about fifteen hundred times larger than a condensation nuclei, and these droplets can condense out of the air in a matter of minutes. As the volume increases, however, the process slows, and

hours and days are required to form the even larger droplets and drops. For comparison to the sizes shown in figure 17.17, consider that the average human hair is about 100 microns in diameter. This is about the same diameter as the large cloud droplet of water. Large raindrops have been observed falling from clouds that formed only a few hours previously, so it must be some process or processes other than the direct condensation of raindrops that form precipitation. These processes are discussed in the section on precipitation.

Fog and Clouds

Fog and clouds are both accumulations of tiny droplets of water that have been condensed from the air. These water droplets are very small, and a very slight upward movement of the air will keep them from falling. If they do fall, they usually evaporate. Fog is sometimes described as a cloud that forms at or near the surface. A fog, as a cloud, forms because air containing water vapor and condensation nuclei has been cooled to the dew point. Some types of fog form under the same "C" night conditions favorable for dew or frost to form, that is, on clear, cool, and calm nights when the relative humidity is high. Sometimes this type of fog forms only in valleys and low-lying areas where cool air accumulates (figure 17.18). This type of fog is typical of inland fogs, those that form away from bodies of water. Other types of fog may form somewhere else, such as in the humid air over an ocean, and then move inland. Many fogs that occur along coastal regions were formed over the ocean and then carried inland by breezes. A third type of fog looks much like a steamy mist rising from melting snow on a street, over a body of water into cold air, or a steamy mist over streets after a summer rain shower. These are examples of a temporary fog that forms as a lot of water vapor is added to cool air. This is a cool fog, like other fogs, and is not hot as the steamlike appearance may lead you to believe.

Damage from a thunderstorm is usually caused by the associated lightning, strong winds, or hail. As illustrated in figure 17.29, the first stage of a thunderstorm begins as convection, mountains, or a dense air mass slightly lifts a mass of warm, moist air in an unstable atmosphere. The lifted air mass expands and cools to the dew point temperature, and a cumulus cloud forms. The latent heat of vaporization released by the condensation process accelerates the upward air motion, called an *updraft,* and the cumulus cloud continues to grow to towering heights. Soon the upward-moving, saturated air reaches the freezing level and ice crystals and snowflakes begin to form. When they become too large to be supported by the updraft, they begin to fall toward the surface, melting into raindrops in the warmer air they fall through. When they reach the surface, this marks the beginning of the mature stage. As the raindrops fall through the air, friction between the falling drops and the cool air produces a downdraft in the region of the precipitation. The cool air accelerates toward the surface at speeds up to 90 km/h (about 55 mi/h), spreading out on the ground when it reaches the surface. In regions where dust is raised by the winds, this spreading mass of cold air from the thunderstorm has the appearance of a small cold front with a steep, bulging leading edge. This miniature cold front may play a role in lifting other masses of warm, moist air in front of the thunderstorm, leading to the development of new cells. This stage in the life of a thunderstorm has the most intense rainfall, winds, and possibly hail. As the downdraft spreads throughout the cloud, the supply of new moisture from the updrafts is cut off and the thunderstorm enters the final, dissipating stage. The entire life cycle, from cumulus cloud to the final stage, lasts for about an hour as the thunderstorm moves across the surface. During the mature stage of powerful updrafts, the top of the thunderstorm may reach all the way to the top of the troposphere, forming a cirrus cloud that is spread into an anvil shape by the strong winds at this high altitude.

The updrafts, downdrafts, and falling precipitation separate tremendous amounts of electric charges that accumulate in different parts of the thundercloud. Large drops of water tend to carry negative charges, and cloud droplets tend to lose them. The upper part of the thunderstorm develops an accumulation of positive charges as cloud droplets are uplifted, and the middle portion develops an accumulation of negative charges from larger drops that fall. The voltage of these charge centers builds to the point that the electrical insulating ability of the air between them is overcome and a giant electrical discharge called *lightning* occurs (figure 17.30). Lightning discharges occur from the cloud to the ground, from the ground to a cloud, from one part of the cloud to another part, or between two different clouds. The discharge takes place in a fraction of a second and may actually consist of a number of strokes rather than one big discharge. The discharge produces an extremely high temperature around the channel, which may be only 6 cm (about 2 in) or so wide. The air it travels through is heated quickly, expanding into a sudden pressure wave that you hear as *thunder.* A nearby lightning strike produces a single, loud crack. Farther away strikes sound more like a rumbling boom as the sound from the separate strokes become separated over distance. Echoing of the thunder produced at farther distances also adds

FIGURE 17.30 Different parts of a thunderstorm cloud develop centers of electric charge. Lightning is a giant electric spark that discharges the accumulated charges.

FIGURE 17.31 These hailstones fell from a thunderstorm in Iowa, damaging automobiles, structures, and crops.
© Telegraph Herald/Photo by Patti Carr

to the rumbling sounds. Lightning can present a risk for people in the open, near bodies of water, or under a single, isolated tree during a thunderstorm. The safest place to be during a thunderstorm is inside a car or a building with a metal frame.

Updrafts are also responsible for **hail,** a frozen form of precipitation that can be very destructive to crops, automobiles, and other property. Hailstones can be irregular, somewhat spherical, or flattened forms of ice that range from the size of a BB to the size of a softball (figure 17.31). Most hailstones, however, are less than 2 cm (about 1 in) in diameter. The larger hailstones have alternating layers of clear and opaque, cloudy ice. These layers are believed to form as the hailstone goes through cycles of falling then being returned to the upper parts of the thundercloud by updrafts. The clear layers are believed to form as the hailstone moves through heavy layers of supercooled water droplets, which accumulate quickly on the hailstone but freeze

Connections . . .

Tornado Damage

slowly because of the release of the latent heat of fusion. The cloudy layers are believed to form as the hailstone accumulates snow crystals or moves through a part of the cloud with less supercooled water droplets. In either case, rapid freezing traps air bubbles, which result in the opaque, cloudy layer. Thunderstorms with hail are most common during the month of May in Colorado, Kansas, and Nebraska.

Tornadoes

A **tornado** is the smallest, most violent weather disturbance that occurs on Earth (figure 17.32). Tornadoes occur with intense thunderstorms and resemble a long, narrow funnel or ropelike structure that drops down from a thundercloud and may or may not touch the ground. This ropelike structure is a rapidly whirling column of air, usually 100 to 400 m (about 330 to 1,300 ft) in diameter. An average tornado will travel 6 to 8 km (about 4 to 5 mi) on the ground, sometimes skipping into the air, then back down again. The bottom of the column moves across the ground at speeds that average about 50 km/h (about 30 mi/h). The speed of the whirling air in the column has been estimated to be up to about 480 km/h (about 300 mi/h), but most tornadoes have winds of less than 180 km/h (112 mi/h). The destruction is produced by the powerful winds, the sudden drop in atmospheric pressure that occurs at the center of the funnel, and the debris that is flung through the air like projectiles. A passing tornado sounds like very loud, continuous rumbling thunder with cracking and hissing noises that are punctuated by the crashing of debris projectiles.

FIGURE 17.32 A tornado might be small, but it is the most violent storm that occurs on Earth. This tornado, moving across an open road, eventually struck Dallas, Texas.

On the average, several hundred tornadoes are reported in the United States every year. These occur mostly during spring and early summer afternoons over the Great Plains states. Texas, Oklahoma, Kansas, and Iowa have such a high occurrence of tornadoes that the region is called "tornado alley." During the spring and early summer, this region has maritime tropical air from the Gulf of Mexico at the surface. Above this warm, moist layer is a layer of dry, unstable air that has just crossed the Rocky Mountains, moved along rapidly by the jet stream. The stage is now set for some event, such as a cold air mass moving in from the north, to shove the warm, moist air upward, and the result will be violent thunderstorms with tornadoes.

Hurricanes

What is the difference between a tropical depression, tropical storm, and a hurricane? In general, they are all storms with strong upward atmospheric motion and a cyclonic surface wind circulation (figure 17.33). They are born over tropical or subtropical waters and are not associated with a weather front.

FIGURE 17.33 This is a satellite photo of hurricane John, showing the eye and counterclockwise motion.

Hurricane Damage

Hurricanes are classified according to category and damage to be expected. Here is the classification scheme:

Category	Damage	Winds	
1	minimal	120–153 km/h	(75–95 mi/h)
2	moderate	154–177 km/h	(96–110 mi/h)
3	extensive	178–210 km/h	(111–130 mi/h)
4	extreme	211–250 km/h	(131–155 mi/h)
5	catastrophic	>250 km/h	(>155 mi/h)

FIGURE 17.34 Cross section of a hurricane.

The varieties of storm intensities are classified according to the *speed* of the maximum sustained surface winds.

A *tropical depression* is an area of low pressure around which the winds are generally moving 55 km/h (about 35 mi/h) or less. The tropical depression might dissolve into nothing, or it might develop into a more intense disturbance. A *tropical storm* is a more intense low pressure area with winds between 56 and 120 km/h (about 35 to 75 mi/h). A **hurricane** is a very intense low-pressure area with winds greater than 120 km/h (about 75 mi/h). A strong storm of this type is called a hurricane if it occurs over the Atlantic Ocean or the Pacific Ocean east of the international date line. It is called a **typhoon** if it occurs over the North Pacific Ocean west of the international date line.

A tropical cyclone is similar to the wave cyclone of the mid-latitudes because both have low-pressure centers with a counterclockwise circulation in the Northern Hemisphere. They are different because a wave cyclone is usually about 2,500 km (about 1,500 mi) wide, has moderate winds, and receives its energy from the temperature differences between two air masses. A tropical cyclone, on the other hand, is often less than 200 km (about 125 mi) wide, has very strong winds, and receives its energy from the latent heat of vaporization released during condensation.

A fully developed hurricane has heavy bands of clouds, showers, and thunderstorms that rapidly rotate around a relatively clear, calm eye (figure 17.34). As a hurricane approaches a location, the air seems unusually calm as a few clouds appear, then thicken as the wind begins to gust. Over the next six hours or so, the overall wind speed increases as strong gusts and intense rain showers occur. Thunderstorms, perhaps with tornadoes, and the strongest winds occur just before the winds suddenly die down and the sky clears with the arrival of the eye of the hurricane. The eye is an average of 10 to 15 km (about 6 to 9 mi) across, and it takes about an hour or so to cross a location. When the eye passes, the intense rain showers, thunderstorms, and hurricane-speed winds begin again, this time blowing from the opposite direction. The whole sequence of events may be over in a day or two, but hurricanes are unpredictable and sometimes stall in one location for days. In general, they move at a rate of 15 to 50 km/h (about 10 to 30 mi/h).

Most of the damage from hurricanes results from strong winds, flooding, and the occasional tornado. Flooding occurs from the intense, heavy rainfall but also from the increased sea level that results from the strong, constant winds blowing seawater toward the shore. The sea level can be raised some 5 m (about 16 ft) above normal, with storm waves up to 15 m (about 50 ft) high on top of this elevated sea level. Overall, large inland areas can be flooded with extensive property damage. A single hurricane moving into a populated coastal region has caused billions of dollars of damage and the loss of hundreds of lives in the past. Today, the National Weather Service tracks hurricanes by weather satellites. Warnings of hurricanes, tornadoes, and severe thunderstorms are broadcast locally over special weather alert stations located across the country.

In August 2005, hurricane Katrina initially struck near Miami, Florida, as a category 1 hurricane. It then moved into the Gulf of Mexico and grew to a strong category 3 hurricane that moved up the Gulf to the eastern Louisiana and western Mississippi coast. This massive storm had hurricane-force winds that extended outward 190 km (120 mi) from the center, resulting in severe storm damage over a wide area as it struck the coast on August 29. Damage resulted from a storm surge that exceeded 7 m (25 ft), heavy rainfall, wind damage, and the failure of the levee system in New Orleans. Overall, this resulted in an estimated $81 billion in damages and more than 1,836 fatalities.

Katrina had sustained winds of 200 km/h (125 mi/h) as it struck the shore, but smaller hurricanes have had stronger sustained winds when they struck the shore. These include

- August 17, 1969: Hurricane Camille hit Mississippi with 306 km/h (190 mi/h) sustained winds.
- August 24, 1992: Hurricane Andrew hit south Florida with 266 km/h (165 mi/h) sustained winds.
- August 13, 2004: Hurricane Charley hit Punta Gorda, Florida, with 240 km/h (150 mi/h) sustained winds.*

WEATHER FORECASTING

Today, weather predictions are based on information about the characteristics, location, and rate of movement of air masses and associated fronts and pressure systems. This information is summarized as average values, then fed into a computer model of the atmosphere. The model is a scaled-down replica of the real atmosphere, and changes in one part of the model result in changes in another part of the model just as they do in the real atmosphere. Underlying the computer model are the basic scientific laws concerning solar radiation, heat, motion, and the gas laws. All these laws are written as a series of mathematical equations, which are applied to thousands of data points in a three-dimensional grid that represents the atmosphere. The computer is given instruc-

*Hurricane data from National Climatic Data Center: www.ncdc.noaa.gov/oa/climate/research/2005/katrina.html.

tions about the starting conditions at each data point, that is, the average values of temperature, atmospheric pressure, humidity, wind speed, and so forth. The computer is then instructed to calculate the changes that will take place at each data point, according to the scientific laws, within a very short period of time. This requires billions of mathematical calculations when the program is run on a worldwide basis. The new calculated values are then used to start the process all over again, and it is repeated some 150 times to obtain a one-day forecast (figure 17.35).

A problem with the computer model of the atmosphere is that small-scale events are inadequately treated, and this introduces errors that grow when predictions are attempted for farther and farther into the future. Small eddies of air, for example, or gusts of wind in a region have an impact on larger-scale atmospheric motions such as those larger than a cumulus cloud. But all of the small eddies and gusts cannot be observed without filling the atmosphere with measuring instruments. This lack of ability to observe small events that can change the large-scale events introduces uncertainties in the data, which, over time, will increasingly affect the validity of a forecast.

To find information about the accuracy of a forecast, the computer model can be run several different times, with each run having slightly different initial conditions. If the results of all the runs are close to each other, the forecasters can feel confident that the atmosphere is in a predictable condition, and this means the forecast is probably accurate. In addition, multiple computer runs can provide forecasts in the form of probabilities. For example, if eight out of ten forecasts indicate rain, the "averaged" forecast might call for an 80 percent chance of rain.

The use of new computer technology has improved the accuracy of next-day forecasts tremendously, and the forecasts up to three days are fairly accurate, too. For forecasts of more than five days, however, the number of calculations and the effect of uncertainties increase greatly. It has been estimated that the reductions of observational errors could increase the range

FIGURE 17.35 Supercomputers make routine weather forecasts possible by solving mathematical equations that describe changes in a mathematical model of the atmosphere. This "fish-eye" view was necessary to show all of this Cray supercomputer.

of accurate forecasting up to two weeks. The ultimate range of accurate forecasting will require a better understanding—and thus, an improved model—of patterns of changes that occur in the ocean as well as in the atmosphere. All of this increased understanding and reduction of errors leads to an estimated ultimate future forecast of three weeks, beyond which any pinpoint forecast would be only slightly better than a wild guess. In the meantime, regional and local daily weather forecasts are fairly accurate, and computer models of the atmosphere now provide the basis for extending the forecasts for up to about a week.

CLIMATE

Changes in the atmospheric condition over a brief period of time, such as a day or a week, are referred to as changes in the *weather*. These changes are part of a composite, larger pattern called **climate.** Climate is the general pattern of the weather that occurs for a region over a number of years. Among other things, the climate determines what types of vegetation grow in a particular region, resulting in characteristic groups of plants associated with the region (figure 17.36). For example, orange, grapefruit, and palm trees grow in a region that has a climate with warm monthly temperatures throughout the year. On the other hand, blueberries, aspen, and birch trees grow in a region that has cool temperature patterns throughout the year. Climate determines what types of plants and animals live in a location, the types of houses that people build, and the lifestyles of people. Climate also influences the processes that shape the landscape, the type of soils that form, the suitability of the region for different types of agriculture, and how productive the agriculture will be in a region. This section is about climate, what determines the climate of a region, and how climate patterns are classified.

Major Climate Groups

Earth's atmosphere is heated directly by incoming solar radiation and by absorption of infrared radiation from the surface. The amount of heating at any particular latitude on the surface depends primarily on two factors: (1) the *intensity* of the incoming solar radiation, which is determined by the angle at which the radiation strikes the surface, and (2) the amount of *time* that the radiation is received at the surface, that is, the number of daylight hours compared to the number of hours of night.

Earth is so far from the Sun that all rays of incoming solar radiation reaching Earth are essentially parallel. Earth, however, has a mostly constant orientation of its axis with respect to the stars as it moves around the Sun in its orbit. Since the inclined axis points toward the Sun part of the year and away from the Sun the other part, radiation reaches different latitudes at different angles during different parts of the year. The orientation of Earth's axis to the Sun during different parts of the year also results in days and nights of nearly equal length in the equatorial region but increasing differences at increasing latitudes to the poles. During the polar winter months, the night is twenty-four hours long, which means no solar radiation is received at all. The equatorial region receives more solar radiation during

Connections . . .

Cities Change the Weather

Using satellite measurements, NASA researchers have confirmed that the presence of a city does influence the weather. First, because of all the concrete, asphalt, and buildings, cities are 0.6°C to 5.6°C (about 1°F to 10°F) warmer than the surrounding country. The added heat of this "urban heat island" tends to make the air less stable. Less stable air brings about the second influence of more rain falling downwind of a city. Evidently, the less stable air and rougher surfaces cause air moving toward a city to rise, and the turbulence mixes in city pollutants that add nuclei for water vapor to condense upon. The result is an average 28 percent increase in the amount of rainfall 30 to 60 km (about 19 to 37 mi) downwind of cities. The NASA research team satellite measurements were verified by data from a large array of ground-based thermometers and rain gauges. The verification of these satellite-based findings are important for urban planning, water resource management, and decisions about where to farm the land. It may also show that local surface environments are more important in computer weather forecast models than had previously been believed.

Myths, Mistakes, and Misunderstandings

Rain Forecast Mistakes

It is a mistake to believe that a 20 percent chance of rain means it will rain over 20 percent of the forecast area. The correct meaning is that it will rain on 20 percent of the days when they have the same atmospheric conditions.

FIGURE 17.36 The climate determines what types of plants and animals live in a location, the types of houses that people build, and the lifestyles of people. This orange tree, for example, requires a climate that is relatively frost-free, yet it requires some cool winter nights to produce a sweet fruit.

a year, and the amount received decreases toward the poles as a result of (1) yearly changes in intensity and (2) yearly changes in the number of daylight hours.

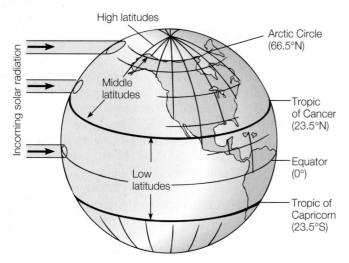

High latitudes
Arctic Circle (66.5°N)
Middle latitudes
Tropic of Cancer (23.5°N)
Equator (0°)
Low latitudes
Tropic of Capricorn (23.5°S)
Incoming solar radiation

FIGURE 17.37   Latitude groups based on incoming solar radiation. The low latitudes receive vertical solar radiation at noon some time of the year, the high latitudes receive no solar radiation at noon during some time of the year, and the middle latitudes are in between.

To generalize about the amount of radiation received at different latitudes, some means of organizing, or grouping, the latitudes is needed (figure 17.37). For this purpose, the latitudes are organized into three groups:

1. The *low latitudes,* those that some time of the year receive *vertical* solar radiation at noon.
2. The *high latitudes,* those that some time of the year receive *no* solar radiation at noon.
3. The *middle latitudes,* which are between the low and high latitudes.

This definition of low, middle, and high latitudes means that the low latitudes are between the tropics of Cancer and Capricorn (between 23.5°N and 23.5°S latitudes) and that the high latitudes are above the Arctic and Antarctic Circles (above 66.5°N and above 66.5°S latitudes).

In general:

1. The low latitudes receive a high amount of incoming solar radiation that varies little during a year. Temperatures are high throughout the year, varying little from month to month.
2. The middle latitudes receive a higher amount of incoming radiation during one part of the year and a lower amount during the other part. Overall temperatures are cooler than in the low latitudes and have a wide seasonal variation.
3. The high latitudes receive a maximum amount of radiation during one part of the year and none during another part. Overall temperatures are low, with the highest range of annual temperatures.

The low, middle, and high latitudes provide a basic framework for describing Earth's climates. These climates are associated with the low, middle, and high latitudes illustrated in figure 17.37, but they are defined in terms of yearly temperature averages. Temperature and moisture are the two most important climate factors, and temperature will be considered first.

The principal climate zones are defined in terms of yearly temperature averages, which occur in broad regions (figure 17.38). They are

1. The *tropical climate zone* of the low latitudes (figure 17.39).
2. The *polar climate zone* of the high latitudes (figure 17.40).
3. The *temperate climate zone* of the middle latitudes (figure 17.41).

FIGURE 17.38   The principal climate zones are defined in terms of yearly temperature averages, which are determined by the amount of solar radiation received at the different latitude groups.

FIGURE 17.39 A wide variety of plant life can grow in a tropical climate, as you can see here.

FIGURE 17.40 Polar climates occur at high elevations as well as high latitudes. This mountain location has a highland polar climate and tundra vegetation but little else.

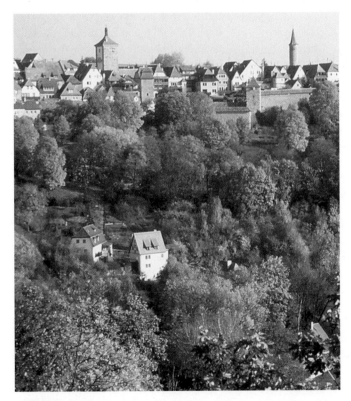

FIGURE 17.41 This temperate-climate deciduous forest responds to seasonable changes in autumn with a show of color.

The tropical climate zone is near the equator and receives the greatest amount of sunlight throughout the year. Overall, the tropical climate zone is hot. Average monthly temperatures stay above 18°C (64°F), even during the coldest month of the year.

The other extreme is found in the polar climate zone, where the sun never sets during some summer days and never rises during some winter days. Overall, the polar climate zone is cold. Average monthly temperatures stay below 10°C (50°F), even during the warmest month of the year.

The temperate climate zone is between the polar and tropical zones, with average temperatures that are neither very cold nor very hot. Average monthly temperatures stay between 10°C and 18°C (50°F and 64°F) throughout the year.

General patterns of precipitation and winds are also associated with the low, middle, and high latitudes. An idealized model of the global atmospheric circulation and pressure patterns was described in the section on weather forecasting. Recall that this model described a huge convective movement of air in the low latitudes, with air being forced upward over the equatorial region. This air expands, cools to the dew point, and produces abundant rainfall throughout most of the year. On the other hand, air is slowly sinking over 30°N and 30°S of the equator, becoming warm and dry as it is compressed. Most of the great deserts of the world are near 30°N or 30°S latitude

A Closer Look

El Niño and La Niña

The term *El Niño* was originally used to describe an occurrence of warm, above-normal ocean temperatures off the South American coast. Fishermen along this coast learned long ago to expect associated changes in fishing patterns about every three to seven years, which usually lasted for about eighteen months. They called this event El Niño, which is Spanish for "the boy child" or "Christ child," because it typically began near Christmas. The El Niño event occurs when the trade winds along the equatorial Pacific become reduced or calm, allowing sea surface temperatures to increase much above normal. The warm water drives the fish to deeper waters or farther away from usual fishing locations.

Today, El Niño is understood to be much more involved than just a warm ocean in the Pacific. It is more than a local event, and the bigger picture is sometimes called the "El Niño–Southern Oscillation," or ENSO. In addition to the warmer tropical Pacific water of El Niño, "the boy," the term *La Niña*, "the girl," has been used to refer to the times when the water of the tropical Pacific is *colder* than normal. The "Southern Oscillation" part of the name comes from observations that atmospheric pressure around Australia seems to be inversely linked to the atmospheric pressure in Tahiti. They seem to be linked because when the pressure is low in Australia, it is high in Tahiti. Conversely, when the atmospheric pressure is high in Australia, it is low in Tahiti. The strength of this Southern Oscillation is measured by the Southern Oscillation Index (SOI), which is defined as the pressure at Darwin, Australia, subtracted from that at Tahiti. Negative values of SOI are usually associated with El Niño events, so the Southern Oscillation and the El Niño are obviously linked. How ENSO can impact the weather in other parts of the world has only recently become better understood.

The atmosphere is a system that responds to incoming solar radiation, the spinning Earth, and other factors, such as the amount of water vapor present. The ocean and atmospheric systems undergo changes by interacting with each other, most visibly in the tropical cyclone. The ocean system supplies water vapor, latent heat, and condensation nuclei, which are the essential elements of a tropical cyclone as well as everyday weather changes and climate. The atmosphere, on the other hand, drives the ocean with prevailing winds, moving warm or cool water to locations where they affect the climate on the land. There is a complex, interdependent relationship between the ocean and the atmosphere, and it is probable that even small changes in one system can lead to bigger changes in the other.

Normally, during non–El Niño times, the Pacific Ocean along the equator has established systems of prevailing wind belts, pressure systems, and ocean currents. In July, these systems push the surface seawater offshore from South America, westward along the equator, and toward Indonesia. During El Niño times, the trade winds weaken and the warm water moves back eastward, across the Pacific to South America, where it then spreads north and south along the coast. Why the trade winds weaken and become calm is unknown and is the subject of ongoing research.

Warmer waters along the coast of South America bring warmer, more humid air and the increased possibility of thunderstorms. Thus, the possibility of towering thunderstorms, tropical storms, or hurricanes increases along the Pacific coast of South America as the warmer waters move north and south. This creates the possibility of weather changes not only along the western coast but elsewhere, too. The towering thunderstorms reach high into the atmosphere, adding tropical moisture and creating changes in prevailing wind belts. These wind belts carry or steer weather systems across the middle latitudes of North America, so typical storm paths are shifted. This shifting can result in

- increased precipitation in California during the fall to spring season;
- a wet winter with more and stronger storms along regions of the southern United States and Mexico;
- a warmer and drier-than-average winter across the northern regions of Canada and the United States;
- a variable effect on central regions of the United States, ranging from reduced snowfall to no effect at all; and
- other changes in the worldwide global complex of ocean and weather events, such as droughts in normally wet climates and heavy rainfall in normally dry climates.

One major problem in these predictions is a lack of understanding of what causes many of the links and a lack of consistency in the links themselves. For example, southern California did not always have an unusually wet season every time an El Niño occurred and in fact experienced a drought during one event.

Scientists have continued to study the El Niño since the mid-1980s, searching for patterns that will reveal consistent cause-and-effect links. Part of the problem may be that other factors, such as a volcanic eruption, may influence part of the linkage but not another part. Another part of the problem may be the influence of unknown factors, such as the circulation of water deep beneath the ocean surface, the track taken by tropical cyclones, or the energy released by tropical cyclones in one year compared to the next.

The results so far have indicated that atmosphere-ocean interactions are much more complex than early theoretical models had predicted. Sometimes a new model will predict some weather changes that occur with El Niño, but no model is yet consistently correct in predicting the conditions that lead to the event and the weather patterns that result. All this may someday lead to a better understanding of how the ocean and the atmosphere interact on this dynamic planet.

Recent years in which El Niño events have occurred are 1951, 1953, 1957–1958, 1965, 1969, 1972–1973, 1976, 1982–1983, 1986–1987, 1991–1992, 1994, 1997–1998, 2002–2003, and 2006–2007.

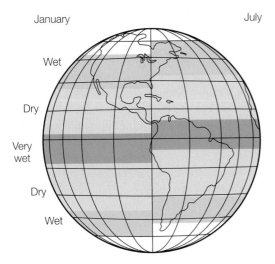

January   July

Wet

Dry

Very
wet

Dry

Wet

FIGURE 17.42 The idealized general rainfall patterns over Earth change with seasonal shifts in the wind and pressure areas of the planet's general atmospheric circulation patterns.

for this reason. There is another wet zone near 60° latitudes and another dry zone near the poles. These wet and dry zones are shifted north and south during the year with the changing seasons. This results in different precipitation patterns in each season. Figure 17.42 shows where the wet and dry zones are in winter and in summer seasons.

Regional Climatic Influence

Latitude determines the basic tropical, temperate, and polar climatic zones, and the wet and dry zones move back and forth over the latitudes with the seasons. If these were the only factors influencing the climate, you would expect to find the same climatic conditions at all locations with the same latitude. This is not what is found, however, because there are four major factors that affect a regional climate. These are (1) altitude, (2) mountains, (3) large bodies of water, and (4) ocean currents. The following describes how these four factors modify the climate of a region.

The first of the four regional climate factors is *altitude*. The atmosphere is warmed mostly by the greenhouse effect from the surface upward, and air at higher altitudes increasingly radiates more and more of its energy to space. Average air temperatures therefore decrease with altitude, and locations with higher altitudes will have lower average temperatures. This is why the tops of mountains are often covered with snow when none is found at lower elevations. St. Louis, Missouri, and Denver, Colorado, are located almost at the same latitude (within 1° of 39°N), so you might expect the two cities to have about the same average temperature. Denver, however, has an altitude of 1,609 m (5,280 ft), and the altitude of St. Louis is 141 m (465 ft). The yearly average temperature for Denver is about 10°C (about 50°F) and for St. Louis it is about 14°C (about 57°F). In general, higher altitude means lower average temperature.

The second of the regional climate factors is the influence of *mountains*. In addition to the temperature change caused by the altitude of the mountain, mountains also affect the conditions of a passing air mass. The western United States has mountainous regions along the coast. When a moist air mass from the Pacific meets these mountains, it is forced upward and cools. Water vapor in the moist air mass condenses, clouds form, and the air mass loses much of its moisture as precipitation falls on the western side of the mountains. Air moving down the eastern slope is compressed and becomes warm and dry. As a result, the western slopes of these mountains are moist and have forests of spruce, redwood, and fir trees. The eastern slopes are dry and have grassland or desert vegetation.

The third of the regional climate factors is the presence of a large body of *water*. Water, as discussed previously, has a higher specific heat than land material, is transparent, and loses energy through evaporation. All of these affect the temperature of a landmass located near a large body of water, making the temperatures more even from day to night and from summer to winter. San Diego, California, and Dallas, Texas, for example, are at about the same latitude (both almost 33°N), but San Diego is at a seacoast and Dallas is inland. Because of its nearness to water, San Diego has an average summer temperature about 7°C (about 13°F) cooler and an average winter temperature about 5°C (about 9°F) warmer than the average temperatures in Dallas. Nearness to a large body of water keeps the temperature more even at San Diego.

The fourth of the regional climate factors is *ocean currents*. In addition to the evenness brought about by being near the ocean, currents in the ocean can bring water that has a temperature different from the land. For example, currents can move warm water northward or they can move cool water southward (figure 17.43). This can influence the temperatures of air masses that move from the water to the land and, thus, the temperatures of the land. For example, the North Pacific current brings warm waters to the western coast of North America, which results in warmer temperatures for cities near the coast.

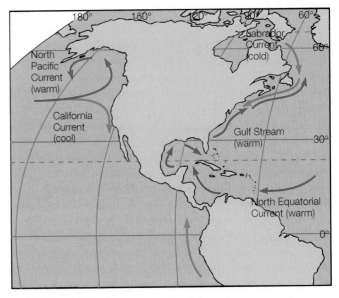

FIGURE 17.43 Ocean currents can move large quantities of warm or cool water to influence the air temperatures of nearby landmasses.

People Behind the Science

Vilhelm Firman Koren Bjerknes (1862–1951)

Vilhelm Bjerknes was the Norwegian scientist who created modern meteorology.

Bjerknes came from a talented family. His father was professor of mathematics at the Christiania (now Oslo) University and a highly influential geophysicist who clearly shaped his son's studies. Bjerknes held chairs at Stockholm and Leipzig before founding the Bergen Geophysical Institute in 1917. Bjerknes made momentous contributions that transformed meteorology into an accepted science. Not least, he showed how weather prediction could be put on a statistical basis, dependent on the use of mathematical models.

During World War I, Bjerknes instituted a network of weather stations throughout Norway; coordination of the findings from such stations led him and his coworkers to develop the highly influential theory of polar fronts, on the basis of the discovery that the atmosphere is made up of discrete air masses displaying dissimilar features. Bjerknes coined the word *front* to delineate the boundaries between such air masses. One of the many contributions of the "Bergen frontal theory" was to explain the generation of cyclones over the Atlantic, at the junction of warm and cold air wedges. Bjerknes's work gave modern meteorology its theoretical tools and methods of investigation.

Source: Modified from the *Hutchinson Dictionary of Scientific Biography*. © RM, 2009. All rights reserved. Helicon Publishing is a division of RM.

Describing Earth's climates presents a problem because there are no sharp boundaries that exist naturally between two adjacent regions with different climates. Even if two adjacent climates are very different, one still blends gradually into the other. For example, if you are driving from one climate zone to another, you might drive for miles before becoming aware that the vegetation is now different than it was an hour ago. Since the vegetation is very different from what it was before, you know that you have driven from one regional climate zone to another. Chapter 23, Ecology and Environment, discusses the typical plants and animals associated with the various climate types.

Actually, no two places on Earth have exactly the same climate. Some plants will grow on the north or south side of a building, for example, but not on the other side. The two sides of the building could be considered as small, local climate zones within a larger, major climate zone.

Climate Change

As stated earlier, *weather* describes changes in the atmospheric condition (rain, wind, etc.) over a brief period of time (day, week). *Climate* describes the general pattern of weather that occurs over a region for a number of months or years. **Climate change** is a departure from the expected average pattern of climate for a region over time.

Scientists measured climate patterns during the past 120 years or so by using thermometers, rain gauges, and other instruments. Climate patterns before 120 years ago are inferred by analyzing evidence found in tree rings, lake-bottom sediments, ice cores drilled from a glacier, and other sources. A natural source used to infer temperature change, rainfall, or some other past climate condition is called **proxy data.**

Proxy data indicates that Earth's climate has undergone major changes in the past, with cold *ice ages* and glaciers dominating the climate for the past several million years (figure 17.44). The most recent ice age covered almost a third of Earth's land surface with up to 4 km (2.5 mi) ice sheets during its maximum extent. About half of the states in the United States were covered by ice, some completely and others partially. A large amount of water was locked up in glaciers, and this caused the sea level to drop some 90 m (about 300 ft), exposing a land bridge between Siberia and Alaska, among other things. The ice sheets of the most recent ice age advanced and retreated at least four times, with the sea level fluctuating with each advance and retreat.

About every 100,000 years Earth enters an *interglacial warming period* before returning to another ice age. We are currently in such a warming period, and the current period began about 18,000 years ago. About 15,000 years ago Earth's climate had warmed enough to stop the advance of glaciers, and since that time the glaciers have been retreating. By about 4000 B.C. the average temperatures had warmed to a few degrees warmer than those of today. The proxy record indicates that such interglacial periods last from 15,000 to 20,000 years before beginning a new glacial period. If the cycle continues as it has in

FIGURE 17.44 Much of Earth's surface is covered with ice sheets and glaciers during an ice age. Both are formed from accumulations of compacted snow.

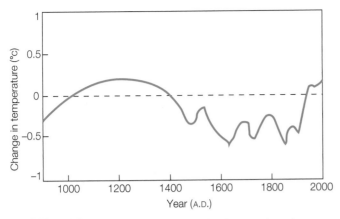

FIGURE 17.45 According to proxy data, Earth has experienced a period of warming and a period of cooling over the past one thousand years.

Source: www.geocraft.com/WVFossils/ice_ages.html

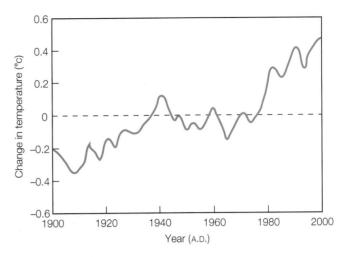

FIGURE 17.46 According to land-based weather stations, Earth has experienced a period of warming, a period of cooling, and most recently, a period of warming over the past one hundred years.

Source: www.ncdc.noaa.gov

the past, we are near the end of the current interglacial period, nearing the beginning of the next ice age.

Data about ice ages and interglacial warming show that periods of warming and cooling occur in large cycles. During the past 1,000 years Earth has also undergone a number of smaller warming and cooling cycles (figure 17.45). For example it was warm—about like today—from 1000 up to 1400. Then from 1400 to 1860 there was a period of cooling called the "little ice age." The little ice age is why the Vikings left Greenland after successfully farming there from the tenth century until the thirteenth.

During the past 100 years, Earth has had two warming cycles separated by one period of cooling. The temperature increased from the early 1800s—before the Industrial Revolution—because we were coming out of the little ice age. Then the temperature decreased from about 1940 to the late 1970s (figure 17.46). And then a slight warming cycle began in the 1980s, continuing through today.

Recording stations began reading small but steady increases in temperatures near the Earth's surface, averaging about 0.25°C (0.4°F) in the past twenty-five years. Sea level has increased 1 to 2 mm per year for the past one hundred years, and this increase is due to thermal expansion of seawater in addition to melting ice.

The recent temperature increase is not uniform around the globe since the southeastern United States has cooled over the past 100 years. Artic sea ice has decreased since the early 1970s, but sea ice in the Antarctic has *increased* during the same period. Nonetheless, some people became concerned that the temperature increase was caused by the release of greenhouse gases—mostly carbon dioxide—that would result in a runaway greenhouse effect (see page 377 for information on the greenhouse effect).

Causes of Global Climate Change

Climate change is brought about by a complex interaction of a number of factors. Some of these factors could be astronomical, and some could be factors occurring in the atmosphere. Solar

FIGURE 17.47 Sunspots appear on the Sun's surface as dark spots because they have a much cooler temperature than their surroundings. Sunspots appear in cycles that vary in number, and cycle 24 began to increase from a minimum on January 4, 2008.

energy is fundamentally responsible for weather and climate, and changes in the Sun's energy output can change the climate. The Sun's output of energy changes with *sunspots,* dark spots that appear to move around the surface of the Sun (figure 17.47).

A Time of the year when Earth is nearest the Sun

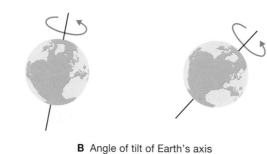

B Angle of tilt of Earth's axis

C Shape of Earth's orbit

FIGURE 17.48 Three variations in Earth's motion that may be responsible for causing ice ages. (*A*) The time of year when Earth is nearest the Sun varies with a period of about twenty-three thousand years. (*B*) The angle of tilt of Earth's axis of rotation varies with a period of about forty-one thousand years. (*C*) The shape of Earth's elliptical orbit varies with a period of about one hundred thousand years. These variations have relatively little effect on the total sunlight reaching Earth but a considerable effect on the sunlight reaching the polar regions in summer. The ellipticity of Earth's orbit is vastly exaggerated here; the orbit is actually less than 2 percent away from a perfect circle.

The number of sunspots varies from year to year, and there are years when sunspots are rare or absent, and years when a peak is reached. The maximum number seems to occur in a cycle that averages 11.1 years. Evidently, the amount of solar energy increases as the number of sunspots increases. More sunspots deliver more energy to Earth's surface, which increases the temperature. Estimates are that the Sun's energy output varies by up to 0.2 percent with each sunspot cycle. If this increase continues into the mid-twenty-first century, Earth's surface temperature will increase by about 0.5°C (about 1°F). However, a relationship between the sunspot cycle *length* and temperatures also comes into play. Higher than normal temperatures tend to occur with shorter cycles, and lower temperatures occur with longer cycles.

In addition to changes in the Sun, changes in the orientation of Earth's tilt, orbital shape, and axis wobble change our orientation to the Sun in predictable cycles. These are called Milankovitch cycles after Milutin Milankovitch (1879–1958), a Yugoslav physicist who calculated how the cycles would affect the climate by altering the amount of solar energy received by Earth (figure

17.48). As it works out, the amount of energy received by high latitudes varies up to 20 percent. A shorter summer allows ice to accumulate, making an ice age. The timing of the ice age and cycles of warmer or cooler average temperatures fits with periods of Earth's orbital variations.

There are also atmospheric factors that can cause climatic changes. The greenhouse gases, for example, can reradiate heat in the atmosphere, producing a warmer climate. The most abundant greenhouse gas is water vapor, which is also the dominate gas in terms of increasing the temperature. Water vapor is followed by carbon dioxide, methane, and then some trace gases (table 17.1). The natural greenhouse effect is a good thing, for without it the average Earth surface temperature would be −18°C (0.4°F). Thanks to the greenhouse effect, the global average is 14°C (57°F).

Carbon dioxide has been increasing in the atmosphere since the late 1800s (figure 17.49) and some believe the burning of fossil fuels is responsible. Precise measurements of atmosphere carbon dioxide concentration have been made since 1958. This found that the year-to-year concentration

TABLE 17.1 Man-made Contribution to the "Greenhouse Effect," expressed as % of Total

Based on concentrations (ppb) adjusted for heat retention characteristics	% of All Greenhouse Gases	% Natural	%Man-made
Water vapor	95.000%	94.999%	0.001%
Carbon Dioxide (CO_2)	3.618%	3.502%	0.117%
Methane (CH_4)	0.360%	0.294%	0.066%
Nitrous Oxide (N_2O)	0.950%	0.903%	0.047%
Misc. gases (CFC's, etc.)	0.072%	0.025%	0.047%
Total	100.00%	99.72	0.28%

*From Internet article *"Global Warming: A Closer Look at the Numbers"* by Monte Hieb. Revised 01/03.

varied but had an average increase of 1.5 parts per million by volume. Carbon dioxide is naturally released and absorbed by plants and animals as well as Earth's oceans. The overall concentration varies because green plants convert carbon dioxide to plant materials. Carbon dioxide is also absorbed by ocean waters, used by marine organisms to make shells, and is converted to mineral deposits such as limestone. Also, note that carbon dioxide can go into solution, and warm ocean water dissolves less carbon dioxide than cool ocean water. Just like a glass of soda, warmer water will release carbon dioxide. Tropical deforestation is reducing one means of removing carbon dioxide from the atmosphere, and this adds up to an estimated 40 percent as much carbon dioxide as the burning of fossil fuels. How much land, the ocean, and plant and animals remove and add carbon dioxide to the atmosphere is still highly uncertain.

example, a 2002 report predicts increased global temperatures during the next one hundred years somewhere between 1.4°C (2.5°F) to 5.8°C (10.4°F) as a result of increased carbon dioxide in the atmosphere. Not all the experts agree with this prediction because it is based on an estimated carbon dioxide concentration that grows exponentially, increasing 1 percent per year. These experts run their own climate model based on the actual increases in carbon dioxide concentration for the past thirty years—which did not grow exponentially—and project additional warming of only 0.5°C (0.9°F)in the next fifty years.

It may be a surprise to find such disagreements between experts. Some argue that changes of the climate are natural but not well understood. Earth has had a history of warming periods, and this was long before activities of people emitted carbon dioxide. Furthermore, not all the "evidence" points to an increase in global temperatures. A retreating glacier, for example, can be a result of less snow in the winter, or the result of a longer summer, not necessarily warmer temperatures.

Do we have increased temperatures and retreating glaciers because of increased carbon dioxide, or are these natural variations in climate cycles that will continue to change? The answer will not come from computer simulation of the global climate because the computer model will always be simpler than the climate itself. Scientists can only weigh the evidence and make a professional judgment, and this can result in disagreement. Today there is no consensus about the cause of the supposed slight warming observed during the past forty years. At least the underlying science of climate change should be understood. Then a critical analysis of climate change should be completed before any future policies about human-made carbon dioxide are implemented.

Global Warming

How do you predict the climate of the future? Scientists use mathematical models to calculate the evolving state of the atmosphere in response to changes in factors. The model is run on a large computer that uses current climatic data from sunlight, land and atmosphere interactions, and interactions with the ocean. A change in some factor is then introduced to calculate a hypothetical future. For example, the changing factor could be an increased concentration of carbon dioxide in the atmosphere. Climate models are imperfect because they do not include all the important factors; for example, they leave out the cycling of carbon dioxide into and out of the atmosphere.

The United Nations established the Intergovernmental Panel on Climate Change (IPCC) in 1988 to address climate change. The IPCC releases reports on the state of the global climate and makes projections about climate change based on changes in climate models. For

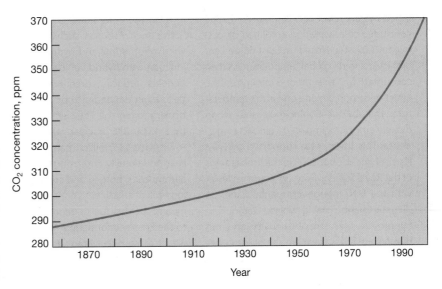

FIGURE 17.49 Carbon dioxide concentration in the atmosphere since 1860, in parts per million (ppm).

Global Warming Lacks a Clear Consensus

By Rep. John Shadegg*

Until the 1890s, it was common for doctors to bleed their patients. The belief was that draining blood removed toxins from the body. Of course, this practice was based on bad science and incomplete knowledge. The loss of blood weakened the patient, often resulting in death. One-dimensional policy proposals aimed only at reducing man-made greenhouse gases, based on what we know now, may well turn out to be the modern-day equivalent of bloodletting. In this case, America is the patient, and the poorest among us will suffer most.

There are those who say we know global warming is occurring, that it is caused by man-made greenhouse gases and is a severe threat. Millions have been spent studying climate change, yet contrary to repeated claims, there remains significant uncertainty and disagreement within the scientific community. In October, Ronald Prinn, a professor of atmospheric chemistry at MIT and director of the Center for Global Change Science, released a study revealing a simultaneous worldwide increase in methane levels, calling into question theories that man is the primary contributor to global warming.

During the December 2007 U.N. climate conference, one hundred scientists wrote a letter to the secretary general calling climate change "a natural phenomenon that has affected humanity through the ages." The groundbreaking "amplifier theory" published by Jan Veizer and Nir Shaviv in 2003 concluded solar variation amplified by cosmic ray flux accounts for at least 66 percent of the variance in global temperatures. When accounted for, this theory minimizes the effect of carbon dioxide on Earth's temperatures, thereby questioning increased carbon dioxide emissions' impact on global temperatures. This theory was supported by Giles Harrison in October 2008 in the Proceedings of the Royal Society. Even an expert reviewer with the U.N. Intergovernmental Panel on Climate Change, David Wojick, stated, "The hypothesis that solar variability and not human activity is warming the oceans goes a long way to explain the puzzling idea that the Earth's surface may be warming while the atmosphere is not."

Consensus on global warming's impact has also not been reached. Soil, the world's largest source of carbon dioxide, produces ten times more than all human activities combined. In November 2008, Johannes Lehmann, a Cornell University professor of biogeochemistry, and others released a study of climate models incorporating increases in carbon dioxide emissions from soil. His study found small changes in soil carbon emissions estimates have large impacts on climate models. Therefore, he concludes predictions about global warming are wide-ranging, and we cannot know the actual effect.

Amid such continuing uncertainty, Congress should move cautiously. At a minimum, we should initially pursue dual-benefit rather than single-benefit strategies. Carbon capture and sequestration, for example, adds cost and provides only a single benefit: limiting greenhouse gas emissions. According to the Pew Center on Global Climate Change, carbon capture alone, excluding sequestration, will cost $390 million to $610 million per coal-fired plant. Similarly, cap-and-trade programs increase cost significantly while only marginally reducing greenhouse gas emissions and doing nothing directly to achieve other energy goals. The European Union's emissions trading scheme has cost $4.7 billion per year and decreased emissions by only 0.8 percent since 2006.

On the other hand, encouraging alternative energy sources such as solar, nuclear, wind, geothermal and hydropower, in conjunction with hybrid vehicles, provides dual benefits: reducing greenhouse gases and decreasing dependence on fossil fuels, particularly foreign oil. The electricity generated by a 1,000 MW nuclear reactor is enough to displace 13.7 million barrels of oil in one year. Additionally, improving the efficiency of automobiles, homes, and commercial buildings will decrease emissions, conserve energy, and cut fossil fuel dependency. According to the Energy Information Administration, if we increase our average fuel economy to forty-five miles per gallon, we can almost eradicate our dependence on foreign oil over the next twenty years. And, by using Energy

Star appliances, Americans can save about 30 percent on their home energy bills.

Should science ultimately conclude global warming is occurring and is human-caused, we will have taken meaningful action to address the problem. Should science, however, determine global warming is not caused by man-made greenhouse gases, we will still have reduced our reliance on, and inefficient use of, fossil fuels. One-dimensional proposals may ultimately be necessary, but they aren't yet. Right now, we can take steps to reduce man-made greenhouse gases while also achieving other important goals.

The single-benefit proposals currently being considered will bleed America's economy. Studies by the Heritage Foundation, the National Association of Manufacturers, the American Council for Capital Formation and others suggest the Climate Security Act, introduced by Senators Joe Lieberman (ID-Conn.) and John Warner (R-Va.), which would establish an economy-wide cap-and-trade program, and the Environmental Protection Agency's regulatory initiative to reduce carbon dioxide emissions will have a catastrophic impact on our nation's economy.

David W. Kreutzer and Karen A. Campbell estimate that the EPA's regulatory initiative will result in a loss of more than 2.8 million manufacturing jobs. And the National Association of Manufacturers (NAM) estimates 1.8 million net jobs lost by 2020 from the Climate Security Act.

Kreutzer and Campbell predict gross domestic product will lose $600 billion in a year. Annual family income will drop $1,170 initially and $4,000 by 2025. Lower-income Americans, the elderly, and those on fixed incomes will be hit the hardest. A state-by-state analysis shows low-income Arizona families, making $14,800 annually, will spend 19 percent of their income on energy by 2020 if the Climate Security Act is implemented.

Proponents of single-benefit strategies talk about "green collar" jobs, but they acknowledge only one side of the equation. They fail to take into account jobs lost. Job losses projected by Kreutzer and Campbell,

---

* Rep. John Shadegg (R-Ariz.) is a member of the Energy and Commerce and Energy Independence and Global Warming Committees.

NAM and the American Council for Capital Formation (ACCF) are net; they account for new "green jobs" as well as jobs lost.

One-dimensional strategies will impose real costs. And we don't even know yet if such policies will slow global warming.

Whether one ascribes to the worst-case global warming scenario or believes global warming is a myth, all can agree it's critical for America to reduce our reliance on fossil fuels and foreign oil and to conserve energy. Until we know more about global warming and its cause, we should pursue dual-benefit

solutions that reduce greenhouse gases and improve our economy. Without more complete information on climate change, insisting upon severe, costly, single-benefit strategies could, like "bleeding," ultimately be worse than the disease.

Questions to Discuss

1. Relative to global warming how is a dual-benefit strategy different from a single-benefit strategy?

2. What one conclusion about global warming would negate the benefits of a single-benefit strategy?

3. Describe several specific examples of dual-benefit strategies.

4. What are the advantages and disadvantages of single-benefit strategies?

5. What are the advantages and disadvantageous of dual-benefit strategies?

SUMMARY

Earth's *atmosphere* thins rapidly with increasing altitude. Pure, dry air is mostly *nitrogen, oxygen,* and *argon,* with traces of *carbon dioxide* and other gases. Atmospheric air also contains a variable amount of *water vapor.* Water vapor cycles into and out of the atmosphere through evaporation and precipitation.

Atmospheric pressure is measured with a *mercury barometer.* At sea level, the atmospheric pressure will support a column of mercury about 76.00 cm (about 29.92 in) tall. This is the average pressure at sea level, and it is called the *standard atmospheric pressure, normal pressure,* or *one atmosphere of pressure.*

Materials on Earth's surface absorb sunlight, emitting more and more *infrared radiation* as they are warmed. *Carbon dioxide* and *molecules* of water vapor in the atmosphere absorb infrared radiation, which then reemit the energy many times before it reaches outer space again. The overall effect warms the lower atmosphere from the bottom up in a process called the *greenhouse effect.*

The layer of the atmosphere from the surface up to where the temperature stops decreasing with height is called the *troposphere.* The *stratosphere* is the layer above the troposphere. Temperatures in the stratosphere increase because of the interaction between ozone (O_3) and ultraviolet radiation from the Sun.

The surface of Earth is not heated uniformly by sunlight. This results in a *differential heating,* which sets the stage for *convection.* The horizontal movement of air on the surface from convection is called *wind.* A generalized model for understanding why the wind blows involves (1) the relationship between *air temperature and air density* and (2) the relationship between *air pressure and the movement of air.* This model explains local wind patterns and wind patterns observed on a global scale.

The amount of water vapor in the air at a particular time is called the *absolute humidity.* The *relative humidity* is a ratio between the amount of water vapor that is in the air and the amount needed to saturate the air at the present temperature.

When the air is saturated, condensation can take place. The temperature at which this occurs is called the *dew point temperature.*

If the dew point temperature is above freezing, *dew* will form. If the temperature is below freezing, *frost* will form. Both dew and frost form directly on objects and do not fall from the air.

Water vapor condenses in the air on *condensation nuclei.* If this happens near the ground, the accumulation of tiny water droplets is called a *fog.* *Clouds* are accumulations of tiny water droplets in the air above the ground. In general, there are three basic shapes of clouds: *cirrus, cumulus,* and *stratus.* These basic cloud shapes have meaning about the atmospheric conditions and about the coming weather conditions.

Water that returns to Earth in liquid or solid form falls from the clouds as *precipitation.* Precipitation forms in clouds through two processes: (1) the *coalescence* of cloud droplets or (2) the *growth of ice crystals* at the expense of water droplets.

Weather changes are associated with the movement of large bodies of air called *air masses,* the leading *fronts* of air masses when they move, and local *high- and low-pressure* patterns that accompany air masses or fronts. Examples of air masses include (1) *continental polar,* (2) *maritime polar,* (3) *continental tropical,* and (4) *maritime tropical.*

When a location is under the influence of an air mass, the location is having *air mass weather* with slow, gradual changes. More rapid changes take place when the *front,* a thin transition zone between two air masses, passes a location.

A *stationary front* often develops a bulge, or *wave,* that forms into a moving cold front and a moving warm front. The faster-moving cold front overtakes the warm front, lifting it into the air to form an *occluded front.* The lifting process forms a low-pressure center called a *cyclone.* Cyclones are associated with heavy clouds, precipitation, and stormy conditions because of the lifting action.

A *thunderstorm* is a brief, intense storm with rain, lightning and thunder, gusty and strong winds, and sometimes hail. A *tornado* is the smallest, most violent weather disturbance that occurs on Earth. A *hurricane* is a *tropical cyclone,* a large, violent circular storm that is born over warm tropical waters near the equator.

The general pattern of the weather that occurs for a region over a number of years is called *climate*. The three principal climate zones are (1) the *tropical climate zone*, (2) the *polar climate zone*, and (3) the *temperate climate zone*. The climate in these zones is influenced by four factors that determine the local climate: (1) *altitude*, (2) *mountains*, (3) *large bodies of water*, and (4) *ocean currents*. The climate for a given location is described by first considering the principal climate zone, then looking at subdivisions within each that result from local influences. Earth's climate has undergone major changes in the past, caused by changes in the Sun, changes in the Earth's orbit, and/or changes in the atmosphere.

KEY TERMS

absolute humidity (p. **384**)	front (p. **391**)	occluded front (p. **392**)	thunderstorm (p. **394**)
air mass (p. **390**)	frost (p. **385**)	precipitation (p. **389**)	tornado (p. **396**)
barometer (p. **376**)	greenhouse effect (p. **377**)	proxy data (p. **404**)	typhoon (p. **397**)
climate (p. **399**)	hail (p. **395**)	relative humidity (p. **384**)	warm front (p. **392**)
climate change (p. **404**)	humidity (p. **384**)	saturated air (p. **383**)	wind (p. **379**)
cold front (p. **392**)	hurricane (p. **397**)	standard atmospheric pressure (p. **377**)	wind chill factor (p. **380**)
condensation nuclei (p. **386**)	ice-forming nuclei (p. **389**)		
dew (p. **385**)	inversion (p. **377**)	stationary front (p. **392**)	
dew point temperature (p. **385**)	jet stream (p. **382**)	storm (p. **394**)	

APPLYING THE CONCEPTS

1. Without adding or removing any water vapor, a sample of air experiencing an increase in temperature will have
   a. a higher relative humidity.
   b. a lower relative humidity.
   c. the same relative humidity.
   d. a changed absolute humidity.

2. Cooling a sample of air results in a (an)
   a. increased capacity to hold water vapor.
   b. decreased capacity to hold water vapor.
   c. unchanged capacity to hold water vapor.

3. On a clear, calm, and cool night, dew or frost is most likely to form
   a. under trees or other shelters.
   b. on bare ground on the side of a hill.
   c. under a tree on the side of a hill.
   d. on grass in an open, low-lying area.

4. Longer periods of drizzle, rain, or snow usually occur from which basic form of a cloud?
   a. stratus
   b. cumulus
   c. cirrus.
   d. None of the above is correct.

5. Brief periods of showers are usually associated with which type of cloud?
   a. stratus
   b. cumulus
   c. cirrus.
   d. None of the above is correct.

6. The type of air mass weather that results from the arrival of polar continental air is
   a. frequent snowstorms with rapid changes.
   b. clear and cold with gradual changes.
   c. unpredictable but with frequent and rapid changes.
   d. much the same from day to day, the conditions depending on the air mass and the local conditions.

7. The appearance of high cirrus clouds followed by thicker, lower stratus clouds and then continuous light rain over several days probably means which of the following air masses has moved to your area?
   a. continental polar
   b. maritime tropical
   c. continental tropical
   d. maritime polar

8. A fully developed cyclonic storm is most likely to form
   a. on a stationary front.
   b. in a high-pressure center.
   c. from differential heating.
   d. over a cool ocean.

9. The basic difference between a tropical storm and a hurricane is
   a. size.
   b. location.
   c. wind speed.
   d. amount of precipitation.

10. Most of the great deserts of the world are located
   a. near the equator.
   b. 30° north or south latitude.
   c. 60° north or south latitude.
   d. anywhere, as there is no pattern to their location.

11. The average temperature of a location is made more even by the influence of
   a. a large body of water.
   b. elevation.
   c. nearby mountains.
   d. dry air.

12. The climate of a specific location is determined by
   a. its latitude.
   b. how much sunlight it receives.
   c. its altitude and nearby mountains and bodies of water.
   d. All of the above are correct.

Answers

1. b **2.** b **3.** d **4.** a **5.** b **6.** d **7.** b **8.** a **9.** c **10.** b **11.** a **12.** d

QUESTIONS FOR THOUGHT

1. Explain why frost is more likely to form on a clear, calm, and cool night than on nights with other conditions.
2. What is a cloud? Describe how a cloud forms.
3. Describe two ways that precipitation may form from the water droplets of a cloud.
4. What is an air mass?
5. What kinds of clouds and weather changes are usually associated with the passing of (a) a warm front? (b) a cold front?
6. Describe the wind direction, pressure, and weather conditions that are usually associated with (a) low-pressure centers and (b) high-pressure centers.
7. In which of the four basic types of air masses would you expect to find afternoon thunderstorms? Explain.
8. Describe the three main stages in the life of a thunderstorm cell, identifying the events that mark the beginning and end of each stage.
9. What is a tornado? When and where do tornadoes usually form?
10. What is a hurricane? Describe how the weather conditions change as a hurricane approaches, passes directly over, then moves away from a location.
11. Describe the average conditions found in the three principal climate zones.
12. Identify the four major factors that influence the climate of a region and explain how each does its influencing.

FOR FURTHER ANALYSIS

1. Describe how you could use a garden hose and a bucket of water to make a barometer. How high a column of water would standard atmospheric pressure balance in a water barometer?
2. If heated air rises, why is there snow on top of a mountain and not at the bottom?
3. According to the U.S. National Oceanic and Atmospheric Administration, the atmospheric concentration of CO_2 has been increasing, global surface temperatures have increased about 0.2°C (0.4°F) over the past twenty-five years, and sea level has been rising 1 to 2 mm/yr since the late nineteenth century. Describe what evidence you would look for to confirm these increases are due to human activity rather than changes in the Sun's output or energy, or changes in Earth's orbit.
4. Evaluate the requirement that differential heating must take place before wind will blow. Do any winds exist without differential heating?
5. Given the current air temperature and relative humidity, explain how you could use the graph in figure 17.15 to find the dew point temperature.
6. Explain why dew is not considered to be a form of precipitation.
7. What are the significant similarities and differences between air mass weather and frontal weather?
8. Analyze and compare the potential damage caused by a hurricane to the potential damage caused by a tornado.
9. Describe several examples of regional climate factors completely overriding the expected weather in a given principal climatic zone. Explain how this happens.

INVITATION TO INQUIRY

Microclimate Experiments

The local climate of a small site or habitat is called a *microclimate*. Certain plants will grow within one microclimate but not another. For example, a north-facing slope of a hill is cooler and loses snow later than a south-facing slope. A different microclimate exists on each side of the hill, and some plants grow better on one side than the other.

Investigate the temperature differences in the microclimate on the north side of a building and on the south side of the same building. Determine the heights, distances, and time of day that you will take temperature readings. Remember to shade the thermometer from direct sunlight for each reading. In your report, discuss the variables that may be influencing the temperature of a microclimate. Design experiments to test your hypotheses about each variable.

18
Earth's Waters

Earth's vast oceans cover more than 70 percent of the surface of Earth. Freshwater is generally abundant on the land because the supply is replenished from ocean waters through the hydrologic cycle.

CORE **CONCEPT**

The hydrologic cycle is water evaporating from the ocean, transport by moving air masses, precipitation on the land, and the movement of water back to the ocean.

OUTLINE

Most *water on Earth is stored in Earth's oceans* (p. 413).

Water fit for human consumption is replenished by the hydrologic cycle (p. 414).

Water on Earth
  Freshwater
  Surface Water
  Groundwater
  Freshwater as a Resource
A Closer Look: Wastewater
    Treatment
Seawater
  Oceans and Seas

The Nature of Seawater
Movement of Seawater
A Closer Look: Rogue Waves
The Ocean Floor
A Closer Look: Key Forecasting Tool
    for the Chesapeake Bay
People Behind the Science: Rachel
    Louise Carson

Precipitation either evaporates, runs across the surface, or soaks in to become groundwater (p. 417).

412

Earth Science

▷ Changes in atmospheric conditions bring precipitation (Ch. 7).

OVERVIEW

Throughout history, humans have diverted rivers and reshaped the land to ensure a supply of freshwater. There is evidence, for example, that ancient civilizations along the Nile River diverted water for storage and irrigation some five thousand years ago. The ancient Greeks and Romans built systems of aqueducts to divert streams to their cities some two thousand years ago. Some of these aqueducts are still standing today. More recent water diversion activities were responsible for the name of Phoenix, Arizona. Phoenix was named after a mythical bird that arose from its ashes after being consumed by fire. The city was given this name because it is built on a system of canals that were first designed and constructed by ancient Native Americans, then abandoned hundreds of years before settlers reconstructed the ancient canal system (figure 18.1). Water is and always has been an essential resource. Where water is in short supply, humans have historically turned to extensive diversion and supply projects to meet their needs.

Precipitation is the basic source of the water supply found today in streams, lakes, and beneath Earth's surface. Much of the precipitation that falls on the land, however, evaporates back into the atmosphere before it has a chance to become a part of this supply. The water that does not evaporate mostly moves directly to rivers and streams, flowing back to the ocean, but some soaks into the land. The evaporation of water, condensation of water vapor, and the precipitation-making processes were introduced in chapter 17 as important weather elements. They are also part of the generalized *hydrologic cycle* of evaporation from the ocean, transport through the atmosphere by moving air masses, precipitation on the land, and movement of water back to the ocean. Only part of this cycle was considered previously, however, and this was the part from evaporation through precipitation. This chapter is concerned with the other parts of the hydrologic cycle, that is, what happens to the water that falls on the land and makes it back to the ocean. It begins with a discussion of how water is distributed on Earth and a more detailed look at the hydrologic cycle. Then the travels of water across and into the land will be considered as streams, wells, springs, and other sources of usable water are discussed as limited resources. The tracing of the hydrologic cycle will be completed as the water finally makes it back to the ocean. This last part of the cycle will consider the nature of the ocean floor, the properties of seawater, and how waves and currents are generated. The water is now ready to evaporate, starting another one of Earth's never-ending cycles.

WATER ON EARTH

Some water is tied up in chemical bonds deep in Earth's interior, but free water is the most abundant chemical compound near the surface. Water is five or six times more abundant than the most abundant mineral in the outer 6 km (about 4 mi) of Earth, so it should be no surprise that water covers about 70 percent of the surface. On average, about 98 percent of this water exists in the liquid state in depressions on the surface and in sediments. Of the remainder, about 2 percent exists in the solid state as snow and ice on the surface in colder locations. Only a fraction of a percent exists as a variable amount of water vapor in the atmosphere at a given time. Water is continually moving back and forth between these "reservoirs," but the percentage found in each is assumed to be essentially constant.

As shown in figure 18.2, over 97 percent of Earth's water is stored in Earth's oceans. This water contains a relatively high level of dissolved salts, which will be discussed in the section on the nature of seawater. These dissolved salts make ocean water unfit for human consumption and for most agricultural purposes. All other water, which is fit for human consumption and agriculture, is called **freshwater.** About two-thirds of Earth's freshwater supply is locked up in the ice caps of Greenland and the Antarctic and in glaciers. This leaves less than 1 percent of all the water found on Earth as available freshwater. There is a generally abundant supply, however, because the freshwater supply is continually replenished by the hydrologic cycle.

Evaporation of water from the ocean is an important process of the hydrologic cycle because (1) water vapor leaves the dissolved salts behind, forming precipitation that is freshwater,

FIGURE 18.1 This is one of the water canals of the present-day system in Phoenix, Arizona. These canals were reconstructed from a system that was built by American Indians, then abandoned. Phoenix is named after a mythical bird that was consumed by fire and then arose from its ashes.

and (2) the gaseous water vapor is easily transported in the atmosphere from one part of Earth to another. Over a year, this natural desalination process produces and transports enough freshwater to cover the entire Earth with a layer about 85 cm (about 33 in) deep. Precipitation is not evenly distributed like this, of course, and some places receive much more, while other places receive almost none. Considering global averages, more water is evaporated from the ocean than returns directly to it by precipitation. On the other hand, more water is precipitated over the land than evaporates from the land surface back to the atmosphere. The net amount evaporated and precipitated over the land and over the ocean is balanced by the return of water to the ocean by streams and rivers. This cycle of evaporation, precipitation, and return of water to the oceans is known as the *hydrologic cycle* (figure 18.3).

Freshwater

The basic source of freshwater is precipitation, but not all precipitation ends up as part of the freshwater supply. Liquid water is always evaporating, even as it falls. In arid climates, rain sometimes evaporates completely before reaching the surface, even from a fully developed thunderstorm. Evaporation continues from the water that does reach the surface. Puddles and standing water on the hard surface of city parking lots and streets, for example, gradually evaporate back to the atmosphere after a rain and the surface is soon dry. Many factors determine how much of a particular rainfall evaporates, but in general, more than two-thirds of the rain eventually returns to the atmosphere. The remaining amount either (1) flows downhill across the surface of the land toward a lower place or (2) soaks into the ground. Water moving across the surface is called **runoff.** Runoff begins as rain accumulates in thin sheets of water that move across the surface of the land. These sheets collect into a

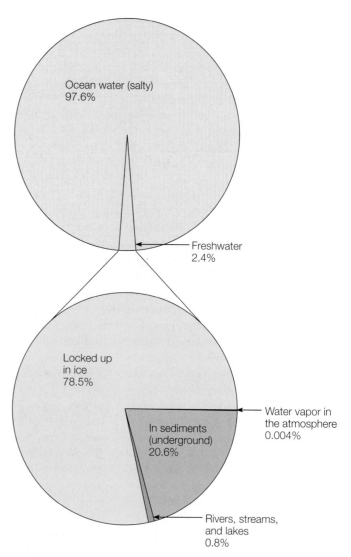

FIGURE 18.2 Estimates of the distribution of all the water found on Earth's surface.

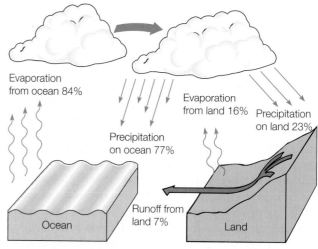

100% is based on a global average of 85 cm/yr precipitation.

FIGURE 18.3 On the average, more water is evaporated from the ocean than is returned by precipitation. More water is precipitated over the land than evaporates. The difference is returned to the ocean by rivers and streams.

small body of running water called a *stream*. A stream is defined as any body of water that is moving across the land, from one so small that you could step across it to the widest river. Water that soaks into the ground moves downward to a saturated zone and is now called *groundwater*. Groundwater moves through sediments and rocks beneath the surface, slowly moving downhill. Streams carry the runoff of a recent rainfall or melting snow, but otherwise most of the flow comes from groundwater that seeps into the stream channel. This explains how a permanent stream is able to continue flowing when it is not being fed by runoff or melting snow (figure 18.4). Where or when the source of groundwater is in low supply, a stream may flow only part of the time, and it is designated as an *intermittent stream*.

The amount of a rainfall that becomes runoff or groundwater depends on a number of factors, including (1) the type of soil on the surface, (2) how dry the soil is, (3) the amount and type of vegetation, (4) the steepness of the slope, and (5) if the rainfall is a long, gentle one or a cloudburst. Different combinations of these factors can result in from 5 percent to almost 100 percent of a rainfall event running off, with the rest evaporating or soaking into the ground. On the average, however, about 70 percent of all precipitation evaporates back into the atmosphere, about 30 percent becomes runoff, and less than 1 percent soaks into the ground.

Surface Water

If you could follow the smallest of streams downhill, you would find that it eventually merges with other streams until they form

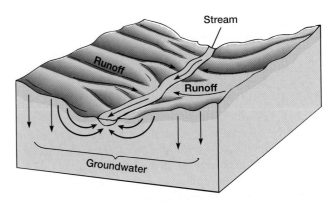

FIGURE 18.4 Some of the precipitation soaks into the ground to become groundwater. Groundwater slowly moves underground, and some of it emerges in streambeds, keeping the streams running during dry spells.

a major river. The land area drained by a stream is known as the stream's drainage basin, or **watershed.** Each stream has its own watershed, but the watershed of a large river includes all the watersheds of the smaller streams that feed into the larger river. Figure 18.5 shows the watersheds of the Columbia River, the Colorado River, and the Mississippi River. Note that the water from the Columbia River and the Colorado River watersheds empties into the Pacific Ocean. The Mississippi River watershed drains into the Gulf of Mexico, which is part of the Atlantic Ocean.

Two adjacent watersheds are separated by a line called a *divide*. Rain that falls on one side of a divide flows into one watershed, and rain that falls on the other side flows into the other

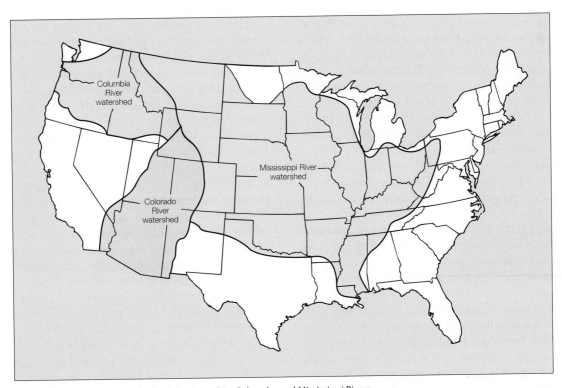

FIGURE 18.5 The approximate watersheds of the Columbia, Colorado, and Mississippi Rivers.

watershed. A *continental divide* separates river systems that drain into opposite sides of a continent. The North American continental divide trends northwestward through the Rocky Mountains. Imagine standing over this line with a glass of water in each hand, then pouring the water to the ground. The water from one glass will eventually end up in the Atlantic Ocean, and the water from the other glass will end up in the Pacific Ocean. Sometimes the Appalachian Mountains are considered to be an eastern continental divide, but water from both sides of this divide ends up on the same side of the continent, in the Atlantic Ocean.

Water moving downhill is sometimes stopped by a depression in a watershed, a depression where water temporarily collects as a standing body of freshwater. A smaller body of standing water is usually called a *pond,* and one of much larger size is called a *lake.* A pond or lake can occur naturally in a depression, or it can be created by building a dam on a stream. A natural pond, a natural lake, or a pond or lake created by building a dam is called a *reservoir* if it is used for (1) water storage, (2) flood control, or (3) generating electricity. A reservoir can be used for one or two of these purposes but not generally for all three. A reservoir built for water storage, for example, is kept as full as possible to store water. This use is incompatible with use for flood control, which would require a low water level in the reservoir in order to catch runoff, preventing waters from flooding the land. In addition, extensive use of reservoir water to generate electricity requires the release of water, which could be incompatible with water storage. The water of streams, ponds, lakes, and reservoirs is collectively called *surface water,* and all serve as sources of freshwater. The management of surface water, as you can see, can present some complicated problems.

Groundwater

Precipitation soaks into the ground, or *percolates* slowly downward until it reaches an area, or zone, where the open spaces between rock and soil particles are completely filled with water. Water from such a saturated zone is called **groundwater.** There is a tremendous amount of water stored as groundwater, which makes up a supply about twenty-five times larger than all the surface water on Earth. Groundwater is an important source of freshwater for human consumption and for agriculture. Groundwater is often found within 100 m (about 330 ft) of the surface, even in arid regions where little surface water is found. Groundwater is the source of water for wells in addition to being the source that keeps streams flowing during dry periods.

Water is able to percolate down to a zone of saturation because sediments contain open spaces between the particles called *pore spaces.* The more pore space a sediment has, the more water it will hold. The total amount of pore spaces in a given sample of sediment is a measure of its *porosity.* Sand and gravel sediments, for example, have grains that have large pore spaces between them, so these sediments have a high porosity. In order for water to move through a sediment, however, the pore spaces must be connected. The ability of a given sample of sediment to transmit water is a measure of its *permeability.* Sand and gravel have a high permeability because the grains

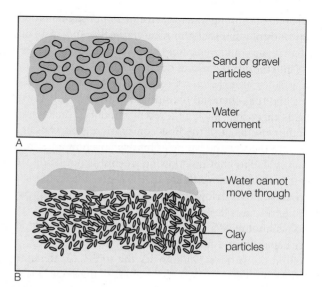

FIGURE 18.6 (A) Sand and gravel have large, irregular particles with large pore spaces, so they have a high porosity. Water can move from one pore space to the next, so they also have a high permeability. (B) Clay has small, flat particles, so it has a low porosity and is practically impermeable because water cannot move from one pore to the next.

do not fit tightly together, allowing water to move from one pore space to the next. Sand and gravel sediments thus have a high porosity as well as a high permeability. Clay sediments, on the other hand, have small, flattened particles that fit tightly together. Clay thus has a low permeability, and when saturated or compressed, clay becomes *impermeable,* meaning water cannot move through it at all (figure 18.6).

The amount of groundwater available in a given location depends on a number of factors, such as the present and past climate, the slope of the land, and the porosity and permeability of the sediments beneath the surface. Generally, sand and gravel sediments, along with solid sandstone, have the best porosity and permeability for transmitting groundwater. Other solid rocks, such as granite, can also transmit groundwater if they are sufficiently fractured by joints and cracks. In any case, groundwater will percolate downward until it reaches an area where pressure and other conditions have eliminated all pores, cracks, and joints. Above this impermeable layer, it collects in all available spaces to form a *zone of saturation.* Water from the zone of saturation is considered to be groundwater. Water from the zone above is not considered to be groundwater. The surface of the boundary between the zone of saturation and the zone above is called the **water table.** The surface of a water table is not necessarily horizontal, but it tends to follow the topography of the surface in a humid climate. A hole that is dug or drilled through the surface to the water table is called a well. The part of the well that is below the water table will fill with groundwater, and the surface of the water in the well is generally at the same level as the water table.

Precipitation falls on the land and percolates down to the zone of saturation, then begins to move laterally, or sideways, to lower and lower elevations until it finds its way back to the

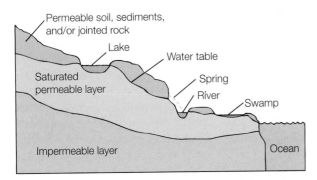

FIGURE 18.7 Groundwater from below the water table seeps into lakes, streams, and swamps and returns to the surface naturally at a spring. Groundwater eventually returns to the ocean, but the trip may take hundreds of years.

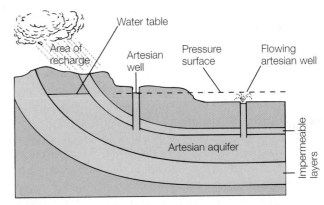

FIGURE 18.8 An artesian aquifer has groundwater that is under pressure because the groundwater is confined between two impermeable layers and has a recharge area at a higher elevation. The pressure will cause the water to rise in a well drilled into the aquifer, becoming a flowing well if the pressure is sufficiently high.

CONCEPTS APPLIED

Table Top

After doing some research, make a drawing to show the depth of the water table in your area and its relationship to local streams, lakes, swamps, and water wells.

surface. This surface outflowing could take place at a stream, pond, lake, swamp, or spring (figure 18.7). Groundwater flows gradually and very slowly through the tiny pore spaces, moving at a rate that ranges from kilometers (miles) per day to meters (feet) per year. Surface streams, on the other hand, move much faster, at rates up to about 30 km per hour (about 20 mi/h).

An **aquifer** is a layer of sand, gravel, sandstone, or other highly permeable material beneath the surface that is capable of producing water in usable quantities. In some places, an aquifer carries water from a higher elevation, resulting in a pressure on water trapped by impermeable layers at lower elevations. Groundwater that is under such a confining pressure is in an *artesian* aquifer. *Artesian* refers to the pressure, and groundwater from an artesian well rises above the top of the aquifer but not necessarily to the surface. Some artesian wells are under sufficient pressure to produce a fountainlike flow or spring (figure 18.8). Some people call groundwater from any deep well "artesian water," which is technically incorrect.

Freshwater as a Resource

Water is an essential resource, not only because it is required for life processes but also because of its role in a modern industrialized society (see chapter 10). Water is used in the home for drinking, cooking, and cleaning, as a carrier to remove wastes, and for maintaining lawns and gardens. These domestic uses lead to an equivalent consumption of about 570 liters per person each day (about 150 gal/person/day), but this is only about 10 percent of the total consumed. Average daily use of water in the United States amounts to some 5,700 liters per person each day (about 1,500 gal/person/day), or about enough water to fill a small swimming pool once a week. The bulk of the water is used by agriculture (about 40 percent), for the production of electricity (about 40 percent), and

for industrial purposes (about 10 percent). These overall percentages of use vary from one region of the country to another.

Most of the water supply is obtained from the surface water resources of streams, lakes, and reservoirs, and 37 percent of the municipal water supply comes from groundwater. If you then add farms, villages, and many suburban areas, the percentage of groundwater used by humans is well above 40 percent. Surface water contains more sediments, bacteria, and possible pollutants than groundwater. This means that surface water requires filtering to remove suspended particles, treatment to kill bacteria, and sometimes processing to remove dissolved chemicals. Groundwater is naturally filtered as it moves through the pore spaces of an aquifer, so it is usually relatively free of suspended particles and bacteria. Thus, the processing or treatment of groundwater is usually not necessary (figure 18.9). But groundwater, on the other hand, will cost more to use as a resource because it must be pumped to the surface. The energy required for this pumping can be very expensive. In addition, groundwater generally contains more dissolved minerals (hard water), which may require additional processing or chemical treatment to remove the troublesome minerals.

The use of surface water as a source of freshwater means that the supply depends on precipitation. When a drought occurs, low river and lake levels may require curtailing water consumption. In some parts of the western United States, such as the Colorado River watershed, *all* of the surface water is already being used, with certain percentages allotted for domestic, industrial, and irrigation uses. Groundwater is also used in this watershed, and in some locations, it is being pumped from the ground faster than it is being replenished by precipitation (figure 18.10). As the population grows and new industries develop, more and more demands are placed on the surface water supply, which has already been committed to other uses, and on the diminishing supply of groundwater. This raises some very controversial issues about how freshwater should be divided among agriculture, industries, and city domestic use. Agricultural interests claim they should have the water because they produce the food and fibers that people must have. Industrial interests claim they should have the water because they create

Science and Society

Water Quality

What do you think of when you see a stream? Do you wonder how deep it is? Do you think of something to do with the water such as swimming or fishing? As you might imagine, not all people look at a stream and think about the same thing. A city engineer, for example, might wonder if the stream has enough water to serve as a source to supplement the city water supply. A rancher or farmer might wonder how the stream could be easily diverted to serve as a source of water for irrigation. An electric utility planner, on the other hand, might wonder if the stream could serve as a source of power.

Water in a stream is a resource that can be used many different ways, but using it requires knowing about the water quality as well as quantity. We need to know if the quality of the water is good enough for the intended use—and different uses have different requirements. Water fit for use in an electric power plant, for example, might not be suitable for use as a city water supply.

Indeed, water fit for use in a power plant might not be suitable for irrigation. Water quality is determined by the kinds and amounts of substances dissolved and suspended in the water and the consequences for users. Whether a source of water can be used for drinking water or not, for example, is regulated by stringent rules and guidelines about what cannot be in the water. These rules are designed to protect human health but do not call for pure water.

The water of even the healthiest stream is not absolutely pure. All water contains many naturally occurring substances such as ions of bicarbonates, calcium, and magnesium. A pollutant is not naturally occurring; it is usually a waste material that contaminates air, soil, or water. There are basically two types of water pollutants: degradable and persistent. Examples of degradable pollutants include sewage, fertilizers, and some industrial wastes. As the term implies, degradable pollutants can be broken down

into simple, nonpolluting substances such as carbon dioxide and nitrogen. Examples of persistent pollutants include some pesticides, petroleum and petroleum products, plastic materials, leached chemicals from landfill sites, oil-based paints, heavy metals and metal compounds such as lead, mercury, and cadmium, and certain radioactive materials. The damage they cause is either irreversible or reparable only over long periods of time.

Questions to Discuss

Discuss with your group the following questions concerning water quality:

1. Which water use requires the purest water? Which requires the least pure?
2. Create a hierarchy of possible water uses; for example, can water used in a power plant later be used for agriculture? For domestic use?
3. What can an individual do to improve water quality?

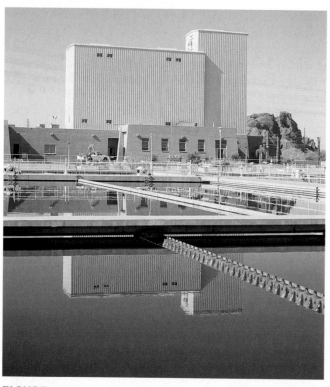

FIGURE 18.9 The filtering beds of a city water treatment facility. Surface water contains more sediments, bacteria, and other suspended materials because it is on the surface and exposed to the atmosphere. This means that surface water must be filtered and treated when used as a domestic resource. Such processing is not required when groundwater is used as the resource.

FIGURE 18.10 This is groundwater pumped from the ground for irrigation. In some areas, groundwater is being removed from the ground faster than it is being replaced by precipitation, resulting in a water table that is falling. It is thus possible that the groundwater resource will soon become depleted in some areas.

the jobs and the products that people must have. Cities, on the other hand, claim that domestic consumption is the most important because people cannot survive without water. Yet, others claim that no group has a right to use water when it is needed to maintain habitats. Who should have the first priority for water use in such cases?

Some have suggested that people should not try to live and grow food in areas that have a short water supply, that plenty of freshwater is available elsewhere. Others have suggested

Wastewater Treatment

One of the most common forms of pollution control in the United States is wastewater treatment. The United States has a vast system of sewer pipes, pumping stations, and treatment plants, and about 74 percent of all Americans are served by such wastewater systems. Sewer pipelines collect the wastewater from homes, businesses, and many industries and deliver it to treatment plants. Most of these plants were designed to make wastewater fit for discharge into streams or other receiving waters.

The basic function of a waste treatment plant is to speed up the natural process of purifying the water. There are two basic stages in the treatment, called the *primary stage* and the *secondary stage*. The primary stage physically removes solids from the wastewater. The secondary stage uses biological processes to further purify wastewater. Sometimes, these stages are combined into one operation.

As raw sewage enters a treatment plant, it first flows through a screen to remove large floating objects such as rags and sticks that might cause clogs. After this initial screening, it passes into a grit chamber where cinders, sand, and small stones settle to the bottom (box figure 18.1). A grit chamber is particularly important in communities with combined sewer systems where sand or gravel may wash into the system along with rain, mud, and other stuff, all with the storm water.

After screening and grit removal, the sewage is basically a mixture of organic and inorganic matter along with other suspended solids. The solids are minute particles that can be removed in a sedimentation tank. The speed of the flow through the larger sedimentation tank is slower, and suspended solids gradually sink to the bottom of the tank. They form a mass of solids called *raw primary sludge,* which is usually removed from the tank by pumping. The sludge may be further treated for use as a fertilizer or disposed of through incineration, if necessary.

Once the effluent has passed through the primary treatment process, which is primarily physical filtering and settling, it enters a secondary stage that involves bio-

Primary Treatment

Screens Grit chamber Sedimentation tank

Box Figure 18.1

logical activities of microorganisms. The secondary stage of treatment removes about 85 percent of the organic matter in sewage by making use of the bacteria that are naturally a part of the sewage. There are two principal techniques used to provide secondary treatment: (1) trickling filters or (2) activated sludge. A trickling filter is simply a bed of stones from 1 to 2 m (3 to 6 ft) deep through which the effluent from the sedimentation tank flows. Interlocking pieces of corrugated plastic or other synthetic media have also been used in trickling beds, but the important part is that it provides a place for bacteria to live and grow. Bacteria grow on the stones or synthetic media and consume most of the organic matter flowing by in the effluent. The now cleaner water trickles out through pipes to another sedimentation tank to remove excess bacteria. Disinfection of the effluent with chlorine is generally used to complete this secondary stage of basic treatment.

The trend today is toward the use of an activated sludge process instead of trickling filters. The activated sludge process speeds up the work of the bacteria by bringing air and sludge heavily laden with bacteria into close contact with the effluent (box figure 18.2). After the effluent leaves the sedimentation tank in the primary stage, it is pumped into an aeration tank, where it is mixed with air and sludge loaded with bacteria and allowed to remain for several hours. During this time, the bacteria break down the organic matter into harmless by-products.

The sludge, now activated with additional millions of bacteria, can be used again by returning it to the aeration tank for mixing with new effluent and ample amounts of air. As with trickling, the final step is generally the addition of chlorine to the effluent, which kills more than 99 percent of the harmful bacteria. Some municipalities are now manufacturing chlorine solution on site to avoid the necessity of transporting and storing large amounts of chlorine, sometimes in a gaseous form. Alternatives to chlorine disinfection, such as ultraviolet light or ozone, are also being used in situations where chlorine in sewage effluents can be harmful to fish and other aquatic life.

New pollution problems have placed additional burdens on wastewater treatment systems. Today's pollutants may be more difficult to remove from water. Increased demands on the water supply only aggravate the problem. These challenges are being met through better and more complete methods of removing pollutants at treatment plants or through prevention of pollution at the source. Pretreatment of industrial waste, for example, removes many troublesome pollutants at the beginning, rather than at the end, of the pipeline.

The increasing need to reuse water calls for better and better wastewater treatment. Every use of water—whether at home, in the factory, or on a farm—results in some change in its quality. New methods for removing pollutants are being developed to return water of more usable quality to receiving lakes and streams. Advanced

—Continued top of next page

Continued—

waste treatment techniques in use or under development range from biological treatment capable of removing nitrogen and phosphorus to physical-chemical separation techniques such as filtration, carbon adsorption, distillation, and reverse osmosis. These activities typically follow secondary treatment and are known as tertiary treatment.

These wastewater treatment processes, alone or in combination, can achieve almost any degree of pollution control desired. As waste effluents are purified to higher degrees by such treatment, the effluent water can be used for industrial, agricultural, or recreational purposes, or even drinking water supplies.

Drawings and some text from *How Wastewater Treatment Works . . . The Basics,* U.S. Environmental Protection Agency, Office of Water, www.epa.gov/owm/featinfo.htm.

Secondary Treatment
Activated sludge process

Aeration tank

Air

Activated sludge

Box Figure 18.2

that humans have historically moved rivers and reshaped the land to obtain water, so perhaps one answer to the problem is to find new sources of freshwater. Possible sources include the recycling of wastewater and turning to the largest supply of water in the world, the ocean. About 90 percent of the water used by industries is presently dumped as a waste product. In some areas, treated city waste water is already being recycled for use in power plants and for watering parks. A practically limitless supply of freshwater could be available by desalting ocean water, something which occurs naturally in the hydrologic cycle. The treatment of seawater to obtain a new supply of freshwater is presently too expensive because of the cost of energy to accomplish the task. New technologies, perhaps ones that use solar energy, may make this more practical in the future. In the meantime, the best sources of extending the supply of freshwater appear to be control of pollution, recycling of wastewater, and conservation of the existing supply.

SEAWATER

More than 70 percent of the surface of Earth is covered by seawater, with an average depth of 3,800 m (about 12,500 ft). The land areas cover 30 percent, less than a third of the surface, with an average elevation of only about 830 m (about 2,700 ft). With this comparison, you can see that humans live on and fulfill most of their needs by drawing from a small part of the total Earth. As populations continue to grow and as resources of the land continue to diminish, the ocean will be looked at more as a resource rather than a convenient place for dumping wastes.

The ocean already provides some food and is a source of some minerals, but it can possibly provide freshwater, new sources of food, new sources of important minerals, and new energy sources in the future. There are vast deposits of phosphorus and manganese minerals on the ocean bottom, for example, that can provide valuable resources. Phosphorus is an element that can be used to manufacture an important fertilizer needed in agriculture, and the land supplies are becoming depleted. Manganese nodules, which occur in great abundance on the ocean bottom, can be a source of manganese, iron, copper, cobalt, and nickel. Seawater contains enough deuterium to make it a feasible source of energy. One gallon of seawater contains about a spoonful of deuterium, with the energy equivalent of 300 gallons of gasoline. It has been estimated there is sufficient deuterium in the oceans to supply power at one hundred times the present consumption for the next ten billion years. The development of controlled nuclear fusion is needed, however, to utilize this potential energy source. The sea may provide new sources of food through *aquaculture,* the farming of the sea the way that the land is presently farmed. Some aquaculture projects have already started with the farming of oysters, clams, lobsters, shrimp, and certain fishes, but these projects have barely begun to utilize the full resources that are possible.

Part of the problem of utilizing the ocean is that the ocean has remained mostly unexplored and a mystery until recent times. Only now are scientists beginning to understand the complex patterns of the circulation of ocean waters, the nature of the chemical processes at work in the ocean, and the interactions of the ocean and the atmosphere, and to chart the topography of the ocean floor.

Oceans and Seas

The vast body of salt water that covers more than 70 percent of Earth's surface is usually called the *ocean* or the *sea*. In general, the **ocean** is a single, continuous body of salt water on the surface of Earth. Although there is really only one big ocean on Earth, specific regions have been given names for convenience in describing locations. For this purpose, three principal regions are recognized: the (1) Atlantic Ocean, (2) Indian Ocean, and (3) Pacific Ocean, as shown in figure 18.11. Specific regions (Atlantic, Indian, and Pacific) are often subdivided further into North Atlantic Ocean, South Atlantic Ocean, and so on.

A *sea* is usually a smaller part of the ocean, a region with characteristics that distinguish it from the larger ocean of which it is a part. Often the term *sea* is used in the name of certain inland bodies of salty water.

The Pacific Ocean is the largest of the three principal ocean regions. It has the largest surface area, covering 180 million km² (about 70 million mi²) and has the greatest average depth of 3.9 km (about 2.4 mi). The Pacific is circled by active converging plate boundaries, so it is sometimes described as being circled by a "rim of fire." It is called this because of the volcanoes associated with the converging plates. The "rim" also has the other associated features of converging plate boundaries such as oceanic trenches, island arcs, and earthquakes. The Atlantic Ocean is second in size, with a surface area of 107 million km² (about 41 million mi²) and the shallowest average depth of only 3.3 km (about 2.1 mi). The Atlantic Ocean is bounded by nearly parallel continental margins with a diverging plate boundary between. It lacks the trench and island arc features of the Pacific, but it does have islands, such as Iceland, that are a part of the Mid-Atlantic Ridge at the plate boundary. The shallow seas of the Atlantic, such as the Mediterranean, Caribbean, and Gulf of Mexico, contribute to the shallow average depth of the Atlantic. The Indian Ocean has the smallest surface area, with 74 million km² (about 29 million mi²) and an average depth of 3.8 km (about 2.4 mi).

As mentioned earlier, a sea is usually a part of an ocean that is identified because some characteristic sets it apart. For example, the Mediterranean, Gulf of Mexico, and Caribbean seas are bounded by land, and they are located in a warm, dry climate. Evaporation of seawater is greater than usual at these locations, which results in the seawater being saltier. Being bounded by land and having saltier seawater characterizes these locations as being different from the rest of the Atlantic. The Sargasso Sea, on the other hand, is a part of the Atlantic that is not bounded by land and has a normal concentration of sea salts. This sea is characterized by having an abundance of floating brown seaweeds that accumulate in this region because of the global wind and ocean current patterns. The Arctic Sea, which is also sometimes called the Arctic Ocean, is a part of the North Atlantic Ocean that is less salty. Thus, the terms *ocean* and *sea* are really arbitrary terms that are used to describe different parts of Earth's one continuous ocean.

The Nature of Seawater

According to one theory, the ocean is an ancient feature of Earth's surface, forming at least three billion years ago as Earth cooled from its early molten state. The seawater and much of the dissolved materials are believed to have formed from the degassing of water vapor and other gases from molten rock materials. The degassed water vapor soon condensed, and over a period of time, it began collecting as a liquid in the depression

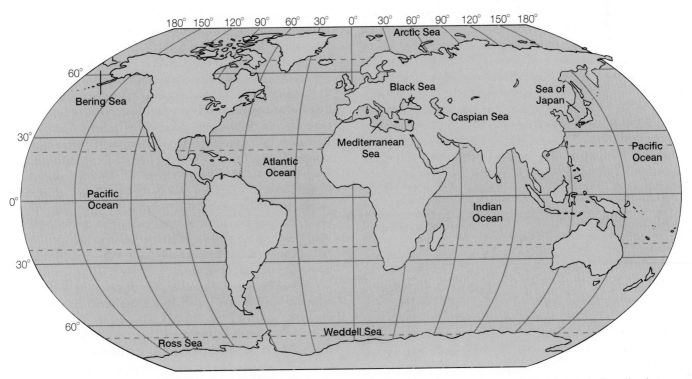

FIGURE 18.11 Distribution of the oceans and major seas on Earth's surface. There is really only one ocean; for example, where is the boundary between the Pacific, Atlantic, and Indian oceans in the Southern Hemisphere?

of the early ocean basin. Ever since, seawater has continuously cycled through the hydrologic cycle, returning water to the ocean through the world's rivers. For millions of years, these rivers have carried large amounts of suspended and dissolved materials to the ocean. These dissolved materials, including salts, stay behind in the seawater as the water again evaporates, condenses, falls on the land, and then brings more dissolved materials much like a continuous conveyor belt.

You might wonder why the ocean basin has not become filled in by the continuous supply of sediments and dissolved materials that would accumulate over millions of years. The basin has not filled in because (1) accumulated sediments have been recycled to Earth's interior through plate tectonics and (2) dissolved materials are removed by natural processes just as fast as they are supplied by the rivers. Some of the dissolved materials, such as calcium and silicon, are removed by organisms to make solid shells, bones, and other hard parts. Other dissolved materials, such as iron, magnesium, and phosphorus, form solid deposits directly and also make sediments that settle to the ocean floor. Hard parts of organisms and solid deposits are cycled to Earth's interior along with suspended sediments that have settled out of the seawater. Studies of fossils and rocks indicate that the composition of seawater has changed little over the past six-hundred million years.

The dissolved materials of seawater are present in the form of ions because of the strong dissolving ability of water molecules. Almost all of the chemical elements are present, but only six ions make up more than 99 percent of any given sample of seawater. As shown in table 18.1, chlorine and sodium are the most abundant ions. These are the elements of sodium chloride, or common table salt. As a sample of seawater evaporates, the positive metal ions join with the different negative ions to form a complex mixture of ionic compounds known as *sea salt*. Sea salt is mostly sodium chloride, but it also contains salts of the four metal ions (sodium, magnesium, calcium, and potassium) combined with the different negative ions of chlorine, sulfate, bicarbonate, and so on.

The amount of dissolved salts in seawater is measured as **salinity.** Salinity is defined as the mass of salts dissolved in 1.0 kg, or 1,000 g, of seawater. Since the salt content is reported in parts per thousand, the symbol ‰ is used (% means parts per hundred). Thus, 35 ‰ means that 1,000 g of seawater contains 35 g of dissolved salts (and 965 g of water). This is the

FIGURE 18.12   Salinity is defined as the mass of salts dissolved in 1.0 kg of seawater. Thus, if a sample of seawater has a salinity of 35 ‰, a 1,000 g sample would evaporate 965 g of water and leave 35 g of sea salts behind.

same concentration as a 3.5 percent salt solution (figure 18.12). Oceanographers use the salinity measure because the mass of a sample of seawater does not change with changes in the water temperature. Other measures of concentration are based on the volume of a sample, and the volume of a liquid does vary as it expands and contracts with changes in the temperature. Thus, by using the salinity measure, any corrections due to temperature differences are eliminated.

The average salinity of seawater is about 35 ‰, but the concentration varies from a low of about 32 ‰ in some locations up to a high of about 36 ‰ in other locations. The salinity of seawater in a given location is affected by factors that tend to increase or decrease the concentration. The concentration is increased by two factors, evaporation and the formation of sea ice. Evaporation increases the concentration because it is water vapor only that evaporates, leaving the dissolved salts behind in a greater concentration. Ice that forms from freezing seawater increases the concentration because when ice forms, the salts are excluded from the crystal structure. Thus, sea ice is freshwater, and the removal of this water leaves the dissolved salts behind in a greater concentration. The salinity of seawater is decreased by three factors: heavy precipitation, the melting of ice, and the addition of freshwater by a large river. All three of these factors tend to dilute seawater with freshwater, which lowers the concentration of salts.

Note that increases or decreases in the salinity of seawater are brought about by the addition or removal of freshwater. This changes only the amount of water present in the solution. The *kinds* or *proportions* of the ions present (table 18.1) in seawater do not change with increased or decreased amounts of freshwater. The same proportion, meaning the same chemical composition, is found in seawater of any salinity of any sample taken from any location anywhere in the world, from any depth of the ocean, or taken any time of the year. Seawater has a remarkably uniform composition that varies only in concentration. This means that the ocean is well mixed and thoroughly stirred around the entire Earth. How seawater becomes so well mixed and stirred on a worldwide basis is discussed in the section on movement of seawater.

If you have ever allowed a glass of tap water to stand for a period of time, you may have noticed tiny bubbles collecting as the water warms. These bubbles are atmospheric gases, such as nitrogen and oxygen, that were dissolved in the water (figure 18.13). Seawater contains dissolved gases in addition to the dissolved salts. Near the surface, seawater contains mostly nitrogen and oxygen in similar proportions to the mixture that is found in the atmosphere. There is more carbon dioxide than

TABLE 18.1   **Major Dissolved Materials in Seawater**

Ion	Percent (by weight)
Chloride (Cl^-)	55.05
Sodium (Na^+)	30.61
Sulfate (SO_4^{-2})	7.68
Magnesium (Mg^{-2})	3.69
Calcium (Ca^{+2})	1.16
Potassium (K^+)	1.10
Bicarbonate (HCO_3^-)	0.41
Bromine (Br^-)	0.19
Total	99.89

FIGURE 18.13 Air will dissolve in water, and cooler water will dissolve more air than warmer water. The bubbles you see here are bubbles of carbon dioxide that came out of solution as the soda became warmer.

you would expect, however, as seawater contains a large amount of this gas. More carbon dioxide can dissolve in seawater because it reacts with water to form carbonic acid, H_2CO_3, the same acid that is found in a bubbly cola. In seawater, carbonic acid breaks down into bicarbonate and carbonate ions, which tend to remain in solution. Water temperature and the salinity have an influence on how much gas can be dissolved in seawater, and increasing either or both will reduce the amount of gases that can be dissolved. Cold, lower salinity seawater in colder regions will dissolve more gases than the warm, higher salinity seawater in tropical locations. Abundant seaweeds (large algae) and phytoplankton (microscopic algae) in the upper, sunlit water tends to reduce the concentration of carbon dioxide and increase the concentration of dissolved oxygen through the process of photosynthesis. With increasing depth, less light penetrates the water, and below about 80 m (about 260 ft), there is insufficient light for photosynthesis. Thus, more algae and phytoplankton and more dissolved oxygen are found above this depth. Below this depth, there is more dissolved carbon dioxide and less dissolved oxygen. The oxygen-poor, deep ocean water does eventually circulate back to the surface, but the complete process may take several thousand years.

Movement of Seawater

Consider the enormity of Earth's ocean, which has a surface area of some 361 million km² (about 139 million mi²) and a volume of 1,370 million km³ (about 328 million mi³) of seawater. There must be a terrific amount of stirring in such an enormous amount of seawater to produce the well-mixed, uniform chemical composition that is found in seawater throughout the world. The amount of mixing required is more easily imagined if you consider the long history of the ocean, the very long period of time over which the mixing has occurred. Based on investigations of the movement of seawater, it has been estimated that there is a complete mixing of all Earth's seawater about every 2,000 years or so. With an assumed age of 3 billion years, this means that

Earth's seawater has been mixed 3,000,000,000 ÷ 2,000, or 1.5 million times. With this much mixing, you would be surprised if seawater were *not* identical all around Earth.

How does seawater move to accomplish such a complete mixing? Seawater is in a constant state of motion, both on the surface and far below the surface. The surface has two types of motion: (1) *waves*, which have been produced by some disturbance, such as the wind, and (2) *currents*, which move water from one place to another. Waves travel across the surface as a series of wrinkles. Waves crash on the shore as booming breakers and make the surf. This produces local currents as water moves along the shore and back out to sea. There are also permanent, worldwide surface currents that move ten thousand times more water across the ocean than all the water moving in all the large rivers on the land. Beneath the surface, there are currents that move water up in some places and move it down in other places. Finally, there are enormous deep ocean currents that move tremendous volumes of seawater. The overall movement of many of the currents on the surface and their relationship to the deep ocean currents are not yet fully mapped or understood. The surface waves are better understood. The general trend and cause of permanent, worldwide currents in the ocean can also be explained. The following is a brief description and explanation of waves, currents, and the deep ocean movements of seawater.

Waves

Any slight disturbance will create ripples that move across a water surface. For example, if you gently blow on the surface of water in a glass, you will see a regular succession of small ripples moving across the surface. These ripples, which look like small, moving wrinkles, are produced by the friction of the air moving across the water surface. The surface of the ocean is much larger, but a gentle wind produces patches of ripples in a similar way. These patches appear, then disappear as the wind begins to blow over calm water. If the wind continues to blow, larger and longer-lasting ripples are made, and the moving air can now push directly on the side of the ripples. A ripple may eventually grow into an **ocean wave,** a moving disturbance that travels across the surface of the ocean. In its simplest form, each wave has a ridge, or mound, of water called a *crest,* which is followed by a depression called a *trough.* Ocean waves are basically repeating series of these crests and troughs that move across the surface like wrinkles (figure 18.14).

The simplest form of an ocean wave can be described by measurements of three distinct characteristics: (1) the *wave height,* which is the vertical distance between the top of a crest and the bottom of the next trough, (2) the *wavelength,* which is the horizontal distance between two successive crests (or other successive parts of the wave), and (3) the *wave period,* which is the time required for two successive crests (or other successive parts) of the wave to pass a given point (figure 18.15).

The characteristics of an ocean wave formed by the wind depend on three factors: (1) the wind speed, (2) the length of time that the wind blows, and (3) the *fetch,* which is the distance the wind blows across the open ocean. As you can imagine, larger waves are produced by strong winds that blow

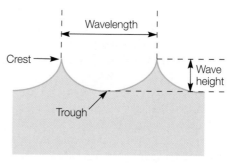

FIGURE 18.15 The simplest form of ocean waves, showing some basic characteristics. Most waves do not look like this representation because most are complicated mixtures of superimposed waves with a wide range of sizes and speeds.

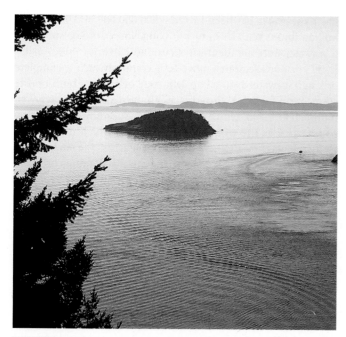

FIGURE 18.14 The surface of the ocean is rarely, if ever, still. Any disturbance can produce a wave, but most waves on the open ocean are formed by a local wind.

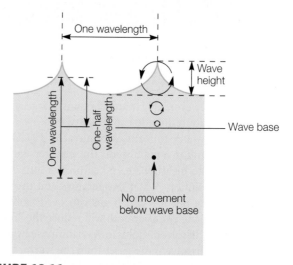

FIGURE 18.16 Water particles are moved in a circular motion by a wave passing in the open ocean. On the surface, a water particle traces out a circle with a diameter that is equal to the wave height. The diameters of the circles traced out by water particles decrease with depth to a depth that is equal to one-half the wavelength of the ocean wave. Bottom sediment cannot be moved by waves when the sediment is below the wave base.

for a longer time over a long fetch. In general, longer-blowing, stronger winds produce waves with greater wave heights, longer wavelengths, and longer periods, but a given wind produces waves with a wide range of sizes and speeds. In addition, the wind does not blow in just one direction, and shifting winds produce a chaotic pattern of waves of many different heights and wavelengths. Thus, the surface of the ocean in the area of a storm or strong wind has a complicated mixture of many sizes and speeds of superimposed waves. The smaller waves soon die out from friction within the water, and the larger ones grow as the wind pushes against their crests. Ocean waves range in height from a few centimeters up to more than 30 m (about 100 ft), but giant waves more than 15 m (about 50 ft) are extremely rare.

The larger waves of the chaotic, superimposed mixture of waves in a storm area last longer than the winds that formed them, and they may travel for hundreds or thousands of kilometers from their place of origin. The longer-wavelength waves travel faster and last longer than the shorter-wavelength waves, so the longer-wavelength waves tend to outrun the shorter-wavelength waves as they die out from energy losses to water friction. Thus, the irregular, superimposed waves created in the area of a storm become transformed as they travel away from the area. They become regular groups of long-wavelength waves with low wave height that are called **swell**. The regular waves of swell that you might observe near a shore may have been produced by a storm that occurred days before thousands of kilometers across the ocean.

The regular crests and troughs of swell carry energy across the ocean, but they do not transport water across the open ocean. If you have ever been in a boat that is floating in swell, you know that you move in a regular pattern of up and forward on each crest, then backward and down on the following trough. The boat does not move along with the waves unless it is

moved along by a wind or by some current. Likewise, a particle of water on the surface moves upward and forward with each wave crest, then backward and down on the following trough, tracing out a nearly circular path through this motion. The particle returns to its initial position, without any forward movement while tracing out the small circle. Note that the diameter of the circular path is equal to the wave height (figure 18.16). Water particles farther below the surface also trace out circular paths as a wave passes. The diameters of these circular paths below the surface are progressively smaller with increasing depth. Below a depth equal to about half the wavelength (the wave base), there is no circular movement of the particles. Thus, you can tell how deeply the passage of a wave disturbs the water below if you measure the wavelength.

As swell moves from the deep ocean to the shore, the waves pass over shallower and shallower water depths. When a depth is reached that is equal to about half the wavelength, the circular motion of the water particles begins to reach the ocean bottom. The water particles now move across the ocean bottom, and the friction between the two results in the waves moving slower as the wave height increases. These important modifications result in a change in the direction of travel and in an increasingly unstable situation as the wave height increases.

Most waves move toward the shore at some angle. As the wave crest nearest the shore starts to slow, the part still over deep water continues on at the same velocity. The slowing at the shoreward side *refracts*, or bends, the wave so it is more parallel to the shore. Thus, waves always appear to approach the shore head-on, arriving at the same time on all parts of the shore.

After the waves reach water that is less than one-half the wavelength, friction between the bottom and the circular motion of the water particles progressively slow the bottom part of the wave. The wave front becomes steeper and steeper as the top overruns the bottom part of the wave. When the wave front becomes too steep, the top part breaks forward and the wave is now called a *breaker* (figure 18.17). In general, this occurs where the water depth is about one and one-third times the wave height. The zone where the breakers occur is called **surf** (figure 18.18).

Waves break in the foamy surf, sometimes forming smaller waves that then proceed to break in progressively shallower water. The surf may have several sets of breakers before the water is finally thrown on the shore as a surging sheet of seawater. The turbulence of the breakers in the surf zone and the final surge expend all the energy that the waves may have brought from thousands of kilometers away. Some of the energy does work in eroding the shoreline, breaking up rock masses into the sands that are carried by local currents back to the ocean. The rest of the energy goes into the kinetic energy of water molecules, which appears as a temperature increase.

Swell does not transport water with the waves over a distance, but small volumes of water are moved as a growing wave is pushed to greater heights by the wind over the open ocean. A strong wind

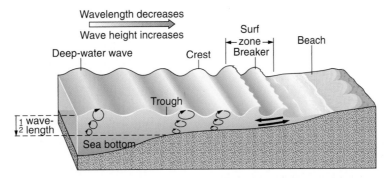

FIGURE 18.17   As a pattern of swell approaches a gently sloping beach, friction between the circular motion of the water particles and the bottom slows the wave and the wave front becomes steeper and steeper. When the depth is about one and one-third times the wave height, the wave breaks forward, moving water toward the beach.

FIGURE 18.18   The white foam is in the surf zone, which is where the waves grow taller and taller, then break forward into a froth of turbulence. Do you see any evidence of rip currents in this picture (see figure 18.19)?

Rogue Waves

A rogue wave is an unusually large wave that appears with smaller waves. The rogue wave has also been called a "freak" wave. Whatever the name, it is generally one wave or a group of two or three waves that are more than twice the size of the normal surrounding waves. The rogue wave is one or several very large "walls of water" and has unpredictable behavior, not following the wind direction, for example. Large rogue waves have been reported from 21 to 35 m (69 to 114 ft) tall, and have been observed to almost capsize large ships.

It is believed that a rogue wave is an extreme storm wave. It probably forms during a storm from constructive interference (see page 114) between smaller waves when the crest and troughs happen to match, coming together to form a mountainous wave that lasts several minutes before subsiding. Other processes, such as wave focusing by the shape of the coast or movement by currents, may play a part in forming rogue waves. The source of rogue waves continues to be a mystery and an active topic of research. This much is known—rogue waves do exist.

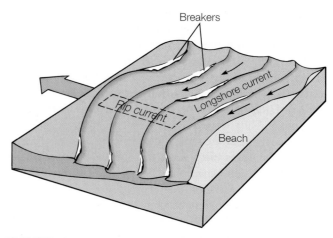

FIGURE 18.19 Breakers result in a buildup of water along the beach that moves as a long-shore current. Where it finds a shore bottom that allows it to return to the sea, it surges out in a strong flow called a *rip current*.

can topple such a wave on the open ocean, producing a foam-topped wave known as a *whitecap*. In general, whitecaps form when the wind is blowing at 30 km/h (about 20 mi/h) or more.

Waves do transport water where breakers occur in the surf zone. When a wave breaks, it tosses water toward the shore, where the water begins to accumulate. This buildup of water tends to move away in currents, or streams, as the water returns to a lower level. Some of the water might return directly to the sea by moving beneath the breakers. This direct return of water forms a weak current known as *undertow*. Other parts of the accumulated water might be pushed along by the waves, producing a *longshore current* that moves parallel to the shore in the surf zone. This current moves parallel to the shore until it finds a lower place or a channel that is deeper than the adjacent bottom. Where the current finds such a channel, it produces a *rip current*, a strong stream of water that bursts out against the waves and returns water through the surf to the sea (figure 18.19). The rip current usually extends beyond the surf zone and then diminishes. A rip current, or where rip currents are occurring, can usually be located by looking for the combination of (1) a lack of surf, (2) darker-looking water, which means a deeper channel, and (3) a turbid, or muddy, streak of water that extends seaward from the channel indicated by the darker water that lacks surf. See chapter 16 for information on how earthquakes can also produce waves.

Ocean Currents

Waves generated by the winds, earthquakes, and tidal forces keep the surface of the ocean in a state of constant motion. Local, temporary currents associated with this motion, such as rip currents or tidal currents, move seawater over a short distance. Seawater also moves in continuous **ocean currents,** streams of water that stay in about the same path as they move

through other seawater over large distances. Ocean currents can be difficult to observe directly since they are surrounded by water that looks just like the water in the current. Wind is likewise difficult to observe directly since the moving air looks just like the rest of the atmosphere. Unlike the wind, an ocean current moves *continuously* in about the same path, often carrying water with different chemical and physical properties than the water it is moving through. Thus, an ocean current can be identified and tracked by measuring the physical and chemical characteristics of the current and the surrounding water. This shows where the current is coming from and where in the world it is going. In general, ocean currents are produced by (1) density differences in seawater and (2) winds that blow persistently in the same direction.

Density Currents. The density of seawater is influenced by three factors: (1) the water temperature, (2) salinity, and (3) suspended sediments. Cold water is generally more dense than warm water, thus sinking and displacing warmer water. Seawater of a high salinity has a higher relative density than less salty water, so it sinks and displaces water of less salinity. Likewise, seawater with a larger amount of suspended sediments has a higher relative density than clear water, so it sinks and displaces clear water. The following describes how these three ways of changing the density of seawater result in the ocean current known as a *density current*, which is an ocean current that flows because of density differences.

Earth receives more incoming solar radiation in the tropics than it does at the poles, which establishes a temperature difference between the tropical and polar oceans. The surface water in the polar ocean is often at or below the freezing point of freshwater, while the surface water in the tropical ocean averages about 26°C (about 79°F). Seawater freezes at a temperature below that of freshwater because the salt content lowers the freezing point. Seawater does not have a set freezing point, however, because as it freezes, the salinity is increased as salt is excluded from the ice structure. Increased salinity lowers the freezing point more, so the more ice that freezes from seawater, the lower the freezing point for the remaining seawater. Cold seawater near the poles is therefore the densest, sinking and creeping slowly as a current across the ocean floor toward the equator. Where and how such a cold, dense bottom current moves is influenced by the shape of the ocean floor, the rotation of Earth, and other factors. The size and the distance that cold bottom currents move can be a surprise. Cold, dense water from the Arctic, for example, moves in a 200 m (about 660 ft) diameter current on the ocean bottom between Greenland and Iceland. This current carries an estimated 5 million m³/s (about 177 million ft³/s) of seawater to the 3.5 km (about 2.1 mi) deep water of the North Atlantic Ocean. This is a flow rate about 250 times larger than that of the Mississippi River. At about 30°N, the cold Arctic waters meet even denser water that has moved in currents all the way from the Antarctic to the deepest part of the North Atlantic Basin (figure 18.20).

A second type of density current results because of differences in salinity. The water in the Mediterranean, for example, has a high salinity because it is mostly surrounded by land in a warm, dry climate. The Mediterranean seawater, with its higher salinity, is more dense than the seawater in the open Atlantic Ocean. This density difference results in two separate currents that flow in opposite directions between the Mediterranean and the Atlantic. The greater density seawater flows from the bottom of the Mediterranean into the Atlantic, while the less dense Atlantic water flows into the Mediterranean near the surface. The dense Mediterranean seawater sinks to a depth of about 1,000 m (about 3,300 ft) in the Atlantic, where it spreads over a large part of the North Atlantic Ocean. This increases the salinity of this part of the ocean, making it one of the more saline areas in the world.

The third type of density current occurs when underwater sediments on a slope slide toward the ocean bottom, producing

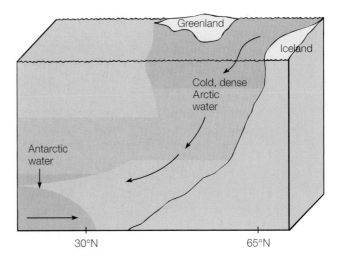

FIGURE 18.20 A cold density current carries about 250 times more water than the Mississippi River from the Arctic and between Greenland and Iceland to the deep Atlantic Ocean. At about 30°N latitude, it meets water that has moved by cold density currents all the way from the Antarctic.

a current of muddy or turbid water called a *turbidity current.* Turbidity currents are believed to be a major mechanism that moves sediments from the continents to the ocean basin. They may also be responsible for some undersea features, such as submarine canyons. Turbidity currents are believed to occur only occasionally, however, and none has ever been directly observed or studied. There is thus no data or direct evidence of how they form or what effects they have on the ocean floor.

Surface Currents. There are broad and deep-running ocean currents that slowly move tremendous volumes of water relatively near the surface. As shown in figure 18.21, each current

FIGURE 18.21 Earth's system of ocean currents.

Thermohaline Circulation

FIGURE 18.22 A summary of the path of the thermohaline circulation. Blue paths represent deep-water currents, while red paths represent surface currents..

is actually part of a worldwide system, or circuit, of currents. This system of ocean currents is very similar to the worldwide system of prevailing winds. This similarity exists because it is the friction of the prevailing winds on the seawater surface that drives the ocean currents. The currents are modified by other factors, such as the rotation of Earth and the shape of the ocean basins, but they are basically maintained by the wind systems.

Each ocean has a great system of moving water called a **gyre** that is centered in the mid-latitudes. The gyres rotate to the right in the Northern Hemisphere and to the left in the Southern Hemisphere. The movement of water around these systems, or gyres, plus some smaller systems, forms the surface circulation system of the world ocean. Each part of the system has a separate name, usually based on its direction of flow. All are called "currents" except one that is called a "stream" (the Gulf Stream) and those that are called "drifts." Both the Gulf Stream and the drifts are currents that are part of the connected system.

The major surface currents are like giant rivers of seawater that move through the ocean near the surface. You know that all the currents are connected, for a giant river of water cannot just start moving in one place, then stop in another. The Gulf Stream, for example, is a current about 100 km (about 60 mi) wide that may extend to a depth of 1 km (about 0.6 mi) below the surface, moving more than 75 million m^3 of water per second (about 2.6 billion ft^3/s). The Gulf Stream carries more than 370 times more water than the Mississippi River. The California Current is weaker and broader, carrying cool water southward at a relatively slow rate. The flow rate of all the currents must be equal, however, since all the ocean basins are connected and the sea level is changing very little, if at all, over long periods of time.

The Great Ocean Conveyor Belt. Density currents and surface currents combine to create **thermohaline circulation,** which occurs deep within the ocean as well as on the surface (figure 18.22). This circulation acts like a conveyer belt as ocean water absorbs and redistributes solar energy, or heat, around Earth, so it is also called "the great ocean conveyer belt." The transport of heat by this conveyer belt influences the climate around the Earth. Ocean water moves slowly on this worldwide transportation system, eventually taking over 2,000 years to complete one trip around Earth.

THE OCEAN FLOOR

Some of the features of the ocean floor were discussed in chapter 16 because they were important in developing the theory of plate tectonics. Many features of the present ocean basins were created from the movement of large crustal plates, according to plate tectonics theory, and in fact, some ocean basins are thought to have originated with the movement of these plates. There is also evidence that some features of the ocean floor were modified during the ice ages of the past. During an ice age, much water becomes locked up in glacial ice, which lowers the sea level. The sea level dropped as much as 140 m (about 460 ft) during the most recent major ice age, exposing the margins of the continents to erosion. Today, these continental margins are flooded with seawater, forming a zone of relatively shallow water called the **continental shelf** (figure 18.23). The continental shelf is considered to be a part of the continent and not the ocean, even though it is covered with an average depth of about 130 m (about 425 ft)

Key Forecasting Tool for the Chesapeake Bay

Submerged aquatic vegetation—SAV, for short—is vital to the Chesapeake Bay ecosystem and a key measuring stick for the Chesapeake's overall health. Eel and wigeon grasses grow in the shallows around the bay and across the bottom, providing food and shelter for baby blue crabs and fish, filtering pollutants, and providing life-sustaining oxygen for the water. So the health of the grasses provides an indication of the overall health of the Chesapeake.

The Chesapeake Bay once had an estimated 600,000 acres of grasses and provided an abundance of oysters, blue crab, shad, haddock, sturgeon, rockfish, and other prized sport fish and seafood. At that time, the bay water was clear, and watermen reported they could clearly see grasses on the bottom some 6 m (about 20 ft) below their boats. Then the water clouded, the grasses began to die, and the ecosystem of the bay began to decline. The low point was reached in 1984, when less than 40,000 acres of grasses could be found in the now murky waters of the Chesapeake.

With the decline of the grasses came the decline of the aquatic species living in the bay. Shad, haddock, and sturgeon once supported large fisheries but are now scarce. The rockfish (striped bass) as well as the prized blue crab were once abundant but seem to decline and recover over the years. The decline of the blue crab population has shocked watermen because the crab is amazingly fertile. The crabs reach sexual maturity in about a year and one female bears millions of eggs. Part of the problem is the loss of the underwater grasses, which shelter the baby crabs from predators. In addition to the loss of habitat, some believe the crab is being overfished.

What happened to the underwater grasses of the Chesapeake? Scientists believe it is a combination of natural erosion and pollutants that wash from farms in the watershed. The pollutants include nutrient-rich fertilizers, chemical residues, and overflow from sewage treatment plants. There are some 6,000 chicken houses around the Chesapeake, raising more than 600 million chickens a year and producing 750,000 tons of manure. The

manure is used as fertilizer, and significant amounts of nitrogen and phosphates wash from these fields. These nutrients accelerate algae growth in the bay, which blocks sunlight. The grass dies and the loss results in muddying of the water, making it impossible for new grasses to begin growing.

There is some evidence to support this idea since the 600,000 acres of underwater grasses died back to a low of 40,000 acres in 1984, then began to rebound when sewage treatment plants were modernized and there were fewer pollutants from industry and farms. The grasses recovered to some 63,500 acres by 1999 but are evidently very sensitive to even slight changes in water quality. Thus, they may be expected to die back with unusual weather conditions that might bring more pollutants or muddy conditions but continue to rebound when the conditions are right. The abundance of blue crabs and other aquatic species can be expected to fluctuate with the health of the grasses. The trends of underwater grass growth do indeed provide a key measure for the health of the Chesapeake Bay.

---

of seawater. The shelf slopes gently away from the shore for an average of 75 km (about 47 mi), but it is much wider on the edge of some parts of continents than other parts.

The continental shelf is a part of the continent that happens to be flooded by seawater at the present time. It still retains some of the general features of the adjacent land that is above water, such as hills, valleys, and mountains, but these features were smoothed off by the eroding action of waves when the sea level was lower. Today, a thin layer of sediments from the adjacent land covers these smoothed-off features.

Beyond the gently sloping continental shelf is a steeper feature called the **continental slope.** The continental slope is the transition between the continent and the deep ocean basin. The water depth at the top of the continental slope is about 120 m (about 390 ft), then plunges to a depth of about 3,000 m (about 10,000 ft) or more. The continental slope is generally 20 to 40 km (about 12 to 25 mi) wide, so the inclination is similar to that encountered driving down a steep mountain road on an interstate highway. At various places around the world, the continental slopes are cut by long, deep, and steep-sided *submarine canyons*. Some of these canyons extend from the top of the slope and down the slope to the ocean basin. Such a submarine canyon can be similar in size and depth to the Grand Canyon on the Colorado River of Arizona. Submarine canyons are believed to have been eroded by turbidity currents, which were discussed in the section on the movement of seawater.

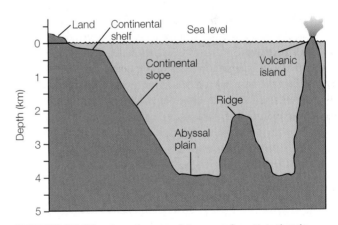

FIGURE 18.23 Some features of the ocean floor. Note that the inclination of the features is not as steep as this vertical exaggeration would suggest.

Beyond the continental slope is the bottom of the ocean floor, the **ocean basin.** Ocean basins are the deepest part of the ocean, covered by about 4 to 6 km (about 2 to 4 mi) of seawater. The basin is mostly a practically level plain called the **abyssal plain** and long, rugged mountain chains called *ridges* that rise thousands of meters above the abyssal plain. The Atlantic Ocean and Indian Ocean basins have ridges that trend north and south near the center of the basin. The Pacific Ocean

People Behind the Science

Rachel Louise Carson (1907–1964)

Rachel Louise Carson was a U.S. biologist, conservationist, and campaigner. Her writings on conservation and the dangers and hazards that many modern practices imposed on the environment inspired the creation of the modern environmental movement.

Carson was born in Springdale, Pennsylvania, on May 27, 1907, and educated at the Pennsylvania College for Women, studying English to achieve her ambition for a literary career. A stimulating biology teacher diverted her toward the study of science, and she went to Johns Hopkins University, graduating in zoology in 1929. She received her master's degree in zoology in 1932 and was then appointed to the department of zoology at the University of Maryland, spending her summers teaching and researching at the Woods Hole Marine Biological Laboratory in Massachusetts. Family commitments to her widowed mother and orphaned nieces forced her to abandon her academic career, and she worked for the U.S. Bureau of Fisheries, writing in her spare time articles on marine life and fish, and producing her first book on the sea just before the Japanese attack on Pearl Harbor. During World War II, she wrote fisheries information bulletins for the U.S. government and reorganized the publications department of what became known after the war as the U.S. Fish and Wildlife Service. In 1949, she was appointed chief biologist and editor of the service. She also became occupied with fieldwork and regularly wrote freelance articles on the natural world. During this period, she was also working on *The Sea Around Us*. Upon its publication in 1951, this book became an immediate best-seller, was translated into several languages, and won several literary awards. Given a measure of financial independence by this success, Carson resigned from her job in 1952 to become a professional writer. Her second book, *The Edge of the Sea* (1955), an ecological exploration of the seashore, further established her reputation as a writer on biological subjects. Her most famous book, *The Silent Spring* (1962), was a powerful indictment of the effects of the chemical poisons, especially DDT, with which humans were destroying the Earth, sea, and sky. Despite denunciations from the influential agrochemical lobby, one immediate effect of Carson's book was the appointment of a presidential advisory committee on the use of pesticides. By this time, Carson was seriously incapacitated by ill health, and she died in Silver Spring, Maryland, on April 14, 1964. On a larger canvas, *The Silent Spring* alerted and inspired a new worldwide movement of environmental concern. While writing about broad scientific issues of pollution and ecological exploitation, Carson also raised important issues about the reckless squandering of natural resources by an industrial world.

Source: Modified from the *Hutchinson Dictionary of Scientific Biography.* © RM, 2009. All rights reserved. Helicon Publishing is a division of RM.

basin has its ridge running north and south near the eastern edge. The Pacific Ocean basin also has more *trenches* than the Atlantic Ocean or the Indian Ocean basins. A trench is a long, relatively narrow, steep-sided trough that occurs along the edges of the ocean basins. Trenches range in depth from about 8 to 11 km (about 5 to 7 mi) deep below sea level.

The ocean basin and ridges of the ocean cover more than half of Earth's surface, accounting for more of the total surface than all the land of the continents. The plain of the ocean basin alone, in fact, covers an area about equal to the area of the land. Scattered over the basin are more than ten thousand steep volcanic peaks called *seamounts*. By definition, seamounts rise more than 1 km (about 0.6 mi) above the ocean floor, sometimes higher than the sea-level surface of the ocean. A seamount that sticks above the water level makes an island. The Hawaiian Islands are examples of such giant volcanoes that have formed islands. Most seamount-formed islands are in the Pacific Ocean. Most islands in the Atlantic, on the other hand, are the tops of volcanoes of the Mid-Atlantic Ridge.

SUMMARY

Precipitation that falls on the land either evaporates, flows across the surface, or soaks into the ground. Water moving across the surface is called *runoff*. Water that moves across the land as a small body of running water is called a *stream*. A stream drains an area of land known as the stream drainage basin or *watershed*. The watershed of one stream is separated from the watershed of another by a line called a *divide*. Water that collects as a small body of standing water is called a *pond*, and a larger body is called a *lake*. A *reservoir* is a natural pond, a natural lake, or a lake or pond created by building a dam for water management or control. The water of streams, ponds, lakes, and reservoirs is collectively called *surface water*.

Precipitation that soaks into the ground *percolates* downward until it reaches a *zone of saturation*. Water from the saturated zone is called *groundwater*. The amount of water that a material will hold depends on its *porosity*, and how well the water can move through the material depends on its *permeability*. The surface of the zone of saturation is called the *water table*.

The *ocean* is the single, continuous body of salt water on the surface of Earth. A *sea* is a smaller part of the ocean with different characteristics. The dissolved materials in seawater are mostly the ions of six substances, but sodium ions and chlorine ions are the most abundant. *Salinity* is a measure of the mass of salts dissolved in 1,000 g of seawater.

An *ocean wave* is a moving disturbance that travels across the surface of the ocean. In its simplest form, a wave has a ridge called a *crest* and a depression called a *trough*. Waves have a characteristic *wave height, wavelength,* and *wave period.* The characteristics of waves made by the wind depend on the wind *speed,* the *time* the wind blows, and the *fetch.* Regular groups of low-profile, long-wavelength waves are called *swell.* When swell approaches a shore, the wave slows and increases in wave height. This slowing *refracts,* or bends, the waves so they approach the shore head-on. When the wave height becomes too steep, the top part breaks forward, forming *breakers* in the *surf zone.* Water accumulates at the shore from the breakers and returns to the sea as *undertow,* as *longshore currents,* or in *rip currents.*

Ocean currents are streams of water that move through other seawater over large distances. Some ocean currents are *density currents,* which are caused by differences in *water temperature, salinity,* or *suspended sediments.* Each ocean has a great system of moving water called a *gyre* that is centered in mid-latitudes. Different parts of a gyre are given different names such as the *Gulf Stream* or the *California Current.*

The ocean floor is made up of the *continental shelf,* the *continental slope,* and the *ocean basin.* The ocean basin has two main parts, the *abyssal plain* and mountain chains called *ridges.*

KEY TERMS

abyssal plain (p. **429**)

aquifer (p. **417**)

continental shelf (p. **428**)

continental slope (p. **429**)

freshwater (p. **413**)

groundwater (p. **416**)

gyre (p. **428**)

ocean (p. **421**)

ocean basin (p. **429**)

ocean currents (p. **426**)

ocean wave (p. **423**)

runoff (p. **414**)

salinity (p. **422**)

surf (p. **425**)

swell (p. **424**)

thermohaline circulation (p. **428**)

watershed (p. **415**)

water table (p. **416**)

APPLYING THE CONCEPTS

1. The most abundant chemical compound at the surface of Earth is
   a. silicon dioxide.
   b. nitrogen gas.
   c. water.
   d. minerals of iron, magnesium, and silicon.

2. Of the total supply, the amount of water that is available for human consumption and agriculture is
   a. 97 percent.
   b. about two-thirds.
   c. about 3 percent.
   d. less than 1 percent.

3. In a region of abundant rainfall, a layer of extensively cracked but otherwise solid granite could serve as a limited source of groundwater because it has
   a. limited permeability and no porosity.
   b. average porosity and average permeability.
   c. no permeability and no porosity.
   d. limited porosity and no permeability.

4. How many different oceans are actually on Earth's surface?
   a. 14
   b. 7
   c. 3
   d. 1

5. The largest of the three principal ocean regions of Earth is the
   a. Atlantic Ocean.
   b. Pacific Ocean.
   c. Indian Ocean.
   d. South American Ocean.

6. The Gulf of Mexico is a shallow sea of the
   a. Atlantic Ocean.
   b. Pacific Ocean.
   c. Indian Ocean.
   d. South American Ocean.

7. Measurement of the salts dissolved in seawater taken from various locations throughout the world show that seawater has a
   a. uniform chemical composition and a variable concentration.
   b. variable chemical composition and a variable concentration.
   c. uniform chemical composition and a uniform concentration.
   d. variable chemical composition and a uniform concentration.

8. The salinity of seawater is *increased* locally by
   a. the addition of water from a large river.
   b. heavy precipitation.
   c. the formation of sea ice.
   d. None of the above is correct.

9. Considering only the available light and the dissolving ability of gases in seawater, more abundant life should be found in a
   a. cool, relatively shallow ocean.
   b. warm, very deep ocean.
   c. warm, relatively shallow ocean.
   d. cool, very deep ocean.

10. The regular, low-profile waves called swell are produced from
    a. constant, prevailing winds.
    b. small, irregular waves becoming superimposed.
    c. longer wavelengths outrunning and outlasting shorter wavelengths.
    d. all wavelengths becoming transformed by gravity as they travel any great distance.

11. If the wavelength of swell is 10.0 m, then you know that the fish below the surface feel the waves to a depth of
    a. 5.0 m.
    b. 10.0 m.
    c. 20.0 m.
    d. however deep it is to the bottom.

Answers

1. c **2.** d **3.** a **4.** d **5.** b **6.** a **7.** a **8.** c **9.** a **10.** c **11.** a

QUESTIONS FOR THOUGHT

1. Describe in general all the things that happen to the water that falls on the land.

2. Explain how a stream can continue to flow even during a dry spell.

3. What is the water table? What is the relationship between the depth to the water table and the depth that a well must be drilled? Explain.

4. Compare the advantages and disadvantages of using (a) surface water and (b) groundwater as a source of freshwater.

5. Prepare arguments for (a) agriculture, (b) industries, and (c) cities each having first priority in the use of a limited water supply. Identify one of these arguments as being the "best case" for first priority, then justify your choice.

6. Discuss some possible ways of extending the supply of freshwater.

7. The world's rivers and streams carry millions of tons of dissolved materials to the ocean each year. Explain why this does not increase the salinity of the ocean.

8. What is swell and how does it form?

9. Why do waves always seem to approach the shore head-on?

10. What factors determine the size of an ocean wave made by the wind?

11. Describe how a breaker forms from swell. What is surf?

12. Describe what you would look for to avoid where rip current occurs at a beach.

FOR FURTHER ANALYSIS

1. Considering the distribution of all the water on Earth, which presently unavailable category would provide the most freshwater at the least cost for transportation, processing, and storage?

2. Describe a number of ways that you believe would increase the amount of precipitation going into groundwater rather than runoff.

3. Some people believe that constructing a reservoir for water storage is a bad idea because (1) it might change the downstream habitat below the dam and (2) the reservoir will eventually fill with silt and sediments. Write a "letter to the editor" that agrees that a reservoir is a bad idea for the reasons given. Now write a second letter that disagrees with the "bad idea" letter and supports the construction of the reservoir.

4. Explain how the average salinity of seawater has remained relatively constant over the past six-hundred million years in spite of the continuous supply of dissolved salts in the river waters of the world.

5. Can ocean waves or ocean currents be used as an energy source? Explain why or why not.

6. What are the significant similarities and differences between a river and an ocean current?

INVITATION TO INQUIRY

Water Use

Investigate the source of water and the amount used by industrial processes, agriculture, and homes in your area. Make pie graphs to compare your area to national averages and develop explanations for any differences. What could be done to increase the supply of water in your area?

19

Organic and Biochemistry

This is a ball-and-stick model of the organic molecule used in many science classrooms to help students understand the 3-D nature of various organic compounds.

CORE **CONCEPT**

The nature of the carbon atom allows for a great variety of organic compounds, many of which play vital roles in living.

OUTLINE

Physics

▷ Organic molecules are involved in energy storage and release (Ch. 3).

Chemistry

▷ Octet rule is followed when bonds are formed (Ch. 9).
▷ Chemistry of the carbon atom is responsible for the traits of organic compounds (Ch. 8).

Earth Science

▷ Fossil fuels were originally the organic components of living things (Ch. 3).
▷ Pollution of air, land, and water may be the result of many types of organic molecules (Ch. 15).

Life Science

▷ The structure of living things is comprised of organic molecules (Ch. 20).
▷ Living things are and utilize organic compounds (Ch. 20).

Astronomy

▷ Extraterrestrial organic compounds are known to exist and may have been the source of the original molecules of life on Earth (Ch. 19).
▷ The exploration for extraterrestrial life centers on the identification of organic compounds (Ch. 19).

OVERVIEW

The impact of ancient Aristotelian ideas on the development of understandings of motion, elements, and matter was discussed in earlier chapters. Historians also trace the "vitalist theory" back to Aristotle. According to Aristotle's idea, all living organisms are composed of the four elements (earth, air, fire, and water) and have in addition an *actuating force,* the life or soul that makes the organism different from nonliving things made of the same four elements. Plants, as well as animals, were considered to have this actuating, or vital, force in the Aristotelian scheme of things.

There were strong proponents of the vitalist theory as recently as the early 1800s. Their basic argument was that organic matter, the materials and chemical compounds recognized as being associated with life, could not be produced in the laboratory. Organic matter could only be produced in a living organism, they argued, because the organism had a vital force that is not present in laboratory chemicals. Then, in 1828, a German chemist named Friedrich Wöhler decomposed a chemical that was *not organic* to produce urea (N_2H_4CO), a known *organic* compound that occurs in urine. Wöhler's production of an organic compound was soon followed by the production of other organic substances by other chemists. The vitalist theory gradually disappeared with each new reaction, and a new field of study, organic chemistry, emerged.

This chapter is an introductory survey of the field of organic chemistry, which is concerned with compounds and reactions of compounds that contain carbon. You will find this an interesting, informative introduction, particularly if you have ever wondered about synthetic materials, natural foods and food products, or any of the thousands of carbon-based chemicals you use every day. The survey begins with the simplest of organic compounds, those consisting of only carbon and hydrogen atoms, compounds known as hydrocarbons. Hydrocarbons are the compounds of crude oil, which is the source of hundreds of petroleum products (figure 19.1).

Most common organic compounds can be considered derivatives of the hydrocarbons, such as alcohols, ethers, fatty acids, and esters. Some of these are the organic compounds that give flavors to foods, and others are used to make hundreds of commercial products, from face cream to margarine. The main groups, or classes, of derivatives will be briefly introduced, along with some interesting examples of each group. A very strong link exists between organic chemistry and the chemistry of living things, which is called biochemistry, or biological chemistry. Some of the important organic compounds of life, including proteins, carbohydrates, and fats, are discussed next. The chapter concludes with an introduction to synthetic polymers, what they are, and how they are related to the fossil fuel supply.

PART I
THE NATURE OF ORGANIC COMPOUNDS

Because there are an infinite variety of organic compounds, this chapter is divided into two parts to help in your understanding. Part I focuses on the features that characterize these compounds, the major groupings into which they are placed, and examples of each. Part II centers on the main classes of biologically important compounds in preparation for your study of living things.

FIGURE 19.1 Refinery and tank storage facilities, such as this one in Texas, are needed to change the hydrocarbons of crude oil into many different petroleum products. The classes and properties of hydrocarbons form one topic of study in organic chemistry.

ORGANIC CHEMISTRY

Today, **organic chemistry** is defined as the study of compounds in which carbon is the principal element, whether the compound was formed by living things or not. The study of compounds that do not contain carbon as a central element is called **inorganic chemistry.** An *organic compound* is thus a compound in which carbon is the principal element, and an *inorganic compound* is any other compound.

The majority of known compounds are organic. Several million are known and thousands of new ones are created every year. You use organic compounds every day, including gasoline, plastics, grain alcohol, foods, flavorings, and many others.

It is the unique properties of carbon that allow it to form so many different compounds. The carbon atom has a valence of four and can combine with one, two, three, or four *other carbon atoms,* in addition to a wide range of other kinds of atoms (figure 19.2). The number of possible molecular combinations is almost limitless, which explains why there are so many organic compounds. Fortunately, there are patterns of groups of carbon atoms and groups of other atoms that lead to similar chemical characteristics, making the study of organic chemistry less difficult.

HYDROCARBONS

Organic compounds consisting of only two elements, carbon and hydrogen, are called **hydrocarbons.** The simplest hydrocarbon has one carbon atom and four hydrogen atoms (figure 19.3), but since carbon atoms can combine with one another, there are thousands of possible structures and arrangements. The carbon-to-carbon bonds are covalent and can be single, double, or triple (figure 19.4). Recall that the dash in a structural means

FIGURE 19.2 Models of a methane molecule. Each of the four bondable electrons in a carbon atom inhabits an area as far away from the other three as possible. Carbon can share with four other atoms at each of these sites. For the sake of simplicity, diagrams of molecules such as the gas methane can be (*A*) two-dimensional drawings, although in reality they are three-dimensional molecules and take up space. Recall from chapter 8 that the higher the atomic number, the larger the space taken up by atoms of that element. The model shown in (*B*) above is called a ball-and-stick model. Part (*C*) is a space-filling model. Each time you see the various ways in which molecules are displayed, try to imagine how much space they actually occupy.

FIGURE 19.3 A molecular formula (*A*) describes the numbers of different kinds of atoms in a molecule, and a structural formula (*B*) represents a two-dimensional model of how the atoms are bonded to each other. Each dash represents a bonding pair of electrons.

FIGURE 19.4 Carbon-to-carbon bonds can be single (*A*), double (*B*), or triple (*C*). Note that in each example, each carbon atom has four dashes, which represent four bonding pairs of electrons, satisfying the octet rule.

People Behind the Science

James Watson (1928–), Francis Crick (1916–2004), and Maurice Wilkins (1916–2004)

James Watson, Francis Crick, and Maurice Wilkins discovered the molecular structure of DNA, for which they shared the Nobel Prize for Physiology or Medicine in 1962. Their discovery of the double helix is considered to be the most important biological advance of the twentieth century.

The American geneticist Watson attended the Horace Mann Grammar School and the South Shore High School in Chicago. At the age of 15, he entered the University of Chicago on a tuition scholarship to study zoology. In 1947 he received a bachelor of science degree in zoology and was awarded a fellowship for graduate study at the Indiana University in Bloomington. In 1950 he achieved his Ph.D. in zoology. It was in the spring of 1951, on a visit to the Zoological Station in Naples, Italy, that he met Maurice Wilkins, who showed him the X-ray picture of crystalline DNA that he had discovered. This inspired his interest in molecular biology, and later that year he moved to the Cavendish Laboratory at the University of Cambridge, England, where he met Francis Crick. They found that they shared an interest in explaining the detailed structure of DNA and started to work together.

English biophysicist Francis Crick went to Northampton Grammar School and when the family moved to London in the 1920s, to Mill Hill School where he developed an interest in physics, chemistry, and mathematics. He then studied physics at University College, London, graduating with a bachelor of science in 1937 and staying on for postgraduate work. However, his studies were interrupted by World War II, and from 1940 he worked for the British Admiralty on the development of radar and magnetic mines for naval warfare. Crick was excited by the idea of applying physics to the study of biology and the investigation of genes at a molecular level. With the financial help of his family and a scholarship from the Medical Research Council, he joined a Cambridge laboratory to study biology, organic chemistry, and X-ray diffraction techniques. By 1949 he was investigating the structure of proteins at the Cavendish Laboratory in Cambridge still driven by the idea of unraveling the mysteries of the genetic code.

New Zealand-born British biophysicist Wilkins moved to England at the age of six. He was sent to King Edward's School,

Birmingham, and later studied physics at St. John's College, Cambridge, graduating in 1938. He was taken on as a research assistant at Birmingham University and gained a Ph.D. from there in 1940. During the war he joined the Ministry of Home Security and Aircraft Production where he worked to improve cathode ray tube screens for radar and studied the separation of uranium isotopes in bombs. In 1946 his research moved him to King's College, London, and the same year he became a member of the Medical Research Council of the Biophysics Research Unit there. His work led to the development of reflecting microscopes for ultraviolet study of nucleic acid. Following the discovery in 1946 that genes consist of DNA, Wilkins began to investigate the structure of the DNA molecule. Studying the X-ray pictures of DNA, he discovered that the molecule has a double helical structure and his colleague Rosalind Franklin also showed that DNA's phosphate groups are situated on the outside of the helix.

By incorporating the results of other researchers (Rosalind Franklin, Edwin Chargaff, and Alexander Todd), Crick and Watson explained that DNA is a double helix consisting of two parallel chains of alternate sugar and phosphate groups linked by pairs of organic bases. They then built a series of molecular models, eventually making one that incorporated all known features of DNA and which gave the same diffraction pattern as that found by Wilkins. They envisioned replication occurring by a parting of the two strands of the double helix, each organic base thus exposed linking with a nucleotide bearing the complementary base. Thus two complete DNA molecules would eventually be formed by this step-by-step linking of nucleotides, with each of the new DNA molecules comprising one strand from the original DNA and one new strand. Their model also explained how genetic information could be coded in a sequence of organic bases. Crick and Watson published their work on the proposed structure of DNA in 1953 and their hypothetical model has been confirmed by thousands of other researchers and accepted as correct.

Source: Modified from the *Hutchinson Dictionary of Scientific Biography.* © RM, 2009. All rights reserved. Helicon Publishing is a division of RM.

---

one shared electron pair has a covalent bond. To satisfy the octet rule, this means that each satisfied carbon atom must have a total of four dashes around it, each attached to another atom, no more and no less. Note that when the carbon atom has double or triple bonds, fewer hydrogen atoms can be attached. One category of hydrocarbons contains only single covalent bonds and are large enough to form chains of carbon atoms with a straight, a branched, or a ring structure (figure 19.5). Examples of these compounds include petroleum and petroleum products.

Members of the *paraffin series*, are not as chemically reactive as the other hydrocarbons. They are called a series because *each higher molecular weight has an additional* CH_2. The simplest is methane, CH_4, and the next highest molecular weight is ethane, C_2H_6. As you can see, C_2H_6 is CH_4 with an additional CH_2.

Note the names of the compounds listed in table 19.1. The names have a consistent prefix and suffix pattern. The prefix and suffix pattern is a code that provides a clue about the compound. The Greek prefix tells you the *number of carbon atoms* in the molecule; for example, *oct-* means eight,

FIGURE 19.5 Carbon-to-carbon chains can be (A) straight, (B) branched, or (C) in a closed ring. (Some carbon bonds are drawn longer but are actually the same length.)

TABLE 19.1   Examples of Straight-Chain Hydrocarbons

Name	Molecular Formula	Structural Formula
Methane	CH_4	
Ethane	C_2H_6	
Propane	C_3H_8	
Butane	C_4H_{10}	
Heptane	C_7H_{16}	
Octane	C_8H_{18}	

so *oct*ane has eight carbon atoms. The suffix *-ane* tells you this hydrocarbon is a member of this series, so it has single bonds only. With the general formula of C_nH_{2n+2}, you can now write the formula when you hear the name. Octane has eight carbon atoms with single bonds and $n = 8$. Two times 8 plus $2(2n + 2)$ is 18, so the formula for octane is C_8H_{18}. Most organic chemical names provide clues like this.

The compounds listed in table 19.1 all have straight chains. Figure 19.6A shows butane with a straight chain and a molecular formula of C_4H_{10}. Figure 19.6B shows a different branched structural formula that has the same C_4H_{10} molecular formula. Compounds with the same molecular formulas with different structures are called **isomers.** The branched isomer is called *isobutane.* The isomers of a particular compound, such as butane, have different physical and chemical properties because they have different structures.

Methane, ethane, and propane can have only one structure each, and butane has two isomers. The number of possible isomers for a particular molecular formula increases rapidly as the number of carbon atoms increase. After butane, hexane has five isomers and octane eighteen isomers. Because they have different structures, each isomer has different physical properties.

FIGURE 19.6 (A) A straight-chain. (B) A branched-chain isomer.

Hydrocarbons with Double or Triple Bonds

Another important category of *hydrocarbons have double covalent carbon-to-carbon bonds*. To denote the presence of a double bond, the suffix *-ane* is changed to *-ene,* as in the compound ethene (table 19.2). Figure 19.4 shows the structural formula for (A) ethane, C_2H_6, and (B) ethene, C_2H_4. These have room for two fewer hydrogen atoms because of the double bond, so the general formula is C_nH_{2n}. Note the simplest of these compounds is called ethene but is commonly known as ethylene.

Ethylene is an important raw material in the chemical industry. Obtained from the processing of petroleum, about half of the commercial ethylene is used to produce the familiar polyethylene plastic. It is also produced by plants to ripen fruit, which explains why unripe fruit enclosed in a sealed plastic bag with ripe fruit will ripen more quickly. The ethylene produced by the ripe fruit acts on the unripe fruit. Commercial fruit packers sometimes use small quantities of ethylene gas to quickly ripen fruit that was picked while green.

Perhaps you have heard the terms *saturated* and *unsaturated* in advertisements for cooking oil or margarine. An organic molecule, such as a hydrocarbon, that does not contain the maximum number of hydrogen atoms is an **unsaturated** hydrocarbon. For example, ethylene can add more hydrogen atoms by reacting with hydrogen gas to form ethane:

Ethene + Hydrogen ⟶ Ethane

The ethane molecule has all the hydrogen atoms possible, so ethane is a **saturated** hydrocarbon. Unsaturated molecules are less stable, which means that they are more chemically reactive than saturated molecules.

Another category of *hydrocarbons has carbon-to-carbon triple bonds* and the general formula of C_nH_{2n-2}. These are highly reactive, and the simplest one, ethyne, has a common

name of acetylene. Acetylene is commonly burned with oxygen gas in a welding torch because the flame reaches a temperature of about 3,000°C. Acetylene is also an important raw material in the production of plastics.

Hydrocarbons That Form Rings

The hydrocarbons discussed up until now have been straight or branched open-ended chains of carbon atoms. Carbon atoms can also bond to each other to form a ring, or *cyclic,* structure (figure 19.7). Because of their ring structure, cyclic versions of hydrocarbons have different chemical properties than similar straight chain hydrocarbons. This shows the importance of structural, rather than simply molecular, formulas in referring to organic compounds.

The six-carbon ring structure shown in figure 19.8A has three double bonds that do not behave as you would expect. In this six-carbon ring, the double bonds are not localized in one place but are spread over the whole molecule. Instead of alternating single and double bonds, all the bonds are something in between. This gives the C_6H_6 molecule increased stability. As a result, the molecule does not behave like other unsaturated compounds; that is, it does not readily react in order to add hydrogen to the ring. The C_6H_6 molecule is the organic compound named *benzene.* To denote the six-carbon ring with delocalized electrons, benzene is represented by the symbol shown in figure 19.8B.

You may have noticed some of the names on labels of paints, paint thinners, and lacquers. Toluene and the xylenes

TABLE 19.2   General Molecular Formulas and Molecular Structures of Different Hydrocarbons

Example Compound	Molecular Structure	General Molecular Formula
Ethane	H—C—C—H (with H above and below each C)	C_nH_{2n+2}
Ethene	C=C (with H's)	C_nH_{2n}
Ethyne	H—C≡C—H	C_nH_{2n-2}

FIGURE 19.7 (A) The "straight" chain has carbon atoms that are able to rotate freely around their single bonds, sometimes linking up in a closed ring. (B) Ring or cyclic compounds.

FIGURE 19.8 (A) The bonds in C_6H_6 are something between single and double, which gives it different chemical properties than double-bonded hydrocarbons. (B) The six-sided symbol with a ring compound known as benzene.

are commonly used in these products as solvents. A benzene ring attached to another molecule is given the name *phenyl*. Phenyl products are good disinfectants and are found in many pine oil products such as Pine Sol.

PETROLEUM

Petroleum is a mixture containing many of the hydrocarbons previously described. While they were once the products of photosynthesis, as time passed, these complex carbohydrates were significantly altered. The origin of petroleum is uncertain, but it is believed to have formed from the slow decomposition of buried marine life, primarily plankton and algae in the absence of oxygen (i.e., anaerobic). Time, temperature, pressure, and perhaps bacteria are considered important in the formation of petroleum. As the petroleum formed, it was forced through porous rock until it reached a rock type or rock structure that stopped it. Here, it accumulated to saturate the porous rock, forming an accumulation called an **oil field.** The composition of petroleum varies from one oil field to the next. The oil from a given field might be dark or light in color. Some oil fields contain oil with a high quantity of sulfur, referred to as "sour crude." Because of such variations, some fields have oil with more desirable qualities than oil from other fields.

Early settlers found oil seeps in the eastern United States and collected the oil for medicinal purposes. One enterprising oil peddler tried to improve the taste by running the petroleum through a whiskey still. He obtained a clear liquid by distilling the petroleum and, by accident, found that the liquid made an excellent lamp oil. This was fortunate timing, for the lamp oil used at that time was whale oil, and whale oil production was declining. This clear liquid obtained by distilling petroleum is today known as *kerosene*.

Wells were drilled, and crude oil refineries were built to produce the newly discovered lamp oil. Gasoline was a by-product and was used primarily as a spot remover. With Henry Ford's automobile production and Thomas Edison's electric light invention, the demand for gasoline increased, and the demand

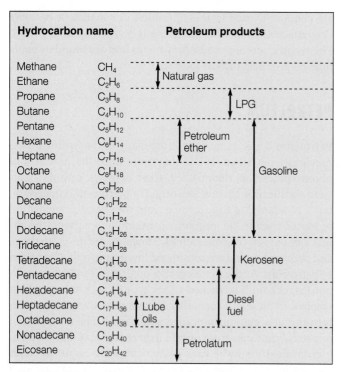

Hydrocarbon name		Petroleum products
Methane	CH_4	
Ethane	C_2H_6	Natural gas
Propane	C_3H_8	
Butane	C_4H_{10}	LPG
Pentane	C_5H_{12}	
Hexane	C_6H_{14}	Petroleum ether
Heptane	C_7H_{16}	
Octane	C_8H_{18}	Gasoline
Nonane	C_9H_{20}	
Decane	$C_{10}H_{22}$	
Undecane	$C_{11}H_{24}$	
Dodecane	$C_{12}H_{26}$	
Tridecane	$C_{13}H_{28}$	
Tetradecane	$C_{14}H_{30}$	Kerosene
Pentadecane	$C_{15}H_{32}$	
Hexadecane	$C_{16}H_{34}$	Diesel fuel
Heptadecane	$C_{17}H_{36}$	Lube oils
Octadecane	$C_{18}H_{38}$	
Nonadecane	$C_{19}H_{40}$	Petrolatum
Eicosane	$C_{20}H_{42}$	

FIGURE 19.9 Petroleum products and the ranges of hydrocarbons in each product.

TABLE 19.3 Petroleum Products

Name	Boiling Range (°C)	Carbon Atoms per Molecule
Natural gas	Less than 0	C_1 to C_4
Petroleum ether	35–100	C_5 to C_7
White "gas"	100–350	C_5 to C_9
Gasoline	35–215	C_5 to C_{12}
Kerosene	35–300	C_{12} to C_{15}
Diesel fuel	300–400	C_{15} to C_{18}
Motor oil, grease	350–400	C_{16} to C_{18}
Paraffin	Solid, melts at about 55	C_{20}
Asphalt	Boiler residue	C_{36} or more

for kerosene decreased. The refineries were converted to produce gasoline, and the petroleum industry grew to become one of the world's largest industries. With the expansion of this industry have come downsides. One of great concern has been climate change associated with the emission of carbon dioxide from a worldwide increase in the number of gasoline-burning engines.

Crude oil is petroleum that is pumped from the ground, a complex and variable mixture of hydrocarbons with an upper limit of about fifty carbon atoms. This thick, smelly black mixture is not usable until it is *refined*, that is, separated into usable groups of hydrocarbons called petroleum products. Petroleum products are separated by distillation—the larger the molecule, the higher the boiling point. Thus, each product has a boiling point range, or "cut," of the distilled vapors. Thus, each product, such as gasoline or heating oil, is made up of hydrocarbons

FIGURE 19.10 The octane rating scale is a description of how rapidly gasoline burns. It is based on (*A*) heptane, with an assigned octane number of 0, and (*B*) trimethylpentane, with an assigned number of 100.

within a range of carbon atoms per molecule (figure 19.9). The products, their boiling ranges, and ranges of carbon atoms per molecule are listed in table 19.3.

The hydrocarbons that have one to four carbon atoms (CH_4 to C_4H_{10}) are gases at room temperature. They can be pumped from certain wells as a gas, but they also can be dissolved in crude oil. *Natural gas* is a mixture of hydrocarbon gases, but it is about 95 percent methane (CH_4). Propane (C_3H_8) and butane (C_4H_{10}) are liquified by compression and cooling and are sold as liquified petroleum gas, or *LPG*. LPG is used where natural gas is not available for cooking or heating and is widely used as a fuel in barbecue grills and camp stoves.

Gasoline is a mixture of hydrocarbons that may have five to twelve carbon atoms per molecule. Gasoline distilled from crude oil consists mostly of straight-chain molecules not suitable for use as an automotive fuel. Straight-chain molecules burn too rapidly in an automobile engine, producing more of an explosion than a smooth burn. You hear these explosions as a knocking or pinging in the engine, and they indicate poor efficiency and could damage the engine. On the other hand, branched chain molecules burn comparatively slower, without the pinging or knocking explosions. The burning rate of gasoline is described by the *octane number* scale. The scale is based on pure heptane, straight-chain molecules that are assigned an octane number of 0, and a multiple branched isomer of octane, trimethylpentane, which is assigned an octane number of 100 (figure 19.10). Most unleaded gasolines have an octane rating of 87, which could be obtained with a mixture that is 87 percent trimethylpentane and 13 percent heptane. Gasoline, however, is a much more complex mixture.

Kerosene is a mixture of hydrocarbons that have from twelve to fifteen carbon atoms. The petroleum product called kerosene is also known by other names, depending on its use. Some of these names are lamp oil (with coloring and fragrance added), jet fuel (with a flash flame retardant added), heating oil, #1 fuel oil, and in some parts of the country, "coal oil."

Diesel fuel is a mixture of a group of hydrocarbons that have from fifteen to eighteen carbon atoms per molecule. Diesel fuel also goes by other names, again depending on its use, that is, distillate fuel oil or #2 fuel oil.

Motor oil and *lubricating oils* have sixteen to eighteen carbon atoms per molecule. Lubricating grease is heavy oil that is thickened with soap. *Petroleum jelly,* also called petrolatum (or Vaseline), is a mixture of hydrocarbons with sixteen to thirty-two carbon atoms per molecule. *Mineral oil* is a light lubricating oil that has been decolorized and purified.

Depending on the source of the crude oil, varying amounts of *paraffin* wax (C_{20} or greater) or *asphalt* (C_{36} or more) may be present. Paraffin is used for candles, waxed paper, and home canning. Asphalt is mixed with gravel and used to surface roads.

HYDROCARBON DERIVATIVES

The hydrocarbons account for only about 5 percent of the known organic compounds, but the other 95 percent can be considered as hydrocarbon derivatives. **Hydrocarbon derivatives** are formed when *one or more hydrogen atoms on a hydrocarbon have been replaced by some element or group of elements other than hydrogen.* For example, the halogens (Fluorine, F_2, Chlorine, Cl_2, and Bromine, Br_2) react in sunlight or when heated, replacing a hydrogen:

In this particular *substitution reaction,* a hydrogen atom on methane is replaced by a chlorine atom to form methyl chloride. Replacement of any number of hydrogen atoms is possible, and a few *organic halides* are illustrated in figure 19.11. Common

Chloroform (CHCl$_3$)

Carbon tetrachloride (CCl$_4$)

Dichlorodifluoromethane (a Freon, CCl$_2$F$_2$)

Vinyl chloride (C$_2$H$_3$Cl)

FIGURE 19.11   Common examples of organic halides.

organic halides include Freons (chlorofluorocarbons) and Teflon (polytetrafluoroethylene).

If a hydrocarbon molecule is unsaturated (has a double or triple bond), a hydrocarbon derivative can be formed by an *addition reaction:*

The bromine atoms add to the double bond on propene, forming dibromopropane. This derivative is used in the synthesis of pharmaceuticals and other organic compounds.

Certain hydrocarbons containing double bonds can combine with each other in an addition reaction to form a very long chain consisting of hundreds of such subunits. A long chain of repeating units is called a **polymer** (*poly* = many; *mer* = segment), individual pieces are called **monomers** (*mono* = single; *mer* = segment or piece), and the reaction is called *addition polymerization.* Ethylene, for example, is heated under pressure with a catalyst to form *polyethylene,* the most popular plastic in the world. It is used to make such products as grocery bags, shampoo bottles, children's toys, and bullet-proof vests. Heating breaks the double bond,

The * indicates that some other atom or series of atoms would be attached at this point.

which provides sites for single covalent bonds to join the ethylene units together,

The * indicates that some other atom or series of atoms would be attached at this point.

which continues the addition polymerization until the chain is hundreds of units long.

Functional Groups Generate Variety

Functional groups are specific combinations of atoms attached to the carbon skeleton that determine specific chemical properties of the organic compound. Functional groups usually have (1) multiple bonds or (2) lone pairs of electrons that cause them

TABLE 19.4   Selected Organic Functional Groups

Name of Functional Group	General Formula	General Structure
Organic halide	RCl	R—C̈l:
Alcohol	ROH	R—Ö—H
Ether	ROR′	R—Ö—R′
Aldehyde	RCHO	R—C—H, ‖ :O:
Ketone	RCOR′	R—C—R′, ‖ :O:
Organic acid	RCOOH	R—C—Ö—H, ‖ :O:
Ester	RCOOR′	R—C—Ö—R′, ‖ :O:
Amino	RNH₂	R—N̈—H, ｜ H
Phosphate	RPO₄	R—O—P=O (with O above and O below)

FIGURE 19.12   Four different alcohols.

to be sites of reactions. Table 19.4 lists some of the common hydrocarbon functional groups. Look over this list and compare the structure of the functional group with the group name. Some of the more interesting examples from a few of these groups will be considered next. Note that the R and R′ (pronounced, "R prime") stand for one or more hydrocarbon groups. For example, in the reaction between methane and chlorine, the product is methyl chloride. In this case, the R in RCl stands for methyl, but it could represent any hydrocarbon group.

Alcohols

An *alcohol* is an organic compound formed by replacing one or more hydrogens on the molecule with a hydroxyl functional group (−OH). The hydroxyl group should not be confused with the hydroxide ion, OH⁻. The hydroxyl group is attached to an organic compound and does not form ions in solution, as the hydroxide ion does. It remains attached to a hydrocarbon group (R), giving the compound its set of properties that are associated with alcohols.

The name of the hydrocarbon group determines the name of the alcohol. If the hydrocarbon group in ROH is methyl, for example, the alcohol is called *methyl alcohol,* or *methanol.* The name of an alcohol has the suffix *-ol* (figure 19.12).

All alcohols have the hydroxyl functional group, and all are chemically similar. Alcohols are toxic to humans except that ethyl alcohol (ethanol) can be consumed in limited quantities. Alcohol has been consumed by people for thousands of years in many cultures. As in the past, there continues to be many health-related problems associated with drinking alcohol, such as fetal alcohol syndrome, liver damage, social disruption, and death. Consumption of small quantities of methanol can result in blindness and death. Ethanol, C_2H_5OH, is produced by the action of yeast or by a chemical reaction of ethylene derived from petroleum refining. Yeast acts on sugars to produce ethanol and CO_2. When beer, wine, and other such beverages are the desired products, the CO_2 escapes during fermentation, and the alcohol remains in solution. In baking, the same reaction utilizes the CO_2 to make dough rise (leavened), and the alcohol is evaporated during baking. Most alcoholic beverages are produced by the yeast fermentation reaction:

sugar → carbon dioxide and ethyl alcohol,

but some are made from ethanol derived from petroleum refining.

Alcohols with six or fewer carbon atoms per molecule are soluble in both certain organic solvents and water. A solution of ethanol and gasoline is called *gasohol*. Today, nearly all gasoline contains ethanol to provide oxygen to cause cleaner burning of hydrocarbons (figure 19.13 and "Connections . . . Biomass to Biofuels").

Alcoholic beverages are a solution of ethanol and water. The *proof* of such a beverage is double the ethanol concentration by volume. Therefore, a solution of 40 percent ethanol by volume in water is 80 proof, and wine that is 12 percent alcohol by volume is 24 proof. Distillation alone will produce a 190 proof concentration, but other techniques are necessary to obtain 200 proof absolute alcohol. *Denatured alcohol* is ethanol with acetone, formaldehyde, and other chemicals in solution that are difficult to separate by distillation. Since these denaturants are toxic, they make consumption impossible, so denatured alcohol is sold without the consumption tax.

Methanol, ethanol, and isopropyl alcohol all have one hydroxyl group per molecule. An alcohol with two hydroxyl groups per molecule is called a *glycol*. Ethylene glycol is perhaps the best-known glycol since it is used as an antifreeze. An alcohol with three hydroxyl groups per molecule is called *glycerol* (or glycerin). Glycerol is a building block of fat molecules and a by-product in the making of soap. It is added to toothpastes, lotions, cosmetics, and some candies to retain moisture and softness. Ethanol, ethylene glycol, and glycerol are compared in figure 19.14.

Glycerol reacts with nitric acid in the presence of sulfuric acid to produce glyceryl trinitrate, commonly known as *nitroglycerine*. Nitroglycerine is a clear oil that is violently explosive, and when warmed, it is extremely unstable. In 1867, Alfred Nobel (for whom the Nobel Prizes were named) discovered that a mixture of nitroglycerine and siliceous earth was more stable than pure nitroglycerine but was nonetheless explosive. Siliceous earth is also known as diatomaceous earth. It is composed of the skeletons of algae known as diatoms and provides an enormous surface area upon which the explosive reaction can take place. The cell walls of diatoms are composed of silicon dioxide (SiO_2). The mixture is packed in a tube and is called *dynamite*. The name *dynamite* comes from the Greek word meaning "power."

Ethers, Aldehydes, and Ketones

An *ether* has a general formula of ROR′, and the best-known ether is diethylether. In a molecule of diethylether, both the R and the R′ are ethyl groups. Diethylether is a volatile, highly flammable liquid that was used as an anesthetic in the past. Today, it is used as an industrial and laboratory solvent.

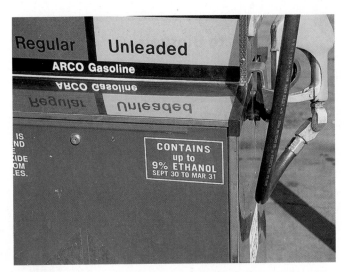

FIGURE 19.13 Gasoline is a mixture of hydrocarbons (C_8H_{18}, for example) that contain no atoms of oxygen. Gasohol contains ethyl alcohol, C_2H_5OH, which does contain oxygen. The addition of alcohol to gasoline, therefore, adds oxygen to the fuel. Since carbon monoxide forms when there is an insufficient supply of oxygen, the addition of alcohol to gasoline helps cut down on carbon monoxide emissions.

FIGURE 19.14 Common examples of alcohols with one, two, and three hydroxyl groups per molecule.

Aldehydes and *ketones* both have a functional group of a carbon atom doubly bonded to an oxygen atom called a *carbonyl group*. The *aldehyde* has a hydrocarbon group, R (or a hydrogen in one case), and a hydrogen attached to the carbonyl group. A *ketone* has a carbonyl group with two hydrocarbon groups attached (figure 19.15).

The simplest aldehyde is *formaldehyde*. Formaldehyde is soluble in water, and a 40 percent concentration called *formalin*

Connections . . .

Biomass to Biofuels

Biomass is any accumulation of organic material produced by living things. The most commonly used biomass sources is wood, agricultural residue from the harvesting of crops, crops grown for their energy content, and animal waste. These traditional, often noncommercial sources of fuel provide more than 10 percent of the world's energy but are not reported in most statistics about global energy. In many developing countries, these sources of fuel are a large proportion of the energy available.

Biomass conversion is the process of obtaining energy from the chemical energy stored in biomass. It is not a new idea; burning wood is a form of biomass conversion that has been used for thousands of years. Biomass can be burned directly as a source of heat for cooking, burned to produce electricity, converted to alcohol, or used to generate methane. China has 500,000 small methane digesters in homes and on farms; India has 100,000; and Korea has 50,000. Brazil is the largest producer of alcohol from biomass. Alcohol provides 50

percent of their automobile fuel. The low price of sugar coupled with the high price of oil has prompted Brazil to use its large crop of sugar cane as a source of energy.

Biomass conversion raises some environmental and economic concerns. Countries that use large amounts of biomass for energy are usually those that have food shortages. Biomass conversion means that fewer nutrients are being returned to the soil, and this compounds the food shortage. If the price of food rises or the price of oil falls, there could be less biomass conversion.

The energy required to produce usable energy stocks from biomass must be taken into account. Growing corn to produce alcohol requires large energy inputs. The amount of energy present in the alcohol produced from the corn is actually about the same as the amount of energy that went into producing the alcohol. Obviously, this makes no sense from an energy point of view. However, the convenience of a liquid fuel may be worth paying for in economic terms.

FIGURE 19.15 The carbonyl group (A) is present in both aldehydes and ketones, as shown in (B). (C) The simplest example of each with the more technical name above and the common name below each formula.

has been used as an embalming agent and to preserve biological specimens. Formaldehyde is also a raw material used to make plastics such as Bakelite. All the aldehydes have odors, and the odors of some aromatic hydrocarbons include the odors of almonds, cinnamon, and vanilla. The simplest ketone is *acetone*. Acetone has a fragrant odor and is used as a solvent in paint removers and nail polish removers.

Organic Acids and Esters

Mineral acids, such as hydrochloric acid and sulfuric acid, are made of inorganic materials. Acids that were derived from organisms are called **organic acids.** Because many of these organic acids can be formed from fats, they are sometimes called *fatty acids.* Chemically, they are known as the *carboxylic acids* because they contain the carboxyl functional group, $-COOH$, and have a general formula of RCOOH.

The simplest carboxylic acid has been known since the Middle Ages, when it was isolated by the distillation of ants. The Latin word *formica* means "ant," so this acid was given the name *formic acid* (figure 19.16A). Formic acid is

$$H - \overset{\displaystyle O}{\underset{\displaystyle \parallel}{C}} - OH$$

A

FIGURE 19.16 This red ant (*Solenopsis invicta*) makes the simplest of the organic acids, formic acid. The sting of bees, ants, and some plants contains formic acid, along with some other irritating materials. Formic acid is HCOOH. The plant known as the stinging nettle (*B*) (*Urtica dioica*) has small, hollow hairs that contain several irritating substances, including formic acid. These hairs have the ability to scratch the skin and mucous membranes, resulting in almost immediate burning, itching, and irritation.

It is formic acid, along with other irritating materials, that causes the sting of bees, ants, and certain plants, for example the stinging nettle (*Urtica dioica*) (figure 19.16B).

Acetic acid, the acid of vinegar, has been known since antiquity. Acetic acid forms from the oxidation of ethanol. An oxidized bottle of wine contains acetic acid in place of the alcohol, which gives the wine a vinegar taste. Before wine is served in a restaurant, the person ordering is customarily handed the bottle cork and a glass with a small amount of wine. You first break the cork in half to make sure it is dry, which tells you that the wine has been sealed from oxygen. The small sip is to taste for vinegar before the wine is served. If the wine has been oxidized, the reaction is

Organic acids are common in many foods. The juice of citrus fruit, for example, contains citric acid, which relieves a thirsty feeling by stimulating the flow of saliva. Lactic acid is found in sour milk, buttermilk, sauerkraut, and pickles. Lactic acid also forms in your muscles as a product of anaerobic carbohydrate metabolism, causing a feeling of fatigue. Anaerobic metabolism of carbohydrates in cells is the breakdown of sugars with the release of usable energy. Citric and lactic acids are small molecules compared to some of the carboxylic acids that are formed from fats. Palmitic acid, for example, is $C_{16}H_{32}O_2$

B

and comes from palm oil. The structure of palmitic acid is a chain of fourteen CH_2 groups with CH_3- at one end and $-COOH$ at the other. Again, it is the functional carboxyl group, $-COOH$, that gives the molecule its acid properties. Organic acids are also raw materials used in the making of polymers of fabric, film, and paint.

CONCEPTS APPLIED

Household Organic Compounds

Pick a household product that has ingredients sounding like they could be organic compounds. On a separate sheet of paper, write the brand name of the product and the type of product (example: Oil of Olay; skin moisturizer), then list the ingredients one under the other (column 1). In a second column, beside each name put the type of compound if you can figure it out from its name or find it in any reference (example: cetyl palmitate—an ester of cetyl alcohol and palmitic acid.) Finally, in a third column, put whatever you know or can find out about the function of that substance in the product. The Internet would be a good resource, as well as the *CRC Handbook* or the *Merck Index*.

Science and Society

A Very Common Organic Compound—Aspirin

Aspirin went on the market as a prescription drug in 1899 after being discovered only two years earlier! However, even primitive humans were familiar with its value as a pain reliever. The bark of the willow tree was known for its "magical" pain relief power thousands of years ago. People in many cultures stripped and chewed the bark for its medicinal effect. It is estimated that more than 80 million tablets of aspirin are consumed in the United States daily. It is the most widely used drug in the world. Just what does it do? Aspirin is really acetylsalicylic acid and is capable of inhibiting the body's production of compounds known as *prostaglandins,* the cause of the pain.

Questions to Discuss

1. What structural parts of aspirin would make it an organic acid?
2. Go to the Internet and find out just how this organic compound acts as a pain reliever.

Esters are organic compounds formed from an alcohol and an organic acid by elimination of water. They are common in both plants and animals, giving fruits and flowers their characteristic odor and taste. For example, the ester amyl acetate has the smell of banana. Esters are also used in perfumes and artificial flavorings. A few of the flavors for which particular esters are responsible are listed in table 19.5. These liquid esters can be obtained from natural sources, or they can be chemically synthesized. Natural flavors are complex mixtures of these esters along with other organic compounds. Lower molecular-weight esters are fragrant smelling liquids, but higher molecular-weight esters are odorless oils and fats.

SYNTHETIC POLYMERS

Natural polymers are huge, chainlike molecules made of hundreds or thousands of smaller, repeating molecular units called monomers. Cellulose and starch are examples of natural polymers made of glucose monomers. *Synthetic polymers* are manufactured from a wide variety of substances. You are familiar with these polymers as synthetic fibers such as nylon and the inexpensive light plastic used for wrappings and containers (figure 19.17).

The first synthetic polymer was a modification of the naturally existing cellulose polymer. Cellulose was chemically modified in 1862 to produce celluloid, the first *plastic*. The term *plastic* means that celluloid could be molded to any desired shape. Celluloid was produced by first reacting cotton with a mixture of nitric and sulfuric acids. The cellulose in the cell walls of the cotton plant reacts with the acids to produce the ester cellulose nitrate. This ester is an explosive compound known as "guncotton," or smokeless gunpowder. When made with ethanol and camphor, the product is less explosive and can be formed and

TABLE 19.5 Flavors and Esters

Ester Name	Formula	Flavor
Amyl acetate	$CH_3 - C - O - C_5H_{11}$ ($\|O$)	Banana
Octyl acetate	$CH_3 - C - O - C_8H_{17}$ ($\|O$)	Orange
Ethyl butyrate	$C_3H_7 - C - O - C_2H_5$ ($\|O$)	Pineapple
Amyl butyrate	$C_3H_7 - C - O - C_5H_{11}$ ($\|O$)	Apricot
Ethyl formate	$H - C - O - C_2H_5$ ($\|O$)	Rum

Name	Chemical unit	Uses	Name	Chemical unit	Uses
Polyethylene		Squeeze bottles, containers, laundry and trash bags, packaging	Polyvinyl acetate		Mixed with vinyl chloride to make vinylite; used as an adhesive and resin in paint
Polypropylene		Indoor-outdoor carpet, pipe valves, bottles	Styrene-butadiene rubber		Automobile tires
Polyvinyl chloride (PVC)		Plumbing pipes, synthetic leather, plastic tablecloths, phonograph records, vinyl tile	Polychloroprene (Neoprene)		Shoe soles, heels
Polyvinylidene chloride (Saran)		Flexible food wrap	Polymethyl methacrylate (Plexiglas, Lucite)		Moldings, transparent surfaces on furniture, lenses, jewelry, transparent plastic "glass"
Polystyrene (Styrofoam)		Coolers, cups, insulating foam, shock-resistant packing material, simulated wood furniture	Polycarbonate (Lexan)		Tough, molded articles such as motorcycle helmets
Polytetrafluoroethylene (Teflon)		Gears, bearings, coating for nonstick surface of cooking utensils	Polyacrylonitrile (Orlon, Acrilan, Creslan)		Textile fibers

FIGURE 19.17   Synthetic polymers, the polymer unit, and some uses of each polymer.

molded into useful articles. This first plastic, celluloid, was used to make dentures, combs, eyeglass frames, and photographic film. Before the discovery of celluloid, many of these articles, including dentures, were made from wood. Today, only Ping-Pong balls are made from cellulose nitrate ("A Closer Look: Nonpersistent and Persistent Organic Pollutants").

Cotton reacted with acetic acid and sulfuric acid produces a cellulose acetate ester. This polymer, through a series of chemical reactions, produces rayon filaments when forced through small holes. The filaments are twisted together to form viscose rayon thread. When forced through a thin slit, a sheet is formed rather than filaments, and the transparent sheet is called *cellophane*. Both rayon and cellophane, as celluloid, are manufactured by modifying the natural polymer of cellulose.

The first truly synthetic polymer was produced in the early 1900s by reacting two chemicals with relatively small molecules rather than modifying a natural polymer. Phenol, an aromatic hydrocarbon, was reacted with formaldehyde, the simplest aldehyde, to produce the polymer named *Bakelite*. Bakelite is a *thermosetting* material that forms cross-links between the polymer chains. Once the links are formed during production, the plastic becomes permanently hardened and cannot be softened or made to flow. Some plastics are *thermoplastic* polymers and soften during heating and harden during cooling because they do not have cross-links.

Polyethylene is a familiar thermoplastic polymer produced by a polymerization reaction of ethylene, which is derived from petroleum. This synthetic was invented just before World War II

A Closer Look

Nonpersistent and Persistent Organic Pollutants

A nonpersistent organic pollutant does not remain in the environment for very long. Most nonpersistent pollutants are *biodegradable*. A biodegradable material is chemically changed by living organisms and often serves as a source of food and energy for decomposer organisms, such as bacteria and fungi. Other nonpersistent pollutants decompose as a result of inorganic chemical reactions. Still others quickly disperse to concentrations that are too low to cause harm.

Nonpersistent toxic materials, such as insecticides, are destroyed by sunlight or reaction with oxygen or water in the atmosphere. For example, organophosphates are commonly

used insecticides (ex. parathion) that usually decompose within several weeks. As a result, they do not accumulate in food chains because they are pollutants for only a short period of time.

Persistent pollutants are those that remain in the environment for many years in an unchanged condition. Most of the persistent pollutants are organic, human-made materials. An estimated 30,000 synthetic chemicals are used in the United States. They are mixed in an endless variety of combinations to produce all types of products used in every aspect of daily life. They are part of our food, transportation, clothing, building materials, home appliances, medicine, recreational equipment, and many other items. Our way of life depends heavily on synthetic materials.

An example of a persistent organic pollutant is polychlorinated biphenyls. PCBs are highly stable organic compounds that resist changes from heat, acids, bases, and oxidation. These characteristics make PCBs desirable for industrial use but also make them persistent pollutants when released into the environment. PCBs have been used in inks, plastics, tapes, paints, glues, waxes, and polishes. Although the manufacture of PCBs in the United States stopped in 1977, these persistent chemicals are still present in the soil and sediments and continue to do harm. PCBs are harmful to fish and other aquatic forms of life because they interfere with reproduction. In humans, PCBs produce liver ailments and skin lesions. In high concentration, they can damage the nervous system, and they are suspected carcinogens.

and was used as an electrical insulating material during the war. Today, there are many variations of polyethylene that are produced by different reaction conditions or by substitution of one or more hydrogen atoms in the ethylene molecule. Polyethylene terephthalate (PET) is used extensively for rigid containers, particularly beverage bottles for carbonated drinks and medicine containers. These containers have the recycle code of number 1. When soft polyethylene near the melting point is rolled in alternating perpendicular directions or expanded and compressed as it is cooled, the polyethylene molecules become ordered in a way that improves the rigidity and tensile strength. One form, high-density polyethylene (HDPE), is used in milk and water jugs, as liners in screw-on jar tops and bottle caps, and as a material for toys. HDPE bottles have the recycling code of number 2 and can be reused to make plastic lumber, motor oil containers, playground equipment, and sheeting. The other form, low-density polyethylene (LDPE), has a recycle code of number 4. Recycled LDPE is used for vegetable, dry cleaning, and grocery bags, and plastic squeeze bottles with a recycle code of number 1.

The properties of polyethylene are also changed by replacing one of the hydrogen atoms in a molecule of ethylene. If the

hydrogen is replaced by a chlorine atom, the compound is called vinyl chloride, and the polymer formed from vinyl chloride is

polyvinyl chloride (PVC). Polyvinyl chloride is used to make plastic water pipes, synthetic leather, and other vinyl products. It differs from the waxy plastic of polyethylene because of the chlorine atom that replaces hydrogen on each monomer.

Replacement of a hydrogen atom with a benzene ring makes a monomer called *styrene*. Styrene is

and polymerization of styrene produces *polystyrene*. Polystyrene is puffed full of air bubbles to produce the familiar Styrofoam coolers, cups, and insulating materials.

If all hydrogens of an ethylene molecule are replaced with atoms of fluorine, the product is polytetrafluoroethylene, a tough plastic that resists high temperatures and acts more like a metal than a plastic. Since it has a low friction, it is used for bearings, gears, and as a nonstick coating on frying pans. You probably know of this plastic by its trade name of *Teflon.*

There are many different polymers in addition to PVC, Styrofoam, and Teflon, and the monomers of some of these are shown in figure 19.17. There are also polymers of isoprene, or synthetic rubber, in wide use. Fibers and fabrics may be polyamides (such as nylon), polyesters (such as Dacron), or polyacrylonitriles (Orlon, Acrilan, Creslan), which have a CN in place of a hydrogen atom on an ethylene molecule and are called acrylic materials. All of these synthetic polymers have added much to practically every part of your life. It would be impossible to list all of their uses here; however, they present problems since (1) they are manufactured from raw materials obtained from coal and a dwindling petroleum supply, and (2) they do not readily decompose when dumped into rivers, oceans, or other parts of the environment. However, research in the polymer sciences is beginning to reflect new understandings learned from research on biological tissues. This could lead to whole new molecular designs for synthetic polymers that will be more compatible with ecosystems.

PART II
ORGANIC COMPOUNDS OF LIFE

ORGANISMS AND THEIR MACROMOLECULES

Aristotle and the later proponents of the vitalist theory were *partly* correct in their concept that living organisms are different from inorganic substances made of the same elements. Living organisms, for example, have the ability to (1) exchange matter and energy with their surroundings and (2) transform matter and energy into different forms as they (3) respond to changes in their surroundings. In addition, living organisms can use the transformed matter and energy to (4) grow and (5) reproduce. Living organisms are able to do these things through a great variety of organic reactions that are catalyzed by enzymes, however, and not through some mysterious "vital force." These enzyme-regulated organic reactions take place because living organisms are highly organized and have an incredible number of relationships among many different chemical processes.

Organic molecules in living things have molecular weights of thousands or millions of atomic mass units (u's) and are therefore referred to as **macromolecules.** Many types of macromolecules are comprised of similar repeated monomers. Monomers are usually combined to form a polymer by a chemical reaction (*dehydration synthesis*) that results in a water molecule being removed from between two monomers. The reverse of this reaction (*hydrolysis*) is the process of splitting a larger polymer into its parts by the addition of water. Digestion of food molecules in the stomach and small intestine is an

important example of this reaction. There are three main types of polymeric macromolecules: (1) carbohydrates, (2) proteins, and (3) nucleic acids. In addition, there is a fourth category, the lipids, which are not true polymers but are macromolecules.

Carbohydrates

One class of organic molecules, **carbohydrates,** is composed of carbon, hydrogen, and oxygen atoms linked together to form monomers called *simple sugars* or *monosaccharides* (*mono* = single; *saccharine* = sweet, sugar) (see table in "Connections. . . How Sweet It Is!"). The name *carbohydrate* literally means "watered carbon," and the empirical formula (CH_2O) for most carbohydrates indicates one carbon (C) atom for each water (H_2O).

The formula for a simple sugar is easy to recognize because there are equal numbers of carbons and oxygens and twice as many hydrogens—for example, $C_3H_6O_3$ or $C_5H_{10}O_5$. We usually describe simple sugars by the number of carbons in the molecule. The ending *-ose* indicates that you are dealing with a carbohydrate. A tri*ose* has three carbons, a pent*ose* has five, and a hex*ose* has six. If you remember that the number of carbons equals the number of oxygen atoms and that the number of hydrogens is double that number, these names tell you the formula for the simple sugar. Carbohydrates play a number of roles in living things. They serve as an immediate source of energy (sugars), provide shape to certain cells (cellulose in plant cell walls), are components of many antibiotics and coenzymes, and are an essential part of the nucleic acids, DNA and RNA.

Simple sugars can be combined with each other to form **complex carbohydrates** (figure 19.18). When two simple sugars bond to each other, a *disaccharide* (*di-* = two) is formed; when three bond together, a *trisaccharide* (*tri-* = three) is formed. Generally, we call a complex carbohydrate that is larger than this a *polysaccharide* (many sugar units). In all cases, the complex carbohydrates are formed by the removal of water from between the sugars. For example, when glucose and fructose are joined together, they form a *disaccharide* (*sucrose*) with the loss of a water molecule,

$$C_6H_{12}O_6 + C_6H_{12}O_6 \rightarrow C_{12}H_{22}O_{11} + H_2O$$
$$\text{glucose} \quad \text{fructose} \quad \text{sucrose} \quad \text{water}$$

Myths, Mistakes, and Misunderstandings

Juice versus Soda Pop

Misconception: Fruit juices have fewer calories from sugar (monosaccharides) than nondiet soft drinks.

In fact, apple juice has more sugar than the same amount of soda pop. Did you also know that it is estimated that people in the United States consume more calories in the form of simple sugar than of meat, chicken, vegetables, and breads combined?

Connections . . .

How Sweet It Is!

Simple sugars, such as glucose, fructose, and galactose, provide the chemical energy necessary to keep organisms alive. *Glucose, $C_6H_{12}O_6$,* is the most abundant carbohydrate and serves as a food and a basic building block for other carbohydrates. Glucose (also called dextrose) is found in the sap of plants, and in the human bloodstream, it is called *blood sugar.* Corn syrup, which is often used as a sweetener, is mostly glucose. Fructose, as its name implies, is the sugar that occurs in fruits, and it is sometimes called *fruit sugar.* Both glucose and fructose have the same molecular formula, but glucose is an aldehyde sugar and fructose is a ketone sugar (box figure 19.2). A mixture of glucose and fructose is found in honey. This mixture also is formed when table sugar (sucrose) is reacted with water in the presence of an acid, a reaction that takes place in the preparation of canned fruit and candies. The mixture of glucose and fructose is called *invert sugar.* Thanks to fructose, invert sugar is about twice as sweet to the taste as the same amount of sucrose (figure 19.18).

Glucose (an aldehyde sugar)

Fructose (a ketone sugar)

Box Figure 19.2   Glucose (blood sugar) is an aldehyde, and fructose (fruit sugar) is a ketone. Both have a molecular formula of $C_6H_{12}O_6$.

Relative Sweetness of Various Sugars and Sugar Substitutes	
Type of Sugar or Artificial Sweetener	**Relative Sweetness**
Lactose (milk sugar)	0.16
Maltose (malt sugar)	0.33
Glucose	0.75
Sucrose (table sugar)	**1.00**
Fructose (fruit sugar)	1.75
Cyclamate	30.00
Aspartame	150.00
Stevia	300.00
Saccharin	350.00
Sucralose	600.00

(a) Cellulose   (b) Plant starches   (c) Glycogen

Amylopectin   Amylose

FIGURE 19.18   Simple sugars are attached to each other by the removal of water from between them. Three common complex carbohydrates are (*A*) cellulose (wood fibers), (*B*) plant starch (amylopectin and amylose), and (*C*) glycogen (sometimes called animal starch). Glycogen is found in muscle cells. Notice how each is similar in that they are all polymers of simple sugars, but each differs from one another in how they are joined together. While many organisms are capable of digesting (hydrolyzing) the bonds that are found in glycogen and plant starch molecules, few are able to break those that link together the monosaccharides of cellulose.

A Closer Look

So You Don't Eat Meat! How to Stay Healthy

Humans require nine amino acids in their diet: threonine, tryptophan, methionine, lysine, phenylalanine, isoleucine, valine, histidine, and leucine. They are called *essential amino acids* because the body is not able to manufacture them. The body uses these *essential amino acids* in the synthesis of the proteins required for good health. For example, the sulfur-containing amino acid methionine is essential for the absorption and transportation of the elements selenium and potassium. It also prevents excess fat buildup in the liver, and it traps heavy metals, such as lead, cadmium, and mercury, bonding with them so that they can be excreted from the body. Essential amino acids are not readily available in most plant proteins and are most easily acquired through meat, fish, and dairy products.

If this is the case, how do people avoid nutritional deficiency if for economic or personal reasons they do not eat meat, poultry, fish, meat products, dairy products, and honey? People who exclude all animal products from their diet are called *vegans*. Those that include only milk are called lacto-vegetarians, those who include eggs are ovo-vegetarians, and those that include both eggs and milk are called lacto-ovo vegetarians. For anyone but a true vegan, the essential amino acids can be provided in even a small amount of milk and eggs. True vegans can get all their essential amino acids by eating certain combinations of plants or plant products. Even though there are certain plants that contain all these amino acids (soy, lupin, hempseed, chia seed, amaranth, buckwheat, and quinoa), most plants contain one or more of the essential amino acids. However, by eating the right combination of different plants, it is possible to get all the essential amino acids in one meal. These combinations are known as complementary foods.

Sucrose, ordinary table sugar, is the most common disaccharide and occurs in high concentrations in sugar cane and sugar beets. It is extracted by crushing the plant materials, then dissolving the sucrose from the materials with water. The water is evaporated and the crystallized sugar is decolorized with charcoal to produce white sugar. Other common disaccharides include *lactose* (milk sugar) and *maltose* (malt sugar). All three disaccharides have similar properties, but maltose tastes only about one-third as sweet as sucrose. Lactose tastes only about one-sixth as sweet as sucrose. No matter which disaccharide sugar is consumed (sucrose, lactose, or maltose), it is converted into glucose and transported by the bloodstream for use by the body.

Some common examples of polysaccharides are **cellulose, starch,** and **glycogen.** Cellulose is an important polysaccharide used in constructing plant cell walls. Humans cannot digest this complex carbohydrate, so we are not able to use it as an energy source. On the other hand, animals known as ruminants (e.g., cows and sheep) and termites have microorganisms within their digestive tracts that do digest cellulose, making it an energy source for them. Plant cell walls add bulk or fiber to our diet, but no calories. Fiber is an important addition to the diet because it helps control weight and reduces the risks of colon cancer. Its large water-holding molecules also help control constipation and diarrhea. Starch is also a plant product digestible by most other organisms. Once broken down, the monosaccharide can be used as an energy source or building materials. A close but structurally different polysaccharide is glycogen. This macromolecule is found in the muscle cells of many animals as a storage form of polysaccharide.

Many types of sugars can be used by cells as components in other, more complex molecules. Sugar molecules are a part of molecules such as DNA (deoxyribonucleic acid), RNA (ribonucleic acid), or ATP (adenosine triphosphate).

Proteins

Proteins play many important roles. Enzymes are catalysts that speed the rate of chemical reactions in living things. Proteins such as hemoglobin serve as carriers of other molecules such as oxygen. Others such as collagen provide shape and support, and several kinds of protein in muscle cells are responsible for movement. Proteins also act as chemical messengers and are called hormones. **Hormones** are chemical messengers secreted by endocrine glands to regulate other parts of the body. Certain other proteins called antibodies help defend the body against dangerous microbes and chemicals. An **antibody** is a globular protein made by the body in response to the presence of a foreign or harmful molecule called an antigen. Antigens are in many cases proteins, too. Chemically, proteins are polymers made up of monomers known as *amino acids*. An **amino acid** is a short carbon skeleton that contains an amino group (a nitrogen and two hydrogens) on one end of the skeleton and a carboxylic acid group at the other end (figure 19.19). In addition, the carbon skeleton may have one of several different *side chains* on it. These vary in their composition and are generally noted as the amino acid's *R-group*. About twenty common amino acids are important to cells, and each differs from one another in the nature of its R-group (see "A Closer Look: So You Don't Eat Meat! How to Stay Healthy").

Amino group **Acid group**

Side chain
"R-group"

FIGURE 19.19 An amino acid is composed of a short carbon skeleton with three functional groups attached: an amino group, a carboxylic acid group (acid group), and an additional variable group (R-group). It is the variable group that determines which specific amino acid is constructed.

Any amino acid can form a bond with any other amino acid. They fit together in a specific way, with the amino group of one bonding to the acid group of the next. You can imagine that by using twenty different amino acids as building blocks, you can construct millions of different combinations. There are over three million possible combinations for a molecule five amino acids long. Each of these combinations is termed a **polypeptide chain.** A specific polypeptide is composed of a specific sequence of amino acids bonded end to end.

The specific sequence of amino acids in a polypeptide is controlled by the genetic information of an organism. Genes are specific messages that tell the cell to link particular amino acids in a specific order; that is, they determine a polypeptide's primary structure. Some sequences of amino acids in a polypeptide are likely to twist (as a coil or a pleated sheet), whereas other sequences remain straight (figure 19.20). For example,

FIGURE 19.20 (A) The structure of a protein molecule can be as simple as a list of its amino acids in the order in which they occur. (B) This shows the next level of complexity (i.e., how one part of the molecule is attached to another part of the same molecule). (C) If already folded parts of a single molecule attach at other places, the molecule displays an even more complex structure. (D) The most complex structural arrangements are displayed by protein molecules that are the result of two or more separate molecules combining into one giant macromolecule.

Protein Structure and Sickle-Cell Anemia

Any changes in the arrangement of amino acids within a protein can have far-reaching effects on its function. For example, normal hemoglobin found in red blood cells consists of two kinds of polypeptide chains called the alpha and beta chains. The beta chain is 146 amino acids long. If just one of these amino acids is replaced by a different one, the hemoglobin molecule may not function properly. A classic example of this results in a condition known as *sickle-cell anemia.* In this case, the sixth amino acid in the beta chain, which is normally glutamic acid, is replaced by valine. This minor change causes the hemoglobin to fold differently, and the red blood cells that contain this altered hemoglobin assume a sickle shape when the body is deprived of an adequate supply of oxygen (see figure 26.16).

some proteins (e.g., hair) take the form of a helix, a shape like that of a coiled spring. Other polypeptides form hydrogen bonds that cause them to make several flat folds that resemble a pleated skirt. This is called a pleated sheet. The way a particular protein folds is important to its function. In bovine spongiform encephalitis (mad cow disease) and Creutzfeldt-Jakob diseases, protein structures are not formed correctly, resulting in characteristic nervous system symptoms.

It is also possible for a single polypeptide to contain one or more coils and pleated sheets along its length. As a result, these different portions of the molecule can interact to form an even more complex globular structure. A good example of a protein with highly complex structure is the myoglobin molecule. This is an oxygen-holding protein found in muscle cells.

Frequently, several different polypeptides, each with its own highly complex structure, twist around each other and chemically combine. The larger, globular structure formed by these interacting polypeptides are displayed by the protein molecules called immunoglobulins or *antibodies,* which are involved in fighting diseases such as influenza and tetanus. The protein portion of the hemoglobin molecule (globin is globular in shape) also demonstrate this highest level of complexity.

Energy in the form of heat or light may break the bonds within protein molecules. When this occurs, the chemical and physical properties of the protein are changed and the protein is said to be **denatured.** (Keep in mind, a protein is a molecule, not a living thing, and therefore cannot be "killed.") A common example of this occurs when the gelatinous, clear portion of an egg is cooked and the protein changes to a white solid.

Some medications are proteins and must be protected from denaturation so as not to lose their effectiveness. Insulin is an example. For protection, such medications may be stored in brown-colored bottles or kept under refrigeration.

The thousands of kinds of proteins can be placed into three categories. Some proteins are important for maintaining the shape of cells and organisms; they are usually referred to as **structural proteins.** The proteins that make up the cell membrane, muscle cells, tendons, and blood cells are examples of structural proteins. The protein collagen is found throughout the human body and gives tissues shape, support, and strength. The second category of proteins, **regulator proteins,** helps determine what activities will occur in the organism. These regulator proteins include *enzymes, chaperones,* and some *hormones.* These molecules help control the chemical activities of cells and organisms. An enzyme is a protein molecule that acts as a catalyst to speed the rate of a reaction. Enzymes can be used over and over again until they are worn out or broken. The production of these protein catalysts is under the direct control of an organism's genetic material (DNA). The instructions for the manufacture of all enzymes are found in the genes of the cell. Organisms make their own enzymes (see "A Closer Look: What Enzymes Are and How They Work"). Enzymes and hormones help control the chemical activities of cells and organisms. Some examples of enzymes are the digestive enzymes in the stomach. The job of a chaperone is to help other proteins fold into their proper shape. For example, some chaperones act as heat shock proteins; that is, they help repair heat damaged proteins. Two hormones that are regulator proteins are insulin and oxytocin. Insulin is produced by the pancreas and controls the amount of glucose in the blood. If insulin production is too low or if the molecule is improperly constructed, glucose molecules are not removed from the bloodstream at a fast enough rate. The excess sugar is then eliminated in the urine. Other symptoms of excess sugar in the blood include excessive thirst and even loss of consciousness. The disease caused by improperly functioning insulin is known as *diabetes.* Oxytocin, a second protein hormone, stimulates the contraction of the uterus during childbirth. It is also an example of an organic molecule that has been produced artificially (e.g., pitocin) and is used by physicians to induce labor. The third category of proteins is **carrier proteins.** Proteins in this category pick up and deliver molecules at one place and transport them to another. For example, proteins regularly attach to cholesterol

What Enzymes Are and How They Work

A *catalyst* is a chemical that speeds a chemical reaction without increasing the temperature and is not used up in the reaction. A protein molecule that acts as a catalyst to speed the rate of a reaction is called an *enzyme*. Enzymes can be used over and over again until they are worn out or broken. The production of these protein catalysts is under the direct control of an organism's genetic material (DNA). Enzymes operate by lowering the amount of energy needed to get a reaction going—the *activation energy*. When this energy is lowered, the nature of the bonds in the substrate is changed so they are more easily broken. Whereas the cartoon (box figure 19.4) shows the breakdown of a single reactant into many end products (as in a hydrolysis reaction), the lowering of activation energy can also result in bonds being broken so that new bonds may be formed in the construction of a single, larger end product from several reactants (as in a synthesis reaction).

During an enzyme-controlled reaction, the enzyme and substrate come together to form a new molecule—the enzyme-substrate complex molecule. This molecule exists for only a very short time. During that time, activation energy is lowered and bonds are changed. The result is the formation of a new molecule or molecules called the end products of the reaction. The number of jobs an enzyme can perform during a particular time period is incredibly large—ranging between a thousand (10^3) and ten thousand trillion (10^{16}) times faster per minute than uncatalyzed reactions! Without the enzyme, perhaps only fifty or one-hundred substrate molecules might be altered in the same time.

Box Figure 19.4

entering the system from the diet, forming molecules called lipoproteins, which are transported through the circulatory system. The cholesterol is released at a distance from the digestive tract, and the proteins return to pick up more dietary cholesterol entering the system.

Nucleic Acids

Nucleic acids are complex polymeric molecules that store and transfer information within a cell. This section is an overview of this important class of organic compounds. More detailed information is presented in chapter 26. There are two types of nucleic acids, DNA and RNA. DNA serves as genetic material determining which proteins will be manufactured, while RNA plays a vital role in the process of protein manufacture. All nucleic acids are constructed of fundamental monomers known as **nucleotides.** Each nucleotide is composed of three parts: (1) a five-carbon simple sugar molecule that may be ribose or deoxyribose, (2) a phosphate group, and (3) a nitrogenous base. There are five types of nitrogenous bases. Two of the bases are

A DNA single strand

A single nucleotide

B RNA

A nucleotide

D = Deoxyribose sugar
R = Ribose sugar
P = Phosphate functional group
G = Guanine nitrogenous base
A = Adenine nitrogenous base
T = Thymine nitrogenous base
C = Cytosine nitrogenous base
U = Uracil nitrogenous base

Backbone

FIGURE 19.21   (A) A single strand of DNA is a polymer composed of nucleotides. Each nucleotide (framed at the top of the molecule) consists of deoxyribose sugar, phosphate, and one of four nitrogenous bases: A,T, G, or C. Notice the backbone of blue-highlighted sugar and phosphate. (B) RNA is also a polymer, but each nucleotide (framed at the top of this RNA molecule) is composed of ribose sugar, phosphate, and one of four nitrogenous bases: A, U, G, or C. The backbone of ribose and phosphate is highlighted in green.

the larger, double-ring molecules, *Adenine* and *Guanine*. The smaller bases are the single-ring bases, *Thymine*, *Cytosine*, and *Uracil*. Nucleotides (monomers) are linked together in long sequences (polymers) so that the sugar and phosphate sequence forms a "backbone" and the nitrogenous bases stick out to the side. DNA has deoxyribose sugar and the bases A, T, G, and C, while RNA has ribose sugar and the bases A, U, G, and C (figure 19.21).

DNA (*deoxyribonucleic acid*) is composed of two strands to form a ladderlike structure thousands of nucleotide bases long. The two strands are attached between their bases according to *the base pair rule*; that is, *Adenine* from one strand always pairs with *Thymine* from the other. *Guanine* always pairs with *Cytosine*.

A T (or A U) and G C

One strand of DNA is called the *coding strand* because it has a meaningful genetic message written using the nitrogenous bases as letters (e.g., the base sequence CATTAGACT) (figure 19.22). This is the basis of the genetic code for all

DNA

FIGURE 19.22   The genetic material is really double-stranded DNA molecules comprised of sequences of nucleotides that spell out an organism's genetic code. The coding strand of the double molecule is the side that can be translated by the cell into meaningful information. The genetic code has the information for telling the cell what proteins to make, which in turn become the major structural and functional components of the cell. The noncoding strand is unable to code for such proteins.

organisms. If these bases are read in groups of three, they make sense to us (i.e.,"cat," "tag," and "act"). The opposite strand is called *noncoding* since it makes no "sense" but protects the coding strand from chemical and physical damage. Both strands are twisted into a helix, that is, a molecule turned around a tubular space like a coiled spring.

The information carried by DNA can be compared to the information in a textbook. Books are composed of words (constructed from individual letters) in particular combinations, organized into chapters. In the same way, DNA is composed of tens of thousands of nucleotides in specific sequences (words) organized into genes (a chapter). Each gene carries the information for producing a protein, just as each chapter carries the information relating to one idea. The order of nucleotides in a gene is directly related to the order of amino acids in a protein. Just as chapters in a book are identified by beginning and ending statements, different genes along a DNA strand have beginning and ending signals. They tell when to start and when to stop reading a particular gene. Human body cells contain forty-six strands (books) of helical DNA, each containing thousands of genes (chapters). These strands are called **chromosomes** when they become supercoiled in preparation for cellular reproduction. Before cell reproduction, the DNA makes copies of the coding and noncoding strands, ensuring that the offspring or *daughter cells* will each receive a full complement of the genes required for their survival. Each chromosome is comprised of a sequence of genes. A **gene** is a segment of DNA that is able to (1) replicate by directing the manufacture of copies of itself; (2) mutate, or chemically change, and transmit these changes to future generations; (3) store information that determines the characteristics of cells and organisms; and (4) use this information to direct the synthesis of structural and regulatory proteins.

Messenger RNA (mRNA) is a single-strand copy of a portion of the coding strand of DNA for a specific gene. When mRNA is formed on the surface of the DNA, the base pair rule applies. However, since **RNA (ribonucleic acid)** does not contain thymine, it uses a U-A pairing instead of the T-A pairing between nucleotides. After mRNA is formed and peeled off, it associates with a cellular structure called the *ribosome* where the genetic message can be translated into a protein molecule. Ribosomes contain another type of RNA, **ribosomal RNA (rRNA)**. rRNA is also an RNA copy of DNA, but after being formed, it becomes twisted and covered in protein to form a ribosome. The third form of RNA, **transfer RNA (tRNA)**, is also made from copies of different segments of DNA, but when peeled off the surface, each takes the form of a cloverleaf. tRNA molecules are responsible for transferring or carrying specific amino acids to the ribosome where all three forms of RNA come together and cooperate in the manufacture of protein molecules.

Lipids

We generally call molecules in this group the **fats.** They are not polymers, as are the previously discussed carbohydrates, proteins, and nucleic acids. However, there are three different types of **lipids:** *true fats* (pork chop fat or olive oil), *phospholipids* (the primary component of cell membranes), and *steroids* (some hormones). In general, lipids are large, nonpolar (do not have a positive end and a negative end), organic molecules that do not easily dissolve in polar solvents such as water. For example, nonpolar vegetable oil molecules do not dissolve in polar water molecules. Lipids are soluble in nonpolar substances such as ether or acetone. Just like carbohydrates, the lipids are composed of carbon, hydrogen, and oxygen. Lipids generally have very small amounts of oxygen in comparison to the amounts of carbon and hydrogen. Simple lipids cannot be digested into smaller, similar subunits. Complex lipids such as true fats and phospholipids can be digested into smaller, similar units.

True (Neutral) Fats

True (neutral) fats are important, complex organic molecules that are used to provide, among other things, energy. The building blocks of a fat are a glycerol molecule and fatty acids. Recall that **glycerol** is a carbon skeleton that has three alcohol groups attached to it. Its chemical formula is $C_3H_5(OH)_3$. At room temperature, glycerol looks like clear, lightweight oil. It is used under the name glycerin as an additive to many cosmetics to make them smooth and easy to spread.

$$
\begin{array}{ccccccc}
 & OH & & OH & & OH & \\
 & | & & | & & | & \\
H- & C & - & C & - & C & -H \\
 & | & & | & & | & \\
 & H & & H & & H &
\end{array}
$$

Glycerol

The **fatty acids** found in fats are a long-chain carbon skeleton that has a carboxylic acid functional group. If the carbon skeleton has as much hydrogen bonded to it as possible, we call it *saturated.* The saturated fatty acid in figure 19.23A is stearic acid, a component of solid meat fats such as bacon. Notice that at every point in this structure, the carbon has as much hydrogen as it can hold. Saturated fats are generally found in animal tissues and tend to be solids at room temperatures.

Some examples of animal products containing saturated fats are butter, whale blubber, suet, and lard.

If the carbons in a fatty acid are double-bonded to each other at one or more points, the fatty acid is said to be *unsaturated*. Unsaturated fats are frequently plant fats or oils, and they are usually liquids at room temperature. Peanut, corn, and olive oil are examples of unsaturated fats that are mixtures of different true fats. When glycerol and three fatty acids are combined, a fat is formed. This reaction is almost exactly the same as the reaction that causes simple sugars to bond together.

Fats are important molecules for storing energy. There is more than twice as much energy in a gram of fat as in a gram of sugar, 9 Calories versus 4 Calories. This is important to an organism because fats can be stored in a relatively small space and still yield a high amount of energy. Fats in animals also provide protection from heat loss. The thick layer of blubber in whales, walruses, and seals prevents the loss of internal body heat to the cold, watery environment in which they live. This same layer of fat, together with the fat deposits around some internal organs—such as the kidneys and heart—serves as a cushion that protects these organs from physical damage. If a fat is formed from a glycerol molecule and three attached fatty acids, it is called a **triglyceride;** if two, a *diglyceride;* and if one, a *monoglyceride* (figure 19.24). Triglycerides account for about 95 percent of the fat stored in human tissue.

Phospholipids

Phospholipids are a class of complex water-insoluble organic molecules that resemble fats but contain a phosphate group (PO_4)

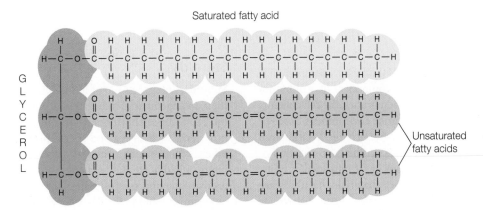

FIGURE 19.23 Structure of saturated and unsaturated fatty acids. (*A*) Stearic acid is an example of a saturated fatty acid. (*B*) Linoleic acid is an example of an unsaturated fatty acid. It is technically an omega-6 fatty acid because the first double bond occurs at carbon number six. (*C*) An omega-3 fatty acid, linolenic acid. Both linoleic and linolenic acids are essential fatty acids for humans.

FIGURE 19.24 The arrangement of the three fatty acids attached to a glycerol molecule is typical of the formation of a fat. The structural formula of the fat appears to be very cluttered until you dissect the fatty acids from the glycerol; then it becomes much more manageable. This example of a triglyceride contains a glycerol molecule, two unsaturated fatty acids (linoleic acid), and a third saturated fatty acid (stearic acid).

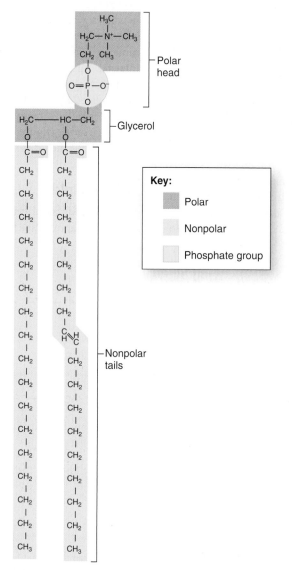

FIGURE 19.25 This molecule is similar to a fat but has a phosphate (yellow) group in its structure. You can think of phospholipid molecules as having a 'head' with two strings dangling down. The head portion is the glycerol and phosphate group, which is polar and soluble in water. The strings are the fatty acid segments of the molecule, which are nonpolar.

in their structure (figure 19.25). One of the reasons phospholipids are important is that they are a major component of membranes in cells. Without these lipids in our membranes, the cell contents would not be separated from the exterior environment. Some of the phospholipids are better known as the *lecithins*. Lecithins are found in cell membranes and also help in the emulsification of fats. They help separate large portions of fat into smaller units. This allows the fat to mix with other materials. Lecithins are added to many types of food for this purpose (chocolate bars, for example). Some people take lecithin as nutritional supplements because they believe it leads to healthier hair and better reasoning ability; but once inside your intestines, lecithins are destroyed by enzymes, just like any other phospholipid.

Steroids

Steroids, another group of lipid molecules, are characterized by their arrangement of interlocking rings of carbon. Many steroid molecules are sex hormones. Some of them regulate reproductive processes, such as egg and sperm production (see chapter 21); others regulate such things as salt concentration in the blood. Figure 19.26 illustrates some of the steroid compounds, such as testosterone and progesterone that are typically manufactured by organisms and in the laboratory as pharmaceuticals. We have already mentioned one steroid molecule: cholesterol. Serum cholesterol (the kind found in your blood and associated with lipoproteins) has been implicated in many cases of athero-sclerosis. However, your body makes this steroid for use as a

Omega Fatty Acids and Your Diet

The occurrence of a double bond in a fatty acid is indicated by the Greek letter ω (omega) followed by a number indicating the location of the first double bond in the molecule. Oleic acid, one of the fatty acids found in olive oil, is comprised of 18 carbons with a single (1) double bond between carbons 9 and 10. Therefore, it is chemically designated C18:1ω9 and is a monounsaturated fatty acid. This fatty acid is commonly referred to as an omega-9 fatty acid. The unsaturated fatty acid in figure 19.23B is linoleic acid, a component of sunflower and safflower oils. Notice that there are two double bonds between

the carbons and fewer hydrogens than in the saturated fatty acid. Linoleic acid is chemically a polyunsaturated fatty acid with two double (2) bonds and is designated C18:2ω6, an omega-6 fatty acid. This indicates that the first double bond of this 18-carbon molecule is between carbons 6 and 7. Since the human body cannot make this fatty acid, it is called an *essential fatty acid* and must be taken in as a part of the diet. The other essential fatty acid, linolenic acid, is C18:3ω3 and has three (3) double bonds. This fatty acid is commonly referred to as an omega-3 fatty acid. One key function of these essential fatty

acids is the synthesis of prostaglandin hormones that are necessary in controlling cell growth and specialization.

Sources of Omega-3 Fatty Acids	Sources of Omega-6 Fatty Acids
Certain fish oil (salmon, sardines, herring)	Corn oil
	Peanut oil
Flaxseed oil	Cottonseed oil
Soybeans	Soybean oil
Soybean oil	Sesame oil
Walnuts	Safflower oil
Walnut oil	Sunflower oil

component of cell membranes. Cholesterol is necessary for the manufacture of vitamin D, which assists in the proper development of bones and teeth. Cholesterol molecules in your skin react with ultraviolet light to produce vitamin D. The body also uses it to make bile acids. These products of the liver are channeled into your intestine to emulsify fats.

Regulating the amount of cholesterol in the body to prevent its negative effects can be difficult, because the body makes it and it is consumed in the diet. Recall that diets high in saturated fats increase the risk for diseases such as atherosclerosis. By watching your diet, you can reduce the amount of cholesterol in your blood serum by about 20 percent, as much as taking a cholesterol-lowering drug. Therefore, it is best to eat foods that are low in cholesterol. Because many foods that claim to be low- or no-cholesterol have high levels of saturated fats, they should also be avoided in order to control serum cholesterol levels.

CONCEPTS APPLIED

Nutrients in Your Pantry

Pick a food product and write down the number of grams of fats, proteins, and carbohydrates per serving according to the information on the label. Multiply the number of grams of proteins and carbohydrates each by 4 Cal/g and the number of grams of fat by 9 Cal/g. Add the total Calories per serving. Does your total agree with the number of Calories per serving given on the label? Also examine the given serving size. Is this a reasonable amount to be consumed at one time, or would you probably eat two or three times this amount? Write the rest of the nutrition information (vitamins, minerals, sodium content, etc.) and then the list of ingredients. Tell what ingredient you think is providing which nutrient. (Example: vegetable oil—source of fat; milk—provides calcium and vitamin A; MSG—source of sodium.)

FIGURE 19.26 (A) Cholesterol is produced by the human body and is found in your cells' membranes. (B) Testosterone increases during puberty, causing the male sex organs to mature. (C) Progesterone is a female sex hormone produced by the ovaries and placenta. Notice the slight structural differences among these molecules.

A Closer Look

Fat and Your Diet

When triglycerides are eaten in fat-containing foods, digestive enzymes hydrolyze them into glycerol and fatty acids. These molecules are absorbed by the intestinal tract and coated with protein to form lipoprotein, as shown in the accompanying diagram. The combination allows the fat to dissolve better in the blood so that it can move throughout the body in the circulatory system.

Five types of lipoproteins found in the body are:

1. Chylomicrons
2. Very-low-density lipoproteins (VLDLs)
3. Low-density lipoproteins (LDLs)
4. High-density lipoproteins (HDLs)
5. Lipoprotein a [Lp(a)]

Chylomicrons are very large particles formed in the intestine; they are 80–95 percent triglycerides. As the chylomicrons circulate through the body, cells remove the triglycerides in order to make sex hormones, store energy, and build new cell parts. When most of the triglycerides have been removed, the remaining portions of the chylomicrons are harmlessly destroyed.

The VLDLs and LDLs are formed in the liver. VLDLs contain all types of lipid, protein, and 10–15 percent cholesterol, whereas the LDLs are about 50 percent cholesterol. As with the chylomicrons, the body uses these molecules for the fats they contain. However, in some people, high levels of LDL and lipoprotein a [Lp(a)] in the blood are associated with atherosclerosis, stroke, and heart attack. It appears that saturated fat disrupts the clearance of LDLs from the bloodstream. Thus, while in the blood, LDLs may stick to the insides of the vessels, forming deposits, which restrict blood flow and contribute to high blood pressure, strokes, and heart attacks. Even though they are 30 percent cholesterol, a high level of HDLs (made in the intestine), compared with LDLs and [Lp(a)], is associated with a lower

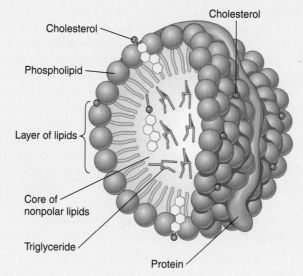

Box Figure 19.5 Diagram of a lipoprotein.

risk for atherosclerosis. One way to reduce the risk of this disease is to lower your intake of LDLs and [Lp(a)]. This can be done by reducing your consumption of saturated fats. An easy way to remember the association between LDLs and HDLs is "L = Lethal" and "H =Healthy" or "Low = Bad" and "High = Good." The federal government's cholesterol guidelines recommend that all adults get a full lipoprotein profile (total cholesterol, HDL, LDL, and triglycerides) once every five years. They also recommend a sliding scale for desirable LDL levels; however, recent studies suggest that one's LDL level should be as low as possible.

Taking certain drugs is one way to control the level of lipoproteins. Statins are a group of medicines (e.g., simvastatin, atorvastatin) that work by blocking the action of enzymes that control the rate of cholesterol production in the body. Their use can lower cholesterol 20-60 percent. They also increase the liver's ability to remove low-density *lipoproteins*. An additional benefit is a slight increase in high-density lipoproteins and a decrease in triglycerides.

Total cholesterol goal values:

- 75–169 mg/dL (milligram per deciliter) for those age twenty and younger
- 100–199 mg/dL for those over age twenty-one

Low density lipoprotein (LDL) goal values:

- Less than 70 mg/dL for those with heart or blood vessel disease and for other patients at very high risk of heart disease (those with metabolic syndrome)
- Less than 100 mg/dL for high-risk patients (for example, some patients who have diabetes or multiple heart disease risk factors)
- Less than 130 mg/dL otherwise

Very low density lipoprotein (VLDL) goal values:

- Less than 40 mg/dL otherwise

High density lipoprotein (HDL) goal value:

- Greater than 45 mg/dL (the higher the better)

Triglyceride goal value:

- Less than 150 mg/dL

SUMMARY

Organic chemistry is the study of compounds that have carbon as the principal element. Such compounds are called *organic compounds,* and all the rest are *inorganic compounds.* There are millions of organic compounds because a carbon atom can link with other carbon atoms as well as atoms of other elements.

A *hydrocarbon* is an organic compound consisting of hydrogen and carbon atoms. The simplest hydrocarbon is one carbon atom and four hydrogen atoms, or CH_4. All hydrocarbons larger than CH_4 have one or more carbon atoms bonded to another carbon atom. The bond can be single, double, or triple, and may be found as chains or rings of carbon.

One category of hydrocarbons has single carbon-to-carbon bonds, a second category has a double carbon-to-carbon bond, and the third has a triple carbon-to-carbon bond. These three categories can have straight- or branched-chain molecules. When the number of carbon atoms is greater than three, different arrangements can occur for a particular number of carbon atoms. Molecules with different arrangements with the same molecular formula are called *isomers.* Isomers have different physical properties, so each isomer is given its own name.

Hydrocarbons that have all the hydrogen atoms possible are *saturated* hydrocarbons. Those that can add more hydrogen to the molecule are called *unsaturated* hydrocarbons. Unsaturated hydrocarbons are more chemically reactive than saturated molecules.

Hydrocarbons that occur in a ring or cycle structure are cyclohydrocarbons. A six-carbon cyclohydrocarbon with three double bonds has different properties than the other cyclohydrocarbons because the double bonds are not localized. This six-carbon molecule is *benzene,* the basic unit of the *aromatic hydrocarbons.*

Petroleum is a mixture of many kinds of hydrocarbons that formed from the slow decomposition of buried marine plankton and algae. Petroleum from the ground, or *crude oil,* is distilled into petroleum products of *natural gas, LPG, petroleum ether, gasoline, kerosene, diesel fuel,* and *motor oils.* Each group contains a range of hydrocarbons and is processed according to use.

In addition to oxidation, hydrocarbons react by *substitution, addition,* and *polymerization* reactions. Reactions take place at sites of multiple bonds or lone pairs of electrons on the *functional groups.* The functional group determines the chemical properties of organic compounds. Changes in functional groups result in the *hydrocarbon derivatives* of *alcohols, ethers, aldehydes, ketones, organic acids, esters,* and *amines.*

Polymers occur naturally in plants and animals, and *synthetic polymers* are made today from variations of the ethylene-derived monomers. Among the more widely used synthetic polymers derived from ethylene are polyethylene, polyvinyl chloride, polystyrene, and Teflon. Problems with the synthetic polymers include that (1) they are manufactured from fossil fuels that are also used as the primary energy supply and (2) they do not readily decompose and tend to accumulate in the environment.

Living organisms have an incredible number of highly organized chemical reactions that are catalyzed by *enzymes,* using food and energy to grow and reproduce. These biochemical processes involve building large *macromolecules* such as proteins, carbohydrates, and lipids.

Carbohydrates are a class of organic molecules composed of CHO. The *monosaccharides* are simple sugars such as *glucose* and *fructose.* Glucose is *blood sugar,* a source of energy. The disaccharides are *sucrose* (table sugar), *lactose* (milk sugar), and *maltose* (malt sugar). The polysaccharides are polymers or glucose in straight or branched chains used as a near-term source of stored energy. Plants store the energy in the form of *starch,* and animals store it in the form of *glycogen. Cellulose* is a polymer similar to starch that humans cannot digest.

Proteins are macromolecular polymers of *amino acids.* There are twenty amino acids that are used in various polymer combinations to build structural, carrier, and functional proteins.

Nucleic acids are polymers composed of nucleotide units. Two forms are recognized, DNA and RNA. In most organisms, DNA serves as the genetic material, while RNA plays crucial roles in the synthesis of proteins.

Lipids are a fourth important class of biochemicals. *True* or *neutral fats* and oils belong in the category known as *lipids* and are esters formed from three fatty acids and glycerol into a *triglyceride.* True fats are usually solid triglycerides associated with animals, and *oils* are liquid triglycerides associated with plant life, but both represent a high-energy storage material. The other two subgroups are the *phospholipids* used in cell membranes, and *steroids,* which primarily serve as hormones.

KEY TERMS

amino acid (p. **451**)
antibody (p. **451**)
carbohydrates (p. **449**)
carrier proteins (p. **453**)
cellulose (p. **451**)
chromosomes (p. **456**)
complex carbohydrate (p. **449**)
denatured (p. **453**)
deoxyribonucleic acid (DNA) (p. **455**)
fats (p. **456**)
fatty acids (p. **456**)

functional group (p. **441**)
gene (p. **456**)
glycerol (p. **456**)
glycogen (p. **451**)
hormone (p. **451**)
hydrocarbon (p. **435**)
hydrocarbon derivatives (p. **441**)
inorganic chemistry (p. **435**)
isomers (p. **437**)
lipid (p. **456**)
macromolecule (p. **449**)

messenger RNA (mRNA) (p. **456**)
monomer (p. **441**)
nucleic acid (p. **454**)
nucleotide (p. **454**)
oil field (p. **439**)
organic acids (p. **444**)
organic chemistry (p. **435**)
petroleum (p. **439**)
phospholipid (p. **457**)
polymer (p. **441**)
polypeptide chain (p. **452**)

proteins (p. **451**)
regulator proteins (p. **453**)
ribonucleic acid (RNA) (p. **456**)
ribosomal RNA (rRNA) (p. **456**)
saturated (p. **438**)
starch (p. **451**)
steroids (p. **458**)
structural proteins (p. **453**)
transfer RNA (tRNA) (p. **456**)
triglyceride (p. **457**)
true (neutral) fats (p. **456**)
unsaturated (p. **438**)

APPLYING THE CONCEPTS

Part I

1. All organic compounds
   a. contain carbon and were formed only by a living organism.
   b. are natural compounds that have not been synthesized.
   c. contain carbon, no matter if they were formed by a living thing or not.
   d. were formed by a plant.

2. There are millions of organic compounds but only thousands of inorganic compounds because
   a. organic compounds were formed by living things.
   b. there is more carbon on Earth's surface than any other element.
   c. atoms of elements other than carbon never combine with themselves.
   d. carbon atoms can combine with up to four other atoms, including other carbon atoms.

3. A hydrocarbon molecule with 4 carbon atoms would be called saturated if it had how many hydrogen atoms?
   a. 4   b. 8
   c. 10   d. 16

4. Isomers are compounds with the same
   a. molecular formula with different structures.
   b. molecular formula with different atomic masses.
   c. atoms but different molecular formulas.
   d. structures but different formulas.

5. The hydrocarbons with a double covalent carbon-carbon bond are called
   a. a polymer.
   b. unsaturated.
   c. petroleum.
   d. None of the above is correct.

6. Petroleum is believed to have formed mostly from the anaerobic decomposition of buried
   a. dinosaurs.
   b. fish.
   c. pine trees.
   d. plankton and algae.

7. Ethylene molecules can add to each other in a reaction to form a long chain called a
   a. monomer.
   b. dimer.
   c. trimer.
   d. polymer.

8. Chemical reactions usually take place on an organic compound at the site of a
   a. double bond.
   b. lone pair of electrons.
   c. functional group.
   d. Any of the above is correct.

9. The R in ROH represents
   a. a functional group.
   b. a hydrocarbon group with a name ending in "-yl."
   c. an atom of an inorganic element.
   d. a polyatomic ion that does not contain carbon.

Part II

10. A protein is a polymer formed from the linking of many
    a. glucose units.
    b. DNA molecules.
    c. amino acid molecules.
    d. monosaccharides.

11. Which of the following is *not* converted to blood sugar by the human body?
    a. lactose
    b. dextrose
    c. cellulose
    d. glycogen

12. Many synthetic polymers become a problem in the environment because they
    a. decompose to nutrients, which accelerates plant growth.
    b. do not readily decompose and tend to accumulate.
    c. do not contain vitamins as natural materials do.
    d. become a source of food for fish, but ruin the flavor of fish meat.

13. What role is played by tRNA in cells?
    a. telling the cell which protein to make.
    b. transferring a specific amino acid to a ribosome for incorporation into a protein.
    c. transferring a copy of the DNA gene to a ribosome for protein synthesis.
    d. terminating protein synthesis.

14. What is the job of a chaperone protein?
    a. Transferring an amino acid to the ribosome for protein synthesis.
    b. Making sure proteins are synthesized.
    c. Helping other proteins fold into their proper shape.
    d. Keeping mutant proteins out of trouble.

15. The basic building block (monomer) of a proteins is a(an)
    a. amino acid.
    b. monosaccharide.
    c. fatty acid.
    d. Nucleotide.

Answers

1. c 2. d 3. c 4. a 5. b 6. d 7. d 8. d 9. b 10. c 11. c 12. b 13. b 14. c 15. a

QUESTIONS FOR THOUGHT

Part I

1. What is an organic compound?
2. What features allow organic molecules to form millions of compounds?
3. Is it possible to have an isomer of ethane? Explain.
4. Suggest a reason that ethylene is an important raw material used in the production of plastics but ethane is not.
5. Give an example of each of the following classes of organic molecules: carbohydrates, proteins, nucleic acids, lipids.
6. What are (a) natural gas, (b) LPG, and (c) diesel fuel?
7. What does the octane number of gasoline describe? On what is the number based?
8. What is a polymer? Give an example of a naturally occurring plant polymer. Give an example of a synthetic polymer.

Part II

9. What is a functional group? List four examples.
10. Draw a structural formula for the fatty acid linoleic acid. What feature makes this a saturated fat?
11. A soft drink is advertised to "contain no sugar." The label lists ingredients of carbonated water, dextrose, corn syrup, fructose, and flavorings. Evaluate the advertising.
12. What are the three subgroups of lipids? What are the differences between saturated and unsaturated fats?
13. What are the key differences between DNA and RNA?
14. List the main roles played by proteins in a cell.
15. What is meant by the term 'denaturation'?

FOR FURTHER ANALYSIS

1. Many people feel that by using a higher-octane gasoline in their vehicles, they will get better performance. What is the difference between high-octane and regular gasoline? Does scientific evidence bear this out? What octane gasoline is recommended for your vehicle?
2. There have been some health concerns about the additives used in gasoline. What are these additives? What purpose do they serve? What do opponents of using such additives propose are their negative impact?
3. The so-called "birth control pill," or "the pill," has been around since the early 1960s. This medication is composed of a variety of organic molecules. What is the nature of these compounds? How do they work to control conception since they do not control birth? What are some of the negative side effects and what are some of the related benefits of taking "the pill"?
4. Many communities throughout the world are involved in recycling. One of the most important classes of materials recycled is plastic. There has been concern that the cost of recycling plastics is higher than the cost of making new plastic products. Go to the Internet and
   a. Identify the kinds of plastics that are commonly recycled.
   b. Explain how a plastic such as HDPE is recycled.
   c. Present an argument for eliminating the recycling of HDPE and for expanding the recycling of HDPE.

INVITATION TO INQUIRY

Alcohols: What Do You Really Know?

Archaeologists, anthropologists, chemists, biologists, and health care professionals agree that the drinking of alcohol dates back thousands of years. Evidence also exists that this practice has occurred in most cultures around the world. Use the Internet to search out answers to the following questions:

1. What is the earliest date for which there is evidence for the production of ethyl alcohol?
2. In which culture did this occur?
3. What is the molecular formula and structure of ethanol?
4. Do alcohol and water mix?
5. How much ethanol is consumed in the form of beverages in the United States each year?
6. What is the legal limit to be considered intoxicated in your state?
7. How is this level measured?
8. Why is there a tax on alcoholic beverages?
9. How do the negative effects of drinking alcohol compare between men and women?
10. Have researchers demonstrated any beneficial effects of drinking alcohol?

Compare what you thought you knew to what is now supported with scientific evidence.

20

The Nature of Living Things

The cell is the simplest structure capable of existing as an individual living unit. Within this unit, many chemical reactions are required for maintaining life. These reactions do not occur at random but are associated with specific parts of the many kinds of cells.

CORE CONCEPT
All living things are composed of cells.

OUTLINE

Physics

▷ Energy in a useful form is essential for all life (Ch. 3).

Chemistry

▷ Molecular movement is required for the transport of atoms and molecules into and out of cells (Ch. 4).

▷ Chemical reactions are the essence of all living things (Ch. 9).

Life Science

▷ The structure of living things is basically the same—a cell (Ch. 20).

▷ The physiological processes of living things are common to all life forms (Ch. 20).

▷ Cell division is required for life to continue (Ch. 20).

▷ All living things have genetic material (Ch. 20).

▷ Many types of diseases are the result of one life form living on another (Ch. 20).

OVERVIEW

What does it mean to be alive? You would think that a science textbook could answer this question easily. However, this is more than just a theoretical question because in recent years, it has been necessary to construct legal definitions of what life is and especially of when it begins and ends. The legal definition of death is important because it may determine whether a person will receive life insurance benefits or if body parts may be used in transplants. In the case of heart transplants, the person donating the heart may be legally "dead," but the heart certainly isn't since it can be removed while it still has "life." In other words, there are different kinds of death. There is the death of the whole living unit and the death of each cell within the living unit. A person actually "dies" before every cell has died. Death, then, is the absence of life, but that still doesn't tell us what life is. At this point, we won't try to define life, but we will describe some of the basic characteristics of living things.

PART I
THE CHARACTERISTICS OF LIFE

This chapter is divided into three parts. The first part focuses on the general structural and functional features displayed by all living things. It is these features that separate them from our nonliving world.

WHAT MAKES SOMETHING ALIVE?

The science of **biology** is, broadly speaking, the study of living things. It draws on chemistry and physics for its foundation and applies these basic physical laws to life processes. Living things have abilities and structures not typically found in things that were never living. The ability to manipulate energy and matter is unique to living things. Developing an understanding of how they modify matter and use energy will help you appreciate how they differ from nonliving objects. Living things show five characteristics that nonliving things do not: (1) metabolic processes, (2) generative processes, (3) responsive processes, (4) control processes, and (5) a unique structural organization (figure 20.1). For something to be "alive," it must display all these characteristics, that is, they must all work together. It is important to recognize that although these characteristics are typical of all living things, they may not all be present in each organism at every point in time. For example, some individuals may reproduce or grow only at certain times. Should a living organism lose one or more of these features for an extended period, it would "die." Something is living when it displays certain unique chemical processes occurring in association with certain unique structures. This section briefly introduces each of these basic characteristics.

1. **Metabolic processes**   All the chemical reactions involving molecules required for a cell to grow, reproduce, and make repairs are referred to as its **metabolism.** Metabolic properties keep a cell alive. The energy that organisms use is stored in the chemical bonds of complex molecules. Even though different kinds of organisms have different ways of metabolizing **nutrients** or food, we are usually talking about three main activities: taking in nutrients, processing them, and eliminating wastes. All living things expend energy to take in nutrients (raw materials) from their environment. Many animals take in these materials by eating or swallowing other organisms. Microorganisms and plants absorb raw materials into their cells to maintain their lives. Once inside, raw materials are used in a series of chemical reactions to manufacture new parts, make repairs, reproduce, and provide energy for essential activities. However, not all the raw materials entering

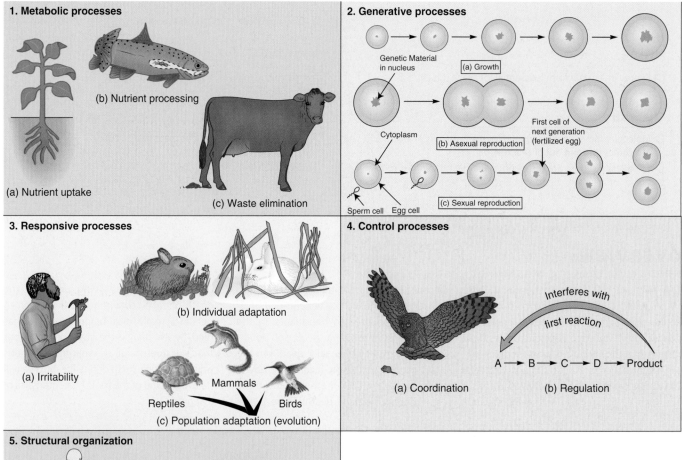

1. Metabolic processes

(a) Nutrient uptake

(b) Nutrient processing

(c) Waste elimination

2. Generative processes

Genetic Material in nucleus

(a) Growth

Cytoplasm

First cell of next generation (fertilized egg)

(b) Asexual reproduction

Sperm cell   Egg cell

(c) Sexual reproduction

3. Responsive processes

(a) Irritability

(b) Individual adaptation

Reptiles   Mammals   Birds

(c) Population adaptation (evolution)

4. Control processes

Interferes with first reaction

A → B → C → D → Product

(a) Coordination

(b) Regulation

5. Structural organization

(a) Organismal organization

(b) Cellular organization

FIGURE 20.1   Living things demonstrate many common characteristics.

a living thing are valuable to it. There may be portions that are useless or even harmful. Organisms eliminate these portions as waste. Metabolic processes also produce unusable heat energy that may also be considered a waste product.

2. The second group of characteristics of life, **generative processes,** are reactions that result in an increase in the size of an individual organism—*growth*—or an increase in the number of individuals in a population of organisms—*reproduction*. During growth, living things add to their structure, repair parts, and store nutrients for later use. Growth and reproduction are directly related to metabolism, since neither can occur without the acquisition and processing of nutrients.

Reproduction of individuals must occur because each organism eventually dies. Thus, reproduction is the only way that living things can perpetuate themselves. There are a number of different ways that various kinds of organisms reproduce and guarantee their continued existence. Some reproductive processes known as sexual reproduction involve two organisms contributing to the creation of a unique, new organism. Asexual (without sex) reproduction occurs when organisms make identical copies of themselves.

3. Organisms also respond to changes within their bodies and in their surroundings in a meaningful way. These **responsive processes** have been organized into three categories: *irritability, individual adaptation,* and *population adaptation* or *evolution.* The term *irritability* is not used in the everyday context of an individual being angry. Here, *irritability* means an individual's rapid response to a stimulus, such as your response to a loud noise, beautiful sunset, or bad smell. This type of response occurs only in the individual receiving the stimulus and is rapid because the mechanism that allows the response to occur (e.g., muscles, bones, and nerves) is already in place. Individual adaptation is also an individual response but is slower since it requires growth or some other fundamental change in an organism. For example, a weasel's fur color will change from its brown summer coat to its white winter coat when genes responsible for the production of brown pigment are "turned off." Or the response of our body to disease organisms requires a change in the way

B

FIGURE 20.9 (B) Ion channels can be opened or closed to allow these sodium ions to be transported to the other side of the membrane. When the signal molecule binds to the ion channel protein, the gate is opened.

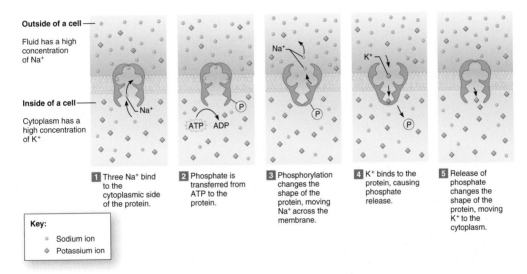

FIGURE 20.10 The action of the carrier protein requires an input of energy (the compound ATP) other than the kinetic energy of the molecules; therefore, this is an *active* (not passive) process termed *active* transport. Active transport mechanisms can transport molecules or ions up a concentration gradient from a low concentration to a higher concentration.

active transport (figure 20.10). Active transport is very specific: Only certain molecules or ions can be moved in this way, and specific proteins in the membrane must carry them. The action of the carrier requires an input of energy other than the kinetic energy of the molecules; therefore, this process is termed *active* transport. For example, some ions, such as sodium and potassium, are actively pumped across cell membranes. Sodium ions are pumped *out of cells* up a concentration gradient. Potassium ions are pumped *into cells* up a concentration gradient.

In addition to active transport, materials can be transported into a cell by *endocytosis* and out by *exocytosis*. **Phagocytosis** is another name for one kind of endocytosis that is

the process cells use to wrap membrane around a particle (usually food) and engulf it (figure 20.11). This is the process leukocytes (white blood cells) use to surround invading bacteria, viruses, and other foreign materials. Because of this, these kinds of cells are called *phagocytes*. When phagocytosis occurs, the material to be engulfed touches the surface of the phagocyte and causes a portion of the outer cell membrane to be indented. The indented cell membrane is pinched off inside the cell to form a sac containing the engulfed material. This sac, composed of a single membrane, is called a **vacuole.** Once inside the cell, the membrane of the vacuole is broken down, releasing its contents inside the cell, or it may combine with another vacuole containing destructive enzymes.

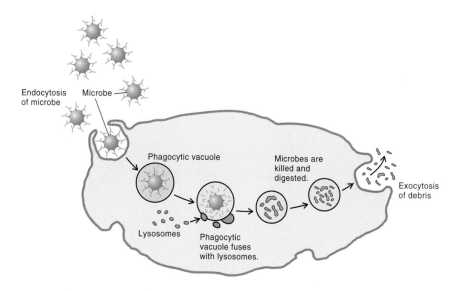

FIGURE 20.11 The sequence illustrates a cell engulfing (endocytosis) a microbe and surrounding it with a membrane. Once encased in a portion of the cell membrane (now called a phagocytic vacuole), a lysosome adds its digestive enzymes to it, which speeds the breakdown of the dangerous microbes. Finally, the hydrolyzed (digested) material moves from the vacuole to the inner surface of the cell membrane, where the contents are discharged (exocytosis).

ORGANELLES COMPOSED OF MEMBRANES

Now that you have some background concerning the structure and the function of membranes, let's turn our attention to the way cells use membranes to build the structural components of their protoplasm. The outer boundary of the cell is termed the cell membrane, or **plasma membrane.** It is associated with one of the characteristics of life, metabolism, including taking up and releasing molecules, sensing stimuli in the environment, recognizing other cell types, and attaching to other cells and nonliving objects. In addition to the cell membrane, many other organelles are composed of membranes. Each of these membranous organelles has a unique shape or structure that is associated with particular functions.

One of the most common organelles found in cells is the *endoplasmic reticulum.* The **endoplasmic reticulum** (ER) is a set of folded membranes and tubes throughout the cell. This system of membranes provides a large surface upon which chemical activities take place (figure 20.12). Since the ER has an enormous surface area, many chemical reactions can be carried out in an extremely small space.

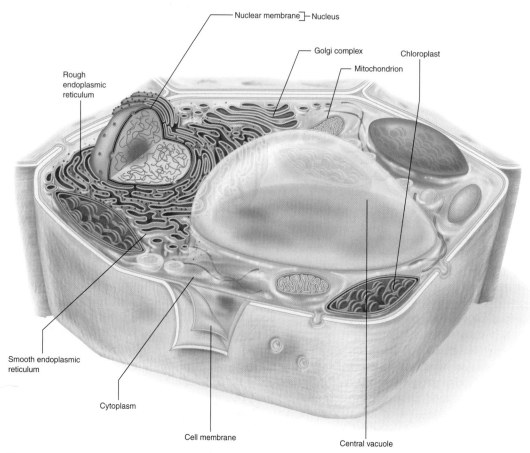

FIGURE 20.12 Certain structures in the cytoplasm are constructed of membranes. Membranes are composed of protein and phospholipids. The membranous cytoplasmic organelles shown are in a typical plant cell.

The **Golgi apparatus** is another organelle composed of membrane. Even though this organelle is also composed of membrane, the way in which it is structured enables it to perform jobs different from those performed by the ER. The typical Golgi is composed of from five to twenty flattened, smooth, membranous sacs, which resemble a stack of pancakes. The Golgi apparatus is the site of the synthesis and packaging of certain molecules produced in the cell. It is also the place where particular chemicals are concentrated prior to their release from the cell or distribution within the cell.

An important group of molecules that is necessary to the cell includes the hydrolytic enzymes. This group of enzymes is capable of destroying carbohydrates, nucleic acids, proteins, and lipids. Since cells contain large amounts of these molecules, these enzymes must be controlled in order to prevent the destruction of the cell. The Golgi apparatus is the site where these enzymes are converted from their inactive to their active forms and packaged in protective membranous sacs. These vesicles are pinched off from the outside surfaces of the Golgi sacs and given the special name **lysosomes,** or "bursting body." Cells use the lysosomes in four major ways:

1. Decompose dead and dying cells.
2. Selectively destroy cells to fashion a developing organism from its immature to mature state (figure 20.13).
3. Digest macromolecules that have been ingested into the cell.
4. Kill dangerous microorganisms that have been taken into the cell by phagocytosis.

A nucleus is a place in a cell—not a solid mass. Just as a room is a place created by walls, a floor, and a ceiling, the nucleus is a place in the cell created by the **nuclear membrane.** This membrane separates the *nucleoplasm,* liquid material in

FIGURE 20.13   This person displays the trait known as syndactylism (*syn* = connected; *dactyl* = finger or toe). In most people, lysosomal enzymes break down the cells between the toes, allowing the toes to separate. In this genetic abnormality, these enzymes fail to do their job.

the nucleus, from the cytoplasm. Because they are separated, the cytoplasm and nucleoplasm can maintain different chemical compositions. If the membrane were not formed around the genetic material, the organelle we call the nucleus would not exist. The nuclear membrane is formed from many flattened sacs fashioned into a hollow sphere around the genetic material, DNA (*d*eoxyribo*n*ucleic *a*cid). It also has large openings called nuclear pores that allow thousands of relatively large molecules to pass into and out of the nucleus each minute. These pores are held open by donut-shaped molecules that resemble the "eyes" in shoes through which the shoelace is strung.

All of the membranous organelles just described can be con-verted from one form to another (figure 20.14). For example,

rough ER
synthesizes proteins and packages them in vesicles.

transport vesicles
from rough ER

Golgi apparatus
modifies lipids and proteins; sorts them and packages them in vesicles.

secretory vesicles
fuses with the plasma membrane as secretion occurs.

smooth ER
synthesizes lipids and performs other functions.

transport vesicles
from smooth ER

lysosomes
digest molecules or old cell parts.

incoming vesicle
bring substances into the cell.

FIGURE 20.14   Eukaryotic cells contain a variety of organelles composed of membranes that consist of two layers of phospholipids and associated proteins. Each organelle has a unique shape and function. Many of these organelles are interconverted from one to another as they perform their essential functions.

phagocytosis results in the formation of vacuolar membrane from cell membrane that fuses with lysosomal membrane, which in turn came from Golgi membrane. Two other organelles composed of membranes are chemically different and are incapable of interconversion. Both types of organelles are associated with energy conversion reactions in the cell. These organelles are the *mitochondrion* and the *chloroplast* (figure 20.15).

The **mitochondrion** is an organelle resembling a small bag with a larger bag inside that is folded back on itself. These inner folded surfaces are known as the *cristae*. Located on the surface of the cristae are particular proteins and enzymes involved in *aerobic cellular respiration*. **Aerobic cellular respiration** is the series of reactions involved in the release of usable energy from food molecules, which requires the participation of oxygen molecules. The average human cell contains upwards of ten thousand mitochondria. Cells that are involved in activities that require large amounts of energy, such as muscle cells, contain many more mitochondria. When properly stained, they can be seen with a compound light microscope. In cells that are functioning aerobically, the mitochondria swell with activity. But when this activity diminishes, they shrink and appear as thread-like structures.

A second energy-converting organelle is the **chloroplast.** Some cells contain only one large chloroplast, while others contain hundreds of smaller chloroplasts. A study of the ultrastructure—that is, the structures seen with an electron microscope—of a chloroplast shows that the entire organelle is enclosed by a membrane, while other membranes are folded and interwoven throughout. As shown in figure 20.15A, in some areas, concentrations of these membranes are stacked up or folded back on themselves. Chlorophyll and other photosynthetic molecules are attached to these membranes. These areas of concentrated chlorophyll are called the *grana* of the chloroplast. The space between the grana, which has no chlorophyll, is known as the *stroma*.

These membranous, saclike organelles are found only in plants and algae. In this organelle, light energy is converted to chemical-bond energy in a process known as *photosynthesis*. **Photosynthesis** is the *trapping of radiant energy and its conversion into the energy of chemical bonds*. Chloroplasts contain a variety of photosynthetic pigments including green *chlorophyll*.

Mitochondria and chloroplasts are different from other kinds of membranous structures in several ways. First, their

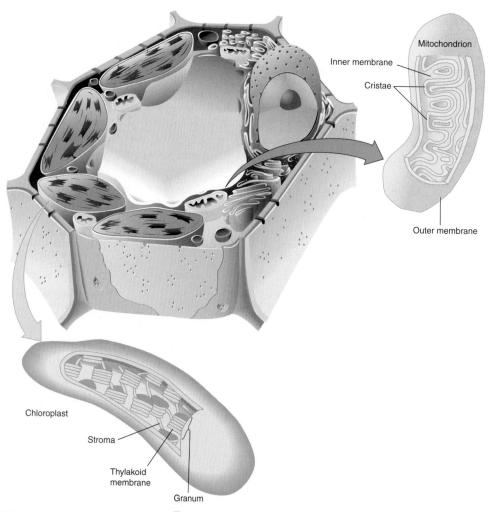

FIGURE 20.15 (*A*) The chloroplast, the container of the pigment chlorophyll, is the site of photosynthesis. The chlorophyll, located in the grana, captures light energy that is used to construct organic molecules in the stroma. (*B*) The mitochondria with their inner folds, called cristae, are the site of aerobic cellular respiration, where food energy is converted to usable cellular energy. Both organelles are composed of phospholipid and protein membranes.

membranes are chemically different from those of other membranous organelles. Second, they are composed of double layers of membrane—an inner and an outer membrane. Third, both of these structures have ribosomes and DNA that are similar to those of bacteria. Finally, these two structures have a certain degree of independence from the rest of the cell—they have a limited ability to reproduce themselves but must rely on DNA from the cell's nucleus for assistance.

The biochemical pathways associated with mitochondria and chloroplasts are respiration and photosynthesis, both oxidation-reduction reactions.

NONMEMBRANOUS ORGANELLES

Suspended in the cytoplasm and associated with the membranous organelles are various kinds of structures that are not composed of phospholipids and proteins arranged in sheets. These are nonmembranous organelles.

In the cytoplasm are many very small structures called **ribosomes** that are composed of ribonucleic acid (RNA) and protein. Ribosomes function in the manufacture of protein. Many ribosomes are found floating freely in the cytoplasm and attached to the endoplasmic reticulum (figure 20.16). Cells that are actively producing protein (e.g., liver cells) have great numbers of free and attached ribosomes.

Among the many types of nonmembranous organelles found there are elongated protein structures that are known as *microtubules, microfilaments,* and *intermediate filaments.* Their various functions are as complex as those provided by the structural framework and cables of a high-rise office building, geodesic dome, or skeletal and muscular systems of a large animal. All three types of organelles interconnect, and some

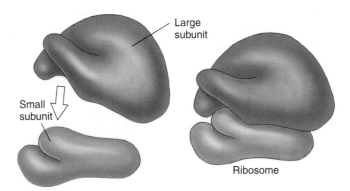

FIGURE 20.16 Each ribosome is constructed of two subunits. Each of the subunits is composed of protein and RNA. These globular organelles are associated with the construction of protein molecules from individual amino acids. The Nobel Prize in Chemistry 2009 was awarded to Drs. Venkatraman Ramakrishan, Thomas A. Steitz, and Ada E. Yonath for determining the structure and function of the ribosomes.

are attached to the inside of the cell membrane, forming what is known as the *cytoskeleton* of the cell (figure 20.17). These cellular components provide the cell with shape, support, and ability to move about the environment. In certain diseases such as ALS (amyotrophic lateral sclerosis, or Lou Gehrig's disease), Alzheimer's, and Down syndrome, there is a buildup of intermediate-type filaments. As a result, researchers are investigating control methods that prevent such buildup.

An arrangement of two sets of microtubules at right angles to each other makes up the structures known as the *centrioles,* and they operate by organizing microtubules into a complex of strings called *spindle fibers.* The *spindle* is the structure upon which chromosomes are attached in order that they may be properly separated during cell division (refer to figure 20.20).

FIGURE 20.17 Microtubules, actin filaments, and intermediate filaments are all interconnected within the cytoplasm of the cell. These structures, along with connections to other cellular organelles, form a cytoskeleton for the cell. The cellular skeleton is not a rigid, fixed-in-place structure but rather changes as the actin and intermediate filaments and microtubule component parts are assembled and disassembled.

Each set is composed of nine groups of short microtubules arranged in a cylinder (figure 20.18). One curious fact about centrioles is that they are present in most animal cells but not in many types of plant cells.

Many cells have microscopic, hairlike structures projecting from their surfaces; these are *cilia* or *flagella* (figure 20.19). In general, we call them flagella if they are long and few in number, and cilia if they are short and more numerous. The cell has the ability to control the action of these microtubular structures, enabling them to be moved in a variety of different ways. Their coordinated actions either propel the cell through the environment or the environment past the cell surface. The flagella of sperm propel them through the reproductive tract toward the egg. The protozoan *Paramecium* is covered with thousands of cilia that actively beat in a rhythmic motion to move the cell through the water. The cilia on the cells that line your trachea move mucus-containing particles from deep within your lungs.

NUCLEAR COMPONENTS

When nuclear structures were first identified, it was noted that certain dyes stained some parts more than others. The parts that stained more heavily were called **chromatin,** which means "colored material." Chromatin is composed of long molecules of deoxyribonucleic acid (DNA) in association with proteins. Chromatin is loosely organized DNA/protein strands in the nucleus. When the chromatin is tightly coiled into shorter, denser structures, we call them **chromosomes.** Chromatin and chromosomes are really the same molecules but differ in structural arrangement. In addition to chromosomes, the nucleus may also contain one, two, or several *nucleoli.* A **nucleolus** is composed of granules and fibers in association with the cell's DNA used in the manufacture of ribosomes.

The final component of the nucleus is its liquid matrix called the *nucleoplasm.* It is a colloidal mixture composed of water and the molecules used in the construction of ribosomes, nucleic acids, and other nuclear material.

MAJOR CELL TYPES

Not all of the cellular organelles we have just described are located in every cell. Some cells typically have combinations of organelles that differ from others. For example, some cells have a nuclear membrane, mitochondria, chloroplasts, ER, and Golgi; others have mitochondria, centrioles, Golgi, ER, and nuclear membrane. Other cells are even more simple and lack the complex membranous organelles described in this chapter. Because of this fact, biologists have been able to classify cells into two major types: prokaryotic and eukaryotic.

The Prokaryotic Cell Structure

Prokaryotic cells, the *bacteria* and *archaea,* do not have a typical nucleus bound by a nuclear membrane, nor do they contain mitochondria, chloroplasts, Golgi, or extensive networks of ER. However, prokaryotic cells contain DNA and enzymes and are able to reproduce and engage in metabolism. They perform all of the basic functions of living things with fewer and more

Microtubule triplet

FIGURE 20.18 These two sets of short microtubules are located just outside the nuclear membrane in many types of cells.

Outer microtubule pair

Plasma membrane

Central microtubule pair

Microtubules

Flagellum

Cilium

Cilia on surface

FIGURE 20.19 Cilia and flagella function like oars or propellers that move the cell through its environment or move the environment past the cell. Cilia and flagella are constructed of groups of microtubules, as in the ciliated protozoan shown on the left and the flagellated alga on the right. Flagella are usually less numerous and longer than cilia.

Antibiotics and Cell Structural Differences

One significant difference between prokaryotic and eukaryotic cells is in the chemical makeup of their ribosomes. The ribosomes of prokaryotic cells contain different proteins from those found in eukaryotic cells. Prokaryotic ribosomes are also smaller. This discovery was important to medicine because many cellular forms of life that cause common diseases are bacterial. As soon as differences in the ribosomes were noted, researchers began to look for ways in which to interfere with the prokaryotic ribosome's function but *not* interfere with the ribosomes of eukaryotic cells. Antibiotics such as streptomycin are the result of this research. This drug combines with prokaryotic ribosomes and causes the death of the prokaryote by preventing the production of proteins essential to its survival. Because eukaryotic ribosomes differ from prokaryotic ribosomes, streptomycin does not interfere with the normal function of ribosomes in human cells.

---

simple organelles. Members of the Archaea are of little concern to the medical profession because none has been identified as disease-causing. They are typically found growing in extreme environments where the pH, salt concentration, or temperatures make it impossible for most other organisms to survive. The other prokaryotic cells are called bacteria, and about 5 percent cause diseases such as tuberculosis, strep throat, gonorrhea, and

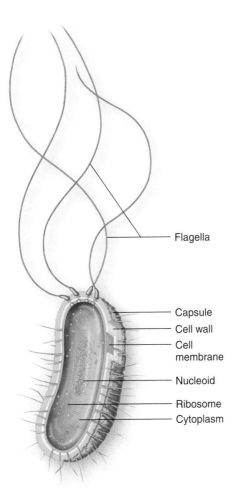

— Flagella

— Capsule
— Cell wall
— Cell membrane
— Nucleoid
— Ribosome
— Cytoplasm

FIGURE 20.20 All bacteria are prokaryotic cells. While smaller and less complex than eukaryotic cells, each is capable of surviving on its own. Most bacteria are involved with decay and decomposition, but a small percentage are pathogens responsible for disease such as strep throat, TB, syphilis, and gas gangrene. The cell illustrated here is a bacillus because it has a rod shape.

acne. Other bacteria are responsible for the decay and decomposition of dead organisms. Although some bacteria have a type of green photosynthetic pigment and carry on photosynthesis, they do so without chloroplasts and use different chemical reactions.

Most prokaryotic cells are surrounded by a *capsule,* slime layer, or spore coat that can be composed of a variety of compounds (figure 20.20). In certain bacteria, this layer is responsible for their ability to stick to surfaces (including host cells) and to resist phagocytosis. Many bacteria also have fimbriae, hairlike protein structures, which help the cell stick to objects. Those with flagella are capable of propelling themselves through the environment. Below the capsule is the rigid cell wall composed of a unique protein-carbohydrate complex called peptidoglycan. This complex provides the cell with the strength to resist osmotic pressure changes and gives the cell shape. Just beneath the wall is the cell membrane. Thinner and with a slightly different chemical composition from eukaryotes, it carries out the same functions as the cell membranes in eukaryotes. Most bacteria are either rods (bacilli), spherical (cocci), or curved (spirilla). The genetic material within the cytoplasm is DNA in the form of a loop.

The Eukaryotic Cell Structure

Eukaryotic cells contain a true nucleus and most of the membranous organelles described earlier. Eukaryotic organisms can be further divided into several categories based on the specific combination of organelles they contain. The cells of plants, fungi, protozoa and algae, and animals are all eukaryotic. The most obvious characteristic that sets the plants and algae apart from other organisms is their green color, which indicates that the cells contain chlorophyll. Chlorophyll is necessary for the process of photosynthesis—the conversion of light energy into chemical-bond energy in food molecules. These cells, then, are different from the other cells in that they contain chloroplasts in their cytoplasm. Another distinguishing characteristic of plants and algae is the presence of cellulose in their cell walls (table 20.2).

The group of organisms that has a cell wall but lacks chlorophyll in chloroplasts is collectively known as *fungi.* They were previously thought to be either plants that had lost their ability to make their own food or animals that had developed cell walls. Organisms that belong in this category of eukaryotic cells include yeasts, molds, mushrooms, and the fungi that cause such human diseases as athlete's foot, "jungle rot," and ringworm. Now we have come to recognize this group as different enough from plants and animals to place them in a separate kingdom.

Eukaryotic organisms that lack cell walls and cannot photosynthesize are placed in separate groups. Organisms that consist of only one cell are called protozoans—examples are *Amoeba* and *Paramecium*. They have all the cellular organelles described in this chapter except the chloroplast; therefore, protozoans must consume food as do the fungi and the multicellular animals.

Although the differences in these groups of organisms may seem to set them worlds apart, their similarity in cellular structure is one of the central themes unifying the field of biology. You can obtain a better understanding of how cells operate in general by studying specific examples. Because the organelles have the same general structure and function regardless of the kind of cell in which they are found, you can learn more about how mitochondria function in plants by studying how mitochondria function in animals. There is a commonality among all living things with regard to their cellular structure and function.

TABLE 20.2 Comparison of General Plant and Animal Cell Structure

Plant Cells	Animal Cells
CELL WALL	————————
Cell membrane	Cell membrane
Cytoplasm	Cytoplasm
Nucleus	Nucleus
Mitochondria	Mitochondria
CENTRAL VACUOLE	
CHLOROPLASTS	————————
	CENTRIOLE
————————	
Golgi apparatus	Golgi apparatus
Endoplasmic reticulum	Endoplasmic reticulum
Lysosomes	Lysosomes
Vacuoles/vesicles	Vacuoles/vesicles
Ribosomes	Ribosomes
Nucleolus	Nucleolus
Inclusions	Inclusions
Cytoskeleton	Cytoskeleton

PART II
ENERGY TRANSFORMATIONS IN CELLS

This part of the chapter explains two of the most fundamental biochemical processes performed by organisms. These processes are the energy transformation functions that keep all living things alive.

RESPIRATION AND PHOTOSYNTHESIS

In living things, oxidation-reduction reactions are metabolic processes that do not take place in one single step but occur in a series of small steps. This allows the potential energy in the oxidized molecule to be released in smaller, more useful amounts. A unique protein catalyst (enzyme) controls each of the steps. Each step begins with a molecule or molecules that serve as the reactant and are called the *substrate*. When the reaction is complete, the substrate is converted to a product, which in turn becomes the new substrate for the next enzyme-controlled reaction. Such a series of enzyme-controlled reactions is often called a *biochemical pathway* or a *metabolic pathway*:

Enzyme 1   Enzyme 2   Enzyme 3   Enzyme 4
Substrate A → Substrate B → Substrate C → Substrate D → End product

Such pathways can be used to create molecules (e.g., photosynthesis), release energy (e.g., aerobic cellular respiration), and perform many other actions. One of the amazing facts of nature is that most organisms use the same basic biochemical or metabolic pathways. So, if you study aerobic cellular respiration in an elephant, it will be essentially the same as that in a petunia, shark, or earthworm. However, since the kinds of enzymes an organism

is able to produce depend on the genes that it has, we should expect some variation in the details of biochemical pathways in different organisms. The fact that so many organisms use essentially the same biochemical processes is a strong argument for the idea of evolution from common ancestors. Once a successful biochemical strategy evolved, the genes and the pathway were retained by evolutionary descendants with slight modifications of the scheme. Two such pathways of importance are aerobic cellular respiration and photosynthesis, both of which involve the transfer of chemical bond energy in the form of ATP.

The Energy Transfer Molecules of Living Things—ATP

To transfer the right amount of chemical-bond energy from energy-releasing to energy-requiring reactions, cells use the molecule ATP. **Adenosine *triphosphate* (ATP)** is a handy source of the right amount of usable chemical-bond energy. Each ATP molecule used in the cell is like a rechargeable AAA battery used to power small toys and electronic equipment. Each contains just the right amount of energy to power the job. When the power has been drained, it can be recharged numerous times before it must be recycled.

Recharging the AAA battery requires getting a small amount of energy from a source of high energy such as a hydro-electric power plant (figure 20.21). Energy from the electric plant is too powerful to directly run a small flashlight or cell phone. If you plug your cell phone directly into the power plant, the cell phone would be destroyed. However, the recharged AAA battery delivers just the right amount of energy at the right time and place. ATP functions in much the same manner. After the chemical-bond energy has been drained by breaking one of its bonds:

FIGURE 20.21 When rechargeable batteries in a flashlight have been drained of their power, they can be recharged by placing them in a specially designed battery charger. This enables the right amount of power from a power plant to be packed into the batteries for reuse. Cells operate in much the same manner. When the cell's "batteries," ATPs, become drained as a result of powering a job such as muscle contraction, these discharged "batteries," ADPs, can be recharged back to full ATP power.

Used to power
chemical reactions

ATP → ADP + P + Energy

Lost as heat to
the environment

the discharged molecule (ADP) is recharged by "plugging it in" to a high-powered energy source. This source may be (1) sunlight (photosynthesis) or (2) chemical-bond energy (released from cellular respiration):

Sunlight (photosynthesis)

Energy + ADP + P → ATP

Chemical-bond energy (cellular respiration)

An ATP molecule is formed from adenine (nitrogenous base), ribose (sugar), and phosphates (figure 20.22). These three are chemically bonded to form AMP, *a*denosine *mono*phosphate (one phosphate). When a second phosphate group is added to the AMP, a molecule of ADP (diphosphate) is formed. The ADP, with the addition of more energy, is able to bond to a third phosphate group and form ATP. The covalent bond that attaches the second phosphate to the AMP molecule is easily broken to release energy for energy-requiring cell processes. Because the energy in this bond is so easy for a cell to use, it is called a *high-energy phosphate bond.* ATP has two high-energy phosphate bonds represented by curved solid lines. Both ADP and ATP, because they contain high-energy bonds, are very unstable molecules and readily lose their phosphates. When this occurs, the energy held in the high-energy bonds of the phosphate can be transferred to another molecule or

FIGURE 20.22 A macromolecule of ATP consists of a molecule of adenine, a molecule of ribose, and three phosphate groups. The two end phosphate groups are bonded together by high-energy bonds. When these bonds are broken, they release an unusually great amount of energy; therefore, they are known as *high-energy bonds*. These bonds are represented by curved, solid lines. The ATP molecule is considered an energy carrier.

released to the environment. It is this ATP energy that is used by cells to perform all the characteristics that keep them alive.

Aerobic Cellular Respiration

Aerobic cellular respiration is a specific series of enzyme-controlled chemical reactions in which oxygen is involved in the breakdown of glucose into carbon dioxide and water, and the chemical-bond energy from glucose is released to the cell in the form of ATP (adenosine triphosphate). While the actual process of aerobic cellular respiration involves many enzyme-controlled steps, the overall process is a reaction between the substrates sugar and oxygen resulting in the formation of the products carbon dioxide and water with the release of energy. The following equation summarizes this process:

$$\text{glucose} + \text{oxygen} \rightarrow \begin{array}{c}\text{carbon}\\\text{dioxide}\end{array} + \text{water} + \text{energy}$$

$$C_6H_{12}O_6 + O_2 \rightarrow CO_2 + H_2O + ATP + heat$$

Covalent bonds are formed by atoms sharing pairs of fast-moving, energetic electrons. Therefore, the covalent bonds in the sugar glucose contain chemical potential energy. Of all the covalent bonds in glucose (O—H, C—H, C—C), those easiest to get at are the C—H and O—H bonds on the outside of the molecule. The chemical activities that remove electrons from glucose result in the glucose being oxidized. When these bonds are broken, several things happen:

1. Some ATP is produced;
2. Hydrogen ions (H^+ or protons) are produced; and,
3. Electrons are released and taken up by special carrier molecules.

The ATP is used to power the metabolic activities of the cell. The chemical activities that remove electrons from glucose result in the glucose being oxidized.

The electrons produced during aerobic cellular respiration must be controlled. If they were allowed to fly about at random, they would quickly combine with other molecules, causing cell death. Electron transfer molecules temporarily hold the electrons and transfer them to other electron transfer molecules. ATP is formed when these transfers take place. Once energy has been removed from electrons for ATP production, the electrons must be placed in a safe location. In *aerobic* cellular respiration, these electrons are ultimately attached to oxygen. Oxygen serves as the final resting place of the less-energetic electrons. When the electrons are added to oxygen, it becomes a negatively charged ion, O=. Since the oxygen has gained electrons, it has been *reduced. So, in the aerobic cellular respiration of glucose, glucose is oxidized and oxygen is reduced.* If something is oxidized (loses electrons), something else must be reduced (gains electrons). A simple way to help identify an oxidation-reduction reaction is to use the mnemonic device "LEO the lion says GER." LEO stands for "Loss of Electrons is Oxidation" and GER stands for "Gain of Electrons is Reduction." A molecule cannot simply lose its electrons—they have to go someplace! Eventually, the positively charged hydrogen ions that were released from the glucose molecule combine with the negatively charged oxygen ion to form water.

Once all the hydrogens have been stripped off the glucose molecule, the remaining carbon and oxygen atoms are rearranged to form individual molecules of CO_2. The oxidation-reduction reaction is complete. All the hydrogen originally a part of the glucose has been moved to the oxygen to form water. All the remaining carbon and oxygen atoms of the original glucose are now in the form of CO_2. The energy released from this process is used to generate ATP.

In cells, these oxidation reactions take place in a particular order and in particular places within the cell. In eukaryotic cells, the process of releasing energy from food molecules begins in the cytoplasm and is completed in the mitochondrion. Three distinct enzymatic pathways are involved (figure 20.23):

1. *Glycolysis* (*glyco* = sugar; *lysis* = to split) is a series of enzyme-controlled reactions that takes place in the cytoplasm of cells and results in the breakdown of glucose with the release of electrons and the formation of ATP. The electrons are sent to the electron-transport system (ETS) for processing.
2. The *Krebs cycle* is a series of enzyme-controlled reactions that takes place inside the mitochondrion and completes the breakdown of the remaining fragments of the original glucose with the release of carbon dioxide, more ATP, and more electrons are sent to the electron-transport system (ETS) for processing.
3. The *electron-transport system (ETS)* occurs in the mitochondria. It is a series of enzyme-controlled reactions that converts the kinetic energy of electrons it receives from glycolysis and the Krebs cycle to ATP. In fact, the majority of ATP produced during aerobic cellular respiration comes from the ETS. The electrons are transferred through a series of oxidation-reduction reactions involving enzymes until eventually, the electrons are accepted by oxygen atoms to form oxygen ions (O=). The negatively charged oxygen atoms attract two positively charged hydrogen ions to form water (H_2O).

Photosynthesis

Ultimately, the energy to power all organisms comes from the Sun's radiant energy. Chlorophyll is a green pigment that absorbs light energy for the process of photosynthesis. Through the process of photosynthesis, plants, algae, and certain bacteria transform light energy to chemical-bond energy in the form of ATP and then use ATP to produce complex organic molecules such as glucose. In algae and the leaves of green plants, photosynthesis occurs in cells that contain chloroplasts.

The following equation summarizes the chemical reactions green plants and many other photosynthetic organisms use to make ATP and organic molecules:

$$\text{light energy} + \begin{array}{c}\text{carbon}\\\text{dioxide}\end{array} + \text{water} \rightarrow \text{glucose} + \text{oxygen}$$

$$\text{light energy} + CO_2 + H_2O \rightarrow C_6H_{12}O_6 + O_2$$

There are three distinct events in the photosynthetic pathway:

1. *Light-capturing events.* In eukaryotic cells, the process of photosynthesis takes place within chloroplasts. Each chloroplast is surrounded by membranes and contains the green pigment, chlorophyll, along with other pigments. Chlorophyll and other *accessory* pigments (yellow, red, and orange) absorb specific wavelengths of light (figure 20.24). After specific amounts of light are absorbed by photosynthetic pigments, their electrons become "excited." With this added energy, these excited electrons are capable of entering into the chemical reactions responsible for the production of ATP.

21

The Origin and Evolution of Life

Earth is the only planet in our solar system with liquid water and the only planet with living things. When seen from space, Earth is blue because of its vast oceans. Early in Earth's history, inorganic materials were converted to organic materials that are thought to have accumulated in the oceans. These organic materials have become combined into living units called cells. Cells are the basic unit of life. While we know a great deal about cells and the organisms they make up, scientists still ask these two questions: What is the nature of life? And how did life originate?

CORE **CONCEPT**
Earth and the life on it have changed over billions of years.

OUTLINE

*Species are genetically
isolated from each other
(p. 516).*

CONNECTIONS

Physics

▷ Newton's laws of motion (Ch. 2).

▷ Measuring certain radioactive isotopes provides a clock to
   measure the age of Earth. (Ch. 11 and 22).

Chemistry

▷ Biochemical reactions evolved with living things (Ch. 19).

▷ Carbon chemistry is the chemistry of living things (Ch. 19).

Astronomy

▷ The origin of the universe (Ch. 12).

▷ Conditions on other planets help us understand what the
   ancient Earth may have been like (Ch. 13).

Earth Science

▷ Atmospheric gases have changed
   over time (Ch. 17).

Life Science

▷ Photosynthesis uses sunlight energy to make organic matter and
   release oxygen (Ch. 20).

▷ Respiration releases usable energy from organic molecules
   (Ch. 20).

▷ DNA stores genetic information that determines the kinds of
   biochemical reactions an organism can perform.
   (Ch. 19 and 26).

OVERVIEW

Understanding how Earth came to have such a complex and diverse combination of living things has challenged thinkers for thousands of years. Has Earth always been filled with living things? Were there kinds of living things that have ceased to exist? Did living things originate on Earth? Do new kinds of living things originate today? As our understanding of the laws of nature and the nature of living things has developed, the way we approach these questions has changed.

Although at one time these questions were just interesting intellectual exercises, today they have many practical applications and affect the decision making of governments, corporations, and individual citizens. For example, we are concerned about the extinction of various species of organisms, and governments spend considerable time and money to prevent extinctions. Understanding how evolutionary processes occur in populations is important for controlling populations of disease organisms that continually develop strains resistant to commonly used antibiotics. People are concerned about the manipulation of the DNA of organisms, which can cause major changes in the nature of living things. And governments are spending vast amounts of money to determine if there is or was life on Mars. However, even from a modern scientific perspective, we are still asking the same basic questions. How did life originate? And how and why do living things change?

PART I
HOW DID LIFE ORIGINATE?

All the ideas in this chapter are related to the origin and modification of living things. However, it may be easier to think about the many aspects of this topic if the chapter is divided into separate parts. Therefore, you will find three parts (Part I, How Did Life Originate?, Part II, The Process of Evolution, and Part III, Speciation), each focusing on one aspect of the general topic of understanding the origin and evolution of life. One of the fundamental questions most ancient societies have sought to answer is, Where did we come from? An extension of this question is, How did life originate? In this part we will look at the various kinds of thinking that apply to this question.

EARLY ATTEMPTS TO UNDERSTAND THE ORIGIN OF LIFE

In earlier times, no one doubted that life originated from non-living things. The concept that life could be generated from nonliving matter is known as **spontaneous generation.** The Greeks, Romans, Chinese, and many other ancient peoples believed that maggots arose from decaying meat; mice developed from wheat stored in dark, damp places; lice formed from sweat; and frogs originated from damp mud. Although these ideas were widely accepted until the seventeenth century, there were some who doubted that spontaneous generation occurred. These people subscribed to an opposing concept that life originates only from preexisting life. The concept that life can come only from other living things is known as **biogenesis.**

One of the earliest challenges to the idea that life can be generated from nonliving matter came in 1668. Francesco Redi, an Italian physician, set up a controlled experiment (figure 21.1). He used two sets of jars that were identical except for one aspect. Both sets of jars contained decaying meat, and both were exposed to the atmosphere; however, one set of jars was covered by gauze, and the other was uncovered. Redi observed that flies settled on the meat in the open jar, but the gauze blocked their access to the meat in the covered jars. When maggots appeared on the meat in the uncovered jars but not on the meat in the covered ones, Redi concluded that the maggots arose from the eggs of the flies and were not produced from the decaying meat.

In 1861, the French chemist Louis Pasteur convinced most scientists that spontaneous generation could not occur. He placed a fermentable sugar solution in a flask that had a long swan neck. The mixture and the flask were boiled for a long time, which would have killed any organisms already in the solution. The long swan neck of the flask was left open to allow oxygen to enter. This was an important experiment because at the time, many thought that oxygen was the "vital element" necessary for the spontaneous generation of living things. Pasteur postulated that it was not the oxygen that caused life to "originate" in such mixtures but that there were organisms in the air that entered nutrient mixtures and caused fermentation. He further postulated that airborne organisms would settle on the bottom of the curved portion of the neck and be unable to reach the sugar-water mixture and thus the sugar solution should not ferment. When he performed the experiment, the solution in the swan-neck flask did not ferment. As a control he cut off the swan neck of other flasks (figure 21.2). This allowed microorganisms from the air to fall into the flask, and within two days, the fermentable solution was supporting a population of microorganisms. In his address to the French Academy, Pasteur stated, "Never will the doctrine of spontaneous generation recover from the mortal blow of this simple experiment. No, there is now no circumstance known in which it can be affirmed that microscopic beings came into the world without germs, without parents similar to themselves."

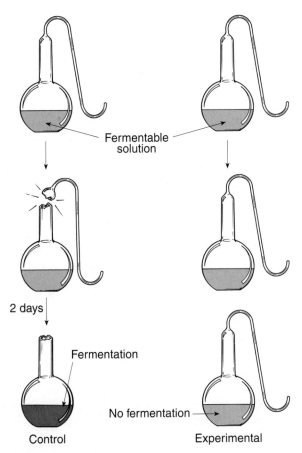

FIGURE 21.2 Louis Pasteur conducted an experiment to test the idea that oxygen was necessary for spontaneous generation to take place. He used swan-neck flasks that allowed oxygen but not airborne organisms to enter the flask. The flasks contained a fermentable solution that had been boiled. One flask was kept intact, but he broke the neck off another flask. The intact flask was the experimental flask and the flask with the broken neck was the control. Within two days, there was growth in the flask with the broken neck but none in the intact flask. Thus, Pasteur demonstrated that it was not oxygen in the air that caused growth in the flasks but living things, which were prevented from entering the flask with the unbroken swan neck. This provided additional evidence against the theory of spontaneous generation.

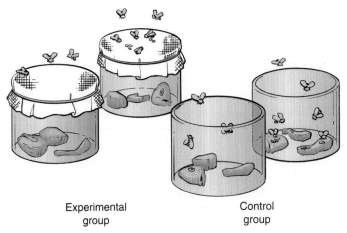

FIGURE 21.1 Francesco Redi performed an experiment in which he prepared two sets of jars that were identical in every way except one. One set of jars had a gauze covering. The uncovered set was the control group; the covered set was the experimental group. Any differences seen between the control and the experimental groups were the result of a single variable—being covered by gauze. In this manner, Redi concluded that the presence of maggots in meat was due to flies laying their eggs on the meat and not spontaneous generation.

CURRENT THINKING ABOUT THE ORIGIN OF LIFE

Extraterrestrial or Earth Origin?

Today, when we look at the question of how life on Earth originated, we still have basically the same two theories. One holds that life arrived on Earth from some extraterrestrial source. This is essentially a variation of the biogenesis argument. The other maintains that life was created on Earth from nonliving material through a process of chemical evolution, which is a variation of the spontaneous generation argument. It is important to recognize that we will probably never know for sure how life on Earth came to be, but it is interesting to speculate and examine the evidence related to this fundamental question.

Extraterrestrial Origin for Life on Earth

Early in the 1900s, Swedish scientist Svante Arrhenius popularized the concept that life arose outside the Earth and that living things were transported to Earth to seed the planet with life. This concept was known as *panspermia*. The argument for extraterrestrial sources of life received some support in 1996 when a meteorite from Antarctica was analyzed. It has been known for many years that meteorites often contain organic molecules, and this suggested that life might have existed elsewhere in the solar system. The chemical makeup of the Antarctic meteorite suggests that it was a portion of the planet Mars that was ejected from Mars as a result of an asteroid colliding with the planet. Analysis of the meteorite shows the presence of complex organic molecules and small structures that resemble those found on Earth that are thought to be the result of the activity of ancient microorganisms. While scientists no longer think these structures are from microorganisms, many still think it is quite possible Mars may have had living things in the past. Since Mars currently has some water as ice and shows features that resemble dried-up river systems and seas, it is highly likely that Mars had much more water in the past. Since water is the most common compound in living things, many consider the presence of water to be a prerequisite for life. For these reasons, many feel it is reasonable to consider that life of a nature similar to that presently found on Earth could have existed on Mars. While previous exploration of

---

CONCEPTS APPLIED

Spontaneous Generation?

At one time, people believed that life could be generated spontaneously from nonliving matter. To get an idea of why they thought this was possible, you can do the following: Take a clear bottle or jar and place a small amount of soil in it. Fill it with water. Place the jar in a warm, well-lighted place. A sunny window is ideal. Observe the changes in it over the next two to three weeks. What evidence do you have that there are living things in the jar? Examine the water under a microscope if you have access to one.

Mars' surface has failed to find evidence for life, many researchers are not satisfied with the results. Thus, current explorations of the Martian surface by the Mars robots—*Spirit* and *Opportunity*—continue to look for evidence of current or past life.

In recent years it has been demonstrated that stars other than our Sun have planets. When people look at the millions of stars and speculate about the number of planets present in the universe, many are convinced that it is possible that life could have arrived on Earth from some other solar system. However, regardless of how life began on Earth, it must have originated somewhere, and many scientists support the idea that life originated on Earth from nonliving material.

Earth Origin for Life on Earth

There has been much research and speculation about the conditions that would have been necessary for life to originate on Earth and the kinds of problems that early living things would have had to solve. Several different kinds of information are important to this discussion.

1. **Water is necessary for life.**   Earth is currently the only planet in our solar system that has a temperature range that allows for water to exist as a liquid on its surface, and water is the most common compound in most kinds of living things.
2. **The atmosphere of the early Earth lacked molecular oxygen.**   Analysis of the atmospheres of other planets shows that they all lack oxygen. The oxygen in Earth's current atmosphere is the result of photosynthetic activity. Therefore, before there was life on Earth, the atmosphere probably lacked oxygen.
3. **Organic molecules can form spontaneously in the absence of oxygen.**   Experiments demonstrate that organic molecules can be generated in an atmosphere that lacks oxygen. Furthermore, these organic molecules could have accumulated in the oceans, since there would have been no oxygen to bring about oxidation to cause their breakdown. (See figure 21.3.)
4. **The early Earth was hot.**   Since it is assumed that all of the planets have been cooling off as they age, it is very likely that Earth was much hotter in the past. The large portions of Earth's surface that are of volcanic origin strongly support this idea.
5. **Sources of energy on the early Earth were different from today.**   The surface of the earth was hotter. A hotter Earth would have contributed to more water vapor in the atmosphere resulting in more storms and lightning. Without oxygen in the atmosphere, much more ultraviolet light energy would have reached Earth's surface.
6. **Simple organisms today often live in conditions similar to those thought to have existed on a primitive Earth.**   Prokaryotic organisms (Bacteria and Archaea) are relatively simple and extremely common. Today, certain prokaryotic organisms live in extreme environments of high temperature, high salinity, low pH, or the absence of oxygen. This suggests that they may have been adapted to life in a world that is very different from today's Earth.

- Carbon
- Nitrogen
- Oxygen
- Hydrogen

FIGURE 21.3 The environment of the primitive Earth was harsh and lifeless. But many scientists believe that it contained the necessary molecules and sources of energy to fashion the first living cell. The energy furnished by volcanoes, lightning, and ultraviolet light could have broken the bonds in the simple inorganic molecules such as carbon dioxide (CO_2), ammonia (NH_3), and methane (CH_4) in the atmosphere. New bonds could have formed as the atoms from the smaller molecules were rearranged and bonded to form simple organic compounds in the atmosphere. Rain and runoff from the land would have carried these chemicals into the oceans. Here, they could have reacted with each other to form more complex organic molecules.

These specialized organisms are found today in unusual locations such as hot springs and around thermal vents in the ocean floor and may be descendants of the first organisms formed on the primitive, more hostile Earth.

7. **Putting it all together.** If we use all these ideas, we can create the following scenario. A hot planet Earth with much ultraviolet light, lightning, and heat to serve as sources of energy and an atmosphere that lacked oxygen could have allowed organic molecules to form from inorganic molecules. These organic molecules could have been washed from the atmosphere by rain and carried by rivers to the oceans and produced a dilute organic "soup" because oxygen was not present to cause them to break down. These organic molecules could have served as "building blocks" for the development of simple organisms. These organic molecules could also have served as a source of energy for newly formed living things. The earliest forms of life would have been adapted to conditions that are very different from those that exist on Earth today.

Meeting Metabolic Needs

Fossil evidence indicates that there were primitive, bacterialike forms of life on Earth at least 3.5 billion years ago. Regardless of how they developed, these first primitive cells would have needed a way to obtain energy and to add new organic molecules to their structure as the organisms grew or as previously existing molecules were lost or destroyed. There are two ways to accomplish this.

Heterotrophs consume organic molecules from their surroundings, which they use to make new molecules and to provide themselves with a source of energy. Today, we recognize all animals, fungi, and most protozoa and Bacteria as being heterotrophs.

Autotrophs use an external energy source such as the energy from inorganic chemical reactions (chemoautotrophs) or sunlight (photoautotrophs) to combine simple inorganic molecules such as water and carbon dioxide to make new organic molecules. These new organic molecules can then be used as building materials for new cells or can be broken down at a later date to provide a source of energy. Today, we recognize eukaryotic plants and algae as being photoautotrophs. Among the Bacteria and Archaea, there are many photoautotrophs and chemoautotrophs.

The Heterotroph Hypothesis

Many scientists support the idea that the first living things produced on Earth were primitive, bacterialike heterotrophs that lived off the organic molecules that would have accumulated in the oceans. Evidence exists to suggest that a wide variety of compounds were present in the early oceans. Since the early heterotrophs are thought to have developed in an atmosphere that lacked oxygen, they would have been, of necessity, anaerobic organisms. Therefore, they did not obtain the maximum amount of energy from the organic molecules they obtained from their environment. At first, this would not have been a problem. The organic molecules that had been accumulating in the ocean for hundreds of millions of years served as an ample source of organic material for the heterotrophs.

Initially these primitive heterotrophs did not have to modify the compounds to meet their needs, and they probably carried out a minimum of enzyme-controlled, metabolic reactions. Those compounds that could be used easily by heterotrophs would have been the first to become depleted from the early environment.

As easily used organic molecules became scarce, it is possible that some of the primitive cells may have had altered (mutated) genetic material that allowed them to convert material that was not directly usable into a compound that could be used. Genetic mutations may have been common because the amount of ultraviolet light, which is known to cause mutations, would have been high. Today, ozone is formed from oxygen in the atmosphere, and ozone screens out much of the ultraviolet light. Before there was oxygen in the atmosphere, there would have been no ozone and much more ultraviolet light would have reached Earth. Heterotrophs with such mutations could have survived, while those without it would have become extinct as the compounds they used for food became scarce. It has been suggested that through a series of mutations in the early heterotrophs, a more complex series of biochemical reactions originated within some of the cells. Such cells could use controlled reactions to modify chemicals from their surroundings and convert them into usable organic compounds.

The Autotroph Hypothesis

As with many areas of science, there are often differences of opinion. Although the heterotroph hypothesis for the origin of living things was the prevailing theory for many years, recent discoveries have caused many scientists to consider an alternative. They propose that the first organism was an autotroph that used the energy released from inorganic chemical reactions to synthesize organic molecules. Such organisms are known as chemoautotrophs. Several kinds of information support this theory. Many kinds of Bacteria and Archaea are chemoautotrophs and live in extremely hostile environments around the world. These organisms are found in hot springs such as those in Yellowstone National Park or near hot thermal vents—areas where hot, mineral-rich water enters seawater from the ocean floor (figure 21.4). They use inorganic chemical reactions as a source of energy to allow them to synthesize organic molecules from inorganic components. The fact that many of these organisms live in very hot environments suggests that they may have originated on an Earth that was much hotter than it is currently. If the first organism was an autotroph, there could have been subsequent evolution of a variety of kinds of cells, both autotrophic and heterotrophic, that could have led to the variety of different prokaryotic cells seen today.

Summary of Ideas About the Origin of Life

As a result of this discussion, you should understand that we do not know how life on Earth originated. Scientists look at many kinds of evidence and continue to explore new avenues of research. So we currently have three competing scientific theories for the origin of life on Earth:

1. Life originated elsewhere in the universe and arrived from some extraterrestrial source.

FIGURE 21.4 Thermal vents are places where hot mineral-rich water enters the ocean water from holes in the ocean bottom. Surrounding these vents are collections of many kinds of animals. Chemoautotrophic members of the Archaea use the hot mineral-rich water as a source of energy with which they synthesize organic molecules. These organisms and the organic molecules they produce are the base of the food chain for a variety of kinds of animals found only in these locations.

2. Life originated on Earth as a heterotroph.
3. Life originated on Earth as an autotroph.

From our modern perspective, we can see that today, all life comes into being as a result of reproduction. Life is generated from other living things, the process of biogenesis. However, reproduction does not answer the question: Where did life come from in the first place? We can speculate, test hypotheses, and discuss various possibilities, but we will probably never know for sure. Life either always was, or it started at some point in the past. If it started, then spontaneous generation of some type had to occur at least once, but it is not happening today.

MAJOR EVENTS IN THE EARLY EVOLUTION OF LIVING THINGS

Regardless of what the first living things were like, once they existed, a process began that resulted in changes in their abilities and structures. Furthermore, the nature of the planet Earth also changed. When we compare the primitive nature of the first living things with the characteristics of current organisms, we recognize that there were several major steps necessary to go from those first organisms to the great diversity and complexity we see today.

Reproduction and the Origin of Genetic Material

In the previous discussion about possible metabolic activities of early life-forms, we made the assumption that there was some kind of genetic material to direct their metabolic activities. There has been considerable thought about the nature of this genetic material. Today, we know that two important *nucleic acid* molecules, DNA and RNA, store and transfer information within a cell. RNA is a simpler molecule than DNA. In most current cells, DNA stores genetic information and RNA assists the DNA in carrying out its instructions. The details of how DNA and RNA function are discussed in chapter 26.

However, it has been discovered that in some viruses, RNA rather than DNA can serve as the genetic material. Other research about the nature of RNA provides interesting information. RNA can be assembled from simpler subunits that could have been present on the early Earth. Scientists have also shown that RNA molecules are able to make copies of themselves without the need for enzymes, and they can do so without being inside cells.

These pieces of evidence suggest that RNA may have been the first genetic material, which helps to solve one of the problems associated with the origin of life: how genetic information was stored in these primitive life-forms. Once a primitive life-form had the ability to copy its genetic material, it would be able to reproduce. Reproduction is one of the most fundamental characteristics of living things.

Once living things existed and had a molecule that stored information, the stage was set for the evolution of new forms of life. Our current knowledge about how DNA and RNA store and use genetic information includes the idea that these molecules are changeable. In other words, *mutations* can occur. Thus, the stage was set for living things to proliferate into a variety of kinds that were adapted to specific environmental conditions.

The Development of an Oxidizing Atmosphere

Ever since its formation, Earth has undergone constant change. In the beginning, it was too hot to support an atmosphere. Later, as it cooled and as gases escaped from volcanoes, a reducing atmosphere (lacking oxygen) was likely to have been formed. The early life-forms would have lived in this reducing atmosphere. However, today we have an oxidizing atmosphere and most organisms use this oxygen as a way to extract energy from organic molecules through a process of aerobic respiration. But what caused the atmosphere to change?

Atmospheric Oxygen Produced by Photosynthesis

Today, it is clear that the oxygen in our atmosphere is the result of the process of photosynthesis. Prokaryotic cyanobacteria are the simplest organisms that are able to photosynthesize. The first fossils of primitive living things show up at about 3.5 billion years ago, and it appears that an oxidizing atmosphere began to develop about 2 billion years ago. Because the fossil record shows that living

things existed before there was oxygen in the atmosphere and that cyanobacteria were present early in the evolution of life, it seemed logical that the first organisms could have accumulated many mutations over time that could have resulted in photosynthetic autotrophs (photoautotrophs). This would have been a significant change because eventually the release of oxygen from photosynthesis would have resulted in an accumulation of oxygen in the atmosphere—an **oxidizing atmosphere.** Once oxygen was present in the atmosphere it would not have been possible to accumulate organic molecules in the oceans as was described earlier in this chapter.

The Significance of Ozone

The presence of oxygen in the atmosphere had an additional significance. Oxygen molecules react with one another to form ozone (O_3). Ozone collects in the upper atmosphere and acts as a screen to prevent most of the ultraviolet light from reaching Earth's surface. The reduction of ultraviolet light reduced the number of mutations in cells.

Evolution of Aerobic Respiration Possible

The presence of oxygen in the atmosphere also allowed for the evolution of aerobic respiration. Since any early heterotrophs were, of necessity, anaerobic organisms, they did not derive large amounts of energy (ATP) from the organic materials available as food. With the evolution of aerobic heterotrophs, there could be a much more efficient conversion of food into usable energy. They could use the oxygen for aerobic respiration and, therefore, generate many more energy-rich ATP molecules from the food molecules they consumed. Because of this, aerobic organisms would have a significant advantage over anaerobic organisms.

The Establishment of Three Major Domains of Life

Although biologists have traditionally divided organisms into kingdoms based on their structure and function, it was very difficult to do this with microscopic organisms. In 1977, Carl Woese published the idea that the "bacteria" (organisms that lack a nucleus), which had been considered a group of similar organisms, were really made up of two very different kinds of organisms, the Bacteria and Archaea. With the newly developed ability to decode the sequence of nucleic acids, it became possible to look at the genetic nature of organisms without being confused by their external structures. Woese studied the sequences of ribosomal RNA and compared similarities and differences. As a result of his studies and those of many others, a new concept of the relationships between various kinds of organisms has emerged. There are three main kinds of living things, Bacteria, Archaea, and Eucarya, that have been labeled **"domains."**

This new picture of living things requires us to reorganize our thinking. It appears that the oldest organisms were members of the Domain Bacteria and that the Domains Archaea and Eucarya are both derived from the Bacteria. Perhaps most interesting is the idea that the Archaea and Eucarya share many characteristics, suggesting that they are more closely related to each other than either is to the Bacteria. It appears that each domain developed specific abilities.

The Bacteria developed many different metabolic abilities. Today, many are able to use organic molecules as a source of energy (heterotrophs). Some of these heterotrophs use anaerobic respiration, whereas others use aerobic respiration. Other Bacteria are autotrophic. Some, such as the cyanobacteria, carry on photosynthesis, whereas others are chemosynthetic and get energy from inorganic chemical reactions.

The Archaea have very diverse metabolic abilities. Some are chemoautotrophic and use inorganic chemical reactions to generate the energy they need to make organic matter. The reactions that provide them with energy often produce methane (CH_4) or hydrogen sulfide (H_2S). In addition, most of these organisms are found in extreme environments such as hot springs or in extremely salty or acidic environments. However, it is becoming clear that they also inhabit soil, the guts of animals, and are particularly abundant in the ocean.

The Eucarya (animals, plants, fungi, protozoa, and algae) are familiar to most people. The cells of eukaryotic organisms are much larger than those of prokaryotic organisms and appear to have incorporated entire cells of other organisms within their own cellular structure. Chloroplasts and mitochondria are both bacterialike structures found inside eukaryotic cells.

The Endosymbiotic Theory and the Origin of Eukaryotic Cells

The earliest fossils occur at about 3.5 billion years ago and appear to be similar in structure to that of present-day Bacteria and Archaea. Therefore, it is likely that the early heterotrophs and autotrophs were probably simple one-celled prokaryotic organisms. The earliest fossils of eukaryotic cells appear at about 1.8 billion years ago.

The **endosymbiotic theory** attempts to explain this evolution. This theory states that eukaryotic cells came into being as a result of the combining of several different types of primitive prokaryotic cells. It is thought that some organelles found in eukaryotic cells originated as free-living prokaryotes. For example, since mitochondria and chloroplasts contain bacterialike DNA and ribosomes, control their own reproduction, and synthesize their own enzymes, it has been suggested that they originated as free-living prokaryotic Bacteria. These bacterial cells could have established a symbiotic relationship with another primitive nuclear-membrane-containing cell type (figure 21.5). When this theory was first suggested, it met with a great deal of criticism and even ridicule. However, continuing research has uncovered several other instances of the probable joining of two different prokaryotic cells to form one. If these cells adapted to one another and were able to survive and reproduce better as a team, it is possible that this relationship may have evolved into present-day eukaryotic cells.

If this relationship had included only a nuclear-membrane-containing cell and aerobic Bacteria, the newly evolved cell would have been similar to present-day heterotrophic protozoa, fungi, and animal cells. If this relationship had included both aerobic Bacteria and photosynthetic Bacteria, the newly formed cell would have been similar to present-day autotrophic algae and plant cells. In addition, it is likely that endosymbiosis occurred

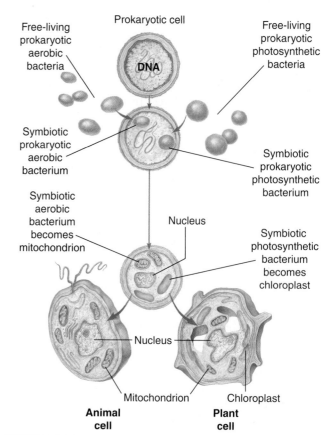

FIGURE 21.5 The endosymbiotic theory proposes that some free-living prokaryotic Bacteria developed symbiotic relationships with a host cell. Some aerobic Bacteria appear to have developed into mitochondria and photosynthetic Bacteria appear to have developed into chloroplasts.

among eukaryotic organisms as well. Several kinds of eukaryotic red and brown algae contain chloroplastlike structures that appear to have originated as free-living eukaryotic cells.

A Summary of the Early Evolution of Life

If we incorporate information about endosymbiosis along with information about the nature of genetic material, we can develop a diagram that helps us see how these various kinds of organisms are related to one another.

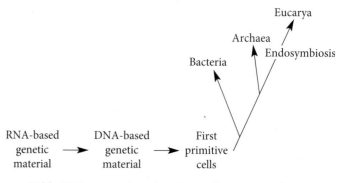

Table 21.1 summarizes the relationships among these three major kinds of organisms. Figure 21.6 summarizes several of the major events in the formation of the planet Earth and the development of life on it.

TABLE 21.1 Major Domains of Life

Bacteria	Archaea	Eucarya
No nuclear membrane	No nuclear membrane	Nuclear membrane present
No membranous structures	No membranous structures	Many kinds of membranous organelles present in cells
Probably ancestral to both Archaea and Eucarya	Probably have a common ancestor with Eucarya	Probably have a common ancestor with Archaea
Chlorophyll-based, oxygen-generating photosynthesis was an invention of the cyanobacteria.	Some have a kind of photosynthesis but do not release oxygen.	Chloroplasts are probably derived from cyanobacteria.
Large number of heterotrophs oxidize organic molecules for energy.	A few are heterotrophs.	Photosynthetic organisms are autotrophs; all others are heterotrophs.
Many use aerobic respiration to obtain energy from organic molecules.	None known to use aerobic respiration.	Mitochondria probably derived from certain aerobic Bacteria.
Some use energy from inorganic chemical reactions to produce organic molecules (chemoautotrophs).	Many are chemoautotrophs.	None is a chemoautotroph.
Some live at high temperatures and may be ancestral to Archaea.	Typically found in extreme environments	
Much metabolic diversity among closely related organisms	None has been identified as pathogenic.	Common evolutionary theme is the development of complex cells through endosymbiosis with other organisms.
Example: *Streptococcus pneumoniae*, one cause of pneumonia	Example: *Pyrolobus fumarii*, deep-sea hydrothermal vents, hot springs, volcanic areas, growth to 113°C (235°F)	Example: *Spirogyra*, a filamentous alga found in freshwater

FIGURE 21.6 Current thinking suggests that several factors shaped the nature of Earth and its atmosphere. Conceivably, the initial atmosphere was primarily carbon dioxide (CO_2) with smaller amounts of nitrogen (N_2), ammonia (NH_3), methane (CH_4), and carbon monoxide (CO). Carbon dioxide decreased as a result of chemical reactions at the surface of Earth. Living organisms have had a significant effect on the atmosphere because once photosynthesis began, carbon dioxide was further reduced and oxygen (O_2) began to increase. The presence of oxygen allowed for the increase in ozone (O_3), which reduces the amount of ultraviolet light penetrating the atmosphere to reach the surface of Earth. Because ultraviolet light causes mutations, a protective layer of ozone probably has reduced the frequency of mutations from that which was present on primitive Earth.

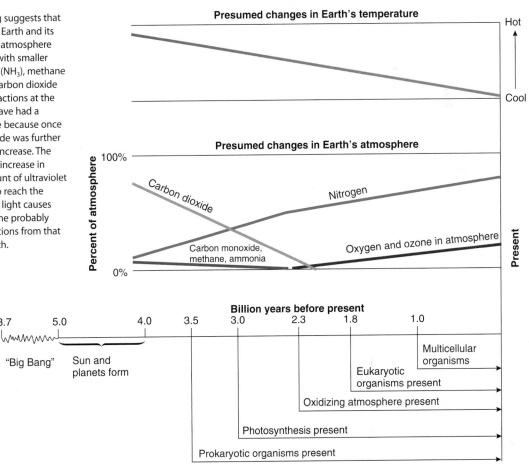

Although we have introduced the idea of evolutionary change while discussing the origin of life, it is important to have a clear understanding what is meant by the evolution of organisms and the mechanisms that drive this important process. This part focuses on the mechanisms that cause evolutionary change.

THE DEVELOPMENT OF EVOLUTIONARY THOUGHT

For centuries, people believed that the various species of plants and animals were fixed and unchanging; that is, they were thought to have remained unchanged from the time of their creation. This was a reasonable assumption because people knew nothing about DNA, how sperm and eggs were formed, or population genetics. Furthermore, the process of evolution is so slow that the results of evolution were usually not evident during a human lifetime. It is even difficult for modern scientists to recognize this slow change in many kinds of organisms.

Charles Darwin (1809–1882) is famous for his contribution of the idea of natural selection as the process that directs evolution. However, he probably would not have recognized the importance of his observations of nature if he had not had the advantage of reading about the ideas of or engaging in discussions with others who had considered the possibility that organisms change over time. The following discussion lists some of the people known to have influenced Darwin's thinking.

Georges-Louis Leclerc, Comte de Buffon (1707–1788)—In his writings, Buffon questioned the established church doctrines of the day that all organisms were created as we currently see them and that they are unchanging. In addition, he questioned the doctrine that Earth was only about six thousand years old. However, he did not suggest any mechanism for evolution. Buffon produced a thirty-six-volume set of books that covered essentially everything known about the natural world. Darwin referred to Buffon as the first person to suggest the idea of natural selection.

Erasmus Darwin (1731–1802)—Erasmus Darwin was Charles Darwin's grandfather. He was a major thinker of his day and had many thoughts about the processes by which organisms could change over time. Although he died before Charles was born, in many ways his thinking anticipated what his grandson would make a basic tenet of biological thought. Many of these ideas are in evidence in the following poem.

Organic life beneath the shoreless waves
Was born and nurs'd in ocean's pearly caves;
First forms minute, unseen by spheric glass,
Move on the mud, or pierce the watery mass;
These, as successive generations bloom,

New powers acquire and larger limbs assume;
Whence countless groups of vegetation spring,
And breathing realms of fin and feet and wing.

Erasmus Darwin, *The Temple of Nature* (1802)

Jean-Baptiste de Lamarck (1744–1829)—In 1809, Lamarck, a student of Buffon's, suggested a process by which evolution could occur. He proposed that acquired characteristics could be transmitted to offspring. For example, he postulated that giraffes originally had short necks. Since giraffes constantly stretched their necks to obtain food, their necks got slightly longer. He thought that the slightly longer neck acquired through stretching could be passed to the offspring, who were themselves stretching their necks, and over time, the necks of giraffes would get longer and longer. Although we now know Lamarck's theory was wrong (because acquired characteristics are not inherited), it stimulated further thought as to how evolution could occur. All during this period, from the mid-1700s to the mid-1800s, lively arguments continued about the possibility of evolutionary change. Darwin was aware of Lamarck's writings and considered him to be an important thinker of the day.

Thomas Malthus (1766–1834)—Malthus' *Essay on the Principle of Population* put forward the concept that people reproduce faster than do resources needed to sustain the population. Therefore, there would be a constant struggle among the members of the population for the limited resources available. Thus, the human population could look forward to poverty and famine unless human reproduction was controlled. Darwin's reading of this essay and its implication of a struggle for limited resources was instrumental in his formulating the theory of natural selection as a mechanism for evolution.

Charles Lyell (1797–1875)—Lyell was an eminent geologist of his day. He is most famous for his elucidation of the idea of uniformitarianism. The central idea of uniformitarianism is that we can interpret evidence of past geologic events by understanding how things are working today. He felt that processes such as sedimentation, volcanic activity, and erosion did not operate differently today from the past. One of the obvious outcomes of this thinking is that the age of Earth must be much greater than a few thousand years. Darwin knew Lyell and Lyell's ideas greatly influenced Darwin's thinking about the time available for the mechanisms of evolution to operate.

Alfred Russel Wallace (1823–1913)—Like Darwin, Wallace traveled extensively and, as a result of his travels, came to the same conclusions about natural selection as did Darwin. He was also influenced by Malthus' ideas, as was Darwin. Although today Wallace is often mentioned as an afterthought, it was his essay concerning his ideas about natural selection that prompted Darwin to publish *On the Origin of Species by Means of Natural Selection, or the Preservation of Favoured Races in the Struggle for Life.*

EVOLUTION AND NATURAL SELECTION

In many cultural contexts, the word *evolution* means progressive change. We talk about the evolution of economies, fashion, or musical tastes. From a biological perspective, the word has a more specific meaning. One of the key characteristics of life is that organisms are able to respond to their surroundings. Evolution is a responsive process that takes place at the population level.

Defining Evolution

Evolution is the change in frequency of genetically determined characteristics within a population over time. There are three key points to this definition. First of all, evolution only occurs in populations. *Populations* are groups of organisms of the same species that are able to interbreed and are thus genetically similar. The second point is that *genes* (specific pieces of DNA) determine the characteristics displayed by organisms. The third point is that the mix of genes (DNA) within populations can change. Thus, evolution involves changes in the genes that are present in a population. By definition, individual organisms are not able to evolve—only populations can.

The Role of the Environment in Evolution

The organism's surroundings determine which characteristics favor survival and reproduction (i.e., which characteristics best fit the organism to its environment). A primary mechanism by which evolution occurs involves the selective passage of genes from one generation to the next through sexual reproduction. This adaptation can take place at different levels and over different time periods. At the smallest level, populations show changes in the genes within a local population. For example, some populations of potato beetles became resistant to certain insecticides and some populations of crabgrass became resistant to specific herbicides when these chemicals became a constant part of their environment. The susceptible members died and the resistant ones lived, leading to populations that consisted primarily of resistant individuals. These populations became genetically adapted to an altered environment. The genes that allowed the organisms to resist the effects of the poisons became more common in these populations. When we look at evolutionary change over millions or billions of years, we recognize that Earth has gone through major changes of climate, sea level, and other conditions. These changes have resulted in changes in the kinds of organisms present as well. Some species went extinct as new species came into being.

Natural Selection Leads to Evolution

The various processes that encourage the passage of beneficial genes to future generations and discourage the passage of harmful or less valuable genes are collectively known as **natural selection.** The idea that some individuals whose gene combinations favor life in their surroundings will be most likely to survive, reproduce, and pass their genes on to the next generation is known as the

FIGURE 21.7 Many kinds of birds, such as this northern flicker, nest in holes in trees. If old and dead trees are not available, they may not be able to breed.

theory of natural selection. The *theory of evolution* states that populations of organisms experience changes in the frequency of genetically determined characteristics over time. Thus, natural selection is a process that brings about evolution by "selecting" which genes will be passed to the next generation. Although natural selection is the major mechanism that drives evolution, there are other processes that can cause changes in genes from generation to generation. "A Closer Look: Other Mechanisms that Cause Evolution" discusses some of these mechanisms.

A theory is a well-established generalization supported by many different kinds of evidence. At one time, the idea that populations of species changed over time was revolutionary. Today, the concepts of evolution and natural selection are central to the study of all of the biological sciences.

The theory of natural selection was first proposed by Charles Darwin and Alfred Wallace and was clearly set forth in 1859 by Darwin in his book *On the Origin of Species by Means of Natural Selection, or the Preservation of Favoured Races in the Struggle for Life.* Since the time it was first proposed, the theory of natural selection has been subjected to countless tests and remains the core concept for explaining how evolution occurs.

There are two common misinterpretations associated with the process of natural selection. The first involves the phrase "survival of the fittest," commonly associated with the theory of natural selection. Individual survival is certainly important because those that do not survive will not reproduce. But the more important factor is the number of descendants an organism leaves. An organism that has survived for hundreds of years but has not reproduced has not contributed any of its genes to the next generation and so has been selected against. The key, therefore, is not survival alone but survival *and reproduction* of the more fit organisms.

Second, the phrase "struggle for life" does not necessarily refer to open conflict and fighting. It is usually much more subtle than that. When a resource such as nesting material, water, sunlight, or food is in short supply, some individuals survive and reproduce more effectively than others. For example, many kinds of birds require holes in trees as nesting places (see figure 21.7).

The Voyage of HMS *Beagle*, 1831–1836

Probably the most significant event in Charles Darwin's life was his opportunity to sail on the British survey ship HMS *Beagle*. Surveys were common at this time; they helped refine maps and chart hazards to shipping. Darwin was twenty-two years old and probably would not have had the opportunity had his uncle not persuaded Darwin's father to allow him to take the voyage. Darwin was to be a gentleman naturalist and companion to the ship's captain, Robert Fitzroy. When the official naturalist left the ship and returned to England, Darwin became the official naturalist for the voyage. The appointment was not a paid position.

The voyage of the *Beagle* lasted nearly five years. During the trip, the ship visited South America, the Galápagos Islands, Australia, and many Pacific Islands. The *Beagle's* entire route is shown on the accompanying map. Darwin suffered greatly from seasickness, and perhaps because of it, he made extensive journeys by mule and on foot some distance inland from wherever the *Beagle* happened to be at anchor. These inland trips provided Darwin the opportunity to make many of his observations. His experience was unique for a man so young

and very difficult to duplicate because of the slow methods of travel used at that time.

Although many people had seen the places that Darwin visited, never before had a student of nature collected volumes of information on them. Also, most other people who had visited these faraway places were military men or adventurers who did not recognize the significance of what they saw. Darwin's notebooks included information on plants, animals, rocks, geography, climate, and the native peoples he encountered. The natural history notes he took during the voyage served as a vast storehouse of information that he used in his writings for the rest of his life. Because Darwin was wealthy, he did not need to work to earn a living and could devote a good deal of his time to the further study of natural history and the analysis of his notes. He was a semi-invalid during much of his later life. Many people think his ill health was caused by a tropical disease he contracted during the voyage of the *Beagle*. As a result of his experiences, he wrote several volumes detailing the events of the voyage, which were first published in 1839 in conjunction with other information related to the voyage of the *Beagle*. His volumes were

revised several times and eventually were entitled *The Voyage of the Beagle*. He also wrote books on barnacles, the formation of coral reefs, how volcanos might have been involved in reef formation, and, finally, *On the Origin of Species by Means of Natural Selection, or the Preservation of Favoured Races in the Struggle for Life*. This last book, written twenty-three years after his return from the voyage, changed biological thinking for all time.

Box Figure 21.1 Young Charles Darwin examining specimens on the Galápagos Islands

The Voyage of HMS *Beagle*, 1831–1836

0 1000 2000 3000
Equatorial scale of miles

Box Figure 21.2

If these are in short supply, some birds will be fortunate and find a top-quality nesting site, others will occupy less suitable holes, and some may not find any. There may or may not be fighting for possession of a site. If a site is already occupied, a bird may not necessarily try to dislodge its occupant but may just continue to search for suitable but less valuable sites. Those that successfully occupy good nesting sites will be much more successful in raising young than will those that must occupy poor sites or those that do not find any.

Similarly, on a forest floor where there is little sunlight, some small plants may grow fast and obtain light while shading out plants that grow more slowly. The struggle for life in this instance involves a subtle difference in the rate at which the plants grow. But the plants are indeed engaged in a struggle, and a superior growth rate is the weapon for survival.

GENETIC DIVERSITY IS IMPORTANT FOR NATURAL SELECTION

Now that we have a basic understanding of how natural selection works, we can look in more detail at factors that influence it. For natural selection to occur there must be genetic differences among the many individuals of an interbreeding population of organisms. If all individuals are identical genetically, it

does not matter which ones reproduce—the same genes will be passed to the next generation and natural selection cannot occur. There are two primary ways in which genetic diversity is generated within the organisms of a species: mutation and genetic recombination during sexual reproduction.

Genetic Diversity Resulting from Mutation

A **mutation** is any change in the genetic information (DNA) of an organism. **Spontaneous mutations** are changes in DNA that cannot be tied to a particular causative agent. It is suspected that cosmic radiation or naturally occurring mutagenic chemicals might be the cause of many of these mutations. It is known that subjecting organisms to high levels of radiation or to certain chemicals increases the rate at which mutations occur. It is for this reason that people who work with radioactive materials or other mutagenic agents take special safety precautions (figure 21.8). It is also known that when cells make copies of DNA, the DNA may not be copied perfectly and small errors can be introduced.

FIGURE 21.8 Because radiation and certain chemicals increase the likelihood of mutations, people who work in hazardous environments receive special training and use protective measures to reduce their exposure to mutagenic agents. It is particularly important to protect the ovaries and testes, since mutations that alter the DNA of eggs and sperm can be passed on to children.

CONCEPTS APPLIED

Simulating Natural Selection

You can simulate the effects of natural selection with a deck of playing cards. Assume the black cards are recessive alleles that make an insect vulnerable to a specific insecticide. Assume that the red cards are dominant alleles that allow the insect to survive exposure to the insecticide. Initially, the dominant resistance allele will be rare in the population. To simulate this situation, keep all the black cards and six of the red cards. The frequency of the red cards is 6/32 = 18.75 percent.

If this insect population were sprayed with the insecticide, what would happen? Use the cards to simulate the effect. Shuffle the cards and draw two cards at random. You are drawing two cards because every individual has two genes (alleles) for each characteristic. If you draw two black cards, discard them. This individual did not have any alleles that allowed it to survive exposure to the insecticide. If you draw two red cards, or a red and a black card, save them. These individuals had a dominant allele that allowed them to resist the insecticide.

After you have drawn all the cards, examine those you did not discard. Calculate the frequency of the red cards (number of red cards/total red and black cards). This change in the frequency of the resistance gene is the result of the different rates of mortality and an example of evolution. This simulation should help you understand why populations of insects can become resistant to insecticides in a few generations.

Naturally occurring mutation rates are low (perhaps one chance in one hundred thousand that a gene will be altered). There are three possible outcomes when a gene is altered. The mutation may be so minor that it has no effect, the mutation may be harmful, or the mutation may be beneficial. Although it is likely that most mutations are neutral or harmful, beneficial mutations do occur. In populations of millions of individuals, each of whom has thousands of genes, over thousands of generations it is highly likely that a new beneficial piece of genetic information could come about as a result of mutation. When we look at the various genes that exist in humans or in any other organism, we should remember that every gene originated as a modification of a previously existing gene. For example, the gene for blue eyes may be a mutated brown-eye gene, or blond hair may have originated as a mutated brown-hair gene. Thus, mutations have been very important for introducing new genetic material into species over time.

In order for mutations to be important in the evolution of organisms, they must be in cells that will become sex cells. Mutations to the cells of the skin or liver will affect only those specific cells and will not be passed to the next generation.

Genetic Diversity Resulting from Sexual Reproduction

A second very important process involved in generating genetic diversity is sexual reproduction. While sexual reproduction does not generate new genetic information the way mutation does, it allows for the recombination of genes into mixtures that did not occur previously. Each individual entering a population by sexual reproduction carries a unique combination of genes; half are donated by the mother and half are donated by the father. During the formation of sex cells, the activities of chromosomes results in new combinations of genes. This means that there are millions of possible combinations of genes in the sex cells of any individual. When fertilization occurs, one of the millions of possible sperm unites with one of the millions of possible eggs, resulting in a genetically unique individual. The gene mixing that occurs during sexual reproduction is known as **genetic recombination.** The new individual has a complete set of genes that is different from that of any other organism that ever existed.

The power of genetic recombination is easily seen when you look at members of your family. Compare yourself with your siblings, parents, and grandparents. Although your genes are traceable back to your grandparents, your specific combination of genes is unique, and this uniqueness shows up in differences in physiology, physical structure, and behavior.

PROCESSES THAT DRIVE NATURAL SELECTION

Several mechanisms allow for selection of certain individuals for survival and successful reproduction. The specific environmental factors that favor certain characteristics are called **selecting agents.** If predators must pursue swift prey organisms, then the faster predators will be selected for, and the selecting agent is the swiftness of available prey. If predators must find prey that are slow but hard to see, then the selecting agent is the camouflage coloration of the prey, and keen eyesight is selected for. If plants are eaten by insects, then the production of toxic materials in the leaves is selected for. All selecting agents influence the likelihood that certain characteristics will be passed to subsequent generations. Three kinds of mechanisms influence which genes are passed to future generations: differential survival, differential reproductive rates, and differential mate selection.

Differential Survival

As stated previously, the phrase "survival of the fittest" is often associated with the theory of natural selection. Although this is recognized as an oversimplification of the concept, survival is an important factor in influencing the flow of genes to subsequent generations. If a population consists of a large number of genetically and structurally different individuals, it is likely that some of them will possess characteristics that make their survival difficult. Therefore, they are likely to die early in life and not have an opportunity to pass their genes on to the next generation.

Charles Darwin described several species of finches on the Galápagos Islands, and scientists have often used these birds in scientific studies of evolution. On one of the islands, scientists studied one of the species of seed-eating ground finches, *Geospiza fortis.* They measured the size of the animals and the size of their bills and related these characteristics to their survival. They found the following: During a drought, the birds ate the smaller, softer seeds more readily than the larger, harder seeds. As the birds consumed the more easily eaten seeds, only the larger, harder seeds remained. During the drought, mortality was extremely high. When scientists looked at ground finch mortality, they found that the larger birds that had stronger, deeper bills survived better than smaller birds with weaker, narrower bills. They also showed that the offspring of the survivors tended to show larger body and bill size as well. The lack of small, easily eaten seeds due to the drought resulted in selection for larger birds with stronger bills that could crack open larger, tougher seeds. Table 21.2 shows data on two of the parameters measured in this study.

TABLE 21.2 **Changes in the Body Structure of *Geospiza fortis***

	Before Drought	After Drought
Average Body Weight	16.06 g	17.13 g
Average Bill Depth	9.21 mm	9.70 mm

As another example of how differential survival can lead to changes in the genes within a population, consider what has happened to many insect populations as we have subjected them to a variety of insecticides. Since there is genetic diversity within all species of insects, an insecticide that is used for the first time on a particular species kills all those that are genetically susceptible. However, the insecticide may not kill individuals with slightly different genetic compositions that give them resistance to the insecticide.

Suppose that in a population of a particular species of insect, 5 percent of the individuals have genes that make them resistant to a specific insecticide. The first application of the insecticide could, therefore, kill 95 percent of the population. However, tolerant individuals would then constitute the majority of the breeding population that survived. This would mean that a larger proportion of the insects in the second generation would be tolerant. The second use of the insecticide on this population would not be as effective as the first. With continued use of the same insecticide, each generation would become more tolerant. Please note in this example that the spraying did not cause mutations that made the population resistant. The use of insecticide killed susceptible individuals, thus leaving resistant individuals to reproduce and pass their resistance genes to the next generation.

Many species of insects produce a new generation each month. In organisms with a short generation time, 99 percent of the population could become resistant to the insecticide in just five years. As a result, the insecticide would no longer be useful in controlling the species. As a new factor (the insecticide) was introduced into the environment of the insect, natural selection resulted in a population that was tolerant of the insecticide.

A recent study of insecticide resistance of houseflies in poultry facilities used in egg production illustrates this point. Poultry facilities are an excellent situation for developing resistance since the flies spend their entire lives within the facility and are subjected to repeated applications of insecticide. Table 21.3 shows the effect of use of the insecticide Cyfluthrin on resistance in a particular housefly population. The standard dose that killed susceptible houseflies did not kill the resistant population. Even at a dosage of one hundred times the standard dose, nearly 40 percent of the houseflies in the resistant population survived.

TABLE 21.3   Insecticide Resistance of Houseflies to the Insecticide Cyfluthrin

	Houseflies from a Susceptible Population		Houseflies from a Poultry Facility Population	
Pesticide Concentration (ng/cm²)	8.3 (standard dose)	8.3 (standard dose)	83 (10 × standard dose)	830 (100 × standard dose)
Percent Surviving	0	100	90	38

Source: Data from Scott, Jeffrey G., et al. "Insecticide resistance in houseflies from caged-layer poultry facilities," *Pest Management Science* 56 (2000):147–53.

Differential Reproductive Rates

Survival alone does not always ensure reproductive success. For a variety of reasons, some organisms may be better able to utilize available resources to produce offspring. If one individual leaves 100 offspring and another leaves only 2, the first organism has passed more copies of its genetic information to the next generation than has the second and the mix of genes in the population of offspring is different from that of their parents.

Scientists have conducted studies of the frequencies of genes for the height of clover plants (figure 21.9). Two identical fields of clover were planted and cows were allowed to graze in one of them. Cows acted as a selecting agent by eating the taller plants first. These tall plants rarely got a chance to reproduce. Only the shorter plants flowered and produced seeds. After some time, seeds were collected from both the grazed and ungrazed fields and grown in a greenhouse under identical conditions. The average height of the plants from the ungrazed field was compared to that of the plants from the grazed field. The seeds from the ungrazed field produced some tall, some short, but mostly medium-sized plants. However, the seeds from the grazed field produced many more shorter plants than medium or tall ones. The cows had selectively eaten the plants that had the genes for tallness. Since the flowers are at the tip of the plant, tall plants were less likely to successfully reproduce, even though they might have been able to survive grazing by cows.

Differential Mate Selection

Within animal populations, some individuals may be chosen as mates more frequently than others. Obviously, those that are frequently chosen have an opportunity to pass on more copies of their genes than those that are rarely chosen. Characteristics of the more frequently chosen individuals may involve general characteristics, such as body size or aggressiveness, or specific conspicuous characteristics attractive to the opposite sex.

For example, male red-winged blackbirds establish territories in cattail marshes where females build their nests. A male will chase out all other males but not females. Some males have large territories, some have small territories, and some are unable to establish territories. Although it is possible for any male to mate, it has been demonstrated that those who have no territory are least likely to mate. Those who defend large territories may have two or more females nesting in their territories and are very likely to mate with those females. It is unclear exactly why females choose one male's territory over another, but the fact is that some males are chosen as mates and others are not.

In other cases, it appears that the females select males that display conspicuous characteristics. Male peacocks have very conspicuous tail feathers. Those with spectacular tails are more likely to mate and have offspring (figure 21.10). Darwin was puzzled by such cases as the peacock, in which the large and conspicuous tail should have been a disadvantage to the bird. Long tails require energy to produce, make it more difficult to fly, and make it more likely that predators will capture the individual. The current theory that seeks to explain this paradox involves female choice. If the females have an innate (genetic) tendency to choose the most elaborately decorated males, genes that favor such plumage will be regularly passed to the next

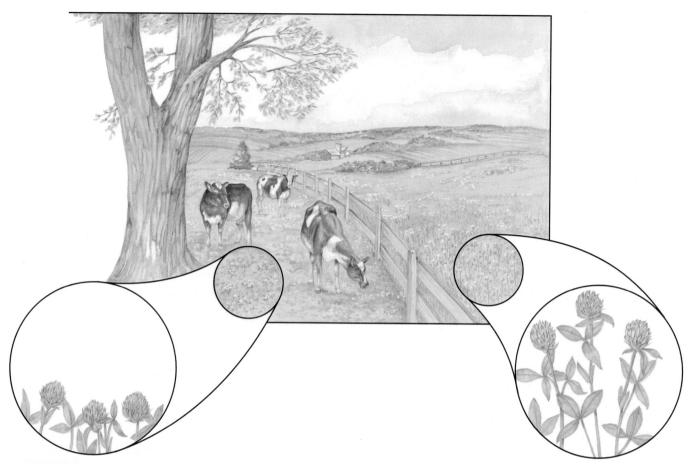

FIGURE 21.9 The clover field to the left of the fence is undergoing natural selection as a result of cattle eating the tall plants and causing them to reproduce less than do the short plants. The other field is not subjected to this selection pressure, so its clover population has more genes for tallness.

FIGURE 21.10 In many animal species, the males display very conspicuous characteristics that are attractive to females. Because the females choose the males they will mate with, those males with the most attractive characteristics will have more offspring, and in future generations, there will be a tendency to enhance the characteristic. With peacocks, those individuals with large colorful displays are more likely to mate. The male in the photograph is displaying to a female. Note that the female is not as highly colored and has a short tail.

generation because females choose the gaudy males as mates. Such special cases in which females choose males with specific characteristics have been called sexual selection.

ACQUIRED CHARACTERISTICS DO NOT INFLUENCE NATURAL SELECTION

Many individual organisms acquire characteristics during their lifetime that are not the result of the genes that they have. For example, a squirrel may learn where a bird feeder is and visit it frequently. It may even learn to defeat the mechanics of special bird feeders designed to prevent it from getting to the food. Such an ability can be very important for survival if food becomes scarce. Such **acquired characteristics** are gained during the life of the organism; they are not genetically determined and, therefore, cannot be passed on to future generations through sexual reproduction.

For example, we often desire a specific set of characteristics in our domesticated animals. The breed of dog known as boxers, for example, is "supposed" to have short tails. However, the genes for short tails are rare in this breed. Consequently, the tails of these dogs are amputated—a procedure called docking.

Similarly, the tails of lambs are also usually amputated. These acquired characteristics are not passed on to the next generation. Removing the tails of these animals does not remove the genes for tail production from their genetic information, and each generation of puppies or sheep is born with tails.

However, before the nature of the gene and mechanisms of heredity were understood it was commonly thought that acquired characteristics could be passed from generation to generation. As mentioned earlier, Lamarck postulated that giraffes could have developed long necks as a result of stretching them to reach food and then these acquired characteristics could be passed to offspring. Darwin's theory of natural selection provides an alternative explanation for long necks in giraffes.

1. In each generation, more giraffes would be born than the food supply could support.
2. In each generation, some giraffes would inherit longer necks, and some would inherit shorter necks.
3. All giraffes would compete for the same food sources.
4. Giraffes with longer necks would obtain more food, have a higher survival rate, and produce more offspring.
5. As a result, succeeding generations would show an increase in the neck length of the giraffe species.

Figure 21.11 contrasts how Lamarck and Darwin provide different explanations for the phenomenon of long necks in giraffes.

THE HARDY-WEINBERG CONCEPT

In the section on evolution and natural selection earlier in this chapter, evolution was described as the change in the genetic makeup of a population over time. We can think of all the genes of all the individuals in a population as a **gene pool** and that a change in the gene pool indicates that evolution is taking place.

In the early 1900s, an English mathematician, G. H. Hardy, and a German physician, Wilhelm Weinberg, recognized that it was possible to apply a simple mathematical relationship to the study of genes in populations if certain conditions were met. An unchanging gene pool over several generations would imply that evolution is *not* taking place. A changing gene pool would indicate that evolution is taking place.

The conditions they felt were necessary for the genetic makeup to *remain constant* (for evolution *not* to occur) are:

1. Mating must be completely random.
2. Mutations must not occur.
3. Migration of individual organisms into and out of the population must not occur.
4. The population must be very large.
5. All genes must have an equal chance of being passed on to the next generation. (Natural selection is not occurring.)

The concept that the genetic makeup will remain constant if these five conditions are met has become known as the **Hardy-Weinberg concept.** The Hardy-Weinberg concept is

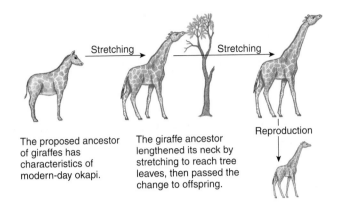

The proposed ancestor of giraffes has characteristics of modern-day okapi.

The giraffe ancestor lengthened its neck by stretching to reach tree leaves, then passed the change to offspring.

Reproduction

A Lamarck's theory: variation is acquired.

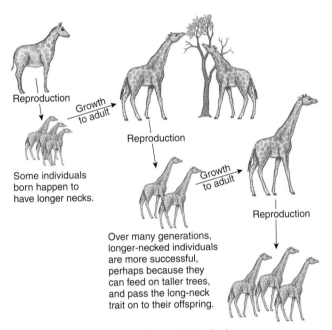

Reproduction

Growth to adult

Reproduction

Some individuals born happen to have longer necks.

Growth to adult

Reproduction

Over many generations, longer-necked individuals are more successful, perhaps because they can feed on taller trees, and pass the long-neck trait on to their offspring.

B Darwin-Wallace theory: variation is inherited.

FIGURE 21.11 (*A*) Lamarck thought that acquired characteristics could be passed on to the next generation. Therefore, he postulated that as giraffes stretched their necks to get food, their necks got slightly longer. This characteristic was passed on to the next generation, which would have longer necks. (*B*) The Darwin-Wallace theory states that there is variation within the population and that those with longer necks would be more likely to survive, reproduce, and pass on their genes for long necks to the next generation.

important because it allows a simple comparison of genes in populations to indicate if genetic changes are occurring within the population. Two different populations of the same species can be compared to see if they have the same genetic makeup, or populations can be examined at different times to see if the genetic makeup is changing.

Although Hardy and Weinberg described the conditions necessary to prevent evolution, an examination of these conditions becomes a powerful argument for why evolution must occur.

A Closer Look

The Development of New Viral Diseases

Mutations are important in the process of evolution. Viruses are particles that are parasites within the cells of organisms. They have either DNA or RNA as their genetic material, and this genetic material only expresses itself, and the virus only displays characteristics of life, when the virus is inside a cell. As parasites and their hosts interact, they are constantly reacting to each other in an evolutionary fashion. Hosts develop new mechanisms to combat parasites, and parasites develop new mechanisms to overcome the defenses of hosts. One of the mechanisms employed by viruses involves high rates of mutation. This is particularly true for RNA viruses because they lack the DNA repair enzymes. Common RNA viruses are HIV, SARS, cold, influenza, polio, and hepatitis viruses. This ability to mutate has resulted in many new, serious human diseases. In addition, many new diseases arise when viruses that cause disease in another animal are able to establish themselves in humans. Many kinds of influenza originated in pigs, ducks, or chickens and were passed to humans through close contact with or eating infected animals. Since in many parts of the world these domesticated animals live in close contact with humans (often in the same building), conditions are favorable for the transmission of animal viruses to humans.

Each year, mutations result in new varieties of influenza and colds that pass through the human population. Occasionally, the new varieties are particularly deadly. In 1918, a new variety of influenza virus originated in pigs in the United States and spread throughout the world. During the 1918–1919 influenza pandemic that followed, twenty million to forty million people died. In 1997 in Hong Kong, a new kind of influenza was identified that killed six of eighteen people infected. Public health officials discovered the virus had come from chickens and ordered the slaughter of all live chickens in Hong Kong. This stopped the spread of the disease. Since 1997 bird flu has been a constant concern and outbreaks have occurred in Vietnam, Indonesia, and Thailand. Isolated cases have occurred in other places. However, most cases appear to be related to human contact with infected birds. In a few cases it is suspected that the disease has been passed from one human to another but so far these cases are rare. If the virus mutates so that human-to-human transmission becomes easy it would become a major world health problem.

In early 2003, an outbreak of a new viral disease known as *severe acute respiratory syndrome* (SARS) originated in China. It is a variation of a coronavirus, a class of virus commonly associated with the common cold, but it caused severe symptoms and resulted in many deaths. In June 2003, the SARS virus was isolated from an animal known as the masked palm civet (*Paguma larvata*). This animal is used for food in China and is the likely origin of this new human disease. Isolation of infected persons prevented the spread, and this new disease was brought under control.

In 2009 a new strain of H1N1 influenza virus resulted in a flu pandemic (flu that spread throughout the world). This new strain of H1N1 virus is commonly called the "swine flu" because it appears to have been transmitted to people from pigs. It is likely that various strains of flu viruses spread through pig populations and that over a number of years, genes from these different strains were recombined to produce a new strain of virus. When it was transmitted from pigs to people it spread readily, since few people had immunity to this new strain of virus. One unusual characteristic of this strain of flu is that it often affects younger people more severely than older people. This may be because older people were exposed to a previous strain of flu, which caused them to have a partial immunity to this new strain.

Random mating rarely occurs. For example, mating between individuals is more likely between those that are nearby than those that are distant. In addition, in animal populations, the possibility exists that females may choose specific males with which to mate.

Mutations occur and alter genetic diversity by either changing one genetic message into another or introducing an entirely new piece of genetic information into the population.

Migration alters the gene pool, since organisms carry their genes with them when they migrate from one place to another. Even plants and fungi migrate when their reproductive stages are carried to new places by wind, water, or animals. Immigration introduces new genetic information, and emigration removes genes from the gene pool.

Population size influences the gene pool. Small populations typically have less genetic diversity than large populations.

Finally, *natural selection* systematically filters some genes from the population, allowing other genes to remain and become more common. The primary mechanisms involved in natural selection are differences in death rates, reproductive rates, and the rate at which individuals are selected as mates. The diagram in figure 21.12 summarizes these ideas.

Common Misconceptions About the Theory of Evolution

1. *Evolution happened only in the past and is not occurring today.* In fact, we see lots of evidence that genetic changes are occurring in the populations of current species (antibiotic resistance, pesticide resistance, and domestication).

2. *Evolution has a specific goal.* Evolution does not move toward a specific goal. Natural selection selects those organisms that best fit the current environment. As the environment changes, so do the characteristics that have value. Random events such as changes in sea level, major changes in climate such as ice ages, or collisions with asteroids have had major influences on subsequent natural selection and evolution. Evolution results in organisms that "fit" the current environment.

3. *Changes in the environment cause mutations that are needed by an organism to survive under the new environmental conditions.* Mutations are random events and are not necessarily adaptive. However, when the environment changes, mutations that were originally detrimental or neutral may have greater value. The gene did not change but the environmental conditions did. In some cases, the mutation rate may increase or there may be more frequent exchanges of genes between individuals when the environment changes, but the mutations are still random. They are not directed to a particular goal.

4. *Individual organisms evolve.* Individuals are stuck with the genes they inherited from their parents. Although individuals may adapt by changing their behavior or physiology, they cannot evolve; only populations can show changes in the gene pool.

5. *Today's species frequently can be shown to be derived from other present-day species. For example, it is commonly stated that apes gave rise to humans.* There are few examples in which it can be demonstrated that one current species gave rise to another. Apes did not become humans, but apes and humans had a common ancestor several million years ago.

(figure 21.13). Since mules are nearly always sterile, there can be no gene flow between horses and donkeys and they are considered to be separate species. Similarly, lions and tigers can be mated in zoos to produce offspring. However, this does not happen in nature, so gene flow does not occur naturally; thus, they are considered to be two separate species.

Still another way to try to determine if two organisms belong to different species is to determine their genetic similarity. The recent advances in molecular genetics allow scientists to examine the structure of the genes present in individuals from a variety of different populations. Those that have a great deal of similarity are assumed to have resulted from populations that have exchanged genes through sexual reproduction in the recent past. If there are significant differences in the genes present in individuals from two populations, they have not exchanged genes recently and are more likely to be members of separate species. Interpretation of the results obtained by examining genetic differences still requires the judgment of experts. It will not unequivocally settle every dispute related to the identification of species, but it is another tool that helps to clarify troublesome situations.

This concept that species can be distinguished from one another by their inability to interbreed or produce fertile offspring is often called the **biological species concept.** Although this concept of a species is useful from a theoretical point of view, it is often not a practical way to distinguish species. Thus, biologists often use specific observable physical, chemical, or behavioral characteristics as guides to distinguishing species. This method of using physical characteristics to identify species is called the **morphological species concept.** Structural differences are useful but not foolproof ways to distinguish species. However, we must rely on such indirect ways to identify species because we cannot possibly test every individual by breeding it with another to see if they will have fertile offspring. Furthermore, many kinds of organisms reproduce primarily by asexual means. Because organisms that reproduce exclusively by asexual methods do not exchange genes with any other individuals, they do not fit our *biological species* definition very well. In addition, the study of fossil species must rely on structural characteristics to make species distinctions since it is impossible to breed extinct organisms.

A Horse

B Donkey

C Mule

FIGURE 21.13 Even though they don't do so in nature, (A) horses and (B) donkeys can be mated. The offspring is called a (C) mule and is sterile. Because the mule is sterile, the horse and the donkey are considered to be of different species.

Connections . . .

Dogs and Wolves Are the Same Species

Scientists compared the mitochondrial DNA sequences from sixty-seven breeds of dogs with that of wolves, coyotes, and jackals. Mitochondrial DNA passes only from the mother to the offspring. All the wolf and dog sequences were similar, and both differed significantly from coyotes and jackals. Careful analysis of the differ-ences in mitochondrial DNA suggests that there were two major domestication events from different parent wolf populations and that there have been incidences of inter-breeding of wolves and dogs several times. Furthermore, the various "breeds" of dogs could not be distinguished by the sequence of DNA. Dogs that superficially resemble one another may have more differences in mitochondrial DNA than those that appear quite different. The breeds of dogs are really rather superficial differences in the basic animal. The unmistakable conclusion is that dogs are simply domesticated wolves.

HOW NEW SPECIES ORIGINATE

The geographic area over which a spe-cies can be found is known as its **range.** The range of the human species is the entire world, while that of a bird known as a snail kite is a small region of southern Florida. As a species expands its range or environmental conditions change in some parts of the range, portions of the popula-tion can become separated from the rest. This means that new colonies or isolated populations have infrequent gene exchange with their geographically distant relatives. Thus, many species consist of partially isolated populations that display charac-teristics that differ significantly from other local populations. Many of the differences observed may be directly related to adap-tations to local environmental conditions. These genetically distinct populations are known as subspecies.

Speciation is the process of generat-ing new species. The process can occur in several ways. One common way that speciation occurs is through geographic isolation.

The Role of Geographic Isolation in Speciation

A portion of a species can become totally isolated from the rest of the gene pool by some geographic change, such as the for-mation of a mountain range, river valley, desert, or ocean. When this happens, the portion of the species is said to be in **geographic isolation** from the rest of the spe-cies. If two populations of a species are geographically isolated, they are also reproductively isolated, and gene exchange is not occurring between them. The geographic features that keep the different portions of the species from exchanging genes are called **geographic barriers.** The uplifting of mountains,

Kaibab squirrel

Aberts squirrel

FIGURE 21.14 These two squirrels are found on opposite sides of the Grand Canyon. Some people consider them to be different species; others consider them to be distinct populations of the same species.

the rerouting of rivers, and the formation of deserts all may separate one portion of a gene pool from another. For example, two kinds of squirrels are found on opposite sides of the Grand Canyon. Some people consider them to be separate species, while others consider them to be different isolated subpopula-tions of the same species (figure 21.14). Even small changes

<hr />

①
No barrier;
one species

②
Barrier
allows differences
to develop in
two populations.

③
Differences
so great that
two species
are evident

④
When barrier is
removed, species
do not interbreed.

Time

FIGURE 21.15 If a single species of snail were to be divided into two different populations by the development of a ridge between them, the two populations could be subjected to different environmental conditions. This could result in a slow accumulation of changes that could ultimately result in two populations that would not be able to interbreed even if the ridge between them were to erode. They would be different species.

differences act as natural selecting agents on the two gene pools and, acting over a long period of time, account for different genetic combinations in the two places. Furthermore, different mutations may occur in the two isolated populations, and each may generate different random combinations of genes as a result of sexual reproduction. This would be particularly true if one of the populations was very small. As a result, the two populations may show differences in color, height, enzyme production, time of seed germination, or many other characteristics.

Reproductive Isolation

Over a long period of time, the genetic differences that accumulate may result in regional populations called **subspecies** that are significantly modified structurally, physiologically, or behaviorally. The differences among some subspecies may be so great that they have reduced reproductive success when the subspecies mate. If enough genetic differences accumulate that two populations are not able to interbreed, then speciation has occurred. Speciation has occurred only if gene flow between isolated populations does not occur even after barriers are removed. In other words, the process of speciation can begin with the geographic isolation of a portion of the species, but new species are generated only if isolated populations become separate from one another genetically. Speciation by this method is really a three-step process. It begins with geographic isolation, is followed by the action of selective agents that choose specific genetic combinations as being valuable, and ends with the genetic differences becoming so great that reproduction between the two groups is impossible.

Speciation Without Isolation

It is also possible to envision ways in which speciation could occur without geographic isolation being necessary. Any process that could result in the reproductive isolation of a portion of a species could lead to the possibility of speciation. For example, within populations, some individuals may breed or flower at a somewhat different time of the year. If the difference in reproductive time is genetically based, different breeding populations could be established, which could eventually lead to speciation. Among animals, variations in the genetically determined behaviors related to courtship and mating could effectively separate one species into two or more separate breeding populations. In plants, genetically determined incompatibility of the pollen

may cause geographic isolation in species that have little ability to move. A fallen tree, a plowed field, or even a new freeway may effectively isolate populations within such species. Snails in two valleys separated by a high ridge have been found to be closely related but different species. The snails cannot get from one valley to the next because of the height and climatic differences presented by the ridge (figure 21.15).

The Role of Natural Selection in Speciation

The separation of a species into two or more isolated subpopulations is not enough to generate new species. Even after many generations of geographic isolation, these separate groups may still be able to exchange genes (mate and produce fertile offspring) if they overcome the geographic barrier, because they have not accumulated enough genetic differences to prevent reproductive success. Differences in environments and natural selection play very important roles in the process of forming new species. Following separation from the main portion of the gene pool by geographic isolation, the organisms within the small, local population are likely to experience different environmental conditions. If, for example, a mountain range has separated a species into two populations, one of them may receive more rain or more sunlight than the other (figure 21.16). These environmental

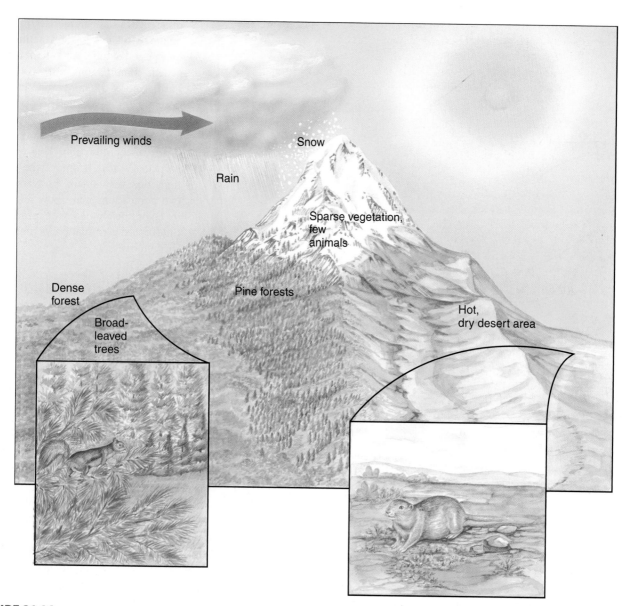

FIGURE 21.16 Most mountain ranges affect the local environment. Because of the prevailing winds, most rain falls on the windward side of the mountain. This supports abundant vegetation. The other side of the mountain receives much less rain and is drier. Often a desert may exist. Both plants and animals must be adapted to the kind of climate typical for their regions. Cactus and ground squirrels would be typical of the desert, and pine trees and tree squirrels would be typical of the windward side of the mountain.

of one population of flowering plants with the flowers of other populations of the same species could lead to separate species.

THE TENTATIVE NATURE OF THE EVOLUTIONARY HISTORY OF ORGANISMS

It is important to understand that thinking about the concept of evolution can take us in several different directions. First, it is clear that genetic changes do occur. Mutations introduce new genes into a species. This has been demonstrated repeatedly with chemicals and radiation. Our recognition of this danger is evident by the ways we protect ourselves against excessive

exposure to mutagenic agents. We also recognize that species can change. We purposely manipulate the genetic constitution of our domesticated plants and animals and change their characteristics to suit our needs. We also recognize that different populations of the same species show genetic differences. Examination of fossils shows that species of organisms that once existed are no longer in existence. We even have historical examples of plants and animals that are now extinct. We can also demonstrate that new species come into existence, and this is easily done with plants. It is clear from this evidence that species are not fixed, unchanging entities.

However, when we try to piece together the evolutionary history of organisms over long periods of time, we must use much indirect evidence, and it becomes difficult to state definitively the specific sequence of steps that the evolution of a species followed. Although it is clear that evolution occurs, it is not possible to state unconditionally that evolution of a particular group of organisms has followed a specific path. There will always be new information that will require changes in thinking, and equally reputable scientists will disagree on the evolutionary processes or the sequence of events that led to a specific group of organisms. But there can be no question that evolution occurred in the past and continues to occur today.

SUMMARY

Current theories about the *origin of life* include the two competing ideas (1) that the primitive Earth environment led to the *spontaneous organization of organic chemicals* into primitive cells or (2) that primitive forms of life *arrived on Earth from space*. Basic units of life were probably similar to present-day *prokaryotes*. These primitive cells could have *changed through time* as a result of *mutation* and *in response to a changing environment*. The presence of living things has affected the nature of Earth's atmosphere. The first organisms would have been *anaerobic* since there was no oxygen in the atmosphere. *Photosynthesis* by organisms such as cyanobacteria would have resulted in oxygen being added to the atmosphere. The presence of oxygen allowed for the development of *aerobic respiration*. Three major kinds of organisms are referred to as *domains: Bacteria, Archaea*, and *Eucarya*. It appears that Bacteria are the oldest group, followed by Archaea and Eucarya. The Eucarya appear to have arisen as a result of the process of *endosymbiosis*.

All sexually reproducing organisms naturally exhibit genetic diversity among the individuals in the population as a result of *mutations* and the *genetic recombination* resulting from sexual reproduction. The *genetic differences* are reflected in physical differences among individu-als. These genetic differences are important for the survival of the species because *natural selection* must have *genetic diversity* to select from. Natural selection by the *environment* results in better-suited individual organisms that produce greater numbers of offspring than those that are less well-off genetically.

Organisms with wide *geographic distribution* often show genetic differences in separate parts of their *range*. A *species* is a group of *organisms* that can interbreed to produce fertile offspring. This definition of a species is also known as the *biological species concept*. People often distinguish species from one another based on structural differences. This is known as a *morphological species concept*. The process of *speciation* usually involves the *geographic separation* of the species into two or more isolated *populations*. While they are separated, natural selection operates to *adapt each population to its environment*.

At one time, people thought that all organisms had remained *unchanged* from the time of their creation. Lamarck suggested that *change did occur* and thought that *acquired characteristics could be passed from generation to generation*. Darwin and Wallace proposed the *theory of natural selection* as the mechanism that drives *evolution*.

KEY TERMS

acquired characteristics (p. **510**)
autotroph (p. **499**)
biogenesis (p. **497**)
biological species concept (p. **517**)
domains (p. **501**)
endosymbiotic theory (p. **502**)
evolution (p. **505**)

gene flow (p. **516**)
gene pool (p. **511**)
genetic recombination (p. **508**)
geographic barriers (p. **518**)
geographic isolation (p. **518**)
Hardy-Weinberg concept
 (p. **511**)

heterotroph (p. **499**)
morphological species concept
 (p. **517**)
mutation (p. **507**)
natural selection (p. **505**)
oxidizing atmosphere (p. **501**)
range (p. **518**)

selecting agents (p. **508**)
speciation (p. **518**)
species (p. **516**)
spontaneous generation (p. **497**)
spontaneous mutation (p. **507**)
subspecies (p. **519**)
theory of natural selection (p. **505**)

APPLYING THE CONCEPTS

Part I

1. Which was *not* a major component of the early Earth's reducing atmosphere?
   a. H_2
   b. CO_2
   c. NH_3
   d. O_2

2. Evidence from fossils shows that prokaryotic cells came into existence approximately ——————— years ago.
   a. 20 billion
   b. 3.5 billion
   c. 4–5 billion
   d. 1.5 billion

3. Oxygen present in our atmosphere is the result of
   a. organisms carrying on respiration.
   b. gases from volcanoes.
   c. the breakdown of carbon dioxide into carbon and oxygen.
   d. organisms carrying on photosynthesis

4. The Domain Eucarya
   a. has the simplest cell structure.
   b. is probably the result of endosymbiosis.
   c. was the first to be produced.
   d. began about 3.5 billion years ago.

Part II

5. Which of the following is involved in generating new genetic combinations in organisms?
   a. acquired characteristics
   b. sexual reproduction
   c. Hardy-Weinberg
   d. natural selection

6. Which of the following could lead to a changed gene frequency in a population?
   a. Individuals with certain characteristics die.
   b. Individuals with certain characteristics are attractive to the opposite sex.
   c. Individuals with certain characteristics have more offspring than others.
   d. All of the above are correct.

7. The Darwin-Wallace theory of natural selection differs from Lamarck's ideas in that
   a. Lamarck understood the role of genes, and Darwin and Wallace did not.
   b. Lamarck assumed that characteristics obtained during an organism's lifetime could be passed to the next generation; Darwin and Wallace did not.
   c. Lamarck did not think that evolution took place; Darwin and Wallace did.
   d. Lamarck developed the basic ideas of speciation, which Darwin and Wallace refined.

8. The Hardy-Weinberg concept requires that
   a. mutations occur   c. migration occur
   b. mating be random   d. natural selection take place

Part III

9. Hybrid animals such as mules are not considered to be a species because they
   a. do not reproduce through many generations.
   b. are not common enough.
   c. can be maintained only by humans.
   d. look different from both parents.

10. Two closely related organisms are not considered to be separate species unless they
    a. look different.
    b. are reproductively isolated.
    c. are able to interbreed.
    d. are in different geographic parts of the world.

11. Which of the following is required for speciation to occur?
    a. genetic isolation from other species
    b. genetic diversity within a species
    c. hundreds of millions of years
    d. reproduction

12. Two groups of organisms belong to different species if
    a. gene flow between the two groups is not possible, even in the absence of physical barriers.
    b. physical barriers separate the two groups, thereby preventing cross matings.
    c. the two groups of organisms have a different physical appearance.
    d. individuals from the two groups, when mated, produce fertile offspring.

Answers
1. d 2. b 3. d 4. b 5. b 6. d 7. b 8. b 9. a 10. b 11. a 12. a

QUESTIONS FOR THOUGHT

Part I

1. In what sequence did the following things happen: living cell, oxidizing atmosphere, eukaryotes developed, reducing atmosphere, first organic molecule?

2. What is meant by *spontaneous generation?* What is meant by *biogenesis?*

3. List two important effects caused by the increase of oxygen in the atmosphere.

4. What evidence supports the theory that eukaryotic cells arose from the development of a symbiotic relationship between primitive prokaryotic cells?

Part II

5. Why are acquired characteristics of little interest to evolutionary biologists?

6. What is natural selection? How does it work?

7. Give two examples of selecting agents and explain how they operate.

8. Why has Lamarck's theory been rejected?

Part III

9. "Evolution is a fact." "Evolution is a theory." Explain how both of these statements can be true.

10. Why is geographic isolation important in the process of speciation?

11. Why aren't mules considered a species?

12. List the series of events necessary for speciation to occur.

FOR FURTHER ANALYSIS

Part I

1. In earlier times, people were certain that horse hairs could turn into worms. There is even a group of simple kinds of worms known as horse-hair worms. Consider the following evidence:

   - Watering troughs were provided for horses to drink from.
   - Hairs were observed to fall from the horse's mane into the water of the watering trough.
   - Long, thin worms were seen to be swimming in the watering troughs.

   Since hair is not alive, if this were to occur, it would be a case of spontaneous generation. Devise an experiment that would determine if nonliving horse hairs become living worms.

Part II

2. People who are hospitalized often develop infections after they have entered the hospital. Often these infections are very difficult to treat with conventional antibiotics. Why do you think this is so?

3. How much diversity is there in hair color in the students of your class? Assume all colors seen are natural. Have the students stand in groups based on distinguishable differences in hair color. How many different categories do you have? What does this tell you about genetic diversity within our species?

4. During the early part of the twentieth century, a major social movement known as the eugenics movement occurred. The purpose of eugenics was to improve the quality of the human gene pool by eliminating "bad genes" from the population. Universities taught courses in eugenics, states passed laws that allowed certain persons to be sterilized against their will, and judging contests were held at state fairs to determine the family with the best characteristics. Hitler's concept of developing a "master race" was an extension of this thinking. Scientists who study evolutionary processes have thoroughly discredited eugenics. However, there are still vestiges of this thinking in society. Describe several examples.

Part III

5. Scientist regularly reorganize how they think specific species of organisms evolved. Some people argue that this is evidence that evolution is "only a theory" and that we can't ever know how change takes place. How would you refute this criticism?

INVITATION TO INQUIRY

The Evolution of Technology

A circular computer disc stores information on its surface in the form of magnetic "spots." This disc spins and is read by a device that detects the different magnetic spots. Assume you are trying to determine the evolutionary process that led to this current level of technology. How would you fit the following "fossils" into your evolutionary scheme: Edison's cylinder phonograph, a magnetic tape machine, an LP record, and a CD. Reread "A Closer Look: The Compact Disc" on page 171 to review how a CD stores information. This is similar to how Edison's cylinder and an LP record store information. Determine the approximate time each technology was developed and describe a logical evolutionary relationship among these technologies. Can you see a direct development from one technology to another? Were there instances of "endosymbiosis" in which one technology was applied within a new system?

22

The History of Life on Earth

These are modern-day stromatolites from Hamlin Pool in Western Australia. The dome-shaped structures shown in the photograph are composed of layers of cyanobacteria and materials they secrete. They grow up to 60 cm (about 2 ft) tall. Some of the oldest fossils are of ancient stromatolites that developed in shallow marine environments about 3.5 billion years ago. When samples from fossil stromatolites are cut into slices, microscopic images can be produced that show the fossil remains of some of the world's oldest cells.

CORE **CONCEPT**

Earth and its kinds of living things changed greatly over billions of years.

OUTLINE

The same geological processes operate today as in the past (p. 527).

Fossils provide evidence for past life and show that the kinds of organisms on Earth have changed (p. 533).

In the past, the positions of the continents changed, the climate changed, and sea level changed (p. 533).

Living things can be organized into logical evolutionary sequences (p. 544).

Humans are a recent addition to Earth (p. 538).

Physics

▷ The nature of radioactivity (Ch. 11).

Chemistry

▷ Fossil fuels are products of past living things (Ch. 20).

Earth Science

▷ Different types of rocks form under specific conditions (Ch. 15).

▷ Plate tectonics results in the movement of continents (Ch. 15).

▷ Weathering processes are important geologic forces (Ch. 16).

Life Science

▷ Evolution of organisms occurs (Ch. 21).

▷ Ecological principles can be applied to past biological communities (Ch. 23).

▷ Endosymbiosis probably led to eukaryotic organisms (Ch. 21).

Astronomy

▷ Meteorite impacts occurred (Ch. 13).

OVERVIEW

At one time, people assumed that Earth and its inhabitants were fixed and unchanging. However, as they began to understand the nature of events such as volcanic eruptions, earthquakes, and erosion, it became obvious that Earth was changeable. Furthermore, the presence of the fossilized remains of extinct organisms showed that the nature of the kinds of things that inhabited Earth had changed as well. These ideas challenged core beliefs and began a long process of modifying the way we look at the history of Earth and life on it. In particular, people needed to develop an understanding that the history of Earth and its life was very long and should be measured in billions of years rather than hundreds or thousands of years.

PART I
THE GEOLOGIC HISTORY OF EARTH

This chapter is divided into two parts. The first part deals with how to interpret the geologic history of Earth and how evidence from geology is used to gain an understanding of how organisms have changed through time.

GEOLOGIC TIME

Reading history from the rocks of Earth's crust requires both a feel for the immensity of geologic time and an understanding of geologic processes. By "geologic time," we mean the age of Earth; the very long span of Earth's history. This span of time is difficult for most of us to comprehend. Human history is measured in units of hundreds and thousands of years, and even the events that can take place in a thousand years are hard to imagine. Geologic time, on the other hand, is measured in units of millions and billions of years.

The understanding of geologic processes has been made possible through the development of various means of measuring ages and time spans in geologic systems. An understanding of geologic time leads to an understanding of geologic processes, which then leads to an understanding of the environmental conditions that must have existed in the past. Thus,

the mineral composition, texture, and sedimentary structure of rocks are clues about past events, events that make up the geological and biological history of Earth.

Early Attempts at Earth Dating

Most human activities are organized by time intervals of minutes, hours, days, months, and years. These time intervals are based on the movements of Earth and the Moon. Short time intervals are measured in minutes and hours, which are typically tracked by watches and clocks. Longer time intervals are measured in days, months, and years, which are typically tracked by calendars. How do you measure and track time intervals for something as old as Earth? First, you would need to know the age of Earth; then you would need some consistent, measurable events to divide the overall age into intervals. Questions about the age of Earth have puzzled people for thousands of years, dating back at least to the time of the ancient Greek philosophers. Many people have attempted to answer this question and understand geologic time but with little success until the last few decades.

Biblical Calculations

One early estimate of the age of Earth was attempted by Archbishop James Ussher of Ireland in the seventeenth century. He painstakingly counted up the generations of people mentioned in biblical history, added some numerological considerations, and arrived at the conclusion that Earth was created on October 23, in the year 4004 B.C. On the authority of biblical scholars, this date was generally accepted for the next century or so, even though some people thought that the geology of Earth seemed to require far longer to develop. The date of 4004 B.C. meant that Earth and all of the surface features on Earth had formed over a period of about six thousand years. In order to account for the location of fossils of marine organisms on mountains, deep river valleys, and other geologic features in such a short time, a model of geologic processes that involved great cataclysmic catastrophes was required. These catastrophes typically included earthquakes and world-wide floods.

Early Scientific Approaches

Near the end of the eighteenth century, James Hutton in Scotland reasoned out the concept that the geologic processes we see today have operated since Earth began. This is known as the *principle of uniformity,* and since many geologic processes such as erosion and sedimentation are slow, people began to assume a much older Earth. Because the principle of uniformity assumed a slow pace for geologic changes, this view of geologic processes became known as gradualism to contrast it with catastrophism, which had been the prevailing view for centuries. (Many of Hutton's ideas were restated and popularized by Charles Lyell, whose writings greatly influenced Charles Darwin.) The problem then became one of finding some uniform change or process that could serve as a geologic clock to measure the age of Earth. To serve as a geologic clock, a process or change would need to meet three criteria: (1) the process must have been operating since Earth began, (2) the process must be uniform or at least subject to averaging, and (3) the process must be measurable.

During the nineteenth century, many attempts were made to find earth processes that would meet the criteria to serve as a geologic clock. Among others, the processes explored were (1) the rate that salt is being added to the ocean, (2) the rate that sediments are being deposited, and (3) the rate that Earth is cooling. Comparing the load of salts being delivered to the ocean by all the rivers, and assuming the ocean was initially pure water, it was calculated that about one-hundred million years would be required for the present salinity to be reached. The calculations did not consider the amount of materials being removed from the ocean by organisms and by chemical sedimentation, however, so this technique was considered to be unacceptable. Even if the amount of materials removed were known, it would actually result in the age of the ocean, not the age of Earth.

A number of separate and independent attempts were made to measure the rate of sediment deposition, then compare that rate to the thickness of sedimentary rocks found on Earth. Dividing the total thickness by the rate of deposition resulted in estimates of an Earth age that ranged from about twenty million to fifteen-hundred million years. The wide differences occurred because there are gaps in many sedimentary rock sequences, periods when sedimentary rocks were being eroded away to be deposited again elsewhere as sediments. There were just too many unknowns for this technique to be considered as acceptable.

The idea of measuring the rate that Earth is cooling for use as a geologic clock assumed that Earth was initially a molten mass that has been cooling ever since. Calculations estimating the temperature that Earth must have been to be molten were compared to Earth's present rate of cooling. This resulted in an estimated age of twenty to forty million years. These calculations were made back in the nineteenth century before it was understood that natural radioactivity is adding heat to Earth's interior, so it has required much longer to cool down to its present temperature.

Modern Techniques for Determining the Age of Earth

Soon after the beginning of the twentieth century, the discovery of the radioactive decay process in the elements of minerals and rocks led to the development of a new, accurate geologic clock. This clock finds the *radiometric age* of rocks in years by measuring how much of an unstable radioactive isotope within the crystals of certain minerals has decayed. Since radioactive decay for each kind of radioactive isotope occurs at a constant, known rate, the ratio of the remaining amount of a radioactive element to the amount of decay products present can be used to calculate the time that the radioactive element has been a part of that crystal (see chapter 11). Certain radioactive isotopes of potassium, uranium, and thorium are often included in the minerals of rocks, so they are often used as "radioactive clocks." By using radiometric aging techniques along with other information, we arrive at a generally accepted age for Earth of about 4.5 billion years. It should be noted that radiometric aging is only useful in aging igneous rocks, since sedimentary rocks are the result of weathering and deposition of other rocky materials.

Table 22.1 lists several radioactive isotopes, their decay products, and half-lives. Often two or more isotopes are used to determine an age for a rock. Agreement between them increases the scientist's confidence in the estimates of the age of the rock. Because there are great differences in the half-lives, some are useful for dating things back to several billion years, while others, such as carbon-14, are only useful for dating things to perhaps fifty thousand years. Carbon-14 is not used to age rock

TABLE 22.1   Radioactive Isotopes and Half-Lives

Radioactive Isotope	Stable Daughter Product	Half-Life
Samarium-147	Neodymium-143	106 billion years
Rubidium-87	Strontium-87	48.8 billion years
Thorium-232	Lead-208	14.0 billion years
Uranium-238	Lead-206	4.5 billion years
Potassium-40	Argon-40	1.25 billion years
Uranium-235	Lead-207	704 million years
Carbon-14	Nitrogen-14	5,730 years

but is very useful in aging materials that are of relatively recent biological origin, since carbon is an important part of all living things. Also, a slightly different method is used to determine carbon-14 dating (See "A Closer Look: What Is Carbon-14 Dating?" on p. 532.)

A recently developed geologic clock is based on the magnetic orientation of magnetic minerals. These minerals become aligned with Earth's magnetic field when the igneous rock crystallizes, making a record of the magnetic field at that time. Earth's magnetic field is global and has undergone a number of reversals in the past. A *geomagnetic time scale* has been established from the number and duration of magnetic field reversals occurring during the past six million years. Combined with radiometric age dating, the geomagnetic time scale is making possible a worldwide geologic clock that can be used to determine local chronologies.

Interpreting the Geologic Record

Features of the geology of an area provide many clues that help us interpret what has happened in the past. However, is important to keep in mind several basic principles of geology. The following is a summary of these basic guiding principles that are used to read a story of geologic events from the rocks.

The *principle of uniformity* is the cornerstone of the logic used to guide thinking about geologic time. As described in chapter 16, this principle is sometimes stated as "the present is the key to the past." This means that the geologic features that you see today were formed in the past by the same processes of crustal movement, erosion, and deposition that are observed today. By studying the processes now shaping Earth, you can understand how it has evolved through time. This principle establishes the understanding that the surface of Earth has been continuously and gradually modified over the immense span of geologic time.

The *principle of original horizontality* is a basic principle that is applied to sedimentary rocks. (See chapter 15 for a discussion of the nature of igneous, sedimentary, and metamorphic rocks.) It is based on the observation that, on a large scale, sediments are commonly deposited in flat-lying layers. Old rocks are continually being changed to new ones in the continuous processes of crustal movement, erosion, and deposition. As sediments are deposited in a basin of deposition, such as a lake or ocean, they accumulate in essentially flat-lying, approximately horizontal layers (figure 22.1). Thus, any layer of sedimentary rocks that is not horizontal has been subjected to forces that have deformed Earth's surface.

The *principle of superposition* is another logical and obvious principle that is applied to sedimentary rocks. Layers of sediments are usually deposited in succession in horizontal layers, which later are compacted and cemented into layers of sedimentary rock. An undisturbed sequence of horizontal layers is thus arranged in chronological order with the oldest layers at the bottom. Each consecutive layer will be younger than the one below it (figure 22.2). This is true, of course, only if the layers have not been turned over by deforming forces.

FIGURE 22.1 (*A*) The principle of original horizontality states that sediments tend to be deposited in horizontal layers. (*B*) Even where the sediments are draped over an irregular surface, they tend toward the horizontal.

FIGURE 22.2 The Grand Canyon, Arizona, provides a majestic cross section of horizontal sedimentary rocks. According to the principle of superposition, traveling deeper and deeper into the Grand Canyon means that you are moving into older and older rocks.

The *principle of crosscutting relationships* is concerned with igneous and metamorphic rock, in addition to sedimentary rock layers. Any geologic feature that cuts across or is intruded (forced) into a rock mass must be younger than the rock mass. Thus, if a fault cuts across a layer of sedimentary rock, the fault is the youngest feature. Faults, folds, and igneous intrusions are always younger than the rocks in which they occur. Often the presence of metamorphic rock provides a further clue to the correct sequence. When hot igneous rock is intruded into preexisting rock, the rock surrounding the hot igneous rock can be "baked," or metamorphosed, into a different form. Thus,

Rocks adjacent to intruding magma may also be metamorphosed by its heat.

A

B

FIGURE 22.3 (*A*) According to the principle of crosscutting relationships, since the granite intrusion is cutting across the sedimentary rock, the granite intrusion is younger than the surrounding sedimentary rock. (*B*) The photo shows two examples of crosscutting relationships. The darker rock was the original rock formed first. Later, the white layers were intruded into the darker rock. Finally, a fault line cuts acoss the white intrusions. So it must have occurred later than the intrusions.

the rock from which the metamorphic rock formed must have preceded the igneous rock (figure 22.3).

Shifting sites of erosion and deposition: The principle of uniformity states that the earth processes going on today have always been occurring. This does not mean, however, that they always occur in the same place. As erosion wears away the rock layers at a site, the sediments produced are deposited someplace else. Later, the sites of erosion and deposition may shift, and the sediments are deposited on top of the eroded area. When the new sediments later are formed into new sedimentary rocks, there will be a time lapse between the top of the eroded layer and the new layers. A time break in the rock record is called an **unconformity.** The unconformity is usually shown by a surface within a sedimentary sequence on which there was a lack of sediment deposition or where active erosion may even have occurred for some period of time. When the rocks are later examined, that time span will not be represented in the record, and if the unconformity is erosional, some of

the record once present will have been lost. An unconformity may occur within a sedimentary sequence of the same kind or between different kinds of rocks. The most obvious kind of unconformity to spot is an *angular unconformity.* An angular unconformity, as illustrated in figure 22.4, is one in which the bedding planes above and below the unconformity are not parallel. An angular unconformity usually implies some kind of tilting or folding, followed by a significant period of erosion, which in turn was followed by a period of deposition (figure 22.5).

The principle of superposition, the principle of crosscutting relationships, and the presence of an unconformity help us interpret the order in which geologic events occurred. This order can be used to unravel a complex sequence of events, such as the one shown in figure 22.6.

1. Deposition

2. Rocks tilted, eroded

3. Subsequent deposition

FIGURE 22.4 An unconformity is a gap in the geologic record caused when the original deeper rocks were subject to erosion and then a subsequent layer of sediments was deposited on top of them. An angular unconformity is easy to identify because the layers of the deeper rocks are not parallel to the horizontal surface rocks.

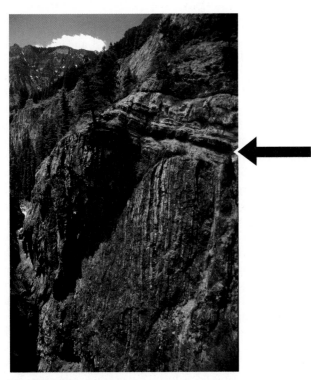

FIGURE 22.5 This photograph shows an angular unconformity. The horizontal sedimentary rock layers overlie almost vertically oriented metamorphic rocks. Metamorphic rocks form deep in Earth, so they must have been uplifted, and the overlying material eroded away before being buried again.

FIGURE 22.6 A complex rock sequence can be deciphered by using geologic principles. The limestones must be oldest (law of superposition), followed by the shales. The granite and basalt must both be younger than the limestone they crosscut (note the metamorphosed zone around the granite). It is not possible to tell whether the igneous rocks (granite and basalt) predate or postdate the shales or to determine whether the sedimentary rocks were tilted before or after the igneous rocks were emplaced. After the limestones and shales were tilted, they were eroded, and then the sandstones were deposited on top. Finally, the lava flow covered the entire sequence.

GEOLOGIC TIME AND THE FOSSIL RECORD

A **fossil** is *any* evidence of former life, so the term means more than fossilized remains such as those pictured in figure 22.7. Evidence can include actual or altered remains of plants, animals, fungi, algae, or even bacteria. It could also be just simple evidence of former life such as the imprint of a leaf, the footprint of a dinosaur, or droppings from bats in a cave.

Early Ideas About Fossils

Our current understanding of fossils requires a great deal of information about geologic time and geologic processes. Without this information, it was difficult for people to appreciate how fossils formed and how to interpret them. The ancient Greek historian Herodotus was among the first to realize that fossil shells found in rocks far from any ocean were remnants of organisms left by a bygone sea. However, one hundred years later, Aristotle still believed that fossils had no relation to living things and were formed inside the rocks. Note that he also believed that living organisms could arise by spontaneous generation from mud. A belief that the fossils must have "grown" in place in rocks would be consistent with a belief in spontaneous generation.

Even when people recognized some similarities between living organisms and certain fossils, they did not make a connection between the fossil and previously living things. Fossils were considered to be the same as quartz crystals or any other mineral crystals that either formed with Earth or grew there later (depending on the philosophical view of the interpreter).

FIGURE 22.7 The hard parts of this fish fossil are beautifully preserved, along with a carbon film, showing a detailed outline of the fish and some of its internal structure.

With the development of Judeo-Christian beliefs, many people came to regard well-preserved fossils that were very similar to living organisms as remains of once-living organisms—that were buried in Noah's flood. Leonardo da Vinci, like other Renaissance scholars, argued that fossils were the remains of organisms that had lived in the past.

By the early 1800s, the true nature of fossils was becoming widely accepted. William "Strata" Smith, an English surveyor, discovered at this time that sedimentary rock strata could be identified by the fossils they contain. Smith grew up in a region of England where fossils were particularly plentiful. He became a collector, keeping careful notes on where he found each fossil and in which type of sedimentary rock layers. During his travels, he discovered that the succession of rock layers on the south coast of England was the same as the succession of rock layers on the east coast. Through his keen observations, Smith found that each kind of sedimentary rock had a distinctive group of fossils that was unlike the group of fossils in other rock layers. Smith amazed his friends by telling them where and in what type of rock they had found their fossils.

Today, the science of discovering fossils, studying the fossil record, and deciphering the history of life from fossils is known as **paleontology.** The word *paleontology* was invented in 1838 by the Scottish geologist Charles Lyell to describe this newly established branch of geology. It is derived from classic Greek roots and means "study of ancient life," and this means a study of fossils.

People sometimes blur the distinction between *paleontology* and *archaeology.* **Archaeology** is the study of past human life and culture from material evidence of artifacts, such as graves, buildings, tools, pottery, landfills, and so on (*artifact* literally means "something made"). The artifacts studied in archaeology can be of any age, from the garbage added to the city landfill yesterday to the fragments of pottery of an ancient tribe that disappeared hundreds of years ago. The word *fossil* originally meant "anything dug up" but today carries the meaning of any *evidence of ancient organisms in the history of life.* Artifacts are therefore not fossils, as you can see from the definitions.

Types of Fossils

Considering all the different things that can happen to the remains of an organism, and considering the conditions needed to form a fossil, it seems amazing that any fossils are formed and then found. For example, the animals you see killed beside the road or the dead trees that fall over in a forest rarely become fossils because scavengers eat the remains of dead animals, and decay organisms break down the organic remains of plants and animals. As a result, very little digestible organic matter escapes destruction.

Thus, a fossil is not likely to form unless there is rapid burial of a recently deceased organism. The presence of hard parts, such as a shell or a skeleton, will also favor the formation of a fossil if there is rapid burial. Several different kinds of fossils can be differentiated based on the kind of material that is preserved and the processes of preservation.

Preserved Whole Organisms

Occasionally, entire organisms are preserved, and in rare cases, even the unaltered remains of an organism's *soft parts* are found. For this to occur, the organism must be quickly protected from scavengers and decomposers following death. Several conditions allow this to occur. The best examples of this uncommon method of fossilization include protection by freezing or entombing in tree resin. Mammoths, for example, have been found frozen and preserved by natural refrigeration in the ices of Alaska and Siberia. The bodies of ice age humans have also been discovered in glaciers, and early inhabitants of South America have been found in ice caves high in mountains. In 1991 the body of a man was discovered in a glacier on the border between Austria and Italy. Subsequent studies of the body showed that he dated from about 3300 B.C. Furthermore, study of the body determined much information about him including what he had recently eaten, that he had been wounded by an arrow, and that he had tattoos. Insects and spiders, complete with delicate appendages, have been found preserved in amber, which is fossilized tree resin (figure 22.8). In each case—ice or resin—the soft parts were protected from scavengers, insects, and bacteria. Additional examples of organisms that have been preserved in a relatively unaltered state include organisms buried in acid bogs or those quickly desiccated following death.

Preserved Hard Parts

Although fossils of soft parts are occasionally formed, the *remains of hard parts* such as shells, bones, and teeth of animals or the pollen and spores of plants, which have hard coverings, are much more common. The mammoth bones shown in figure 22.9 are a good example of fossils of hard parts of organisms. The bones were covered with sediment and preserved.

Because hard parts are more easily preserved, organisms that have such parts are more often found as fossils than those

FIGURE 22.8 This fly was stuck in and covered over with plant pitch. When the pitch fossilized to form amber, the fly was preserved as well. The entire fly is preserved in the amber.

fossil evidence of life of any kind living on the land during the Cambrian. The dominant life-forms of the Cambrian ocean were echinoderms, mollusks, trilobites, and brachiopods. The trilobites, now extinct, made up more than half of the kinds of living things during the Cambrian.

During the Ordovician and Silurian periods of the Paleozoic era, most living things were still marine organisms with various kinds of jawless fish becoming prominent. Also, by the end of the Silurian, some primitive plants were found on land. The Devonian period saw the further development of different kinds of fish, including those that had jaws, and many kinds of land plants and animals. Coral reefs were also common in the Devonian. The Carboniferous period was a time of vast swamps of ferns, horsetails, and other primitive nonseed plants that would form great coal deposits. Fossils of the first reptiles and the first winged insects are found in rocks from this age. The Paleozoic era closed with the extinction at the end of the Permian period of about 90 percent of plant and animal life of that time.

3. The Mesozoic era was a time when the development of life on land flourished. The dinosaurs first appeared in the Triassic period, outnumbering all the other reptiles until the close of the Mesozoic. Fossils of the first mammals and modern forms of gymnosperms (cone-bearing plants) developed in the Triassic. The first flowering plants, the first deciduous trees, and the first birds appeared in the Cretaceous period. The Cretaceous is the final period of the Mesozoic era and is characterized by the dominance of dinosaurs and extensive evolution of flowering plants and the insects that pollinate them. Birds and mammals also increased in variety. Like the close of the Paleozoic, the Mesozoic era ended with a great dying of land and marine life that resulted in the extinction of many species, including the dinosaurs.

4. As the Cenozoic era opened, the dinosaurs were extinct and the mammals became the dominant vertebrate life-form. The Cenozoic is thus called the Age of the Mammals. However, there were also major increases in the kinds of insects, flowering plants (particularly grasses), and birds. Finally, toward the end of this period of time, humans arrived on the scene, and many other kinds of large mammals such as mammoths, mastodons, giant ground sloths, and saber-toothed tigers went extinct.

Mass Extinctions

When we look at the fossil record, there is evidence of several mass extinctions. Five are recognized for causing the extinction of 50 percent or more of the species present. It is important to understand that although these extinctions were "sudden" in geologic time, they occurred over millions of years. Each resulted in a change in the kinds of organisms present with major groups going extinct and the evolution of new kinds of organisms. The boundaries between many of the geologic time designations are defined by major extinction events. Geologists have developed theories about the causes of each of these mass extinctions. Many of these theories involve changes in the size and location of continents as a result of plate tectonics.

1. The mass extinction at the end of the Ordovician period resulted in the extinction of 60 percent of genera of organisms. At that time, most organisms lived in the oceans. It is thought that the large continent of Gondwana migrated to the South Pole and this resulted in the development of large glaciers and a drastic drop in sea level along with a cooling of the waters.

2. At the end of the Devonian period, there was a mass extinction that affected primarily marine organisms. Approximately 60 percent of genera went extinct. Since many of the organisms that went extinct were warm-water, marine organisms, glaciation along with a cooling of the oceans is a widely held theory for the cause of this extinction.

3. The mass extinction at the end of the Permian period is unusual in several ways. It resulted in the extinction of about 90 percent of organisms and took place over a very short time—less than a million years. Because of this, it is often referred to as the "Great Dying." Both marine and terrestrial organisms were affected. Because this extinction event occurred over a short time and affected all species, it is assumed that a major, worldwide event was responsible. However, at this time, there is no clearly identifiable cause. Suggestions include a meteorite impact, a supernova, massive volcanic activity, or a combination of factors.

4. The extinction at the end of the Triassic period was relatively mild compared to others. About 50 percent of species appear to have gone extinct. There is no clear cause for this extinction.

5. The mass extinction at the end of the Cretaceous period resulted in the extinction of about 60 percent of species. Based on evidence of a thin clay layer marking the boundary between the Cretaceous and Tertiary periods, one theory proposes that a huge (16 km diameter and 10^{15} kg mass) meteorite struck Earth (figure 22.15). The impact

FIGURE 22.15 A widely held theory is that a large meteorite was responsible for the mass extinction at the end of the Cretaceous period. Several of the other mass extinctions may have also been caused by meteorite impacts. This is an artist's depiction of the kinds of impact a large meteorite could have had on Earth.

TABLE 22.2 Geologic Time and the History of Life (To trace the history of life, start with the oldest time period, the Hadean.)

Eon	Era	Period	Epoch	Million Years Ago*	Major Physical Events	Major Biological Events
Phanerozoic	Cenozoic	Quaternary	Holocene (Recent)	0.01 to present	—Much of Earth has been modified by human activity	—Dominance of modern humans
			Pleistocene	0.01 ← 1.8	—Extinction of many kinds of large mammals in late Pleistocene, perhaps due to human activity; —Most recent periods of glaciation	—Origin of modern humans; —Many kinds of large mammals present in early Pleistocene
		Tertiary	Pliocene	1.8 ← 5.3	—Cool, dry period; —North and South America join; —The Indian Plate collides with Asia to form the Himalaya Mountains	—Origin of ancestors of humans; —Grasslands and grazing mammals widespread; —All modern groups of mammals present
			Miocene	5.3 ← 23.8	—Warm, moist period; —Current arrangement of continents present	—Grasslands and grazers common; —Mammals widespread; —Marine kelp forests common
			Oligocene	23.8 ← 33.7	—Cool period; —India, South America, and Australia are separate continents	—Tropics diminish; —Woodlands and grasslands expand; —Many new grazing mammals
			Eocene	33.7 ← 54.8	—Australia and South America separate from Antarctica; —North America separates from Europe	—All modern forms of flowering plants present; —Forests common; —All major groups of mammals present; —Grasses found for the first time; —First primates
			Paleocene	54.8 ← 65.0	—Warm period	—Many new kinds of mammals; —Many new kinds of birds
	Mesozoic	Cretaceous		65 ← 144	—Major extinction at the end of Cretaceous affected both land and ocean organisms, probably because of meteorite impact; —Warm, moist period; —Continents continue to separate	—Dinosaurs dominant; —Evolution of birds; —Coevolution of flowering plants and their insect pollinators; —The three modern mammal groups found; —First flowering plants
		Jurassic		144 ← 206	—Pangaea begins to split up; —North and South America split; —South America and Africa split; —Climate cooler and wetter as Pangaea splits	—Dinosaurs dominant; —First birds; —Many small primitive mammals
		Triassic		206 ← 248	—Major extinction of about 50% of species; —No clear reason for extinction; —Supercontinent, Pangaea, near equator; —Warm, dry climate	—Many modern insects; —Explosion in reptile diversity; —Evolution of modern gymnosperms; —Evolution of modern corals; —First dinosaurs; —First mammals; —Mollusks are dominant aquatic invertebrates
	Paleozoic	Permian		248 ← 290	—Largest of all extinctions occurred at the end of the Permian (90% of all species) but no clear cause for this extinction; —New giant continent of Pangaea forms; —Much of Pangaea was probably desert; —Modern levels of oxygen in atmosphere	—Gymnosperms dominant; —Many new species of insects; —Amphibians and reptiles dominant

Eon	Era	Period	Epoch	Million Years Ago*	Major Physical Events	Major Biological Events
Phanerozoic (continued)	Paleozoic (continued)	Carboniferous		290 ← 354	—Major coal deposits formed —Two major continents moving toward each other to form Pangaea —Warm climate	—Gymnosperms (cone-bearing plants) present by the end of the period —Reptiles present by the end of the period —Vast swamps of club mosses, horsetails, and ferns common —Many kinds of sharks —First winged insects, some very large —Amphibians common —Armored fish extinct
		Devonian		354 ← 417	—Glaciation probable cause of mass extinction at end of Devonian —About 60% of genera extinct —Extinction primarily affected warm-water, marine life —Warm, shallow seas common for much of this period —Two large continents present	—First seed plants (seed ferns) —Both jawless and jawed fish abundant —Coral reefs abundant —Land plants abundant (first forests) —Many new kinds of wingless insects —Amphibians on land
		Silurian		417 ← 443	—Melting of glaciers led to higher sea level —A large continent, Gondwana, is present at South Pole, but there are many smaller equatorial continents	—Numerous coral reefs —Fungi present —First vascular plants —Widespread nonvascular plants —First land animals—Arthropods (spiders and centipedes) —First jawed fish —Jawless armored fish abundant —Some jawless fish found in freshwater
		Ordovician		443 ← 490	—Giant continent, Gondwana, located at South Pole resulted in formation of glaciers —Formation of glaciers led to drop in sea level, which is the probable cause of mass extinction at the end of Ordovician (60% of genera go extinct)	—Primitive, nonvascular land plants present —Jawless, armored fish common —Diverse marine invertebrates
		Cambrian		490 ← 543	—Rodinia breaks up into pieces —Glaciation probable cause of mass extinction at the end of Cambrian	—All major groups of animals present —Mollusks, brachiopods, echinoderms, trilobites, and nautiloids common
Proterozoic		This period of time is also known as the Precambrian		543 ← 2,500	—Single large supercontinent (Rodinia) existed about 1.1 billion years ago —Oxidation of terrestrial iron deposits forms red beds that indicate presence of atmospheric oxygen (oxidized iron is red) at about 2.5 billion years ago —About 1.8 billion years ago, the atmospheric oxygen concentration was probably about 15% of current level	—Primitive multicellular, soft-bodied marine animals present about 1 billion years ago —Algae present about 1 billion years ago —Fossil record poorly preserved —First eukaryotic cells present about 1.8 billion years ago
Archean				2,500 ← 3,800	—Formation of continents —Atmosphere lacked oxygen —Oldest rocks 3.8 billion years ago	—Fossil stromatolites (layers of cyanobacteria) common —Cyanobacteria carried on photosynthesis, releasing oxygen —Fossil cyanobacteria present 3.5 billion years ago —Origin of life about 3.8 billion years ago
Hadean				3,800 ← 4,500	—Crust of Earth in process of solidifying —No rocks of this age —Origin of Earth	

* Time scale is based on *1999 Geologic Time Scale* of the Geological Society of America.

would have thrown a tremendous amount of dust into the atmosphere, obscuring the Sun and significantly changing the climate and thus the conditions of life on Earth. The resulting colder climate may have led to the extinction of many plant and animal species, including the dinosaurs. This theory is based on the clay layer, which theoretically formed as the dust settled, and its location in the rock record at the time of the extinctions. The layer is enriched with a rare metal, iridium, which is not found on Earth in abundance but occurs in certain meteorites in greater abundance.

6. During the Quaternary period at the end of the last ice age, there was a relatively minor extinction of the many species of large mammals. This is thought to be due to either a major change in climate at the end of the last ice age or hunting by humans as they expanded their range from Africa to Europe, Asia, and the Americas.

7. Many people are convinced that we are currently experiencing a human-caused mass extinction because of our ability to alter the face of Earth and destroy the habitats needed by plants and animals.

Interpreting Geologic History—A Summary

When interpreting the fossil record in any part of the world, there are several things to keep in mind:

1. *We are dealing with long periods of time.* The history of life on Earth goes back to 3.5 billion years ago, and the evolution of humans took place over a period of several million years. It is important to understand that many of the processes of sedimentation, continental drift, and climate change took place slowly over many millions of years.

2. *Earth has changed greatly over its history.* There have been repeated periods of warming and cooling. During periods of cooling, glaciers formed, which resulted in lower sea level, which in turn exposed more land and changed the climate of any particular continent. In addition, the continents were not fixed in position. Changes in position affected the climate that the continents experienced. For example, at one time, what is now North America was attached to Antarctica near the South Pole.

3. *There have been many periods in the history of Earth when most of the organisms went extinct.* Cooling climates, changes in sea level, and meteorite impacts are all suspected of causing mass extinctions. However, it should not be thought that these were sudden extinctions. Most took place over millions of years. Some of the extinctions affected as much as 90 percent of the things living at the time.

4. *New forms of life evolved that replaced those that went extinct.* The earliest organisms we see in the fossil record were prokaryotic marine organisms similar to present-day bacteria. The oldest fossils of these organisms date to about 3.5 billion years ago. The next major step was the development of eukaryotic marine organisms about 1.8 billion years ago. The first eukaryotic multicellular organisms were present by about 1 billion years ago. The development of multicellular organisms ultimately led to

the colonization of land by plants and animals, with plants colonizing about 500 million years ago and animals at about 450 million years ago.

5. *Although there were massive extinctions, there are many examples of the descendants of early life-forms present today.* Bacteria and many kinds of simple eukaryotic organisms are extremely common today, as are various kinds of algae and primitive forms of plants. In the oceans, many kinds of marine animals such as starfish, jelly fish, and clams are descendants of earlier forms.

6. *The kinds of organisms present have changed the nature of Earth.* The oxygen in the atmosphere is the result of the process of photosynthesis. Its presence has altered the amount of ultraviolet radiation reaching Earth. Plants tend to reduce the erosive effects of running water and humans have significantly changed the surface of Earth.

Table 22.2 lists the various geologic time periods and major events that occurred during those time periods.

PALEONTOLOGY, ARCHAEOLOGY, AND HUMAN EVOLUTION

There is intense curiosity about how our species (*Homo sapiens*) came to be, and we recognize that as animals, humans are subject to the same evolutionary forces as other organisms. Since all of our close evolutionary relatives are extinct, however, it is difficult for us to visualize our evolutionary development, and we tend to think we are unique and not subject to the laws of nature. Thus, the evolution of the human species remains an interesting and controversial topic.

We use several kinds of evidence to try to sort out our evolutionary history. Fossils of various kinds of prehuman and ancient human ancestors have been found, but these are often only fragments of skeletons that are often hard to date. Stone tools of various kinds have also been found associated with prehuman and early human sites. Finally, other aspects of the culture of our human ancestors have been found in burial sites, cave paintings, and the creation of ceremonial objects. Various methods have been used to age these findings. Some can be dated quite accurately, while others are more difficult to pinpoint.

When fossils are examined, anthropologists can identify differences in the structures of bones that are consistent with changes in species. Based on the amount of change they see and the ages of the fossils, these scientists make judgments about the species to which the fossil belongs. As new discoveries are made, opinions of experts will change and our evolutionary history may become clearer as old ideas are replaced. It is also clear from the fossil record that humans are relatively recent additions to the forms of life.

As evidence of how new discoveries change the way we think, the term *hominin* is a collective term currently used to refer to humans and their humanlike ancestors. Previously, the term *hominid* was used to refer to this group but is now used to refer to the broader group that includes all humanlike organisms plus the great apes. Consequently, when you read material

A Closer Look

Neandertals Were Probably a Different Species

An ongoing controversy surrounds the relationship between Neandertals and other forms of prehistoric humans. One position is that Neandertals were a small separate race or subspecies of human that could have interbred with other humans and may have disappeared because their subspecies was eliminated by interbreeding with more populous, successful groups. (Many small, remote tribes have been lost as distinct cultural and genetic entities in the same way in recent history.)

Others maintain that the Neandertals showed such great difference from other early humans that they must have been a different species and became extinct because they could not compete with more successful *Homo sapiens* immigrants from Africa. (The names of these ancient people typically are derived from the place where the fossils were first discovered. For example, Neandertals were first found in the Neander Valley of Germany and Cro-Magnons [considered to be *Homo sapiens*]

were initially found in the Cro-Magnon caves in France.)

The use of molecular genetic technology has shed some light on the relationship of Neandertals to other kinds of humans. Examination of the mitochondrial DNA obtained from the bones of a Neandertal individual reveals significant differences between Neandertals and other kinds of early humans. This greatly strengthens the argument that Neandertals were a separate species, *Homo neanderthalensis*.

---

about this topic, you will need to determine how the terms are being used. Although we do not have a clear picture of how humans evolved, several points are well accepted.

1. Recently published analyses of fossils of *Ardipithecus ramidus* suggest that it was an early ancestor in the human line of evolution.
2. Fossils of *Orrorin* and *Sahelanthropus* are older than *Ardipithecus* and are poorly understood. However, these organisms may have been similar to *Ardipithecus*.
3. There is a great deal of fossil evidence that several bipedal species of the genera *Australopithecus* and *Paranthropus* were common in Africa beginning about four million years ago. These organisms are often referred to collectively as australopiths.
4. Based on fossil evidence, it appears that the climate of Africa was becoming drier during much of the time when the evolution of humans was occurring.
5. Several species of the genus *Homo* became prominent in Africa beginning about 2.2 million years ago and appear to have made a change from a primarily herbivorous diet to a carnivorous or omnivorous diet.
6. All members of the genus *Homo* have relatively large brains compared to australopiths. The cranial capacity ranges from about 650 cm^3 for early fossils of *Homo* to about 1,450 cm^3 for modern humans. These organisms are also associated with various degrees of stone tool construction and use. It is possible that some of the australopiths also may have constructed stone tools.
7. The fossils of australopiths, several early species of *Homo*, and the fossils of earlier organisms such as *Ardipithecus*, *Orrorin*, and *Sahelanthropus* are found only in Africa. Fossils of several later species of the genus *Homo* have larger brains and are found in Africa, Europe, and Asia but not in Australia or the Americas. Only *Homo sapiens* is found in Australia and the Americas.
8. Since the fossils of *Homo* species found in Asia and Europe are generally younger than the early *Homo* species found in Africa, it is assumed that species of *Homo* migrated from Africa to Europe and Asia.

9. Differences in size between males and females are less prominent in members of the genus *Homo*, so perhaps there were fewer sexual differences in activities than with the australopiths.

When we try to put all of these bits of information together, we can construct the following plausible scenario for the evolution of our species.

Ardipithecus

Ardipithecus ramidus lived in woodlands and was able to walk upright when on the ground as well as navigate on branches of trees. The fossils from the best-studied site have been aged at about 4.4 million years ago. Studies of these fossils indicate that *Ardipithecus ramidus* was about 120 cm (4 ft) tall when standing, weighed about 50 kg (110 lb), and had a brain capacity of less than 400 cm^3. It was able to walk and run on its hind limbs but had an opposable big toe and was able to climb and walk on the tops of branches on all fours. Males and females were about the same size. They ate a variety of plant and animal materials in their woodland habitat.

The Australopiths

Various species of *Australopithecus* and *Paranthropus* were present in Africa from over four million years ago until about one million years ago. It is important to recognize that there are few fossils of these early humanlike organisms and that often they are fragments of the whole organism. This has led to much speculation and argument among experts about the specific position each fossil has in the evolutionary history of humans. They were sexually dimorphic with the males much larger than the females. Males were 1.5 m (5 ft) or less and females about 1.1 m (3.5 ft). They were stocky and walked upright like humans. They had relatively small brains (cranial capacity 530 cm^3 or less—about the size of a standard softball). Anthropologists recognize several different species of *Australopithecus* and *Paranthropus*.

An upright posture had several advantages in a world that was becoming drier. It allowed for more rapid movement over long distances, the ability to see longer distances, and reduced the amount of heat gained from the Sun. In addition, upright posture freed the arms for other uses such as carrying and manipulating objects, and using tools. The various species of *Australopithecus* and *Paranthropus* shared these characteristics and, based on the structure of their skulls, jaws, and teeth, appeared to have been herbivores with relatively small brains.

The Genus *Homo*

About 2.5 million years ago, the first members of the genus *Homo* appeared on the scene. There is considerable disagreement about how many species there were, but *Homo habilis* is one of the earliest. *H. habilis* had a larger brain (650 cm³) and smaller teeth than australopiths and made much more use of stone tools. Some people feel that it was a direct descendant of *Australopithecus africanus*. Many experts feel that *H. habilis* was a scavenger that made use of group activities, tools, and higher intelligence to take over the kills made by other carnivores. The higher-quality diet would have supported the metabolic needs of the larger brain.

About 1.8 million years ago, *Homo ergaster* appeared on the scene. It was much larger, up to 1.6 m (6 ft) tall, than *H. habilis*,

who was about 1.3 m (4 ft) tall, and also had a much larger brain (cranial capacity of 850 cm³). A little later, a similar species (*Homo erectus*) appears in the fossil record. Some people consider *H. ergaster* and *H. erectus* to be variations of the same species. Along with their larger brain and body size, *H. ergaster* and *H. erectus* are distinguished from earlier species by their extensive use of stone tools. Hand axes were manufactured and used to cut the flesh of prey and crush the bones to obtain the fatty marrow. These organisms appear to have been predators, whereas *H. habilis* was a scavenger. The use of meat as food allows animals to move about more freely, because appropriate food is available almost everywhere. By contrast, herbivores are often confined to places that have foods appropriate to their use: fruits for fruit eaters, grass for grazers, forests for browsers, etc. In fact, fossils of *H. erectus* have been found in the Middle East and Asia as well as Africa. Most experts feel that *H. erectus* originated in Africa and migrated through the Middle East to Asia.

At about eight hundred thousand years ago, another hominin classified as *Homo heidelbergensis* appears in the fossil record. Since fossils of this species are found in Africa, Europe, and Asia, it appears that they constitute a second wave of migration of early *Homo* from Africa to other parts of the world.

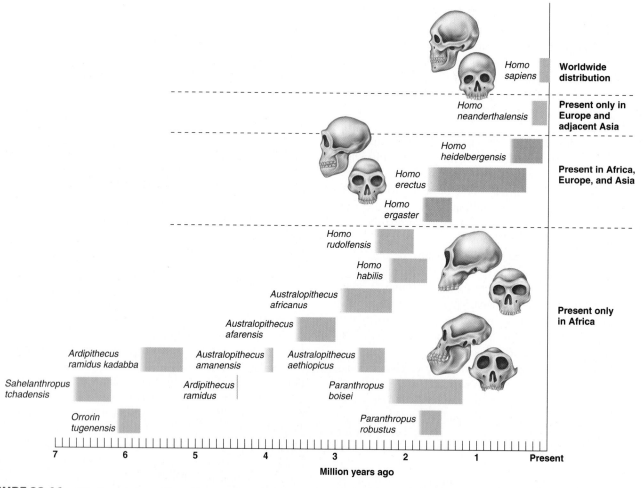

FIGURE 22.16   This diagram shows the various organisms thought to be relatives of humans. The bars represent approximate times the species are thought to have existed. Notice that (1) all species are extinct today except for modern humans; (2) several different species of organisms coexisted for extensive periods; (3) all the older species are found only in Africa; and (4) more recent species of *Homo* are found in Europe and Asia as well as Africa.

Both *H. erectus* and *H. heidelbergensis* disappear from the fossil record as two new species (*Homo neanderthalensis* and *Homo sapiens*) become common.

The Neandertals were primarily found in Europe and adjoining parts of Asia and were not found in Africa. Since Neandertals were common in Europe, many people feel Neandertals are descendants of *H. heidelbergensis,* which also was common in Europe and preceded Neandertals.

The Origin of *Homo sapiens*

Homo sapiens is found throughout the world and is now the only hominin species remaining of a long line of ancestors. There are two different theories that seek to explain the origin of *Homo sapiens.* One theory, known as the **out-of-Africa hypothesis,** states that modern humans, *H. sapiens,* originated in Africa as had several other hominin species and migrated from Africa to Asia and Europe and displaced species such as *H. erectus* and *H. heidelbergensis* that had migrated into these areas previously. The other theory, known as the **multiregional hypothesis,** states that *H. erectus* evolved into *H. sapiens.* During a period of about 1.7 million years, fossils of *H. erectus* showed a progressive increase in the size of the cranial capacity and reduction in the size of the jaw, so that it becomes difficult to distinguish *H. erectus* from *H. heidelbergensis* and *H. heidelbergensis* from *H. sapiens.* Proponents of this hypothesis believe that *H. heidelbergensis* is not a distinct species but an intermediate between the earlier *H. erectus* and *H. sapiens.* According to this theory, various subgroups of *H. erectus* existed throughout Africa, Asia, and Europe, and interbreeding among the various groups gave rise to the various races of humans we see today.

Another continuing puzzle is the relationship of humans that clearly belong to the species *Homo sapiens* with a contemporary group known as Neandertals. Some people consider Neandertals to be a race of *H. sapiens* specially adapted to life in the harsh conditions found in postglacial Europe. Others consider them to be a separate species (*Homo neanderthalensis*). The Neandertals were muscular, had a larger brain capacity than modern humans, and had many elements of culture, including burials. The cause of their disappearance from the fossil record at about twenty-five thousand years ago remains a mystery. Perhaps climate change to a warmer climate was responsible. Perhaps contact with *H. sapiens* resulted in their elimination either through hostile interactions or, if they were able to interbreed with *H. sapiens,* they could have been absorbed into the larger *H. sapiens* population.

The discovery of fossils of a small human in the sediments of a cave on an Indonesian island in 2003 has led to much speculation about a possible new species of *Homo*; *Homo floresiensis.* The fossils were dated to about 18,000 years ago. The most interesting feature of this discovery is that the individuals were of small size; perhaps a little over 1 meter (3.5 feet) tall and with a correspondingly small skull. There has been much discussion among scientists about the significance of this discovery. Some feel that the small size indicates a developmental abnormality and that the remains could be that of an abnormally small *H. sapiens.* Others suggest that there are many examples of island animals that show small size and that these fossils could be adapted to island conditions. Others are convinced that the discovery represents a new species. Research continues and a more complete understanding will develop as more information is gathered.

Large numbers of fossils of prehistoric humans have been found in all parts of the world. Many of these show evidence of a collective group memory we call *culture.* Cave paintings, carvings in wood and bone, tools of various kinds, and burials are examples. These are also evidence of a capacity to think and invent, and "free time" to devote to things other than gathering food and other necessities of life. We may never know how we came to be, but we will always be curious and will continue to search and speculate about our beginnings. Figure 22.16 summarizes the current knowledge of the historical record of humans and their relatives.

PART II
KINDS OF ORGANISMS

Part II is a brief description of the kinds of organisms that live on Earth and the processes involved in organizing this vast biodiversity into logical arrangements.

THE CLASSIFICATION OF ORGANISMS

To talk about items in our world, we must have names for them. As new items come into being or are discovered, new words are devised to describe them. For example, the words *laptop, palm pilot,* and *text message* describe new technology that did not exist thirty years ago. Similarly, in the biological world, people have given names to newly discovered organisms so they can communicate with others about the organism.

The Problem with Common Names

The common names used by people of distinct cultures are usually different. A *dog* in English is *chien* in French, *perro* in Spanish, and *cane* in Italian. It is even possible that different names can be used to identify the same organism in different regions within a country. For example, the common garter snake may be called a garden snake or gardner snake depending on where you live (figure 22.17). Actually, there are several different species of "garter snakes" that have been identified as distinct from one another. Thus, common names can be confusing, and scientists sought a more acceptable way to give organisms names that would eliminate confusion and would be used by all scientists.

The naming of organisms is a technical process, but it is extremely important. When biologists are describing their research, common names such as robin, maple tree, or garter snake are not good enough. They must be able to accurately

FIGURE 22.17 Depending on where you live you may call this organism a garter snake, a garden snake, or a gardner snake. These common names can lead to confusion about what kind of snake a person might be talking about. However, the scientific name (*Thamnophis sirtalis*) is recognized worldwide by the scientific community.

identify the organisms involved so that everyone who reads the report, wherever they live in the world, will be able to know what organism is being discussed. The scientific identification of organisms really involves two different but related activities. One, *taxonomy*, involves the naming of organisms, and the other, *phylogeny*, involves showing how organisms are related evolutionarily. In reality no taxonimic decisions are made without considering the evolutionary history of the organism.

Taxonomy

Taxonomy is the science of naming organisms and grouping them into logical categories. Various approaches have been used to classify organisms. The Greek philosopher Aristotle (384–322 B.C.) had an interest in nature and was the first person to attempt a logical classification system. The root word for *taxonomy* is the Greek word *taxis,* which means "arrangement." Aristotle used the size of plants to divide them into the categories of trees, shrubs, and herbs.

During the Middle Ages, Latin was widely used as the scientific language. As new species were identified, they were given Latin names, often using as many as fifteen words to describe a single organism. Although using Latin meant that most biologists, regardless of their native language, could understand a species name, it did not completely do away with duplicate names. Because many of the organisms could be found over wide geographic areas and communication was slow, there could still be two or more Latin names for a species. To make the situation even more confusing, ordinary people used common local names.

Binomial System of Nomenclature

The modern system of classification began in 1758, when Carolus Linnaeus (1707–1778), a Swedish doctor and botanist, published his tenth edition of *Systema Naturae* (figure 22.18). (Linnaeus's original name was Karl von Linne, which he "latinized" to Carolus Linnaeus.) In the previous editions, Linnaeus had used the

FIGURE 22.18 Carolus Linnaeus (1707–1778), a Swedish doctor and botanist, originated the modern system of taxonomy.

polynomial (many-names) Latin system. However, in the tenth edition, he introduced the *binomial system of nomenclature.* The **binomial system of nomenclature** uses only two Latin names—the genus name and the specific epithet (*epithet* = descriptive word)—for each species of organism. Recall that a species is a population of organisms capable of interbreeding and producing fertile offspring. Individual organisms are members of a species. A **genus** (plural, *genera*) is a group of closely related species of organisms; the specific epithet is a word added to the genus name to identify which one of several species within the genus we are discussing. This is similar to the naming system we use with people. When you look in the phone book, you look for the last name (surname), which gets you in the correct general category. Then you look for the first name (given name) to identify the individual you wish to call. The unique name given to a particular type of organism is called its species name or scientific name. To clearly identify the scientific name from other words, binomial names are either *italicized* or <u>underlined</u>. The first letter of the genus name is capitalized. The specific epithet is always written in lowercase. For example, *Thamnophis sirtalis* is the binomial name for the common garter snake.

In addition to assigning a specific name to each species, Linnaeus recognized a need for placing organisms into meaningful groups. In his system, he divided all forms of life into two *kingdoms,* Plantae and Animalia, and then further subdivided the kingdoms into other smaller units.

Organizing Species into Logical Groups

Since Linnaeus's initial attempts to place all organisms into categories, there have been many changes. One of the most fundamental is the recent recognition that there are three major categories of organisms that have been called *domains*.

A **domain** is the largest category into which organisms are classified. There are three domains: Bacteria, Archaea, and Eucarya (figure 22.19.). Organisms are separated into these three domains based on specific structural and biochemical features of their cells. The Bacteria and Archaea are prokaryotic, and the Eucarya are eukaryotic.

A **kingdom** is a subdivision of a domain. There are several kingdoms within the Bacteria and Archaea that are based primarily on differences in the metabolism and genetic composition of the organisms. Within the domain Eucarya, there are four kingdoms: Plantae, Animalia, Fungi, and Protista (protozoa and algae). Figure 22.20 provides examples of the three domains. However, no clear consensus exists at this time about how many kingdoms there are in the Bacteria and Archaea.

A **phylum** is a subdivision of a kingdom. However, microbiologists and botanists often use the term *division* in place of the term *phylum*.

All kingdoms have more than one phylum. For example, the kingdom Plantae contains several phyla, including flowering plants, conifer trees, mosses, ferns, and several other groups. Organisms are placed in phyla based on careful investigation of the specific nature of their structure, metabolism, and biochemistry. An attempt is made to identify natural groups rather than artificial or haphazard arrangements. For example, while nearly all plants are green and carry on photosynthesis, only flowering plants have flowers and produce seeds; conifers lack flowers but have seeds in cones; ferns lack flowers, cones, and seeds; and mosses are so simple in structure that they even lack tissues for transporting water.

A **class** is a subdivision within a phylum. For example, within the phylum Chordata, there are seven classes: mammals, birds, reptiles, amphibians, and three classes of fishes.

An **order** is a category within a class. Carnivora is an order of meat-eating animals within the class Mammalia. There are several other orders of mammals, including horses and their relatives, cattle and their relatives, rodents, rabbits, bats, seals, whales, humans, and many others.

A **family** is a subdivision of an order and consists of a group of closely related genera, which in turn are composed of groups of closely related species. The cat family, Felidae, is a subgrouping of the order Carnivora and includes many species in several genera, including the Canada lynx and bobcat (genus *Lynx*), the cougar (genus *Puma*), the leopard, tiger, jaguar, and lion (genus *Panthera*), the house cat (genus *Felis*), and several others. Thus, in the present-day science of taxonomy, each organism that has been classified has its own unique binomial

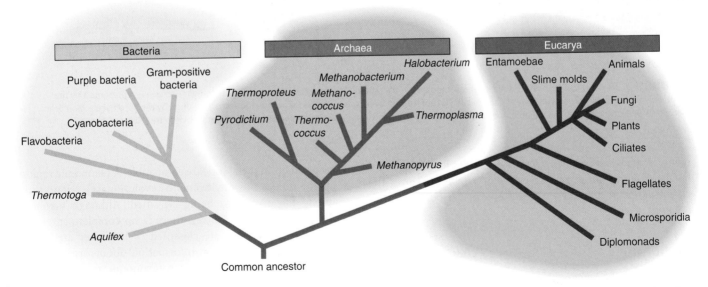

FIGURE 22.19 This diagram shows the three domains of living things and the way they are related to one another evolutionarily. The domain Bacteria is the oldest group. The domains Archaea and Eucarya are derived from the Bacteria.

A *Streptococcus pyogenes*

B *Methanopyrus*

C *Morchella esculenta*

D *Amoeba proteus*

E *Homo sapiens*

F *Acer saccharum*

FIGURE 22.20 (*A*) The domain Bacteria is represented by the bacterium *Streptococcus pyogenes*. (*B*) The domain Archaea is represented by the methane-producing *Methanopyrus*. The domain Eucarya is represented by members of the following kingdoms: (*C*) Fungi, by the ascomycete *Morchella esculenta*; (*D*) Protista, by the one-celled *Amoeba proteus*; (*E*) Animalia, by the animal *Homo sapiens*; and (*F*) Plantae, by the tree *Acer saccharum*.

name. In turn, it is assigned to larger groupings that are thought to have a common evolutionary history. Table 22.3 uses the classification of humans to show how the various categories are used.

Phylogeny

Phylogeny is the science that explores the evolutionary relationships among organisms and seeks to reconstruct evolutionary history. Taxonomists and phylogenists work together so that the products of their work are compatible. A taxonomic ranking should reflect the phylogenetic (evolutionary) relationships among the organisms being classified. Although taxonomy and phylogeny are sciences, there is no complete agreement as to how organisms are classified or how they are related. New organisms and new information about organisms are discovered constantly. Therefore, taxonomic and phylogenetic relationships are constantly being revised. During this revision process, scientists often have differences of opinion about the significance of new information.

herbivores. *Homo habilis* had a larger brain and appears to have been a scavenger. Several other species of the genus *Homo* arose in Africa. These forms appear to have been carnivores. Some of these migrated to Europe and Asia. The origin of *Homo sapiens* is not completely understood. Two alternative views are that *H. sapiens* arose in Africa and migrated throughout the world or that *H. sapiens* evolved from earlier ancestors found throughout Africa, Asia, and Europe.

To facilitate accurate *communication*, biologists assign a *specific name* to each *species* that is cataloged. The various species are cataloged into larger groups on the basis of similar traits. *Taxonomy* is the science of classifying and naming organisms. *Phylogeny* is the science of trying to figure out the evolutionary history of a particular organism. The *taxonomic ranking of organisms* reflects their *evolutionary relationships*. *Fossil evidence, comparative anatomy, developmental stages,* and *biochemical evidence* are employed in the sciences of taxonomy and phylogeny.

The first organisms thought to have evolved were single-celled, prokaryotic organisms. These simple organisms gave rise to two different prokaryotic domains, Bacteria and Archaea. The Bacteria are a very diverse group of organisms. Some are autotrophs, while others are heterotrophs. The cyanobacteria were probably the first organisms to carry on photosynthesis. Many kinds of heterotrophic Bacteria are decomposers, while some are parasites, mutualistic, or commensal organisms. The Archaea are distinguished from Bacteria by differences in their cell walls, cell membranes and the nature of their DNA. Many of the Archaea live in very hostile environments. Eukaryotic organisms were probably derived from the Bacteria and Archaea and constitute the domain Eucarya. There are four kingdoms in the Eucarya: *Protista, Fungi, Plantae,* and *Animalia*. The Protista includes organisms such as protozoa and algae that consist of single cells or small colonies of cells. The Fungi kingdom includes multicellular, heterotrophic organisms with cell walls made of chitin. The Plantae are complex, multicellular, autotrophic organisms that have cell walls made of cellulose and are primarily terrestrial. The Animalia are complex, multicellular, heterotrophic organisms that lack cell walls and typically have the ability to move.

KEY TERMS

archaeology (p. **530**)

binomial system of nomenclature (p. **542**)

class (p. **543**)

domain (p. **543**)

eon (p. **534**)

epoch (p. **534**)

era (p. **534**)

family (p. **543**)

fossil (p. **529**)

fungus (p. **550**)

genus (p. **542**)

geologic time scale (p. **533**)

index fossils (p. **533**)

kingdom (p. **543**)

multiregional hypothesis (p. **541**)

obligate intracellular parasite (p. **555**)

order (p. **543**)

out-of-Africa hypothesis (p. **541**)

paleontology (p. **530**)

parasite (p. **549**)

period (p. **534**)

phylogeny (p. **544**)

phylum (p. **543**)

prions (p. **556**)

taxonomy (p. **542**)

unconformity (p. **528**)

viroid (p. **556**)

virus (p. **555**)

APPLYING THE CONCEPTS

Part I

1. Some of the oldest fossils are about how many years old?
   a. 4.55 billion
   b. 3.5 billion
   c. 250 million
   d. 10,000

2. A fossil of a jellyfish, if found, would most likely be
   a. a preserved fossil.
   b. one formed by mineralization.
   c. a carbon film.
   d. a cast or a mold.

3. In any sequence of sedimentary rock layers that has not been subjected to stresses, you would expect to find
   a. essentially horizontal stratified layers.
   b. the oldest layers at the bottom and the youngest at the top.
   c. younger faults, folds, and intrusions in the rock layers.
   d. All of the above are correct.

4. You would expect to find the least number of fossils in rocks from which geologic time?
   a. Cenozoic
   b. Mesozoic
   c. Paleozoic
   d. Precambrian

5. The numerical dates associated with events on the geologic time scale were determined by
   a. relative dating of the rate of sediment deposition.
   b. radiometric dating using radioactive decay.
   c. the temperature of Earth.
   d. the rate that salt is being added to the ocean.

6. In the evolution of humans, the fossil evidence suggests that
   a. the earliest humanlike primates were present in Africa, Europe, and North America.
   b. *Homo sapiens* has been around for about two million years.
   c. there were several different species of human ancestors in Africa.
   d. large numbers of fossils clearly show the way humans evolved.

Part II

7. The three main kinds of living things, Bacteria, Archaea, and Eucarya, have been labeled as
   a. species.
   b. kingdoms.
   c. domains.
   d. families.

8. Which is commonly used as sources of information when developing a phylogeny?
   a. fossil evidence
   b. biochemical information
   c. embryological development
   d. All of the above are correct.

9. Which of the following kingdoms contain members that are autotrophs?
   a. Protista
   b. Plantae
   c. Bacteria
   d. All of the above are correct.
10. Plants and fungi differ in that
    a. plants are autotrophic and fungi are heterotrophic.
    b. plants have cell walls of cellulose and fungi have cell walls of chitin.
    c. plants carry on photosynthesis and fungi do not.
    d. All of the above are correct.
11. Viruses
    a. can be free-living or parasitic.
    b. belong to the domain Archaea.
    c. can function and reproduce only inside a host cell.
    d. All of the above are correct.
12. The Archaea
    a. are ancestors of Bacteria.
    b. are eukaryotic.
    c. lack a nucleus.
    d. None of the above is correct.

Answers

1. b **2.** c **3.** d **4.** d **5.** b **6.** c **7.** c **8.** d **9.** d **10.** d **11.** c **12.** c

QUESTIONS FOR THOUGHT

Part I

1. Why does the rock record go back only 3.8 billion years? If this missing record were available, what do you think it would show? Explain.

2. What major event marked the end of the Paleozoic and Mesozoic eras according to the fossil record? Describe one theory that proposes to account for this.

3. Describe how the principles of superposition, horizontality, and faunal succession are used in the relative dating of sedimentary rock layers.

4. Describe four different kinds of fossils and how they were formed.

5. Describe four things that fossils can tell you about Earth's history.

6. What are some of the major steps thought to have been involved in the evolution of humans?

Part II

7. How do the domains Bacteria and Archaea differ?

8. What is the value of taxonomy?

9. How do viruses reproduce?

10. Why are Latin names used for genus and species?

11. List two characteristics typical for each of the following kingdoms: Protista, Fungi, Plantae, Animalia.

12. Describe four kinds of evidence used to establish phylogenetic relationships among organisms.

FOR FURTHER ANALYSIS

Part I

1. Take a long narrow strip of paper. Tear it in half crossways. Tear one of the remaining pieces in half. Continue until you have made five tears. Measure the length of the piece you have remaining. Use this information and the number of times you "halved" the paper to determine the length of the original strip of paper. This is analogous to the way in which the measurement of radioactive isotopes can be used to determine the age of rocks.

2. Consider the following questions:

   Could our early hominin ancestors have seen living dinosaurs?

   What did the animals in the Cambrian period eat?

   What organisms have dominated most of the history of Earth?

   How might viruses and viroids be related?

Part II

3. Develop a "phylogeny" for motor vehicles and draw a diagram of your phylogeny. Consider the following questions among those you include in determining your phylogeny:

   How many "kinds" currently exist?

   Are there "fossils" that may have been precursors to motor vehicles?

   What was the ancestor of motor vehicles?

   What was the first motor vehicle?

   What motor vehicles have gone extinct?

   What major evolutionary changes have occurred?

   What "environmental factors" shaped the evolution of motor vehicles?

   Present your phylogeny to your class and have them critique your effort.

INVITATION TO INQUIRY

Understanding Geological Time

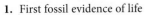

It is often difficult to appreciate the extremely long period of time involved in the history of Earth and life. It is also often difficult to appreciate that the kinds of living things on Earth have changed significantly. From table 22.2, obtain the approximate dates for the following significant events and place them on the time line.

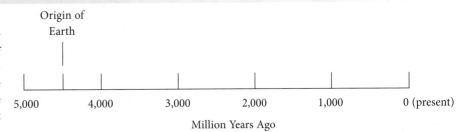

1. First fossil evidence of life
2. First evidence of eukaryotic cells
3. First animals
4. First terrestrial plants
5. First dinosaurs
6. First mammals
7. Dinosaurs go extinct
8. First humans

Now do the following:

1. Circle the period of time in which only prokaryotic organisms were present.
2. Circle the period of time during which dinosaurs were present.
3. Circle the period of time in which humans have been present.
4. Circle the period of time in which only marine organisms were present.

23

Ecology and Environment

Organisms interact in many ways.

CORE CONCEPT
Everything is interconnected.

OUTLINE

Energy and matter flow from one organism to another in food chains (p. 562).

A Definition of Environment
The Organization of Ecological Systems
People Behind The Science: Eugene Odum
Energy Flow in Ecosystems
Community Interactions

Climate affects the kinds of organisms that live in an area (p. 567).

Types of Terrestrial Communities
  Temperate Deciduous Forest
  Temperate Grassland or Prairie
  Savanna
  Desert
  Boreal Coniferous Forest
  Mediterranean Shrublands (Chaparral)
  Temperate Rainforest
  Tundra
  Tropical Rainforest
  Tropical Dry Forest
Types of Aquatic Communities
  Marine Communities
  Freshwater Communities
  Estuaries
A Closer Look: The Importance of Habitat Size

Individual Species Requirements: Habitat and Niche
  Habitat
  Niche
A Closer Look: Zebra Mussels: Invaders from Europe
Kinds of Organism Interactions
  Predation
  Parasitism
  Commensalism
  Mutualism
  Competition
  Competition and Natural Selection
The Cycling of Materials in Ecosystems
  The Carbon Cycle
  The Nitrogen Cycle
A Closer Look: Carbon Dioxide and Global Warming
  The Phosphorus Cycle
  Nutrient Cycles and Geologic Time
  Bioaccumulation and Biomagnification

Populations are limited by environmental factors (p. 589).

Human activities alter natural systems and processes (p. 592).

CONNECTIONS

Physics
▷ Second law of thermodynamics (Ch. 4).
▷ Energy flows (Ch. 3).

Astronomy
▷ Energy comes from the Sun (Ch. 12 and 13).
▷ Seasons occur because Earth orbits the Sun (Ch. 17).

Chemistry
▷ Biochemistry is the chemistry of living things (Ch. 19).
▷ Chemical reactions cause material and energy changes (Ch. 9).

Earth Science
▷ Atmospheric gases affect living things (Ch. 17).
▷ Climate is different at different places on Earth (Ch. 17).
▷ The greenhouse effect warms Earth (Ch. 17).

Life Science
▷ Evolutionary processes lead to adaptation (Ch. 21).
▷ Photosynthesis traps sunlight energy as chemical energy (Ch. 20).
▷ Respiration releases chemical bond energy for organisms (Ch. 20).

OVERVIEW

Today, we recognize that environmental problems are a worldwide concern. Poor agricultural practices in Africa and China result in dust storms that affect the local people and are also carried to the rest of the world. Concern about global warming has led to the convening of international conferences and the drafting of treaties whose goal is to reduce the impact of energy use. Many are concerned about protecting endangered species and the forests, grasslands, and oceans that provide habitats for them. In many ways, these problems are the result of an incredible increase in the size of the human population. As the human population has increased, humans have sought to use land for agriculture and other purposes. Converting land to agricultural use results in its loss for other purposes such as habitat and watershed protection. To understand the nature of environmental problems and steps we can take to solve them, we need to be familiar with some of the central ideas of the science of ecology.

A DEFINITION OF ENVIRONMENT

Ecology is the branch of biology that studies the relationships between organisms and their environments. This is a very simple definition for a very complex branch of science. Most ecologists define the word **environment** very broadly as anything that affects an organism during its lifetime. Environmental influences can be divided into two categories: **biotic factors** are other living things that affect an organism, and **abiotic factors** are nonliving influences that affect an organism (figure 23.1). If we consider a fish in a stream, we can identify many environmental factors that are important to its life. The temperature of the water is extremely important as an abiotic factor that may be influenced by the presence of trees (biotic factor) along the stream bank that shade the stream and prevent the Sun from heating it. Obviously, the kind and number of food organisms in the stream are important biotic factors as well. The type of material that makes up the stream bottom and the amount of oxygen dissolved in the water are other important abiotic factors, both of which are related to how rapidly the water is flowing.

Characterizing the environment of an organism is a complex and challenging process; everything seems to be influenced or modified by other factors. A plant is influenced by many different factors during its lifetime: the types and amounts of minerals in the soil, the amount of sunlight hitting the plant, the animals that eat the plant, and the wind, water, and temperature. Each item on this list can be further subdivided into other areas of study. For instance, water is important in the life of plants, so rainfall is studied in plant ecology. But even the study of rainfall is not simple. In some areas of the world, it rains only during one part of

FIGURE 23.1 (A) The sticks and branches bald eagles use to build their nest are part of their biotic environment. The tree in which this nest is built is also part of the eagle's biotic environment. (B) The irregular shape of the trees is the result of wind and snow, both abiotic factors. Snow driven by these prevailing winds tends to "sandblast" one side of the tree and prevent limb growth.

A Biotic Factor (nesting material and tree)

B Abiotic Factor (wind-driven snow)

the year, while in others, rain is evenly distributed throughout the year. The rainfall could be hard and driving, or it could come as gentle, misty showers of long duration. The water could soak into the soil for later use, or it could run off into streams and be carried away.

The animals in an area are influenced as much by environmental factors as are the plants. If environmental factors do not favor the growth of plants, there will be little food and few hiding places for animal life. Two types of areas that support only small numbers of living animals are deserts and polar regions. Near the poles, the low temperature and short growing season inhibit growth; therefore, there are relatively few species of animals with relatively small numbers of individuals. Deserts receive little rainfall and therefore have poor plant growth and low concentrations of animals. On the other hand, tropical rainforests have high rates of plant growth and large numbers of animals of many kinds.

THE ORGANIZATION OF ECOLOGICAL SYSTEMS

Ecologists can study ecological relationships at several different levels. The smallest living unit is the individual **organism,** which is composed of atoms and molecules arranged in a highly organized manner. Each kind of organism has a particular functional role in the community, which is known as its niche. Groups of organisms of the same species are called populations. Interacting groups of populations of different species are called **communities.** An **ecosystem** consists of all the interacting organisms in an area and their interactions with their abiotic surroundings. Figure 23.2 shows how these different levels of organization are related to one another.

All living things require a continuous supply of energy to maintain life. Therefore, many people like to organize living systems by the energy relationships that exist among the different kinds of organ-

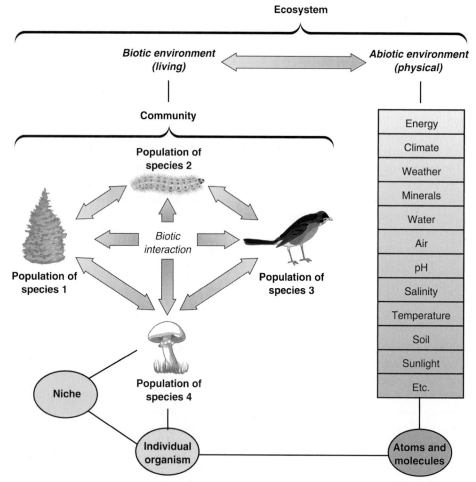

FIGURE 23.2 Ecology is the science that deals with the interactions between organisms and their environment. This study can take place at several different levels, from the broad ecosystem level through community interactions and population studies to the study of the niche of individual organisms. Ecology also involves study of the physical environment and the atoms and molecules that make up both the living and nonliving parts of an ecosystem. The same organism can be viewed in many ways. We can study it as an individual, as a member of a population, or as a participant in a community or ecosystem.

Eugene Odum (1913–2002)

Eugene Odum was born in 1913 to Howard W. and Anna K. Odum, the first of three children. His father was a professor of sociology at the University of North Carolina. His mother was an urban planner. As a teenager, he wrote about bird life for a local newspaper. He continued to be interested in birds for the rest of his life. He received his bachelor's and master's degrees from the University of North Carolina and his doctorate from the University of Illinois.

In 1939, he married Martha Ann Huff. They had two sons. He joined the faculty of the University of Georgia in 1940 and retired in 1984.

One of his major contributions to science was the popularization of the ecosystem concept. Earlier ecologists had looked primarily at individual organisms and how they reacted with their surroundings or at small systems such as ponds that were easy to quantify.

In 1953, Eugene Odum and his brother, Howard T. Odum, published *Fundamentals of Ecology,* which clearly described ecology as an integrated discipline that could be approached from a systems point of view. Furthermore, it was possible to look at ecology from a regional point of view. For ten years, this was the only book of its type available.

Eugene Odum's interest in ecosystem-level processes quite naturally led to concerns about how humans interact with their surroundings. Thus, his thinking was an important catalyst of the environmental movement of the 1970s.

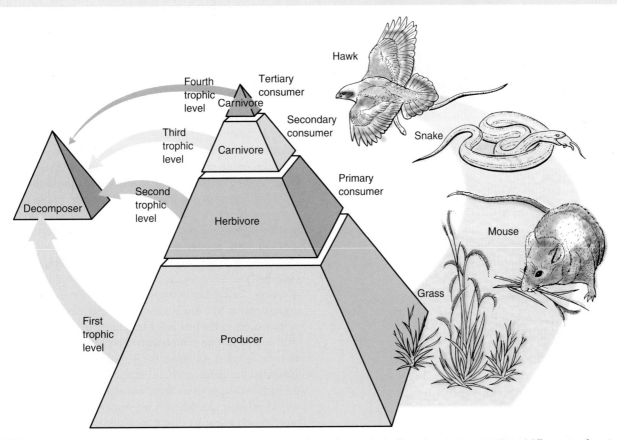

FIGURE 23.3 Organisms within ecosystems can be divided into various trophic levels on the basis of how they obtain energy. Several different sets of terminology are used to identify these different roles. This illustration shows how the different sets of terminology are related to one another. Each colored block in this diagram represents the amount of energy or living material at each trophic level. The amount of energy and the amount of living matter decreases sharply as they are passed to higher trophic levels.

isms present. An ecosystem contains several different kinds of organisms. Those that trap sunlight for photosynthesis, resulting in the production of organic material from inorganic material, are called **producers.** Green plants and other photosynthetic organisms such as algae and cyanobacteria are, in effect, converting sunlight energy into the energy contained within the chemical bonds of organic compounds. There is a flow of energy from the Sun into the living matter of plants.

The energy that plants trap can be transferred through a number of other organisms in the ecosystem. Since all organisms, other than producers, must obtain energy in the form of organic matter, they are called **consumers.** Consumers cannot capture energy from the Sun as plants do. All animals are consumers. They either eat plants directly or eat other sources of organic matter derived from plants. Each time the energy enters a different organism, it is said to enter a different **trophic level,** which is a step, or stage, in the flow of energy through an ecosystem (figure 23.3). The plants (producers) receive their energy directly from the Sun and are said to occupy the *first trophic level.*

Various kinds of consumers can be divided into several categories, depending on how they fit into the flow of energy through an ecosystem. Animals that feed directly on plants

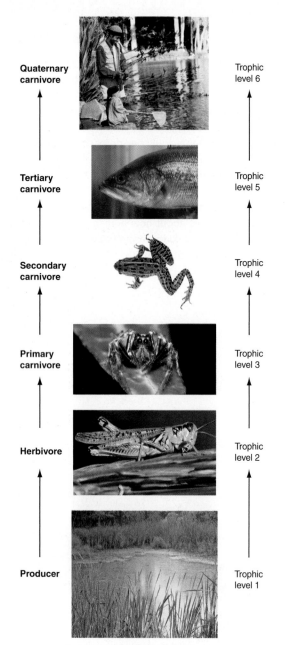

	Trophic level 6
Quaternary carnivore	
Tertiary carnivore	Trophic level 5
Secondary carnivore	Trophic level 4
Primary carnivore	Trophic level 3
Herbivore	Trophic level 2
Producer	Trophic level 1

FIGURE 23.4 As one organism feeds on another organism, there is a flow of energy from one trophic level to the next. This illustration shows six trophic levels.

are called **herbivores,** or **primary consumers,** and occupy the *second trophic level.* Animals that eat other animals are called **carnivores,** or **secondary consumers,** and can be subdivided into different trophic levels depending on what animals they eat. Animals that feed on herbivores occupy the *third trophic level* and are known as **primary carnivores.** Animals that feed on the primary carnivores are known as **secondary carnivores** and occupy the *fourth trophic level.*

This sequence of organisms feeding on one another is known as a **food chain.** Often food chains are very complex and can involve many trophic levels. For example, a human may eat a fish that ate a frog that ate a spider that ate an insect that consumed plants for food. Figure 23.4 shows the six different trophic levels in this food chain. Obviously, there can be higher categories, and some organisms don't fit neatly into this theoretical scheme. Some animals are carnivores at some times and herbivores at others; they are called **omnivores.** They are classified into different trophic levels depending on what they happen to be eating at the moment.

If an organism dies, the energy contained within the organic compounds of its body is finally released to the environment as heat by organisms that decompose the dead body into carbon dioxide, water, ammonia, and other simple inorganic molecules. Organisms of decay, called **decomposers,** are things such as bacteria, fungi, and other organisms that use dead organisms as sources of energy. This group of organisms efficiently converts nonliving organic matter into simple inorganic molecules that can be reused by producers in the process of trapping energy. Decomposers are thus very important components of ecosystems that cause materials to be recycled. As long as the Sun supplies the energy, elements are cycled through ecosystems repeatedly. Table 23.1 summarizes the various categories of organisms within an ecosystem. Now that we have a better idea of how ecosystems are organized, we can look more closely at energy flow through ecosystems.

TABLE 23.1   Roles in an Ecosystem

Classification	Description	Examples
Producers	Organisms that convert simple inorganic compounds into complex organic compounds by photosynthesis	Trees, flowers, grasses, ferns, mosses, algae, cyanobacteria
Consumers	Organisms that rely on other organisms as food; animals that eat plants or other animals	
Herbivore	Eats plants directly	Deer, goose, cricket, vegetarian human, many snails
Carnivore	Eats meat	Wolf, pike, dragonfly
Omnivore	Eats plants and meat	Rat, most humans
Scavenger	Eats food left by others	Coyote, skunk, vulture, crayfish
Parasite	Lives in or on another organism, using it for food	Tick, tapeworm, many insects
Decomposers	Organisms that return organic compounds to inorganic compounds; important components in recycling elements	Bacteria, fungi

ENERGY FLOW IN ECOSYSTEMS

All living things need a constant source of energy. Two fundamental physical laws of energy are important when looking at ecological systems from an energy point of view. The *first law of thermodynamics*—also known as the *law of conservation of energy*—states that energy is neither created nor destroyed. This means that we should be able to describe the amounts of energy in each trophic level and follow energy as it flows through successive trophic levels. The *second law of thermodynamics* states that natural processes proceed toward a state of greater disorder. This disordered condition is known as entropy. Another way to state the second law of thermodynamics is to say that when energy is converted from one form to another, useful energy is lost, which is to say there is an increase in entropy. This means that as energy passes from one trophic level to the next, there will be a reduction in the amount of available energy in living things and an increase in the amount of heat in their surroundings.

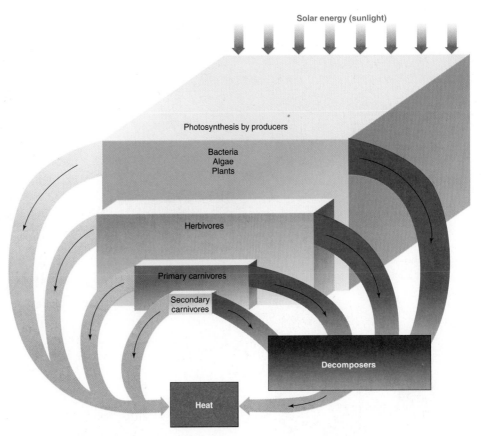

FIGURE 23.5 As energy flows from one trophic level to the next, approximately 90 percent of it is lost. This means that the amount of energy at the producer level is ten times larger than the amount of energy at the herbivore level.

The total energy of an ecosystem can be measured in several ways. An indirect way to get an idea of the amount of energy at a trophic level is to determine its **biomass,** which is the weight of the organisms at that trophic level. Since organic matter will burn, it is possible to convert the energy in biomass into heat and light. For example, the total producer trophic level can be harvested and burned. The number of kilocalories of heat energy produced by burning is equivalent to the energy content of the organic material of the plants. This is what you do when you burn wood at a campfire. Another way of determining the energy present is to measure the rate of photosynthesis and respiration, and calculate the amount of energy being trapped in the living material of the plants. However, this is technically difficult. It requires careful measurements of tiny changes in the amount of carbon dioxide and oxygen in the atmosphere as a result of photosynthesis and respiration or the use of techniques to follow traceable atoms, such as radioisotopes, from one organism to another.

Since only the plants, algae, and cyanobacteria in the producer trophic level are capable of capturing energy from the Sun, all other organisms are directly or indirectly dependent on the producer trophic level. The second trophic level consists of herbivores that eat the producers. This trophic level has significantly less energy and biomass in it for several reasons. *In general, there is about a 90 percent loss of available energy as we proceed from one trophic level to the next higher level.* Actual measurements will vary from one ecosystem to another, but 90 percent is a good rule of thumb. This loss in energy content at

the second and subsequent trophic levels is predicted by the second law of thermodynamics. When the energy in producers is converted to the energy of herbivores, much of the energy is lost as heat to the surroundings.

In addition to this loss of available energy, there is an additional loss involved in the capture and processing of food material by herbivores. Although herbivores don't need to chase their food, they do need to travel to where food is available, then gather, chew, digest, and metabolize it. All these processes require energy.

Just as the herbivore trophic level experiences a 90 percent loss in energy content, the higher trophic levels of primary carnivores, secondary carnivores, and tertiary carnivores also experience a reduction in the energy available to them. This energy loss is reflected in the number of organisms at each trophic level. A field may have millions of producers (grass and other plants), thousands of herbivores that eat grass (grasshoppers, prairie dogs, etc.), and a few carnivores (birds, weasels, etc.) that eat the herbivores. Finally, decomposers act on dead organisms with this same kind of available energy loss. Figure 23.5 shows a diagram in which the energy content decreases by 90 percent as we pass from one trophic level to the next.

COMMUNITY INTERACTIONS

In the section on energy flow in ecosystems, we looked at ecological relationships from the point of view of ecosystems and the way energy flows through them. But we can also study

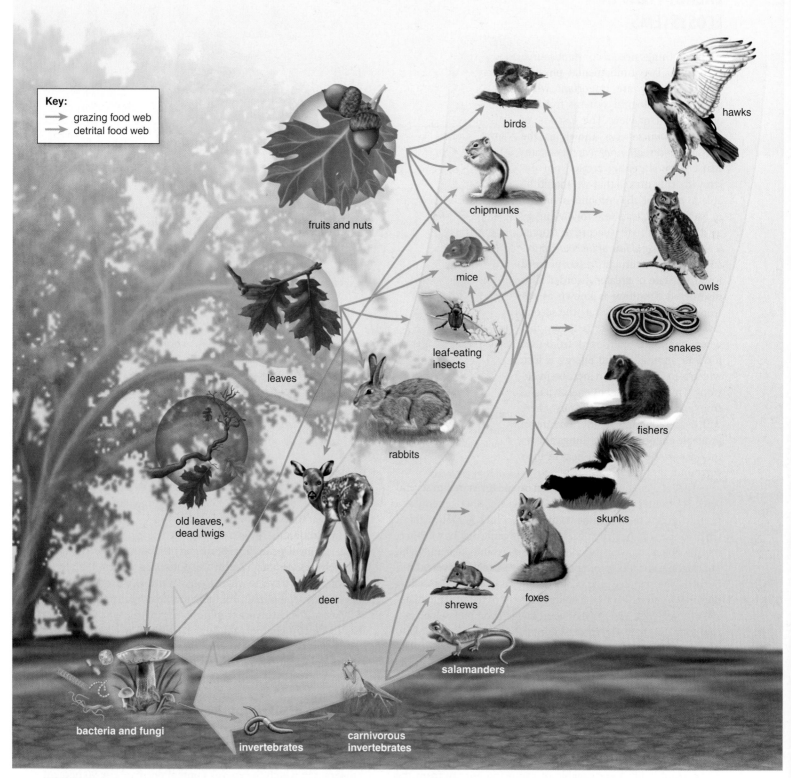

Key:
→ grazing food web
→ detrital food web

fruits and nuts
leaves
old leaves, dead twigs
birds
chipmunks
mice
leaf-eating insects
rabbits
deer
shrews
salamanders
foxes
bacteria and fungi
invertebrates
carnivorous invertebrates
hawks
owls
snakes
fishers
skunks

FIGURE 23.6 As organisms feed on one another they establish a web of relationships known as a food web. This illustration shows the interactions between grazing and detrital food webs. In grazing food webs, photosynthesis by plants provides the energy for grazing animals that eat plants, which in turn provide energy for carnivores. Since all organisms die they ultimately become part of a detrital food web in which dead organic matter and waste products supply the energy for a series of bacteria, fungi, and animals.

relationships at the community level and focus on the kinds of interactions that take place among organisms.

As you know, one of the ways that organisms interact is by feeding on one another. A community includes many different food chains, and many organisms may be involved in several of the food chains at the same time, so the food chains become interwoven into a *food web* (figure 23.6). In general, communities are relatively stable in terms of the kinds of organisms present and how they interact. Some communities, such as tropical rainforests, have large numbers of different kinds of organisms present. Such communities have high **biodiversity.** Others, such as tundra communities, have low biodiversity. People also talk about a loss of biodiversity when a specific kind of organism is eliminated from a region. For example, when wolves were eliminated from much of North America, there was a loss of biodiversity. Although communities are relatively stable, we need to recognize that they are also dynamic collections of organisms. Although the kinds of organisms present may not change, the numbers of each kind may change significantly throughout the year or over several years. For example, a drought will reduce the survival of plants, and an epidemic of disease will reduce the survival of many birds.

If numbers of a particular kind of organism in a community increase or decrease significantly, there will be a ripple effect through the community. For example, the populations of many kinds of small mammals fluctuate from year to year. This results in changes in the numbers of their predators, and often the predators switch to other prey species, which impacts other parts of the community.

Not all organisms within a community have the same level of importance. Obviously, plants are key organisms that supply the energy for all other organisms. However, even certain animal species can have a very high impact on the nature of a community. For example, sea otters feed on sea urchins that feed on a seaweed called kelp. When sea otter populations were reduced significantly, sea urchin populations increased, they ate much of the kelp, and other species of animals had fewer places to hide. Thus, the sea otters played a key role in shaping the nature of the community. Such organisms are often referred to as **keystone species.**

TYPES OF TERRESTRIAL COMMUNITIES

Although each community of organisms is unique, similar communities can be combined into broad categories based on specific characteristics of the physical environment and the kinds of organisms that live in the area. The kind of terrestrial community that develops in an area is determined primarily by climatic factors of precipitation patterns and temperature ranges. These large regional communities are known as **biomes.** The map in figure 23.7 shows the distribution of the major terrestrial biomes of the world.

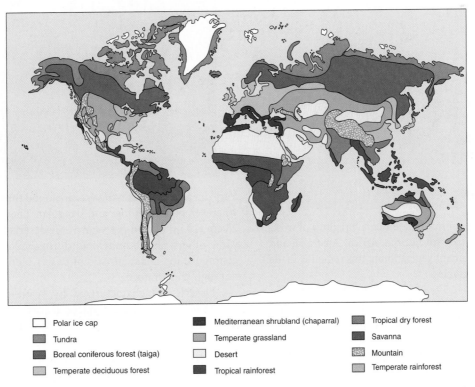

☐ Polar ice cap	◼ Mediterranean shrubland (chaparral)	◼ Tropical dry forest
◼ Tundra	◼ Temperate grassland	◼ Savanna
◼ Boreal coniferous forest (taiga)	☐ Desert	▨ Mountain
◼ Temperate deciduous forest	◼ Tropical rainforest	◼ Temperate rainforest

FIGURE 23.7 Major climatic differences determine the kind of vegetation that can live in a region of the world. Associated with specialized groups of plants are particular kinds of animals. These regional ecosystems are called biomes.

Temperate Deciduous Forest

The *temperate deciduous forest* biome is found in parts of the world (primarily eastern North America, parts of Europe, Japan, Korea, and parts of China) that have 75 to 100 cm (30 to 40 in) of rainfall and cold weather for a significant part of the year. Precipitation is spread throughout the year. The predominant plants are large trees that lose their leaves more or less completely during the fall of the year and are therefore called *deciduous.* Aspen, birch, cottonwood, oak, hickory, beech, and maple are typical trees found in this geographic region. Typical animals of this biome are skunks, porcupines, deer, frogs, opossums, owls, mosquitoes, and beetles.

In much of this region, the natural vegetation has been removed to allow for agriculture, so the original character of the biome is gone except where farming is not practical or the original forest has been preserved.

Temperate Grassland or Prairie

Temperate grasslands are common in temperate regions of western North America and parts of Eurasia, Africa, Australia, and South America. The dominant vegetation in this region is made up of various species of grasses.

The annual rainfall is 25 to 75 cm (10 to 30 in), an amount that is not adequate to support the growth of dense forests. Trees in grasslands grow primarily along streams where they can obtain sufficient water. Animals found in this area include the prairie dog, pronghorn antelope, prairie chicken, grasshopper, rattlesnake, and meadowlark. Most of the original

grasslands, like the temperate deciduous forest, have been converted to agricultural uses.

Savanna

Savannas are found in tropical regions of central Africa, northern Australia, and parts of South America that have pronounced rainy and dry seasons. Although these regions may receive 100 cm (40 in) of rain per year, there is an extended dry season of three months or more.

Therefore, savannas consist of grasses with scattered trees. Grazing animals such as antelope, zebra, and buffalo are common, as are termites and many other insects, which serve as food for many birds and reptiles.

Desert

Very dry areas known as *deserts* are found throughout the world wherever rainfall is low and irregular. They receive less than 25 cm (10 in) of rain per year. Some deserts are extremely hot, while others can be quite cold during much of the year. The distinguishing characteristic of desert biomes is low rainfall, not high temperature.

Deserts are characterized by scattered, thorny plants that lack leaves or have reduced leaves. Since leaves tend to lose water rapidly, the lack of leaves is an adaptation to dry conditions. Although a desert is a very harsh environment, many kinds of insects, reptiles, birds, and mammals can live in this biome. The animals usually avoid the hottest part of the day by staying in burrows or other shaded, cooler areas. Staying underground or in the shade also allows the animals to conserve water.

Boreal Coniferous Forest

The *boreal coniferous forest* (also known as the *northern coniferous forest,* or *taiga*) is found through parts of southern Canada, extending southward along the mountains of the United States, and in much of northern Europe and Asia. This region has short, cool summers and long, cold winters with abundant snowfall. Precipitation ranges from 25 to 100 cm (10 to 40 in) per year. However, even if precipitation is low, the extensive snowmelt in spring and low evaporation rate result in humid summers. The dominant vegetation consists of evergreen trees that are especially adapted to withstand long, cold winters with abundant snowfall. Spruces and firs are common tree species that are intermingled with many other kinds of vegetation. Small lakes and bogs are common. Characteristic animals in this biome include mice, bears, wolves, squirrels, moose, midges, and flies.

Mediterranean Shrublands (Chaparral)

Mediterranean shrublands are located near an ocean and are dominated by shrubby plants. As the name implies, this biome is typical of the Mediterranean coast and is also found in coastal southern California, the southern tip of Africa, a portion of the west coast of Chile, and southern Australia. The climate has wet, cool winters and hot, dry summers. Rainfall is 40 to 100 cm (15 to 40 in) per year. The vegetation is dominated by woody shrubs that are adapted to withstand the hot, dry summer. Often the plants are dormant during the summer. Fire is a common feature of this biome, and the shrubs are adapted to withstand occasional fires. The kinds of animals vary widely in the different regions of the world with this biome. Many kinds of insects, reptiles, birds, and mammals are found in these areas. In the chaparral of California, rattlesnakes, spiders, coyotes, lizards, and rodents are typical inhabitants.

Temperate Rainforest

The coastal areas of northern California, Oregon, Washington, British Columbia, and southern Alaska contain an unusual set of environmental conditions that supports a *temperate rainforest*. There are also small areas of temperate rainforest in southern Chile and on the west coast of New Zealand. The temperate rainforests occur along the coast where the prevailing winds from the west bring moisture-laden air. As the air meets the coastal mountains and is forced to rise, it cools and the moisture falls as rain or snow. Most of these areas receive 200 cm (80 in) or more of precipitation per year. This abundance of water, along with fertile soil and mild temperatures, results in a lush growth of plants.

Sitka spruce, Douglas fir, and western hemlock are typical evergreen coniferous trees in the temperate rainforest of North America. Undisturbed (old growth) forests of this region have trees as old as eight hundred years that are nearly 100 m (about 300 ft) tall. Deciduous trees of various kinds (red alder, big leaf maple, black cottonwood) also exist in open areas where they can get enough light. All trees are covered with mosses, ferns, and other plants that grow on the surface of the trees. The kinds of animals present are similar to those of a temperate deciduous forest.

Tundra

North of the coniferous forest biome in North America and northern Europe and Asia is an area known as the *tundra*. It is characterized by extremely long, severe winters and short, cool summers. The deeper layers of the soil remain permanently frozen, forming a layer called the *permafrost*. Although precipitation is typically less than 25 cm (10 in) per year, the melting of snow during the brief summer (about one hundred days) results in moist conditions. Summer temperatures may be near freezing at night and rise to about 10°C (50°F) during the day. Because the deeper layers of the soil are frozen, when the surface thaws, the water forms puddles.

Under these conditions of low temperature and short growing season, very few kinds of animals and plants can survive. No trees can live in this region. Typical plants and animals of the area are dwarf willow and some other shrubs, reindeer moss (actually a lichen), some flowering plants, caribou, wolves, musk oxen, fox, snowy owls, mice, and many kinds of insects. Many kinds of birds are summer residents only.

Tropical Rainforest

Tropical rainforests are found primarily near the equator in Central and South America, Africa, parts of southern Asia, and some Pacific Islands. The temperature is quite warm and constant, and rain falls nearly every day. Most areas receive over 200 cm (80 in) of rain per year and some exceed 500 cm (200 in).

These areas have high biodiversity. Balsa (a very light wood), teak (used in furniture), and ferns the size of trees are examples of plants from the tropical rainforest. Typically, every plant has other plants growing on it. Tree trunks are likely to be covered with orchids, many kinds of vines, and mosses. Tree frogs, bats, lizards, birds, monkeys, and an almost infinite variety of insects inhabit the rainforest. These forests are very dense and little sunlight reaches the forest floor. When the forest is opened up (by a hurricane or the death of a large tree) and sunlight reaches the forest floor, the opened area is rapidly overgrown with vegetation.

Tropical Dry Forest

Tropical dry forests are found in parts of Central and South America, Australia, Africa, and Asia (particularly India and Myanmar). A major characteristic of this biome is seasonal rainfall. Many of the tropical dry forests have a monsoon climate in which several months of heavy rainfall are followed by extensive periods without rain. Some tropical dry forests have as much as eight months without rain. The rainfall may be as low as 50 cm (20 in) or as high as 200 cm (80 in). Since the rainfall is highly seasonal, many of the plants have special adaptations for enduring drought. In many of the regions that have extensive dry periods, many of the trees drop their leaves during the dry period. Many of the species of animals found here are also found in more moist tropical forests of the region. However, there are fewer kinds in dry forests than in rainforests.

TYPES OF AQUATIC COMMUNITIES

Terrestrial biomes are determined by the amount and kind of precipitation and by temperature ranges. Other factors, such as soil type and wind, also play a part. Aquatic ecosystems also are shaped by key environmental factors. Several important factors are the ability of the Sun's rays to penetrate the water, the depth of the water, the nature of the bottom substrate, the water temperature, and the amount of dissolved salts. Aquatic systems are often divided into **marine communities,** which have a high salt content, and **freshwater communities,** which have low salt concentrations.

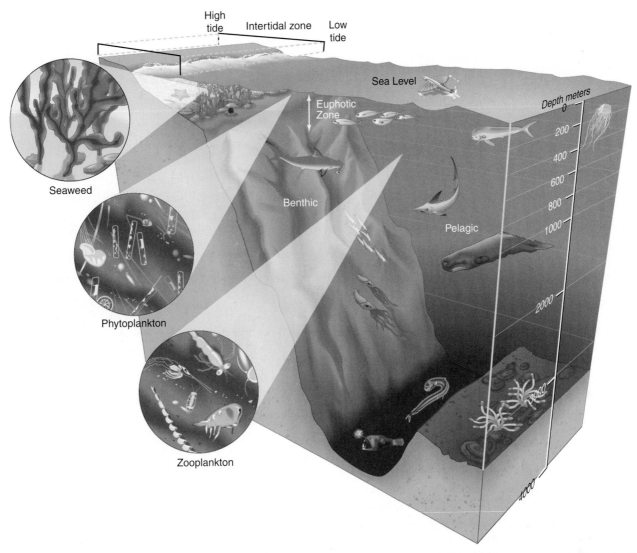

High tide   Intertidal zone   Low tide

Sea Level

Euphotic Zone

Benthic

Pelagic

Depth meters
0
200
400
600
800
1000
2000
4000

Seaweed

Phytoplankton

Zooplankton

FIGURE 23.8   All of the photosynthetic activity of the ocean occurs in shallow water called the euphotic zone, either by attached algae near the shore or by minute phytoplankton in the upper levels of the open ocean. Consumers are either free-swimming pelagic organisms or benthic organisms that live on the bottom. Small animals that feed on phytoplankton are known as zooplankton.

Marine Communities

In the open ocean, many kinds of organisms float or swim actively and are called **pelagic.** Among the pelagic organisms are two types: larger organisms that actively swim and smaller, weakly swimming organisms. The term **plankton** is used to describe aquatic organisms that are so small and weakly swimming that they are simply carried by currents. The planktonic organisms that carry on photosynthesis are called *phytoplankton.* In the open ocean, a majority of these organisms are microscopic floating algae and cyanobacteria. The upper layer of the ocean, where the Sun's rays penetrate, is known as the *euphotic zone.* It is in this euphotic zone where phytoplankton are most common. Small, weakly swimming animals of many kinds, known as *zooplankton,* feed on the phytoplankton. The zooplankton are in turn eaten by larger animals such as fish and larger shrimp, which are eaten by larger fish such as salmon, tuna, sharks, and mackerel (figure 23.8).

Those marine organisms that live on the bottom are called **benthic.** Many seaweeds, some fish, clams, oysters, various

crustaceans, sponges, sea anemones, and many other kinds of organisms live on the bottom. In shallow water, sunlight can penetrate to the bottom, and a variety of attached seaweeds trap this energy as they carry on photosynthesis. Many of the benthic animals graze on the seaweeds and are in turn eaten by other larger animals. Some benthic animals, such as clams, filter water to obtain plankton for food.

As with terrestrial communities, environmental conditions influence the kinds of marine communities that exist. Coral reefs are found in warm, tropical, shallow seas, and "forests" of large seaweeds are found along cool, rocky shores. Sandy shores with lots of wave action typically have few plants, while shallow, protected, muddy areas may support mangrove swamps or salt marshes.

Freshwater Communities

Freshwater ecosystems differ from marine ecosystems in several ways. In comparison to marine communities, the amount of salt

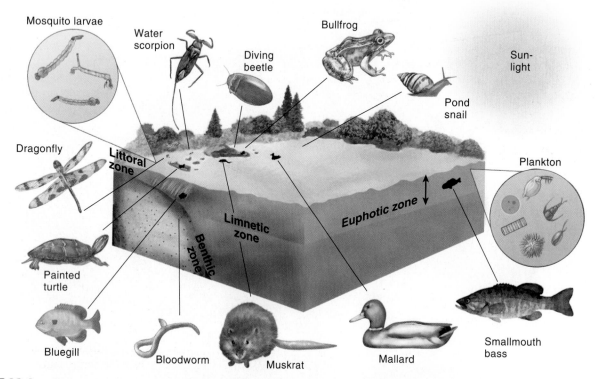

Mosquito larvae

Water scorpion

Diving beetle

Bullfrog

Sun-light

Pond snail

Dragonfly

Littoral zone

Plankton

Limnetic zone

Benthic zone

Euphotic zone

Painted turtle

Bluegill

Bloodworm

Muskrat

Mallard

Smallmouth bass

FIGURE 23.9 Lakes are similar in structure to oceans except that the species are different because most marine organisms cannot live in freshwater. Insects are common organisms in freshwater lakes, as are many kinds of fish, zooplankton, and phytoplankton.

present is much less, the temperature of the water can change greatly, the water is in the process of moving to the ocean, oxygen can often be in short supply, and the organisms that inhabit freshwater systems are different. Freshwater ecosystems can be divided into two categories: those in which the water is relatively stationary, such as lakes, ponds, and reservoirs, and those in which the water is running downhill, such as streams and rivers.

Large lakes have many of the same characteristics as the ocean. If the lake is deep, there is a euphotic zone at the top, with many kinds of phytoplankton, and zooplankton that feed on the phytoplankton. Small fish feed on the zooplankton, which are in turn eaten by larger fish. The species of organisms found in freshwater lakes are different from those found in the ocean, but the roles played are similar, so the same terminology is used.

Along the shore and in the shallower parts of lakes, many kinds of flowering plants are rooted in the bottom. Some are completely submerged, while others have leaves that float on the surface or protrude above the water. Cattails, bulrushes, arrowhead plants, and water lilies are examples. In addition, many kinds of freshwater algae are present.

Associated with the plants and algae are a large number of different kinds of animals. Fish, crayfish, clams, and many kinds of aquatic insects are common inhabitants of this mixture of plants and algae. This region, with rooted vegetation, is known as the *littoral zone,* and the portion of the lake that does not have rooted vegetation is called the *limnetic zone* (figure 23.9).

Streams and rivers present a different set of conditions. Since the water is moving, planktonic organisms are less important than are attached organisms. Most algae grow attached to rocks and other objects on the bottom. Since the water is

shallow, light can penetrate easily to the bottom (except for large or extremely muddy rivers). Even so, it is difficult for photosynthetic organisms to accumulate the nutrients necessary for growth, and most streams are not very productive. As a matter of fact, the major input of nutrients is from organic matter that falls into the stream from terrestrial sources. These are primarily the leaves from trees and other vegetation, as well as the bodies of living and dead insects.

Within the stream is a community of organisms specifically adapted to use the debris as a source of food. Bacteria and fungi colonize the organic matter, and many kinds of insects shred and eat this organic matter as well as the fungi and bacteria living on it. The feces (intestinal wastes) of these insects and the tiny particles produced during the eating process become food for other insects that build nets to capture the tiny bits of organic matter that drift their way. These herbivorous insects are in turn eaten by carnivorous insects and fish.

Estuaries

An **estuary** is a special category of aquatic ecosystem that consists of shallow, partially enclosed areas where freshwater enters the ocean. The saltiness of the water in the estuary changes with tides and the flow of water from rivers. The organisms that live here are specially adapted to this set of physical conditions, and the number of species is less than in the ocean or in freshwater. Estuaries are especially productive ecosystems because of the large quantity of nutrients introduced into the basin from the rivers that run into them. This productivity is further enhanced by the fact that the shallow water allows light

The Importance of Habitat Size

Many people interested in songbird populations have documented a significant decrease in the numbers of certain songbird species. Species particularly affected are those that migrate between North America and South America and require relatively large areas of undisturbed forest in both their northern and southern homes. Many of these species are being hurt by human activities that fragment large patches of forest into many smaller patches, creating more edges between different habitat types.

Bird and other animal species that thrive in edge habitats replace the songbirds, which require large patches of undisturbed forest. Cowbirds that normally live in open areas are a particular problem. Cowbirds do not build nests but lay their eggs in the nests of other birds after removing the eggs of the host species. When forests are broken into small patches, cowbirds reach a larger percentage of forest-nesting birds because the forest birds must nest closer to the edge.

The species most severely affected are those that have both their northern and southern habitats disturbed. A study of migrating sharp-shinned hawks indicated that their numbers have also been greatly reduced in recent years. It is thought that since sharp-shinned hawks use small songbirds as their primary source of food, the reduction in hawks is directly related to the reduction in migratory songbirds.

to penetrate to most of the water in the basin. Phytoplankton and attached algae and plants are able to use the sunlight and the nutrients for rapid growth. This photosynthetic activity supports many kinds of organisms in the estuary. Estuaries are especially important as nursery sites for fish and crustaceans such as flounder and shrimp. The adults enter these productive, sheltered areas to reproduce and then return to the ocean. The young spend their early life in the estuary and eventually leave as they get larger and are more able to survive in the ocean.

INDIVIDUAL SPECIES REQUIREMENTS: HABITAT AND NICHE

People approach the study of organism interactions in two major ways. Many people look at interrelationships from the broad ecosystem point of view, while others focus on individual organisms and the specific things that affect them in their daily lives. The first approach involves the study of all the organisms that interact with one another—the community—and usually looks at general relationships among them. For example, organisms are lumped into categories such as producers, consumers, and decomposers because they perform different functions in a community. These ideas were described earlier in this chapter. The second way of looking at interrelationships is to study in detail the ecological relationships of a certain species of organism. Two important concepts related to the ecological study of individual species are the concepts of habitat and niche.

Habitat

Each organism has particular requirements for life and lives where the environment provides what it needs. The environmental requirements of a whale include large expanses of ocean but with seasonally important feeding areas and protected locations used for giving birth. The kind of place, or part of an ecosystem, occupied by an organism, is known as its **habitat.** Habitats are usually described in terms of conspicuous or particularly

significant features in the area where the organism lives. For example, the habitat of a prairie dog is usually described as a grassland, while the habitat of a tuna is described as the open ocean. The key thing to keep in mind when you think of habitat is that it is the *place* a particular kind of organism lives. When describing the habitats of organisms, we sometimes use the terminology of the major biomes of the world, such as desert, grassland, or savanna. It is also possible to describe the habitat of the bacterium *Escherichia coli* as the gut of humans and other mammals or the habitat of a fungus as a rotting log.

Niche

Each species has particular requirements for life and places specific demands on the habitat in which it lives. The specific functional role of an organism is its **niche.** Its niche is the way it goes about living its life. Just as the word *place* is the key to understanding the concept of habitat, the word *function* is the key to understanding the concept of a niche. Understanding the niche of an organism involves a detailed understanding of the impacts an organism has on its biotic and abiotic surroundings as well as all of the factors that affect the organism. For example, the niche of an earthworm includes abiotic items such as soil particle size, soil texture, and the moisture, pH, and temperature of the soil. The earthworm's niche also includes biotic impacts such as serving as food for birds, moles, and shrews; as bait for anglers; or as a consumer of dead plant organic matter (figure 23.10). In addition, an earthworm serves as a host for a variety of parasites, transports minerals and nutrients from deeper soil layers to the surface, incorporates organic matter into the soil, and creates burrows that allow air and water to penetrate the soil more easily. And this is only a limited sample of all of the aspects of its niche.

Some organisms have rather broad niches; others, with very specialized requirements and limited roles to play, have niches that are quite narrow. The opossum (figure 23.11A) is an animal with a very broad niche. It eats a wide variety of plant and animal foods, can adjust to a wide variety of climates, is used as food

Zebra Mussels: Invaders from Europe

In the mid-1980s, a clamlike organism called the zebra mussel, *Dresssenia polymorpha,* was accidentally introduced into the waters of the Great Lakes. It probably arrived in the ballast water of a ship from Europe. Ballast water is pumped into empty ships to make them more stable when crossing the ocean. Immature stages of the zebra mussel were probably emptied into Lake St. Clair, near Detroit, Michigan, when a ship discharged its ballast water to take on cargo. This organism has since spread to many areas of the Great Lakes and smaller inland lakes. It has also been discovered in other parts of the United States, including the mouth of the Mississippi River. Zebra mussels attach to any hard surface and reproduce rapidly. Densities of more than twenty thousand individuals per square meter have been documented in Lake Erie.

These invaders are of concern for several reasons. First, they coat the intake pipes of municipal water plants and other facilities, requiring expensive measures to clean the pipes. Second, they coat any suitable surface, preventing native organisms from using the space. Third, they introduce a new organism into the food chain. Zebra mussels filter small aquatic organisms from the water very efficiently and may remove the food organisms required by native species. Their filtering activity has significantly increased the clarity of the water in several areas in the Great Lakes. This can affect the kinds of fish present, because greater clarity allows predator fish to find

Box Figure 23.1

prey more easily. In addition, the clarity of the water has allowed for greater production of aquatic vegetation and algae, because light penetrates the water better and to a greater depth. There is concern that zebra mussels have significantly changed the ecological organization of the Great Lakes.

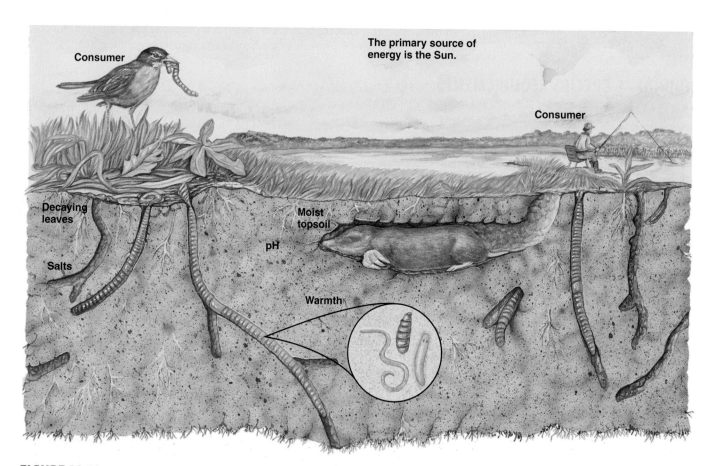

FIGURE 23.10 The niche of an earthworm involves a great many factors. It includes the fact that the earthworm is a consumer of dead organic matter, a source of food for other animals, host to parasites, and bait for an angler. Furthermore, the earthworm loosens the soil by its burrowing and "plows" the soil when it deposits materials on the surface. In addition, the pH, texture, and moisture content of the soil have an impact on the earthworm. Keep in mind that this is but a small part of what the niche of the earthworm includes.

A The oppossum has a broad niche.

FIGURE 23.11 (A) The opossum has a very broad niche. It eats a variety of foods, is able to live in a variety of habitats, and has a large reproductive capacity. It is generally extending its range in the United States. (B) The koala has a narrow niche. It feeds on the leaves of only a few species of *Eucalyptus* trees, is restricted to relatively warm, forested areas of Australia, and is generally endangered in much of its habitat.

B The koala has a narrow niche.

by many kinds of carnivores (including humans), and produces large numbers of offspring. By contrast, the koala of Australia (figure 23.11B) has a very narrow niche. It can live only in areas of Australia with specific species of *Eucalyptus* trees, because it eats the leaves of only a few kinds of these trees. Furthermore, it cannot tolerate low temperatures and does not produce large numbers of offspring. As you might guess, the opossum is expanding its range and the koala is endangered in much of its range.

It is often easy to overlook important roles played by some organisms. For example, when Europeans introduced cattle into Australia—a continent where there had previously been no large, hoofed mammals—they did not think about the impact of cow manure or the significance of the niche of a group of beetles called *dung beetles*. These beetles rapidly colonize fresh dung and cause it to be broken down. No such beetles existed in Australia; therefore, in areas where cattle were raised, a significant amount of land became covered with accumulated cow dung. This reduced the area where grass could grow and reduced productivity. The problem was eventually solved by the importation of several species of dung beetles from Africa, where large, hoofed mammals are common. The dung beetles made use of what the cattle did not digest, returning it to a form that plants could more easily recycle into plant biomass.

KINDS OF ORGANISM INTERACTIONS

When organisms encounter one another, they can influence one another in numerous ways. Some interactions are harmful to one or both of the organisms. Others are beneficial. Ecologists have classified kinds of interactions between organisms into several broad categories.

Predation

Predation occurs when one animal captures, kills, and eats another animal. The organism that is killed is called the *prey,* and the one that does the killing is called the *predator.* The predator obviously benefits from the relationship, while the prey organism is harmed. Most predators are relatively large compared to their prey and have specific adaptations that aid them in catching prey. Many spiders build webs that serve as nets to catch flying insects. The prey are quickly paralyzed by the spider's bite and wrapped in a tangle of silk threads. Other rapidly moving spiders, such as wolf spiders and jumping spiders, have large eyes that help them find prey without using webs. Dragonflies patrol areas where they can capture flying insects. Hawks and owls have excellent eyesight that allows them to find their prey. Many predators, such as leopards, lions, and cheetahs, use speed to run down their prey, while others such as frogs, toads, and many kinds of lizards blend in with their surroundings and strike quickly when a prey organism happens by (figure 23.12).

Parasitism

Another kind of interaction in which one organism is harmed and the other aided is the relationship of parasitism. In fact, there are more species of parasites in the world than there are nonparasites, making this a very common kind of relationship. **Parasitism** involves one organism living in or on another living

A Lions use strength and speed to capture prey.

B Spiders use nets to trap prey.

FIGURE 23.12 (*A*) Many predators, such as lions, capture prey by making use of speed and strength. (*B*) Other predators, such as many kinds of spiders, use nets to trap prey which they then bite and kill. Obviously, predators benefit from the food they obtain to the detriment of the prey organism.

A Tapeworm—an internal parasite

B Tick—an external parasite

C Indian Pipe—a parasitic plant

FIGURE 23.13 Parasites benefit from the relationship because they obtain nourishment from the host. Tapeworms (*A*) are internal parasites in the guts of their host, where they absorb food from the host's gut. The tick (*B*) is an external parasite that sucks blood from its host. Indian pipe (*Monotropa uniflora*) (*C*) is a flowering plant that lacks chlorophyll and is parasitic on fungi that have a mutualistic relationship with tree roots.

A A remora on a shark.

FIGURE 23.14 In the relationship called commensalism, one organism benefits and the other is not affected. (*A*) The remora fish shown here hitchhikes a ride on the shark. It eats scraps of food left over by the messy eating habits of the shark. The shark does not seem to be hindered in any way. (*B*) The grey Spanish moss hanging on this oak tree is a good example of an epiphyte. The Spanish Moss does not harm the tree but benefits from using the tree surface as a place to grow.

B Spanish moss is an epiphyte.

organism from which it derives nourishment. The *parasite* derives the benefit and harms the *host,* the organism it lives in or on (figure 23.13). Many kinds of fungi live on trees and other kinds of plants, including those that are commercially valuable. Dutch elm disease is caused by a fungus that infects the living, sap-carrying parts of the tree. Mistletoe is a common plant that is a parasite on other plants. The mistletoe plant invades the tissues of the tree it is living on and derives nourishment from the tree. There are even parasitic plants that have lost their ability to produce chlorophyll.

Many kinds of worms, protozoa, bacteria, and viruses are important parasites. Parasites that live on the outside of their hosts are called *external parasites.* For example, ticks live on the outside of the bodies of mammals such as rats, dogs, cats, and humans, where they suck blood and do harm to their host. At the same time, the host could also have a tapeworm in its intestine. Since the tapeworm lives inside the host, it is called an *internal parasite.* Bacterial and viral parasites are common causes of disease in humans and other organisms.

Commensalism

Predation and parasitism both involve one organism benefiting while the other is harmed. Another common relationship is one in which one organism benefits while the other is not affected. This is known as **commensalism.** For example, sharks often have another fish, the remora, attached to them. The remora has a sucker on the top side of its head that allows it to attach to the shark and get a free ride (figure 23.14A). While the remora benefits from the free ride and by eating leftovers from the shark's meals, the shark does not appear to be troubled by this uninvited guest, nor does it benefit from the presence of the remora.

Another example of commensalism is the relationship between trees and epiphytic plants. Epiphytes are plants that live on the surface of other plants but do not derive nourishment from them (figure 23.14B). Many kinds of plants (e.g., orchids, ferns, Spanish moss, and mosses) use the surface of trees as a place to live. These kinds of organisms are particularly common in tropical rainforests. Many epiphytes derive benefit from the relationship because they are able to be located in the top of the tree, where they receive more sunlight and moisture. The trees derive no benefit from the relationship, nor are they harmed; they simply serve as a support surface for epiphytes.

Mutualism

So far in our examples, only one species has benefited from the association of two species. There are also many situations in which two species live in close association with one another, and both benefit. This is called **mutualism.** One interesting example of mutualism involves digestion in animals such as rabbits, cows, and termites. These animals eat plant material that is high in cellulose, even though they do not produce the enzymes capable of breaking down cellulose molecules into simple sugars. They manage to get energy out of these cellulose molecules with the help of special microorganisms living in their digestive tracts. The microorganisms produce cellulose-digesting enzymes, called *cellulases,* which break down cellulose into smaller carbohydrate molecules that the host's digestive enzymes can break down into smaller glucose molecules. The microorganisms benefit because the host's gut provides them with a moist, warm, nourishing environment in which to live. The hosts benefit because the microorganisms provide them with a source of food.

Another kind of mutualistic relationship exists between flowering plants and bees. Undoubtedly, you have observed bees and other insects visiting flowers to obtain nectar from the blossoms. Usually the flowers are constructed in such a manner that the bees pick up pollen (sperm-containing packages) on their hairy bodies and transfer it to the female part of the next flower they visit (figure 23.15). Because bees normally visit many individual flowers of the same species for several minutes and ignore other species of flowers, they can serve as pollen carriers between two flowers of the same species. Plants pollinated in this manner produce less pollen than do plants that rely on the wind to transfer pollen. This saves the plant energy because it doesn't need to produce huge quantities of pollen. It does, however, need to transfer some of its energy savings into the production of showy flowers and nectar to attract the bees. The bees benefit from both the nectar and pollen; they use both for food.

Symbiosis is a close physical relationship between two kinds of organisms. This term is sometimes used to mean the same thing as mutualism. At other times, it is used to include mutualism, parasitism, and commensalism as different kinds of symbiotic relationships. So you will need to determine the meaning from the context in which the word is used.

FIGURE 23.15 Mutualism is an interaction between two organisms in which both benefit. The plant benefits because cross-fertilization (exchange of gametes from a different plant) is more probable; the bee benefits by acquiring nectar and pollen for food.

Competition

So far in our discussion of organism interactions we have left out the most common one. It is reasonable to envision every organism on the face of Earth being involved in *competitive* interactions. **Competition** is a kind of interaction between organisms in which both organisms are harmed to some extent. Competition occurs whenever two organisms both need a vital resource that is in short supply (figure 23.16). The vital resource

B Interspecific competition for food.

FIGURE 23.16 Whenever a needed resource is in limited supply, organisms compete for it. When competition occurs between members of the same species it is called *intraspecific competition*. (*A*) Intraspecific competition for sunlight among these pine trees has resulted in the tall, straight trunks. Those trees that did not grow fast enough died. Competition between different species is called *interspecific competition*. (*B*) The lion and vultures are competing for the lion's zebra kill.

A Intraspecific competition for light.

could be food, shelter, nesting sites, water, mates, or space. It may involve a snarling tug-of-war between two dogs over a scrap of food, or it can be a silent struggle between plants for access to available light. If you have ever started tomato seeds (or other garden plants) in a garden and failed to eliminate the weeds, you have witnessed competition. If the weeds are not removed, they compete with the garden plants for available sunlight, water, and nutrients, resulting in poor growth of both the garden plants and the weeds.

Competition and Natural Selection

It is important to recognize that although competition results in harm to both organisms, there can still be winners and losers. The two organisms may not be harmed to the same extent, which results in one having greater access to the limited resource. Furthermore, even the loser can continue to survive if it migrates to an area where competition is less intense. Over many generations, when individuals of two different species compete, it is possible that the species may evolve through natural selection to exploit different niches. Thus, competition provides a major mechanism for natural selection. With the development of slight differences between niches, the intensity of competition is reduced. For example, many birds catch flying insects as food. However, they do not compete directly with each other because some feed at night, some feed high in the air, some feed only near the ground, and still others perch on branches and wait for insects to fly past. The insect-eating niche can be further subdivided by specializing on particular sizes or kinds of insects.

Many of the relationships just described involve the transfer of nutrients from one organism to another (predation, parasitism, mutualism). Another important way scientists look at ecosystems is to look at how materials are cycled from organism to organism.

THE CYCLING OF MATERIALS IN ECOSYSTEMS

Although some new atoms are being added to Earth from cosmic dust and meteorites, this amount is not significant in relation to the entire mass of Earth. Therefore, Earth can be considered to be a closed ecosystem as far as matter is concerned. Only sunlight energy comes to Earth in a continuous stream, and even this is ultimately returned to space as radiant energy. However, it is this flow of energy that drives all biological processes. Living systems have evolved ways of using this energy to continue life through growth and reproduction and the continual reuse of existing atoms. In this recycling process, inorganic molecules are combined to form the organic compounds of living things. If there were no way of recycling this organic matter back into its inorganic forms, organic material would build up as the bodies of dead organisms. This is thought to have occurred millions of years ago when the present deposits of coal, oil, and natural gas were formed. However, under most conditions, decomposers are available to break down

organic material to inorganic material that can then be reused by other organisms to rebuild organic material.

Carbon, hydrogen, phosphorus, potassium, nitrogen, sulfur, oxygen, calcium, and many other kinds of atoms are involved in the structure of living things. These atoms are constantly recycled. One way to get an appreciation of how various kinds of organisms interact to cycle materials is to look at a specific kind of atom and follow its progress through an ecosystem.

The Carbon Cycle

The **carbon cycle** is the series of activities of organisms that result in the cycling of carbon atoms between the atmosphere and living things. All living things contain organic molecules that are composed of long chains of carbon atoms. Carbon and oxygen atoms combine to form the molecule carbon dioxide (CO_2), which is a gas found in small quantities (about 0.04 percent) in the atmosphere. Producers combine carbon dioxide (CO_2) and water (H_2O) to form complex organic molecules such as sugar ($C_6H_{12}O_6$) during photosynthesis. At the same time, oxygen molecules (O_2) are released. Terrestrial producers obtain CO_2 from and release O_2 to the atmosphere. Aquatic producers obtain CO_2 from and release O_2 to the water in which they live.

The organic matter in the bodies of plants may be used by herbivores as food. When an herbivore eats a plant, it breaks down the complex organic molecules into simpler organic molecules, such as simple sugars, amino acids, glycerol, and fatty acids. These can be used as building blocks in the construction of its own body. Thus, the atoms in the body of the herbivore can be traced back to the plants that were eaten. Similarly, when carnivores eat herbivores, these same atoms are transferred to them. Finally, the waste products of plants and animals and the remains of dead organisms are used by decomposer organisms as sources of carbon and other atoms they need for survival.

In addition, all the organisms in this cycle—plants, herbivores, carnivores, and decomposers—obtain energy (ATP) from the process of respiration, in which oxygen (O_2) is used to break down organic compounds into carbon dioxide (CO_2) and water (H_2O). Thus, the carbon atoms that started out as components of carbon dioxide (CO_2) molecules have passed through the bodies of living organisms as parts of their organic molecules and returned as carbon dioxide as a result of respiration, ready to be cycled again. Similarly, the oxygen atoms (O) released as oxygen molecules (O_2) during photosynthesis have been used during the process of respiration (figure 23.17).

The Nitrogen Cycle

Another important element for living things is nitrogen (N). Nitrogen is essential in the formation of amino acids, which are needed to form proteins, and in the formation of nitrogenous bases, which are a part of ATP and the nucleic acids DNA and RNA. Nitrogen (N) is found as molecules of nitrogen gas (N_2) in the atmosphere. Although nitrogen gas (N_2) makes up approximately 80 percent of Earth's atmosphere, only

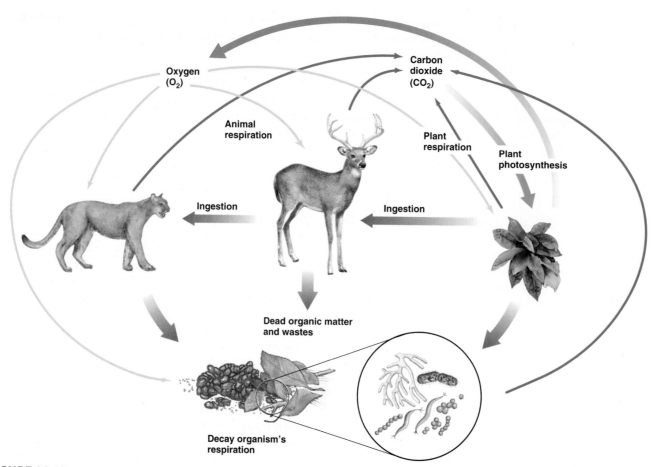

FIGURE 23.17 Carbon atoms are cycled through ecosystems. Carbon dioxide (green arrows) produced by respiration is the source of carbon that plants incorporate into organic molecules when they carry on photosynthesis. These carbon-containing organic molecules (black arrows) are passed to animals when they eat plants and other animals. Organic molecules in waste or dead organisms are consumed by decomposer organisms in the soil when they break down organic molecules into carbon dioxide and water. All organisms (plants, animals, and decomposers) return carbon atoms to the atmosphere or water as carbon dioxide when they carry on cellular respiration. Oxygen (blue arrows) is being cycled at the same time that carbon is. The oxygen is released into the atmosphere and into the water during photosynthesis and taken up during cellular respiration.

a few kinds of bacteria are able to convert it into nitrogen compounds that other organisms can use. Therefore, in most terrestrial ecosystems, the amount of nitrogen available limits the amount of plant biomass that can be produced. (Most aquatic ecosystems are limited by the amount of phosphorus rather than the amount of nitrogen.) Plants utilize several different nitrogen-containing compounds to obtain the nitrogen atoms they need to make amino acids and other compounds (figure 23.18).

Symbiotic nitrogen-fixing bacteria live in the roots of certain kinds of plants, where they convert nitrogen gas molecules into compounds that the plants can use to make amino acids and nucleic acids. The most common plants that enter into this mutualistic relationship with bacteria are the legumes, such as beans, clover, peas, alfalfa, and locust trees. Some other organisms, such as alder trees and even a kind of aquatic fern, can also participate in this relationship. There are also **free-living nitrogen-fixing bacteria** in the soil that provide nitrogen compounds that can be taken up through the roots, but the bacteria do not live in a close physical union with plants. In addition to these soil nitrogen-fixing bacteria, cyanobacteria are important in providing nitrogen in aquatic environments, and recent studies have shown that several kinds of Archaea may also be involved in the nitrogen cycle as nitrogen-fixers.

Another way plants get usable nitrogen compounds involves a series of different bacteria. *Decomposer bacteria* convert organic nitrogen-containing compounds into ammonia (NH_3). *Nitrifying bacteria* can convert ammonia (NH_3) into nitrite-containing (NO_2^-) compounds, which in turn can be converted into nitrate-containing (NO_3^-) compounds. Many kinds of plants can use either ammonia (NH_3) or nitrate (NO_3^-) from the soil as building blocks for amino acids and nucleic acids.

All animals obtain their nitrogen from the food they eat. The ingested proteins are broken down into their component amino acids during digestion. These amino acids can then be reassembled into new proteins characteristic of the animal. All dead organic matter and waste products of plants and animals are acted upon by decomposer organisms, and the nitrogen is released as ammonia (NH_3), which can be taken up by plants or acted upon by nitrifying bacteria to make nitrate (NO_3^-).

Finally, other kinds of bacteria called *denitrifying bacteria* are capable of converting nitrite (NO_2^-) to nitrogen gas (N_2), which is released into the atmosphere. Thus, in the **nitrogen cycle,** nitrogen from the atmosphere is passed through a

Chemical Fertilizers and the Nitrogen Cycle

Modern agriculture depends on the use of fertilizers to replace the elements removed from the soil by the plants that are harvested. The three most important elements in fertilizers are nitrogen, phosphorus, and potassium. The numbers on bags of fertilizer tell you how much of each of these three components is present. For example, a 6-12-12 fertilizer has 6 percent nitrogen, 12 percent phosphorus, and 12 percent potassium compounds. Although molecular nitrogen makes up 80 percent of the atmosphere, it is not available to plants. From a chemical point of view, it is very difficult to get molecular nitrogen, which consists of two nitrogen atoms bonded to each other, to react with other atoms. Although some bacteria are able to attach hydrogen to nitrogen to form ammonia, a chemical method for producing ammonia was not determined until 1909, when Fritz Haber determined the conditions under which nitrogen and hydrogen would combine to form ammonia. The process of attaching hydrogen to ammonia requires a great deal of energy. This was an important discovery because it led to a practical way to produce a variety of nitrogen-containing compounds that can be used for fertilizer. Farmers apply ammonia directly into the soil to provide nitrogen for many kinds of plants. Ammonia can be reacted with other materials to produce a variety of compounds that are used as nitrogen fertilizers, including urea, ammonium nitrate, and ammonium sulfate.

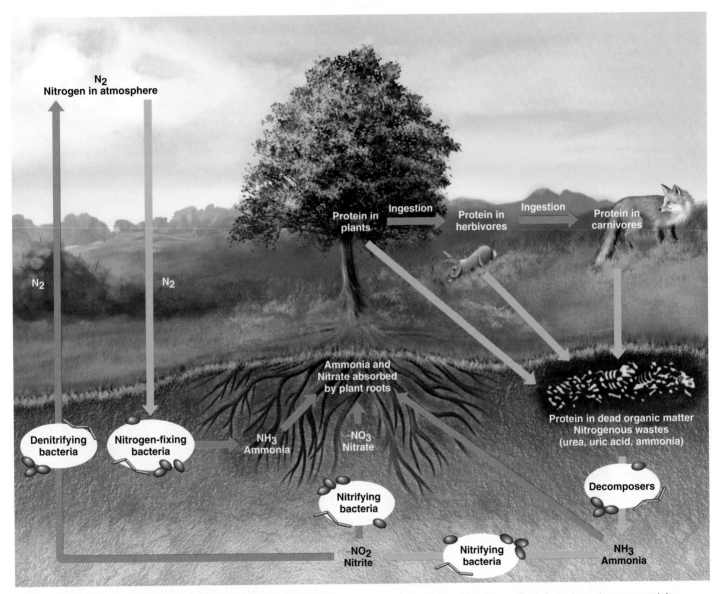

FIGURE 23.18 Nitrogen atoms are cycled through ecosystems. Atmospheric nitrogen is converted by nitrogen-fixing bacteria to nitrogen-containing compounds that plants can use to make proteins and other compounds. Proteins are passed to other organisms when one organism is eaten by another. Dead organisms and their waste products are acted upon by decomposer organisms to form ammonia, which may be reused by plants and converted to other nitrogen compounds by nitrifying bacteria. Denitrifying bacteria return nitrogen as a gas to the atmosphere.

A Closer Look

Carbon Dioxide and Global Warming

Humans have significantly altered the carbon cycle. As we burn fossil fuels, the amount of carbon dioxide in the atmosphere continually increases. It has risen over 22 percent, from about 317 parts per million (0.0317 percent) in 1960 to about 387 parts per million (0.0387 percent) in 2009. Although present in tiny amounts, it has a significant effect. Carbon dioxide, methane, and chlorofluorocarbons are often referred to as greenhouse gases because they have the effect of warming the Earth. When solar radiation hits the Earth, the Earth is warmed and infrared radiation is radiated away from the Earth. Molecules of carbon dioxide and other greenhouse gases absorb some of the infrared radiation. These molecules, in turn, radiate energy in all directions, some of it back to the Earth. Although the mechanisms are different, many people have compared the warming of the Earth by these gases to the warming that occurs in a greenhouse and the warming of the Earth is often referred to as the greenhouse effect. (See box figure 23.2.)

Many are concerned that increased carbon dioxide levels are causing a warming of the planet that will cause major changes in weather and climate, leading to the flooding of coastal cities and major changes in agricultural production.

Others maintain that weather and climate cycles are normal and that the recent period of warming is part of a natural cycle. Because major disagreements have arisen over the significance of global warming, the United Nations Environment Programme established an Intergovernmental Panel on Climate Change (IPCC), which included scientists, political leaders, economists, and others, to study the issue and make recommendations. A main activity of the IPCC is to provide at regular intervals an assessment of the state of knowledge about climate change.

The fourth Assessment Report, *Climate Change 2007*, presented several important conclusions:

1. Evidence of increased temperature is clear.
   - The average temperature of the Earth has increased 0.56–0.92°C (1.0–1.7 °F) in the past 100 years.
   1998 was the warmest year on record and 2005 was the second warmest. Eight of the warmest years on record have occurred in the ten-year period between 1998 and 2007.
   - Sea level is rising about 1.8mm/year. This equates to a rise of 18 cm (7 inches) over 100 years.
   - Sea ice has decreased.
   - The arrival of spring is earlier in many parts of the world.
2. A strong correlation exists between the increase in temperature and the concentration of greenhouse gases in the atmosphere.
3. Human activity has greatly increased the amounts of these greenhouse gases. Greenhouse gas emissions increased 70 percent between 1970 and 2004.

Concern about the possibility of global warming caused by increased carbon dioxide in the atmosphere has resulted in a series of international meetings in which governmental leaders of many nations have pledged to reduce the release of

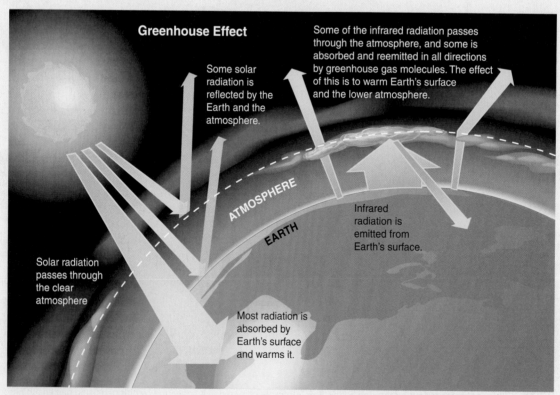

Box Figure 23.2 The greenhouse effect naturally warms Earth's surface. Without it, Earth would be 33°C (60°F) cooler than it is today— uninhabitable for life as we know it.

Source: Data from Climate Change—State of Knowledge. October 1997, Office of Science and Technology Policy, Washington, D.C.

carbon dioxide and other greenhouse gases. However, emissions of carbon dioxide are directly related to economic activity and the energy usage that fuels it. Therefore, it remains to be seen if countries will meet their goals or will succumb to economic pressures to allow continued use of large amounts of fossil fuels. Of special concern are the major economically developing countries, particularly China and India. As they seek to grow their economies and improve the economic well-being of their people, they must inevitably increase their energy consumption and increase their release of carbon dioxide.

Some countries have sought to offset their contribution to the amount of carbon dioxide in the atmosphere by planting millions of trees or by preventing the destruction of forests. The thought is that the trees carry on photosynthesis, grow, and store carbon in their bodies, leading to reduced carbon dioxide levels. At the same time, people in other parts of the world continue to destroy forests at a rapid rate. When the two activities are compared, tree planting does not offset deforestation. In addition, the trees that have been planted will ultimately die and decompose, releasing carbon dioxide back into the atmosphere, so it is not clear that this is an effective means of reducing atmospheric carbon dioxide over the long term. (See Chapter 17 for discussion of other aspects of climate change.)

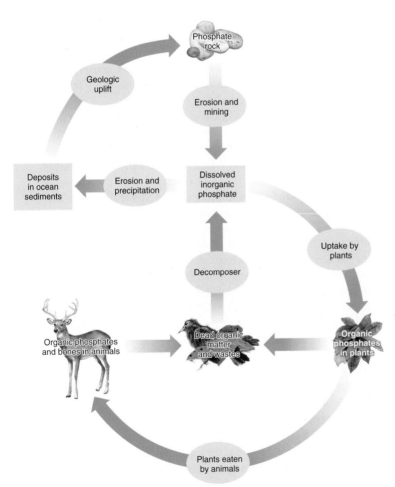

FIGURE 23.19 The source of phosphorus is rock that, when dissolved, provides phosphorus used by plants and animals.

series of organisms, many of which are bacteria, and ultimately returns to the atmosphere to be cycled again. However, the nitrogen cycle differs from the carbon cycle in a fundamental way. Inorganic nitrogen compounds are not necessarily returned to the atmosphere but are returned to plants by way of decomposer, nitrate, and nitrite bacteria.

The Phosphorus Cycle

Phosphorus is another atom common in the structure of living things. It is present in many important biological molecules, such as DNA, and in the membrane structure of cells. In addition, the bones and teeth of vertebrate animals contain significant quantities of phosphorus. The ultimate source of phosphorus atoms is rock. In nature, new phosphorus compounds are released by the erosion of rock and dissolved in water. Plants use the dissolved phosphorus compounds to construct the molecules they need. Animals obtain the phosphorus they need when they consume plants or other animals. When an organism dies or excretes waste products, decomposer organisms recycle the phosphorus compounds back into the soil. Phosphorus compounds that are dissolved in water are ultimately precipitated as deposits. Geologic processes elevate these deposits and expose them to erosion, thus making these deposits available to organisms (figure 23.19). Waste products of animals often have significant amounts of phosphorus. In places where large numbers of seabirds or bats have congregated for hundreds of years, the thickness of their droppings (called *guano*) can be a significant source of phosphorus for fertilizer. In many soils, phosphorus is in short supply and must be provided to crop plants to get maximum yields. Phosphorus is also in short supply in aquatic ecosystems, and the addition of phosphorus from detergents in sewage and runoff from lawns and farmland can cause undesirable growth of algae and plants in lakes, rivers, estuaries, and ocean waters near land.

Nutrient Cycles and Geologic Time

The nutrient cycles we have just discussed act on a short-term basis in which elements are continually being reused among organisms and on a long-term basis in which certain elements are tied up for long time periods and are not part of the active nutrient cycle. In our discussion of the phosphorus cycle, it was mentioned that the source of phosphorus is rock. While phosphorus moves rapidly through organisms in food chains, phosphorus ions are not very soluble in water and tend to precipitate in the oceans to form sediments that eventually become rock on the ocean floor.

Once this has occurred it takes the process of geologic uplift followed by erosion to make phosphorus ions available to terrestrial ecosystems. Thus we can think of the ocean as a place where phosphorus is removed from the active nutrient cycle (this situation is known as a *sink*).

There are also long-term aspects to the carbon cycle. Organic matter in soil and sediments are the remains of once living organisms. Thus, these compounds constitute a sink for carbon, particularly in ecosystems in which decomposition is slow (tundra, northern forests, grasslands, swamps, marine sediments). These materials can tie up carbon for hundreds to thousands of years. Fossil fuels (coal, petroleum, and natural gas), which were also formed from the remains of organisms, are a longer-term sink that involves hundreds of millions of years. These carbon atoms in fossil fuels at one time were part of the active carbon cycle but were removed from the active cycle when the organisms accumulated without decomposing. The organisms that formed petroleum and natural gas are thought to be the remains of marine organisms that got covered by sediments. Coal was formed from the remains of plants that were buried by sediments. Once the organisms were buried, their decomposition would be slowed, and heat from the Earth and pressure from the sediments helped to transform the remains of living things into fossil fuels. The carbon atoms in fossil fuels have been locked up for hundreds of millions of years. Thus the formation of fossil fuels was a sink for carbon atoms.

Oceans are a major carbon sink. Carbon dioxide is highly soluble in water. Many kinds of carbonate sedimentary rock are formed from the precipitation of carbonates from solution in oceans. In addition, many marine organisms form skeletons or shells of calcium carbonate. These materials accumulate on the ocean floor as sediments that over time can be converted to limestone. Limestone typically contains large numbers of fossils. The huge amount of carbonate rock is an indication that there must have been higher amounts of carbon dioxide in the Earth's atmosphere in the past.

Since fossil fuels are the remains of once living things and living things have nitrogen as a part of protein, nitrogen that was once part of the active nitrogen cycle was removed when the fossil fuels were formed. In ecosystems in which large amounts of nonliving organic matter accumulates (swamps, humus in forests, and marine sediments) nitrogen can be tied up for relatively long time periods. In addition, some nitrogen may be tied up in sedimentary rock and in some cases is released with weathering. However, it appears that the major sink for nitrogen is as nitrogen in the atmosphere. Nitrogen compounds are very soluble in water so when sedimentary rock is exposed to water, these materials are dissolved and reenter the active nitrogen cycle.

Bioaccumulation and Biomagnification

Bioaccumulation occurs when organisms capture and store materials. Carbon, nitrogen, and phosphorus are examples of elements that are temporarily bioaccumulated by living things. These materials are released from organisms as part of their metabolism or when they die. However, it is also possible for organisms to bioaccumulate toxins that are taken in by organisms but not metabolized or excreted by them. Two kinds of materials that have received attention because they bioaccumulate are methylmercury (CH_3Hg^+) and persistent organic pollutants.

Methylmercury is formed by certain kinds of Bacteria that live in low-oxygen environments. Although elemental mercury is present in rocks, the actions of humans have increased the amount of mercury available for action by these Bacteria. Mining activity and the burning of coal are major sources of mercury. Persistent organic pollutants result from the manufacture of certain plastics, fire retardants, and pesticides. Since these chemicals are not a product of nature, most organisms, including decomposers, do not have the necessary metabolic pathways to cause their decomposition. Therefore, some of these materials are taken up by organisms and bioaccumulate. The more methylmercury or persistent organic compounds an organism is exposed to, the more they accumulate, and the more likely they will affect the health of the organism.

Compounding the problem is the phenomenon of biomagnification in food chains. When an herbivore eats plants, it accumulates the persistent materials already accumulated by the plant. Similarly, a carnivore further concentrates the materials accumulated by the herbivores it eats. Thus, as one proceeds to higher trophic levels, the concentration found in animals increases—it is biomagnified.

The well-documented case of DDT is an example of how biomagnification occurs. At one time DDT was a commonly used insecticide. DDT is not very soluble in water but dissolves in oil or fatty compounds. When DDT falls on an insect or is consumed by the insect, the DDT is accumulated in the insect's fatty tissue. Large doses kill insects but small doses do not, and their bodies may contain as much as one part per billion of DDT. This is not very much, but it can have a tremendous effect on the animals that feed on the insects.

When an aquatic habitat was sprayed with a small concentration of DDT or received DDT from runoff, small aquatic organisms accumulated a concentration that was up to 250 times greater than the concentration of DDT in the surrounding water. These organisms were eaten by shrimp, clams, and small fish, which were in turn eaten by larger fish. DDT concentrations of large fish were as much as 2000 times the original concentration sprayed on the area. What was a very small initial concentration had become so high that it could be fatal to animals at higher trophic levels. This was of particular concern for birds, since DDT interferes with the production of eggshells, making them much more fragile. This problem was more common in carnivorous birds because they are at the top of the food chain. Although all birds of prey probably were affected to some degree, those that relied on fish for food were affected most severely. Eagles, osprey, cormorants, and pelicans were particularly susceptible species. (See figure 23.20.) Because it was linked to the decline of many populations of fish-eating birds, the manufacture and use of DDT was prohibited in the United States in the early 1970s and is now prohibited for most uses throughout the world.

The populations of several species of birds, including the brown pelican, bald eagle, osprey, and cormorant, all of which feed primarily on fish, were severely affected because of biomagnification of persistent organic chemicals. With the control

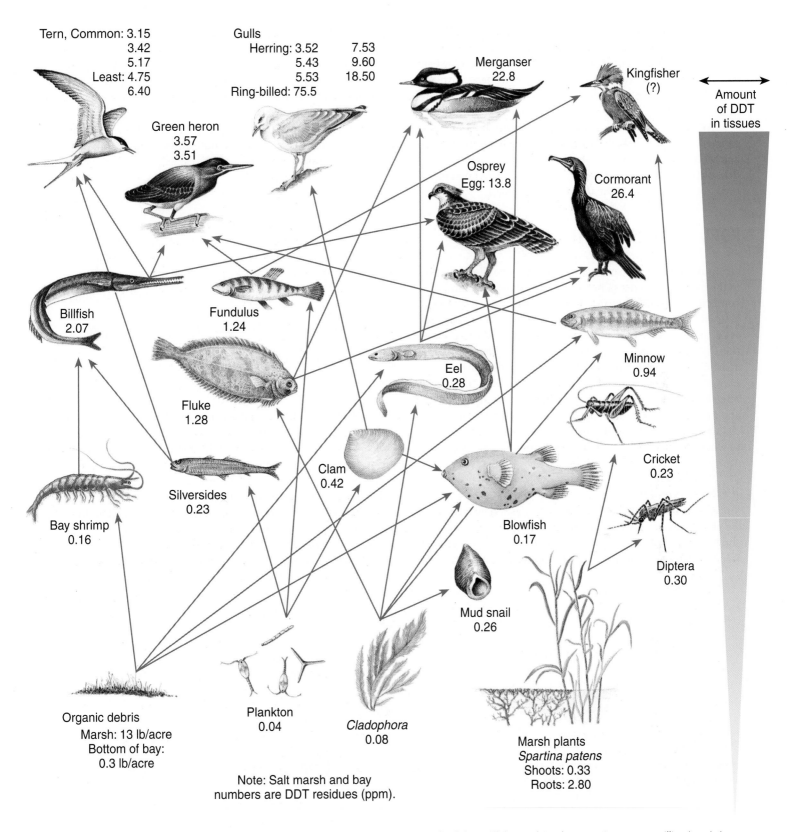

Tern, Common: 3.15
3.42
5.17
Least: 4.75
6.40

Gulls
Herring: 3.52   7.53
5.43   9.60
5.53   18.50
Ring-billed: 75.5

Merganser
22.8

Kingfisher
(?)

Amount
of DDT
in tissues

Green heron
3.57
3.51

Osprey
Egg: 13.8

Cormorant
26.4

Billfish
2.07

Fundulus
1.24

Minnow
0.94

Fluke
1.28

Eel
0.28

Cricket
0.23

Clam
0.42

Silversides
0.23

Blowfish
0.17

Diptera
0.30

Bay shrimp
0.16

Mud snail
0.26

Organic debris
Marsh: 13 lb/acre
Bottom of bay:
0.3 lb/acre

Plankton
0.04

Cladophora
0.08

Marsh plants
Spartina patens
Shoots: 0.33
Roots: 2.80

Note: Salt marsh and bay
numbers are DDT residues (ppm).

FIGURE 23.20   This illustration shows the biomagnification of DDT in an aquatic food chain. All the numbers shown are in parts per million (ppm). A concentration of one part per million means that in a million equal parts of the organism, one of the parts would be DDT. Notice how the amount of DDT in the bodies of the organisms increases as we go from producers to herbivores to carnivores. Because DDT is persistent, it builds up in the top trophic levels of the food chain.

or restriction of most persistent pesticides and several other chemicals, the levels of these chemicals in their body tissues have declined, and their populations have rebounded. The bald eagle has been removed from the endangered species list, and the pelican has been removed from the endangered species list in part of its range.

Today we still see evidence of the problem of biomagnifications in advisories that caution people (particularly children and women of childbearing age) from eating certain kinds of fish. This is primarily because of the level of methylmercury in predator fish that are at the top of the food chain.

POPULATION CHARACTERISTICS

A **population** is a group of organisms of the same species located in the same place at the same time. Examples are the number of dandelions in your front yard, the rat population in the sewers of your city, or the number of people in New York City. On a larger scale, all the people of the world constitute the human population. The terms *species* and *population* are interrelated because a species is a population—the largest possible population of a particular kind of organism. The term *population*, however, is often used to refer to portions of a species by specifying a space and time. For example, the size of the human population in a city changes from hour to hour during the day and varies according to where you set the boundaries of the city.

Genetic Differences

Since each local population is a small portion of a species, we should expect distinct populations of the same species to show differences. Some of these differences will be genetic. *Gene frequency* is a measure of how often a specific gene for a

characteristic shows up in the individuals of a population. Two populations of the same species often have quite different gene frequencies. For example, many populations of mosquitoes have high frequencies of insecticide-resistant genes, whereas others do not. The frequency of the genes for tallness in humans is greater in certain African tribes than in any other human population. Figure 23.21 shows that the frequency of the genetic information for type B blood differs significantly from one human population to another.

Age Structure

Another feature of a population is its *age distribution,* which is the number of organisms of each age in the population. In addition, organisms are often grouped into the following categories:

- prereproductive juveniles—insect larvae, plant seedlings, or babies;
- reproductive adults—mature insects, plants producing seeds, or humans in early adulthood; or
- postreproductive adults no longer capable of reproduction—annual plants that have shed their seeds, salmon that have spawned, and many elderly humans.

A population is not necessarily divided into equal thirds (figure 23.22). In some situations, a population may be made up of a majority of one age group. If the majority of the population is prereproductive, then a period of rapid population growth should be anticipated in the future as the prereproductive individuals reach sexual maturity. If a majority of the population is reproductive, the population should be growing rapidly. If the majority of the population is postreproductive, a population decline should be anticipated. Many organisms that only live a short time and have high reproductive rates can have age

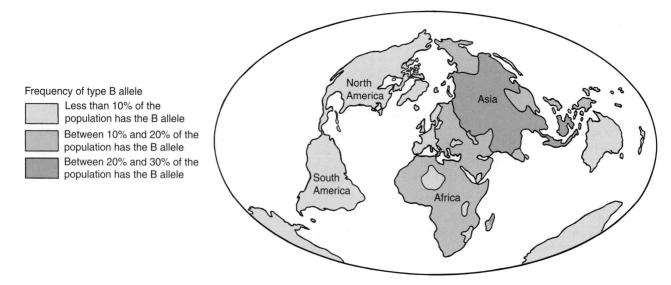

Frequency of type B allele

Less than 10% of the population has the B allele

Between 10% and 20% of the population has the B allele

Between 20% and 30% of the population has the B allele

FIGURE 23.21 The allele for type B blood is not evenly distributed in the world. This map shows that the type B allele is most common in parts of Asia and has been dispersed to the Middle East and parts of Europe and Africa. There has been very little flow of the allele to the Americas.

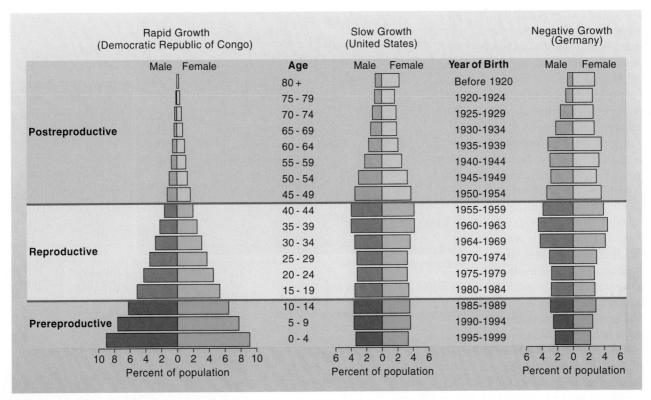

		Age			Year of Birth		
Rapid Growth (Democratic Republic of Congo) Male / Female			**Slow Growth** (United States) Male / Female			**Negative Growth** (Germany) Male / Female	
		80 +			Before 1920		
		75 - 79			1920-1924		
		70 - 74			1925-1929		
Postreproductive		65 - 69			1930-1934		
		60 - 64			1935-1939		
		55 - 59			1940-1944		
		50 - 54			1945-1949		
		45 - 49			1950-1954		
		40 - 44			1955-1959		
		35 - 39			1960-1963		
		30 - 34			1964-1969		
Reproductive		25 - 29			1970-1974		
		20 - 24			1975-1979		
		15 - 19			1980-1984		
		10 - 14			1985-1989		
Prereproductive		5 - 9			1990-1994		
		0 - 4			1995-1999		

Rapid Growth: 10 8 6 4 2 0 2 4 6 8 10 Percent of population
Slow Growth: 6 4 2 0 2 4 6 Percent of population
Negative Growth: 6 4 2 0 2 4 6 Percent of population

FIGURE 23.22 The relative numbers of individuals in each of the three categories (prereproductive, reproductive, and postreproductive) are good clues to the future growth of a population. The Democratic Republic of Congo has a large number of young individuals who will become reproducing adults. Therefore, this population is likely to grow rapidly. The United States has a large proportion of reproductive individuals and a moderate number of prereproductive individuals. Therefore, this population is likely to grow slowly. Germany has a declining number of reproductive individuals and a very small number of prereprodutive individuals. Therefore, its population has begun to decline.

Source: Data from Population Reference Bureau, Inc.

distributions that change significantly in a matter of weeks or months. For example, many birds have a flurry of reproductive activity during the summer months. Therefore, if you sample the population of a specific species of bird at different times during the summer, you would have widely different proportions of reproductive and prereproductive individuals.

Sex Ratio

Populations can also differ in their sex ratios. The *sex ratio* is the number of males in a population compared to the number of females. In bird and mammal species where strong pair-bonding occurs, the sex ratio may be nearly one to one (1:1). Among mammals and birds that do not have strong pair-bonding, sex ratios may show a larger number of females than males. This is particularly true among species that are hunted, because more males than females are shot, resulting in a higher proportion of females surviving. Since one male can fertilize several females, the population can remain large even though the females outnumber the males. Many species of animals such as sea lions, horses, and elk have mating systems in which one male maintains a harem of females. The sex ratio in these small groups is quite different from a 1:1 ratio (figure 23.23). Some male sea lions can maintain a harem of up to fifteen females. There are very few situations in which the number of males exceeds the number of females.

FIGURE 23.23 Some male animals, such as this Stellar sea lion, defend a harem of females; therefore, the sex ratio in these groups is several females per male.

Population Density

Regardless of the sex ratio of a population, most species can generate large numbers of offspring, producing a concentration of organisms in an area. *Population density* is the number of organisms of a species per unit area. Some populations are extremely concentrated into a limited space, while others are well dispersed. As the population density increases, competition among members of the population for the necessities of life increases. This increases the likelihood that some individuals will explore new habitats and migrate to new areas. Increases in the intensity of competition that cause changes in the environment and lead to dispersal are often referred to as *population pressure*. The dispersal of individuals to new areas can relieve the pressure on the home area and lead to the establishment of new populations. Among animals, it is often the juveniles who participate in this dispersal process.

In plant populations, dispersal is not able to relieve population pressure; instead, the death of weaker individuals usually results in reduced population density. In the lodgepole pine, seedlings become established in areas following fire and dense thickets of young trees are established. As the stand ages, many small trees die. The remaining trees grow larger as the population density drops (figure 23.24).

A Dense seedlings

FIGURE 23.24 This population of lodgepole pine seedlings consists of a large number of individuals very close to one another (*A*). As the trees grow, many of the weaker trees will die, the distance between individuals will increase, and the population density will be reduced (*B*).

B Less dense adult trees

THE POPULATION GROWTH CURVE

Because most species of organisms have a high reproductive capacity, there is a tendency for populations to grow if environmental conditions permit. For example, if the usual litter size for a pair of mice is four, the four would produce eight, which in turn would produce sixteen, and so forth. Figure 23.25 shows a graph of change in population size over time known as a **population growth curve.** This kind of curve is typical for situations where a species is introduced into a previously unutilized area.

The change in the size of a population depends on the rate at which new organisms enter the population compared to the rate at which they leave. The number of new individuals added to the population by reproduction per thousand individuals is called **natality.** The number of individuals leaving the population by death per thousand individuals is called **mortality.** When a small number of organisms (two mice) first invades an area, there is a period of time before reproduction takes place when the population remains small and relatively constant. This part of the population growth curve is known as the **lag phase.** During the lag phase, both natality and mortality are low. The lag phase occurs because reproduction is not an instantaneous event. Even after animals enter an area, they must mate and produce young. In organisms that take a long time to mature and produce young, such as elephants, deer, and many kinds of plants, the lag phase may be measured in years. With the mice in our example, it will be measured in weeks. The first litter of young will be produced and

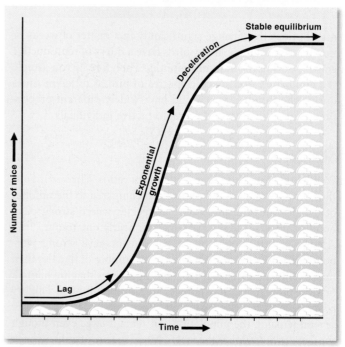

FIGURE 23.25 In this mouse population, the period of time in which there is little growth is known as the lag phase. This is followed by a rapid increase in population as the offspring of the originating population begin to reproduce themselves; this is known as the exponential growth phase. Eventually, the population growth rate begins to slow during the deceleration phase, and the population reaches a stable equilibrium phase, during which the birthrate equals the death rate.

will themselves be able to reproduce in a matter of weeks. Furthermore, the original parents will probably produce an additional litter or two during this time period. Now we have several pairs of mice reproducing more than just once. With several pairs of mice reproducing, natality increases while mortality remains low; therefore, the population begins to grow at an ever-increasing (accelerating) rate. This portion of the population growth curve is known as the **exponential growth phase.**

The number of mice (or any other organism) cannot continue to increase at a faster and faster rate because, eventually, something in the environment will become limiting and cause an increase in the number of deaths. For animals, food, water, or nesting sites may be in short supply, or predators or disease may kill many individuals. Plants may lack water, soil nutrients, or sunlight. As environmental pressures increase, the number of individuals entering the population by birth or migration begins to decline and the number leaving by death or migration begins to increase. This results in a slowing of the rate at which the population is growing, known as the **deceleration phase.** Eventually, the number of individuals entering the population by reproduction or immigration will come to equal the number of individuals leaving it by death or migration, and the population size becomes stable. Often there is both a decrease in natality and an increase in mortality at this point. This portion of the population growth curve is known as the **stable equilibrium phase.** It is important to recognize that this is still a population with births, deaths, migration, and a changing mix of individuals; however, the size of the population is stable.

POPULATION-SIZE LIMITATIONS

Populations cannot continue to increase indefinitely; eventually, some factor or set of factors acts to limit the size of a population, leading to the development of a stable equilibrium phase or even to a reduction in population size. The identifiable factors that prevent unlimited population growth are known as **limiting factors.** All of the different limiting factors that act on a population are collectively known as **environmental resistance,** and the maximum population that an area can support is known as the **carrying capacity** of the area. In general, organisms that are small and have short life spans tend to have fluctuating populations and do not reach a carrying capacity, whereas large organisms that live a long time tend to reach an optimum population size that can be sustained over an extended period (figure 23.26). A forest ecosystem contains populations of many insect species that fluctuate widely and rarely reach a carrying capacity, but the number of specific tree species or large animals such as owls or deer is relatively constant.

However, the carrying capacity is not an inflexible number. Often environmental changes such as climate variations, disease epidemics, forest fires, or floods can change the carrying capacity of an area for specific species. In aquatic ecosystems, one of the major factors that determines the carrying capacity is the amount of nutrients in the water. In areas where nutrients are abundant,

FIGURE 23.26 A number of factors in the environment, such as food, oxygen supply, diseases, predators, and space, determine the number of organisms that can survive in a given area—the carrying capacity of that area. The environmental factors that limit populations are collectively known as environmental resistance.

the numbers of various kinds of organisms are high. Often nutrient levels fluctuate with changes in current or runoff from the land, and plant and animal populations fluctuate as well. In addition, a change that negatively affects the carrying capacity for one species may increase the carrying capacity for another. For example, the cutting down of a mature forest followed by the growth of young trees increases the carrying capacity for deer and rabbits, which use the new growth for food, but decreases the carrying capacity for squirrels, which need mature, fruit-producing trees as a source of food and old, hollow trees for shelter.

LIMITING FACTORS TO HUMAN POPULATION GROWTH

Today, we hear differing opinions about the state of the world's human population. On one hand, we hear that the population is growing rapidly. In contrast, we hear that some European countries have shrinking populations. Other countries are concerned about the aging of their populations because birthrates and death rates are low. In magazines and on television, we see that there are starving people in the world. At the same time, we hear discussions about the problem of food surpluses and obesity in

TABLE 23.2   Comparison of Population Characteristics of More and Less Developed Nations (2008)

Population Characteristic	More Developed Nations	Less Developed Nations	Significance
Total population (2008)	1.227 billion	5.479 billion	Less developed nations constitute nearly 82 percent of world's population.
Expected 2025 population	1.269 billion	6.731 billion	Less developed nations are growing much faster.
Birthrate	12 births per 1,000 in population	23 births per 1,000 in population	Less developed nations have high birthrates.
Death rate	10 deaths per 1,000 in population	8 deaths per 1,000 in population	More developed nations have higher death rates because their populations are older.
Total fertility rate	1.6 births per woman per lifetime	2.8 births per woman per lifetime	Women in less developed nations have nearly twice as many children as those in more developed nations.
Infant mortality rate	6 deaths per 1,000 births	54 deaths per 1,000 births	Less developed nations have much higher infant mortality rates.
Percent under 15 years of age	17	30	Less developed nations have young populations.
Percent over 65 years of age	16	6	More developed nations have aging populations.
Gross national income per person	US$ 31,200	US$ 4,760	Incomes in less developed nations are about 15 percent of those in developed nations.

Source: Data from the Population Reference Bureau, *2008 World Population Data Sheet.*

many countries. Some have even said that the most important problem in the world today is the rate at which the human population is growing; others maintain that the growing population will provide markets for goods and be an economic boon. How do we reconcile this mass of conflicting information?

The world population is over 6.7 billion people and is expected to grow to 8 billion by 2025. This is nearly a 20 percent increase in less than twenty years. The distribution of this growth, however, will not be even. The human population is divided into two rather distinct segments: the more economically developed countries of the world (Europe, North America, Australia, New Zealand, and Japan) and the less developed nations. Sharp

contrasts exist between the populations of these two groups of nations. Table 23.2 highlights some of these differences.

Regardless of whether we are considering the developed world, the less developed world, or the total world population, it is important to realize that human populations follow the same patterns of growth and are acted upon by the same kinds of limiting factors as are populations of other organisms. When we look at the curve of population growth over the past several thousand years, estimates are that the human population remained low and constant for thousands of years but has increased explosively (from 1 billion to 6.7 billion) in the past two hundred years (figure 23.27). For example, it has

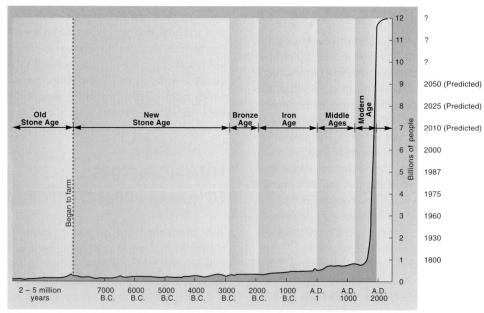

FIGURE 23.27   The number of humans doubled from A.D. 1800 to 1930 (from 1 billion to 2 billion), doubled again by 1975 (4 billion), and is projected to double again by the year 2025. How long can the human population continue to double before Earth's ultimate carrying capacity is reached?

Source: Data from Jean Van Der Tak, et al., "Our Population Predicament: A New Look," in *Population Bulletin,* vol. 34, no. 5, December 1979, Population Reference Bureau, Washington, D.C., and more recent data from the Population Reference Bureau.

Government Policy and Population Control

The actions of government can have a significant impact on the population growth patterns of nations. Some countries have policies that encourage couples to have children. The U.S. tax code indirectly encourages births by providing tax advantages to the parents of children. Some countries in Europe are concerned about the lack of working-age people in the future and are considering ways to encourage births.

China and India are the two most populous countries in the world. Both have over 1 billion people. However, China has taken steps to control its population and now has a total fertility rate of 1.6 children per woman, while India has a total fertility rate of 2.8. The total fertility rate is the average number of children born to a woman during her lifetime. The difference in the total fertility rate between these two countries is the result of different policy decisions over the last fifty years. The history of China's population policy is an interesting study of how government policy affects reproductive activity among its citizens. When the People's Republic of China was established in 1949, the official policy of the government was to encourage births because more Chinese would be able to produce more goods and services, and production was the key to economic prosperity. The population grew from 540 million to 614 million between 1949 and 1955 while economic progress was slow. Consequently, the government changed its policy and began to promote population control.

The first family-planning program began in 1955, as a means of improving maternal and child health. Birthrates fell, but other social changes resulted in widespread famine and increased death rates and low birthrates in the late 1950s and early 1960s.

The present family-planning policy began in 1971 with the launching of the *wan xi shao* campaign. Translated, this phrase means "later" (marriages), "longer" (intervals between births), and "fewer" (children). As part of this program, the legal ages for marriage were raised. For women and men in rural areas, the ages were raised to twenty-three and twenty-five, respectively; for women and men in urban areas, the ages were raised to twenty-five and twenty-eight, respectively. These policies resulted in a reduction of birthrates by nearly 50 percent between 1970 and 1979.

An even more restrictive "one child" campaign was begun in 1978–1979. The program offered incentives for couples to restrict their family size to one child. Couples enrolled in the program would receive free medical care, cash bonuses for their work, special housing treatment, and extra old-age benefits. Those who broke their pledge were penalized by the loss of these benefits as well as other economic penalties. By the mid-1980s, less than 20 percent of the eligible couples were signing up for the program. Rural couples, particularly, desired more than one child. In fact, in a country where about 60 percent of the population is rural, the rural total fertility rate was 2.5 children per woman. In 1988, a second child was sanctioned for rural couples if their first child was a girl, which legalized what had been happening anyway.

The programs appear to have had an effect because the current total fertility rate has fallen to 1.6 children per woman. Replacement fertility, the total fertility rate at which the population would eventually stabilize, is 2.1 children per woman per lifetime. Furthermore, over 90 percent of couples use contraception. Abortion is also an important aspect of this program, with a ratio of more than six hundred abortions per thousand live births.

Although the program has been effective in reducing the birthrate, there has been one unexpected outcome: the sex ratio shifting in favor of males. (Among children the ratio is about 120 boys to 100 girls.) It is not clear what is causing the change; however, as in many societies, there is a strong social desire for sons. Because of this desire for sons, many speculate that one or more of the following may be the cause: female fetuses may be aborted more frequently than males; there may be a higher infant mortality rate among female children due to neglect; or female off-spring may be given up for adoption more frequently than males. Recently the Chinese government has responded to this inequality with policy changes. Public information sources, advice from health care workers, and small financial incentives are all aimed at making the raising of girl children socially acceptable.

By contrast, during the same fifty years, India has had little success in controlling its population. In 2000, a new plan was unveiled that includes the goal of bringing the total fertility rate from 3.2 children per woman to 2 (replacement rate) by 2010. In the past, the emphasis of government programs was on meeting goals of sterilization and contraceptive use, but this has not been successful. Today, less than 50 percent of couples use contraceptives. This new plan emphasizes improvements in the quality of life of the people. The major thrusts are to reduce infant and maternal death, immunize children against preventable disease, and encourage girls to attend school. It is hoped that improved health will remove the perceived need for large numbers of births. Currently, less than 50 percent of the women in India can read and write. The emphasis on improving the educational status of women is related to the experiences of other developing countries. In many other countries, it has been shown that an increase in the education level of women has been linked to lower fertility rates.

It seems overly optimistic to think that the total fertility rate can be reduced from 3.2 to 2 in just ten years, but it had declined to 2.28 by 2008. This is an encouriging sign and programs that emphasize improvements in maternal and child health and increasing the educational level of women have been very effective in reducing total fertility in other countries.

Questions to Discuss

1. Under what conditions should governments interfere with the reproductive desires of their people?
2. Compare the population control plans of China and India. List the advantages and disadvantages of each.
3. Assume the U.S. government needs to develop specific population control measures. Develop a list of at least three population control measures you would support.

Source: Data from H. Yuan Tien, "China's Demographic Dilemmas," in *Population Bulletin, 1992,* Population Reference Bureau, Washington, D.C., and Natural Family Planning Commission of China; and more recent data taken from Population Reference Bureau.

been estimated that when Columbus discovered America, the Native American population was about 1 million and was at or near the carrying capacity. Today, the population of the United States and Canada is about 338 million people. Does this mean that humans are different from other animal species? Can the human population continue to grow forever?

The human species is no different from other animals. It has an upper limit set by the carrying capacity of the environment, but the human population has been able to increase astronomically because technological changes and displacement of other species have allowed us to shift the carrying capacity upward. Much of the exponential growth phase of the human population can be attributed to improvements in sanitation, control of diseases, improvement in agricultural methods, and replacement of natural ecosystems with artificial agricultural ecosystems. But even these conditions have their limits. There will be some limiting factors that eventually cause a leveling off of our population growth curve. We cannot increase beyond our ability to get raw materials and energy, nor can we ignore the waste products we produce or the other organisms with which we interact. Probably the ultimate limiting factor will be the ability to produce food. Improvements in agricultural production have not kept pace with population growth. Indeed, in many parts of the world, food is in short supply despite surpluses in much of the developed world.

Humans are different from most other organisms in a fundamental way: We are able to predict the outcome of a specific course of action. Current technology and medical knowledge are available to control human population and improve the health and well-being of the people of the world. Why then does the human population continue to grow, resulting in human suffering and stressing the environment in which we live? Because we are social animals that have freedom of choice, we frequently do not do what is considered "best" from an unemotional, unselfish, biological point of view. People make decisions based on historical, social, cultural, ethical, and personal considerations. What is best for the population as a whole may be bad for you as an individual. The biggest problems associated with control of the human population are not biological problems but rather require the efforts of philosophers, theologians, politicians, and sociologists. If our population increases, so will political, social, and biological problems; there will be less individual freedom, there will be intense competition for resources, and famine and starvation will become even more common. The knowledge and technology necessary to control the human population are available, but the will is not. What will eventually limit the size of our population? Will it be lack of resources, lack of energy, accumulated waste products, competition among ourselves, or rational planning of family size?

HUMAN POPULATION GROWTH AND THE GLOBAL ECOSYSTEM

The growth of the human population has had major consequences for the global ecosystem. As the size of the human population has increased, more and more resources have been diverted to human use and fewer are available for other organisms. Nearly Earth's entire surface has been affected by human activity. Over 80 percent of the land surface has been modified for human uses such as agricultural and grazing land, managed forests, urban space, and transportation corridors. Essentially all of the land that can support agriculture has been converted from forests or grasslands to raising crops. Oceans have been significantly impacted as well. Most fish populations are either overfished or are being fished at capacity.

When humans convert terrestrial or aquatic ecosystems to their use, these ecosystems are unavailable for their original inhabitants. This habitat destruction has resulted in a loss of biodiversity due to extinction and endangerment of species that compete with humans for resources. For example, the bison population of North America is restricted to a few protected areas, since nearly all the grasslands have been converted to agriculture and grazing. Habitat destruction in the ocean also occurs. The use of trawls—weighted nets that are pulled along the ocean bottom—to catch fish has greatly modified the ocean bottom in areas of high fishing activity.

Other human activities affect Earth globally. Global climate change is linked to increased amounts of carbon dioxide, methane, and other greenhouse gases in the atmosphere. Increased greenhouse gases lead to a general warming of Earth that results in the melting of glaciers, affects precipitation patterns, and causes a rise in sea level. These changes have the potential to cause significant change in biomes. The increase in atmospheric carbon dioxide is a direct result of the use of fossil fuels and the destruction of forests. The increases in methane can be traced to, in large part, domesticated livestock that release methane from their guts and decomposition that occurs in flooded rice paddies.

In addition, persistent organic pollutants such as DDT, polychlorinated biphenyls (PCBs), and many other compounds from industry and mercury from the burning of coal in electric power plants have a global circulation and are found in humans and other animals throughout the world. Many kinds of ocean fish have elevated levels of mercury.

Lifestyles affect the degree of impact. Consumption of energy and other resources is much higher in the nations of the developed world than in those of the less developed world. The developed world, with less than 20 percent of the world's human population, consumes about 60 percent of the world's energy. The United States, with less than 5 percent of the world's population, consumes about 25 percent of the world's energy. The consumption of goods and services by the people of the developed world has a global impact. Even the food consumption habits of people have wide-ranging effects. Rainforests in Brazil are burned to create grazing land to raise beef to sell to developed nations.

As the human population continues to increase, so will the pressure to raise more food, produce more consumer products, and supply more energy. These needs will place increased pressure on the natural resources of Earth.

SUMMARY

Ecology is the study of how organisms interact with their environment. The *environment* consists of *biotic* and *abiotic* components that are interrelated in an *ecosystem*. All *ecosystems* must have a constant *input of energy* from the Sun. *Producer organisms* are capable of trapping the Sun's energy and converting it into *biomass*. *Herbivores* feed on producers and are in turn eaten by *carnivores*, which may be eaten by *other carnivores*. Each level in the *food chain* is known as a *trophic level*. Other kinds of organisms involved in food chains are *omnivores*, which eat both plant and animal food, and *decomposers*, which break down dead organic matter and waste products. All *ecosystems* have a large *producer* base with successively smaller amounts of energy at the *herbivore, primary carnivore,* and *secondary carnivore trophic levels*. This is because each time *energy* passes from *one trophic level* to the next, about 90 percent of the energy is lost from the *ecosystem*. A *community* consists of the interacting *populations of organisms* in an area. The organisms are interrelated in many ways in food chains that interlock to create *food webs*. Because of this interlocking, changes in one part of the *community* can have effects elsewhere.

Major land-based regional *ecosystems* are known as *biomes*. The *temperate deciduous forest, boreal coniferous forest, temperate rainforest, tropical rainforest, tropical dry forest, Mediterranean shrublands, temperate grasslands, desert, savanna,* and *tundra* are examples of biomes. Aquatic communities can be divided into marine, freshwater, and estuarine communities.

Each organism in a community occupies a specific space known as its *habitat* and has a specific functional role to play known as its *niche*. An *organism's habitat* is usually described in terms of some conspicuous element of its surroundings. The *niche* is difficult to describe because it involves so many interactions with the physical environment and other living things.

Interactions between organisms fit into several categories. *Predation* is one organism benefiting (*predator*) at the expense of the organism killed and eaten (*prey*). *Parasitism* is one organism benefiting (*parasite*) by living in or on another organism (*host*) and deriving nourishment from it. *Commensal* relationships exist when one organism is helped but the other is not affected. *Mutualistic* relationships benefit both organisms. *Competition* causes harm to both of the organisms involved, although one may be harmed more than the other and may become extinct, evolve into a different niche, or be forced to migrate.

Many atoms are cycled through ecosystems. The *carbon atoms* of living things are trapped by *photosynthesis*, passed from organism to organism as food, and released to the atmosphere by *respiration*. *Nitrogen* originates in the atmosphere, is trapped by *nitrogen-fixing bacteria*, passes through a series of organisms, and is ultimately released to the atmosphere by *denitrifying bacteria*. *Phosphorus* originates in rock and is used by organisms and eventually deposited as sediments.

A *population* is a group of organisms of the same species in a particular place at a particular time. *Populations* differ from one another in *gene frequency, age distribution, sex ratio,* and *population density*. A typical population growth curve consists of a *lag phase* in which population rises very slowly, followed by an *exponential growth phase* in which the population increases at an accelerating rate, followed by a leveling off during the *deceleration phase* until population growth stops in the *stable equilibrium phase* as the *carrying capacity* of the environment is reached.

Humans as a species have the same limits and influences that other organisms do. However, humans can reason and predict, thus offering the possibility of population control through *conscious population limitation*. Human activities have had many impacts on the global ecosystem. These include: habitat destruction, release of pollutants, climate change, and reduced biodiversity.

KEY TERMS

abiotic factors (p. **561**)
benthic (p. **571**)
biodiversity (p. **567**)
biomass (p. **565**)
biomes (p. **567**)
biotic factors (p. **561**)
carbon cycle (p. **579**)
carnivore (p. **564**)
carrying capacity (p. **589**)
commensalism (p. **577**)
community (p. **562**)
competition (p. **578**)
consumer (p. **563**)
deceleration phase (p. 589)

decomposers (p. **564**)
ecology (p. **561**)
ecosystem (p. **562**)
environment (p. **561**)
environmental resistance (p. **589**)
estuary (p. **572**)
exponential growth phase (p. **589**)
food chain (p. **564**)
free-living nitrogen-fixing
    bacteria (p. **580**)
freshwater community (p. **570**)
habitat (p. **573**)
herbivore (p. **564**)
keystone species (p. **567**)

lag phase (p. **588**)
limiting factors (p. **589**)
marine community (p. **570**)
mortality (p. **588**)
mutualism (p. **577**)
natality (p. **588**)
niche (p. **573**)
nitrogen cycle (p. **583**)
omnivore (p. **564**)
organism (p. **562**)
parasitism (p. **575**)
pelagic (p. **571**)
plankton (p. **571**)
population (p. **586**)

population growth curve (p. **588**)
predation (p. **575**)
primary carnivore (p. **564**)
primary consumer (p. **564**)
producer (p. **563**)
secondary carnivore (p. **564**)
secondary consumer (p. **564**)
stable equilibrium phase
    (p. **589**)
symbiosis (p. **578**)
symbiotic nitrogen-fixing
    bacteria (p. **580**)
trophic level (p. **563**)

APPLYING THE CONCEPTS

1. As energy is passed from one trophic level to the next in a food chain, about _____ percent is lost at each transfer.
   a. 10
   b. 50
   c. 75
   d. 90

2. The primary factor that determines whether a geographic area will support temperate deciduous forest or prairie is
   a. the amount of rainfall.
   b. the severity of the winters.
   c. the depth of the soil.
   d. the kinds of animals present.

3. Which one of the following organisms is most likely to be at the second trophic level?
   a. a maple tree
   b. a snake
   c. a fungus
   d. an elephant

4. Which one of the following processes is necessary for the production of organic molecules from inorganic molecules?
   a. respiration by animals
   b. decomposer organisms releasing carbon dioxide
   c. carbon dioxide uptake by plants
   d. herbivores consuming plants

5. If two species of organisms occupy the same niche,
   a. mutualism will result.
   b. competition will be very intense.
   c. both organisms will become extinct.
   d. both will need to enlarge their habitat.

6. If nitrogen-fixing bacteria were to become extinct
   a. life on Earth would stop immediately since there would be no source of nitrogen.
   b. life on Earth would continue indefinitely.
   c. life on Earth would slowly dwindle as useful nitrogen became less available.
   d. life on Earth would be unchanged except that proteins would be less common.

7. You obtain nitrogen for the organic molecules in your body from
   a. the air you breathe.
   b. the water you drink.
   c. carbohydrates in the food you eat.
   d. proteins in the food you eat.

8. Many plants (flowers) provide nectar for insects. The insects, in turn, pollinate the flower. This relationship between the insect and plant represents
   a. parasitism.
   b. commensalism.
   c. mutualism.
   d. predation.

9. As population density increases, which one of the following is likely to occur?
   a. Natality will increase.
   b. Mortality will decrease.
   c. The population will experience exponential growth.
   d. Individuals will migrate from the area of highest density.

10. A population made up primarily of prereproductive individuals will
    a. increase rapidly in the future.
    b. become extinct.
    c. rarely occur.
    d. remain stable for several generations.

11. Listed below are the sex ratios for four populations. All other things being equal (including current population size), which population should experience the greatest future growth?
    a. 1 male:1 female
    b. 2 male:1 female
    c. i male: 2 female
    d. 3 male: 1 female

12. The current human population of the world is experiencing
    a. a population decline.
    b. slow steady growth.
    c. stable equilibrium.
    d. rapid growth.

Answers
1. d **2.** a **3.** d **4.** c **5.** b **6.** c **7.** d **8.** c **9.** d **10.** a **11.** c **12.** d

QUESTIONS FOR THOUGHT

1. Describe the flow of energy through an ecosystem.
2. What role does each of the following play in an ecosystem: sunlight, plants, the second law of thermodynamics, consumers, decomposers, herbivores, carnivores, and omnivores?
3. Why is there usually a larger herbivore biomass than a carnivore biomass?
4. List a predominant abiotic factor in each of the following biomes: temperate deciduous forest, boreal coniferous forest, desert, tundra, tropical rainforest, and savanna.
5. Can energy be recycled through an ecosystem? Explain why or why not.
6. Describe your niche.
7. What do parasites, commensal organisms, and mutualistic organisms have in common? How are they different?
8. Trace the flow of carbon atoms through a community that contains plants, herbivores, decomposers, and parasites.
9. Describe four different roles played by bacteria in the nitrogen cycle.
10. What kinds of organisms have their population size limited by the carrying capacity of their environment?
11. If competition is intense, how can both kinds of competing organisms continue to exist?
12. How does the population growth curve of humans compare with that of other kinds of animals?

FOR FURTHER ANALYSIS

1. Many cities have serious problems with large populations of deer. They eat the plants in people's yards and cause serious traffic accidents. You are a consultant hired to help a city solve its deer population problem. Using what you have learned about the principles of population growth, describe two alternative plans that would work.

2. Sulfur is present in all organisms, particularly in proteins. Based on what you know about the carbon and nitrogen cycles, develop an outline of what you think the sulfur cycle would look like.

3. Many people advocate the use of biofuels (biodiesel, ethanol, etc.) as a substitute for fossil fuels. They suggest that these fuels cause less environmental damage and are renewable. Based on what you know about energy flow through ecosystems and the effect of agriculture on biodiversity, what kinds of questions would you like to ask these advocates of biofuels production?

4. Biomes are regional ecological systems. What is the predominant biome where you live? Develop a list of ten characteristics you consider important in describing your biome.

5. Compare your social interactions with other people to the kinds of ecological interactions that occur among organisms. List examples of social interactions that could be considered to be:

    mutualism

    parasitism

    commensalism

    competition

INVITATION TO INQUIRY

World Population Characteristics

The Population Reference Bureau maintains a website that contains the World Population Data Sheet. It produces a new data sheet each year. Go to that website and access the World Population Data Sheet. Determine which countries have each of the following:

1. The highest infant mortality rate
2. The highest total fertility rate
3. The lowest life expectancy

Compare the countries you have listed. In what ways are they similar?

Now determine which countries have each of the following:

1. The lowest infant mortality rate
2. The lowest total fertility rate
3. The highest life expectancy

Compare the countries you have listed. In what ways are they similar?

24

Human Biology: Materials Exchange and Control Mechanisms

Because many different systems each perform a specific set of tasks in the human body, it is necessary to coordinate these various activities with one another.

CORE **CONCEPT**
The human body consists of many interacting systems.

OUTLINE

Physics

▷ Kinetic molecular theory (Ch. 4).
▷ Pressure is a force (Ch. 4).
▷ Gas laws (Ch. 4).

Chemistry

▷ pH is a measure of hydrogen ion concentration (Ch. 10).
▷ Many large organic molecules are polymers (Ch. 19).
▷ Ions are positively or negatively charged (Ch. 10).

Earth Science

▷ Minerals are inorganic materials (Ch. 15).

Life Science

▷ Enzymes control biological reactions (Ch. 20).
▷ Aerobic cellular respiration requires oxygen (Ch. 20).
▷ Cell membranes are selectively permeable (Ch. 20).

OVERVIEW

All cells must have nutrients, waste products, gases, and other molecules pass through their cell membranes. In single-celled organisms, the cell membrane is in direct contact with the cell's surroundings so diffusion and other processes accomplish this exchange. However, large, multicellular organisms consist of trillions of individual cells, most of which are not located near the surface of the organism yet must exchange materials in the same manner as in single-celled creatures. Therefore, large organisms consist of many different systems that assist in the exchange of materials such as food, wastes, and gases between the cells of the organisms and the external environment. The primary systems involved are the circulatory, respiratory, digestive, and excretory systems.

These systems must be coordinated and regulated in such a way that they supply the needs of the individual cells. Two other systems, the nervous and endocrine systems, are involved in managing these functions. This chapter will look at the systems involved in the exchange of materials and coordination of activities within the human body.

HOMEOSTASIS

Living things are like complex machines with many parts that must function in a coordinated fashion. All systems are integrated and affect one another in many ways. For example, when you run up a hill, your leg and arm muscles move in a coordinated way to provide power. They burn fuel (glucose) for energy and produce carbon dioxide and lactic acid as waste products, which tend to lower the pH of the blood. Your heart beats faster to provide oxygen and nutrients to the muscles, you breathe faster to supply the muscles with oxygen and get rid of carbon dioxide, and the blood vessels in the muscles dilate to allow more blood to flow to them. As you run, you generate excess heat. As a result, more blood flows to the skin to get rid of the heat and sweat glands begin to secrete, thus cooling the skin. All of these automatic internal adjustments help to regulate the pH of the blood; the concentration of oxygen, carbon dioxide, and glucose in the blood; and body temperature within narrow ranges. These processes can be summed up in the concept of *homeostasis*. **Homeostasis** is the process of maintaining a constant internal environment as a result of monitoring and modifying the functioning of various systems.

EXCHANGING MATERIALS: BASIC PRINCIPLES

In order to maintain homeostasis, organisms must exchange materials with their surroundings. For materials to enter or leave an organism, they must pass through a membrane. The ability to transport materials into or out of an organsim is determined by the surface area where exchange takes place. As you increase the size of an object, the volume increases faster than the surface area. (See chapter 1, page 8, for a discussion of surface-area-to-volume ratio.) So, the larger an organism or cell becomes, the more difficult it is to satisfy its needs.

A major mechanism for exchange of materials between organisms and their environment is diffusion. In addition to the limitation that surface area presents to the transport of materials, cells also have a problem with diffusion of materials within the cell. The molecular process of diffusion is quite rapid over short distances but becomes very slow over longer distances. Diffusion is generally insufficient to handle the needs of cells if it must take place over a distance of more than 1 mm. The center of the cell would die before it received the molecules it needed if the distance were greater. Because of this, it is understandable that the basic unit of life, the cell, must remain small.

A Marine flatworm

B Marine polychaete with gills

C Horseshoe crab with book gills

FIGURE 24.1   As multicellular organisms evolved, special structures with increased surface area developed to assist in the exchange of gases between the organism and its surroundings. (*A*) Flatworm have no special structures, but their flat shape assires that no cell is far from the surface. (*B*) Many kinds of polychaete worms have feathery gills that provide a large surface area for the exchange of gases. The same structure with its large surface area is used to capture tiny food organisms. (*C*) The horseshoe crab has folds on the undersurface that serve as gills. Because they are like the pages in a book, they are called *book gills*.

When we look at the evolution of organisms, we find that the first organisms were small, single-celled, and marine. Under these conditions, most exchanges could simply occur through their cell membranes. As more complex, multicellular organisms evolved, they had many cells that were not in contact with their surroundings and developed other ways to allow for efficient exchange of materials. The evolution of organs and organ systems allowed them to exchange and distribute materials to all cells even if many cells were not exposed to the sea. Often it is possible to see a pattern of evolutionary development of the structures involved in the exchange of materials. For example, many tiny multicellular marine animals still can get sufficient oxygen through their surfaces without special structures. Larger marine animals have some sort of gill (a folding of the body surface) to assist gas exchange (figure 24.1).

TRANSPORTING MATERIALS: THE CIRCULATORY SYSTEM

Large, multicellular organisms such as humans consist of trillions of cells. Since many of these cells are buried within the organism far from the body surface, there must be some sort of distribution system to assist them in solving their materials exchange problems. The primary mechanism used is the circulatory system.

The circulatory system consists of several fundamental parts. **Blood** is the fluid medium in animals that assists in the transport of materials and heat. The **heart** is a pump that forces the fluid blood from one part of the body to another. The heart pumps blood into **arteries,** which distribute blood to organs. It flows into successively smaller arteries until it reaches tiny vessels called **capillaries,** where materials are exchanged between the blood and tissues through the walls of the capillaries. The blood flows from the capillaries into **veins** that combine into larger veins that ultimately return the blood to the heart from the tissues.

The Nature of Blood

Blood is a fluid that consists of several kinds of cells suspended in a watery matrix called **plasma.** Table 24.1 lists the chemical composition of the plasma and the variety of cells found in blood. The plasma contains many kinds of dissolved molecules. The primary function of the blood is to transport molecules, cells, and heat from one part of the body to another. The major kinds of molecules that are distributed by the blood are respiratory gases (oxygen and carbon dioxide), nutrients of various kinds, waste products, disease-fighting antibodies, and chemical messengers (hormones). Blood has special characteristics that allow it to distribute respiratory gases very efficiently. Although little oxygen is carried as free, dissolved oxygen molecules in the plasma, *red blood cells* (*RBCs*)—also known as *erythrocytes*—contain **hemoglobin,** an iron-containing molecule, to which oxygen molecules readily bind. This allows for much more oxygen to be carried than could be possible if it were simply dissolved in the blood. Thus, the circulating blood is constantly picking up and delivering molecules to where they are needed. This is important for maintaining homeostasis.

Another important regulatory function of the blood involves the various kinds of cells involved in immunity. The *white blood cells* (*WBCs*) carried in the blood help the body resist many diseases. They constitute the core of the *immune system*. The various WBCs participate in providing immunity

TABLE 24.1   The Composition of Blood

Component	Quantity Present	Function
Plasma	55%	Maintain fluid nature of blood
Water	91.5%	
Protein	7.0%	
Other materials	1.5%	
Cellular material	45%	
Red blood cells (erythrocytes)	4.3–5.8 million/mm^3	Carry oxygen and carbon dioxide
White blood cells (leukocytes)	5–9 thousand/mm^3	Immunity
Lymphocytes	25%–30% of white cells present	
Monocytes	3%–7% of white cells present	
Neutrophils	57%–67% of white cells present	
Eosinophils	1%–3% of white cells present	
Basophils	Less than 1% of white cells present	
Platelets	250–400 thousand/mm^3	Clotting

Neutrophils   Eosinophils   Basophils

Lymphocytes   Monocytes   Platelets   Erythrocytes

in several ways. First of all, immune system cells are able to recognize cells and molecules that are harmful to the body. If a molecule is recognized as harmful, certain white blood cells called *lymphocytes* produce *antibodies* (*immunoglobulins*) that attach to the dangerous materials. The dangerous molecules that stimulate the production of antibodies are called *antigens* (*immunogens*). When harmful microorganisms (e.g., bacteria, viruses, fungi) or toxic molecules enter the body or cancer cells are produced, a combination of several kinds of lymphocytes (1) recognize, (2) boost their abilities to respond to,

(3) move toward, and (4) destroy the problem causers. *Neutrophils* and *monocytes* are specific kinds of WBCs capable of engulfing harmful material, a process called phagocytosis. Thus, they are often called *phagocytes*. While most can move from the bloodstream into the surrounding tissue, monocytes undergo such a striking increase in size that they are given a different name—*macrophages*. Macrophages can be found throughout the body and are the most active of the phagocytes. *Eosinophils* are primarily involved in destroying multicellular parasites such as worms, and *basophils* appear to be primarily involved in the

inflammation response, which results in increased blood flow to an injured area, causing swelling and reddening.

Another kind of cellular particle in the blood is the *platelet*. These are fragments of specific kinds of white blood cells and are important in blood clotting. They collect at the site of a wound, where they break down, releasing specific molecules, which begin a series of reactions that results in the formation of fibers that trap blood cells and form a plug in the opening of the wound. This mixture of fibers and trapped blood cells dries and hardens to form a scab.

The Heart

Blood can perform its transportation function only if it moves. The organ responsible for providing the energy to pump the blood is the heart. The heart is a muscular pump that provides the pressure necessary to propel the blood throughout the body. It must continue its cycle of contraction and relaxation, or blood stops flowing and body cells are unable to obtain nutrients or get rid of wastes. Some cells, such as brain cells, are extremely sensitive to having their flow of blood interrupted because they require a constant supply of glucose and oxygen. Others, such as muscle cells or skin cells, are much better able to withstand temporary interruptions of blood flow.

The hearts of humans, other mammals, and birds consist of four chambers and four sets of valves that work together to ensure that blood flows in one direction only. Two of these chambers, the right and left *atria* (singular, *atrium*), are relatively thin-walled structures that collect blood from the major veins and empty it into the larger, more muscular ventricles (figure 24.2).

The right and left *ventricles* are chambers that have powerful muscular walls whose contraction forces blood to flow through the arteries to all parts of the body. The valves between the atria and ventricles, known as *atrioventricular valves*, are important one-way valves that allow the blood to flow from the atria to the ventricles but prevent flow in the opposite direction. Similarly, there are valves in the aorta and pulmonary artery, known as *semilunar valves*. The **aorta** is the large artery that carries blood from the left ventricle to the body, and the **pulmonary artery** carries blood from the right ventricle to the lungs. The semilunar valves prevent blood from flowing back into the ventricles. If the atrioventricular or semilunar valves are damaged or function improperly, the efficiency of the heart as a pump is diminished, and the person may develop an enlarged heart or other symptoms.

The right and left sides of the heart have slightly different jobs, because they pump blood to different parts of the body. The right side of the heart receives blood from the general body and pumps it through the pulmonary arteries to the lungs, where exchange of oxygen and carbon dioxide takes place, and the blood returns from the lungs to the left atrium. This is called **pulmonary circulation.** The larger, more powerful left side of the heart receives oxygenated blood from the lungs, delivers it

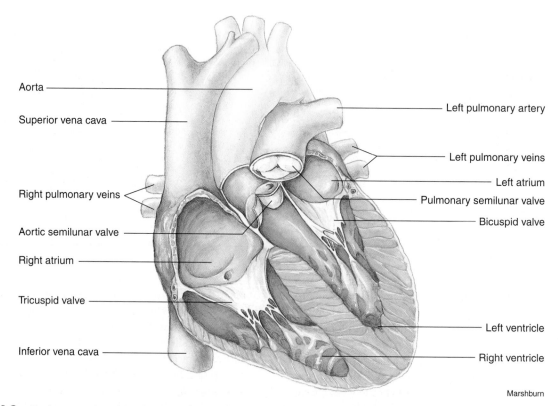

Marshburn

FIGURE 24.2 The heart consists of two thin-walled chambers called atria that contract to force blood into the two ventricles. When the ventricles contract, the atrioventricular valves (bicuspid and tricuspid) close, and blood is forced into the aorta and pulmonary artery. Semilunar valves in the aorta and pulmonary artery prevent the blood from flowing back into the ventricles when they relax.

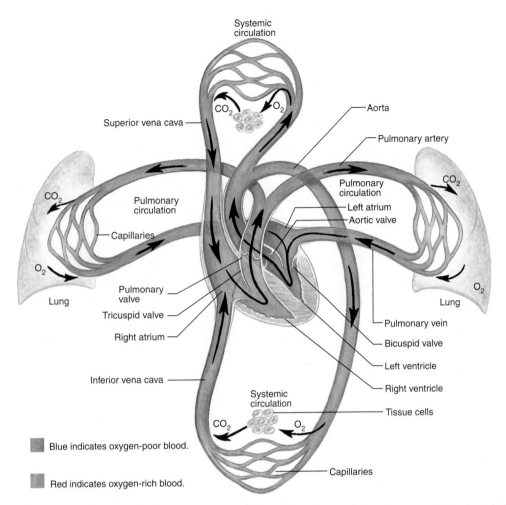

Systemic
circulation

Superior vena cava

CO₂   O₂

Aorta

Pulmonary artery

Pulmonary
circulation

Pulmonary
circulation

Left atrium

Aortic valve

CO₂

Capillaries

CO₂

Lung

Pulmonary
valve

Tricuspid valve

Right atrium

O₂

Lung

O₂

Pulmonary vein

Bicuspid valve

Left ventricle

Inferior vena cava

Right ventricle

Systemic
circulation

Tissue cells

CO₂   O₂

Blue indicates oxygen-poor blood.

Red indicates oxygen-rich blood.

Capillaries

FIGURE 24.3   The right ventricle pumps blood that is poor in oxygen to the two lungs by way of the pulmonary arteries, where the blood receives oxygen and turns bright red. The blood is then returned to the left atrium by way of four pulmonary veins. This part of the circulatory system is known as pulmonary circulation. The left ventricle pumps oxygen-rich blood by way of the aorta to all parts of the body except the lungs. This blood returns to the right atrium, depleted of its oxygen, by way of the superior vena cava from the head region and the inferior vena cava from the rest of the body. This portion of the circulatory system is known as systemic circulation.

through the aorta to all parts of the body, and returns it to the right atrium by way of large veins known as the *superior vena cava* and *inferior vena cava*. This is known as **systemic circulation.** Both circulatory pathways are shown in figure 24.3. The systemic circulation is responsible for gas, nutrient, and waste exchange in all parts of the body except the lungs.

Arteries, Veins, and Capillaries

Arteries and veins are the tubes that transport blood from one place to another within the body.

Blood Pressure

Arteries carry blood away from the heart because the contraction of the walls of the ventricles increases the pressure in the arteries compared to that in the capillaries and veins downstream of the heart. A typical pressure recorded in a large artery while the heart is contracting is about 120 millimeters of mercury. This is known as the **systolic blood pressure.** The pressure recorded while the heart is relaxing is about 80

millimeters of mercury. This is known as the **diastolic blood pressure.** A blood pressure reading includes both of these numbers and is recorded as 120/80. Most physicians consider a systolic pressure above 140 or a diastolic pressure above 90 to be cause for concern and will recommend treatment for high blood pressure.

Structure and Function of Arteries and Veins

The walls of arteries are relatively thick and muscular yet elastic. Healthy arteries have the ability to expand as blood is pumped into them and return to normal as the pressure

---

CONCEPTS APPLIED

Go with the Flow

Look at figure 24.3. Start with blood in the left lung and trace the flow of blood until it arrives back at the left lung. Make a list of the structures encountered.

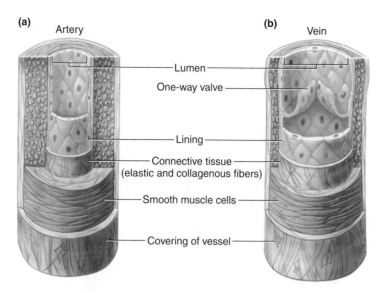

(a) Artery

(b) Vein

Lumen

One-way valve

Lining

Connective tissue
(elastic and collagenous fibers)

Smooth muscle cells

Covering of vessel

FIGURE 24.4 The walls of arteries are much thicker than the walls of veins. (The pressure in arteries is much higher than the pressure in veins.) The pressure generated by the ventricles of the heart forces blood through the arteries. Veins often have very low pressure. The valves in the veins prevent the blood from flowing backward, away from the heart.

drops. This ability to expand absorbs some of the pressure and reduces the peak pressure within the arteries, thus reducing the likelihood that they will burst. If arteries become hardened and less resilient, the peak blood pressure rises and they are more likely to rupture. The elastic nature of the arteries is also responsible for assisting the flow of blood. When they return to normal from their stretched condition, they give a little push to the blood that is flowing through them.

Blood is distributed from the large aorta through smaller and smaller blood vessels to millions of tiny capillaries. Some of the smaller arteries, called **arterioles,** may contract or relax to regulate the flow of blood to specific parts of the body. Major parts of the body that receive differing amounts of blood, depending on need, are the digestive system, muscles, and skin.

Veins collect blood from the capillaries and return it to the heart. The pressure in these blood vessels is very low. Some of the largest veins may have a blood pressure of 0.0 mmHg for brief periods. Since pressure in veins is so low, muscular movements of the body are important in helping to return blood to the heart. When muscles of the body contract, they compress nearby veins, and this pressure pushes blood along in the veins. Because valves in the veins allow blood to flow only toward the heart, this activity acts as an additional pump to help return blood to the heart. Figure 24.4 compares the structure and function of arteries and veins.

Structure and Function of Capillaries

Although the arteries are responsible for distributing blood to various parts of the body and arterioles regulate where blood goes, it is the function of capillaries to assist in the exchange of materials between the blood and cells. Capillaries are tiny, thin-walled tubes that receive blood from arterioles. They are so small

that red blood cells must go through them in single file. They are so numerous that each cell in the body has a capillary located near it. It is estimated that there are about 1,000 m² of surface area represented by the capillary surface in a typical human. Each capillary wall consists of a single layer of cells and therefore presents only a thin barrier to the diffusion of materials between the blood and the cells that surround the capillaries. Figure 24.5 shows the close physical relationship between capillaries and body cells. The flow of blood through these smallest blood vessels is relatively slow. This allows time for the diffusion of such materials as oxygen, glucose, and water from the blood to surrounding cells and for the movement of such materials as carbon dioxide, lactic acid, and ammonia from the cells into the blood.

In addition to molecular exchange by diffusion, considerable amounts of water and dissolved materials leak through the small holes in the capillaries. This liquid is known as **lymph.** Lymph is produced when the blood pressure forces water and some small dissolved molecules through the walls of the capillaries. Lymph bathes the cells but must eventually be returned to the circulatory system by lymph vessels or swelling will occur. Return is accomplished by the *lymphatic system,* a collection of thin-walled tubes that branch throughout the body. These tubes collect lymph that left the circulatory system and ultimately empty it into major blood vessels near the heart. Figure 24.6 shows the structure of the lymphatic system.

SKIN: THE BODY'S CONTAINER

The skin is the outer covering of the body. It is the largest organ in the body, has a relatively complex structure, and serves several important functions.

FIGURE 24.5 Capillaries are tiny blood vessels. Exchange of cells and molecules can occur between blood and tissues through their thin walls. Molecules diffuse in and out of the blood, and cells such as monocytes can move from the blood through the thin walls into the surrounding tissue. There is also a flow of liquid through holes in the capillary walls. This liquid, called lymph, bathes the cells and eventually enters small lymph vessels that return lymph to the circulatory system near the heart.

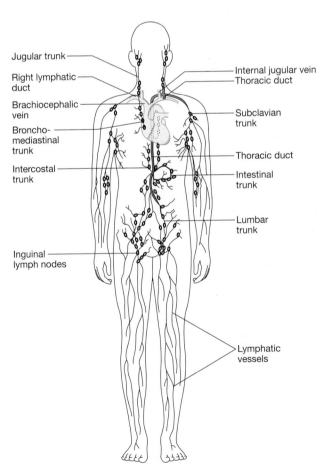

FIGURE 24.6 The lymphatic system consists of many ducts that transport lymph fluid back toward the heart. Along the way the lymph is filtered in the lymph nodes, and bacteria and other foreign materials are removed before the lymph is returned to the circulatory system near the heart.

Primary Functions of the Skin

Although it serves several important functions, the importance of the skin is often overlooked. It acts as an impermeable barrier between us and our surroundings. While aquatic animals such as fish and amphibians can get by with a water-permeable skin, terrestrial animals such as reptiles, birds, and mammals must prevent the loss of water from their body surface. An impermeable skin also is a barrier to the entry of microorganisms and chemicals.

The skin is actually the largest organ of the body. For an average human, it constitutes about 7,600 cm^2 (3,000 in^2). This is about the surface area of a 4-by-5 ft rug. The large surface area of the skin is important in getting rid of heat. Shunting blood to the skin results in heat loss (as long as the environmental temperature is less than the body temperature), and the secretions of sweat glands spread over this large surface provide evaporative cooling.

The Structure of the Skin

The skin is divided into three distinct layers: the epidermis, the dermis, and the subcutaneous layers. The epidermis is the outer surface of the skin that you can see. It consists of a layer of cells at its base, next to the dermis, that undergoes mitosis to produce new cells. This layer of cells is called the germinating layer. The cells produced by the germinating layer migrate toward the surface. As they migrate to the surface, they produce a protein called keratin, lose their nuclei and other structures, and become flattened disks. This outer layer of the epidermis is called the cornified layer and consists of several layers of these flattened, scalelike, keratin-filled cell remnants. The cornified layer provides toughness to the epidermis and is impermeable to most chemicals and microorganisms. Thus, the cornified layer prevents the loss of water from and the entry of microorganisms to the underlying tissues of the body.

Since the cornified layer is in direct contact with the environment, it is constantly worn away by friction with surroundings. The constant replenishment of this layer from below, as a result of the germinating layer, is essential to maintain this function of the skin. If the cornified layer is destroyed, people lose fluids very rapidly and are subject to invasion by various kinds of disease microorganisms. This is probably most dramatically illustrated by the problems experienced by victims of heat or chemical burns in which the cornified layer is destroyed. They must receive intravenous fluids to replace the liquids lost and often succumb to generalized infections because they have no barrier to microorganisms.

The dermis is immediately below the epidermis. It contains a great many connective tissue fibers that provide the flexible but tough texture of the skin. There are many blood vessels and nerve endings in the dermis. The overlying epidermis does not have blood vessels or nerve endings. The cells of the epidermis receive nutrition from the blood vessels in the dermis, and pain, touch, and temperature are sensed by the nerve endings in the dermis.

The subcutaneous layer lies below the dermis and contains a great many connective tissue fibers and fat cells. The fibers in this layer bind the skin to underlying tissues, primarily muscle. There are also blood vessels and nerves in this layer.

Other Features of the Skin

Several other features of the skin deserve comment. Melanin is a brown-black pigment produced by special cells in the germinating layer. Dark-skinned people have melanin-producing cells that are more active than those of light-skinned people. Melanin is a protection against the damaging effects of ultraviolet light. When people are exposed to ultraviolet light, the melanin-producing cells are stimulated to produce more melanin and tanning occurs. Though tanning is most noticeable in light-skinned people, the melanin-producing cells of all people increase activity in response to ultraviolet light.

Hair and nails (and hooves, claws, and horns in other mammals) are composed of keratin. Therefore, they are specialized accumulations produced by the epidermis of the skin.

Several kinds of glands are derived from the epidermis but project down into the underlying dermis. Oil glands associated with hair follicles secrete oily material that is released onto the skin surface where it lubricates the skin and hair. Scent glands are also associated with hair follicles and produce a watery secretion that contains particular odor molecules. In humans, scent glands are found primarily in the armpit and pubic area. Other mammals may have scent glands near their eyes, around the anus, on their feet, or on their belly. Humans and a few other mammals have sweat glands over much of the skin surface. As mentioned earlier, evaporation of sweat from the skin surface cools the body. Finally, mammary glands, which are responsible for milk production, are specialized sweat glands. Figure 24.7 illustrates the various parts of the skin.

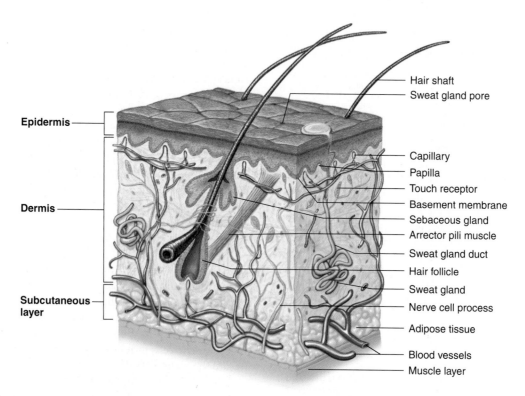

FIGURE 24.7 The skin is the largest organ of the body and consists of several layers and structures that prevent water loss, protect against infection, provide for cooling, and serve as sense organs.

EXCHANGING GASES: THE RESPIRATORY SYSTEM

In terrestrial vertebrates, the primary structures involved in gas exchange are the lungs. Since the lungs are internal structures, the amount of water lost is much less than if these moist surfaces were on the outer surface of the body, as are the gills of fish.

Structure and Function of Lungs

The **lungs** are organs of the body that allow gas exchange to take place between the air and blood. Associated with the lungs is a set of tubes that conducts air from outside the body to the lungs. The single large-diameter **trachea** is supported by rings of cartilage that prevent its collapse. It branches into two major **bronchi** (singular, *bronchus*) that deliver air to smaller and smaller branches. Bronchi are also supported by cartilage. The smallest tubes, known as **bronchioles,** contain smooth muscle and are therefore capable of constricting. Finally, the bronchioles deliver air to clusters of tiny sacs, known as **alveoli** (singular, *alveolus*), where the exchange of gases takes place between the air and blood. The alveoli are very thin-walled and surrounded by capillaries. This close association makes gas exchange by diffusion relatively easy. Figure 24.8 illustrates the parts of the respiratory system and the close relationship between the alveoli of the lungs and the capillaries of the circulatory system.

The nose, mouth, and throat are also important parts of the air-transport pathway because they modify the humidity and temperature of the air and clean the air as it passes. The lining of the trachea contains cells that have cilia that beat in a direction that moves mucus and foreign materials from the lungs. The foreign matter may then be expelled by coughing or "clearing your throat," thus keeping the air passage clear.

The Mechanism of Breathing

Breathing is the process of moving air in and out of the lungs. It is accomplished by the movement of a muscular organ known as the **diaphragm,** which separates the chest cavity, which contains the lungs, from the abdominal cavity. In addition, muscles located between the ribs (*intercostal* muscles) assist in breathing. During inhalation, the diaphragm moves downward, and the external intercostal muscles of the chest wall contract, causing the chest wall to move outward and upward. Both of these muscular contractions cause the volume of the chest cavity to increase. This increase in volume results in a lower pressure in the chest cavity compared to the outside air pressure. Consequently, air flows from the outside, higher-pressure area through the trachea, bronchi, and bronchioles to the alveoli (figure 24.9). During normal relaxed breathing, exhalation is accomplished by the chest wall and diaphragm simply returning to their normal, relaxed position. Muscular contraction is not involved. During vigorous inhalation, additional muscles of the chest wall are involved, making the

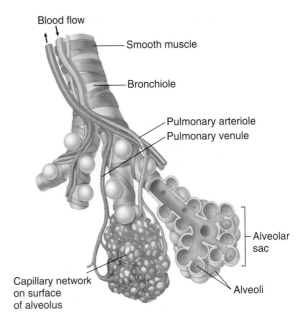

FIGURE 24.8   The exchange of gases takes place between the air-filled alveolus and the blood-filled capillary. The capillaries form a network around the saclike alveoli. The thin walls of the alveolus and capillary are in direct contact with one another; their combined thickness is usually less than 1 micron (a thousandth of a millimeter). This close relationship and the thinness of the walls allow for the easy diffusion of gases.

Cigarette Smoking and Your Health

Cigarette smoking is becoming less and less acceptable in our society. The banning of smoking in public buildings and on domestic air flights attests to this fact. Yet in spite of social pressure to quit smoking, research linking smoking with lung and heart disease, and evidence that even secondhand smoke can be harmful, about 20 percent of American adults are smokers. The same is true for high school students—about 20 percent are smokers.

Hazards of Cigarette Smoking

Bronchitis. Cigarette smoking is the leading cause of chronic bronchitis, which involves the inflammation of the bronchi. A common symptom of bronchitis is a harsh cough that expels a greenish-yellow mucus.

Emphysema. Emphysema is a progressive disease in which some of the alveoli are lost. People afflicted with this disease have less and less respiratory surface area and experience greater difficulty getting adequate oxygen, even though they may be breathing more rapidly. It may be caused by cigarette smoke and other air pollutants that damage alveoli. This damage reduces the capacity of the lungs to exchange gases with the bloodstream. A common symptom of emphysema is difficulty exhaling. Several years of an emphysema sufferer's forced breathing can increase the size of the chest and give it a barrel appearance.

Asthma. Cigarette smoke is one of many environmental factors that may trigger an asthma attack. Asthma is an allergic reaction that results in the narrowing of the bronchioles of the lungs and the excess production of fluids that limit the amount of air that can enter the lungs. Symptoms of asthma include coughing, wheezing, and difficulty breathing.

Lung cancer. Lung cancer develops twenty times more frequently in heavy smokers than in nonsmokers. Typically, lung cancer starts in the bronchi. Cigarette smoke and other pollutants cause cells below the surface of the bronchi to divide at an abnormally high rate. This malignant growth may spread through the lung and move into other parts of the body. Occurrence of cancers of the mouth, larynx, esophagus, pancreas, and bladder is also significantly greater in smokers than in nonsmokers.

Pneumonia. Cigarette smokers have an increased risk of developing pneumonia. Pneumonia involves the infection or inflammation of alveoli, which leads to fluid filling the alveolar sacs. Pneumonia is typically caused by the bacterium *Streptococcus pneumoniae* but in some cases is caused by other bacteria, fungi, protists, or viruses.

Smoking during pregnancy. Cigarette smoking during pregnancy has been linked to low birth weight and higher rates of fetal and infant death. Children of mothers who smoke during pregnancy have a higher incidence of heart abnormalities, cleft palate, and sudden infant death syndrome (SIDS). Nursing infants of smoking mothers have higher than normal rates of intestinal problems, and infants exposed to secondhand smoke have an increased incidence of respiratory disorders.

Heart disease. Smoking is a major contributor to heart disease. The action of nicotine from cigarette smoke results in constriction of blood vessels and the reduction of blood flow.

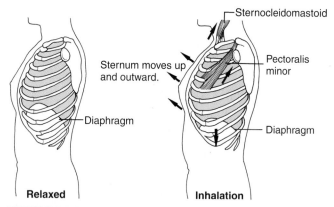

FIGURE 24.9 During inhalation, the diaphragm and external intercostal muscles between the ribs contract, causing the volume of the chest cavity to increase. During a normal exhalation, these muscles relax, and the chest volume returns to normal.

chest cavity larger. In vigorous exhalation, internal intercostal muscles of the chest wall and muscles of the abdominal wall contract to compress the chest cavity and more completely empty the lungs of air.

Homeostasis and Breathing

Exercising causes an increase in the amount of carbon dioxide in the blood because muscles are oxidizing glucose more rapidly. When carbon dioxide dissolves in water, it forms a weak acid (carbonic acid). This lowers the pH of the blood. Certain brain cells and specialized cells in the aortic arch and carotid arteries are sensitive to changes in blood pH. When they sense a lower pH, nerve impulses are sent more frequently to the diaphragm and intercostal muscles. These muscles contract more rapidly and more forcefully, resulting in more rapid, deeper breathing. Because more air is being exchanged per minute, carbon dioxide is lost from the lungs more rapidly. When exercise stops, the carbon dioxide level drops, blood pH rises, and breathing eventually returns to normal.

OBTAINING NUTRIENTS: THE DIGESTIVE SYSTEM

All cells must have a continuous supply of nutrients to provide the energy they require and the building blocks needed to construct the macromolecules typical of living things. This section will deal with the processing and distribution of different kinds of nutrients.

The digestive system consists of a muscular tube with several specialized segments. In addition, there are glands that secrete digestive juices into the tube. Several different kinds of activities are involved in getting nutrients to the cells that need them.

Processing Food

The digestive system is designed as a disassembly system. Its purpose is to take large chunks of food and break them down to small molecules that can be taken up by the circulatory system and distributed to cells. The first steps in this process take place in the mouth.

Digestive Activities in the Mouth

It is important to grind large particles into small pieces by chewing in order to increase their surface area and allow for more efficient chemical reactions. It is also important to add water to the food, which further disperses the particles and provides the watery environment needed for these chemical reactions. Materials must also be mixed, so that all the molecules that need to interact with one another have a good chance of doing so. The mouth and the stomach are the major body regions involved in reducing the size of food particles. The teeth are involved in cutting and grinding food to increase its surface area. The watery mixture that is added to the food in the mouth is known as *saliva,* and the three pairs of glands that produce saliva are known as *salivary glands.* Saliva contains the enzyme *salivary amylase,* which initiates the chemical breakdown of starch. Saliva also lubricates the mouth and helps to bind food before swallowing. Figure 24.10 shows the structures of the digestive system.

Swallowing

Once the food has been chewed, it is swallowed and passes down the esophagus to the stomach. The process of swallowing involves a complex series of events. First, a ball of food, known as a *bolus,* is formed by the tongue and moved to the back of

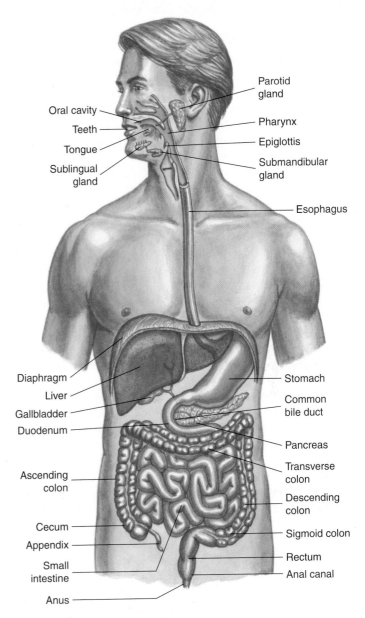

FIGURE 24.10 In the digestive system, the teeth, tongue, and enzymes from the salivary glands modify the food before it is swallowed. The stomach adds acid and enzymes and further changes the texture of the food. The food is eventually emptied into the duodenum, where the liver and pancreas add their secretions. The small intestine also adds enzymes and is involved in absorbing nutrients. The large intestine is primarily involved in removing water.

the mouth cavity. Here it stimulates the walls of the throat, also known as the *pharynx.* Nerve endings in the lining of the pharynx are stimulated, causing a reflex contraction of the walls of the esophagus, which transports the bolus to the stomach.

Since both food and air pass through the pharynx, it is important to prevent food from getting into the lungs. During swallowing, the larynx is pulled upward. This causes a flap of tissue called the *epiglottis* to cover the opening to the trachea and prevent food from entering the trachea.

People Behind the Science

William Beaumont (1785–1853)

William Beaumont was born in 1785 in Lebanon, Connecticut, to Samuel and Lucretia Beaumont. He was the third of nine children. William attended school in Lebanon and, at the age of twenty-two, went to live in Champlain, New York, near the border with Vermont, just south of the Canadian border. He had relatives living in the area. In 1807, he became the schoolmaster of Champlain. In 1809, he began to study medicine. As was the custom of the time, he began reading about medical subjects and served an apprenticeship under the guidance of two local doctors in St. Albans, Vermont, about 55 km (35 mi) from Champlain. In June 1812, he was licensed to practice medicine. In September 1812, he joined the U.S. Army as a surgeon's mate and was assigned to the Sixth Infantry Regiment in Plattsburgh, New York. During the War of 1812, he treated the various illnesses and wounds related to the poor living conditions and battle injuries of the soldiers.

In 1815, after the war, he entered private practice in Plattsburgh. But in 1819, he reenlisted in the Army and was sent to Fort Mackinac on Mackinac Island in Lake Huron near the Canadian border. While living in Plattsburgh, he met Deborah Green, who was divorced from Nathanial Platt. In 1821, Beaumont returned to Plattsburgh from Fort Mackinac and married Green, who returned with him to Fort Mackinac. They had four children, two of whom were born while the Beaumonts were at Fort Mackinac.

At Fort Mackinac, Dr. Beaumont had an experience that would make him an expert on the physiology of digestion.

On the morning of June 6, 1822, a nineteen-year-old French-Canadian fur trapper named Alexis St. Martin was shot in the stomach by an accidental discharge from a shotgun. Dr. Beaumont was called to attend the wounded man. Part of the stomach and body wall had been shot away and parts of St. Martin's clothing were imbedded in the wound. Although Dr. Beaumont did not expect St. Martin to live, he quickly cleaned the wound, pushed portions of the lung and stomach that were protruding back into the cavity, and dressed the wound. The next day, Beaumont was surprised to find St. Martin alive and was encouraged to do what he could to extend the young man's life. In fact, Beaumont cared for St. Martin for two years. When the wound was completely healed, the stomach had fused to the body wall and a hole allowed Beaumont to look into the stomach. Fortunately for St. Martin, a flap of tissue from the lining of the stomach closed off the opening so that what he ate did not leak out.

Beaumont found that he could look through the opening and observe the activities in the stomach. He recognized that this presented an opportunity to study the function of the stomach in a way that had not been done before. He gathered gastric juice, had its components identified, introduced food into the hole with a string attached so that he could retrieve the food particles that were partially digested for examination, and observed the effect of emotion on digestion. He discovered many things that were new to science and contrary to the teachings of the time. He recounted many of his observations and experiments in his journal.

I consider myself but a humble inquirer after truth—a simple experimenter. And if I have been led to conclusions opposite to the opinions of many who have been considered luminaries of physiology, and, in some instances, from all the professors of this science, I hope the claim of sincerity will be conceded to me, when I say that such difference of opinion has been forced upon me by the convictions of experiment, and the fair deductions of reasoning.

Digestive Activities in the Stomach

In the stomach, additional liquid, called *gastric juice,* is added to the food. Gastric juice contains enzymes and hydrochloric acid. The major enzyme of the stomach is *pepsin,* which initiates the chemical breakdown of protein. The pH of gastric juice is very low, generally around pH 2. The entire mixture is churned by the contractions of the three layers of muscle in the stomach wall. The combined activities of enzymatic breakdown, chemical breakdown by hydrochloric acid, and mechanical processing by muscular movement result in a thoroughly mixed liquid called *chyme.* Chyme eventually leaves the stomach through a valve known as the pyloric sphincter and enters the small intestine.

Digestive Activities in the Small Intestine

The first part of the **small intestine** is known as the **duodenum.** In addition to producing enzymes, the duodenum secretes several kinds of hormones that regulate the release of food from the stomach and the release of secretions from the pancreas and liver. The **pancreas** produces a number of different digestive enzymes and also secretes large amounts of bicarbonate ions, which neutralize stomach acid so that the pH of the duodenum is about pH 8. The **liver** is a large organ in the upper abdomen that performs several functions. One of its functions is the secretion of *bile.* When bile leaves the liver, it is stored in the *gallbladder* prior to being released into the duodenum. When bile is released from the gallbladder, it assists mechanical mixing by breaking large fat globules into smaller particles. This process is called *emulsification.*

Digestive Activities in the Large Intestine

Along the length of the small intestine, additional watery juices are added until the mixture reaches the **large intestine.** The large intestine is primarily involved in reabsorbing the water that has been added to the food tube along with saliva, gastric juice, bile, pancreatic secretions, and intestinal juices. The large intestine is also home to a variety of different kinds of bacteria.

In 1833, Beaumont published *Experiments and Observations on the Gastric Juice and the Physiology of Digestion*, which contained descriptions of some 240 experiments. The following are some of his important discoveries.

1. He measured the temperature of the stomach and found that it does not heat up when food is introduced, as was thought at the time. *"But from the result of a great number of experiments and examinations, made with a view to asserting the truth of this opinion, in the empty and full state of the organ, . . . I am convinced that there is no alteration of temperature. . . ."*

2. He found that pure gastric juice contains large amounts of hydrochloric acid. This was contrary to the prevailing opinion that gastric juice was simply water. *"I think I am warranted, from the result of all the experiments, in saying, that the gastric juice, so far from being 'inert as water,' as some authors assert, is the most general solvent in nature, of alimentary matter—even the hardest bone cannot withstand its action."*

3. He observed that gastric juice is not stored in the stomach but is secreted when food is eaten. *"The gastric juice does not accumulate in the cavity of the stomach, until alimentary matter is received, and excite its vessels to discharge their contents, for the immediate purpose of digestion."*

4. He realized that digestion begins immediately when food enters the stomach. The prevailing opinion of the day was that nothing happened for an hour or more. *"At 2 o'clock* P.M.*—twenty minutes after having eaten an ordinary dinner of boiled, salted beef, bread, potatoes, and turnips, and drank [sic] a gill of water, I took from his stomach, through the artificial opening, a*

gill of the contents. . . . Digestion had evidently commenced, and was perceptually progressing, at the time."

5. He discovered that food in the stomach satisfies hunger even though it is not eaten. *"To ascertain whether the sense of hunger would be allayed without food being passed through the oesophagus [sic], he fasted from breakfast time, 'til 4 o'clock* P.M.*, and became quite hungry. I then put in at the aperture, three and a half drachms of lean, boiled beef. The sense of hunger immediately subsided, and stopped the borborygmus, or croaking noise, caused by the motion of the air in the stomach and intestines, peculiar to him since the wound, and almost always observed when the stomach is empty."*

Dr. Beaumont was eventually posted to St. Louis, Missouri, where he left the military in 1839 and went into private practice. In 1853, he slipped coming out of a patient's home, fell, and hit his head. The resulting injury became infected and resulted in his death at the age of sixty-eight. Today, his role as an army surgeon and his major contribution to an understanding of gastric physiology are recognized in the naming of the William Beaumont Army Medical Center in El Paso, Texas.

St. Martin did not take kindly to the probings and experiments of Beaumont and twice ran away from Beaumont's care back to Canada, where he married, had two children, and resumed his former life as a voyageur, fur trapper, and wood chopper. He did not die until 1880 at the age of eighty-six, having lived over sixty years with a hole in his stomach.

Most live on the undigested food that makes it through the small intestine. Some provide additional benefit by producing vitamins that can be absorbed from the large intestine. A few kinds may cause disease.

Nutrient Uptake

The process of digestion has resulted in a variety of simple organic molecules that are available for absorption from the tube of the gut into the circulatory system. Nearly all nutrients are absorbed in the small intestine. Many features of the small intestine provide a large surface area for the absorption of nutrients. First of all, the small intestine is a very long tube; the longer the tube, the greater the internal surface area. In a typical adult human, it is about 6 m (about 20 ft) long. In addition to length, the lining of the intestine consists of millions of tiny projections called *villi*, which increase the surface area. When we examine the cells that make up the villi, we find that they also have folds in their surface membranes. All of these characteristics increase

the surface area available for the transport of materials from the gut into the circulatory system (figure 24.11).

Scientists estimate that the cumulative effect of all of these features produces a total intestinal surface area of about 250 m^2 (about 2,600 ft^2). That is equivalent to about the area of a basketball court.

The surface area by itself would be of little value if it were not for the intimate contact of the circulatory system with this lining. Each villus contains several capillaries and a branch of the lymphatic system called a *lacteal*. The close association between the intestinal surface and the circulatory and lymphatic systems allows for the efficient uptake of nutrients from the cavity of the gut into the circulatory system.

Several different kinds of processes are involved in the transport of materials from the intestine to the circulatory system. Some molecules, such as water and many ions, simply diffuse through the wall of the intestine into the circulatory system. Other materials, such as amino acids and simple sugars, are assisted across the membrane by carrier molecules. Fatty

A Photo of villi in small intestine

B Photo of microvilli in intestinal cell

Capillaries

Villi

Lacteals

C Relationship of villi, capillaries, and lacteals

FIGURE 24.11 (*A*) The surface area of the intestinal lining is increased by the many fingerlike projections known as villi. (*B*) The cells that make up the surface of the villi have many tiny projections on them known as microvilli. Both the villi and microvilli provide a large surface area for the exchange of materials between the gut and the circulatory system. (*C*) Within each villus are capillaries and lacteals. Most kinds of materials enter the capillaries, but most fat-soluble substances enter the lacteals, giving them a milky appearance. Lacteals are part of the lymphatic system. Because the lymphatic system empties into the circulatory system, fat-soluble materials also eventually enter the circulatory system. The close relationship between these vessels and the epithelial lining of the villus allows for efficient exchange of materials from the intestinal cavity to the circulatory system.

acids and glycerol are absorbed into the intestinal lining cells where they are resynthesized into fats and enter lacteals in the villi. Since the lacteals are part of the lymphatic system, which eventually empties its contents into the circulatory system, fats are also transported by the blood. They just reach the blood by a different route.

NUTRITION

Organisms maintain themselves by constantly processing molecules to provide building blocks for new living material and energy to sustain themselves. The word **nutrition** is used in two related contexts. First of all, nutrition is a branch of science that seeks to understand food, its nutrients, how the nutrients are used by the body, and how inappropriate combinations or quantities of nutrients lead to ill health. The word *nutrition*

is also used in a slightly different context to refer to all the processes by which we take in food and utilize it, including *ingestion, digestion, absorption,* and *assimilation. Ingestion* involves the process of taking food into the body through eating. *Digestion* involves the breakdown of complex food molecules to simpler molecules. *Absorption* involves the movement of simple molecules from the digestive system to the circulatory system for dispersal throughout the body. *Assimilation* involves the modification and incorporation of absorbed molecules into the structure of the organism.

Kinds of Nutrients

Nutrients are the kinds of molecules that must be consumed to maintain the body. They can be organized into several categories based on their chemical structure and the functions they serve in the body. Commonly accepted categories are: carbohydrates, fats, proteins, vitamins, minerals, and water. The chemical nature of organic molecules is discussed in chapter 19.

1. *Carbohydrates* are organic molecules that contain carbon, hydrogen, and oxygen atoms in the following ratio: CH_2O. Common carbohydrates that serve as nutrients are various kinds of sugars and starches. The primary function of these molecules is to provide energy. Each gram of carbohydrate provides about 4 kilocalories of energy. The sugar glucose undergoes aerobic cellular respiration to provide carbon dioxide, water, and ATP. See chapter 20 for a discussion of the process of aerobic cellular respiration. Starch and some other kinds of sugars can be converted to glucose and thus are able to undergo aerobic cellular respiration.

2. *Fats* are also organic molecules that consist of carbon, hydrogen, and oxygen, but fats have fewer oxygen atoms than carbohydrates. Fats are excellent sources of energy since they can be broken down into smaller units that enter the Krebs cycle to provide carbon dioxide, water, and ATP. Fats provide about 9 kilocalories of energy per gram—more than twice the energy of carbohydrates.

3. *Proteins* are polymers of amino acids that contain carbon, hydrogen, oxygen, and nitrogen, and some contain sulfur. Proteins are important in the structure of cells and the body. All cells and parts of the body contain protein. Thus, protein is needed as a nutrient to provide for growth and repair of cells, tissues, and organs. Proteins also contain energy (about 4 kilocalories per gram) but yield less energy than fats or carbohydrates because they are more difficult to break down in aerobic cellular respiration.

4. *Vitamins* are organic molecules that cannot be manufactured by the body but are needed in small amounts to assist enzymes in their functions.

5. *Minerals* are inorganic molecules needed for a variety of purposes. Some such as calcium are needed as a component of bone. Others are needed as components of enzymes or other important molecules. Most minerals are needed in relatively small amounts.

6. *Water* is often not even classified as a nutrient but is vitally important to sustain life. Water is the most common molecule in the body. Since it is constantly lost in urine, feces, and by evaporation, it must be replaced.

All of these nutrients must be obtained in adequate amounts on a daily basis or ill health results.

CONCEPTS APPLIED

Check Out the Nutrition Labels

The next time you're in the kitchen or grocery store, read and compare some food labels. Compare the list of nutrients and their quantities found in one product from each of the basic food groups to determine which has the highest amount of your chosen nutrient. Now select three foods from the same food group and make a similar comparison.

Connections . . .

Measuring the Caloric Value of Foods

A *bomb calorimeter* is an instrument that is used to determine the heat of combustion or caloric value of foods. To operate the instrument, a small food sample is formed into a pellet and sealed inside a strong container called the bomb. The bomb is filled with 30 atmospheres of oxygen and then placed in a surrounding jacket that is filled with water. The sample is electrically ignited, and as it aerobically reacts with the oxygen, the bomb releases heat to the surrounding water jacket. As the heat is released, the temperature of the water rises. Recall that a calorie is the amount of energy (or heat) needed to increase the temperature of one gram of water one degree Celsius. Therefore, if one gram of food results in a water temperature increase of 13°C for each gram of water, that food has the equivalent of 13 calories.

Guidelines for Obtaining Adequate Nutrients

The food and drink consumed from day to day constitutes a person's **diet.** A good understanding of nutrition can promote good health and involves an understanding of the energy and nutrient content in various foods. In an attempt to give people some guidelines for planning a diet that provides adequate amounts of the six classes of nutrients, nutritional scientists and government officials in the United States and many other countries developed nutrient standards.

Dietary Reference Intakes

The U.S. Department of Agriculture has over the years published various guidelines as aids to maintaining good nutritional health. The current guidelines are called **Dietary Reference Intakes** and provide recommendations on the amounts of different kinds of nutrients people should receive. These guidelines are very detailed and include specific guidelines for males and females of various ages and for pregnant and nursing mothers. They also provide information about the maximum amount of certain nutrients that people should not exceed. In other words, you can have too much of a good thing.

TABLE 24.2   Dietary Reference Intakes for Some Common Nutrients

Nutrient	Women 19–30 Years Old	Men 19–30 Years Old	Maximum for Persons 19–30 Years Old	Value of Nutrient
Carbohydrates	130 g/day (45–65% of kilocalories)	130 g/day (45–65% of kilocalories)	No maximum set, but refined sugars should not exceed 25% of total kilocalories.	A source of energy
Proteins	46 g/day (10–35% of kilocalories)	56 g/day (10–35% of kilocalories)	No maximum set, but high protein diets stress kidneys.	Proteins are structural components of all cells. There are ten essential amino acids that must be obtained in the diet.
Fats	20–35% of kilocalories	20–35% of kilocalories	Up to 35% of total kilocalories	Energy source and building blocks needed for many molecules
Saturated and trans fatty acids	No reference amount determined	No reference amount determined	As low as possible	
Linoleic acid	12 g/day	17 g/day	No maximum set	Essential fatty acid needed for enzyme function and to maintain epithelial cells
Linolenic acid	1.1 g/day	1.6 g/day	No maximum set	Essential fatty acid needed to reduce risk of coronary heart disease
Cholesterol	No reference amount determined	No reference amount determined	As low as possible	None needed, since the liver makes cholesterol
Total fiber	25 g/day	38 g/day	No maximum set	Improve gut function
Water	2.7 L/day	3.7 L/Day	No maximum set	Improve gut and kidney function
Calcium	1 g/day	1 g/day	2.5 g/day	Needed for the structure of bones and many other functions
Iron	18 mg/day	8 mg/day	45 mg/day	Needed to build hemoglobin of red blood cells
Sodium	1.5 g/day	1.5 g/day	2.3 g/day (most people exceed this limit)	Needed for normal cell function
Vitamin A	700 ug/day	900 ug/day	3,000 ug/day	Maintains skin and intestinal lining
Vitamin C	75 mg/day	90 mg/day	2,000 mg/day	Maintains connective tissue and skin
Vitamin D	5 ug/day	5 ug/day	50 ug/day	Needed to absorb calcium for bones

Dietary Reference Intakes are used when preparing product labels. Federal law requires that labels list ingredients from the greatest to the least in quantity. These labels typically provide information about the amounts of various nutrients present and the number of kilocalories present. In addition to carbohydrates, fats, proteins, and fiber, there are about twenty-five vitamins and minerals for which Dietary Reference Intakes standards are set. Table 24.2 give examples of some of the more common nutrients and their reference amounts for young adults. There are special recommendations for pregnant women and nursing mothers in this age group.

Counting Calories

One of the primary pieces of information on the nutritional facts label on foods you purchase is the energy content of the food item. The unit used to measure the amount of energy in foods is the **kilocalorie (kcal).** One kilocalorie is the amount of heat needed to increase the temperature of one *kilogram* of water one degree Celsius. Remember that the prefix *kilo*-means "1,000 times" the value listed. Therefore, a kilocalorie is 1,000 times more heat energy than a **calorie,** which is the amount of heat needed to raise the temperature of one *gram* of water one degree Celsius. Although the amount of energy in food is measured in kilocalories, for nutritional purposes a kilocalorie is usually designated as Calorie with a capital C. This is unfortunate because it is easy to confuse a Calorie, which is really a

kilocalorie, with a calorie. Most books on nutrition and dieting use the term *Calorie* to refer to *food Calories.* Thus, the information on a nutrition facts label can be used to help a person determine the number of Calories they are consuming per day. A commonly used standard for a daily diet is 2,000 Calories. However, the number of Calories needed by an individual is highly variable and depends on a person's specific metabolism and level of physical activity.

The Food Guide Pyramid

Using Dietary Reference Intakes and product labels or counting kilocalories is a pretty complicated way for a person to plan a diet. Planning a diet around basic food groups is generally easier. The federal government adopted the **Food Guide Pyramid** of the Department of Agriculture as one of its primary tools to help the general public plan for good nutrition. Over the years, as new nutritional information became available, nutritional suggestions have been modified. The current Food Guide Pyramid, released in 2005, is a modification of earlier versions (see figure 24.12). It contains six basic groups of foods with guidelines for the amounts one needs daily from each group for ideal nutritional planning. The Food Guide Pyramid is designed to encourage people to reduce the amount of fat, sugar, and salt in the diet, while increasing daily servings of fruits and vegetables. It also includes guidelines on the amount of daily exercise people should get.

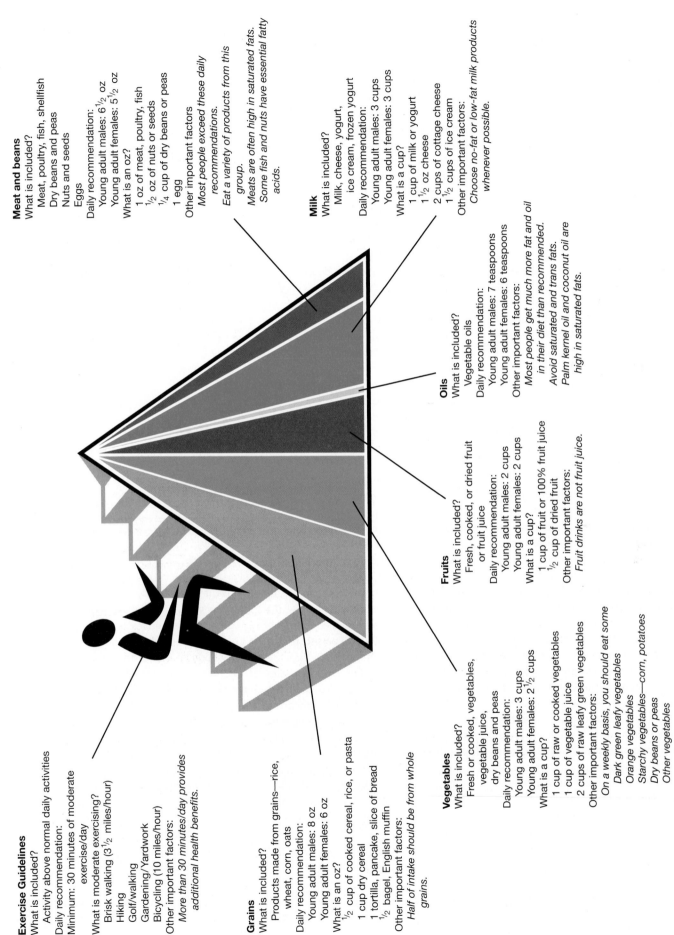

Exercise Guidelines

What is included?
Activity above normal daily activities

Daily recommendation:
Minimum: 30 minutes of moderate exercise/day

What is moderate exercising?
Brisk walking (3½ miles/hour)
Hiking
Golf/walking
Gardening/Yardwork
Bicycling (10 miles/hour)

Other important factors:
More than 30 minutes/day provides additional health benefits.

Grains

What is included?
Products made from grains—rice, wheat, corn, oats

Daily recommendation:
Young adult males: 8 oz
Young adult females: 6 oz

What is an oz?
½ cup of cooked cereal, rice, or pasta
1 cup dry cereal
1 tortilla, pancake, slice of bread
½ bagel, English muffin

Other important factors:
Half of intake should be from whole grains.

Vegetables

What is included?
Fresh or cooked, vegetables, vegetable juice, dry beans and peas

Daily recommendation:
Young adult males: 3 cups
Young adult females: 2½ cups

What is a cup?
1 cup of raw or cooked vegetables
1 cup of vegetable juice
2 cups of raw leafy green vegetables

Other important factors:
On a weekly basis, you should eat some
Dark green leafy vegetables
Orange vegetables
Starchy vegetables—corn, potatoes
Dry beans or peas
Other vegetables

Fruits

What is included?
Fresh, cooked, or dried fruit or fruit juice

Daily recommendation:
Young adult males: 2 cups
Young adult females: 2 cups

What is a cup?
1 cup of fruit or 100% fruit juice
½ cup of dried fruit

Other important factors:
Fruit drinks are not fruit juice.

Oils

What is included?
Vegetable oils

Daily recommendation:
Young adult males: 7 teaspoons
Young adult females: 6 teaspoons

Other important factors:
Most people get much more fat and oil in their diet than recommended.
Avoid saturated and trans fats.
Palm kernel oil and coconut oil are high in saturated fats.

Milk

What is included?
Milk, cheese, yogurt, ice cream, frozen yogurt

Daily recommendation:
Young adult males: 3 cups
Young adult females: 3 cups

What is a cup?
1 cup of milk or yogurt
1½ oz cheese
2 cups of cottage cheese
1½ cups of ice cream

Other important factors:
Choose no-fat or low-fat milk products whenever possible.

Meat and beans

What is included?
Meat, poultry, fish, shellfish
Dry beans and peas
Nuts and seeds
Eggs

Daily recommendation:
Young adult males: 6½ oz
Young adult females: 5½ oz

What is an oz?
1 oz of meat, poultry, fish
½ oz of nuts or seeds
¼ cup of dry beans or peas
1 egg

Other important factors:
Most people exceed these daily recommendations.
Eat a variety of products from this group.
Meats are often high in saturated fats.
Some fish and nuts have essential fatty acids.

FIGURE 24.12 In 2005, the U.S. Department of Agriculture published a new food guide pyramid. It includes guidelines for people of different ages. Only the guidelines for young adults nineteen to thirty years of age are included here. For more information see the website http://www.mypyramid.gov/.

Source: U.S. Department of Agriculture, 2005.

The Dynamic Skeleton

Most people think of the skeleton as a nonliving framework that supports the more "soft" body parts. However, the skeleton is anything but inert and serves functions in addition to that of support. Box figure 24.2 shows the primary parts of the human skeleton.

A primary function of the skeleton is to provide support while allowing motion. At joints, tough, fibrous cords called ligaments hold bones together, and muscles are attached to the bones by tough, fibrous cords called *tendons*. Muscles are typically attached to two different bones, and muscular contraction results in one bone moving with respect to another, causing a bending or rotation at the joint. The actions of muscles are also important in providing support because they generate tension that provides rigidity to the skeletal framework, like the cables that hold up the mast of a ship.

Protection is another major function of the skeleton. The brain, spinal cord, and organs of the chest are enclosed by and protected by the bony structures of the skull, vertebrae, and ribs and associated structures, respectively.

Chemically, the skeleton is a combination of an inorganic mineral, hydroxyapatite, and organic fibers of various kinds of protein. Hydroxyapatite contains calcium and phosphorus and has the chemical formula $Ca_{10}(PO_4)_6(OH)_2$. There is constant turnover of both hydroxyapatite and protein in the bones of the skeleton. For example, calcium and phosphorus have several important functions in the body and are distributed throughout the body by the circulatory system. If the blood does not contain adequate calcium and phosphorus from the diet, these minerals are taken from the skeleton to supply immediate needs. These substances can be replaced later, when they are present in the diet. This function is particularly important in pregnant and nursing mothers, who must provide calcium and phosphorus for bone development of embryos and as a source of calcium in the mother's milk for nursing infants. Without adequate calcium and phosphorus in the diet, the mother can experience significant bone loss and skeletal development in the infant is impaired.

Nowhere is the dynamic nature of the skeleton more evident than in the process of growth. Because long bones are typically connected to other bones at a joint, it is important that the joint continue to function while the bone is growing. Long bones of the body have special growth areas near their ends but not on the very end. Thus, new bone is added near the ends so that the bones become longer without interfering with the function of the joint. Bones of the skull and certain other bones grow directly from their edges, eventually growing together and fusing into a solid structure. The bones of the skull of newborn babies are not joined and are capable of being moved around during the birth process. As the individual bones of the skull grow, the bones eventually knit together to form a solid skull, but a "soft spot" persists on the top of the head for a time after birth.

One of the activities that stimulates change in the nature of the skeleton is the stress and strain placed on it. When the skeleton is "challenged," it gets stronger by adding material. When it is not challenged, material is lost and the bones get weaker. For example, astronauts in weightless conditions lose bone mass. Therefore, exercise programs are very important parts of the schedule of astronauts. Since exercise strengthens bones, it is particularly important in elderly persons who tend to have reduced bone mass for age-related reasons.

The skeleton is even involved in the production of blood. The red bone marrow, found in the core of certain bones of the skeleton (vertebrae, ribs, breast bone, and upper arm and leg bones), is the site of production of red and white blood cells and platelets. The lymph nodes, tonsils, spleen, and thymus also produce some kinds of white blood cells. Bone marrow transplants are often provided for people whose blood cell production capacity has been destroyed by chemotherapy or radiation.

Box Figure 24.2 Major bones of the skeleton.

1. Grain Products Group   Grains include vitamin-enriched or whole-grain cereals and products such as breads, bagels, buns, crackers, dry and cooked cereals, pancakes, pasta, and tortillas. Because whole grain foods contain more nutrients than those that are not whole grain, it is recommended that half of all grains consumed be whole grain products. The Food Guide Pyramid recommends that people consume 5–8 ounces of grain products per day depending on their age and sex. This is relatively easy to meet, since a slice of bread, tortilla, half a bagel, or a cup of dry cereal is equivalent to an ounce. Grains should provide most of your kilocalorie requirements in the form of complex carbohydrates such as starch, which is the main ingredient in most grain products.

Significant nutritional components: carbohydrate, fiber, several B vitamins, vitamin E, and the minerals iron, selenium, and magnesium.

2. Vegetables Group   Vegetables include either raw or cooked nonsweet plant materials, such as broccoli, carrots, cabbage, corn, green beans, tomatoes, potatoes, lettuce, and spinach. The Food Guide Pyramid recommends 1–3 cups of items from this food group each day depending on a person's age and sex. Since different vegetables contain different amounts of vitamins, it is wise to include as much variety as possible in this group. Therefore, vegetables are divided into five subgroups: dark green vegetables, orange vegetables, dry beans and peas, starchy vegetables, and other vegetables. On a weekly basis, people should be sure to consume some vegetables from each of the following three groups: dark green, orange, and dry beans and peas. The amount of starchy vegetables can vary considerably depending on a person's activity level and caloric needs.

Significant nutritional components: carbohydrate, fiber, several B vitamins, vitamins A, C, E, and K, and the minerals iron, potassium, and magnesium.

3. Fruits Group   Fruits include such sweet plant products as melons, berries, apples, oranges, and bananas. The Food Guide Pyramid recommends 1–2 cups of fruit per day depending on a person's age and sex. Fruit may be fresh or preserved or fruit juice. Many individual pieces of fruit (apple, orange, banana) are about 1 cup. A half cup of dried fruit is equivalent to 1 cup.

However, since these foods tend to be high in natural sugars, consumption of large amounts of fruits can add significant amounts of kilocalories to your diet. Fruit drinks are not fruit juices and have high amounts of sugars added to them.

Significant nutritional components: carbohydrate, fiber, water, and vitamin C.

4. Dairy Products Group   Dairy products include all cheeses, ice cream, yogurt, and milk. Two to three cups from this food group are recommended each day depending on a person's age. Vitamin D fortified dairy products are the primary dietary source of vitamin D. Vitamin D is also manufactured in the skin exposed to sunlight. Because milk products contain saturated fats, choose no-fat or low-fat milk products. Many cheeses contain large amounts of cholesterol and fat for each serving. The cholesterol in low fat cheeses is significantly reduced.

Significant nutritional components: protein, carbohydrate, fat, some B vitamins and vitamin D, and the minerals calcium and potassium.

5. Meat, Poultry, Fish, and Dry Beans Group   This group contains most of the things we eat as a source of protein. In addition to meat, poultry, fish and beans, nuts, peas, tofu, and eggs are considered a part of this group. The Food Guide Pyramid recommends that adults receive 5–6½ ounces of protein per day depending on a person's age and sex. Since the amount needed per day is so small, most people eat many times what is needed.

Daily protein intake is essential since protein is not stored in the body like fats or carbohydrates. Furthermore, of the amino acids that combine to form proteins, there are about ten that the body cannot manufacture. These are called essential amino acids since it is essential that we receive them in the diet. Animal sources of protein have all the amino acids present and are called complete proteins. Many plant proteins may lack specific amino acids and are called incomplete proteins. Vegetarians must pay particular attention to acquiring adequate sources of protein because they have eliminated animal flesh from their diet. They can get all the essential amino acids if they eat proper combinations of plant materials.

Since many sources of protein (particularly meat and nuts) also include significant fat and health recommendations suggest reduced fat, choosing low-fat sources of protein is important. Selecting foods that have less fat, broiling rather than frying, or removing the fat before cooking reduces fat.

Significant nutritional components: protein, fat, several B vitamins and vitamin E from seeds and nuts, and the minerals iron, zinc, and magnesium.

6. Oils   Oils are fats that are liquids at room temperature. The Food Guide Pyramid recommends 3–7 teaspoons per day depending on age and sex. This is a very small amount and most people get more than they need in salad dressing and oils used in cooking and baking. Small amounts of oils are important in the diet because certain essential fatty acids cannot be manufactured by the body and must be obtained in the diet. However, since oils are fats, they have a high caloric content.

Significant nutritional components: Essential fatty acids and vitamin E.

7. Exercise   Since over 65 percent of adult Americans are overweight and over 30 percent are obese, exercise is important to improving nutritional health. Therefore, exercise is included as a part of the Food Guide Pyramid. Although it is not directly related to nutrition, the amount of exercise people get affects the number of food calories they can consume on a daily basis without gaining weight. Exercise has other health benefits as well. The pyramid recommends at least thirty minutes of moderate exercise per day, and longer periods or more vigorous exercise has additional health benefits. Moderate exercise elevates the heart rate significantly. Activities such as doing household chores or walking while shopping do not elevate heart rate and therefore do not count as moderate exercise.

Eating Disorders

The three most common eating disorders are over-eating leading to obesity, bulimia, and anorexia nervosa. All three disorders involve behaviors that lead to ill health. The causes of these behaviors are complex and often involve a strong psychological component. Metabolic imbalances may also contribute to the development of these disorders, particularly obesity. (Some people have high metabolic rates, while others have low metabolic rates for a variety of reasons.) In some cases people are strongly influenced by prevailing cultural perceptions and values. (They want to be thin, or muscular, or have a certain body shape.) Some studies have shown that there is a tendency for members of a family to develop the same kinds of eating disorders. Therefore, it has been suggested that genetic factors influence the risk of developing eating disorders. On the other hand, members of the same family are likely to develop patterns of eating related to culture and other factors. Eating disorders are often associated with psychological depression.

Obesity

Obesity is the condition of being overweight to the extent that a person's health and life span are adversely affected. Obesity occurs when people consistently take in more food energy than is necessary to meet their daily requirements. About 30 percent of the U.S. population is considered to be obese based on their having a *body mass index* (BMI) above 30.

$$\text{BMI} = \frac{\text{weight in kilograms}}{\text{height in meters}^2}$$

BMI is used as an indirect measurement of the amount of fat a person has stored. For most people, this is a reasonably good tool. Thin people have a low BMI, and obese people have a high BMI. However, since a given volume of muscle tissue weights more than an equal volume of fat, extremely muscular people may have a high BMI and actually have a low amount of fat. This is only likely to be true in people who engage in heavy muscular exercise. Figure 24.13 shows the body mass index for various heights and weights. Although the height and weight in the chart are in feet and pounds, the BMI is given in kg/m².

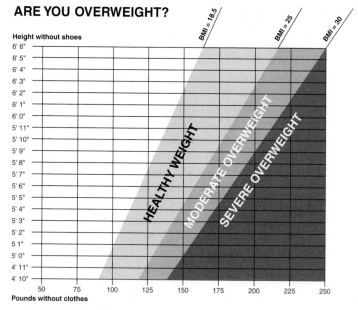

ARE YOU OVERWEIGHT?

FIGURE 24.13 Find your height and weight and determine your approximate BMI.

At first glance, it would appear that obesity is a simple problem to solve. To lose weight, all that people must do is consume fewer kilocalories, exercise more, or do both. While all obese people have an imbalance between their need for kilocalories and the amount of food they eat, the reasons for this imbalance are complex and varied.

1. *Metabolic Differences*   Even at rest, energy is required to maintain breathing, heart rate, and other normal body functions. The **basal metabolic rate** is the rate at which a person uses energy when at rest. Some people have a low basal metabolic rate and, therefore, use fewer kilocalories for maintenance than other people.
2. *Inactive Lifestyle*   Often exercise is all that is needed to control a weight problem. In addition to increasing metabolism during the exercise itself, exercise tends to raise the basal metabolic rate for a period of time following exercise. Table 24.3 lists several common activities and the amount of energy needed to sustain the activity.

TABLE 24.3   Typical Energy Requirements for Common Activities

Light Activities (120–150 kcal/h)	Light to Moderate Activities (150–300 kcal/h)	Moderate Activities (300–400 kcal/h)	Heavy Activities (420–600 kcal/h)
Dressing	Sweeping floors	Pulling weeds	Chopping wood
Typing	Painting	Walking behind a lawnmower	Shoveling snow
Slow walking	Walking 2–3 mi/h	Walking 3.5–4 mi/h on level surface	Walking or jogging 5 mi/h
Standing	Bowling	Golf (no cart)	Walking up hills
Studying	Store clerking	Doubles tennis	Cross-country skiing
Sitting activities	Canoeing 2.5–3 mi/h	Canoeing 4 mi/h	Swimming
	Bicycling on level surface at 5.5 mi/h	Volleyball	Bicycling 11–12 mi/h or up and down hills

Diet and Nutrition

There is probably no area of modern life in which there is more misinformation, pseudoscience, and myth than nutrition. Diet books abound, each purporting to provide the answer to nutritional and weight control problems. It is helpful to look at the science behind these various statements. The list below provides several examples of the mismatch between claims and scientific reality.

Myth or Misunderstanding	Scientific Basis
1. Exercise burns calories.	Calories are not molecules, so they cannot be burned in the physical sense, but you do oxidize (burn) the fuels to provide yourself with the energy needed to perform various activities.
2. Active people who are increasing their fitness need more protein.	The amount of protein needed per day is very small—about 50 g. Most people get many times the amount of protein required.
3. Vitamins give you energy.	Most vitamins assist enzymes in bringing about chemical reactions, some of which may be energy-yielding, but vitamins are not sources of energy.
4. Large amounts of protein are needed to build muscle.	A person can build only a few grams of new muscle per day. Therefore, consuming large amounts of protein will not increase the rate of muscle growth.
5. Large quantities (mega doses) of vitamins will fight disease, build strength, and increase length of life.	Quantities of vitamins that greatly exceed recommendations have not been shown to be beneficial. Large doses of some vitamins (vitamins A, D, B_3) are toxic.
6. Special protein supplements are more quickly absorbed and can build muscle faster.	Adequate protein is obtained in nearly all diets. The supplements may be absorbed faster, but that does not mean that they will be incorporated into muscle mass faster.
7. Vitamins prevent cancer, heart disease, and other health problems.	Vitamins are important to health. However, it is a gross oversimplification to suggest that consumption of excess amounts of specific vitamins will prevent certain diseases. Many factors contribute to the cause of disease.

3. *Cultural Influences* Our culture encourages food consumption. Social occasions and business meetings frequently involve eating, and restaurants and other food preparers have increased the size of portions significantly over the past fifty years. For example, in the 1950s, a fast-food serving of French fries was 2.4 ounces. Today, that size is still available, but it is the small size, and medium and large sizes contain two to three times the quantity of the small size.

The abundance of snack foods high in sugars and fats is another important cultural influence. Comsumption of these foods at other than meal times increases the total number of calories consumed during the day. Furthermore, many of these foods are said to provide "empty calories"—calories are provided by sugar or fat but there is little or no other nutritional benefit (protein, vitamins, minerals, or fiber).

4. *Psychological factors* Overeating is associated with a variety of psychological factors. Eating is a pleasurable activity, and with all pleasurable activities it is sometimes difficult to determine when to stop. Conversely, overeating is also often associated with depression.

Bulimia

Bulimia ("hunger of an ox" in Greek) is an eating disorder in which the person engages in a cycle of eating large amounts of food (binges) followed by purging the body of the food by inducing vomiting or using laxatives. Many bulimics also use diuretics that cause the body to lose water and therefore reduce weight. It is often called the silent killer because it is difficult to detect. Bulimics are usually of normal body weight or overweight. The behavior has a strong psychological component and is often associated with depression or other psychological problems.

Vomiting may be induced physically or by the use of some nonprescription drugs. Case studies have shown that bulimics may take forty to sixty laxatives a day to rid themselves of food. For some, the laxative becomes addictive. The binge-purge cycle and associated use of diuretics results in a variety of symptoms that can be deadly.

Anorexia Nervosa

Anorexia nervosa is an eating disorder characterized by severe, prolonged weight loss as a result of a voluntary severe restriction in food intake. It is most common among adolescent and preadolescent females. An anorexic person's fear of becoming overweight is so intense that even though weight loss occurs, it does not lessen the fear of obesity, and the person continues to diet, often even refusing to maintain the optimum body weight for his or her age, sex, and height. Persons who have this disorder have a distorted perception of their bodies. They see themselves as fat when in fact they are starving to death. Society's preoccupation with body image, particularly among young people, may contribute to the incidence of this disorder.

A Closer Look

Nutrition During Pregnancy and Lactation

The period of pregnancy and milk production (lactation) requires that special attention be paid to the diet to ensure proper fetal development, a safe delivery, and a healthy milk supply.

During pregnancy and lactation there are additional metabolic demands on the body of the mother to support the growth of the fetus and produce milk. Therefore, the amount of essential nutrients and kilocalories consumed must increase. Kilocalories should be increased by 300 per day to meet the needs of an increased basal metabolic rate; the development of the uterus, breasts, and placenta; and the

work required for fetal growth. Some of these kilocalories can be obtained by drinking milk, which simultaneously supplies calcium needed for fetal bone development. Individuals who cannot tolerate milk should use supplementary sources of calcium. In addition, the daily intake of protein should be at least 65 g per day. As was mentioned earlier, most people consume much more than this amount of protein per day. Two essential nutrients, folic acid and iron, should be obtained through prenatal supplements because they are so essential to cell division and development of the fetal blood supply.

The mother's nutritional status affects the developing baby in several ways. If she is under fifteen years of age or has had two or more pregnancies in a two-year period, her nutritional stores often are inadequate to support a successful pregnancy. The use of drugs such as alcohol, caffeine, nicotine, and "hard" drugs (e.g., heroin) can result in decreased nutrient exchange between the mother and fetus. In particular, heavy smoking can result in low birth weights, and alcohol abuse is responsible for fetal alcohol syndrome.

WASTE DISPOSAL: THE EXCRETORY SYSTEM

Because cells are modifying molecules during metabolic processes, harmful waste products are constantly being formed. Urea is a common waste product formed during the metabolic breakdown of amino acids, and hydrogen ions are produced by many metabolic reactions. Other molecules, such as water and salts, may be consumed in excessive amounts and must be removed. The primary organs involved in regulating the level of toxic or unnecessary molecules are the **kidneys.** The urine they produce flows to the urinary bladder through tubes known as *ureters*. The urine leaves the bladder through a tube known as the urethra (figure 24.14).

The kidneys consist of about 2.4 million tiny units called **nephrons.** At one end of a nephron is a cup-shaped structure called *Bowman's capsule*, which surrounds a knot of capillaries known as a *glomerulus*. In addition to Bowman's capsule, a nephron consists of three distinctly different regions: the *proximal convoluted tubule*, the *loop of Henle*, and the *distal convoluted tubule*. The distal convoluted tubule of a nephron is connected to a collecting duct that transports fluid to the ureters and ultimately to the urinary bladder, where it is stored until it can be eliminated (figure 24.15).

As in the other systems, the excretory system involves a close connection between the circulatory system and a surface. In this case, the large surface is provided by the walls of the millions of nephrons, which are surrounded by capillaries. Three major activities occur at these surfaces: filtration, reabsorption, and secretion.

Filtration occurs through the glomerulus. The glomerulus presents a large surface for the filtering of material from the blood to Bowman's capsule. Blood that enters the glomerulus is under pressure from the muscular contraction of the heart. The capillaries of the glomerulus are quite porous and provide a large surface area for the movement of water and small dissolved

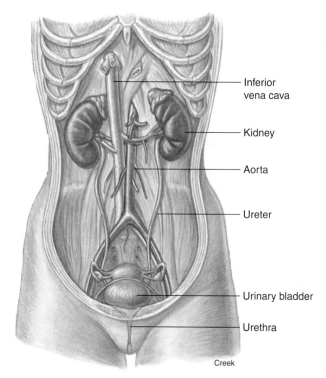

FIGURE 24.14 The primary organs involved in removing materials from the blood are the kidneys. The urine produced by the kidneys is transported by the ureters to the urinary bladder. From the bladder, the urine is emptied to the outside of the body by way of the urethra.

molecules from the blood into Bowman's capsule. Normally, only the smaller molecules, such as glucose, amino acids, and ions, are able to pass through the glomerulus into Bowman's capsule at the end of the nephron. The various kinds of blood cells and larger molecules such as proteins do not pass out of the blood into the nephron. This physical filtration process allows many kinds of molecules to leave the blood and enter the nephron. The volume

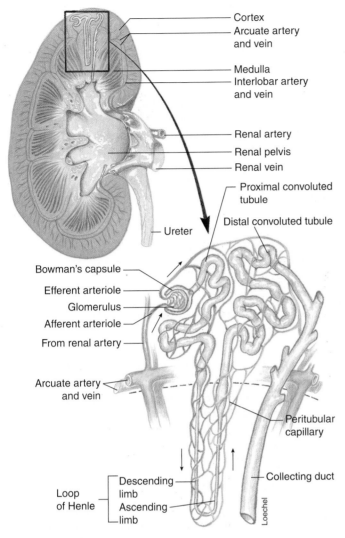

Cortex
Arcuate artery and vein
Medulla
Interlobar artery and vein
Renal artery
Renal pelvis
Renal vein
Proximal convoluted tubule
Distal convoluted tubule
Ureter
Bowman's capsule
Efferent arteriole
Glomerulus
Afferent arteriole
From renal artery
Arcuate artery and vein
Peritubular capillary
Collecting duct
Descending limb
Ascending limb
Loop of Henle
Loechel

FIGURE 24.15 The nephron and the closely associated blood vessels create a system that allows for the passage of materials from the circulatory system to the nephron by way of the glomerulus and Bowman's capsule. Materials are added to and removed from the fluid in the nephron via the tubular portions of the nephron.

of material filtered in this way is about 7.5 liters per hour. Since your entire blood supply is about 5 to 6 liters, there must be some method of recovering much of this fluid.

Reabsorption occurs in several parts of the nephron. The proximal convoluted tubule reabsorbs most of the useful molecules (water, phosphate, salts, sugars, amino acids) as the material filtered from the glomerulus passes through the nephron. The loop of Henle is involved in the reabsorption of water.

The distal convoluted tubule is involved in both reabsorption and secretion and is involved in the regulation of ions, such as hydrogen, sodium, potassium, and calcium ions.

Some molecules that pass through the nephron are relatively unaffected by the various activities going on in the kidney. One of these is urea, which is filtered through the glomerulus into Bowman's capsule. As it passes through the nephron, much of it stays in the tubule and is eliminated in the urine. Many other kinds of molecules, such as minor metabolic waste products and some drugs, are also lost in the urine.

CONTROL MECHANISMS

The nervous and endocrine systems are the major systems of the body that integrate stimuli and generate appropriate responses necessary to maintain homeostasis. There are many kinds of sense organs located within organs and on the surface of the body that respond to specific kinds of stimuli. A **stimulus** is any change in the environment that the organism can detect. Some stimuli, such as light or sound, are typically external to the organism; others, such as the pain generated by an infection, are internal. The reaction of the organism to a stimulus is known as a **response** (figure 24.16).

The *nervous system* consists of a network of cells with fibrous extensions that carry information throughout the body. The *endocrine system* consists of a number of glands that communicate with one another and with other tissues through chemicals distributed throughout the organism. **Glands** are organs that manufacture molecules that are secreted either through ducts or into surrounding tissue, where they are picked up by the circulatory system. *Endocrine glands* have no ducts and secrete their products, called **hormones,** into the circulatory system; other glands, such as the digestive glands and sweat glands, empty their contents through ducts. These kinds of glands are called *exocrine glands.*

Although the functions of the nervous and endocrine systems can overlap and be interrelated, these two systems have quite different methods of action. The nervous system functions very much like a computer. A message is sent along established pathways from a specific initiating point to a specific end point, and the transmission is very rapid. The endocrine system functions in a manner analogous to a radio broadcast system. Radio stations send their signals in all directions, but only those radio receivers that are tuned to the correct frequency can receive the message. Messenger molecules (hormones) are typically distributed throughout the body by the circulatory system, but only those cells that have the proper receptor sites can receive and respond to the molecules.

The Structure of the Nervous System

The basic unit of the nervous system is a specialized cell called a *neuron,* or *nerve cell.* A typical neuron consists of a central body called the *soma,* or *cell body,* that contains the nucleus, and several long, protoplasmic extensions called nerve fibers. There are two kinds of fibers: *axons,* which carry information away from the cell body, and *dendrites,* which carry information toward the cell body. Most nerve cells have one axon and several dendrites.

Neurons are arranged into two major systems. The *central nervous system,* which consists of the brain and spinal cord, is surrounded by the skull and the vertebrae of the spinal column. It receives input from sense organs, interprets information, and generates responses. The *peripheral nervous system* is located outside the skull and spinal column and consists of bundles of long axons and dendrites called nerves. There are two different sets of neurons in the peripheral nervous system. *Motor neurons* carry messages from the central nervous system to muscles and

Stimulus	Integration	Response

FIGURE 24.16 A stimulus is any detectable change in the surroundings of an organism. When an organism receives a stimulus, it processes the information and may ignore the stimulus or generate a response to it.

Sensory neuron
- Cell body
- Myelin sheath
- Direction of conduction
- Axon
- Dendrites

Interneuron
- Cell body

Motor neuron
- Axon
- Dendrites
- Cell body
- Myelin sheath
- Neurofibril node
- Axon
- Axon terminal

glands, and *sensory neurons* carry input from sense organs to the central nervous system. Motor neurons typically have one long axon that runs from the spinal cord to a muscle or gland, while sensory neurons have long dendrites that carry input from the sense organs to the central nervous system (figure 24.17).

The Nature of the Nerve Impulse

Since most nerve cells have long fibrous extensions, it is possible for information to be passed along the nerve cell from one end to the other. The message that travels along a neuron is known as a **nerve impulse.** A nerve impulse is not a simple electric current but involves a specific sequence of chemical events involving activities at the cell membrane.

Since all cell membranes are differentially permeable, it is difficult for some ions to pass through the membrane, and the combination of ions inside the membrane is different from that on the outside. Cell membranes also contain proteins that actively transport specific ions from one side of the membrane to the other. Active transport involves the cell's use of ATP to move materials from one side of the cell membrane to the other. One of the ions that is actively transported from cells is the sodium ion (Na^+). At the same time sodium ions (Na^+) are being transported out of cells, potassium ions (K^+) are being transported into the normal resting cells.

FIGURE 24.17 Nerve cells consist of a nerve cell body that contains the nucleus and several fibrous extensions. The fibers that carry impulses to the nerve cell body are dendrites. The long fiber that carries the impulse away from the cell body is the axon. Sensory neurons typically have long fibers running to and from the cell body. Motor neurons have short dendrites and long axons leaving the cell body.

However, there are more sodium ions (Na^+) transported out than potassium ions (K^+) transported in.

Because a normal resting cell has more positively charged Na^+ ions on the outside of the cell than on the inside, a small but measurable voltage exists across the membrane of the cell. The voltage difference between the inside and outside of a cell membrane is about 70 millivolts (0.07 volt). The two sides of the cell membrane are, therefore, polarized in the same sense that a battery is polarized, with a positive and negative pole. A resting neuron has its positive pole on the outside of the cell membrane and its negative pole on the inside of the membrane.

When a cell is stimulated at a specific point on the cell membrane, the cell membrane changes its permeability and lets sodium ions (Na^+) pass through it from the outside to the inside and potassium ions (K^+) diffuse from the inside to the outside. The membrane is thus *depolarized;* it loses its difference in charge as sodium ions (Na^+) diffuse into the cell from the outside. Sodium ions diffuse into the cell because initially they are in greater concentration outside the cell than inside. When the membrane becomes more permeable, they are able to diffuse into the cell, toward the area of lower concentration.

The depolarization of one point on the cell membrane causes the adjacent portion of the cell membrane to change its permeability as well, and it also depolarizes. Thus, a wave of depolarization passes along the length of the neuron from one end to the other (figure 24.18). The depolarization and passage of an impulse along any portion of the neuron is a momentary event. As soon as a section of the membrane has been depolarized, a series of events begins to reestablish the original polarized state, and the membrane is said to be *repolarized.* When the nerve impulse reaches the end of the axon, it stimulates the release of a molecule that stimulates depolarization of the next neuron in the chain.

Activities at the Synapse

Neurons are arranged end-to-end to form long chains of cells. Between the fibers of adjacent neurons is a space called the **synapse.** Many chemical events occur in the synapse that are important in the function of the nervous system. When a neuron is stimulated, an impulse passes along its length from one end to the other. When the impulse reaches a synapse, a molecule called a **neurotransmitter** is released into the synapse from the axon. It diffuses across the synapse and binds to specific receptor sites on the dendrite of the next neuron. When enough neurotransmitter molecules have bound to the second neuron, an impulse is initiated in it as well. Several kinds of neurotransmitters are produced by specific neurons. These include dopamine, epinephrine, acetylcholine, and several other molecules. The first neurotransmitter identified was *acetylcholine.*

As long as a neurotransmitter is bound to its receptor, it continues to stimulate the nerve cell. Thus, if acetylcholine continues to occupy receptors, the neuron continues to be stimulated again and again. An enzyme called *acetylcholinesterase* destroys acetylcholine and prevents this from happening. The breakdown products of the acetylcholine can be used to remanufacture new acetylcholine molecules at the end of the axon. The destruction of acetylcholine allows the second neuron in the chain to return to normal. Thus, it will be ready to

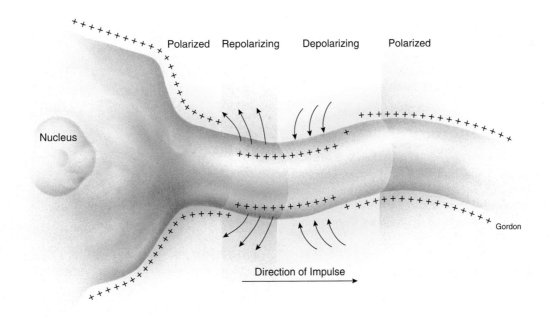

FIGURE 24.18 When a nerve cell is stimulated, a small portion of the cell membrane depolarizes as Na^+ flows into the cell through the membrane. This encourages the depolarization of an adjacent portion of the membrane, and it depolarizes a short time later. In this way a wave of depolarization passes down the length of the nerve cell. Shortly after a portion of the membrane is depolarized, the ionic balance is reestablished. It is repolarized and ready to be stimulated again.

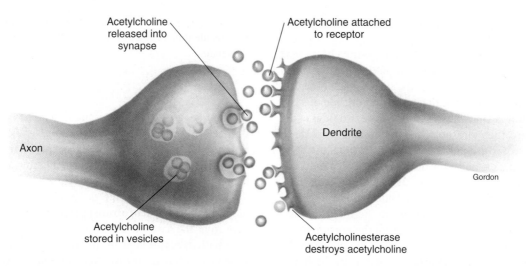

Acetylcholine released into synapse

Acetylcholine attached to receptor

Axon

Dendrite

Gordon

Acetylcholine stored in vesicles

Acetylcholinesterase destroys acetylcholine

FIGURE 24.19 When a nerve impulse reaches the end of an axon, it releases a neurotransmitter into the synapse. In this illustration, the neurotransmitter is acetylcholine. When acetylcholine is released into the synapse, acetylcholine molecules diffuse across the synapse and bind to receptors on the dendrite, initiating an impulse in the next neuron. Acetylcholinesterase is an enzyme that destroys acetylcholine, preventing continuous stimulation of the dendrite.

accept another burst of acetylcholine from the first neuron a short time later. Neurons must also constantly manufacture new acetylcholine molecules, or they will exhaust their supply and be unable to conduct an impulse across a synapse (figure 24.19).

Caffeine and nicotine are considered stimulants because they make it easier for impulses to pass through the synapse. The nerve cells are sensitized and will respond to smaller amounts of acetylcholine than normal. Drinking coffee, taking caffeine pills, or smoking tobacco will increase the sensitivity of the nervous system to stimulation. However, as with most kinds of drugs, continual use or abuse tends to lead to a loss of the effect as the nervous system adapts to the constant presence of the drugs.

Because of the way the synapse works, impulses can go in only one direction: Only axons secrete acetylcholine, and only dendrites have receptors. This explains why there are separate sensory and motor neurons to carry messages to and from the central nervous system.

Endocrine System Function

As mentioned previously, the endocrine system is basically a broadcasting system in which glands secrete messenger molecules, called hormones, which are distributed throughout the body by the circulatory system (figure 24.20). However, each kind of hormone attaches only to appropriate receptor molecules on the surfaces of certain cells.

How Cells Respond to Hormones

The cells that receive the message typically respond in one of three ways: (1) Some cells release products that have been previously manufactured, (2) other cells are stimulated to synthesize molecules or to begin metabolic activities, and (3) some are stimulated to divide and grow.

These different kinds of responses mean that some endocrine responses are relatively rapid, while others are very slow.

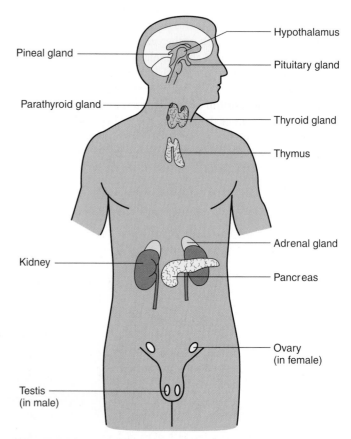

Pineal gland

Hypothalamus

Pituitary gland

Parathyroid gland

Thyroid gland

Thymus

Kidney

Adrenal gland

Pancreas

Ovary (in female)

Testis (in male)

FIGURE 24.20 The endocrine glands are located at various places within the body and cause their effects by secreting hormones.

For example, the release of the hormones *epinephrine* and *norepinephrine* (also known as adrenaline and noradrenaline) from the adrenal medulla, located near the kidney or certain nerve cells, causes a rapid change in the behavior of an organism. The heart rate increases, blood pressure rises, blood is shunted to muscles, and the breathing rate increases. You have certainly experienced this reaction many times in your lifetime as when you nearly have an automobile accident or slip and nearly fall.

Another hormone, called *antidiuretic hormone (ADH),* acts more slowly. It is released from the posterior pituitary gland at the base of the brain and regulates the rate at which the body loses water through the kidneys. It does this by allowing the reabsorption of water from the kidney back into the bloodstream. The effects of this hormone can be noticed in a matter of minutes to hours. Insulin is another hormone whose effects are quite rapid. Insulin is produced by the pancreas, located near the stomach, and stimulates cells—particularly muscle, liver, and fat cells—to take up glucose from the blood. After a meal that is high in carbohydrates, the level of glucose in the blood begins to rise, stimulating the pancreas to release insulin. The increased insulin causes glucose levels to fall as the sugar is taken up by cells. People with diabetes have insufficient or improperly acting insulin or lack the receptors to respond to the insulin and therefore have difficulty regulating glucose levels in their blood.

The responses that result from the growth of cells may take weeks or years to occur. For example, *growth-stimulating hormone (GSH)* is produced by the anterior pituitary gland over a period of years and results in typical human growth. After sexual maturity, the amount of this hormone generally drops to very low levels and body growth stops. Sexual development is also largely the result of the growth of specific tissues and organs. The male sex hormone *testosterone,* produced by the testes, causes the growth of male sex organs and a change to the adult body form. The female counterpart, *estrogen,* results in the development of female sex organs and the adult body form. In all of these cases, it is the release of hormones over long periods, continually stimulating the growth of tissues sensitive to the hormones, that results in the normal developmental pattern. The absence or inhibition of any of these hormones early in life changes the normal growth process.

Interactions of the Nervous and Endocrine Systems

Because the pituitary is constantly receiving information from the hypothalamus of the brain, many kinds of sensory stimuli to the body can affect the functioning of the endocrine system. One example is the way in which the nervous system and endocrine system interact to influence the menstrual cycle. At least three different hormones are involved in the cycle of changes that affect the ovary and the lining of the uterus. It is well documented that stress caused by tension or worry can interfere with the normal cycle of hormones and delay or stop menstrual cycles. In addition, young women living in groups, such as in college dormitories, often find that their menstrual cycles become synchronized. Although the exact mechanism involved in this phenomenon is unknown, it is suspected that input from the nervous system causes this synchronization. (Odors and sympathetic feelings have been suggested as causes.)

Although we still tend to think of the nervous and endocrine systems as being separate and different, it is becoming clear that they are interconnected. These two systems cooperate to bring about appropriate responses to environmental challenges. The nervous system is specialized for receiving and sending short-term messages, whereas activities that require long-term, growth-related actions are handled by the endocrine system.

SENSORY INPUT

The activities of the nervous and endocrine systems are often responses to some kind of input received from the sense organs. Sense organs of various types are located throughout the body. Many of them are located on the surface, where environmental changes can be easily detected. Hearing, sight, and touch are good examples of such senses. Other sense organs are located within the body and indicate to the organism how its various parts are changing. For example, pain and pressure are often used to monitor internal conditions. The sense organs detect changes, but the brain is responsible for perception—the recognition that a stimulus has been received. Sensory abilities involve many different kinds of mechanisms, including chemical recognition, the detection of energy changes, and the monitoring of forces.

Chemical Detection

All cells have receptors on their surfaces that can bind selectively to molecules they encounter. This binding process can cause changes in the cell in several ways. In some cells, it causes depolarization. When this happens, the cells can stimulate neurons and cause messages to be sent to the central nervous system, informing it of some change in the surroundings. In other cases, a molecule binding to the cell surface may cause certain genes to be expressed, and the cell responds by changing the molecules it produces. This is typical of the way the endocrine system receives and delivers messages.

Taste

Most cells have specific binding sites for particular molecules. Others, such as the taste buds on the tongue, appear to respond to classes of molecules. Traditionally we have distinguished four kinds of tastes: sweet, sour, salt, and bitter. However, recently, a fifth kind of taste, *umami* (meaty), has been identified that responds to the amino acid glutamate, which is present in many kinds of foods and can be added as a flavor enhancer (monosodium glutamate). Sourness appears to result from taste buds that respond to the presence of hydrogen ions (H^+), because acids taste sour. Saltiness is sensed when sodium chloride stimulates taste buds. However, the sensations of sweetness, bitterness, and umami occur when molecules bind to specific surface receptors on the cell. Sweetness can be stimulated by many kinds of organic molecules, including sugars and artificial sweeteners, and also by inorganic lead compounds. When a molecule binds to a sweetness receptor, a molecule is split, and its splitting stimulates an enzyme that leads to the depolarization of the cell. The sweet taste of lead salts in

Connections . . .

Negative Feedback Control

old paints partly explains why children sometimes eat paint chips. Because the lead interferes with normal brain development, this behavior can have disastrous results. Many kinds of compounds of diverse structures give the bitter sensation. The cells that respond to bitter sensations have a variety of receptor molecules on their surface. When a substance binds to one of the receptors, the cell depolarizes. In the case of umami, it is the glutamate molecule that binds to receptors on the cells of the taste buds.

It is also important to understand that much of what we often refer to as taste involves such inputs as temperature, texture, and smell. Cold coffee has a different taste than hot coffee even though they are chemically the same. Lumpy, cooked cereal and smooth cereal have different tastes. If you are unable to smell food, it doesn't taste as it should, which is why you sometimes lose your appetite when you have a stuffy nose. We still have much to learn about how the tongue detects chemicals and the role that other associated senses play in modifying taste.

Smell

The other major chemical sense, the sense of smell, is much more versatile; it can detect thousands of different molecules at very low concentrations. The cells that make up the *olfactory epithelium*, the cells that line the nasal cavity and respond to smells, apparently bind molecules to receptors on their surface. Exactly how this can account for the large number of recognizably different odors is unknown, but the receptor cells are extremely sensitive. In some cases, a single molecule of a substance is sufficient to cause a receptor cell to send a message to the brain, where the sensation of odor is perceived. These sensory cells for smell also fatigue rapidly. You have probably noticed that when you first walk into a room, specific odors are readily detected, but after a few minutes, you are unable to

detect them. Most perfumes and aftershaves are undetectable after fifteen minutes of continuous stimulation.

Other Chemical Senses

Many internal sense organs also respond to specific molecules. For example, the brain and aorta contain cells that respond to concentrations of hydrogen ions, carbon dioxide, and oxygen in the blood. Remember, too, that the endocrine system relies on the detection of specific messenger molecules to trigger its activities.

Light Detection

The eyes primarily respond to changes in the flow of light energy. The structure of the eye is designed to focus light on a light-sensitive layer of the back of the eye known as the **retina** (figure 24.21). There are two kinds of receptors in the retina of the eye. The cells called *rods* respond to a broad range of wavelengths of light and are responsible for black-and-white vision. Since rods are very sensitive to light, they are particularly useful in dim light. Rods are located over most of the retinal surface except for the area of most acute vision known as the *fovea centralis*. The other receptor cells, called *cones,* are found throughout the retina but are particularly concentrated in the fovea centralis. Cones are not as sensitive to light, but they can detect light of different wavelengths (color). This combination of receptors gives us the ability to detect color when light levels are high, but we rely on black-and-white vision at night. There are three different varieties of cones: One type responds best to red light, another responds best to green light, and the third responds best to blue light. Stimulation of various combinations of these three kinds of cones allows us to detect different shades of color.

Rods and the three different kinds of cones each contain a pigment that decomposes when struck by light of the proper wavelength and sufficient strength. The pigment found in rods is called *rhodopsin*. This change in the structure of rhodopsin causes the rod to stimulate the nerve cell connected to it and send messages to the brain. Cone cells have a similar mechanism of action, and each of the three kinds of cones has a different pigment. Thus, the pattern of color and light intensity recorded on the retina is detected by rods and cones and converted into a series of nerve impulses that are received and interpreted by the brain.

Sound Detection

The ears respond to changes in sound waves. Sound is produced by the vibration of molecules. Consequently, the ears are detecting changes in the quantity of energy and the quality of sound waves. Sound has several characteristics. Loudness, or volume, is a measure of the intensity of sound energy that arrives at the ear. Very loud sounds will literally vibrate your body and can cause hearing loss if they are too intense. Pitch is a quality of sound that is determined by the frequency of the sound vibrations. High-pitched sounds have short wavelengths; low-pitched sounds have long wavelengths.

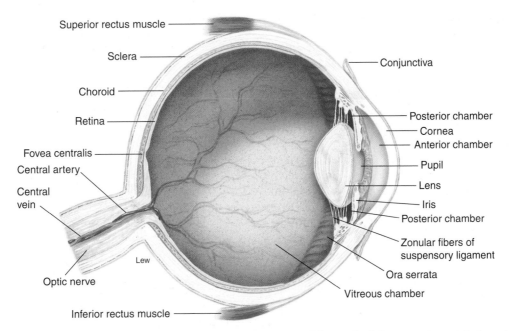

FIGURE 24.21   The eye contains a cornea and lens that focus the light on the retina of the eye. The light causes pigments in the rods and cones of the retina to decompose. This leads to the stimulation of neurons that send messages to the brain.

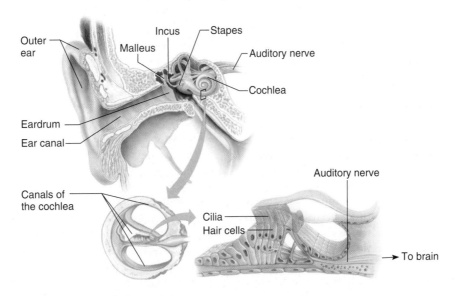

FIGURE 24.22   The ear consists of an external cone that directs sound waves to the tympanum (eardrum). Vibrations of the tympanum move the ear bones and vibrate the oval window of the cochlea, where the sound is detected. The semicircular canals monitor changes in the position of the head, helping us maintain balance.

Figure 24.22 shows the anatomy of the ear. The sound that arrives at the ear is first funneled by the external ear to the **tympanum,** also known as the *eardrum.* The cone-shaped nature of the external ear focuses sound on the tympanum and causes it to vibrate at the same frequency as the sound waves reaching it. Attached to the tympanum are three tiny bones known as the *malleus* (hammer), *incus* (anvil), and *stapes* (stirrup). The malleus is attached to the tympanum, the incus is attached to the malleus and stapes, and the stapes is attached to a small, membrane-covered opening called the oval window in a snail-shaped structure known as the cochlea. The vibration of the tympanum causes the tiny bones (malleus, incus,

and stapes) to vibrate, and they in turn cause a corresponding vibration in the membrane of the *oval window.*

The *cochlea* of the ear is the structure that detects sound and consists of a snail-shaped set of fluid-filled tubes. When the oval window vibrates, the fluid in the cochlea begins to move, causing a membrane in the cochlea, called the *basilar membrane,* to vibrate. High-pitched, short-wavelength sounds cause the basilar membrane to vibrate at the base of the cochlea near the oval window. Low-pitched, long-wavelength sounds vibrate the basilar membrane far from the oval window. Loud sounds cause the basilar membrane to vibrate more vigorously than do faint sounds. Cells on this membrane depolarize when they are stimulated by its vibrations. Because they synapse with neurons, messages can be sent to the brain.

Because sounds of different wavelengths stimulate different portions of the cochlea, the brain is able to determine the pitch of a sound. Most sounds consist of a mixture of pitches that are heard. Louder sounds stimulate the membrane more forcefully, causing the sensory cells in the cochlea to send more nerve impulses per second. Thus, the brain is able to perceive the loudness of various sounds, as well as the pitch.

Associated with the cochlea are two fluid-filled chambers and a set of fluid-filled tubes called the *semicircular canals*. These structures are not involved in hearing but are involved in maintaining balance and posture. In the walls of these canals and chambers are cells similar to those found on the basilar membrane. These cells are stimulated by movements of the head and by the position of the head with respect to the force of gravity. The constantly changing position of the head results in sensory input that is important in maintaining balance.

Touch

What we normally call the sense of *touch* consists of a variety of different kinds of input. Some receptors respond to pressure, others to temperature, and others, which we call *pain receptors*, usually respond to cell damage. When these receptors are appropriately stimulated, they send a message to the brain. Because receptors are stimulated in particular parts of the body, the brain is able to localize the sensation. However, not all parts of the body are equally supplied with these receptors. The tips of the fingers, lips, and external genitals have the highest density of these nerve endings, while the back, legs, and arms have far fewer receptors.

Some internal receptors, such as pain and pressure receptors, are important in allowing us to monitor our internal activities. Many pains generated by the internal organs are often perceived as if they were somewhere else. For example, the pain associated with heart attack is often perceived to be in the left arm.

Pressure receptors in joints and muscles are important in providing information about the degree of stress being placed on a portion of the body. This is also important information to send back to the brain so that adjustments can be made in movements to maintain posture. If you have ever had your foot "go to sleep" because the nerve stopped functioning, you have experienced what it is like to lose this constant input of nerve messages from the pressure sensors to assist in guiding the movements you make. Your movements become uncoordinated until the nerve function returns to normal.

OUTPUT MECHANISMS

The nervous system and endocrine system cause changes in several ways. Both systems can stimulate muscles to contract and glands to secrete. The endocrine system is also able to alter the metabolic activities of cells and regulate the growth

Creek

FIGURE 24.23 Because muscles cannot actively lengthen, it is necessary to have sets of muscles that oppose one another. The contraction and shortening of one muscle cause the stretching of a relaxed muscle.

Triceps extends arm.

Biceps flexes arm.

of tissues. The nervous system acts upon two kinds of organs: muscles and glands. The actions of muscles and glands are simple and direct: muscles contract and glands secrete.

Muscles

The ability to move is one of the fundamental characteristics of animals. Through the coordinated contraction of many muscles, the intricate, precise movements of a dancer, basketball player, or writer are accomplished. It is important to recognize that muscles can pull only by contracting; they are unable to push by lengthening. The work of any muscle is done during its contraction. Relaxation is the passive state of the muscle. There must always be some force available that will stretch a muscle after it has stopped contracting and relaxes. Therefore, the muscles that control the movements of the skeleton are present in antagonistic sets—for every muscle's action, there is another muscle that has the opposite action. For example, the biceps muscle causes the arm to flex (bend) as the muscle shortens. The contraction of its antagonist, the triceps muscle, causes the arm to extend (straighten) and at the same time stretches the relaxed biceps muscle (figure 24.23).

There are three major types of muscle: *skeletal, smooth,* and *cardiac*. These differ from one another in several ways. *Skeletal muscle* is voluntary muscle; it is under the control of the nervous system. The brain or spinal cord sends a message to skeletal muscles, and they contract to move the legs, fingers, and other parts of the body. This does not mean that you must make a conscious decision every time you want to move a muscle. Many of the

Which Type of Exercise Do You Do?

Aerobic exercise occurs when the muscles being contracted are supplied with sufficient oxygen to continue aerobic cellular respiration:

$$C_6H_{12}O_6 + 6O_2 \rightarrow 6CO_2 + 6H_2O + energy$$

This type of exercise involves long periods of activity with elevated breathing and heart rate. It results in strengthened chest muscles, which enable more complete exchange of air during breathing. It also improves the strength of the heart, enabling it to pump more efficiently. Flow of blood to the muscles also improves. All these changes increase endurance.

Anaerobic exercise takes place when insufficient amounts of oxygen reach the contracting muscle cells and they shift to anaerobic cellular respiration:

$$C_6H_{12}O_6 \rightarrow lactic\ acid + less\ energy$$

The buildup of lactic acid results in muscle pain and eventually prevents further contraction. Anaerobic exercise involves explosive bouts of activity, as in sprints or jumping. This kind of exercise increases muscle strength but does little to improve endurance.

Resistance exercise occurs when muscles contract against an object that does not allow the muscles to move that object. This type of exercise does not improve the ability of your body to deliver oxygen to your muscles, nor does it increase your endurance. However, it does stimulate your muscle cells to manufacture more contractile protein fibers.

Make a list of your activities that would be considered aerobic, anaerobic, or resistance exercise. Which activities would result in weight reduction? Weight gain? Improved cardiovascular fitness?

---

movements we make are learned initially but become automatic as a result of practice. For example, walking, swimming, or riding a bicycle required a great amount of practice originally, but now you probably perform these movements without thinking about them. They are, however, still considered voluntary actions.

Smooth muscles make up the walls of muscular internal organs, such as the gut, blood vessels, and reproductive organs. They have the property of contracting as a response to being stretched. Since much of the digestive system is being stretched constantly, the responsive contractions contribute to the normal rhythmic movements associated with the digestive system. These are involuntary muscles; they can contract on their own without receiving direct messages from the nervous system. Although they are involuntary, they can be stimulated to contract by the presence of certain hormones (the hormone oxytocin causes the smooth muscle of the uterus to contract), or their degree of contraction can be modified by nervous stimulation.

Cardiac muscle is the muscle that makes up the heart. It has the ability to contract rapidly like skeletal muscle but does not require nervous stimulation to do so. Nervous stimulation can, however, cause the heart to speed or slow its rate of contraction. Hormones, such as epinephrine and norepinephrine, also influence the heart by increasing its rate and strength of contraction. Cardiac muscle also has the characteristic of being unable to stay contracted. It will contract quickly but must have a short period of relaxation before it will be able to contract a second time. This makes sense in light of its continuous, rhythmic, pumping function.

Glands

The glands of the body are of two different kinds. Those that secrete into the bloodstream are called **endocrine glands.** We have already talked about several of these: the pituitary, thyroid, ovary, and testis are examples. The **exocrine glands** are those

that secrete to the surface of the body or into one of the tubular organs of the body, such as the gut or reproductive tract. Examples are the salivary glands, intestinal mucous glands, and sweat glands. Some of these glands, such as salivary glands and sweat glands, are under nervous control. When stimulated by the nervous system, they secrete their contents.

Many other exocrine glands are under hormonal control. Many of the digestive enzymes of the stomach and intestine are secreted in response to local hormones produced in the gut. These are circulated through the blood to the digestive glands, which respond by secreting the appropriate digestive enzymes and other molecules. Endocrine glands also respond to nervous stimulation or stimulation by hormones.

Growth Responses

The hormones produced by the endocrine system can have a variety of effects. As mentioned earlier, hormones can stimulate smooth muscle to contract and can influence the contraction of cardiac muscle as well. Many kinds of glands, both endocrine and exocrine, are caused to secrete as a result of a hormonal stimulus. However, the endocrine system has one major effect that is not equaled by the nervous system: Hormones regulate growth. *Growth-stimulating hormone (GSH)* is produced over a period of years to bring about the increase in size of most of the structures of the body. A low level of this hormone results in a person with small body size. It is important to recognize that the amount of growth-stimulating hormone (GSH) present varies from time to time. It is present in fairly high amounts throughout childhood and results in steady growth. It also appears to be present at higher levels at certain times, resulting in growth spurts. Finally, as adulthood is reached, the level of this hormone falls, and growth stops.

Similarly, testosterone produced during adolescence influences the growth of bone and muscle to provide men with larger, more muscular bodies than those of women. In addition, there is growth of the penis, growth of the larynx, and increased growth of hair on the face and body. The primary female hormone, estrogen, causes growth of reproductive organs and development of breast tissue. It is also involved, along with other hormones, in the cyclic growth and sloughing of the wall of the uterus.

SUMMARY

The body's various *systems* must be integrated in such a way that the *internal environment* stays relatively constant. This concept is called *homeostasis*.

The *circulatory, respiratory, digestive,* and *excretory systems* are involved in the exchange of materials across cell membranes. All of these systems have special features that provide *large surface* areas to allow for necessary *exchanges*.

The *circulatory system* consists of a pump, the *heart,* and *blood vessels* that distribute the blood to all parts of the body. The *blood* is a carrier fluid that transports molecules and heat. The exchange of materials between the *blood* and body *cells* takes place through the walls of the *capillaries*. Hemoglobin in *red blood cells* is very important in the transport of *oxygen*.

The *skin* is the outer covering of the body and performs several functions, including protecting underlying tissues, preventing water loss, and regulating temperature.

The *respiratory system* consists of the *lungs* and associated tubes that allow air to enter and leave the lungs. The *diaphragm* and *muscles* of the chest wall are important in the process of *breathing*. In the lungs, tiny sacs called *alveoli* provide a large surface area in association with capillaries, which allows for rapid exchange of oxygen and carbon dioxide.

The *digestive system* is involved in disassembling food molecules. This involves several processes: *grinding* by the teeth and stomach, *emulsification* of fats by bile from the liver, *addition of water* to dissolve molecules, and *enzymatic action* to break complex molecules into simpler molecules for *absorption*. The intestine provides a *large surface area* for the *absorption of nutrients* because it is long and its wall contains many tiny projections that increase surface area.

To maintain good health, people must receive *nutrient molecules* that can enter the cells and function in the metabolic processes. The *proper quantity and quality of nutrients* are essential to good health. An important measure of the amount of energy required to sustain a human at rest is the *basal metabolic rate*. To meet this and all additional requirements, the United States has established recommendations for each nutrient. Should there be metabolic or psychological problems associated with a person's *normal metabolism*, a variety of disorders may occur, including *obesity, anorexia nervosa,* and *bulimia*.

The *excretory system* is a filtering system of the body. The *kidneys* consist of *nephrons* into which the *circulatory system* filters fluid. Most of this fluid is useful and is reclaimed by the cells that make up the walls of these *tubules*. Materials that are present in excess or those that are harmful are allowed to escape.

A *nerve impulse* is caused by sodium ions entering the cell as a result of a change in the permeability of the cell membrane. Thus, a *wave of depolarization* passes down the length of a *neuron* to the *synapse*. The *axon* of a neuron secretes a *neurotransmitter,* such as *acetylcholine,* into the *synapse,* where these molecules bind to the dendrite of the next cell in the chain, resulting in an impulse in it as well. The *acetylcholinesterase* present in the synapse destroys acetylcholine so that it does not repeatedly stimulate the dendrite.

Several kinds of *sensory inputs* are possible. Many kinds of chemicals can bind to cell surfaces and be recognized. This is probably how the *sense of taste* and the *sense of smell* function. Light energy can be detected because light causes certain molecules in the *retina* of the eye to decompose and stimulate neurons. *Sound* can be detected because fluid in the *cochlea* of the *ear* is caused to vibrate, and special cells detect this movement and stimulate neurons. The sense of *touch* consists of a variety of receptors that respond to *pressure, cell damage,* and *temperature*.

There are two major kinds of organs that respond to nervous and hormonal stimulation: *muscles* and *glands*. There are three kinds of muscle: *skeletal muscle,* which moves parts of the skeleton; *cardiac muscle*—the heart; and *smooth muscle,* which makes up the muscular walls of internal organs.

Glands are of two types: *exocrine glands,* which secrete through ducts into the cavity of an organ or to the surface of the skin, and *endocrine glands,* which release their secretions into the circulatory system. It is becoming clear that the endocrine system and the nervous system are interrelated. Actions of the endocrine system can change how the nervous system functions, and the reverse is also true.

KEY TERMS

alveoli (p. **605**)	arteries (p. **598**)	blood (p. **598**)	bulimia (p. **617**)
anorexia nervosa (p. **617**)	arterioles (p. **602**)	bronchi (p. **605**)	calorie (p. **612**)
aorta (p. **600**)	basal metabolic rate (p. **616**)	bronchioles (p. **605**)	capillaries (p. **598**)

APPLYING THE CONCEPTS

1. The primary structures involved in pumping blood are the
   a. veins.
   b. atria.
   c. capillaries.
   d. ventricles.

2. The fluid portion of the blood that leaves the capillaries and surrounds the cells is
   a. hemoglobin.
   b. edema.
   c. lymph.
   d. lacteal.

3. Blood is carried through vessels to all parts of the body except the lungs by
   a. pulmonary circulation.
   b. the pulmonary artery.
   c. systemic circulation.
   d. the lymphatic system.

4. As air passes through the lungs, it follows the path:
   a. trachea → bronchi → bronchioles → alveoli
   b. trachea → bronchioles → bronchi → alveoli
   c. bronchi → trachea → alveoli → bronchioles
   d. bronchioles → alveoli → bronchi → trachea

5. The levels of water, hydrogen ions, salts, and urea in the blood are regulated by the
   a. liver.
   b. kidneys.
   c. bladder.
   d. rectum.

6. An organism's reaction to a change in the environment is a(n)
   a. stimulus.
   b. impulse.
   c. response.
   d. perception.

7. A light stimulus is received by the nervous system, which results in growth. The growth is the direct result of
   a. hormones stimulating cells.
   b. activating muscles.
   c. the endocrine system stimulating the nervous system.
   d. increasing nervous activity.

8. When a nerve cell is stimulated,
   a. acetylcholine is destroyed.
   b. potassium ions enter the neuron.
   c. sodium ions enter the neuron.
   d. calcium attaches to the dendrites.

9. The central nervous system consists of the
   a. brain only.
   b. brain and spinal cord.
   c. brain, spinal cord, and nerves.
   d. motor neurons and sensory neurons.

10. Olfactory senses detect
    a. light.
    b. sound.
    c. chemicals.
    d. pain.

11. Strict vegetarians do not eat animals or their products (milk, eggs). Which one of the following nutritional needs would they have the greatest difficulty meeting from food alone?
    a. carbohydrates
    b. essential amino acids
    c. essential fatty acids
    d. vitamin A

12. Large food molecules are chemically broken down to smaller molecules by
    a. bile salts.
    b. enzymes.
    c. dissolving.
    d. chewing.

Answers

1. d **2.** c **3.** c **4.** a **5.** b **6.** c **7.** a **8.** c **9.** b **10.** c **11.** b **12.** b

QUESTIONS FOR THOUGHT

1. What are the functions of the heart, arteries, veins, arterioles, the blood, and capillaries?

2. Describe three ways in which the digestive system increases its ability to absorb nutrients.

3. Describe the mechanics of breathing.

4. Describe how changing permeability of the cell membrane and the movement of sodium ions cause a nerve impulse.

5. What is the role of acetylcholine in a synapse?

6. What is actually detected by the nasal epithelium, taste buds, cochlea of the ear, and retina of the eye?

7. List the differences between the central and peripheral nervous systems.

8. List two differences in the way the endocrine and nervous systems function.

9. Why do large, multicellular organisms need a circulatory system?

10. List the six groups of the Food Guide Pyramid, along with a major contribution of each to the diet.

11. Describe how villi and the length of the small intestine are related to the ability to absorb nutrients.

12. Describe two ways in which smooth muscle and skeletal muscle differ in the way they function.

FOR FURTHER ANALYSIS

1. The heart can be replaced with a mechanical pump. Major blood vessels can be replaced with plastic tubing. Can you envision artificial capillaries? What technical problems would need to be overcome before these would be possible?

2. The skin is an excellent organ for getting rid of heat. What role does surface area play in this function?

3. If you were allergic to dairy products, what are the particular nutrients for which you would need to find alternative sources?

4. Describe the ways in which the functions of the nervous and endocrine systems are similar. In what ways are they different?

5. How is the structure of each of the following organs able to provide a large surface area for the exchange of molecules: lungs, small intestine, and kidney?

INVITATION TO INQUIRY

Monitor Your Diet

Many people have a poor idea of their normal dietary intake. For twenty-four hours, keep track of what you eat. Record the food and quantity eaten. Include all food and drink (including alcoholic beverages) consumed, not just those foods eaten at mealtime. If you are eating packaged foods, save the package because it has the quantity contained and the nutritional facts about the food. For other items, estimate the size. You can use the following chart to record your food intake.

Meal	Item	Quantity	Nutritional Information
Breakfast	1. 2. 3. 4.		
Snacks	1. 2. 3. 4.		
Lunch	1. 2. 3. 4.		
Snacks	1. 2. 3. 4.		
Dinner	1. 2. 3. 4.		
Snacks	1. 2. 3. 4.		

Now compare your daily diet with the Food Guide Pyramid. What categories were consumed in excess? What categories do you need to consume more of? Did you get a good mixture of vitamins and minerals?

25

Human Biology: Reproduction

This is a photograph of a human embryo at approximately eight weeks of development. The length of the embryo is approximately 35–40mm CR (Crown-Rump). This is the embryonic period and the beginning of the fetal period.

CORE **CONCEPT**

Humans are sexual beings and possess the structural, functional, and behavioral capabilities of transmitting necessary genetic material to the next generation.

OUTLINE

The process of meiosis reduces the number of chromosomes during sex cell formation and introduces genetic variation to the offspring (p. 632).

Sexuality involves distinct hereditary, anatomical, and behavioral aspects (p. 635).

Hormones play important roles in sexual development and function (p. 642).

Life Science

▷ A specialized form of cell division plays a vital role in sexual reproduction (Ch. 25).

▷ One's sexuality is a complex of anatomy, physiology, behavior, and culture (Ch. 25).

▷ Males and females have specialized anatomical features for reproduction (Ch. 25).

▷ Gamete formation in males and females differs (Ch. 25).

▷ Reproduction is under the control of many different hormones (Ch. 25).

▷ Understanding sexuality requires knowledge of the process from gamete production to old age (Ch. 25).

OVERVIEW

If both parents contribute equally to the genetic information of the child, how can the chromosome number in humans remain constant for generation after generation? There is a specialized kind of cell division that results in the formation of sex cells. In this chapter, we will discuss the mechanics of this process. Knowing the mechanics is essential to understanding how genetic variety can occur in sex cells. This variety ultimately shows up as differences in offspring as a result of sexual reproduction. Sex and your sexuality influence you in many ways throughout life. Before birth, sex-determining chromosomes direct the formation of hormones that control the development of sex organs, after which the effects of these hormones diminish. With the start of puberty, renewed hormonal activity causes major structural and behavioral changes that influence you for the remainder of your life.

SEXUAL REPRODUCTION

Nearly all organisms have a method of shuffling and exchanging genetic information. Humans and many other organisms have two sets of genetic data, one inherited from each of their parents. **Sexual reproduction** is the formation of a new individual by the union of two sex cells, one donated by each parent. The sex cells produced by male organisms are called **sperm,** and those produced by females are called **eggs.** A general term sometimes used to refer to either eggs or sperm is **gamete** (sex cell). During this process, the two sets of genetic information found in each body cell must be reduced to one set, that is, one of each kind of chromosome. The cellular process that is responsible for forming gametes is called **gametogenesis.** This is somewhat similar to shuffling a deck of cards and dealing out hands; the shuffling and dealing ensure that each hand will be different. An organism with two sets of chromosomes can produce many combinations of chromosomes when it produces sex cells, just as many different hands can be dealt from one pack of cards. When a sperm unites with an egg (*fertilization),* a new organism containing two sets of genetic information is formed. This new organism's genetic information might have survival advantages over the information found in either parent; this is the value of sexual reproduction.

The life cycle of a sexually reproducing organism begins with the fertilization of an egg by a sperm. This new cell, called a **zygote,** divides by mitosis, as do the resulting cells, and growth occurs. When the organism reaches sexual maturity, it is able to produce gametes by another form of cell division, *meiosis* and the life cycle is complete. Notice in figure 25.1 that the zygote and its descendants have two sets of chromosomes. However, male gametes and female gametes each contain only one set of chromosomes. These sex cells are said to be **haploid.** The haploid number of chromosomes is noted as *n.* A zygote contains two sets and is said to be **diploid.**

Meiosis

Sex cell formation

Single adult cell → Single gamete (sperm or egg)

Contains 2 sets of chromosomes → Contains 1 set of chromosomes

$$2n \rightarrow n$$

The diploid number of chromosomes is noted as *2n* (*n* + *n* = *2n*). Diploid cells have two sets of chromosomes, one set from each parent. Remember, a chromosome is composed of two chromatids, each containing double-stranded DNA. These two chromatids are attached to each other at a point called the *centromere.* In a diploid nucleus, the chromosomes occur as *homologous chromosomes*—a pair of chromosomes in a diploid cell that contain similar (not necessarily identical) genes throughout their length. One of the chromosomes of a homologous pair was donated by the father, the other by the mother (figure 25.2).

It is necessary for organisms that reproduce sexually to form gametes having only one set of chromosomes. If gametes contained two sets of chromosomes, the zygote resulting from their union would have four sets of chromosomes. The number of chromosomes would continue to double with each new generation. However, this does not usually happen; the

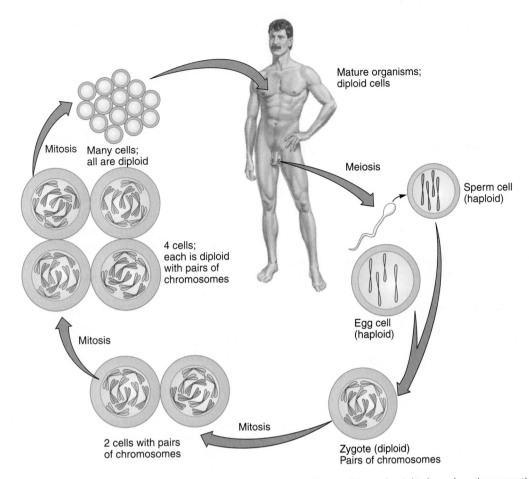

FIGURE 25.1 The cells of this adult human actually have forty-six chromosomes in their nuclei, not the eight shown here. In preparation for sexual reproduction, the number of chromosomes must be reduced by half so that fertilization will result in the original number of forty-six chromosomes in the new individual. The offspring will grow and produce new cells by mitosis, completing the life cycle.

FIGURE 25.2 Pairs of chromosomes that are of similar size and shape that have genes for the same traits are said to be homologous. Notice that members of a gene pair may not be identical but code for the same general type of information. Homologous chromosomes are of the same length, have the same types of genes in the same sequence, and have their centromeres in the same location—one came from the male parent and the other was contributed by the female parent.

FIGURE 25.3 Humans produce cells with a haploid number of chromosomes in organs called gonads, the ovaries and the testes.

number of chromosomes remains constant generation after generation. How are sperm and egg cells formed so that they get only half the chromosomes of the diploid cell? The answer lies in the process of **meiosis,** the specialized pair of cell divisions that reduce the chromosome number from diploid ($2n$) to haploid (n). One of the major functions of meiosis is to produce cells that have one set of genetic information. Therefore, when fertilization occurs between a haploid sperm (n) and a haploid egg (n), the resulting zygote will have two sets of chromosomes ($n + n = 2n$), as did each parent.

Not every cell in the body goes through the process of meiosis. Only specialized organs are capable of producing haploid cells (figure 25.3). In humans, the organs in which meiosis occurs are called **gonads.** The female gonads that produce eggs are called **ovaries.** The male gonads that produce sperm are called **testes.**

Connections . . .

Why Am I Different?

With the possible exception of identical twins, every human who has ever been born is genetically unique. The formation of haploid cells by meiosis and the combination of sex cells from two parents to form a diploid offspring by sexual reproduction result in genetic diversity in the offspring. The five factors that influence genetic diversity in offspring are mutations, crossing-over, segregation and independent assortment, and fertilization.

Mutations

Two types of mutations will be discussed in chapter 26: point mutations and chromosomal aberrations. In point mutations, a change in a DNA nucleotide results in the production of a different protein. In chromosomal aberrations, genes are rearranged when parts of chromosomes are duplicated, deleted, or inverted. By causing the production of different proteins, both types of mutations increase diversity.

Box Figure 25.1

Crossing-Over

When crossing-over occurs during gametogenesis in your ovaries or testes, pieces of genetic material are exchanged between the chromosomes you received from your mother and father. This exchange results in a chromosome with a new combination of genes, *combinations* of traits that were probably never before found in the family.

Segregation and Independent Assortment

The process of segregation relates to how the chromosomes donated by your father line up or assort themselves in pairs with those donated by your mother. How one pair of chromosomes lines up during metaphase I does not affect how the other pair lines up; i.e., they assort themselves independently of the others (box figure 25.1). If all those donated by your father line up on the right side of the cell during metaphase I of meiosis and all those donated by your mother line up on the left, when the chromosomes separate or *segregate,* the resulting gametes will be of two kinds: half of your gametes will contain only your father's chromosomes and the other half of the gametes you form will contain only your mother's chromosomes. If however, some of the chromosomes from your father line up on the right side and some of your mother's line up on the left, the gametes you form will contain some of your mother's and some of your father's chromosomes. There are millions of ways (2^{23}) in which these chromosomes can line up and divide.

Fertilization

Because of the large number of possible gametes resulting from independent assortment, segregation, mutations, and crossing-over, an incredibly large number of types of offspring can result. Because human males and females can produce millions of genetically unique sperm and eggs, the number of kinds of offspring possible is infinite.

THE MECHANICS OF MEIOSIS

The whole point of meiosis is to distribute the chromosomes and the genes they carry so that each daughter cell gets one member of each homologous pair. In this way, each daughter cell gets one complete set of genetic information. Meiosis consists of two divisions known as meiosis I and meiosis II. During meiosis I, the number of chromosomes is reduced from the diploid number to the haploid number. Therefore, it is often referred to as a *reduction division.* The sequence of events in meiosis I is artificially divided into four phases: prophase I, metaphase I, anaphase I, and telophase I. Meiosis II includes four phases: prophase II, metaphase II, anaphase II, and telophase II. The two daughter cells formed during meiosis I continue through meiosis II, so that four cells usually result from the two divisions. Table 25.1 is a review of the stages of meiosis as they would occur in a cell with diploid number of chromosomes equal to 8, (i.e., $n = 4$).

CONCEPTS APPLIED

Meiosis Models

Models can be very useful in helping us to understand complex biological events such as meiosis. You can create model chromosomes very easily by using various lengths of colored strings, threads, or yarns to simulate the twenty-three pairs of homologous chromosomes in a human cell. Each homologous pair should be different from the other pairs, either in color, length, or both. Begin your modeling with each chromosome in its replicated form (i.e., two chromatids per chromosome). Attach the two chromatids with a loose twist. Manipulate your twenty-three pairs of model chromosomes through the stages of meiosis I and II. If you have performed the actions properly, you should end up with four cells, each haploid ($n = 23$).

TABLE 25.1   A Review of the Stages of Meiosis

Stage		Description
Prophase I		During prophase I, the cell is preparing itself for division. The chromosomes are composed of two chromatids and coil so that they become shorter and thicker. Homologous chromosomes pair with each other. This pairing is called *synapsis*. While homologous chromosomes are paired, they can exchange equivalent sections of DNA. This process is called *crossing-over* and is important in mixing up the genes that will be divided into separate gametes. Long fibers known as *spindle fibers* become attached to the chromosomes.
Metaphase I		The synapsed homologous chromosomes move to the equator of the cell as single units. How they are arranged on the equator (which one is on the left and which one is on the right) is determined by chance. In the drawing, three green chromosomes from the father and one purple chromosome from the mother are lined up on the left. Similarly, one green chromosome from the father and three purple chromosomes from the mother are on the right. They could have aligned themselves in several other ways.
Anaphase I		During this stage, homologous chromosomes separate and their number is reduced from diploid to haploid. The two members of each pair of homologous chromosomes move away from each other toward opposite poles. The direction each takes is determined by how each pair was originally arranged on the spindle. Each chromosome is independently attached to a spindle fiber at its centromere. Each chromosome still consists of two chromatids. Because the homologous chromosomes and the genes they carry are being separated from one another, this process is called *segregation*. The way in which a single pair of homologous chromosomes segregates does not influence how other pairs of homologous chromosomes segregate; that is, each pair segregates independently of other pairs. This is known as *independent assortment* of chromosomes.
Telophase I		The two newly forming daughter cells are now haploid (*n*) since each contains only one of each pair of homologous chromosomes.
Prophase II		The chromosomes of each of the two haploid daughter cells shorten and become attached to spindle fibers by their centromeres.
Metaphase II		Chromosomes move to the equator of the cell.
Anaphase II		Centromeres divide, allowing the chromatids to separate toward the poles.
Telophase II		Four haploid (*n*) cells are formed from the division of the two meiosis I cells. Each of these sex cells is genetically different from all the others. These cells become the sex cells (egg or sperm) of a higher organism.

HUMAN SEXUALITY FROM DIFFERENT POINTS OF VIEW

Probably nothing interests people more than sex and sexuality. **Sex** is the nature of the biological differences between males and females. By **sexuality,** we mean all the factors that contribute to one's female or male nature. These include the structure and function of the sex organs, the behaviors that involve these structures, psychological components, and the role culture plays in manipulating our sexual behavior. Males and females have different behavior patterns for a variety of reasons. Some behavioral differences are learned (patterns of dress, use of facial makeup), whereas others appear to be less dependent on culture (degree of aggressiveness, frequency of

Speculation on the Evolution of Human Sexual Behavior

Much speculation has occurred about how human sexual behavior evolved. It is important to recognize that this speculation is not fact but an attempt to evaluate human sexual behaviors from an evolutionary perspective.

When we compare human sexuality with that of other mammals, human sexuality differs in several ways. Whereas most mammals are sexually active during specific periods of the year, humans may engage in sexual intercourse at any time throughout the year. The sex act appears to be important as a pleasurable activity rather than a purely reproductive act. Associated with this difference is the fact that human females do not display changes that indicate they are releasing eggs (ovulating).

All other female mammals display changes in odor, appearance, or behavior that clearly indicate to the males of the species that the female is ovulating and sexually responsive. This is referred to as "being in heat." This is not true for humans. Human males are unable to differentiate ovulating females from those who are not ovulating.

In other mammals with few exceptions, infants grow to sexual maturity in a year or less. Although extremely long-lived mammals (elephants or whales) do not reach sexual maturity in a year, their young have well-developed muscles that allow them to move about with a high degree of independence. While the young of these species still rely on their mothers for milk and protection, they are capable of obtaining other food for themselves as well. This is not true for human infants, who are extremely immature when born and develop walking skills slowly. They also require several years of training before they are able to function independently (box figure 25.2).

Perhaps the extremely immature condition in which human infants are born is related to human brain size. The size of the head is very large and just fits through the birth canal in the mother's pelvis. One way to accommodate a large brain size and not need to redesign the basic anatomy of the female pelvis would be to have the young be born in a very immature condition while the brain is still small and growing. Having the young born in an immature condition can solve one problem but creates another. The immature condition of human infants is associated with a need to provide extensive care for them.

Box Figure 25.2

Raising young requires a considerable investment of time and resources. Females invest significant resources in the pregnancy itself. Fat stores provide energy necessary to a successful pregnancy. Female mammals, including humans, that have little stored fat often have difficulty becoming pregnant in the first place and are more likely to die of complications resulting from the pregnancy. Nutritional counseling is an important part of modern prenatal care because it protects the health of both mother and developing fetus. The long duration of pregnancy in humans requires good nutrition over an extended period. Once the child is born, the mother continues to require good nutrition because she provides the majority of food for the infant through her breast milk. As the child grows, other food items are added to its diet. Since the young child is unable to find and prepare its own food, the mother or father or both must expend energy to feed the child.

With all these ideas in mind, we can speculate about how human sexual behavior may have evolved. Imagine a primitive Stone Age human culture. Females have a great deal invested in each child produced. They will only be able to produce a few children during their lifetimes, and many children will die because of malnutrition, disease, and accidents. Those females who have genes that will allow more of their offspring to survive will be selected for. Human males, on the other hand, have very little invested in each child and can impregnate many different females. Males who have many children who survive are selected for. How might these different male and female goals fit together to provide insight into the sexual behaviors we see in humans today?

The males of most mammals contribute little toward the raising of young. In many species, males meet with females only for mating (deer, cats, rabbits, mice). In some species, the male and female form short-term pair-bonds for one season and the males share the burden of raising the young (foxes). Only a few (wolves) form pair-bonds lasting for years in which males and females cooperate in the raising of young. However, pair bonding in humans is usually a long-term relationship. The significance of this relationship can be evaluated from an evolutionary perspective. This long-term pair-bond can serve the interests of both males and females. When males form long-term relationships with females, the females gain an additional source of nutrition for their offspring, who will be completely dependent on their parents for food and protection for several years, thus increasing the likelihood that the young will survive. Human males benefit from the long-term pair-bond as well. Because human females do not display the fact that they are ovulating, the only way a male can be assured that the child he is raising is his is to have exclusive mating rights with a specific female. The establishment of bonding involves a great deal of sexual activity, much more than is necessary to just bring about reproduction. It is interesting to speculate that sexual behavior in humans is as much involved in maintaining pair-bonds as it is in creating new humans.

Much has been written about the differences in sexual behavior between men and women and that men and women look for different things when assessing individuals as potential mates. It is very difficult to distinguish behaviors that are truly biologically determined and those that are culturally determined. However, some differences may have biological roots. Females benefit from bonding with males who have access to resources that are shared in the raising of young. Do women look for financial security and a willingness to share in a mate? Because pregnancy and nursing young require a great deal of nutrition, it is in the male's interest to choose a mate who is healthy, young, and in good nutritional condition. Since the breasts and buttocks are places of fat storage in women, do men look for youth and appropriate amounts of nutritionally important fat stored in the breasts and buttocks? If these differences between men and women really exist, are they purely cultural, or is there an evolutionary input from our primitive ancestors?

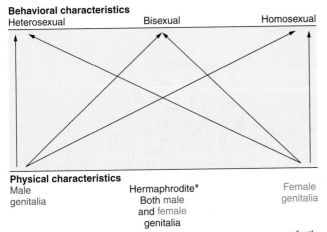

Behavioral characteristics

Heterosexual   Bisexual   Homosexual

Physical characteristics

Male genitalia   Hermaphrodite*   Female genitalia
Both male and female genitalia

*Humans with both male and female sex characteristics are rare; most are not fertile, but that does not rule out sexual interest or potential for enjoyment of sexual activity.

FIGURE 25.4 This figure shows how behavioral and physical sexual characteristics interrelate. At the ends of the behavioral spectrum, individuals can be strongly heterosexual and homosexual. They might also find that their physical traits do not match this behavior, but they might be attracted to both sexes or to a person of the opposite sex.

sexual thoughts). We have an intense interest in the facts about our own sexual nature and the sexual behavior of members of the opposite sex and that of peoples of other cultures.

There are several different ways to look at human sexuality. The behavioral sciences tend to focus on the behaviors associated with being male and female and what is considered appropriate and inappropriate sexual behavior. Sex is considered a strong drive, appetite, or urge by psychologists. They describe the sex drive as a basic impulse to satisfy a biological, social, or psychological need. Other social scientists (sociologists, cultural anthropologists) are interested in sexual behavior as it occurs in different cultures and subcultures. When a variety of cultures is examined, it becomes very difficult to classify various kinds of sexual behavior as normal or abnormal. What is considered abnormal in one culture may be normal in another. For example, public nudity is considered abnormal in many cultures but not in others.

Biologists have studied the sexual behavior of nonhuman animals for centuries. They have long considered the function of sex and sexuality in light of its value to the population or species. Sexual reproduction results in new combinations of genes that are important in the process of natural selection. Many biologists today are attempting to look at human sexual behavior from an evolutionary perspective and speculate on why certain sexual behaviors are common in humans (see "A Closer Look: Speculation on the Evolution of Human Sexual Behavior" in this chapter). The behaviors of courtship, mating, rearing of the young, and the division of labor between the sexes are complex in all social animals, including humans. These are demonstrated in the elaborate social behaviors surrounding mate selection and the establishment of families. It is difficult to draw the line between the biological development of sexuality and the social establishment of customs related to the sexual aspects of human life. However, the biological mechanism that determines whether an individual will develop into a female or male has been well documented.

Sexuality ranges from strongly heterosexual to strongly homosexual. Human sexual orientation is a complex trait, and evidence suggests that there is no one gene that determines where a person falls on the sexuality spectrum. It is most likely a combination of various genes acting together and interacting with environmental factors (figure 25.4) The primary biological goal of sexual intercourse (coitus, mating) is the union of sperm and egg to form offspring. However, this interaction can also improve the health of and bring pleasure to willing partners.

CHROMOSOMAL DETERMINATION OF SEX AND EARLY DEVELOPMENT

Two of the forty-six chromosomes in a human adult are the **sex-determining chromosomes.** The other twenty-two pairs that do not determine the sex of an individual are called **autosomes.** There are two kinds of sex-determining chromosomes: the **X chromosome** and the **Y chromosome** (see box figure 25.3 in "A Closer Look: Karyotyping and Down Syndrome"). The two sex-determining chromosomes, X and Y, do not carry equivalent amounts of information, nor do they have equal functions. The X chromosome carries typical genetic information about the production of specific proteins in addition to its function in determining sex. For example, the X chromosome carries information on blood clotting, color vision, and many other characteristics. The Y chromosome, however, appears to be primarily concerned with determining male sexual differentiation and has few other genes on it.

When a human sperm cell is produced, it carries twenty-two autosomes and a sex-determining chromosome. Unlike eggs, which always carry an X chromosome, half the sperm cells carry an X chromosome and the other half carry a Y chromosome. If an X-carrying sperm cell fertilizes an X-containing egg cell, the resultant embryo will develop into a female. If a Y-carrying sperm cell fertilizes the egg, a male embryo develops. It is the presence or absence of the *SRY* (**s**ex-determining **r**egion **Y**) gene typically located on the short arm of the Y chromosome that determines the sex of the developing individual. The SRY gene produces a chemical called the testes determining factor (TDF) and acts as a master switch that triggers the events that converts the embryo into a male. Without a functioning SRY gene, the embryo would become female.

Chromosomal Abnormalities and Sex

Evidence that the Y chromosome and SRY gene controls male development comes as a result of many kinds of studies including individuals who have an abnormal number of chromosomes. An abnormal meiotic division that results in sex cells with too many or too few chromosomes is called *nondisjunction*. If nondisjunction affects the X and Y chromosomes, a gamete might be produced that has only twenty-two chromosomes and lacks a sex-determining chromosome, or it might have twenty-four, with two sex-determining chromosomes. If a cell with too few or too many sex chromosomes is fertilized, sexual

A Closer Look

Karyotyping and Down Syndrome

It is possible to examine cells and count chromosomes. Among the easiest cells to examine in this way are white blood cells. They are dropped onto a microscope slide so that the cells are broken open and the chromosomes are separated. Photographs are taken of chromosomes from cells in the metaphase stage of mitosis. The chromosomes in the photographs can then be cut and arranged for comparison to known samples (box figure 25.3). This picture of an individual's chromosomal makeup is referred to as that person's *karyotype.*

In the normal process of meiosis, diploid cells have their number of chromosomes reduced to haploid. This involves segregating homologous chromosomes into separate cells during the first meiotic division. Occasionally, a pair of homologous chromosomes does not segregate properly during gametogenesis and both chromosomes of a pair end up in the same gamete. This kind of division is known as *nondisjunction.* One example of the effects of nondisjunction is the condition known as Down syndrome. If an abnormal gamete with two number 21 chromosomes has been fertilized by another containing the typical one copy of chromosome number 21, the resulting zygote would have forty-seven chromosomes (e.g., twenty-four from the female plus twenty-three from the male parent) (box figure 25.4). Since the zygote divides by mitosis, the child who develops from this fertilization would have forty-seven chromosomes in

every cell of his or her body and would have the symptoms characteristic of Down syndrome. These may include thickened eyelids, some mental impairment, and faulty speech (box figure 25.5). Premature aging is probably the most significant impact of this genetic disease. On the other hand, a child born with only one chromosome 21 rarely survives.

It is known that the mother's age at childbirth plays an important part in the occurrence of *trisomies* (three of one of the kinds of chromosomes instead of the normal two), such as Down syndrome. In women, gametogenesis begins early in life, but cells destined to become eggs are put on hold during meiosis I. Beginning at puberty and ending at menopause, one of these cells completes meiosis I monthly. This means that cells released for fertilization later in life are older than those released earlier in life. Therefore, it was believed that the chances of abnormalities such as nondisjunction increase as the age of the egg increases. However, the evidence no longer supports this age-egg link. Currently, the increase in frequency of trisomies with age has been correlated

A

B

C

Box Figure 25.3

development is usually affected. If a normal egg cell is fertilized by a sperm cell with no sex chromosome, the offspring will have only one X chromosome. These people are designated as XO. They develop a collection of characteristics known as *Turner's syndrome* (figure 25.5). An individual with this condition is female, is short for her age, and fails to mature sexually, resulting in sterility. In addition, she may have a thickened neck (termed *webbing*), hearing impairment, and some abnormalities in the cardiovascular system. When the condition is diagnosed, some of the physical conditions can be modified with treatment. Treatment involves the use of growth-stimulating hormones

to increase growth rate and female sex hormones to stimulate sexual development, although sterility is not corrected.

An individual who has XXY chromosomes is basically male (figure 25.6). This genetic condition is termed *Klinefelter's syndrome* and is probably the most common chromosomal variation found in humans. The largest percentage of these men leads healthy, normal lives and is indistinguishable from normal males. However, those with Klinefelter's syndrome may be sterile and show breast enlargement, incomplete masculine body form, lack of facial hair, and have some minor learning problems. These characteristics vary greatly in degree, and

Box Figure 25.4

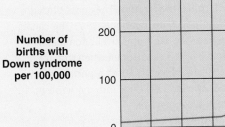

Number of births with Down syndrome per 100,000

Mother's age (years)

Box Figure 25.6

Box Figure 25.5

with a decrease in the activity of a woman's immune system. As she ages, her immune system is less likely to recognize the difference between an abnormal and a normal embryo. This means that she is more likely to carry an abnormal fetus to full term.

Box figure 25.6 illustrates the frequency of occurrence of Down syndrome at different ages in women. Notice that the frequency increases very rapidly after age thirty-seven. For this reason, many physicians encourage couples to have their children in their early to mid-twenties and not in their late thirties or early forties. Physicians normally encourage older women who are pregnant to have the cells of their fetus checked to see if they have the normal chromosome number. It is important to know that the male parent can also contribute the extra chromosome 21. However, it appears that this occurs less than 30 percent of the time.

Sometimes a portion of chromosome 14 may be cut out and joined to chromosome 21. A person with this 14/21 transfer is monosomic and has only forty-five chromosomes; one 14 and one 21 are missing and replaced by the translocated 14/21. Statistically, about 15 percent of the children of carrier mothers inherit the 14/21 chromosome and have Down syndrome. Fewer of the children born to fathers with the 14/21 translocation inherit the abnormal chromosome and are Downic.

Whenever an individual is born with a chromosomal abnormality, it is recommended that both parents have a karyotype evaluation in an attempt to identify the possible source of the problem. This is not to fix blame but to provide information on the likelihood that another pregnancy would also result in a child with a chromosomal abnormality.

many men are diagnosed with Klinefelter's syndrome only after they undergo testing to determine why they are infertile. This condition is present in about 1 in 500 to 1,000 men. Treatment may involve testosterone therapy and breast-reduction surgery in males who have significant breast development.

Because both conditions involve abnormal numbers of X or Y chromosomes, they provide strong evidence that these chromosomes are involved in determining sexual development. The early embryo resulting from fertilization and cell division is neither male nor female but becomes female or male later in development—based on the sex-determining chromosomes that control the specialization of the cells of the undeveloped, embryonic gonads into female ovaries or male testes. This specialization of embryonic cells is termed **differentiation.** The embryonic gonads begin to differentiate into testes about five to seven weeks after **conception** (fertilization) if the SRY gene is present and functioning. The Y chromosome seems to control this differentiation process in males because the gonads do not differentiate into female sex organs until later and then only if two X chromosomes are present. It is the absence of the Y chromosome that determines female sexual differentiation.

A Closer Look

Cryptorchidism—Hidden Testes

At about the seventh month of pregnancy (gestation), in normal male fetuses each testis moves from a position in the abdominal cavity to an external sac, called the scrotum. They pass through an opening called the **inguinal canal** (box figure 25.7). This canal closes off but continues to be a weak area in the abdominal wall, and it may rupture later in life. This can happen when strain (e.g., from improperly lifting heavy objects) causes a portion of the intestine to push through the inguinal canal into the scrotum, a condition known as an inguinal hernia.

Occasionally, the testes do not descend, resulting in a condition known as **cryptorchidism** (*crypt*=hidden; *orchidos*= testes). Sometimes, the descent occurs during puberty; if not, there is a twenty-five to fifty times increased risk for testicular cancer. Because of this increased risk, surgery can be done to allow the undescended testes to be moved into the scrotum. Sterility will result if the testes remain in the abdomen. This happens because normal sperm cell development cannot occur in a very warm environment. The temperature in the abdomen is higher than the temperature in the scrotum. Normally, the temperature of the testes is very carefully regulated by muscles that control their distance from the body. Physicians have even diagnosed cases of male infertility as being caused by tight-fitting pants that hold the testes so close to the body that the temperature increase interferes with normal sperm development. Recent evidence has also suggested that teenage boys and young men working with computers in the laptop position for extended periods may also be at risk for lowered sperm counts.

Box Figure 25.6

FIGURE 25.5 Turner's syndrome individuals have forty-five chromosomes. They have only one of the sex chromosomes and it is an X chromosome. Individuals with this condition are females, have delayed growth, and fail to develop sexually. This woman is less than 150 cm (5 ft) tall and lacks typical secondary sexual development for her age. She also has a "webbed neck" that is common among Turner's syndrome individuals.

FIGURE 25.6 Klinefelter's syndrome individuals have two X chromosomes and a Y chromosome; they are male, are sterile, and often show some degree of breast development and female body form. They are typically tall. The two photos show a Klinefelter's individual before and after receiving testosterone hormone therapy.

Fetal Development and Sex

Development of embryonic gonads begins very early during fetal growth. First, a group of cells begins to differentiate into primitive gonads at about week 5 (figure 25.7). By week 5 or 7 if a Y chromosome is present, a gene product (TDF) from the chromosome will begin the differentiation of these gonads into testes; they will develop into ovaries beginning about week 12 if two X chromosomes are present (Y chromosome is absent).

As soon as the gonad has differentiated into an embryonic testis at about week 8, it begins to produce testosterone. The presence of testosterone results in the differentiation of male sexual anatomy, and the absence of testosterone results in the differentiation into female sexual anatomy.

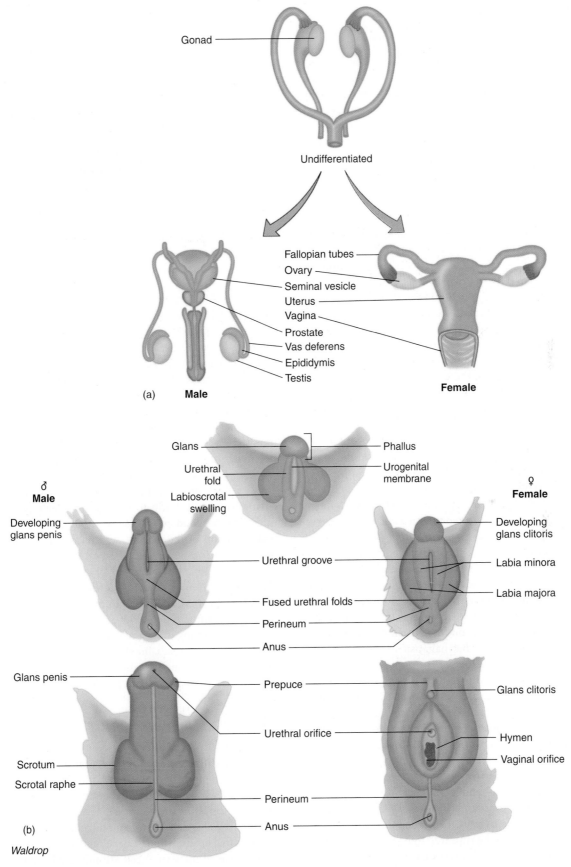

Gonad

Undifferentiated

Fallopian tubes
Ovary
Seminal vesicle
Uterus
Vagina
Prostate
Vas deferens
Epididymis
Testis

(a) **Male** **Female**

Glans
Urethral fold
Labioscrotal swelling

Phallus
Urogenital membrane

♂
Male

♀
Female

Developing glans penis

Urethral groove

Fused urethral folds

Perineum

Anus

Developing glans clitoris

Labia minora

Labia majora

Glans penis

Prepuce

Scrotum
Scrotal raphe

Urethral orifice

Glans clitoris

Hymen

Vaginal orifice

Perineum

Anus

(b)

Waldrop

FIGURE 25.7 The early embryo grows without showing any sexual characteristics. The male and female sexual organs eventually develop from common basic structures. (*A*) shows the development of the internal anatomy. (*B*) shows the development of external anatomy.

SEXUAL MATURATION OF YOUNG ADULTS

Following birth, sexuality plays only a small part in physical development for several years. Culture and environment shape the responses that the individual will come to recognize as normal behavior. During **puberty** (the developmental period when the body changes and becomes able to reproduce), normally between eleven and fourteen years of age, increased production of sex hormones causes major changes as the individual reaches sexual maturity. Generally, females reach puberty six months to two years before males. After puberty, humans are sexually mature and have the capacity to produce offspring.

The Maturation of Females

Female children typically begin to produce quantities of sex hormones from the hypothalamus portion of the brain and the pituitary gland, ovaries, and adrenal glands at nine to twelve years of age. This marks the onset of puberty. The **hypothalamus** controls the functioning of many other glands throughout the body, including the **pituitary gland.** At puberty, the hypothalamus begins to release a hormone known as **gonadotropin-releasing hormone (GnRH),** which stimulates the pituitary to release luteinizing hormone (LH) and **follicle-stimulating hormone (FSH).** Increased levels of FSH stimulate the development of **follicles,** saclike structures that produce oocytes in the ovary, and the increased luteinizing hormone stimulates the ovary to produce larger quantities of **estrogens.** The increasing supply of estrogen is responsible for the many changes in sexual development that can be noted at this time. These changes include breast growth, changes in the walls of the uterus and vagina, increased blood supply to the clitoris, and changes in the pelvic bone structure. The **clitoris** is a small, elongated erectile structure located between and at the head of the labia; it is equivalent to the penis.

Estrogen also stimulates the female adrenal gland to produce **androgens,** male sex hormones. The androgens are responsible for the production of pubic hair, and they seem to have an influence on the female sex drive. Those features that are not primarily involved in sexual reproduction but are characteristic of a sex are called **secondary sexual characteristics.** In women, the distribution of body hair, patterns of fat deposits, and a higher voice are examples.

A major development during this time is the establishment of the **menstrual cycle.** This involves the periodic growth and shedding of the lining of the uterus. These changes are under the control of a number of hormones produced by the pituitary and ovaries. The ovaries are stimulated to release their hormones by the pituitary gland, which is in turn influenced by the ovarian hormones. Both follicle-stimulating hormone (FSH) and luteinizing hormone (LH) are produced by the pituitary gland. FSH causes the maturation and development of the ovaries. LH is important in causing ovulation and converting the ruptured follicle into a structure known as the *corpus luteum*. The corpus luteum produces the hormone **progesterone,** which is important in maintaining

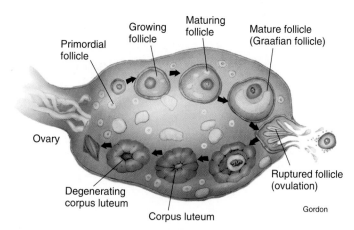

FIGURE 25.8 In the ovary, the egg begins development inside a sac of cells known as a follicle. Each month, one of these follicles develops and releases its product. This release through the wall of the ovary is known as ovulation.

the lining of the uterus. Changes in the levels of progesterone result in a periodic buildup and shedding of the lining of the uterus known as the menstrual cycle. Table 25.2 summarizes the activities of these various hormones. Associated with the menstrual cycle is the periodic release of sex cells from the surface of the ovary, called **ovulation** (figure 25.8). Initially, these two cycles, menstruation and ovulation, may be irregular, which is normal during puberty. Eventually, hormone production becomes regulated so that ovulation and menstruation take place on a regular monthly basis in most women, although normal cycles may vary from twenty-one to forty-five days.

As girls progress through puberty, curiosity about the changing female body form and new feelings lead to self-investigation. Studies have shown that sexual activity such as manipulation of the clitoris, which causes pleasurable sensations, is performed by a large percentage of young women. Self-stimulation, frequently to orgasm, is a common result. This stimulation is termed *masturbation,* and it should be stressed that it is considered a normal part of sexual development. **Orgasm** is the peak of sexual arousal. During orgasm, whether occurring during masturbation or sexual intercourse, the heart rate increases, breathing quickens, and blood pressure rises; muscles throughout the body spasm but mainly those in the vagina, uterus, anus, and pelvic floor.

The Maturation of Males

Males typically reach puberty about two years later (ages eleven to fourteen) than females. Puberty in males also begins with a change in hormone levels. At puberty, the hypothalamus releases increased amounts of gonadotropin-releasing hormone (GnRH), resulting in increased levels of follicle-stimulating hormone (FSH) and luteinizing hormone. These are the same changes that occur in female development. Luteinizing hormone is often called interstitial cell-stimulating hormone (ICSH) in males. LH stimulates the testes to produce **testosterone,** the

TABLE 25.2   Human Reproductive Hormones

Hormone	Production Site	Target Organ	Function
Gonadotropin-releasing hormone (GnRH)	Hypothalamus	Anterior pituitary	Stimulates the release of FSH and LH from anterior pituitary
Luteinizing hormone (LH) or interstitial cell-stimulating hormone (ICSH)	Anterior pituitary	Ovary, testes	Stimulates ovulation in females and sex-hormone (estrogens and testosterone) production in both males and females
Follicle-stimulating hormone (FSH)	Anterior pituitary	Ovary, testes	Stimulates ovary and testis development; stimulates egg production in females and sperm production in males
Estrogens	Ovary	Entire body	Stimulates development of female reproductive tract and secondary sexual characteristics
Testosterone	Testes	Entire body	Stimulates development of male reproductive tract and secondary sexual characteristics
Progesterone	Corpus luteum of ovary	Uterus, breasts	Causes uterine thickening and maturation; maintains pregnancy; contributes to milk production
Androgens	Testes, adrenal glands	Entire body	Stimulates development of male reproductive tract and secondary sexual characteristics in males and females
Oxytocin	Posterior pituitary	Uterus, breasts	Causes uterus to contract and breasts to release milk
Prolactin, lactogenic, or luteotropic hormone	Anterior pituitary	Breasts, ovary	Stimulates milk production; also helps maintain normal ovarian cycle
Human chorionic gonadotropin	Placenta	Corpus luteum	Maintains corpus luteum so that it continues to secrete progesterone and maintain pregnancy

primary sex hormone in males. The testosterone produced by the embryonic testes caused the differentiation of internal and external genital anatomy in the male embryo. At puberty, the increased amount of testosterone is responsible for sexual maturation and the development of male secondary sexual characteristics, and is important in the maturation and production of sperm.

The major changes during puberty include growth of the testes and scrotum, pubic-hair development, and increased size of the penis. Secondary sex characteristics also begin to become apparent. Facial hair, underarm hair, and chest hair are some of the most obvious. The male voice changes as the larynx (voice box) begins to change shape. Body contours also change, and a growth spurt increases height. In addition, the proportion of the body that is muscle increases and the proportion of body fat decreases. At this time, a boy's body begins to take on the characteristic adult male shape, with broader shoulders and heavier muscles.

In addition to these external changes, increased testosterone causes the production of seminal fluid by the accessory glands. FSH stimulates the production of sperm cells. The release of sperm cells and seminal fluid begins during puberty and is termed *ejaculation*. This release is generally accompanied by the pleasurable sensations of orgasm. The sensations associated with ejaculation may lead to self-stimulation, or masturbation. Masturbation is a common and normal activity as a boy goes through puberty. Studies of adult sexual behavior have shown that nearly all men masturbate at some time during their lives.

SPERMATOGENESIS

The biological reason for sexual activity is the production of offspring. The process of producing gametes includes meiosis and is called gametogenesis (gamete formation). The term **spermatogenesis** is used to describe gametogenesis that takes place in the testes of males. The two bean-shaped testes are composed of many small sperm-producing tubes, or **seminiferous tubules,** and collecting ducts that store sperm. These are held together by a thin covering membrane. The seminiferous tubules join together and eventually become the epididymis, a long, narrow convoluted tube in which sperm cells are stored and mature before ejaculation (figure 25.9).

Leading from the epididymis is the vas deferens, or sperm duct; this empties into the urethra, which conducts the sperm out of the body through the **penis** (figure 25.10). Before puberty, the seminiferous tubules are packed solid with diploid cells called spermatogonia. These cells, which are found just inside the tubule wall, undergo *mitosis* and produce more spermatogonia. Beginning about age eleven, some of the spermatogonia specialize and begin the process of *meiosis,* whereas others continue to divide by mitosis, ensuring a constant and continuous supply of spermatogonia. Spermatogenesis needs to occur below body temperature, which is why the testicles are in a sack, the scrotum, outside the body. Once spermatogenesis begins, the seminiferous tubules become hollow and can transport the mature sperm.

Spermatogenesis involves several steps. It begins when some of the spermatogonia in the walls of the seminiferous tubules differentiate and enlarge to become *primary spermatocytes.*

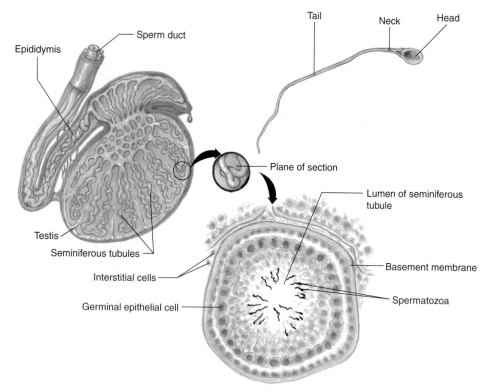

FIGURE 25.9 The testis consists of many tiny tubes called seminiferous tubules. The walls of the tubes consist of cells that continually divide, producing large numbers of sperm. The sperm leave the seminiferous tubules and enter the epididymis, where they are stored prior to ejaculation through the sperm duct. Sperm cells have a head region, with an enzyme-containing cap and the DNA. They also have a neck region, with ATP-generating mitochondria, and a tail flagellum, which propels the sperm.

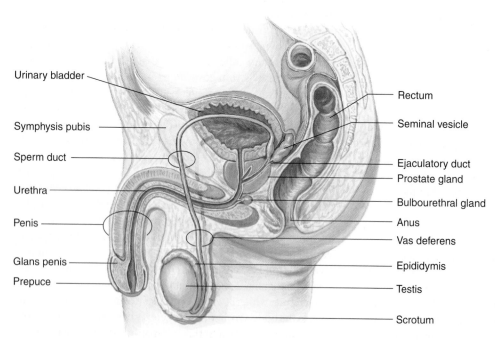

FIGURE 25.10 The male reproductive system consists of two testes that produce sperm, ducts that carry the sperm, and various glands. Muscular contractions propel the sperm through the vas deferens past the seminal vesicles, prostate gland, and bulbourethral gland, where most of the liquid of the semen is added. The semen passes through the urethra of the penis to the outside of the body.

These diploid cells undergo the first meiotic division, which produces two haploid *secondary spermatocytes.* The secondary spermatocytes go through the second meiotic division, resulting in four haploid **spermatids,** which lose much of their cytoplasm and develop long tails to mature into sperm. The sperm have only a small amount of food reserves. Therefore, once they are released and become active swimmers, they live no more than seventy-two hours. However, if the sperm are placed in a special protective solution, the temperature can be lowered drastically to −196°C. Under these conditions, the sperm become deactivated, freeze, and can survive for years outside the testes. This has led to the development of sperm banks. Artificial insemination (placing stored sperm into the reproductive tract of a female) or *in vivo* fertilization of cattle, horses, and other domesticated animals with sperm from sperm banks is common. The same techniques can be used in humans but are most often used when men produce insufficient sperm so that the sperm must be collected and stored to increase the number of sperm to enhance the likelihood of fertilization.

Semen, also known as *seminal fluid,* is a mixture of sperm and secretions from three *accessory glands.* These are the *seminal vesicles, prostate, and bulbourethral glands.* They produce secretions that nourish and activate the spermatozoa, and clear the urethral tract before ejaculation. They also serve to lubricate the tract and act as the vehicle to help propel the spermatozoa.

Seminal vesicles secrete an alkaline fluid that contains fructose and hormones. The alkaline nature of the fluid helps to neutralize the acid environment in the female reproductive tract. This improves the chances that sperm will make their way through the tract to the egg. The fructose provides energy for the sperm. The hormones stimulate contractions in the male's system that help to propel sperm. Seminal vesicle secretions make up about 60 percent of the total

Radioactive Seeds and Prostate Cancer Therapy

Radioactive isotopes of iodine-125 (I^{125}) are placed within thin metal capsules only 4.5 mm long and 0.8 mm in diameter or a rigid, absorbable suture. It only takes about forty-five minutes using a hypodermic needlelike apparatus to insert them through the skin between the scrotum and rectum into a prostate tumor. No incisions are necessary. Radiation released from these "seeds" is absorbed within the tumor and the patient is not considered to be radioactive. Depending on the size of the tumor, twenty, thirty, forty, or more seeds might be inserted. The energy released by the small amount of I^{125} acts locally and lasts only long enough to destroy the tumor cells. Patients are usually up and about in a matter of days and return to their normal activities after two to four weeks.

Myths, Mistakes, and Misunderstandings

Mumps and Sterility

Misunderstanding: Both boys and girls might become sterile if they get the mumps.

In fact, mumps is a viral infection primarily of the salivary glands. It causes them to swell, and the swelling reaches its peak in about forty-eight hours and may last two weeks. Inflammation and swelling of the testes (called orchitis) can also occur in about 15 to 25 percent of the male cases. In rare cases, the swollen testes rupture their thin covering membrane, leading to sterility. In females, inflammation of the ovaries (called oophoritis) occurs in only 5 percent of the mumps cases and does not cause sterility, since the ovaries are not bound by a membrane and can swell without rupturing.

seminal fluid. Bulbourethral gland secretions are also alkaline and help neutralize the vaginal tract making fertilization more likely. The prostate gland (note spelling) produces a thin milky fluid with a characteristic odor that makes up about 25 percent of semen and contains sperm-activating enzymes.

Prostate cancer can be a significant problem for men. Age appears to be the greatest risk factor of this disease. It has been estimated that 75 percent of men over sixty-five years of age are diagnosed with this form of cancer. This may be because prostate cancer is a slow, progressive disease. Prostate cancer is more common in African-American men in the United States than in any other group. Treatment of the disease may be watchful waiting, changes in diet, surgery, radiation therapy, hormonal therapy, or a combination of these. Surgery, if not done very carefully, can lead to problems with urination or an inability to obtain an erection. A recent study found that 30 percent of men who had their prostate removed and experienced erectile dysfunction regained their ability to have an erection after taking erectile dysfunction medications for nine months. The prostate also uses zinc when functioning properly. Low levels of dietary zinc can lead to enlargement, which may pinch the urethra, interfering with urination. Dietary supplements of zinc, selenium, and vitamin E are currently being researched as ways of combating this swelling.

Spermatogenesis in human males takes place continuously throughout a male's reproductive life, although the number of sperm produced decreases as a man ages. Sperm counts can be taken and used to determine the probability of successful fertilization. A healthy male probably releases about 150 million sperm with each ejaculation. A man must be able to release at least 100 million sperm per milliliter to be fertile. Men with sperm counts under 50 million/ml are often infertile, and those with sperm counts below 20 million/ml are clinically infertile. These vast numbers of sperm are necessary because so many die during their journey that large numbers are needed for the few survivors to reach the egg. In addition, each sperm contains enzymes in its head that are needed to digest through mucus and protein found in the female reproductive tract. Millions of sperm contribute in this way to the process of fertilization, but only one is involved in fertilizing the egg.

OOGENESIS

The term **oogenesis** refers to the production of egg cells. This process starts before a girl is born during prenatal development of the ovary. It occurs when diploid oogonia cease dividing by *mitosis* and enlarge to become *primary oocytes.* All of the primary oocytes that a woman will ever have are already formed before her birth. At this time, they number approximately two million, but that number is reduced by cell death to between three-hundred and four-hundred thousand cells by the time of puberty. Primary oocytes begin to undergo meiosis and pause in prophase I. Oogenesis pauses at this point, and all the primary oocytes remain just under the surface of the ovary.

At puberty and on a regular basis thereafter, the sex hormones stimulate a primary oocyte to continue its maturation process. It completes the first meiotic division of a single primary oocyte, which began before the woman's birth, about once a month. But in telophase I, the two cells that form receive unequal portions of cytoplasm. You might think of it as a lopsided division. The smaller of the two cells is called a *polar body,* and the larger haploid cell is the *secondary oocyte* (people commonly refer to this cell as an *egg,* or *ovum,* although technically it is not). The other primary oocytes remain in the ovary. Ovulation begins when the soon-to-be-released secondary oocyte, encased in a saclike structure known as a follicle, grows and moves near the surface of the ovary. When this maturation is complete, the follicle ruptures and the secondary oocyte is released. It is swept into the **oviduct** (fallopian tube) by ciliated

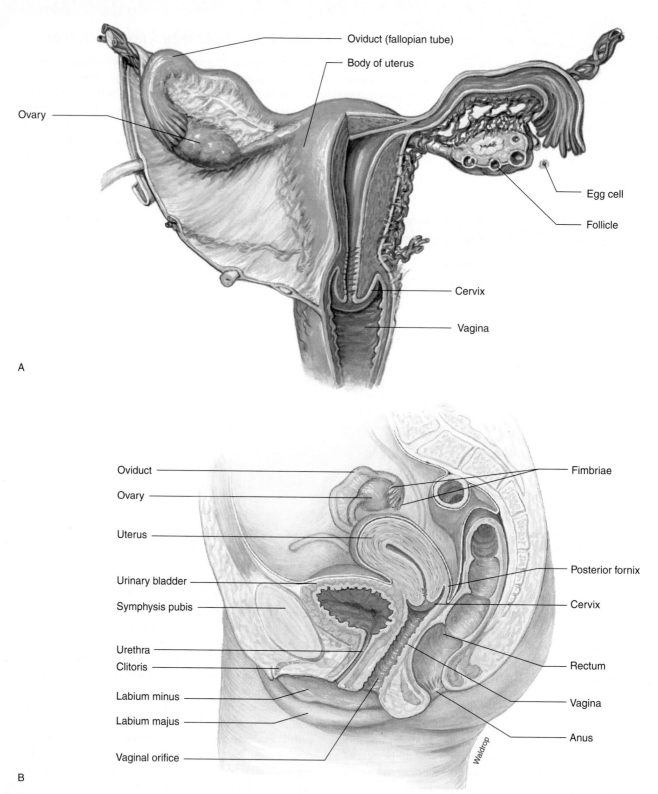

Oviduct (fallopian tube)

Body of uterus

Ovary

Egg cell

Follicle

Cervix

Vagina

A

Oviduct

Fimbriae

Ovary

Uterus

Urinary bladder

Posterior fornix

Symphysis pubis

Cervix

Urethra

Clitoris

Rectum

Labium minus

Vagina

Labium majus

Anus

Vaginal orifice

Waldrop

B

FIGURE 25.11   In the human female reproductive system (A), after ovulation the cell travels down the oviduct to the uterus. If it is not fertilized, it is shed when the uterine lining is lost during menstruation. (B) The human female reproductive system, side view.

cells and travels toward the **uterus** (figure 25.11). Because of the action of the luteinizing hormone, the follicle from which the oocyte ovulated develops into a glandlike structure, the corpus luteum, which produces hormones (progesterone and estrogen) that prevent the release of other secondary oocytes.

If the secondary oocyte is fertilized, it completes meiosis with the sperm DNA inside, and the haploid egg nucleus and sperm nucleus unite to form the zygote. If the secondary oocyte is not fertilized, it passes through the **vagina** to the outside during menstruation. During her lifetime, a female releases about three

A Link Between Menstruation and Endometriosis?

Most of the endometrium, along with its blood and capillaries, is released from the uterus when a woman has her period. It passes out through the cervix and the vagina. However, medical exams often reveal that a small amount of blood and other tissues from the uterine lining has moved up through the oviducts. If these tissues come in contact with the ovary and top of the oviduct, they may cause inflammation and the tissues may grow together. Some physicians believe that this is a likely cause of a condition called endometriosis, in which the endometrial tissue enters the body cavity through the oviduct and grows in other parts of the body. Thus, tissues that normally slide over one another grow together, forming adhesions. Because they are stuck together, the movement of these tissues may cause abdominal pain during bowel movements, sexual intercourse, or other activities. In some women, the pain is so bad that they choose not to engage in sexual intercourse. It can also cause irregular menstrual bleeding and, because the oviducts may be blocked, infertility. Endometriosis is very common in women who experience infertility.

hundred to five hundred secondary oocytes. Obviously, few of these cells are fertilized.

One distinguishing characteristics is the relative age of male and female sex cells. In males, sperm production is continuous throughout life. Sperm do not remain in the tubes of the male reproductive system for very long. They are either released shortly after they form or die and are harmlessly absorbed.

In females, meiosis begins before birth, but the oogenesis process is not completed, and the cell is not released for many years. A secondary oocyte released when a woman is thirty-seven years old began meiosis thirty-seven years before! During that time, the cell was exposed to many influences, a number of which may have damaged the DNA or interfered with the meiotic process. The increased risk of abnormal births in older mothers may be related to the age of their eggs. Such alterations are less likely to occur in males because new gametes are being produced continuously. Also, defective sperm appear to be much less likely to be involved in fertilization.

Hormonal Control of Female Sexual Cycles

Hormones control the cycle of changes in breast tissue, in the ovaries, and in the uterus. In particular, estrogen and progesterone stimulate milk production by the breasts and cause the lining of the uterus to become thicker and filled with blood vessels before ovulation. This ensures that if fertilization occurs, the resultant embryo will be able to attach itself to the wall of the uterus and receive nourishment. If the cell is not fertilized, the lining of the uterus, *endometrium,* is shed. This is known as *menstruation, menstrual flow,* the *menses,* or a *period.*

The activities of the ovulatory cycle and the menstrual cycle are coordinated. During the first part of the menstrual cycle, increased amounts of FSH cause the follicle to increase in size. Simultaneously, the follicle secretes increased amounts of estrogen that cause the lining of the uterus to increase in thickness. When ovulation occurs, the remains of the follicle are converted into a corpus luteum by the action of LH. The corpus luteum begins to secrete progesterone, and the nature of the uterine lining changes by becoming more vascularized. This is choreographed so that if an embryo arrives in the uterus shortly after ovulation, it meets with a uterine lining prepared to accept it. If pregnancy does not occur, the corpus luteum degenerates, resulting in a reduction in the amount of progesterone needed to maintain the lining of the uterus, and the lining is shed. At the same time that hormones are regulating ovulation and the menstrual cycle, changes are taking place in the breasts. The same hormones that prepare the uterus to receive the embryo also prepare the breasts to produce milk. These changes in the breasts, however, are relatively minor unless pregnancy occurs.

HORMONAL CONTROL OF FERTILITY

An understanding of how various hormones manipulate the menstrual cycle, ovulation, milk production, and sexual behavior has led to the medical use of certain hormones. Some women are unable to have children because they do not release oocytes from their ovaries or they release them at the wrong time. Physicians can now regulate the release of oocytes from the ovary using certain hormones, commonly called *fertility drugs.* These hormones can be used to stimulate the release of oocytes for capture and use in what is called *in vitro* fertilization (*IVF* or *test-tube* fertilization) or to increase the probability of natural conception; that is, *in vivo* fertilization (*in-life* fertilization).

Unfortunately, the use of these techniques often results in multiple embryos being implanted in the uterus. This is likely to occur because the drugs may cause too many secondary oocytes to be released at one time. In the case of *in vitro* fertilization, because there is a high rate of failure and the process is expensive, typically several early-stage embryos are inserted into the uterus to increase the likelihood that one will implant. If several are successful, multiple embryos implant. The implantation of multiple embryos makes it difficult for one embryo to develop properly and be carried through the entire nine-month gestation period. When we understand the action of hormones better, we may be able to control the effects of fertility drugs and eliminate the problem of multiple implantations.

A second medical use of hormones is in the control of conception by the use of birth-control pills—oral contraceptives. Birth-control pills have the opposite effect of fertility drugs. They raise the levels of estrogen and progesterone, which suppresses

TABLE 25.3 Common Causes of Infertility

Females

Endometriosis	A condition in which endometrial tissue grows outside the uterus, forming adhesions between organs
Ovulation problems	Usually a hormonal condition that prevents the release of a mature egg from an ovary
Poor egg quality	Eggs that are damaged or develop chromosomal abnormalities; usually age-related
Polycystic ovary syndrome	Ovaries that contain many small cysts causing hormonal imbalances and therefore do not ovulate regularly
Female tube blockages	Causes can include pelvic inflammatory disease (PID), sexually transmitted diseases, and previous sterilization surgery

Males

Male tube blockages	Varicose veins in the testicles are the most common cause of male tube blockages. Sexually transmitted diseases also cause blockage
Sperm problems	Low or no sperm counts, poor sperm motility (the ability to move), and abnormally shaped sperm
Sperm allergy	An immune reaction to sperm causing them to be killed by antibodies; occurs in fewer than 10 percent of infertility cases
Chronic stress	Decreased LH levels, which cause lower testosterone levels and decreased spermatogenesis

the production of FSH and LH, preventing the release of secondary oocytes from the ovary. Hormonal control of fertility is not as easy to achieve in men because there is no comparable cycle of gamete release. However, a new, reversible conception control method for males has been developed. It relies on using an implant containing the male sex hormone testosterone and a three-monthly injection of a progestin, a hormone used in female contraceptive pills. The combination of the two hormones temporarily turns off the normal signals from the brain that stimulate sperm production. An extra dose of testosterone may be given to help maintain sex drive. The use of drugs and laboratory procedures to help infertile couples have children has also raised the technical possibility of cloning in humans (Table 25.3).

FERTILIZATION, PREGNANCY, AND BIRTH

In most women, a secondary oocyte is released from the ovary about fourteen days before the next menstrual period. The menstrual cycle is usually said to begin on the first day of menstruation. Therefore, if a woman has a regular twenty-eight-day cycle, the cell is released approximately on day 14 (figure 25.12). If a woman normally has a regular twenty-one-day menstrual cycle, ovulation would occur about day 7 in the cycle. If a

woman has a regular forty-day cycle, ovulation would occur about day 26 of her menstrual cycle. Some women, however, have very irregular menstrual cycles, and it is difficult to determine just when the oocyte will be released to become available for fertilization. Once the cell is released, it is swept into the oviduct and moved toward the uterus. If sperm are present, they swarm around the secondary oocyte as it passes down the oviduct, but only one sperm penetrates the outer layer to fertilize it and cause it to complete meiosis II. The other sperm contribute enzymes that digest away the protein and mucus barrier between the egg and the successful sperm.

During this second meiotic division, the second polar body is pinched off and the *ovum* (egg) is formed. Because chromosomes from the sperm are already inside, they simply intermingle with those of the ovum, forming a diploid zygote or fertilized egg. As the zygote continues to travel down the oviduct, it begins to divide by mitosis into smaller and smaller cells without having the mass of cells increase in size (figure 25.13). This division process is called *cleavage*. Eventually, a solid ball of cells is produced, known as the morula stage of embryological development. The morula travels down the oviduct and continues to divide by mitosis. The result is called a *blastocyst*. The blastocyst becomes embedded, or implanted, when it reaches the lining of the uterus.

The next stage in the development is known as the gastrula stage because the gut is formed during this time (*gastro* = stomach). In many kinds of animals, the gastrula is formed by a complex folding of the blastocyst walls. The embryo eventually develops a tube that becomes the gut. The formation of the primitive gut is just one of a series of changes that result in an embryo that is recognizable as a miniature human being.

Most of the time during its development, the embryo is enclosed in a water-filled membrane, the amnion, which protects it from blows and keeps it moist. Two other membranes, the chorion and allantois, fuse with the lining of the uterus to form the **placenta** (figure 25.14). A fourth sac, the yolk sac, is well developed in birds, fish, amphibians, and reptiles. The yolk sac in these animals contains a large amount of food used by the developing embryo. Although a yolk sac is present in mammals, it is small and does not contain yolk. The nutritional needs of the embryo are met through the placenta. The placenta also produces the hormone chorionic gonadotropin, which stimulates the corpus luteum to continue producing progesterone and thus prevents menstruation and ovulation during gestation.

As the embryo's cells divide and grow, some of them become differentiated into nerve cells, bone cells, blood cells, or other specialized cells. To divide, grow, and differentiate, cells must receive nourishment. This is provided by the mother through the placenta, in which both fetal and maternal blood vessels are abundant, allowing for the exchange of substances between the mother and embryo. The materials diffusing across the placenta include oxygen, carbon dioxide, nutrients, and a variety of waste products. The materials entering the embryo travel through blood vessels in the umbilical cord. The diet and behavior of the mother are extremely important. Any molecules consumed by the mother can affect the embryo. Cocaine, alcohol, heroin,

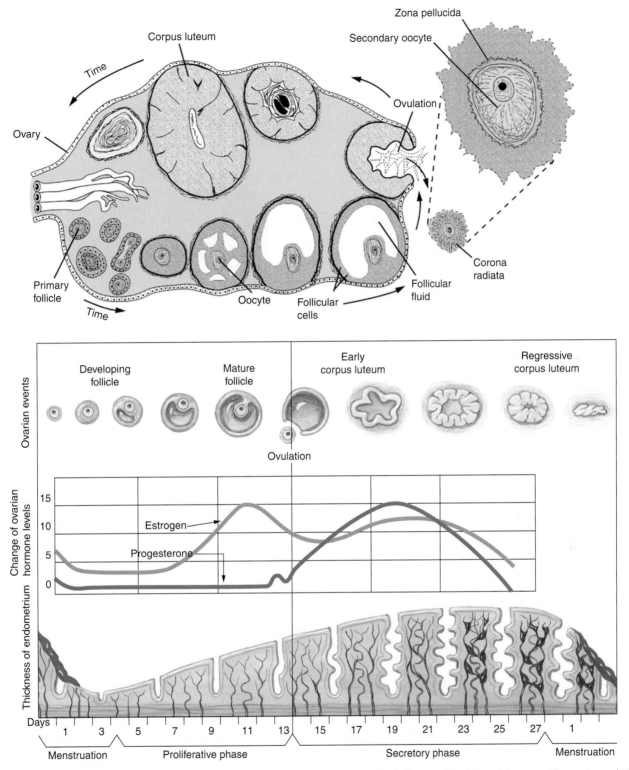

FIGURE 25.12 The release of a secondary oocyte (ovulation) is timed to coincide with the thickening of the lining of the uterus. The uterine cycle in humans involves the preparation of the uterine wall to receive the embryo if fertilization occurs. Knowing how these two cycles compare, it is possible to determine when pregnancy is most likely to occur.

and chemicals in cigarette smoke can all cross the placenta and affect the development of the embryo. The growth of the embryo results in the development of major parts of the body by the tenth week of pregnancy. After this time, the embryo continues to increase in size, and the structure of the body is refined.

Twins

In the United States, women giving birth have a 1 in 40 chance of delivering twins and a 1 in 650 chance of triplets or other multiple births. Twins happens in two ways. In the case of identical twins (approximately one-third of twins), during cleavage

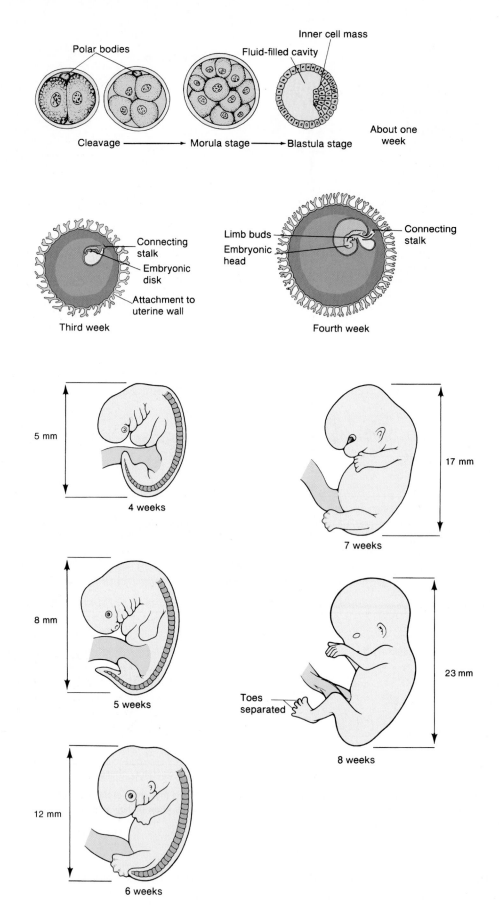

FIGURE 25.13 During the period of time between fertilization and birth, many changes take place in the embryo. Here we see some of the changes that take place during the first eight weeks.

Robert Geoffrey Edwards (1925–) and Patrick Christopher Steptoe (1913–1988)

Robert Geoffrey Edwards and Patrick Christopher Steptoe were British researchers—a physiologist and an obstetric surgeon, respectively—who devised a technique for fertilizing a human egg outside the body and transferring the fertilized embryo to the uterus of a woman. A child born following the use of this technique is popularly known as a "test-tube baby."

Robert Edwards was educated at the University of Wales and the University College of North Wales, Bangor, and from 1951 to 1957 at the University of Edinburgh. In 1957, Edwards went to the California Institute of Technology but returned to England the following year to the National Institute of Medical Research at Mill Hill. He remained there until 1962, when he took an appointment at Glasgow University. A year later, he moved again to the Department of Physiology at Cambridge University. From 1985 to 1989,

Box Figure 25.8

he was professor of human reproduction at Cambridge.

During his research in Edinburgh, Edwards successfully replanted mouse embryos into the uterus of a mouse and wondered if the same process could be used to replant a human embryo into the uterus of a woman. Edwards was able to obtain human eggs from pieces of ovarian tissue removed during surgery. He found that the ripening process was very slow, the first division beginning only after twenty-four hours.

He studied the maturation of eggs of different species of mammals and in 1965 attempted the fertilization of human eggs. He left mature eggs with sperm overnight and found just one where a sperm had passed through the outer membrane, but it had failed to fertilize the egg. In 1967, Edwards read a paper by Patrick Steptoe describing the use of a new instrument, known as the *laparoscope,* to view the internal organs, which he saw had a possible application to his own research.

Patrick Christopher Steptoe was educated at Kings's College and St. George's Hospital Medical School, London. He was appointed chief assistant obstetrician and gynecologist at St. George's Hospital, London, in 1947, and senior registrar at the Whittington Hospital, London, in 1949. From 1951 to 1978, he was senior obstetrician and gynecologist at Oldham General Hospital, and from 1969, director of the Centre for Human Reproduction.

The paper describing laparoscopy that interested Edwards dealt with Steptoe's method of exploring the interior of the abdomen without a major operation. Steptoe inserted the laparoscope through a small incision near the navel, and by means of this telescopelike instrument, with its object lens inside the body and its eyepiece outside, he was able to examine the ovaries and other internal organs.

Early in 1968, Edwards and Steptoe repeated experiments on the fertilization of human eggs. Steptoe treated volunteer patients with fertility drug to stimulate maturation of the eggs in the ovary. Edwards devised a simple piece of apparatus to be used with the laparoscope for collecting mature eggs from human ovaries. The mature eggs were removed and Edwards then prepared them for fertilization using sperm provided by the patient's husband. For a year, they continued experiments of this kind until they were sure that the fertilized eggs were developing normally. The next step was to see if an eight-celled embryo would develop.

In 1971, Edwards and Steptoe were ready to introduce an eight-celled embryo into the uterus of a volunteer patient. In 1975, an embryo did implant but in the stump of a fallopian tube, where it could not develop properly and was a danger to the mother. It was removed, but it did demonstrate that the basic technique was sound. In 1977, it was decided to abandon the use of the fertility drug and remove the egg at precisely the right natural stage of maturity, an egg was fertilized and then reimplanted (a process called *in vitro fertilization*) in the mother two days later. The patient became pregnant; twelve weeks later, the position of the baby was found to be satisfactory, and its heartbeat could be heard. During the last eight weeks of pregnancy, the mother was kept under close medical supervision, and a healthy girl—Louise Brown—was delivered by cesarean section on July 25, 1978.

In vitro fertilization has also been used to overcome the infertility in men that is due to a low sperm count. Edwards' research has further added to knowledge of the development of the human egg and young embryo, and Steptoe's laparoscope is a valuable instrument capable of wider application.

Science and Society

Multiple Births, IVF, and Hormones

The worldwide incidence of multiple pregnancies (twins, triplets, quadruplets, and more) is increasing as advanced types of infertility treatments such as in vitro fertilization (IVF) have become more ordinary. *In vitro* fertilization is a method that uses hormones to stimulate egg production; removing the oocyte from the ovary and fertilizing it with donated sperm. The fertilized egg

is incubated to stimulate cell division in a laboratory dish and then placed in the uterus.

In the United States, the number of twin births has risen more than 50 percent in the past thirty years. The incidence of higher-order multiple pregnancies (triplets or greater) has increased about one hundred times. According to the National Center for Health Statistics, multiple pregnancies in the United States has seen the greatest increase among women in their thirties and forties. This trend is due in part to the fact that older women are less able to get pregnant naturally and are more likely to undergo infertility treatment.

One of the reasons why IVF is so successful is that it usually involves implanting more than one embryo, increasing the chance of a pregnancy. IVF babies are twenty times more likely to be born as multiple birth babies; one study shows that about 45 percent of all IVF newborns are born as multiple birth babies.

The use of hormones to stimulate super-ovulation, coupled with insemination, is a more common method of infertility treatment. Super-ovulation involves using the hormone gonadotropin to stimulate ovulation of more than one egg. The more eggs released, the greater the chance that a multiple pregnancy will result. Because it is a more commonly used method, and because it is so potent, super-ovulation accounts for as many if not more multiple births than IVF.

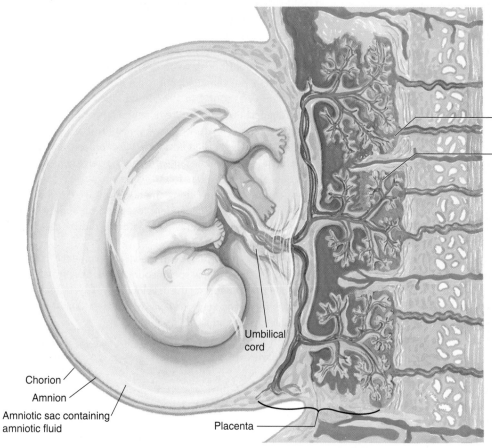

Decidua basalis

Chorion frondosum

Umbilical cord

Chorion

Amnion

Amniotic sac containing amniotic fluid

Placenta

FIGURE 25.14 The embryonic blood vessels that supply the developing embryo with nutrients and remove the metabolic wastes are separate from the blood vessels of the mother. Because of this separation, the placenta can selectively filter many types of incoming materials and microorganisms.

the embryo splits into two separate groups of cells. Each develops into an independent embryo. Because they come from the same single fertilized ovum, they have the same genes and are of the same sex. Should separation be incomplete, the twins would be born attached to one another, a condition referred to as conjoined twins. Conjoined twins occur once in every seventy thousand to one-hundred thousand live births.

Fraternal twins do not contain the same genetic information and may be of different sexes. They result from the fertilization of two separate oocytes by different sperm. Therefore, they no more resemble each other than do ordinary brothers and sisters.

Birth

The process of giving birth is also known as *parturition* or birthing. At the end of about nine months, hormone changes in the mother's body stimulate contractions of the muscles of the uterus during a period before birth called *labor*. These contractions are stimulated by the hormone oxytocin, which is released from the posterior pituitary. The contractions normally move the baby headfirst through the vagina, or birth canal. One of the first effects of these contractions may be bursting of the amnion (bag of water) surrounding the baby. Following this, the uterine contractions become stronger, and shortly thereafter, the baby is born. In some cases, the baby becomes turned in the uterus before labor. If this occurs, the feet or buttocks appear first. Such a birth is called a *breech birth*. This can be a dangerous situation because the baby's source of oxygen is being cut off as the placenta begins to separate from the mother's body.

If for any reason the baby does not begin to breathe on its own, it will not receive enough oxygen to prevent the death of nerve cells; thus, brain damage or death can result.

Occasionally, a baby may not be able to be born normally because of its position in the uterus, the location of the placenta on the uterine wall, the size of the birth canal, the number of babies in the uterus, or many other reasons. A common procedure to resolve this problem is the surgical removal of the baby through the mother's abdomen. This procedure is known as a cesarean, or C-section. The procedure was apparently named after the Roman emperor Julius Caesar, who was said to have been the first child to be delivered by this method. Currently, over 20 percent of births in the United States are by cesarean section. While C-sections are known to have been performed before Caesar, the name stuck. This rate reflects the fact that many women who are prone to problem pregnancies are having children rather than forgoing pregnancy. In addition, changes in surgical techniques have made the procedure much safer. Finally, many physicians who are faced with liability issues related to problem pregnancy may encourage cesarean section rather than normal birth.

Following the birth of the baby, the placenta, also called the *afterbirth,* is expelled. Once born, the baby begins to function on its own. The umbilical cord collapses, and the baby's lungs, kidneys, and digestive system must now support all its bodily needs. This change is quite a shock, but the baby's loud protests fill the lungs with air and stimulate breathing.

Over the next few weeks, the mother's body returns to normal, with one major exception. The breasts, which have undergone changes during the period of pregnancy, are ready to produce milk to feed the baby. Following birth, prolactin, a hormone from the pituitary gland, stimulates the production of milk, and oxytocin stimulates its release. If the baby is breastfed, the stimulus of the baby's sucking will prolong the time during which milk is produced. This response involves both the nervous and endocrine systems. The sucking stimulates nerves in the nipple and breast, which results in the release of prolactin and oxytocin from the pituitary.

In some cultures, breast-feeding continues for two to three years, and the continued production of milk often delays the reestablishment of the normal cycles of ovulation and menstruation. Many people believe that a woman cannot become pregnant while she is nursing a baby. However, because there is so much variation among women, relying on this as a natural conception-control method is not a good choice. Many women have been surprised to find themselves pregnant again a few months after delivery.

CONTRACEPTION

Throughout history, people have tried various methods of conception control (figure 25.15). In ancient times, conception control was encouraged during times of food shortage or when tribes were on the move from one area to another in search of a new home. Writings as early as 1500 B.C. indicate that the Egyptians used a form of tampon medicated with the ground powder of a shrub to prevent fertilization. This may sound primitive, but we use the same basic principle today to destroy sperm in the vagina. As you read about the various methods of contraception, remember that no method described is 100 percent effective for avoiding pregnancy and preventing sexually transmitted diseases (STDs) except abstinence.

Chemical Methods

Contraceptive jellies and foams make the environment of the vagina more acidic, which diminishes the sperm's chances of survival. The spermicidal (sperm-killing) foam or jelly is placed in the vagina before intercourse. When the sperm make contact with the acidic environment, they stop swimming and soon die. Aerosol foams are an effective method of conception control, but interfering with the hormonal regulation of ovulation is more effective. Contraceptive foams and jellies provide very little protection from sexually transmitted diseases.

Hormonal Control Methods

The first successful method of hormonal control was "the pill," or "birth-control pill." However, the way the pill works is by preventing ovulation and secondarily by interfering with implantation. The quantity and balance of hormones (estrogen and progesterone) in the pill serve to fool the ovary into functioning as if the woman was already pregnant. Therefore, conception is highly unlikely. The emergency contraceptive pill (ECP), Plan B®, or "morning-after pill," uses high doses

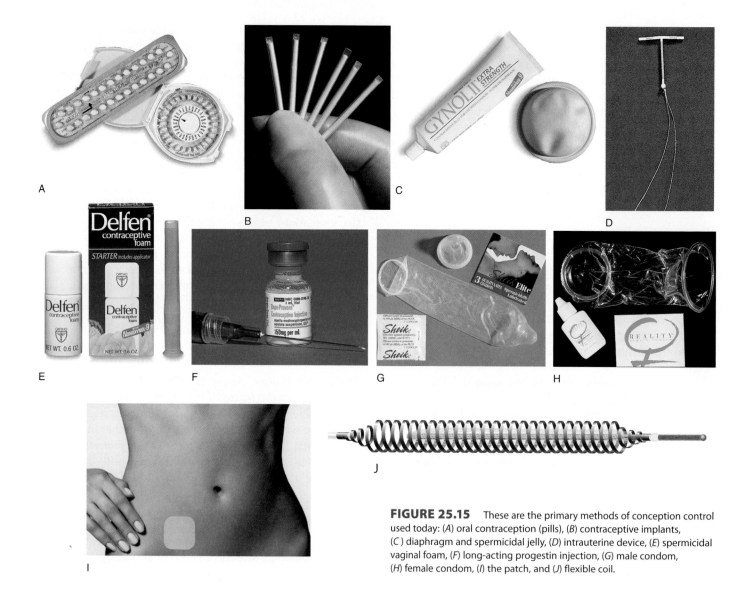

FIGURE 25.15 These are the primary methods of conception control used today: (A) oral contraception (pills), (B) contraceptive implants, (C) diaphragm and spermicidal jelly, (D) intrauterine device, (E) spermicidal vaginal foam, (F) long-acting progestin injection, (G) male condom, (H) female condom, (I) the patch, and (J) flexible coil.

of the same hormones found in oral contraceptives, which prevents the woman from becoming pregnant in the first place. In fact, "the pill" in higher dosages can be used as an ECP. ECPs prevent conception in several ways: (1) delay or inhibit ovulation; (2) inhibit transport of the egg or sperm; (3) prevent fertilization; (4) inhibit implantation of a fertilized egg; or (5) stimulate an autoimmune response that kills the sperm. This method is effective if used within seventy-two hours after unprotected intercourse. The failure rate for hormonal methods of control is between 1 and 2 per 100 women per year; the procedure provides no protection from sexually transmitted disease.

Another method of conception control that may again become available in the United States also involves hormones. The hormones are contained within small rods or capsules known as implants that are placed under a woman's skin. These rods, when properly implanted, slowly release hormones and prevent the maturation and release of oocytes from the follicle. The major advantage of the implant is its convenience. Once the implant has been inserted, the woman can forget

about contraceptive protection for several years. If she wants to become pregnant, the implants are removed, and her normal menstrual and ovulation cycles return over a period of weeks.

The vaginal ring is another method that utilizes hormones. The ring is inserted into the vagina and releases a continuous low dose of estrogen and progestin for twenty-one days. At the end of the twenty-one days, the ring is removed for seven days to allow the menstrual period to occur. A new ring is inserted monthly.

Contraceptive hormones can also be used to prevent menstrual periods, thereby preventing the symptoms of premenstrual syndrome and improving a woman's sex life. The birth-control pill, Seasonale, limits periods to four a year. The first continuous-use birth-control pill, Lybrel, may soon be available in the United States.

Still another method is the hormone-infused contraceptive patch. It is an adhesive square that can be attached to the abdomen, buttocks, upper arm, or upper torso. The patch works by slowly releasing through the skin a combination of estrogen and

Sexually Transmitted Diseases

Diseases currently referred to as *sexually transmitted diseases (STDs)* were formerly called *venereal diseases (VDs)*. The term *venereal* is derived from the name of the Roman goddess for love, Venus. Although these kinds of illnesses are most frequently transmitted by sexual activity, many can also be spread by other methods of direct contact, such as hypodermic needles, blood transfusions, and blood-contaminated materials. Currently, the Centers for Disease Control and Prevention (CDC) in Atlanta, Georgia, recognizes over twenty diseases as being sexually transmitted (Table 25.A).

The United States has the highest rate of sexually transmitted disease among industrially developed countries—65 million people (nearly one-fifth of the population) have incurable sexually transmitted diseases. The CDC estimates there are 15 million new cases of sexually transmitted diseases each year, nearly 4 million among teenagers. Table 25.B lists the most common STDs and estimates of the number of new cases each year. The portions of the public that are most at risk are teenagers, minorities, and women. Some of the most important STDs are described here because of their high incidence in the population and our inability to bring some of them under control. For example, there is no known cure for the HIV virus that is responsible for AIDS. There has also been a sharp rise in the number of gonorrhea cases in the United States caused by a form of the bacterium *Neisseria gonorrhoeae* that has become resistant to the drug penicillin by producing an enzyme that actually destroys the antibiotic. However, most of the infectious agents can be controlled if diagnosis occurs early and treatment programs are carefully followed by the patient.

The spread of STDs during sexual intercourse is significantly diminished by the use of condoms. Other types of sexual contact (i.e., hand, oral, anal) and congenital transmission (i.e., from the mother to the fetus during pregnancy) help maintain some of these diseases in the population at high enough levels to warrant attention by public health officials, the U.S. Public Health Service, the CDC, and state and local public health agencies. All of these agencies are involved in attempts to raise the general public health to a higher level. Their investigations have resulted in the successful control of many diseases and the identification of special problems, such as those associated with the STDs. Because the United States has an incidence rate of STDs that is fifty to one-hundred times higher than other industrially developed countries, there is still much that needs to be done.

High-risk behaviors associated with contracting STDs include sex with multiple partners and failing to use condoms. While some STDs are simply inconvenient or annoying, others severely compromise health and can result in death. As one health official stated, "We should be knowledgeable enough about our own sexuality and the STDs to answer the question, Is what I'm about to do worth dying for?"

Table 25.A

Sexually Transmitted Diseases	
Disease	**Agent**
Genital herpes	Virus
Gonorrhea	Bacterium
Syphilis	Bacterium
Acquired immunodeficiency syndrome (AIDS)	Virus
Candidiasis	Yeast
Chancroid	Bacterium
Genital warts	Virus
Gardnerella vaginalis	Bacterium
Genital *Chlamydia* infection	Bacterium
Genital cytomegalovirus infection	Virus
Genital *Mycoplasma* infection	Bacterium
Group B *Streptococcus* infection	Bacterium
Nongonococcal urethritis	Bacterium
Pelvic inflammatory disease (PID)	Bacterium
Molluscum contagiosum	Virus
Crabs	Body lice
Scabies	Mite
Trichomoniasis	Protozoan
Hepatitis B	Virus

Table 25.B

Diagnosed Cases of Sexually Transmitted Diseases United States		
Reported cases by state health departments		
	Cases 2003	Cases 2007
Syphilis	34,270	11,466
Chlamydia	877,478	1,108,374
Gonorrhea	335,104	355,991
Chancroid	54	23
HIV/AIDS	39,666	44,084

Source: Data from the Centers for Disease Control and Prevention publication, TRACKING THE HIDDEN EPIDEMICS, TRENDS IN STDS IN THE UNITED STATES 2000.

progestin that prevents ovulation. The hormones also cause the cervical mucus to thicken, creating a barrier to prevent sperm from entering the uterus. When used correctly, it is about 99 percent effective. It does not protect against reproductive tract infections such as HIV/AIDS.

Timing Method

Killing sperm and preventing ovulation or sperm production are not the only methods of preventing conception. Any method that prevents the sperm from reaching the oocyte prevents conception. One method is to avoid intercourse during those times of the month when a secondary oocyte may be present. This is known as the *rhythm method* of conception control. Although at first glance it appears to be the simplest and least expensive, determining just when a secondary oocyte is likely to be present can be very difficult. A woman with a regular twenty-eight-day menstrual cycle will typically ovulate about fourteen days before the onset of the next menstrual flow. To avoid pregnancy, couples need to abstain from intercourse a few days before and after this date. However, if a woman has an irregular menstrual cycle, there may be only a few days each month for intercourse without the possibility of pregnancy. In addition to calculating safe days based on the length of the menstrual cycle, a woman can better estimate the time of ovulation by keeping a record of changes in her body temperature and vaginal pH. Both changes are tied to the menstrual cycle and can therefore help a woman predict ovulation. In particular, at about the time of ovulation, a woman has a slight rise in body temperature—less than 1°C. Thus, one should use an extremely sensitive thermometer. A digital-readout thermometer on the market spells out the word "yes" or "no." The failure rate for the rhythm method is less than 9 per 100 women per year; the procedure provides no protection from sexually transmitted disease.

Barrier Methods

Other methods of conception control that prevent the sperm from reaching the secondary oocyte include the diaphragm, cap, sponge, and condom. The diaphragm is a specially fitted membranous shield that is inserted into the vagina before intercourse and positioned so that it covers the cervix, which contains the opening of the uterus. Because of anatomical differences among females, diaphragms must be fitted by a physician. The effectiveness of the diaphragm is increased if spermicidal foam or jelly is also used. The vaginal cap functions in a similar way. The contraceptive sponge, as the name indicates, is a small amount of absorbent material that is soaked in a spermicide. The sponge is placed within the vagina and chemically and physically prevents the sperm cells from reaching the oocyte. The contraceptive sponge is no longer available for use in the United States but is still available in many other parts of the world. The failure rate for diaphragms, cervical caps, and sponges including the use of spermicides is between 14 and 50 per 100 women per year; the methods are not an effective protection against sexually transmitted disease.

The male condom is probably the most popular contraceptive device. It is a thin sheath that is placed over the erect penis before intercourse. In addition to preventing sperm from reaching the secondary oocyte, this physical barrier also helps prevent the transmission of the microbes that cause STDs, such as syphilis, gonorrhea, and AIDS, from being passed from one person to another during sexual intercourse. The most desirable condoms are made of a thin layer of latex that does not reduce the sensitivity of the penis. Latex condoms have also been determined to be the most effective in preventing transmission of the AIDS virus. The condom is most effective if it is prelubricated with a spermicidal material such as nonoxynol-9. This lubricant also has the advantage of providing some protection against the spread of the HIV virus. The failure rate for latex condoms is 11 per 100 women per year. Except for abstinence, latex condoms are the best protection against sexually transmitted diseases.

Recently developed condoms for women are now available. One called the Femidom is a polyurethane sheath that, once inserted, lines the contours of the woman's vagina. It has an inner ring that sits over the cervix and an outer ring that lies flat against the labia. Research shows that this device protects against STDs and is as effective a contraceptive as the condom used by men. The failure rate for such barrier protection is 21 per 100 women per year; the method may give some protection against sexually transmitted diseases but is not as effective as latex condoms.

The intrauterine device (IUD) is not a physical barrier that prevents the gametes from uniting. How this mechanical device works is not completely known. It may in some way interfere with the implantation of the embryo. The IUD must be fitted and inserted into the uterus by a physician, who can also remove it if pregnancy is desired. IUDs are used successfully in many countries. Current research with new and different intrauterine implants indicates that they are able to prevent pregnancy, and one is currently available in the United States. The IUD can also be used for "emergency contraception"—in cases of unprotected sex (forced sex) or failure of a conception control method (a condom slips or breaks). The IUD must be inserted within seven days of unprotected sex. The failure rate for the IUD is less than 1 per 100 women per year; the device provides no protection from sexually transmitted disease.

Another barrier method is tubal implant sterilization. In this conception control method, a soft, flexible coil is inserted into both fallopian tubes (figure 25.15 J). After being inserted the coil stimulates a tissue growth resulting in blockage of the fallopian tube. This device is greater than 99 percent efficient and should be considered a form of sterilization. It is considered to be permanent and irreversible and does not protect against STDs.

Surgical Methods

Two contraceptive methods that require surgery are tubal ligation and vasectomy (figure 25.16). Tubal ligation is the cutting and tying off of the oviducts and can be done on an outpatient

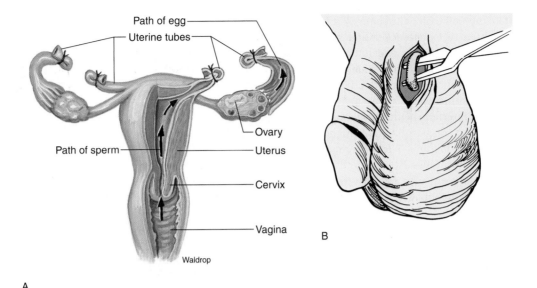

FIGURE 25.16 Two very effective contraceptive methods require surgery. Tubal ligation (*A*) involves severing the oviducts and suturing or sealing the cut ends. This prevents the sperm cell and the secondary oocyte from meeting. This procedure is generally considered ambulatory surgery, or at most requires a short hospitalization period. Vasectomy (*B*) requires minor surgery, usually in a clinic under local anesthesia. Following the procedure, minor discomfort may be experienced for several days. The severing and sealing of the vas deferens prevents the release of sperm cells from the body by ejaculation.

basis in most cases. An alternative to sterilization by tubal ligation involves the insertion of small, flexible devices called micro-inserts into each fallopian tube. Once inserted, tissue grows into the inserts, blocking the tubes. Ovulation continues as usual, but the sperm and egg cannot unite. Vasectomy is not the same as castration. Castration is the surgical removal of testes. Vasectomy can be performed in a physician's office and does not require hospitalization. A small opening is made above the scrotum, and the spermatic cord (vas deferens) is cut and tied. This prevents sperm from moving through the ducts to the outside. Because most of the sperm-carrying fluid (semen) is produced by the seminal vesicles, prostate gland, and bulbourethral glands, a vasectomy does not interfere with normal ejaculation. The sperm that are still being produced die and are reabsorbed in the testes. Neither tubal ligation nor vasectomy interferes with normal sex drives. However, these medical procedures are generally not reversible and should not be considered by those who may want to have children at a future date. The failure rate for sterilization is less than 1 per 100 women per year; the procedure provides no protection from sexually transmitted disease.

TERMINATION OF PREGNANCY—ABORTION

Another medical procedure often associated with birth control is abortion, which has been used throughout history. Abortion involves various medical procedures that cause the death and removal of the developing embryo. Abortion is obviously not a method of conception control; rather, it prevents the normal development of the embryo and causes its death. Abortion is a highly charged subject. Some people believe that abortion should be prohibited by law in all cases. Others think

that abortion should be allowed in certain situations, such as in pregnancies that endanger the mother's life or that are the result of rape or incest. Still others think that abortion should be available to any woman under any circumstances. Regardless of the moral and ethical issues that surround abortion, it is still a common method of terminating unwanted pregnancies throughout the world.

The abortion techniques used in the United States today all involve the possibility of infections, particularly if done by poorly trained personnel. The three most common techniques are scraping the inside of the uterus with special instruments (called a *D and C* or *dilation and curettage*), injecting a saline solution into the uterine cavity, or using a suction device to remove the embryo from the uterus. In the future, abortion may be accomplished by a medication prescribed by a physician. One drug, RU-486, is currently used in about 15 percent or more of the elective abortions in France. It has received approval for use in the United States. RU-486 is an antiprogestin and works by blocking progesterone receptors on cells. The medication is administered orally under the direction of a physician, and several days later, a hormone is administered. This usually results in the onset of contractions that expel the fetus. A follow-up examination of the woman is made after several weeks to ensure that there are no serious side effects of the medication.

CHANGES IN SEXUAL FUNCTION WITH AGE

Although there is a great deal of variation, somewhere around the age of fifty, a woman's hormonal balance begins to change because of changes in the production of hormones by the ovaries. At this time, the menstrual cycle becomes less regular and ovulation is often unpredictable. The changes in hormone levels cause many women to experience mood swings and physical symptoms, including cramps and hot flashes. This period when the ovaries stop producing viable secondary oocytes and the body becomes nonreproductive is known as the **menopause.** Occasionally, the physical impairment becomes so severe that it interferes with normal life and the enjoyment of sexual activity, and a physician might recommend *hormone replacement therapy (HRT)* to augment the natural production of the hormones estrogen and progestin. Normally, the sexual enjoyment of a healthy woman continues during the time of menopause and for many years thereafter.

Hormone Therapy

The U.S. Preventive Services Task Force (USPSTF) was set up by the U.S. Public Health Service to determine the merits of hormone replacement therapy. In 1996, the USPSTF recommended that physicians advise all women about the potential benefits and problems of HT, while stating that the evidence was insufficient to recommend for or against HT for all women after menopause. The task force concluded in 1996 that HT might have long-term benefits for women with osteoporosis and heart disease but that it might increase the risk for breast cancer.

Although recent evidence confirmed that HT reduces the risk for bone fractures and may reduce the likelihood of developing colorectal cancer, it has also confirmed that HT increases the risk for blood clots, breast cancer, and stroke. Most important, new evidence indicates that HT does not reduce the risk for coronary heart disease and that combined estrogen and progestin may actually increase the risk. In 2003, the USPSTF recommended against the routine use of combined estrogen and progestin for the prevention of chronic conditions in postmenopausal women because the problems are likely to outweigh the benefits for most women. So far, the evidence isn't strong enough to recommend for or against the use of estrogen alone for the prevention of chronic conditions in postmenopausal women who have had a hysterectomy.

Impotence

Human males do not experience a relatively abrupt change in their reproductive or sexual lives. The word *impotence* is used to describe problems that interfere with sexual intercourse and reproduction. These may include a lack of sexual desire, problems with ejaculation or orgasm, and erectile dysfunction. Erectile dysfunction (ED) is the recurring inability to get or keep an erection firm enough for sexual intercourse. Most incidences of ED at any age are physical, not psychological. In older men, this is usually the result of disease, injury, or a side effect of medication. Blood pressure drugs, antihistamines, antidepressants, tranquilizers, appetite suppressants, and certain ulcer drugs have been associated with ED. Damage to nerves, arteries, smooth muscles, and other tissues is the most common cause of ED. Diseases linked with ED include: diabetes, kidney disease, chronic alcoholism, multiple sclerosis, arteriosclerosis, vascular disease, and neurologic disease. Other possible causes are smoking, which reduces blood flow in veins and arteries, and lowered amounts of testosterone.

ED is frequently treated with psychotherapy, behavior modifications, oral or locally injected drugs, vacuum devices, and surgically implanted devices. Oral medications (e.g., Viagra and Levitra) work by enhancing the effects of nitric oxide (NO). NO relaxes smooth muscles in the penis during sexual stimulation and allows increased blood flow. It is increased blood in the penis that causes an erection.

ED and loss of sexual desire are not an inevitable part of aging. Rather, sexual desires tend to wane slowly as a man ages. Men also produce fewer sperm cells and less seminal fluid with age. Nevertheless, healthy individuals can experience a satisfying sex life during aging. Recent evidence indicates that men also experience hormonal and emotional changes similar to those seen as women go through menopause. Human sexual behavior is quite variable. The same is true of older persons. The whole range of responses to sexual partners continues but generally in a diminished form. People who were very active sexually when young continue to be active but less so as they reach middle age. Those who were less active tend to decrease their sexual activity also. It is reasonable to state that one's sexuality continues from before birth until death.

SUMMARY

The human sex drive is a powerful motivator for many activities in our lives. Although it provides for reproduction and improvement of the gene pool, it also has a nonbiological, sociocultural dimension. Sexuality begins before birth, as sexual anatomy is determined by the sex-determining chromosome complement that we receive at fertilization. Females receive two X chromosomes. Males receive one X and one Y *sex-determining chromosome.* It is the presence and activity of the *SRY gene* that causes male development and its absence or inactivity that allows female development.

At *puberty,* hormones influence the development of secondary sex characteristics and the functioning of gonads. As the *ovaries* and

testes begin to produce gametes, fertilization becomes possible. Sexual reproduction involves the production of gametes by meiosis in the ovaries and testes. The production and release of these gametes is controlled by the interaction of hormones. In males, each cell that undergoes *spermatogenesis* results in four sperm; in females, each cell that undergoes *oogenesis* results in one oocyte and two polar bodies. Successful sexual reproduction depends on proper hormone balance, proper meiotic division, fertilization, placenta formation, proper diet of the mother, and birth. *Hormones* regulate ovulation and menstruation and may also be used to encourage or discourage ovulation. *Fertility drugs, the patch* and *birth-control pills,* for example, involve hormonal control. A number of contraceptive methods have been developed, including the diaphragm, condom, IUD, spermicidal jellies and foams, contraceptive implants, the sponge, tubal ligation, and vasectomy.

Hormones continue to direct our sexuality throughout our lives. Even after menopause, when fertilization and pregnancy are no longer possible for a female, normal sexual activity can continue in both women and men.

KEY TERMS

androgens (p. **642**)
autosomes (p. **637**)
clitoris (p. **642**)
conception (p. **639**)
cryptorchidism (p. **640**)
differentiation (p. **639**)
diploid (p. **632**)
egg (p. **632**)
estrogens (p. **642**)
follicle (p. **642**)
follicle-stimulating hormone (FSH) (p. **642**)
gamete (p. **632**)
gametogenesis (p. **632**)

gonad (p. **633**)
gonadotropin-releasing hormone (GnRH) (p. **642**)
haploid (p. **632**)
hypothalamus (p. **642**)
inguinal canal (p. **640**)
meiosis (p. **633**)
menopause (p. **658**)
menstrual cycle (p. **642**)
oogenesis (p. **645**)
orgasm (p. **642**)
ovaries (p. **633**)
oviduct (p. **645**)
ovulation (p. **642**)

penis (p. **643**)
pituitary gland (p. **642**)
placenta (p. **648**)
progesterone (p. **642**)
puberty (p. **642**)
secondary sexual characteristics (p. **642**)
semen (p. **644**)
seminiferous tubules (p. **643**)
sex (p. **635**)
sex-determining chromosome (p. **637**)
sexual reproduction (p. **632**)
sexuality (p. **635**)

sperm (p. **632**)
spermatids (p. **644**)
spermatogenesis (p. **643**)
testes (p. **633**)
testosterone (p. **642**)
uterus (p. **646**)
vagina (p. **646**)
X chromosome (p. **637**)
Y chromosome (p. **637**)
zygote (p. **632**)

APPLYING THE CONCEPTS

1. Crossing-over between segments of homologous chromosomes results in
   a. new gene combinations.
   b. zygotes.
   c. diploid cells.
   d. segregation of genes.

2. An event unique to prophase I is
   a. segregation.
   b. synapsis.
   c. reduction division.
   d. independent assortment.

3. The fact that each homologous pair of chromosomes in humans separates and moves to the poles without being influenced by the other pairs is
   a. segregation.
   b. disintegration.
   c. independent assortment.
   d. fertilization.

4. An organism having a diploid number of 12 forms gametes having
   a. 6 chromosomes.
   b. 12 chromosomes.
   c. 18 chromosomes.
   d. 24 chromosomes.

5. Gametogenesis produces
   a. sex cells.
   b. gonads.
   c. zygotes.
   d. testes.

6. Human females typically are fertile during which point of their ovarian cycle?
   a. only after sexual intercourse
   b. immediately following their period
   c. about day 13
   d. continuously

7. Which of the following sexually transmitted diseases is caused by a virus?
   a. gonorrhea
   b. scabies
   c. candidiasis
   d. genital herpes

8. Which pituitary hormone stimulates the release of LH and FSH, stimulating the development of a follicle?
   a. estrogen
   b. gonadotropin-releasing hormone (GnRH)
   c. testosterone
   d. androgen

9. Without this, the embryo would become female.
   a. FSH gene
   b. estrogen
   c. SRY gene
   d. x-bodies

10. Which of these is not an accessory gland of males?
    a. follicle
    b. seminal vesicle
    c. bulbourethral gland
    d. prostate gland

11. Fertilization normally occurs in the
    a. uterus.
    b. ovary.
    c. oviduct.
    d. vagina.

12. Which is not a hormonal conception control method?
    a. the pill
    b. condom
    c. the patch
    d. morning-after pill

Answers

1. a **2.** b **3.** a **4.** a **5.** a **6.** c **7.** d **8.** b **9.** c **10.** a **11.** c **12.** b

QUESTIONS FOR THOUGHT

1. How do haploid cells differ from diploid cells?
2. What is unique about metaphase I?
3. How does mitosis differ from meiosis?
4. What are advantages and disadvantages of the rhythm method of conception control?
5. Diagram fertilization as it would occur between a sperm and an egg with the haploid number of 3.
6. What is HRT?
7. What structures are associated with the human female reproductive system? What are their functions?
8. What structures are associated with the human male reproductive system? What are their functions?
9. What are the differences between oogenesis and spermatogenesis in humans?
10. List the hormones associated with the functioning of the male and female reproductive systems. Describe their function.
11. What is the difference between the origin of fraternal and identical twins?
12. List three likely causes of infertility in males and three in females.

FOR FURTHER ANALYSIS

Sexual orientation is a complex concept because there is so much confusion and a lack of information among a large part of society. Some very knowledgeable professionals lead into an explanation of this topic by describing intersex conditions and the complex intersections of what many refer to as "nature and nurture." The intersex concept is one that describes one's sexual orientation as being at some point on a continuum from strongly heterosexual at one end to strongly homosexual on the other. Each person's orientation is somewhere on this continuum. Research this issue by exploring the following:

1. The biological bases of sexual orientation
2. How various cultural biases influence sexual behavior
3. The influence various religious organizations have on an individual's thinking

Lately, there has been news on circumstances surrounding surgically changing one's sex. Thinking critically about this subject, answer the following questions:

1. What genetic reasons are associated with such a decision?
2. What cultural reasons support or rebuff such a decision?
3. What would be your reaction if a relative or close personal friend were to make the decision to change their sex?

INVITATION TO INQUIRY

Sex and Society: How Do Cultures Differ?

Archaeologists, anthropologists, and biologists have studied sexuality in many cultures. They have discovered that there are many differences in attitude and behaviors. Use the Internet to search out answers to the following questions:

1. Is the maturation of males and females the same in all cultures?
2. What differences in attitude are there among cultures with regard to public nudity?
3. What differences exist among cultures with regard to homosexual behavior?
4. Do all cultures show sexual dominance of the male over the female?
5. What methods of conception control are favored in other cultures?
6. What do various religions have to say about sexual behaviors? Which behaviors are acceptable and which are not?

26

Mendelian and Molecular Genetics

The expression of many genes is influenced by the environment. The gene for dark hair in the cat is sensitive to temperature and expresses itself only in the parts of the body that stay cool.

CORE **CONCEPT**

All living things survive and reproduce only as a result of maintaining molecular genetic information that is passed from one generation to th next.

OUTLINE

Chemistry

▷ Hydrogen bonds help hold the genetic material together (Ch. 9).

▷ Covalent bonds are needed to stabilize nucleic acids and proteins (Ch. 9).

▷ Chemical reactions are responsible for DNA replication and protein synthesis (Ch. 9).

▷ Enzymes control all the reaction involved in genetic mechanisms (Ch. 20).

Life Science

▷ Mendel's laws still explain the transmission of genetic traits from one generation to another (Ch. 25).

▷ There are some simple but many more complex mechanisms used to express a genetic trait (Ch. 26).

▷ Mutations are not all bad, but many result in genetic abnormalities (Ch. 26).

▷ The central dogma explains how genetic information is used to generate physical traits (Ch. 26).

OVERVIEW

Why do you have a particular blood type or hair color? Why do some people have the same skin color as their parents, while others have a skin color different from that of their parents? Why do flowers show such a wide variety of colors? Why is it that generation after generation of plants, animals, and microbes look so much like members of their own kind? What are genetically modified organisms, and how are they produced? These questions and many others can be better answered if you have an understanding of Mendelian and molecular genetics.

PART I
MENDELIAN GENETICS UPDATED

This chapter is divided into two parts. The first presents the fundamental concepts needed to understand Mendelian genetics, and Part II covers topics related to molecular genetics.

GENETICS, MEIOSIS, AND CELLS

A **gene** is a portion of DNA that determines a characteristic. Through meiosis and reproduction, genes can be transmitted from one generation to another. The study of genes, how genes produce characteristics, and how the characteristics are inherited is the field of biology called **genetics.** The first person to systematically study inheritance and formulate laws about how characteristics are passed from one generation to the next was an Augustinian monk named Gregor Mendel (1822–1884). Mendel's work was not generally accepted until 1900, when three men, working independently, rediscovered some of the ideas that Mendel had formulated more than thirty years earlier. Because of his early work, the study of the pattern of inheritance that follows the laws formulated by Gregor Mendel is often called **Mendelian genetics.**

To understand this chapter, you need to know some basic terminology. One term that you have already encountered is *gene.* Mendel thought of a gene as a particle that could be passed from the parents to the *offspring* (children, descendants, or progeny). Today, we know that genes are actually composed of specific sequences of DNA nucleotides. The particle concept is not entirely inaccurate, because a particular gene is located at a specific place on a chromosome called its *locus (loci* = pl.; location). When we study genetics, we study how these particles or genes are passed from parents to their offspring or children.

Another important idea to remember is that all sexually reproducing organisms have a diploid ($2n$) stage. Since gametes are haploid (n) and most organisms are diploid, the conversion of diploid to haploid cells during meiosis is an important process.

$$2(n) \rightarrow \text{meiosis } (n) \rightarrow \text{gametes}$$

The diploid cells have two sets of chromosomes—one set inherited from each parent.

$$n + n \text{ gametes} \rightarrow \text{fertilization} \rightarrow 2n$$

Therefore, each individual has two chromosomes of each kind and two genes for each characteristic (refer to figure 25.1), one from the mother and one from the father. When sex cells are produced by meiosis, reduction division occurs, and the diploid number is reduced to haploid. Therefore, the sex cells produced by meiosis have only one chromosome of each of the homologous pairs that were in the diploid cell that began meiosis. Diploid organisms usually result from the fertilization of a haploid egg by a haploid sperm. Therefore, they inherit one gene of each type from each parent. For example, each of us has two genes for earlobe shape: one came with our father's sperm, the other with our mother's egg (figure 26.1).

A   **B**

FIGURE 26.1 Whether your earlobe is free (*A*) or attached (*B*) depends on the genes you have inherited. As genes express themselves, their actions affect the development of various tissues and organs. Some people's earlobes do not separate from the sides of their heads in the same manner as do those of others. How genes control this complex growth pattern and why certain genes function differently from others is yet to be clarified. How much variation do you see among fellow class members?

SINGLE-GENE INHERITANCE PATTERNS

In diploid organisms, there may be two different forms of the gene. In fact, there may be *several* alternative forms or **alleles** of each gene within a population. The word *gene* is used to refer to the genetic material in general, whereas the term *allele* more specifically refers to the alternative forms of the genetic material for a certain characteristic. In people, for example, there are two alleles for earlobe shape. One allele produces an earlobe that is fleshy and hangs free, while the other allele produces a lobe that is attached to the side of the face and does not hang free. The type of earlobe that is present is determined by the type of allele (gene) received from each parent and the way in which these alleles interact with one another. Alleles are located on the same pair of homologous chromosomes—one allele on each chromosome. These alleles are also at the same specific location, or locus (figure 26.2).

The **genome** is a set of all the genes necessary to specify an organism's complete list of characteristics. The term *genome* is used in two ways. It may refer to the diploid (*2n*) or haploid (*n*) number of chromosomes in a cell. Be sure to clarify how your instructor uses this term. The **genotype** of an organism is a listing of the genes present in that organism. It consists of the cell's DNA code; therefore, you cannot see the genotype of an organism. It is not yet possible to know the complete genotype of most organisms, but it is often possible to figure out the genes present that determine a particular characteristic. For example, there are three possible genotypic combinations of the two alleles for earlobe shape. Genotypes are typically represented by upper- and lowercase letters. In the case of the earlobe trait, the allele for free earlobes is designated "*E*"while that for attached earlobes is "*e.*"

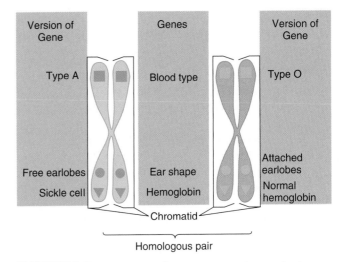

FIGURE 26.2 Homologous chromosomes contain genes for the same characteristics at the same place. Note that the attached-earlobe allele is located at the ear-shape locus on one chromosome and the free-earlobe allele is located at the ear-shape locus on the other member of the homologous pair of chromosomes. The other two genes are for hemoglobin structure (alleles for normal and sickled cells) and blood type (alleles for blood types A and O).The examples presented here are for illustrative purposes only. We do not really know if these particular genes are on these chromosomes.

A person's genotype could be (1) two alleles for attached earlobes (*ee*), (2) one allele for attached earlobes and one allele for free earlobes (*Ee*), or (3) two alleles for free earlobes (*EE*).

How would individuals with each of these three genotypes look? The observable characteristics of an organism are known as its **phenotype** and are determined by the ways in which each combination of alleles expresses (shows) itself. A gene *expresses* itself by producing product molecules (proteins) that control and direct chemical activities or become structural components of the organism. The way combinations of alleles express themselves is also influenced by the environment of the organism. The phrase *gene expression* refers to the degree to which a gene goes through protein synthesis to show itself as a physical feature of the individual.

A person with two alleles for attached earlobes will have earlobes that do not hang free. A person with one allele for attached earlobes and one allele for free earlobes will have a phenotype that exhibits free earlobes. An individual with two alleles for free earlobes will also have free earlobes. Notice that there are three genotypes but only two phenotypes. The individuals with the free-earlobe phenotype can have different genotypes.

Alleles	Genotypes	Phenotypes
E = free earlobes	*EE*	free earlobes
e = attached earlobes	*Ee*	free earlobes
	ee	attached earlobes

The expression of some alleles is directly influenced by the presence of other alleles of the same gene. For any particular pair of alleles in an individual, the two alleles from the two parents are either identical or not identical. A person is **homozygous** for a trait when they have the combination of two identical alleles for that particular characteristic, for example, *EE* and *ee*.

A Closer Look

Cystic Fibrosis—What Is the Probability?

Cystic fibrosis is among the most common lethal genetic disorders that affect Caucasians in the United States. Cystic fibrosis affects nearly thirty thousand people in North America. One in every twenty persons has a defective recessive allele that causes cystic fibrosis, but most of these individuals display no cystic fibrosis symptoms, because the recessive allele is masked by a normal dominant allele. Only those with two copies of the defective recessive gene develop symptoms. About one thousand new cystic fibrosis cases are identified in the United States each year. The gene for cystic fibrosis occurs on chromosome 7; it is responsible for the manufacture of **c**ystic **f**ibrosis **t**ransmembrane **r**egulator (CFTR) protein. The CFTR protein controls the movement of chloride ions across the cell membrane.

There are many possible types of mutations in the CFTR gene. The most common mutation results in a CFTR protein with a deletion of a single amino acid. As a result, CFTR protein is unable to control the movement of chloride ions across the cell membrane. The major result is mucus filling the bronchioles, resulting in blocked breathing and frequent respiratory infections. It is also responsible for other symptoms:

1. A malfunction of sweat glands in the skin and the secretion of excess chloride ions
2. Bile duct clogging, which interferes with digestion and liver function
3. Mucus clogging the pancreas ducts, preventing the flow of digestive enzymes into the intestinal tract
4. Bowel obstructions caused by thickened stools
5. Sterility in males due to the absence of vas deferens development and, on occasion, female sterility due to dense mucus blocking sperm from reaching eggs

Consider the facts about the frequency of the cystic fibrosis gene in the population. What is the probability that any set of parents will have a child with cystic fibrosis? What is the probability that a person who carries the cystic fibrosis allele will marry someone who also has the allele?

A person with two alleles for freckles is said to be homozygous for that trait. A person with two alleles for no freckles is also homozygous. If an organism is homozygous, the characteristic expresses itself in a specific manner. A person homozygous for free earlobes has free earlobes, and a person homozygous for attached earlobes has attached earlobes.

An individual is designated as **heterozygous** when they have two different allelic forms of a particular gene, for example, *Ee*. The heterozygous individual received one allelic form of the gene from one parent and a different allele from the other parent. For instance, a person with one allele for freckles and one allele for no freckles is heterozygous. If an organism is heterozygous, these two different alleles interact to determine a characteristic. The phenotype produced by the alleles can vary, depending on how the two different alleles interact.

TABLE 26.1

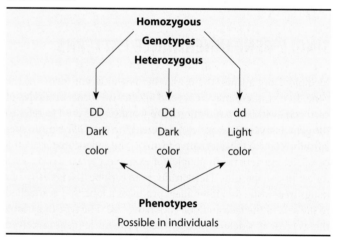

A SIMPLE MODEL OF INHERITANCE— DOMINANT AND RECESSIVE ALLELES

Phenotypes are determined by the type of allele (gene) received from each parent and the way in which these alleles interact with one another. The simplest rule to use to predict phenotype from genotype describes when one allele in the pair expresses itself and hides the other. This is an inheritance pattern called *dominance* and *recessiveness*. A **dominant allele** (denoted by a capital letter, such as *E* in the earlobe example) masks the effect of other alleles for the trait (table 26.1). For example, if a person has one allele for free earlobes and one allele for attached earlobes, that person has a phenotype of free earlobes. We say the allele for free earlobes is dominant. A **recessive allele** (denoted by a lowercase letter, here *e*) is one that, when present with another allele, has its actions overshadowed by the other; it is masked by the effect of the other allele. For example, if a person is heterozygous (has one allele for free earlobes and one allele for attached earlobes, *Ee*), the person will have a phenotype of free earlobes. The allele for free earlobes is dominant. While *EE* or *Ee* may show the free earlobe phenotype, recessive phenotypes are only observable when two recessive alleles (*ee*) are present.

Many people have some common misconceptions about recessive alleles. One is that recessive alleles are all bad or harmful. Don't think that recessive genes are necessarily bad. *The term recessive has nothing to do with the significance or value of the gene—it simply describes how it can be expressed. Recessive alleles are not less likely to be inherited but must be present in a homozygous condition to express themselves. This may make recessive alleles less likely to be observed. A second misconception is that recessive alleles occur less frequently in a population.*

This is not always the case. The number of fingers a person has is determined by a gene. The recessive allele for this gene is five fingers. The dominant allele is six. Clearly, the five-finger allele is more frequent in our population because we see many more five-fingered individuals than individuals with six fingers.

MENDEL'S LAWS OF HEREDITY

Heredity problems are concerned with determining which alleles are passed from the parents to the offspring and how likely it is that various types of offspring will be produced.

The Augustinian monk Gregor Mendel developed a method of predicting the outcome of inheritance. He performed experiments that showed the inheritance patterns of certain characteristics in common garden pea plants (*Pisum sativum*). From his work, Mendel concluded which traits were dominant and which were recessive. Some of his results are shown in table 26.2.

What made Mendel's work unique was that he studied only one trait at a time. Previous investigations had tried to follow numerous traits at the same time. When this was attempted, the total set of characteristics was so cumbersome to work with that no clear idea could be formed of how the offspring inherited traits. Mendel used traits with clear-cut alternatives, such as purple or white flower color, yellow or green seed pods, and tall or dwarf pea plants. He was very lucky to have chosen pea plants in his study because they naturally self-pollinate. When self-pollination occurs in pea plants over many generations, it is possible to develop a population of plants that is homozygous for a number of characteristics. Such a population is known as a *pure line*.

Mendel took a pure line of pea plants having purple flower color, removed the male parts (anthers), and discarded them so that the plants could not self-pollinate. He then took anthers from a pure-breeding white-flowered plant and pollinated the antherless purple flowers.

When the pollinated flowers produced seeds, Mendel collected, labeled, and planted them.

When these seeds germinated and grew, they eventually produced flowers.

You might be surprised to learn that all of the plants resulting from this cross had purple flowers. One of the prevailing hypotheses of Mendel's day would have predicted that the purple and white colors would have blended, resulting in flowers that were lighter than the parental purple flowers. Another hypothesis would have predicted that the offspring would have

had a mixture of white and purple flowers. The unexpected result—all of the offspring produced flowers like those of one parent and no flowers like those of the other—caused Mendel to examine other traits as well and formed the basis for much of the rest of his work. He repeated his experiments using pure strains for other traits. Pure-breeding tall plants were crossed with pure-breeding dwarf plants. Pure-breeding plants with yellow pods were crossed with pure-breeding plants with green pods. The results were all the same: the offspring showed the characteristic of one parent and not the other.

Next, Mendel crossed the offspring of the white-purple cross (all of which had purple flowers) with each other to see what the third generation would be like. Had the characteristic of the original white-flowered parent been lost completely? This second-generation cross was made by pollinating these purple flowers that had one white parent among themselves. The seeds produced from this cross were collected and grown.

When these plants flowered, three-fourths of them produced purple flowers and one-fourth produced white flowers.

After analyzing his data, Mendel formulated several genetic laws to describe how characteristics are passed from one generation to the next and how they are expressed in an individual.

Mendel's law of dominance—When an organism has two different alleles for a given trait, the allele that is expressed, overshadowing the expression of the other allele, is said to be dominant. The gene whose expression is overshadowed is said to be recessive.

Mendel's law of segregation—When gametes are formed by a diploid organism, the alleles that control a trait separate from one another into different gametes, retaining their individuality. Parental alleles of a gene separate at meiosis into gametes:

Parental cell
Bb

B   b

Gamete type one   Gamete type two

Mendel's law of independent assortment—Members of one *pair of alleles* separate from each other independently of the members of the other pair of alleles. In other words, if two different genes are separate chromosomes, where the one allele goes is not influenced by where the other goes.

At the time of Mendel's research, biologists knew nothing of chromosomes or DNA or of the processes of mitosis and meiosis. Mendel assumed that each gene was separate from other genes. It was fortunate for him that most of the characteristics he picked to study were found on separate chromosomes. If two or more of these genes had been located on the same chromosome (*linked genes*), he probably would not have been able to formulate his laws. The discovery of chromosomes and DNA has led to modifications in Mendel's laws, but it was Mendel's work that formed the foundation for the science of genetics.

TABLE 26.2   **Dominant and Recessive Traits in Pea Plants**

Characteristic	Dominant Allele	Recessive Allele
Plant height	Tall	Dwarf
Pod shape	Full	Constricted
Pod color	Green	Yellow
Seed surface	Round	Wrinkled
Seed color	Yellow	Green
Flower color	Purple	White

People Behind the Science

Gregor Johann Mendel (1822–1884)

Gregor Johann Mendel was an Austrian monk who discovered the basic laws of heredity, thereby laying the foundation of modern genetics—although the importance of his work was not recognized until after his death.

Mendel was born Johann Mendel on July 22, 1822, in Heinzendorf, Austria, the son of a peasant farmer. He studied for two years at the Philosophical Institute in Olomouc, after which, in 1843, he entered the Augustinian monastery in Brünn, Moravia, taking the name Gregor. In 1847, he was ordained a priest. During his religious training, Mendel taught himself a certain amount of science. In 1850, he tried to pass an examination to obtain a teaching license but failed, and in 1851, he was sent by his abbot to the University of Vienna to study physics, chemistry, mathematics, zoology, and botany. Mendel left the university in 1853 and returned to the monastery in Brünn in 1854. He then taught natural science in the local Technical High School until 1868, during which period he again tried, and failed, to gain a teaching certificate that would have enabled him to teach in more advanced institutions. It was also in the period 1854 to 1868 that Mendel performed most of his scientific work on heredity. He was elected abbot of his monastery at Brünn in 1868, and the administrative duties involved left him little time for further scientific investigations. Mendel remained abbot at Brünn until his death on January 6, 1884.

Mendel began the experiments that led to his discovery of the basic laws of heredity in 1856. Much of his work was performed on the edible pea (*Pisum* sp.), which he grew in the monastery garden. He carefully self-pollinated and wrapped (to prevent accidental pollination by insects) each individual plant, collected the seeds produced by the plants, and studied the offspring of these seeds. He found that dwarf plants produced only dwarf offspring and that the seeds produced by this second

Box Figure 26.1

generation also produced only dwarf offspring. With tall plants, however, he found that both tall and dwarf offspring were produced and that only about one-third of the tall plants bred true, from which he concluded that there were two types of tall plants, those that bred true and those that did not. Next, he crossbred dwarf plants with true-breeding tall plants, planted the resulting seeds, and then self-pollinated each plant from this second generation. He found that all the offspring in the first generation were tall but that the offspring from the self-pollination of this first generation were a mixture of about one-quarter true-breeding dwarf plants, one-quarter true-breeding tall plants, and one-half nontrue-breeding tall plants. Mendel also studied other characteristics in pea plants, such as flower color, seed shape, and flower position, finding that, as with height, simple laws governed the inheritance of these traits. From his findings, Mendel concluded that each parent plant contributes a factor that determines a particular trait and that the pairs of factors in the offspring do not give rise to a merger of traits. These conclusions, in turn, led him to formulate his famous law of segregation and law of independent assortment of characters, which are now recognized as two of the fundamental laws of heredity.

Mendel reported his findings to the Brünn Society for the Study of Natural Science in 1865, and in the following year, he published *Experiments with Plant Hybrids*, a paper that summarized his results. But the importance of his work was not recognized at the time. It was not until 1900, when his work was rediscovered by Hugo De Vries, Carl Erich Correns, and Erich Tschermak von Seysenegg, that Mendel achieved fame—sixteen years after his death.

Source: Modified from the *Hutchinson Dictionary of Scientific Biography*. © RM, 2009. All rights reserved. Helicon Publishing is a division of RM.

STEPS IN SOLVING HEREDITY PROBLEMS: SINGLE-FACTOR CROSSES

People have long been interested in understanding why the offspring of plants and animals resemble or don't resemble their parents. Mendel's work helped formulate a way to answer such questions. The first type of problem we will consider is the easiest type, a single-factor cross. A **single-factor cross** (sometimes called a monohybrid cross: *mono* = one; *hybrid* = combination) is a genetic cross or mating in which a single characteristic is followed from one generation to the next.

For example, in humans, the allele for Tourette syndrome (TS) is inherited as a dominant allele. For centuries, people displaying this genetic disorder were thought to be possessed by the devil since they displayed such unusual behaviors. These motor and verbal behaviors or tics are involuntary and range from mild (e.g., leg tapping, eye blinking, face twitching) to the more violent forms such as the shouting of profanities, head

jerking, spitting, compulsive repetition of words, or even barking like a dog. The symptoms result from an excess production of the brain messenger, dopamine.

A Single-Factor Problem: If both parents are heterozygous (have one allele for Tourette syndrome and one allele for no Tourette syndrome), what is the probability that they can have a child without Tourette syndrome? With Tourette syndrome?

Step 1: Assign a symbol for each allele.

Usually a capital letter is used for a dominant allele and a small letter for a recessive allele. Use the symbol *T* for Tourette and *t* for no Tourette.

Allele	Genotype	Phenotype
T = Tourette	*TT*	Tourette syndrome
t = normal	*Tt*	Tourette syndrome
	tt	Normal

Myths, Mistakes, and Misunderstandings

You Don't Always See It Even If You Have It

Misconception: If you've got a dominant gene, you'll show the trait.

In fact, just because you have the dominant allele for a trait doesn't mean you will show it. *Penetrance* is the frequency with which the allele expresses itself when present. Understanding penetrance helps answer the question, "If all these people have this gene, why don't I see more people with the condition?" For example, heterozygous males show a 90 percent penetrance of Tourette syndrome; that is, 90 percent of men with the dominant allele for TS express the trait enough to be diagnosed with TS. There is only 60 to 70 percent penetrance in the female population.

Step 2: Determine the genotype of each parent and indicate a mating.

Because both parents are heterozygous, the male genotype is *Tt*. The female genotype is also *Tt*. The × between them is used to indicate a mating.

$$Tt \times Tt$$

Step 3: Determine all the possible kinds of gametes each parent can produce.

Remember that gametes are haploid; therefore, they can have only one allele instead of the two present in the diploid cell. Because the male has both the Tourette syndrome allele and the normal allele, half of his gametes will contain the Tourette syndrome allele and the other half will contain the normal allele. Because the female has the same genotype, her gametes will be the same as his.

For genetic problems, a *Punnett square* is used. A **Punnett square** is a box figure that allows you to determine the probability of genotypes and phenotypes of the progeny of a particular cross. Remember, because of the process of meiosis, each gamete receives only one allele for each characteristic listed. Therefore, the male will produce sperm with either a *T* or a *t;* the female will produce ova with either a *T* or a *t*. The possible gametes produced by the male parent are listed on the left side of the square and the female gametes are listed on the top. In our example, the Punnett square would show a single dominant allele and a single recessive allele from the male on the left side. The alleles from the female would appear on the top.

		Female genotype *Tt* Possible female gametes *T* and *t*	
Male genotype *Tt*		*T*	*t*
Possible male gametes *T* and *t*	*T*		
	t		

Muscular Dystrophy and Genetics

Because most genes are comprised of thousands of nucleotide base pairs, there are many different changes in these sequences that can result in the formation of multiple "bad" gene products and, therefore, an abnormal phenotype. The same gene can have many different "bad" forms. The fact that a phenotypic characteristic can be determined by many different alleles for a particular characteristic is called *genetic heterogeneity*. For example, two of the best-known forms of *muscular dystrophy* (*MD*) are Duchenne's and Becker's. Duchenne's (DMD) is characterized by a severe, progressive weakening of the muscles, the appearance of a false muscle atrophy (degeneration) in the calves of the legs, onset in early childhood, and a high likelihood of death in the thirties. DMD is caused by a mutation in the dystrophin gene located on the X chromosome and is a recessive trait. Becker's (BMD) is caused by a mutation in the same dystrophin gene but at a different location. BMD is a milder form of MD.

Myths, Mistakes, and Misunderstandings

Another Fatal Mistake?

Misconception: All mutations are fatal.

In fact, this is not true! (1) A mutation may have no effect; that is, it may replace the intended amino acid with the same one. (2) The mutation may result in a new protein that does not interfere with metabolism enough to be lethal. (3) The mutation may be in a gene that is not used.

Step 4: Determine all the gene combinations that can result when these gametes unite.

To determine the possible combinations of alleles that could occur as a result of this mating, simply fill in each of the empty squares with the alleles that can be donated from each parent. Determine all the gene combinations that can result when these gametes unite.

	T	*t*
T	*TT*	*Tt*
t	*Tt*	*tt*

Step 5: Determine the phenotype of each possible gene combination.

In this problem, three of the offspring, *TT, Tt,* and *Tt,* have Tourette syndrome. One progeny, *tt,* is normal. Therefore, the answer to the problem is that the probability of having offspring with Tourette syndrome is 3/4; for no Tourette syndrome, it is 1/4.

Take the time to learn these five steps. All single-factor problems can be solved using this method; the only variation in the problems will be the types of alleles and the number of possible types of gametes the parents can produce.

Are You a "Taster"?

Some humans inherit the ability to taste the chemical phenylthiocarbamide (PTC), and others are unable to taste PTC. Ask your instructor to furnish you with a supply of paper impregnated with PTC. Place a piece of this paper on your tongue. If you experience a bitter taste, you are a taster. If you do not, you are a nontaster. Take enough strips to test members of your family you can readily contact. These may include your grandparents, parents, siblings, children, aunts, and uncles. After you have tested the members of your family, construct a pedigree as described below. Is it possible for you to determine the genotypes of your family members? Is tasting PTC a dominant or a recessive trait? Could it be a case of lack of dominance? Might it be linked to another trait?

In a pedigree, circles represent females and squares represent males. Symbols of parents are connected by a horizontal mating line, and the offspring are shown on a horizontal line below the parents. Individuals who are tasters are represented by a solid symbol, nontasters by an open symbol. A typical pedigree might look something like the one below, which shows three generations. Create your own pedigree. For this analysis, you can use only "blood" (biological) relatives.

MORE COMPLEX MODELS OF INHERITANCE

So far, we have considered phenotypes that are determined by dominance and recessiveness between two alleles. In fact, dominant/recessive inheritance is probably the least common inheritance model since most inheritance patterns do not fit this simple pattern. In most situations, phenotypes are not only determined by the type of allele (gene) received but also by the alternative ways in which they interact with each other and how the environment influences their expression. There are six generally recognized patterns: X-linked characteristics, codominance, incomplete dominance, multiple alleles, polygenic inheritance, and pleiotropy.

X-Linked Genes

Alleles located on the same chromosome tend to be inherited together. They are said to be linked. The process of crossing-over, which occurs during prophase I of meiosis I, may split up these linkage groups. Crossing-over happens between homologous chromosomes donated by the mother and the father, and results in a mixing of genes. The closer two genes are to each other on a chromosome, the more probable it is that they will be inherited together.

People and many other organisms have two types of chromosomes. Autosomes (22 pairs) are not involved in sex deter-

mination and have alleles for the same traits on both members of the homologous pair of chromosomes. *Sex chromosomes* are a pair of chromosomes that control the sex of an organism. In humans and some other animals, there are two types of sex chromosomes—the X chromosome and the Y chromosome. The Y chromosome is much shorter than the X chromosome and has few genes for traits found on the X chromosome in humans. One genetic trait that is located on the Y chromosome contains the testis-determining gene—SRY. Females are normally produced when two X chromosomes are present because they do not have the SRY gene. Males are usually produced when one X chromosome and one Y chromosome are present.

Genes found together on the X chromosomes are said to be **X-linked.** Because the Y chromosome is shorter than the X chromosome, it does not have many of the alleles that are found on the comparable portion of the X chromosome. Therefore, in a man, the presence of a single allele on his only X chromosome will be expressed, regardless of whether it is dominant or recessive because there is no corresponding allele on the Y chromosome. A Y-linked trait in humans is the SRY gene. This gene controls the differentiation of the embryonic gonad to a male testis. By contrast, more than one hundred genes are on the X chromosome. Some of these X-linked genes can result in abnormal traits such as *color deficiency* (formerly referred to as color blindness), *hemophilia, brown teeth,* and at least two forms of *muscular dystrophy* (Becker's and Duchenne's).

Codominance

In some inheritance situations, alleles are neither dominant nor recessive to one another, and both express themselves in the heterozygous individual. This inheritance pattern is called codominance. In a case of codominance, the phenotype of each allele is expressed in the heterozygous condition. Consequently, a person with the heterozygous genotype can have a phenotype very different from either of their homozygous parents. For codominant alleles, all uppercase symbols are used and superscripts are added to represent the different alleles. The uppercase letters call attention to the fact that each allele can be detected phenotypically to some degree even when in the presence of its alternative allele. For example, the coat colors (C) of horses, cattle, and other animals may be phenotypically roan. The roan gene adds white hairs among the other body hairs, whatever their color. The roan gene can be applied to any color. The most common base roan color is red. The color is the result of individual hairs that are two different colors. A red roan would have both red and white hairs intermingled. Animals that are true red would have the genotype $C^R C^R$, red roan would have $C^R C^W$, and true white would have $C^W C^W$ (figure 26.3).

Incomplete Dominance

In the inheritance pattern known as partial or **incomplete dominance**, alleles lack dominant relationships. The results of a cross are heterozygous offspring that have an intermediate phenotype that is not distinct from either homozygous parent.

Another example of incomplete dominance occurs in certain colors of horses. Palomino and buckskin horses show a cream-

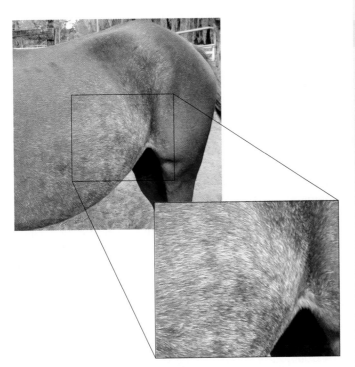

FIGURE 26.3 The color of this horse, an Arab, displays the codominant color pattern called roan. Notice that there are places on the body where both white and red hairs are found side by side.

color allele for the basic color gene (C^{cr}). In these horses, the C^{cr} allele dilutes the hair pigment to yellow. In the homozygous condition ($C^{cr}C^{cr}$), the horse is called cremello; both black and red pigments are intermingled in a single hair to produce yellow.

A classic example of incomplete dominance in plants involves the color of the petals of snapdragons. The gene products interact to produce a blended result. There are two alleles for the color of these flowers. Because neither allele is recessive, we cannot use the traditional capital and small letters as symbols for these alleles. Instead, the allele for white petals is given the symbol F^W, and the one for red petals is given the symbol F^R (figure 26.4).

There are three possible combinations of these two alleles:

Genotype	Phenotype
F^WF^W	White flower
F^RF^R	Red flower
F^RF^W	Pink flower

Notice that there are only two different alleles, red and white, but there are three phenotypes, red, white, and pink. Both the red-flower allele and the white-flower allele partially express themselves when both are present, and this results in pink.

Multiple Alleles

So far, we have discussed only traits that are determined by two alleles, for example, *A, a*. However, there can be more than two different alleles for a single trait. All the various forms of the same gene (alleles) that control a particular trait are referred to as **multiple alleles.** However, one person can have only a maximum

Blame That Trait on Your Mother!

Within a eukaryotic cell, the bulk of DNA is located in the nucleus. It is this genetic material that controls the majority of biochemical processes of the cell. It has long been recognized that the mitochondria also contain DNA, mtDNA. This genetic material works in conjunction with that located in the nucleus. However, an interesting thing happens when a sperm is formed as a result of meiosis. All the mitochondria are packed into the "necks" and not the "heads" of the sperm. When a sperm penetrates a secondary oocyte, the head enters but the majority of the mitochondria in the neck remain outside. Therefore, fathers rarely, if ever, contribute mitochondrial genetic information to their children. Should a mutation occur in the mother's mitochondrial DNA, she will pass the abnormality on to her children. Should a mutation occur in the father's mitochondrial DNA, he will not transmit the abnormality. This unusual method of transmission is called *mitochondrial* (or *maternal*) *inheritance.* Most abnormalities transmitted through mitochondrial genes are associated with muscular weakness because the mitochondria are the major source of ATP in eukaryotic cells. People with one form of this rare type of genetic abnormality, *Leber's hereditary optic neuropathy,* show symptoms of sudden loss of central vision due to optic nerve death in young adults with onset at about age twenty. Both males and females may be affected.

A $F^R F^R$ B $F^W F^W$

C $F^W F^R$

FIGURE 26.4 The colors of these snapdragons are determined by two alleles for petal color, F^W and F^R. There are three different phenotypes because of the way in which the alleles interact with one another. In the heterozygous condition, neither of the alleles dominates the other.

of two of the alleles for the characteristic. A good example of a characteristic that is determined by multiple alleles is the ABO blood type. There are three alleles for blood type:

Allele*

I^A = blood has type A antigens on red blood cell surface
I^B = blood has type B antigens on red blood cell surface
i = blood type O has neither type A nor type B antigens on red blood cell surface

*The symbols I^A and I^B stand for the technical term for the antigenic carbohydrates attached to red blood cells, the *Immunogens*. These alleles are located on human chromosome 9. The ABO system is not the only one used to type blood. Others include the Rh, MNS, and Xg systems.

Note in the midst of this example of multiple alleles that in the ABO system, A and B show *codominance* when they are together in the same individual. However, both A and B are dominant over the O allele. These three alleles can be combined as pairs in six different ways, resulting in four different phenotypes:

Genotype		Phenotype
$I^A I^A$	=	Blood type A
$I^A i$	=	Blood type A
$I^B I^B$	=	Blood type B
$I^B i$	=	Blood type B
$I^A I^B$	=	Blood type AB
ii	=	Blood type O

Multiple-allele problems are worked as single-factor problems. Some examples are in the practice problems at the end of this chapter.

Polygenic Inheritance

Thus far, we have considered phenotypic characteristics that are determined by alleles at a single place and chromosome. However, some characteristics are determined by the interaction of genes at several different loci (on different chromosomes or at different places on a single chromosome). This is called polygenic inheritance. A number of different pairs of alleles may combine their efforts to determine a characteristic. Skin color in humans is a good example of this inheritance pattern. According to some experts, genes for skin color are located at a minimum of three loci. At each of these loci, the allele for dark skin is dominant over the allele for light skin. Therefore, a wide variety of skin colors is possible depending on how many dark-skin alleles are present (figure 26.5).

Polygenic inheritance is very common in determining characteristics that appear on a gradient in nature. In the skin-color example, and in many other traits as well, the characteristics cannot be categorized in terms of *either/or*, but the variation in phenotypes can be classified as *how much* or *by what amount*.

For instance, people show great variations in height. There are not just tall and short people—there is a wide range. Some people are as short as 1 m, and others are taller than 2 m. This quantitative trait is probably determined by a number of different genes. Intelligence also varies significantly, from those who are severely retarded to those who are geniuses. Many of these traits may be influenced by outside environmental factors such as diet, disease, accidents, and social factors. These are just a few examples of polygenic inheritance patterns.

Pleiotropy

Even though a single gene produces only one product, it often has a variety of effects on the phenotype of the person. This is called *pleiotropy*. **Pleiotropy** (*pleio-* =changeable) is a term used to describe the multiple effects that a gene may have on the phenotype. A good example of pleiotropy is *Marfan syndrome* (figure 26.6), a disease suspected to have occurred in U.S. President Abraham Lincoln. Marfan is a disorder of the body's connective tissue but can also have effects in many other organs, including the eyes, heart, blood, skeleton, and lungs. Symptoms generally appear as a tall, lanky body with long arms and spiderfingers, scoliosis (curvature of the spine), and depression of the chest wall. Many people with Marfan syndrome are near-

Locus 1	$d^1 d^1$	$d^1 D^1$	$d^1 D^1$	$D^1 D^1$	$D^1 d^1$	$D^1 d^1$	$D^1 D^1$
Locus 2	$d^2 d^2$	$d^2 d^2$	$d^2 D^2$	$D^2 d^2$	$D^2 d^2$	$D^2 D^2$	$D^2 D^2$
Locus 3	$d^3 d^3$	$d^3 d^3$	$d^3 d^3$	$d^3 d^3$	$D^3 D^3$	$D^3 D^3$	$D^3 D^3$
Total number of dark-skin genes	0	1	2	3	4	5	6

FIGURE 26.5 Skin color in humans is an example of polygenic inheritance. The darkness of the skin is determined by the number of dark-skin genes a person inherits from his or her parents.

The Inheritance of Eye Color

It is commonly thought that eye color is inherited in a simple dominant/recessive manner. Brown eyes are considered to be dominant over blue eyes. The real pattern of inheritance, however, is considerably more complicated than this. Eye color is determined by the amount of a brown pigment, known as melanin, present in the iris of the eye. If there is a large quantity of melanin present on the anterior surface of the iris, the eyes are dark. Black eyes have a greater quantity of melanin than brown eyes.

If there is not a large amount of melanin present on the anterior surface of the iris, the eyes will appear blue, not because of a blue pigment but because blue light is returned from the iris (see box figure 26.2). The iris appears blue for the same reason that deep bodies of water tend to appear blue. There is no blue pigment in the water, but blue wavelengths of light are returned to the eye from the water. People appear to have blue eyes because the blue wavelengths of light are reflected from the iris.

Just as black and brown eyes are determined by the amount of pigment present, colors such as green, gray, and hazel are produced by the various amounts of melanin in the iris. If a very small amount of brown melanin is present in the iris, the eye tends to appear green, whereas relatively large amounts of melanin produce hazel eyes.

Several different genes are probably involved in determining the quantity and placement of the melanin and, therefore, in determining eye color. These genes interact in such a way that a wide range of eye color is possible. Eye color is probably determined by polygenic inheritance, just as skin color and height are. Some newborn babies have blue eyes that later become brown. This is because they have not yet begun to produce melanin in their irises at the time of birth.

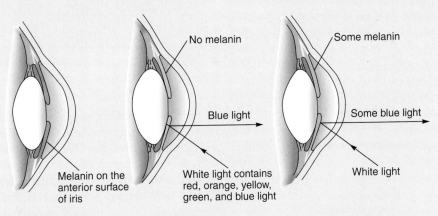

Box Figure 26.2

Iris of the eye is dark colored. — Melanin on the anterior surface of iris

Iris of the eye appears blue. — No melanin; Blue light; White light contains red, orange, yellow, green, and blue light

Iris of the eye appears green or hazel. — Some melanin; Some blue light; White light

A

B

C

D

FIGURE 26.6 It is estimated that about 40,000 (1 out of 10,000) people in the United States have this autosomal dominant abnormality. Notice the common lanky appearance to the body and face of (A) this person with Marfan syndrome and (B) former U.S. President Abraham Lincoln. Photos (C) and (D) illustrate their unusually long fingers.

Inheritance Pattern of Spina Bifida

Many polygenic traits are influenced by outside environmental factors, such as diet, drugs, disease, accidents, and social factors. The degree to which the environment impacts the genes can determine the nature of phenotype. Patterns of inheritance that are the result of polygenic traits influenced by environmental factors are called multifactorial inheritance. In fact, most polygenic traits are multifactorial. *Multifactorial inheritance* is not yet completely understood and is hard to trace since it does not follow the typical single-gene inheritance pattern. The first generation (children) to follow individuals with such diseases (parents) shows a markedly high risk of acquiring the gene, whereas the second generation (grandchildren) has a much lower risk. Examples of multifactorial traits include neural tube defects (e.g., spina bifida), insulin-dependent diabetes mellitus, cleft palate, different fingerprints in identical twins, and congenital heart disease. In the case of spina bifida, the arches of the vertebrae fail to enclose the spinal cord (typically in the lower back region), and in some cases, the cord and nerves may be exposed at birth. This abnormality is a leading cause of stillbirths and death during early infancy and, if unable to be surgically corrected, is a handicap in survivors. One factor that appears to influence this disease is the amount of the vitamin folic acid the mother has in her system prior to conception. Other factors include chromosome disorders, single-gene mutations, and teratogens (chemicals that cause developmental abnormalities).

FIGURE 26.7 PKU is an recessive disorder located on chromosome 12. The diagram on the left shows how the normal pathway works. The diagram on the right shows an abnormal pathway. If the enzyme phenylalanine hydroxylase is not produced because of a mutated gene, the amino acid phenylalanine cannot be converted to tyrosine and is converted into phenylpyruvic acid, which accumulates in body fluids. The buildup of phenylpyruvic acid causes the death of nerve cells and ultimately results in mental retardation. Because phenylalanine is not converted to tyrosine, subsequent reactions in the pathway are also affected.

In this genetic abnormality, people are unable to convert the amino acid phenylalanine into the amino acid tyrosine. The buildup of phenylalanine in the body prevents the normal development of the nervous system. Such individuals suffer from PKU and may become mentally retarded. There is a concern in the medical profession that there may be many adults with PKU who do not know they have the condition. Children diagnosed with PKU are not tracked throughout their life, and many parents do not tell their children they have the condition. After these children become adults, they could easily be unaware of their medical history. The problem comes if a female PKU patient becomes pregnant and does not have proper prenatal care because neither she nor her physician is aware of her medical history.

Environmental Influences on Gene Expression

One of the complicating aspects of genetics is that expression of an allele's phenotype depends on so much more than just the genotype of the organism. Sometimes the physical environment determines whether or not dominant or recessive alleles are expressed. For example, in humans, genes for freckles do not express themselves fully unless the person's skin is exposed to sunlight (figure 26.8). Another example is the allele for six fingers (*polydactylism*), which is dominant over the allele for five fingers in humans. Some people who have received the allele for six fingers have a fairly complete sixth finger; in others, it may appear as a little stub. In another case, a dominant allele causes the formation of a little finger that cannot be bent as a normal little finger. However, not all people who are believed to have inherited that allele will have a stiff little finger. In some cases, this dominant characteristic is not expressed or perhaps only shows on one hand. Thus, there may be variation in the

sighted because they have a dislocation of the lens of the eye. The white of the eye (sclera) may appear bluish. Heart problems include dilation of the aorta and abnormalities of the heart valves. Death may be caused by a dissection (tear) in the aorta from the rupture in a weakened and dilated area of the aorta called an aortic *aneurysm*. Thus, the alleles that produce Marfan syndrome have multiple effects on height, eyesight, and other traits.

Another example of pleiotropy is the genetic abnormality PKU (phenylketonuria). In this example, a single gene affects many different chemical reactions that depend on the way a cell metabolizes the amino acid phenylalanine, commonly found in many foods (refer to figure 26.7).

A   B   C

FIGURE 26.8   The expression of many genes is influenced by the environment. The allele for dark hair in the cat (*A*) is sensitive to temperature and expresses itself only in the parts of the body that stay cool. (*B* and *C*) The allele for freckles expresses itself more fully when a person is exposed to sunlight.

FIGURE 26.9   Neurofibromatosis I is seen in many forms, including benign fibromatous skin tumors, "café-au-lait" spots, nodules in the iris, and possible malignant tumors. It is extremely variable in its expressivity; that is, the traits may be almost unnoticeable or extensive. An autosomal dominant trait, it is the result of a mutation and the production of a protein (neurofibromin) that normally would suppress the activity of a gene that causes tumor formation.

degree to which an allele expresses itself in an *individual*. Geneticists refer to this as *variable expressivity*. A good example of this occurs in the genetic abnormality *neurofibromatosis type I* (NF1) (figure 26.9). In some cases, it may not be expressed in the *population* at all. This is referred to as a *lack of penetrance*. Other genes may be interacting with these dominant alleles, causing the variation in expression.

Both internal and external environmental factors can influence the expression of genes. For example, at conception, a male receives genes that will eventually determine the pitch of his voice. However, these genes are expressed differently after puberty. At puberty, male sex hormones are released. This internal environmental change results in the deeper male voice. A male who does not produce these hormones retains a higher-pitched voice in later life. Comparable changes can occur in females when an abnormally functioning adrenal gland causes the release of large amounts of male hormones.

Diet is an external environmental factor that can influence the phenotype of an individual. *Diabetes mellitus,* a metabolic disorder in which glucose is not properly metabolized and is passed out of the body in the urine, has a genetic basis. Some people who have a family history of diabetes are thought to have inherited the trait for this disease. Evidence indicates that they can delay the onset of the disease by reducing the amount of sugar in their diet.

Epigenetics and Gene Expression

Epigenetics is the study of changes in gene expression caused by factors other than alterations in a cell's DNA. The term actually means "in addition to genetics," (i.e., non-genetic factors that cause the cell's genes to express themselves differently). When an epigenetic change occurs, it may last for the life of the cell and may even be passed on to the next generation. This is what happens when a cell (e.g., stem cell) undergoes the process of differentiation. Stem cells are called pluripotent because they have the potential to be any kind of cell found in the body (e.g., muscle, bone, or skin cell). However, once they become differentiated they lose the ability to become other kinds of cells and so do the cells they produce by cell division. For example, if a pluripotent cell were to express muscle protein genes and not insulin protein genes, it would differentiate into a muscle cell, not an insulin-producing cell.

The following are four examples of events that can cause an epigenetic effect:

1. Adding a methyl group to a cytosine in the gene changes it to methylcytosine. Since methylcytosine cannot be read during translation, the gene is turned off.
2. Altering the shape of the histones around the gene. Modifying histones ensures that a differentiated cell would stay differentiated and not convert back into being a pluripotent cell.
3. Having the protein that has already been transcribed return to the gene and keep it turned on.
4. Splicing RNA into sequences not originally determined by the gene.

Some compounds are considered epigenetic carcinogens; that is, they are able to cause cells to form tumors, but they do not change the nucleotide sequence of a gene. Examples include certain chlorinated hydrocarbons used as fungicides and some nickel-containing compounds.

PART II
THE MOLECULAR BASIS OF GENETICS

While molecular genetics is only about fifty years old, this science has progressed rapidly. Our understanding of DNA structure and functions contributes greatly to our ability to control the genetics of organisms to our advantage.

THE CENTRAL DOGMA

As scientists began to understand the chemical makeup of the **nucleic acids,** an attempt was made to understand how DNA and RNA relate to inheritance, cell structure, and cell activities. The concept that resulted is known as the *central dogma.* The belief can be written in this form:

What this concept map says is that at the center of it all is DNA, the genetic material of the cell, and (going to the left) it is capable of reproducing itself, a process called **DNA replication**. Going to the right, DNA is capable of supervising the manufacture of RNA (a process known as **transcription),** which in turn is involved in the production of protein molecules, a process known as **translation.**

DNA replication occurs in cells in preparation for the cell division processes of mitosis and meiosis. Without replication, daughter cells would not receive the library of information required to sustain life. The transcription process results in the formation of a strand of RNA that is a copy of a segment of the DNA on which it is formed. Some of the RNA molecules become involved in various biochemical processes, while others are used in the translation of the RNA information into proteins. Structural proteins are used by the cell as building materials (feathers, collagen, hair), while others are used to direct and control chemical reactions (enzymes or hormones) or carry molecules from place to place (hemoglobin).

It is the processes of transcription and translation that result in the manufacture of all enzymes. Each unique enzyme molecule is made from a blueprint in the form of a DNA nucleotide sequence, or *gene.* Some of the thousands of enzymes manufactured in the cell are the tools required for transcription and translation to take place. *The enzymes made by the process carry out the process of making more enzymes!*

Tools are made to make more tools! The same is true for DNA replication. Enzymes made from the DNA blueprints by transcription and translation are used as tools to make exact copies of the genetic material! More blueprints are made so that future generations of cells will have the genetic materials necessary for them to manufacture their own regulatory and structural

proteins. Without DNA, RNA, and enzymes functioning in the proper manner, life as we know it would not occur.

DNA has four properties that enable it to function as genetic material. It is able to (1) *replicate* by directing the manufacture of copies of itself; (2) *mutate,* or chemically change, and transmit these changes to future generations; (3) *store* information that determines the characteristics of cells and organisms; and (4) use this information to *direct* the synthesis of structural and regulatory proteins essential to the operation of the cell or organism.

The Structure of DNA and RNA

Nucleic acid molecules are enormous and complex polymers made up of monomers called **nucleotides.** Each nucleotide is composed of a sugar molecule (S) containing five carbon atoms, a phosphate group (P), and a molecule containing nitrogen that will be referred to as a *nitrogenous base* (B) (figure 26.10). It is possible to classify nucleic acids into two main groups based on the kinds of sugars and nitrogenous bases used in the nucleotides (i.e., DNA and RNA).

In cells, DNA is the nucleic acid that functions as the original blueprint for the synthesis of proteins. It contains the sugar *deoxyribose;* phosphates; and the nitrogenous bases adenine, guanine, cytosine, and thymine (A, G, C, T). RNA is a type of nucleic acid that is directly involved in the synthesis of protein. It contains the sugar *ribose;* phosphates; and adenine, guanine, cytosine, and uracil (A, G, C, U). There is no thymine in RNA and no uracil in DNA.

DNA and RNA differ in one other respect. DNA is actually a double molecule. It consists of two flexible strands held together between their protruding bases. The two strands are twisted about each other in a coil or double helix that resembles a twisted ladder (figure 26.11). The two strands of the molecule are held together because they "fit" each other like two jigsaw puzzle pieces that interlock with one another and are stabilized by weak chemical forces, hydrogen bonds. The four kinds of bases always pair in a definite way: adenine (A) with thymine (T), and guanine (G) with cytosine (C). Notice that the large molecules (A and G) pair with the small ones (T and C), thus keeping the two complementary (matched) strands parallel. The bases that pair are said to be **complementary bases.**

$$A : T$$
$$\text{and}$$
$$G : C$$

You can "write" a message in the form of a stable DNA molecule by combining the four different DNA nucleotides (A, T, G, C) in particular sequences. The four DNA nucleotides are being used as an alphabet to construct three-letter words. To make sense out of such a code, it is necessary to read in one direction. Reading the sequence in reverse does not always make sense, just as reading this paragraph in reverse would not make sense.

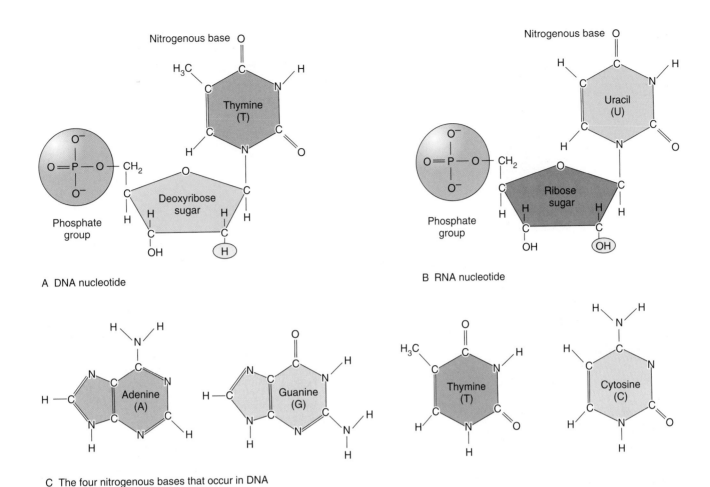

A DNA nucleotide

B RNA nucleotide

C The four nitrogenous bases that occur in DNA

FIGURE 26.10 (*A*) The nucleotide is the basic structural unit of all nucleic acid molecules. A thymine nucleotide of DNA is comprised of phosphate, deoxyribose sugar, and the nitrogenous base, thymine (T). Notice in the nucleotides that the phosphate group is written in "shorthand" form as a P inside a circle. (*B*) The RNA uracil nucleotide is comprised of a phosphate, ribose sugar, and the nitrogenous base, uracil (U). Notice the difference between the sugars and how the bases differ from one another. (*C*) Using these basic components (phosphate, sugars, and bases), the cell can construct eight common types of nucleotides. Can you describe all eight?

The genetic material of humans and other eukaryotic organisms is *strands* of coiled *double-stranded DNA*, which has histone proteins attached along its length. When eukaryotic chromatin fibers coil into condensed, highly knotted bodies, they are seen easily through a microscope after staining with dye. Condensed like this, a chromatin fiber is referred to as a **chromosome** (figure 26.12B). The genetic material in bacteria is also double-stranded DNA, but the ends of the molecule are connected to form a *loop* and they do not form condensed chromosomes (figure 26.13).

Each chromatin DNA strand is different because each strand has a different chemical code. Coded DNA serves as a central cell library. Tens of thousands of messages are in this storehouse of information. This information tells the cell such things as (1) how to produce enzymes required for the digestion of nutrients, (2) how to manufacture enzymes that will metabolize the nutrients and eliminate harmful wastes, (3) how to repair and assemble cell parts, (4) how to reproduce healthy offspring, (5) when and how to react to favorable and unfavorable changes in the environment, and (6) how to coordinate and regulate all of life's essential functions. If any of these functions is not performed properly, the cell may die. The importance of

maintaining essential DNA in a cell becomes clear when we consider cells that have lost their DNA. For example, human red blood cells lose their nuclei as they mature and become specialized for carrying oxygen and carbon dioxide throughout the body. Without DNA, they are unable to manufacture the essential cell components needed to sustain themselves. They continue to exist for about 120 days, functioning only on enzymes manufactured earlier in their lives. When these enzymes are gone, the cells die. Because these specialized cells begin to die the moment they lose their DNA, they are more accurately called *red blood corpuscles (RBCs):* "little dying red bodies."

DNA Replication

Because all reproducing cells must maintain a complete set of genetic material, there must be a doubling of DNA in order to have enough to pass on to the offspring (refer to chapter 20 and "The Importance of Cell Division"). DNA replication is the process of duplicating the genetic material prior to its distribution to daughter cells. When a cell divides into two daughter cells, each new cell must receive a complete copy of the parent cell's genetic information, or it will not be able to manufacture all the proteins

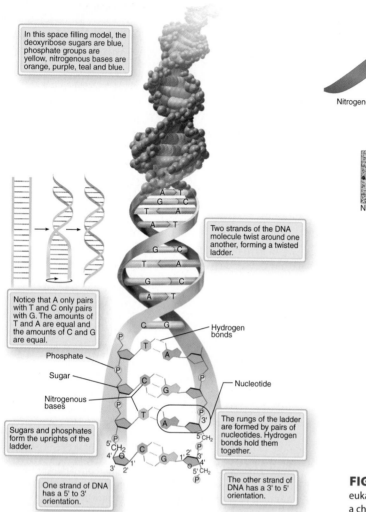

In this space filling model, the deoxyribose sugars are blue, phosphate groups are yellow, nitrogenous bases are orange, purple, teal and blue.

Two strands of the DNA molecule twist around one another, forming a twisted ladder.

Notice that A only pairs with T and C only pairs with G. The amounts of T and A are equal and the amounts of C and G are equal.

Phosphate
Sugar
Nitrogenous bases

Hydrogen bonds

Nucleotide

Sugars and phosphates form the uprights of the ladder.

The rungs of the ladder are formed by pairs of nucleotides. Hydrogen bonds hold them together.

One strand of DNA has a 5' to 3' orientation.

The other strand of DNA has a 3' to 5' orientation.

FIGURE 26.11 Polymerized deoxyribonucleic acid (DNA) is a helical molecule. The nucleotides within each strand are held together by covalent bonds. The two parallel strands are linked by hydrogen bonds between the base paired nitrogen bases.

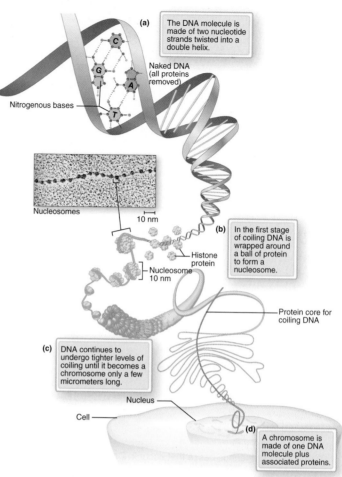

(a) The DNA molecule is made of two nucleotide strands twisted into a double helix.

Naked DNA (all proteins removed)

Nitrogenous bases

Nucleosomes

10 nm

(b) In the first stage of coiling DNA is wrapped around a ball of protein to form a nucleosome.

Histone protein
Nucleosome 10 nm

Protein core for coiling DNA

(c) DNA continues to undergo tighter levels of coiling until it becomes a chromosome only a few micrometers long.

Nucleus
Cell

(d) A chromosome is made of one DNA molecule plus associated proteins.

FIGURE 26.12 During certain stages in the life cycle of a eukaryotic cell, the DNA is tightly coiled to form a chromosome. To form a chromosome, the DNA molecule is wrapped around a group of several histone proteins. Together the histones and the DNA form a structure called the nucleosome. The nucleosomes are stacked together in coils to form a chromosome.

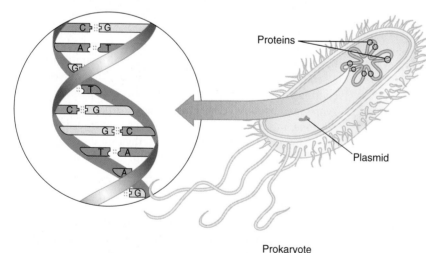

Proteins

Plasmid

Prokaryote

FIGURE 26.13 The nucleic acid of prokaryotic cells (the bacteria) has a different kind of protein compared to eukaryotic cells. In addition, the ends of the giant nucleoprotein molecule overlap and bind with one another to form a loop. The additional small loop of DNA is the plasmid, which contains genes that are not essential to the daily life of the cell.

vital to its existence. Accuracy of duplication is essential to guarantee the continued existence of that type of cell. Should the daughters not receive exact copies, they would most likely die.

1. The DNA replication process begins as an enzyme breaks the attachments between the two strands of DNA. In eukaryotic cells, this occurs in hundreds of different spots along the length of the DNA (figure 26.14).

2. Moving along the DNA, the enzyme "unzips" the halves of the DNA (figure 26.14A). Hydrogen bonds hold each in position (AT, GC), while the new nucleotide is covalently bonded between the sugar and phosphate of the new backbone.

3. Proceeding in opposite directions on each side, the enzyme *DNA polymerase* moves down the length of the DNA, attaching new DNA nucleotides into position. Relatively weak hydrogen bonds hold each in position (AT, GC), while the new nucleotide is

FIGURE 26.14 (*A*) *Helicase* enzymes bind to the DNA molecule. (*B*) The enzymes separate the two strands of DNA. (*C, D*) As the DNA strands are separated, new DNA nucleotides are added to the new strands by DNA polymerase. The new DNA strands are synthesized according to base-pairing rules for nucleic acids. (*E*) By working in two directions at once along the DNA strand, the cell us able to replicate the DNA more quickly. Each new daughter cell receives one of these copies.

bonded to its neighboring nucleotide more strongly with covalent bonds between the sugar and phosphate of the new backbone (figure 26.14B).

4. The enzyme that speeds the addition of new nucleotides to the growing chain works along with another enzyme to make sure that no mistakes are made. If the wrong nucleotide appears to be headed for a match, the enzyme will reject it in favor of the correct nucleotide. If a mistake is made and a wrong nucleotide is paired into position, specific enzymes have the ability to replace it with the correct one (figure 26.14C).

5. Replication proceeds in both directions and at multiple sites along the chromatin, appearing as "bubbles" (figure 26.14D).

6. The complementary molecules (AT, GC) pair with the exposed nitrogenous bases of both DNA strands by forming new hydrogen bonds (figure 26.14C).

7. Once properly aligned, a bond is formed between the sugars and phosphates of the newly positioned nucleotides. A strong sugar and phosphate backbone is formed in the process (figure 26.14C).

8. This process continues until all the replication "bubbles" join (figure 26.14D).

A new complementary strand of DNA forms on each of the old DNA strands, resulting in the formation of two double-stranded DNA molecules. In this way, the exposed nitrogenous bases of the original DNA serve as a *template,* or pattern, for

the formation of the new DNA. As the new DNA is completed, it twists into its double-helix shape.

The completion of the DNA replication process yields two double helices that are identical in their nucleotide sequences. Half of each is new, half is the original parent DNA molecule. The DNA replication process is highly accurate. It has been estimated that there is only one error made for every 2×10^9 nucleotides. A human cell contains forty-six chromosomes consisting of about 3,000,000,000 (3 billion) base pairs. This averages to about 1.5 errors per cell! Don't forget that this figure is an estimate. While some cells may have five errors per replication, others may have more, and some may have no errors at all. It is also important to note that some errors may be major and deadly, while others are insignificant. Because this error rate is so small, DNA replication is considered by most to be essentially error-free. Following DNA replication, the cell now contains twice the amount of genetic information and is ready to begin the process of distributing one set of genetic information to each of its two daughter cells.

The distribution of DNA involves splitting the cell and distributing a set of genetic information to the two new daughter cells. In this way, each new cell has the necessary information to control its activities. The mother cell ceases to exist when it divides its contents between the two smaller daughter cells. A cell does not really die when it reproduces itself; it merely starts over again.

A   Newly forming RNA

RNA polymerase

RNA nucleotides

Noncoding DNA strand

Coding DNA strand

B   Newly forming RNA

C   New RNA

FIGURE 26.15   This figure illustrates the basic events that occur during transcription. (*A*) An enzyme, RNA polymerase, attaches to the DNA at the (see *D*) promoter sequence and then separates the complementary strands. The enzyme then proceeds along the DNA strand in the correct direction to find the (*D*) protein coding region of the gene. (*B*) As RNA polymerase moves down the coding strand, new complementary RNA nucleotides are base-paired to one of the exposed DNA strands. The base-paired RNA nucleotides are linked together by RNA polymerase to form a new RNA molecule that is complementary to the nucleotide sequence of the DNA. The (*D*) termination sequence signals the RNA polymerase to end mRNA transcription, so that the RNA can leave the nucleus to aid in translation. (*C*) The newly formed (transcribed) RNA is then separated from the DNA molecule and used by the cell.

D   Promoter sequence   Protein code   Termination sequence

DNA Transcription

DNA functions in the manner of a reference library that does not allow its books to circulate. Information from the originals must be copied for use outside the library. The second major function of DNA is the making of these single-stranded, complementary RNA copies of DNA. This operation is called transcription (*scribe* = to write), which means to transfer data from one form to another. In this case, the data are copied from DNA language to RNA language. The same base-pairing rules that control the accuracy of DNA replication apply to the process of transcription. Using this process, the genetic information stored as a DNA chemical code is carried in the form of an RNA copy to other parts of the cell. It is RNA that is used to guide the assembly of amino acids into structural and regulatory proteins. Without the process of transcription, genetic information would be useless in directing cell functions.

Although many types of RNA are synthesized from the genes, the three most important are *messenger RNA (mRNA), transfer RNA (tRNA),* and *ribosomal RNA (rRNA).*

Transcription begins in a way that is similar to DNA replication. The DNA strands are separated by an enzyme, exposing the nitrogenous-base sequences of the two strands. However, unlike DNA replication, transcription occurs only on one of the two DNA strands, which serves as a template, or pattern, for the synthesis of RNA (see figure 26.15). This side is also referred to as the genetic *coding strand* of the DNA. But which strand is copied? Where does it start and when does it stop? Where along the sequence of thousands of nitrogenous bases does the chemical code for the manufacture of a particular enzyme begin and where does it end? If transcription begins randomly, the resulting RNA may not be an accurate copy of the code, and the enzyme product may be useless or deadly to the cell. To

Basic Steps of Translation

1. An mRNA molecule is placed in the smaller of the two portions of a ribosome so that six nucleotides (two codons) are locked into position.

2. The larger ribosomal unit is added to the ribosome/mRNA combination.

3. A tRNA with bases that match the second mRNA codon attaches to the mRNA. The tRNA is carrying a specific amino acid. Once attached, a second tRNA carrying another specific amino acid moves in and attaches to its complementary mRNA codon right next to the first tRNA/amino acid complex.

4. The two tRNAs properly align their two amino acids so that the amino acids may be chemically attached to one another (box figure 26.3).

5. Once the two amino acids are connected to one another by a covalent peptide bond, the first tRNA detaches from its amino acid and mRNA codon.

6. The ribosome moves along the mRNA to the next codon (the first tRNA is set free to move through the cytoplasm to attach to and transfer another amino acid).

7. The next tRNA/amino acid unit enters the ribosome and attaches to its codon next to the first set of amino acids.

8. The tRNAs properly align their amino acids so that they may be chemically attached to one another, forming a chain of three amino acids.

9. Once three amino acids are connected to one another, the second tRNA is released from its amino acid and

mRNA (this tRNA is set free to move through the cytoplasm to attach to and transfer another amino acid).

10. The ribosome moves along the mRNA to the next codon, and the fourth tRNA arrives (box figure 26.4).

11. This process repeats until all the amino acids needed to form the protein have attached to one another in the proper sequence. This amino acid sequence was encoded by the DNA gene.

12. Once the final amino acid is attached to the growing chain of amino acids, all the molecules (mRNA, tRNA, and newly formed protein) are released from the ribosome. The "stop" mRNA codon signals this action.

13. The ribosome is again free to become involved in another protein-synthesis operation.

14. The newly synthesized chain of amino acids (the new protein) leaves the ribosome to begin its work. However, the protein may need to be altered by the cell before it will be ready for use (box figure 26.5).

Box Figure 26.3

Box Figure 26.4

Box Figure 26.5

answer these questions, it is necessary to explore the nature of the genetic code itself.

We know that genetic information is in chemical-code form in the DNA molecule. When the coded information is used or *expressed,* it guides the assembly of particular amino acids into structural and regulatory polypeptides and proteins. If DNA is molecular language, then each nucleotide in this language can be

thought of as a letter within a four-letter alphabet. Each word, or code, is always three letters (nucleotides) long, and only three-letter words can be written. A **DNA code** is a triplet nucleotide sequence that codes for one of the twenty common amino acids. The number of codes in this language is limited because there are only four different nucleotides, which are used only in groups of three. The order of these three letters is just as important in DNA language as it is in our language. We recognize that CAT is not the same as TAC. If all the possible three-letter codes were written using only the four DNA nucleotides for letters, there would be a total of sixty-four combinations.

$$4^3 = 4 \times 4 \times 4 = 64$$

When codes are found at a particular place along a coding strand of DNA and the sequence has meaning, the sequence is a gene. "Meaning" in this case refers to the fact that the gene can be transcribed into an RNA molecule, which in turn may control the assembly of individual amino acids into a polypeptide.

When a gene is transcribed into RNA, the double-stranded DNA is "unzipped," and an enzyme known as *RNA polymerase* attaches to the DNA at the promoter region. It is from this region that the enzymes will begin to assemble RNA nucleotides into a complete, single-stranded copy of the gene, including initiation and termination codes. Triplet RNA nucleotide sequences complementary to DNA codes are called **codons**. Remember that there is no thymine in RNA molecules; it is replaced with uracil. Therefore, the initiation code in DNA (TAC) would be base-paired by RNA polymerase to form the RNA codon AUG. When transcription is complete, the newly assembled RNA is separated from its DNA template and made available for use in the cell; the DNA recoils into its original double-helix form.

In summary:

1. The process begins as one portion of the enzyme RNA polymerase breaks the attachments between the two strands of DNA; the enzyme "unzips" the two strands of the DNA.
2. A second portion of the enzyme RNA polymerase attaches at a particular spot on the DNA. It proceeds in one direction along one of the two DNA strands, attaching new RNA nucleotides into position until it reaches the end of the gene. The enzymes then assemble RNA nucleotides into a complete, singlestranded RNA copy of the gene. There is no thymine in RNA molecules; it is replaced by uracil. Therefore, the start codon in DNA (TAC) would be paired by RNA polymerase to form the RNA codon AUG.
3. The enzyme that speeds the addition of new nucleotides to the growing chain works along with another enzyme to make sure that no mistakes are made.
4. When transcription is complete, the newly assembled RNA is separated from its DNA template and made available for use in the cell; the DNA recoils into its original double-helix form.

As previously mentioned, three general types of RNA are produced by transcription: messenger RNA, transfer RNA, and ribosomal RNA. Each kind of RNA is made from a specific gene and performs a specific function in the synthesis of polypeptides from individual amino acids at ribosomes. **Messenger RNA (mRNA)** is a mature, straight-chain copy of a gene that describes the exact sequence in which amino acids should be bonded together to form a polypeptide.

Transfer RNA (tRNA) molecules are responsible for picking up particular amino acids and transferring them to the ribosome for assembly into the polypeptide (refer to

TABLE 26.3   **Twenty Common Amino Acids and Their Codes**

Amino Acid	Three-Letter Code
glycine	Gly
alanine	Ala
valine	Val
leucine	Leu
isoleucine	Ile
methionine	Met
phenylalanine	Phe
tryptophan	Trp
proline	Pro
serine	Ser
threonine	Thr
cysteine	Cys
tyrosine	Tyr
asparagine	Asn
glutamine	Gln
aspartic acid	Asp
glutamic acid	Glu
lysine	Lys
arginine	Arg
histidine	His

These are the twenty common amino acids used in the protein synthesis operation of a cell. Each has a known chemical structure.

"A Closer Look: Basic Steps of Translation"). All tRNA molecules are shaped like a cloverleaf. This shape is formed when they fold, and some of the bases form hydrogen bonds that hold the molecule together. One end of the tRNA is able to attach to a specific amino acid. Toward the midsection of the molecule, a triplet nucleotide sequence can base-pair with a codon on mRNA. This triplet nucleotide sequence on tRNA that is complementary to a codon of mRNA is called an **anticodon**. **Ribosomal RNA (rRNA)** is a highly coiled molecule and is used, along with protein molecules, in the manufacture of all ribosomes, the cytoplasmic organelles where tRNA, mRNA, and rRNA come together to help in the synthesis of proteins.

Translation or Protein Synthesis

The mRNA molecule is a coded message written in the biological world's universal nucleic acid language. The code is read in one direction, starting at the initiator. The information is used to assemble amino acids into protein by a process called translation. The word *translation* refers to the fact that nucleic acid language is being changed to protein language. To translate mRNA language into protein language, a dictionary is necessary. Remember, the four letters in the nucleic acid alphabet yield sixty-four possible three-letter words. The protein language has twenty words in the form of twenty common amino acids (table 26.3). Thus, there are more than enough nucleotide words for the twenty amino acid molecules because each nucleotide triplet codes for an amino acid.

TABLE 26.4   Amino Acid–mRNA Nucleic Acid Dictionary

Second letter

	U	C	A	G	
U	UUU ⎫ Phe UUC ⎬ UUA ⎫ Leu UUG ⎬	UCU ⎫ UCC ⎬ Ser UCA UCG ⎭	UAU ⎫ Tyr UAC ⎬ UAA Stop UAG Stop	UGU ⎫ Cys UGC ⎬ UGA Stop UGG Try	U C A G
C	CUU ⎫ CUC ⎬ Leu CUA CUG ⎭	CCU ⎫ CCC ⎬ Pro CCA CCG ⎭	CAU ⎫ His CAC ⎬ CAA ⎫ Gln CAG ⎬	CGU ⎫ CGC ⎬ Arg CGA CGG ⎭	U C A G
A	AUU ⎫ AUC ⎬ Ile AUA ⎭ AUG Met or start	ACU ⎫ ACC ⎬ Thr ACA ACG ⎭	AAU ⎫ ASN AAC ⎬ AAA ⎫ Lys AAG ⎬	AGU ⎫ Ser AGC ⎬ AGA ⎫ Arg AGG ⎬	U C A G
G	GUU ⎫ GUC ⎬ Val GUA GUG ⎭	GCU ⎫ GCC ⎬ Ala GCA GCG ⎭	GAU ⎫ Asp GAC ⎬ GAA ⎫ Glu GAG ⎬	GGU ⎫ GGC ⎬ Gly GGA GGG ⎭	U C A G

First letter · Third letter

Table 26.4 is an amino acid–nucleic acid dictionary. Notice that more than one codon may code for the same amino acid. Some would contend that this is needless repetition, but such "synonyms" can have survival value. If, for example, the gene or the mRNA becomes damaged in a way that causes a particular nucleotide base to change to another type, the chances are still good that the proper amino acid will be read into its proper position. But not all such changes can be compensated for by the codon system, and an altered protein may be produced. Changes can occur that cause great harm. Some damage is so extensive that the entire strand of DNA is broken, resulting in improper protein synthesis or a total lack of synthesis. Any change in DNA is called a **mutation.** Recall that some mutations may be so minor that they have no effect, while others may be harmful or beneficial. Although most mutations are likely neutral or harmful, beneficial mutations do occur.

The construction site of the protein molecules (i.e., the translation site) is on the ribosome, a cellular organelle that serves as the meeting place for mRNA and the tRNA that is carrying amino acid building blocks. Ribosomes can be found free in the cytoplasm or attached to the endoplasmic reticulum.

Thus, the mRNA moves through the ribosomes with the mRNA's codon sequence allowing for the chemical bonding of a specific sequence of amino acids. Remember that the sequence was originally determined by the DNA.

A protein's three-dimensional shape is determined by its amino acid sequence. This shape determines the activity of the protein molecule. The protein may be a structural component

of a cell or a regulatory protein, such as an enzyme. Any change in amino acids or their order changes the action of the protein molecule. The protein insulin, for example, has a different amino acid sequence than the digestive enzyme trypsin. Both proteins are essential to human life and must be produced constantly and accurately. The amino acid sequence of each is determined by a different gene. Each gene is a particular sequence of DNA nucleotides. Any alteration of that sequence can directly alter the protein structure and, therefore, the survival of the organism.

Alterations of DNA

Several kinds of changes to DNA may result in mutations. Phenomena that are either known or suspected causes of DNA damage are called **mutagenic agents.** Agents known to cause damage to DNA are certain viruses (e.g., papillomavirus), weak or "fragile" spots in the DNA, X radiation (X rays), and chemicals found in foods and other products, such as chemicals from burning tobacco. All have been studied extensively, and there is little doubt that they cause mutations. *Chromosomal aberrations* is the term used to describe major changes in DNA. Four types of aberrations are inversions, translocations, duplications, and deletions. An *inversion* occurs when a chromosome is broken and this piece becomes reattached to its original chromosome but in reverse order. It has been cut out and flipped around. A *translocation* occurs when one broken segment of DNA becomes integrated into a different chromosome. *Duplication* occurs when a portion of a chromosome is replicated and attached to the original section

Connections . . .

Astronauts, Cells, and DNA

Astronauts are regularly exposed to cosmic radiation and on occasion their DNA is damaged (box figure 26.6). Because DNA carries a cell's genetic information, damage may result in cell death, cancers, or other abnormalities. Research has shown that cells have the ability to repair damaged DNA. However, while cells can often fix minor damage successfully, they sometimes botch major repairs that can make a cell even more prone to becoming cancerous. So rather than attempt to fix itself, the repair mechanisms may be blocked by enzymes forcing a severely damaged cell to self-destruct. This actually keeps the astronaut healthier overall.

One hypothesis on how damaged DNA is repaired suggests that the repair happens right where the damage occurs. New research shows that some strands of DNA with minor damage are repaired on the spot. A second hypothesis proposes that cells move the most damaged DNA to special "repair shops" inside the cell. Scientist at NASA's Space Radiation Program suggest that rather than trying to gather the repair enzymes at the damage site, it might be more efficient to keep all these enzymes in "shops" near the chromosomes and take damaged DNA to them.

Should exposure to radiation increase for earth-bound organisms, it will be important for scientists to understand the molecular biology of DNA repair. This better understanding could enable medical professionals to limit or control radiation-induced illness.

A   B

FIGURE 26.16   (A) A normal red blood cell is shown in comparison with (B) a cell having the sickle shape. This sickling is the result of a single amino acid change in the hemoglobin molecule.

TABLE 26.5   Types of Chromosomal Aberrations

Normal Sequence	THE ONE BIG FLY HAD ONE RED EYE*
KIND OF MUTATION	**SEQUENCE CHANGE**
Missense	THQ ONE BIG FLY HAD ONE RED EYE
Nonsense	THE ONE BIG
Frameshift	THE ONE QBI GFL YHA DON ERE DEY
Deletion	THE ONE BIG HAD ONE RED EYE
Duplication	THE ONE BIG FLY FLY HAD ONE RED EYE
Insertion	THE ONE BIG WET FLY HAD ONE RED EYE
Expanding mutation:	
Parents	THE ONE BIG FLY HAD ONE RED EYE
Children	THE ONE BIG FLY FLY FLY HAD ONE RED EYE
Grandchildren	THE ONE BIG FLY FLY FLY FLY FLY FLY HAD ONE RED EYE

Based on R. Lewis, *Human Genetics, Concepts and Applications,* 2d ed., Dubuque, IA: Wm. C. Brown Publishers, 1997.

*A sentence composed of three-letter words can provide an analogy to the effect of mutations on a gene's nucleotide sequence.

in sequence. *Deletion* aberrations result when the broken piece becomes lost or is destroyed before it can be reattached.

In some individuals, a single nucleotide of the gene may be changed. This type of mutation is called a *point mutation.* An example of the effects of altered DNA may be seen in human red blood cells. Red blood cells contain the oxygen-transport molecule, hemoglobin. Normal hemoglobin molecules are composed of 150 amino acids in four chains—two alpha and two beta. The nucleotide sequence of the gene for the beta chain is known, as is the amino acid sequence for this chain. In normal individuals, the sequence begins like this:

Val-His-Leu-Thr-Pro-Glu-Glu-Lys . . . .

A single nucleotide change in the DNA sequence results in a mutation that produces a new amino acid sequence in all the red blood cells:

Val-His-Leu-Thr-Pro-Val-Glu-Lys . . . .

This single nucleotide change (known as a *missense point mutation*), which causes a single amino acid to change, may seem minor. However, it is the cause of **sickle-cell anemia,** a disease that affects the red blood cells by changing them from a circular to a sickle shape when oxygen levels are low (figure 26.16). When this sickling occurs, the red blood cells do not flow smoothly through capillaries. Their irregular shapes cause them to clump, clogging the blood vessels. This prevents them from delivering their oxygen load to the oxygen-demanding tissues. A number of physical disabilities may result, including physical weakness, brain damage, pain and stiffness of the joints, kidney damage, rheumatism, and, in severe cases, death.

Changes in the structure of DNA may have harmful effects on the next generation if they occur in the sex cells. Some damage to DNA is so extensive that the entire strand of DNA is broken, resulting in the synthesis of abnormal proteins or a total lack of protein synthesis. A number of experiments indicate that many street drugs, such as LSD (lysergic acid diethylamide), are mutagenic agents and cause DNA to break. Abnormalities have also been identified that are the result of changes in the number or sequence of bases. One way to illustrate these various kinds of mutations is seen in table 26.5.

USING DNA TO OUR ADVANTAGE

Advances in the field of biotechnology have allowed scientists to do many exciting things. To understand how these accomplishments are achieved by the scientists, we need to explore two key strategies: (1) genetic modification of an organism and (2) DNA sequencing.

Genetic Modification of Organisms

Since biblical times, civilizations have attempted to improve the quality of their livestock and agricultural stock. Cows that produced more milk or more tender meat were valued over those that were dry or tough. Initial attempts to develop improved agricultural stock were limited to selective breeding programs where organisms with the desired characteristics were allowed to breed. With a greater understanding of the genetics, scientists irradiated cells to produce mutations that were then screened to determine if they were desirable. While this approach is a very informative way to learn about the genetics of an organism, it lacked the ability to create a specific change. Creating mutations is a very haphazard process. However, today the results are achieved in a much more directed manner that uses biotechnology's ability to transfer DNA from one organism to another. Techniques such as gene cloning and transformation (introducing new DNA sequences into an organism to alter their genetic makeup) allow scientists to introduce very specific characteristics to an organism. These techniques use enzymes that manipulate DNAs to form complementary base pairs. Once the new DNA sequences are stable and transferred into the host cell, the cell is genetically altered and begins to read the new DNA and produce new cell products such as enzymes and other important products. The resulting new form of DNA is called **recombinant DNA.** Organisms that contain these genetic changes are referred to as **genetically modified (GM).** Genetic engineers identify and isolate sequences of nucleotides from a living or dead cell. Organisms such as viruses, bacteria, fungi, plants, and animals or their offspring have been engineered so that they contain genes from at least one unrelated organism.

As this highly sophisticated procedure has been refined, it has become possible to quickly and accurately splice genes from a variety of species into host bacteria or other host cells by a process called *gene cloning.* This makes possible the synthesis of large quantities of important products. For example, recombinant DNA procedures are responsible for the production of human insulin, used in the control of diabetes; nutritionally enriched "golden rice," capable of supplying underdeveloped nations with missing essential amino acids; interferon, used as an antiviral agent; human growth hormone, used to stimulate growth in children lacking this hormone; and somatostatin, a brain hormone also implicated in growth. Over two hundred such products have been manufactured using these methods.

Although some of these chemicals have been produced in small amounts from genetically engineered microorganisms, crops such as turnips, rice, soybeans, potatoes, cotton, corn, and tobacco can generate tens or hundreds of kilograms of specialty chemicals per year. Many of these GM crops also have increased nutritional value and yet can be cultivated using traditional methods. Such crops have the potential of supplying the essential amino acids, fatty acids, and other nutrients now lacking in the diets of people in underdeveloped or developing nations. Researchers have also shown, for example, that turnips can produce interferon (an antiviral agent), tobacco can create antibodies to fight human disease, oilseed rape plants can serve as a source of human brain hormones, and potatoes can synthesize human serum albumin that is indistinguishable from the genuine human blood protein.

Some of the likely rewards of biotechnology are (1) production of additional, medically useful proteins; (2) mapping of the locations of genes on human chromosomes; (3) a more complete understanding of how genes are regulated; (4) production of crop plants with increased yields; and (5) development of new species of garden plants (figure 26.17).

Gene Therapy

The field of biotechnology allows scientists and medical doctors to work together and potentially cure genetic disorders. Unlike contagious diseases, genetic diseases cannot be caught or transmitted because they are caused by a genetic predisposition for a particular disorder—not separate, disease-causing organisms such as a bacterium or virus. Gene therapy involves inserting genes, deleting genes, or manipulating the action of genes in order to cure or lessen the effect of genetic diseases. These activities are very new and experimental. While these lines of investigation create hope for many, there are many problems that must be addressed before gene therapy becomes a realistic treatment for many disorders.

The strategy for treating someone with gene therapy varies depending on the individual's disorder. When designing a gene therapy treatment, scientists have to ask themselves, "Exactly what is the problem?" Is the mutant gene not working at all? Is it working normally, but is there too little activity? Is there too much protein being made? Or possibly, is the problem that the gene is acting in a unique and new manner? If there is no gene activity or too little gene activity, the scientists need to somehow introduce a more active version of the gene. If there is too

FIGURE 26.17 Using our understanding of how DNA codes for proteins allows scientists to develop treatments for disease as well as develop crops with greater yield and nutritional value.

much activity or if the problem is caused by the gene having a new activity, this excess activity must first be stopped and then the normal activity restored.

To stop a mutant gene from working, scientists must change it. This typically involves inserting a mutation into the proteincoding region of the gene or the region that is necessary to activate the gene. Scientists have used some types of viruses to do this in organisms other than humans for some time now. The difficulty in this technique is to mutate only that one gene without disturbing the other genes of the cells and creating more mutations in other genes. Developing reliable methods to accomplish this is a major focus of gene therapy. Once the mutant gene is silenced, the scientists begin the work of introducing a

"good" copy of the gene. Again, there are many difficulties in this process:

- Scientists must find a way of returning the corrected DNA to the cell.
- The corrected DNA must be made a part of the cell's DNA so that it is passed on with each cell division, it doesn't interfere with other genes, and it can be transcribed by the cell as needed.
- Finally, cells containing the corrected DNA must be reintroduced to the patient.

Many of these techniques are experimental, and the medical community is still evaluating their usefulness in treating many

Stem cells are cells that are self-renewing and have not yet completed determination or differentiation, so they have the potential to develop into many different cell types. Scientists can generate stem cells by nuclear transfer techniques but occur naturally throughout the body. They are involved in many activities including tissue regeneration, wound healing, and cancer.

If scientists had the ability to control determination and differentiation, it may allow the manipulation of an organism's cells or the insertion of cells into an organism to allow the regrowth of damaged tissues and organs in humans. This could aid in the cure or treatment of many medical problems, such as the repair of damaged knee cartilage, heart tissue from a heart attack, or damaged nerve tissue from spinal or head injuries. Some kinds of degenerative diseases occur because specific kinds of cells die or cease to function properly. Parkinson's disease results from malfunctioning brain cells, and many forms of diabetes are caused by malfunctioning cells in the pancreas. If stem cells could be used to replace these malfunctioning cells, normal function could be restored and the diseases cured.

A cloning experiment has great scientific importance, because it represents an advance in scientists understanding of the processes of *determination* and *differentiation*. Recall that determination is the process a cell goes through to select which genes it will express. A differentiated cell has become a particular cell type because of the proteins that it expresses. Differentiation is more or less a permanent condition. The techniques that produced Dolly and other cloned animals use a differentiated cell and reverse the determination process, so that this cell is able to express all the genes necessary to create an entirely new organism. Until this point, scientists were not sure that this was possible.

Embryonic and Adult Stem Cells

Because embryonic stem cells have not undergone determination and differentiation and have the ability to become *any* tissue in the body, they are of great interest to scientists. As an embryo develops, its stem cells go through the process of determination and differentiation to create all the necessary tissues. To study embryonic stem cells, scientists must remove them from embryos, destroying the embryos (box figure 26.7).

Scientists have also explored other methods of obtaining stem cells. Embryonic stem cells reach an intermediary level of determination at which they are committed to becoming a particular *tissue* type, but not necessarily a particular *cell* type. An example of this intermediate determination occurs when stems cells become determined to be any one of several types of nerve cells but have not yet committed to becoming any one nerve cell. Scientists call these partially determined stem cells

"tissue-specific." These types of stem cells can be found in adults. One example is hematopoietic stem cells. These cells are able to become the many different types of cells found in blood—red blood cells, white blood cells, and platelets. The disadvantage of using these types of stem cells is that they have already become partially determined and do not have the potential to become every cell type.

Personalized Stem Cell Lines

Scientists hope that eventually it will be possible to produce embryonic stem cells from somatic cells by using somatic cell nuclear transfer techniques similar to that used for cloning a sheep. This technique would involve transferring a nucleus from the patient's cell to a human egg that has had its original nucleus removed. The human egg would be allowed to grow and develop to produce embryonic stem cells. If the process of determination and differentiation can be controlled, new tissues or even organs could be developed through what is termed regenerative medicine.

Under normal circumstances, organ transplant patients must always worry about rejecting their transplant and take strong immunosuppressant drugs to avoid organ rejection. Tissue and organs grown from customized stem cells would have the benefit of being immunologically compatible with the patient; thus, organ rejection would not be a concern (box figure 26.8).

(–continued on page 686)

Egg

Inner cell mass

Embryonic stem-cell culture

Sperm

Blastocyst

Once sperm cell and egg cell have joined, cell cleavage produces a blastocyst. The inner cell mass of the blastocyst develops into the human embryo.

Embryo

Embrionic stem cells can also be harvested from the part of the embryo that will develop into gonads.

Box Figure 26.7 After fertilization of an egg with sperm, the cell begins to divide and form a mass of cells. Each of these cells has the potential to become any cell in the embryo. Embryonic stem cells may be harvested at this point or at other points in the determination process.

(–continued from page 685)

The potential therapeutic value of stem cells has resulted in the founding of many clinics around the world. These clinics offer stem cell-based therapies to patients with a variety of medical conditions. However, the benefits of these therapies are as yet unproven and in fact have the potential for serious harm. "Stem cell tourism" is a phrase being used to describe this industry since desperate patients travel to these clinics in hopes that such therapies will save their lives. This new industry is teeming with 'medical tourist traps' offering unproven medical treatments to unsuspecting consumers. Unfortunately, the days of customized stem cells, stem cell therapies, and organ culture are still in the future.

Box Figure 26.8 One potential use of biotechnology is the production of customized stem cell lines. In this application a somatic cell from a patient would be inserted into a human egg from which the nucleus has been removed. The egg would divide and generate stem cells. These cells could then be cultured and used for therapy. In this example the stem cells could be used to create pancreatic cells to treat a diabetic patient.

disorders, as well as the risks that these techniques pose to the patient.

Recently, the first efforts to determine the human genome, the entire human DNA sequence, were completed. Many scientists feel that advances in medical treatments will occur more quickly by having this information available. Three new fields of biology have grown out of these efforts: *genomics, transcriptomics,* and *proteomics.* Genomics is the study of the DNA sequence and looks at the significance of how different genes and DNA sequences are related to each other. When genes are identified from the DNA sequence, transcriptomics looks at when, where, and how much a gene is expressed. Finally, proteomics examines the proteins that are predicted from the DNA sequence. From these types of studies, we are able to identify gene families that can be used to determine how humans have evolved on a molecular level, how genes are used in an organism throughout its body and over its life span, and how to identify common themes from one protein to the next.

Strategy Two: Sequencing

The second strategy of biotechnology works with comparing of nucleotide sequences from different cells. The closer organisms are genetically, the more closely their DNA sequence will resemble one another. Base sequence analysis can be done two ways. Researchers can look at the DNA sequences directly for comparison. However, this is a very tedious and time-consuming process. A frequently used alternative is to use enzymes to cut the DNA in

Cloning—What This Term Really Means

The term **clone** (Greek word *klön* for "twig") refers to exact copies of biological entities such as genes, cells, or organisms. The term refers to the outcome, not the way that the results are achieved. Many whole organisms "clone" themselves (e.g., bacteria undergo asexual binary fission, and many plants reproduce "clonally" through asexual means, sometimes referred to as *vegetative regeneration*). One process of making exact copies of the cells of a cat, sheep, or other organism in the laboratory is technically known as "*somatic cell nuclear transfer.*" The purpose of somatic cell nuclear transfer is not the "cloning" of a human or other adult but the establishment of a clonal cell line. In this technique, the nucleus of a somatic (body) cell of an adult donor is transferred into an oocyte, or "egg cell," whose own nucleus has been removed. Should such hybrid cells be cultured *in vitro* (*vitro* = glass; e.g., in test tubes) for a few weeks, the result is the production of a line of cloned cells (e.g., human embryonic stem cells). These can be used for a variety of medically valuable therapies or research (e.g., investigating certain genetic or other diseases). Using the other technique, *reproductive cloning*, the hybrid cell can be implanted into the uterus of a host or surrogate mother, where it may develop into a genetic copy or clone of the whole individual who donated the nucleus. A great variety of animals has already been cloned in this fashion, including cats, sheep, mice, monkeys, pigs, cattle, rabbits, mules, deer, horses, fish, and frogs.

In the film *Godsend*, starring Robert De Niro, Greg Kinnear, and Rebecca Romijn-Stamos, parents agree to clone their accidentally killed, eight-year-old son but wish they hadn't when the clone starts behaving erratically.

Questions to Discuss

1. Based on what you have learned, do you think cloning as depicted in *Godsend* would be possible?
2. What are the ethical and moral implications of taking such action?
3. Does it make a difference if the organism cloned is a human or a cat?

A

B

Box Figure 26.9 "CC," the cloned kitty, with surrogate mother and her genetic mother, "Rainbow." Out of 87 implanted cloned embryos, CC (*Copy Cat!*) is the only one to survive—comparable to the success rate in sheep, mice, cows, goats, and pigs. (*A*) Notice that she is completely unlike her tabby surrogate mother. (*B*) "Rainbow" is her genetic donor, and both are female calico domestic shorthair cats.
Source: Nature, AOP, published online: 14 February 2002; DOI: 10.1038/nature723.

specific places to create DNA fragments of different lengths, separate out DNA fragments by size, and analyze these for similarities and differences. This second approach scans bigger stretches of DNA more quickly but not as closely. This method can be used by biologists to better understand evolutionary relationships between what at first glance might be unrelated species. This second category of research involves directly manipulating DNA using the more sophisticated techniques such as the polymerase chain reaction (PCR), genetic fingerprinting, and cloning.

The PCR Reaction

Both the cloning and sequencing strategies of biotechnology have been greatly aided by the development of a technique called PCR. In 1989, the American Association for the Advancement of Science named DNA polymerase "Molecule of the Year." The value of this enzyme in the *polymerase chain reaction (PCR)* is so great that it could not be ignored. Just what is the PCR, how does it work, and what can you do with it?

PCR is a laboratory procedure for copying selected segments of DNA. A single cell can provide enough DNA for analysis and identification! Having a large number of copies of a "target sequence" of nucleotides enables biochemists to more easily work with DNA. This is like increasing the one "needle in the haystack" to such large numbers (one-hundred billion in only a matter of hours) that they are not hard to find, recognize, and work with. The types of specimens that can be used include semen, hair, blood, bacteria, protozoa, viruses, mummified tissue, and frozen cells. The process requires the DNA specimen, free DNA nucleotides, synthetic "primer" DNA, DNA polymerase, and simple lab equipment such as a test tube and a source of heat.

Having decided which target sequence of nucleotides (which "needle") is to be replicated, scientists heat the specimen of DNA to separate the coding and noncoding strands. Molecules of synthetic "primer" DNA are added to the specimen. These primer molecules are specifically designed to attach to the ends of the target sequence. Next, a mixture of nucleotides is added so that they can become the newly replicated DNA. The presence of the primer, attached to the DNA and added nucleotides, serves as the substrate for the DNA polymerase. Once added, the polymerase begins making its way down the length of the DNA from one attached primer end to the other. The enzyme bonds the new DNA nucleotides to the strand, replicating the molecule as it goes. It stops when it reaches the other end, having produced a new copy of the target sequence. Because the DNA polymerase will continue to operate as long as enzymes and substrates are available, the process continues, and in a short time, there are billions of small pieces of DNA, all replicas of the target sequence.

So what, you say? Well, consider the following. Using the PCR, scientists have been able to:

1. More accurately diagnose such diseases as sickle-cell anemia, cancer, Lyme disease, AIDS, and Legionnaires' disease;
2. Perform highly accurate tissue typing for matching organ transplant donors and recipients;
3. Help resolve criminal cases of rape, murder, assault, and robbery by matching a suspect's DNA to that found at the crime scene;
4. Detect specific bacteria in environmental samples;
5. Monitor the spread of genetically engineered microorganisms in the environment;
6. Check water quality by detecting bacterial contamination from feces;
7. Identify viruses in water samples;
8. Identify disease-causing protozoa in water;
9. Determine specific metabolic pathways and activities occurring in microorganisms;
10. Determine races, distribution patterns, kinships, migration patterns, evolutionary relationships, and rates of evolution of long-extinct species;
11. Accurately settle paternity suits;
12. Confirm identity in amnesia cases;
13. Identify a person as a relative for immigration purposes;
14. Provide the basis for making human antibodies in specific bacteria;
15. Possibly provide the basis for replicating genes that could be transplanted into individuals suffering from genetic diseases; and
16. Identify nucleotide sequences peculiar to the human genome (an application currently underway as part of the Human Genome Project).

Genetic Fingerprinting

With another genetic engineering accomplishment, *genetic fingerprinting*, it is possible to show the nucleotide sequence differences between individuals since no two people have the same

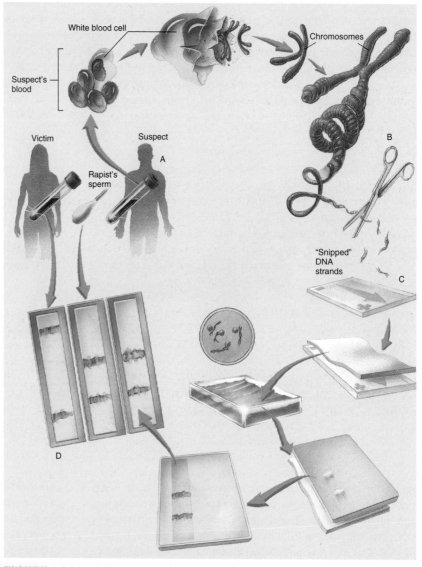

FIGURE 26.18 Because every person's DNA is unique (*A*), when samples of an individual's DNA are subjected to restriction enzymes, the cuts will occur in different places and DNA chunks of different sizes will result. (*B*) Restriction enzymes have the ability to cut DNA at places where specific sequences of nucleotides occur. When the chunks are caused to migrate across an electrophoresis gel (*C*), the smaller fragments migrate farther than the larger fragments, producing a pattern known as a "DNA fingerprint" (*D*). Because of individual differences in DNA sequences, these sites vary from person to person. As a result, the DNA fingerprint that separates DNA fragments on the basis of size can appear different from one person to another. Several controls are done in this type of experiment. One of the controls is the presence of the victim's DNA. Why is this important for the interpretation of this test?

nucleotide sequences. While this sounds like an easy task, the presence of many millions of base pairs in a person's chromosomes makes this a lengthy and sometimes impractical process. Therefore, scientists don't really do a complete fingerprint but focus only on certain shorter, repeating patterns in the DNA. By focusing on these shorter, repeating nucleotide sequences, it is possible to determine whether samples from two individuals have these same repeating segments. Scientists use a small number of sequences that are known to vary a great deal among individuals and compare those to get a certain probability of a match. The more similar the sequences, the more likely the two samples are from the same person. The less similar the sequences, the less likely the two samples are from the same person. In criminal cases, DNA samples from the crime site can be compared to those taken from suspects. If 100 percent of the short, repeating sequence matches, it is highly probable that the suspect was at the scene of the crime and may be the guilty party. This same procedure can also be used to confirm the identity of a person, as in cases of amnesia, murder, or accidental death (figure 26.18).

SUMMARY

Genes are units of heredity composed of specific lengths of *DNA* that determine the characteristics an organism displays. Specific genes are at specific loci on specific chromosomes. The *phenotype* displayed by an organism is the result of the effect of the environment on the ability of the genes to *express* themselves.

Diploid organisms have two genes for each characteristic. The alternative forms of genes for a characteristic are called *alleles*. There may be many different alleles for a particular characteristic. Organisms with two identical alleles are *homozygous* for a characteristic; those with different alleles are *heterozygous*. Some alleles are *dominant* over other alleles that are said to be *recessive*.

Sometimes two alleles express themselves, and often a gene has more than one recognizable effect on the phenotype of the organism. Some characteristics may be determined by several different pairs of alleles. In humans and some other organisms, phenotypes are determined not only by the type of allele (gene) received but also by the alternative ways in which they interact with each other and how the environment influences their expression and epigenetic factors. There are six generally recognized patterns: *multiple allelic, polygenic, pleiotropic, codominant, incomplete dominant,* and *X-linked characteristics.*

The successful operation of a living cell depends on its ability to accurately use the genetic information found in its DNA. The *enzymes* that can be synthesized using the information in DNA are responsible for the efficient control of a cell's metabolism. However, the production of protein molecules is under the control of the *nucleic acids,* the primary control molecules of the cell. The structure of the nucleic acids, *DNA* and *RNA,* determines the structure of the proteins, whereas the structure of the proteins determines their function in the cell's life cycle. *Protein synthesis* involves the decoding of the DNA into specific protein molecules and the use of the intermediate molecules, mRNA and tRNA, at the ribosome. Errors in any of the codons of these molecules may produce observable changes in the cell's functioning and can lead to cell death.

DNA replication results in an exact doubling of the genetic material. The process virtually guarantees that identical strands of DNA will be passed on to the next generation of cells.

Methods of manipulating DNA have led to the controlled transfer of genes from one kind of organism to another. This has made it possible for bacteria to produce a number of human gene products.

KEY TERMS

alleles (p. **663**)
anticodon (p. **680**)
chromosome (p. **675**)
clone (p. **687**)
codon (p. **680**)
complementary base (p. **674**)
DNA code (p. **679**)
DNA replication (p. **674**)
dominant allele (p. **664**)
epigenetics (p. **673**)
gene (p. **662**)
genetically modified (GM)
   organisms (p. **683**)

genetics (p. **662**)
genome (p. **663**)
genotype (p. **663**)
heterozygous (p. **664**)
homozygous (p. **663**)
incomplete dominance
   (p. **668**)
law of dominance (p. **665**)
law of independent assortment
   (p. **665**)
law of segregation (p. **665**)
Mendelian genetics (p. **662**)
messenger RNA (mRNA) (p. **680**)

multiple alleles (p. **669**)
mutagenic agent (p. **681**)
mutation (p. **681**)
nucleic acids (p. **674**)
nucleotides (p. **674**)
phenotype (p. **663**)
pleiotropy (p. **670**)
polygenic inheritance (p. **670**)
Punnett square (p. **667**)
recessive allele (p. **664**)
recombinant DNA (p. **683**)
ribosomal RNA (rRNA)
   (p. **680**)

sickle-cell anemia (p. **682**)
single-factor cross (p. **666**)
stem cell (p. **685**)
transcription (p. **674**)
transfer RNA (tRNA)
   (p. **680**)
translation (p. **674**)
X-linked gene (p. **668**)

APPLYING THE CONCEPTS

1. An example of a phenotype is
   a. AB type blood.
   b. an allele for type A and B blood.
   c. a sperm with an allele for A blood.
   d. lack of iron in the diet causing anemia.

2. A person's height is determined by the interaction of numerous alleles. This is an example of
   a. autosomes.
   b. pleiotropy.
   c. single-factor crosses.
   d. polygenic inheritance.

3. In order for a recessive X-linked trait to appear in a female, she must inherit a recessive allele from
   a. neither parent.
   b. both parents.
   c. her father only.
   d. her mother only.

4. It was noticed that in certain flowers, if a flower had red petals, they were usually small petals. If the petals were white, pink, or orange, they were usually large petals. This could be the result of
   a. codominance.
   b. polygenic inheritance.
   c. incomplete dominance.
   d. None of these is correct.

5. A woman with blood type O and a man with blood type AB have a child together. What are the possible blood types of this child?
   a. AB or O
   b. A, B, or O
   c. A or B
   d. AB, A, B, or O

6. "She has really long fingers and toes and is exceptionally tall." This statement expresses the _____ of an individual.
   a. genotype
   b. phenotype
   c. monohybridization
   d. locus placement

7. One difference between mRNA and mature tRNA is the:
   a. kinds of polypeptide components.
   b. mRNA is straight while tRNA is cloverleaf-shaped.
   c. part of RNA from which they were coded.
   d. age of the molecules only.

8. While one strand of double-stranded DNA is being transcribed to mRNA:
   a. the complementary strand makes tRNA.
   b. the complementary strand is inactive.
   c. the complementary strand at this point is replicating.
   d. mutations are impossible during this short period.

9. A major difference between the genetic data of prokaryotic and eukaryotic cells is that in prokaryotes, the
   a. genes are RNA not DNA.
   b. histones are arranged differently.
   c. double-stranded DNA is circular.
   d. double-stranded DNA is absent in bacteria.

10. If the DNA gene strand has the base sequence CCA, which of the following DNA nitrogenous bases would pair with them during DNA replication?
    a. CCA
    b. GGU
    c. GGU
    d. GGT

11. Using the Amino Acid–mRNA Nucleic Acid Dictionary (table 26.4), the codon ACG would control the placement of which amino acid in the synthesis of a new protein?
    a. proline
    b. isoleucine
    c. glycine
    d. threonine

12. This new field of biotechnology examines the proteins that are predicted from the DNA sequence.
    a. human genome
    b. codon technology
    c. proteomics
    d. hormone therapy

Answers

1. a 2. d 3. b 4. c 5. c 6. b 7. b 8. b 9. c 10. d 11. d 12. c

QUESTIONS FOR THOUGHT

1. What are the differences between DNA and RNA?

2. List the sequence of events that takes place when a DNA message is translated into protein.

3. Chromosomal and point mutations both occur in DNA. In what ways do they differ? How is this related to recombinant DNA?

4. How does DNA replication differ from the manufacture of an RNA molecule?

5. If a DNA nucleotide sequence is CATAAAGCA, what is the mRNA nucleotide sequence that would base-pair with it?

6. What amino acids would occur in the protein chemically coded by the sequence of nucleotides in question 5?

7. What is the probability of each of the following sets of parents producing the given genotypes in their offspring?

Parents	Offspring	Genotype
(a) AA	× aa	Aa
(b) Aa	× Aa	Aa
(c) Aa	× Aa	aa

8. If an offspring has the genotype Aa, what possible combinations of parental genotypes can exist?

9. In humans, the allele for albinism is recessive to the allele for normal skin pigmentation.
   a. What is the probability that a child of a heterozygous mother and father will be an albino?
   b. If a child is normal, what is the probability that it is a carrier of the recessive albino allele?

10. Dwarfism (achondroplasia) is the result of a mutation in the gene for the production of fibroblast growth factor. It is inherited as an autosomal dominant disorder and often occurs as a new mutation. What might be the explanation if two people of normal height and body stature have five children, one of whom displays achondroplasia?

11. A woman with color-deficient vision marries a man with normal vision. They have ten children—six boys and four girls.
    a. How many are expected to have normal vision?
    b. How many are expected to have color-deficient vision?

12. Hemophilia is a disease that prevents the blood from clotting normally. It is caused by a recessive allele located on the X chromosome. A boy has the disease; neither his parents nor his grandparents have the disease. What are the genotypes of his parents and grandparents?

FOR FURTHER ANALYSIS

1. There are two approaches to gene therapy: somatic cell and germ cell line therapy. Compare and contrast these two approaches. Give examples of genetic abnormalities that might be controlled with each method.

2. It is important to remember that Mendel's work was introduced only a little over a hundred years ago. But in that time, science has progressed from his most basic hypothesis to a level of understanding that has already enabled the control of some of our most notorious genetic abnormalities. List five such diseases. List the kinds of genetic medicine that are proposed to help with each.

3. What is the genetic basis of the following diseases and what are their symptoms?
   a. Kearns-Sayre syndrome
   b. Leber's hereditary optic neuropathy (LHON)
   c. Myoclonic epilepsy and red ragged fibers (MERRF)

4. Polydactyly is the most common genetic abnormality of the human hand and is classified into three types.
   a. What are these types?
   b. Diagram how such phenotypes would arise beginning with a mutation in the DNA, following through the central dogma and ending with how this phenotype would be produced.

5. It is known that the difference between the genomes of humans and chimpanzees is about 1.5 percent. If there is only this small difference between what appears to be two strikingly different organisms, what percentage difference do you think exists among humans? Based on this information, comment on the validity of separating humans of the world into separate races.

INVITATION TO INQUIRY

Smoking and Mutations: What Do You Really Know?

For decades, cancer has been empirically (based on commonplace experiences) linked to the use of tobacco, but only recently have researchers showed a cause-and-effect relationship between the two. Studies of patients with lung cancer revealed that 60 percent exhibited mutations in a gene known as p53. Use the Internet to answer the following questions:

1. What is the p53 gene?
2. What is apoptosis or "programmed cell death"?
3. What does the compound in tobacco (benzopyrene) have to do with this gene?
4. What is known about the p53 gene and lung cancer?

Appendix A

Mathematical Review

WORKING WITH EQUATIONS

Many of the problems of science involve an equation, a short-hand way of describing patterns and relationships that are observed in nature. Equations are also used to identify properties and to define certain concepts, but all uses have well-established meanings, symbols that are used by convention, and allowed mathematical operations. This appendix will assist you in better understanding equations and the reasoning that goes with the manipulation of equations in problem-solving activities.

Background

In addition to a knowledge of rules for carrying out mathematical operations, an understanding of certain quantitative ideas and concepts can be very helpful when working with equations. Among these helpful concepts are (1) the meaning of inverse and reciprocal, (2) the concept of a ratio, and (3) fractions.

The term *inverse* means the opposite, or reverse, of something. For example, addition is the opposite, or inverse, of subtraction, and division is the inverse of multiplication. A *reciprocal* is defined as an inverse multiplication relationship between two numbers. For example, if the symbol n represents any number (except zero), then the reciprocal of n is $1/n$. The reciprocal of a number $(1/n)$ multiplied by that number (n) always gives a product of 1. Thus, the number multiplied by 5 to give 1 is $1/5(5 \times 1/5 = 5/5 = 1)$. So 1/5 is the reciprocal of 5, and 5 is the reciprocal of 1/5. Each number is the *inverse* of the other.

The fraction 1/5 means 1 divided by 5, and if you carry out the division, it gives the decimal 0.2. Calculators that have a $1/x$ key will do the operation automatically. If you enter 5, then press the $1/x$ key, the answer of 0.2 is given. If you press the $1/x$ key again, the answer of 5 is given. Each of these numbers is a reciprocal of the other.

A *ratio* is a comparison between two numbers. If the symbols m and n are used to represent any two numbers, then the ratio of the number m to the number n is the fraction m/n. This expression means to divide m by n. For example, if m is 10 and n is 5, the ratio of 10 to 5 is 10/5, or 2:1.

Working with *fractions* is sometimes necessary in problem-solving exercises, and an understanding of these operations is needed to carry out unit calculations. It is helpful in many of these operations to remember that a number (or a unit) divided by itself is equal to 1; for example,

$$\frac{5}{5} = 1 \qquad \frac{\text{inch}}{\text{inch}} = 1 \qquad \frac{5 \text{ inches}}{5 \text{ inches}} = 1$$

When one fraction is divided by another fraction, the operation commonly applied is to "invert the denominator and multiply." For example, 2/5 divided by 1/2 is

$$\frac{\frac{2}{5}}{\frac{1}{2}} = \frac{2}{5} \times \frac{2}{1} = \frac{4}{5}$$

What you are really doing when you invert the denominator of the larger fraction and multiply is making the denominator (1/2) equal to 1. Both the numerator (2/5) and the denominator (1/2) are multiplied by 2/1, which does not change the value of the overall expression. The complete operation is

$$\frac{\frac{2}{5}}{\frac{1}{2}} \times \frac{\frac{2}{1}}{\frac{2}{1}} = \frac{\frac{2}{5} \times \frac{2}{1}}{\frac{1}{2} \times \frac{2}{1}} = \frac{\frac{4}{5}}{\frac{2}{2}} = \frac{\frac{4}{5}}{1} = \frac{4}{5}$$

Symbols and Operations

The use of symbols seems to cause confusion for some students because it seems different from their ordinary experiences with arithmetic. The rules are the same for symbols as they are for numbers, but you cannot do the operations with the symbols until you know what values they represent. The operation signs, such as $+$, $\div$, $\times$, and $-$, are used with symbols to indicate the operation that you *would* do if you knew the values. Some of the mathematical operations are indicated several ways. For example, $a \times b$, $a \cdot b$, and ab all indicate the same thing, that a is to be multiplied by b. Likewise, $a \div b$, a/b, and $a \times 1/b$ all indicate that a is to be divided by b. Since it is not possible to carry out the operations on symbols alone, they are called *indicated operations*.

Operations in Equations

An equation is a shorthand way of expressing a simple sentence with symbols. The equation has three parts: (1) a left side, (2) an equal sign (=), which indicates the equivalence of the two sides, and (3) a right side. The left side has the same value and units as the right side, but the two sides may have a very different appearance. The two sides may also have the symbols that indicate mathematical operations ($+$, $-$, $\times$, and so forth) and may be in certain forms that indicate operations (a/b, ab, and so forth). In any case, the equation is a complete expression that states the left side has the same value and units as the right side.

Equations may contain different symbols, each representing some unknown quantity. In science, the expression "solve the equation" means to perform certain operations with one symbol (which represents some variable) by itself on one side of the equation. This single symbol is usually, but not necessarily, on the left side and is not present on the other side. For example, the equation $F = ma$ has the symbol F on the left side. In science, you would say that this equation is solved for F. It could also be solved for m or for a, which will be considered shortly. The equation $F = ma$ is solved for F, and the *indicated operation* is to multiply m by a because they are in the form ma, which means the same thing as $m \times a$. This is the only indicated operation in this equation.

A solved equation is a set of instructions that has an order of indicated operations. For example, the equation for the relationship between a Fahrenheit and Celsius temperature, solved for °C, is $C = 5/9(F - 32)$. A list of indicated operations in this equation is as follows:

1. Subtract 32° from the given Fahrenheit temperature.
2. Multiply the result of (1) by 5.
3. Divide the result of (2) by 9.

Why are the operations indicated in this order? Because the bracket means 5/9 of the *quantity* $(F - 32°)$. In its expanded form, you can see that $5/9(F - 32°)$ actually means $5/9(F) - 5/9(32°)$. Thus, you cannot multiply by 5 or divide by 9 until you have found the quantity of $(F - 32°)$. Once you have figured out the order of operations, finding the answer to a problem becomes almost routine as you complete the needed operations on both the numbers and the units.

Solving Equations

Sometimes it is necessary to rearrange an equation to move a different symbol to one side by itself. This is known as solving an equation for an unknown quantity. But you cannot simply move a symbol to one side of an equation. Since an equation is a statement of equivalence, the right side has the same value as the left side. If you move a symbol, you must perform the operation in a way that the two sides remain equivalent. This is accomplished by "canceling out" symbols until you have the unknown on one side by itself. One key to understanding the canceling operation is to remember that a fraction with the same number (or unit) over itself is equal to 1. For example, consider the equation $F = ma$, which is solved for F. Suppose you are considering a problem in which F and m are given, and the unknown is a. You need to solve the equation for a so it is on one side by itself. To eliminate the m, you do the *inverse* of the indicated operation on m, dividing both sides by m. Thus,

Solve an Equation

The equation for finding the kinetic energy of a moving body is $KE = 1/2mv^2$. You need to solve this equation for the velocity, v.

Answer

The order of indicated operations in the equation is as follows:

1. Square v.
2. Multiply v^2 by m.
3. Divide the result of (2) by 2.

To solve for v, this order is *reversed* as the "canceling operations" are used:

Step 1: Multiply both sides by 2

$$KE = \frac{1}{2}mv^2$$

$$2\,KE = \frac{2}{2}mv^2$$

$$2\,KE = mv^2$$

Step 2: Divide both sides by m

$$\frac{2\,KE}{m} = \frac{mv^2}{m}$$

$$\frac{2\,KE}{m} = v^2$$

Step 3: Take the square root of both sides

$$\sqrt{\frac{2\,KE}{m}} = \sqrt{v^2}$$

$$\sqrt{\frac{2\,KE}{m}} = v$$

or

$$v = \sqrt{\frac{2\,KE}{m}}$$

The equation has been solved for v, and you are now ready to substitute quantities and perform the needed operations.

$$F = ma$$

$$\frac{F}{m} = \frac{ma}{m}$$

$$\frac{F}{m} = a$$

Since m/m is equal to 1, the a remains by itself on the right side. For convenience, the whole equation may be flipped to move the unknown to the left side,

$$a = \frac{F}{m}$$

Thus, a quantity that indicated a multiplication (ma) was removed from one side by an inverse operation of dividing by m.

Consider the following inverse operations to "cancel" a quantity from one side of an equation, moving it to the other side:

If the Indicated Operation of the Symbol You Wish to Remove Is:	Perform This Inverse Operation on Both Sides of the Equation
multiplication	division
division	multiplication
addition	subtraction
subtraction	addition
squared	square root
square root	square

SIGNIFICANT FIGURES

The numerical value of any measurement will always contain some uncertainty. Suppose, for example, that you are measuring one side of a square piece of paper as shown in figure A.1. You could say that the paper is *about* 3.5 cm wide and you would be correct. This measurement, however, would be unsatisfactory for many purposes. It does not approach the true value of the length and contains too much uncertainty. It seems clear that the paper width is larger than 3.4 cm but shorter than 3.5 cm. But how much larger than 3.4 cm? You cannot be certain if the paper is 3.44, 3.45, or 3.46 cm wide. As your best estimate, you might say that the paper is 3.45 cm wide. Everyone would agree that you can be certain about the first two numbers (3.4) and they should be recorded. The last number (0.05) has been estimated and is not certain. The two certain numbers, together with one uncertain number, represent the greatest accuracy possible with the ruler being used. The paper is said to be 3.45 cm wide.

A *significant figure* is a number that is believed to be correct with some uncertainty only in the last digit. The value of the width of the paper, 3.45 cm, represents three significant figures. As you can see, the number of significant figures can be determined by the degree of accuracy of the measuring instrument being used. But suppose you need to calculate the area of the paper. You would multiply 3.45 cm $\times$ 3.45 cm and the product for the area

would be 11.9025 cm^2. This is a greater accuracy than you were able to obtain with your measuring instrument. The result of a calculation can be no more accurate than the values being treated. Because the measurement had only three significant figures (two certain, one uncertain), then the answer can have only three significant figures. The area is correctly expressed as 11.9 cm^2.

There are a few simple rules that will help you determine how many significant figures are contained in a reported measurement:

1. All digits reported as a direct result of a measurement are significant.
2. Zero is significant when it occurs between nonzero digits. For example, 607 has three significant figures, and the zero is one of the significant figures.
3. In figures reported as *larger than the digit one,* the digit zero is not significant when it follows a nonzero digit to indicate place. For example, in a report that "23,000 people attended the rock concert," the digits 2 and 3 are significant but the zeros are not significant. In this situation, the 23 is the measured part of the figure, and the three zeros tell you an estimate of how many attended the concert, that is, 23 thousand. If the figure is a measurement rather than an estimate, then it is written *with a decimal point after the last zero* to indicate that the zeros *are* significant. Thus, 23,000 has *two* significant figures (2 and 3), but 23,000. has *five* significant figures. The figure 23,000 means "about 23 thousand," but 23,000. means 23,000. and not 22,999 or 23,001.
4. In figures reported as *smaller than the digit 1,* zeros after a decimal point that come before nonzero digits *are not* significant and serve only as place holders. For example, 0.0023 has two significant figures, 2 and 3. Zeros alone after a decimal point or zeros after a nonzero digit indicate a measurement, however, so these zeros *are* significant. The figure 0.00230, for example, has three significant figures since the 230 means 230 and not 229 or 231. Likewise, the figure 3.000 cm has four significant figures because the presence of the three zeros means that the measurement was actually 3.000 and not 2.999 or 3.001.

Multiplication and Division

When multiplying or dividing measurement figures, the answer may have no more significant figures than the *least* number of significant figures in the figures being multiplied or divided. This simply means that an answer can be no more accurate than the least accurate measurement entering into the calculation and that you cannot improve the accuracy of a measurement by doing a calculation. For example, in multiplying 54.2 mi/h $\times$ 4.0 h to find out the total distance traveled, the first figure (54.2) has three significant figures but the second (4.0) has only two significant figures. The answer can contain only two significant figures since this is the weakest number of those involved in the calculation. The correct answer is therefore 220 mi, not 216.8 mi. This may seem strange since multiplying the two numbers together gives the answer of 216.8 mi. This answer, however, means a greater accuracy than is possible, and the accuracy cannot be improved over the weakest number involved in the calculation. Since the weakest number (4.0) has

FIGURE A.1

only two significant figures the answer must also have only two significant figures, which is 220 mi.

The result of a calculation is *rounded* to have the same least number of significant figures as the least number of a measurement involved in the calculation. When rounding numbers, the last significant figure is increased by 1 if the number after it is 5 or larger. If the number after the last significant figure is 4 or less, the nonsignificant figures are simply dropped. Thus, if two significant figures are called for in the answer of the previous example, 216.8 is rounded up to 220 because the last number after the two significant figures is 6 (a number larger than 5). If the calculation result had been 214.8, the rounded number would be 210 miles.

Note that *measurement figures* are the only figures involved in the number of significant figures in the answer. Numbers that are counted or **defined** are not included in the determination of significant figures in an answer. For example, when dividing by 2 to find an average, the 2 is ignored when considering the number of significant figures. Defined numbers are defined exactly and are not used in significant figures. Since 1 kilogram is *defined* to be exactly 1,000 grams, such a conversion is not a measurement.

Addition and Subtraction

Addition and subtraction operations involving measurements, as with multiplication and division, cannot result in an answer that implies greater accuracy than the measurements had before the calculation. Recall that the last digit to the right in a measurement is uncertain; that is, it is the result of an estimate. The answer to an addition or subtraction calculation can have this uncertain number *no farther from the decimal place than it was in the weakest number involved in the calculation.* Thus, when 8.4 is added to 4.926, the weakest number is 8.4, and the uncertain number is .4, one place to the right of the decimal. The sum of 13.326 is therefore rounded to 13.3, reflecting the placement of this weakest doubtful figure.

The rules for counting zeros tell us that the numbers 203 and 0.200 both have three significant figures. Likewise, the numbers 230 and 0.23 only have two significant figures. Once you remember the rules, the counting of significant figures is straightforward. On the other hand, sometimes you find a number that seems to make it impossible to follow the rules. For example, how would you write 3,000 with two significant figures? There are several special systems in use for taking care of problems such as this, including the placing of a little bar over the last significant digit. One of the convenient ways of showing significant figures for difficult numbers is to use scientific notation, which is also discussed in the "Scientific Notation" section in this appendix. The convention for writing significant figures is to display one digit to the left of the decimal. The exponents are not considered when showing the number of significant figures in scientific notation. Thus, if you want to write three thousand showing one significant figure, you would write 3×10^3. To show two significant figures, it is 3.0×10^3, and for three significant figures it becomes 3.00×10^3. As you can see, the correct use of scientific notation leaves little room for doubt about how many significant figures are intended.

CONVERSION OF UNITS

The measurement of most properties results in both a numerical value and a unit. The statement that a glass contains 50 cm^3 of a liquid conveys two important concepts: the numerical value of 50 and the referent unit of cubic centimeters. Both the numerical value and the unit are necessary to communicate correctly the volume of the liquid.

When working with calculations involving measurement units, *both* the numerical value and the units are treated mathematically. As in other mathematical operations, there are general rules to follow.

1. Only properties with *like units* may be added or subtracted. It should be obvious that adding quantities such as 5 dollars and 10 dimes is meaningless. You must first convert to like units before adding or subtracting.
2. Like or unlike units may be multiplied or divided and treated in the same manner as numbers. You have used this rule when dealing with area (length $\times$ length = length^2, for example, cm $\times$ cm = cm^2) and when dealing with volume (length $\times$ length $\times$ length = length^3, for example, cm $\times$ cm $\times$ cm = cm^3).

You can use these two rules to create a *conversion ratio* that will help you change one unit to another. Suppose you need to convert 2.3 kg to grams. First, write the relationship between kilograms and grams:

$$1,000 \text{ g} = 1 \text{ kg}$$

Next, divide both sides by what you wish to convert *from* (kilograms in this example):

$$\frac{1,000 \text{ g}}{1 \text{ kg}} = \frac{1 \text{ kg}}{1 \text{ kg}}$$

One kilogram divided by 1 kg equals 1, just as 10 divided by 10 equals 1. Therefore, the right side of the relationship becomes 1 and the equation is:

$$\frac{1,000 \text{ g}}{1 \text{ kg}} = 1$$

CONCEPTS APPLIED

Unit Conversion

A distance is reported as 100.0 km, and you want to know how far this is in miles.

Answer

First, you need to obtain a *conversion factor* from a textbook or reference book, which usually lists the conversion factors by properties in a table. Such a table will show two conversion factors for kilometers and miles: (1) 1 km = 0.621 mi and (2) 1 mi = 1.609 km. You select the factor that is in the same form as your problem; for example, your problem is 100.0 km = ? mi. The conversion factor in this form is 1 km = 0.621 mi.

Second, you convert this conversion factor into a *conversion ratio* by dividing the factor by what you wish to convert *from:*

conversion factor :	1 km = 0.621 mi
divide factor by what you want to convert from :	$\dfrac{1\ km}{1\ km} = \dfrac{0.621\ mi}{1\ km}$
resulting conversion rate :	$\dfrac{0.621\ mi}{km}$

Note that if you had used the 1 mi = 1.609 km factor, the resulting units would be meaningless. The conversion ratio is now multiplied by the numerical value *and unit* you wish to convert:

$$100.0\ km \times \frac{0.621\ mi}{km}$$

$$(100.0)(0.621)\frac{km \cdot mi}{km}$$

$$0.621\ mi$$

The 1 is usually understood, that is, not stated, and the operation is called *canceling*. Canceling leaves you with the fraction 1,000 g/1 kg, which is a conversion ratio that can be used to convert from kilograms to grams. You simply multiply the conversion ratio by the numerical value and unit you wish to convert:

$$= 2.3\ kg \times \frac{1,000\ g}{1\ kg}$$

$$= \frac{2.3 \times 1,000}{1}\ \frac{kg \times g}{kg}$$

$$= \boxed{2,300\ g}$$

The kilogram units cancel. Showing the whole operation with units only, you can see how you end up with the correct unit of grams:

$$kg \times \frac{g}{kg} = \frac{kg \cdot g}{kg} = g$$

Since you did obtain the correct unit, you know that you used the correct conversion ratio. If you had blundered and used an inverted conversion ratio, you would obtain

$$2.3\ kg \times \frac{1\ kg}{1,000\ g} = .0023\ \frac{kg^2}{g}$$

which yields the meaningless, incorrect units of kg^2/g. Carrying out the mathematical operations on the numbers and the units will always tell you whether or not you used the correct conversion ratio.

SCIENTIFIC NOTATION

Most of the properties of things that you might measure in your everyday world can be expressed with a small range of numerical values together with some standard unit of measure. The range of numerical values for most everyday things can be dealt with by using units (1s), tens (10s), hundreds (100s), or perhaps thousands (1,000s). But the actual universe contains some objects of incredibly large size that require some very big numbers to describe. The Sun, for example, has a mass of about 1,970,000,000,000,000,000,000,000,000,000 kg. On the other hand, very small numbers are needed to measure the size and parts of an atom. The radius of a hydrogen atom, for example, is about 0.00000000005 m. Such extremely large and small numbers are cumbersome and awkward since there are so many zeros to keep track of, even if you are successful in carefully counting all the zeros. A method does exist to deal with extremely large or small numbers in a more condensed form. The method is called *scientific notation*, but it is also sometimes called *powers of ten* or *exponential notation*, since it is based on exponents of 10. Whatever it is called, the method is a compact way of dealing with numbers that not only helps you keep track of zeros but provides a simplified way to make calculations as well.

In algebra you save a lot of time (as well as paper) by writing $(a \times a \times a \times a \times a)$ as a^5. The small number written to the right and above a letter or number is a superscript called an *exponent*. The exponent means that the letter or number is to be multiplied by itself that many times; for example, a^5 means a multiplied by itself five times; or $a \times a \times a \times a \times a$. As you can see, it is much easier to write the exponential form of this operation than it is to write it out in the long form. Scientific notation uses an exponent to indicate the power of the base 10. The exponent tells how many times the base, 10, is multiplied by itself. For example,

$$10,000 = 10^4$$
$$1,000 = 10^3$$
$$100 = 10^2$$
$$10 = 10^1$$
$$1 = 10^0$$
$$0.1 = 10^{-1}$$
$$0.01 = 10^{-2}$$
$$0.001 = 10^{-3}$$
$$0.0001 = 10^{-4}$$

This table could be extended indefinitely, but this somewhat shorter version will give you an idea of how the method works. The symbol 10^4 is read as "ten to the fourth power" and means $10 \times 10 \times 10 \times 10$. Ten times itself four times is 10,000, so 10^4 is the scientific notation for 10,000. It is also equal to the number of zeros between the 1 and the decimal point; that is, to write the longer form of 10^4, you simply write 1, then move the decimal point four places to the *right*—10 to the fourth power is 10,000.

Scientific Notation Expression

What is 26,000,000 in scientific notation?

Answer

Count how many times you must shift the decimal point until one digit remains to the left of the decimal point. For numbers larger than the digit 1, the number of shifts tells you how much the exponent is increased, so the answer is

$$2.6 \times 10^7$$

which means the coefficient 2.6 is multiplied by 10 seven times.

The power of ten table also shows that numbers smaller than 1 have negative exponents. A negative exponent means a reciprocal:

$$10^{-1} = \frac{1}{10} = 0.1$$

$$10^{-2} = \frac{1}{100} = 0.01$$

$$10^{-3} = \frac{1}{1000} = 0.001$$

To write the longer form of 10^{-4}, you simply write 1, then move the decimal point four places to the *left;* 10 to the negative fourth power is 0.0001.

Scientific notation usually, but not always, is expressed as the product of two numbers: (1) a number between 1 and 10 that is called the *coefficient* and (2) a power of ten that is called the *exponent.* For example, the mass of the Sun that was given in long form earlier is expressed in scientific notation as

$$1.97 \times 10^{30} \text{ kg}$$

and the radius of a hydrogen atom is

$$5.0 \times 10^{-11} \text{ m}$$

In these expressions, the coefficients are 1.97 and 5.0, and the power of ten notations are the exponents. Note that in both of these examples, the exponent tells you where to place the decimal point if you wish to write the number all the way out in the long form. Sometimes scientific notation is written without a coefficient, showing only the exponent. In these cases, the coefficient of 1.0 is understood, that is, not stated. If you try to enter a scientific notation in your calculator, however, you will need to enter the understood 1.0, or the calculator will not be able to function correctly. Note also that 1.97×10^{30} kg and the expressions 0.197×10^{31} kg and 19.7×10^{29} kg are all correct expressions of the mass of the Sun. By convention, however, you will use the form that has one digit to the left of the decimal.

It was stated earlier that scientific notation provides a compact way of dealing with very large or very small numbers, but it provides a simplified way to make calculations as well. There are a few mathematical rules that will describe how the use of scientific notation simplifies these calculations.

To *multiply* two scientific notation numbers, the coefficients are multiplied as usual, and the exponents are *added*

Scientific Notation Calculation

Solve the following problem concerning scientific notation:

$$\frac{(2 \times 10^4) \times (8 \times 10^{-6})}{8 \times 10^4}$$

Answer

First, separate the coefficients from the exponents,

$$\frac{2 \times 8}{8} \times \frac{10^4 \times 10^{-6}}{10^4}$$

then multiply and divide the coefficients and add and subtract the exponents as the problem requires,

$$2 \times 10^{\{[(4) + (-6)] = [4]\}}$$

Solving the remaining additions and subtractions of the coefficients gives

$$2 \times 10^{-6}$$

---

algebraically. For example, to multiply (2×10^2) by (3×10^3), first separate the coefficients from the exponents,

$$(2 \times 3) \times (10^2 \times 10^3),$$

then multiply the coefficients and add the exponents,

$$6 \times 10^{(2+3)} = 6 \times 10^5$$

Adding the exponents is possible because $10^2 \times 10^3$ means the same thing as $(10 \times 10) \times (10 \times 10 \times 10)$, which equals $(100) \times (1,000)$, or 100,000, which is expressed as 10^5 in scientific notation. Note that two negative exponents add algebraically, for example $10^{-2} \times 10^{-3} = 10^{[(-2) + (-3)]} = 10^{-5}$. A negative and a positive exponent also add algebraically, as in $10^5 \times 10^{-3} = 10^{[(+5) + (-3)]} = 10^2$.

If the result of a calculation involving two scientific notation numbers does not have the conventional one digit to the left of the decimal, move the decimal point so it does, changing the exponent according to which way and how much the decimal point is moved. Note that the exponent increases by one number for each decimal point moved to the left. Likewise, the exponent decreases by one number for each decimal point moved to the right. For example, $938. \times 10^3$ becomes 9.38×10^5 when the decimal point is moved two places to the left.

To *divide* two scientific notation numbers, the coefficients are divided as usual and the exponents are *subtracted.* For example, to divide (6×10^6) by (3×10^2), first separate the coefficients from the exponents,

$$(6 \div 3) \times (10^6 \div 10^2)$$

then divide the coefficients and subtract the exponents,

$$2 \times 10^{(6-2)} = 2 \times 10^4$$

Note that when you subtract a negative exponent, for example, $10^{[(3) - (-2)]}$, you change the sign and add, $10^{(3 + 2)} = 10^5$.

Appendix B

Solubilities Chart

	Acetate	Bromide	Carbonate	Chloride	Fluoride	Hydroxide	Iodide	Nitrate	Oxide	Phosphate	Sulfate	Sulfide
Aluminum	S	S	—	S	s	i	S	S	i	i	S	d
Ammonium	S	S	S	S	S	S	S	S	—	S	S	S
Barium	S	S	i	S	s	S	S	S	S	i	i	d
Calcium	S	S	i	S	i	s	S	S	s	i	s	d
Copper(I)	—	s	i	s	i	—	i	—	i	—	d	i
Copper(II)	S	S	i	S	S	i	S	S	i	i	S	i
Iron(II)	S	S	i	S	s	i	S	S	i	i	S	i
Iron(III)	S	S	i	S	s	i	S	S	i	i	S	d
Lead	S	s	i	s	i	i	s	S	i	i	i	i
Magnesium	S	S	i	S	i	i	S	S	i	i	S	d
Mercury(I)	s	i	i	i	d	d	i	S	i	i	i	i
Mercury(II)	S	s	i	S	d	i	i	S	i	i	i	i
Potassium	S	S	S	S	S	S	S	S	S	S	S	i
Silver	s	i	i	i	S	—	i	S	i	i	i	i
Sodium	S	S	S	S	S	S	S	S	d	S	S	S
Strontium	S	S	s	S	i	s	S	S	—	i	i	i
Zinc	S	S	i	S	S	i	S	S	i	i	S	i

S = soluble
i = insoluble
s = slightly soluble
d = decomposes

698

Relative Humidity Table

Dry-Bulb Temperature (°C)	Difference Between Wet-Bulb and Dry-Bulb Temperatures (°C)																			
	1	2	3	4	5	6	7	8	9	10	11	12	13	14	15	16	17	18	19	20
0	81	64	46	29	13															
1	83	66	49	33	17															
2	84	68	52	37	22	7														
3	84	70	55	40	26	12														
4	86	71	57	43	29	16														
5	86	72	58	45	33	20	7													
6	86	73	60	48	35	24	11													
7	87	74	62	50	38	26	15													
8	87	75	63	51	40	29	19	8												
9	88	76	64	53	42	32	22	12												
10	88	77	66	55	44	34	24	15	6											
11	89	78	67	56	46	36	27	18	9											
12	89	78	68	58	48	39	29	21	12											
13	89	79	69	59	50	41	32	23	15	7										
14	90	79	70	60	51	42	34	26	18	10										
15	90	80	71	61	53	44	36	27	20	13	6									
16	90	81	71	63	54	46	38	30	23	15	8									
17	90	81	72	64	55	47	40	32	25	18	11									
18	91	82	73	65	57	49	41	34	27	20	14	7								
19	91	82	74	65	58	50	43	36	29	22	16	10								
20	91	83	74	66	59	51	44	37	31	24	18	12	6							
21	91	83	75	67	60	53	46	39	32	26	20	14	9							
22	92	83	76	68	61	54	47	40	34	28	22	17	11	6						
23	92	84	76	69	62	55	48	42	36	30	24	19	13	8						
24	92	84	77	69	62	56	49	43	37	31	26	20	15	10	5					
25	92	84	77	70	63	57	50	44	39	33	28	22	17	12	8					
26	92	85	78	71	64	58	51	46	40	34	29	24	19	14	10	5				
27	92	85	78	71	65	58	52	47	41	36	31	26	21	16	12	7				
28	93	85	78	72	65	59	53	48	42	37	32	27	22	18	13	9	5			
29	93	86	79	72	66	60	54	49	43	38	33	28	24	19	15	11	7			
30	93	86	79	73	67	61	55	50	44	39	35	30	25	21	17	13	9	5		
31	93	86	80	73	67	61	56	51	45	40	36	31	27	22	18	14	11	7		
32	93	86	80	74	68	62	57	51	46	41	37	32	28	24	20	16	12	9	5	
33	93	87	80	74	68	63	57	52	47	42	38	33	29	25	21	17	14	10	7	
34	93	87	81	75	69	63	58	53	48	43	39	35	30	28	23	19	15	12	8	5
35	94	87	81	75	69	64	59	54	49	44	40	36	32	28	24	20	17	13	10	7

Appendix D

Solutions for Second Example Exercises

Note: Solutions that involve calculations of measurements are rounded up or down to conform to the rules for significant figures as described in appendix A.

Chapter 1

1.2

$m = 15.0\text{ g}$
$V = 4.50\text{ cm}^3$
$\rho = ?$

$$\rho = \frac{m}{V}$$

$$= \frac{15.0\,\text{g}}{4.50\,\text{cm}^3}$$

$$= \boxed{3.33\,\frac{\text{g}}{\text{cm}^3}}$$

Chapter 2

2.2

$\bar{v} = 8.00\text{ km/h}$
$t = 10.0\text{ s}$
$d = ?$

The bicycle has a speed of 8.00 km/h, and the time factor is 10.0 s, so km/h must be converted to m/s:

$$\bar{v} = \frac{0.2778\,\frac{\text{m}}{\text{s}}}{\frac{\text{km}}{\text{h}}} \times 8.00\,\frac{\text{km}}{\text{h}}$$

$$= (0.2778)(8.00)\frac{\text{m}}{\text{s}} \times \frac{\text{h}}{\text{km}} \times \frac{\text{km}}{\text{h}}$$

$$= 2.22\,\frac{\text{m}}{\text{s}}$$

$$\bar{v} = \frac{d}{t}$$

$$\bar{v}t = \frac{dt}{t}$$

$$d = \bar{v}t$$

$$= \left(2.22\,\frac{\text{m}}{\text{s}}\right)(10.0\,\text{s})$$

$$= (2.22)(10.0)\frac{\text{m}}{\text{s}} \times \frac{\text{s}}{1}$$

$$= \boxed{22.2\,\text{m}}$$

2.4

$v_i = 0\,\frac{\text{m}}{\text{s}}$

$v_f = ?$

$a = 5\,\frac{\text{m}}{\text{s}^2}$

$t = 6\,\text{s}$

$$a = \frac{v_f - v_i}{t} \quad \therefore \quad v_f = at + v_i$$

$$= \left(5\,\frac{\text{m}}{\text{s}^2}\right)(6\,\text{s}) + 0$$

$$= (5)(6)\,\frac{\text{m}}{\text{s}^2} \times \frac{\text{s}}{1}$$

$$= \boxed{30\,\frac{\text{m}}{\text{s}}}$$

2.6

$m = 20\text{ kg}$
$F = 40\text{ N}$
$a = ?$

$$F = ma \quad \therefore \quad a = \frac{F}{m}$$

$$= \frac{40\,\frac{\text{kg}\cdot\text{m}}{\text{s}^2}}{20\,\text{kg}}$$

$$= \frac{40}{20}\,\frac{\text{kg}\cdot\text{m}}{\text{s}^2} \times \frac{1}{\text{kg}}$$

$$= \boxed{2\,\frac{\text{m}}{\text{s}^2}}$$

2.8

$m = 60.0\text{ kg}$
$w = 100.0\text{ N}$
$g = ?$

$$w = mg \quad \therefore \quad g = \frac{w}{m}$$

$$= \frac{100.0\,\frac{\text{kg}\cdot\text{m}}{\text{s}^2}}{60.0\,\text{kg}}$$

$$= \frac{100.0}{60.0}\,\frac{\text{kg}\cdot\text{m}}{\text{s}^2} \times \frac{1}{\text{kg}}$$

$$= \boxed{1.67\,\frac{\text{m}}{\text{s}^2}}$$

2.10

$m = 0.25$ kg
$r = 0.25$ m
$v = 2.0$ m/s
$F = ?$

$$F = \frac{mv^2}{r}$$

$$= \frac{(0.25\,\text{kg})\left(2.0\,\dfrac{\text{m}}{\text{s}}\right)^2}{0.25\,\text{m}}$$

$$= \frac{(0.25\,\text{kg})\left(4.0\,\dfrac{\text{m}^2}{\text{s}^2}\right)}{0.25\,\text{m}}$$

$$= \frac{(0.25)(4.0)}{0.25}\,\frac{\text{kg}\cdot\text{m}^2}{\text{s}^2}\times\frac{1}{\text{m}}$$

$$= 4.0\,\frac{\text{kg}\cdot\text{m}}{\text{s}^2}$$

$$= \boxed{4.0\,\text{N}}$$

2.12

$G = 6.67\times10^{-11}\,\text{N}\cdot\text{m}^2/\text{kg}^2$
$m_e = 6.0\times10^{24}\,\text{kg}$
$d = 12.8\times10^6\,\text{m}$
$g = ?$

$$g = \frac{Gm_e}{d^2}$$

$$= \frac{\left(6.67\times10^{-11}\,\dfrac{\text{N}\cdot\text{m}^2}{\text{kg}^2}\right)(6.0\times10^{24}\,\text{kg})}{(12.8\times10^6\,\text{m})^2}$$

$$= \frac{\left(6.67\times10^{-11}\,\dfrac{\text{N}\cdot\text{m}^2}{\text{kg}^2}\right)(6.0\times10^{24}\,\text{kg})}{1.64\times10^{14}\,\text{m}^2}$$

$$= \frac{(6.67\times10^{-11})(6.0\times10^{24})}{1.64\times10^{14}}\,\frac{\text{N}\cdot\text{m}^2}{\text{kg}^2}\times\frac{\text{kg}}{1}\times\frac{1}{\text{m}^2}$$

$$= 2.44\,\frac{\dfrac{\text{kg}\cdot\text{m}}{\text{s}^2}}{\text{kg}}$$

$$= \boxed{2.44\,\frac{\text{m}}{\text{s}^2}}$$

Chapter 3

3.2

$w = 50$ lb
$d = 2$ ft
$W = ?$

$$W = Fd$$

$$= (50\,\text{lb})\,(2\,\text{ft})$$

$$= (50)\,(2)\quad\text{ft}\times\text{lb}$$

$$= \boxed{100\,\text{ft}\cdot\text{lb}}$$

3.4

$w = 150$ lb
$h = 15$ ft
$t = 10.0$ s
$P = ?$

$$P = \frac{wh}{t}$$

$$= \frac{(150\,\text{lb})(15\,\text{ft})}{10.0\,\text{s}}$$

$$= \frac{(150)(15)}{10.0}\,\frac{\text{ft}\cdot\text{lb}}{\text{s}}$$

$$= 225\,\frac{\text{ft}\cdot\text{lb}}{\text{s}}$$

$$= \frac{225\,\dfrac{\text{ft}\cdot\text{lb}}{\text{s}}}{550\,\dfrac{\dfrac{\text{ft}\cdot\text{lb}}{\text{s}}}{\text{hp}}}$$

$$= 0.41\frac{\text{ft}\cdot\text{lb}}{\text{s}}\times\frac{\text{s}}{\text{ft}\cdot\text{lb}}\times\text{hp}$$

$$= \boxed{0.41\,\text{hp}}$$

3.6

$m = 5.00$ kg
$g = 9.8$ m/s^2
$h = 5.00$ m
$W = ?$

$$W = Fd$$
$$W = mgh$$
$$= (5.00 \, \text{kg})\left(9.8\frac{\text{m}}{\text{s}^2}\right)(5.00 \, \text{m})$$
$$= (5.00)(9.8)(5.00) \, \frac{\text{kg} \cdot \text{m}}{\text{s}^2} \times \text{m}$$
$$= 245 \, \text{N} \cdot \text{m}$$
$$= \boxed{250 \, \text{J}}$$

3.8

$m = 100.0$ kg
$v = 6.0$ m/s
$W = ?$

$$W = K.E.$$
$$K.E. = \frac{1}{2}mv^2$$
$$= \frac{1}{2}(100.0 \, \text{kg})\left(6.0\frac{\text{m}}{\text{s}}\right)^2$$
$$= \frac{1}{2}(100.0 \, \text{kg})\left(36\frac{\text{m}^2}{\text{s}^2}\right)$$
$$= \frac{1}{2}(100.0)(36)\frac{\text{kg} \cdot \text{m}}{\text{s}^2} \times \text{m}$$
$$= 1,800 \, \text{N} \cdot \text{m}$$
$$= \boxed{1,800 \, \text{J}}$$

Chapter 4

4.2

$T_\text{C} = 20°$
$T_\text{F} = ?$

$$T_\text{F} = \frac{9}{5}T_\text{C} + 32°$$
$$= \frac{9}{5}20° + 32°$$
$$= \frac{180°}{5} + 32°$$
$$= 36° + 32°$$
$$= \boxed{68°}$$

4.6

$m = 2$ kg
$Q = 1.2$ kcal
$\Delta T = 20°$ C
$c = ?$

$$Q = mc\Delta T \quad \therefore \quad c = \frac{Q}{m\Delta T}$$
$$= \frac{1.2 \, \text{kcal}}{(2 \, \text{kg})(20.0 \text{C}°)}$$
$$= \frac{1.2}{(2)(20.0)}\frac{\text{kcal}}{\text{kgC}°}$$
$$= \boxed{0.03\frac{\text{kcal}}{\text{kgC}°}}$$

Chapter 5

5.3

$f = 2,500$ Hz
$v = 330$ m/s
$\lambda = ?$

$$v = \lambda f \quad \therefore \quad \lambda = \frac{v}{f}$$
$$= \frac{330\frac{\text{m}}{\text{s}}}{2,500\frac{1}{\text{s}}}$$
$$= \frac{330}{2,500}\frac{\text{m}}{\text{s}} \times \frac{\text{s}}{1}$$
$$= 0.13 \, \text{m} \quad \text{or} \quad \boxed{13 \, \text{cm}}$$

5.5

$t = 1.00$ s
$v = 1,147$ ft/s
$d = ?$

$$v = \frac{d}{t} \quad \therefore \quad d = vt$$
$$= \left(1147\frac{\text{ft}}{\text{s}}\right)(1.00 \, \text{s})$$
$$= (1147)(1.00)\frac{\text{ft}}{\text{s}} \times \frac{\text{s}}{1}$$
$$= 1147 \, \text{ft}$$

Sound traveled from the source, to a reflecting surface, then back to the source, so the distance is

$$1147 \, \text{ft} \times \frac{1}{2} = \boxed{574 \, \text{ft}}$$

Chapter 6

6.3

$$V = 120 \text{ V}$$
$$R = 30 \ \Omega$$
$$I = ?$$

$$V = IR \quad \therefore \quad I = \frac{V}{R}$$
$$= \frac{120 \text{ V}}{30 \dfrac{\text{V}}{\text{A}}}$$
$$= \frac{120}{30} \ \frac{\text{V}}{1} \times \frac{\text{A}}{\text{V}}$$
$$= \boxed{4 \text{ A}}$$

6.5

$$I = 0.5 \text{ A}$$
$$V = 120 \text{ V}$$
$$P = ?$$

$$P = IV$$
$$= (0.5 \text{ A})(120 \text{ V})$$
$$= (0.5)(120) \ \frac{\text{C}}{\text{s}} \times \frac{\text{J}}{\text{C}}$$
$$= 60 \frac{\text{J}}{\text{s}}$$
$$= \boxed{60 \text{ W}}$$

6.7

$$I = 0.5 \text{ A}$$
$$V = 120 \text{ V}$$
$$P = IV = 60 \text{ W}$$
$$Rate = \$0.10/\text{kWh}$$
$$Cost = ?$$

$$\text{cost} = \frac{(\text{watts})\,(\text{time})\,(\text{rate})}{1{,}000 \dfrac{\text{W}}{\text{kW}}}$$

$$= \frac{(60\text{W})\,(1.00\text{h})\left(\$0.10\dfrac{\text{W}}{\text{kW}}\right)}{1{,}000 \dfrac{\text{W}}{\text{kW}}}$$

$$= \frac{(60)\,(1.00)\,(0.10)}{1{,}000} \ \frac{\text{W}}{1} \times \frac{\text{h}}{1} \times \frac{\$}{\text{kWh}} \times \frac{\text{kW}}{\text{W}}$$

$$= \$0.006 \quad \text{or} \quad \boxed{0.6 \text{ of a cent per hour}}$$

Chapter 7

7.4

$$f = 7.00 \times 10^{14} \text{ Hz}$$
$$h = 6.63 \times 10^{-34} \text{ J} \cdot \text{s}$$
$$E = ?$$

$$E = hf$$
$$= (6.63 \times 10^{-34}\text{J} \cdot \text{s})\left(7.00 \times 10^{14} \frac{1}{\text{s}}\right)$$
$$= (6.63 \times 10^{-34})\,(7.00 \times 10^{14}) \ \text{J} \cdot \text{s} \times \frac{1}{\text{s}}$$
$$= \boxed{4.64 \times 10^{-19}\text{J}}$$

Chapter 8

8.2

$$f = 7.30 \times 10^{14} \text{ Hz}$$
$$h = 6.63 \times 10^{-34} \text{ J} \cdot \text{s}$$
$$E = ?$$

$$E = hf$$
$$= (6.63 \times 10^{-34}\text{J} \cdot \text{s})\left(7.30 \times 10^{14} \frac{1}{\text{s}}\right)$$
$$= (6.63 \times 10^{-34})\,(7.30 \times 10^{14}) \ \text{J} \cdot \text{s} \times \frac{1}{\text{s}}$$
$$= \boxed{4.84 \times 10^{-19}\text{J}}$$

Appendix E

Solutions for Group A Parallel Exercises

Note: Solutions that involve calculations of measurements are rounded up or down to conform to the rules for significant figures described in appendix A.

Chapter 1

1.1 Answers will vary but should have the relationship of 100 cm in 1 m, for example, 178 cm = 1.78 m.

1.2 Since density is given by the relationship $\rho = m/V$, then

$$\rho = \frac{m}{V} = \frac{272\text{ g}}{20.0\text{ cm}^3}$$

$$= \frac{272}{20.0}\ \frac{\text{g}}{\text{cm}^3}$$

$$= \boxed{13.6\ \frac{\text{g}}{\text{cm}^3}}$$

1.3 The volume of a sample of lead is given and the problem asks for the mass. From the relationship of $\rho = m/V$, solving for the mass (m) tells you that the density (ρ) times the volume (V) equals the mass, or $m = \rho V$. The density of lead, 11.4 g/cm³, can be obtained from table 1.3, so

$$\rho = \frac{m}{V}$$

$$V\rho = \frac{m\cancel{V}}{\cancel{V}}$$

$$m = \rho V$$

$$m = \left(11.4\ \frac{\text{g}}{\text{cm}^3}\right)(10.0\text{ cm}^3)$$

$$11.4 \times 10.0\ \frac{\text{g}}{\text{cm}^3} \times \text{cm}^3$$

$$114\ \frac{\text{g} \cdot \cancel{\text{cm}^3}}{\cancel{\text{cm}^3}}$$

$$= \boxed{114\text{ g}}$$

1.4 Solving the relationship $\rho = m/V$ for volume gives $V = m/\rho$, and

$$\rho = \frac{m}{V}$$

$$V\rho = \frac{m\cancel{V}}{\cancel{V}}$$

$$\frac{V\cancel{\rho}}{\cancel{\rho}} = \frac{m}{\rho}$$

$$V = \frac{m}{\rho}$$

$$V = \frac{600\text{ g}}{3.00\ \dfrac{\text{g}}{\text{cm}^3}}$$

$$= \frac{600}{3.00}\ \frac{\text{g}}{1} \times \frac{\text{cm}^3}{\text{g}}$$

$$= 200\ \frac{\cancel{\text{g}} \cdot \text{cm}^3}{\cancel{\text{g}}}$$

$$= \boxed{200\text{ cm}^3}$$

1.5 A 50.0 cm³ sample with a mass of 34.0 grams has a density of

$$\rho = \frac{m}{V} = \frac{34.0\text{ g}}{50.0\text{ cm}^3}$$

$$= \frac{34.0}{50.0}\ \frac{\text{g}}{\text{cm}^3}$$

$$= \boxed{0.680\ \frac{\text{g}}{\text{cm}^3}}$$

According to table 1.3, 0.680 g/cm³ is the density of gasoline, so the substance must be gasoline.

1.6 The problem asks for a mass and gives a volume, so you need a relationship between mass and volume. Table 1.3 gives the density of water as 1.00 g/cm³, which is a density that is easily remembered. The volume is given in liters (L), which should first be converted to cm³ because this is the unit in which density is expressed. The relationship of $\rho = m/V$ solved for mass is ρV, so the solution is

$$\rho = \frac{m}{V} \therefore m = \rho V$$

$$m = \left(1.00 \, \frac{g}{cm^3}\right)(40,000 \, cm^3)$$

$$= 1.00 \times 40,000 \, \frac{g}{cm^3} \times cm^3$$

$$= 40,000 \, \frac{g \cdot \cancel{cm^3}}{\cancel{cm^3}}$$

$$= 40,000 \, g$$

$$= \boxed{40 \, kg}$$

1.7 From table 1.3, the density of aluminum is given as 2.70 g/cm³. Converting 2.1 kg to the same units as the density gives 2,100 g. Solving $\rho = m/V$ for the volume gives

$$V = \frac{m}{\rho} = \frac{2,100 \, g}{2.70 \, \dfrac{g}{cm^3}}$$

$$= \frac{2,100}{2.70} \, \frac{g}{1} \times \frac{cm^3}{g}$$

$$= 777.78 \, \frac{\cancel{g} \cdot cm^3}{\cancel{g}}$$

$$= \boxed{780 \, cm^3}$$

1.8 The length of one side of the box is 0.1 m. Reasoning: Since the density of water is 1.00 g/cm³, then the volume of 1,000 g of water is 1,000 cm³. A cubic box with a volume of 1,000 cm³ is 10 cm (since 10 × 10 × 10 = 1,000). Converting 10 cm to m units, the cube is 0.1 m on each edge.

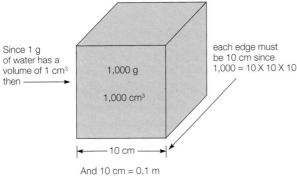

Since 1 g of water has a volume of 1 cm³ then ⟶ 1,000 g / 1,000 cm³

each edge must be 10 cm since 1,000 = 10 × 10 × 10

⟵ 10 cm ⟶

And 10 cm = 0.1 m

FIGURE A.4

1.9 The relationship between mass, volume, and density is $\rho = m/V$. The problem gives a volume but not a mass. The mass, however, can be assumed to remain constant during the compression of the bread, so the mass can be obtained from the original volume and density, or

$$\rho = \frac{m}{V} \therefore m = \rho V$$

$$m = \left(0.2 \, \frac{g}{cm^3}\right)(3,000 \, cm^3)$$

$$= 0.2 \times 3,000 \, \frac{g}{cm^3} \times cm^3$$

$$= 600 \, \frac{g \cdot \cancel{cm^3}}{\cancel{cm^3}}$$

$$= 600 \, g$$

A mass of 600 g and the new volume of 1,500 cm³ means that the new density of the crushed bread is

$$\rho = \frac{m}{V}$$

$$= \frac{600 \, g}{1,500 \, cm^3}$$

$$= \frac{600}{1,500} \, \frac{g}{cm^3}$$

$$= \boxed{0.4 \, \frac{g}{cm^3}}$$

1.10 According to table 1.3, lead has a density of 11.4 g/cm³. Therefore, a 1.00 cm³ sample of lead would have a mass of

$$\rho = \frac{m}{V} \therefore m = \rho V$$

$$m = \left(11.4 \, \frac{g}{cm^3}\right)(1.00 \, cm^3)$$

$$= 11.4 \times 1.00 \, \frac{g}{cm^3} \times cm^3$$

$$= 11.4 \, \frac{g \cdot \cancel{cm^3}}{\cancel{cm^3}}$$

$$= 11.4 \, g$$

Also according to table 1.3, copper has a density of 8.96 g/cm³. To balance a mass of 11.4 g of lead, a volume of this much copper would be required:

$$\rho = \frac{m}{V} \therefore V = \frac{m}{\rho}$$

$$V = \frac{11.4 \, g}{8.96 \, \dfrac{g}{cm^3}}$$

$$= \frac{11.4}{8.96} \, \frac{g}{1} \times \frac{cm^3}{g}$$

$$= 1.27232 \, \frac{\cancel{g} \cdot cm^3}{\cancel{g}}$$

$$= \boxed{1.27 \, cm^3}$$

Chapter 2

2.1 Listing the quantities with their symbols, we can see the problem involves the quantities found in the definition of average speed:

$$\bar{v} = 350.0 \, m/s$$

$$t = 5.00 \, s$$

$$d = ?$$

$$\bar{v} = \frac{d}{t} \therefore d = \bar{v}t$$

$$d = \left(350.0 \, \frac{m}{s}\right)(5.00 \, s)$$

$$= (350.0)(5.00) \, \frac{m}{s} \times s$$

$$= \boxed{1,750 \, m}$$

2.2 The initial velocity, final velocity, and time are known and the problem asked for the acceleration. Listing these quantities with their symbols, we have

$$v_i = 0 \text{ m/s}$$
$$v_f = 15.0 \text{ m/s}$$
$$t = 10.0 \text{ s}$$
$$a = ?$$

These are the quantities involved in the acceleration equation, which is already solved for the unknown:

$$a = \frac{v_f - v_i}{t}$$
$$a = \frac{15.0 \text{ m/s} - 0 \text{ m/s}}{10.0 \text{ s}}$$
$$= \frac{15.0}{10.0} \frac{\text{m}}{\text{s}} \times \frac{1}{\text{s}}$$
$$= \boxed{1.50 \frac{\text{m}}{\text{s}^2}}$$

2.3 The distance (d) and the time (t) quantities are given in the problem, and

$$\bar{v} = \frac{d}{t}$$
$$= \frac{160 \text{ km}}{2.0 \text{ h}}$$
$$= \frac{160}{2.0} \frac{\text{km}}{\text{h}}$$
$$= \boxed{80 \text{ km/h}}$$

The units cannot be simplified further. Note two significant figures in the answer, which is the least number of significant figures involved in the division operation.

2.4 Listing the known and unknown quantities:

$$m = 40.0 \text{ kg}$$
$$a = 2.4 \text{ m/s}^2$$
$$F = ?$$

These are the quantities found in Newton's second law of motion, $F = ma$, which is already solved for force (F). Thus,

$$F = (40.0 \text{ kg})\left(2.4 \frac{\text{m}}{\text{s}^2}\right)$$
$$= 40.0 \times 2.4 \frac{\text{kg} \cdot \text{m}}{\text{s}^2}$$
$$= \boxed{96 \text{ N}}$$

2.5 List the known and unknown quantities for the first situation, using an unbalanced force of 18.0 N to give the object an acceleration of 3 m/s²:

$$F_1 = 18 \text{ N}$$
$$a_1 = 3 \text{ m/s}^2$$

For the second situation, we are asked to find the force needed for an acceleration of 10 m/s²:

$$a_2 = 10 \text{ m/s}^2$$
$$F = ?$$

These are the quantities of Newton's second law of motion, $F = ma$, but the mass appears to be missing. The mass can be found from

$$F = ma \therefore m_1 = \frac{F_1}{a_1}$$
$$= \frac{18 \dfrac{\text{kg} \cdot \text{m}}{\text{s}^2}}{3 \dfrac{\text{m}}{\text{s}^2}}$$
$$= \frac{18}{3} \frac{\text{kg} \cdot \text{m}}{\text{s}^2} \times \frac{\text{s}^2}{\text{m}}$$
$$= 6 \text{ kg}$$

Now that we have the mass, we can easily find the force needed for an acceleration of 10 m/s²:

$$F_2 = m_2 a_2$$
$$= (6 \text{ kg})\left(10 \frac{\text{m}}{\text{s}^2}\right)$$
$$= 6 \times 10 \frac{\text{kg} \cdot \text{m}}{\text{s}^2}$$
$$= \boxed{60 \text{ N}}$$

2.6 Listing the known and unknown quantities:

$$m = 70.0 \text{ kg}$$
$$g = 9.8 \text{ m/s}^2$$
$$w = ?$$
$$w = mg$$
$$= (70.0 \text{ kg})\left(9.8 \frac{\text{m}}{\text{s}^2}\right)$$
$$= 70.0 \times 9.8 \text{ kg} \times \frac{\text{m}}{\text{s}^2}$$
$$= 686 \frac{\text{kg} \cdot \text{m}}{\text{s}^2}$$
$$= \boxed{690 \text{ N}}$$

2.7 Listing the known and unknown quantities:

$$m = 100 \text{ kg}$$
$$v = 6 \text{ m/s}$$
$$p = ?$$

These are the quantities found in the equation for momentum, $p = mv$, which is already solved for momentum (p). Thus,

$$p = mv$$
$$= (100 \text{ kg})\left(6 \frac{\text{m}}{\text{s}}\right)$$
$$= \boxed{600 \frac{\text{kg} \cdot \text{m}}{\text{s}}}$$

(Note the lowercase p is the symbol used for momentum. This is one of the few cases where the English letter does not provide a clue about what it stands for. The units for momentum are also somewhat unusual for metric units since they do not have a name or single symbol to represent them.)

2.8 Listing the known and unknown quantities:

$$w = 13,720 \text{ N}$$
$$v = 91 \text{ km/h}$$
$$p = ?$$

The equation for momentum is $p = mv$, which is already solved for momentum (p). The weight unit must be first converted to a mass unit:

$$w = mg \therefore m = \frac{w}{g}$$

$$= \frac{13,720 \dfrac{\text{kg} \cdot \text{m}}{\text{s}^2}}{9.8 \dfrac{\text{m}}{\text{s}^2}}$$

$$= \frac{13,720}{9.8} \frac{\text{kg} \cdot \text{m}}{\text{s}^2} \times \frac{\text{s}^2}{\text{m}}$$

$$= 1,400 \text{ kg}$$

The km/h unit should next be converted to m/s. Using the conversion factor from inside the front cover:

$$\frac{0.2778 \dfrac{\text{m}}{\text{s}}}{1 \dfrac{\text{km}}{\text{h}}} \times 91 \frac{\text{km}}{\text{h}}$$

$$0.2778 \times 91 \frac{\text{m}}{\text{s}} \times \frac{\text{h}}{\text{km}} \times \frac{\text{km}}{\text{h}}$$

$$25.2798 \frac{\text{m}}{\text{s}}$$

$$25 \frac{\text{m}}{\text{s}}$$

Now, listing the converted known and unknown quantities:

$$m = 1,400 \text{ kg}$$
$$v = 25 \text{ m/s}$$
$$p = ?$$

and solving for momentum (p),

$$p = mv$$

$$= (1,400 \text{ kg}) \left(25 \frac{\text{m}}{\text{s}} \right)$$

$$= \boxed{35,000 \; \frac{\text{kg} \cdot \text{m}}{\text{s}}}$$

2.9 Listing the known and unknown quantities:

Bullet $\rightarrow m = 0.015 \text{ kg}$ Rifle $\rightarrow m = 6 \text{ kg}$
Bullet $\rightarrow v = 200 \text{ m/s}$ Rifle $\rightarrow v = ? \text{ m/s}$

Note the mass of the bullet was converted to kilograms. This is a conservation of momentum question, where the

bullet and rifle can be considered as a system of interacting objects:

$$\text{Bullet momentum} = -\text{rifle momentum}$$
$$(mv)_b = -(mv)_r$$
$$(mv)_b - (mv)_r = 0$$

$$(0.015 \text{ kg}) \left(200 \frac{\text{m}}{\text{s}} \right) - (6 \text{ kg}) v_r = 0$$

$$\left(3 \text{ kg} \cdot \frac{\text{m}}{\text{s}} \right) - (6 \text{ kg} \cdot v_r) = 0$$

$$\left(3 \text{ kg} \cdot \frac{\text{m}}{\text{s}} \right) = (6 \text{ kg} \cdot v_r)$$

$$v_r = \frac{3 \text{ kg} \cdot \dfrac{\text{m}}{\text{s}}}{6 \text{ kg}}$$

$$= \frac{3}{6} \frac{\text{kg}}{1} \times \frac{1}{\text{kg}} \times \frac{\text{m}}{\text{s}}$$

$$= \boxed{0.5 \; \frac{\text{m}}{\text{s}}}$$

The rifle recoils with a velocity of 0.5 m/s.

2.10 A unit conversion is needed:

$$\left(90.0 \frac{\text{km}}{\text{h}} \right) \left(0.2778 \frac{\dfrac{\text{m}}{\text{s}}}{\dfrac{\text{km}}{\text{h}}} \right) = 25.0 \text{ m/s}$$

a. $F = ma \therefore m = \dfrac{F}{a}$ and $a = \dfrac{v_f - v_i}{t}$, so

$$m = \frac{F}{\dfrac{v_f - v_i}{t}} = \frac{5,000.0 \dfrac{\text{kg} \cdot \text{m}}{\text{s}^2}}{\dfrac{25.0 \text{ m/s} - 0}{5.0 \text{ s}}}$$

$$= \frac{5,000.0 \dfrac{\text{kg} \cdot \text{m}}{\text{s}^2}}{5.0 \dfrac{\text{m}}{\text{s}^2}}$$

$$= \frac{5,000.0}{5.0} \frac{\text{kg} \cdot \text{m}}{\text{s}^2} \times \frac{\text{s}^2}{\text{m}}$$

$$= 1,000 \frac{\text{kg} \cdot \cancel{\text{m}} \cdot \cancel{\text{s}^2}}{\cancel{\text{m}} \cdot \cancel{\text{s}^2}}$$

$$= \boxed{1.0 \times 10^3 \text{ kg}}$$

b. $w = mg$

$$= (1.0 \times 10^3 \text{ kg}) \left(9.8 \frac{\text{m}}{\text{s}^2} \right)$$

$$= (1.0 \times 10^3)(9.8) \text{ kg} \times \frac{\text{m}}{\text{s}^2}$$

$$= 9.8 \times 10^3 \frac{\text{kg} \cdot \text{m}}{\text{s}^2}$$

$$= \boxed{9.8 \times 10^3 \text{ N}}$$

2.11

$$F = \frac{mv^2}{r}$$

$$= \frac{(0.20 \text{ kg})\left(3.0 \dfrac{\text{m}}{\text{s}}\right)^2}{1.5 \text{ m}}$$

$$= \frac{(0.20 \text{ kg})\left(9.0 \dfrac{\text{m}^2}{\text{s}^2}\right)}{1.5 \text{ m}}$$

$$= \frac{0.20 \times 9.0}{1.5} \frac{\text{kg} \cdot \text{m}^2}{\text{s}^2} \times \frac{1}{\text{m}}$$

$$= 1.2 \frac{\text{kg} \cdot \text{m} \cdot \cancel{\text{m}}}{\text{s}^2 \cdot \cancel{\text{m}}}$$

$$= \boxed{1.2 \text{ N}}$$

2.12 Newton's laws of motion consider the resistance to a change of motion, or mass, and not weight. The astronaut's mass is

$$w = mg \therefore m = \frac{w}{g}$$

$$= \frac{1,960.0 \dfrac{\text{kg} \cdot \text{m}}{\text{s}^2}}{9.8 \dfrac{\text{m}}{\text{s}^2}}$$

$$= \frac{1,960.0}{9.8} \frac{\text{kg} \cdot \text{m}}{\text{s}^2} \times \frac{\text{s}^2}{\text{m}}$$

$$= 200 \text{ kg}$$

a. From Newton's second law of motion, you can see that the 100 N rocket gives the 200 kg astronaut an acceleration of:

$$F = ma \therefore a = \frac{F}{m}$$

$$= \frac{100 \dfrac{\text{kg} \cdot \text{m}}{\text{s}^2}}{200 \text{ kg}}$$

$$= \frac{100 \text{ kg} \cdot \text{m}}{200 \text{ s}^2} \times \frac{1}{\text{kg}}$$

$$= 0.5 \text{ m/s}^2$$

b. An acceleration of 0.5 m/s² for 2.0 s will result in a final velocity of

$$a = \frac{v_f - v_i}{t} \therefore v_f = at + v_i$$

$$= (0.5 \text{ m/s}^2)(2.0 \text{ s}) + 0 \text{ m/s}$$

$$= \boxed{1 \text{ m/s}}$$

Chapter 3

3.1 Listing the known and unknown quantities:

$$F = 200 \text{ N}$$
$$d = 3 \text{ m}$$
$$W = ?$$

These are the quantities found in the equation for work, $W = Fd$, which is already solved for work (W). Thus,

$$W = Fd$$

$$= \left(200 \frac{\text{kg} \cdot \text{m}}{\text{s}^2}\right)(3 \text{ m})$$

$$= (200)(3) \text{ N} \cdot \text{m}$$

$$= \boxed{600 \text{ J}}$$

3.2 Listing the known and unknown quantities:

$$F = 440 \text{ N}$$
$$d = 5.0 \text{ m}$$
$$w = 880 \text{ N}$$
$$W = ?$$

These are the quantities found in the equation for work, $W = Fd$, which is already solved for work (W). As you can see in the equation, the force exerted and the distance the box was moved are the quantities used in determining the work accomplished. The weight of the box is a different variable and one that is not used in this equation. Thus,

$$W = Fd$$

$$= \left(440 \frac{\text{kg} \cdot \text{m}}{\text{s}^2}\right)(5.0 \text{ m})$$

$$= 2,200 \text{ N} \cdot \text{m}$$

$$= \boxed{2,200 \text{ J}}$$

3.3 Note that 10.0 kg is a mass quantity and not a weight quantity. Weight is found from $w = mg$, a form of Newton's second law of motion. Thus, the force that must be exerted to lift the backpack is its weight, or (10.0 kg) × (9.8 m/s²), which is 98 N. Therefore, a force of 98 N was exerted on the backpack through a distance of 1.5 m, and

$$W = Fd$$

$$= \left(98 \frac{\text{kg} \cdot \text{m}}{\text{s}^2}\right)(1.5 \text{ m})$$

$$= 147 \text{ N} \cdot \text{m}$$

$$= \boxed{150 \text{ J}}$$

3.4 Weight is defined as the force of gravity acting on an object, and the greater the force of gravity, the harder it is to lift the object. The force is proportional to the mass of the object, as the equation $w = mg$ tells you. Thus, the force you exert when lifting is $F = w = mg$, so the work you do on an object you lift must be $W = mgh$.

You know the mass of the box and you know the work accomplished. You also know the value of the acceleration due to gravity, g, so the list of known and unknown quantities is:

$$m = 102 \text{ kg}$$
$$g = 9.8 \text{ m/s}^2$$
$$W = 5,000 \text{ J}$$
$$h = ?$$

The equation $W = mgh$ is solved for work, so the first thing to do is to solve it for h, the unknown height in this problem (note that height is also a distance):

$$W = mgh \therefore h = \frac{W}{mg}$$

$$= \frac{5{,}000 \frac{kg \cdot m}{s^2} \times m}{(102 \text{ kg}) \left(9.8 \frac{m}{s^2} \right)}$$

$$= \frac{5{,}000.0}{102 \times 9.8} \frac{kg \cdot m}{s^2} \times \frac{m}{1} \times \frac{1}{kg} \times \frac{s^2}{m}$$

$$= \frac{5{,}000}{999.6} m$$

$$= \boxed{5 \text{ m}}$$

3.5 A student running up the stairs has to lift herself, so her weight is the required force needed. Thus, the force exerted is $F = w = mg$, and the work done is $W = mgh$. You know the mass of the student, the height, and the time. You also know the value of the acceleration due to gravity, g, so the list of known and unknown quantities is:

$$m = 60.0 \text{ kg}$$
$$g = 9.8 \text{ m/s}^2$$
$$h = 5.00 \text{ m}$$
$$t = 3.92 \text{ s}$$
$$P = ?$$

The equation $P = \dfrac{mgh}{t}$ is already solved for power, so:

a.
$$P = \frac{mgh}{t}$$

$$= \frac{(60.0 \text{ kg}) \left(9.8 \frac{m}{s^2} \right) (5.00 \text{ m})}{3.92 \text{ s}}$$

$$= \frac{(60.0)(9.8)(5.00) \left(\frac{kg \cdot m}{s^2} \right) \times m}{(3.92) \quad s}$$

$$= \frac{2940}{3.92} \frac{N \cdot m}{s}$$

$$= 750 \frac{J}{s}$$

$$= \boxed{750 \text{ W}}$$

b. A power of 750 watts is almost one horsepower.

3.6 Listing the known and unknown quantities:

$$m = 2{,}000 \text{ kg}$$
$$v = 72 \text{ km/h}$$
$$KE = ?$$

These are the quantities found in the equation for kinetic energy, $KE = 1/2mv^2$, which is already solved. However, note that the velocity is in units of km/h, which must be changed to m/s before doing anything else (it must be m/s because all energy and work units are in units of the joule [J]. A joule is a newton-meter, and a newton is a $kg \cdot m/s^2$).

Using the conversion factor from inside the front cover of your text,

$$\frac{0.2778 \frac{m}{s}}{1.0 \frac{km}{h}} \times 72 \frac{km}{h}$$

$$(0.2778)(72) \frac{m}{s} \times \frac{h}{km} \times \frac{km}{h}$$

$$20 \frac{m}{s}$$

and

$$KE = \frac{1}{2}mv^2$$

$$= \frac{1}{2}(2{,}000 \text{ kg}) \left(20 \frac{m}{s} \right)^2$$

$$= \frac{1}{2}(2{,}000 \text{ kg}) \left(400 \frac{m^2}{s^2} \right)$$

$$= \frac{1}{2} \times 2{,}000 \times 400 \frac{kg \cdot m^2}{s^2}$$

$$= 400{,}000 \frac{kg \cdot m}{s^2} \times m$$

$$= 400{,}000 \text{ N} \cdot m$$

$$= \boxed{4 \times 10^5 \text{ J}}$$

Scientific notation is used here to simplify a large number and to show one significant figure.

3.7 Recall the relationship between work and energy—that you do work on an object when you throw it, giving it kinetic energy, and the kinetic energy it has will do work on something else when stopping. Because of the relationship between work and energy, you can calculate (1) the work you do, (2) the kinetic energy a moving object has as a result of your work, and (3) the work it will do when coming to a stop, and all three answers should be the same. Thus, you do not have a force or a distance to calculate the work needed to stop a moving car, but you can simply calculate the kinetic energy of the car. Both answers should be the same.

Before you start, note that the velocity is in units of km/h, which must be changed to m/s before doing anything else (it must be m/s because all energy and work units are in units of the joule [J]. A joule is a newton-meter, and a newton is a $kg \cdot m/s^2$). Using the conversion factor from inside the front cover,

$$\frac{0.2778 \frac{m}{s}}{1.0 \frac{km}{h}} \times 54.0 \frac{km}{h}$$

$$0.2778 \times 54.0 \frac{m}{s} \times \frac{h}{km} \times \frac{km}{h}$$

$$15.0 \frac{m}{s}$$

and

$$KE = \frac{1}{2}mv^2$$

$$= \frac{1}{2}(1{,}000.0 \text{ kg})\left(15.0 \frac{\text{m}}{\text{s}}\right)^2$$

$$= \frac{1}{2}(1{,}000.0 \text{ kg})\left(225 \frac{\text{m}^2}{\text{s}^2}\right)$$

$$= \frac{1}{2} \times 1{,}000.0 \times 225 \frac{\text{kg} \cdot \text{m}^2}{\text{s}^2}$$

$$= 112{,}500 \frac{\text{kg} \cdot \text{m}}{\text{s}^2} \times \text{m}$$

$$= 112{,}500 \text{ N} \cdot \text{m}$$

$$= \boxed{1.13 \times 10^5 \text{ J}}$$

Scientific notation is used here to simplify a large number and to easily show three significant figures. The answer could likewise be expressed as 113 kJ.

3.8 a. How much energy was used by a 1,000 kg car climbing a hill 51.02 m high is answered by how much work the car did. In this case, $W = Fd$ and the force exerted is the weight of the car, $w = mgh$. Thus,

$$w = mgh$$

$$= (1{,}000 \text{ kg})\left(9.8 \frac{\text{m}}{\text{s}^2}\right)(51.02 \text{ m})$$

$$= 1{,}000 \times 9.8 \times 51.02 \text{ kg} \times \frac{\text{m}}{\text{s}^2} \times \text{m}$$

$$= 499{,}996 \frac{\text{kg} \cdot \text{m}}{\text{s}^2} \times \text{m}$$

$$= 500{,}000 \text{ N} \cdot \text{m}$$

$$= 5 \times 10^5 \text{ J (or 500 kJ)}$$

b. How much potential energy the car has is found in the potential energy equation, $PE = mgh$. Note the potential energy is path independent, that is, depends only on the vertical height of the hill. Thus,

$$PE = mgh$$

$$= (1{,}000 \text{ kg})\left(9.8 \frac{\text{m}}{\text{s}^2}\right)(51.02 \text{ m})$$

$$= 1{,}000 \times 9.8 \times 51.02 \text{ kg} \times \frac{\text{m}}{\text{s}^2} \times \text{m}$$

$$= 499{,}996 \frac{\text{kg} \cdot \text{m}}{\text{s}^2} \times \text{m}$$

$$= 500{,}000 \text{ N} \cdot \text{m}$$

$$= \boxed{5 \times 10^5 \text{ J (or 500 kJ)}}$$

As you can see, the potential energy of the car is exactly the same as the amount of energy used to climb the hill.

3.9 a.

$$W = Fd$$

$$= (10 \text{ lb})(5 \text{ ft})$$

$$= (10)(5) \text{ ft} \times \text{lb}$$

$$= 50 \text{ ft} \cdot \text{lb}$$

b. The distance of the bookcase from some horizontal reference level did not change, so the gravitational potential energy does not change.

3.10 The force (F) needed to lift the book is equal to the weight (w) of the book, or $F = w$. Since $w = mg$, then $F = mg$. Work is defined as the product of a force moved through a distance, or $W = Fd$. The work done in lifting the book is therefore $W = mgh$, and:

a.

$$W = mgh$$

$$= (2.0 \text{ kg})(9.8 \text{ m/s}^2)(2.00 \text{ m})$$

$$= (2.0)(9.8)(2.00) \frac{\text{kg} \cdot \text{m}}{\text{s}^2} \times \text{m}$$

$$= 39.2 \frac{\text{kg} \cdot \text{m}^2}{\text{s}^2}$$

$$= 39.2 \text{ J} = \boxed{39 \text{ J}}$$

b.

$$PE = mgh = \boxed{39 \text{ J}}$$

c.

$$PE_{\text{lost}} = KE_{\text{gained}} = mgh = \boxed{39 \text{ J}}$$

(or)

$$v = \sqrt{2gh} = \sqrt{(2)(9.8 \text{ m/s}^2)(2.00 \text{ m})}$$

$$= \sqrt{39.2 \text{ m}^2/\text{s}^2}$$

$$= 6.26 \text{ m/s}$$

$$KE = \frac{1}{2}mv^2 = \left(\frac{1}{2}\right)(2.0 \text{ kg})(6.26 \text{ m/s})^2$$

$$= \left(\frac{1}{2}\right)(2.0 \text{ kg})(39.2 \text{ m}^2/\text{s}^2)$$

$$= (1.0)(39.2) \frac{\text{kg} \cdot \text{m}^2}{\text{s}^2}$$

$$= \boxed{39 \text{ J}}$$

3.11

$$KE = \frac{1}{2}mv^2$$

$$= \frac{1}{2}(60.0 \text{ kg})\left(2.0 \frac{\text{m}}{\text{s}}\right)^2$$

$$= \frac{1}{2}(60.0 \text{ kg})\left(4.0 \frac{\text{m}^2}{\text{s}^2}\right)$$

$$= 30.0 \times 4.0 \text{ kg} \times \left(\frac{\text{m}^2}{\text{s}^2}\right)$$

$$= \boxed{120 \text{ J}}$$

$$KE = \frac{1}{2}mv^2$$

$$= \frac{1}{2}(60.0 \text{ kg})\left(4.0 \frac{\text{m}}{\text{s}}\right)^2$$

$$= \frac{1}{2}(60.0 \text{ kg})\left(16 \frac{\text{m}^2}{\text{s}^2}\right)$$

$$= 30.0 \times 16 \text{ kg} \times \left(\frac{\text{m}^2}{\text{s}^2}\right)$$

$$= \boxed{480 \text{ J}}$$

Thus, doubling the speed results in a four-fold increase in kinetic energy.

3.12 a. The force needed is equal to the weight of the student. The English unit of a pound is a force unit, so

$$W = Fd$$
$$= (170.0 \text{ lb})(25.0 \text{ ft})$$
$$= \boxed{4{,}250 \text{ ft·lb}}$$

b. Work (W) is defined as a force (F) moved through a distance (d), or $W = Fd$. Power (P) is defined as work (W) per unit of time (t), or $P = W/t$. Therefore,

$$P = \frac{Fd}{t}$$
$$= \frac{(170.0 \text{ lb})(25.0 \text{ ft})}{10.0 \text{ s}}$$
$$= \frac{(170.0)(25.0)}{10.0} \frac{\text{ft·lb}}{\text{s}}$$
$$= 425 \frac{\text{ft·lb}}{\text{s}}$$

One hp is defined as $550 \dfrac{\text{ft·lb}}{\text{s}}$ and

$$\frac{425 \text{ ft·lb/s}}{550 \dfrac{\text{ft·lb/s}}{\text{hp}}} = \boxed{0.77 \text{ hp}}$$

(Note that the student's power rating [425 ft·lb/s] is less than the power rating defined as 1 horsepower [550 ft·lb/s]. Thus, the student's horsepower must be *less* than 1 horsepower. A simple analysis such as this will let you know if you inverted the ratio or not.)

Chapter 4

4.1 Listing the known and unknown quantities:
body temperature $\quad T_F = 98.6^\circ$
$$T_C = ?$$

These are the quantities found in the equation for conversion of Fahrenheit to Celsius, $T_C = \dfrac{5}{9}(T_F - 32^\circ)$, where T_F is the temperature in Fahrenheit and T_C is the temperature in Celsius. This equation describes a relationship between the two temperature scales and is used to convert a Fahrenheit temperature to Celsius. The equation is already solved for the Celsius temperature, T_C. Thus,

$$T_C = \frac{5}{9}(T_F - 32^\circ)$$
$$= \frac{5}{9}(98.6^\circ - 32^\circ)$$
$$= \frac{333^\circ}{9}$$
$$= \boxed{37^\circ C}$$

4.2 $\quad Q = mc\,\Delta T$
$$= (221 \text{ g})\left(0.093 \frac{\text{cal}}{\text{gC}^\circ}\right)(38.0^\circ C - 20.0^\circ C)$$
$$= (221)(0.093)(18.0)\,\text{g} \times \frac{\text{cal}}{\text{gC}^\circ} \times {}^\circ C$$
$$= 370 \frac{\text{g·cal·}^\circ\!C}{\text{gC}^\circ}$$
$$= \boxed{370 \text{ cal}}$$

4.3 First, you need to know the energy of the moving bike and rider. Since the speed is given as 36.0 km/h, convert to m/s by multiplying times 0.2778 m/s per km/h:

$$\left(36.0 \frac{\text{km}}{\text{h}}\right)\left(0.2778 \frac{\text{m/s}}{\text{km/h}}\right)$$
$$= (36.0)(0.2778) \frac{\text{km}}{\text{h}} \times \frac{\text{h}}{\text{km}} \times \frac{\text{m}}{\text{s}}$$
$$= 10.0 \text{ m/s}$$

Then,

$$KE = \frac{1}{2}mv^2$$
$$= \frac{1}{2}(100.0 \text{ kg})(10.0 \text{ m/s})^2$$
$$= \frac{1}{2}(100.0 \text{ kg})(100 \text{ m}^2/\text{s}^2)$$
$$= \frac{1}{2}(100.0)(100) \frac{\text{kg·m}^2}{\text{s}^2}$$
$$= 5{,}000 \text{ J}$$

Second, this energy is converted to the calorie heat unit through the mechanical equivalent of heat relationship, that 1.0 kcal = 4,184 J, or that 1.0 cal = 4.184 J. Thus,

$$\frac{5{,}000 \text{ J}}{4{,}184 \dfrac{\text{J}}{\text{kcal}}}$$
$$1.195 \frac{\text{J}}{1} \times \frac{\text{kcal}}{\text{J}}$$
$$\boxed{1.20 \text{ kcal}}$$

4.4 First, you need to find the energy of the falling bag. Since the potential energy lost equals the kinetic energy gained, the energy of the bag just as it hits the ground can be found from

$$PE = mgh$$
$$= (15.53 \text{ kg})(9.8 \text{ m/s}^2)(5.50 \text{ m})$$
$$= (15.53)(9.8)(5.50) \frac{\text{kg·m}}{\text{s}^2} \times \text{m}$$
$$= 837 \text{ J}$$

In calories, this energy is equivalent to

$$\frac{837 \text{ J}}{4{,}184 \dfrac{\text{J}}{\text{kcal}}} = 0.200 \text{ kcal}$$

Second, the temperature change can be calculated from the equation giving the relationship between a quantity of heat (Q), mass (m), specific heat of the substance (c), and the change of temperature:

$$Q = mc\,\Delta T \ \therefore\ \Delta T = \frac{Q}{mc}$$
$$= \frac{0.200 \text{ kcal}}{(15.53 \text{ kg})\left(0.200 \dfrac{\text{kcal}}{\text{kgC}^\circ}\right)}$$
$$= \frac{0.200}{(15.53)(0.200)} \frac{\text{kcal}}{1} \times \frac{1}{\text{kg}} \times \frac{\text{kgC}^\circ}{\text{kcal}}$$
$$= 0.064 \frac{\text{kcal·kgC}^\circ}{\text{kcal·kg}}$$
$$= \boxed{6.4 \times 10^{-2}\,^\circ C}$$

4.5 The Calorie used by dietitians is a kilocalorie; thus, 250.0 Cal is 250.0 kcal. The mechanical energy equivalent is 1 kcal = 4,184 J, so (250.0 kcal)(4,184 J/kcal) = 1,046,250 J.

Since $W = Fd$ and the force needed is equal to the weight (mg) of the person, $W = mgh = (75.0 \text{ kg})(9.8 \text{ m/s}^2)(10.0 \text{ m}) = 7,350$ J for each stairway climbed.

A total of 1,046,250 J of energy from the french fries would require 1,046,250 J/7,350 J per climb, or 142.3 trips up the stairs.

4.6 For unit consistency,

$$T_C = \frac{5}{9}(T_F - 32°) = \frac{5}{9}(68° - 32°) = \frac{5}{9}(36°) = 20°C$$

$$= \frac{5}{9}(32° - 32°) = \frac{5}{9}(0°) = 0°C$$

Glass bowl:

$$Q = mc\Delta T$$

$$= (0.5 \text{ kg})\left(0.2 \frac{\text{kcal}}{\text{kg°C}}\right)(20°C)$$

$$= (0.5)(0.2)(20) \frac{\text{kg}}{1} \times \frac{\text{kcal}}{\text{kgC°}} \times \frac{°C}{1}$$

$$= \boxed{2 \text{ kcal}}$$

Iron pan:

$$Q = mc\Delta T$$

$$= (0.5 \text{ kg})\left(0.11 \frac{\text{kcal}}{\text{kgC°}}\right)(20°C)$$

$$= (0.5)(0.11)(20) \text{ kg} \times \frac{\text{kcal}}{\text{kgC°}} \times °C$$

$$= \boxed{1 \text{ kcal}}$$

4.7 Note that a specific heat expressed in cal/gC° has the same numerical value as a specific heat expressed in kcal/kgC° because you can cancel the k units. You could convert 896 cal to 0.896 kcal, but one of the two conversion methods is needed for consistency with other units in the problem.

$$Q = mc\Delta T \therefore m = \frac{Q}{c\Delta T}$$

$$= \frac{896 \text{ cal}}{\left(0.056 \frac{\text{cal}}{\text{gC°}}\right)(80.0°C)}$$

$$= \frac{896}{(0.056)(80.0)} \frac{\text{cal}}{1} \times \frac{\text{gC°}}{\text{cal}} \times \frac{1}{C°}$$

$$= 200 \text{ g}$$

$$= \boxed{0.20 \text{ kg}}$$

4.8 Since a watt is defined as a joule/s, finding the total energy in joules will tell the time:

$$Q = mc\Delta T$$

$$= (250.0 \text{ g})\left(1.00 \frac{\text{cal}}{\text{gC°}}\right)(60.0°C)$$

$$= (250.0)(1.00)(60.0) \text{ g} \times \frac{\text{cal}}{\text{gC°}} \times °C$$

$$= 1.50 \times 10^4 \text{ cal}$$

This energy in joules is $(1.50 \times 10^4 \text{ cal})\left(4.184 \frac{\text{J}}{\text{cal}}\right) = 62,800$ J

A 300 watt heater uses energy at a rate of $300 \frac{\text{J}}{\text{s}}$, so

$$\frac{62,800 \text{ J}}{300 \text{ J/s}} = 209 \text{ s} \quad \text{is required, which is} \quad \frac{209 \text{ s}}{60 \frac{\text{s}}{\text{min}}} = 3.48 \text{ min},$$

or

$$\boxed{\text{about } 3\frac{1}{2} \text{ min}}$$

4.9
$$Q = mc\Delta \therefore c = \frac{Q}{m\Delta T}$$

$$= \frac{60.0 \text{ cal}}{(100.0 \text{ g})(20.0°C)}$$

$$= \frac{60.0}{(100.0)(20.0)} \frac{\text{cal}}{\text{gC°}}$$

$$= \boxed{0.0300 \frac{\text{cal}}{\text{gC°}}}$$

4.10 To change water at 80.0°C to steam at 100.0°C requires two separate quantities of heat that can be called Q_1 and Q_2. The quantity Q_1 is the amount of heat needed to warm the water from 80.0°C to the boiling point, which is 100.0°C at sea level pressure ($\Delta T = 20.0°C$). The relationship between the variables involved is $Q = mc\Delta T$. The quantity Q_2 is the amount of heat needed to take 100.0°C water through the phase change to steam (water vapor) at 100.0°C. The phase change from a liquid to a gas (or gas to liquid) is concerned with the latent heat of vaporization. For water, the latent heat of vaporization is given as 540.0 cal/g.

$m = 250.0$ g

$L_{v(water)} = 540.0$ cal/g

$Q = ?$

$$Q_1 = mc\Delta T$$

$$= (250.0 \text{ g})\left(1.00 \frac{\text{cal}}{\text{gC°}}\right)(20.0°C)$$

$$= (250.0)(1.00)(20.0) \text{ g} \times \frac{\text{cal}}{\text{gC°}} \times °C$$

$$= 5,000 \frac{\text{g·cal·C°}}{\text{g·C°}}$$

$$= 5,000 \text{ cal}$$

$$= 5.00 \text{ kcal}$$

$$Q_2 = mL_v$$

$$= (250.0 \text{ g})\left(540.0 \frac{\text{cal}}{\text{g}}\right)$$

$$= 250.0 \times 540.0 \frac{\text{g·cal}}{\text{g}}$$

$$= 135,000 \text{ cal}$$

$$= 135.0 \text{ kcal}$$

$$Q_{Total} = Q_1 + Q_2$$

$$= 5.00 \text{ kcal} + 135.0 \text{ kcal}$$

$$= \boxed{140.0 \text{ kcal}}$$

4.11 To change 20.0°C water to steam at 125.0°C requires three separate quantities of heat. First, the quantity Q_1 is the amount of heat needed to warm the water from 20.0°C to 100.0°C ($\Delta T = 80.0$°C). The quantity Q_2 is the amount of heat needed to take 100.0°C water to steam at 100.0°C. Finally, the quantity Q_3 is the amount of heat needed to warm the steam from 100.0° to 125.0°C. According to table 4.3, the c for steam is 0.480 cal/g°C.

$$m = 100.0 \text{ g}$$
$$\Delta T_{\text{water}} = 80.0\text{°C}$$
$$\Delta T_{\text{steam}} = 25.0\text{°C}$$
$$L_{v(\text{water})} = 540.0 \text{ cal/g}$$
$$c_{\text{steam}} = 0.480 \text{ cal/gC°}$$

$$Q_1 = mc\,\Delta T$$
$$= (100.0 \text{ g})\left(1.00\,\frac{\text{cal}}{\text{gC°}}\right)(80.0\text{°C})$$
$$= (100.0)(1.00)(80.0) \text{ g} \times \frac{\text{cal}}{\text{gC°}} \times \text{°C}$$
$$= 8,000\,\frac{\text{g} \cdot \text{cal} \cdot \text{°C}}{\text{g} \cdot \text{C°}}$$
$$= 8,000 \text{ cal}$$
$$= 8.00 \text{ kcal}$$

$$Q_2 = mL_v$$
$$= (100.0 \text{ g})\left(540.0\,\frac{\text{cal}}{\text{g}}\right)$$
$$= 100.0 \times 540.0\,\frac{\text{g} \cdot \text{cal}}{\text{g}}$$
$$= 54,000 \text{ cal}$$
$$= 54.00 \text{ kcal}$$

$$Q_3 = mc\,\Delta T$$
$$= (100.0 \text{ g})\left(0.480\,\frac{\text{cal}}{\text{gC°}}\right)(25.0\text{°C})$$
$$= (100.0)(0.480)(25.0) \text{ g} \times \frac{\text{cal}}{\text{gC°}} \times \text{°C}$$
$$= 1,200\,\frac{\text{g} \cdot \text{cal} \cdot \text{°C}}{\text{g} \cdot \text{C°}}$$
$$= 1,200 \text{ cal}$$
$$= 1.20 \text{ kcal}$$

$$Q_{\text{total}} = Q_1 + Q_2 + Q_3$$
$$= 8.00 \text{ kcal} + 54.00 \text{ kcal} + 1.20 \text{ kcal}$$
$$= \boxed{63.20 \text{ kcal}}$$

4.12 a. Step 1: Cool the water from 18.0°C to 0°C.

$$Q_1 = mc\,\Delta T$$
$$= (400.0 \text{ g})\left(1.00\,\frac{\text{cal}}{\text{gC°}}\right)(18.0\text{°C})$$
$$= (400.0)(1.00)(18.0) \text{ g} \times \frac{\text{cal}}{\text{gC°}} \times \text{°C}$$
$$= 7,200\,\frac{\text{g} \cdot \text{cal} \cdot \text{°C}}{\text{g} \cdot \text{C°}}$$
$$= 7,200 \text{ cal}$$
$$= 7.20 \text{ kcal}$$

Step 2: Find the energy needed for the phase change of water at 0°C to ice at 0°C.

$$Q_2 = mL_f$$
$$= (400.0 \text{ g})\left(80.0\,\frac{\text{cal}}{\text{g}}\right)$$
$$= 400.0 \times 80.0\,\frac{\text{g} \cdot \text{cal}}{\text{g}}$$
$$= 32,000 \text{ cal}$$
$$= 32.0 \text{ kcal}$$

Step 3: Cool the ice from 0°C to ice at -5.00°C.

$$Q_3 = mc\,\Delta T$$
$$= (400.0 \text{ g})\left(0.50\,\frac{\text{cal}}{\text{gC°}}\right)(5.00\text{°C})$$
$$= 400.0 \times 0.50 \times 5.00 \text{ g} \times \frac{\text{cal}}{\text{gC°}} \times \text{°C}$$
$$= 1,000\,\frac{\text{g} \cdot \text{cal} \cdot \text{°C}}{\text{g} \cdot \text{C°}}$$
$$= 1,000 \text{ cal}$$
$$= 1.0 \text{ kcal}$$

$$Q_{\text{total}} = Q_1 + Q_2 + Q_3$$
$$= 7.20 \text{ kcal} + 32.0 \text{ kcal} + 1.0 \text{ kcal}$$
$$= \boxed{40.2 \text{ kcal}}$$

Chapter 5

5.1
$$v = f\lambda$$
$$= \left(10\,\frac{1}{\text{s}}\right)(0.50 \text{ m})$$
$$= 5\,\frac{\text{m}}{\text{s}}$$

5.2 The distance between two *consecutive* condensations (or rarefactions) is one wavelength, so $\lambda = 3.00$ m and

$$v = f\lambda$$
$$= \left(112.0\,\frac{1}{\text{s}}\right)(3.00 \text{ m})$$
$$= 336\,\frac{\text{m}}{\text{s}}$$

5.3 a. One complete wave every 4 s means that $T = 4.00$ s. (Note that the symbol for the *time of a cycle* is T. Do not confuse this symbol with the symbol for temperature.)

b.
$$f = \frac{1}{T}$$
$$= \frac{1}{4.0 \text{ s}}$$
$$= \frac{1}{4.0}\,\frac{1}{\text{s}}$$
$$= 0.25\,\frac{1}{\text{s}}$$
$$= \boxed{0.25 \text{ Hz}}$$

5.4 The distance from one condensation to the next is one wavelength, so

$$v = f\lambda \therefore \lambda = \frac{v}{f}$$

$$= \frac{330\ \frac{\text{m}}{\text{s}}}{260\ \frac{1}{\text{s}}}$$

$$= \frac{330}{260}\ \frac{\text{m}}{\text{s}} \times \frac{\text{s}}{1}$$

$$= \boxed{1.3\ \text{m}}$$

5.5 a. $v = f\lambda = \left(256\ \frac{1}{\text{s}}\right)(1.34\ \text{m}) = \boxed{343\ \text{m/s}}$

b. $= \left(440.0\ \frac{1}{\text{s}}\right)(0.780\ \text{m}) = \boxed{343\ \text{m/s}}$

c. $= \left(750.0\ \frac{1}{\text{s}}\right)(0.457\ \text{m}) = \boxed{343\ \text{m/s}}$

d. $= \left(2,500.0\ \frac{1}{\text{s}}\right)(0.137\ \text{m}) = \boxed{343\ \text{m/s}}$

5.6 The speed of sound and time are given and you are looking for a distance.

$$v = 1,100\ \text{ft/s}$$
$$t = 4.80\ \text{s}$$
$$d = ?$$

Calculating the total distance the sound traveled,

$$v = \frac{d}{t} \therefore d = vt$$

$$= \left(1,100\ \frac{\text{ft}}{\text{s}}\right)(4.80\ \text{s})$$

$$= (1,100)(4.80)\frac{\text{ft}}{\text{s}} \times \text{s}$$

$$= 4,800\ \text{ft}$$

Since the sound travels from you to the cliff, then back to you, the cliff must be half the total distance:

$$\frac{4,800\ \text{ft}}{2} = \boxed{2,400\ \text{ft}}$$

5.7 The speed of the sound and the time between the lightning and thunder are given:

$$v = 1,140\ \text{ft/s}$$
$$t = 4.63\ \text{s}$$
$$d = ?$$

The distance that a sound with this velocity travels in the given time is

$$v = \frac{d}{t} \therefore d = vt$$

$$= \left(1,140\ \frac{\text{ft}}{\text{s}}\right)(4.63\ \text{s})$$

$$= (1,140)(4.63)\frac{\text{ft}}{\text{s}} \times \text{s}$$

$$= \boxed{5,280\ \text{ft or 1 mile}}$$

5.8 a. $v = f\lambda \therefore \lambda = \frac{v}{f}$

$$= \frac{1,125\ \frac{\text{ft}}{\text{s}}}{440\ \frac{1}{\text{s}}}$$

$$= \frac{1,125}{440}\ \frac{\text{ft}}{\text{s}} \times \frac{\text{s}}{1}$$

$$= 2.56\ \frac{\text{ft}\cdot\text{s}}{\text{s}}$$

$$= \boxed{2.6\ \text{ft}}$$

b. $v = f\lambda \therefore \lambda = \frac{v}{f}$

$$= \frac{5,020}{440}\ \frac{\text{ft}}{\text{s}} \times \frac{\text{s}}{1}$$

$$= 11.4\ \text{ft} = \boxed{11\ \text{ft}}$$

Chapter 6

6.1 First, recall that a negative charge means an excess of electrons. Second, the relationship between the total charge (q), the number of electrons (n), and the charge of a single electron (e) is $q = ne$. The fundamental charge of a single ($n = 1$) electron (e) is $1.60 \times 10^{-19}\ \text{C}$. Thus

$$q = ne \therefore n = \frac{q}{e}$$

$$= \frac{1.00 \times 10^{-14}\ \text{C}}{1.60 \times 10^{-19}\ \frac{\text{C}}{\text{electron}}}$$

$$= \frac{1.00 \times 10^{-14}}{1.60 \times 10^{-19}}\ \frac{\text{C}}{1} \times \frac{\text{electron}}{\text{C}}$$

$$= 6.25 \times 10^4\ \frac{\text{C}\cdot\text{electron}}{\text{C}}$$

$$= \boxed{6.25 \times 10^4\ \text{electron}}$$

6.2

$$\frac{\text{electric}}{\text{current}} = \frac{\text{charge}}{\text{time}}$$

or

$$I = \frac{q}{t}$$

$$= \frac{6.00\ \text{C}}{2.00\ \text{s}}$$

$$= 3.00\ \frac{\text{C}}{\text{s}}$$

$$= \boxed{3.00\ \text{A}}$$

6.3

$$R = \frac{V}{I}$$

$$= \frac{120.0\ \text{V}}{4.00\ \text{A}}$$

$$= 30.0\ \frac{\text{V}}{\text{A}}$$

$$= \boxed{30.0\ \Omega}$$

6.4
$$R = \frac{V}{I} \therefore I = \frac{V}{R}$$
$$= \frac{120.0 \text{ V}}{60.0 \frac{\text{V}}{\text{A}}}$$
$$= \frac{120.0}{60.0} \text{V} \times \frac{\text{A}}{\text{V}}$$
$$= \boxed{2.00 \text{ A}}$$

6.5 a.
$$R = \frac{V}{I} \therefore V = IR$$
$$= (1.20 \text{ A})\left(10.0 \frac{\text{V}}{\text{A}}\right)$$
$$= \boxed{12.0 \text{ V}}$$

b. Power = (current)(potential difference)

or

$$P = IV$$
$$= \left(1.20 \frac{\text{C}}{\text{s}}\right)\left(12.0 \frac{\text{J}}{\text{C}}\right)$$
$$= (1.20)(12.0) \frac{\text{C}}{\text{s}} \times \frac{\text{J}}{\text{C}}$$
$$= 14.4 \frac{\text{J}}{\text{s}}$$
$$= \boxed{14.4 \text{ W}}$$

6.6 Note that there are two separate electrical units that are rates: (1) the amp (coulomb/s) and (2) the watt (joule/s). The question asked for a rate of using energy. Energy is measured in joules, so you are looking for the power of the radio in watts. To find watts ($P = IV$), you will need to calculate the current (I) since it is not given. The current can be obtained from the relationship of Ohm's law:

$$I = \frac{V}{R}$$
$$= \frac{3.00 \text{ V}}{15.0 \frac{\text{V}}{\text{A}}}$$
$$= 0.200 \text{ A}$$
$$P = IV$$
$$= (0.200 \text{ C/s})(3.00 \text{ J/C})$$
$$= \boxed{0.600 \text{ W}}$$

6.7
$$\frac{(1,200 \text{ W})(0.25 \text{ h})(\$0.10/\text{kWh})}{1,000 \frac{\text{W}}{\text{kW}}}$$
$$\frac{(1,200)(0.25)(0.10)}{1,000} \frac{\text{W}}{1} \times \frac{\text{h}}{1} \times \frac{\$}{\text{kWh}} \times \frac{\text{kW}}{\text{W}}$$
$$\boxed{\$0.03} \text{ (3 cents)}$$

6.8 The relationship between power (P), current (I), and voltage (V) will provide a solution. Since the relationship considers

power in watts, the first step is to convert horsepower to watts. One horsepower is equivalent to 746 watts, so:

$$(746 \text{ W/hp})(2.00 \text{ hp}) = 1,492 \text{ W}$$
$$P = IV \therefore I = \frac{P}{V}$$
$$= \frac{1,492 \frac{\text{J}}{\text{s}}}{12.0 \frac{\text{J}}{\text{C}}}$$
$$= \frac{1,492}{12.0} \frac{\text{J}}{\text{s}} \times \frac{\text{C}}{\text{J}}$$
$$= 124.3 \frac{\text{C}}{\text{s}}$$
$$= \boxed{124 \text{ A}}$$

6.9 a. The rate of using energy is joule/s, or the watt. Since 1.00 hp = 746 W,
inside motor: (746 W/hp)(1/3 hp) = 249 W
outside motor: (746 W/hp)(1/3 hp) = 249 W
compressor motor: (746 W/hp)(3.70 hp) = 2,760 W

$$249 \text{ W} + 249 \text{ W} + 2,760 \text{ W} = \boxed{3,258 \text{ W}}$$

b. $\dfrac{(3,258 \text{ W})(1.00 \text{ h})(\$0.10/\text{kWh})}{1,000 \text{ W/kW}} = \$0.33$ per hour

c. ($0.33/h)(12 h/day)(30 day/mo) = $\boxed{\$118.80}$

6.10 The solution is to find how much current each device draws and then to see if the total current is less or greater than the breaker rating:

Toaster: $I = \dfrac{V}{R} = \dfrac{120 \text{ V}}{15 \text{ V/A}} = 8.0 \text{ A}$

Motor: (0.20 hp)(746 W/hp) = 150 W

$$I = \frac{P}{V} = \frac{15 \text{ J/s}}{120 \text{ J/C}} = 1.3 \text{ A}$$

Three 100 W bulbs: 3 × 100 W = 300 W

$$I = \frac{P}{V} = \frac{300 \text{ J/s}}{120 \text{ J/C}} = 2.5 \text{ A}$$

Iron: $I = \dfrac{P}{V} = \dfrac{600 \text{ J/s}}{120 \text{ J/C}} = 5.0 \text{ A}$

The sum of the currents is 8.0A + 1.3A + 2.5A + 5.0A = 16.8A, so the total current is greater than 15.0 amp and the circuit breaker will trip.

6.11 a.
$$V_p = 1,200 \text{ V}$$
$$N_p = 1 \text{ loop}$$
$$N_s = 200 \text{ loops}$$
$$V_s = ?$$
$$\frac{V_p}{N_p} = \frac{V_s}{N_s} \therefore V_s = \frac{V_p N_s}{N_p}$$
$$V_s = \frac{(1,200 \text{ V})(200 \text{ loops})}{1 \text{ loop}}$$
$$= \boxed{240,000 \text{ V}}$$

b.
$$I_p = 40 \text{ A} \quad V_p I_p = V_s I_s \therefore I_s = \frac{V_p I_p}{V_s}$$

$$I_s = ? \qquad I_s = \frac{1,200 \text{ V} \times 40 \text{ A}}{240,000 \text{ V}}$$

$$= \frac{1,200 \times 40}{240,000} \frac{\text{V} \cdot \text{A}}{\text{V}}$$

$$= \boxed{0.2 \text{ A}}$$

6.12 a.
$$V_s = 12 \text{ V}$$
$$I_s = 0.5 \text{ A} \qquad \frac{V_p}{N_p} = \frac{V_s}{N_s} \therefore \frac{N_p}{N_s} = \frac{V_p}{V_s}$$
$$V_p = 120 \text{ V} \qquad \frac{N_p}{N_s} = \frac{120 \text{ V}}{12 \text{ V}} = \frac{10}{1}$$
$$\frac{N_p}{N_s} = ?$$

or

$$\boxed{10 \text{ primary to 1 secondary}}$$

b.
$$I_p = ? \quad V_p I_p = V_s I_s \therefore I_p = \frac{V_s I_s}{V_p}$$

$$I_p = \frac{(12 \text{ V})(0.5 \text{ A})}{120 \text{ V}}$$

$$= \frac{12 \times 0.5}{120} \frac{\text{V} \cdot \text{A}}{\text{V}}$$

$$= \boxed{0.05 \text{ A}}$$

c.
$$P_s = ? \quad P_s = I_s V_s$$
$$= (0.5 \text{ A})(12 \text{ V})$$
$$= 0.5 \times 12 \frac{\text{C}}{\text{s}} \times \frac{\text{J}}{\text{C}}$$
$$= 6 \frac{\text{J}}{\text{s}}$$
$$= \boxed{6 \text{ W}}$$

Chapter 7

7.1 The relationship between the speed of light in a transparent material (v), the speed of light in a vacuum ($c = 3.00 \times 10^8$ m/s), and the index of refraction (n) is $n = c/v$. According to table 7.1, the index of refraction for water is $n = 1.33$ and for ice is $n = 1.31$.

a.
$$c = 3.00 \times 10^8 \text{ m/s}$$
$$n = 1.33$$
$$v = ?$$
$$n = \frac{c}{v} \therefore v = \frac{c}{n}$$
$$v = \frac{3.00 \times 10^8 \text{ m/s}}{1.33}$$
$$= \boxed{2.26 \times 10^8 \text{ m/s}}$$

b.
$$c = 3.00 \times 10^8 \text{ m/s}$$
$$n = 1.31$$
$$v = ?$$
$$v = \frac{3.00 \times 10^8 \text{ m/s}}{1.31}$$
$$= \boxed{2.29 \times 10^8 \text{ m/s}}$$

7.2
$$d = 1.50 \times 10^8 \text{ km}$$
$$= 1.50 \times 10^{11} \text{ m}$$
$$c = 3.00 \times 10^8 \text{ m/s}$$
$$t = ?$$
$$v = \frac{d}{t} \therefore t = \frac{d}{v}$$
$$t = \frac{1.50 \times 10^{11} \text{ m}}{3.00 \times 10^8 \frac{\text{m}}{\text{s}}}$$
$$= \frac{1.50 \times 10^{11}}{3.00 \times 10^8} \text{ m} \times \frac{\text{s}}{\text{m}}$$
$$= 5.00 \times 10^2 \frac{\text{m} \cdot \text{s}}{\text{m}}$$
$$= \frac{5.00 \times 10^2 \text{ s}}{60.0 \frac{\text{s}}{\text{min}}}$$
$$= \frac{5.00 \times 10^2}{60.0} \text{ s} \times \frac{\text{min}}{\text{s}}$$
$$= \boxed{8.33 \text{ min}}$$

7.3
$$d = 6.00 \times 10^9 \text{ km}$$
$$= 6.00 \times 10^{12} \text{ m}$$
$$c = 3.00 \times 10^8 \text{ m/s}$$
$$t = ?$$
$$v = \frac{d}{t} \therefore t = \frac{d}{v}$$
$$t = \frac{6.00 \times 10^{12} \text{ m}}{3.00 \times 10^8 \frac{\text{m}}{\text{s}}}$$
$$= \frac{6.00 \times 10^{12}}{3.00 \times 10^8} \text{ m} \times \frac{\text{s}}{\text{m}}$$
$$= 2.00 \times 10^4 \text{ s}$$
$$= \frac{2.00 \times 10^4 \text{ s}}{3,600 \frac{\text{s}}{\text{h}}}$$
$$= \frac{2.00 \times 10^4}{3.600 \times 10^3} \text{ s} \times \frac{\text{h}}{\text{s}}$$
$$= \boxed{5.56 \text{ h}}$$

7.4 From equation 7.1, note that both angles are measured from the normal and that the angle of incidence (θ_i) equals the angle of reflection (θ_r), or

$$\theta_i = \theta_r \therefore \boxed{\theta_i = 10°}$$

7.5
$$v = 2.20 \times 10^8 \text{ m/s}$$
$$c = 3.00 \times 10^8 \text{ m/s}$$
$$n = ?$$
$$n = \frac{c}{v}$$
$$= \frac{3.00 \times 10^8 \frac{\text{m}}{\text{s}}}{2.20 \times 10^8 \frac{\text{m}}{\text{s}}}$$
$$= 1.36$$

According to table 7.1, the substance with an index of refraction of 1.36 is ethyl alcohol.

7.6 a. From equation 7.3:

$$\lambda = 6.00 \times 10^{-7} \text{ m} \qquad c = \lambda f \therefore f = \frac{c}{\lambda}$$

$$c = 3.00 \times 10^8 \text{ m/s}$$

$$f = ?$$

$$f = \frac{3.00 \times 10^8 \frac{m}{s}}{6.00 \times 10^{-7} \text{ m}}$$

$$= \frac{3.00 \times 10^8}{6.00 \times 10^{-7}} \frac{m}{s} \times \frac{1}{m}$$

$$= 5.00 \times 10^{14} \frac{1}{s}$$

$$= \boxed{5.00 \times 10^{14} \text{ Hz}}$$

b. From equation 7.4:

$$f = 5.00 \times 10^{14} \text{ Hz}$$

$$h = 6.63 \times 10^{-34} \text{ J·s}$$

$$E = ?$$

$$E = hf$$

$$= (6.63 \times 10^{-34} \text{ J·s})\left(5.00 \times 10^{14} \frac{1}{s}\right)$$

$$= (6.63 \times 10^{-34})(5.00 \times 10^{14}) \text{ J·s} \times \frac{1}{s}$$

$$= \boxed{3.32 \times 10^{-19} \text{ J}}$$

7.7 First, you can find the energy of one photon of the peak intensity wavelength (5.60×10^{-7} m) by using equation 7.3 to find the frequency, then equation 7.4 to find the energy:

Step 1:
$$c = \lambda f \therefore f = \frac{c}{\lambda}$$

$$= \frac{3.00 \times 10^8 \frac{m}{s}}{5.60 \times 10^{-7} \text{ m}}$$

$$= 5.36 \times 10^{14} \text{ Hz}$$

Step 2: $E = hf$

$$= (6.63 \times 10^{-34} \text{ J·s})(5.36 \times 10^{14} \text{ Hz})$$

$$= 3.55 \times 10^{-19} \text{ J}$$

Step 3: Since one photon carries an energy of 3.55×10^{-19} J and the overall intensity is 1,000.0 W, each square meter must receive an average of

$$\frac{1,000.0 \frac{J}{s}}{3.55 \times 10^{-19} \frac{J}{\text{photon}}}$$

$$\frac{1.000 \times 10^3}{3.55 \times 10^{-19}} \frac{J}{s} \times \frac{\text{photon}}{J}$$

$$\boxed{2.82 \times 10^{21} \frac{\text{photon}}{s}}$$

7.8 a. $f = 4.90 \times 10^{14}$ Hz $\qquad c = \lambda f \therefore \lambda = \frac{c}{f}$

$$c = 3.00 \times 10^8 \text{ m/s}$$

$$\lambda = ?$$

$$\lambda = \frac{3.00 \times 10^8 \frac{m}{s}}{4.90 \times 10^{14} \frac{1}{s}}$$

$$= \frac{3.00 \times 10^8}{4.90 \times 10^{14}} \frac{m}{s} \times \frac{s}{1}$$

$$= \boxed{6.12 \times 10^{-7} \text{ m}}$$

b. According to table 7.2, this is the frequency and wavelength of orange light.

7.9
$$f = 5.00 \times 10^{20} \text{ Hz}$$

$$h = 6.63 \times 10^{-34} \text{ J·s}$$

$$E = ?$$

$$E = hf$$

$$= (6.63 \times 10^{-34} \text{ J·s})\left(5.00 \times 10^{20} \frac{1}{s}\right)$$

$$= (6.63 \times 10^{-34})(5.00 \times 10^{20}) \text{ J·s} \times \frac{1}{s}$$

$$= \boxed{3.32 \times 10^{-13} \text{ J}}$$

7.10
$$\lambda = 1.00 \text{ mm}$$

$$= 0.001 \text{ m}$$

$$f = ?$$

$$c = 3.00 \times 10^8 \text{ m/s}$$

$$h = 6.63 \times 10^{-34} \text{ J·s}$$

$$E = ?$$

Step 1:
$$c = \lambda f \therefore f = \frac{c}{\lambda}$$

$$f = \frac{3.00 \times 10^8 \frac{m}{s}}{1.00 \times 10^{-3} \text{ m}}$$

$$= \frac{3.00 \times 10^8}{1.00 \times 10^{-3}} \frac{m}{s} \times \frac{1}{m}$$

$$= 3.00 \times 10^{11} \text{ Hz}$$

Step 2: $E = hf$

$$= (6.63 \times 10^{-34} \text{ J·s})\left(3.00 \times 10^{11} \frac{1}{s}\right)$$

$$= (6.63 \times 10^{-34})(3.00 \times 10^{11}) \text{ J·s} \times \frac{1}{s}$$

$$= \boxed{1.99 \times 10^{-22} \text{ J}}$$

Chapter 8

8.1 Energy is related to the frequency and Planck's constant in equation 8.1, $E = hf$.

$$\text{For } n = 6, E_H = 6.05 \times 10^{-20} \text{ J}$$
$$\text{For } n = 2, E_L = 5.44 \times 10^{-19} \text{ J}$$
$$E = ? \text{ J}$$
$$E = E_H - E_L$$
$$= (-6.05 \times 10^{-20} \text{ J}) - (-5.44 \times 10^{-19} \text{ J})$$
$$= \boxed{4.84 \times 10^{-19} \text{ J}}$$

8.2
$$\text{For } n = 6, E_H = -6.05 \times 10^{-20} \text{ J}$$
$$\text{For } n = 2, E_L = -5.44 \times 10^{-19} \text{ J}$$
$$h = 6.63 \times 10^{-34} \text{ J·s}$$
$$f = ?$$
$$hf = E_H - E_L \therefore f = \frac{E_H - E_L}{h}$$
$$f = \frac{(-6.05 \times 10^{-20} \text{ J}) - (-5.44 \times 10^{-19} \text{ J})}{6.63 \times 10^{-34} \text{ J·s}}$$
$$= \frac{4.84 \times 10^{-19}}{6.63 \times 10^{-34}} \frac{\cancel{J}}{\cancel{J}\text{·s}}$$
$$= 7.29 \times 10^{14} \frac{1}{s}$$
$$= \boxed{7.29 \times 10^{14} \text{ Hz (violet)}}$$

8.3 $(n = 1) = -13.6 \text{ eV}$

Since the energy of the electron is -13.6 eV, it will require 13.6 eV (or 2.17×10^{-18} J) to remove the electron.

8.4
$$q/m = -1.76 \times 10^{11} \text{ C/kg}$$
$$q = -1.60 \times 10^{-19} \text{ C}$$
$$m = ?$$
$$\text{mass} = \frac{\text{charge}}{\text{charge}/\text{mass}}$$
$$= \frac{-1.60 \times 10^{-19} \text{ C}}{-1.76 \times 10^{11} \dfrac{\text{C}}{\text{kg}}}$$
$$= \frac{-1.60 \times 10^{-19}}{-1.76 \times 10^{11}} \cancel{C} \times \frac{\text{kg}}{\cancel{C}}$$
$$= \boxed{9.09 \times 10^{-31} \text{ kg}}$$

Chapter 9

9.1 Recall that the number of outer energy level electrons is the same as the family number for the representative elements:

a. Li: 1
b. N: 5
c. F: 7
d. Cl: 7
e. Ra: 2
f. Be: 2

9.2 The same information that was used in question 1 can be used to draw the dot notation:

(a) $\dot{\text{B}}\cdot$

(c) $\text{Ca}\colon$

(e) $\cdot\ddot{\ddot{\text{O}}}\cdot$

(b) $\cdot\ddot{\ddot{\text{Br}}}\cdot$

(d) $\text{K}\cdot$

(f) $\cdot\ddot{\ddot{\text{S}}}\cdot$

9.3 The charge is found by identifying how many electrons are lost or gained in achieving the noble gas structure:

a. Boron 3+
b. Bromine 1−
c. Calcium 2+
d. Potassium 1+
e. Oxygen 2−
f. Nitrogen 3−

9.4 The name of some common polyatomic ions are in box table 9.2. Using this table as a reference, the names are

a. hydroxide
b. sulfite
c. hypochlorite
d. nitrate
e. carbonate
f. perchlorate

9.5 The Roman numeral tells you the charge on the variable charge elements. The charges for the polyatomic ions are found in box table 9.2. The charges for metallic elements can be found in tables 9.2 and 9.3. Using these resources and the crossover technique, the formulas are as follows:

a. $Fe(OH)_3$
b. $Pb_3(PO_4)_2$
c. $ZnCO_3$
d. NH_4NO_3
e. $KHCO_3$
f. K_2SO_3

9.6 Box table 9.3 has information about the meaning of prefixes and stem names used in naming covalent compounds. (a), for example, asks for the formula of carbon tetrachloride. Carbon has no prefixes, so there is one carbon atom, and it comes first in the formula because it comes first in the name. The *tetra*-prefix means four, so there are four chlorine atoms. The name ends in *-ide,* so you know there are only two elements in the compound. The symbols can be obtained from the list of elements on the inside back cover of this text. Using all this information from the name, you can think out the formula for carbon tetrachloride. The same process is used for the other compounds and formulas:

a. CCl_4
b. H_2O
c. MnO_2
d. SO_3
e. N_2O_5
f. As_2S_5

9.7 Again using information from box table 9.3, this question requires you to reverse the thinking procedure you learned in question 6.

  a. carbon monoxide

  b. carbon dioxide

  c. carbon disulfide

  d. dinitrogen monoxide

  e. tetraphosphorus trisulfide

  f. dinitrogen trioxide

9.8 **a.** $2\,SO_2 + O_2 \rightarrow 2\,SO_3$

  b. $4\,P + 5\,O_2 \rightarrow 2\,P_2O_5$

  c. $2\,Al + 6\,HCl \rightarrow 2\,AlCl_3 + 3\,H_2$

  d. $2\,NaOH + H_2SO_4 \rightarrow Na_2SO_4 + 2\,H_2O$

  e. $Fe_2O_3 + 3\,CO \rightarrow 2\,Fe + 3\,CO_2$

  f. $3\,Mg(OH)_2 + 2\,H_3PO_4 \rightarrow Mg_3(PO_4)_2 + 6\,H_2O$

9.9 **a.** General form of $XY + AZ \rightarrow XZ + AY$ with precipitate formed: Ion exchange reaction.

  b. General form of $X + Y \rightarrow XY$: Combination reaction.

  c. General form of $XY \rightarrow X + Y + \ldots$: Decomposition reaction.

  d. General form of $X + Y \rightarrow XY$: Combination reaction.

  e. General form of $XY + A \rightarrow AY + X$: Replacement reaction.

  f. General form of $X + Y \rightarrow XY$: Combination reaction.

9.10 **a.** $C_5H_{12(g)} + 8\,O_{2(g)} \rightarrow 5\,CO_{2(g)} + 6\,H_2O_{(g)}$

  b. $HCl_{(aq)} + NaOH_{(aq)} \rightarrow NaCl_{(aq)} + H_2O_{(l)}$

  c. $2\,Al_{(s)} + Fe_2O_{3(s)} \rightarrow Al_2O_{3(s)} + 2\,Fe_{(l)}$

  d. $Fe_{(s)} + CuSO_{4(aq)} \rightarrow FeSO_{4(aq)} + Cu_{(s)}$

  e. $MgCl_{(aq)} + Fe(NO_3)_{2(aq)} \rightarrow$ No reaction (all possible compounds are soluble and no gas or water was formed)

  f. $C_6H_{10}O_{5(s)} + 6\,O_{2(g)} \rightarrow 6\,CO_{2(g)} + 5\,H_2O_{(g)}$

Chapter 11

11.1 **a.** cobalt-60: 27 protons, 33 neutrons

  b. potassium-40: 19 protons, 21 neutrons

  c. neon-24: 10 protons, 14 neutrons

  d. lead-208: 82 protons, 126 neutrons

11.2 **a.** $^{60}_{27}Co$

  b. $^{40}_{19}K$

  c. $^{24}_{10}Ne$

  d. $^{204}_{82}Pb$

11.3 **a.** cobalt-60: Radioactive because odd numbers of protons (27) and odd numbers of neutrons (33) are usually unstable.

  b. potassium-40: Radioactive, again having an odd number of protons (19) and an odd number of neutrons (21).

  c. neon-24: Stable, because even numbers of protons and neutrons are usually stable.

  d. lead-208: Stable, because even numbers of protons and neutrons, *and* because 82 is a particularly stable number of nucleons.

11.4 **a.** $^{56}_{26}Fe \rightarrow \,^{0}_{-1}e + \,^{56}_{27}Co$

  b. $^{7}_{4}Be \rightarrow \,^{0}_{-1}e + \,^{7}_{5}B$

  c. $^{64}_{29}Cu \rightarrow \,^{0}_{-1}e + \,^{64}_{30}Zn$

  d. $^{24}_{11}Na \rightarrow \,^{0}_{-1}e + \,^{24}_{12}Mg$

  e. $^{214}_{82}Pb \rightarrow \,^{0}_{-1}e + \,^{214}_{83}Bi$

  f. $^{32}_{15}P \rightarrow \,^{0}_{-1}e + \,^{32}_{16}S$

11.5 **a.** $^{235}_{92}U \rightarrow \,^{4}_{2}He + \,^{231}_{90}Th$

  b. $^{226}_{88}Ra \rightarrow \,^{4}_{2}He + \,^{222}_{86}Rn$

  c. $^{239}_{94}Pu \rightarrow \,^{4}_{2}He + \,^{235}_{92}U$

  d. $^{214}_{83}Bi \rightarrow \,^{4}_{2}He + \,^{210}_{81}Tl$

  e. $^{230}_{90}Th \rightarrow \,^{4}_{2}He + \,^{226}_{88}Ra$

  f. $^{210}_{84}Po \rightarrow \,^{4}_{2}He + \,^{206}_{82}Pb$

11.6 Thirty-two days is four half-lives. After the first half-life (8 days), 1/2 oz will remain. After the second half-life (8 + 8, or 16 days), 1/4 oz will remain. After the third half-life (8 + 8 + 8, or 24 days), 1/8 oz will remain. After the fourth half-life (8 + 8 + 8 + 8, or 32 days), 1/16 oz will remain, or 6.3×10^{-2} oz.

11.7 The Fe-56 nucleus has a mass of 55.9206 u, but the individual masses of the nucleons are

$$26 \text{ protons} \times 1.00728 \text{ u} = 26.1893 \text{ u}$$
$$30 \text{ neutrons} \times 1.00867 \text{ u} = \frac{30.2601 \text{ u}}{56.4494 \text{ u}}$$

The mass defect is thus

$$
\begin{array}{r}
56.4494 \text{ u} \\
-55.9206 \text{ u} \\
\hline
0.5288 \text{ u}
\end{array}
$$

The atomic mass unit (u) is equal to the mass of a mole (g), therefore 0.5288 u = 0.5288 g. The mass defect is equivalent to the binding energy according to $E = mc^2$. For a molar mass of Fe-56, the mass defect is

$$E = (5.29 \times 10^{-4} \text{ kg})\left(3.00 \times 10^8 \, \frac{m}{s}\right)^2$$
$$= (5.29 \times 10^{-4} \text{ kg})\left(9.00 \times 10^{16} \, \frac{m^2}{s^2}\right)$$
$$= 4.76 \times 10^{13} \, \frac{kg \cdot m^2}{s^2}$$
$$= 4.76 \times 10^{13} \text{ J}$$

For a single nucleus,

$$\frac{4.76 \times 10^{13} \text{ J}}{6.02 \times 10^{23} \text{ nuclei}} = 7.90 \times 10^{-11} \text{ J/nuclei}$$

Appendix F

Problem Solving

Students are sometimes apprehensive when assigned problem exercises. This apprehension comes from a lack of experience in problem solving and not knowing how to proceed. This reading is concerned with a basic approach and procedures that you can follow to simplify problem-solving exercises. Thinking in terms of quantitative ideas is a skill that you can learn. Actually, no mathematics beyond addition, subtraction, multiplication, and division is needed. What is needed is knowledge of certain techniques. If you follow the suggested formatting procedures and seek help from the mathematical review appendix as needed, you will find that problem solving is a simple, fun activity that helps you to learn to think in a new way. Here are some more considerations that will prove helpful.

1. Read the problem carefully, perhaps several times, to understand the problem situation. If possible, make a sketch to help you visualize and understand the problem in terms of the real world.

2. Be alert for information that is not stated directly. For example, if a moving object "comes to a stop," you know that the final velocity is zero, even though this was not stated outright. Likewise, questions about "how far?" are usually asking about distance, and questions about "how long?" are usually asking about time. Such information can be very important in procedure step 1 (see procedure steps in right column), the listing of quantities and their symbols. Overlooked or missing quantities and symbols can make it difficult to identify the appropriate equation.

3. Understand the meaning and concepts that an equation represents. An equation represents a relationship that exists between variables. Understanding the relationship helps you to identify the appropriate equation or equations by inspection of the list of known and unknown quantities (see procedure step 2). You will find a list of the equations being considered at the end of each chapter. Information about the meaning and the concepts that an equation represents is found within each chapter.

4. Solve the equation *before* substituting numbers and units for symbols (see procedure step 3). A helpful discussion of the mathematical procedures required, with examples, is in appendix A.

5. Note if the quantities are in the same units. A mathematical operation requires the units to be the same; for example, you cannot add nickels, dimes, and quarters until you first convert them all to the same unit of money. Likewise, you cannot correctly solve a problem if one time quantity is in seconds and another time quantity is in hours. The quantities must be converted to the same units before anything else is done (see procedure step 4). There is a helpful section on how to use conversion ratios in appendix A.

6. Perform the required mathematical operations on the numbers and the units as if they were two separate problems (see procedure step 6). You will find that following this step will facilitate problem-solving activities because the units you obtain will tell you if you have worked the problem correctly. If you just write the units that you *think* should appear in the answer, you have missed this valuable self-check.

7. Be aware that not all learning takes place in a given time frame and that solutions to problems are not necessarily arrived at "by the clock." If you have spent a half an hour or so unsuccessfully trying to solve a particular problem, move on to another problem or do something entirely different for a while. Problem solving often requires time for something to happen in your brain. If you move on to some other activity, you might find that the answer to a problem that you have been stuck on will come to you "out of the blue" when you are not even thinking about the problem. This unexpected revelation of solutions is common to many real-world professions and activities that involve thinking.

Until you are comfortable in a problem-solving situation, you should follow a formatting procedure that will help you organize your thinking. Here is an example of such a formatting procedure:

Step 1: List the quantities involved together with their symbols on the left side of the page, including the unknown quantity with a question mark.

Step 2: Inspect the given quantities and the unknown quantity as listed, and *identify* the equation that expresses a relationship between these quantities. A list of the equations that express the relationships discussed in each chapter is found at the end of that chapter. *Write* the identified equation on the right side of your paper, opposite the list of symbols and quantities.

Step 3: If necessary, *solve* the equation for the variable in question. This step must be done before substituting any numbers or units in the equation. This simplifies things and keeps down the confusion that may otherwise result. If you need help solving an equation, see the section on this topic in appendix A.

Step 4: If necessary, *convert* any unlike units so they are all the same. If the equation involves a time in seconds, for example, and a speed in kilometers per hour, you should convert the km/h to m/s. Again, this step should be done at this point to avoid confusion and incorrect operations in a later step. If you need help converting units, see the section on this topic in appendix A.

Step 5: *Substitute* the known quantities in the equation, replacing each symbol with both the number value and units represented by the symbol.

Step 6: *Perform* the required mathematical operations on the numbers and on the units. This performance is less confusing if you first separate the numbers and units, as shown in the following example and in the examples throughout this text, and then perform the mathematical operations on the numbers and units as separate steps, showing all work.

Step 7: *Draw a box* around your answer (numbers and units) to communicate that you have found what you were looking for. The box is a signal that you have finished your work on this problem.

EXAMPLE PROBLEM

Mercury is a liquid metal with a mass density of 13.6 g/cm^3. What is the mass of 10.0 cm^3 of mercury?

SOLUTION

The problem gives two known quantities, the mass density (ρ) of mercury and a known volume (V), and identifies an unknown quantity, the mass (m) of that volume. Make a list of these quantities:

$$\rho = 13.6 \, \text{g/cm}^3$$
$$V = 10.0 \, \text{cm}^3$$
$$m = ?$$

The appropriate equation for this problem is the relationship between mass density (ρ), mass (m), and volume (V):

$$\rho = \frac{m}{V}$$

The unknown in this case is the mass, m. Solving the equation for m, by multiplying both sides by V, gives:

$$V\rho = \frac{m\cancel{V}}{\cancel{V}}$$
$$V\rho = m, \text{ or}$$
$$m = V\rho$$

Now you are ready to substitute the known quantities in the equation and perform the mathematical operations on the numbers and on the units:

$$m = \left(13.6 \, \frac{\text{g}}{\text{cm}^3}\right)(10.0 \, \text{cm}^3)$$
$$= (13.6)(10.0)\left(\frac{\text{g}}{\text{cm}^3}\right)(\text{cm}^3)$$
$$= 136 \, \frac{\text{g} \cdot \cancel{\text{cm}^3}}{\cancel{\text{cm}^3}}$$
$$= \boxed{136 \, \text{g}}$$

Glossary

A

abiotic factors nonliving parts of an organism's environment

absolute humidity a measure of the actual amount of water vapor in the air at a given time—for example, in grams per cubic meter

absolute magnitude a classification scheme to compensate for the distance differences to stars; calculations of the brightness that stars would appear to have if they were all at a defined, standard distance

absolute scale temperature scale set so that zero is at the theoretical lowest temperature possible, which would occur when all random motion of molecules has ceased

absolute zero the theoretical lowest temperature possible, which occurs when all random motion of molecules has ceased

abyssal plain the practically level plain of the ocean floor

acceleration a change in velocity per change in time; by definition, this change in velocity can result from a change in speed, a change in direction, or a combination of changes in speed and direction

accretion disk fat bulging disk of gas and dust from the remains of the gas cloud that forms around a protostar

acetylcholine a neurotransmitter secreted into the synapse by many axons and received by dendrites

acetylcholinesterase an enzyme present in the synapse that destroys acetylcholine

achondrites homogeneously textured stony meteorites

acid any substance that is a proton donor when dissolved in water; generally considered a solution of hydronium ions in water that can neutralize a base, forming a salt and water

acid-base indicator a vegetable dye used to distinguish acid and base solutions by a color change

acquired characteristics characteristics an organism gains during its lifetime that are not genetically determined and therefore cannot be passed on to future generations

active transport use of a carrier molecule to move molecules through a cell membrane in a direction opposite that of the concentration gradient; the carrier requires an input of energy other than the kinetic energy of the molecules

adenine a double-ring nitrogenous-base molecule in DNA and RNA; the complementary base of thymine or uracil

adenosine triphosphate (ATP) a molecule formed from the building blocks of adenine, ribose, and phosphates; it functions as the primary energy carrier in the cell

aerobic cellular respiration the biochemical pathway that requires oxygen and converts food, such as carbohydrates, to carbon dioxide and water; during this conversion, it releases the chemical-bond energy as ATP molecules

air mass a large, more or less uniform body of air with nearly the same temperature and moisture conditions throughout

air mass weather the weather experienced within a given air mass; characterized by slow, gradual changes from day to day

alcohol an organic compound with a general formula of ROH, where R is one of the hydrocarbon groups; for example, methyl or ethyl

aldehyde an organic molecule with the general formula RCHO, where R is one of the hydrocarbon groups; for example, methyl or ethyl

alkali metals members of family IA of the periodic table, having common properties of shiny, low-density metals that can be cut with a knife and that react violently with water to form an alkaline solution

alkaline earth metals members of family IIA of the periodic table, having common properties of soft, reactive metals that are less reactive than alkali metals

alleles alternative forms of a gene for a particular characteristic (e.g., attached-earlobe and free-earlobe are alternative alleles for ear shape)

alpha particle the nucleus of a helium atom (two protons and two neutrons) emitted as radiation from a decaying heavy nucleus; also known as an alpha ray

alpine glaciers glaciers that form at high elevations in mountainous regions

alternating current an electric current that first moves one direction, then the opposite direction with a regular frequency

alternation of generations a term used to describe that aspect of the life cycle in which there are two distinctly different forms of an organism; each form is involved in the production of the other, and only one form is involved in producing gametes

alveoli tiny sacs that are part of the structure of the lungs where gas exchange takes place

amino acids organic molecules that join to form polypeptides and proteins

amp unit of electric current; equivalent to C/s

ampere full name of the unit amp

amplitude the extent of displacement from the equilibrium condition; the size of a wave from the rest (equilibrium) position

anaphase the third stage of mitosis, characterized by dividing of the centromeres and movement of the chromosomes to the poles

androgens male sex hormones produced by the testes that cause the differentiation of typical internal and external genital male anatomy

angle of incidence angle of an incident (arriving) ray or particle to a surface; measured from a line perpendicular to the surface (the normal)

angle of reflection angle of a reflected ray or particle from a surface; measured from a line perpendicular to the surface (the normal)

angular momentum quantum number in the quantum mechanics model of the atom, one of four descriptions of the energy state of an electron wave; this quantum number describes the energy sublevels of electrons within the main energy levels of an atom

annular eclipse occurs when the penumbra reaches the surface of Earth; as seen from Earth, the Sun forms a bright ring around the disk of the Moon

anorexia nervosa a nutritional deficiency disease characterized by severe, prolonged weight loss for fear of becoming obese

Antarctic Circle parallel identifying the limit toward the equator where the Sun appears above the horizon all day for six months during the summer; located at 66.5°S latitude

anther the sex organ in plants that produces the pollen that contains the sperm

antibody a globular protein molecule made by the body in response to the presence of a foreign or harmful molecule called an antigen; these molecules are capable of combining with the foreign molecules or microbes to inactivate or kill them

anticline an arch-shaped fold in layered bed rock

anticodon a sequence of three nitrogenous bases on a tRNA molecule capable of forming hydrogen bonds with three complementary bases on an mRNA codon during translation

anticyclone a high-pressure center with winds flowing away from the center; associated with clear, fair weather

antinode region of maximum amplitude between adjacent nodes in a standing wave

aorta the large blood vessel that carries blood from the left ventricle to the majority of the body

apogee the point at which the Moon's elliptical orbit takes the Moon farthest from Earth

apoptosis also known as "programmed cell death"; death of specific cells that has a genetic basis and is not the result of injury

apparent local noon the instant when the Sun crosses the celestial meridian at any particular longitude

apparent local solar time the time found from the position of the Sun in the sky; the shadow of the gnomon on a sundial

apparent magnitude a classification scheme for different levels of brightness of stars that you see; brightness values range from one to six with the number one (first magnitude) assigned to the brightest star and the number six (sixth magnitude) assigned to the faintest star that can be seen

apparent solar day the interval between two consecutive crossings of the celestial meridian by the Sun

aquifer a layer of sand, gravel, or other highly permeable material beneath the surface that is saturated with water and is capable of producing water in a well or spring

Archaea the domain in which are found prokaryotic organisms that live in extreme habitats

archaeology the study of past human life and culture from material evidence

Arctic Circle parallel identifying the limit toward the equator where the Sun appears above the horizon all day for one day up to six months during the summer; located at 66.5°N latitude

area the extent of a surface; the surface bounded by three or more lines

arid dry climate classification; receives less than 25 cm (10 in) precipitation per year

aromatic hydrocarbon organic compound with at least one benzene ring structure; cyclic hydrocarbons and their derivatives

arteries the blood vessels that carry blood away from the heart

arterioles small arteries located just before capillaries that can expand and contract to regulate the flow of blood to parts of the body

artesian term describing the condition where confining pressure forces groundwater from a well to rise above the aquifer

asbestos the common name for any one of several incombustible fibrous minerals that will not melt or ignite and can be woven into a fireproof cloth or used directly in fireproof insulation; about six different commercial varieties of asbestos are used, one of which has been linked to cancer under heavy exposure

assimilation the physiological process that takes place in a living cell as it converts nutrients in food into specific molecules required by the organism

asteroids small rocky bodies left over from the formation of the solar system; most are accumulated in a zone between the orbits of Mars and Jupiter

asthenosphere a plastic, mobile layer of Earth's structure that extends around Earth below the lithosphere

astronomical unit the radius of Earth's orbit is defined as one astronomical unit (AU)

atom the smallest unit of an element that can exist alone or in combination with other elements

atomic mass unit relative mass unit (u) of an isotope based on the standard of the carbon-12 isotope, which is defined as a mass of exactly 12.00 u; one atomic mass unit (1 u) is 1/12 the mass of a carbon-12 atom

atomic number the number of protons in the nucleus of an atom

atomic weight weighted average of the masses of stable isotopes of an element as they occur in nature, based on the abundance of each isotope of the element and the atomic mass of the isotope compared to carbon-12

atria thin-walled sacs of the heart that receive blood from the veins of the body and empty into the ventricles; singular, *atrium*

atrioventricular valves located between the atria and ventricles of the heart preventing blood from flowing backward from the ventricles into the atria

autosomes chromosomes that typically carry genetic information used by the organism for characteristics other than the primary determination of sex

autotroph an organism that is able to make its food molecules from inorganic raw materials by using basic energy sources such as sunlight

autumnal equinox one of two times a year that daylight and night are of equal length; occurs on or about September 23 and identifies the beginning of the fall season

axis the imaginary line about which a planet or other object rotates

axon a neuronal fiber that carries information away from the nerve cell body

B

background radiation ionizing radiation (alpha, beta, gamma, etc.) from natural sources; between 100 and 500 millirems/yr of exposure to natural radioactivity from the environment

Balmer series a set of four line spectra, narrow lines of color emitted by hydrogen atom electrons as they drop from excited states to the ground state

band of stability a region of a graph of the number of neutrons versus the number of protons in nuclei; nuclei that have the neutron to proton ratios located in this band do not undergo radioactive decay

barometer an instrument that measures atmospheric pressure, used in weather forecasting and in determining elevation above sea level

basal metabolic rate (BMR) the amount of energy required to maintain normal body activity while at rest

base any substance that is a proton acceptor when dissolved in water; generally considered a solution that forms hydroxide ions in water that can neutralize an acid, forming a salt and water

basilar membrane a membrane in the cochlea containing sensory cells that are stimulated by the vibrations caused by sound waves

basin a large, bowl-shaped fold in the land into which streams drain; also a small enclosed or partly enclosed body of water

beat rhythmic increases and decreases of volume from constructive and destructive interference between two sound waves of slightly different frequencies

benign tumor an abnormal cell mass that does not fragment and spread beyond its original area of growth

benthic aquatic organisms that live on the bottom

beta particle high-energy electron emitted as ionizing radiation from a decaying nucleus; also known as a beta ray

big bang theory current model of galactic evolution in which the universe was created from an intense and brilliant explosion from a primeval fireball

bile a product of the liver, stored in the gallbladder, which is responsible for the emulsification of fats

binding energy the energy required to break a nucleus into its constituent protons and neutrons; also the energy equivalent released when a nucleus is formed

binomial system of nomenclature a naming system that uses two Latin names, genus and specific epithet, for each type of organism

biodiversity the number of different kinds of organisms found in an area

biogenesis the concept that life originates only from preexisting life

biological species concept the concept that species can be distinguished from one another by their inability to interbreed

biology the science that deals with the study of living things and how living things interact with things around them

biomass (energy) any material formed by photosynthesis, including plants, trees, and crops, and any garbage, crop residue, or animal waste; the mass of living material in an area

biomes large regional communities primarily determined by climate

biotic factors living parts of an organism's environment

black hole the theoretical remaining core of a supernova with an intense gravitational field

blackbody radiation electromagnetic radiation emitted by an ideal material (the blackbody) that perfectly absorbs and perfectly emits radiation

blood the fluid medium consisting of cells and plasma that assists in the transport of materials and heat

Bohr model model of the structure of the atom that attempted to correct the deficiencies of the solar system model and account for the Balmer series

boiling point the temperature at which a phase change of liquid to gas takes place through boiling; the same temperature as the condensation point

boundary the division between two regions of differing physical properties

breaker a wave whose front has become so steep that the top part has broken forward of the wave, breaking into foam, especially against a shoreline

British thermal unit the amount of energy or heat needed to increase the temperature of 1 pound of water 1 degree Fahrenheit (abbreviated Btu)

bronchi major branches of the trachea that ultimately deliver air to bronchioles in the lungs

bronchioles small tubes that deliver air to the alveoli in the lung; they are capable of contracting

bulimia a nutritional deficiency disease characterized by a binge-and-purge cycle of eating

C

calorie the amount of energy (or heat) needed to increase the temperature of 1 gram of water 1 degree Celsius

Calorie the nutritionist's "calorie"; equivalent to 1 kilocalorie

cancer any abnormal growth of cells that has a malignant potential

capillaries tiny blood vessels through the walls of which exchange between cells and the blood takes place

carbohydrates organic compounds that include sugars, starches, and cellulose; carbohydrates are used by plants and animals for structure, protection, and food

carbon cycle the series of processes and organisms through which carbon atoms pass in ecological systems

carbon film a type of fossil formed when the volatile and gaseous constituents of a buried organic structure are distilled away, leaving a carbon film as a record

carbonation in chemical weathering, a reaction that occurs naturally between carbonic acid (H_2CO_3) and rock minerals

carcinogens agents responsible for causing cancer

carnivores animals that eat other animals

carrier protein proteins that pick up molecules from one place in a cell or multicellular organism and transport them to another; for example, certain blood proteins

carrying capacity (ecosystem) the *optimum* population size an area can support over an extended period of time

cast sediments deposited by groundwater in a mold, taking the shape and external features of the organism that was removed to form the mold, then gradually changing to sedimentary rock

cathode rays negatively charged particles (electrons) that are emitted from a negative terminal in an evacuated glass tube

cell the basic structural unit that makes up all living things

cell membrane the outer-boundary membrane of the cell; also known as the *plasma membrane*

cell nucleus the central part of a cell that contains the genetic material

cell plate a plant-cell structure that begins to form in the center of the cell and proceeds to the cell membrane, resulting in cytokinesis

cell wall an outer covering on some cells; may be composed of cellulose, chitin, or peptidoglycan depending on the kind of organism

cellulose a polysaccharide abundant in plants that forms the fibers in cell walls that provide structure for plant materials

Celsius scale referent scale that defines numerical values for measuring hotness or coldness, defined as degrees of temperature; based on the reference points of the freezing point of water and the boiling point of water at sea-level pressure, with 100 degrees between the two points

cementation process by which spaces between buried sediment particles under compaction are filled with binding chemical deposits, binding the particles into a rigid, cohesive mass of a sedimentary rock

Cenozoic the most recent geologic era; the time of recent life, meaning the fossils of this era are identical to the life found on Earth today

central nervous system the portion of the nervous system consisting of the brain and spinal cord

centrifugal force an apparent outward force on an object following a circular path that is a consequence of the third law of motion

centrioles organelles composed of microtubules located just outside the nucleus

centripetal force the force required to pull an object out of its natural straight-line path and into a circular path; *centripetal* means "center seeking"

centromere the unreplicated region where two chromatids are joined

Cepheid variables a bright variable star that can be used to measure distance

cerebellum a region of the brain connected to the medulla oblongata that receives many kinds of sensory stimuli and coordinates muscle movement

cerebrum a region of the brain that surrounds most of the other parts of the brain and is involved in consciousness and thought

chain reaction a self-sustaining reaction where some of the products are able to produce more reactions of the same kind; in a nuclear chain reaction, neutrons are the products that produce more nuclear reactions in a self-sustaining series

chemical bond an attractive force that holds atoms together in a compound

chemical change a change in which the identity of matter is altered and new substances are formed

chemical energy a form of energy involved in chemical reactions associated with changes in internal potential energy; a kind of potential energy that is stored and later released during a chemical reaction

chemical equation concise way of describing what happens in a chemical reaction

chemical reaction a change in matter where different chemical substances are created by forming or breaking chemical bonds

chemical weathering the breakdown of minerals in rocks by chemical reactions with water, gases of the atmosphere, or solutions

chemistry the science concerned with the study of the composition, structure, and properties of substances and the transformations they undergo

Chinook a warm wind that has been warmed by compression; also called Santa Ana

chloroplast an energy-converting, membranous, saclike organelle in plant cells containing the green pigment chlorophyll

chondrites subdivision of stony meteorites containing small spherical lumps of silicate minerals or glass

chondrules small spherical lumps of silicate minerals or glass found in some meteorites

chromatid one of two component parts of a chromosome formed by replication and attached at the centromere

chromatin areas or structures within the nucleus of a cell composed of long, loosely arranged molecules of deoxyribonucleic acid (DNA) in association with proteins

chromatin fibers see *nucleoproteins*

chromosomal aberrations changes in the gene arrangement in chromosomes; for example, translocation and duplication mutations

chromosomes complex tightly coiled structures within the nucleus composed of various kinds of histone proteins and DNA that contains the cell's genetic information

class a group of closely related orders found within a phylum

cleavage furrow an indentation of the cell membrane of an animal cell that pinches the cytoplasm into two parts

climate the general pattern of weather that occurs in a region over a number of years

climate change departure from the expected average pattern of climate for a given region over time

clitoris a small, elongated erectile structure located between and at the head of the labia

clones genetically identical individuals that were reproduced asexually

coalescence process (meteorology) the process by which large raindrops form from the merging and uniting of millions of tiny water droplets

cochlea the part of the ear that converts sound into nerve impulses

coding strand one of two DNA strands that serves as a template, or pattern, for the synthesis of RNA

codon a sequence of three nucleotides of an mRNA molecule that directs the placement of a particular amino acid during translation

cold front the front that is formed as a cold air mass moves into warmer air

combination chemical reaction a synthesis reaction in which two or more substances combine to form a single compound

comets celestial objects originating from the outer edges of the solar system that move about the Sun in highly elliptical orbits; solar heating and pressure from the solar wind form a tail on the comet that points away from the Sun

commensalism a relationship between two organisms in which one organism is helped and the other is not affected

community all of the kinds of interacting organisms within an ecosystem

compaction the process of pressure from a depth of overlying sediments squeezing together the deeper sediments and squeezing out water

competition a kind of interaction between organisms in which both organisms are harmed to some extent

complementary base a nitrogenous base in nucleic acids that can form hydrogen bonds with another base of a specific nucleotide

complex carbohydrates macromolecules formed as a result of the chemical combining of simple sugars (monomers) to form polysaccharides; for example, starch

compound a pure chemical substance that can be decomposed by a chemical change into simpler substances with a fixed mass ratio

compressive stress a force that tends to compress the surface as Earth's plates move into each other

concentration an arbitrary description of the relative amounts of solute and solvent in a solution;

a larger amount of solute makes a concentrated solution, and a small amount of solute makes a dilute concentration

conception fertilization

condensation (sound) a compression of gas molecules; a pulse of increased density and pressure that moves through the air at the speed of sound

condensation (water vapor) when more vapor or gas molecules are returning to the liquid state than are evaporating

condensation nuclei tiny particles such as tiny dust, smoke, soot, and salt crystals that are suspended in the air on which water condenses

condensation point the temperature at which a gas or vapor changes back to a liquid

conduction the transfer of heat from a region of higher temperature to a region of lower temperature by increased kinetic energy moving from molecule to molecule

cones light-sensitive cells in the retina of the eye that respond to different colors of light

consistent law principle the laws of physics are the same in all reference frames that are moving at a constant velocity with respect to each other

constancy of speed principle the speed of light in empty space has the same value for all observers regardless of their velocity

constructive interference the condition in which two waves arriving at the same place, at the same time and in phase, add amplitudes to create a new wave

consumers organisms that must obtain energy in the form of organic matter

continental air mass dry air masses that form over large land areas

continental climate a climate influenced by air masses from large land areas; hot summers and cold winters

continental drift a concept that continents shift positions on Earth's surface, moving across the surface rather than being fixed, stationary landmasses

continental glaciers glaciers that cover a large area of a continent; for example, Greenland and the Antarctic

continental shelf a feature of the ocean floor; the flooded margins of the continents that form a zone of relatively shallow water adjacent to the continents

continental slope a feature of the ocean floor; a steep slope forming the transition between the continental shelf and the deep ocean basin

control group the situation used as the basis for comparison in a controlled experiment

control processes mechanisms that ensure that an organism will carry out all metabolic activities in the proper sequence (coordination) and at the proper rate (regulation)

control rods rods inserted between fuel rods in a nuclear reactor to absorb neutrons and thus control the rate of the nuclear chain reaction

controlled experiment an experiment that allows for a comparison of two events that are identical in all but one respect

convection transfer of heat from a region of higher temperature to a region of lower temperature by the displacement of high-energy molecules; for example, the displacement of warmer, less dense air (higher kinetic energy) by cooler, more dense air (lower kinetic energy)

convection cell complete convective circulation pattern; also, slowly turning regions in the plastic asthenosphere that might drive the motion of plate tectonics

convection zone (of a star) part of the interior of a star according to a model; the region directly above the radiation zone where gases are heated by the radiation zone below and move upward by convection to the surface, where they emit energy in the form of visible light, ultraviolet radiation, and infrared radiation

convergent boundaries boundaries that occur between two plates moving toward each other

core (of Earth) the center part of Earth, which consists of a solid inner part and liquid outer part; makes up about 15 percent of Earth's total volume and about one-third of its mass

core (of a star) dense, very hot region of a star where nuclear fusion reactions release gamma and X-ray radiation

Coriolis effect the apparent deflection due to the rotation of Earth; it is to the right in the Northern Hemisphere

corpus luteum a glandlike structure produced in the ovary that produces hormones (progesterone and estrogen) that prevent the release of other eggs

correlation the determination of the equivalence in geologic age by comparing the rocks in two separate locations

coulomb unit used to measure quantity of electric charge; equivalent to the charge resulting from the transfer of 6.24 billion particles such as the electron

Coulomb's law relationship between charge, distance, and magnitude of the electrical force between two bodies

covalent bond a chemical bond formed by the sharing of a pair of electrons

covalent compound chemical compound held together by a covalent bond or bonds

creep the slow downhill movement of soil down a steep slope

crest the high mound of water that is part of a wave; also refers to the condensation, or high-pressure part, of a sound wave

critical angle limit to the angle of incidence when all light rays are reflected internally

critical mass mass of fissionable material needed to sustain a chain reaction

crossing-over the process that occurs when homologous chromosomes exchange equivalent sections of DNA during meiosis

crude oil petroleum pumped from the ground that has not yet been refined into usable products

crust the outermost part of Earth's interior structure; the thin, solid layer of rock that rests on top of the Mohorovicic discontinuity

cryptorchidism a developmental condition in which the testes do not migrate from the abdomen through the inguinal canal to the scrotum

curie unit of nuclear activity defined as 3.70×10^{10} nuclear disintegrations per second

cycle a complete vibration

cyclone a low-pressure center where the winds move into the low-pressure center and are forced upward; a low-pressure center with clouds, precipitation, and stormy conditions

cytokinesis division of the cytoplasm of one cell into two new cells

cytoplasm the more fluid portion of the protoplasm that surrounds the nucleus

cytosine a single-ring nitrogenous-base molecule in DNA and RNA; complementary to guanine

D

dark energy a recently discovered mystery force that is apparently accelerating the expanding universe

dark matter missing matter of the universe that is believed to exist but is invisible and unseen

data measurement information used to describe something

data points points that may be plotted on a graph to represent simultaneous measurements of two related variables

daughter cells two cells formed by cell division

daughter chromosomes chromosomes produced by DNA replication that contain identical genetic information; formed after chromosome division in anaphase

daughter nuclei two nuclei formed by mitosis

daylight saving time setting clocks ahead one hour during the summer to more effectively utilize the longer days of summer, then setting the clocks back in the fall

deceleration phase a period in a population growth curve following exponential growth in which the rate of population growth begins to slow

decibel scale a nonlinear scale of loudness based on the ratio of the intensity level of a sound to the intensity at the threshold of hearing

decomposers organisms that use dead organic matter as a source of energy

decomposition chemical reaction a chemical reaction in which a compound is broken down into the elements that make up the compound into simpler compounds or into elements and simpler compounds

deflation the widespread removal of base materials from the surface by the wind

delta a somewhat triangular deposit at the mouth of a river formed where a stream flowing into a body of water slowed and lost its sediment-carrying ability

denature a change in the chemical and physical properties of a molecule as a result of the breaking of chemical bonds within the molecule; for example, the change in egg white as a result of cooking

dendrites neuronal fibers that receive information from axons and carry it toward the nerve-cell body

denitrifying bacteria bacteria capable of converting nitrite (NO_2^-) to nitrogen gas (N_2), which is released into the atmosphere

density the compactness of matter described by a ratio of mass per unit volume

deoxyribonucleic acid (DNA) a polymer of nucleotides that serves as genetic information; in prokaryotic cells, it is a double-stranded DNA loop and contains attached HU proteins, and in eukaryotic cells, it is found in strands with attached histone proteins; when tightly coiled, it is known as a chromosome

deoxyribose a 5-carbon sugar molecule; a component of DNA

depolarized having lost the electrical difference existing between two points or objects

destructive interference the condition in which two waves arriving at the same point at the same time out of phase add amplitudes to create zero total disturbance

dew condensation of water vapor into droplets of liquid on surfaces

dew point temperature the temperature at which condensation begins

diaphragm a muscle separating the lung cavity from the abdominal cavity that is involved in exchanging the air in the lungs

diastolic blood pressure the blood pressure recorded while the heart is relaxing

diastrophism all-inclusive term that means any and all possible movements of Earth's plates, including drift, isostatic adjustment, and any other process that deforms or changes Earth's surface by movement

diet the food and drink consumed by a person from day to day

Dietary Reference Intakes current U.S. Department of Agriculture guidelines that provide information on the amounts of certain nutrients members of the public should receive

differentiation the process of forming specialized cells within a multicellular organism

diffuse reflection light rays reflected in many random directions, as opposed to the parallel rays reflected from a perfectly smooth surface such as a mirror

diffusion net movement of a kind of molecule from an area of higher concentration to an area of lesser concentration

diploid having two sets of chromosomes: one set from the maternal parent and one set from the paternal parent

direct current an electrical current that always moves in one direction

direct proportion when two variables increase or decrease together in the same ratio (at the same rate)

disaccharides two monosaccharides joined together with the loss of a water molecule; examples are sucrose (table sugar), lactose, and maltose

dispersion the effect of spreading colors of light into a spectrum with a material that has an index of refraction that varies with wavelength

divergent boundaries boundaries that occur between two plates moving away from each other

divide line separating two adjacent watersheds

DNA code a sequence of three nucleotides of a DNA molecule

DNA polymerase an enzyme that bonds DNA nucleotides together when they base pair with an existing DNA strand

DNA replication the process by which the genetic material (DNA) of the cell reproduces itself prior to its distribution to the next generation of cells

domain a classification group above that of kingdoms

dome a large, upwardly bulging, symmetrical fold that resembles a dome

dominant allele an allele that expresses itself and masks the effects of other alleles for the trait

Doppler effect an apparent shift in the frequency of sound or light due to relative motion between the source of the sound or light and the observer

double bond covalent bond formed when two pairs of electrons are shared by two atoms

dune a hill, low mound, or ridge of windblown sand or other sediments

duodenum the first part of the small intestine, which receives food from the stomach and secretions from the liver and pancreas

dwarf planet an object that is orbiting the Sun, is nearly spherical, but has not cleared matter from its orbital zone and is not a satellite

E

earthquake a quaking, shaking, vibrating, or upheaval of Earth's surface

earthquake epicenter point on Earth's surface directly above an earthquake focus

earthquake focus place where seismic waves originate beneath the surface of Earth

echo a reflected sound that can be distinguished from the original sound, which usually arrives 0.1 s or more after the original sound

eclipse when the shadow of a celestial body falls on the surface of another celestial body

ecology the branch of biology that studies the relationships between organisms and their environment

ecosystem an interacting unit consisting of a collection of organisms and abiotic factors

egg the haploid sex cell produced by sexually mature females

ejaculation the release of sperm cells and seminal fluid through the penis

El Niño changes in the atmospheric pressure systems, ocean currents, water temperatures, and wind patterns that seem to be linked to worldwide changes in the weather

elastic rebound the sudden snap of stressed rock into new positions; the recovery from elastic strain that results in an earthquake

elastic strain an adjustment to stress in which materials recover their original shape after a stress is released

electric charge a fundamental property of electrons and protons; electrons have a negative electric charge and protons have a positive electric charge

electric circuit consists of a voltage source that maintains an electrical potential, a continuous conducting path for a current to follow, and a device where work is done by the electrical potential; a switch in the circuit is used to complete or interrupt the conducting path

electric current the flow of electric charge

electric field force field produced by an electric charge

electric field lines a map of an electric field representing the direction of the force that a test charge would experience; the direction of an electric field shown by lines of force

electric generator a mechanical device that uses wire loops rotating in a magnetic field to produce electromagnetic induction in order to generate electricity

electrical energy a form of energy from electromagnetic interactions; one of five forms of energy—mechanical, chemical, radiant, electrical, and nuclear

electrical force a fundamental force that results from the interaction of electrical charge and is billions and billions of times stronger than the gravitational force; sometimes called the "electromagnetic force" because of the strong association between electricity and magnetism

electrical insulators electrical nonconductors, or materials that obstruct the flow of electric current

electrical nonconductors materials that have electrons that are not moved easily within the material—for example, rubber; electrical nonconductors are also called electrical insulators

electrolyte water solution of ionic substances that conducts an electric current

electromagnet a magnet formed by a solenoid that can be turned on and off by turning the current on and off

electromagnetic force one of four fundamental forces; the force of attraction or repulsion between two charged particles

electromagnetic induction process in which current is induced by moving a loop of wire in a magnetic field or by changing the magnetic field

electron subatomic particle that has the smallest negative charge possible and usually found in an orbital of an atom, but gained or lost when atoms become ions

electron configuration the arrangement of electrons in orbitals and suborbitals about the nucleus of an atom

electron dot notation notation made by writing the chemical symbol of an element with dots around the symbol to indicate the number of outer orbital electrons

electron pair a pair of electrons with different spin quantum numbers that may occupy an orbital

electrostatic charge an accumulated electric charge on an object from a surplus or deficiency of electrons; also called "static electricity"

element a pure chemical substance that cannot be broken down into anything simpler by chemical or physical means; there are over one hundred known elements, the fundamental materials of which all matter is made

endocrine glands secrete chemical messengers into the circulatory system

endocrine system a number of glands that communicate with one another and other tissues through chemical messengers transported throughout the body by the circulatory system

endoplasmic reticulum (ER) folded membranes and tubes throughout the eukaryotic cell that provide a large surface upon which chemical activities take place

endosymbiotic theory a theory suggesting that some organelles found in eukaryotic cells may have originated as free-living prokaryotes

energy the ability to do work

English system a system of measurement that originally used sizes of parts of the human body as referents

entropy the measure of disorder in thermodynamics

environment the surroundings; anything that affects an organism during its lifetime

environmental resistance the collective set of factors that limit population growth

enzymes protein molecules, produced by organisms, that are able to control the rate at which chemical reactions occur

eon the largest blocks of time in Earth's geologic history: the Hadean, Archean, Proterozoic, and Phanerozoic eons

epigenetics the study of changes in gene expression caused by factors other than alterations in a cell's DNA.

epinephrine a hormone produced by the adrenal medulla that increases heart rate, blood pressure, and breathing rate

epiphyte a plant that lives on the surface of another plant without doing harm

epochs subdivisions of geologic periods

equation a statement that describes a relationship in which quantities on one side of the equal sign are identical to quantities on the other side

equation of time the cumulative variation between the apparent local solar time and the mean solar time

equinoxes Latin meaning "equal nights"; time when daylight and night are of equal length, which occurs during the spring equinox and the autumnal equinox

eras the major blocks of time in Earth's geologic history: the Cenozoic, Mesozoic, Paleozoic, and Precambrian eras

erosion the process of physically removing weathered materials; for example, rock fragments are physically picked up by an erosion agent such as a stream or a glacier

essential amino acids those amino acids that cannot be synthesized by the human body and must be part of the diet (e.g., lysine, tryptophan, and valine)

esters class of organic compounds with the general structure of RCOOR′, where R is one of the hydrocarbon groups—for example, methyl or ethyl; esters make up fats, oils, and waxes, and some give fruit and flowers their taste and odor

estrogen one of the female sex hormones that cause the differentiation of typical female internal and external genital anatomy; responsible for the changes in breasts, vagina, uterus, clitoris, and pelvic bone structure at puberty

estuary a special category of aquatic ecosystem that consists of shallow, partially enclosed areas where freshwater enters the ocean

ether class of organic compounds with the general formula ROR′, where R is one of the hydrocarbon groups—for example, methyl or ethyl; mostly used as industrial and laboratory solvents

Eubacteria the domain in which are found organisms commonly known as bacteria

Eucarya the domain in which are found all eukaryotic organisms

eukaryotic cells one of the two major types of cells; characterized by cells that have a true nucleus, as in plants, fungi, protists, and animals

euphotic zone the upper layer of the ocean, where the Sun's rays penetrate

evaporation process of more molecules leaving a liquid for the gaseous state than returning from the gas to the liquid; can occur at any given temperature from the surface of a liquid

evolution the continuous genetic adaptation of a population of organisms to its environment over time

excited state as applied to an atom, describes the energy state of an atom that has electrons in a state above the minimum energy state for that atom; as applied to a nucleus, describes the energy state of a nucleus that has particles in a state above the minimum energy state for that nuclear configuration

exfoliation the fracturing and breaking away of curved, sheetlike plates from bare rock surfaces via physical or chemical weathering, resulting in dome-shaped hills and rounded boulders

exocrine glands use ducts to secrete to the surface of the body or into hollow organs of the body; for example, sweat glands or pancreas

exosphere the outermost layer of the atmosphere, where gas molecules merge with the diffuse vacuum of space

experiment a re-creation of an event in a way that enables a scientist to gain valid and reliable empirical evidence

experimental group the group in a controlled experiment that is identical to the control group in all respects but one

exponential growth phase a period of time during population growth when the population increases at an accelerating rate

external energy the total potential and kinetic energy of an everyday-sized object

external parasite a parasite that lives on the outside of its host

F

facilitated diffusion diffusion assisted by carrier molecules

Fahrenheit scale referent scale that defines numerical values for measuring hotness or coldness, defined as degrees of temperature; based on the reference points of the freezing point of water and the boiling point of water at sea-level pressure, with 180 degrees between the two points

family (elements) vertical columns of the periodic table consisting of elements that have similar properties

family (organisms) a group of closely related genera within an order

fats organic compounds of esters formed from glycerol and three long-chain carboxylic acids that are also called triglycerides; called fats in animals and oils in plants

fatty acid one of the building blocks of true fats and phospholipids; composed of a longchain carbon skeleton with a carboxylic acid functional group at one end; for example, linoleic acid

fault a break in the continuity of a rock formation along which relative movement has occurred between the rocks on either side

fault plane the surface along which relative movement has occurred between the rocks on either side; the surface of the break in continuity of a rock formation

ferromagnesian silicates silicates that contain iron and magnesium; examples include the dark-colored minerals olivine, augite, hornblende, and biotite

fertilization the joining of haploid nuclei, usually from an egg and a sperm cell, resulting in a diploid cell called the zygote

fiber natural (plant) or industrially produced polysaccharides that are resistant to hydrolysis by human digestive enzymes

first law of motion every object remains at rest or in a state of uniform straight-line motion unless acted on by an unbalanced force

first law of thermodynamics a statement of the law of conservation of energy in the relationship between internal energy, work, and heat

first quarter the moon phase between the new phase and the full phase when the Moon is perpendicular to a line drawn through Earth and the Sun; one-half of the lighted Moon can be seen from Earth, so this phase is called the first quarter

floodplain the wide, level floor of a valley built by a stream; the river valley where a stream floods when it spills out of its channel

fluids matter that has the ability to flow or be poured; the individual molecules of a fluid are able to move, rolling over or by one another

folds bends in layered bedrock as a result of stress or stresses that occurred when the rock layers were in a ductile condition, probably under considerable confining pressure from deep burial

foliation the alignment of flat crystal flakes of a rock into parallel sheets

follicle the saclike structure near the surface of the ovary that encases the soon-to-be released secondary oocyte

follicle-stimulating hormone (FSH) the pituitary secretion that causes the ovaries to begin to produce larger quantities of estrogen and to develop the follicle and prepare the egg for ovulation

food chain a sequence of organisms that feed on one another, resulting in a flow of energy from a producer through a series of consumers

Food Guide Pyramid a tool developed by the U.S. Department of Agriculture to help the general public plan for good nutrition; guidelines for required daily intake from each of five food groups

food web a system of interlocking food chains

force a push or pull capable of changing the state of motion of an object; since a force has magnitude (strength) as well as direction, it is a vector quantity

formula describes what elements are in a compound and in what proportions

formula weight the sum of the atomic weights of all the atoms in a chemical formula

fossil any evidence of former life

fossil fuels organic fuels that contain the stored radiant energy of the Sun converted to chemical energy by plants or animals that lived millions of years ago; coal, petroleum, and natural gas are the common fossil fuels

Foucault pendulum a heavy mass swinging from a long wire that can be used to provide evidence about the rotation of Earth

fovea centralis the area of sharpest vision on the retina, containing only cones, where light is sharply focused

fracture strain an adjustment to stress in which materials crack or break as a result of the stress

free fall when objects fall toward Earth with no forces acting upward; air resistance is neglected when considering an object to be in free fall

free-living nitrogen-fixing bacteria soil bacteria that provide nitrogen compounds that can be taken up by plants through their roots

freezing point the temperature at which a phase change of liquid to solid takes place; the same temperature as the melting point for a given substance

frequency the number of cycles of a vibration or of a wave occurring in one second, measured in units of cycles per second (hertz)

freshwater water that is not saline and is fit for human consumption

freshwater communities aquatic communities that have low salt concentrations

front the boundary, or thin transition zone, between air masses of different temperatures

frost ice crystals formed by water vapor condensing directly from the vapor phase; frozen water vapor that forms on objects

frost wedging the process of freezing and thawing water in small rock pores and cracks that become larger and larger, eventually forcing pieces of rock to break off

fuel rod long zirconium alloy tubes containing fissionable material for use in a nuclear reactor

full moon the moon phase when Earth is between the Sun and the Moon and the entire side of the Moon facing Earth is illuminated by sunlight

functional group the atom or group of atoms in an organic molecule that is responsible for the chemical properties of a particular class or group of organic chemicals

fundamental charge smallest common charge known; the magnitude of the charge of an electron and a proton

fundamental forces forces that cannot be explained in terms of any other force; gravitational, electromagnetic, weak, and strong nuclear force

fundamental frequency the lowest frequency (longest wavelength) that can set up standing waves in an air column or on a string

fundamental properties a property that cannot be defined in simpler terms other than to describe how it is measured; the fundamental properties are length, mass, time, and charge

fungus the common name for the kingdom Fungi; examples, yeast and mold

G

g symbol representing the acceleration of an object in free fall due to the force of gravity

galactic clusters gravitationally bound subgroups of as many as 1,000 stars that move together within the Milky Way galaxy

galaxy group of billions and billions of stars that form the basic unit of the universe; for example, Earth is part of the solar system, which is located in the Milky Way galaxy

gallbladder an organ attached to the liver that stores bile

galvanometer a device that measures the size of an electric current from the size of the magnetic field produced by the current

gamete a haploid sex cell

gametogenesis the generating of gametes; the meiotic cell-division process that produces sex cells; oogenesis and spermatogenesis

gametophyte stage a life-cycle stage in plants in which a haploid sex cell is produced by mitosis

gamma ray very short wavelength electromagnetic radiation emitted by decaying nuclei

gases a phase of matter composed of molecules that are relatively far apart moving freely in a constant, random motion and have weak cohesive forces acting between them, resulting in the characteristic indefinite shape and indefinite volume of a gas

gasohol solution of ethanol and gasoline

gastric juice the secretions of the stomach that contain enzymes and hydrochloric acid

Geiger counter a device that indirectly measures ionizing radiation (beta and/or gamma) by detecting "avalanches" of electrons that are able to move because of the ions produced by the passage of ionizing radiation

gene a unit of heredity located on a chromosome and composed of a sequence of DNA nucleotides

gene flow the movement of genes from one generation to another or from one place to another

gene frequency a measure of how often a specific gene for a characteristic shows up in the individuals of a population

gene pool all of the genes of a population of organisms

genetically modified (GM) organisms that contain genetic changes

general theory of relativity geometric theory of gravity from an analysis of space and time and how it interacts with a mass

generative processes actions that increase the size of an individual organism (growth) or increase the number of individuals in a population (reproduction)

genetic recombination the gene mixing that occurs as a result of sexual reproduction

genetics the study of genes, how genes produce characteristics, and how the characteristics are inherited

genome a set of all the genes necessary to specify an organism's complete list of characteristics

genotype the catalog of genes of an organism, whether or not these genes are expressed

genus (plural, *genera*) a group of closely related species within a family

geographic barriers geographic features that keep different portions of a species from exchanging genes

geographic isolation a condition in which part of the gene pool is separated by geographic barriers from the rest of the population

geologic time scale a "calendar" of geologic history based on the appearance and disappearance of particular fossils in the sedimentary rock record

geomagnetic time scale time scale established from the number and duration of magnetic field reversals during the past 6 million years

geosynchronous satellite a satellite with a period of one day, turning with Earth and appearing to move around a fixed point in the sky

geothermal energy energy from beneath Earth's surface

giant planets the large outer planets of Jupiter, Saturn, Uranus, and Neptune, which all have similar densities and compositions

glacier a large mass of ice on land that formed from compacted snow and slowly moves under its own weight

gland an organ that manufactures and secretes a material either through ducts or directly into the circulatory system

globular clusters symmetrical and tightly packed clusters of as many as a million stars that move together as subgroups within the Milky Way galaxy

glycerol an alcohol with three hydroxyl groups per molecule; for example, glycerin (1, 2, 3-propanetriol)

glycogen a highly branched polysaccharide synthesized by the human body and stored in the muscles and liver; serves as a direct reserve source of energy

glycol an alcohol with two hydroxyl groups per molecule; for example, ethylene glycol, which is used as an antifreeze

Golgi apparatus a stack of flattened, smooth, membranous sacs; the site of synthesis and packaging of certain molecules in eukaryotic cells

gonad a generalized term for organs in which meiosis occurs to produce gametes; ovary or testis

gonadotropin-releasing hormone (GnRH) a hormone released at puberty by the hypothalamus that stimulates the pituitary gland to release luteinizing hormone (LH) and follicle-stimulating hormone (FSH)

granite light-colored, coarse-grained igneous rock common on continents; igneous rocks formed by blends of quartz and feldspars, with small amounts of micas, hornblende, and other minerals

greenhouse effect the process of increasing the temperature of the lower parts of the atmosphere through redirecting energy back toward the surface; the absorption and reemission of infrared radiation by carbon dioxide, water vapor, and a few other gases in the atmosphere

ground state energy state of an atom with electrons at the lowest energy state possible for that atom

groundwater water from a saturated zone beneath the surface; water from beneath the surface that supplies wells and springs

growth-stimulating hormone (GSH) a hormone produced by the anterior pituitary gland that stimulates tissues to grow

guanine a double-ring nitrogenous-base molecule in DNA and RNA; the complementary base of cytosine

gyre the great circular systems of moving water in each ocean

H

habitat the place or part of an ecosystem occupied by an organism

hail a frozen form of precipitation, sometimes with alternating layers of clear and opaque, cloudy ice

hair hygrometer a device that measures relative humidity from changes in the length of hair

half-life the time required for one-half of the unstable nuclei in a radioactive substance to decay into a new element

halogen member of family VIIA of the periodic table, having common properties of very reactive nonmetallic elements common in salt compounds

haploid having a single set of chromosomes resulting from the reduction division of meiosis

hard water water that contains relatively high concentrations of dissolved salts of calcium and magnesium

Hardy-Weinberg concept the concept that population must be infinitely large, have random mating, no mutations, no migration, and no selection for specific characteristics in order to prevent evolution from taking place

heart the muscular pump that forces the blood through the blood vessels of the body

heat total internal energy of molecules, which is increased by gaining energy from a temperature difference (conduction, convection, radiation) or by gaining energy from a form conversion (mechanical, chemical, radiant, electrical, nuclear)

heat of formation energy released in a chemical reaction

Heisenberg uncertainty principle you cannot measure both the exact momentum and the exact position of a subatomic particle at the same time—when the more exact of the two is known, the less certain you are of the value of the other

hemoglobin an iron-containing molecule found in red blood cells, to which oxygen molecules bind

herbivores animals that feed directly on plants

hertz unit of frequency; equivalent to one cycle per second

Hertzsprung-Russell diagram diagram to classify stars with a temperature-luminosity graph

heterotroph an organism that requires a source of organic material from its environment; it cannot produce its own food

heterozygous describes a diploid organism that has two different alleles for a particular characteristic

high short for high-pressure center (anticyclone), which is associated with clear, fair weather

high latitudes latitudes close to the poles; those that for a period of time during the winter months receive no solar radiation at noon

high-pressure center another term for *anticyclone*

homeostasis the process of maintaining a constant internal environment as a result of constant monitoring and modification of the functioning of various systems

homologous chromosomes a pair of chromosomes in a diploid cell that contain similar genes at corresponding loci throughout their length

homozygous describes a diploid organism that has two identical alleles for a particular characteristic

hormones chemical messengers secreted by endocrine glands to regulate other parts of the body

horsepower measurement of power defined as a power rating of 550 ft·lb/s

host an organism that a parasite lives in or on

hot spots sites on Earth's surface where plumes of hot rock materials rise from deep within the mantle

humid moist climate classification; receives more than 50 cm (20 in) precipitation per year

humidity the amount of water vapor in the air; see *relative humidity*

hurricane a tropical cyclone with heavy rains and winds exceeding 120 km/h

hydration the attraction of water molecules for ions; a reaction that occurs between water and minerals that make up rocks

hydrocarbon an organic compound consisting of only the two elements hydrogen and carbon

hydrocarbon derivatives organic compounds that can be thought of as forming when one or more hydrogen atoms on a hydrocarbon have been replaced by an element or a group of elements other than hydrogen

hydrogen bonding a strong bond that occurs between the hydrogen end of one molecule and the fluorine, oxygen, or nitrogen end of another molecule

hydrologic cycle water vapor cycling into and out of the atmosphere through continuous evaporation of liquid water from the surface and precipitation of water back to the surface

hydronium ion a molecule of water with an attached hydrogen ion, H_3O

hypothalamus a region of the brain located in the floor of the thalamus and connected to the pituitary gland that is involved in sleep and arousal; emotions such as anger, fear, pleasure, hunger, sexual response, and pain; and automatic functions such as temperature, blood pressure, and water balance

hypothesis a tentative explanation of a phenomenon that is compatible with the data and provides a framework for understanding and describing that phenomenon

I

ice-crystal process a precipitation-forming process that brings water droplets of a cloud together through the formation of ice crystals

ice-forming nuclei small, solid particles suspended in air; ice can form on the suspended particles

igneous rocks rocks that formed from magma, which is a hot, molten mass of melted rock materials

immune system a system of white blood cells specialized to provide the body with resistance to disease; there are two types, antibody-mediated immunity and cell-mediated immunity

impulse a change of motion is brought about by an impulse; the product of the size of an applied force and the time the force is applied

incandescent matter emitting visible light as a result of high temperature; for example, a lightbulb, a flame from any burning source, and the Sun are all incandescent sources because of high temperature

incident ray line representing the direction of motion of incoming light approaching a boundary

incomplete dominance describes a situation in which the phenotype of a heterozygote is intermediate between the two homozygotes on a phenotypic gradient; that is, the phenotypes appear to be "blended" in heterozygotes

incus the ear bone that is located between the malleus and the stapes

independent assortment the segregation, or assortment, of one pair of homologous chromosomes independently of the segregation, or assortment, of any other pair of chromosomes

index fossils distinctive fossils of organisms that lived only a brief time; used to compare the age of rocks exposed in two different locations

index of refraction the ratio of the speed of light in a vacuum to the speed of light in a material

inertia a property of matter describing the tendency of an object to resist a change in its state of motion; an object will remain in unchanging motion or at rest in the absence of an unbalanced force

inferior vena cava a major vein that returns blood to the heart from the lower body

infrasonic sound waves having too low a frequency to be heard by the human ear; sound having a frequency of less than 20 Hz

inguinal canal opening in the floor of the abdominal cavity through which the testes in a human male fetus descend into the scrotum

initiation code the code on DNA with the base sequence TAC that begins the process of transcription

inorganic chemistry the study of all compounds and elements in which carbon is not the principal element

insulators materials that are poor conductors of heat—for example, heat flows slowly through materials with air pockets because the molecules making up air are far apart; also, materials that are poor conductors of electricity—for example, glass or wood

intensity a measure of the energy carried by a wave

interference phenomenon of light where the relative phase difference between two light waves produces light or dark spots, a result of light's wavelike nature

intermediate-focus earthquakes earthquakes that occur in the upper part of the mantle, between 70 to 350 km below the surface of Earth

intermolecular forces forces of interaction between molecules

internal energy sum of all the potential energy and all the kinetic energy of all the molecules of an object

internal parasite a parasite that lives inside its host

international date line the 180° meridian is arbitrarily called the international date line; used to compensate for cumulative time zone changes by adding or subtracting a day when the line is crossed

interphase the stage between cell divisions in which the cell is engaged in metabolic activities

interstitial cell-stimulating hormone (ICSH) the chemical messenger molecule released from the pituitary gland that causes the testes to produce testosterone, the primary male sex hormone; same as follicle-stimulating hormone in females

intertropical convergence zone a part of the lower troposphere in a belt from 10°N to 10°S of the equator where air is heated, expands, and becomes less dense and rises around the belt

intrusive igneous rocks coarse-grained igneous rocks formed as magma cools slowly deep below the surface

inverse proportion the relationship in which the value of one variable increases while the value of the second variable decreases at the same rate (in the same ratio)

inversion a condition of the troposphere when temperature increases with height rather than decreasing with height; a cap of cold air over warmer air that results in trapped air pollution

ion an atom or a particle that has a net charge because of the gain or loss of electrons; polyatomic ions are groups of bonded atoms that have a net charge

ion exchange reaction a reaction that takes place when the ions of one compound interact with the ions of another, forming a solid that comes out of solution, a gas, or water

ionic bond chemical bond of electrostatic attraction between negative and positive ions

ionic compounds chemical compounds that are held together by ionic bonds; that is, bonds of electrostatic attraction between negative and positive ions

ionization process of forming ions from molecules

ionization counter a device that measures ionizing radiation (alpha, beta, gamma, etc.) by indirectly counting the ions produced by the radiation

ionized an atom or a particle that has a net charge because it has gained or lost electrons

ionosphere refers to that part of the atmosphere—parts of the thermosphere and upper mesosphere—where free electrons and ions reflect radio waves around Earth and where the northern lights occur

iron meteorites meteorite classification group whose members are composed mainly of iron

island arcs curving chains of volcanic islands that occur over belts of deep-seated earthquakes; for example, the Japanese and Indonesian islands

isomers chemical compounds with the same molecular formula but different molecular structure; compounds that are made from the same numbers of the same elements but have different molecular arrangements

isotope atoms of an element with identical chemical properties but with different masses; isotopes are atoms of the same element with different numbers of neutrons

J

jet stream a powerful, winding belt of wind near the top of the troposphere that tends to extend all the way around Earth, moving generally from the west in both hemispheres at speeds of 160 km/h or more

joint a break in the continuity of a rock formation without a relative movement of the rock on either side of the break

joule metric unit used to measure work and energy; can also be used to measure heat; equivalent to newton-meter

K

Kelvin scale a temperature scale that does not have arbitrarily assigned referent points, and zero means nothing; the zero point on the Kelvin scale (also called absolute scale) is the lowest limit of temperature, where all random kinetic energy of molecules ceases

ketone an organic compound with the general formula RCOR′, where R is one of the hydrocarbon groups; for example, methyl or ethyl

keystone species species in an ecosystem that affects many aspects of the ecosystem and whose removal causes significant alteration of the ecosystem

kidneys the primary organs involved in regulating blood levels of water, hydrogen ions, salts, and urea

kilocalorie the amount of energy required to increase the temperature of 1 kilogram of water 1 degree Celsius; equivalent to 1,000 calories

kilogram the fundamental unit of mass in the metric system of measurement

kinetic energy the energy of motion; can be measured from the work done to put an object in motion, from the mass and velocity of the object while in motion, or from the amount of work the object can do because of its motion

kinetic molecular theory the collection of assumptions that all matter is made up of tiny atoms and molecules that interact physically, that explain the various states of matter, and that have an average kinetic energy that defines the temperature of a substance

kingdom a category within a domain used in the classification of organisms

Kuiper Belt a disk-shaped region of small icy bodies some 30 to 100 AU from the Sun; the source of short-period comets

L

lack of dominance the condition of two unlike alleles both expressing themselves, neither being dominant

lag phase a stage of a population growth curve during which both natality and mortality are low

lake a large inland body of standing water

large intestine the last portion of the food tube; primarily involved in reabsorbing water

last quarter the moon phase between the full phase and the new phase when the Moon is perpendicular to a line drawn through Earth and the Sun; one half of the lighted Moon can be seen from Earth, so this phase is called the last quarter

latent heat refers to the heat "hidden" in phase changes

latent heat of fusion the heat absorbed when 1 gram of a substance changes from the solid to the liquid phase, or the heat released by 1 gram of a substance when changing from the liquid phase to the solid phase

latent heat of vaporization the heat absorbed when 1 gram of a substance changes from the liquid phase to the gaseous phase, or the heat released when 1 gram of gas changes from the gaseous phase to the liquid phase

latitude the angular distance from the equator to a point on a parallel that tells you how far north or south of the equator the point is located

lava magma, or molten rock, that is forced to the surface from a volcano or a crack in Earth's surface

law of conservation of energy energy is never created or destroyed; it can only be converted from one form to another as the total energy remains constant

law of conservation of mass same as law of conservation of matter; mass, including single atoms, is neither created nor destroyed in a chemical reaction

law of conservation of momentum the total momentum of a group of interacting objects remains constant in the absence of external forces

law of dominance when an organism has two different alleles for a trait, the allele that is expressed and overshadows the expression of the other allele is said to be dominant; the allele whose expression is overshadowed is said to be recessive

law of independent assortment members of one allelic pair will separate from each other independently of the members of other allele pairs

law of segregation when gametes are formed by a diploid organism, the alleles that control a trait separate from one another into different gametes, retaining their individuality

light ray model model using lines to show the direction of motion of light to describe the travels of light

light-year the distance that light travels through empty space in one year, approximately 9.5×10^{12} km

limiting factors the identifiable factors that prevent unlimited population growth

limnetic zone the region in a lake that does not have rooted vegetation

line spectrum narrow lines of color in an otherwise dark spectrum; these lines can be used as "fingerprints" to identify gases

linear scale a scale, generally on a graph, where equal intervals represent equal changes in the value of a variable

lines of force lines drawn to make an electric field strength map, with each line originating on a positive charge and ending on a negative charge; each line represents a path on which a charge would experience a constant force and lines closer together mean a stronger electric field

linkage group genes located on the same chromosome that tend to be inherited together

lipid large, nonpolar, organic molecules that do not easily dissolve in polar solvents such as water; there are three different types of lipids: *true fats* (pork chop fat or olive oil), *phospholipids* (the primary component of cell membranes), and *steroids* (most hormones)

liquids a phase of matter composed of molecules that have interactions stronger than those found in a gas but not strong enough to keep the molecules near the equilibrium positions of a solid, resulting in the characteristic definite volume but indefinite shape of a liquid

liter a metric system unit of volume usually used for liquids

lithosphere solid layer of Earth's structure that is above the asthenosphere and includes the entire crust, the Moho, and the upper part of the mantle

littoral zone the region in a lake with rooted vegetation

liver an organ of the body responsible for secreting bile, filtering the blood, detoxifying molecules, and modifying molecules absorbed from the gut

locus the spot on a chromosome where an allele is located

loess very fine dust or silt that has been deposited by the wind over a large area

longitude angular distance of a point east or west from the prime meridian on a parallel

longitudinal wave a mechanical disturbance that causes particles to move closer together and farther apart in the same direction that the wave is traveling

longshore current a current that moves parallel to the shore, pushed along by waves that move accumulated water from breakers

loudness a subjective interpretation of a sound that is related to the energy of the vibrating source, related to the condition of the transmitting medium, and related to the distance involved

low latitudes latitudes close to the equator; those that receive vertical solar radiation at noon during some part of the year

luminous an object or objects that produce visible light; for example, the Sun, stars, lightbulbs, and burning materials are all luminous

lunar eclipse occurs when the Moon is full and the Sun, Moon, and Earth are lined up so the shadow of Earth falls on the Moon

lung a respiratory organ in which air and blood are brought close to one another and gas exchange occurs

L-wave seismic waves that move on the solid surface of Earth much as water waves move across the surface of a body of water

lymph liquid material that leaves the circulatory system to surround cells

lymphatic system a collection of thin-walled tubes that collects, filters, and returns lymph from the body to the circulatory system

lysosome a specialized, submicroscopic organelle that holds a mixture of hydrolytic enzymes

M

macromolecule very large molecule, with a molecular weight of thousands or millions of atomic mass units, that is made up of a combination of many smaller, similar molecules

magma a mass of molten rock material either below or on Earth's crust from which igneous rock is formed by cooling and hardening

magnetic field model used to describe how magnetic forces on moving charges act at a distance

magnetic poles the ends, or sides, of a magnet about which the force of magnetic attraction seems to be concentrated

magnetic quantum number from the quantum mechanics model of the atom, one of four descriptions of the energy state of an electron wave; this quantum number describes the energy of an electron orbital as the orbital is oriented in space by an external magnetic field, a kind of energy sub-sublevel

magnetic reversal the flipping of polarity of Earth's magnetic field as the north magnetic pole and the south magnetic pole exchange positions

main sequence stars normal, mature stars that use their nuclear fuel at a steady rate; stars on the Hertzsprung-Russell diagram in a narrow band that runs from the top left to the lower right

malignant tumor an abnormal, nonencapsulated growth of tumor cells that may spread to other parts of the body

malleus the ear bone that is attached to the tympanum

manipulated variable in an experiment, a quantity that can be controlled or manipulated; also known as the independent variable

mantle middle part of Earth's interior; a 2,870 km (about 1,780 mi) thick shell between the core and the crust

marine climate a climate influenced by air masses from over an ocean, with mild winters and cool summers compared to areas farther inland

marine communities aquatic communities that have a high salt content

maritime air mass a moist air mass that forms over the ocean

mass a measure of inertia, which means a resistance to a change of motion

mass defect the difference between the sum of the masses of the individual nucleons forming a nucleus and the actual mass of that nucleus

mass movement erosion caused by the direct action of gravity

mass number the sum of the number of protons and neutrons in a nucleus defines the mass number of an atom; used to identify isotopes; for example, uranium 238

masturbation stimulation of one's own sex organs

matter anything that occupies space and has mass

meanders winding, circuitous turns or bends of a stream

mean solar day is 24 hours long and is averaged from the mean solar time

mean solar time a uniform time averaged from the apparent solar time

measurement the process of comparing a property of an object to a well-defined and agreed-upon referent

mechanical energy the form of energy associated with machines, objects in motion, and objects having potential energy that results from gravity

mechanical weathering the physical breaking up of rocks without any changes in their chemical composition

medulla oblongata a region of the more primitive portion of the brain connected to the spinal cord that controls such automatic functions as blood pressure, breathing, and heart rate

meiosis the specialized pair of cell divisions that reduces the chromosome number from diploid $(2n)$ to haploid (n)

melting point the temperature at which a phase change of solid to liquid takes place; the same temperature as the freezing point for a given substance

Mendelian genetics the pattern of inheriting characteristics that follows the laws formulated by Gregor Mendel

menopause the period when the ovaries stop producing viable secondary oocytes and the body becomes nonreproductive

menstrual cycle (menses, menstrual flow, period) the repeated building up and shedding of the lining of the uterus

meridians north-south running arcs that intersect at both poles and are perpendicular to the parallels

Mesozoic a geologic era; the time of middle life, meaning some of the fossils for this time period are similar to the life found on Earth today, but many are different from anything living today

messenger RNA (mRNA) a molecule composed of ribonucleotides that functions as a copy of the gene and is used in the cytoplasm of the cell during protein synthesis

metabolism the total of all the chemical reactions and energy changes that take place in an organism

metabolic processes the total of all chemical reactions within an organism; for example, nutrient uptake and processing, and waste elimination

metal matter having the physical properties of conductivity, malleability, ductility, and luster

metamorphic rocks previously existing rocks that have been changed into a distinctly different rock by heat, pressure, or hot solutions

metaphase the second stage in mitosis, characterized by alignment of the chromosomes at the equatorial plane

metastasize the process of cells of tumors moving from the original site and establishing new colonies in other regions of the body

meteor the streak of light and smoke that appears in the sky when a meteoroid is made incandescent by compression of Earth's atmosphere

meteor shower event when many meteorites fall in a short period of time

meteorite the solid iron or stony material of a meteoroid that survives passage through Earth's atmosphere and reaches the surface

meteoroids remnants of comets and asteroids in space

meteorology the science of understanding and predicting weather

meter the fundamental metric unit of length

metric system a system of referent units based on invariable referents of nature that have been defined as standards

microclimate a local, small-scale pattern of climate; for example, the north side of a house has a different microclimate than the south side

microvilli tiny projections from the surfaces of cells that line the intestine

middle latitudes latitudes equally far from the poles and equator; between the high and low latitudes

mineral (geology) a naturally occurring, inorganic solid element or chemical compound with a crystalline structure

minerals (biology) inorganic elements that cannot be manufactured by the body but are required in low concentrations; essential to metabolism

miscible fluids two or more kinds of fluids that can mix in any proportion

mitochondrion a membranous organelle resembling a small bag with a larger bag inside that is folded back on itself; serves as the site of aerobic cellular respiration

mitosis a process that results in equal and identical distribution of replicated chromosomes into two newly formed nuclei

mixture matter composed of two or more kinds of matter that has a variable composition and can be separated into its component parts by physical means

model a mental or physical representation of something that cannot be observed directly that is usually used as an aid to understanding

moderator a substance in a nuclear reactor that slows fast neutrons so the neutrons can participate in nuclear reactions

mold the preservation of the shape of an organism by the dissolution of the remains of a buried organism, leaving an empty space where the remains were

molecule from the chemical point of view, a particle composed of two or more atoms held together by an attractive force called a chemical bond; from the kinetic theory point of view, smallest particle of a compound or gaseous element that can exist and still retain the characteristic properties of a substance

momentum the product of the mass of an object times its velocity

monomer individual, repeating units or segments of complex molecules that chemically combine to form long chainlike molecules called polymers; for example, monosaccharides link to form polysaccharides

monosaccharides simple sugars containing 3 to 6 carbon atoms; the most common kinds are 6-carbon molecules such as glucose and fructose

moraines deposits of rocks and other mounded materials bulldozed into position by a glacier and left behind when the glacier melted

morphological species concept the concept that species can be distinguished from one another by structural characteristics

mortality the number of individuals leaving the population by death per thousand individuals in the population

motor neurons those neurons that carry information from the central nervous system to muscles or glands

motor unit all of the muscle cells stimulated by a single neuron

mountain a natural elevation of Earth's crust that rises above the surrounding surface

multiple alleles a term used to refer to conditions in which there are several different alleles for a particular characteristic, not just two

multiregional hypothesis states that *Homo erectus* evolved into *Homo sapiens*

mutagen any chemical, object, or energy that causes a change in the genetic information of an organism

mutagenic agent anything that causes permanent change in DNA

mutation any change in the genetic information of a cell

mutualism a relationship between two organisms in which both organisms benefit

N

natality the number of individuals entering the population by reproduction per thousand individuals in the population

natural frequency the frequency of vibration of an elastic object that depends on the size, composition, and shape of the object

natural selection the processes that encourage the passage of beneficial genes and discourage the passage of harmful or unfavorable genes from one generation to the next

neap tide period of less-pronounced high and low tides; occurs when the Sun and Moon are at right angles to one another

nebulae a diffuse mass of interstellar clouds of hydrogen gas or dust that may develop into a star

negative electric charge one of the two types of electric charge; repels other negative charges and attracts positive charges

negative ion atom or particle that has a surplus, or imbalance, of electrons and, thus, a negative charge

nephrons millions of tiny tubular units that make up the kidneys, which are responsible for filtering the blood

nerve cell see *neuron*

nerve impulse a series of changes that take place in the neuron, resulting in a wave of depolarization that passes from one end of the neuron to the other

nerves bundles of neuronal fibers

nervous system a network of neurons that carry information from sense organs to the central nervous system and from the central nervous system to muscles and glands

net force the resulting force after all forces have been added; if a net force is zero, all the forces have canceled each other and there is not an unbalanced force

neuron the cellular unit consisting of a cell body and fibers that makes up the nervous system; also called nerve cell

neurotransmitter a molecule released by the axons of neurons that stimulates other cells

neutralized acid or base properties have been lost through a chemical reaction

neutron neutral subatomic particle usually found in the nucleus of an atom

neutron star very small superdense remains of a supernova with a center core of pure neutrons

new crust zone zone of a divergent boundary where new crust is formed by magma upwelling at the boundary

new moon the moon phase when the Moon is between Earth and the Sun and the entire side of the Moon facing Earth is dark

newton a unit of force defined as $kg \cdot m/s^2$; that is, a 1 newton force is needed to accelerate a 1 kg mass $1 \ m/s^2$

niche the functional role of an organism

nitrifying bacteria bacteria that can convert ammonia (NH_3) into nitrite-containing (NO_2^-) compounds, which in turn can be converted into nitrate-containing (NO_3^-) compounds

nitrogen cycle nitrogen in the atmosphere is acted on and used by many bacteria and other organisms and ultimately returned to the atmosphere

nitrogenous base a category of organic molecules found as components of the nucleic acids; there are five common types: thymine, guanine, cytosine, adenine, and uracil

noble gas members of family VIII of the periodic table, having common properties of colorless, odorless, chemically inert gases; also known as rare gases or inert gases

node regions on a standing wave that do not oscillate

noise sounds made up of groups of waves of random frequency and intensity

nonelectrolytes water solutions that do not conduct an electric current; covalent compounds that form molecular solutions and cannot conduct an electric current

nonferromagnesian silicates silicates that do not contain iron or magnesium ions; examples include the minerals of muscovite (white mica), the feldspars, and quartz

nonmetal an element that is brittle (when a solid), does not have a metallic luster, is a poor conductor of heat and electricity, and is not malleable or ductile

nonsilicates minerals that do not have the silicon-oxygen tetrahedra in their crystal structure

noon the event of time when the Sun moves across the celestial meridian

norepinephrine a hormone produced by the adrenal medulla that increases heart rate, blood pressure, and breathing rate

normal a line perpendicular to the surface of a boundary

normal fault a fault where the hanging wall has moved downward with respect to the foot wall

north pole (of a magnet) short for "north seeking"; the pole of a magnet that points northward when it is free to turn

nova a star that explodes or suddenly erupts and increases in brightness

nuclear energy the form of energy from reactions involving the nucleus, the innermost part of an atom

nuclear fission nuclear reaction of splitting a massive nucleus into more stable, less massive nuclei with an accompanying release of energy

nuclear force one of four fundamental forces, a strong force of attraction that operates over very short distances between subatomic particles; this force overcomes the electric repulsion of protons in a nucleus and binds the nucleus together

nuclear fusion nuclear reaction of low-mass nuclei fusing together to form more stable and more massive nuclei with an accompanying release of energy

nuclear membrane the structure surrounding the nucleus that separates the nucleoplasm from the cytoplasm

nuclear reactor steel vessel in which a controlled chain reaction of fissionable materials releases energy

nucleic acids complex molecules that store and transfer genetic information within a cell; constructed of fundamental monomers known as nucleotides; the two common forms are DNA and RNA

nucleoli (singular, *nucleolus*) nuclear structures composed of completed or partially completed ribosomes and the specific parts of chromosomes that contain the information for their construction

nucleons name used to refer to both the protons and neutrons in the nucleus of an atom

nucleoproteins the double-stranded DNA with attached proteins; also called chromatin fibers

nucleosomes histone clusters with their encircling DNA

nucleotide the building block of the nucleic acids, composed of a 5-carbon sugar, a phosphate, and a nitrogenous base

nucleus (atom) tiny, relatively massive and positively charged center of an atom containing protons and neutrons; the small, dense center of an atom

numerical constant a constant without units; a number

nutrients molecules required by the body for growth, reproduction, or repair

nutrition collectively, the processes involved in taking in, assimilating, and utilizing nutrients

O

obesity the condition of being 15 percent to 20 percent above the individual's ideal weight

obligate intracellular parasites infectious particles (viruses) that can function only when inside a living cell

observed lapse rate the rate of change in temperature compared to change in altitude

occluded front a front that has been lifted completely off the ground into the atmosphere, forming a cyclonic storm

ocean the single, continuous body of salt water on the surface of Earth

ocean basin the deep bottom of the ocean floor, which starts beyond the continental slope

ocean currents streams of water within the ocean that stay in about the same path as they move over large distances; steady and continuous onward movement of a channel of water in the ocean

ocean wave a moving disturbance that travels across the surface of the ocean

oceanic ridges long, high, continuous, suboceanic mountain chains; for example, the Mid-Atlantic Ridge in the center of the Atlantic Ocean Basin

oceanic trenches long, narrow, deep troughs with steep sides that run parallel to the edge of continents

octet rule a generalization that helps keep track of the valence electrons in most representative elements; atoms of the representative elements (A families) attempt to acquire an outer orbital with eight electrons through chemical reactions

offspring descendants of a set of parents

ohm unit of resistance; equivalent to volts/amps

Ohm's law the electric potential difference is directly proportional to the product of the current times the resistance

oil field petroleum accumulated and trapped in extensive porous rock structure or structures

oils organic compounds of esters formed from glycerol and three long-chain carboxylic acids that are also called triglycerides; called fats in animals and oils in plants

olfactory epithelium the cells of the nasal cavity that respond to chemicals

omnivores animals that are carnivores at some times and herbivores at others

oogenesis the specific name given to the gametogenesis process that leads to the formation of eggs

Oort cloud a spherical "cloud" of small, icy bodies from 30,000 AU out to a light-year from the Sun; the source of long-period comets

opaque materials that do not allow the transmission of any light

orbital the region of space around the nucleus of an atom where an electron is likely to be found

order a group of closely related families within a class

ore mineral mineral deposits with an economic value

organ a structure composed of two or more kinds of tissues that perform a particular function

organ system a group of organs that performs a particular function

organelles cellular structures that perform specific functions in the cell; the function of an organelle is directly related to its structure

organic acids organic compounds with a general formula of RCOOH, where R is one of the hydrocarbon groups; for example, methyl or ethyl

organic chemistry the study of compounds in which carbon is the principal element

organism an independent living unit

orgasm a complex series of responses to sexual stimulation that results in intense frenzy of sexual excitement

osmosis the net movement of water molecules through a selectively permeable membrane

osteoporosis a disease condition resulting from the demineralization of the bone, resulting in pain, deformities, and fractures; related to a loss of calcium

out-of-Africa hypothesis states that modern humans (*Homo sapiens*) originated in Africa, as had several other hominid species, and migrated from Africa to Asia and Europe and displaced species such as *H. erectus* and *H. heidelbergensis* that had migrated into these areas previously

oval window the membrane-covered opening of the cochlea, to which the stapes is attached

ovaries the female sex organs that produce haploid sex cells—the eggs or ova

overtones higher resonant frequencies that occur at the same time as the fundamental frequency, giving a musical instrument its characteristic sound quality

oviduct the tube (fallopian tube) that carries the oocyte to the uterus

ovulation the release of a secondary oocyte from the surface of the ovary

oxbow lake a small body of water, or lake, that formed when two bends of a stream came together and cut off a meander

oxidation the process of a substance losing electrons during a chemical reaction; a reaction between oxygen and the minerals making up rocks

oxidation-reduction reaction a chemical reaction in which electrons are transferred from one atom to another; sometimes called "redox" for short

oxidizing agents substances that take electrons from other substances

oxidizing atmosphere an atmosphere that contains molecular oxygen

oxytocin a hormone released from the posterior pituitary that causes contraction of the uterus

P

paleontology the science of discovering fossils, studying the fossil record, and deciphering the history of life

Paleozoic a geologic era; time of ancient life, meaning the fossils from this time period are very different from anything living on Earth today

pancreas an organ of the body that secretes many kinds of digestive enzymes into the duodenum

parallels reference lines on Earth used to identify where in the world you are northward or southward from the equator; east and west running circles that are parallel to the equator on a globe with the distance from the equator called the latitude

parasite an organism that lives in or on another organism and derives nourishment from it

parasitism a relationship which involves one organism living in or on another living organism from which it derives nourishment

parts per billion concentration ratio of parts of solute in every 1 billion parts of solution (ppb); could be expressed as ppb by volume or as ppb by weight

parts per million concentration ratio of parts of solute in every 1 million parts of solution (ppm); could be expressed as ppm by volume or as ppm by weight

Pauli exclusion principle no two electrons in an atom can have the same four quantum numbers; thus, a maximum of two electrons can occupy a given orbital

pelagic describes aquatic organisms that float or swim actively

penis the portion of the male reproductive system that deposits sperm in the female reproductive tract

penumbra the zone of partial darkness in a shadow

pepsin an enzyme produced by the stomach that is responsible for beginning the digestion of proteins

percent by volume the volume of solute in 100 volumes of solution

percent by weight the weight of solute in 100 weight units of solution

perception recognition by the brain that a stimulus has been received

perigee when the Moon's elliptical orbit brings the Moon closest to Earth

perihelion the point at which an orbit comes closest to the Sun

period (geologic time) subdivisions of geologic eras

period (periodic table) horizontal rows of elements with increasing atomic numbers; runs from left to right on the element table

period (wave) the time required for one complete cycle of a wave

periodic law similar physical and chemical properties recur periodically when the elements are listed in order of increasing atomic number

peripheral nervous system the fibers that communicate between the central nervous system and other parts of the body

permeability the ability to transmit fluids through openings, small passageways, or gaps

permineralization the process that forms a fossil by alteration of an organism's buried remains by circulating groundwater depositing calcium carbonate, silica, or pyrite

petroleum oil that comes from oil-bearing rock, a mixture of hydrocarbons that is believed to have formed from ancient accumulations of buried organic materials such as remains of algae

pH scale scale that measures the acidity of a solution with numbers below 7 representing acids, 7 representing neutral, and numbers above 7 representing bases

phagocytosis the process by which the cell wraps around a particle and engulfs it

pharynx the region at the back of the mouth cavity; the throat

phase change the action of a substance changing from one state of matter to another; a phase change always absorbs or releases internal potential energy without a temperature change

phases of matter the different physical forms that matter can take as a result of different molecular arrangements, resulting in characteristics of the common phases of a solid, liquid, or gas

phenotype the physical, chemical, and behavioral expression of the genes possessed by an organism

phospholipids a class of complex water-insoluble organic molecules that resemble fats but contain a phosphate group (PO_4) in their structure; a component of cellular membranes

photoelectric effect the movement of electrons in some materials as a result of energy acquired from absorbed light

photon a quantum of energy in a light wave; the particle associated with light

photosynthesis a series of reactions that takes place in chloroplasts and results in the storage of sunlight energy in the form of chemical-bond energy

phylogeny the science that explores the evolutionary relationships among organisms and seeks to reconstruct evolutionary history

phylum a subdivision of a kingdom

physical change a change of the state of a substance but not the identity of the substance

phytoplankton planktonic organisms that carry on photosynthesis

pistil the sex organ in plants that produces eggs or ova

pitch the frequency of a sound wave

pituitary gland the gland at the base of the brain that controls the functioning of other glands throughout the organism

placenta an organ made up of tissues from the embryo and the uterus of the mother that allows for the exchange of materials between the mother's bloodstream and the embryo's bloodstream; it also produces hormones

Planck's constant proportionality constant in the relationship between the energy of vibrating molecules and their frequency of vibration; a value of 6.63×10^{-34} Js

plane of the ecliptic the plane of Earth's orbit

planet an object that is orbiting the Sun, is nearly spherical, and is large enough to clear all matter from its orbital zone

plankton aquatic organisms that are so small and weakly swimming that they are simply carried by currents.

plasma (biology) the watery matrix that contains the molecules and cells of the blood

plasma (physics) a phase of matter; a very hot gas consisting of electrons and atoms that have been stripped of their electrons because of high kinetic energies

plasma membrane the outer-boundary membrane of the cell; see *cell membrane*

plastic strain an adjustment to stress in which materials become molded or bent out of shape under stress and do not return to their original shape after the stress is released

plate tectonics the theory that Earth's crust is made of rigid plates that float on the asthenosphere

pleiotropy the multiple effects that a gene may have on the phenotype of an organism

point mutation a change in the DNA of a cell as a result of a loss or change in a single nitrogenous-base

polar air mass cold air mass that forms in cold regions

polar body the smaller of two cells formed by unequal meiotic division during oogenesis

polar climate zone climate zone of the high latitudes; average monthly temperatures stay below 10°C (50°F), even during the warmest month of the year

polar molecule a molecule that has a negative charge on one part and a positive charge on another part

polarized light whose constituent transverse waves are all vibrating in the same plane; also known as plane-polarized light

Polaroid a film that transmits only polarized light

polyatomic ion ion made up of many atoms

polygenic inheritance the concept that a number of different pairs of alleles may combine their efforts to determine a characteristic

polymers huge, chainlike molecules made of hundreds or thousands of smaller repeating molecular units called monomers

polypeptide chain polymers of amino acids; sometimes called proteins

polysaccharides polymers consisting of monosaccharide units joined together in straight or branched chains; starches, glycogen, or cellulose

pond a small body of standing water, smaller than a lake

pons a region of the brain immediately anterior to the medulla oblongata that connects to the cerebellum and higher regions of the brain and controls several sensory and motor functions of the head and face

population a group of organisms of the same species located in an area

population growth curve a graph of the change in population size over time

porosity the ratio of pore space to the total volume of a rock or soil sample, expressed as a percentage; freely admitting the passage of fluids through pores or small spaces between parts of the rock or soil

positive electric charge one of the two types of electric charge; repels other positive charges and attracts negative charges

positive ion atom or particle that has a net positive charge due to an electron or electrons being torn away

potential energy energy due to position; energy associated with changes in position (e.g., gravitational potential energy) or changes in shape (e.g., compressed or stretched spring)

power the rate at which energy is transferred or the rate at which work is performed; defined as work per unit of time

Precambrian the time before the time of ancient life, meaning the rocks for this time period contain very few fossils

precession the slow wobble of the axis of Earth similar to the wobble of a spinning top

precipitation water that falls to the surface of Earth in the solid or liquid form

predation a relationship between two organisms that involves the capturing, killing, and eating of one by the other

predator an organism that captures, kills, and eats another animal

pressure defined as force per unit area; for example, pounds per square inch (lb/in^2)

prey an organism captured, killed, and eaten by a predator

primary carnivores carnivores that eat herbivores and are therefore on the third trophic level

primary coil part of a transformer; a coil of wire that is connected to a source of alternating current

primary consumers organisms that feed directly on plants—herbivores

primary loop part of the energy-converting system of a nuclear power plant; the closed pipe system that carries heated water from the nuclear reactor to a steam generator

primary oocyte the diploid cell of the ovary that begins to undergo the first meiotic division in the process of oogenesis

primary spermatocyte the diploid cell in the testes that undergoes the first meiotic division in the process of spermatogenesis

prime meridian the referent meridian (0°) that passes through the Greenwich Observatory in England

principal quantum number from the quantum mechanics model of the atom, one of four descriptions of the energy state of an electron wave; this quantum number describes the main energy level of an electron in terms of its most probable distance from the nucleus

principle of uniformity a frame of reference that the same processes that changed the landscape in the past are the same processes you see changing the landscape today

prions proteins that can be passed from one individual to another and cause disease

probability the chance that an event will happen, expressed as a percent or fraction

producers organisms that produce new organic material from inorganic material with the aid of sunlight

progesterone a hormone released by the corpus luteum that is important in maintaining the lining of the uterus

prokaryotic cells one of the two major types of cells; they do not have a typical nucleus bound by a nuclear membrane and lack many of the other membranous cellular organelles; for example, members of the Eubacteria and Archaea

promoter a region of DNA at the beginning of each gene, just ahead of an initiator code

proof a measure of ethanol concentration of an alcoholic beverage; proof is double the concentration by volume; for example, 50 percent by volume is 100 proof

properties qualities or attributes that, taken together, are usually unique to an object; for example, color, texture, and size

prophase the first phase of mitosis during which individual chromosomes become visible

proportionality constant a constant applied to a proportionality statement that transforms the statement into an equation

protein synthesis the process whereby the tRNA utilizes the mRNA as a guide to arrange the amino acids in their proper sequence according to the genetic information in the chemical code of DNA

proteins macromolecular polymers made of smaller molecules of amino acids, with molecular weight from about six thousand to 50 million; proteins are amino acid polymers with roles in biological structures or functions; without such a function, they are known as polypeptides

proton subatomic particle that has the smallest possible positive charge, usually found in the nucleus of an atom

protoplanet nebular model a model of the formation of the solar system that states that the planets formed from gas and dust left over from the formation of the Sun

protoplasm the living portion of a cell as distinguished from the nonliving cell wall

protostar an accumulation of gases that will become a star

proxy data data from natural sources used to infer temperature change, rainfall change, or some other climate condition

pseudoscience use of the appearance of science to mislead; the assertions made are not valid or reliable

psychrometer a two-thermometer device used to measure the relative humidity

puberty a time in the life of a developing individual characterized by the increasing production of sex hormones, which cause it to reach sexual maturity

pulmonary artery the major blood vessel that carries blood from the right ventricle to the lungs

pulmonary circulation the flow of blood through certain chambers of the heart and blood vessels to the lungs and back to the heart

pulsar the source of regular, equally spaced pulsating radio signals believed to be the result of the magnetic field of a rotating neutron star

Punnett square a method used to determine the probabilities of allele combinations in an offspring

pure substance materials that are the same throughout and have a fixed definite composition

pure tone sound made by very regular intensities and very regular frequencies from regular repeating vibrations

P-wave a pressure, or compressional, wave in which a disturbance vibrates materials back and forth in the same direction as the direction of wave movement

P-wave shadow zone a region on Earth between 103° and 142° of arc from an earthquake where no P-waves are received; believed to be explained by P-waves being refracted by the core

pyloric sphincter a valve located at the end of the stomach that regulates the flow of food from the stomach to the duodenum

Q

quad 1 quadrillion Btu (10^{15} Btu); used to describe very large amounts of energy

quanta fixed amounts; usually referring to fixed amounts of energy absorbed or emitted by matter (singular, *quantum*)

quantities measured properties; includes the numerical value of the measurement and the unit used in the measurement

quantum mechanics model of the atom based on the wave nature of subatomic particles, the mechanics of electron waves; also called wave mechanics

quantum numbers numbers that describe energy states of an electron; in the Bohr model of the atom, the orbit quantum numbers could be any whole number 1, 2, 3, and so on out from the nucleus; in the quantum mechanics model of the atom, four quantum numbers are used to describe the energy state of an electron wave

R

rad a measure of radiation received by a material (radiation absorbed dose)

radiant energy the form of energy that can travel through space; for example, visible light and other parts of the electromagnetic spectrum

radiation the transfer of heat from a region of higher temperature to a region of lower temperature by greater emission of radiant energy from the region of higher temperature

radiation zone part of the interior of a star according to a model; the region directly above the core where gamma and X rays from the core are absorbed and reemitted, with the radiation slowly working its way outward

radioactive decay the natural spontaneous disintegration or decomposition of a nucleus

radioactive decay constant a specific constant for a particular isotope that is the ratio of the rate of nuclear disintegration per unit of time to the total number of radioactive nuclei

radioactive decay series series of decay reactions that begins with one radioactive nucleus that decays to a second nucleus that decays to a third nucleus and so on until a stable nucleus is reached

radioactivity spontaneous emission of particles or energy from an atomic nucleus as it disintegrates

radiometric age age of rocks determined by measuring the radioactive decay of unstable elements within the crystals of certain minerals in the rocks

range the geographical distribution of a species

rarefaction a thinning or pulse of decreased density and pressure of gas molecules

ratio a relationship between two numbers, one divided by the other; the ratio of distance per time is speed

real image an image generated by a lens or mirror that can be projected onto a screen

recessive allele an allele that, when present with a dominant allele, does not express itself and is masked by the effect of the dominant allele

recombinant DNA DNA that has been constructed by inserting new pieces of DNA into the DNA of another organism, such as a bacterium

red giant stars one of two groups of stars on the Hertzsprung-Russell diagram that have a different set of properties than the main sequence stars; bright, low-temperature giant stars that are enormously bright for their temperature

redox (chemical) reaction short name for oxidation-reduction reaction

reducing agent supplies electrons to the substance being reduced in a chemical reaction

reduction division (also **meiosis**) a type of cell division in which daughter cells get only half the chromosomes from the parent cell

referent referring to or thinking of a property in terms of another, more familiar object

reflected ray a line representing direction of motion of light reflected from a boundary

reflection the change when light, sound, or other waves bounce backward off a boundary

refraction a change in the direction of travel of light, sound, or other waves crossing a boundary

regulator protein proteins that help determine the activities that will occur in a cell or multicellular organism; for example, enzymes and some hormones

relative dating dating the age of a rock unit or geological event relative to some other unit or event

relative humidity ratio (times 100 percent) of how much water vapor is in the air to the maximum amount of water vapor that could be in the air at a given temperature

reliable a term used to describe results that remain consistent over successive trials

rem measure of radiation that considers the biological effects of different kinds of ionizing radiation

replacement chemical reaction reaction in which an atom or polyatomic ion is replaced in a compound by a different atom or polyatomic ion

replacement (fossil formation) process in which an organism's buried remains are altered by circulating groundwaters carrying elements in solution; the removal of original materials by dissolutions and the replacement of new materials an atom or molecule at a time

representative elements name given to the members of the A-group families of the periodic table; also called the main-group elements

reservoir a natural or artificial pond or lake used to store water, control floods, or generate electricity; a body of water stored for public use

resonance when the frequency of an external force matches the natural frequency of a material and standing waves are set up

response the reaction of an organism to a stimulus

responsive processes those abilities to react to external and internal changes in the environment; for example, irritability, individual adaptation, and evolution

retina the light-sensitive region of the eye

reverberation apparent increase in volume caused by reflections, usually arriving within 0.1 second after the original sound

reverse fault a fault where the hanging wall has moved upward with respect to the foot wall

revolution the motion of a planet as it orbits the Sun

rhodopsin a light-sensitive pigment found in the rods of the retina

ribonucleic acid (RNA) a polymer of nucleotides formed on the template surface of DNA by transcription; three forms that have been identified are mRNA, rRNA, and tRNA

ribose a 5-carbon sugar molecule that is a component of RNA

ribosomal RNA (rRNA) a globular form of RNA; a part of ribosomes

ribosomes small structures composed of two protein and ribonucleic acid subunits; involved in the assembly of proteins from amino acids

Richter scale expresses the intensity of an earthquake in terms of a scale with each higher number indicating ten times more ground movement and about thirty times more energy released than the preceding number

ridges long, rugged mountain chains rising thousands of meters above the abyssal plains of the ocean basin

rift a split or fracture in a rock formation, land formation, or in the crust of Earth

rip current strong, brief current that runs against the surf and out to sea

RNA polymerase an enzyme that attaches to the DNA at the promoter region of a gene and assists in combining RNA nucleotides when the genetic information is transcribed into RNA

rock a solid aggregation of minerals or mineral materials that have been brought together into a cohesive solid

rock cycle understanding of igneous, sedimentary, or metamorphic rock as a temporary state in an ongoing transformation of rocks into new types; the process of rocks continually changing from one type to another

rock flour rock pulverized by a glacier into powdery, silt-sized sediment

rods light-sensitive cells in the retina of the eye that respond to low-intensity light but do not respond to different colors of light

rotation the spinning of a planet on its axis

runoff water moving across the surface of Earth as opposed to soaking into the ground

S

salivary amylase an enzyme present in saliva that breaks starch molecules into smaller molecules

salivary glands glands that produce saliva

salt any ionic compound except one with hydroxide or oxide ions

San Andreas fault in California, the boundary between the North American Plate and the Pacific Plate that runs north-south for some 1,300 km (800 miles) with the Pacific Plate moving northwest and the North American Plate moving southeast

saturated air air in which an equilibrium exists between evaporation and condensation; the relative humidity is 100 percent

saturated molecule an organic molecule that has the maximum number of hydrogen atoms possible

saturated solution the apparent limit to dissolving a given solid in a specified amount of water at a given temperature; a state of equilibrium that exists between dissolving solute and solute coming out of solution

scientific law a relationship between quantities, usually described by an equation in the physical sciences; is more important and describes a wider range of phenomena than a scientific principle

scintillation counter a device that indirectly measures ionizing radiation (alpha, beta, gamma, etc.) by measuring the flashes of light produced when the radiation strikes a phosphor

sea a smaller part of the ocean with characteristics that distinguish it from the larger ocean

sea breeze cool, dense air from over water moving over land as part of convective circulation

seafloor spreading the process by which hot, molten rock moves up from the interior of Earth to emerge along mid-oceanic rifts, flowing out in both directions to create new rocks and spread apart the seafloor

seamounts steep, submerged volcanic peaks on the abyssal plain

second the standard unit of time in both the metric and English systems of measurement

second law of motion the acceleration of an object is directly proportional to the net force acting on that object and inversely proportional to the mass of the object

second law of thermodynamics a statement that the natural process proceeds from a state of higher order to a state of greater disorder

secondary carnivores carnivores that feed on primary carnivores and are therefore at the fourth trophic level

secondary coil part of a transformer, a coil of wire in which the voltage of the original alternating current in the primary coil is stepped up or down by way of electromagnetic induction

secondary consumers animals that eat other animals—carnivores

secondary loop part of a nuclear power plant; the closed pipe system that carries steam from a steam generator to the turbines, then back to the steam generator as feedwater

secondary oocyte the larger of the two cells resulting from the unequal cytoplasmic division of a primary oocyte in meiosis I of oogenesis

secondary sexual characteristics characteristics of the adult male or female, including the typical shape that develops at puberty: broader shoulders, heavier long-bone muscles, development of facial hair, axillary hair, and chest hair, and changes in the shape of the larynx in the male; rounding of the pelvis and breasts and changes in deposition of fat in the female

secondary spermatocyte cells in the seminiferous tubules that go through the second meiotic division, resulting in four haploid spermatids

sedimentary rocks rocks formed from particles or dissolved minerals from previously existing rocks that were deposited from air or water

sediments accumulations of silt, sand, or gravel that settled out of the atmosphere or out of water

segregation the separation and movement of homologous chromosomes to the opposite poles of the cell

seismic waves vibrations that move as waves through any part of Earth, usually associated with earthquakes, volcanoes, or large explosions

seismograph an instrument that measures and records seismic wave data

selecting agents specific environmental factors that favor the passage of certain characteristics from one generation to the next and discourage others

selectively permeable the property of a membrane that allows certain molecules to pass through it, but interferes with the passage of others

semen the sperm-carrying fluid produced by the seminal vesicles, prostate glands, and bulbourethral glands of males

semiarid climate classification between arid and humid; receives between 25 and 50 cm (10 and 20 in) precipitation per year

semicircular canals a set of tubular organs associated with the cochlea that sense changes in the movement or position of the head

semiconductors elements that have properties between those of a metal and those of a nonmetal, sometimes conducting an electric current and sometimes acting like an electrical insulator depending on the conditions and their purity; also called metalloids

semilunar valves pulmonary artery and aorta valves that prevent the flow of blood backward into the ventricles

seminal vesicle a part of the male reproductive system that produces a portion of the semen

seminiferous tubules sperm-producing tubes in the testes

sensory neurons those neurons that send information from sense organs to the central nervous system

sex the nature of biological differences between males and females

sex chromosomes a pair of chromosomes that determines the sex of an organism; X and Y chromosomes

sex-determining chromosome the chromosomes X and Y that are primarily responsible for determining if an individual will develop as a male or female

sexual reproduction the propagation of organisms involving the union of gametes from two parents

sexuality a term used in reference to the totality of the aspects—physical, psychological, and cultural—of our sexual nature

shallow-focus earthquakes earthquakes that occur from the surface down to 70 km deep

shear stress produced when two plates slide past one another or by one plate sliding past another plate that is not moving

shell model of the nucleus model of the nucleus that has protons and neutrons moving in energy levels or shells in the nucleus (similar to the shell structure of electrons in an atom)

shells the layers that electrons occupy around the nucleus

shield volcano a broad, gently sloping volcanic cone constructed of solidified lava flows

shock wave a large, intense wave disturbance of very high pressure; the pressure wave created by an explosion, for example

sickle-cell anemia a disease caused by a point mutation; this malfunction produces sickle-shaped red blood cells

sidereal day the interval between two consecutive crossings of the celestial meridian by a particular star

sidereal month the time interval between two consecutive crossings of the Moon across any star

sidereal year the time interval required for Earth to move around its orbit so that the Sun is again in the same position against the stars

silicates minerals that contain silicon-oxygen tetrahedra either isolated or joined together in a crystal structure

sill a tabular-shaped intrusive rock that formed when magma moved into the plane of contact between sedimentary rock layers

simple harmonic motion the vibratory motion that occurs when there is a restoring force opposite to and proportional to a displacement

single bond covalent bond in which a single pair of electrons is shared by two atoms

single-factor cross a genetic study in which one characteristic is followed from the parental generation to the offspring

small intestine the portion of the digestive system immediately following the stomach; responsible for digestion and absorption

small solar system bodies all objects orbiting the Sun that are not planets or dwarf planets

soil a mixture of unconsolidated weathered earth materials and humus, which is altered, decay-resistant organic matter

solenoid a cylindrical coil of wire that becomes electromagnetic when a current runs through it

solids a phase of matter with molecules that remain close to fixed equilibrium positions due to strong interactions between the molecules, resulting in the characteristic definite shape and definite volume of a solid

solstices times when the Sun is at its maximum or minimum altitude in the sky, known as the summer solstice and the winter solstice

solubility dissolving ability of a given solute in a specified amount of solvent, the concentration that is reached as a saturated solution is achieved at a particular temperature

solute the component of a solution that dissolves in the other component; the solvent

solution a homogeneous mixture of ions or molecules of two or more substances

solvent the component of a solution present in the larger amount; the solute dissolves in the solvent to make a solution

soma the cell body of a neuron, which contains the nucleus

sonic boom sound waves that pile up into a shock wave when a source is traveling at or faster than the speed of sound

sound quality characteristic of the sound produced by a musical instrument; determined by the presence and relative strengths of the overtones produced by the instrument

south pole (of a magnet) short for "south seeking"; the pole of a magnet that points southward when it is free to turn

special theory of relativity analysis of how space and time are affected by motion between an observer and what is being measured

speciation the process of generating new species

species a group of organisms that can interbreed naturally to produce fertile offspring

specific heat each substance has its own specific heat, which is defined as the amount of energy (or heat) needed to increase the temperature of 1 gram of a substance 1 degree Celsius

speed a measure of how fast an object is moving—the rate of change of position per change in time; speed has magnitude only and does not include the direction of change

sperm the haploid sex cells produced by sexually mature males

spermatids haploid cells produced by spermatogenesis that change into sperm

spermatogenesis the specific name given to the gametogenesis process that leads to the formation of sperm

spin quantum number from the quantum mechanics model of the atom, one of four descriptions of the energy state of an electron wave; this quantum number describes the spin orientation of an electron relative to an external magnetic field

spinal cord the portion of the central nervous system located within the vertebral column that carries both sensory and motor information between the brain and the periphery of the body

spindle fibers an array of microtubules extending from pole to pole; used in the movement of chromosomes

spontaneous generation the theory that living organisms arose from nonliving material

spontaneous mutation a change in the DNA of an organism for which there is no known cause

spring equinox one of two times a year that daylight and night are of equal length; occurs on or about March 21 and identifies the beginning of the spring season

spring tides unusually high and low tides that occur every two weeks because of the relative positions of Earth, Moon, and Sun

stable equilibrium phase a stage in a population growth curve following rapid growth in which there is both a decrease in natality and an increase in mortality so that the size of the population is stable

standard atmospheric pressure the average atmospheric pressure at sea level, which is also known as normal pressure; the standard pressure is 29.92 in or 760.0 mm of mercury (1,013.25 millibar)

standard time zones 15° wide zones defined to have the same time throughout the zone, with the time defined as the mean solar time at the middle of each zone

standard unit a measurement unit established as the standard upon which the value of the other referent units of the same type are based

standing waves condition where two waves of equal frequency traveling in opposite directions meet and form stationary regions of maximum displacement due to constructive interference and stationary regions of zero displacement due to destructive interference

stapes the ear bone that is attached to the oval window

starch a group of complex carbohydrates (polysaccharides) that plants use as a stored food source and that serves as an important source of food for animals

stationary front occurs when the edge of a weather front is not advancing

steam generator part of a nuclear power plant; the heat exchanger that heats feedwater from the secondary loop to steam with the very hot water from the primary loop

stem cells cells that have not yet completed determination or differentiation, so they have the potential to develop into many different cell types

step-down transformer a transformer that decreases the voltage

step-up transformer a transformer that increases the voltage

steroids a group of lipid molecules characterized by an arrangement of interlocking rings of carbon; they serve as hormones that aid in regulating body processes

stimulus any change in the internal or external environment of an organism that it can detect

stony meteorites meteorites composed mostly of silicate minerals that usually make up rocks on Earth

stony-iron meteorites meteorites composed of silicate minerals and metallic iron

storm a rapid and violent weather change with strong winds, heavy rain, snow, or hail

strain adjustment to stress; a rock unit might respond to stress by changes in volume or shape, or by breaking

stratosphere the layer of the atmosphere above the troposphere where temperature increases with height

stream body of running water

stress a force that tends to compress, pull apart, or deform rock; stress on rocks in Earth's solid outer crust results as Earth's plates move into, away from, or alongside each other

structural protein protein molecules whose function is to provide support and shape to a cell or multicellular organism; for example, muscle protein fibers

structural similarities one of the characteristics of living things; describes the fact that all living things are composed of cells, either prokaryotic or eukaryotic

subduction zone the region of a convergent boundary where the crust of one plate is forced under the crust of another plate into the interior of Earth

sublimation the phase change of a solid directly into a vapor or gas

submarine canyons a feature of the ocean basin; deep, steep-sided canyons that cut through the continental slopes

subspecies regional groups within a species that are significantly different structurally, physiologically, or behaviorally yet are capable of exchanging genes by interbreeding with other members of the species

summer solstice in the Northern Hemisphere, the time when the Sun reaches its maximum altitude in the sky, which occurs on or about June 22 and identifies the beginning of the summer season

superconductors some materials in which, under certain conditions, the electrical resistance approaches zero

supercooled water in the liquid phase when the temperature is below the freezing point

superior vena cava a major vein that returns blood to the heart from the head and upper body

supernova a rare catastrophic explosion of a star into an extremely bright but short-lived phenomenon

supersaturated containing more than the normal saturation amount of a solute at a given temperature

surf the zone where breakers occur; the water zone between the shoreline and the outermost boundary of the breakers

surface wave a seismic wave that moves across Earth's surface, spreading across the surface like water waves spread on the surface of a pond from a disturbance

S-wave a sideways or shear wave in which a disturbance vibrates materials from side to side, perpendicular to the direction of wave movement

S-wave shadow zone a region of Earth more than 103° of arc away from the epicenter of an earthquake where S-waves are not recorded; believed to be the result of the core of Earth that is a liquid, or at least acts like a liquid

swell regular groups of low-profile, long wavelength waves that move continuously

symbiosis a close physical relationship between two kinds of organisms

symbiotic nitrogen-fixing bacteria bacteria that live in the roots of certain kinds of plants, where they convert nitrogen gas molecules into compounds that the plants can use to make amino acids and nucleic acids

synapse the space between the axon of one neuron and the dendrite of the next, where chemicals are secreted to cause an impulse to be initiated in the second neuron

synapsis the condition in which the two members of a pair of homologous chromosomes come to lie close to one another

syncline a trough-shaped fold in layered bedrock

synodic month the interval of time from new moon to new moon (or any two consecutive identical phases)

systemic circulation the flow of blood through certain chambers of the heart and blood vessels to the general body and back to the heart

systolic blood pressure the blood pressure recorded in a large artery while the heart is contracting

T

talus steep, conical, or apronlike accumulations of rock fragments at the base of a slope

taxonomy the science of classifying and naming organisms

telophase the last phase in mitosis, characterized by the formation of daughter nuclei

temperate climate zone climate zone of the middle latitudes; average monthly temperatures stay between 10°C and 18°C (50°F and 64°F) throughout the year

temperature how hot or how cold something is; a measure of the average kinetic energy of the molecules making up a substance

tensional stress the opposite of compressional stress; occurs when one part of a plate moves away from another part that does not move

termination code the DNA nucleotide sequence at the end of a gene with the code ATT, ATC, or ACT that signals "stop here"

terrestrial planets planets Mercury, Venus, Earth, and Mars that have similar densities and compositions as compared to the outer giant planets

testes the male sex organs that produce haploid cells called sperm

testosterone the male sex hormone produced in the testes that controls male sexual development

thalamus a region of the brain that relays information between the cerebrum and lower portions of the brain, providing some level of awareness in that it determines pleasant and unpleasant stimuli and is involved in sleep and arousal

theory a broad, detailed explanation that guides the development of hypotheses and interpretations of experiments in a field of study

theory of natural selection the idea that some individuals within a population will have favorable combinations of genes that make it very likely that those individuals will survive, reproduce, and pass their genes to the next generation

thermohaline circulation the movement of the waters of the oceans caused by surface currents and changes in density due to differences in temperature or salinity

thermometer a device used to measure the hotness or coldness of a substance

third law of motion whenever two objects interact, the force exerted on one object is equal in size and opposite in direction to the force exerted on the other object; forces always occur in matched pairs that are equal and opposite

thrust fault a reverse fault with a low-angle fault plane

thunderstorm a brief, intense electrical storm with rain, lightning, thunder, strong winds, and sometimes hail

thymine a single-ring nitrogenous-base molecule in DNA but not in RNA; it is complementary to adenine

thyroid-stimulating hormone (TSH) a hormone secreted by the pituitary gland that stimulates the thyroid to secrete thyroxine

thyroxine a hormone produced by the thyroid gland that speeds up the metabolic rate

tidal bore a strong tidal current, sometimes resembling a wave, produced in very long, very narrow bays as the tide rises

tidal currents a steady and continuous onward movement of water produced in narrow bays by the tides

tides periodic rise and fall of the level of the sea from the gravitational attraction of the Moon and Sun

tissue a group of specialized cells that work together to perform a specific function

tornado a long, narrow, funnel-shaped column of violently whirling air from a thundercloud that moves destructively over a narrow path when it touches the ground

total internal reflection condition where all light is reflected back from a boundary between materials; occurs when light arrives at a boundary at the critical angle or beyond

total solar eclipse eclipse that occurs when Earth, the Moon, and the Sun are lined up so the Moon completely covers the disk of the Sun; the umbra of the Moon's shadow falls on the surface of Earth

trachea a major tube supported by cartilage that carries air to the bronchi; also known as the windpipe

transcription the process of manufacturing RNA from the template surface of DNA; three forms of RNA that may be produced are mRNA, rRNA, and tRNA

transfer RNA (tRNA) a molecule composed of ribonucleic acid; it is responsible for transporting a specific amino acid into a ribosome for assembly into a protein

transform boundaries in plate tectonics, boundaries that occur between two plates sliding horizontally by each other along a long, vertical fault; sudden jerks along the boundary result in the vibrations of earthquakes

transformer a device consisting of a primary coil of wire connected to a source of alternating current and a secondary coil of wire in which electromagnetic induction increases or decreases the voltage of the source

transition elements members of the B-group families of the periodic table

translation the assembly of individual amino acids into a polypeptide

transparent term describing materials that allow the transmission of light; for example, glass and clear water are transparent materials

transportation the movement of eroded materials by agents such as rivers, glaciers, wind, or waves

transverse wave a mechanical disturbance that causes particles to move perpendicular to the direction that the wave is traveling

triglyceride organic compound of esters formed from glycerol and three long-chain carboxylic acids; also called fats in animals and oils in plants

triple bond covalent bond formed when three pairs of electrons are shared by two atoms

trophic level a step in the flow of energy through an ecosystem

tropic of Cancer parallel identifying the northern limit where the Sun appears directly overhead; located at 23.5°N latitude

tropic of Capricorn parallel identifying the southern limit where the Sun appears directly overhead; located at 23.5°S latitutde

tropical air mass a warm air mass from warm regions

tropical climate zone climate zone of the low latitudes; average monthly temperatures stay above 18°C (64°F), even during the coldest month of the year

tropical cyclone a large, violent circular storm that is born over the warm, tropical ocean near the equator; also called hurricane (Atlantic and eastern Pacific) and typhoon (in western Pacific)

tropical year the time interval between two consecutive spring equinoxes; used as standard for the common calendar year

troposphere layer of the atmosphere from the surface to where the temperature stops decreasing with height

trough the low mound of water that is part of a wave; also refers to the rarefaction, or low-pressure part of a sound wave

true fats also known as *neutral fats*; a category of lipids composed of glycerol and fatty acids; for example, pork chop fat or olive oil

tsunami very large, fast, and destructive ocean wave created by an undersea earthquake, landslide, or volcanic explosion; a seismic sea wave

tumor a mass of undifferentiated cells not normally found in a certain portion of the body

turbidity current a muddy current produced by underwater landslides

tympanum the eardrum

typhoon the name for hurricanes in the western Pacific

U

ultrasonic sound waves too high in frequency to be heard by the human ear; frequencies above 20,000 Hz

umbra the inner core of a complete shadow

unconformity a time break in the rock record

undertow a current beneath the surface of the water produced by the return of water from the shore to the sea

unit in measurement, a well-defined and agreed-upon referent

universal law of gravitation every object in the universe is attracted to every other object with a force directly proportional to the product of their masses and inversely proportional to the square of the distance between the centers of the two masses

unpolarized light light consisting of transverse waves vibrating in all conceivable random directions

unsaturated molecule an organic molecule that does not contain the maximum number of hydrogen atoms; a molecule that can add more hydrogen atoms because of the presence of double or triple bonds

uracil a single-ring nitrogenous-base molecule in RNA but not in DNA; it is complementary to adenine

uterus the organ in female mammals in which the embryo develops

V

vacuole a large sac within the cytoplasm of a cell, composed of a single membrane

vagina the passageway between the uterus and outside of the body; the birth canal

valence the number of covalent bonds an atom can form

valence electrons electrons of the outermost shell; the electrons that determine the chemical properties of an atom and the electrons that participate in chemical bonding

valid a term used to describe meaningful data that fit into the framework of scientific knowledge

Van Allen belts belts of radiation caused by cosmic-ray particles becoming trapped and following Earth's magnetic field lines between the poles

vapor the gaseous state of a substance that is normally in the liquid state

variables changing quantities usually represented by a letter or symbol

veins the blood vessels that return blood to the heart

velocity describes both the speed and direction of a moving object; a change in velocity is a change in speed, in direction of travel, or both

ventifacts rocks sculpted by wind abrasion

ventricles the powerful muscular chambers of the heart whose contractions force blood to flow through the arteries to all parts of the body

vernal equinox another name for the spring equinox, which occurs on or about March 21 and marks the beginning of the spring season

vibration a back-and-forth motion that repeats itself

villi tiny fingerlike projections in the lining of the intestine that increase the surface area for absorption

viroids infectious particles that are composed solely of a small, single strand of RNA

virtual image an image where light rays appear to originate from a mirror or lens; this image cannot be projected on a screen

virus a nucleic acid particle coated with protein that functions as an obligate intracellular parasite

vitamins organic molecules that cannot be manufactured by the body but are required in very low concentrations for good health

volcanism volcanic activity; the movement of magma

volcano a hill or mountain formed by the extrusion of lava or rock fragments from a mass of magma below

volt the ratio of work done to move a quantity of charge

voltage source a device that does work in moving a quantity of charge

volume how much space something occupies

vulcanism volcanic activity; the movement of magma

W

warm front the front that forms when a warm air mass advances against a cool air mass

water table the boundary below which the ground is saturated with water

watershed the region or land area drained by a stream; a stream drainage basin

watt metric unit for power; equivalent to J/s

wave a disturbance or oscillation that moves through a medium

wave equation the relationship of the velocity of a wave to the product of the wavelength and frequency of the wave

wave front a region of maximum displacement in a wave; a condensation in a sound wave

wave height the vertical distance of an ocean wave between the top of the wave crest and the bottom of the next trough

wave mechanics alternate name for quantum mechanics derived from the wavelike properties of subatomic particles

wave period the time required for two successive crests or other successive parts of the wave to pass a given point

wavelength the horizontal distance between successive wave crests or other successive parts of the wave

weak acid acids that only partially ionize because of an equilibrium reaction with water

weak base a base only partially ionized because of an equilibrium reaction with water

weathering slow changes that result in the breaking up, crumbling, and destruction of any kind of solid rock

white dwarf stars one of two groups of stars on the Hertzsprung-Russell diagram that have a different set of properties than the main sequence stars; faint, white-hot stars that are very small and dense

wind a horizontal movement of air that moves along or parallel to the ground, sometimes in currents or streams

wind abrasion the natural sand-blasting process that occurs when wind-driven particles break off small particles of rock and polish the rock they strike

wind chill factor a factor that compares heat loss from bodies in still air with those in moving air; moving air removes heat more rapidly and causes a person to feel that the air is colder than its actual temperature; the cooling power of wind

winter solstice in the Northern Hemisphere, the time when the Sun reaches its minimum altitude, which occurs on or about December 22 and identifies the beginning of the winter season

work the magnitude of applied force times the distance through which the force acts; can be thought of as the process by which one form of energy is transformed to another

X

X chromosome the chromosome in a human female egg (and in one-half of sperm cells) that is associated with the determination of sexual characteristics

X-linked gene a gene located on one of the sex-determining X chromosomes

Y

Y chromosome the sex-determining chromosome in one-half of the sperm cells of human males responsible for determining maleness

Z

zone of saturation zone of sediments beneath the surface in which water has collected in all available spaces

zooplankton small, weakly swimming animals of many kinds

zygote a diploid cell that results from the union of an egg and a sperm; a fertilized egg

Credits

Photographs

Chapter 1 Opener: © Werner H. Muller/ Peter Arnold; 1.1: © Spencer Grant/Photo Researchers; 1.2: © Rafael Macia/Photo Researchers; 1.3: © Bill W. Tillery; 1.12: © BananaStock/PunchStock; 1.13a: © David R. Frazier/Photo Researchers; Box 1.2: © Klaus Hackenberg/Corbis; Box 1.3: © USGS Photo Library.

Chapter 2 Opener: © Douglas Faulkner/Photo Researchers; 2.1: © Brand X Pictures; 2.5: © Bill W. Tillery; 2.7: © Scala/Art Resource, NY; Box 2.1: © Eunice Harris/Photo Researchers; 2.14, 2.19: Arizona State University Media Relations Office; Box 2.2: © Image Club; 2.21: NASA; 2.25: © Stock Trek/ Getty; 2.26: © NASA.

Chapter 3 Opener: © Hank Morgan/Science Source/Photo Researchers; 3.1, 3.9, 3.10a: © Bill W. Tillery; 3.10b: © Alan Oddie/PhotoEdit; 3.11, 3.13: Arizona Public Service Company; Box 3.1: © Archive Photos/Getty; 3.18: © Bill W. Tillery.

Chapter 4 Opener: © Lawrence Livermore National Laboratory/SPL/Photo Researchers; 4.1: Bethlehem Steel Corporation; 4.5: Reprinted with permission from Honeywell, Inc.; 4.13: © Stockbyte/SuperStock; 4.19: © Bill W. Tillery; Box 4.4: North Wind Picture Archives.

Chapter 5 Opener: © Clyde H. Smith/Peter Arnold; 5.1: © Patrick Coughlin/Stone/Getty; 5.5: © Patrick Clark/Getty Images; Box 5.1B: Jones & Childers; 5.13: © The McGraw-Hill Companies, Inc./Bob Coyle, Photographer; 5.24: NASA; 5.25: NASA.

Chapter 6 Opener: © Keith Kent/Peter Arnold; 6.1: © Bill W. Tillery; Box 6.2: © The Corcoran Gallery of Art/Corbis; 6.11 (all): © Bill W. Tillery; Box 6.4a: © 1989 Paul Silverman/Fundamental Photographs; 6.12: © Steve Cole/PhotoDisc/Getty; 6.24 a–b, Box 6.6 a–b: © Bill W. Tillery.

Chapter 7 Opener: © James L. Amos/Peter Arnold; 7.1: © John Kieffer/Peter Arnold; 7.8, 7.16: © Bill W. Tillery; 7.23: © Brock May/ Photo Researchers; Box 7.9: © Baldwin H. Ward & Kathryn C. Ward/Corbis.

Chapter 8 Opener: © Astrid & Hanns-Frieder Michler/SPL/Photo Researchers; 8.1: © Ken Edward/Science Source/Photo Researchers; Box 8.1: © Scala/Art Resource, NY.

Chapter 9 Opener: © David Weintraub/Photo Researchers; 9.1: © Bill W. Tillery; 9.3 a–b: © Leonard Lessin/Peter Arnold; 9.4: © Bill W. Tillery; 9.5: © Richard Megna/Fundamental Photographs; 9.9: © Charles M. Falco/Photo Researchers; 9.11, 9.12: © Bill W. Tillery; 9.13: Ryan McVay/Getty Images; 9.15, 9.16, 9.18. Box 9.2: © Bill W. Tillery; p. 220: © Roger Ressmeyer/Corbis.

Chapter 10 Opener: © Blair Seitz/Photo Researchers; 10.1: © Craig Tuttle/Corbis; 10.10 a–b, 10.11, 10.13, Box 10.1: © Bill W. Tillery.

Chapter 11 Opener: © Yann Arthus-Bertrand/Peter Arnold; 11.1: Arizona Public Service Company; 11.2 a–b: © Fundamental Photographs; 11.14 a–b, 11.16, 11.17: Arizona Public Service Company; Box 11.3: © Bill W. Tillery; p. 264: © Bettman/Corbis.

Chapter 12 Opener: NASA; 12.1a: © UC Regents/Lick Observatory; 12.2: © J. Fuller/ Visuals Unlimited; 12.8, 12.11, 12.13: © UC Regents/Lick Observatory; Box 12.1: © Hencoup Enterprises/SPL/Photo Researchers.

Chapter 13 Opener: NASA; 13.1, 13.4, 13.5: NASA; 13.6: © UC Regents/Lick Observatory; 13.7, 13.8: NASA/JPL/Cornell; 13.10 a–b, 13.11: NASA; 13.12 a–b: © Space Telescope Science Institute/NASA/SPL/Photo Researchers; 13.13, 13.14: NASA; 13.16: "Oort Cloud" by Don Davis reproduced with permission of Sky Publishing Corp.; 13.17: © UC Regents/Lick Observatory; 13.19 a–b: Center for Meteorite Studies, Arizona State University.

Chapter 14 Opener: Jacques Descloitres, MODIS Rapid Response Team, NASA/GSFC; 14.1: © Jason Reed/Photodisc/Getty; 14.3: NASA; 14.18: © Bill W. Tillery; p. 328: © Bettman/Corbis.

Chapter 15 Opener: © Brian Parker/Tom Stack & Associates; 15.1: © Bill W. Tillery; 15.4: © Charles Falco/Photo Researchers; 15.5: © Bill W. Tillery; Box 15.1: © Charles C. Plummer; 15.8, 15.9: © Bill W. Tillery; 15.10: © Charles C. Plummer.

Chapter 16 Opener: © Jeff Foot/Tom Stacks & Associates; 16.1: Photo by A. Post, USGS Photo Library, Denver, CO; 16.2: © Bill W. Tillery; 16.4a: © Robert W. Northrop; 16.4b: © Bill W. Tillery; 16.5: © Charles C. Plummer; 16.8b: © Frank M. Hanna; 16.12d: Photo by University of Colorado; courtesy National Geophysical Data Center, Boulder; 16.15: NASA; 16.18: © Bill W. Tillery; 16.20: National Park Service, Photo by Wm. Belnap, Jr.; 16.22a: © A.J. Copley/Visuals Unlimited; 16.22b: © L. Linkhard/Visuals Unlimited; 16.23: © Ken Wagner/Visuals Unlimited; 16.24 a–b: © Bill W. Tillery; 16.26: © William J. Weber/Visuals Unlimited; 16.27: © Doug Sherman/Geofile; 16.29: © Burt Amundson; 16.30a: Photo by D.A. Rahm, courtesy of Rahm Memorial Collections, Western Washington; 16.31: © Charles C. Plummer.

Chapter 17 Opener: © Royalty-Free/Corbis; 17.1, 17.16, 17.18 a–b, 17.19 a–f: © Bill W. Tillery; 17.20, 17.22: NOAA; 17.31: © Telegraph Herald/Photo by Patti Carr; 17.32, 17.33: NOAA; 17.35: © David Parker/SPL/Photo Researchers; 17.36: © Bill W. Tillery; 17.39: Dr. Charles Hogue, Curator of Entomology, Los Angeles County; 17.40: © Bob Wallen; 17.41: © Elizabeth Wallen; 17.44: © Glen Allison/Getty Images; 17.47: © Digital Vision/PunchStock.

Chapter 18 Opener: © James H. Karales/ Peter Arnold; 18.1: Salt River Project; 18.9: City of Tempe, Arizona; 18.10: Salt River Project; 18.13, 18.14: © Bill W. Tillery; 18.18: © The McGraw-Hill Companies, Inc./John A. Karachewski, photographer.

Chapter 19 Opener: © Royalty-Free/Corbis; 19.1: American Petroleum Institute; p. 438: © The McGraw-Hill Companies, Inc./Pat Watson, photographer; p. 439: Texas Energy Museum, Beaumont, Texas; 19.13: © Bill W. Tillery; p. 444: © Eldon Enger; 19.16a: © R.J. Erwin/Photo Researchers; 19.16b: © Michael Rosenfeld/Photographer's Choice/Getty; p. 448: Photo by Tim McCabe, courtesy of USDA Natural Resources Conservation Service; 19.18 (left & middle): © Royalty-Free/Corbis; 19.18 (right): © PhotoLink/Getty; p. 451: © Jonelle Weaver/getty; p. 453: © I. Rozenbaum/F. Cirou/ Photo Alto.

Chapter 20 Opener: © Melba Photo Agency/ PunchStock; 20.2 (bacteria): CDC/Janice Carr; 20.2 (amoeba): © Melba Photo Agency/ PunchStock; 20.2 (humans): © Stockbyte/ PunchStock; 20.2 (orchid): © The McGraw-Hill Companies, Inc./Steven P. Lynch, photographer; Box 20.1: The Granger Collection, NY; Box 20.2: © Mary Evans Picture Library/Alamy; Box 20.3a: © Fred Ross; Box 20.3b: © Creatas Images/Jupiterimages; 20.6: Courtesy of Tampa Hyperbaric Enterprise, Inc., www.oxytank.com; Box 20.4b, 20.13, 20.24: © Fred Ross; Box 20.5: © BananaStock/PunchStock.

Chapter 21 Opener: © Brand X Pictures/ Punchstock; 21.4: NOAA; p. 506: © Walt Anderson/Visuals Unlimited; 21.7: © U.S. Fish and Wildlife Service/Dave Menke photographer; 21.8: © D.O.E./Science Source/ Photo Researchers; 21.10: © Carl & Ann Purcell/ Corbis; p. 515: © Stockbyte; 21.13a: © Fred Ross; 21.13b: © Eldon Enger; 21.13c: © Fred Ross; 21.14 (both): © Tom & Pat Leeson.

Index

Table of Atomic Weights (Based on Carbon-12)

Name	Symbol	Atomic Number	Atomic Weight	Name	Symbol	Atomic Number	Atomic Weight
Actinium	Ac	89	(227)	Mendelevium	Md	101	258.10
Aluminum	Al	13	26.9815	Mercury	Hg	80	200.59
Americium	Am	95	(243)	Molybdenum	Mo	42	95.94
Antimony	Sb	51	121.75	Neodymium	Nd	60	144.24
Argon	Ar	18	39.948	Neon	Ne	10	20.179
Arsenic	As	33	74.922	Neptunium	Np	93	(237)
Astatine	At	85	(210)	Nickel	Ni	28	58.71
Barium	Ba	56	137.34	Niobium	Nb	41	92.906
Berkelium	Bk	97	(247)	Nitrogen	N	7	14.0067
Beryllium	Be	4	9.0122	Nobelium	No	102	259.101
Bismuth	Bi	83	208.980	Osmium	Os	76	190.2
Bohrium	Bh	107	264	Oxygen	O	8	15.9994
Boron	B	5	10.811	Palladium	Pd	46	106.4
Bromine	Br	35	79.904	Phosphorus	P	15	30.9738
Cadmium	Cd	48	112.40	Platinum	Pt	78	195.09
Calcium	Ca	20	40.08	Plutonium	Pu	94	244.064
Californium	Cf	98	242.058	Polonium	Po	84	(209)
Carbon	C	6	12.0112	Potassium	K	19	39.098
Cerium	Ce	58	140.12	Praseodymium	Pr	59	140.907
Cesium	Cs	55	132.905	Promethium	Pm	61	144.913
Chlorine	Cl	17	35.453	Protactinium	Pa	91	(231)
Chromium	Cr	24	51.996	Radium	Ra	88	(226)
Cobalt	Co	27	58.933	Radon	Rn	86	(222)
Copper	Cu	29	63.546	Rhenium	Re	75	186.2
Curium	Cm	96	(247)	Rhodium	Rh	45	102.905
Dubnium	Db	105	(262)	Rubidium	Rb	37	85.468
Dysprosium	Dy	66	162.50	Ruthenium	Ru	44	101.07
Einsteinium	Es	99	(254)	Rutherfordium	Rf	104	(261)
Erbium	Er	68	167.26	Samarium	Sm	62	150.35
Europium	Eu	63	151.96	Scandium	Sc	21	44.956
Fermium	Fm	100	257.095	Seaborgium	Sg	106	(266)
Fluorine	F	9	18.9984	Selenium	Se	34	78.96
Francium	Fr	87	(223)	Silicon	Si	14	28.086
Gadolinium	Gd	64	157.25	Silver	Ag	47	107.868
Gallium	Ga	31	69.723	Sodium	Na	11	22.989
Germanium	Ge	32	72.59	Strontium	Sr	38	87.62
Gold	Au	79	196.967	Sulfur	S	16	32.064
Hafnium	Hf	72	178.49	Tantalum	Ta	73	180.948
Hassium	Hs	108	(269)	Technetium	Tc	43	(99)
Helium	He	2	4.0026	Tellurium	Te	52	127.60
Holmium	Ho	67	164.930	Terbium	Tb	65	158.925
Hydrogen	H	1	1.0079	Thallium	Tl	81	204.37
Indium	In	49	114.82	Thorium	Th	90	232.038
Iodine	I	53	126.904	Thulium	Tm	69	168.934
Iridium	Ir	77	192.2	Tin	Sn	50	118.69
Iron	Fe	26	55.847	Titanium	Ti	22	47.90
Krypton	Kr	36	83.80	Tungsten	W	74	183.85
Lanthanum	La	57	138.91	Uranium	U	92	238.03
Lawrencium	Lr	103	260.105	Vanadium	V	23	50.942
Lead	Pb	82	207.19	Xenon	Xe	54	131.30
Lithium	Li	3	6.941	Ytterbium	Yb	70	173.04
Lutetium	Lu	71	174.97	Yttrium	Y	39	88.905
Magnesium	Mg	12	24.305	Zinc	Zn	30	65.38
Manganese	Mn	25	54.938	Zirconium	Zr	40	91.22
Meitnerium	Mt	109	(268)				

Note: A value given in parentheses denotes the number of the longest-lived or best-known isotope.